Power System Optimization

THIRD EDITION

D.P. KOTHARI

Former Director-in-Charge
Indian Institute of Technology Delhi

Former Principal
VNIT, Nagpur

Former Vice-Chancellor
VIT University, Vellore

J.S. DHILLON

Professor
Department of Electrical and Instrumentation Engineering
Sant Longowal Institute of Engineering and Technology
(Deemed-to-be-University)
Longowal, Punjab

PHI Learning Private Limited

Delhi-110092
2025

In fond memory of ***Shri Asoke K. Ghosh*** *(October 1942 – February 2024), Founder Chairman and Managing Director of PHI Learning, whose vision endlessly inspires.*

The Legacy Continues....

Published by Pushpita Ghosh, PHI Learning Private Limited, Rimjhim House, 111, Patparganj Industrial Estate, Delhi-110092 and Printed by Syndicate Binders, A-20, Hosiery Complex, Noida, Phase-II Extension, Noida-201305 (N.C.R. Delhi).

₹1495.00

POWER SYSTEM OPTIMIZATION, Third Edition
D.P. Kothari and J.S. Dhillon

ISBN-978-93-5443-992-6 (Print Book)
ISBN-978-93-5443-924-7 (e-Book)

To my grandchildren, Aditi, Anushri and Anirudh

— D.P. Kothari

To my father, Sh. Harcharan Singh Dhillon

— J.S. Dhillon

Contents

Preface

In recent years, myriad meta-heuristic optimization techniques have been applied to power systems. These commonly used algorithms include ant colony optimization, genetic algorithm, differential evolution algorithm, particle swarm optimization, artificial bee colony, bacterial foraging optimization algorithm, gravitational search algorithm, intelligent water drops, river formation dynamics, charged system search, harmony search algorithm, group search optimizer, and fruit fly optimization and many more. Optimization algorithms have advantages and disadvantages compared to each other and may show different performances when solving discrete and continuous problems. Countless variations and combinations of such meta-heuristic optimization techniques are proposed to improve the performance in terms of convergence rate and solution accuracy. In general, heuristic computation is applied to solve optimization problems as well as multiobjective optimization problems that cannot be solved with other tools because of their intrinsic difficulty, high dimensionality, or incomplete definition. Hence, meta-heuristic algorithms require sometimes a massive computational effort to yield efficient and competitive solutions to real-size power system engineering problems, which would, otherwise rest unsolved today.

This book focuses on the aspects related to the soft computation, such as genetic algorithm, evolutionary programming, particle swarm optimization, and differential evolution procedures' implementation on power system applications. Chapter 9 is added to include Biogeography-based optimization for economic dispatch with prohibited operating zones and ramp rate, multi-fuel economic dispatch using artificial bee colony algorithm and combined heat and power economic dispatch using teaching learning based optimization. The chapter also presents short-range variable head hydrothermal generation scheduling using real coded genetic algorithm and short-term multi-chain hydrothermal generation scheduling using predator prey optimization.

Unit commitment is a very significant optimization task that plays important role in operation planning of power systems. Chapter 10 is added to include unit commitment problem. Enumerated method, priority list method, dynamic programming, Lagrange relaxation and heuristic methods are elaborated to solve unit commitment problem. The edition gives less space for computation of loss formula with conventional method.

Still, the intent of this volume has been to offer a wide spectrum of sample works developed in leading research throughout the world, about search techniques at the heart of computational intelligence to deal with power system optimization problems. The style of writing is specially adapted to self-study. The book should be useful for both beginners and experienced researchers in the field of power system optimization.

The authors would like to acknowledge all our graduate and postgraduate students whose thesis work on various topics of power system operations contributed to the broadening of our horizon on the subject. The recent work of Dr. Jarnail Singh Dhillon and Dr. Nitin Narang has formed the basis for the material of Chapter 9. Authors are grateful to the authorities of Sant Longowal Institute of Engineering & Technology (SLIET), Longowal and Visvesvaraya National Institute of Technology (VNIT), Nagpur for their encouragement in writing this latest edition. The authors appreciate the suggestions and help offered by professors who have used the first and second editions and our students. We hope that the reader will find the edition useful.

Finally, we would like to thank all our families, colleagues, and friends for their love and support.

D.P. KOTHARI [E-mail: dpk0710@yahoo.com]
J.S. DHILLON [E-mail: jsdhillonp@yahoo.com]

D.P. KOTHARI
J.S. DHILLON

Preface to the First Edition

In response to increasing public awareness of the environmental situation and the plea for clean air, many engineers came up with new methods to reduce air pollution in parallel with the pursuit of economy. Engineers are devoting considerable time to handle such conflicting situations through multiobjective optimization. The aim of multiobjective optimization is to help engineers (or decision makers) take the right decision in conflicting situations bedevilled with several objectives to be satisfied simultaneously. Further for large-scale integrated electric power systems, there is no other alternative but to use the digital computer as a computation tool for fast, accurate, and robust solution procedures.

This book is intended to serve as an introductory text to the topic of multiobjective optimization in electric power systems. It may also be used for self-study by practising professionals engaged in planning and operation of thermal as well as integrated hydrothermal electric power systems. It has been the endeavour of the authors to provide simple and understandable basic computational algorithms so that students or practising engineers can develop their own programs in any high level language or improve the existing ones. Solved examples are given for better understanding of each power system problem discussed. The reader is expected to have a prior knowledge of basics of electric power system, optimization techniques, numerical methods, and matrix operations. The first chapter introduces the power system components, planning and operation problems, potential application of fuzzy theory and artificial neural networks in power systems.

Chapter 2 elaborates on power network modelling and important techniques of ac load flow analysis like Gauss–Seidel, Newton–Raphson, and decoupled load flow. To reduce the computation burden, initial guess for load flow is also explained. This chapter also deals with modelling and solution procedure for ac–dc load flow.

Chapter 3 is devoted to economic dispatch of thermal power systems. Newton–Raphson and approximations to Newton–Raphson method are discussed to solve the classical economic dispatch. The economic dispatch procedures are elaborated here to consider the exact loss formula as well as the real and reactive power balance. Rigorous economic dispatch techniques such as penalty factor method, gradient method, and Newton–Raphson method are discussed. The chapter also deals with the evaluation of B-coefficients by the classical method, Y-bus method, and the sensitivity factor method. It also explains the development of exact transmission loss formula.

Chapter 4 deals with the foundations of hydrothermal scheduling such as fixed-head, variable-head for short-term and long-term problems. It elaborates upon the classical Newton–Raphson and approximate Newton–Raphson methods to solve the fixed-head, short-range hydrothermal problem. Classical and approximate Newton–Raphson methods for short-range, variable-head, hydrothermal problems are also discussed in this chapter. It also deals with hydro plant modelling for long-range operations like hydro plants on different streams, cascaded hydro plants multi-chain hydro plants, and pumped storage hydro plants. This chapter also discusses the solution procedure for long-range generation scheduling of hydrothermal plants.

Chapter 5 provides the necessary background of multiobjective optimization and explains the various methods, namely weighting, ε-constraint, min-max, utility function and global criteria methods. Basic fuzzy set theory is also discussed in this chapter as required for decision making. The chapter elaborates on Surrogate Worth Trade-off approach for multiobjective thermal power dispatch and weighting method for (i) multiobjective thermal power dispatch, (ii) multiobjective thermal power dispatch considering active and reactive power balance, and (iii) multiobjective short-term hydrothermal scheduling.

Chapter 6 deals with multiobjective stochastic optimal power dispatch problems such as (i) multiobjective stochastic optimal thermal power dispatch using the ε-constraint method, (ii) multiobjective stochastic optimal thermal power dispatch using the Surrogate Worth Trade-off method, (iii) multiobjective stochastic optimal thermal power dispatch using the weighting method, (iv) stochastic economic-emission load dispatch, (v) multiobjective thermal power dispatch using the risk/dispersion method, (vi) stochastic multiobjective short-term hydrothermal scheduling, (vii) stochastic multiobjective long-term hydrothermal scheduling, and (viii) multiobjective thermal power dispatch using artificial neural networks (ANNs).

Chapter 7 provides an introduction to evolutionary programming technique for generation scheduling. The basics of genetic algorithms such as coding, genetic operators, random number generation are discussed in this chapter. The step-wise procedure to solve the economic dispatch problem using the genetic algorithm is also presented. Necessary appendices have been provided covering topics such as evaluation of expected values of used functions, evaluation of coefficient of variance of generator output, Kuhn–Tucker theorem, Newton–Raphson, Gauss elimination, Gauss–Seidel methods and Primal-Dual Interior Point method to solve optimization problems.

We are indebted to our colleagues at Giani Zail Singh College of Engineering & Technology, Bathinda, and the Indian Institute of Technology Delhi for their encouragement and various useful suggestions. We express our gratitude to Dr. S.C. Parti, Professor (Retd.), TIET (Thapar Institute of Engineering & Technology), Patiala for his constant interest and support. We hope this book will challenge the readers to delve into an insightful understanding of multiobjective optimization in power systems. We will welcome constructive criticism and appraisal by readers.

D.P. KOTHARI
J.S. DHILLON

CHAPTER 1

Introduction

1.1 A PERSPECTIVE

Electric power today plays an exceedingly important role in the life of the community and in the development of various sectors of economy. In fact, the modern economy is totally dependent on the electricity as a basic input. This in turn has led to the increase in the number of power stations and their capacities and the consequent increase in the power transmission lines which connect the generating stations to the load centres. Interconnections between systems are also on the increase to enhance reliability and economy. The transmission voltage, while dependent on the quantum of power transmitted, should fit in with the long-term system requirement as well as provide flexibility in system operation.

Conventionally, electric energy is obtained by conversion from fossil fuels, namely coal, oil, natural gas, and also from nuclear and hydro sources. Heat energy released by burning fossil fuels or by fission of nuclear material is converted to electricity by first converting it to the mechanical form through a thermo cycle and then converting the mechanical energy through generators to the electrical form. A thermo cycle is basically a low efficiency process—going up to 40% in the case of modern large plants while smaller plants may have considerably lower efficiencies. The earth has fixed non-replenishable resources of fossil fuels and nuclear materials with certain countries over endowed by nature while others are deficient. Hydro energy, though replenishable, is also limited in terms of power. The world's increasing power requirements can only be met by hydro sources.

With the ever increasing per capita energy consumption and exponentially rising population, the technologist already sees the end of the earth's non-replenishable fuel resources. The oil crises of the 1970s dramatically drew attention to this fact. In fact, we can no longer afford the use of oil as a fuel for generation of electricity. In terms of bulk electric energy generation, a distinct shift is taking place across the world in favour of coal and in particular nuclear sources. Also the problems of air and thermal pollution caused by power generation have to be efficiently tackled to avoid ecological disasters. A coordinated worldwide action plan is, therefore, necessary to ensure that energy supply to humanity at large is assured for a long time and at low economic cost.

The ecological and environmental aspects relating to various types of power stations and also transmission lines have to receive full consideration before the long-term power perspective is evolved. Such a study presupposes the availability of a system for investigation of alternative locations of generation, determination of an appropriate order of priorities for their construction and relative costs, identification of demand centres and the possible routing of transmission and interconnection facilities, over a given time frame.

Environmental restrictions or compliance costs can lead to options other than those indicated by traditional engineering approaches. Public safety has never before been so politicized nor have interpretations of what is safe and what is unsafe varied so widely, both within and outside the engineering community. Thus influences external to the electric utility have now come to dominate the system planning activity. Therefore, a broader set of qualifications for the planning engineering has been necessitated for dealing with the expanded aspects of the planning function.

Therefore for meeting the future energy demand much importance needs to be placed on electricity generation from three types of primary energy, namely hydro, coal and nuclear in the best technical, economic and environmental conditions. Of these three sources, even after going for hydroelectric development and expansion of nuclear power capacity, there would have to be a substantial amount of new capacity to be obtained from coal.

1.2 THE COMPONENTS OF A POWER SYSTEM

Electrical engineering is an essential ingredient for the industrial and all-round development of any country. It is coveted form of energy, since it can be generated centrally in bulk and transmitted economically over long distances. Also, it can be adapted easily and efficiently to various applications in both industries and domestic fields. The system which generates, controls, transmits, and finally consumes electrical energy is called an electrical power system. Figure 1.1 shows the structure of a typical power system. Electric energy is produced in generators, transformed to an appropriate voltage level in transformers and then despatched via the buses on the transmission lines for final distribution to the loads. Through tie-lines, the system is connected to neighbouring systems belonging to the same pool (grid). For most system studies, it is sufficient to use lumped or composite type representations of the loads.

The circuit breakers allow the tripping of faulty elements and also sectionalizing of the system. High voltage is now being generated, transformed, transmitted and distributed as three-phase AC power. Collectively all the power system components, namely generators, transformers, buses, lines and loads, form the power network or grid.

The generating plant consists of generating units comprising boiler-turbine-alternator complete with necessary accessories. Exciters and voltage regulators, and step-up transformers also form part of the generating subsystem. Regulating transformers are present in the transmission subsystem to control the active and reactive powers. Static/rotating VAR generators are also used for voltage control.

Electricity cannot be stored economically and the electric utility can exercise little control over the load or power demand at any time. The system, therefore, should be capable of matching the output from the generators to the demand at any time at the specified voltage and frequency. The generating subsystem converts the energy available from the natural sources into the electrical form in a most efficient way.

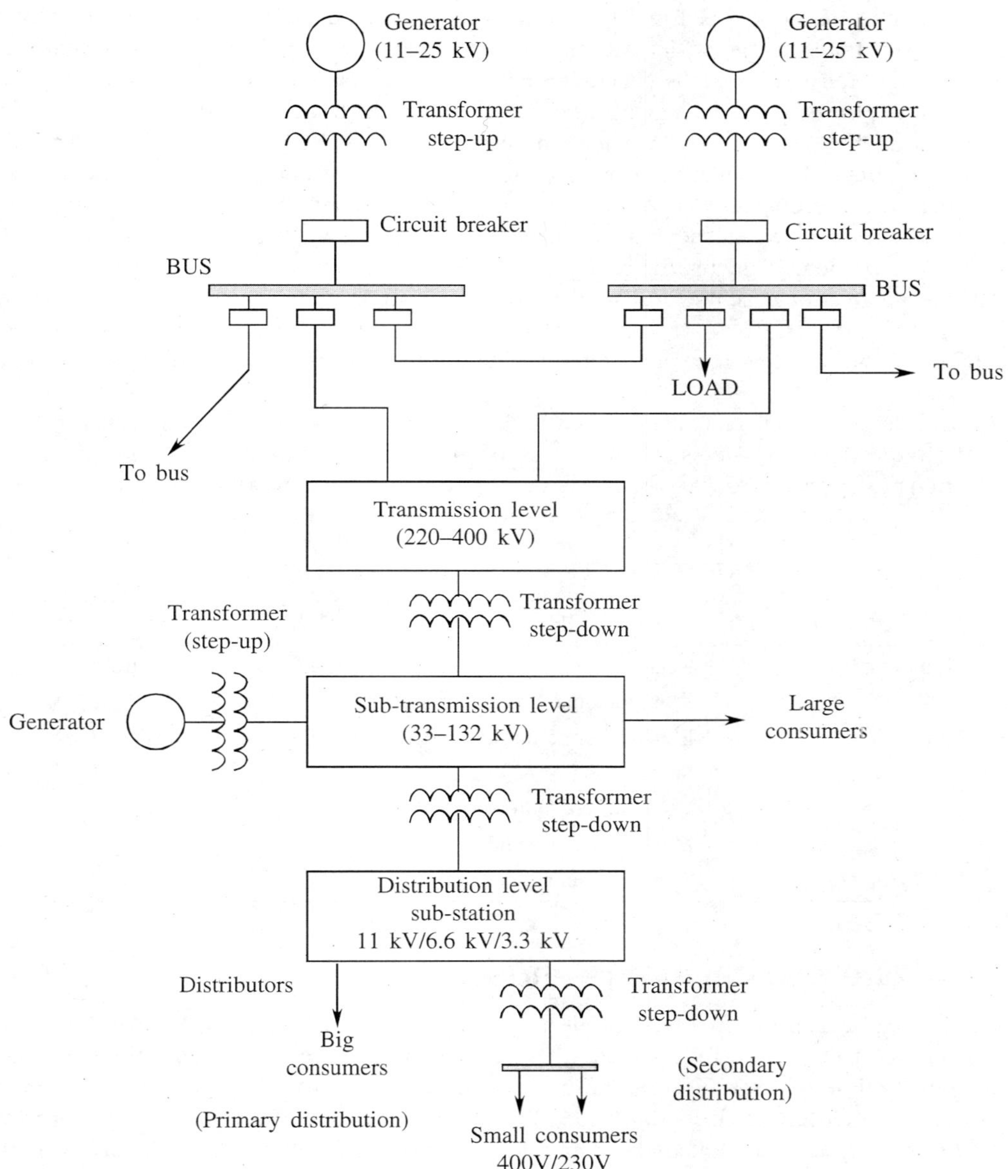

FIGURE 1.1 Typical structure of a power system.

1.3 POWER SYSTEM AND COMPUTERS

The methods first used for solving various power system problems were AC and DC network analyzers developed in early 1930s. AC analyzers were used for load flow and stability studies whereas DC analyzers were preferred for short-circuit studies.

Analog computers developed in 1940s were used in conjunction with AC network analyzers to solve various problems for offline studies. In 1950s many analog devices were developed to control the online functions such as generation control, frequency, and tie-line control.

The 1950s also saw the advent of digital computers which were first used to solve a load flow problem in 1956. Power system studies by computers provided greater flexibility, accuracy, speed, and economy. Till 1970s, there was a wide spread use of computers in system analysis. With the entry of microprocessors in the arena, now besides mainframe computers, mini, micro and personal computers are all increasingly being used to carry out various power system studies and solve power system problems for offline and online applications.

Offline applications include research, routine evaluation of system performance and data assimilation and retrieval. They are mainly used for planning and analyzing some new aspects of the system. Online and real-time applications include data logging and the monitoring of the system state.

A large central computer is used in central load despatch centres for economical and secure control of large integrated systems. Microprocessors and computers installed in generating stations control various local processes such as the starting up of a generator from the cold state. Table 1.1 depicts the time scale of various hierarchical control problems to be solved by computers.

TABLE 1.1 Hierarchical control problems

Time scale	*Control problem*
Millisecond	Relaying and system voltage control and excitation control
2 s–5 min	AGC (automatic generation control)
10 min–few hours	ED (Economic dispatch/MED (Minimum emission dispatch))
–do–	Security analysis
Few hours–1 week	UC (Unit commitment)
1–6 months	Maintenance scheduling
1–10 years	System planning

1.4 PLANNING AND OPERATING PROBLEMS

The operating and expansion strategies of electric utilities have been developed under the premise that all loads must be met in full, as and when they occur and with very high reliability. Since there are few facilities to store energy, the net production of a utility (generation plus the inflow over its ties) must closely track its total load. The phenomenon that affects the ability to perform this tracking span a large time interval—from microseconds, for fast transients, to many years into the future for planning decisions. In order to avoid having to deal with the entire interval all at once, the expansion and operating strategies are organized in a hierarchy as given below.

1.4.1 Resource and Equipment Planning

Generation planning and production costing

The forecasts of the load demand and fuel prices are given and it is required to find the least cost-effective additions to generation capacity to meet the load within the prescribed margins. The horizon of time is a compromise between the need to look past the life span of equipment (about 40 years) and difficulty in forecasting load for more than a decade into the future. The approximate horizon is 20 years ahead.

Long range fuel planning

The generating plants are known a priori. It is required to find the least cost-effective source of fuel and schedule deliveries. Constraints imposed are regulatory policies, i.e. limits on the amount of fuel like oil that can be burnt and environmental policy, i.e. limits on pollutants like sulphur content of coal. It is required to plan the strategy 20 years in advance.

Transmission and distribution planning

The load forecasts and planned generation additions are known well in advance. The planning problem is to find the necessary transmission and distribution additions so that the area load demands can be met economically, reliably and in an environmentally acceptable way. The strategy should be planned 5 to 15 years ahead.

1.4.2 Operation Planning

Maintenance and production scheduling

It should be planned 2 to 5 years ahead, when load forecasts and the equipment to meet the load, inter-utility sales of energy and routine maintenance for the equipment are known. Major equipment need to be taken out of service for periods of one to two weeks for maintenance at intervals of the order of a year. The objective of planning is to maintain the prescribed capacity margins at all times or failing this, to minimize the risk of energy interruption to the customer while minimizing production cost.

Fuel scheduling

Considering the limitations imposed by long-term yearly fuel contracts, the objective is to schedule fuel deliveries and storage to meet plant requirements.

1.4.3 Real-Time Operation

Unit commitment

When load forecasts and generators available for power generation are given, then there is a need to decide when each generator would be started up and shut down as fixed costs are involved in starting and stopping generators. So, the main objective is to minimize the operating cost while having enough capacity online to track the load during changes and cover for random generator failures.

Dispatching

The objective is to schedule the committed generators to meet the load, maintain voltages and frequency within prescribed tolerances and minimize operating cost without unduly stressing the equipment.

Automatic protection

It is required to design protection schemes to minimize damage to equipment and interruptions of service to customers resulting from random failures.

1.5 ARTIFICIAL INTELLIGENCE AND NEURAL NETWORKS

Artificial intelligence (AI) is the study of how to make computers do things which at the moment, people do better. If people are more intelligent than computers and if AI tries to improve the performance of computers in activities that people do better, then the goal of AI is to make computers more intelligent. Researchers in AI have used many different techniques to determine the process used by human beings to produce a particular type of intelligent behaviour, and then to simulate that process on computers. This AI technique is called *modelling* or *simulation*.

Currently, the most well-known area of AI research is the expert system, wherein programs include expert level knowledge of a particular field in order to assist experts in that field. AI research in new areas of power engineering applications such as power system planning, fault diagnosis, protection and monitoring, and control has been conducted in the last decade. An expert system approach has been utilized in the area of scheduling generators online to meet the daily varying load demand. The expert system acts as a preprocessor for a dynamic programming type program for unit commitment. An AI-based algorithm has been exploited to schedule thermal generators in a 24-hour scheduling horizon.

The latest trend in AI is the *resurrection* of neural networks (NN). Although still in an evolutionary stage, these networks have been used in a wide range of real-world applications such as pattern classification, function approximation, automatic control, and optimization. The current interest in the development of ANN is largely due to their brain-like organizational structure and learning ability. These networks have multiple layers of neurons that process information in parallel, and act asynchronously in real time through feedforward and feedback interconnections.

There is a long history of application of neural networks to various power system problems listed below:

- Load forecasting
- Security assessment
- Contingency analysis
- Alarm processing and diagnosis
- Control and observability
- Modelling and identifications

However, pioneering efforts in NN application areas have been for solving power system optimization problems such as multiple criterion decision-making problems. In the book, this area has been further explored.

1.6 FUZZY THEORY IN POWER SYSTEMS

With the penetration of fuzzy set theory into manufacturing and computer products, the application of fuzzy set theory in power systems is beginning to receive attention from power systems researchers. Fuzzy sets were first introduced in solving power systems long-range decision-making problems more than a decade ago. However, substantial interest in its applications to power area is fairly recent.

Analytical solution methods exist for many power systems operation, planning, and control problems. However, the mathematical formulations of real-world problems are derived under certain restrictive assumptions and even with these assumptions, the solution of large-scale power system problems is not trivial. On the other hand, there are many uncertainties in various power systems problems because power systems are large, complex, geographically widely distributed systems and influenced by unexpected events. These facts make it difficult to effectively deal with many power system problems through strict mathematical formulations alone. Therefore, expert system approaches, as one area of artificial intelligence, emerged in recent years in power systems as a complement to mathematical approaches and proved to be effective when properly coupled together.

There are problems in power systems that contain conflicting objectives. In power system operation, economy and security, maximum load supply and minimum generating cost are conflicting objectives. The combination of these objectives by weighting coefficients is the traditional approach for solving this problem. Fuzzy set theory offers a better compromise to obtain solutions which cannot be easily found by weighting methods.

Power system components have physical and operational limits which are usually described as hard inequality constraints in mathematical formulations. The elimination of minor violations of some constraints usually greatly increases the computational burden and decreases the efficiency and may even prevent finding a feasible solution. In practice, certain slight violations of the inequality constraints are permissible. This means that there is not a clear constraint boundary and the constraints can be made soft. Traditionally, this problem has been managed either by modifying the objective function or by modifying the underlying iterative process. The fuzzy set approach inherently incorporates soft constraints and thus simplifies the implementation of such considerations. The application of fuzzy set theory has emerged in more common areas of power systems such as planning, operation, control, and diagnosis and is being widely accepted.

1.7 EVOLUTIONARY ALGORITHMS

Evolutionary algorithms mimic natural evolutionary principles to constitute random search and optimization procedures. Evolutionary algorithms are different from direct search and optimization procedures in a variety of ways. Evolutionary optimization techniques find and maintain multiple solutions in one single simulation run. However, direct search and optimization algorithms use a single solution update during iterations and use a deterministic transition rule. (Figure 1.2 illustrates the classification of optimization techniques).

The concept of a genetic algorithm was first conceived by John Holland of the University of Michigan, Ann Arbor. Genetic algorithms (GA) are search and optimization procedures that are motivated by the principle of natural genetics and natural selection. For higher precision, larger string length is required. The population size requirement is also large for large strings. Thereby the computational complexity of the algorithm increases. Since a fixed coding scheme is applied to code the decision variables. Variable bounds must be such that they bracket the optimum variable values. In many problems, this information is not usually known a priori. Then this may cause some difficulty in using binary-coded GAs in such problems. Furthermore, a careful thinking of the schema processing in binary strings reveals that not all Holland's schemata are equally important in most problems having a continuous search space. To a continuous search space, the meaningful schemata are those that represent the contiguous regions of the search space. Thus, the crossover

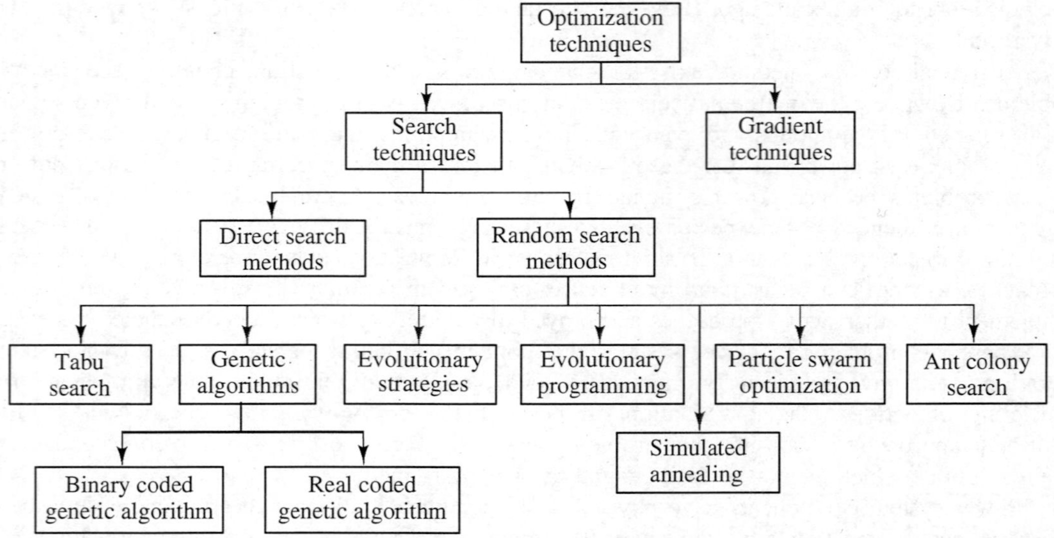

FIGURE 1.2 Classification of optimization techniques.

operator used in the binary coding needs to be redesigned in order to increase the propagation of more meaningful schemata pertaining to a continuous search space.

Real-parameter GA applies crossover and mutation operators (Figure 1.3) directly to real parameter values. Since real parameters are used directly, without any string coding, solving real-parameter optimization problems is a step easier when compared to the binary-coded GAs. Unlike in the binary-coded GAs, decision variables can be directly used to compute the fitness values. Since the selection operator works with the fitness value, any selection operator used with binary-coded GAs can be used in real-parameter GAs. However, the difficulty arises with the search operators. In the binary-coded GAs, decision variables are coded into finite-length strings and exchanging portions of two parent strings is easier to implement and visualize. Simply flipping a bit to perform mutation is also convenient and resembles a natural mutation event. In real-parameter GAs, the main challenge is how to use a pair of real-parameter decision variable vectors to create a new pair of offspring vectors or how to perturb a decision variable vector to a mutated vector in a meaningful manner. The term 'crossover' is not that meaningful and can be best described as blending operators. However, most blending operators in real-parameter GAs are known as crossover operators.

Evolution strategy (ES) is suggested during the early Sixties by P. Bienert, I. Rechenberg and H.-P. Schwefel of Technical University of Berlin. Since the evaluation of a solution in each of these problems was difficult and time-consuming, a simple two-membered ES was used in all of the early studies. However, Schwefel was the first to simulate a different version of the ES on a computer in 1965. Thereafter, multi-membered ESs, recombinative ESs, and-self-adaptive ESs were all suggested. However, the early ES procedure is fundamentally different from binary GAs in mainly two ways:

(i) ESs use real parameter values and
(ii) early ESs do not use any crossover-like operator.

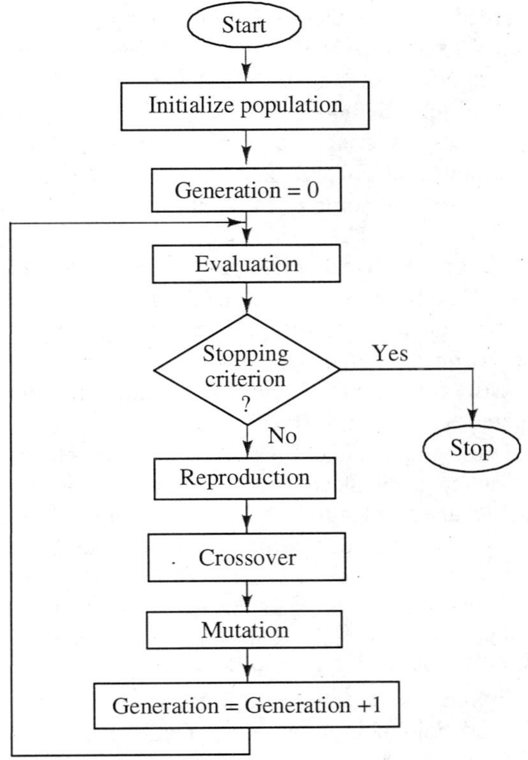

FIGURE 1.3 Genetic algorithm.

However, the working principle of an ES is similar to that of a real-parameter GA used with selection and mutation operators only. ES studies have introduced crossover-like operators.

Evolutionary programming (EP) is a mutation-based evolutionary algorithm applied to discrete search spaces. David Fogel extended the initial work of his father Larry Fogel for real-parameter optimization problems. Real-parameter EP is similar in principle to evolution strategy (ES). Normally distributed mutations are performed in both algorithms. Both algorithms encode mutation strength or variance of the normal distribution for each decision variable. Self-adapting rule is used to update the mutation strengths. EP begins its search with a set of solutions initialized randomly in a given bounded space. Thereafter, EP is allowed to search anywhere in the real space, similar to the real-parameter GAs. Each solution is evaluated to calculate its objective function value.

Although evolutionary programming (EP) was ûrst proposed as an approach to artificial intelligence, it has been recently applied with success to many numerical and combinatorial optimization problems. Optimization by EP can be summarized into two major steps:

(1) mutate the solutions in the current population,
(2) select the next generation from the mutated and the current solutions.

These two steps can be regarded as a population-based version of the classical generate-and-test method, where mutation is used to generate new solutions (offspring) and selection is used to test which of the newly generated solutions should survive to the next generation. Formulating EP as a special case of the generate-and-test method establishes a bridge between EP and other search algorithms, such as evolution strategies, genetic algorithms, simulated annealing (SA), tabu search (TS), and others, and thus facilitates cross-fertilization among different research areas.

One disadvantage of EP in solving some of the multimodal optimization problems is its slow convergence to a good near-optimum. The generate-and-test formulation of EP indicates that mutation is a key search operator which generates new solutions from the current ones. A new mutation operator based on Cauchy random numbers is proposed and tested on a set of 23 functions. The new EP with Cauchy mutation significantly outperforms the classical EP (CEP), which uses Gaussian mutation, on a number of multimodal functions with many local minima while being comparable to CEP for unimodal and multimodal functions with only a few local minima. The new EP is denoted as "fast EP" (FEP).

The process to heat up a metal and to cool it down slowly is referred to as annealing. On the other hand, the process to heat up a metal and cool it down fast is referred to as quenching or hardening. Since according to the molecule's behaviour, internal energy is large when its temperature is satisfactorily high and state or internal energy of metal is determined stochastically. From high temperature state, if temperature is cooled down slowly, its internal thermal energy decreases slowly. Since a cooling process of metal is ruled by stochastic thermal dynamics, the final state of molecules is determined randomly according to the behaviour of the molecules or its cooling speed. By simulating a property of annealing process in which the smallest internal energy can be reached to find the minimum of an objective function.

Tabu search is a restricted neighbourhood search technique, and is an iteration algorithm. The fundamental idea of TS is the use of flexible memory of search history which thus guides the search process to surmount local optimal solutions. The basic components of the Tabu search are the moves, Tabu list and aspiration level.

Ant colony search algorithms, to some extent, mimic the behaviour of real ants. As is well known, real ants are capable of finding the shortest path from food sources to the nest without using visual cues. They are also capable of adapting to changes in the environment, for example, finding a new shortest path once the old one is no longer feasible due to a new obstacle. The studies by ethnologists reveal that such capabilities ants have are essentially due to what is called "pheromone trails" which ants use to communicate information among individuals regarding path and to decide where to go. Ants deposit a certain amount of pheromone while walking, and each ant probabilistically prefers to follow a direction rich in pheromone rather than a poorer one.

In 1995, Kennedy and Eberhart first introduced the Particle Swarm Optimization (PSO) method, motivated by social behaviour of organisms such as fish schooling and bird flocking. PSO, as an optimization tool, provides a population-based search procedure in which individuals called particles change their positions (states) with time. In a PSO system, particles fly around in a multidimensional search space. During flight, each particle adjusts its position according to its own experience, and the experience of neighbouring particles, making use of the best position encountered by itself and its neighbours. The swarm direction of a particle is defined by the set of particles neighbouring the particle and its history experience.

REFERENCES

Books

Deb, K., *Multiobjective Optimization using Evolutionary Algorithm*, John Wiley & Sons, New York, 2001.

Goldberg, D.E., *Genetic Algorithm in Search, Optimization and Machine Learning*, Addison-Wesley, 1989.

Kothari, D.P. and I.J. *Nagrath, Modern Power System Analysis*, 3rd ed., Tata McGraw-Hill, New Delhi, 2003.

Nagrath, I.J. and D.P. Kothari, *Power System Engineering*, Tata McGraw-Hill, New Delhi, 1994.

Papers

Carpentier, J.L., Optimal power flow: Uses, methods and developments, *Proceedings of IFAC Conference*, RJ Brazil, pp. 11–25, 1985.

Chowdhury B.H. and S. Rahman, A review of recent advances in economic dispatch, *IEEE Trans. on Power Systems*, Vol. **PWRS-5 (4)**, pp. 1248–1259, 1990.

Happ, H.H., Optimal power dispatch: A comprehensive survey, *IEEE Trans. on Power Apparatus and Systems*, Vol. **PAS-96 (3)**, pp. 841–854, 1977.

Huneault, M. and F.D. Galiana, A survey of the optimal power flow literature, *IEEE Trans. on Power Systems*, Vol. **PWRS-6 (2)**, pp. 762–770, 1991.

IEEE working group, Description and bibliography of major economic security function, Part II and III—Bibliography (1959–1972 and 1973–1979), *IEEE Trans. on Power Apparatus and Systems*, Vol. **PAS-100 (1)**, pp. 215–235, 1981.

Sasson, A.M. and H.M. Merrill, Some applications of optimization techniques to power system problems, *Proc. IEEE*, Vol. **62 (7)**, pp. 959–972, 1974.

Talukdar, S.N. and F.F. Wu, Computer-aided dispatch for electric power systems, *Proceedings of IEEE*, Vol. **69 (10)**, pp. 1212–1231, 1981.

CHAPTER 2

Load Flow Studies

2.1 INTRODUCTION

Electrical transmission systems operate in their steady-state mode under normal conditions. Three major problems encountered in steady-state mode of operations are listed below in their hierarchical order of importance:

1. Load flow problem
2. Optimal load dispatch problem
3. Systems control problem

The computational procedure required to determine the steady-state operating characteristics of a power system network is termed *load flow* (or *power flow*). The aim of power flow calculations is to determine the steady-state operating characteristics of a power generation/transmission system for a given set of bus bar loads. Active power generations are specified according to economic dispatching. The magnitude of generation voltage is maintained at the specified level by an automatic voltage regulator acting on the machine excitation. Loads are specified by their constant active and reactive power requirements. The loads are assumed to be unaffected by the small variations of voltage and frequency expected during normal steady-state operation.

The direct analysis of the network is not possible, as the loads are given in terms of complex powers rather than impedances. The generators behave more like power sources than voltage sources. The main information obtained from the load flow study consists of:

- Magnitudes and phase angles of load bus voltages
- Reactive powers and voltage phase angles at generator buses
- Real and reactive power flow on transmission lines
- Power at the reference bus

This information is essential for the continuous monitoring of the current state of the system. The information is also important for analyzing the effectiveness of the alternative plans for the future, such as adding new generator sites, meeting increased load demand and locating new transmission sites.

The single-line diagram of a power system having four buses is shown in Figure 2.1. In the power system, the variables defined on each bus are:

- Complex powers supplied by generators S_{g_1} and S_{g_2}
- Complex powers drawn by loads, S_{d_1}, S_{d_3} and S_{d_4}
- Complex voltages, V_1, V_2, V_3 and V_4

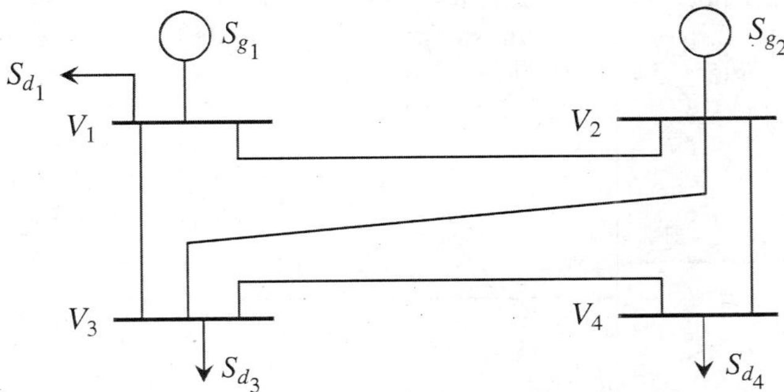

FIGURE 2.1 Single-line diagram of a four-bus system.

There results a net injection of power into the transmission system. The transmission system may be a primary transmission system or sub-transmission system. The primary transmission system transmits bulk power from the generators to the bulk power substations. The sub-transmission system transmits power from the substations or some old generators to the distribution substations. The transmission system has to be designed in such a manner that the power system operation is reliable and economic and no difficulties are encountered in its operation. The difficulties involved, however, are:

- One or more transmission lines becoming overloaded
- Generator(s) becoming overloaded
- The stability margins for a transmission link being too small

There may be emergencies, such as:

- The loss of one or more transmission links
- Shutdown of some generators which gives rise to overloading of other generators and transmission links.

In system operation and planning, the bus voltages and powers are kept within certain limits. The power system networks of today are highly complicated consisting of hundreds of buses and transmission links. Thus, the load flow study involves extensive calculations. With the advent of fast digital computers with huge memory, all kinds of power system studies including the load flow study can now be carried out conveniently. The type of solution for a load flow also determines the method used which can be:

- Accurate or approximate
- Unadjusted or adjusted
- Offline or online
- Single case or multiple cases

2.2 NETWORK MODEL FORMULATION

A real-life power system comprises a large number of buses for a load flow study. These buses are interconnected by means of transmission lines. Power is injected into a bus from generators, while the loads are tapped from it. There may be some buses with no generation facilities and some buses may have VAR generators attached to them. The surplus power at some of the buses is transported via transmission lines to buses deficit in power. The single-line diagram of a three-bus system is shown in Figure 2.2. Normally a transmission line is modelled by a nominal-Π, while the line resistance is always neglected in load flow analysis.

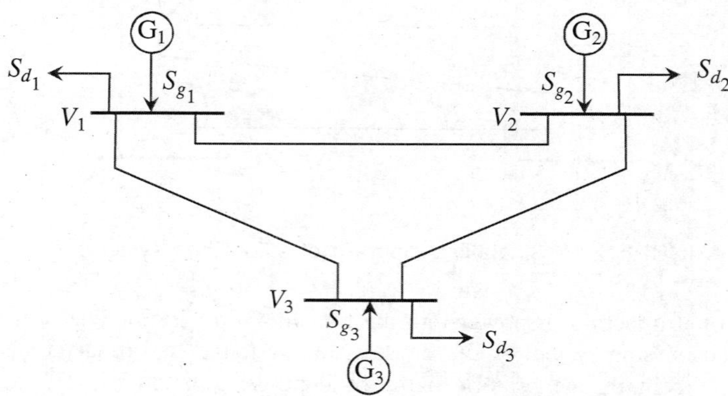

FIGURE 2.2 Single-line diagram of a three-bus system.

The loads are considered negative generators and lump together the generator and load power at the buses. The power at the ith bus injected into the transmission system is called the *bus power* and is defined as

$$S_i = S_{g_i} - S_{d_i} \tag{2.1}$$

where

the complex power supplied by generators is, $S_{g_i} = P_{g_i} + jQ_{g_i}$
the complex power drawn by loads is, $S_{d_i} = P_{d_i} + jQ_{d_i}$
the complex power injected is, $S_i = P_i + jQ_i$

Equation (2.1) can be rewritten as

$$P_i + jQ_i = (P_{g_i} + jQ_{g_i}) - (P_{d_i} + jQ_{d_i})$$

or

$$P_i + jQ_i = (P_{g_i} - P_{d_i}) + j(Q_{g_i} - Q_{d_i})$$

The real and reactive powers injected into the ith bus are

$$P_i = P_{g_i} - P_{d_i} \tag{2.2a}$$

$$Q_i = Q_{g_i} - Q_{d_i} \tag{2.2b}$$

So the 'bus current' at the ith bus is defined as

$$I_i = I_{g_i} - I_{d_i} \tag{2.3}$$

The equivalent power source at the ith bus injects current I_i into the bus. All the sources are always connected to a common ground node. The transmission lines are replaced by their nominal-Π equivalent (Figure 2.3). The line admittance between the nodes i and k is depicted by y_{ik} and $y_{ik} = y_{ki}$. The mutual admittances between the lines is assumed to be zero. Applying the Kirchhoff's current law (KCL) at nodes 1, 2, and 3, respectively (Figure 2.4):

$$I_1 = y_{10}V_1 + y_{12}(V_1 - V_2) + y_{13}(V_1 - V_3)$$

$$I_2 = y_{20}V_2 + y_{12}(V_2 - V_1) + y_{23}(V_2 - V_3)$$

$$I_3 = y_{30}V_3 + y_{13}(V_3 - V_1) + y_{23}(V_3 - V_2)$$

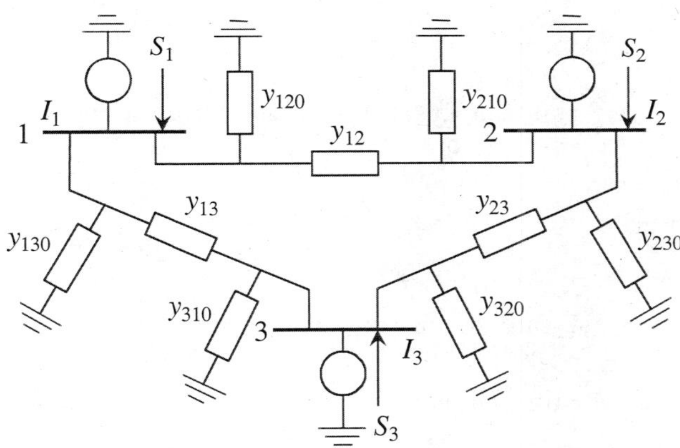

FIGURE 2.3 Equivalent circuit of the power system of Figure 2.2.

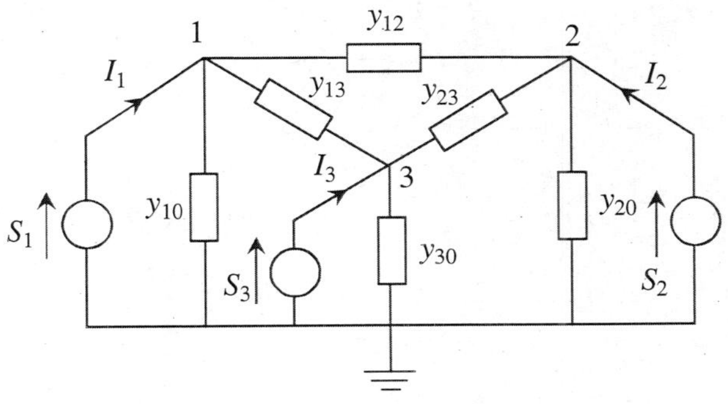

$y_{10} = y_{120} + y_{130}$, $\quad y_{20} = y_{210} + y_{230}$ and $\quad y_{30} = y_{310} + y_{320}$

FIGURE 2.4 Reduced circuit diagram of Figure 2.3.

Rearranging the preceding equations,

$$I_1 = (y_{10} + y_{12} + y_{13})V_1 - y_{12}V_2 - y_{13}V_3 \tag{2.4}$$

$$I_2 = -y_{12}V_1 + (y_{20} + y_{12} + y_{23})V_2 - y_{23}V_3 \tag{2.5}$$

$$I_3 = -y_{13}V_1 - y_{23}V_2 + (y_{30} + y_{13} + y_{23})V_3 \tag{2.6}$$

Writing in the standard matrix form,

$$\begin{bmatrix} I_1 \\ I_2 \\ I_3 \end{bmatrix} = \begin{bmatrix} Y_{11} & Y_{12} & Y_{13} \\ Y_{21} & Y_{22} & Y_{23} \\ Y_{31} & Y_{32} & Y_{33} \end{bmatrix} \begin{bmatrix} V_1 \\ V_2 \\ V_3 \end{bmatrix} \tag{2.7}$$

The diagonal elements of the Y matrix are self-admittances,

$$Y_{ii} = \frac{I_i}{V_i} \quad \text{(all } V = 0 \text{ except } V_i\text{)}$$

= short-circuit driving point admittance or self-admittance at the ith bus

$$Y_{11} = y_{10} + y_{12} + y_{13}$$

$$Y_{22} = y_{20} + y_{12} + y_{23}$$

$$Y_{33} = y_{30} + y_{13} + y_{23}$$

The off-diagonal elements of Y_{BUS} are the transfer admittances,

$$Y_{ik}(i \neq k) = \frac{I_i}{V_k} \quad \text{(all } V = 0 \text{ except } V_k\text{)}$$

= short-circuit transfer admittance between the ith and kth buses.

$$Y_{12} = Y_{21} = -y_{12}$$

$$Y_{13} = Y_{31} = -y_{13}$$

$$Y_{23} = Y_{32} = -y_{23}$$

Using the index notation, Eq. (2.7) can be rewritten as

$$I_i = \sum_{k=1}^{NB} Y_{ik}V_k; \quad i = 1, 2, ..., \text{NB (number of buses)} \tag{2.8}$$

In matrix notation

$$I_{BUS} = Y_{BUS}V_{BUS} \tag{2.9}$$

where

I_{BUS} is NB × 1 column vector of bus currents
V_{BUS} is NB × 1 column vector of bus voltages
Y_{BUS} is NB × NB matrix of bus admittance matrix.

It follows that:

- Y_{BUS} is a symmetric matrix $[Y_{ik} = Y_{ki}(k \neq i)]$, if there is no regulating/phase transformer. So only NB(NB + 1)/2 terms are to be stored for the NB bus system.
- $Y_{ik}(i \neq k) = 0$ if ith and kth buses are not connected.

Since in a power network each bus is connected only to a few other buses, the Y_{BUS} of a large network is very sparse, i.e. it has a large number of zero elements. The sparsity feature reduces numerical computations in load flow studies and minimizes the memory requirement, as only nonzero elements are stored.

Equation (2.9) can also be written in the form,

$$V_{BUS} = Z_{BUS} I_{BUS} \tag{2.10}$$

where $Z_{BUS} = Y_{BUS}^{-1}$.

Z_{BUS} is known as the Bus Impedance Matrix. It may be noted that:

- Size of Z_{BUS} is (NB × NB)
- The diagonal elements are the short-circuit driving point impedances and the off-diagonal elements are the short-circuit transfer impedances.
- Symmetric Y_{BUS} yields symmetrical Z_{BUS}.
- Z_{BUS} is a full matrix, i.e. zero elements in Y_{BUS} become nonzero elements in the corresponding Z_{BUS}.

Y_{BUS} is used to solve load flow problems. It has gained widespread application owing to its simplicity in data preparation and the ease with which it can be formed and modified for network changes.

2.3 Y_{BUS} FORMULATION

2.3.1 No Mutual Coupling between Transmission Lines

Initially all the elements of Y_{BUS} are set to zero. Addition of admittance y between buses i and j affects four entries in Y_{BUS}, namely Y_{ii}, Y_{ij}, Y_{ji}, and Y_{jj} as follows:

$$Y_{ii}^{new} = Y_{ii}^{old} + y \tag{2.11a}$$

$$Y_{ij}^{new} = Y_{ij}^{old} - y \tag{2.11b}$$

$$Y_{ji}^{new} = Y_{ji}^{old} - y \tag{2.11c}$$

$$Y_{jj}^{new} = Y_{jj}^{old} + y \tag{2.11d}$$

Addition of an element of admittance y from bus to ground will only affect,

$$Y_{ii}^{new} = Y_{ii}^{old} + y \tag{2.12}$$

The detailed algorithm is outlined next:

Algorithm 2.1: To Build Y_{BUS}

1. Read NL (number of lines); NB (number of buses).
2. Initialize $Y_{ij}^{0} = 0$ ($i = 1, 2, ..., \text{NB}; j = 1, 2, ..., \text{NB}$).
3. Set the line number counter $i = 1$.
4. Read the admittance y between buses j to k and shunt admittance y^{sh} between the bus and reference.
   SB_i— stores the index 'i' of the jth bus which is linked to the kth bus.
   EB_i— stores the index 'i' of the kth bus which is linked to the jth bus.
   y_i— admittance between the jth and kth buses.
   y_i^{sh} — admittance between the jth and reference buses.
5. Assign the values
   $l = \text{SB}_i$
   $m = \text{EB}_i$

   $$Y_{ll}^{\text{new}} = Y_{ll}^{\text{old}} + y_i + y_i^{\text{sh}}$$

   $$Y_{mm}^{\text{new}} = Y_{mm}^{\text{old}} + y_i + y_i^{\text{sh}}$$

   $$Y_{lm}^{\text{new}} = Y_{lm}^{\text{old}} - y_i$$

   $$Y_{ml}^{\text{new}} = Y_{ml}^{\text{old}} - y_i$$

6. Check $i \geq \text{NL}$
   if 'yes' then GOTO Step 7

   else $i = i + 1$ and GOTO Step 4 and repeat.
7. Write the matrix and stop.

2.3.2 Mutual Coupling between Transmission Lines

The equivalent circuit of mutually coupled transmission lines is shown in Figure 2.5. Shunt elements are omitted for simplicity. The mutual impedance between the transmission lines is z_{m} and the series impedances are z_{s1} and z_{s2}.

From Figure 2.5

$$V_i = z_{\text{s1}} I_i + z_{\text{m}} I_k + V_j$$

$$V_k = z_{\text{s2}} I_k + z_{\text{m}} I_i + V_n$$

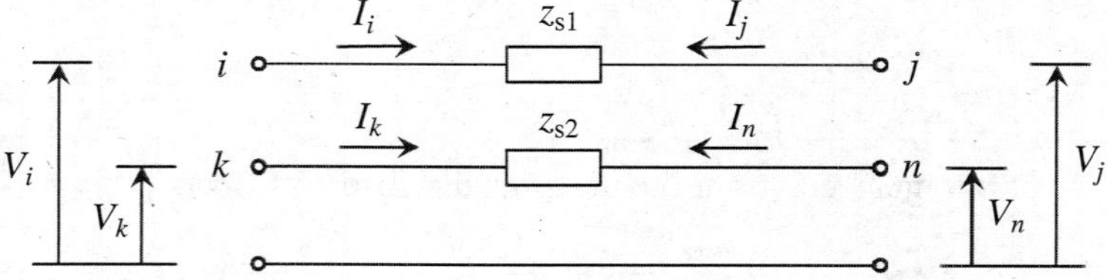

FIGURE 2.5 Mutually coupled transmission lines.

or

$$\begin{bmatrix} V_i \\ V_k \end{bmatrix} - \begin{bmatrix} V_j \\ V_n \end{bmatrix} = \begin{bmatrix} z_{s1} & z_m \\ z_m & z_{s2} \end{bmatrix} \begin{bmatrix} I_i \\ I_k \end{bmatrix} \tag{2.13}$$

or

$$\begin{bmatrix} I_i \\ I_k \end{bmatrix} = \begin{bmatrix} y_{s1} & y_m \\ y_m & y_{s2} \end{bmatrix} \begin{bmatrix} V_i - V_j \\ V_k - V_n \end{bmatrix} \tag{2.14}$$

where

$$\begin{bmatrix} y_{s1} & y_m \\ y_m & y_{s2} \end{bmatrix} = \begin{bmatrix} z_{s1} & z_m \\ z_m & z_{s2} \end{bmatrix}^{-1}$$

Similarly,

$$\begin{bmatrix} I_j \\ I_n \end{bmatrix} = \begin{bmatrix} y_{s1} & y_m \\ y_m & y_{s2} \end{bmatrix} \begin{bmatrix} V_j - V_i \\ V_n - V_k \end{bmatrix} \tag{2.15}$$

From Eqs. (2.14) and (2.15), the elements of Y_{BUS} become

$$Y_{ii}^{new} = Y_{ii}^{old} + y_{s1} \tag{2.16a}$$

$$Y_{jj}^{new} = Y_{jj}^{old} + y_{s1} \tag{2.16b}$$

$$Y_{kk}^{new} = Y_{kk}^{old} + y_{s2} \tag{2.16c}$$

$$Y_{nn}^{new} = Y_{nn}^{old} + y_{s2} \tag{2.16d}$$

$$Y_{ij}^{new} = Y_{ji}^{new} = Y_{ij}^{old} - y_{s1} \tag{2.17a}$$

$$Y_{kn}^{new} = Y_{nk}^{new} = Y_{kn}^{old} - y_{s2} \tag{2.17b}$$

$$Y_{ik}^{new} = Y_{ki}^{new} = Y_{ik}^{old} + y_m \tag{2.18a}$$

$$Y_{jn}^{new} = Y_{nj}^{new} = Y_{jn}^{old} + y_m \tag{2.18b}$$

$$Y_{in}^{new} = Y_{ni}^{new} = Y_{in}^{old} - y_m \tag{2.19a}$$

$$Y_{jk}^{new} = Y_{kj}^{new} = Y_{jk}^{old} - y_m \tag{2.19b}$$

EXAMPLE 2.1 Obtain Y_{BUS} for the 4-bus sample system given in Figure 2.6. The series impedances and the shunt admittances of each line are given in Table 2.1.

FIGURE 2.6 Sample network.

TABLE 2.1 Line impedances for sample network of Figure 2.6

Line no.	*Bus code (p–q)*	*Shunt Admittance (y)*	*Series Impedance (z)*
1	1–2	$j0.0$	$0.08 + j0.20$
2	1–4	$j0.0$	$0.05 + j0.10$
3	2–3	$j0.0$	$0.04 + j0.12$
4	3–4	$j0.0$	$0.04 + j0.14$

Solution Y_{BUS} elements are initialized to zero. The number of buses are 4, so the size of the matrix is 4 × 4.

$$\left.\begin{aligned} Y_{11} = 0.0,\ Y_{12} = 0.0,\ Y_{13} = 0.0,\ Y_{14} = 0.0 \\ Y_{21} = 0.0,\ Y_{22} = 0.0,\ Y_{23} = 0.0,\ Y_{24} = 0.0 \\ Y_{31} = 0.0,\ Y_{32} = 0.0,\ Y_{33} = 0.0,\ Y_{34} = 0.0 \\ Y_{41} = 0.0,\ Y_{42} = 0.0,\ Y_{43} = 0.0,\ Y_{44} = 0.0 \end{aligned}\right\} \quad Y_{ij} = 0\ (i = 1, 2, 3, 4;\ j = 1, 2, 3, 4)$$

Branch y_1 is added between the bus numbers 1 and 2 and the elements of Y_{BUS} are modified as given below:

$$\begin{aligned} y_1 &= 1/z_1 = 1.724138 - j4.310345 \\ Y_{11} &= Y_{11} + y_1 = \ \ 1.724138 - j4.310345 \\ Y_{22} &= Y_{22} + y_1 = \ \ 1.724138 - j4.310345 \\ Y_{12} &= Y_{12} - y_1 = -1.724138 + j4.310345 \\ Y_{21} &= Y_{21} - y_1 = -1.724138 + j4.310345 \end{aligned}$$

Branch y_2 is added between the bus numbers 1 and 4 and the elements of Y_{BUS} are modified as given below:

$$\begin{aligned} y_2 &= 1/z_2 = 4.0 - j8.0 \\ Y_{11} &= Y_{11} + y_2 = 5.724138 - j12.310345 \\ Y_{44} &= Y_{44} + y_2 = \ \ 4.0 - j8.0 \\ Y_{14} &= Y_{14} - y_2 = -4.0 + j8.0 \\ Y_{41} &= Y_{41} - y_2 = -4.0 + j8.0 \end{aligned}$$

Branch y_3 is added between the bus numbers 2 and 3 and the elements of Y_{BUS} are modified as given below:

$$y_3 = 1/z_3 = 2.5 - j7.5$$

$$Y_{22} = Y_{22} + y_3 = 4.224138 - j11.810345$$

$$Y_{33} = Y_{33} + y_3 = 2.5 - j7.5$$

$$Y_{23} = Y_{23} - y_3 = -2.5 + j7.5$$

$$Y_{32} = Y_{32} - y_3 = -2.5 + j7.5$$

Branch y_4 is added between the bus numbers 3 and 4 and the elements of Y_{BUS} are modified as given below:

$$y_4 = 1/z_4 = 1.886792 - j6.603774$$

$$Y_{33} = Y_{33} + y_4 = 4.386792 - j14.103774$$

$$Y_{44} = Y_{44} + y_4 = 5.886792 - j14.603774$$

$$Y_{34} = Y_{34} - y_4 = -1.886792 + j6.603774$$

$$Y_{43} = Y_{43} - y_4 = -1.886792 + j6.603774$$

The elements of Y_{BUS} are obtained as

$$Y_{11} = 5.724138 - j12.310345$$

$$Y_{12} = -1.724138 + j4.310345$$

$$Y_{13} = 0.0 + j0.0$$

$$Y_{14} = -4.0 + j8.0$$

$$Y_{21} = -1.724138 + j4.310345$$

$$Y_{22} = 4.224138 - j11.810345$$

$$Y_{23} = -2.5 + j7.5$$

$$Y_{24} = 0.0 + j0.0$$

$$Y_{31} = 0.0 + j0.0$$

$$Y_{32} = -2.5 + j7.5$$

$$Y_{33} = 4.386792 - j14.103774$$

$$Y_{34} = -1.886792 + j6.603774$$

$$Y_{41} = -4.0 + j8.0$$

$$Y_{42} = 0.0 + j0.0$$

$$Y_{43} = -1.886792 + j6.603774$$

$$Y_{44} = 5.886792 - j14.603774$$

2.4 NODE ELIMINATION IN Z_{BUS}

Equation (2.10) is considered here

$$V_{BUS} = Z_{BUS}\, I_{BUS}$$

Equation (2.10) can be written in the partitioned form as

$$\begin{bmatrix} V_{\text{BUS}} \\ e_L \end{bmatrix} = \begin{bmatrix} Z_{\text{BUS}} & Z_{iL} \\ Z_{Lj} & Z_{LL} \end{bmatrix} \begin{bmatrix} I_{\text{BUS}} \\ I_L \end{bmatrix} \tag{2.20}$$

where

e_L is the voltage element to be eliminated
I_L is the current element to be eliminated
V_{BUS} is the (NB × 1) vector of voltages to be retained
I_{BUS} is the (NB × 1) vector of currents to be retained
Z_{BUS} is the (NB × NB) sub-matrix of impedances to be retained
Z_{iL} is the off-diagonal (NB × 1) sub-matrix
Z_{Lj} is the transpose on Z_{iL}
Z_{LL} is the element of the impedance matrix to be eliminated.

The matrices are arranged in such a way that the nodes to be eliminated are in the bottom rows, i.e.

$$E_i = \sum_{j=1}^{\text{NB}} Z_{ij} I_j + Z_{iL} I_L \qquad (i = 1, 2, ..., \text{NB}) \tag{2.21}$$

$$e_L = \sum_{j=1}^{\text{NB}} Z_{Lj} I_j + Z_{LL} I_L \tag{2.22}$$

To eliminate bus L, short-circuit voltage source ($e_L = 0$). Therefore, from Eq. (2.22),

$$e_L = \sum_{j=1}^{\text{NB}} Z_{Lj} I_j + Z_{LL} I_L = 0$$

or

$$I_L = -\sum_{j=1}^{\text{NB}} \frac{Z_{Lj}}{Z_{LL}} I_j \tag{2.23}$$

Substituting Eq. (2.23) into Eq. (2.21), the following is obtained:

$$E_i = \sum_{j=1}^{\text{NB}} Z_{ij} I_j - \sum_{j=1}^{\text{NB}} \frac{Z_{iL} Z_{Lj}}{Z_{LL}} I_j$$

$$= \sum_{j=1}^{\text{NB}} \left[Z_{ij} - \frac{Z_{iL} Z_{Lj}}{Z_{LL}} \right] I_j \tag{2.24}$$

The modified elements can be represented as

$$Z_{ij}^{\text{new}} = Z_{ij}^{\text{old}} - \frac{Z_{iL} Z_{Lj}}{Z_{LL}} \qquad (i = 1, 2, ..., \text{NB}; j = 1, 2, ..., \text{NB}) \tag{2.25}$$

2.5 Z_{BUS} FORMULATION

Equation (2.10) can be rewritten as

$$I_{BUS} = Y_{BUS}V_{BUS}$$

or

$$V_{BUS} = Y_{BUS}^{-1}I_{BUS} = Z_{BUS}I_{BUS} \tag{2.26}$$

or

$$Z_{BUS} = Y_{BUS}^{-1}$$

The sparsity of Y_{BUS} may be retained by using an efficient inversion technique [Tinney and Walker, 1967], and the nodal impedance matrix can then be calculated directly from the factorized admittance matrix. Equation (2.26) can be written in expanded form as

$$V_i = \sum_{k=1}^{NB} Z_{ik}I_k \tag{2.27}$$

It follows from Eq. (2.27) that

$$Z_{ik} = \left.\frac{V_i}{I_k}\right|_{\substack{I_1 = I_2 = \dots = I_{NB} = 0 \\ I_k \neq 0}} \tag{2.28}$$

Also, $Z_{ik} = Z_{ki}$ (Z_{BUS} is a symmetrical matrix).

As per Eq. (2.28) if a unit current is injected at bus (node) k, while the other buses are kept open-circuited, the bus voltages yield the values of the kth column of Z_{BUS}. The Z_{BUS} algorithm is a step-by-step programmable technique which proceeds branch by branch. It has the advantage that any modification of the network does not require complete rebuilding of Z_{BUS}.

2.5.1 No Mutual Coupling between Transmission Lines

Consider that Z_{BUS} has been formulated up to a certain stage and another branch is now added. Upon adding a new branch, one of the following situations is presented.

1. Z_b is added from a new bus to the reference bus. A new branch is added and the dimension of Z_{BUS} goes up by one. This is considered as type-1 modification.
2. Z_b is added from a new bus to an old bus. A new branch is added and the dimension of Z_{BUS} goes up by one. This is considered as type-2 modification.
3. Z_b is added from an old bus to the reference bus. A new loop is formed but the dimension of Z_{BUS} does not change. This is considered as type-3 modification.
4. Z_b connects two old buses. A new loop is formed but the dimension of Z_{BUS} does not change. This is considered as type-4 modification.

Type-1 modification

Figure 2.7 shows a passive (linear) 2-bus network.

$$\begin{bmatrix} V_1 \\ V_2 \end{bmatrix} = \begin{bmatrix} Z_{11} & Z_{12} \\ Z_{21} & Z_{22} \end{bmatrix} \begin{bmatrix} I_1 \\ I_2 \end{bmatrix} \tag{2.29}$$

A branch with impedance Z_b is added between the new bus 3 and the reference bus.

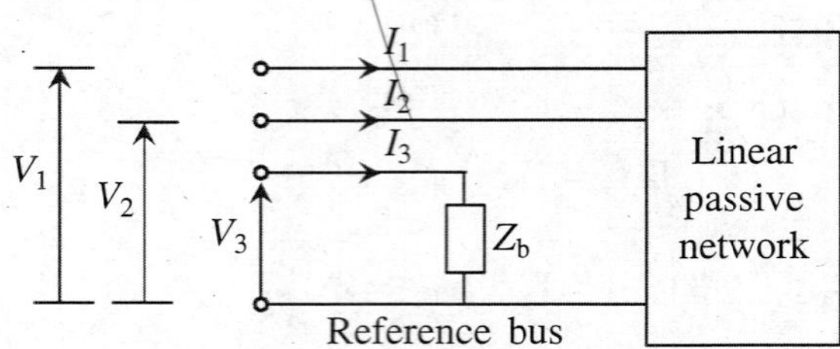

FIGURE 2.7 Type-1 modification.

Now,

$$V_3 = Z_b I_3$$

Equation (2.29) can be modified as

$$\begin{bmatrix} V_1 \\ V_2 \\ V_3 \end{bmatrix} = \begin{bmatrix} Z_{11} & Z_{12} & Z_{13} \\ Z_{21} & Z_{22} & Z_{23} \\ Z_{31} & Z_{32} & Z_{33} \end{bmatrix} \begin{bmatrix} I_1 \\ I_2 \\ I_3 \end{bmatrix} \tag{2.30}$$

where $Z_{31} = Z_{32} = Z_{13} = Z_{23} = 0$ and $Z_{33} = Z_b$. In generalized form, it is rewritten for the added branch with impedance Z_b between the new bus k and the reference bus. Thus,

$$Z_{ki} = 0 \qquad (i = 1, 2, ..., n; \quad i \neq k) \tag{2.31}$$

$$Z_{ik} = Z_{ki}$$

$$Z_{kk} = Z_b \tag{2.32}$$

where n is size of the previous Z_{BUS}.

Type-2 modification

Z_b is added from the new bus 3 to the old bus 2. It follows from Figure 2.8 that

$$V_3 = V_2 + Z_b I_3 \tag{2.33}$$

$$V_1 = Z_{11} I_1 + Z_{12}(I_2 + I_3) \tag{2.34}$$

$$V_2 = Z_{21} I_1 + Z_{22}(I_2 + I_3) \tag{2.35}$$

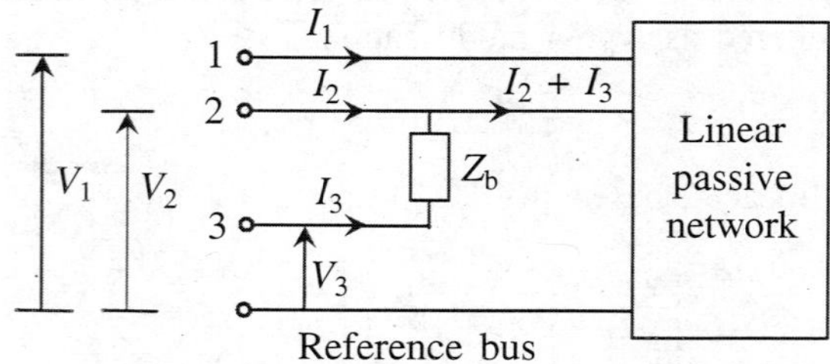

FIGURE 2.8 Type-2 modification.

In matrix form,

$$\begin{bmatrix} V_1 \\ V_2 \end{bmatrix} = \begin{bmatrix} Z_{11} & Z_{12} \\ Z_{21} & Z_{22} \end{bmatrix} \begin{bmatrix} I_1 \\ I_2 + I_3 \end{bmatrix} \tag{2.36}$$

Substituting Eq. (2.35) into Eq. (2.33), we have

$$V_3 = Z_{21}I_1 + Z_{22}(I_2 + I_3) + Z_b I_3$$

Rearranging the above equation, we obtain

$$V_3 = Z_{21}I_1 + Z_{22}I_2 + (Z_{22} + Z_b)I_3 \tag{2.37}$$

Equations (2.34), (2.35) and (2.37) give the new matrix

$$\begin{bmatrix} V_1 \\ V_2 \\ V_3 \end{bmatrix} = \begin{bmatrix} Z_{11} & Z_{12} & Z_{13} \\ Z_{21} & Z_{22} & Z_{23} \\ Z_{31} & Z_{32} & Z_{33} \end{bmatrix} \begin{bmatrix} I_1 \\ I_2 \\ I_3 \end{bmatrix} \tag{2.38}$$

where $Z_{13} = Z_{12}$, $Z_{31} = Z_{21}$, $Z_{23} = Z_{22}$, $Z_{32} = Z_{22}$, $Z_{33} = Z_{22} + Z_b$

In generalized form the above equation can be rewritten for the added branch with impedance Z_b between the new bus k and the old bus j. Thus,

$$Z_{ki} = Z_{ji} \qquad (i = 1, 2, \ldots, n; \quad i \neq k) \tag{2.39}$$

$$Z_{ik} = Z_{ki}$$

$$Z_{kk} = Z_b + Z_{jj} \tag{2.40}$$

where n is size of the previous Z_{BUS}.

Type-3 modification

Here Z_b connects an old bus 2 to the reference bus as shown in Figure 2.9. This follows from Figure 2.8 by connecting bus 3 to the reference bus, i.e. by setting $V_3 = 0$. From Eq. (2.37), the following is obtained.

$$Z_{21}I_1 + Z_{22}I_2 + (Z_{22} + Z_b)I_3 = 0$$

or

$$I_3 = \frac{-1}{Z_{22} + Z_b}(Z_{21}I_1 + Z_{22}I_2) \tag{2.41}$$

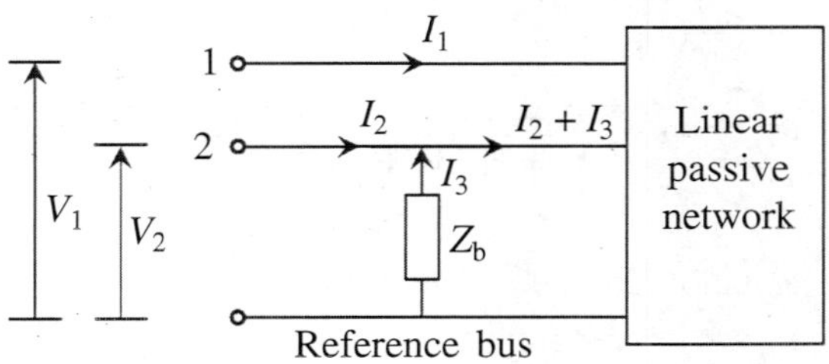

FIGURE 2.9 Type-3 modification.

Substituting the value of I_3 obtained from Eq. (2.41) into Eq. (2.34),

$$V_1 = Z_{11}I_1 + Z_{12}\left(I_2 - \frac{1}{Z_{22} + Z_b}(Z_{21}I_1 + Z_{22}I_2)\right)$$

Rearranging the above equation, we obtain

$$V_1 = \left(Z_{11} - \frac{Z_{12}Z_{21}}{Z_{22} + Z_b}\right)I_1 + \left(Z_{12} - \frac{Z_{12}Z_{22}}{Z_{22} + Z_b}\right)I_2 \tag{2.42}$$

Substituting the value of I_3 obtained from Eq. (2.41) into Eq. (2.35), we get

$$V_2 = Z_{21}I_1 + Z_{22}\left(I_2 - \frac{1}{Z_{22} + Z_b}(Z_{21}I_1 + Z_{22}I_2)\right)$$

Rearranging the above equation, we have

$$V_2 = \left(Z_{21} - \frac{Z_{22}Z_{21}}{Z_{22} + Z_b}\right)I_1 + \left(Z_{22} - \frac{Z_{22}Z_{22}}{Z_{22} + Z_b}\right)I_2 \tag{2.43}$$

Equations (2.42) and (2.43) can be written in matrix form,

$$\begin{bmatrix} V_1 \\ V_2 \end{bmatrix} = \begin{bmatrix} Z_{11} & Z_{12} \\ Z_{21} & Z_{22} \end{bmatrix}\begin{bmatrix} I_1 \\ I_2 \end{bmatrix} - \frac{1}{Z_b + Z_{22}}\begin{bmatrix} Z_{12}Z_{21} & Z_{12}Z_{22} \\ Z_{22}Z_{21} & Z_{22}Z_{22} \end{bmatrix}\begin{bmatrix} I_1 \\ I_2 \end{bmatrix}$$

or

$$\begin{bmatrix} V_1 \\ V_2 \end{bmatrix} = \begin{bmatrix} Z_{11} & Z_{12} \\ Z_{21} & Z_{22} \end{bmatrix}\begin{bmatrix} I_1 \\ I_2 \end{bmatrix} - \frac{1}{Z_b + Z_{22}}\begin{bmatrix} Z_{12} \\ Z_{22} \end{bmatrix}[Z_{21} \quad Z_{22}]\begin{bmatrix} I_1 \\ I_2 \end{bmatrix}$$

or

$$Z_{\text{BUS}}^{\text{new}} = Z_{\text{BUS}}^{\text{old}} - \frac{1}{Z_b + Z_{22}}\begin{bmatrix} Z_{12} \\ Z_{22} \end{bmatrix}[Z_{21} \quad Z_{22}]$$

The above equation can be written in generalized form. If the branch with impedance Z_b is added between the old bus k and the reference bus, then

$$Z_{\text{BUS}}^{\text{new}} = Z_{\text{BUS}}^{\text{old}} - \frac{1}{Z_b + Z_{kk}}\begin{bmatrix} Z_{1k} \\ \vdots \\ Z_{nk} \end{bmatrix}[Z_{k1} \; \ldots \; Z_{kn}] \tag{2.44a}$$

In the index form the above equation can be rewritten as

$$Z_{ij}^{\text{new}} = Z_{ij}^{\text{old}} - \frac{Z_{ik}Z_{kj}}{Z_{kk} + Z_b}; \qquad (i = 1, 2, \ldots, n;\ j = 1, 2, \ldots, n) \tag{2.44b}$$

Type-4 modification

Here Z_b connects an old bus 2 to another old bus 3 as shown in Figure 2.10. It thus follows that:

$$\begin{bmatrix} V_1 \\ V_2 \\ V_3 \end{bmatrix} = \begin{bmatrix} Z_{11} & Z_{12} & Z_{13} \\ Z_{21} & Z_{22} & Z_{23} \\ Z_{31} & Z_{32} & Z_{33} \end{bmatrix} \begin{bmatrix} I_1 \\ I_2 + I_4 \\ I_3 - I_4 \end{bmatrix} \tag{2.45}$$

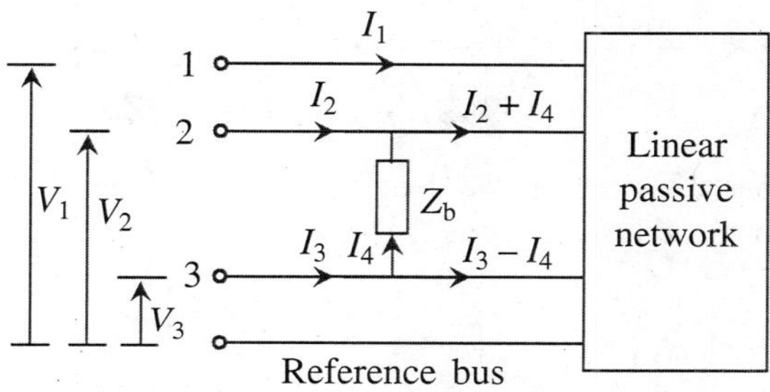

FIGURE 2.10 Type-4 modification.

The voltages of the buses 2 and 3 are constrained by the equation

$$V_3 = V_2 + Z_b I_4 \tag{2.46}$$

or

$$Z_{31}I_1 + Z_{32}(I_2 + I_4) + Z_{33}(I_3 - I_4) = Z_{21}I_1 + Z_{22}(I_2 + I_4) + Z_{23}(I_3 - I_4) + Z_b I_4$$

Rearranging, we obtain

$$0 = (Z_{21} - Z_{31})I_1 + (Z_{22} - Z_{32})I_2 + (Z_{23} - Z_{33})I_3 + (Z_{22} + Z_{33} - Z_{23} - Z_{32} + Z_b)I_4$$

or

$$I_4 = -\frac{1}{P}\,[(Z_{21} - Z_{31})I_1 + (Z_{22} - Z_{32})I_2 + (Z_{23} - Z_{33})I_3]$$

where $P = Z_b + Z_{22} + Z_{33} - Z_{23} - Z_{32}$.

The above equation can be rewritten in matrix form as

$$I_4 = -\frac{1}{P} \begin{bmatrix} Z_{21} - Z_{31} \\ Z_{22} - Z_{32} \\ Z_{23} - Z_{33} \end{bmatrix}^T \begin{bmatrix} I_1 \\ I_2 \\ I_3 \end{bmatrix} \tag{2.47}$$

Equation (2.45) can be rewritten in the form,

$$\begin{bmatrix} V_1 \\ V_2 \\ V_3 \end{bmatrix} = \begin{bmatrix} Z_{11} & Z_{12} & Z_{13} \\ Z_{21} & Z_{22} & Z_{23} \\ Z_{31} & Z_{32} & Z_{33} \end{bmatrix} \begin{bmatrix} I_1 \\ I_2 \\ I_3 \end{bmatrix} + \begin{bmatrix} Z_{12} - Z_{13} \\ Z_{22} - Z_{23} \\ Z_{32} - Z_{33} \end{bmatrix} I_4 \tag{2.48}$$

Substituting the value of I_4 from Eq. (2.47) into Eq. (2.48), we have

$$\begin{bmatrix} V_1 \\ V_2 \\ V_3 \end{bmatrix} = \begin{bmatrix} Z_{11} & Z_{12} & Z_{13} \\ Z_{21} & Z_{22} & Z_{23} \\ Z_{31} & Z_{32} & Z_{33} \end{bmatrix} \begin{bmatrix} I_1 \\ I_2 \\ I_3 \end{bmatrix} - \frac{1}{P} \begin{bmatrix} Z_{12} - Z_{13} \\ Z_{22} - Z_{23} \\ Z_{32} - Z_{33} \end{bmatrix} \begin{bmatrix} Z_{21} - Z_{31} \\ Z_{22} - Z_{32} \\ Z_{23} - Z_{33} \end{bmatrix}^T \begin{bmatrix} I_1 \\ I_2 \\ I_3 \end{bmatrix}$$

$$Z_{\text{BUS}}^{\text{new}} = Z_{\text{BUS}}^{\text{old}} - \frac{1}{P} \begin{bmatrix} Z_{12} - Z_{13} \\ Z_{22} - Z_{23} \\ Z_{32} - Z_{33} \end{bmatrix} \begin{bmatrix} Z_{21} - Z_{31} \\ Z_{22} - Z_{32} \\ Z_{23} - Z_{33} \end{bmatrix}^T \tag{2.49a}$$

The above equation can be written in the generalized form. If the branch with impedance Z_b is added between the old bus k and the old bus l, then

$$Z_{\text{BUS}}^{\text{new}} = Z_{\text{BUS}}^{\text{old}} - \frac{1}{P} \begin{bmatrix} Z_{1k} - Z_{1l} \\ Z_{2k} - Z_{2l} \\ \vdots \\ Z_{nk} - Z_{nl} \end{bmatrix} \begin{bmatrix} Z_{k1} - Z_{l1} \\ Z_{k2} - Z_{l2} \\ \vdots \\ Z_{kn} - Z_{ln} \end{bmatrix}^T \tag{2.49b}$$

where $P = Z_b + Z_{kk} + Z_{ll} - Z_{lk} - Z_{kl}$.

In the index form the above equation can be rewritten as

$$Z_{ij}^{\text{new}} = Z_{ij}^{\text{old}} - \frac{(Z_{ik} - Z_{il})(Z_{kj} - Z_{lj})}{Z_b + Z_{kk} + Z_{ll} - Z_{kl} - Z_{lk}} \qquad (i = 1, 2, \ldots, n; \quad j = 1, 2, \ldots, n) \tag{2.49c}$$

The opening of a line (Z_{ij}) is equivalent to adding a branch in parallel to it with impedance $-Z_{ij}$.

Algorithm 2.2: To Build Z_{BUS}

1. Input data—NL (no. of lines) and NB (no. of buses).
2. Initialize $Z_{ij} = 0$ (i = 1, 2, ..., NB; j = 1, 2, ..., NB).
3. Set the line number and bus number counter ($i = 1$ and $m = 0$).
4. Input the impedance Z_b between the buses k to l and type modification.
5. Check type modification
   If type modification is 1 then GOTO Step 6
   If type modification is 2 then GOTO Step 7
   If type modification is 3 then GOTO Step 8
   If type modification is 4 then GOTO Step 9.
6. Type modification 1 means that a branch is added from the new bus k to the reference bus

$$Z_{kk} = Z_b$$

$$Z_{kj} = Z_{jk} = 0.0 \qquad (j = 1, 2, \ldots, m)$$

   modify, $m = m + 1$ and GOTO Step 10.

7. Type modification 2 means that a branch is added from the new bus k to the old bus l

$$Z_{kk} = Z_{ll} + Z_b$$
$$Z_{kj} = Z_{lj}; \quad j = 1, 2, ..., m$$
$$Z_{jk} = Z_{kj}; \quad j = 1, 2, ..., m$$

modify, $m = m + 1$ and GOTO Step 10.

8. Type modification 3 means that a link is added from the old bus k to the reference bus

$$Z_{ij}^{\text{new}} = Z_{ij}^{\text{old}} - \frac{Z_{ik}Z_{kj}}{Z_{kk} + Z_b} \qquad (i = 1, 2, ..., m; \; j = 1, 2, ..., m)$$

GOTO Step 10.

9. Type modification 4 means that link is added from the old bus k to the old bus l

$$Z_{ij}^{\text{new}} = Z_{ij}^{\text{old}} - \frac{(Z_{ik} - Z_{il})(Z_{kj} - Z_{lj})}{Z_b + Z_{kk} + Z_{ll} - Z_{lk} - Z_{kl}} \qquad (i = 1, 2, ..., m; \; j = 1, 2, ..., m)$$

GOTO Step 10.

10. Check $i \geq$ NL
if 'yes' then GOTO Step 11, else $i = i + 1$ and GOTO Step 4 and repeat.

11. Write the matrix and stop.

2.5.2 Mutual Coupling between Transmission Lines

Addition of a branch

In this case, a new branch p–q is added as shown in Figure 2.11 to a partial (already assembled) network.

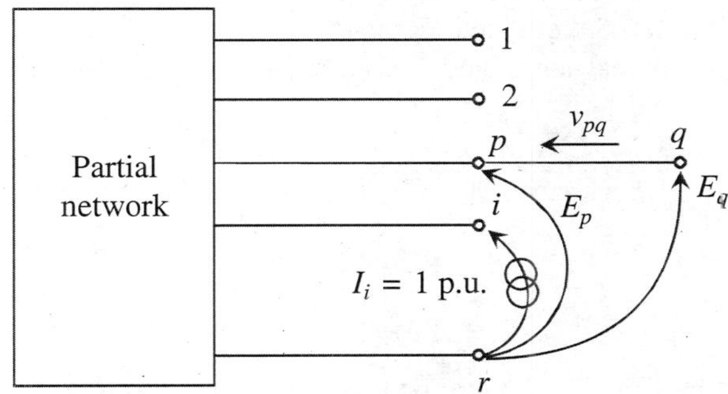

FIGURE 2.11 Addition of a branch to partial network.

The performance equation for the already assembled network with an added branch p–q is

$$\begin{bmatrix} E_1 \\ E_2 \\ \vdots \\ E_p \\ \vdots \\ E_m \\ E_q \end{bmatrix} = \begin{bmatrix} Z_{11} & Z_{12} & \cdots & Z_{1p} & \cdots & Z_{1m} & Z_{1q} \\ Z_{21} & Z_{22} & \cdots & Z_{2p} & \cdots & Z_{2m} & Z_{2q} \\ \vdots & \vdots & \vdots & \vdots & & \vdots & \vdots \\ Z_{p1} & Z_{p2} & \cdots & Z_{pp} & \cdots & Z_{pm} & Z_{pq} \\ \vdots & \vdots & \vdots & \vdots & & \vdots & \vdots \\ Z_{m1} & Z_{m2} & \cdots & Z_{mp} & \cdots & Z_{mm} & Z_{mq} \\ Z_{q1} & Z_{q2} & \cdots & Z_{qp} & \cdots & Z_{qm} & Z_{qq} \end{bmatrix} \begin{bmatrix} I_1 \\ I_2 \\ \vdots \\ I_p \\ \vdots \\ I_m \\ I_q \end{bmatrix} \tag{2.50}$$

The network consists of bilateral passive elements, so $Z_{qi} = Z_{iq}$ (q = 1, 2, ..., m). The elements Z_{iq} can be determined by injecting current $I_i = 1$ pu at the ith bus and calculating the voltage at the qth bus with respect to the reference node r as shown in Figure 2.11. Since all other bus currents are zero, then

$$E_i = Z_{qi} \qquad (i = 1, 2, ..., m) \tag{2.51}$$

Assume that the element pq is mutually coupled with a group of elements indicated by rs. The currents in the elements of the network can be written in terms of primitive admittances and the voltages across the elements. Thus,

$$\begin{bmatrix} i_{pq} \\ i_{rs} \end{bmatrix} = \begin{bmatrix} y_{pq} & y_{pq,\,rs} \\ y_{rs,\,pq} & y_{rs} \end{bmatrix} \begin{bmatrix} v_{pq} \\ v_{rs} \end{bmatrix} \tag{2.52}$$

where

i_{pq} is the current through the added element
v_{pq} is the voltage across the added element
i_{rs} is the current vector of the already built network
v_{rs} is the voltage vector of the already built network
y_{pq} is the self-admittance of the added element
$y_{pq,\,rs}$ is the vector of mutual admittance between p–q and all other elements with which it has coupling.

From Figure 2.11,

$$E_q = E_p - v_{pq} \tag{2.53}$$

Further, from Eq. (2.52)

$$i_{pq} = y_{pq}v_{pq} + y_{pq,rs}v_{rs} \tag{2.54}$$

The current in the added branch, i_{pq}, is zero, because the current source is connected between the bus i and the reference. But the voltage across pq, v_{pq}, is not zero due to mutual coupling. Thus from Eq. (2.54),

$$y_{pq}v_{pq} + y_{pq,rs}v_{rs} = 0$$

or

$$v_{pq} = -\frac{y_{pq,\,rs}v_{rs}}{y_{pq}} \tag{2.55}$$

Substituting for v_{pq} in Eq. (2.53), we get

$$E_q = E_p + \frac{y_{pq,\,rs}v_{rs}}{y_{pq}}$$

or

$$E_q = E_p + \frac{y_{pq,\,rs}(E_r - E_s)}{y_{pq}}$$

Using Eq. (2.51)

$$Z_{qi} = Z_{pi} + \frac{y_{pq,\,rs}(Z_{ri} - Z_{si})}{y_{pq}} \qquad (i = 1, 2, ..., m;\ i \neq q) \tag{2.56}$$

To calculate Z_{qq}, inject $I_q = 1$ p.u. current and measure E_q with respect to reference. Since all other currents are zero, using Eq. (2.50)

$$E_j = Z_{jq} \qquad (j = 1, 2, ..., m)$$

$$i_{pq} = -I_q = -1 \tag{2.57}$$

Substituting for i_{pq} in Eq. (2.54), we obtain

$$y_{pq}v_{pq} + y_{pq,\,rs}v_{rs} = -1$$

or

$$v_{pq} = -\frac{1 + y_{pq,\,rs}\,v_{rs}}{y_{pq}}$$

Substituting for v_{pq} in Eq. (2.53), we obtain

$$E_q = E_p + \frac{1 + y_{pq,\,rs}\,v_{rs}}{y_{pq}}$$

or

$$E_q = E_p + \frac{1 + y_{pq,\,rs}\,(E_r - E_s)}{y_{pq}}$$

Using Eq. (2.57), we get

$$Z_{qq} = Z_{pq} + \frac{1 + y_{pq,\,rs}\,(Z_{rp} - Z_{sp})}{y_{pq}} \tag{2.58}$$

Addition of a link

Let the added element p–q be a link. A voltage source e_l is connected in series with the added element for recalculating the elements of the bus impedance matrix, as shown in Figure 2.12. This creates a new fictitious node l which will be eventually eliminated. The value of e_l is such that the current through the added element is zero.

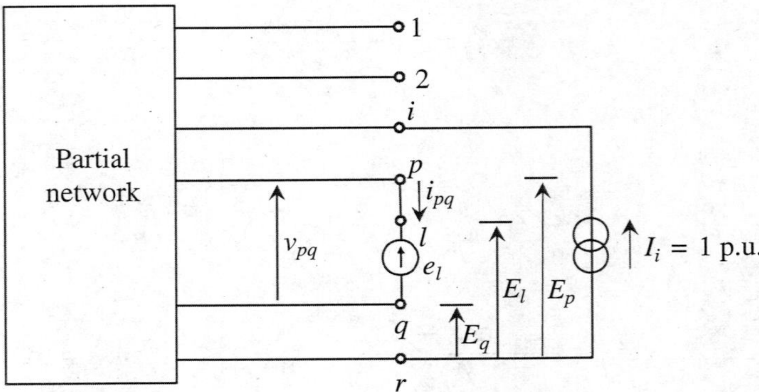

FIGURE 2.12 Addition of a link to a partial network.

The performance equation for the partial network with the added element p–l and the series voltage source e_l is

$$\begin{bmatrix} E_1 \\ E_2 \\ \vdots \\ E_p \\ \vdots \\ E_m \\ E_l \end{bmatrix} = \begin{bmatrix} Z_{11} & Z_{12} & \cdots & Z_{1p} & \cdots & Z_{1m} & Z_{1q} \\ Z_{21} & Z_{22} & \cdots & Z_{2p} & \cdots & Z_{2m} & Z_{2q} \\ \vdots & \vdots & \vdots & \vdots & & \vdots & \vdots \\ Z_{p1} & Z_{p2} & \cdots & Z_{pp} & \cdots & Z_{pm} & Z_{pq} \\ \vdots & \vdots & \vdots & \vdots & & \vdots & \vdots \\ Z_{m1} & Z_{m2} & \cdots & Z_{mp} & \cdots & Z_{mm} & Z_{mq} \\ Z_{l1} & Z_{l2} & \cdots & Z_{lp} & \cdots & Z_{lm} & Z_{ll} \end{bmatrix} \begin{bmatrix} I_1 \\ I_2 \\ \vdots \\ I_p \\ \vdots \\ I_m \\ I_l \end{bmatrix} \tag{2.59}$$

In the above equation, E_1, E_2, ..., E_m are with respect to the reference bus. But e_l is with respect to bus q. So,

$$e_l = E_l - E_q \tag{2.60}$$

The elements Z_{li}, $i \neq l$ can be determined by injecting 1 pu current at the bus i as shown in Figure 2.12 and calculating e_l with respect to the q bus. Since all other bus currents are zero, from Eq. (2.59)

$$E_k = Z_{ki} \qquad (k = 1, 2, ..., m)$$

$$e_l = Z_{li} \tag{2.61}$$

The voltage of the series source is

$$e_l = E_p - E_q - v_{pl} \tag{2.62}$$

Since $i_{pq} = 0$, the elements p–l can be treated as branch, and also $i_{pl} = 0$. Thus,

$$i_{pl} = y_{pl}v_{pl} + y_{pl,\,rs}v_{rs}$$

Therefore,

$$v_{pl} = -\frac{y_{pl,\,rs}v_{rs}}{y_{pl}}$$

or

$$v_{pl} = -\frac{y_{pq,\,rs}v_{rs}}{y_{pq}} \qquad (\because\ y_{pl,\,rs} = y_{pq,\,rs},\ y_{pl} = y_{pq}) \tag{2.63}$$

Substituting Eq. (2.63) in Eq. (2.62),

$$e_l = E_p - E_q + \frac{y_{pq,\,rs}v_{rs}}{y_{pq}}$$

or

$$e_l = E_p - E_q + \frac{y_{pq,\,rs}(E_r - E_s)}{y_{pq}}$$

Substituting Eq. (2.61) into the above equation,

$$Z_{lj} = Z_{pj} - Z_{qj} + \frac{y_{pq,\,rs}(Z_{rj} - Z_{sj})}{y_{pq}} \qquad (i = 1,\ 2,\ \ldots,\ m;\ j \neq l) \tag{2.64}$$

To calculate Z_{ll}, there is need to inject 1 pu current at bus l with bus q as reference. Determine voltage at bus l with respect to bus q.

Then from Eq. (2.59),

$$E_k = Z_{kl} \qquad (k = 1,\ 2,\ \ldots,\ m)$$

$$e_l = Z_{ll} \tag{2.65}$$

where E_k are the voltages with respect to the reference bus. Z_{ll} can be found out directly by computing e_l. The current in the element p–l is

$$i_{pl} = -I_i = -1$$

The current in terms of primitive admittances and the voltages across the element are

$$i_{pl} = y_{pl}v_{pl} + y_{pl,\,rs}v_{rs} = -1$$

Since $y_{pl,\,rs} = y_{pq,\,rs}$ and $y_{pl} = y_{pq}$, then

$$v_{pl} = -\frac{1 + y_{pl,rs}v_{rs}}{y_{pl}}$$

or

$$v_{pl} = -\frac{1 + y_{pq,\,rs} v_{rs}}{y_{pq}} \tag{2.66}$$

Substituting Eq. (2.66) in Eq. (2.62), we get

$$e_l = E_p - E_q + \frac{1 + y_{pq,\,rs} v_{rs}}{y_{pq}}$$

or

$$e_l = E_p - E_q + \frac{1 + y_{pq,\,rs}(E_r - E_s)}{y_{pq}}$$

Substituting Eq. (2.65) into the above equation,

$$Z_{ll} = Z_{pl} - Z_{ql} + \frac{1 + y_{pq,\,rs}(Z_{rl} - Z_{sl})}{y_{pq}} \tag{2.67}$$

Here, the summary of equations for formation of the bus impedance matrix is given as per type of modification defined earlier in Section 2.5.1.

Type modification 1: A branch is added from a new bus q to the reference bus having mutual coupling with link from bus r to bus s.

$$Z_{qi} = \frac{y_{pq,\,rs}(Z_{ri} - Z_{si})}{y_{pq}} \qquad (i = 1, 2, \ldots, m)$$

$$Z_{qq} = \frac{1 + y_{pq,\,rs}(Z_{rq} - Z_{sq})}{y_{pq}}$$

Type modification 2: A branch is added from a new bus q to an old bus p having mutual coupling with link from bus r to bus s.

$$Z_{qi} = Z_{pi} + \frac{y_{pq,\,rs}(Z_{ri} - Z_{si})}{y_{pq}} \qquad (i = 1, 2, \ldots, m)$$

$$Z_{qq} = Z_{pq} + \frac{1 + y_{pq,\,rs}(Z_{rq} - Z_{sq})}{y_{pq}}$$

Type modification 3: A link is added from an old bus q to the reference bus having mutual coupling with link from bus r to bus s.

$$Z_{li} = -Z_{qi} + \frac{y_{pq,\,rs}(Z_{ri} - Z_{si})}{y_{pq}} \qquad (i = 1, 2, \ldots, m)$$

$$Z_{ll} = -Z_{ql} + \frac{1 + y_{pq,\,rs}(Z_{rl} - Z_{sl})}{y_{pq}}$$

$$Z_{ij}^{\text{new}} = Z_{ij}^{\text{old}} + \frac{Z_{il}Z_{lj}}{Z_{ll}} \qquad (i = 1, 2, ..., m; \quad j = 1, 2, ..., m)$$

Type modification 4: A link is added from an old bus q to the another old bus p having mutual coupling with link from bus r to bus s.

$$Z_{li} = Z_{pi} - Z_{qi} + \frac{y_{pq,\,rs}(Z_{ri} - Z_{si})}{y_{pq}} \qquad (i = 1, 2, ..., m)$$

$$Z_{ll} = Z_{pl} - Z_{ql} + \frac{1 + y_{pq,\,rs}(Z_{rl} - Z_{sl})}{y_{pq}}$$

$$Z_{ij}^{\text{new}} = Z_{ij}^{\text{old}} + \frac{Z_{il}Z_{lj}}{Z_{ll}} \qquad (i = 1, 2, ..., m; \quad j = 1, 2, ..., m)$$

Removal of elements

Let the equation of the system be

$$E_{\text{BUS}} = Z_{\text{BUS}}I_{\text{BUS}}$$

$$E_i = \sum_{k=1}^{\text{NB}} Z_{ik}I_k \qquad (i = 1, 2, ..., \text{NB}) \tag{2.68}$$

On removing an element, there may be change in the impedance matrix of network, Eq. (2.68) can then be written as

$$E_i^{\text{new}} = \sum_{k=1}^{\text{NB}} Z_{ik}^{\text{new}} I_k \qquad (i = 1, 2, ..., \text{NB}) \tag{2.69}$$

where E_i^{new} is the new bus voltages of the ith bus.

Now, if the original impedance matrix is retained then to maintain the new bus voltages, appropriate changes in the bus current of affected elements are required. Then, Eq. (2.69) becomes

$$E_i^{\text{new}} = \sum_{k=1}^{\text{NB}} Z_{ik}(I_k + \Delta I_k) \qquad (i = 1, 2, ..., \text{NB}) \tag{2.70}$$

Suppose an element p–q is mutually coupled to an element r–s. Element p–q is to be removed. Following Eq. (2.70), it is obtained as

$$\begin{aligned} \Delta I_k &= \Delta I_{pq} && \text{for } k = p \\ \Delta I_k &= -\Delta I_{pq} && \text{for } k = q \\ \Delta I_k &= \Delta I_{rs} && \text{for } k = r \\ \Delta I_k &= -\Delta I_{rs} && \text{for } k = s \end{aligned} \tag{2.71}$$

and $\Delta I_k = 0$ for the remaining bus.

A current source of 1 p.u. at the jth bus is connected (i.e. I_j = 1 p.u.) and other buses are kept open-circuited.

$$I_j = 1 \text{ pu}$$

$$I_k = 0 \qquad (\text{for } k = 1, 2, \ldots, n; \quad k \neq j) \tag{2.72}$$

Substituting Eqs. (2.71) and (2.72) in Eq. (2.70), we obtain

$$E_i^{\text{new}} = Z_{ij} + Z_{ip}\Delta I_{pq} - Z_{iq}\Delta I_{pq} + Z_{ir}\Delta I_{rs} - Z_{is}\Delta I_{rs}$$

or

$$E_i^{\text{new}} = Z_{ij} + (Z_{ip} - Z_{iq})\Delta I_{pq} + (Z_{ir} - Z_{is})\,\Delta I_{rs} \tag{2.73}$$

Let $\alpha = [p \quad r]^T$ and $\beta = [q \quad s]^T$.

Substituting the above defined variables into Eq. (2.73), we get

$$E_i^{\text{new}} = Z_{ij} + (Z_{i\alpha} - Z_{i\beta})\,\Delta I_{\alpha\beta} \tag{2.74}$$

In view of the primitive admittance matrix, the performance equation can be written as

$$\Delta I_{\alpha\beta} = \Delta Y_s (E_\gamma^{\text{new}} - E_\delta^{\text{new}}) \tag{2.75}$$

where

$$\Delta Y_s = Y_s - Y_s^{\text{m}}$$

and

Y_s is primitive admittance matrix before removing the element

Y_s^{m} is primitive admittance matrix after removing the element

$E_\gamma^{\text{new}}, E_\delta^{\text{new}}$ are the new voltages.

From Eq. (2.74), the new voltages can be written as

$$E_\gamma^{\text{new}} = Z_{\gamma j} + (Z_{\gamma\alpha} - Z_{\gamma\beta})\Delta I_{\alpha\beta} \tag{2.76}$$

$$E_\delta^{\text{new}} = Z_{\delta j} + (Z_{\delta\alpha} - Z_{\delta\beta})\Delta I_{\alpha\beta} \tag{2.77}$$

Substituting Eqs. (2.76) and (2.77) in Eq. (2.75), we get

$$\Delta I_{\alpha\beta} = \Delta Y_s[Z_{\gamma j} - Z_{\delta j} + (Z_{\gamma\alpha} - Z_{\gamma\beta} - Z_{\delta\alpha} + Z_{\delta\beta})\,\Delta I_{\alpha\beta}]$$

Rearranging, we obtain

$$[I - \Delta Y_s(Z_{\gamma\alpha} - Z_{\gamma\beta} - Z_{\delta\alpha} + Z_{\delta\beta})]\Delta I_{\alpha\beta} = \Delta Y_s(Z_{\gamma j} - Z_{\delta j})$$

On simplification,

$$\Delta I_{\alpha\beta} = M^{-1}\Delta Y_s(Z_{\gamma j} - Z_{\delta j}) \tag{2.78}$$

where $M = I - \Delta Y_s(Z_{\gamma\alpha} - Z_{\gamma\beta} - Z_{\delta\alpha} + Z_{\delta\beta})$.

Substituting Eq. (2.78) in Eq. (2.74), we get

$$E_i^{\text{new}} = Z_{ij} + (Z_{i\alpha} - Z_{i\beta})M^{-1}\Delta Y_s(Z_{\gamma j} - Z_{\delta j})$$

From the above equation the elements of the impedance matrix can be obtained as

$$Z_{ij}^{\text{new}} = Z_{ij} + (Z_{i\alpha} - Z_{i\beta})M^{-1}\Delta Y_s(Z_{\gamma j} - Z_{\delta j}) \qquad (i = 1, 2, ..., \text{NB}; \; j = 1, 2, ..., \text{NB}) \quad (2.79)$$

EXAMPLE 2.2 Obtain Z_{BUS} for the 4-bus sample system given in Figure 2.13. The series impedances of each line are given in Table 2.2. Reference bus is represented by R. No line is mutually coupled.

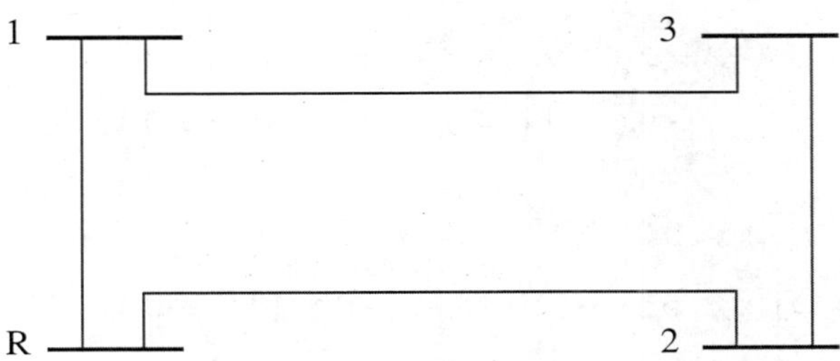

FIGURE 2.13 Sample network.

TABLE 2.2 Line impedance data of a sample network

Line no.	*Bus code (p–q)*	*Impedance (Z)*	*Type of modification*
1	R–1	0.6	1 (New bus to reference)
2	R–2	0.5	1 (New bus to reference)
3	2–3	0.5	2 (New bus to old bus)
4	1–3	0.25	4 (Old bus to old bus)

Solution

Step 1: Add branch $Z_{1\text{R}} = 0.6$. It is type-1 modification, i.e. from bus 1 to reference bus R. Thus,

$$Z_{\text{BUS}} = [0.6]$$

Step 2: Add branch $Z_{2\text{R}} = 0.5$. It is type-1 modification, i.e. from bus 2 to reference bus R. Thus,

$$Z_{\text{BUS}} = \begin{bmatrix} 0.6 & 0.0 \\ 0.0 & 0.5 \end{bmatrix}$$

Step 3: Add branch $Z_{23} = 0.5$. It is type-2 modification, i.e. from new bus 3 to old bus 2. Thus,

$$Z_{\text{BUS}} = \begin{bmatrix} 0.6 & 0.0 & 0.0 \\ 0.0 & 0.5 & 0.5 \\ 0.0 & 0.5 & 1.0 \end{bmatrix}$$

Step 4: Add branch $Z_{13} = 0.25$. It is type-4 modification, i.e. from old bus 1 to old bus 3. Thus,

$$Z_{BUS} = \begin{bmatrix} 0.6 & 0.0 & 0.0 \\ 0.0 & 0.5 & 0.5 \\ 0.0 & 0.5 & 1.0 \end{bmatrix} - \frac{1}{P}\begin{bmatrix} 0.6 - 0.0 \\ 0.0 - 0.5 \\ 0.0 - 1.0 \end{bmatrix}\begin{bmatrix} 0.6 - 0.0 \\ 0.0 - 0.5 \\ 0.0 - 1.0 \end{bmatrix}^T$$

where

$$P = Z_{11} + Z_{33} - Z_{13} - Z_{31} + 0.25$$
$$= 0.6 + 1.0 - 0.0 - 0.0 + 0.25 = 1.85$$

Thus,

$$Z_{BUS} = \begin{bmatrix} 0.6 & 0.0 & 0.0 \\ 0.0 & 0.5 & 0.5 \\ 0.0 & 0.5 & 1.0 \end{bmatrix} - \frac{1}{1.85}\begin{bmatrix} 0.6 \\ -0.5 \\ -1.0 \end{bmatrix}[0.6 \quad -0.5 \quad -1.0]$$

or

$$Z_{BUS} = \begin{bmatrix} 0.405405 & 0.162162 & 0.324324 \\ 0.162162 & 0.364865 & 0.229730 \\ 0.324324 & 0.229730 & 0.459459 \end{bmatrix}$$

EXAMPLE 2.3 Find the modified bus impedance matrix of Example 2.2, if the line from bus 1 to bus 3 is removed.

Solution This is equivalent to connecting an impedance of – 0.25 between old bus 1 and old bus 3, i.e. a type-4 modification. Therefore,

$$Z_{BUS} = \begin{bmatrix} 0.405405 & 0.162162 & 0.324324 \\ 0.162162 & 0.364865 & 0.229730 \\ 0.324324 & 0.229730 & 0.459459 \end{bmatrix} - \frac{1}{P}\begin{bmatrix} 0.081081 \\ -0.067568 \\ -0.135135 \end{bmatrix}\begin{bmatrix} 0.081081 \\ -0.067568 \\ -0.135135 \end{bmatrix}^T$$

where

$$P = Z_{11} + Z_{33} - Z_{13} - Z_{31} - 0.25$$
$$= 0.405405 + 0.459459 - 0.324324 - 0.324324 - 0.25$$
$$= -\ 0.033784$$

Therefore,

$$Z_{BUS} = \begin{bmatrix} 0.6 & 0.0 & 0.0 \\ 0.0 & 0.5 & 0.5 \\ 0.0 & 0.5 & 1.0 \end{bmatrix}$$

EXAMPLE 2.4 Obtain Z_{BUS} for the 4-bus sample system given in Figure 2.14. The self and mutual impedances of each line are given in Table 2.3. Reference bus is represented by R.

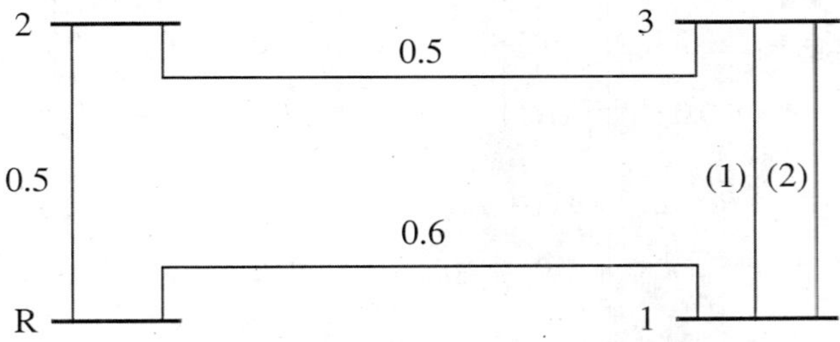

FIGURE 2.14 Sample network.

TABLE 2.3 Line impedance data of a sample network

Line no.	*Self*		*Mutual*		*Modification type*
	Bus code $(p–q)$	*Impedance* (Z_{pq})	*Bus code* $(r–s)$	*Impedance* $(Z_{pq,rs})$	
1	R–1	0.60			1
2	R–2	0.50			1
3	2–3	0.50			2
4	1–3(1)	0.25	1–3(2)	0.10	4
5	1–3(2)	0.20			4

Solution

Step 1: Add branch $Z_{1R} = 0.60$. It is type-1 modification, i.e. from bus 1 to reference bus R. Thus,

$$Z_{BUS} = [0.6]$$

Step 2: Add branch $Z_{2R} = 0.50$. It is type-1 modification, i.e. from bus 2 to reference bus R. Thus,

$$Z_{BUS} = \begin{bmatrix} 0.6 & 0.0 \\ 0.0 & 0.5 \end{bmatrix}$$

Step 3: Add branch $Z_{23} = 0.50$. It is type-2 modification, i.e. from new bus 3 to old bus 2. Thus,

$$Z_{BUS} = \begin{bmatrix} 0.6 & 0.0 & 0.0 \\ 0.0 & 0.5 & 0.5 \\ 0.0 & 0.5 & 1.0 \end{bmatrix}$$

Step 4: Add branch $Z_{1-3(1)} = 0.25$. It is type-4 modification, i.e. from old bus 1 to old bus 3.

Thus,

$$Z_{BUS} = \begin{bmatrix} 0.6 & 0.0 & 0.0 \\ 0.0 & 0.5 & 0.5 \\ 0.0 & 0.5 & 1.0 \end{bmatrix} - \frac{1}{P} \begin{bmatrix} 0.6 - 0.0 \\ 0.0 - 0.5 \\ 0.0 - 1.0 \end{bmatrix} \begin{bmatrix} 0.6 - 0.0 \\ 0.0 - 0.5 \\ 0.0 - 1.0 \end{bmatrix}^T$$

where

$$P = Z_{11} + Z_{33} - Z_{13} - Z_{31} + 0.25$$

$$= 0.6 + 1.0 - 0.0 - 0.0 + 0.25 = 1.85$$

Therefore,

$$Z_{BUS} = \begin{bmatrix} 0.6 & 0.0 & 0.0 \\ 0.0 & 0.5 & 0.5 \\ 0.0 & 0.5 & 1.0 \end{bmatrix} - \frac{1}{1.85} \begin{bmatrix} 0.6 \\ -0.5 \\ -1.0 \end{bmatrix} [0.6 \quad -0.5 \quad -1.0]$$

or

$$Z_{BUS} = \begin{bmatrix} 0.405405 & 0.162162 & 0.324324 \\ 0.162162 & 0.364865 & 0.229730 \\ 0.324324 & 0.229730 & 0.459459 \end{bmatrix}$$

Step 5: Add branch $Z_{1-3(2)} = 0.2$ which is mutually coupled to branch $Z_{1-3(1)}$. $p = 1$, $q = 3$ and $r = 1$, $s = 3$.

The primitive impedance matrix of the coupled elements is

$$Z_{pq, rs} = \begin{bmatrix} 0.25 & 0.1 \\ 0.1 & 0.2 \end{bmatrix}$$

$$y_{pq, rs} = Z_{pq,rs}^{-1} = \begin{bmatrix} 5.0 & -2.5 \\ -2.5 & 6.25 \end{bmatrix}$$

From the above admittance matrix,

$$y_{1-3(2), 1-3(2)} = 6.25, \quad y_{1-3(1), 1-3(2)} = -2.5$$

$$Z_{lk} = Z_{pk} - Z_{qk} + \frac{y_{pq, rs}(Z_{rk} - Z_{sk})}{y_{pq, rs}} \qquad (k = 1, 2,..., \text{NB})$$

or

$$Z_{ll} = Z_{pl} - Z_{ql} + \frac{1 + y_{pq,rs}(Z_{rl} - Z_{sl})}{y_{pq,rs}}$$

When $k = 1$

$$Z_{l1} = Z_{11} - Z_{31} + \frac{y_{13, 13}(Z_{11} - Z_{31})}{y_{1-3(1), 1-3(2)}}$$

or

$$Z_{l1} = 0.405405 - 0.324324 + \frac{-2.5(0.405405 - 0.324324)}{6.25} = 0.0486486$$

When $k = 2$

$$Z_{l2} = Z_{12} - Z_{32} + \frac{y_{13,13}\,(Z_{12} - Z_{32})}{y_{1-3(1),\,1-3(2)}}$$

or

$$Z_{l2} = 0.162162 - 0.229730 + \frac{-2.5(0.162162 - 0.229730)}{6.25} = -0.040541$$

When $k = 3$

$$Z_{l3} = Z_{13} - Z_{33} + \frac{y_{13,13}\,(Z_{13} - Z_{33})}{y_{1-3(1),\,1-3(2)}}$$

or

$$Z_{l3} = 0.324324 - 0.459459 + \frac{-2.5(0.324324 - 0.459459)}{6.25} = -0.081081$$

Now,

$$Z_{ll} = Z_{1l} - Z_{3l} + \frac{1 + y_{13,13}(Z_{1l} - Z_{3l})}{y_{1-3(1),\,1-3(2)}}$$

$$Z_{ll} = 0.0486486 + 0.081081 + \frac{1.0 - 2.5(0.0486486 - 0.081081)}{6.25} = 0.237838$$

$$Z_{BUS} = \begin{bmatrix} 0.405405 & 0.162162 & 0.324324 & 0.0486486 \\ 0.162162 & 0.364865 & 0.229730 & -0.040541 \\ 0.324324 & 0.229730 & 0.459459 & -0.081081 \\ 0.0486486 & -0.040541 & -0.081081 & 0.237838 \end{bmatrix}$$

Eliminating the node. Suppose the fourth node is eliminated.

$$Z_{ij}^{\text{new}} = Z_{ij}^{\text{old}} - \frac{Z_{ik}^{\text{old}} Z_{kj}^{\text{old}}}{Z_{kk}^{\text{old}}} \qquad (i = 1, 2, 3; \quad j = 1, 2, 3; \quad k = 4)$$

When $i = 1, j = 1$ and $k = 4$

$$Z_{11}^{\text{new}} = Z_{11}^{\text{old}} - \frac{Z_{14}^{\text{old}} Z_{41}^{\text{old}}}{Z_{44}^{\text{old}}}$$

or

$$Z_{11}^{\text{new}} = 0.405405 - \frac{0.0486486 \times 0.0486486}{0.237838} = 0.395454$$

When $i = 1$, $j = 2$ and $k = 4$

$$Z_{12}^{new} = Z_{12}^{old} - \frac{Z_{14}^{old} Z_{42}^{old}}{Z_{44}^{old}}$$

or

$$Z_{12}^{new} = 0.162162 - \frac{0.0486486 \times (-0.040541)}{0.237838} = 0.170454$$

When $i = 1$, $j = 3$ and $k = 4$

$$Z_{13}^{new} = Z_{13}^{old} - \frac{Z_{14}^{old} Z_{43}^{old}}{Z_{44}^{old}}$$

or

$$Z_{13}^{new} = 0.324324 - \frac{0.0486486 \times (-0.081081)}{0.237838} = 0.340909$$

When $i = 2$, $j = 2$ and $k = 4$

$$Z_{22}^{new} = Z_{22}^{old} - \frac{Z_{24}^{old} Z_{42}^{old}}{Z_{44}^{old}}$$

or

$$Z_{22}^{new} = 0.364865 - \frac{(-0.040541) \times (-0.040541)}{0.237838} = 0.357955$$

When $i = 2$, $j = 3$ and $k = 4$

$$Z_{23}^{new} = Z_{23}^{old} - \frac{Z_{24}^{old} Z_{43}^{old}}{Z_{44}^{old}}$$

or

$$Z_{23}^{new} = 0.229730 - \frac{(-0.040541) \times (-0.081081)}{0.237838} = 0.215909$$

When $i = 3$, $j = 3$ and $k = 4$

$$Z_{33}^{new} = Z_{33}^{old} - \frac{Z_{34}^{old} Z_{43}^{old}}{Z_{44}^{old}}$$

or

$$Z_{33}^{new} = 0.459459 - \frac{(-0.081081) \times (-0.081081)}{0.237838} = 0.431818$$

Finally, the modified impedance matrix becomes

$$Z_{BUS} = \begin{bmatrix} 0.395454 & 0.170454 & 0.340909 \\ 0.170454 & 0.357955 & 0.215909 \\ 0.340909 & 0.215909 & 0.431818 \end{bmatrix}$$

EXAMPLE 2.5 Find the modified bus impedance, Z_{BUS}, of Example 2.4 if the element from bus 1 to 3 (Figure 2.14) is removed.

Solution Removal of the network element 1–3(2) mutually coupled to network element 1–3(1) is obtained from

$$Z_{ij}^{new} = Z_{ij} + (\bar{Z}_{i\alpha} - \bar{Z}_{i\beta})[M]^{-1}[\Delta y_s](\bar{Z}_{\gamma j} - \bar{Z}_{\delta j})$$

where p–q is 1–3 and r–s is also 1–3. The indices $\alpha = 1$, $\gamma = 1$, $\beta = 3$, and $\delta = 3$.

The primitive impedance matrix is

$$Z_{1-3(1),\,1-3(2)} = \begin{bmatrix} 0.25 & 0.1 \\ 0.1 & 0.2 \end{bmatrix}$$

$$y_s = Z_{pq,\,rs}^{-1} = \begin{bmatrix} 5.0 & -2.5 \\ -2.5 & 6.25 \end{bmatrix}$$

The modified primitive admittance submatrix is

$$y_s' = \begin{bmatrix} 4.0 & 0.0 \\ 0.0 & 0.0 \end{bmatrix}$$

Thus,

$$\Delta y_s = y_s - y_s' = \begin{bmatrix} 1.0 & -2.5 \\ -2.5 & 6.25 \end{bmatrix}$$

Now, the M matrix is defined by

$$[M] = \{I - [\Delta y_s]([Z_{\gamma\alpha}] - [Z_{\delta\alpha}] - [Z_{\gamma\beta}] + [Z_{\delta\beta}])\}$$

where

$$[Z_{\gamma\alpha}] = \begin{bmatrix} Z_{11} & Z_{11} \\ Z_{11} & Z_{11} \end{bmatrix} = \begin{bmatrix} 0.395454 & 0.395454 \\ 0.395454 & 0.395454 \end{bmatrix}$$

$$[Z_{\delta\alpha}] = \begin{bmatrix} Z_{31} & Z_{31} \\ Z_{31} & Z_{31} \end{bmatrix} = \begin{bmatrix} 0.340909 & 0.340909 \\ 0.340909 & 0.340909 \end{bmatrix}$$

$$[Z_{\gamma\beta}] = \begin{bmatrix} Z_{13} & Z_{13} \\ Z_{13} & Z_{13} \end{bmatrix} = \begin{bmatrix} 0.340909 & 0.340909 \\ 0.340909 & 0.340909 \end{bmatrix}$$

$$[Z_{\delta\beta}] = \begin{bmatrix} Z_{33} & Z_{33} \\ Z_{33} & Z_{33} \end{bmatrix} = \begin{bmatrix} 0.431818 & 0.431818 \\ 0.431818 & 0.431818 \end{bmatrix}$$

The elements of the M matrix are obtained as

$$[Z_{\gamma\alpha}] - [Z_{\delta\alpha}] - [Z_{\gamma\beta}] + [Z_{\delta\beta}] = \begin{bmatrix} 0.145454 & 0.145454 \\ 0.145454 & 0.145454 \end{bmatrix}$$

$$[M] = \begin{bmatrix} 1 & 0 \\ 0 & 1 \end{bmatrix} - \begin{bmatrix} 1.0 & -2.5 \\ -2.5 & 6.25 \end{bmatrix} \begin{bmatrix} 0.145454 & 0.145454 \\ 0.145454 & 0.145454 \end{bmatrix}$$

or

$$[M] = \begin{bmatrix} 1 & 0 \\ 0 & 1 \end{bmatrix} - \begin{bmatrix} -0.218181 & -0.218181 \\ 0.545453 & 0.545453 \end{bmatrix} = \begin{bmatrix} 1.218181 & 0.218181 \\ -0.545453 & 0.454547 \end{bmatrix}$$

The inverse of M matrix is

$$[M]^{-1} = \begin{bmatrix} 0.675677 & -0.324323 \\ 0.810808 & 1.810808 \end{bmatrix}$$

Also,

$$[M]^{-1}[\Delta y_s] = \begin{bmatrix} 0.675677 & -0.324323 \\ 0.810808 & 1.810808 \end{bmatrix} \begin{bmatrix} 1 & -2.5 \\ -2.5 & 6.25 \end{bmatrix} = \begin{bmatrix} 1.486485 & -3.716211 \\ -3.716212 & 9.290529 \end{bmatrix}$$

The elements of the impedance matrix are obtained as

$$Z_{ij}^{\text{new}} = Z_{ij} + \begin{bmatrix} Z_{i1} - Z_{i3} \\ Z_{i1} - Z_{i3} \end{bmatrix}^T \begin{bmatrix} 1.486485 & -3.716211 \\ -3.716212 & 9.290529 \end{bmatrix} \begin{bmatrix} Z_{1j} - Z_{3j} \\ Z_{1j} - Z_{3j} \end{bmatrix} \qquad (i = 1, 2, 3; \quad j = 1, 2, 3)$$

$$Z_{\text{BUS}} = \begin{bmatrix} 0.405405 & 0.162162 & 0.324324 \\ 0.162162 & 0.364865 & 0.229730 \\ 0.324324 & 0.229730 & 0.459459 \end{bmatrix}$$

2.6 LOAD FLOW PROBLEM

The complex power injected by the source into the ith bus of a power system is

$$S_i = P_i + jQ_i = V_i I_i^* \qquad (i = 1, 2, ..., \text{NB})$$

Substituting the value of $I_i = \sum_{k=1}^{\text{NB}} Y_{ik} V_k$ in the above power equation,

$$P_i + jQ_i = V_i\left(\sum_{k=1}^{NB} Y_{ik}^* V_k^*\right) \qquad (k = 1, 2, ..., NB) \tag{2.80}$$

where

V_i is the voltage of the ith bus

Y_{ik} is the element of the admittance bus.

Equating the real and imaginary parts,

$$P_i = \text{Re}\left(V_i\left(\sum_{k=1}^{NB} Y_{ik}^* V_k^*\right)\right) \qquad (k = 1, 2, ..., NB) \tag{2.81a}$$

$$Q_i = \text{Im}\left(V_i\left(\sum_{k=1}^{NB} Y_{ik}^* V_k^*\right)\right) \qquad (k = 1, 2, ..., NB) \tag{2.81b}$$

where

P_i is the real power

Q_i is the reactive power.

Let $V_i = |V_i| e^{j\delta_i}$, $V_k = |V_k| e^{j\delta_k}$, $Y_{ik} = |Y_{ik}| e^{j\theta_{ik}}$

where

$|V_i|$ is the magnitude of the voltage

δ_i is the angle of the voltage

θ_{ik} is the load angle.

Substituting for V_i, V_k and Y_{ik} in Eq. (2.80)

$$P_i + jQ_i = |V_i| e^{j\delta_i} \sum_{k=1}^{NB} |V_k|\, e^{-j\delta_k} |Y_{ik}| e^{-j\theta_{ik}} \qquad (i = 1, 2, ..., NB)$$

or

$$P_i + jQ_i = |V_i| \sum_{k=1}^{NB} |V_k|\, |Y_{ik}| e^{j(\delta_i - \delta_k - \theta_{ik})} \qquad (i = 1, 2, ..., NB)$$

or

$$P_i + jQ_i = |V_i| \sum_{k=1}^{NB} |V_k|\, |Y_{ik}| \angle(\delta_i - \delta_k - \theta_{ik}) \qquad (i = 1, 2, ..., NB)$$

or

$$P_i + jQ_i = |V_i| \sum_{k=1}^{NB} |V_k|\, |Y_{ik}| (\cos(\delta_i - \delta_k - \theta_{ik}) + j\sin(\delta_i - \delta_k - \theta_{ik})) \qquad (i = 1, 2, ..., NB)$$

Separating the real and imaginary parts of above equation to get real and reactive powers,

$$P_i = |V_i| \sum_{k=1}^{NB} |V_k|\, |Y_{ik}| \cos(\delta_i - \delta_k - \theta_{ik}) \qquad (i = 1, 2, ..., NB) \tag{2.82a}$$

$$Q_i = |V_i| \sum_{k=1}^{NB} |V_k|\, |Y_{ik}| \sin(\delta_i - \delta_k - \theta_{ik}) \qquad (i = 1, 2, ..., NB) \tag{2.82b}$$

Equation (2.80) can also be written as

$$P_i + jQ_i = |V_i| e^{j\delta_i} \sum_{k=1}^{NB} (G_{ik} - jB_{ik}) |V_k| e^{-j\delta_k} \qquad (i = 1, 2, ..., NB)$$

where

$$Y_{ik} = G_{ik} + jB_{ik}$$

$$Y_{ik}^* = G_{ik} - jB_{ik}$$

$$P_i + jQ_i = |V_i| \sum_{k=1}^{NB} (G_{ik} - jB_{ik}) |V_k| e^{j(\delta_i - \delta_k)} \qquad (i = 1, 2, ..., NB)$$

or

$$P_i + jQ_i = |V_i| \sum_{k=1}^{NB} (G_{ik} - jB_{ik}) |V_k| \angle(\delta_i - \delta_k) \qquad (i = 1, 2, ..., NB)$$

$$P_i + jQ_i = |V_i| \sum_{k=1}^{NB} |V_k| (G_{ik} - jB_{ik}) (\cos(\delta_i - \delta_k) + j\sin(\delta_i - \delta_k)) \qquad (i = 1, 2, ..., NB)$$

$$P_i + jQ_i = |V_i| \sum_{k=1}^{NB} |V_k| \{[G_{ik} \cos(\delta_i - \delta_k) + B_{ik} \sin(\delta_i - \delta_k)]$$

$$+ j[G_{ik} \sin(\delta_i - \delta_k) - B_{ik} \cos(\delta_i - \delta_k)]\} \qquad (i = 1, 2, ..., NB)$$

Separating the real and imaginary parts of the above equation to get real and reactive powers,

$$P_i = |V_i| \sum_{k=1}^{NB} |V_k| [G_{ik} \cos(\delta_i - \delta_k) + B_{ik} \sin(\delta_i - \delta_k)]; \qquad (i = 1, 2, ..., NB) \quad (2.83a)$$

$$Q_i = |V_i| \sum_{k=1}^{NB} |V_k| [G_{ik} \sin(\delta_i - \delta_k) - B_{ik} \cos(\delta_i - \delta_k)]; \qquad (i = 1, 2, ..., NB) \quad (2.83b)$$

Equations (2.81a) and (2.81b), Eqs. (2.82a) and (2.82b), Eqs. (2.83a) and (2.83b) are called power flow equations. These are NB real and NB reactive power flow equations giving a total of 2 × NB power flow equations. At each bus there are four variables, namely $|V_i|$, δ_i, P_i and Q_i, giving a total of 4 × NB variables (for NB buses). If at every bus two variables are specified (thus specifying a total of 2 × NB variables), the remaining two variables at every bus (a total of 2 × NB remaining variables) can be found by solving 2 × NB power flow equations. In a physical system, the variables are specified depending on what kind of devices are connected to that bus. In general, four types of buses are defined.

2.6.1 Slack Bus/Swing Bus/Reference Bus

Voltage magnitude and voltage phase angle are specified. Normally, voltage magnitude is set to 1 pu and voltage angle is set to zero. The real and reactive powers are not specified.

The known parameters are voltage magnitude $|V_i|$, and voltage angle δ_i.
The unknown parameters are real power P_i and reactive power Q_i.

In a load flow study, real and reactive powers cannot be fixed a priori at all the buses as the net complex power flow into the network is not known in advance. So, the system power losses are unknown till the load flow study is complete. It is, therefore, necessary to have one bus at which complex power is unspecified so that it supplies the difference in the total system load plus losses and the sum of the complex powers specified at the remaining buses. Such a bus is known as slack bus and must be a generator bus. If slack bus is not specified, the bus connected to the largest generating station is normally selected as the slack bus.

2.6.2 *PQ* Bus/Load Bus

A pure load bus is a *PQ* bus. A load bus has no generating facility (i.e. $P_{g_i} = Q_{g_i} = 0$). At this type of bus, the net real power P_i and the reactive power Q_i are known as

$$P_i = P_{g_i} - P_{d_i} \quad \text{and} \quad Q_i = Q_{g_i} - Q_{d_i}$$

where

P_{g_i}, Q_{g_i} are the real and reactive power generations at the bus respectively.

P_{d_i}, Q_{d_i} are the real and reactive power demands at the bus respectively.

P_{d_i} and Q_{d_i} are known from load forecasting and P_{g_i} and Q_{g_i} are specified.

The known variables on bus are real power P_i and the reactive power Q_i.

The unknowns are voltage magnitude $|V_i|$ and voltage angle δ_i.

The *PQ* buses are the most common comprising almost 85% of all the buses in a given power system.

2.6.3 *PV* Bus/Generator Bus

A generator is always connected to a *PV* bus. Hence the net power P_i, known as P_{d_i} is known from load forecasting.

The knowns are real power P_i and the voltage magnitude $|V_i|$.

The unknowns are reactive power Q_i and the voltage angle δ_i.

PV buses comprise about 15% of all the buses in a power system.

2.6.4 Voltage-Controlled Buses

Generally the *PV* buses and the voltage-controlled buses are grouped together but these buses have physical difference. The voltage-controlled bus has also voltage control capabilities, and uses a tap-adjustable transformer and/or a static VAR compensator instead of a generator. Hence, $P_{g_i} = Q_{g_i} = 0$ at these buses. Thus $P_i = -P_{d_i}$ and $Q_i = -Q_{d_i}$ at these buses.

The knowns are real power P_i, reactive power Q_i, and voltage magnitude $|V_i|$. The voltage angle δ_i is an unknown parameter.

2.6.5 Limits

For static load flow equations (SLFE) solution to have practical significance, all the state and control variables must be within the specified practical limits. These limits are represented by specifications of power system hardware and operating constraints, and are described as follows:

- Voltage magnitude $|V_i|$ must satisfy the inequality

$$|V_i|^{min} \le |V_i| \le |V_i|^{max}$$

This limit arises due to the fact that the power system equipment is designed to operate at fixed voltages within the allowable variations of ± (5–10)% of rated values.

- Certain of the voltage angles δ_i (state variables) must satisfy

$$\left|\delta_i - \delta_k\right| \le \left|\delta_i - \delta_k\right|^{max}$$

This constraint limits the maximum permissible power angle of transmission line connecting buses i and k and is imposed by considerations of stability.

- Owing to physical limitations of P and/or Q, generation sources are constrained.

$$P_{gi}^{min} \le P_{g_i} \le P_{gi}^{max} \qquad (i = 1, 2, ..., NB)$$

$$Q_{gi}^{min} \le Q_{g_i} \le Q_{gi}^{max} \qquad (i = 1, 2, ..., NB)$$

Also, the equality constraints are

$$\sum_{i=1}^{NB} P_{g_i} = \sum_{i=1}^{NB} P_{d_i} + P_L$$

$$\sum_{i=1}^{NB} Q_{g_i} = \sum_{i=1}^{NB} Q_{d_i} + Q_L$$

where P_L and Q_L are system real and reactive power losses respectively.

The load flow equations are nonlinear algebraic equations and have to be solved through iterative numerical techniques such as Gauss Elimination, Gauss–Seidel and Newton–Raphson techniques, etc. At the cost of solution accuracy, it is possible to linearize load flow equations by making suitable assumptions and approximations so that fast and explicit solutions become possible.

2.7 COMPUTATION OF LINE FLOWS

Consider the line connecting the buses i and k (Figure 2.15).
where

y_{ik} is the series admittance

y_{ik0}, y_{ki0} are the shunt admittances respectively

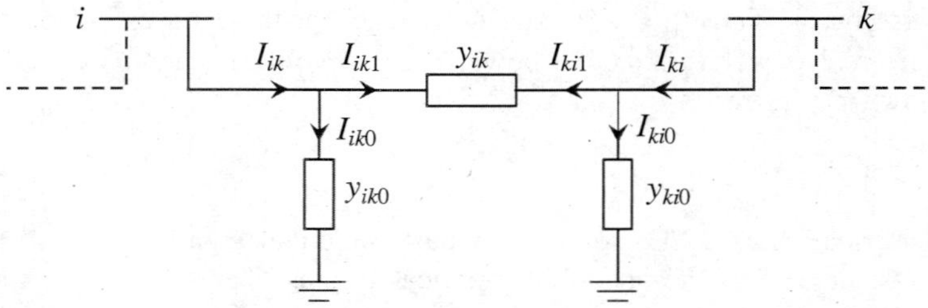

FIGURE 2.15 Transmission line connecting *i* and *k* buses.

S_{ik} is the power injected into the line from the ith bus
I_{ik} is the current injected by the ith bus
V_i is voltage at the ith bus.

The current fed by the bus i into the line can be expressed as

$$I_{ik} = I_{ik1} + I_{ik0}$$

$$I_{ik} = (V_i - V_k)y_{ik} + V_i y_{ik0} \tag{2.84}$$

where

$$I_{ik1} = (V_i - V_k)y_{ik}$$

$$I_{ik0} = V_i y_{ik0}$$

The power injected into the line from bus i to bus k is

$$S_{ik} = P_{ik} + jQ_{ik}$$

$$S_{ik} = V_i I_{ik}^*$$

Substituting the conjugate of Eq. (2.84) into the above equation

$$S_{ik} = V_i [(V_i^* - V_k^*) y_{ik}^* + V_i^* y_{ik0}^*] \tag{2.85}$$

Similarly, the power injected into the line from bus k to bus i is

$$S_{ki} = V_k [(V_k^* - V_i^*) y_{ik}^* + V_k^* y_{ki0}^*] \tag{2.86}$$

The power loss in the $(i–k)$th line is the sum of the power flows in the $(i–k)$th line from the ith bus and the kth bus, respectively, i.e.

$$P_{L_{ik}} = S_{ik} + S_{ki} \tag{2.87}$$

Total transmission losses can be computed by summing all the line flows of the power system.

$$P_{\text{Loss}} = \sum_{l=1}^{\text{NL}} S_l \tag{2.88}$$

where $S_l = S_{ik} + S_{ki}$.

The slack bus power can also be obtained by summing the line flows on the lines terminating at the slack bus.

2.8 MODELLING OF REGULATING TRANSFORMERS

Transformers provide a convenient means of controlling real power, and reactive power flow along a transmission line. Real power can be controlled by means of shifting the phase of voltage, and reactive power by changing its magnitude. Voltage magnitude can be changed by transformers provided with tap changing under load (TCUL) gear. Transformers specially designed to adjust voltage magnitude or phase angle through small values are called *regulating transformers*. The presence of regulating transformers in lines modifies the Y_{BUS} matrix thereby modifying the load flow solution. Consider a line, connecting two buses, having a regulating transformer with off-nominal turns (tap) ratio a included at one end as shown in Figure 2.16.

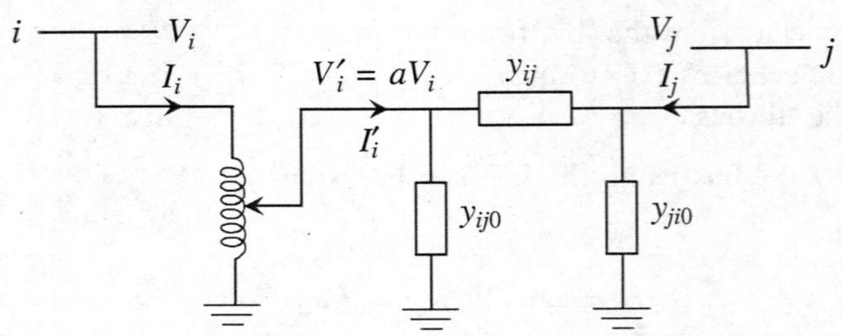

FIGURE 2.16 Transmission line with regulating transformer.

Since the transformer is assumed ideal, complex power output from it equals complex power input, i.e.

$$S_i = V_i I_i^* = V_i'(I_i')^*$$

or

$$V_i I_i^* = aV_i (I_i')^* \qquad (\because V_i' = aV_i)$$

or

$$I_i = a^* I_i' \tag{2.89}$$

For transmission line,

$$I_i' = y_{ij0} V_i' + y_{ij}(V_i' - V_j)$$

or

$$I_i' = a y_{ij0} V_i + y_{ij}(aV_i - V_j)$$

or

$$I_i/a^* = a y_{ij0} V_i + a y_{ij} V_i - y_{ij} V_j$$

or

$$I_i = a^2 (y_{ij0} + y_{ij}) V_i - a^* y_{ij} V_j \tag{2.90}$$

Similarly,

$$I_j = y_{ij0} V_j + y_{ij}(V_j - V_i')$$

or

$$I_j = y_{ij0} V_j + y_{ij}(V_j - aV_i)$$

or

$$I_j = y_{ij0} V_j + y_{ij} V_j - a y_{ij} V_i$$

or

$$I_j = -a y_{ij} V_i + (y_{ij0} + y_{ij}) V_j \tag{2.91}$$

Equations (2.90) and (2.91) can be written in matrix form,

$$\begin{bmatrix} I_i \\ I_j \end{bmatrix} = \begin{bmatrix} a^2 (y_{ij0} + y_{ij}) & -a^* y_{ij} \\ -a y_{ij} & (y_{ij0} + y_{ij}) \end{bmatrix} \begin{bmatrix} V_i \\ V_j \end{bmatrix} \tag{2.92}$$

Alternative method

From Figure 2.16,

$$\begin{bmatrix} I_i' \\ I_j \end{bmatrix} = \begin{bmatrix} (y_{ij0} + y_{ij}) & -y_{ij} \\ -y_{ij} & (y_{ij0} + y_{ij}) \end{bmatrix} \begin{bmatrix} V_i' \\ V_j \end{bmatrix} \tag{2.93}$$

where

$$\begin{bmatrix} V_i' \\ V_j \end{bmatrix} = \begin{bmatrix} a & 0 \\ 0 & 1 \end{bmatrix} \begin{bmatrix} V_i \\ V_j \end{bmatrix} \tag{2.94a}$$

$$\begin{bmatrix} I_i' \\ I_j \end{bmatrix} = \begin{bmatrix} \dfrac{1}{a^*} & 0 \\ 0 & 1 \end{bmatrix} \begin{bmatrix} I_i \\ I_j \end{bmatrix} \tag{2.94b}$$

Substituting Eqs. (2.94a) and (2.94b) into Eq. (2.93),

$$\begin{bmatrix} \dfrac{1}{a^*} & 0 \\ 0 & 1 \end{bmatrix} \begin{bmatrix} I_i \\ I_j \end{bmatrix} = \begin{bmatrix} (y_{ij0} + y_{ij}) & - y_{ij} \\ - y_{ij} & (y_{ij0} + y_{ij}) \end{bmatrix} \begin{bmatrix} a & 0 \\ 0 & 1 \end{bmatrix} \begin{bmatrix} V_i \\ V_j \end{bmatrix}$$

or

$$\begin{bmatrix} I_i \\ I_j \end{bmatrix} = \begin{bmatrix} \dfrac{1}{a^*} & 0 \\ 0 & 1 \end{bmatrix}^{-1} \begin{bmatrix} a(y_{ij0} + y_{ij}) & - y_{ij} \\ - ay_{ij} & (y_{ij0} + y_{ij}) \end{bmatrix} \begin{bmatrix} V_i \\ V_j \end{bmatrix}$$

or

$$\begin{bmatrix} I_i \\ I_j \end{bmatrix} = \begin{bmatrix} a^* & 0 \\ 0 & 1 \end{bmatrix} \begin{bmatrix} a(y_{ij0} + y_{ij}) & - y_{ij} \\ - ay_{ij} & (y_{ij0} + y_{ij}) \end{bmatrix} \begin{bmatrix} V_i \\ V_j \end{bmatrix}$$

or

$$\begin{bmatrix} I_i \\ I_j \end{bmatrix} = \begin{bmatrix} a^2 (y_{ij0} + y_{ij}) & - a^* y_{ij} \\ - ay_{ij} & (y_{ij0} + y_{ij}) \end{bmatrix} \begin{bmatrix} V_i \\ V_j \end{bmatrix} \tag{2.95}$$

π-network representation of an off-nominal transformer

For an off-nominal turns ratio transformer, a is real, i.e. $a^* = a$. Therefore, Eq. (2.90) can be rewritten as

$$I_i = a^2(y_{ij0} + y_{ij})V_i - ay_{ij}V_j$$

or

$$I_i = (a^2y_{ij0} + a^2y_{ij})V_i - ay_{ij}V_i + ay_{ij}V_i - ay_{ij}V_j$$

or

$$I_i = [a^2y_{ij0} + a(a - 1)y_{ij}]V_i + ay_{ij}(V_i - V_j) \tag{2.96}$$

Equation (2.91) can be rewritten as

$$I_j = ay_{ij}V_j - ay_{ij}V_i + (y_{ij0} + y_{ij})V_j - ay_{ij}V_j$$

or

$$I_j = ay_{ij}(V_j - V_i) + [y_{ij0} + (1 - a)y_{ij}]V_j \tag{2.97}$$

Equations (2.96) and (2.97), can be represented by a π-network as shown in Figure 2.17.

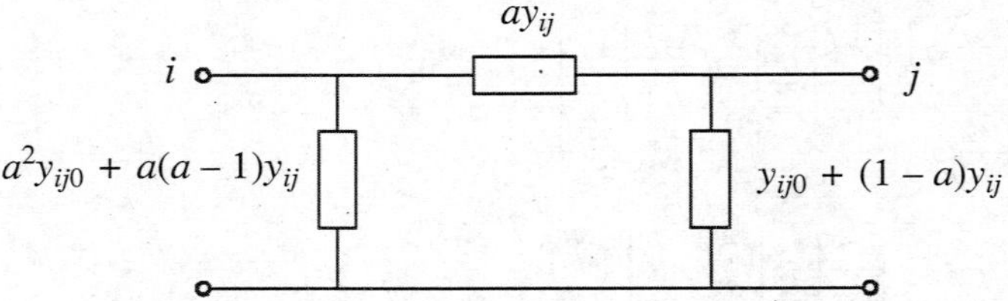

FIGURE 2.17 π-representation of a transmission line.

Off-nominal transformers at both line ends

With off-nominal tap-setting transformers at each end as shown in Figure 2.18, we have

$$\begin{bmatrix} I_i' \\ I_j' \end{bmatrix} = \begin{bmatrix} (y_{ij0} + y_{ij}) & -y_{ij} \\ -y_{ij} & (y_{ij0} + y_{ij}) \end{bmatrix} \begin{bmatrix} V_i' \\ V_j' \end{bmatrix} \tag{2.98}$$

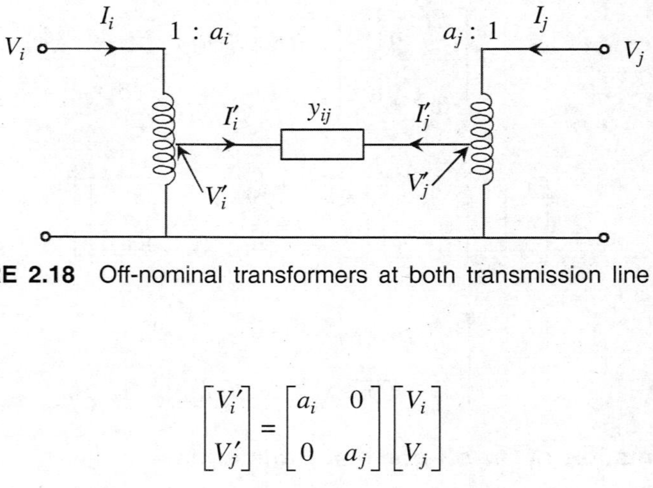

FIGURE 2.18 Off-nominal transformers at both transmission line ends.

where

$$\begin{bmatrix} V_i' \\ V_j' \end{bmatrix} = \begin{bmatrix} a_i & 0 \\ 0 & a_j \end{bmatrix} \begin{bmatrix} V_i \\ V_j \end{bmatrix} \tag{2.99a}$$

$$\begin{bmatrix} I_i' \\ I_j' \end{bmatrix} = \begin{bmatrix} \dfrac{1}{a_i^*} & 0 \\ 0 & \dfrac{1}{a_j^*} \end{bmatrix} \begin{bmatrix} I_i \\ I_j \end{bmatrix} \tag{2.99b}$$

Substituting Eqs. (2.99a) and (2.99b) into Eq. (2.98),

$$\begin{bmatrix} \frac{1}{a_i^*} & 0 \\ 0 & \frac{1}{a_j^*} \end{bmatrix} \begin{bmatrix} I_i \\ I_j \end{bmatrix} = \begin{bmatrix} (y_{ij0} + y_{ij}) & -y_{ij} \\ -y_{ij} & (y_{ij0} + y_{ij}) \end{bmatrix} \begin{bmatrix} a_i & 0 \\ 0 & a_j \end{bmatrix} \begin{bmatrix} V_i \\ V_j \end{bmatrix}$$

or

$$\begin{bmatrix} I_i \\ I_j \end{bmatrix} = \begin{bmatrix} \frac{1}{a_i^*} & 0 \\ 0 & \frac{1}{a_j^*} \end{bmatrix}^{-1} \begin{bmatrix} a_i(y_{ij0} + y_{ij}) & -a_j y_{ij} \\ -a_i y_{ij} & a_j(y_{ij0} + y_{ij}) \end{bmatrix} \begin{bmatrix} V_i \\ V_j \end{bmatrix}$$

or

$$\begin{bmatrix} I_i \\ I_j \end{bmatrix} = \begin{bmatrix} a_i^* & 0 \\ 0 & a_j^* \end{bmatrix} \begin{bmatrix} a_i(y_{ij0} + y_{ij}) & -a_j y_{ij} \\ -a_i y_{ij} & a_j(y_{ij0} + y_{ij}) \end{bmatrix} \begin{bmatrix} V_i \\ V_j \end{bmatrix}$$

or

$$\begin{bmatrix} I_i \\ I_j \end{bmatrix} = \begin{bmatrix} a_i a_i^*(y_{ij0} + y_{ij}) & -a_i^* a_j y_{ij} \\ -a_i a_j^* y_{ij} & a_j a_j^*(y_{ij0} + y_{ij}) \end{bmatrix} \begin{bmatrix} V_i \\ V_j \end{bmatrix} \quad (2.100)$$

2.9 GAUSS–SEIDEL METHOD

Let there be NB buses. Power system is comprised of three types of buses, namely slack bus, *PQ* buses and *PV* buses. The slack bus is numbered one, and the remaining (NB – 1) buses may be *PQ* buses or *PV* buses. Let 2 to NP buses be *PQ* buses and NP + 1 to NB buses be *PV* buses.

PQ buses

At *PQ* buses, real power, $P_i (i = 2, 3, ..., \text{NP})$ and reactive power, $Q_i (i = 2, 3, ..., \text{NP})$ are specified or known whereas voltage magnitudes, $|V_i|$ $(i = 2, 3, ..., \text{NP})$ and voltage angles, δ_i $(i = 2, 3, ..., \text{NP})$ are to be calculated. So, V_i is assumed initially and updated in every iteration.

The complex power injected into the *i*th bus is

$$S_i = P_i + jQ_i = V_i I_i^*$$

or

$$P_i - jQ_i = V_i^* I_i$$

or

$$I_i = \frac{P_i - jQ_i}{V_i^*} \quad (2.101a)$$

Current injected at each bus is

$$I_i = \sum_{k=1}^{\text{NB}} Y_{ik} V_k \qquad (i = 2, 3, ..., \text{NP}) \quad (2.101b)$$

Substituting the value of I_i into Eq. (2.101a), we get

$$P_i - jQ_i = V_i^* \sum_{k=1}^{NB} Y_{ik} V_k \qquad (i = 2, 3, ..., NP) \tag{2.101c}$$

From Eq. (2.101b), which can be rewritten as

$$I_i = Y_{ii} V_i + \sum_{\substack{k=1 \\ k \neq i}}^{NB} Y_{ik} V_k \qquad (i = 2, 3, ..., NP)$$

or

$$V_i = \frac{1}{Y_{ii}} \left\{ I_i - \sum_{\substack{k=1 \\ k \neq i}}^{NB} Y_{ik} V_k \right\} \qquad (i = 2, 3, ..., NP)$$

Substituting Eq. (2.101a) in the above equation,

$$V_i = \frac{1}{Y_{ii}} \left\{ \frac{P_i - jQ_i}{V_i^*} - \sum_{\substack{k=1 \\ k \neq i}}^{NB} Y_{ik} V_k \right\} \qquad (i = 2, 3, ..., NP) \tag{2.102}$$

Now, for the $(r + 1)$th iteration, the voltage becomes

$$V_i^{r+1} = \frac{1}{Y_{ii}} \left\{ \frac{P_i - jQ_i}{(V_i^*)^r} - \sum_{k=1}^{i-1} Y_{ik} V_k^{r+1} - \sum_{k=i+1}^{NB} Y_{ik} V_k^r \right\} \qquad (i = 2, 3, ..., NP) \tag{2.103}$$

For the $(r + 1)$th iteration, the updated values of V_k^{r+1} $(k = 1, 2, ..., i - 1)$ are used and for the rest of voltages previous values, i.e. V_k^r $(k = i + 1, i + 2, ..., NB)$ are used. The iterative process is continued till no further improvement in voltage is achieved.

$$\max_i \left| V_i^{r+1} - V_i^r \right| \le \varepsilon \qquad (i = 2, 3, ..., NP) \tag{2.104}$$

The limits of voltage magnitude can be checked and fixed as

$$V_i = \begin{cases} |V_i|^{min} & \text{if } |V_i| \le |V_i|^{min} \\ |V_i|^{max} & \text{if } |V_i| \ge |V_i|^{max} \\ |V_i| & \text{if } |V_i|^{min} < |V_i| < |V_i|^{max} \end{cases} \qquad (i = 2, 3, ..., NP) \tag{2.105}$$

PV buses

Real power, $P_i(i = NP + 1, NP + 2, ..., NB)$ and voltage magnitude, $|V_i|$ $(i = NP + 1, NP + 2, ..., NB)$ are known at *PV* buses. Reactive power, $Q_i(i = NP + 1, NP + 2, ..., NB)$ and voltage angle, $\delta_i(i = NP + 1, NP + 2, ..., NB)$ are unknown at *PV* buses. So, Q_i and δ_i are updated in every iteration.

From Eq. (2.101c),

$$Q_i = -\text{Im}\left\{V_i^* \sum_{k=1}^{\text{NB}} Y_{ik} V_k\right\} \qquad (i = \text{NP} + 1, \text{NP} + 2, ..., \text{NB}) \tag{2.106a}$$

The revised value of Q_i is obtained from the above equation by substituting the most updated values of voltages. Thus,

$$Q_i = -\text{Im}\left\{(V_i^*)^r \sum_{k=1}^{i-1} Y_{ik} V_k^{r+1} + (V_i^*)^r \sum_{k=i}^{\text{NB}} Y_{ik} V_k^r\right\} \qquad (i = \text{NP} + 1, \text{NP} + 2, ..., \text{NB}) \tag{2.106b}$$

The revised value of δ_i is obtained from Eq. (2.103) as

$$\delta_i^{r+1} = \angle\left[\frac{1}{Y_{ii}}\left\{\frac{P_i - jQ_i^{r+1}}{(V_i^*)^r} - \sum_{k=1}^{i-1} Y_{ik} V_k^{r+1} - \sum_{k=i+1}^{\text{NB}} Y_{ik} V_k^r\right\}\right] \qquad (i = \text{NP} + 1, \text{NP} + 2, ..., \text{NB}) \tag{2.106c}$$

Limits of reactive power can be checked and fixed as given below:

$$Q_i = \begin{cases} Q_i^{\min} & \text{if } Q_i \le Q_i^{\min} \\ Q_i^{\max} & \text{if } Q_i \ge Q_i^{\max} \\ Q_i & \text{if } Q_i^{\min} < Q_i < Q_i^{\max} \end{cases} \qquad (i = \text{NP} + 1, \text{NP} + 2, ..., \text{NB}) \tag{2.107}$$

If any limit (either maximum or minimum) is violated, then that bus is treated as the *PQ* bus. But if in a subsequent computation iteration, Q_i comes within the limits then the bus is converted back to the *PV* bus.

Slack bus

At slack bus, voltage magnitude $|V_1|$ and voltage angle δ_1 are specified or known, and real power P_1 and reactive power Q_1 are to be calculated. To calculate power, the following equations can be used as

$$P_1 = \text{Re}\left\{V_1^* \sum_{k=1}^{\text{NB}} Y_{1k} V_k\right\} \tag{2.108a}$$

$$Q_1 = -\text{Im}\left\{V_1^* \sum_{k=1}^{\text{NB}} Y_{1k} V_k\right\} \tag{2.108b}$$

Acceleration convergence

Convergence in the Gauss–Seidel method can be speeded up by the use of the acceleration factor

$$V_i^{r+1} = V_i^r + \alpha(V_i^{r+1} - V_i^r) \tag{2.109}$$

where α is a real number called the *acceleration factor*.

A suitable value of α for any system can be obtained by trial load flow studies. A general recommended value is 1.6. A wrong choice of α may slow down convergence or even cause the method to diverge. The Gauss–Seidel method requires the smallest number of arithmetic operations to complete an iteration. A detailed stepwise procedure is explained here.

Algorithm 2.3: Gauss–Seidel Method to Perform the Load Flow

1. Read data
   NB (number of buses); NP (number of PQ buses).
   V_1, δ_1 for slack bus, $P_i(i = 2, 3, ..., \text{NB})$ for PQ and PV buses.
   $Q_i(i = 2, 3, ..., \text{NP})$ for PQ buses, $V_i^S(i = \text{NP} + 1, \text{NP} + 2, ..., \text{NB})$ for PV buses.
   $V_i^{\min}$, $V_i^{\max}$ $(i = 2, 3, ..., \text{NP})$ for PQ buses.
   $Q_i^{\min}$, $Q_i^{\max}$ $(i = \text{NP} + 1, \text{NP} + 2, ..., \text{NB})$ for PV buses. α (step length), R (no. of iterations), ε (convergence tolerance)
2. Form Y_{BUS} using Algorithm 2.1.
3. Assume initially

$$\left|V_i\right| (i = 2, 3, ..., \text{NP}) \quad \text{and} \quad \delta_i(i = 2, 3, ..., \text{NB})$$

4. Set iteration count, $r = 0$ and $|\Delta V^{\max}| = 0$
5. Set bus count $i = 0$
6. If BUS is PQ-bus then
   6.1. Compute V_i^{new} from Eq. (2.102) as

$$V_i^{\text{new}} = \frac{1}{Y_{ii}} \left\{ \frac{P_i - jQ_i}{V_i^*} - \sum_{\substack{k=1 \\ k \neq i}}^{\text{NB}} Y_{ik} V_k \right\}$$

   6.2. Update the voltage according to Eq. (2.109) as

$$V_i^{\text{new}} = V_i + \alpha(V_i^{\text{new}} - V_i)$$

   6.3. Check the limits of $|V_i^{\text{new}}|$ and set according to Eq. (2.105), i.e.

$$\left|V_i^{\text{new}}\right| = \begin{cases} \left|V_i\right|^{\min} & \text{if } \left|V_i^{\text{new}}\right| \le \left|V_i\right|^{\min} \\ \left|V_i\right|^{\max} & \text{if } \left|V_i^{\text{new}}\right| \ge \left|V_i\right|^{\max} \\ \left|V_i\right| & \text{if } \left|V_i\right|^{\min} < \left|V_i^{\text{new}}\right| < \left|V_i\right|^{\max} \end{cases}$$

   6.4. Compute $\Delta V_i = V_i^{\text{new}} - V_i$

   if $|\Delta V_i| > |\Delta V^{\max}|$ then $|\Delta V^{\max}| = |\Delta V_i|$

   6.5. Assign new voltage to old

$$V_i = V_i^{\text{new}}$$

7. If BUS is the *PV*-bus then

   7.1. Compute Q_i^{new} for the *PV* bus using Eq. (2.106a)

$$Q_i^{\text{new}} = -\text{Im}\left\{V_i^* \sum_{k=1}^{\text{NB}} Y_{ik} V_k\right\}$$

   7.2. Check the limits of Q_i and set according to Eq. (2.107a)

$$Q_i^{\text{new}} = \begin{cases} Q_i^{\min} & \text{if } Q_i^{\text{new}} \le Q_i^{\min} \\ Q_i^{\max} & \text{if } Q_i^{\text{new}} \ge Q_i^{\max} \\ Q_i & \text{if } Q_i^{\min} < Q_i^{\text{new}} < Q_i^{\max} \end{cases}$$

   If no limit is violated then set $Q^{\text{limit}} = 0$
   If any limit is violated then set $Q^{\text{limit}} = 1$

   7.3. Compute the voltage angle for the *PV* bus using Eq. (2.106c)

$$V_i^{\text{new}} = \left|V_i^{\text{new}}\right| \angle\delta_i^{\text{new}} = \left[\frac{1}{Y_{ii}}\left\{\frac{P_i - jQ_i^{\text{new}}}{V_i^*} - \sum_{\substack{k=1 \\ k \ne i}}^{\text{NB}} Y_{ik} V_k\right\}\right]$$

   7.4. Update the voltage according to Eq. (2.109)

$$V_i^{\text{new}} = V_i + \alpha(V_i^{\text{new}} - V_i)$$

   7.5. If $Q^{\text{limit}} = 0$ then

$$\delta_i^{\text{new}} = \tan^{-1}\left(\frac{\text{Im}\,(V_i^{\text{new}})}{\text{Re}\,(V_i^{\text{new}})}\right)$$

$$V_i^{\text{new}} = V_i^{\text{S}} \angle\delta_i^{\text{new}}$$

$$V_i^{\text{new}} = V_i^{\text{S}} (\cos \delta_i^{\text{new}} + j \sin \delta_i^{\text{new}})$$

   7.6. Assign new voltage to old

$$V_i = V_i^{\text{new}}$$

8. Increment the bus counter, $i = i + 1$
9. Check that all voltages of *PQ* and *PV* buses have been modified
   if $i \le \text{NB}$ then GOTO Step 6 and repeat.
10. Check convergence
    $r = r + 1$
    If $|\Delta V^{\max}| \ge \varepsilon$ and $r \le R$ then GOTO Step 5 and repeat.

11. Compute powers on slack bus

$$P_1 - jQ_1 = \left\{V_1^* \sum_{k=1}^{NB} Y_{1k} V_k\right\}$$

12. Calculate line flows using Eqs. (2.85) and (2.86)

$$S_{ik} = V_i \left[(V_i^* - V_k^*)\, y_{ik}^* + V_i^* y_{ik0}^*\right]$$

$$S_{ki} = V_k \left[(V_k^* - V_i^*)\, y_{ik}^* + V_k^* y_{ki0}^*\right]$$

13. Stop.

EXAMPLE 2.6 For the sample system of Figure 2.19, the generators are connected at all the four buses, while loads are at buses 2, 3, and 4. The values of real and reactive powers are listed in Table 2.4. All buses other than slack are of *PQ*-type. Line data are given in Table 2.5. Find the voltages and the bus angles at the three buses using the GS (Gauss–Seidel) iteration.

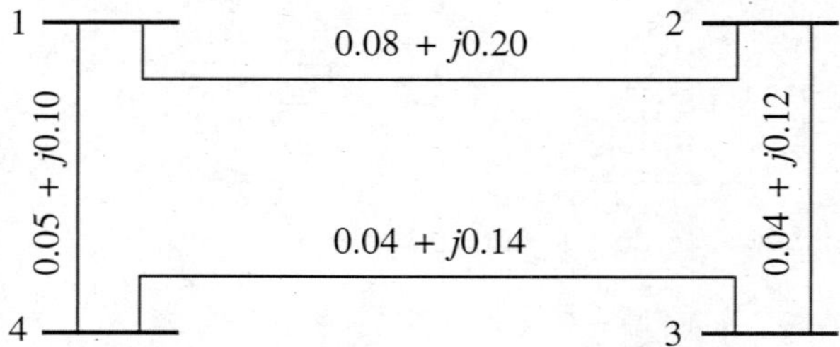

FIGURE 2.19 Sample system.

TABLE 2.4 Input data

Bus	P_i (p.u.)	Q_i (p.u.)	V_i (p.u.)	*Type of bus*
1	–	–	$1.05\angle 0$	Slack
2	– 0.45	– 0.15	–	*PQ*
3	– 0.51	– 0.25	–	*PQ*
4	– 0.60	– 0.30	–	*PQ*

TABLE 2.5 Line data

Line no.	*Bus code (p–q)*	*Line impedance*
1	1–2	0.08 + *j*0.20
2	1–4	0.05 + *j*0.10
3	2–3	0.04 + *j*0.12
4	3–4	0.04 + *j*0.14

Solution Y-bus is calculated using Algorithm 2.1. The elements of Y_{BUS} are given below:

Y_{11} = 5.724138 – j12.310340 $\quad$ Y_{12} = – 1.724138 + j4.310345

Y_{13} = 0.0 + j0.0 $\quad$ Y_{14} = – 4.0 + j8.0

Y_{21} = –1.724138 + j4.310345 $\quad$ Y_{22} = 4.224138 – j11.810340

Y_{23} = –2.5 + j7.5 $\quad$ Y_{24} = 0.0 + j0.0

Y_{31} = 0.0 + j0.0 $\quad$ Y_{32} = – 2.5 + j7.50

Y_{33} = 4.386792 – j14.103770 $\quad$ Y_{34} = – 1.886792 + j6.603774

Y_{41} = – 4.0 + j8.0 $\quad$ Y_{42} = 0.0 + j0.0

Y_{43} = – 1.886792 + j6.603774 $\quad$ Y_{44} = 5.886792 – j14.603770

Assuming flat start (means set the voltage to 1 p.u. value)

$$V_1 = 1.05 + j0.0 \text{ p.u.} \qquad V_2 = 1.00 + j0.0 \text{ p.u.}$$

$$V_3 = 1.00 + j0.0 \text{ p.u.} \qquad V_4 = 1.00 + j0.0 \text{ p.u.}$$

Set iteration counter, $r = 0$

Let $$|\Delta V^{max}| = 0$$

Using Eq. (2.102), V_2^{new} is computed as

$$V_2^{new} = \frac{1}{Y_{22}}\left[\frac{P_2 - jQ_2}{V_2^*} - \sum_{\substack{j=1\\ j\neq 2}}^{4} Y_{2j}V_j\right]$$

or

$$V_2^{new} = \frac{1}{Y_{22}}\left[\frac{P_2 - jQ_2}{V_2^*} - Y_{21}V_1 - Y_{23}V_3 - Y_{24}V_4\right]$$

or

$$V_2^{new} = 0.9951507 - j0.0290685 \text{ p.u.}$$

Using Eq. (2.109), V_2^{new} is accelerated

$$V_2^{new} = V_2 + \alpha(V_2^{new} - V_2)$$

where α =1.2 (assumed value). Therefore,

$$V_2^{new} = 0.9941808 - j0.03488218 \text{ p.u.}$$

Change in voltage is computed as

$$|\Delta V| = \left|V_2^{new} - V_2\right| = 0.0353624$$

Maximum change is recorded to check the convergence, i.e.

$$|\Delta V| > \left|\Delta V^{max}\right|, \text{ so } \left|\Delta V^{max}\right| = 0.0353624$$

The calculated value of V_2^{new} is assigned to V_2, i.e.

$$V_2 = V_2^{new} = 0.9941808 - j0.03488218 \text{ p.u.}$$

Now to calculate V_3^{new}, the following voltage values are considered in which the value of voltage V_2 is the updated value.

$$V_1 = 1.05 + j0.0 \text{ p.u.}$$
$$V_2 = 0.9941808 - j0.03488218 \text{ p.u.}$$
$$V_3 = 1.00 + j0.0 \text{ p.u.}$$
$$V_4 = 1.00 + j0.0 \text{ p.u.}$$

Using Eq. (2.102), V_3^{new} is computed as

$$V_3^{new} = \frac{1}{Y_{33}}\left[\frac{P_3 - jQ_3}{V_3^*} - \sum_{\substack{j=1\\j\neq 3}}^{4} Y_{3j}V_j\right]$$

or

$$V_3^{new} = \frac{1}{Y_{33}}\left[\frac{P_3 - jQ_3}{V_3^*} - Y_{31}V_1 - Y_{32}V_2 - Y_{34}V_4\right]$$

or

$$V_3^{new} = 0.9708458 - j0.04667337 \text{ p.u.}$$

Using Eq. (2.109), V_3^{new} is accelerated, i.e.

$$V_3^{new} = V_3 + \alpha(V_3^{new} - V_3)$$

or

$$V_3^{new} = 0.9650149 - j0.05600805 \text{ p.u.}$$

Change in voltage is computed as

$$|\Delta V| = |V_3^{new} - V_3| = 0.06603678$$

Maximum change is recorded to check the convergence, i.e.

$$|\Delta V| > |\Delta V^{max}|, \text{ so } |\Delta V^{max}| = 0.06603678$$

The calculated value of V_3^{new} is assigned to V_3. Thus,

$$V_3 = V_3^{new} = 0.9650149 - j0.05600805 \text{ p.u.}$$

To calculate V_4^{new}, the following voltage values are used in which the value of voltage V_3 is the updated value.

$$V_1 = 1.05 + j0.0 \text{ p.u.}$$
$$V_2 = 0.9941808 - j0.03488218 \text{ p.u.}$$
$$V_3 = 0.9650149 - j0.05600805 \text{ p.u.}$$
$$V_4 = 1.00 + j0.0 \text{ p.u.}$$

Using Eq. (2.102), V_4^{new} is computed as

$$V_4^{new} = \frac{1}{Y_{44}}\left[\frac{P_4 - jQ_4}{V_4^*} - \sum_{\substack{j=1\\ j\neq 4}}^{4} Y_{4j}V_j\right]$$

or

$$V_4^{new} = \frac{1}{Y_{44}}\left[\frac{P_4 - jQ_4}{V_4^*} - Y_{41}V_1 - Y_{42}V_2 - Y_{43}V_3\right]$$

or

$$V_4^{new} = 0.9786592 - j0.04863431 \text{ p.u.}$$

Using Eq. (2.109), V_4^{new} is accelerated, i.e.

$$V_4^{new} = V_4 + \alpha(V_4^{new} - V_4)$$

or

$$V_4^{new} = 0.974391 - j0.05836118 \text{ p.u.}$$

Change in voltage is computed as

$$\left|\Delta V\right| = \left|V_4^{new} - V_4\right| = 0.06373263$$

Maximum change is recorded to check the convergence, i.e.

$$\left|\Delta V\right| < \left|\Delta V^{max}\right|, \text{ so } \left|\Delta V^{max}\right| = 0.06603678$$

The calculated value of V_4^{new} is assigned to V_4, i.e.

$$V_4 = V_4^{new} = 0.9650149 - j0.05600805 \text{ p.u.}$$

The convergence is checked as below:

$$\text{Is } \left|\Delta V^{max}\right| < \varepsilon(0.0001)$$

If the convergence criterion is not satisfied, go for the next iteration by incrementing the counter r, i.e. $r = r + 1$.

To start the new iteration, the following voltage values are used in which the value of voltage V_4 is the updated value.

$$V_1 = 1.05 + j0.0 \text{ p.u.}$$

$$V_2 = 0.9941808 - j0.03488218 \text{ p.u.}$$

$$V_3 = 0.9650149 - j0.05600805 \text{ p.u.}$$

$$V_4 = 0.9650149 - j0.05600805 \text{ p.u.}$$

After 13 iterations, the final voltage values, when $\left|\Delta V^{max}\right| = 5.584756 \times 10^{-06}$, are given as

$$V_1 = 1.05 + j0.0$$

$$V_2 = 0.9286948 - j0.0970232 \text{ p.u.}$$

$$V_3 = 0.9046143 - j0.1109611 \text{ p.u.}$$

$$V_4 = 0.9461540 - j0.0687263 \text{ p.u.}$$

Voltage magnitudes and angles are given below:

$$|V_1| = 1.05 \text{ p.u.} \qquad \delta_1 = 0.0 \text{ rad}$$

$$|V_2| = 0.9337450 \text{ p.u.} \qquad \delta_2 = -0.1040950 \text{ rad}$$

$$|V_3| = 0.9113891 \text{ p.u.} \qquad \delta_3 = -0.1220516 \text{ rad}$$

$$|V_4| = 0.9486451 \text{ p.u.} \qquad \delta_4 = -0.0725102 \text{ rad}$$

The slack bus real and reactive powers are computed using the following equation.

$$P_1 - jQ_1 = V_1^* \sum_{j=1}^{4} Y_{1j} V_j = 1.6721720 - j0.9570209 \text{ p.u.}$$

The line flows are tabulated below in Table 2.6.

TABLE 2.6 Line flows

Line no.	*Bus code* (*i–k*)	S_{ik}
1	1–2	0.6587177 + j0.3733650
1	2–1	–0.6171169 – j0.2693631
2	1–4	1.0134540 + j0.5836556
2	4–1	–0.9514246 – j0.4595973
3	2–3	0.1671311 + j0.1193821
3	3–2	–0.1651958 – j0.1135760
4	3–4	–0.3448024 – j0.1364204
4	4–3	0.3514238 + j0.1595952

EXAMPLE 2.7 For the sample system of Figure 2.19, the generators are connected at all the four buses, while loads are at buses 2, 3 and 4. The values of real and reactive powers are listed in Table 2.7 along with the type of bus. Line data are given in Table 2.5. Find the voltage and the bus angles at the three buses using GS (Gauss–Seidel) iteration.

TABLE 2.7 Input data

Bus	P_i (p.u.)	Q_i (p.u.)	V_i (p.u.)	*Type of bus*
1	–	–	$1.05\angle 0$	Slack
2	– 0.45	– 0.15	$1.00\angle 0$	*PV*
3	– 0.51	– 0.25	–	*PQ*
4	– 0.60	– 0.30	–	*PQ*

Solution Y-bus is calculated using Algorithm 2.1 and the values of Y_{BUS} elements are the same as those given in Example 2.6.

Assuming flat start, means that the voltage is set to 1 p.u. value.

$V_1 = 1.05 + j0.0$ p.u., $V_3 = 1.00 + j0.0$ p.u., and $V_4 = 1.00 + j0.0$ p.u.

Set iteration counter, $r = 0$.

Let
$$|\Delta V^{\max}| = 0$$

Q_2 is computed using Eq. (2.106a), i.e.

$$Q_2 = -\text{Im}\left\{V_2^* \sum_{k=1}^{4} Y_{2k}V_k\right\}$$

or

$$Q_2 = -\text{Im}\{V_2^*(Y_{21}V_1 + Y_{22}V_2 + Y_{23}V_3 + Y_{24}V_4)\} = -0.215517 \text{ p.u.}$$

Using Eq. (2.102), V_2^{new} is computed as

$$V_2^{\text{new}} = \frac{1}{Y_{22}}\left[\frac{P_2 - jQ_2}{V_2^*} - \sum_{\substack{j=1\\ j\neq 2}}^{4} Y_{2j}V_j\right]$$

or

$$V_2^{\text{new}} = \frac{1}{Y_{22}}\left[\frac{P_2 - jQ_2}{V_2^*} - Y_{21}V_1 - Y_{23}V_3 - Y_{24}V_4\right]$$

or

$$V_2^{\text{new}} = 0.9902325 - j0.0273094 \text{ p.u.}$$

Using Eq. (2.109), V_2^{new} is accelerated.

$$V_2^{\text{new}} = V_2 + \alpha(V_2^{\text{new}} - V_2)$$

Assuming $\alpha = 1.2$

$$V_2^{\text{new}} = 0.988279 - j0.03277127 \text{ p.u.}$$

The voltage angle is separated, i.e.

$$\delta_2^{\text{new}} = \tan^{-1}\left(\frac{-0.03277127}{0.988279}\right) = -0.0331478$$

The voltage angle is attached with the specified voltage of the PV bus.

$$V_2^{\text{new}} = 1.0 \times [\cos(-0.0331478) + j\sin(-0.0331478)]$$

or

$$V_2^{\text{new}} = 0.9994507 - j0.03314173 \text{ p.u.}$$

The computed value of V_2^{new} is assigned to V_2, i.e.

$$V_2 = V_2^{\text{new}} = 0.9994507 - j0.03314173 \text{ p.u.}$$

Now to calculate V_3^{new}, the following voltage values are used in which the value of voltage V_2 is the updated value.

$$V_1 = 1.05 + j0.0 \text{ p.u.}$$

$$V_2 = 0.9994507 - j0.03314173 \text{ p.u.}$$

$$V_3 = 1.00 + j0.0 \text{ p.u.}$$

$$V_4 = 1.00 + j0.0 \text{ p.u.}$$

Using Eq. (2.102), V_3^{new} is computed as

$$V_3^{\text{new}} = \frac{1}{Y_{33}}\left[\frac{P_3 - jQ_3}{V_3^*} - \sum_{\substack{j=1 \\ j\neq 3}}^{4} Y_{3j}V_j\right]$$

or

$$V_3^{\text{new}} = \frac{1}{Y_{33}}\left[\frac{P_3 - jQ_3}{V_3^*} - Y_{31}V_1 - Y_{32}V_2 - Y_{34}V_4\right]$$

or

$$V_3^{\text{new}} = 0.9736471 - j0.04568502 \text{ p.u.}$$

Using Eq. (2.109), V_3^{new} is accelerated, i.e.

$$V_3^{\text{new}} = V_3 + \alpha(V_3^{\text{new}} - V_3)$$

or

$$V_3^{\text{new}} = 0.9683765 - j0.05482203 \text{ p.u.}$$

Change in voltage is computed as

$$\left|\Delta V\right| = \left|V_3^{\text{new}} - V_3\right| = 0.06328905$$

Maximum change is recorded to check the convergence, i.e.

$$\text{If} \quad \left|\Delta V\right| > \left|\Delta V^{\max}\right|, \quad \text{then} \quad \left|\Delta V^{\max}\right| = 0.06328905$$

The computed value of V_3^{new} is assigned to V_3, i.e.

$$V_3 = V_3^{\text{new}} = 0.9683765 - j0.05482203 \text{ p.u.}$$

To calculate V_4^{new}, the following voltage values are used in which the value of voltage V_3 is the updated value.

$$V_1 = 1.05 + j0.0 \text{ p.u.}$$

$$V_2 = 0.9994507 - j0.03314173 \text{ p.u.}$$

$$V_3 = 0.9683765 - j0.05482203 \text{ p.u.}$$

$$V_4 = 1.00 + j0.0 \text{ p.u.}$$

Using Eq. (2.102), V_4^{new} is computed as

$$V_4^{\text{new}} = \frac{1}{Y_{44}}\left[\frac{P_4 - jQ_4}{V_4^*} - \sum_{\substack{j=1 \\ j\neq 4}}^{4} Y_{4j}V_j\right]$$

or

$$V_4^{\text{new}} = \frac{1}{Y_{44}}\left[\frac{P_4 - jQ_4}{V_4^*} - Y_{41}V_1 - Y_{42}V_2 - Y_{43}V_3\right]$$

or

$$V_4^{\text{new}} = 0.9801714 - j0.04827332 \text{ p.u.}$$

Using Eq. (2.109), V_4^{new} is accelerated, i.e.

$$V_4^{\text{new}} = V_4 + \alpha(V_4^{\text{new}} - V_4)$$

or

$$V_4^{\text{new}} = 0.9762057 - j0.05792799 \text{ p.u.}$$

Change in voltage is computed as

$$|\Delta V| = \left|V_4^{\text{new}} - V_4\right| = 0.06262444$$

Maximum change is recorded to check the convergence, i.e.

$$\text{If} \quad |\Delta V| < \left|\Delta V^{\max}\right|, \quad \text{then} \quad \left|\Delta V^{\max}\right| = 0.06328905$$

The computed value of V_4^{new} is assigned to V_4, i.e.

$$V_4 = V_4^{\text{new}} = 0.9762057 - j0.05792799 \text{ p.u.}$$

The convergence is checked as

$$\text{Is}\left|\Delta V^{\max}\right| < \varepsilon(0.0001)$$

Convergence criterion is not satisfied, so we go for the next iteration by updating the counter by 1.

$$r = r + 1$$

To start new iteration, the following voltage values are used in which the value of voltage V_4 is the updated value.

$$V_1 = 1.05 + j0.0 \text{ p.u.}$$

$$V_2 = 0.9994507 - j0.03314173 \text{ p.u.}$$

$$V_3 = 0.9683765 - j0.05482203 \text{ p.u.}$$

$$V_4 = 0.9762057 - j0.05792799 \text{ p.u.}$$

After 11 iterations the final voltage values are given below, when $\left|\Delta V^{\max}\right| = 9.300476 \times 10^{-06}$.

$$V_1 = 1.05 + j0.0$$

$$V_2 = 0.9925385 - j0.1219321 \text{ p.u.}$$

$$V_3 = 0.9492246 - j0.1257839 \text{ p.u.}$$

$$V_4 = 0.9653854 - j0.0764109 \text{ p.u.}$$

Voltage magnitudes and angles are mentioned here:

$$|V_1| = 1.05 \text{ p.u.} \qquad \delta_1 = 0.0 \text{ rad}$$

$$|V_2| = 1.00 \text{ p.u.} \qquad \delta_2 = -0.1222363 \text{ rad}$$

$$|V_3| = 0.9575223 \text{ p.u.} \qquad \delta_3 = -0.1317447 \text{ rad}$$

$$|V_4| = 0.9684047 \text{ p.u.} \qquad \delta_4 = -0.0789860 \text{ rad}$$

The slack bus real and reactive powers are computed and are given below:

$$P_1 - jQ_1 = V_1^* \sum_{j=1}^{4} Y_{1j} V_j$$

or

$$P_1 - jQ_1 = 1.6531060 - j0.429160 \text{ p.u.}$$

The line flows are tabulated below in Table 2.8.

TABLE 2.8 Line flows

Line no.	*Bus code (i–k)*	S_{ik}
1	1–2	0.6558731 + *j*0.0393236
1	2–1	–0.6245468 + *j*0.0389922
2	1–4	0.9972326 + *j*0.3898367
2	4–1	–0.9452397 – *j*0.2858507
3	2–3	0.1745856 + *j*0.2961466
3	3–2	–0.1698582 – *j*0.2819647
4	3–4	–0.3401432 + *j*0.0319703
4	4–3	0.3452354 – *j*0.0141476

2.10 NEWTON–RAPHSON METHOD

The load flow solution must satisfy the following nonlinear algebraic equations, i.e.

$$f_i(V, \delta) = 0 \text{ for all buses} \tag{2.110}$$

At each bus, except the slack bus, real and reactive powers need to be evaluated.

$$f_i(V, \delta) = P_i^S - P_i(V, \delta) = 0 \quad \text{for all } PQ \text{ and } PV \text{ buses} \qquad (2.111a)$$

$$f_i(V, \delta) = Q_i^S - Q_i(V, \delta) = 0 \quad \text{for all } PQ \text{ buses.} \qquad (2.111b)$$

where

P_i^S is specified real power of the ith bus
Q_i^S is specified reactive power of the ith bus
V_i is magnitude of voltage at the ith bus
δ_i is angle of voltage at the ith bus
NB is the number of buses
NV is the number of PV buses.

Consider that number one bus of the power system is the slack bus. Assume that initial values of the unknown variables are as under:

$$V_i^0 \ (i = \text{NV} + 1, \text{NV} + 2, ..., \text{NB} \qquad \text{for all } PQ \text{ buses})$$

$$\delta_i^0 \ (i = 2, 3, ..., \text{NB} \qquad \text{for all } PQ \text{ and } PV \text{ buses})$$

Let ΔV_i (i = NV + 1, NV + 2, ..., NB) and $\Delta\delta_i$ (i = 2, 3, ..., NB) be the corrections, which on being added to the initial guess, give the actual solution. Equation (2.110) can be written as

$$f_i(V_k^0 + \Delta V_k, \delta_k^0 + \Delta\delta_k) = 0 \qquad (i = 2, 3, ..., \text{NB}) \qquad (2.112)$$

Expanding the above equation using Taylor's series around the initial guess, and ignoring the higher order terms

$$f_i(V_k^0 + \Delta V_k, \delta_k^0 + \Delta\delta_k) = f_i(V_k^0, \delta_i^0) + \sum_{k=\text{NV}+1}^{\text{NB}} \frac{\partial f_i}{\partial V_k} \Delta V_k + \sum_{k=2}^{\text{NB}} \frac{\partial f_i}{\partial \delta_k} \Delta\delta_k$$

Partial derivatives are evaluated around the initial points (V_k^0, δ_k^0). Using Eq. (2.112), the above equation can be rewritten as

$$\sum_{k=\text{NV}+1}^{\text{NB}} \frac{\partial f_i}{\partial V_k} \Delta V_k + \sum_{k=2}^{\text{NB}} \frac{\partial f_i}{\partial \delta_k} \Delta\delta_k = -f_i(V_k^0, \delta_k^0) \qquad (2.113)$$

After expanding Eq. (2.111a) using Taylor's series around initial guess, the equation is written in the form of Eq. (2.113) as

$$\sum_{k=2}^{\text{NB}} \frac{\partial P_i}{\partial \delta_k} \Delta\delta_k + \sum_{k=\text{NV}+1}^{\text{NB}} \frac{\partial P_i}{\partial V_k} \Delta V_k = \Delta P_i \qquad (i = 2, 3, ..., \text{NB}) \qquad (2.114)$$

where $\Delta P_i = P_i(V_k^0, \delta_k^0) - P_i^S$.

Similarly, after expanding Eq. (2.111b) using Taylor's series around the initial guess, the equation is written in the form of Eq. (2.113) as

$$\sum_{k=2}^{\text{NB}} \frac{\partial Q_i}{\partial \delta_k} \Delta\delta_k + \sum_{k=\text{NV}+1}^{\text{NB}} \frac{\partial Q_i}{\partial V_k} \Delta V_k = \Delta Q_i \qquad (i = \text{NV} + 1, \text{NV} + 2, ..., \text{NB}) \qquad (2.115)$$

where $\Delta Q_i = Q_i(V_k^0, \delta_k^0) - Q_i^S$.

Equations (2.114) and (2.115) can be written in matrix form as

$$\begin{bmatrix} \dfrac{\partial P}{\partial \delta} & \dfrac{\partial P}{\partial V} \\ \dfrac{\partial Q}{\partial \delta} & \dfrac{\partial Q}{\partial V} \end{bmatrix} \begin{bmatrix} \Delta\delta \\ \Delta V \end{bmatrix} = \begin{bmatrix} \Delta P \\ \Delta Q \end{bmatrix} \tag{2.116}$$

Size of matrix = NB + (NB – (NV+1)) – 1 = 2(NB – 1) – NV

where

NB – (NV + 1) is the number of *PQ* buses
NB is number of total buses
NV is number of *PV* buses.

Equations (2.114) and (2.115) are solved iteratively, till ΔP and ΔQ become almost zero. Let

$$H_{ik} = \frac{\partial P_i}{\partial \delta_k}, \quad N_{ik} = V_k \frac{\partial P_i}{\partial V_k}, \quad J_{ik} = \frac{\partial Q_i}{\partial \delta_k}, \quad L_{ik} = V_k \frac{\partial Q_i}{\partial V_k}$$

Equation (2.116) can be rewritten as

$$\begin{bmatrix} H & N \\ J & L \end{bmatrix} \begin{bmatrix} \Delta\delta \\ \dfrac{\Delta V}{V} \end{bmatrix} = \begin{bmatrix} \Delta P \\ \Delta Q \end{bmatrix} \tag{2.117}$$

Equation (2.114) can be rewritten in terms of *H* and *N* as

$$\sum_{k=2}^{\text{NB}} H_{ik}\Delta\delta_k + \sum_{k=\text{NV}+1}^{\text{NB}} N_{ik}\frac{\Delta V_k}{V_k} = \Delta P_i \qquad (i = 2, 3, ..., \text{NB}) \tag{2.118}$$

Equation (2.115) can be rewritten in terms of *J* and *L* as

$$\sum_{k=2}^{\text{NB}} J_{ik}\Delta\delta_k + \sum_{k=\text{NV}+1}^{\text{NB}} L_{ik}\frac{\Delta V_k}{V_k} = \Delta Q_i \qquad (i = \text{NV} + 1, \text{NV} + 2, ..., \text{NB}) \tag{2.119}$$

Evaluation of Jacobian matrix elements

Real and reactive powers are represented by Eq. (2.83a) and Eq. (2.83b), respectively, and are enlisted below:

$$P_i = \sum_{k=1}^{\text{NB}} V_i V_k \,[G_{ik} \cos(\delta_i - \delta_k) + B_{ik} \sin(\delta_i - \delta_k)]$$

$$Q_i = \sum_{k=1}^{\text{NB}} V_i V_k \,[G_{ik} \sin(\delta_i - \delta_k) - B_{ik} \cos(\delta_i - \delta_k)]$$

The above equations can be rewritten as

$$P_i = G_{ii}V_i^2 + \sum_{\substack{k=1 \\ k\neq i}}^{\text{NB}} V_i V_k \,[G_{ik} \cos(\delta_i - \delta_k) + B_{ik} \sin(\delta_i - \delta_k)] \tag{2.120a}$$

$$Q_i = -B_{ii}V_i^2 + \sum_{\substack{k=1 \\ k \neq i}}^{NB} V_i V_k \, [G_{ik} \sin(\delta_i - \delta_k) - B_{ik} \cos(\delta_i - \delta_k)] \tag{2.120b}$$

Differentiating Eq. (2.120a) w.r.t. δ_i to find H_{ii}

$$\frac{\partial P_i}{\partial \delta_i} = \sum_{\substack{k=1 \\ k \neq i}}^{NB} V_i V_k \, [-G_{ik} \sin(\delta_i - \delta_k) + B_{ik} \cos(\delta_i - \delta_k)] \tag{2.121}$$

Adding Eqs. (2.121) and (2.120b),

$$Q_i + \frac{\partial P_i}{\partial \delta_i} = -B_{ii}V_i^2$$

Rearranging the above equation,

$$H_{ii} = \frac{\partial P_i}{\partial \delta_i} = -Q_i - B_{ii}V_i^2 \tag{2.122a}$$

Differentiating Eq. (2.120a) with respect to δ_k $(k \neq i)$ to find H_{ik} $(k \neq i)$

$$H_{ik} = \frac{\partial P_i}{\partial \delta_k} = V_i V_k [G_{ik} \sin(\delta_i - \delta_k) - B_{ik} \cos(\delta_i - \delta_k)] \tag{2.122b}$$

Differentiating Eq. (2.120a) with respect to V_i to find N_{ii}

$$\frac{\partial P_i}{\partial V_i} = 2G_{ii}V_i + \sum_{\substack{k=1 \\ k \neq i}}^{NB} V_k [G_{ik} \cos(\delta_i - \delta_k) + B_{ik} \sin(\delta_i - \delta_k)]$$

Multiplying the above equation by V_i,

$$V_i \frac{\partial P_i}{\partial V_i} = 2G_{ii}V_i^2 + \sum_{\substack{k=1 \\ k \neq i}}^{NB} V_i V_k [G_{ik} \cos(\delta_i - \delta_k) + B_{ik} \sin(\delta_i - \delta_k)] \tag{2.123}$$

Subtracting Eq. (2.120a) from Eq. (2.123),

$$V_i \frac{\partial P_i}{\partial \delta_i} - P_i = G_{ii}V_i^2$$

Rearranging the above equation,

$$N_{ii} = V_i \frac{\partial P_i}{\partial \delta_i} = P_i + G_{ii}V_i^2 \tag{2.124a}$$

Differentiating Eq. (2.120a) with respect to V_k $(k \neq i)$ to find N_{ik} $(k \neq i)$,

$$\frac{\partial P_i}{\partial V_k} = V_i [G_{ik} \cos(\delta_i - \delta_k) + B_{ik} \sin(\delta_i - \delta_k)]$$

Multiplying by V_k,

$$N_{ik} = V_k \frac{\partial P_i}{\partial V_k} = V_i V_k [G_{ik} \cos(\delta_i - \delta_k) + B_{ik} \sin(\delta_i - \delta_k)] \tag{2.124b}$$

Differentiating Eq. (2.120b) with respect to δ_i to find J_{ii}

$$\frac{\partial Q_i}{\partial \delta_i} = \sum_{\substack{k=1 \\ k\neq 1}}^{NB} V_i V_k [G_{ik} \cos(\delta_i - \delta_k) + B_{ik} \sin(\delta_i - \delta_k)] \tag{2.125}$$

Subtracting Eq. (2.125) from Eq. (2.120a),

$$P_i - \frac{\partial Q_i}{\partial \delta_i} = G_{ii} V_i^2$$

Rearranging the above equation,

$$J_{ii} = \frac{\partial Q_i}{\partial \delta_i} = P_i - G_{ii} V_i^2 \tag{2.126a}$$

Differentiating Eq. (2.120b) with respect to δ_k $(k \neq i)$ to find J_{ik} $(k \neq i)$

$$J_{ik} = \frac{\partial Q_i}{\partial \delta_k} = V_i V_k \; [-G_{ik} \cos(\delta_i - \delta_k) - B_{ik} \sin(\delta_i - \delta_k)] \tag{2.126b}$$

Differentiating Eq. (2.120b) with respect to V_i to find L_{ii}

$$\frac{\partial Q_i}{\partial V_i} = -2 B_{ii} V_i + \sum_{\substack{k=1 \\ k\neq i}}^{NB} V_k [G_{ik} \sin(\delta_i - \delta_k) - B_{ik} \cos(\delta_i - \delta_k)]$$

Multiplying the above equation by V_i,

$$V_i \frac{\partial Q_i}{\partial V_i} = -2 B_{ii} V_i^2 + \sum_{\substack{k=1 \\ k\neq i}}^{NB} V_i V_k [G_{ik} \sin(\delta_i - \delta_k) - B_{ik} \cos(\delta_i - \delta_k)] \tag{2.127}$$

Subtracting Eq. (2.120b) from Eq. (2.127),

$$V_i \frac{\partial Q_i}{\partial V_i} - Q_i = B_{ii} V_i^2$$

Rearranging the above equation,

$$L_{ii} = V_i \frac{\partial Q_i}{\partial V_i} = Q_i - B_{ii} V_i^2 \tag{2.128a}$$

Differentiating Eq. (2.120b) with respect to $V_k (k \neq i)$ to find $L_{ik} (k \neq i)$

$$\frac{\partial Q_i}{\partial V_k} = V_i [G_{ik} \sin(\delta_i - \delta_k) - B_{ik} \cos(\delta_i - \delta_k)]$$

Multiplying by V_k,

$$L_{ik} = V_k \frac{\partial Q_i}{\partial V_k} = V_i V_k [G_{ik} \sin(\delta_i - \delta_k) - B_{ik} \cos(\delta_i - \delta_k)] \tag{2.128b}$$

Limits on the controllable variable Q_i of PV buses

Q_i is computed using Eq. (2.120b). The limits are set as following and the PV bus is considered as the PQ bus on violation of limits.

$$Q_i = \begin{cases} Q_i^{\min} & \text{if } Q_i < Q_i^{\min} \\ Q_i^{\max} & \text{if } Q_i > Q_i^{\max} \\ Q_i & \text{if } Q_i^{\min} \le Q_i \le Q_i^{\max} \end{cases} \tag{2.129}$$

The bus which has been changed from the PV bus to the PQ bus on violation of Q limit needs the calculation of change in voltage magnitude. It is calculated from Eq. (2.119).

$$\sum_{k=2}^{\text{NB}} J_{ik}\Delta\delta_k + L_{ii}\frac{\Delta V_i}{V_i} + \sum_{\substack{k=\text{NV}+1 \\ k \ne i}}^{\text{NB}} L_{ik}\frac{\Delta V_k}{V_k} = \Delta Q_i$$

Rearranging the above equation,

$$\Delta V_i = \frac{V_i}{L_{ii}}\left[\Delta Q_i - \sum_{k=2}^{\text{NB}} J_{ik}\Delta\delta_k - \sum_{\substack{k=\text{NV}+1 \\ k \ne i}}^{\text{NB}} L_{ik}\frac{\Delta V_k}{V_k}\right] \tag{2.130}$$

and specified voltage magnitude of the PV bus is updated as

$$V_i = V_i^{\text{S}} + \Delta V_i$$

where V_i^{S} is scheduled voltage magnitude of the ith bus.

With the new value of V_i, the bus is restored to PV bus and iteration is continued.

Limits on the voltage magnitude, V_i

The power system equipment is designed to operate at fixed voltages with allowable variations of ±(5–10)% of the rated values. If the voltage magnitude limits are violated, then the voltage is fixed as follows:

$$V_i = \begin{cases} V_i^{\min} & \text{if } V_i < V_i^{\min} \\ V_i^{\max} & \text{if } V_i > V_i^{\max} \\ V_i & \text{if } V_i^{\min} \le V_i \le V_i^{\max} \end{cases} \tag{2.131}$$

The Newton–Raphson (NR) method is useful for large systems. The NR method requires more memory when rectangular coordinates are used. Hence polar coordinates are preferred for the NR method. To avoid time consuming sine and cosine terms in the Jacobian elements in the polar version of the NR method, the elements of the Jacobian are calculated by the rectangular version. The rectangular version is faster in convergence, but slightly less reliable than the polar version. With the NR method, the power differences and elements of the Jacobian are to be computed per iteration and triangularization has also to be done per iteration, so that the time taken per iteration is considerably longer as compared to the Gauss–Seidel method. However, the NR method gives the accurate results and convergence is guaranteed. The choice of slack bus does not affect the solution. Moreover, it works with the regulating transformers etc. equally well. A detailed stepwise procedure is explained here.

Algorithm 2.4: Newton–Raphson Method to Perform Load Flow

1. Read data
   NB (number of buses); NV (the number of *PV* buses).
   V_1, δ_1 for slack bus, P_i^S (i = 2, 3, ..., NB) for *PQ* and *PV* buses.
   Q_i^S (i = NV + 1, NV + 2, ..., NB for PQ buses), V_i^S (i = 2, 3, ..., NV for *PV* buses),
   V_i^{min}, V_i^{max} (i = NV + 1, NV + 2, ..., NB for *PQ* buses).
   Q_i^{min}, Q_i^{max} (i = 2, 3, ..., NV for *PV* buses).
   R(maximum number of iterations), ε (tolerance in convergence).
2. Form Y_{BUS} as explained in Section 2.3.
3. Assume initially
   $|V_i|$ (i = NV + 1, NV + 2, ..., NB) and δ_i (i = 2, 3, ..., NB)
4. Set iteration count, r = 0.
5. Compute P_i, ΔP_i using Eq. (2.120a) or Eq. (2.83a)

$$P_i = \sum_{k=1}^{NB} V_i V_k [G_{ik} \cos(\delta_i - \delta_k) + B_{ik} \sin(\delta_i - \delta_k)] \quad (i = 2, 3, ..., NB)$$

$$\Delta P_i = P_i^S - P_i \quad (i = 2, 3, ..., NB)$$

   Compute Q_i, ΔQ_i using Eq. (2.120b) or Eq. (2.83b)

$$Q_i = \sum_{k=1}^{NB} V_i V_k [G_{ik} \sin(\delta_i - \delta_k) - B_{ik} \cos(\delta_i - \delta_k)] \quad (i = NV+1, NV+2, ..., NB)$$

$$\Delta Q_i = Q_i^S - Q_i \quad (i = NV + 1, NV + 2, ..., NB)$$

6. If maximum $|\Delta P_i|$ (i = 2, 3, ..., NB) and $|\Delta Q_i|$ (i = NV + 1, NV + 2, ..., NB} $\le \varepsilon$ then GOTO Step 15.
7. Compute Jacobian matrix elements using Eqs. (2.109).
   When $i = k$

$$H_{ii} = \frac{\partial P_i}{\partial \delta_i} = -Q_i - B_{ii} V_i^2, \; N_{ii} = V_i \frac{\partial P_i}{\partial \delta_i} = P_i + G_{ii} V_i^2$$

$$J_{ii} = \frac{\partial Q_i}{\partial \delta_i} = P_i - G_{ii} V_i^2, \; L_{ii} = V_i \frac{\partial Q_i}{\partial \delta_i} = Q_i - B_{ii} V_i^2$$

   When $i \neq k$

$$H_{ik} = \frac{\partial P_i}{\partial \delta_k} = V_i V_k [G_{ik} \sin(\delta_i - \delta_k) - B_{ik} \cos(\delta_i - \delta_k)]$$

$$N_{ik} = V_k \frac{\partial P_i}{\partial V_k} = V_i V_k [G_{ik} \cos(\delta_i - \delta_k) + B_{ik} \sin(\delta_i - \delta_k)]$$

$$J_{ik} = \frac{\partial Q_i}{\partial \delta_k} = V_i V_k [-G_{ik} \cos(\delta_i - \delta_k) - B_{ik} \sin(\delta_i - \delta_k)]$$

$$L_{ik} = V_k \frac{\partial Q_i}{\partial V_k} = V_i V_k [G_{ik} \sin(\delta_i - \delta_k) - B_{ik} \cos(\delta_i - \delta_k)]$$

8. Compute $\Delta\delta_i$ (i = 2, 3, ..., NB) and $\Delta V_i/V_i$ (i = NV + 1, NV + 2, ..., NB) using Eq. (2.117),

$$\begin{bmatrix} H & N \\ J & L \end{bmatrix} \begin{bmatrix} \Delta\delta \\ \dfrac{\Delta V}{V} \end{bmatrix} = \begin{bmatrix} \Delta P \\ \Delta Q \end{bmatrix}$$

9. Modify δ_i and V_i,

$$\delta_i = \delta_i + \Delta\delta_i \qquad (i = 2, 3, ..., \text{NB})$$

$$V_i = V_i + \frac{\Delta V_i}{V_i} V_i \qquad (i = \text{NV} + 1, \text{NV} + 2, ..., \text{NB})$$

10. Set bus count i = 2.
11. If *PQ* bus then check the limits of V_i and set according to Eq. (2.131), i.e.

$$V_i = V_i^{\min} \quad \text{if } V_i \le V_i^{\min}$$

$$V_i = V_i^{\max} \quad \text{if } V_i \ge V_i^{\max}$$

12. If *PV* bus then compute Q_i using Eq. (2.120b) and check the limits of Q_i and set according to Eq. (2.129), i.e.

$$Q_i = Q_i^{\min} \quad \text{if } Q_i \le Q_i^{\min}$$

$$Q_i = Q_i^{\max} \quad \text{if } Q_i \ge Q_i^{\max}$$

If limits are violated then *PV* bus is temporarily converted to *PQ* bus. So, compute J_{ik} and L_{ik} with updated values of Q_i, V_i, and δ_i. Using Eq. (2.130), calculate the change in voltage, i.e.

$$\Delta V_i = \frac{V_i}{L_{ii}} \left[\Delta Q_i - \sum_{k=2}^{\text{NB}} J_{ik} \Delta\delta_k - \sum_{\substack{k=\text{NV}+1 \\ k \ne i}}^{\text{NB}} L_{ik} \frac{\Delta V_k}{V_k} \right]$$

and specified voltage magnitude of *PV* bus is updated as

$$V_i = V_i^{\text{S}} + \Delta V_i$$

13. Increment the bus count, $i = i + 1$
    If $i \le$ NB, then GOTO Step 11.

14. Advance the count, $r = r + 1$
    If $r \le R$ then GOTO Step 5 and repeat.

15. Compute the active and reactive power on slack bus using Eq. (2.83a) and Eq. (2.83b), i.e.

$$P_1 = \sum_{k=1}^{NB} V_1 V_k [G_{1k} \cos(\delta_1 - \delta_k) + B_{1k} \sin(\delta_1 - \delta_k)]$$

$$Q_1 = \sum_{k=1}^{NB} V_1 V_k [G_{1k} \sin(\delta_i - \delta_k) - B_{1k} \cos(\delta_1 - \delta_k)]$$

16. Calculate line flows using Eqs. (2.85) and (2.86), i.e.

$$S_{ik} = V_i^c [\{(V_i^c)^* - (V_k^c)^*\} \overset{*}{y}_{ik} + (V_i^c)^* \overset{*}{y}_{ik0}]$$

$$S_{ki} = V_k^c [\{(V_k^c)^* - (V_i^c)^*\} \overset{*}{y}_{ki} + (V_k^c)^* \overset{*}{y}_{ki0}]$$

where $V_i^c = V_i(\cos\delta_i + j\sin\delta_i)$

17. Stop.

EXAMPLE 2.8 For the sample system of Figure 2.19, the generators are connected at all the four buses, while the loads are at buses 2, 3, and 4. The values of real and reactive powers are listed in Table 2.4. All buses other than slack are of *PQ*-type. Line data are given in Table 2.5. Find the voltages and the bus angles at the three buses using the NR (Newton–Raphson) method.

Solution *Y*-bus is formed using Algorithm 2.1. The real and imaginary parts are separated to obtain the *G* and *B* matrices.

$$G = \begin{bmatrix} 5.724138 & -1.724138 & 0.0 & -4.0 \\ -1.724138 & 4.224138 & -2.5 & 0.0 \\ 0.0 & -2.5 & 4.386792 & -1.886792 \\ -4.0 & 0.0 & -1.886792 & 5.886792 \end{bmatrix}$$

$$B = \begin{bmatrix} -12.31034 & 4.310345 & 0.0 & 8.0 \\ 4.310345 & -11.81034 & 7.5 & 0.0 \\ 0.0 & 7.5 & -14.10377 & 6.603774 \\ 8.0 & 0.0 & 6.603774 & -14.60377 \end{bmatrix}$$

To start iteration, choose the initial values of V_i and δ_i.

$$\delta_2 = \delta_3 = \delta_4 = 0$$

$$V_2 = V_3 = V_4 = 1.0 \text{ p.u.}$$

Specified real and reactive powers are taken as

$$P_2^S = -0.45 \text{ p.u.}, \quad P_3^S = -0.51 \text{ p.u.}, \quad P_4^S = -0.60 \text{ p.u.}$$

$$Q_2^S = -0.15 \text{ p.u.}, \quad Q_3^S = -0.25 \text{ p.u.}, \quad Q_4^S = -0.30 \text{ p.u.}$$

The real powers (P_2, P_3, and P_4) are computed from Eq. (2.120a) as

$$P_2 = \sum_{k=1}^{4} V_2 V_k [G_{2k} \cos(\delta_2 - \delta_k) + B_{2k} \sin(\delta_2 - \delta_k)] = -8.620691 \times 10^{-2} \text{ p.u.}$$

$$P_3 = \sum_{k=1}^{4} V_3 V_k [G_{3k} \cos(\delta_3 - \delta_k) + B_{3k} \sin(\delta_3 - \delta_k)] = -2.384186 \times 10^{-7} \text{ p.u.}$$

$$P_4 = \sum_{k=1}^{4} V_4 V_k [G_{4k} \cos(\delta_4 - \delta_k) + B_{4k} \sin(\delta_4 - \delta_k)] = -1.999998 \times 10^{-1} \text{ p.u.}$$

Real power residuals are calculated as

$$\Delta P_2 = P_2^S - P_2 = -0.3638 \text{ p.u.}$$

$$\Delta P_3 = P_3^S - P_3 = -0.5100 \text{ p.u.}$$

$$\Delta P_4 = P_4^S - P_4 = -0.4000 \text{ p.u.}$$

Using Eq. (2.120b), the reactive powers Q_2, Q_3, and Q_4 are computed as

$$Q_2 = \sum_{k=1}^{4} V_2 V_k [G_{2k} \sin(\delta_2 - \delta_k) - B_{2k} \cos(\delta_2 - \delta_k)] = -2.155170 \times 10^{-1} \text{ p.u.}$$

$$Q_3 = \sum_{k=1}^{4} V_3 V_k [G_{3k} \sin(\delta_3 - \delta_k) - B_{3k} \cos(\delta_3 - \delta_k)] = -4.768372 \times 10^{-7} \text{ p.u.}$$

$$Q_4 = \sum_{k=1}^{4} V_4 V_k [G_{4k} \sin(\delta_4 - \delta_k) - B_{4k} \cos(\delta_4 - \delta_k)] = -3.999996 \times 10^{-1} \text{ p.u.}$$

Reactive power residuals are calculated as

$$\Delta Q_2 = Q_2^S - Q_2 = 0.0655 \text{ p.u.}$$

$$\Delta Q_3 = Q_3^S - Q_3 = -0.2500 \text{ p.u.}$$

$$\Delta Q_4 = Q_4^S - Q_4 = 0.1000 \text{ p.u.}$$

The convergence criterion is checked to stop the iterations, i.e.

$$\text{maximum } \{|\Delta P_i| \ (i = 2, 3, 4) \text{ and } |\Delta Q_i| \ (i = 2, 3, 4)\} \le \varepsilon \ (0.001)$$

$$\text{maximum } \{0.3636, 0.51, 0.40, 0.0655, 0.25, 0.1\} = 0.51 > 0.001$$

The convergence criterion is not satisfied, so the changes in variables at the end of the first iteration are obtained as follows:

$$\begin{bmatrix} H_{22} & H_{23} & H_{24} & N_{22} & N_{23} & N_{24} \\ H_{32} & H_{33} & H_{34} & N_{32} & N_{33} & N_{34} \\ H_{42} & H_{43} & H_{44} & N_{42} & N_{43} & N_{44} \\ J_{22} & J_{23} & J_{24} & L_{22} & L_{23} & L_{24} \\ J_{32} & J_{33} & J_{34} & L_{32} & L_{33} & L_{34} \\ J_{42} & J_{43} & J_{44} & L_{42} & L_{43} & L_{44} \end{bmatrix} \begin{bmatrix} \Delta\delta_2 \\ \Delta\delta_3 \\ \Delta\delta_4 \\ \Delta V_2/V_2 \\ \Delta V_3/V_3 \\ \Delta V_4/V_4 \end{bmatrix} = \begin{bmatrix} \Delta P_2 \\ \Delta P_3 \\ \Delta P_4 \\ \Delta Q_2 \\ \Delta Q_3 \\ \Delta Q_4 \end{bmatrix}$$

where

$$H_{22} = \frac{\partial P_2}{\partial \delta_2} = -Q_2 - B_{22}V_2^2 = 12.0259$$

$$H_{23} = \frac{\partial P_2}{\partial \delta_3} = V_2V_3[G_{23}\sin(\delta_2 - \delta_3) - B_{23}\cos(\delta_2 - \delta_3)] = -7.5$$

$$H_{24} = \frac{\partial P_2}{\partial \delta_4} = V_2V_4[G_{24}\sin(\delta_2 - \delta_4) - B_{24}\cos(\delta_2 - \delta_4)] = 0.0$$

$$H_{32} = \frac{\partial P_3}{\partial \delta_2} = V_3V_2[G_{32}\sin(\delta_3 - \delta_2) - B_{32}\cos(\delta_3 - \delta_2)] = -7.5$$

$$H_{33} = \frac{\partial P_3}{\partial \delta_3} = -Q_3 - B_{33}V_3^2 = 14.1038$$

$$H_{34} = \frac{\partial P_3}{\partial \delta_4} = V_3V_4[G_{34}\sin(\delta_3 - \delta_4) - B_{34}\cos(\delta_3 - \delta_4)] = -6.6038$$

$$H_{42} = \frac{\partial P_4}{\partial \delta_2} = V_4V_2[G_{42}\sin(\delta_4 - \delta_2) - B_{42}\cos(\delta_4 - \delta_2)] = 0.0$$

$$H_{43} = \frac{\partial P_4}{\partial \delta_3} = V_4V_3[G_{43}\sin(\delta_4 - \delta_3) - B_{43}\cos(\delta_4 - \delta_3)] = -6.6038$$

$$H_{44} = \frac{\partial P_4}{\partial \delta_4} = -Q_4 - B_{44}V_4^2 = 15.0038$$

$$N_{22} = V_2\frac{\partial P_2}{\partial V_2} = P_2 + G_{22}V_2^2 = 4.1379$$

$$N_{23} = V_3 \frac{\partial P_2}{\partial V_3} = V_2 V_3 [G_{23} \cos(\delta_2 - \delta_3) + B_{23} \sin(\delta_2 - \delta_3)] = -2.5$$

$$N_{24} = V_4 \frac{\partial P_2}{\partial V_4} = V_2 V_4 [G_{24} \cos(\delta_2 - \delta_4) + B_{24} \sin(\delta_2 - \delta_4)] = 0.0$$

$$N_{32} = V_2 \frac{\partial P_3}{\partial V_2} = V_3 V_2 [G_{32} \cos(\delta_3 - \delta_2) + B_{32} \sin(\delta_3 - \delta_2)] = -2.5$$

$$N_{33} = V_3 \frac{\partial P_3}{\partial V_3} = P_3 + G_{33} V_3^2 = 4.3868$$

$$N_{34} = V_4 \frac{\partial P_3}{\partial V_4} = V_3 V_4 [G_{34} \cos(\delta_3 - \delta_4) + B_{34} \sin(\delta_3 - \delta_4)] = -1.8868$$

$$N_{42} = V_2 \frac{\partial P_4}{\partial V_2} = V_4 V_2 [G_{42} \cos(\delta_4 - \delta_2) + B_{42} \sin(\delta_4 - \delta_2)] = 0.0$$

$$N_{43} = V_3 \frac{\partial P_4}{\partial V_3} = V_4 V_3 [G_{43} \cos(\delta_4 - \delta_3) + B_{43} \sin(\delta_4 - \delta_3)] = -1.8868$$

$$N_{44} = V_4 \frac{\partial P_4}{\partial V_4} = P_4 + G_{44} V_4^2 = 5.6868$$

$$J_{22} = \frac{\partial Q_2}{\partial \delta_2} = P_2 - G_{22} V_2^2 = -4.3103$$

$$J_{23} = \frac{\partial Q_2}{\partial \delta_3} = V_2 V_3 [-G_{23} \cos(\delta_2 - \delta_3) - B_{23} \sin(\delta_2 - \delta_3)] = 2.5$$

$$J_{24} = \frac{\partial Q_2}{\partial \delta_4} = V_2 V_4 [-G_{24} \cos(\delta_2 - \delta_4) - B_{24} \sin(\delta_2 - \delta_4)] = 0.0$$

$$J_{32} = \frac{\partial Q_3}{\partial \delta_2} = V_3 V_2 [-G_{32} \cos(\delta_3 - \delta_2) - B_{32} \sin(\delta_3 - \delta_2)] = 2.5$$

$$J_{33} = \frac{\partial Q_3}{\partial \delta_3} = P_3 - G_{33} V_3^2 = -4.3868$$

$$J_{34} = \frac{\partial Q_3}{\partial \delta_4} = V_3 V_4 [-G_{34} \cos(\delta_3 - \delta_4) - B_{34} \sin(\delta_3 - \delta_4)] = 1.8868$$

$$J_{42} = \frac{\partial Q_4}{\partial \delta_2} = V_4 V_2 [-G_{42} \cos(\delta_4 - \delta_2) - B_{42} \sin(\delta_4 - \delta_2)] = 0.0$$

$$J_{43} = \frac{\partial Q_4}{\partial \delta_3} = V_4 V_3 [-G_{43} \cos(\delta_4 - \delta_3) - B_{43} \sin(\delta_4 - \delta_3)] = 1.8868$$

$$J_{44} = \frac{\partial Q_4}{\partial \delta_4} = P_4 - G_{44} V_4^2 = -6.0868$$

$$L_{22} = V_2 \frac{\partial Q_2}{\partial V_2} = Q_2 - B_{22} V_2^2 = 11.5948$$

$$L_{23} = V_3 \frac{\partial Q_2}{\partial V_3} = V_2 V_3 [G_{23} \sin(\delta_2 - \delta_3) - B_{23} \cos(\delta_2 - \delta_3)] = -7.5$$

$$L_{24} = V_4 \frac{\partial Q_2}{\partial V_4} = V_2 V_4 [G_{24} \sin(\delta_2 - \delta_4) - B_{24} \cos(\delta_2 - \delta_4)] = 0.0$$

$$L_{32} = V_2 \frac{\partial Q_3}{\partial V_2} = V_3 V_2 [G_{32} \sin(\delta_3 - \delta_2) - B_{32} \cos(\delta_3 - \delta_2)] = -7.5$$

$$L_{33} = V_3 \frac{\partial Q_3}{\partial V_3} = Q_3 - B_{33} V_3^2 = 14.1038$$

$$L_{34} = V_4 \frac{\partial Q_3}{\partial V_4} = V_3 V_4 [G_{34} \sin(\delta_3 - \delta_4) - B_{34} \cos(\delta_3 - \delta_4)] = -6.6038$$

$$L_{42} = V_2 \frac{\partial Q_4}{\partial V_2} = V_4 V_2 [G_{42} \sin(\delta_4 - \delta_2) - B_{42} \cos(\delta_4 - \delta_2)] = 0.0$$

$$L_{43} = V_3 \frac{\partial Q_4}{\partial V_3} = V_4 V_3 [G_{43} \sin(\delta_4 - \delta_3) - B_{43} \cos(\delta_4 - \delta_3)] = -6.6038$$

$$L_{44} = V_4 \frac{\partial Q_4}{\partial V_4} = Q_4 - B_{44} V_4^2 = 14.2038$$

Using the gauss elimination method, in which triangularization and back substitution process is performed to compute the changes in δ and V, we have

$$\Delta\delta_2 = -0.09696, \quad \Delta\delta_3 = -0.11217, \quad \Delta\delta_4 = -0.06928$$

$$\Delta V_2/V_2 = -0.05504, \quad \Delta V_3/V_3 = -0.07549, \quad \Delta V_4/V_4 = -0.04285$$

Modified values of δ_i and V_i are computed as

$$\delta_2 = \delta_2 + \Delta\delta_2 = 0.0 - 0.09696 = -0.09696 \text{ rad}$$

$$\delta_3 = \delta_3 + \Delta\delta_3 = 0.0 - 0.11217 = -0.11217 \text{ rad}$$

$$\delta_4 = \delta_4 + \Delta\delta_4 = 0.0 - 0.06928 = -0.06928 \text{ rad}$$

$$V_2 = V_2 + \frac{\Delta V_2}{V_2} V_2 = 1.0 - \frac{0.05504}{1.0} \times 1.0 = 0.094496 \text{ p.u.}$$

$$V_3 = V_3 + \frac{\Delta V_3}{V_3} V_3 = 1.0 - \frac{0.07549}{1.0} \times 1.0 = 0.92451 \text{ p.u.}$$

$$V_4 = V_4 + \frac{\Delta V_4}{V_4} V_4 = 1.0 - \frac{0.04285}{1.0} \times 1.0 = 0.95715 \text{ p.u.}$$

After seven iterations, the results obtained are tabulated in (Table 2.9).

TABLE 2.9 Results after seven iterations

Bus	P^S (p.u.)	Q^S (p.u.)	P (p.u.)	Q (p.u.)	V (p.u.)	δ (rad)
1	0.0	0.00	1.5694500	0.7529464	1.05	0.0
2	−0.45	−0.15	−0.4499942	−0.1499812	0.9337445	−0.10409630
3	−0.51	−0.25	−0.5100103	−0.2500389	0.9113876	−0.12205260
4	−0.60	−0.30	−0.5999956	−0.2999793	0.9486454	−0.07251044

EXAMPLE 2.9 For the sample system of Figure 2.19, the generators are connected at all the four buses, while the loads are at buses 2, 3, and 4. The values of real and reactive powers are listed in Table 2.7 along with the type of buses. Line data are given in Table 2.5. Find the voltages and the bus angles at the three buses using the NR method.

Solution Y-bus is formed using Algorithm 2.1. The real and imaginary parts are separated to obtain the G and B matrices. The obtained values of G and B matrices are given in Example 2.8.

To start iteration, choose initial values of V_i and δ_i as

$$\delta_2 = \delta_3 = \delta_4 = 0$$

$$V_3 = V_4 = 1.0 \text{ p.u.}$$

Specified real and reactive powers are taken as

$$P_2^S = -0.45 \text{ p.u.}, \qquad P_3^S = -0.51 \text{ p.u.}, \qquad P_4^S = -0.60 \text{ p.u.},$$

$$Q_3^S = -0.25 \text{ p.u.}, \qquad Q_4^S = -0.30 \text{ p.u.},$$

The real powers (P_2, P_3, and P_4) are computed from Eq. (2.120a) as

$$P_2 = \sum_{k=1}^{4} V_2 V_k \left[G_{2k} \cos(\delta_2 - \delta_k) + B_{2k} \sin(\delta_2 - \delta_k)\right] = -8.620691 \times 10^{-2} \text{ p.u.}$$

$$P_3 = \sum_{k=1}^{4} V_3 V_k \left[G_{3k} \cos(\delta_3 - \delta_k) + B_{3k} \sin(\delta_3 - \delta_k)\right] = -2.384186 \times 10^{-7} \text{ p.u.}$$

$$P_4 = \sum_{k=1}^{4} V_4 V_k \,[G_{4k} \cos (\delta_4 - \delta_k) + B_{4k} \sin (\delta_4 - \delta_k) = -1.999998 \times 10^{-1} \text{ p.u.}$$

Real power residuals are calculated as

$$\Delta P_2 = P_2^S - P_2 = -0.3638 \text{ p.u.}$$

$$\Delta P_3 = P_3^S - P_3 = -0.5100 \text{ p.u.}$$

$$\Delta P_4 = P_4^S - P_4 = -0.4000 \text{ p.u.}$$

Using Eq. (2.120b), the reactive powers Q_3 and Q_4 are computed as

$$Q_3 = \sum_{k=1}^{4} V_3 V_k \,[G_{3k} \sin (\delta_3 - \delta_k) - B_{3k} \cos (\delta_3 - \delta_k)] = -4.768372 \times 10^{-7} \text{ p.u.}$$

$$Q_4 = \sum_{k=1}^{4} V_4 V_k \,[G_{4k} \sin (\delta_4 - \delta_k) - B_{4k} \cos (\delta_4 - \delta_k)] = -3.999996 \times 10^{-1} \text{ p.u.}$$

Reactive power residuals are calculated as

$$\Delta Q_3 = Q_3^S - Q_3 = -0.2500 \text{ p.u.}$$

$$\Delta Q_4 = Q_4^S - Q_4 = 0.1000 \text{ p.u.}$$

The convergence criterion is checked to stop the iterations, i.e.

$$\text{maximum } \{|\Delta P_i| \; (i = 2, 3, 4) \text{ and } |\Delta Q_i| \; (i = 3, 4)\} \le \varepsilon \; (0.001)$$

$$\text{maximum } \{0.3636, 0.51, 0.40, 0.25, 0.1\} = 0.51 > 0.001$$

The convergence criterion has not been satisfied, so the change in variables at the end of first iteration are obtained as follows:

$$\begin{bmatrix} H_{22} & H_{23} & H_{24} & N_{23} & N_{24} \\ H_{32} & H_{33} & H_{34} & N_{33} & N_{34} \\ H_{42} & H_{43} & H_{44} & N_{43} & N_{44} \\ J_{32} & J_{33} & J_{34} & L_{33} & L_{34} \\ J_{42} & J_{43} & J_{44} & L_{43} & L_{44} \end{bmatrix} \begin{bmatrix} \Delta\delta_2 \\ \Delta\delta_3 \\ \Delta\delta_4 \\ \Delta V_3/V_3 \\ \Delta V_4/V_4 \end{bmatrix} = \begin{bmatrix} \Delta P_2 \\ \Delta P_3 \\ \Delta P_4 \\ \Delta Q_3 \\ \Delta Q_4 \end{bmatrix}$$

where

$$H_{22} = \frac{\partial P_2}{\partial \delta_2} = -Q_2 - B_{22} V_2^2 = 12.0259$$

$$H_{23} = \frac{\partial P_2}{\partial \delta_3} = V_2 V_3 [G_{23} \sin (\delta_2 - \delta_3) - B_{23} \cos (\delta_2 - \delta_3)] = -7.5$$

$$H_{24} = \frac{\partial P_2}{\partial \delta_4} = V_2 V_4 [G_{24} \sin (\delta_2 - \delta_4) - B_{24} \cos (\delta_2 - \delta_4)] = 0.0$$

$$H_{32} = \frac{\partial P_3}{\partial \delta_2} = V_3 V_2 [G_{32} \sin(\delta_3 - \delta_2) - B_{32} \cos(\delta_3 - \delta_2)] = -7.5$$

$$H_{33} = \frac{\partial P_3}{\partial \delta_3} = -Q_3 - B_{33} V_3^2 = 14.1038$$

$$H_{34} = \frac{\partial P_3}{\partial \delta_4} = V_3 V_4 [G_{34} \sin(\delta_3 - \delta_4) - B_{34} \cos(\delta_3 - \delta_4)] = -6.6038$$

$$H_{42} = \frac{\partial P_4}{\partial \delta_2} = V_4 V_2 [G_{42} \sin(\delta_4 - \delta_2) - B_{42} \cos(\delta_4 - \delta_2)] = 0.0$$

$$H_{43} = \frac{\partial P_4}{\partial \delta_3} = V_4 V_3 [G_{43} \sin(\delta_4 - \delta_3) - B_{43} \cos(\delta_4 - \delta_3)] = -6.6038$$

$$H_{44} = \frac{\partial P_4}{\partial \delta_4} = -Q_4 - B_{44} V_4^2 = 15.0038$$

$$N_{23} = V_3 \frac{\partial P_2}{\partial V_3} = V_2 V_3 [G_{23} \cos(\delta_2 - \delta_3) + B_{23} \sin(\delta_2 - \delta_3)] = -2.5$$

$$N_{24} = V_4 \frac{\partial P_2}{\partial V_4} = V_2 V_4 [G_{24} \cos(\delta_2 - \delta_4) + B_{24} \sin(\delta_2 - \delta_4)] = 0.0$$

$$N_{33} = V_3 \frac{\partial P_3}{\partial V_3} = P_3 + G_{33} V_3^2 = 4.3868$$

$$N_{34} = V_4 \frac{\partial P_3}{\partial V_4} = V_3 V_4 [G_{34} \cos(\delta_3 - \delta_4) + B_{34} \sin(\delta_3 - \delta_4)] = -1.8868$$

$$N_{43} = V_3 \frac{\partial P_4}{\partial V_3} = V_4 V_3 [G_{43} \cos(\delta_4 - \delta_3) + B_{43} \sin(\delta_4 - \delta_3)] = -1.8868$$

$$N_{44} = V_4 \frac{\partial P_4}{\partial V_4} = P_4 + G_{44} V_4^2 = 5.6868$$

$$J_{32} = \frac{\partial Q_3}{\partial \delta_2} = V_3 V_2 [-G_{32} \cos(\delta_3 - \delta_2) - B_{32} \sin(\delta_3 - \delta_2)] = 2.5$$

$$J_{33} = \frac{\partial Q_3}{\partial \delta_3} = P_3 - G_{33} V_3^2 = -4.3868$$

$$J_{34} = \frac{\partial Q_3}{\partial \delta_4} = V_3 V_4 [-G_{34} \cos(\delta_3 - \delta_4) - B_{34} \sin(\delta_3 - \delta_4)] = 1.8868$$

$$J_{42} = \frac{\partial Q_4}{\partial \delta_2} = V_4 V_2 [-G_{42} \cos(\delta_4 - \delta_2) - B_{42} \sin(\delta_4 - \delta_2)] = 0.0$$

$$J_{43} = \frac{\partial Q_4}{\partial \delta_3} = V_4 V_3 [-G_{43} \cos(\delta_4 - \delta_3) - B_{43} \sin(\delta_4 - \delta_3)] = 1.8868$$

$$J_{44} = \frac{\partial Q_4}{\partial \delta_4} = P_4 - G_{44}V_4^2 = -6.0868$$

$$L_{33} = V_3 \frac{\partial Q_3}{\partial V_3} = Q_3 - B_{33}V_3^2 = 14.1038$$

$$L_{34} = V_4 \frac{\partial Q_3}{\partial V_4} = V_3V_4\,[G_{34} \sin(\delta_3 - \delta_4) - B_{34} \cos(\delta_3 - \delta_4)] = -6.6038$$

$$L_{43} = V_3 \frac{\partial Q_4}{\partial V_3} = V_4V_3\,[G_{43} \sin(\delta_4 - \delta_3) - B_{43} \cos(\delta_4 - \delta_3)] = -6.6038$$

$$L_{44} = V_4 \frac{\partial Q_4}{\partial V_4} = Q_4 - B_{44}V_4^2 = 14.2038$$

Using the gauss elimination method, in which triangularization and back substitution process is followed, changes in δ and V are found out as given below:

$$\Delta\delta_2 = -0.11664, \qquad \Delta\delta_3 = -0.12567, \qquad \Delta\delta_4 = -0.07658$$

$$\Delta V_3/V_3 = -0.03854, \qquad \Delta V_4/V_4 = -0.02700$$

Modified values of δ_i and V_i are computed as

$$\delta_2 = \delta_2 + \Delta\delta_2 = 0.0 - 0.11664 = -0.11664 \text{ rad}$$

$$\delta_3 = \delta_3 + \Delta\delta_3 = 0.0 - 0.12567 = -0.12567 \text{ rad}$$

$$\delta_4 = \delta_4 + \Delta\delta_4 = 0.0 - 0.07658 = -0.07658 \text{ rad}$$

$$V_3 = V_3 + \frac{\Delta V_3}{V_3} V_3 = 1.0 - \frac{0.03854}{1.0} \times 1.0 = 0.96146 \text{ p.u.}$$

$$V_4 = V_4 + \frac{\Delta V_4}{V_4} V_4 = 1.0 - \frac{0.02700}{1.0} \times 1.0 = 0.97300 \text{ p.u.}$$

This procedure is repeated and the results after five iterations are given below (Table 2.10).

TABLE 2.10 Results after five iterations

Bus	P^S (p.u.)	Q^S (p.u.)	P (p.u.)	Q (p.u.)	V (p.u.)	δ (rad)
1	0.0	0.00	1.5511550	0.2742737	1.05	0.0
2	−0.45	−0.15	−0.4499938	0.3351800	1.00	−0.122242
3	−0.51	−0.25	−0.5100070	−0.2500268	0.9575192	−0.131748
4	−0.60	−0.30	−0.6000010	−0.2999982	0.9684032	−0.078987

2.11 DECOUPLED NEWTON METHOD

An intrinsic characteristic of any practical electric power system operating in steady state is strong inter-reliance between real power and bus voltage angles and between reactive powers and voltage magnitudes. The property of feeble coupling between P–δ and Q–V variables results in developing decoupled load flow (DLF) method. P–δ and Q–V are solved separately. In view of the above, Eqs. (2.116)–(2.119) can be modified as given below:

$$\begin{bmatrix} \frac{\partial P}{\partial \delta} & 0 \\ 0 & \frac{\partial Q}{\partial V} \end{bmatrix} \begin{bmatrix} \Delta\delta \\ \Delta V \end{bmatrix} = \begin{bmatrix} \Delta P \\ \Delta Q \end{bmatrix} \tag{2.132}$$

$$\begin{bmatrix} H & 0 \\ 0 & L \end{bmatrix} \begin{bmatrix} \Delta\delta \\ \frac{\Delta V}{V} \end{bmatrix} = \begin{bmatrix} \Delta P \\ \Delta Q \end{bmatrix} \tag{2.133}$$

$$\sum_{k=2}^{NB} H_{ik}\,\Delta\delta_k = \Delta P_i \qquad (i = 2, 3, ..., NB) \tag{2.134a}$$

$$\sum_{k=NV+1}^{NB} L_{ik}\frac{\Delta V_k}{V_k} = \Delta Q_i \qquad (i = NV + 1, NV + 2, ..., NB) \tag{2.134b}$$

where

$$H_{ii} = \frac{\partial P_i}{\partial \delta_i} = -\,Q_i - B_{ii}V_i^2 \qquad (i = k)$$

$$H_{ik} = \frac{\partial P_i}{\partial \delta_k} = V_iV_k\,[G_{ik}\sin(\delta_i - \delta_k) - B_{ik}\cos(\delta_i - \delta_k)] \qquad (i \neq k)$$

$$L_{ii} = V_i\frac{\partial Q_i}{\partial \delta_i} = Q_i - B_{ii}V_i^2 \qquad (i = k)$$

$$L_{ik} = V_k\frac{\partial Q_i}{\partial V_k} = V_iV_k\,[G_{ik}\sin(\delta_i - \delta_k) - B_{ik}\cos(\delta_i - \delta_k)] \qquad (i \neq k)$$

Equation (2.134a) is solved to find $\Delta\delta$. The updated δ is then used to solve Eq. (2.134b) to compute ΔV.

Limits on the controllable variable Q_i of PV buses

Q_i is computed using Eq. (2.134b). The limits are set as following and the PV bus is considered as a PQ bus on violation of limits.

$$Q_i = \begin{cases} Q_i^{\min} & \text{if } Q_i < Q_i^{\min} \\ Q_i^{\max} & \text{if } Q_i > Q_i^{\max} \\ Q_i & \text{if } Q_i^{\min} \le Q_i \le Q_i^{\max} \end{cases} \tag{2.135}$$

The bus which has been changed from *PV* to *PQ* on violation, needs the calculation of change in voltage magnitude. It is calculated from Eq. (2.134b)

$$L_{ii}\frac{\Delta V_i}{V_i} + \sum_{\substack{k=\text{NV}+1 \\ k \neq i}}^{\text{NB}} L_{ik}\frac{\Delta V_k}{V_k} = \Delta Q_i$$

Rearranging the above equation

$$\Delta V_i = \frac{V_i}{L_{ii}}\left[\Delta Q_i - \sum_{\substack{k=\text{NV}+1 \\ k \neq i}}^{\text{NB}} L_{ik}\frac{\Delta V_k}{V_k}\right] \tag{2.136}$$

and the specified voltage magnitude of the *PV* bus is updated as

$$V_i = V_i^{\text{S}} + \Delta V_i$$

where V_i^{S} is scheduled voltage magnitude of the *i*th bus.

With the new value of V_i, the bus is restored to *PV* bus and iteration is continued.

Limits on the voltage magnitude, V_i

If the voltage magnitude limits are violated then the voltage is fixed as per the following:

$$V_i = \begin{cases} V_i^{\min} & \text{if } V_i < V_i^{\min} \\ V_i^{\max} & \text{if } V_i > V_i^{\max} \\ V_i & \text{if } V_i^{\min} \le V_i \le V_i^{\max} \end{cases} \tag{2.137}$$

The reliability of the decoupled Newton method is comparable to the formal Newton method for ill-conditioned problems. But, the decoupled method is simple and computationally efficient than the formal Newton method. The storage of Jacobian elements and triangularization is less. But computation time per iteration is less in the Newton method. A detailed stepwise procedure is explained here.

Algorithm 2.5: Decoupled Newton–Raphson Method for Load Flow Calculations

1. Read data

   NB (number of buses), NV (number of *PV* buses).
   V_1, δ_1 for slack bus, P_i^{S} (i = 2, 3, ..., NB) for *PQ* and *PV* buses.
   Q_i^{S}(i = NV + 1, NV + 2, ..., NB for *PQ* buses), V_i^{S}(i = 2, 3, ..., NV for *PV* buses), $V_i^{\min}$, $V_i^{\max}$ (i = NV + 1, NV + 2, ..., NB for *PQ* buses).
   $Q_i^{\min}$, $Q_i^{\max}$ (i = 2, 3, ..., NV for *PV* buses), R(maximum number of iterations), ε (tolerance in convergence)

2. Form Y_{BUS} as explained in Section 2.3.

3. Assume initially voltage magnitude

   $|V_i|$ (i = NV + 1, NV + 2, ..., NB) and angle of voltage δ_i (i = 2, 3, ..., NB)

4. Set iteration count, $r = 0$.
5. Compute P_i, ΔP_i using Eq. (2.120a) as

$$P_i = \sum_{k=1}^{NB} V_i V_k [G_{ik} \cos(\delta_i - \delta_k) + B_{ik} \sin(\delta_i - \delta_k)] \qquad (i = 2, 3, ..., NB)$$

$$\Delta P_i = P_i^S - P_i \qquad (i = 2, 3, ..., NB)$$

6. If maximum $\{|\Delta P_i|\ (i = 2, 3, ..., NB)\} \le \varepsilon$ then GOTO Step 10.
7. Compute the elements of the Jacobian matrix H as

$$H_{ii} = \frac{\partial P_i}{\partial \delta_i} = -Q_i - B_{ii} V_i^2 \qquad (i = k)$$

$$H_{ik} = \frac{\partial P_i}{\partial \delta_k} = V_i V_k\ [G_{ik} \sin(\delta_i - \delta_k) - B_{ik} \cos(\delta_i - \delta_k)] \qquad (i \ne k)$$

8. Compute $\Delta\delta_i$ $(i = 2, 3, ..., NB)$ using Eq. (2.134a) as

$$[H]\ [\Delta\delta] = [\Delta P]$$

9. Modified value of δ_i is computed as

$$\delta_i = \delta_i + \Delta\delta_i \qquad (i = 2, 3, ..., NB)$$

10. Compute Q_i, $|\Delta Q_i|$ using Eq. (2.120b), i.e.

$$Q_i = \sum_{k=1}^{NB} V_i V_k [G_{ik} \sin(\delta_i - \delta_k) - B_{ik} \cos(\delta_i - \delta_k)] \quad (i = NV + 1, NV + 2, ..., NB)$$

$$\Delta Q_i = Q_i^S - Q_i \qquad (i = NV + 1, NV + 2, ..., NB)$$

11. If maximum $\{|\Delta Q_i|\ (i = NV + 1, NV + 2, ..., NB)\} \le \varepsilon$ then GOTO Step 20.
12. Compute the elements of the Jacobian matrix L as

$$L_{ii} = V_i \frac{\partial Q_i}{\partial \delta_i} = Q_i - B_{ii} V_i^2 \qquad (i = k)$$

$$L_{ik} = V_k \frac{\partial Q_i}{\partial V_k} = V_i V_k [G_{ik} \sin(\delta_i - \delta_k) - B_{ik} \cos(\delta_i - \delta_k)] \qquad (i \ne k)$$

13. Compute $\Delta V_i/V_i$ $(i = NV + 1, NV + 2, ..., NB)$ using Eq. (2.134b), i.e.

$$[L]\left[\frac{\Delta V}{V}\right] = [\Delta Q]$$

14. The modified values of voltage magnitude, V_i are computed as

$$V_i = V_i + \frac{\Delta V_i}{V_i} V_i \qquad (i = NV + 1, NV + 2, ..., NB)$$

15. Set bus count $i = 2$.
16. If bus is PQ then check the limits of V_i and set according to Eq. (2.137), i.e.

$$V_i = V_i^{\min} \quad \text{if } V_i \le V_i^{\min}$$

$$V_i = V_i^{\max} \quad \text{if } V_i \ge V_i^{\max}$$

17. If bus is PV, then compute Q_i using Eq. (2.120b) and check the limits of Q_i and set according to Eq. (2.135), i.e.

$$Q_i = Q_i^{\min} \quad \text{if } Q_i \le Q_i^{\min}$$

$$Q_i = Q_i^{\max} \quad \text{if } Q_i \ge Q_i^{\max}$$

If any limit is violated then PV bus is temporarily converted to PQ bus. So, compute L_{ik} with updated values of Q_i, V_i and δ_i. Using Eq. (2.136), calculate the change in voltage as

$$\Delta V_i = \frac{V_i}{L_{ii}} \left[\Delta Q_i - \sum_{\substack{k = NV+1 \\ k \ne i}}^{NB} L_{ik} \frac{\Delta V_k}{V_k} \right]$$

and the specified voltage magnitude of the PV bus is updated as

$$V_i = V_i^S + \Delta V_i$$

18. Increment the bus count $i = i + 1$
If $i \le NB$, then GOTO Step 16.
19. Advance the count $r = r + 1$
If $r \le R$ then GOTO Step 5 and repeat.
20. Compute slack bus active and reactive powers as

$$P_1 = \sum_{k=1}^{NB} V_1 V_k [G_{1k} \cos(\delta_1 - \delta_k) + B_{1k} \sin(\delta_1 - \delta_k)]$$

$$Q_1 = \sum_{k=1}^{NB} V_1 V_k [G_{1k} \sin(\delta_1 - \delta_k) - B_{1k} \cos(\delta_1 - \delta_k)]$$

21. Calculate line flows using Eqs. (2.85) and (2.86), i.e.

$$S_{ik} = V_i^c [\{(V_i^c)^* - (V_k^c)^*\} y_{ik}^* + (V_i^c)^* y_{ik0}^*]$$

$$S_{ki} = V_k^c [\{(V_k^c)^* - (V_i^c)^*\} y_{ki}^* + (V_k^c)^* y_{ki0}^*]$$

where $V_i^c = V_i(\cos\delta_i + j \sin\delta_i)$
22. Stop.

EXAMPLE 2.10 For the sample system of Figure 2.19, the generators are connected at all the four buses, while loads are at buses 2, 3, and 4. The values of real and reactive powers are listed in Table 2.4. All buses other than slack are of the *PQ*-type. Line data is given in Table 2.5. Find the voltages and the bus angles at the three buses using the decoupled Newton–Raphson method.

Solution The *G* and *B* matrices are given below:

$$G = \begin{bmatrix} 5.724138 & -1.724138 & 0.0 & -4.0 \\ -1.724138 & 4.224138 & -2.5 & 0.0 \\ 0.0 & -2.5 & 4.386792 & -1.886792 \\ -4.0 & 0.0 & -1.886792 & 5.886792 \end{bmatrix}$$

$$B = \begin{bmatrix} -12.31034 & 4.310345 & 0.0 & 8.0 \\ 4.310345 & -11.81034 & 7.5 & 0.0 \\ 0.0 & 7.5 & -14.10377 & 6.603774 \\ 8.0 & 0.0 & 6.603774 & -14.60377 \end{bmatrix}$$

Assume that bus number one is slack bus. To start iteration, choose initial values as

$$\delta_2 = \delta_3 = \delta_4 = 0$$

$$V_2 = V_3 = V_4 = 1.0 \text{ p.u.}$$

Specified real power values are

$$P_2^S = -0.45 \text{ p.u.}, \qquad P_3^S = -0.51 \text{ p.u.}, \qquad P_4^S = -0.60 \text{ p.u.}$$

The real powers (P_2, P_3, and P_4) are computed from Eq. (2.120a) as

$$P_2 = \sum_{k=1}^{4} V_2 V_k \left[G_{2k} \cos(\delta_2 - \delta_k) + B_{2k} \sin(\delta_2 - \delta_k)\right] = -8.620691 \times 10^{-2} \text{ p.u.}$$

$$P_3 = \sum_{k=1}^{4} V_3 V_k \left[G_{3k} \cos(\delta_3 - \delta_k) + B_{3k} \sin(\delta_3 - \delta_k)\right] = -2.384186 \times 10^{-7} \text{ p.u.}$$

$$P_4 = \sum_{k=1}^{4} V_4 V_k \left[G_{4k} \cos(\delta_4 - \delta_k) + B_{4k} \sin(\delta_4 - \delta_k)\right] = -1.999998 \times 10^{-1} \text{ p.u.}$$

Real power residuals are calculated as

$$\Delta P_2 = P_2^S - P_2 = -0.3638 \text{ p.u.}$$

$$\Delta P_3 = P_3^S - P_3 = -0.5100 \text{ p.u.}$$

$$\Delta P_4 = P_4^S - P_4 = -0.4000 \text{ p.u.}$$

If maximum $\{|\Delta P_i|\ (i = 2, 3, 4)\} > \varepsilon\ (0.001)$ then calculate the change in voltage angles, i.e.

$$\begin{bmatrix} H_{22} & H_{23} & H_{24} \\ H_{32} & H_{33} & H_{34} \\ H_{42} & H_{43} & H_{44} \end{bmatrix} \begin{bmatrix} \Delta\delta_2 \\ \Delta\delta_3 \\ \Delta\delta_4 \end{bmatrix} = \begin{bmatrix} \Delta P_2 \\ \Delta P_3 \\ \Delta P_4 \end{bmatrix}$$

where

$$H_{22} = \frac{\partial P_2}{\partial \delta_2} = -Q_2 - B_{22}V_2^2 = 12.0259$$

$$H_{23} = \frac{\partial P_2}{\partial \delta_3} = V_2V_3[G_{23}\sin(\delta_2 - \delta_3) - B_{23}\cos(\delta_2 - \delta_3)] = -7.5$$

$$H_{24} = \frac{\partial P_2}{\partial \delta_4} = V_2V_4[G_{24}\sin(\delta_2 - \delta_4) - B_{24}\cos(\delta_2 - \delta_4)] = 0.0$$

$$H_{32} = \frac{\partial P_3}{\partial \delta_2} = V_3V_2[G_{32}\sin(\delta_3 - \delta_2) - B_{32}\cos(\delta_3 - \delta_2)] = -7.5$$

$$H_{33} = \frac{\partial P_3}{\partial \delta_3} = -Q_3 - B_{33}V_3^2 = 14.1038$$

$$H_{34} = \frac{\partial P_3}{\partial \delta_4} = V_3V_4[G_{34}\sin(\delta_3 - \delta_4) - B_{34}\cos(\delta_3 - \delta_4)] = -6.6038$$

$$H_{42} = \frac{\partial P_4}{\partial \delta_2} = V_4V_2[G_{42}\sin(\delta_4 - \delta_2) - B_{42}\cos(\delta_4 - \delta_2)] = 0.0$$

$$H_{43} = \frac{\partial P_4}{\partial \delta_3} = V_4V_3[G_{43}\sin(\delta_4 - \delta_3) - B_{43}\cos(\delta_4 - \delta_3)] = -6.6038$$

$$H_{44} = \frac{\partial P_4}{\partial \delta_4} = -Q_4 - B_{44}V_4^2 = 15.0038$$

or

$$\begin{bmatrix} 12.0259 & -7.5000 & 0.0000 \\ -7.5000 & 14.1038 & -6.6038 \\ 0.0000 & -6.6038 & 15.0038 \end{bmatrix} \begin{bmatrix} \Delta\delta_2 \\ \Delta\delta_3 \\ \Delta\delta_4 \end{bmatrix} = \begin{bmatrix} -0.3638 \\ -0.5100 \\ -0.4000 \end{bmatrix}$$

After triangularization the above matrix becomes

$$\begin{bmatrix} 12.0259 & -7.5000 & 0.0000 \\ 0.0000 & 9.4264 & -6.6038 \\ 0.0000 & 0.0000 & 10.3774 \end{bmatrix} \begin{bmatrix} \Delta\delta_2 \\ \Delta\delta_3 \\ \Delta\delta_4 \end{bmatrix} = \begin{bmatrix} -0.3638 \\ -0.7369 \\ -0.9162 \end{bmatrix}$$

Change in voltage angle at each bus, $\Delta\delta_i (i = 2, 3, 4)$ is obtained as

$$\Delta\delta_2 = -0.11758, \qquad \Delta\delta_3 = -0.14003, \qquad \Delta\delta_4 = -0.08829$$

Modified δ_i, voltage angle, is computed as

$$\delta_2 = \delta_2 + \Delta\delta_2 = 0.0 - 0.11758 = -0.11758 \text{ rad}$$

$$\delta_3 = \delta_3 + \Delta\delta_3 = 0.0 - 0.14003 = -0.14003 \text{ rad}$$

$$\delta_4 = \delta_4 + \Delta\delta_4 = 0.0 - 0.08829 = -0.08829 \text{ rad}$$

Specified reactive powers are given as

$$Q_2^S = -0.15 \text{ p.u.}, \; Q_3^S = -0.25 \text{ p.u.}, \; Q_4^S = -0.30 \text{ p.u.}$$

Compute Q_2, Q_3, and Q_4 using Eq. (2.120b), i.e.

$$Q_2 = \sum_{k=1}^{4} V_2 V_k \left[G_{2k} \sin(\delta_2 - \delta_k) - B_{2k} \cos(\delta_2 - \delta_k)\right] = -2.612371 \times 10^{-2} \text{ p.u.}$$

$$Q_3 = \sum_{k=1}^{4} V_3 V_k \left[G_{3k} \sin(\delta_3 - \delta_k) - B_{3k} \cos(\delta_3 - \delta_k)\right] = 1.644072 \times 10^{-1} \text{ p.u.}$$

$$Q_4 = \sum_{k=1}^{4} V_4 V_k \left[G_{4k} \sin(\delta_4 - \delta_k) - B_{4k} \cos(\delta_4 - \delta_k)\right] = -8.567333 \times 10^{-2} \text{ p.u.}$$

Reactive power residuals are calculated as

$$\Delta Q_2 = Q_2^S - Q_2 = -0.1239 \text{ p.u.}$$

$$\Delta Q_3 = Q_3^S - Q_3 = -0.4144 \text{ p.u.}$$

$$\Delta Q_4 = Q_4^S - Q_4 = -0.2143 \text{ p.u.}$$

Convergence is checked. If maximum $\{|\Delta P_i|\ (i = 2, 3, 4) \text{ and } |\Delta Q_i|\ (i = 2, 3, 4)\} \le \varepsilon\ (0.001)$ then stop, otherwise compute change in voltage magnitudes.

$$\begin{bmatrix} L_{22} & L_{23} & L_{24} \\ L_{32} & L_{33} & L_{34} \\ L_{42} & L_{43} & L_{44} \end{bmatrix} \begin{bmatrix} \Delta V_2/V_2 \\ \Delta V_3/V_3 \\ \Delta V_4/V_4 \end{bmatrix} = \begin{bmatrix} \Delta Q_2 \\ \Delta Q_3 \\ \Delta Q_4 \end{bmatrix}$$

where

$$L_{22} = V_2 \frac{\partial Q_2}{\partial V_2} = Q_2 - B_{22} V_2^2 = 11.7842$$

$$L_{23} = V_3 \frac{\partial Q_2}{\partial V_3} = V_2 V_3 \left[G_{23} \sin(\delta_2 - \delta_3) - B_{23} \cos(\delta_2 - \delta_3)\right] = -7.5542$$

$$L_{24} = V_4 \frac{\partial Q_2}{\partial V_4} = V_2V_4 [G_{24} \sin(\delta_2 - \delta_4) - B_{24} \cos(\delta_2 - \delta_4)] = 0.0$$

$$L_{32} = V_2 \frac{\partial Q_3}{\partial V_2} = V_3V_2 [G_{32} \sin(\delta_3 - \delta_2) - B_{32} \cos(\delta_3 - \delta_2)] = -7.4420$$

$$L_{33} = V_3 \frac{\partial Q_3}{\partial V_3} = Q_3 - B_{33}V_3^2 = 14.2682$$

$$L_{34} = V_4 \frac{\partial Q_3}{\partial V_4} = V_3V_4 [G_{34} \sin(\delta_3 - \delta_4) - B_{34} \cos(\delta_3 - \delta_4)] = -6.4974$$

$$L_{42} = V_2 \frac{\partial Q_4}{\partial V_2} = V_4V_2 [G_{42} \sin(\delta_4 - \delta_2) - B_{42} \cos(\delta_4 - \delta_2)] = 0.0$$

$$L_{43} = V_3 \frac{\partial Q_4}{\partial V_3} = V_4V_3 [G_{43} \sin(\delta_4 - \delta_3) - B_{43} \cos(\delta_4 - \delta_3)] = -6.6925$$

$$L_{44} = V_4 \frac{\partial Q_4}{\partial V_4} = Q_4 - B_{44}V_4^2 = 14.5181$$

The elements are represented in matrix form as

$$\begin{bmatrix} 11.7842 & -7.5542 & 0.0000 \\ -7.4420 & 14.2682 & -6.4974 \\ 0.0000 & -6.6925 & 14.5181 \end{bmatrix} \begin{bmatrix} \Delta V_2/V_2 \\ \Delta V_3/V_3 \\ \Delta V_4/V_4 \end{bmatrix} = \begin{bmatrix} -0.1239 \\ -0.4144 \\ -0.2143 \end{bmatrix}$$

After the triangularization process, the above matrix becomes

$$\begin{bmatrix} 11.7842 & -7.5542 & 0.0000 \\ 0.0000 & 9.4975 & -6.4974 \\ 0.0000 & 0.0000 & 9.9397 \end{bmatrix} \begin{bmatrix} \Delta V_2/V_2 \\ \Delta V_3/V_3 \\ \Delta V_4/V_4 \end{bmatrix} = \begin{bmatrix} -0.1239 \\ -0.4926 \\ -0.5615 \end{bmatrix}$$

Employing back substitution, change in voltage magnitudes, $\Delta V_i/V_i$ (i = 2, 3, 4) is computed as

$$\Delta V_2/V_2 = -0.06854, \qquad \Delta V_3/V_3 = -0.09051, \qquad \Delta V_4/V_4 = -0.05649$$

Modified voltage, V_i (i = 2, 3, 4) is computed as

$$V_2 = V_2 + \frac{\Delta V_2}{V_2} V_2 = 1.0 - \frac{0.06854}{1.0} \times 1.0 = 0.93146 \text{ p.u.}$$

$$V_3 = V_3 + \frac{\Delta V_3}{V_3} V_3 = 1.0 - \frac{0.09051}{1.0} \times 1.0 = 0.90949 \text{ p.u.}$$

$$V_4 = V_4 + \frac{\Delta V_4}{V_4} V_4 = 1.0 - \frac{0.05649}{1.0} \times 1.0 = 0.94351 \text{ p.u.}$$

This procedure is repeated and results after seven iterations are given below in Table 2.11 and line flows are given in Table 2.12.

TABLE 2.11 Results after seven iterations

Bus	P^S (p.u.)	Q^S (p.u.)	P (p.u.)	Q (p.u.)	V (p.u.)	δ (rad)
1	0.0	0.00	1.6721890	0.9570484	1.05	0.0
2	– 0.45	– 0.15	– 0.4499998	– 0.1500000	0.9337450	– 0.1040963
3	– 0.51	– 0.25	– 0.5099999	– 0.2500001	0.9113911	– 0.1220526
4	– 0.60	– 0.30	0.6000003	– 0.3000002	0.9486455	– 0.0725104

TABLE 2.12 Line flows

Line no.	*Bus code (i–k)*	S_{ik} (p.u.)
1	1–2	0.6587290 + *j*0.3733827
1	2–1	– 0.6171262 – *j*0.2693757
2	1–4	1.0134600 + *j*0.5836660
2	4–1	– 0.9514301 – *j*0.4596054
3	2–3	0.1671262 + *j*0.1193755
3	3–2	– 0.1651910 – *j*0.1135699
4	3–4	– 0.3448088 – *j*0.1364298
4	4–3	0.3514306 + *j*0.1596059

EXAMPLE 2.11 For the sample system of Figure 2.19, the generators are connected at all the four buses, while the loads are at buses 2, 3, and 4. The values of real and reactive powers are listed in Table 2.7. Line data are given in Table 2.5. Find the voltages and the bus angles at the three buses using the decoupled Newton–Raphson method.

Solution The G and B matrices are as in Example 2.10.

Consider that bus 1 is the slack bus. To start iteration, choose the initial values as

$$\delta_2 = \delta_3 = \delta_4 = 0$$

$$V_3 = V_4 = 1.0 \text{ p.u.}$$

Specified real power values are given as

$$P_2^S = -0.45 \text{ p.u.}, \qquad P_3^S = -0.51 \text{ p.u.}, \qquad P_4^S = -0.60 \text{ p.u.}$$

The real powers (P_2, P_3, and P_4) are computed from Eq. (2.120a) as

$$P_2 = \sum_{k=1}^{4} V_2 V_k \left[G_{2k} \cos(\delta_2 - \delta_k) + B_{2k} \sin(\delta_2 - \delta_k)\right] = -8.620691 \times 10^{-2} \text{ p.u.}$$

$$P_3 = \sum_{k=1}^{4} V_3 V_k \left[G_{3k} \cos(\delta_3 - \delta_k) + B_{3k} \sin(\delta_3 - \delta_k)\right] = -2.384186 \times 10^{-7} \text{ p.u.}$$

$$P_4 = \sum_{k=1}^{4} V_4 V_k \left[G_{4k} \cos(\delta_4 - \delta_k) + B_{4k} \sin(\delta_4 - \delta_k)\right] = -1.999998 \times 10^{-1} \text{ p.u.}$$

Real power residuals are calculated as

$$\Delta P_2 = P_2^S - P_2 = -0.3638 \text{ p.u.}$$

$$\Delta P_3 = P_3^S - P_3 = -0.5100 \text{ p.u.}$$

$$\Delta P_4 = P_4^S - P_4 = -0.4000 \text{ p.u.}$$

If maximum $\{|\Delta P_i| \; (i = 2, 3, 4)\} > \varepsilon \; (0.001)$ then calculate the change in voltage angle at *PV* and *PQ* buses.

$$\begin{bmatrix} H_{22} & H_{23} & H_{24} \\ H_{32} & H_{33} & H_{34} \\ H_{42} & H_{43} & H_{44} \end{bmatrix} \begin{bmatrix} \Delta\delta_2 \\ \Delta\delta_3 \\ \Delta\delta_4 \end{bmatrix} = \begin{bmatrix} \Delta P_2 \\ \Delta P_3 \\ \Delta P_4 \end{bmatrix}$$

where

$$H_{22} = \frac{\partial P_2}{\partial \delta_2} = -Q_2 - B_{22} V_2^2 = 12.0259$$

$$H_{23} = \frac{\partial P_2}{\partial \delta_3} = V_2 V_3 \left[G_{23} \sin(\delta_2 - \delta_3) - B_{23} \cos(\delta_2 - \delta_3)\right] = -7.5$$

$$H_{24} = \frac{\partial P_2}{\partial \delta_4} = V_2 V_4 \left[G_{24} \sin(\delta_2 - \delta_4) - B_{24} \cos(\delta_2 - \delta_4)\right] = 0.0$$

$$H_{32} = \frac{\partial P_3}{\partial \delta_2} = V_3 V_2 \left[G_{32} \sin(\delta_3 - \delta_2) - B_{32} \cos(\delta_3 - \delta_2)\right] = -7.5$$

$$H_{33} = \frac{\partial P_3}{\partial \delta_3} = -Q_3 - B_{33} V_3^2 = 14.1038$$

$$H_{34} = \frac{\partial P_3}{\partial \delta_4} = V_3 V_4 \left[G_{34} \sin(\delta_3 - \delta_4) - B_{34} \cos(\delta_3 - \delta_4)\right] = -6.6038$$

$$H_{42} = \frac{\partial P_4}{\partial \delta_2} = V_4 V_2 [G_{42} \sin(\delta_4 - \delta_2) - B_{42} \cos(\delta_4 - \delta_2)] = 0.0$$

$$H_{43} = \frac{\partial P_4}{\partial \delta_3} = V_4 V_3 [G_{43} \sin(\delta_4 - \delta_3) - B_{43} \cos(\delta_4 - \delta_3)] = -6.6038$$

$$H_{44} = \frac{\partial P_4}{\partial \delta_4} = -Q_4 - B_{44} V_4^2 = 15.0038$$

The above elements are represented in matrix form as

$$\begin{bmatrix} 12.0259 & -7.5000 & 0.0000 \\ -7.5000 & 14.1038 & -6.6038 \\ 0.0000 & -6.6038 & 15.0038 \end{bmatrix} \begin{bmatrix} \Delta\delta_2 \\ \Delta\delta_3 \\ \Delta\delta_4 \end{bmatrix} = \begin{bmatrix} -0.3638 \\ -0.5100 \\ -0.4000 \end{bmatrix}$$

After triangularization the above matrix becomes

$$\begin{bmatrix} 12.0259 & -7.5000 & 0.0000 \\ 0.0000 & 9.4264 & -6.6038 \\ 0.0000 & 0.0000 & 10.3774 \end{bmatrix} \begin{bmatrix} \Delta\delta_2 \\ \Delta\delta_3 \\ \Delta\delta_4 \end{bmatrix} = \begin{bmatrix} -0.3638 \\ -0.7369 \\ -0.9162 \end{bmatrix}$$

Employing back substitution, the change in voltage angles, $\delta_i (i = 2, 3, 4)$ is computed as

$$\Delta\delta_2 = -0.11758, \qquad \Delta\delta_3 = -0.14003, \qquad \Delta\delta_4 = -0.08829$$

Modified values of $\delta_i (i = 2, 3, 4)$, are computed as

$$\delta_2 = \delta_2 + \Delta\delta_2 = 0.0 - 0.11758 = -0.11758 \text{ rad}$$

$$\delta_3 = \delta_3 + \Delta\delta_3 = 0.0 - 0.14003 = -0.14003 \text{ rad}$$

$$\delta_4 = \delta_4 + \Delta\delta_4 = 0.0 - 0.08829 = -0.08829 \text{ rad}$$

Specified reactive powers are

$$Q_3^S = -0.25 \text{ p.u.}, \qquad Q_4^S = -0.30 \text{ p.u.}$$

Compute Q_3 and Q_4 using Eq. (2.120b)

$$Q_3 = \sum_{k=1}^{4} V_3 V_k [G_{3k} \sin(\delta_3 - \delta_k) - B_{3k} \cos(\delta_3 - \delta_k)] = 1.644072 \times 10^{-1} \text{ p.u.}$$

$$Q_4 = \sum_{k=1}^{4} V_4 V_k [G_{4k} \sin(\delta_4 - \delta_k) - B_{4k} \cos(\delta_4 - \delta_k)] = -8.567333 \times 10^{-2} \text{ p.u.}$$

Reactive power residuals are calculated as

$$\Delta Q_3 = Q_3^S - Q_3 = -0.4144 \text{ p.u.}$$

$$\Delta Q_4 = Q_4^S - Q_4 = -0.2143 \text{ p.u.}$$

Convergence is checked as

$$\text{maximum } \{|\Delta P_i| \ (i = 2, 3, 4) \text{ and } |\Delta Q_i| \ (i = 3, 4)\} \le \varepsilon \ (0.001) \text{ then stop}$$

$$\text{maximum } \{0.3638, 0.5100, 0.40, 0.4144, 0.2143\} = 0.51 > 0.001$$

Convergence is not met, so the change in voltage magnitude at *PQ* buses is given by

$$\begin{bmatrix} L_{33} & L_{34} \\ L_{43} & L_{44} \end{bmatrix} \begin{bmatrix} \Delta V_3/V_3 \\ \Delta V_4/V_4 \end{bmatrix} = \begin{bmatrix} \Delta Q_3 \\ \Delta Q_4 \end{bmatrix}$$

where

$$L_{33} = V_3 \frac{\partial Q_3}{\partial V_3} = Q_3 - B_{33}V_3^2 = 14.2682$$

$$L_{34} = V_4 \frac{\partial Q_3}{\partial V_4} = V_3V_4 \, [G_{34} \sin (\delta_3 - \delta_4) - B_{34} \cos (\delta_3 - \delta_4)] = -6.4974$$

$$L_{43} = V_3 \frac{\partial Q_4}{\partial V_3} = V_4V_3 \, [G_{43} \sin (\delta_4 - \delta_3) - B_{43} \cos (\delta_4 - \delta_3)] = -6.6925$$

$$L_{44} = V_4 \frac{\partial Q_4}{\partial V_4} = Q_4 - B_{44}V_4^2 = 14.5181$$

Elements are arranged in matrix form as

$$\begin{bmatrix} 14.2682 & -6.4974 \\ -6.6925 & 14.5181 \end{bmatrix} \begin{bmatrix} \Delta V_3/V_3 \\ \Delta V_4/V_4 \end{bmatrix} = \begin{bmatrix} -0.4144 \\ -0.2143 \end{bmatrix}$$

Triangularizing the above matrix,

$$\begin{bmatrix} 11.2682 & -6.4974 \\ 0.0000 & 11.4705 \end{bmatrix} \begin{bmatrix} \Delta V_3/V_3 \\ \Delta V_4/V_4 \end{bmatrix} = \begin{bmatrix} -0.4144 \\ -0.4087 \end{bmatrix}$$

After back substitution, we get

$$\Delta V_3/V_3 = -0.04527, \qquad \Delta V_4/V_4 = -0.03563,$$

Modified voltage is computed as

$$V_3 = V_3 + \frac{\Delta V_3}{V_3} V_3 = 1.0 - \frac{0.04527}{1.0} \times 1.0 = 0.95473 \text{ p.u.}$$

$$V_4 = V_4 + \frac{\Delta V_4}{V_4} V_4 = 1.0 - \frac{0.03563}{1.0} \times 1.0 = 0.96437 \text{ p.u.}$$

This procedure is repeated and the results after nine iterations are given in Table 2.13. Line flows are given in Table 2.14.

TABLE 2.13 Results after nine iterations

Bus	P^S (p.u.)	Q^S (p.u.)	P (p.u.)	Q (p.u.)	V (p.u.)	δ (rad)
1	0.0	0.00	1.6531390	0.4291533	1.0500000	0.0
2	– 0.45	– 0.15	– 0.4500009	– 0.3351609	1.0	– 0.1222418
3	– 0.51	– 0.25	– 0.5100001	– 0.2499999	0.9575218	– 0.1317479
4	– 0.60	– 0.30	– 0.5999997	– 0.3000001	0.9684043	– 0.0789868

TABLE 2.14 Line flows

Line no.	*Bus code* (*i–k*)	S_{ik} (p.u.)
1	1–2	0.6558992 + *j*0.0393169
1	2–1	– 0.6245705 + *j*0.0390050
2	1–4	0.9972406 + *j*0.3898371
2	4–1	– 0.9452469 – *j*0.2858496
3	2–3	0.1745696 + *j*0.2961555
3	3–2	– 0.1698423 – *j*0.2819736
4	3–4	– 0.3401575 + *j*0.0319743
4	4–3	0.3452502 – *j*0.0141501

2.12 FAST DECOUPLED LOAD FLOW (FDLF)

The Fast Decoupled Load Flow (FDLF) was developed by B. Stott in 1974. The assumptions which are valid in normal power system operation, are made as follows:

(i) Under normal loading conditions, angle differences, $(\delta_i - \delta_k)$, across transmission lines are small, i.e. $\cos(\delta_i - \delta_k) \cong 1$, $\sin(\delta_i - \delta_k) \cong 0$

(ii) For a transmission line, its reactance is more than its resistance. In other words, $X/R >> 1$. So, G_{ik} can be ignored because $G_{ik} << B_{ik}$.

In view of the above, $G_{ik} \sin(\delta_i - \delta_k) << B_{ik}$ and $Q_i << B_{ii} V_i^2$.

With these assumptions, the elements of H and L sub-matrices become considerably simplified as

$$H_{ii} = \frac{\partial P_i}{\partial \delta_i} = -B_{ii}V_i^2 \qquad (i = k) \tag{2.138a}$$

$$H_{ik} = \frac{\partial P_i}{\partial \delta_k} = -V_i V_k B_{ik} \qquad (i \neq k) \tag{2.138b}$$

$$L_{ii} = V_i \frac{\partial Q_i}{\partial \delta_i} = -B_{ii}V_i^2 \qquad (i = k) \tag{2.139a}$$

$$L_{ik} = V_k \frac{\partial Q_i}{\partial V_k} = -\ V_i V_k B_{ik} \qquad (i \neq k) \tag{2.139b}$$

Substituting Eqs. (2.138a) and (2.138b) in Eq. (2.134a), we have

$$\sum_{k=2}^{NB} [-V_i V_k B_{ik}]\, \Delta\delta_k = \Delta P_i \qquad (i = 2, 3, ..., NB) \tag{2.140a}$$

or

$$\sum_{k=2}^{NB} [-V_k B_{ik}]\, \Delta\delta_k = \frac{\Delta P_i}{V_i} \qquad (i = 2, 3, ..., NB) \tag{2.140b}$$

Setting V_k = 1 p.u. on right-hand side of Eq. (2.140a),

$$\sum_{k=2}^{NB} [-B_{ik}]\, \Delta\delta_k = \frac{\Delta P_i}{V_i} \qquad (i = 2, 3, ..., NB) \tag{2.140c}$$

Substituting Eqs. (2.139a) and (2.139b) in Eq. (2.134b), we get

$$\sum_{k=NV+1}^{NB} [-V_i V_k B_{ik}] \frac{\Delta V_k}{V_k} = \Delta Q_i \qquad (i = NV + 1, NV + 2, ..., NB) \tag{2.141a}$$

or

$$\sum_{k=NV+1}^{NB} [-B_{ik}]\, \Delta V_k = \frac{\Delta Q_i}{V_i} \qquad (i = NV + 1, NV + 2, ..., NB) \tag{2.141b}$$

Setting V_k = 1 p.u. on right-hand side of Eq. (2.140a),

$$\sum_{k=NV+1}^{NB} [-B_{ik}]\, \Delta V_k = \frac{\Delta Q_i}{V_i} \qquad (i = NV + 1, NV + 2, ..., NB) \tag{2.141c}$$

Above equations can be written in matrix form as

$$[B'] [\Delta\delta] = \left[\frac{\Delta P}{V}\right] \tag{2.142a}$$

$$[B''] [\Delta V] = \left[\frac{\Delta Q}{V}\right] \tag{2.142b}$$

where

B' is the matrix having elements $-B_{ik}$ (i = 2, 3, ..., NB and k = 2, 3, ..., NB)

B'' is the matrix having elements $-B_{ik}$ (i = NV + 1, NV + 2, ..., NB and k = NV + 1, NV + 2, ..., NB).

Further simplification of the FDLF algorithm is achieved by:

1. Omitting the elements of $[B']$ that predominately affect reactive power flows, i.e. shunt reactances and transformer off-nominal in-phase taps.

2. Omitting from $[B'']$ the angle shifting effect of the phase shifter that predominately affects reactive power flows.
3. Ignoring the series resistance in calculating the elements of $[B']$, which then becomes the dc approximation of the power flow matrix.

Equations (2.142a) and (2.142b) are solved alternatively, always employing the most recent voltage values.

A detailed stepwise procedure is explained here.

Algorithm 2.6: Fast Decoupled Load Flow Method

1. Read data
   NB (number of buses); NV (number of *PV* buses).
   V_1, δ_1 for slack bus, $P_i^S (i = 2, 3, ..., \text{NB})$ for *PQ* and *PV* buses.
   Q_i^S $(i = \text{NV} + 1, \text{NV} + 2, ..., \text{NB})$ for *PQ* buses, V_i^S $(i = 2, 3, ..., \text{NV}$ for *PV* buses), $V_i^{\min}$, $V_i^{\max}$ $(i = \text{NV} + 1, \text{NV} + 2, ..., \text{NB})$ for *PQ* buses.
   $Q_i^{\min}$, $Q_i^{\max}$ $(i = 2, 3, ..., \text{NV})$ for *PV* buses, R (maximum number of iterations), ε (tolerance in convergence)
2. Form Y_{BUS} as explained in Section 2.3 and form B' and B'' matrices.
3. Assume initially, voltage magnitude and voltage angles
   $$|V_i| \ (i = \text{NV} + 1, \text{NV} + 2, ..., \text{NB}) \quad \text{and} \quad \delta_i \ (i = 2, 3, ..., \text{NB})$$
4. Set iteration count, $r = 0$.
5. Compute P_i, ΔP_i using Eq. (2.120a)
   $$P_i = \sum_{k=1}^{\text{NB}} V_i V_k [G_{ik} \cos(\delta_i - \delta_k) + B_{ik} \sin(\delta_i - \delta_k)] \qquad (i = 2, 3, ..., \text{NB})$$
   $$\Delta P_i = P_i^S - P_i \qquad (i = 2, 3, ..., \text{NB})$$
6. Compute $\Delta P^{\max} = \text{maximum}\ \{|\Delta P_i|\ (i = 2, 3, ..., \text{NB})\}$.
   If $\Delta P^{\max} \le \varepsilon_p$ then GOTO Step 9.
7. Compute $\Delta\delta_i$ $(i = 2, 3, ..., \text{NB})$ using Eq. (2.142a), i.e.
   $$[B'][\Delta\delta] = \left[\frac{\Delta P}{V}\right]$$
8. Modified δ_i is computed as
   $$\delta_i = \delta_i + \Delta\delta_i \qquad (i = 2, 3, ..., \text{NB})$$
9. Compute Q_i, ΔQ_i using Eq. (2.120b), i.e.
   $$Q_i = \sum_{k=1}^{\text{NB}} V_i V_k [G_{ik} \sin(\delta_i - \delta_k) - B_{ik} \cos(\delta_i - \delta_k)] \qquad (i = \text{NV} + 1, \text{NV} + 2, ..., \text{NB})$$
   $$\Delta Q_i = Q_i^S - Q_i \qquad (i = \text{NV} + 1, \text{NV} + 2, ..., \text{NB})$$

10. Compute $\Delta Q^{\max}$ = maximum $\{|\Delta Q_i|\ (i = \text{NV} + 1, \text{NV} + 2, ..., \text{NB})\}$.

    If ($\Delta Q^{\max} \le \varepsilon_q$ and $\Delta P^{\max} \le \varepsilon_p$) then GOTO Step 14.

11. Compute ΔV_i (i = NV + 1, NV + 2, ..., NB) using Eq. (2.142b), i.e.

$$[B''][\Delta V] = \left[\frac{\Delta Q}{V}\right]$$

12. Modify V_i as

$$V_i = V_i + \frac{\Delta V_i}{V_i} V_i \qquad (i = \text{NV} + 1, \text{NV} + 2, ..., \text{NB})$$

13. Advance the count $r = r + 1$.

    Is $r \le R$ then GOTO Step 5 and repeat.

14. Compute the active and reactive powers on slack bus as

$$P_1 = \sum_{k=1}^{\text{NB}} V_1 V_k [G_{1k} \cos(\delta_1 - \delta_k) + B_{1k} \sin(\delta_1 - \delta_k)]$$

$$Q_1 = \sum_{k=1}^{\text{NB}} V_1 V_k [G_{1k} \sin(\delta_1 - \delta_k) + B_{1k} \cos(\delta_1 - \delta_k)]$$

15. Calculate line flows using Eqs. (2.85) and (2.86), i.e.

$$S_{ik} = V_i^c [\{(V_i^c)^* - (V_k^c)^*\} \overset{*}{y}_{ik} + (V_i^c)^* \overset{*}{y}_{ik0}]$$

$$S_{ki} = V_k^c [\{(V_k^c)^* - (V_i^c)^*\} \overset{*}{y}_{ki} + (V_k^c)^* \overset{*}{y}_{ki0}]$$

    where $V_i^c = V_i(\cos\delta_i + j\sin\delta_i)$

16. Stop.

EXAMPLE 2.12 For the sample system of Figure 2.19, the generators are connected at all the four buses, while loads are at buses 2, 3, and 4. The values of real and reactive powers are listed in Table 2.4. All buses other than slack are of the *PQ*-type. Line data is given in Table 2.5. Find the voltages and the bus angles at the three buses using the fast decoupled load flow (FDLF) method.

Solution

$$B = \begin{bmatrix} -12.31034 & 4.310345 & 0.0 & 8.0 \\ 4.310345 & -11.81034 & 7.5 & 0.0 \\ 0.0 & 7.5 & -14.10377 & 6.603774 \\ 8.0 & 0.0 & 6.603774 & -14.60377 \end{bmatrix}$$

Consider bus 1 as the slack bus. To start iteration, choose initial values of δ_i and V_i as

$$\delta_2 = \delta_3 = \delta_4 = 0 \text{ rad}$$

$$V_2 = V_3 = V_4 = 1.0 \text{ p.u.}$$

Specified real power values are

$$P_2^S = -0.45 \text{ p.u.}, \qquad P_3^S = -0.51 \text{ p.u.}, \qquad P_4^S = -0.60 \text{ p.u.}$$

The real powers (P_2, P_3, and P_4) are computed from Eq. (2.120a) as

$$P_2 = \sum_{k=1}^{4} V_2 V_k \,[G_{2k} \cos(\delta_2 - \delta_k) + B_{2k} \sin(\delta_2 - \delta_k)] = -8.620691 \times 10^{-2} \text{ p.u.}$$

$$P_3 = \sum_{k=1}^{4} V_3 V_k \,[G_{3k} \cos(\delta_3 - \delta_k) + B_{3k} \sin(\delta_3 - \delta_k)] = -2.384186 \times 10^{-7} \text{ p.u.}$$

$$P_4 = \sum_{k=1}^{4} V_4 V_k \,[G_{4k} \cos(\delta_4 - \delta_k) + B_{4k} \sin(\delta_4 - \delta_k)] = -1.999998 \times 10^{-1} \text{ p.u.}$$

Real power residuals are calculated as

$$\Delta P_2 = P_2^S - P_2 = -0.3638 \text{ p.u.}$$

$$\Delta P_3 = P_3^S - P_3 = -0.5100 \text{ p.u.}$$

$$\Delta P_4 = P_4^S - P_4 = -0.4000 \text{ p.u.}$$

Convergence is checked as

$$\text{maximum } \{|\Delta P_i| \ (i = 2, 3, 4)\} \le \varepsilon \ (0.001)$$

$$\text{maximum } \{0.3638, 0.51, 0.4\} \doteq 0.51 > \varepsilon \ (0.001)$$

Convergence is not obtained, so calculate the change in voltage angles

$$\begin{bmatrix} -B_{22} & -B_{23} & -B_{24} \\ -B_{32} & -B_{33} & -B_{34} \\ -B_{42} & -B_{43} & -B_{44} \end{bmatrix} \begin{bmatrix} \Delta\delta_2 \\ \Delta\delta_3 \\ \Delta\delta_4 \end{bmatrix} = \begin{bmatrix} \Delta P_2/V_2 \\ \Delta P_3/V_3 \\ \Delta P_4/V_4 \end{bmatrix}$$

or

$$\begin{bmatrix} 11.8103 & -7.5000 & 0.0000 \\ -7.5000 & 14.1038 & -6.6038 \\ 0.0000 & -6.6038 & 14.6038 \end{bmatrix} \begin{bmatrix} \Delta\delta_2 \\ \Delta\delta_3 \\ \Delta\delta_4 \end{bmatrix} = \begin{bmatrix} -0.3638/1.0 \\ -0.5100/1.0 \\ -0.4000/1.0 \end{bmatrix}$$

Triangularizing the above matrix, we have

$$\begin{bmatrix} 11.8103 & -7.5000 & 0.0000 \\ 0.0000 & 9.3410 & -6.6038 \\ 0.0000 & 0.0000 & 9.9351 \end{bmatrix} \begin{bmatrix} \Delta\delta_2 \\ \Delta\delta_3 \\ \Delta\delta_4 \end{bmatrix} = \begin{bmatrix} -0.3638 \\ -0.7410 \\ -0.9239 \end{bmatrix}$$

Back substitution gives the change in voltage angles as

$$\Delta\delta_2 = -0.12293, \qquad \Delta\delta_3 = -0.14507, \qquad \Delta\delta_4 = -0.09299$$

Modify $\delta_i (i = 2, 3, 4)$ as

$$\delta_2 = \delta_2 + \Delta\delta_2 = 0.0 - 0.12293 = -\ 0.12293 \text{ rad}$$

$$\delta_3 = \delta_3 + \Delta\delta_3 = 0.0 - 0.14507 = -\ 0.14507 \text{ rad}$$

$$\delta_4 = \delta_4 + \Delta\delta_4 = 0.0 - 0.09299 = -\ 0.09299 \text{ rad}$$

Specified reactive powers are

$$Q_2^S = -0.15 \text{ p.u.}, \qquad Q_3^S = -0.25 \text{ p.u.}, \qquad Q_4^S = -0.30 \text{ p.u.}$$

Compute Q_2, Q_3, and Q_4 using Eq. (2.120b) as

$$Q_2 = \sum_{k=1}^{4} V_2 V_k \,[G_{2k} \sin(\delta_2 - \delta_k) - B_{2k} \cos(\delta_2 - \delta_k)] = -1.289478 \times 10^{-2} \text{ p.u.}$$

$$Q_3 = \sum_{k=1}^{4} V_3 V_k \,[G_{3k} \sin(\delta_3 - \delta_k) - B_{3k} \cos(\delta_3 - \delta_k)] = 1.643656 \times 10^{-1} \text{ p.u.}$$

$$Q_4 = \sum_{k=1}^{4} V_4 V_k \,[G_{4k} \sin(\delta_4 - \delta_k) - B_{4k} \cos(\delta_4 - \delta_k)] = -6.297398 \times 10^{-2} \text{ p.u.}$$

Reactive power residuals are calculated as

$$\Delta Q_2 = Q_2^S - Q_2 = -0.1371 \text{ p.u.}$$

$$\Delta Q_3 = Q_3^S - Q_3 = -0.4144 \text{ p.u.}$$

$$\Delta Q_4 = Q_4^S - Q_4 = -0.2370 \text{ p.u.}$$

Convergence criterion is considered as

$$\text{maximum } \{|\Delta P_i| \ (i = 2, 3, 4) \text{ and } |\Delta Q_i| \ (i = 2, 3, 4)\} \le \varepsilon \ (0.001)$$

$$\text{maximum } \{0.3638, 0.51, 0.4, 0.1371, 0.4144, 0.2370\} = 0.51 > \varepsilon$$

Convergence has not been obtained, so the change in voltage magnitudes is computed as

$$\begin{bmatrix} -B_{22} & -B_{23} & -B_{24} \\ -B_{32} & -B_{33} & -B_{34} \\ -B_{42} & -B_{43} & -B_{44} \end{bmatrix} \begin{bmatrix} \Delta V_2 \\ \Delta V_3 \\ \Delta V_4 \end{bmatrix} = \begin{bmatrix} \Delta Q_2 / V_2 \\ \Delta Q_3 / V_3 \\ \Delta Q_4 / V_4 \end{bmatrix}$$

or

$$\begin{bmatrix} 11.8103 & -7.5000 & 0.0000 \\ -7.5000 & 14.1038 & -6.6038 \\ 0.0000 & -6.6038 & 14.6038 \end{bmatrix} \begin{bmatrix} \Delta V_2 \\ \Delta V_3 \\ \Delta V_4 \end{bmatrix} = \begin{bmatrix} -0.1371/1.0 \\ -0.4144/1.0 \\ -0.2370/1.0 \end{bmatrix}$$

Triangularizing the above matrix, we have

$$\begin{bmatrix} 11.8103 & -7.5000 & 0.0000 \\ 0.0000 & 9.3410 & -6.6038 \\ 0.0000 & 0.0000 & 9.9351 \end{bmatrix} \begin{bmatrix} \Delta V_2 \\ \Delta V_3 \\ \Delta V_4 \end{bmatrix} = \begin{bmatrix} -0.1371 \\ -0.5014 \\ -0.5915 \end{bmatrix}$$

Back substitution gives the change in voltage magnitude as

$$\Delta V_2 = -0.07243, \qquad \Delta V_3 = -0.09577, \qquad \Delta V_4 = -0.05954$$

Modified voltages are

$$V_2 = 1.0 - 0.07243 = 0.92757 \text{ p.u.}$$
$$V_3 = 1.0 - 0.09577 = 0.90423 \text{ p.u.}$$
$$V_4 = 1.0 - 0.05954 = 0.94046 \text{ p.u.}$$

This procedure is repeated and results after 14 iterations are given in Table 2.15. Line flows are given in Table 2.16.

TABLE 2.15 Results after 14 iterations

Bus	P^S (p.u.)	Q^S (p.u.)	P (p.u.)	Q (p.u.)	V (p.u.)	δ (rad)
1	0.0	0.00	1.6721970	0.9570510	1.50	0.0
2	−0.45	−0.15	−0.4500004	−0.1499991	0.9337449	−0.1040965
3	−0.51	−0.25	−0.5099981	−0.2500028	0.9113908	−0.1220529
4	−0.60	−0.30	0.6000080	−0.2999880	0.9486451	−0.0725110

TABLE 2.16 Line flows

Line no.	*Bus code* (*i–k*)	S_{ik} (p.u.)
1	1–2	0.6815881 + *j*0.3377037
1	2–1	− 0.6396030 − *j*0.2327412
2	1–4	1.0347830 + *j*0.5468349
2	4–1	− 0.9726602 − *j*0.4225897
3	2–3	0.1907119 + *j*0.0947944
3	3–2	− 0.1886566 − *j*0.0886283
4	3–4	− 0.3808545 − *j*0.0905160
4	4–3	0.3881041 + *j*0.1158895

EXAMPLE 2.13 For the sample system of Figure 2.19, the generators are connected at all the four buses, while the loads are at buses 2, 3, and 4. The values of real and reactive powers are listed in Table 2.7. Line data are given in Table 2.5. Find the voltages and the bus angles at the three buses using the fast decoupled load flow method.

Solution

$$B = \begin{bmatrix} -12.31034 & 4.310345 & 0.0 & 8.0 \\ 4.310345 & -11.81034 & 7.5 & 0.0 \\ 0.0 & 7.5 & -14.10377 & 6.603774 \\ 8.0 & 0.0 & 6.603774 & -14.60377 \end{bmatrix}$$

To start iteration choose initial values of δ_i and V_i as

$$\delta_2 = \delta_3 = \delta_4 = 0$$

$$V_3 = V_4 = 1.0 \text{ p.u.}$$

Specified real power values are

$$P_2^S = -0.45 \text{ p.u.}, \qquad P_3^S = -0.51 \text{ p.u.}, \qquad P_4^S = -0.60 \text{ p.u.}$$

The real powers (P_2, P_3, and P_4) are computed from Eq. (2.120a) as

$$P_2 = \sum_{k=1}^{4} V_2 V_k \,[G_{2k} \cos (\delta_2 - \delta_k) + B_{2k} \sin (\delta_2 - \delta_k)] = -8.620691 \times 10^{-2} \text{ p.u.}$$

$$P_3 = \sum_{k=1}^{4} V_3 V_k \,[G_{3k} \cos (\delta_3 - \delta_k) + B_{3k} \sin (\delta_3 - \delta_k)] = -2.384186 \times 10^{-7} \text{ p.u.}$$

$$P_4 = \sum_{k=1}^{4} V_4 V_k \,[G_{4k} \cos (\delta_4 - \delta_k) + B_{4k} \sin (\delta_4 - \delta_k)] = -1.999998 \times 10^{-1} \text{ p.u.}$$

Real power residuals are calculated as

$$\Delta P_2 = P_2^S - P_2 = -0.3638 \text{ p.u.}$$

$$\Delta P_3 = P_3^S - P_3 = -0.5100 \text{ p.u.}$$

$$\Delta P_4 = P_4^S - P_4 = -0.4000 \text{ p.u.}$$

Convergence is checked as

$$\text{maximum } \{|\Delta P_i| \ (i = 2, 3, 4)\} \le \varepsilon \ (0.001)$$

$$\text{maximum } \{0.3638, 0.51, 0.4\} = 0.51 > \varepsilon \ (0.001)$$

Since convergence has not been achieved, changes in voltage angles are calculated as

$$\begin{bmatrix} -B_{22} & -B_{23} & -B_{24} \\ -B_{32} & -B_{33} & -B_{34} \\ -B_{42} & -B_{43} & -B_{44} \end{bmatrix} \begin{bmatrix} \Delta\delta_2 \\ \Delta\delta_3 \\ \Delta\delta_4 \end{bmatrix} = \begin{bmatrix} \Delta P_2/V_2 \\ \Delta P_3/V_3 \\ \Delta P_4/V_4 \end{bmatrix}$$

$$\begin{bmatrix} 11.8103 & -7.5000 & 0.0000 \\ -7.5000 & 14.1038 & -6.6038 \\ 0.0000 & -6.6038 & 14.6038 \end{bmatrix} \begin{bmatrix} \Delta\delta_2 \\ \Delta\delta_3 \\ \Delta\delta_4 \end{bmatrix} = \begin{bmatrix} -0.3638/1.0 \\ -0.5100/1.0 \\ -0.4000/1.0 \end{bmatrix}$$

Triangularizing the above matrix,

$$\begin{bmatrix} 11.8103 & -7.5000 & 0.0000 \\ 0.0000 & 9.3410 & -6.6038 \\ 0.0000 & 0.0000 & 9.9351 \end{bmatrix} \begin{bmatrix} \Delta\delta_2 \\ \Delta\delta_3 \\ \Delta\delta_4 \end{bmatrix} = \begin{bmatrix} -0.3638 \\ -0.7410 \\ -0.9239 \end{bmatrix}$$

Back substitution gives the change in voltage angles as

$$\Delta\delta_2 = -0.12293, \qquad \Delta\delta_3 = -0.14507, \qquad \Delta\delta_4 = -0.09299$$

Modified values of δ_i (i = 2, 3, 4) are computed as

$$\delta_2 = \delta_2 + \Delta\delta_2 = 0.0 - 0.12293 = -0.12293 \text{ rad}$$

$$\delta_3 = \delta_3 + \Delta\delta_3 = 0.0 - 0.14507 = -0.14507 \text{ rad}$$

$$\delta_4 = \delta_4 + \Delta\delta_4 = 0.0 - 0.09299 = -0.09299 \text{ rad}$$

Specified reactive powers are

$$Q_3^S = -0.25 \text{ p.u.}, \qquad Q_4^S = -0.30 \text{ p.u.},$$

Compute Q_3 and Q_4 using Eq. (2.120b),

$$Q_3 = \sum_{k=1}^{4} V_3 V_k \, [G_{3k} \sin(\delta_3 - \delta_k) - B_{3k} \cos(\delta_3 - \delta_k)] = 1.643656 \times 10^{-1} \text{ p.u.}$$

$$Q_4 = \sum_{k=1}^{4} V_4 V_k \, [G_{4k} \sin(\delta_4 - \delta_k) - B_{4k} \cos(\delta_4 - \delta_k)] = -6.297398 \times 10^{-2} \text{ p.u.}$$

Reactive power residuals are calculated as

$$\Delta Q_3 = Q_3^S - Q_3 = -0.4144 \text{ p.u.}$$

$$\Delta Q_4 = Q_4^S - Q_4 = -0.2370 \text{ p.u.}$$

Convergence criterion is considered as

$$\text{maximum } \{|\Delta P_i| \ (i = 2, 3, 4) \text{ and } |\Delta Q_i| \ (i = 3, 4)\} \leq \varepsilon \ (0.001)$$

or

$$\text{maximum } \{0.3638, 0.51, 0.4, 0.4144, 0.2370\} = 0.51 > \varepsilon$$

Convergence has not been achieved, so change in voltage magnitudes is computed.

$$\begin{bmatrix} -B_{33} & -B_{34} \\ -B_{43} & -B_{44} \end{bmatrix} \begin{bmatrix} \Delta V_3 \\ \Delta V_4 \end{bmatrix} = \begin{bmatrix} \Delta Q_3/V_3 \\ \Delta Q_4/V_4 \end{bmatrix}$$

or

$$\begin{bmatrix} 14.1038 & -6.6038 \\ -6.6038 & 14.6038 \end{bmatrix} \begin{bmatrix} \Delta V_3 \\ \Delta V_4 \end{bmatrix} = \begin{bmatrix} -0.4144/1.0 \\ -0.2370/1.0 \end{bmatrix}$$

Triangularizing the above matrix, we have

$$\begin{bmatrix} 14.1038 & -6.6038 \\ 0.0000 & 11.5117 \end{bmatrix} \begin{bmatrix} \Delta V_3 \\ \Delta V_4 \end{bmatrix} = \begin{bmatrix} -0.4144 \\ -0.4310 \end{bmatrix}$$

Back substitution gives the change in voltage magnitudes as

$$\Delta V_3 = -0.04691, \qquad \Delta V_4 = -0.03744$$

Modified voltage at each PQ bus is computed as

$$V_3 = 1.0 - 0.04691 = 0.95309 \text{ p.u.}$$

$$V_4 = 1.0 - 0.03744 = 0.96256 \text{ p.u.}$$

This procedure is repeated and results after twelve iterations are given in Table 2.17, when $|\Delta P| = 1.452649 \times 10^{-5}$ and $|\Delta Q| = 8.388814 \times 10^{-6}$.

TABLE 2.17 Results after twelve iterations

Bus	P^S (p.u.)	Q^S (p.u.)	P (p.u.)	Q (p.u.)	V (p.u.)	δ (rad)
1	0.0	0.00	1.6531440	0.4291472	1.05	0.0
2	−0.45	−0.15	−0.4500015	0.3351632	1.00	−0.1222420
3	−0.51	−0.25	−0.5100006	−0.2499992	0.9575220	−0.1317483
4	−0.60	−0.30	−0.5999855	−0.2999916	0.9684048	−0.0789874

Line flows are given in Table 2.18.

TABLE 2.18 Line flows

Line no.	*Bus code* (*i–k*)	S_{ik} (p.u.)
1	1–2	0.6437674 + *j*0.0049928
1	2–1	– 0.6136931 + *j*0.0701930
2	1–4	1.0061940 + *j*0.3536526
2	4–1	– 0.9546064 – *j*0.2504784
3	2–3	0.2135317 + *j*0.2731219
3	3–2	– 0.2087948 – *j*0.2589113
4	3–4	– 0.3496354 + *j*0.0751152
4	4–3	0.3551111 – *j*0.0559500

2.13 INITIAL GUESS FOR LOAD FLOW

Leoniopoulos [1994] derived the method to calculate the initial guess instead of flat voltage start. The real power is represented by the equation

$$P_i = V_i \sum_{k=1}^{NB} V_k [G_{ik} \cos(\delta_i - \delta_k) + B_{ik} \sin(\delta_i - \delta_k)] \tag{2.143}$$

where

P_i is the real power of the *i*th bus
V_i is voltage magnitude of the *i*th bus
δ_i is voltage angle of the *i*th bus
$Y_{ik} = G_{ik} + jB_{ik}$ is the element of admittance bus matrix
NB is number of buses.

Expanding the following around zero,

$$\sin(\delta_i - \delta_k) = (\delta_i - \delta_k) - \frac{(\delta_i - \delta_k)^3}{3!} + \cdots$$

$$\cos(\delta_i - \delta_k) = 1 - \frac{(\delta_i - \delta_k)^2}{2!} + \cdots$$

If angles are small, then the higher order terms can be ignored, i.e.

$$\sin(\delta_i - \delta_k) \approx (\delta_i - \delta_k) \tag{2.144}$$

$$\cos(\delta_i - \delta_k) \approx 1 \tag{2.145}$$

Assume $V_i = 1.0$ ($i = 2, 3, ..., NB$)

Substituting the above assumptions in Eq. (2.143),

$$P_i = V_i \sum_{k=1}^{NB} V_k [G_{ik} + B_{ik} (\delta_i - \delta_k)]$$

or

$$P_i = \sum_{k=1}^{NB} G_{ik} + \sum_{k=1}^{NB} B_{ik} (\delta_i - \delta_k)$$

or

$$P_i - \sum_{k=1}^{NB} G_{ik} = \sum_{\substack{k=1 \\ k \neq i}}^{NB} B_{ik} \delta_i - \sum_{\substack{k=1 \\ k \neq i}}^{NB} B_{ik} \delta_k$$

The above equation can be rewritten as

$$P_i - \sum_{k=1}^{NB} G_{ik} = \sum_{k=1}^{NB} - B_{ik} \delta_k \tag{2.146}$$

where

$$- B_{ii} = \sum_{\substack{k=1 \\ k \neq i}}^{NB} B_{ik}$$

The slack bus voltage is known a priori, so Eq. (2.146) can be modified as

$$P_i - \sum_{k=1}^{NB} G_{ik} + B_{iS} \delta_S = \sum_{\substack{k=1 \\ k \neq S}}^{NB} - B_{ik} \delta_k \tag{2.147}$$

From Eq. (2.147), voltage angles of *PV* and *PQ* buses can be obtained by using the Gauss elimination method or any other method used to solve the linear simultaneous equations.

The reactive power is represented by the equation

$$Q_i = V_i \sum_{k=1}^{NB} V_k [G_{ik} \sin (\delta_i - \delta_k) - B_{ik} \cos (\delta_i - \delta_k)] \tag{2.148}$$

where Q_i is the reactive power of the ith bus.

Let

$$V_i = 1 \text{ p.u.}$$

$$\sin (\delta_i - \delta_k) \approx (\delta_i - \delta_k)$$

$$\cos (\delta_i - \delta_k) \approx 1$$

$$G_{ik} \sin (\delta_i - \delta_k) << B_{ik}$$

Substituting the above assumptions in Eq. (2.148),

$$Q_i = \sum_{k=1}^{NB} V_k (-B_{ik})$$

or

$$Q_i - \sum_{k=1}^{NV} (-B_{ik})\, V_k = \sum_{k=NV+1}^{NB} V_k(-B_{ik}) \qquad (2.149)$$

where NV is the number of slack and *PV* buses.

From the above equation, voltage magnitudes of *PQ* buses can be obtained by using the Gauss elimination method or any other method to solve the linear simultaneous equations.

EXAMPLE 2.14 Line data consisting of line charging and line impedance of a power system are given in Table 2.19. The scheduled generation, load and specified voltages on various types of buses are given in Table 2.20. Find the starting point for the load flow solution.

Solution Bus 1 is the slack bus. Bus 2 is the *PV* bus. The *B* matrix is given below.

$$B = \begin{bmatrix} -8.6275 & 2.3529 & 0.0 & 1.5686 & 4.7059 \\ 2.3529 & -9.4118 & 4.7059 & 2.3529 & 0.0 \\ 0.0 & 4.7059 & -9.4118 & 0.0 & 4.7059 \\ 1.5686 & 2.3529 & 0.0 & -3.9217 & 0.0 \\ 4.7059 & 0.0 & 4.7059 & 0.0 & -9.4118 \end{bmatrix}$$

TABLE 2.19 Line data

Line no.	*Bus code (p–q)*	*Line charging* Y_{pq}	*Impedance* Z_{pq}
1	1–2	$j0.0$	$0.10 + j0.4$
2	1–4	$j0.0$	$0.15 + j0.6$
3	1–5	$j0.0$	$0.05 + j0.2$
4	2–3	$j0.0$	$0.05 + j0.2$
5	2–4	$j0.0$	$0.10 + j0.4$
6	3–5	$j0.0$	$0.05 + j0.2$

TABLE 2.20 Generation, load, and voltage at buses

Bus no.	*Generation*		*Load*		*Voltage*		*Type of bus*
	P_{gi} (p.u.)	Q_{gi} (p.u.)	P_{d_i} (p.u.)	Q_{d_i} (p.u.)	V_i (p.u.)	δ_i (rad)	
1	–	–	0.00	0.00	1.02	0.00	Slack
2	0.0	0.0	1.00	0.00	1.00	–	*PV*
3	0.0	0.0	0.60	0.30	–	–	*PQ*
4	0.0	0.0	0.40	0.10	–	–	*PQ*
5	0.0	0.0	0.60	0.20	–	–	*PQ*

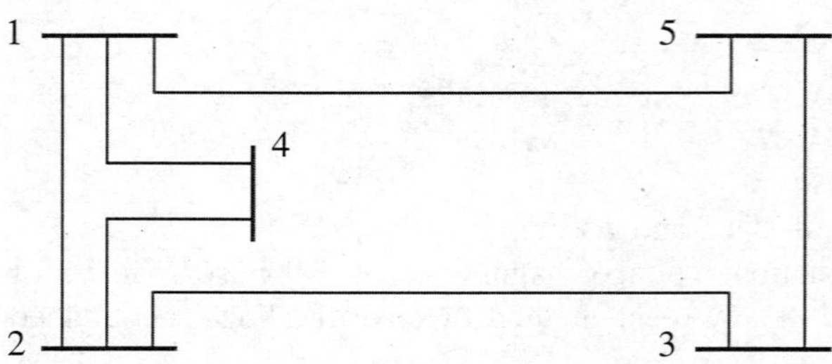

FIGURE 2.20 Sample system.

Equation (2.147) can be written in matrix form as

$$\begin{bmatrix} -B_{22} & -B_{23} & -B_{24} & -B_{25} \\ -B_{32} & -B_{33} & -B_{34} & -B_{35} \\ -B_{42} & -B_{43} & -B_{44} & -B_{45} \\ -B_{52} & -B_{53} & -B_{54} & -B_{55} \end{bmatrix} \begin{bmatrix} \delta_2 \\ \delta_3 \\ \delta_4 \\ \delta_5 \end{bmatrix} = \begin{bmatrix} P_2 - G_{T2} + B_{21}\delta_1 \\ P_3 - G_{T3} + B_{31}\delta_1 \\ P_4 - G_{T4} + B_{41}\delta_1 \\ P_5 - G_{T5} + B_{51}\delta_1 \end{bmatrix}$$

where

$$-B_{22} = B_{12} + B_{32} + B_{42} + B_{52} \quad \text{and} \quad G_{T2} = \sum_{k=1}^{5} G_{2k}$$

$$-B_{33} = B_{13} + B_{23} + B_{43} + B_{53} \quad \text{and} \quad G_{T3} = \sum_{k=1}^{5} G_{3k}$$

$$-B_{44} = B_{41} + B_{42} + B_{43} + B_{45} \quad \text{and} \quad G_{T4} = \sum_{k=1}^{5} G_{4k}$$

$$-B_{55} = B_{51} + B_{52} + B_{53} + B_{54} \quad \text{and} \quad G_{T5} = \sum_{k=1}^{5} G_{5k}$$

$$\begin{bmatrix} -9.4118 & -4.7059 & -2.3529 & 0.0 \\ -4.7059 & -9.4118 & 0.0 & -4.7059 \\ -2.3529 & 0.0 & -3.9216 & 0.0 \\ 0.0 & -4.7059 & 0.0 & -9.4118 \end{bmatrix} \begin{bmatrix} \delta_2 \\ \delta_3 \\ \delta_4 \\ \delta_5 \end{bmatrix} = \begin{bmatrix} -1.0 \\ -0.6 \\ -0.4 \\ -0.6 \end{bmatrix}$$

The above matrix is triangularized as

$$\begin{bmatrix} -9.4118 & -4.7059 & -2.3529 & 0.0 \\ 0.0 & -7.0588 & 1.1765 & -4.7059 \\ 0.0 & 0.0 & -3.1373 & -0.7843 \\ 0.0 & 0.0 & 0.0 & -6.0784 \end{bmatrix} \begin{bmatrix} \delta_2 \\ \delta_3 \\ \delta_4 \\ \delta_5 \end{bmatrix} = \begin{bmatrix} -1.0 \\ -0.1 \\ -0.1667 \\ -0.4917 \end{bmatrix}$$

By employing back substitution, the voltage angles are obtained as

$$\delta_2 = 0.11516 \text{ rad}, \quad \delta_3 = -0.03427 \text{ rad}, \quad \delta_4 = 0.03290 \text{ rad}, \quad \delta_5 = 0.08089 \text{ rad}$$

Equation (2.149) can be written in matrix form as

$$\begin{bmatrix} -B_{33} & -B_{34} & -B_{35} \\ -B_{43} & -B_{44} & -B_{45} \\ -B_{53} & -B_{54} & -B_{55} \end{bmatrix} \begin{bmatrix} V_3 \\ V_4 \\ V_5 \end{bmatrix} = \begin{bmatrix} Q_3 + B_{31}V_1 + B_{32}V_2 \\ Q_4 + B_{41}V_1 + B_{42}V_2 \\ Q_5 + B_{51}V_1 + B_{52}V_2 \end{bmatrix}$$

or

$$\begin{bmatrix} 9.4118 & 0.0 & -4.7059 \\ 0.0 & 3.9216 & 0.0 \\ -4.7059 & 0.0 & 9.4118 \end{bmatrix} \begin{bmatrix} V_3 \\ V_4 \\ V_5 \end{bmatrix} = \begin{bmatrix} 4.5941 \\ 3.9471 \\ 4.6000 \end{bmatrix}$$

Solving we get

$$V_3 = 0.97667 \text{ p.u.}, \qquad V_4 = 1.0065 \text{ p.u.}, \qquad V_5 = 0.97708 \text{ p.u.}$$

2.14 DC SYSTEM MODEL

To analyze the steady-state dc converter operation, the following assumptions are made:

- At the terminal busbar, the three-phase ac supply is balanced and sinusoidal
- The converter operation is balanced
- The converter transformer is lossless and the magnetizing admittance is ignored.

A basic dc converter system is shown in Figure 2.21(a) representing the fundamental frequency. An equivalent circuit for the converter is shown in Figure 2.21(b).

$$I_s = k\frac{3\sqrt{2}}{\pi} I_{dc} \tag{2.150}$$

where

k is very close to unity

I_s is the fundamental current on secondary of converter transformer

I_{dc} is the converter dc current.

The fundamental current magnitudes on both sides of the lossless transformer are related by the following equation.

$$I_i = aI_s \tag{2.151}$$

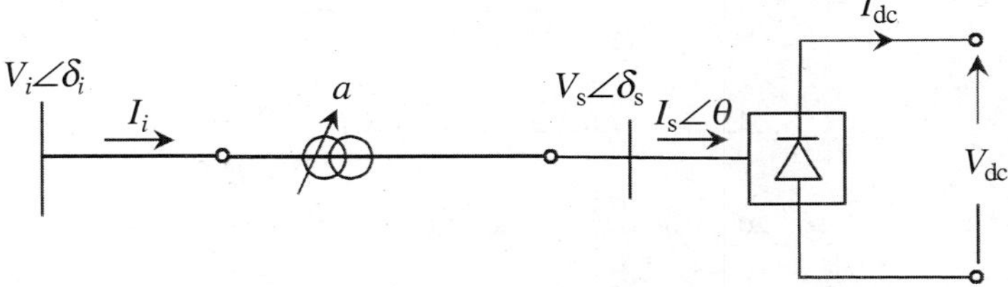

FIGURE 2.21(a) Schematic diagram of a dc converter.

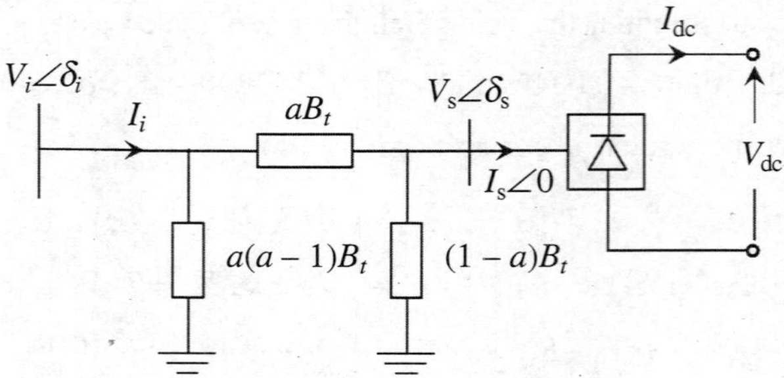

FIGURE 2.21(b) Single-phase equivalent circuit for the basic converter.

where

I_i is the fundamental current on primary of the converter transformer
a is the off-nominal tap ratio.

Therefore,

$$I_i = ak\frac{3\sqrt{2}}{\pi}I_{dc} \tag{2.152}$$

The real power of harmonic frequencies at the bus bars is presented. The power on the secondary side of the transformer is equal to the dc power, i.e.

$$V_{dc}I_{dc} = V_sI_s \cos \delta_s \tag{2.153}$$

where

V_s is the magnitude of the voltage at the converter transformer secondary, i.e. on the rectifier side (fundamental frequency component is taken)
δ_s is the angle of voltage at the converter transformer secondary (on the rectifier side)
V_{dc} is the average dc voltage
I_{dc} is the converter dc current.

It is assumed that transformer is lossless, the real power may be equated to the dc power, i.e.

$$V_{dc}I_{dc} = V_iI_i \cos \delta_i \tag{2.154}$$

where

V_i is the magnitude of the voltage at the converter transformer on the bus bar side
δ_i is the angle of voltage at the converter transformer on the bus bar side.

Substituting the value of I_i from Eq. (2.152) into Eq. (2.154) and rearranging,

$$V_{dc}I_{dc} - V_ia\left(k\frac{3\sqrt{2}}{\pi}\right)I_{dc} \cos \delta_i = 0$$

On simplification,

$$V_{dc}I_{dc} - k_1aV_iI_{dc} \cos \delta_i = 0 \tag{2.155}$$

where $k_1 = k(3\sqrt{2})/\pi$.

The dc voltage is expressed in terms of the ac source voltage as

$$V_{dc} = k_1aV_i \cos \alpha - k_2I_{dc}X_c \tag{2.156}$$

where

α is the firing delay angle of thyristors in rectifier
X_c is commutation reactance of the converter at dc bus bar
k_2 is a constant (= $3/\pi$).

The dc current and voltage satisfy the relation

$$f(V_{dc}, I_{dc}) = 0 \tag{2.157}$$

From Figure 2.21(b), the following relation can be established using the Kirchhoff's voltage law

$$V_{dc} - I_{dc}R_{dc} = 0 \tag{2.158}$$

where R_{dc} is the resistance in the dc circuit.

The dc model may be summarized as follows:

$$R(x, V_i)_k = 0 \tag{2.159}$$

where

$$R(1) = V_{dc} - k_1 a V_i \cos \delta_i \tag{2.159a}$$

$$R(2) = V_{dc} - k_1 a V_i \cos \alpha + k_2 X_c I_{dc} \tag{2.159b}$$

$$R(3) = f(V_{dc}, I_{dc}) = V_{dc} - R_{dc} I_{dc} \tag{2.159c}$$

$$R(j) = \text{control equations}$$

$$x = [V_{dc}, I_{dc}, a, \cos \alpha, \delta_i]^T$$

Control equations may be written as

$$R(4) = a - a^{sp}$$

$$R(5) = V_{dc} I_{dc} - P_{dc}^{sp}$$

$$R(6) = \cos \alpha - \cos \alpha^{sp}$$

where

a^{sp} is the specified converter transformer tap ratio
P_{dc}^{sp} is the specified dc power
α^{sp} is the specified minimum firing angle.

The dc real and reactive powers are expressed as

$$P_{dc} = V_i I_i \cos \delta_i = k_1 a V_i I_{dc} \cos \delta_i \tag{2.160a}$$

$$Q_{dc} = V_i I_i \sin \delta_i = k_1 a V_i I_{dc} \sin \delta_i \tag{2.160b}$$

The equivalent inverter model

The equivalent inverter circuit is shown in Figure 2.22. The residual equations are similar to those of the rectifier with the condition that

$$\alpha + \gamma = 180°$$

where γ is the extinction advance angle in the case of inverter.

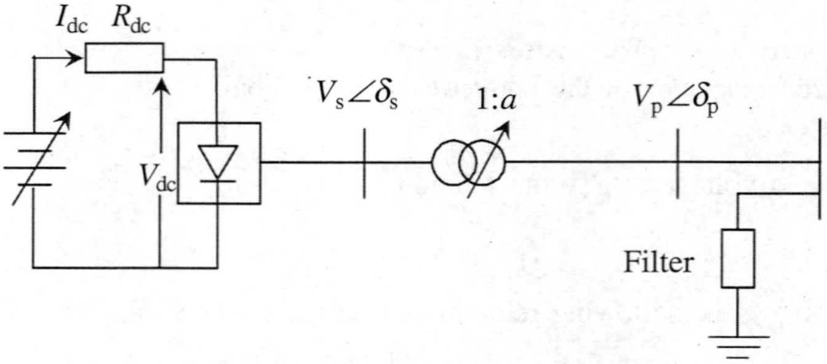

FIGURE 2.22 The equivalent inverter.

$$R(1) = V_{dc} - k_1 a V_i \cos \delta_i \quad (2.161a)$$

$$R(2) = V_{dc} - k_1 a V_i \cos (\pi - \gamma) - k_2 X_c I_{dc} \quad (2.161b)$$

$$R(3) = V_{dc} + R_{dc} I_{dc} - f(I_{dc}, \cos \alpha) \quad (2.161c)$$

$$R(j) = \text{control equations}$$

where $x = [V_{dc}, I_{dc}, a, \cos \alpha, \delta_i, \cos \gamma]^T$.

The control equations may be written as

$$R(4) = a - a^{sp}$$

$$R(5) = V_{dc} I_{dc} - P_{dc}^{sp}$$

$$R(6) = \cos \alpha - \cos \alpha^{sp}$$

$$R(7) = \cos (\pi - \gamma) - \cos (\pi - \gamma^{sp})$$

where

a^{sp} is the specified converter transformer tap ratio
P_{dc}^{sp} is the specified dc power
α^{sp} is the specified minimum firing angle
γ^{sp} is the specified minimum for minimum reactance power consumption of the inverter.

The residual equation for the rectifier and the inverter model can be generalized as given below:

$$R(1) = V_{dc} - k_1 a V_i \cos \delta_i \quad (2.162a)$$

$$R(2) = V_{dc} - k_1 a V_i \cos \alpha + S_i k_2 X_c I_{dc} \quad (2.162b)$$

$$R(3) = V_{dc} - S_i R_{dc} I_{dc} - f(I_{dc}, \cos \alpha) \quad (2.162c)$$

$$R(j) = \text{control equations}$$

where

$x = [V_{dc}, I_{dc}, a, \cos \alpha, \delta_i]^T$

$\alpha = \pi - \gamma$ for inverter operation

$$S_i = \begin{cases} +1 \text{ for rectifier operation} \\ -1 \text{ for inverter operation} \end{cases}$$

Control equations may be considered as

$$R(4) = a - a^{sp}$$

$$R(5) = V_{dc}I_{dc} - P_{dc}^{sp}$$

$$R(6) = \cos \alpha - \cos \alpha^{sp}$$

2.15 AC–DC LOAD FLOW

The load flow equations relating to ac system variables are derived from the specified ac system operating conditions.

$$\Delta P_i = \sum_{k=2}^{NB} \frac{\partial P_i}{\partial \delta_k} \Delta\delta_k + \sum_{k=NV+1}^{NB} \frac{\partial P_i}{\partial V_k} \Delta V_k \tag{2.163a}$$

$$\Delta Q_i = \sum_{k=2}^{NB} \frac{\partial Q_i}{\partial \delta_k} \Delta\delta_k + \sum_{k=NV+1}^{NB} \frac{\partial Q_i}{\partial V_k} \Delta V_k \tag{2.163b}$$

where

P_i is the real power of the ith bus
Q_i is the reactive power of the ith bus
V_i is the magnitude of voltage of the ith bus
δ_i is the angle of voltage of the ith bus
NB is the total number of buses in the system
NV is the number of the PV buses.

The modification required to the usual real and reactive powers mismatches occur at the converter terminal bus bar, i.

$$P^{sp} - P_i^{ac} - P_i^{dc} = 0 \tag{2.164a}$$

$$Q^{sp} - Q_i^{ac} - Q_i^{dc} = 0 \tag{2.164b}$$

where

$$P_i^{ac} = \sum_{k=1}^{NB} V_i V_k [G_{ik} \cos(\delta_i - \delta_k) + B_{ik} \sin(\delta_i - \delta_k)]$$

$$Q_i^{ac} = \sum_{k=1}^{NB} V_i V_k [G_{ik} \sin(\delta_i - \delta_k) - B_{ik} \cos(\delta_i - \delta_k)]$$

$$P_i^{dc} = V_i I_i \cos \delta_i = a k_1 I_{dc_i} \cos \delta_i$$

$$Q_i^{dc} = V_i I_i \sin \delta_i = a k_1 I_{dc_i} \sin \delta_i$$

P_i^{ac} is the real ac power at the ith bus
Q_i^{ac} is the reactive ac power at the ith bus
P_i^{dc} is the real dc power at the ith bus
Q_i^{dc} is the reactive dc power at the ith bus.

Small variations about the initial values of Eqs. (2.164a) and (2.164b) can be obtained by the following equations:

$$\Delta P_i = \Sigma \frac{\partial P_i^{ac}}{\partial \delta_k} \Delta \delta_k + \Sigma \frac{\partial P_i^{ac}}{\partial V_k} \Delta V_k + \frac{\partial P_i^{dc}}{\partial x_k} \Delta x_k \tag{2.165a}$$

$$\Delta Q_i = \Sigma \frac{\partial Q_i^{ac}}{\partial \delta_k} \Delta \delta_k + \Sigma \frac{\partial Q_i^{ac}}{\partial V_k} \Delta V_k + \frac{\partial Q_i^{dc}}{\partial x_k} \Delta x_k \tag{2.165b}$$

$$\Delta R = \Sigma \frac{\partial R}{\partial x} \Delta x \tag{2.165c}$$

where $x = [V_{dc}, I_{dc}, a, \cos \alpha, \delta_i]^T$.

The dc system equations are made independent of the ac system angle δ by selecting a separate angle reference for the dc system variables. This decouples the angle dependance of ac and dc systems. The real power is more related to δ and reactive power to V in the case of ac.

$$\Delta P_i = \Sigma \frac{\partial P_i^{ac}}{\partial \delta_k} \Delta \delta_k \tag{2.166a}$$

$$\Delta Q_i = \Sigma \frac{\partial Q_i^{ac}}{\partial V_k} \Delta V_k \tag{2.166b}$$

$$\Delta R = \Sigma \frac{\partial R}{\partial x} \Delta x \tag{2.166c}$$

For ac network,

$$H_{ii} = \frac{\partial P_i}{\partial \delta_i} = -Q_i - B_{ii} V_i^2 \qquad (i = k) \tag{2.167a}$$

$$H_{ik} = \frac{\partial P_i}{\partial \delta_k} = V_i V_k [G_{ik} \sin (\delta_i - \delta_k) - B_{ik} \cos (\delta_i - \delta_k)] \qquad (i \neq k) \tag{2.167b}$$

$$L_{ii} = V_i \frac{\partial Q_i}{\partial \delta_i} = Q_i - B_{ii} V_i^2 \qquad (i = k) \tag{2.167c}$$

$$L_{ik} = V_k \frac{\partial Q_i}{\partial V_k} = V_i V_k [G_{ik} \sin (\delta_i - \delta_k) - B_{ik} \cos (\delta_i - \delta_k)] \qquad (i \neq k) \tag{2.167d}$$

For the rectifier model,

$$\Delta R(1) = \frac{\partial R(1)}{\partial V_{dc}} \Delta V_{dc} + \frac{\partial R(1)}{\partial I_{dc}} \Delta I_{dc} + \frac{\partial R(1)}{\partial a} \Delta a + \frac{\partial R(1)}{\partial (\cos \alpha)} \Delta (\cos \alpha) + \frac{\partial R(1)}{\partial \delta_i} \Delta \delta_i$$

From Eq. (2.159a),

$$\Delta R(1) = \Delta V_{dc} - (k_1 V_i \cos \delta_i) \Delta a + (k_1 a V_i \sin \delta_i) \Delta \delta_i \tag{2.168a}$$

Similarly, from Eq. (2.159b),

$$\Delta R(2) = \Delta V_{dc} - (k_2 X_c) \Delta I_{dc} - (k_1 V_i \cos \delta_i) \Delta a + (k_1 a V_i) \Delta (\cos \alpha) \tag{2.168b}$$

$$\Delta R(3) = \Delta V_{dc} - R_{dc} \Delta I_{dc} \tag{2.168c}$$

$$\Delta R(4) = \Delta a \tag{2.168d}$$

$$\Delta R(5) = I_{dc}\Delta V_{dc} - V_{dc}\Delta I_{dc} \quad (2.168e)$$

$$\Delta R(6) = \Delta(\cos \alpha) \quad (2.168f)$$

For the inverter model,

$$\Delta R(1) = \Delta V_{dc} - (k_1 V_i \cos \delta_i)\Delta a + (k_1 a V_i \sin \delta_i)\Delta\delta_i \quad (2.169a)$$

$$\Delta R(2) = \Delta V_{dc} - (k_2 X_c)\Delta I_{dc} - (k_1 V_i \cos(\pi - \gamma))\Delta a + (k_1 a V_i)\Delta(\cos \gamma) \quad (2.169b)$$

$$\Delta R(3) = \Delta V_{dc} + R_{dc}\Delta I_{dc} + \frac{\delta V_{dc}}{\delta I_{dc}}\Delta I_{dc} + \frac{\delta V_{dc}}{\delta I_{dc}}\Delta(\cos \alpha) \quad (2.169c)$$

$$\Delta R(4) = \Delta a \quad (2.169d)$$

$$\Delta R(5) = I_{dc}\Delta V_{dc} - V_{dc}\Delta I_{dc} \quad (2.169e)$$

$$\Delta R(6) = \Delta(\cos \alpha) \quad (2.169f)$$

$$\Delta R(7) = -\Delta(\cos \gamma) \quad (2.169g)$$

Equations (2.165a), (2.165b) and (2.165c) can be solved using the Gauss elimination method.

Algorithm 2.7: Decoupled Method for ac–dc Load Flow

1. Read data (ac)
   NB (number of buses); NV (number of PV buses).
   V_1, δ_1 for slack bus, $P_i^S (i = 2, 3, ..., \text{NB})$ for PQ and PV buses.
   Q_i^S $(i = \text{NV} + 1, \text{NV} + 2, ..., \text{NB})$ for PQ buses, V_i^S $(i = 2, 3, ..., \text{NV})$ for PV buses, V_i^{min}, V_i^{max} $(i = \text{NV} + 1, \text{NV} + 2, ..., \text{NB})$ for PQ buses.
   Q_i^{min}, Q_i^{max} $(i = 2, 3, ..., \text{NV})$ for PV buses, R(maximum number of iterations), $\in$ (tolerance in convergence)
   Read data (dc)
   HVDC link between bus bars, X_c (Commutation reactance), R_{dc} (dc link resistance), Control parameters P_{dc}^S(dc link power), α^S(firing angle), γ^S(Extinction advance angle), etc.
2. Form Y_{BUS} as explained in Section 2.3 for ac bus bars
3. Assume initially magnitude and angle of voltage

$$|V_i| \; (i = \text{NV} + 1, \text{NV} + 2, ..., \text{NB}) \quad \text{and} \quad \delta_i^0 \quad (i = 2, 3, ..., \text{NB})$$

4. Set iteration count $r = 0$.
5. Compute P_i, ΔP_i using Eq. (2.120a), i.e.

$$P_i = \sum_{k=1}^{\text{NB}} V_i V_k [G_{ik} \cos(\delta_i - \delta_k) + B_{ik} \sin(\delta_i - \delta_k)] \qquad (i = 2, 3, ..., \text{NB})$$

$$\Delta P_i = P_i^S - P_i \qquad (i = 2, 3, ..., \text{NB})$$

6. If maximum $\{|\Delta P_i| \; (i = 2, 3, ..., \text{NB})\} \le \varepsilon$ then GOTO Step 10.
7. Compute the elements of Jacobian matrix H, i.e.

$$H_{ii} = \frac{\partial P_i}{\partial \delta_i} = -Q_i - B_{ii}V_i^2 \quad i = k$$

$$H_{ik} = \frac{\partial P_i}{\partial \delta_k} = V_i V_k [G_{ik} \sin(\delta_i - \delta_k) - B_{ik} \cos(\delta_i - \delta_k)] \qquad (i \neq k)$$

8. Compute $\Delta\delta_i$ (i = 2, 3, ..., NB) using Eq. (2.134a), i.e.

$$[H]\,[\Delta\delta] = [\Delta P]$$

9. Modified value of δ_i is computed as

$$\delta_i = \delta_i + \Delta\delta_i \qquad (i = 2, 3, ..., \text{NB})$$

10. Compute Q_i, ΔQ_i using Eq. (2.120b), i.e.

$$Q_i = \sum_{k=1}^{\text{NB}} V_i V_k [G_{ik} \sin(\delta_i - \delta_k) - B_{ik} \cos(\delta_i - \delta_k)] \qquad (i = \text{NV} + 1, \text{NV} + 2, ..., \text{NB})$$

$$\Delta Q_i = Q_i^{\text{S}} - Q_i \qquad (i = \text{NV} + 1, \text{NV} + 2, ..., \text{NB})$$

11. If maximum $\{|\Delta Q_i|$ (i = NV + 1, NV + 2, ..., NB)$\} \le \varepsilon$ then GOTO Step 22.
12. Compute elements of Jacobian matrix L as

$$L_{ii} = V_i \frac{\partial Q_i}{\partial \delta_i} = Q_i - B_{ii} V_i^2 \qquad (i = k)$$

$$L_{ik} = V_k \frac{\partial Q_i}{\partial V_k} = V_1 V_k [G_{ik} \sin(\delta_i - \delta_k) - B_{ik} \cos(\delta_i - \delta_k)] \qquad (i \neq k)$$

13. Compute $\Delta V_i/V_i$ (i = NV + 1, NV + 2, ..., NB) using Eq. (2.134b), i.e.

$$[L]\left[\frac{\Delta V}{V}\right] = [\Delta Q]$$

14. Modified value of voltage magnitude V_i is computed as

$$V_i = V_i + \frac{\Delta V_i}{V_i} V_i \qquad (i = \text{NV} + 1, \text{NV} + 2, ..., \text{NB})$$

15. Set bus count i = 2.
16. If the bus is PQ, then check the limits of V_i and set according to Eq. (2.137), i.e.

$$V_i = V_i^{\min} \quad \text{if } V_i \ge V_i^{\min}$$

$$V_i = V_i^{\max} \quad \text{if } V_i \ge V_i^{\max}$$

17. If the bus is PV, then compute Q_i using Eq. (2.120b) and check the limits of Q_i and set according to Eq. (2.135), i.e.

$$Q_i = Q_i^{\min} \quad \text{if } Q_i \le Q_i^{\min}$$

$$Q_i = Q_i^{\max} \quad \text{if } Q_i \ge Q_i^{\max}$$

If any limit is violated then the *PV* bus is temporarily converted to *PQ* bus. So, compute L_{ik} with the updated values of Q_i, V_i, and δ_i. Using Eq. (2.136), calculate change in voltage, i.e.

$$\Delta V_i = \frac{V_i}{L_{ii}} \left[\Delta Q_i - \sum_{\substack{k = NV+1 \\ k \neq i}}^{NB} L_{ik} \frac{\Delta V_k}{V_k} \right]$$

and specified voltage magnitude of the *PV* bus is updated as

$$V_i = V_i^S + \Delta V_i$$

18. Increment the bus count $i = i + 1$

    If $i \leq$ NB, then GOTO Step 16.

19. Calculate dc residuals using Eq. (2.159) or (2.161).

    Calculate the dc Jacobian, $J_{dc} = \partial R / \partial X$ using Eq. (2.168) or Eq. (2.169).

20. Compute ΔX and update X

    $$X = X + \Delta X \text{ where } X \text{ is a vector of } X = [V_{dc}, I_{dc}, a, \cos \alpha, \delta_i]^T$$

21. Advance the count $r = r + 1$

    If $r \leq R$ then GOTO Step 5 and repeat.

22. Compute powers on slack bus, i.e.

$$P_1 = \sum_{k=1}^{NB} V_1 V_k [G_{1k} \cos(\delta_1 - \delta_k) + B_{1k} \sin(\delta_1 - \delta_k)]$$

$$Q_1 = \sum_{k=1}^{NB} V_1 V_k [G_{1k} \sin(\delta_1 - \delta_k) - B_{1k} \cos(\delta_1 - \delta_k)]$$

23. Stop.

2.16 CONCLUSION

In this chapter, load flow study of power systems has been discussed. Important methods, namely Gauss–Seidel, Newton–Raphson (NR), Decoupled Newton (DLF) and Fast Decoupled Load Flow (FDLF), have been described. It is difficult to say which one of the existing methods is dictated by the types and sizes of the problems to be solved, as well as the precise details of implementations. Choice of a particular method in any given situation is normally a compromise between the various criteria of goodness of the load flow methods. It would not be incorrect to say that among the existing methods, no single method meets all the desirable requirements of an ideal load flow method, namely high speed, low storage, reliability for ill-conditioned problems, versatility in handling various adjustments and simplicity in programming. Fortunately, not all the desirable features of a load flow method are needed in all situations.

Despite a large number of load flow methods available, it is easy to see that only the NR and FDLF load flow methods are the most important ones for general purpose load flow analysis. The FDLF method is clearly superior to the NR method from the point of view of speed as well as storage. Yet, the NR method is still in use because of its high versatility, accuracy and reliability and, as such, is being widely used for a variety of system optimization calculations.

Sample System 2.1

Line data consisting of line charging and line impedance of a power system are given in Table 2.21. The scheduled generation, load, and specified voltage on various types of buses are given in Table 2.22.

TABLE 2.21 Line data

Line no.	*Branch p–q*	*Line charging* Y_{pq}	*Impedance* Z_{pq}
1	1–2	$j0.0$	$0.10 + j0.4$
2	1–4	$j0.0$	$0.15 + j0.6$
3	1–5	$j0.0$	$0.05 + j0.2$
4	2–3	$j0.0$	$0.05 + j0.2$
5	2–4	$j0.0$	$0.10 + j0.4$
6	3–5	$j0.0$	$0.05 + j0.2$

TABLE 2.22 Generation, load, and voltage at buses

Bus	*Generation*		*Load*		*Voltage*		*Type*
no.	P_{gi} (p.u.)	Q_{gi} (p.u.)	P_{d_i} (p.u.)	Q_{d_i} (p.u.)	V_i (p.u.)	δ_i (rad)	*of bus*
1	–	–	0.00	0.00	1.02	0.00	Slack
2	0.0	0.0	1.00	0.00	1.00	–	*PV*
3	0.0	0.0	0.60	0.30	–	–	*PQ*
4	0.0	0.0	0.40	0.10	–	–	*PQ*
5	0.0	0.0	0.60	0.20	–	–	*PQ*

Reactive power limit of the *PV*-bus is given as

$$0.0 \le Q_2 \le 0.2$$

Decoupled Newton–Raphson method is employed and convergence ($\varepsilon = 0.00001$) is achieved in 11 iterations. Voltage magnitudes, voltage angles and power injected at all buses are tabulated in Table 2.23. Line flows are given in Table 2.24.

TABLE 2.23 Voltage and power injected at buses

Bus no.	V_i (p.u.)	δ_i (rad)	P_i (p.u.)	Q_i (p.u.)
1	1.02	0.0	2.9175840	1.3705670
2	0.8612135	−0.4462322	−0.9994068	0.5000000
3	0.7862771	−0.4350828	−0.5996794	−0.2986889
4	0.8385913	−0.3690717	−0.4000205	−0.0991933
5	0.8380031	−0.2634329	−0.6002641	−0.1998347

TABLE 2.24 Line flows

Line no.	*Branch i–k*	S_{ik}	*Line no.*	*Branch i–k*	S_{ik}
1	1–2	1.0378870 + *j*0.3604762	6	3–5	−0.5660844 − *j*0.0134186
1	2–1	−0.9218592 + *j*0.1036347	6	5–3	0.5920158 + *j*0.1171440
2	1–4	0.5791855 + *j*0.2595949			
2	4–1	−0.5211051 − *j*0.0272735			
3	1–5	1.3005110 + *j*0.7504954			
3	5–1	−1.1921600 − *j*0.3170906			
4	2–3	0.0404465 + *j*0.3127801			
4	3–2	−0.0337410 − *j*0.2859583			
5	2–4	−0.1182650 + *j*0.0836448			
5	4–2	0.1210941 − *j*0.0723285			

Sample System 2.2

Line data consisting of line charging and line impedance of a power system are given in Table 2.25. The scheduled generation, load, and specified voltage on various types of buses are given in Table 2.26.

TABLE 2.25 Line data

Line no.	*Branch p–q*	*Line charging* Y_{pq}	*Impedance* Z_{pq}
1	1–2	*j*0.06	0.02 + *j*0.06
2	1–3	*j*0.05	0.08 + *j*0.24
3	2–3	*j*0.04	0.06 + *j*0.18
4	2–4	*j*0.04	0.06 + *j*0.18
5	2–5	*j*0.03	0.04 + *j*0.12
6	3–4	*j*0.02	0.01 + *j*0.03
7	4–5	*j*0.05	0.08 + *j*0.24

TABLE 2.26 Generation, load, and voltage at buses

Bus	*Generation*		*Load*		*Voltage*		*Type*
no.	P_{gi} (p.u.)	Q_{gi} (p.u.)	P_{d_i} (p.u.)	Q_{d_i} (p.u.)	V_i (p.u.)	δ_i (rad)	*of bus*
1	–	–	0.00	0.00	1.06	0.00	Slack
2	0.2	0.2	0.00	0.00	–	–	*PQ*
3	0.0	0.0	0.45	0.15	–	–	*PQ*
4	0.0	0.0	0.40	0.05	–	–	*PQ*
5	0.0	0.0	0.60	0.10	–	–	*PQ*

Decoupled Newton–Raphson method is employed and convergence ($\varepsilon = 0.0001$) is achieved in five iterations. Voltage magnitudes, voltage angles and power injected at all buses are tabulated in Table 2.27. Line flows are given in Table 2.28.

TABLE 2.27 Voltage and power injected at buses

Bus no.	V_i (p.u.)	δ_i (rad)	P_i (p.u.)	Q_i (p.u.)
1	1.0600000	0.0	1.2958920	−0.3998962
2	1.0586840	−0.0519997	0.1999401	0.2000005
3	1.0400140	−0.0906581	−0.4499944	−0.1500008
4	1.0403200	−0.0965536	−0.3999603	−0.0500075
5	1.0347560	−0.1104136	−0.5999067	−0.1000094

TABLE 2.28 Line flows

Line no.	*Branch i–k*	S_{ik}	*Line no.*	*Branch i–k*	S_{ik}
1	1–2	0.8894789 − *j*0.3153776	5	2–5	0.5476638 + *j*0.0104937
1	2–1	−0.8743017 + *j*0.2262446	5	5–2	−0.5368901 − *j*0.0439186
2	1–3	0.4064125 − *j*0.0845181	6	3–4	0.1883627 − *j*0.0943952
2	3–1	−0.3945952 + *j*0.0097086	6	4–3	−0.1879857 + *j*0.0522482
3	2–3	0.2470838 − *j*0.0128177	7	4–5	0.0633137 − *j*0.0506671
3	3–2	−0.2437607 − *j*0.0653108	7	5–4	−0.0630165 − *j*0.0560906
4	2–4	0.2794943 − *j*0.0239181			
4	4–2	−0.2752891 − *j*0.0515893			

Sample System 2.3

Line data consisting of line charging, line impedance and off-nominal turn-ratio of a power system is given in Table 2.29. The scheduled generation, load, and specified voltage on various types of buses are given in Table 2.30.

TABLE 2.29 Line data

Line no.	*Branch p–q*	*Line charging* Y_{pq}	*Impedance* Z_{pq}	*Turn-ratio a*
1	1–4	+*j*0.015	0.080 + *j*0.370	1.0
2	1–6	+*j*0.021	0.123 + *j*0.518	1.0
3	2–3	+*j*0.000	0.723 + *j*1.050	1.0
4	2–5	+*j*0.000	0.282 + *j*0.640	1.0
5	4–3	+*j*0.000	0.000 + *j*0.133	0.909
6	4–6	+*j*0.015	0.097 + *j*0.407	1.0
7	6–5	+*j*0.000	0.000 + *j*0.300	0.975

TABLE 2.30 Generation, load, and voltage at buses

Bus no.	Generation		Load		Voltage		Type of bus
	P_{gi} (p.u.)	Q_{gi} (p.u.)	P_{d_i} (p.u.)	Q_{d_i} (p.u.)	V_i (p.u.)	δ_i (rad)	
1	–	–	0.00	0.00	1.05	0.0	Slack
2	0.5	0.0	0.00	0.00	1.10	–	*PV*
3	0.0	0.0	0.55	0.13	–	–	*PQ*
4	0.0	0.0	0.00	0.00	–	–	*PQ*
5	0.0	0.0	0.30	0.18	–	–	*PQ*
6	0.0	0.0	0.50	0.05	–	–	*PQ*

Reactive power limit of *PV*-bus is given as

$$0.0 \le Q_2 \le 0.2$$

Decoupled Newton–Raphson method is employed and convergence (ε = 0.000001) is achieved in eight iterations. Voltage magnitudes, voltage angles and power injected at all buses are tabulated in Table 2.31. Line flows are given in Table 2.32.

TABLE 2.31 Voltage and power injected at buses

Bus no.	V_i (p.u.)	δ_i (rad)	P_i (p.u.)	Q_i (p.u.)
1	1.05	0.0	0.9521295	0.4326028
2	1.10	−0.0584521	0.4999997	0.1844000
3	1.0007640	−0.2231306	−0.5500001	−0.1300005
4	0.9297400	−0.1716732	0.0000002	0.0000004
5	0.9197770	−0.2152802	−0.3000000	−0.1799997
6	0.9191896	−0.2136118	−0.5000000	−0.0500002

TABLE 2.32 Line flows

Line no.	Branch i–k	S_{ik}	Line no.	Branch i–k	S_{ik}
1	1–4	0.5091050 + *j*0.2534483	5	4–3	0.3598305 − *j*0.4872383
1	4–1	−0.4850085 − *j*0.1715056	5	3–4	−0.3598305 + *j*0.5436867
2	1–6	0.4430244 + *j*0.1791546	6	4–6	0.0891556 − *j*0.0082673
2	6–1	−0.4165614 − *j*0.1086044	6	6–4	−0.0882611 − *j*0.0136196
3	2–3	0.1717787 − *j*0.0001369	7	6–5	0.0047020 − *j*0.0017959
3	3–2	−0.1541471 + *j*0.0257429	7	5–6	−0.0047020 + *j*0.0018049
4	2–5	0.3282209 + *j*0.1845368			
4	5–2	−0.2951773 − *j*0.1095443			

Sample System 2.4

Line data consisting of line charging and line impedance of a power system are given in Table 2.33. The scheduled generation, load, and specified voltage on various types of buses are given in Table 2.34.

TABLE 2.33 Line data

Line no.	*Branch* p–q	*Line charging* Y_{pq}	*Impedance* Z_{pq}
1	1–9	$j0.030$	$0.15 + j0.50$
2	1–11	$j0.010$	$0.05 + j0.16$
3	2–3	$j0.030$	$0.15 + j0.50$
4	2–7	$j0.020$	$0.10 + j0.28$
5	2–10	$j0.010$	$0.05 + j0.16$
6	3–4	$j0.015$	$0.08 + j0.24$
7	4–6	$j0.020$	$0.10 + j0.28$
8	4–8	$j0.020$	$0.10 + j0.28$
9	4–9	$j0.030$	$0.15 + j0.50$
10	5–6	$j0.025$	$0.12 + j0.36$
11	5–9	$j0.010$	$0.05 + j0.16$
12	7–8	$j0.010$	$0.05 + j0.16$
13	7–10	$j0.015$	$0.08 + j0.24$
14	8–9	$j0.025$	$0.12 + j0.36$
15	8–10	$j0.015$	$0.08 + j0.24$
16	8–11	$j0.020$	$0.10 + j0.28$
17	10–11	$j0.025$	$0.12 + j0.36$

TABLE 2.34 Generation, load, and voltage at buses

Bus	*Generation*		*Load*		*Voltage*		*Type*
no.	P_{gi} (p.u.)	Q_{gi} (p.u.)	P_{d_i} (p.u.)	Q_{d_i} (p.u.)	V_i (p.u.)	δ_i (rad)	*of bus*
1	–	–	0.00	0.00	1.070	0.0	Slack
2	0.6625	–	0.00	0.00	1.088	–	*PV*
3	0.6625	–	0.00	0.00	1.095	–	*PV*
4	0.4778	–	0.00	0.00	1.062	–	*PV*
5	0.4778	–	0.00	0.00	1.046	–	*PV*
6	0.0	0.0	0.10	0.02	–	–	*PQ*
7	0.0	0.0	0.40	0.10	–	–	*PQ*
8	0.0	0.0	0.90	0.45	–	–	*PQ*
9	0.0	0.0	0.70	0.35	–	–	*PQ*
10	0.0	0.0	0.25	0.05	–	–	*PQ*
11	0.0	0.0	0.25	0.05	–	–	*PQ*

Reactive power limits of *PV*-buses are given as

$$-0.1 \leq Q_2 \leq 0.5$$

$$-0.1 \leq Q_3 \leq 0.5$$

$$-0.1 \leq Q_4 \leq 0.5$$

$$-0.1 \leq Q_5 \leq 0.5$$

Decoupled Newton–Raphson method is employed and convergence ($\varepsilon = 0.00001$) is achieved in seven iterations. Voltage magnitudes, voltage angles and power injected at all buses are tabulated in Table 2.35. Line flows are given in Table 2.36.

TABLE 2.35 Voltage and power injected at buses

Bus no.	V_i (p.u.)	δ_i (rad)	P_i (p.u.)	Q_i (p.u.)
1	1.070	0.0	0.4296081	0.1998982
2	1.088	0.0314549	0.6625003	0.3650117
3	1.095	0.1498241	0.6625000	−0.0680686
4	1.062	0.0704448	0.4777996	0.0597918
5	1.046	0.0405004	0.4778001	0.0833654
6	1.0538600	0.0418368	−0.1000002	−0.0200001
7	1.0160200	−0.0532843	−0.4000000	−0.0999998
8	0.9949282	−0.0592053	−0.8999985	−0.4500006
9	1.0033960	−0.0272797	−0.7000003	−0.3500005
10	1.0361120	−0.0349548	−0.2500006	−0.0500000
11	1.0357370	−0.0427044	−0.2500002	−0.0500000

TABLE 2.36 Line flows

Line no.	*Branch i–k*	S_{ik}	*Line no.*	*Branch i–k*	S_{ik}
1	1–9	0.0931827 + *j*0.0810305	10	5–6	−0.0105336 − *j*0.0466769
1	9–1	−0.0903010 − *j*0.1359760	10	6–5	0.0105867 − *j*0.0082821
2	1–11	0.3364255 + *j*0.1188675	11	5–9	0.4883338 + *j*0.1300431
2	11–1	−0.3307410 − *j*0.1228535	11	9–5	−0.4765276 − *j*0.1132725
3	2–3	−0.2577526 + *j*0.0432544	12	7–8	0.0722417 + *j*0.1011454
3	3–2	0.2669573 − *j*0.0840549	12	8–7	−0.0713871 − *j*0.1186324
4	2–7	0.3894258 + *j*0.1311059	13	7–10	−0.0976511 − *j*0.0672555
4	7–2	−0.3745907 − *j*0.1338884	13	10–7	0.0985978 + *j*0.0385082
5	2–10	0.5308269 + *j*0.1906501	14	8–9	−0.0862617 − *j*0.0179814
5	10–2	−0.5171931 − *j*0.1695946	14	9–8	0.0871694 − *j*0.0292129

(Contd.)

TABLE 2.36 Line flows (*Contd.*)

Line no.	Branch i–k	S_{ik}	Line no.	Branch i–k	S_{ik}
6	3–4	0.3955428 + j0.0159869	15	8–10	−0.1445756 − j0.1361212
6	4–3	−0.3850270 − j0.0193427	15	10–8	0.1474535 + j0.1138036
7	4–6	0.1116981 − j0.0299398	16	8–11	−0.0996269 − j0.1287218
7	6–4	−0.1105870 − j0.0117185	16	11–8	0.1018282 + j0.0936327
8	4–8	0.5232996 + j0.0766157	17	10–11	0.0211412 − j0.0327170
8	8–4	−0.4981474 − j0.0485441	17	11–10	−0.0210874 − j0.0207785
9	4–9	0.2278291 + j0.0324603			
9	9–4	−0.2203412 − j0.0715401			

Sample System 2.5

Line data consisting of line charging, line impedance and off-nominal turn-ratio of a power system is given in Table 2.37. The scheduled generation, load, and specified voltage on various types of buses are given in Table 2.38.

TABLE 2.37 Line data

Line no.	Branch p–q	Line charging Y_{pq}	Impedance Z_{pq}	Turn-ratio a
1	1–2	j0.0264	0.01938 + j0.05917	1.0
2	1–5	j0.0264	0.05403 + j0.22304	1.0
3	2–3	j0.0219	0.04699 + j0.19797	1.0
4	2–4	j0.0187	0.05811 + j0.17632	1.0
5	2–5	j0.0170	0.05695 + j0.17388	1.0
6	3–4	j0.0173	0.06701 + j0.17103	1.0
7	4–5	j0.0064	0.01335 + j0.04211	1.0
8	4–7	j0.0	0.0 + j0.20912	0.932
9	4–9	j0.0	0.0 + j0.55618	0.978
10	5–6	j0.0	0.0 + j0.25202	1.0
11	6–11	j0.0	0.09498 + j0.19890	0.969
12	6–12	j0.0	0.12291 + j0.25581	1.0
13	6–13	j0.0	0.06615 + j0.13027	1.0
14	7–8	j0.0	0.0 + j0.17615	1.0
15	7–9	j0.0	0.0 + j0.11001	1.0
16	9–10	j0.0	0.03181 + j0.08450	1.0
17	9–14	j0.0	0.19711 + j0.27038	1.0
18	10–11	j0.0	0.08205 + j0.19207	1.0
19	12–13	j0.0	0.22092 + j0.19988	1.0
20	13–14	j0.0	0.17093 + j0.34802	1.0

TABLE 2.38 Generation, load, and voltage at buses

Bus no.	*Generation*		*Load*		*Voltage*		*Type of bus*
	P_{gi} (p.u.)	Q_{gi} (p.u.)	P_{d_i} (p.u.)	Q_{d_i} (p.u.)	V_i (p.u.)	δ_i (rad)	
1	–	–	0.0	0.0	1.06	0.0	Slack
2	0.0	0.0	–0.183	0.297	–	–	*PQ*
3	0.0	0.0	0.942	–0.044	–	–	*PQ*
4	0.0	0.0	0.478	–0.039	–	–	*PQ*
5	0.0	0.0	0.076	0.016	–	–	*PQ*
6	0.0	0.0	0.112	–0.0474	–	–	*PQ*
7	0.0	0.0	0.0	0.0	–	–	*PQ*
8	0.0	0.0	0.0	–0.174	–	–	*PQ*
9	0.0	0.0	0.295	–0.046	–	–	*PQ*
10	0.0	0.0	0.090	0.058	–	–	*PQ*
11	0.0	0.0	0.035	0.018	–	–	*PQ*
12	0.0	0.0	0.060	0.016	–	–	*PQ*
13	0.0	0.0	0.135	0.058	–	–	*PQ*
14	0.0	0.0	0.149	0.050	–	–	*PQ*

Decoupled Newton–Raphson method is employed and convergence ($\varepsilon = 0.00001$) is achieved in 20 iterations. Voltage magnitudes, voltage angles and power injected at all buses are tabulated in Table 2.39. Line flows are given in Table 2.40.

TABLE 2.39 Voltage and power injected at buses

Bus no.	V_i (p.u.)	δ_i (rad)	P_i (p.u.)	Q_i (p.u.)
1	1.06	0.0	2.3303080	0.4607337
2	1.0150030	–0.0800955	0.1829994	–0.2969987
3	0.9813243	–0.2235264	–0.9419996	0.0439997
4	0.9932188	–0.1802138	–0.4779998	0.0389987
5	0.9970710	–0.1524515	–0.0760006	–0.0160021
6	1.0433390	–0.2518152	–0.1119988	0.0473992
7	1.0366250	–0.2362034	–0.0000001	0.0000001
8	1.0653940	–0.2362034	0.0000000	0.1739998
9	1.0310100	–0.2652050	–0.2949985	0.0460000
10	1.0255730	–0.2680940	–0.0900005	–0.0580009
11	1.0307980	–0.2623896	–0.0350000	–0.0179997
12	1.0278830	–0.2673156	–0.0599925	–0.0159998
13	1.0225750	–0.2686943	–0.1350083	–0.0579995
14	1.0051830	–0.2840871	–0.1490010	–0.0500003

TABLE 2.40 Line flows

Line no.	*Branch i–k*	S_{ik}	*Line no.*	*Branch i–k*	S_{ik}
1	1–2	1.5695780 + *j*0.3206484	11	6–11	0.0722568 + *j*0.0315819
1	2–1	−1.5249690 − *j*0.2413125	11	11–6	−0.0717143 − *j*0.0304457
2	1–5	0.7607304 + *j*0.1400845	12	6–12	0.0775966 + *j*0.0262592
2	5–1	−0.7315167 − *j*0.0753964	12	12–6	−0.0768389 − *j*0.0246822
3	2–3	0.7312234 + *j*0.0282107	13	6–13	0.1775005 + *j*0.0773352
3	3–2	−0.7067181 + *j*0.0313790	13	13–6	−0.1752225 − *j*0.0728490
4	2–4	0.5612764 − *j*0.0502135	14	7–8	0.0000000 − *j*0.1693013
4	4–2	−0.5434532 + *j*0.0665812	14	8–7	0.0000000 + *j*0.1739998
5	2–5	0.4154675 − *j*0.0336869	15	7–9	0.2817171 + *j*0.0569965
5	5–2	−0.4059112 + *j*0.0284499	15	9–7	−0.2817171 − *j*0.0485391
6	3–4	−0.2352813 + *j*0.0126209	16	9–10	0.0535518 + *j*0.0462236
6	4–3	0.2391930 − *j*0.0363631	16	10–9	−0.0534020 − *j*0.0458258
7	4–5	−0.6167516 + *j*0.1074161	17	9–14	0.0944612 + *j*0.0303033
7	5–4	0.6220743 − *j*0.1033027	17	14–9	−0.0926364 − *j*0.0278001
8	4–7	0.2755194 − *j*0.1984432	18	10–11	−0.0365981 − *j*0.0121743
8	7–4	−0.2755194 + *j*0.2228831	18	11–10	0.0367141 + *j*0.0124460
9	4–9	0.1562945 − *j*0.0608410	19	12–13	0.0168462 + *j*0.0086827
9	9–4	−0.1562945 + *j*0.0767004	19	13–12	−0.0167711 − *j*0.0086147
10	5–6	0.4094770 − *j*0.1626914	20	13–14	0.0569854 + *j*0.0234643
10	6–5	−0.4094770 + *j*0.2119064	20	14–13	−0.0563646 − *j*0.0222002

Sample System 2.6

Line data consisting of line charging, line impedance and off-nominal turn-ratio of a power system is given in Table 2.41. The scheduled generation, load, and specified voltage on various types of buses are given in Table 2.42.

TABLE 2.41 Line data

Line no.	*Branch p–q*	*Line charging* Y_{pq}	*Impedance* Z_{pq}	*Turn-ratio* *a*
1	1–3	*j*0.0179	0.0720 + *j*0.2876	1.0
2	1–16	*j*0.0337	0.0290 + *j*0.1379	1.0
3	1–17	*j*0.0148	0.1012 + *j*0.2794	1.0
4	1–19	*j*0.0224	0.1487 + *j*0.3897	1.0

(*Contd.*)

TABLE 2.41 Line data (*Contd.*)

Line no.	*Branch p–q*	*Line charging* Y_{pq}	*Impedance* Z_{pq}	*Turn-ratio a*
5	1–23	j0.0573	0.1085 + j0.2245	1.0
6	1–25	j0.0873	0.0753 + j0.3593	1.0
7	2–6	j0.0186	0.0617 + j0.2935	1.0
8	2–7	j0.0155	0.0511 + j0.2442	1.0
9	2–8	j0.0175	0.0579 + j0.2763	1.0
10	3–13	j0.0085	0.0564 + j0.1487	1.0
11	3–14	j0.0185	0.1183 + j0.3573	1.0
12	4–19	j0.0113	0.0196 + j0.0514	1.0
13	4–20	j0.0220	0.0382 + j0.1007	1.0
14	4–21	j0.0558	0.0970 + j0.2547	1.0
15	5–10	j0.0557	0.0497 + j0.2372	1.0
16	5–17	j0.1335	0.0144 + j0.1269	1.0
17	5–19	j0.0140	0.0929 + j0.2442	1.0
18	6–13	j0.0040	0.0263 + j0.0691	1.0
19	7–8	j0.0078	0.0529 + j0.1465	1.0
20	7–12	j0.0110	0.0364 + j0.1736	1.0
21	8–9	j0.0118	0.0387 + j0.1847	1.0
22	8–17	j0.0572	0.0497 + j0.2372	1.0
23	9–10	j0.0085	0.0973 + j0.2691	1.0
24	10–11	j0.0135	0.0898 + j0.2359	1.0
25	11–17	j0.0161	0.1068 + j0.2807	1.0
26	12–17	j0.0135	0.0460 + j0.2196	1.0
27	14–15	j0.0044	0.0281 + j0.0764	1.0
28	15–16	j0.0148	0.0256 + j0.0673	1.0
29	17–18	j0.0122	0.0806 + j0.2119	1.0
30	18–19	j0.0132	0.0872 + j0.2294	1.0
31	20–21	j0.0354	0.0615 + j0.1613	1.0
32	21–22	j0.0238	0.0414 + j0.1087	1.0
33	22–23	j0.0169	0.2250 + j0.3559	1.0
34	22–24	j0.0567	0.0970 + j0.2595	1.0
35	24–25	j0.0317	0.0472 + j0.1458	1.0

TABLE 2.42 Generation, load, and voltage at buses

Bus no.	*Generation* P_{gi} (p.u.)	Q_{gi} (p.u.)	*Load* P_{d_i} (p.u.)	Q_{d_i} (p.u.)	*Voltage* V_i (p.u.)	δ_i (rad)	*Type of bus*
1	–	–	0.00	0.00	1.030	0.0	Slack
2	0.936	–	0.10	0.03	1.002	–	*PV*
3	1.513	–	0.50	0.17	1.050	–	*PV*
4	0.480	–	0.30	0.10	1.015	–	*PV*
5	1.784	0.5	0.25	0.08	1.007	–	*PV*
6	0.0	0.0	0.15	0.05	–	–	*PQ*
7	0.0	0.0	0.15	0.05	–	–	*PQ*
8	0.0	0.0	0.25	0.00	–	–	*PQ*
9	0.0	0.0	0.15	0.05	–	–	*PQ*
10	0.0	0.0	0.15	0.05	–	–	*PQ*
11	0.0	0.0	0.05	0.00	–	–	*PQ*
12	0.0	0.0	0.10	0.00	–	–	*PQ*
13	0.0	0.0	0.25	0.08	–	–	*PQ*
14	0.0	0.0	0.20	0.07	–	–	*PQ*
15	0.0	0.0	0.30	0.10	–	–	*PQ*
16	0.0	0.0	0.30	0.10	–	–	*PQ*
17	0.0	0.0	0.60	0.20	–	–	*PQ*
18	0.0	0.0	0.15	0.05	–	–	*PQ*
19	0.0	0.0	0.15	0.05	–	–	*PQ*
20	0.0	0.0	0.25	0.08	–	–	*PQ*
21	0.0	0.0	0.20	0.07	–	–	*PQ*
22	0.0	0.0	0.20	0.07	–	–	*PQ*
23	0.0	0.0	0.15	0.05	–	–	*PQ*
24	0.0	0.0	0.15	0.05	–	–	*PQ*
25	0.0	0.0	0.25	0.08	–	–	*PQ*

Decoupled Newton–Raphson method is employed and convergence ($\varepsilon = 0.0001$) is achieved in eight iterations. Voltage magnitudes, voltage angles and power injected at all buses are tabulated in Table 2.43.

TABLE 2.43 Voltage and power injected at buses

Bus no.	V_i (p.u.)	δ_i (rad)	P_i (p.u.)	Q_i (p.u.)
1	1.0300000	0.0000000	0.7240468	0.0779707
2	1.0020000	0.1317326	0.8360000	−0.1865165
3	1.0500000	0.0973678	1.0130000	0.1948333
4	1.0150000	−0.0479114	0.1800069	0.1539479
5	1.0070000	0.1150286	1.5340000	−0.4440076
6	1.0083200	0.0777504	−0.1500006	−0.0500004
7	0.9975652	0.0487078	−0.1499998	−0.0500001
8	0.9995552	0.0380560	−0.2499998	0.0000006
9	0.9926407	0.0249825	−0.1500001	−0.0500001
10	1.0026400	0.0444859	−0.1500002	−0.0500006
11	1.0054140	0.0266034	−0.0499998	0.0000004
12	1.0014450	0.0268849	−0.1000001	−0.0000003
13	1.0139090	0.0739145	−0.2499987	−0.0799992
14	0.9862505	−0.0377593	−0.1999998	−0.0699996
15	0.9830649	−0.0540977	−0.3000011	−0.1000007
16	0.9936035	−0.0498755	−0.2999990	−0.0999990
17	1.0058740	0.0224340	−0.5999998	−0.2000007
18	0.9983927	−0.0145077	−0.1500000	−0.0499995
19	1.0098490	−0.0223375	−0.1500007	−0.0500009
20	1.0016820	−0.0844541	−0.2500004	−0.0800000
21	1.0004880	−0.1049661	−0.2000009	−0.0699991
22	0.9978289	−0.1199480	−0.2000258	−0.0700003
23	1.0118770	−0.0665098	−0.1499694	−0.0500006
24	1.0037240	−0.1374826	−0.1499978	−0.0499999
25	1.0089770	−0.1235014	−0.2499990	−0.0800005

Sample System 2.7

Line data consisting of line charging, line impedance and off-nominal turn-ratio of a power system is given in Table 2.44. The scheduled generation, load, and specified voltage on various types of buses are given in Table 2.45.

TABLE 2.44 Line data

Line no.	*Branch p–q*	*Line charging* Y_{pq}	*Impedance* Z_{pq}	*Turn-ratio a*
1	1–11	*j*0.050	0.060 + *j*0.206	1.0
2	11–12	*j*0.014	0.160 + *j*0.524	1.0
3	11–17	*j*0.018	0.210 + *j*0.694	1.0
4	11–24	*j*0.032	0.084 + *j*0.280	1.0
5	12–17	*j*0.006	0.066 + *j*0.236	1.0
6	24–13	*j*0.021	0.066 + *j*0.208	1.0
7	13–14	*j*0.021	0.060 + *j*0.200	1.0
8	13–16	*j*0.020	0.268 + *j*0.802	1.0
9	14–15	*j*0.000	0.065 + *j*0.194	1.0
10	14–4	*j*0.038	0.078 + *j*0.350	1.0
11	15–9	*j*0.021	0.074 + *j*0.212	1.0
12	15–10	*j*0.026	0.192 + *j*0.950	1.0
13	15–18	*j*0.010	0.066 + *j*0.186	1.0
14	9–16	*j*0.006	0.084 + *j*0.256	1.0
15	18–20	*j*0.003	0.038 + *j*0.108	1.0
16	18–10	*j*0.018	0.168 + *j*0.680	1.0
17	10–19	*j*0.018	0.174 + *j*0.704	1.0
18	19–20	*j*0.005	0.058 + *j*0.164	1.0
19	20–2	*j*0.026	0.054 + *j*0.242	1.0
20	20–4	*j*0.052	0.118 + *j*0.334	1.0
21	20–23	*j*0.026	0.094 + *j*0.256	1.0
22	2–21	*j*0.000	0.000 + *j*0.146	1.0
23	21–22	*j*0.000	0.146 + *j*0.248	1.0
24	22–23	*j*0.000	0.000 + *j*0.064	1.0
25	23–3	*j*0.000	0.000 + *j*0.058	1.0
26	4–5	*j*0.038	0.108 + *j*0.358	1.0
27	5–26	*j*0.028	0.078 + *j*0.266	1.0
28	5–27	*j*0.024	0.066 + *j*0.220	1.0
29	26–25	*j*0.016	0.046 + *j*0.154	1.0
30	26–27	*j*0.025	0.066 + *j*0.216	1.0
31	25–17	*j*0.085	0.212 + *j*0.806	1.0
32	27–6	*j*0.017	0.024 + *j*0.076	1.0
33	27–7	*j*0.019	0.066 + *j*0.184	1.0
34	7–28	*j*0.023	0.066 + *j*0.224	1.0
35	28–8	*j*0.014	0.038 + *j*0.128	1.0

TABLE 2.45 Generation, load, and voltage at buses

Bus no.	Generation		Load		Voltage		Type of bus
	P_{gi} (p.u.)	Q_{gi} (p.u.)	P_{d_i} (p.u.)	Q_{d_i} (p.u.)	V_i (p.u.)	δ_i (rad)	
1	–	–	0.0000	0.0000	1.012	0.0	Slack
2	0.758	0.075	0.2550	0.1250	1.002	–	*PV*
3	0.201	0.123	0.0000	0.0000	1.001	–	*PV*
4	1.190	0.129	0.4050	0.1940	1.040	–	*PV*
5	1.425	−0.128	0.0255	0.0155	1.050	–	*PV*
6	0.580	0.205	0.5850	0.1400	1.036	–	*PV*
7	0.229	0.150	0.4200	0.0515	1.028	–	*PV*
8	0.235	0.160	0.2190	0.0200	1.040	–	*PV*
9	0.0	0.0	0.0520	0.0075	–	–	*PQ*
10	0.0	0.0	0.0555	0.2150	–	–	*PQ*
11	0.0	0.0	0.2365	0.1590	–	–	*PQ*
12	0.0	0.0	0.0895	0.0760	–	–	*PQ*
13	0.0	0.0	0.0750	0.0400	–	–	*PQ*
14	0.0	0.0	0.1255	0.0750	–	–	*PQ*
15	0.0	0.0	0.1150	0.0650	–	–	*PQ*
16	0.0	0.0	0.0850	0.0400	–	–	*PQ*
17	0.0	0.0	0.1910	0.1250	–	–	*PQ*
18	0.0	0.0	0.0845	0.0145	–	–	*PQ*
19	0.0	0.0	0.1500	0.0250	–	–	*PQ*
20	0.0	0.0	0.6350	0.4300	–	–	*PQ*
21	0.0	0.0	0.0000	0.0000	–	–	*PQ*
22	0.0	0.0	0.0000	0.0000	–	–	*PQ*
23	0.0	0.0	0.2375	0.1230	–	–	*PQ*
24	0.0	0.0	0.1600	0.0750	–	–	*PQ*
25	0.0	0.0	0.1410	0.0500	–	–	*PQ*
26	0.0	0.0	0.2670	0.1000	–	–	*PQ*
27	0.0	0.0	0.1455	0.0800	–	–	*PQ*
28	0.0	0.0	0.2750	0.1245	–	–	*PQ*

Decoupled Newton–Raphson method is employed and convergence ($\varepsilon = 0.00001$) is achieved in eight iterations. Voltage magnitudes, voltage angles and power injected at all buses are tabulated in Table 2.46.

TABLE 2.46 Voltage and power injected at buses

Bus no.	V_i (p.u.)	δ_i (rad)	P_i (p.u.)	Q_i (p.u.)
1	1.012	0.0000000	0.6380979	0.1490656
2	1.002	−0.1044606	0.5029998	0.2108777
3	1.001	−0.1604965	0.2009998	0.2940157
4	1.040	−0.0250462	0.7849998	0.3107586
5	1.050	0.0402329	1.3995000	−0.1395139
6	1.036	−0.1026045	−0.0049999	0.0714689
7	1.028	−0.1969713	−0.1910000	0.1014132
8	1.040	−0.2702140	−0.0140001	0.1395656
9	0.9089335	−0.2274111	−0.0520002	−0.0075003
10	0.8576921	−0.2101920	−0.0554945	−0.2149965
11	0.9408321	−0.1257688	−0.2365004	−0.1590001
12	0.9066542	−0.1856025	−0.0894994	−0.0759997
13	0.9293590	−0.1904876	−0.0750007	−0.0400000
14	0.9434978	−0.1676937	−0.1255015	−0.0750012
15	0.9146234	−0.2051601	−0.1149993	−0.0650001
16	0.9035132	−0.2368435	−0.0849994	−0.0399996
17	0.9147613	−0.1927515	−0.1910003	−0.1250004
18	0.9155408	−0.1991368	−0.0845001	−0.0145009
19	0.9054530	−0.2111868	−0.1499998	−0.0250006
20	0.9279057	−0.1837509	−0.6350048	−0.4300030
21	1.0025250	−0.1255868	0.0000004	0.0000005
22	0.9833152	−0.1629218	−0.0000006	−0.0000006
23	0.9840331	−0.1723321	−0.2375002	−0.1230007
24	0.9235564	−0.1835407	−0.1599999	−0.0749996
25	0.9940771	−0.1268076	−0.1409999	−0.0499998
26	1.0095570	−0.0918859	−0.2669999	−0.1000003
27	1.0295370	−0.1002295	−0.1454994	−0.0800003
28	1.0214970	−0.2629933	−0.2749998	−0.1245002

REFERENCES

Books

Arrillaga, J. and C.P. Arnold, *Computer Analysis of Power Systems*, John Wiley & Sons, Singapore, 1990.

Arrillaga, J., C.P. Arnold and B.J. Harker, *Computer Modelling of Electrical Power Systems*, Wiley, New York, 1983.

Brown, H.E., *Solution of Large Networks by Matrix Methods*, Wiley, New York, 1975.

Elgerd, O.I., *Electric Energy Systems Theory: An Introduction*, 2nd ed., Tata McGraw-Hill, New Delhi, 1983.

El-Hawary, M.E. and G.S. Christensen, *Optimal Economic Operation of Power Systems*, Academic Press, New York, 1979.

Grainger, J.J. and W.D. Stevenson, *Power System Analysis*, McGraw-Hill, New York, 1994.

Gross, C.A., *Power System Analysis*, Wiley, New York, 1979.

Kirchmayer, L.K., *Economic Operation of Power Systems*, Wiley Eastern, New Delhi, 1958.

Knight, U.G., *Power System Engineering and Mathematics*, Pergamon Press, New York, 1972.

Kothari, D.P. and I.J. *Nagrath, Modern Power System Analysis*, 3rd ed., Tata McGraw-Hill, New Delhi, 2003.

Kusic, G.L., *Computer-Aided Power Systems Analysis*, Prentice Hall of India, New Delhi, 1986.

Mahalanabis, A.K., D.P. Kothari and S.I. Ahson, *Computer-Aided Power System Analysis and Control*, Tata McGraw-Hill, New Delhi, 1991.

Nagrath, I.J. and D.P. Kothari, *Power System Engineering*, Tata McGraw-Hill, New Delhi, 1994.

Shipley, R.B., *Introduction to Matrices and Power Systems*, Wiley, New York, 1976.

Singh, L.P., *Advanced Power System Analysis and Dynamics*, 2nd ed., Wiley Eastern, New Delhi, 1993.

Stagg, G.W. and A.H. Ei-Abiad, *Computer Methods in Power Systems Analysis*, McGraw-Hill, New Delhi, 1968.

Stevenson, W.D., *Elements of Power System Analysis*, 4th ed., McGraw-Hill, New York, 1982.

Weedy, B.M. and B.J. Cory, *Electrical Power Systems*, 4th ed., Wiley, New York, 1998.

Wood, A.J. and B. Wollenberg, *Power Generation, Operation and Control*, 2nd ed., John Wiley, New York, 1996.

Papers

Arrillaga, J., C.P. Arnold, J.R. Camacho and S. Sankar, AC-DC Load Flow with Unit-Connected Generator-Converter Infeeds, *IEEE Trans. on Power Systems*, Vol. **PWRS-8,** No. 2, pp. 701–706, 1993.

Carpentier, J.L., Optimal power flow: Uses, methods and developments, *Proceedings of IFAC Conference*, R.J. Brazil, pp. 11–25, 1985.

Dommel, H.W. and W.F. Tinney, Optimal power flow solutions, *IEEE Trans. on Power Apparatus and Systems*, Vol. **PAS-87,** No. 10, pp. 1866–1876, 1968.

Ekwue, A.O. and J.F. Macqueen, Comparison of load flow solution methods, *Electric Power Systems Research*, Vol. **22,** No. 3, pp. 213–222, 1991.

Iwamoto, S. and Y. Tamura, A fast load flow method retaining non-linearity, *IEEE Trans. on Power Apparatus and Systems,* Vol. **PAS-97,** No. 5, pp. 1586–1599, Sep./Oct., 1978.

Iwamoto, S. and Y. Tamura, A load flow calculation method for ill-conditioned power systems, *IEEE Trans. on Power Apparatus and Systems,* Vol. **PAS-100,** No. 4, pp. 1736–1743, April 1981.

Leoniopoulos, G., Efficient starting point of load-flow equations, *Int. J. Electrical Power and Energy Systems*, Vol. **16,** No. 6, pp. 419–422, Dec. 1994.

Nagendra Rao, K.S. Prakasa Rao and J. Nanda, A novel hybrid load flow methods, *IEEE Trans. on Power Apparatus and Systems*, Vol. **PAS-100,** No. 1, pp. 303–308, 1981.

Nanda, J., D.P. Kothari and S.C. Srivastava, A novel second order fast decoupled load flow method in polar coordinates, *Electric Machines and Power Systems*, Vol. **14,** No. 5, pp. 339–351, 1989.

Nanda, J., D.P. Kothari and S.C. Srivastava, Some important observations on FDF algorithms, *Proceedings of IEEE*, pp. 732–33, May 1987.

Nanda, J., P.R. Bijwe, D.P. Kothari and D.I. Shenoy, Second order decoupled load flow, *Electric Machines and Power Systems*, Vol. **12,** No. 5, pp. 301–302, 1987.

Sachdev, M.S. and T.K.P. Medicherla, A second order load flow technique, *IEEE Trans. on Power Apparatus and Systems*, Vol. **PAS-96,** No. 1, pp. 189–197, 1977.

Sasson, A.M., Combined use of the Powell and Fletcher Powell nonlinear programming methods for optimal load flows, *IEEE Trans. on Power Apparatus and Systems*, Vol. **PAS-88,** No. 10, pp. 1530–1535, 1969.

Sasson, A.M., Non-linear programming solutions for load flow, minimum loss and economic dispatching problems, *IEEE Trans. on Power Apparatus and Systems*, Vol. **PAS-88,** No. 4, pp. 399–409, 1969.

Sato, N. and W.F. Tinney, Techniques for exploiting sparsity of network admittance matrix, *IEEE Trans. on Power Apparatus and Systems,* Vol. **PAS-82**, pp. 44, 1963.

Sinha, A.K., L. Roy and H.N.P. Srivastava, A decoupled second order state estimator for ac-dc power systems, *IEEE Trans. on Power Systems*, Vol. **PWRS-9,** No. 3, pp. 1485–1493, August 1994.

Stott, B. and O. Alsac, Fast decoupled load flow, *IEEE Trans. on Power Apparatus and Systems*, Vol. **PAS-83,** No. 3, pp. 859–869, 1974.

Stott, B., Decoupled Newton load flow, *IEEE Trans. on Power Apparatus and Systems*, Vol. **PAS-91,** No. 5, pp. 1955–1959, 1972.

Stott, B., Review of load-flow calculation method, *Proceedings of IEEE*, pp. 916, 1974.

Tinney, W.F. and C.E. Hart, Power flow solution by Newton's method, *IEEE Trans. on Power Apparatus and Systems*, Vol. **PAS-86, No. 11**, pp. 1449–1460, Nov. 1967.

Tinney, W.F. and W.S. Mayer, Solution of large sparse systems by ordered triangular factorization, *IEEE Trans. on Automatic Control*, Vol. **18**, pp. 333, August 1973.

Tinney, W.F. and J.W. Walker, Direct solutions of sparse network equations by optimally ordered triangular factorizations, *Proceedings of IEEE*, Vol. **55**, pp. 1801, 1967.

Wang, L. and X. Rong Li, Robust fast decoupled power flow, *IEEE Trans. on Power Systems*, Vol. **PWRS-15,** No. 1, pp. 208–215, 2000.

Ward, J.B. and H.W. Hale, Digital computer solution of power problems, *AIEE Trans. Part-III: Power Apparatus and Systems*, Vol. **75,** No. 3, pp. 398–404, June 1956.

CHAPTER 3

Economic Load Dispatch of Thermal Generating Units

3.1 INTRODUCTION

Electrical energy cannot be stored, but is generated from natural sources and delivered as demand arises. A transmission system is used for the delivery of bulk power over considerable distances, and a distribution system is used for local deliveries. As depicted in Figure 3.1, an interconnected power system consists of mainly three parts: the generators which produce the electrical energy, the transmission lines which transmit it to far away places, and the loads which use it. Such a configuration applies to all interconnected networks (regional, national, international), where the number of elements may vary. The transmission networks are interconnected through ties so that utilities can exchange power, share reserves and render assistance to one another in times of need.

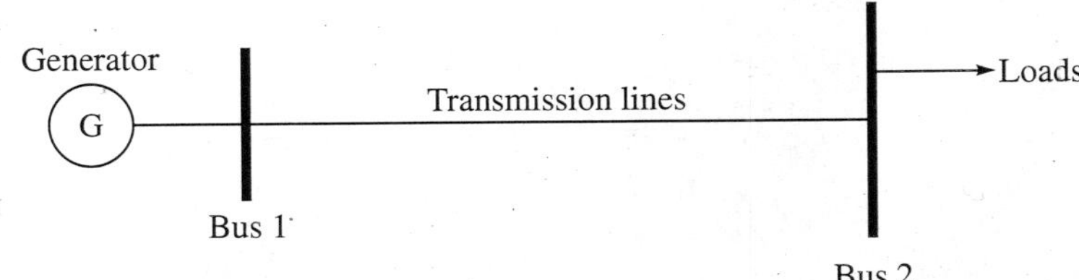

FIGURE 3.1 A simplified configuration of an interconnected power system.

Since the sources of energy are so diverse (coal, oil or gas, river water, marine tide, a radioactive matter, sun power), the choice of one or the other is made on economic, technical or geographic basis. As there are few facilities to store electrical energy, the net production of a utility (generation plus the inflows over its ties) must clearly track its total load. For an interconnected system, the fundamental problem is one of minimizing the source expenses. The economic dispatching problem is to define the production level of each plant so that the total cost of generation and transmission is minimum for a prescribed schedule of loads.

- Forecasting includes determining the peak rate of supply (power demand and volume), i.e. energy demand for both long-term investment decisions and short-term operating decisions.

- Operating applications include allocation of output (dispatching), unit start-up selection, (unit commitment), hydrothermal coordination, and maintenance scheduling.
- The investment planning applications cover the generation and the transmission systems.

3.2 GENERATOR OPERATING COST

The majority of generators in extant systems are of three types—nuclear, hydro, and fossil (coal, oil or gases). Nuclear plants tend to be operated at constant output levels and hydro plants have essentially no variable operating costs. Therefore, the components of cost that fall under the category of dispatching procedures are the costs of the fuel burnt in the fossil plants. The total cost of operation includes the fuel cost, costs of labour, supplies and maintenance. Generally, costs of labour, supplies and maintenance are fixed percentages of incoming fuel costs. Figure 3.2 shows a simple model of a fossil plant. The power output of the fossil plant is increased sequentially by opening a set of valves at the inlet to its steam turbine. The throttling losses in a valve are large when it is just opened and small when it is fully opened.

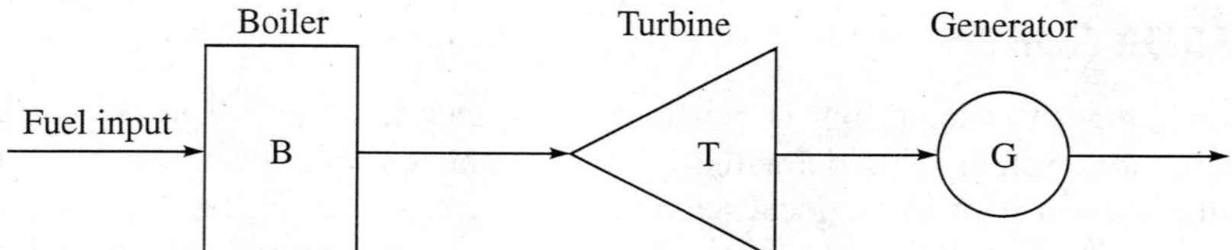

FIGURE 3.2 Simple model of a fossil plant.

As a result, the operating cost of the plant has the form shown in Figure 3.3. For dispatching purposes, this cost is usually approximated by one or more quadratic segments. So, the fuel cost curve is modelled as a quadratic in the active power generation

$$F(P_{g_i}) = a_i P_{g_i}^2 + b_i P_{g_i} + c_i \ \text{Rs/h} \tag{3.1}$$

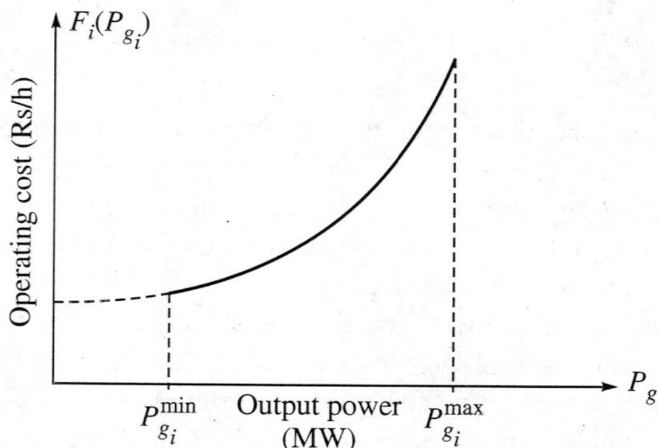

FIGURE 3.3 Operating costs of a fossil-fired generator. $P_{g_i}^{\min}$ and $P_{g_i}^{\max}$ are the lower and the upper limits on its output.

The $P_{g_i}^{\min}$ is the minimum loading limit below which it is uneconomical (or may be technically infeasible) to operate the unit and $P_{g_i}^{\max}$ is the maximum output limit.

The fuel cost curve may have a number of discontinuities. The discontinuities occur when the output power has to be extended by using additional boilers, steam condensers, or other equipment. Discontinuities also appear if the cost represents the operation of an entire power station, so that cost has discontinuities on paralleling of generators. Within the continuity range the incremental fuel cost may be expressed by a number of short line segments or piece-wise linearizations.

3.3 ECONOMIC DISPATCH PROBLEM ON A BUS BAR

Let us assume that it is known a priori which generators are to be run to meet a particular load demand on the station. Suppose there is a station with NG generators committed and the active power load demand P_D is given, the real power generation P_{g_i} for each generator has to be allocated so as to minimize the total cost. The optimization problem can therefore be stated as

Minimize
$$F(P_{gi}) = \sum_{i=1}^{NG} F_i(P_{gi}) \tag{3.2a}$$

subject to (i) the energy balance equation

$$\sum_{i=1}^{NG} P_{g_i} = P_D \tag{3.2b}$$

(ii) and the inequality constraints

$$P_{g_i}^{\min} \le P_{g_i} \le P_{gi}^{\max} \qquad (i = 1, 2, \ldots, \text{NG}) \tag{3.2c}$$

where

P_{g_i} is the decision variable, i.e. real power generation
P_D is the real power demand
NG is the number of generation plants
$P_{g_i}^{\min}$ is the lower permissible limit of real power generation
$P_{gi}^{\max}$ is the upper permissible limit of real power generation
$F_i(P_{g_i})$ is the operating fuel cost of the ith plant and is given by the quadratic equation

$$F_i(P_{g_i}) = a_i P_{g_i}^2 + b_i P_{g_i} + c_i \ \text{Rs/h} \tag{3.2d}$$

The above constrained optimization problem is converted into an unconstrained optimization problem. Lagrange multiplier method is used in which a function is minimized (or maximized) with side conditions in the form of equality constraints. Using this method, an augmented function is defined as

$$L(P_{g_i}, \lambda) = F(P_{g_i}) + \lambda\left(P_D - \sum_{i=1}^{NG} P_{g_i}\right) \tag{3.3}$$

where λ is the Lagrangian multiplier.

A necessary condition for a function $F(P_{g_i})$, subject to energy balance constraint to have a relative minimum at point $P_{g_i}^*$ is that the partial derivative of the Lagrange function defined by $L = L(P_{g_i}, \lambda)$ with respect to each of its arguments must be zero. So, the necessary conditions for the optimization problem are

$$\frac{\partial L(P_{g_i}, \lambda)}{\partial P_{g_i}} = \frac{\partial F(P_{g_i})}{\partial P_{g_i}} - \lambda = 0 \qquad (i = 1, 2, ..., \text{NG}) \tag{3.4}$$

and

$$\frac{\partial L(P_{g_i}, \lambda)}{\partial \lambda} = P_D - \sum_{i=1}^{\text{NG}} P_{g_i} = 0 \tag{3.5}$$

From Eq. (3.4),

$$\frac{\partial F(P_{g_i})}{\partial P_{g_i}} = \lambda \qquad (i = 1, 2, ..., \text{NG}) \tag{3.6}$$

where $\partial F(P_{g_i})/\partial P_{g_i}$ is the incremental fuel cost of the *i*th generator (₹/MWh).

Optimal loading of generators corresponds to the equal incremental cost point of all the generators. Equation (3.6), called the coordination equations numbering NG are solved simultaneously with the load demand to yield a solution for Lagrange multiplier λ and the optimal generation of NG generators. Considering the cost function given by Eq. (3.2d), the incremental cost can be defined as

$$\frac{\partial F(P_{g_i})}{\partial P_{g_i}} = 2a_i P_{g_i} + b_i \tag{3.7}$$

Substituting the incremental cost into Eq. (3.6), this equation becomes

$$2a_i P_{g_i} + b_i = \lambda \qquad (i = 1, 2, ..., \text{NG}) \tag{3.8}$$

Rearranging Eq. (3.8) to get P_{g_i}

$$P_{g_i} = \frac{\lambda - b_i}{2a_i} \qquad (i = 1, 2, ..., \text{NG}) \tag{3.9}$$

Substituting the value of P_{g_i} in Eq. (3.5), we get

$$\sum_{i=1}^{\text{NG}} \frac{\lambda - b_i}{2a_i} = P_D$$

or

$$\lambda = \frac{P_D + \sum_{i=1}^{\text{NG}} \frac{b_i}{2a_i}}{\sum_{i=1}^{\text{NG}} \frac{1}{2a_i}} \tag{3.10}$$

Thus, λ can be calculated using Eq. (3.10) and P_{g_i} can be calculated using Eq. (3.9). Now consider the effect of the generator limits given by the inequality constraint of Eq. (3.2c). If a particular generator loading P_{g_i} reaches the limit $P_{g_i}^{min}$ or $P_{g_i}^{max}$, its loading is held fixed at this value and the balance load is shared between the remaining generators on an equal incremental cost basis.

3.3.1 Limit Constraint Fixing

To fix up the limits the following strategy can be applied [Famideh-Vojdani, 1980].

Let

$$\Delta h = \sum_{i=1}^{R_1} h_i^{max} - \sum_{i=1}^{R_2} h_i^{min} \tag{3.11}$$

where

$$h_i^{max} = P_{g_i} - P_{g_i}^{max} \quad (i = 1, 2, ..., R_1 \text{ upper bound violations})$$

$$h_i^{min} = P_{g_i}^{min} - P_{g_i} \quad (i = 1, 2, ..., R_2 \text{ lower bound violations})$$

(i) If $\Delta h > 0$, fix all R_1 upper bound violations to the upper limits, i.e. $P_{g_i}^{max}$

(ii) If $\Delta h < 0$, fix all R_2 lower bound violations to the lower limits, i.e. $P_{g_i}^{min}$

(iii) On the other side, if $\Delta h = 0$, fix both R_1 upper and R_2 lower bound violations to their respective upper $P_{g_i}^{max}$ and lower $P_{g_i}^{min}$ limits.

Determine the new demand which is original P_D minus the sum of fixed generation levels, i.e.

$$P_D^{new} = P_D - \sum_{i=1}^{R_1+R_2} P_{g_i}$$

The new demand is allocated to other committed generators on an equal incremental cost basis.

EXAMPLE 3.1 Two units of the system have the following cost curves:

$$F(P_{g_1}) = 0.05P_{g_1}^2 + 22P_{g_1} + 120 \quad ₹/\text{h}$$

$$F(P_{g_2}) = 0.06P_{g_2}^2 + 16P_{g_2} + 120 \quad ₹/\text{h}$$

where P_g is in MW. Both the units operate at all times and the maximum and minimum loads on each unit are 100 MW and 20 MW, respectively. Determine the economic operating schedule of the plants for loads of 80 MW, 120 MW, and 180 MW, neglecting the transmission line losses.

Solution Using Eq. (3.10), compute incremental cost, λ

$$\lambda = \frac{P_D + \dfrac{22}{2 \times 0.05} + \dfrac{16}{2 \times 0.06}}{\dfrac{1}{2 \times 0.05} + \dfrac{1}{2 \times 0.06}} \quad ₹/\text{MWh}$$

$$= \frac{P_D + 353.333}{18.333} \quad ₹/\text{MWh} \tag{i}$$

When $P_D = 80$ MW
Substituting the value of P_D in Eq. (i) and solving for λ, we get

$$\lambda = 23.6368 \text{ ₹/MWh}$$

Using Eq. (3.9) to calculate generations

$$P_{g_1} = \frac{23.6368 - 22}{2 \times 0.05} = 16.36 \text{ MW}; \qquad P_{g_2} = \frac{23.6368 - 16}{2 \times 0.06} = 63.64 \text{ MW}$$

But $P_{g_1} = 16.36$ MW < 20 MW, so fix P_{g_1} at the lower limit 20 MW and the rest of the demand will be met by the second generator.

$$\text{So} \qquad P_{g_1} = 20 \text{ MW} \quad \text{and} \quad P_{g_2} = 80 - 20 = 60 \text{ MW}$$

When $P_D = 120$ MW
Substitute the value of P_D in Eq. (i) and solve for λ, i.e.

$$\lambda = 25.81818 \text{ ₹/MWh}$$

From Eq. (3.9),

$$P_{g_1} = \frac{25.81818 - 22}{2 \times 0.05} = 38.1818 \text{ MW}; \qquad P_{g_2} = \frac{25.81818 - 16}{2 \times 0.06} = 81.8182 \text{ MW}$$

When $P_D = 180$ MW
Substitute the value of P_D in Eq. (i) and solve for λ, i.e.

$$\lambda = 29.0914 \text{ ₹/MWh}$$

From Eq. (3.9),

$$P_{g_1} = \frac{29.0914 - 22}{2 \times 0.05} = 70.9142 \text{ MW}; \qquad P_{g_2} = \frac{29.0914 - 16}{2 \times 0.06} = 109.095 \text{ MW}$$

But $P_{g_2} = 109.095$ MW > 100 MW, so fix P_{g_2} at the upper limit 100 MW and the rest of the demand will be met by the first generator.

$$\text{So} \qquad P_{g_2} = 100 \text{ MW} \quad \text{and} \quad P_{g_1} = 180 - 100 = 80 \text{ MW}$$

EXAMPLE 3.2 Incremental fuel costs in rupees per MWh for a plant consisting of two units are

$$\frac{\partial F_1}{\partial P_{g_1}} = 0.20\, P_{g_1} + 40 \quad \text{and} \quad \frac{\partial F_2}{\partial P_{g_2}} = 0.40\, P_{g_2} + 30$$

and the generator limits are

$$30 \text{ MW} \le P_{g_1} \le 175 \text{ MW} \quad \text{and} \quad 20 \text{ MW} \le P_{g_2} \le 125 \text{ MW}$$

Assume that both units are operating at all times. How will the load be shared between the two units as the system load varies over the full range of the load values? What are the corresponding values of the plant incremental costs?

Solution From Eq. (3.10),

$$\lambda = \frac{P_D + \dfrac{40}{0.2} + \dfrac{30}{0.4}}{\dfrac{1}{0.2} + \dfrac{1}{0.4}} \text{ ₹/MWh}$$

$$= \frac{P_D + 275}{7.5} \text{ ₹/MWh}$$

or

$$P_D = (7.5\lambda - 275) \text{ MW} \tag{i}$$

The values of λ at minimum and maximum operating limits can be obtained as follows:

When unit 1 is operating at minimum limit, $P_{g_1}^{\min} = 30$ MW, then

$$\lambda_1^{\min} = 0.2 \times 30 + 40 = 46 \text{ ₹/MWh}$$

When unit 2 is operating at minimum limit, $P_{g_2}^{\min} = 20$ MW, then

$$\lambda_2^{\min} = 0.4 \times 20 + 30 = 38 \text{ ₹/MWh}$$

When unit 1 is operating at maximum limit, $P_{g_1}^{\max} = 175$ MW, then

$$\lambda_1^{\max} = 0.2 \times 175 + 40 = 75 \text{ ₹/MWh}$$

When unit 2 is operating at maximum limit, $P_{g_2}^{\max} = 125$ MW, then

$$\lambda_2^{\max} = 0.4 \times 125 + 30 = 80 \text{ ₹/MWh}$$

So, there are three operating conditions for λ, which are obtained from minimum and maximum operating values, i.e.

(i) $38 \le \lambda \le 46$
(ii) $46 \le \lambda \le 75$
(iii) $75 \le \lambda \le 80$

The range of load demand can be obtained as given below:

(i) For $38 \le \lambda \le 46$, the lower power limit of unit 1 is violated because $\lambda = 46$ ₹/MWh when $P_1^{\min} = 30$ MW.

The minimum demand of the system that can be met will be, $P_D = P_1^{\min} + P_2^{\min}$, i.e. $P_D = 30 + 20 = 50$ MW when $\lambda = 38$ ₹/MWh. For $\lambda = 46$, the value of P_D can be determined from Eq. (i), i.e.

$$P_D = 7.5 \times 46 - 275 = 70 \text{ MW}$$

So, the range of demand is $50 \le P_D \le 70$.

As P_D increases beyond 50 MW the load increments are placed on unit 2 because unit 1 is fixed at 30 MW (minimum operating limit). So, the incremental cost equation of unit 2 can be written in terms of P_D as

$$\lambda = 0.4 \times (P_D - 30) + 30$$

On simplification

$$\lambda = (0.4P_D + 18) \text{ ₹/MWh for } 50 \le P_D \le 70$$

(ii) For $46 \le \lambda \le 75$, the operation of generators will be within the operating limits. When $\lambda = 75$, the value of P_D can be determined from Eq. (i), i.e.

$$P_D = 7.5 \times 75 - 275 = 287.5 \text{ MW}$$

So, the range of demand is $70 \le P_D \le 287.5$, and

$$\lambda = \frac{P_D + 275}{7.5} \text{ ₹/MWh for } 70 \le P_D \le 287.5$$

(iii) For $75 \le \lambda \le 80$, the upper limit of unit 1 is violated because $\lambda = 75$ ₹/MWh when $P_1^{\max} = 175$ MW. The maximum demand of the system that can be met when $\lambda = 80$ ₹/MWh, is

$$P_D = P_1^{\max} + P_2^{\max}, \text{ i.e } P_D = 175 + 125 = 300 \text{ MW}$$

So, the range of demand is $287.5 \le P_D \le 300$.

As P_D increases beyond 287.5 MW, the load increments are placed on unit 2, because unit 1 is fixed at 175 MW. So, the incremental cost equation of unit 2 can be written in terms of P_D as

$$\lambda = 0.4 \times (P_D - 175) + 30 = 0.4P_D - 40 \quad \text{₹/MWh for } 287.5 \le P_D \le 300$$

The generation schedule is given in Table 3.1.

3.4 OPTIMAL GENERATION SCHEDULING

From the unit commitment table of a given plant, the fuel cost curve of the plant can be determined in the form of a polynomial of suitable degree by the method of least squares fit. If the transmission losses are neglected, the total system load can be optimally divided among the various generating plants using the equal incremental cost criterion of Eq. (3.6). It is, however, unrealistic to neglect transmission losses particularly when long distance transmission of power is involved. A modern electric utility serves over a vast area of relatively low load density. The transmission losses may vary from 5 to 15 per cent of total load. Therefore, it is essential to account for transmission losses while developing an economic load dispatch policy.

The economic dispatch problem is defined as that which minimizes the total operating cost of a power system while meeting the total load plus transmission losses within generator limits. Mathematically, the problem is defined as

$$\text{Minimize} \qquad F(P_{g_i}) = \sum_{i=1}^{NG} (a_i P_{g_i}^2 + b_i P_{g_i} + c_i) \text{ ₹/h} \tag{3.12a}$$

TABLE 3.1 Generation schedule (Example 3.2)

λ (₹/MWh)	Unit 1, P_{g_1} (MW)	Unit 2, P_{g_2} (MW)	$P_D = P_{g_1} + P_{g_2}$ (MW)
38	30	20	50
40	30	25	55
42	30	30	60
44	30	35	65
46	30	40	70
48	40	45	85
50	50	50	100
60	100	75	175
75	175	112.5	287.5
78	175	120	295
80	175	125	300

subject to (i) the energy balance equation

$$\sum_{i=1}^{NG} P_{g_i} = P_D + P_L \tag{3.12b}$$

(ii) and the inequality constraints

$$P_{g_i}^{\min} \le P_{gi} \le P_{g_i}^{\max} \qquad (i = 1, 2, ..., \text{NG}) \tag{3.12c}$$

where

a_i, b_i, and c_i are the cost coefficients
P_D is the load demand
P_{gi} is the real power generation and will act as decision variable
NG is the number of generation buses
P_L is the transmission power loss.

One of the most important, simple but approximate method of expressing transmission loss as a function of generator powers is through *B*-coefficients. This method uses the fact that under normal operating conditions, the transmission loss is quadratic in the injected bus real powers. The general form of the loss formula (derived later in this section) using *B*-coefficients is

$$P_L = \sum_{i=1}^{NG} \sum_{j=1}^{NG} P_{gi} B_{ij} P_{gj} \quad \text{MW} \tag{3.13}$$

where

P_{gi} and P_{gj} are the real power generations at the *i*th and *j*th buses, respectively
B_{ij} are the loss coefficients which are constant under certain assumed conditions
NG is number of generation buses.

The transmission loss formula of Eq. (3.13) is known as the George's formula.

Another more accurate form of transmission loss expression, frequently known as the Kron's loss formula is

$$P_L = B_{00} + \sum_{i=1}^{NG} B_{i0} P_{g_i} + \sum_{i=1}^{NG} \sum_{j=1}^{NG} P_{g_i} B_{ij} P_{g_j} \quad \text{MW} \tag{3.14}$$

where

P_{g_i} and P_{g_j} are the real power generation at *i*th and *j*th buses, respectively.

B_{00}, B_{i0}, and B_{ij} are the loss coefficients which are constant under certain assumed conditions NG is number of generation buses.

The above constrained optimization problem is converted into an unconstrained optimization problem. Lagrange multiplier method is used in which the function is minimized (or maximized) with side conditions in the form of equality constraints. Using Lagrange multipliers, an augmented function is defined as

$$L(P_{g_i}, \lambda) = F(P_{g_i}) + \lambda \left(P_D + P_L - \sum_{i=1}^{NG} P_{g_i} \right) \tag{3.15}$$

where λ is the Lagrangian multiplier.

Necessary conditions for the optimization problem are

$$\frac{\partial L(P_{g_i}, \lambda)}{\partial P_{g_i}} = \frac{\partial F(P_{g_i})}{\partial P_{g_i}} + \lambda \left(\frac{\partial P_L}{\partial P_{g_i}} - 1 \right) = 0 \qquad (i = 1, 2, ..., \text{NG})$$

Rearranging the above equation,

$$\frac{\partial F(P_{g_i})}{\partial P_{g_i}} = \lambda \left(1 - \frac{\partial P_L}{\partial P_{g_i}} \right) \qquad (i = 1, 2, ..., \text{NG}) \tag{3.16}$$

where

$\frac{\partial F(P_{g_i})}{\partial P_{g_i}}$ is the incremental operating cost of the *i*th generator (₹/MWh)

$\frac{\partial P_L}{\partial P_{g_i}}$ is the incremental transmission losses.

Equation (3.16) is known as the exact coordination equation, and

$$\frac{\partial L(P_{g_i}, \lambda)}{\partial \lambda} = P_D + P_L - \sum_{i=1}^{NG} P_{g_i} = 0 \tag{3.17}$$

Equation (3.16), the so-called coordination equation, numbering NG is solved simultaneously with Eq. (3.17) to yield a solution for Lagrange multiplier λ and the optimal generation of NG generators. By differentiating the transmission loss equation, Eq. (3.14) with respect to P_{g_i}, the incremental transmission loss can be obtained as

$$\frac{\partial P_L}{\partial P_{g_i}} = B_{i0} + \sum_{j=1}^{NG} 2 B_{ij} P_{g_j} \qquad (i = 1, 2, ..., \text{NG}) \tag{3.18}$$

and by differentiating cost function Eq. (3.12a), with respect to P_{g_i}, the incremental cost can be obtained as

$$\frac{\partial F(P_{g_i})}{\partial P_{g_i}} = 2a_i P_{g_i} + b_i \qquad (i = 1, 2, ..., \text{NG}) \tag{3.19}$$

Equation (3.16) can be rewritten as

$$\frac{\dfrac{\partial F(P_{g_i})}{\partial P_{g_i}}}{1 - \dfrac{\partial P_L}{\partial P_{g_i}}} = \lambda$$

or

$$\left(\frac{\partial F(P_{g_i})}{\partial P_{g_i}}\right) L_i = \lambda \qquad (i = 1, 2, ..., \text{NG}) \tag{3.20}$$

where $L_i = \dfrac{1}{1 - \dfrac{\partial P_L}{\partial P_{g_i}}}$ is called the *penalty factor* of the ith plant.

To obtain the solution, substitute Eqs. (3.18) and (3.19) into Eq. (3.16)

$$2a_i P_{g_i} + b_i = \lambda\left(1 - B_{i0} - \sum_{j=1}^{\text{NG}} 2B_{ij}P_{g_j}\right) \qquad (i = 1, 2, ..., \text{NG})$$

Rearranging the above equation to get P_{g_i}, we have

$$2(a_i + \lambda B_{ii})P_{g_i} = \lambda\left(1 - B_{i0} - \sum_{\substack{j=1 \\ j \neq i}}^{\text{NG}} 2B_{ij}P_{g_j}\right) - b_i \qquad (i = 1, 2, ..., \text{NG})$$

The value of P_{g_i} can be obtained as

$$P_{g_i} = \frac{\lambda\left(1 - B_{i0} - \sum_{\substack{j=1 \\ j \neq i}}^{\text{NG}} 2B_{ij}P_{g_j}\right) - b_i}{2(a_i + \lambda B_{ii})} \qquad (i = 1, 2, ..., \text{NG}) \tag{3.21}$$

If the initial values of P_{g_i} $(i = 1, 2, ..., \text{NG})$ and λ are known, the above equation can be solved iteratively until Eq. (3.17) is satisfied by modifying λ. This technique is known as successive approximation. The stepwise procedure is explained below. For simplicity it is considered that the solution remains within limits.

Algorithm 3.1: Economic Dispatch (Classical Method)

1. Read data, namely cost coefficients, a_i, b_i, c_i; B-coefficients, B_{ij}, B_{i0}, B_{00} (i = 1, 2, ..., NG; j = 1, 2, ..., NG); convergence tolerance, ε; step size α; and maximum iterations allowed, ITMAX, etc.
2. Compute the initial values of P_{g_i}(i = 1, 2, ..., NG) and λ by assuming that the transmission losses are zero, i.e. $P_L = 0$. Then the problem can be stated by Eqs. (3.2a) and (3.2b) and the solution can be obtained directly using Eqs. (3.10) and (3.9).
3. Set iteration counter, IT = 1.
4. Compute P_{g_i} (i = 1, ..., NG) using Eq. (3.21).
5. Compute transmission loss using Eq. (3.14).
6. Compute $\Delta P = P_D + P_L - \sum_{i=1}^{NG} P_{g_i}$.
7. Check $|\Delta P| \le \varepsilon$, if 'yes', then GOTO Step 10.

   Check IT $\ge$ ITMAX, if 'yes' then GOTO Step 10. (It means program terminated without obtaining the required convergence.)
8. Update $\lambda^{new} = \lambda + \alpha\,\Delta P$, where α is the step-size used to increase or decrease the value of λ in order to meet the Step 6.
9. IT = IT + 1, $\lambda = \lambda^{new}$ and GOTO Step 4 and repeat.
10. Compute optimal total cost from Eq. (3.12a) and transmission loss from Eq. (3.14).
11. Stop.

Consider now the effect of the generator limits given by the inequality constraint of Eq. (3.12c). If a particular generator loading P_{g_i} reaches the lower limit, $P_{g_i}^{min}$, or the upper limit, $P_{g_i}^{max}$, its loading from then on is held fixed at this value and the balance load is shared between the remaining generators. A stepwise procedure to obtain the optimal generation schedule when the operating generation limits are imposed is given below.

Algorithm 3.2: Economic Dispatch Considering Limits (Classical Method)

1. Read data, namely cost coefficients, a_i, b_i, c_i; B-coefficients, B_{ij}, B_{i0}, B_{00} (i = 1, 2, ..., NG; j = 1, 2, ..., NG); convergence tolerance, ε; step size α; and maximum allowed iterations, ITMAX, etc.
2. Compute the initial values of P_{g_i} (i = 1, 2, ..., NG) and λ by assuming that $P_L = 0$. Then the problem can be stated by Eqs. (3.2a) and (3.2b) and the values of λ and P_{g_i} (i = 1, 2, ..., NG) can be obtained directly using Eqs. (3.10) and (3.9), respectively.
3. Assume no generator has been fixed at either lower limit or at upper limit.
4. Set iteration counter, IT = 1.
5. Compute P_{g_i}(i = 1, 2, ..., R) of generators which are not fixed at either upper or lower limits, using Eq. (3.21), where R is the number of participating generators.
6. Compute the transmission loss using Eq. (3.14).
7. Compute $\Delta P = P_D + P_L - \sum_{i=1}^{NG} P_{g_i}$.

8. Check $|\Delta P| \le \varepsilon$, if 'yes' then GOTO Step 11.

   Check IT $\ge$ ITMAX, if 'yes', then GOTO Step 11. (It means the program moves forward without obtaining required convergence.)
9. Modify $\lambda^{new} = \lambda + \alpha\Delta P$, where α is the step-size used to increase or decrease the value of λ in order to meet the Step 7.
10. IT = IT + 1, $\lambda = \lambda^{new}$ and GOTO Step 5 and repeat.
11. Check the limits of generators, if no more violations then GOTO Step 13, else fix as following.

    If $P_{g_i} < P_{g_i}^{\min}$ then $P_{g_i} = P_{g_i}^{\min}$

    If $P_{g_i} > P_{g_i}^{\max}$ then $P_{g_i} = P_{g_i}^{\max}$.
12. GOTO Step 4.
13. Compute the optimal total cost from Eq. (3.12a) and transmission loss from Eq. (3.14).
14. Stop.

The above strategy is demonstrated in Example 3.4.

EXAMPLE 3.3 The fuel inputs per hour of two plants are given as

$$F_1(P_{g_1}) = (0.00889\,P_{g_1}^2 + 10.333\,P_{g_1} + 200)\ ₹/\text{h}$$

$$F_2(P_{g_2}) = (0.00741\,P_{g_2}^2 + 10.833\,P_{g_2} + 240)\ ₹/\text{h}$$

Determine the economic schedule to meet the demand of 150 MW and the corresponding cost of generation. The transmission losses are given by

$$P_L = 0.001P_{g_1}^2 + 0.002P_{g_2}^2 - 2 \times 0.0002P_{g_1}P_{g_2}$$

Solution Follow the stepwise procedure mentioned in Algorithm 3.1.

Step 1: Recognize the data and assume $\alpha = 0.05$, $\varepsilon = 0.0001$, and ITMAX = 15

Step 2: Compute λ using Eq. (3.10), i.e.

$$\lambda = \frac{150 + \dfrac{10.333}{2 \times 0.00889} + \dfrac{10.833}{2 \times 0.00741}}{\dfrac{1}{2 \times 0.00889} + \dfrac{1}{2 \times 0.00741}} = 11.81812\ ₹/\text{MWh}$$

Compute P_{g_1} and P_{g_2}, using Eq. (3.9)

$$P_{g_1} = \frac{11.81812 - 10.333}{2 \times 0.00889} = 83.5276\ \text{MW}$$

$$P_{g_2} = \frac{11.81812 - 10.833}{2 \times 0.00741} = 66.47239\ \text{MW}$$

Step 3: Set IT = 1

Step 4: Compute P_{g_1} and P_{g_2}, using Eq. (3.21)

$$P_{g_1} = \frac{11.81812\,[1 - 2.0 \times (-0.0002) \times 66.47239] - 10.333}{2[0.00889 + (11.81812 \times 0.001)]} = 43.44557 \text{ MW}$$

$$P_{g_2} = \frac{11.81812[1 - 2.0 \times (-0.0002) \times 83.5276] - 10.833}{2[0.00741 + (11.81812 \times 0.002)]} = 22.22453 \text{ MW}$$

Step 5: Compute transmission loss,

$$P_L = (0.001)(43.44557)^2 + (0.002)(22.22453)^2 - 2(0.0002)(43.44557)(22.22453) = 2.489154 \text{ MW}$$

Step 6: Compute the energy balance requirement to be met

$$\Delta P = P_D + P_L - \sum_{i=1}^{2} P_{g_i} = 150.0 + 2.489154 - 65.6701 = +86.81905 \text{ MW}$$

Step 7: If $|\Delta P| \le 0.0001$, the criterion is not met then GOTO Step 8.

Step 8: Modify λ = 11.81812 + 0.05(+86.81905) = 16.15907 ₹/MWh

IT = IT + 1 and GOTO Step 4 and repeat.

The iteration-wise obtained results of P_{g_1}, P_{g_2}, λ, ΔP, and P_L are given in Table 3.2.

TABLE 3.2 Generation schedule (Example 3.3)

IT	P_{g_1} (MW)	P_{g_2} (MW)	λ (₹/MWh)	ΔP (MW)	P_L (MW)
1	43.4456	22.2245	11.81812	+86.81905	2.48915
2	119.1606	70.5657	16.15907	–18.93145	20.79482
3	110.1316	67.4586	15.21250	–9.331604	18.25856
4	101.7692	61.8196	14.74592	1.894872	15.48376
5	102.7077	52.1681	14.84067	+0.848730	15.72457
6	103.4826	62.6953	14.88311	–0.203037	15.97491
7	103.3739	62.6494	14.87295	–0.077766	15.94553
8	103.3012	62.5993	14.86907	+0.021338	15.92185
9	103.3133	62.6048	14.87013	+0.007057	15.92521
10	103.3201	62.6096	14.87049	–0.002229	15.92743
11	103.3188	62.6089	14.87037	–0.000624	15.92705
12	103.3181	62.6085	14.87034	+0.000239	15.92684
13	103.3183	62.6086	14.87035	+0.000041	15.92689

The final optimal schedule is:

P_{g_1} = 103.3183 MW, P_{g_2} = 62.6086 MW, λ = 14.87035 ₹/MWh, and cost = 2309.77 ₹/h

EXAMPLE 3.4 For a three-generator system, the fuel cost coefficients and the operating generator limits are given in Table 3.3(a). The B-coefficients for transmission loss are given in Table 3.3(b). Determine the economic schedule for loads 160 MW and 210 MW.

Solution Algorithm 3.2 is followed to get the optimal generation schedule. The achieved generation schedule is given in Table 3.4. The number of iterations taken by the algorithm are given in Table 3.5. The values of step length α chosen, and the achieved convergence are given in Table 3.5.

The method is very sensitive to the value of α, i.e. the step size. The number of iterations depend upon the assumed value of α. An incorrect value of α, sets the solution procedure in the oscillations. To avoid this problem the Newton–Raphson method can be implemented to get the solution.

TABLE 3.3(a) Fuel cost coefficients and operating generator limits (Example 3.4)

Generator i	a_i (₹/MW²h)	b_i (₹/MWh)	c_i (₹/h)	$P_{g_i}^{min}$ (MW)	$P_{g_i}^{max}$ (MW)
1	0.006085	10.04025	136.9125	5.0	150.0
2	0.005915	9.760576	59.1550	15.0	100.0
3	0.005250	8.662500	328.1250	50.0	250.0

TABLE 3.3(b) B-coefficients MW^{-1} (Example 3.4)

0.0001363	0.0000175	0.0001839
0.0000175	0.0001545	0.0002828
0.0001839	0.0002828	0.0016147

TABLE 3.4 Optimal generation schedule (Example 3.4)

P_D (MW)	P_{g1} (MW)	P_{g2} (MW)	P_{g3} (MW)	λ (₹/MWh)	F (₹/h)	P_L (MW)
No generation limits imposed						
160.0	57.5577	70.5238	37.9172	11.09701	2176.023	5.998648
210.0	83.4010	95.6169	39.4862	11.52315	2741.473	8.503935
Generation limits imposed						
160.0	53.3906	64.6094	50.0000	11.08013	2179.159	7.999945
210.0	79.9043	90.5531	50.0000	11.51164	2743.905	10.457380

3.5 ECONOMIC DISPATCH USING NEWTON–RAPHSON METHOD

The economic dispatch problem is expressed by Eqs. (3.12a), (3.12b), and (3.12c) and is converted into an unconstrained optimization problem as in Eq. (3.15). Necessary conditions for the optimization problems [Eq. (3.15)] are given by Eqs. (3.16) and Eq. (3.17). The solution of nonlinear Eq. (3.16) can be obtained using the Newton–Raphson method in which any change in

TABLE 3.5 Optimal schedule (Example 3.4)

P_D (MW)	*Iterations*	$\mid\Delta P\mid$ (MW)	α
No generation limits imposed			
160.0	20	0.7009506×10^{-4}	0.005
210.0	20	0.7820129×10^{-4}	0.005
Generation limits imposed			
160.0	13	0.5531311×10^{-4}	0.005
210.0	14	0.2193451×10^{-4}	0.005

control variables, about their initial values can be obtained using Taylor's expansion. Taylor's expansion to second order of Eq. (3.16) and Eq. (3.17) can be written as

$$\frac{\partial^2 L}{\partial P_{g_i}^2}\Delta P_{g_i} + \sum_{\substack{j=1\\ j\neq i}}^{NG}\frac{\partial^2 L}{\partial P_{g_i}\partial P_{g_j}}\Delta P_{g_j} + \frac{\partial^2 L}{\partial P_{g_i}\partial\lambda}\Delta\lambda = -\frac{\partial L}{\partial P_{g_i}} \quad (i = 1, 2, ..., NG) \tag{3.22}$$

$$\sum_{j=1}^{NG}\frac{\partial^2 L}{\partial\lambda\partial P_{g_j}}\Delta P_{g_j} + \frac{\partial^2 L}{\partial\lambda^2}\Delta\lambda = -\frac{\partial L}{\partial\lambda} \tag{3.23}$$

The above equations can be rewritten in matrix form as

$$\begin{bmatrix}\nabla_{P_gP_g}L & \nabla_{P_g\lambda}L\\ (\nabla_{P_g\lambda}L)^T & \nabla_{\lambda\lambda}L\end{bmatrix}\begin{bmatrix}\Delta P_g\\ \Delta\lambda\end{bmatrix} = \begin{bmatrix}-\nabla_{P_g}L\\ -\nabla_{\lambda}L\end{bmatrix} \tag{3.24}$$

Derivatives can be obtained as follows:

$$\frac{\partial L}{\partial P_{g_i}} = \frac{\partial F_i}{\partial P_{g_i}} + \lambda\left(\frac{\partial P_L}{\partial P_{g_i}} - 1\right) = (2a_iP_{g_i} + b_i) + \lambda\left(B_{i0} + \sum_{j=1}^{NG}2B_{ij}P_{g_j} - 1\right) \quad (i = 1, 2, ..., NG) \tag{3.25a}$$

$$\frac{\partial L}{\partial\lambda} = P_D + P_L - \sum_{i=1}^{NG}P_{g_i} \tag{3.25b}$$

Taking derivatives of Eq. (3.25a) with respect to P_{g_i},

$$\frac{\partial^2 L}{\partial P_{g_i}^2} = \frac{\partial^2 F_i}{\partial P_{g_i}^2} + \lambda\frac{\partial^2 P_L}{\partial P_{g_i}^2} = 2a_i + 2\lambda B_{ii} \qquad (i = 1, 2, ..., NG) \tag{3.26a}$$

$$\frac{\partial^2 L}{\partial P_{g_i}\partial P_{g_j}} = \lambda\frac{\partial^2 P_L}{\partial P_{g_i}\partial P_{g_j}} = 2\lambda B_{ij} \qquad (i = 1, 2, ..., NG;\ \ j = 1, 2, ..., NG;\ \ i \neq j) \tag{3.26b}$$

Taking derivatives of Eqs. (3.25a) and (3.25b) with respect to λ,

$$\frac{\partial^2 L}{\partial \lambda \partial P_{g_i}} = \frac{\partial^2 L}{\partial P_{g_i} \partial \lambda} = \frac{\partial P_L}{\partial P_{g_i}} - 1 = B_{i0} + \sum_{j=1}^{NG} 2B_{ij} P_{g_j} - 1 \qquad (i = 1, 2, ..., NG) \qquad (3.26c)$$

$$\frac{\partial^2 L}{\partial \lambda^2} = 0 \qquad (3.26d)$$

Equations (3.22) and (3.23) [or Eq. (3.24)] are iterated till no further improvement is obtained, or single derivatives with respect to control variables become zero. The stepwise procedure is outlined here.

Algorithm 3.3: Economic Dispatch (Newton–Raphson Method)

1. Read data, namely a_i, b_i, c_i (cost coefficients); B_{ij}, B_{i0}, B_{00} (B-coefficients) (i = 1, 2, ..., NG; j = 1, 2, ..., NG) convergence tolerance, ε; and ITMAX (maximum allowed iterations), etc.
2. Compute the initial values of P_{g_i} (i = 1, 2, ..., NG) and λ by presuming that $P_L = 0$. The values of λ and P_{g_i}(i = 1, 2, ..., NG) can be computed directly using Eqs. (3.10) and (3.9), respectively.
3. Assume that no generator has been fixed either at lower limit or at upper limit.
4. Set iteration counter IT = 1.
5. Compute Hessian and Jacobian matrix elements using Eqs. (3.25) and (3.26).

$$[H]\begin{bmatrix} \Delta P_g \\ \Delta \lambda \end{bmatrix} = -[J]$$

   Deactivate row and column of Hessian matrix and row of Jacobian matrix representing the generator whose generation is fixed either at lower limit or at upper limit. This is done so that fixed generators cannot participate in allocation.
6. Gauss elimination method is employed in which triangularization and back-substitution processes are performed to find P_{g_i}(i = 1, 2, ..., R and $\Delta\lambda$). Here R is the number of generators which can participate in allocation.
7. Check either $\sqrt{\sum_{i=1}^{R} (\Delta P_{g_i})^2 + (\Delta \lambda)^2} \le \varepsilon$ or $\sqrt{\sum_{i=1}^{R} \left(\frac{\partial L}{\partial P_{g_i}}\right)^2 + \left(\frac{\partial L}{\partial \lambda}\right)^2} \le \varepsilon$

   If convergence condition is 'yes' then GOTO Step 10.

   Check IT > ITMAX, if condition is 'yes', GOTO Step 10. (It means the procedure proceeds without obtaining required convergence.).
8. Modify control variables,

$$P_{g_i}^{\text{new}} = P_{g_i} + \Delta P_{g_i} \qquad (i = 1, 2, ..., R \text{ and } \lambda^{\text{new}} = \lambda + \Delta\lambda)$$

9. IT = IT + 1, $P_{g_i} = P_{g_i}^{\text{new}}$, $\lambda = \lambda^{\text{new}}$ and GOTO Step 5 and repeat.
10. If no more violations then GOTO Step 12, else check the limits of generators and fix up as follows:

    If $P_{g_i} < P_{g_i}^{\min}$ then $P_{g_i} = P_{g_i}^{\min}$

    If $P_{g_i} > P_{g_i}^{\max}$ then $P_{g_i} = P_{g_i}^{\max}$
11. GOTO Step 4 and repeat.
12. Compute the optimal total cost and transmission loss.
13. Stop.

EXAMPLE 3.5 Determine the economic schedule to meet the demand of 150 MW using the Newton–Raphson method. Use the data of Example 3.3.

Solution Given P_D = 150 MW. Initial values are presented below as calculated in Example 3.3.

$$P_{g_1} = 83.5276 \text{ MW}; \; P_{g_2} = 66.47239 \text{ MW}; \quad \text{and} \quad \lambda = 11.81812 \text{ ₹/MWh}$$

The Hessian Matrix elements are obtained and are given below:

$$\frac{\partial^2 L}{\partial P_{g_1}^2} = 2a_1 + 2\lambda B_{11} = 0.041416 \text{ ₹/MW}^2\text{h}$$

$$\frac{\partial^2 L}{\partial P_{g_2}^2} = 2a_2 + 2\lambda B_{22} = 0.062092 \text{ ₹/MW}^2\text{h}$$

$$\frac{\partial^2 L}{\partial P_{g_1} \partial P_{g_2}} = \frac{\partial^2 L}{\partial P_{g_2} \partial P_{g_1}} = 2\lambda B_{12} = -0.004727 \text{ ₹/MW}^2\text{h}$$

$$\frac{\partial^2 L}{\partial \lambda \partial P_{g_1}} = \frac{\partial^2 L}{\partial P_{g_1} \partial \lambda} = 2B_{11} P_{g_1} + 2B_{12} P_{g_2} = -1.0 = -0.859534$$

$$\frac{\partial^2 L}{\partial \lambda \partial P_{g_2}} = \frac{\partial^2 L}{\partial P_{g_2} \partial \lambda} = 2B_{21} P_{g_1} + 2B_{22} P_{g_2} - 1.0 = -0.767522$$

$$\frac{\partial^2 L}{\partial \lambda^2} = 0$$

The Jacobian matrix elements are computed and are shown below:

$$\frac{\partial L}{\partial P_{g_1}} = 2a_1 P_{g_1} + b_1 + \lambda(2B_{11} P_{g_1} + 2B_{21} P_{g_2} - 1) = 1.660047 \text{ ₹/MWh}$$

$$\frac{\partial L}{\partial P_{g_2}} = 2a_2 P_{g_2} + b_2 + \lambda(2B_{21} P_{g_1} + 2B_{22} P_{g_2} - 1) = 2.747459 \text{ ₹/MWh}$$

$$P_L = B_{11}P_{g_1}^2 + B_{22}P_{g_2}^2 + 2B_{12}P_{g_1}P_{g_2} = 13.59311 \text{ MW}$$

$$\frac{\partial L}{\partial \lambda} = P_D + P_L - (P_{g_1} + P_{g_2}) = 13.59311 \text{ MW}$$

The above elements of Hessian and Jacobian matrices are written in matrix form as

$$\begin{bmatrix} 0.041416 & -0.004727 & -0.859534 \\ -0.004727 & 0.062092 & -0.767522 \\ -0.859534 & -0.767522 & 0 \end{bmatrix} \begin{bmatrix} \Delta P_{g_1} \\ \Delta P_{g_2} \\ \Delta \lambda \end{bmatrix} = \begin{bmatrix} -1.660047 \\ -2.747459 \\ -13.59311 \end{bmatrix}$$

Using the gauss-elimination method,

$$\begin{bmatrix} -0.859534 & -0.767522 & 0 \\ 0 & 0.066314 & -0.767522 \\ 0 & 0 & -1.342289 \end{bmatrix} \begin{bmatrix} \Delta P_{g_1} \\ \Delta P_{g_2} \\ \Delta \lambda \end{bmatrix} = \begin{bmatrix} -13.59311 \\ -2.672700 \\ -3.996098 \end{bmatrix}$$

Using back substitution,

$$\Delta\lambda = 2.977077 \text{ ₹/MWh}, \ \Delta P_{g_2} = -5.846895 \text{ MW}, \ \Delta P_{g_1} = 21.035510 \text{ MW}$$

$$P_{g_1} = 83.5276 + 21.035510 = 104.563110 \text{ MW}$$

$$P_{g_2} = 66.47239 - 5.846895 = 60.6255 \text{ MW}$$

$$\lambda = 11.81812 + 2.977077 = 14.7952 \text{ ₹/MWh}$$

Check the convergence,

$$[(\Delta P_{g_1})^2 + (\Delta P_{g_2})^2 + (\Delta\lambda)^2]^{1/2} = 22.035014$$

Compute $\Delta P = |P_{g_1} + P_{g_2} - P_L - P_D| = 0.560071 > 0.001$

The Iteration-wise schedule is given in Table 3.6.

TABLE 3.6 Schedule of generation during iterations (Example 3.5)

IT	P_{g1} (MW)	P_{g2} (MW)	λ (₹/MWh)	ΔP (MW)	P_L (MW)
2	103.2330	62.56359	14.86603	0.1043829	15.90201
3	103.3140	62.60577	14.87012	0.0578308	15.92552
4	103.3183	62.60843	14.87034	0.0031089	15.92682
5	103.3183	62.60858	14.87035	0.0001812	15.92690

Cost = 2309.771 ₹/h

3.6 ECONOMIC DISPATCH USING THE APPROXIMATE NEWTON–RAPHSON METHOD

The economic dispatch problem is solved by solving Eq. (3.22) and Eq. (3.23) iteratively. Equation (3.22) can be approximated without much loss in the accuracy of solution by neglecting the term $\sum_{\substack{j=1 \\ j\neq i}}^{NG}\left(\frac{\partial^2 L}{\partial P_{g_i}\partial P_{g_j}}\right)$. So, Eq. (3.22) can be rewritten as

$$\frac{\partial^2 L}{\partial P_{g_i}^2}\Delta P_{g_i} + \frac{\partial^2 L}{\partial P_{g_i}\partial \lambda}\Delta\lambda = -\frac{\partial L}{\partial P_{g_i}} \quad (i = 1, 2, ..., \text{NG}) \tag{3.27}$$

In view of Eq. (3.26d), Eq. (3.23) is rewritten as

$$\sum_{j=1}^{NG}\frac{\partial^2 L}{\partial \lambda\,\partial P_{g_j}}\Delta P_{g_j} = -\frac{\partial L}{\partial \lambda} \tag{3.28}$$

Substituting Eqs. (3.25a), (3.26a) and (3.26c) in Eq. (3.27),

$$\left(\frac{\partial^2 F_i}{\partial P_{g_i}^2}+\lambda\frac{\partial^2 P_L}{\partial P_{g_i}^2}\right)\Delta P_{g_i} + \left(\frac{\partial P_L}{\partial P_{g_i}} - 1\right)\Delta\lambda = -\left(\frac{\partial F_i}{\partial P_{g_i}} + \lambda\left(\frac{\partial P_L}{\partial P_{g_i}} - 1\right)\right)$$

or

$$\left(\frac{\partial^2 F_i}{\partial P_{g_i}^2}+\lambda\frac{\partial^2 P_L}{\partial P_{g_i}^2}\right)\Delta P_{g_i} = \left(1 - \frac{\partial P_L}{\partial P_{g_i}}\right)(\lambda + \Delta\lambda) - \left(\frac{\partial F_i}{\partial P_{g_i}}\right) \tag{3.29}$$

Let

$$X_i = \frac{\partial F_i}{\partial P_{g_i}} = 2a_i P_{g_i} + b_i \text{ ₹/MWh} \tag{3.30}$$

$$Y_i = \frac{\partial^2 F_i}{\partial P_{g_i}^2} + \lambda\frac{\partial^2 P_L}{\partial P_{g_i}^2} = 2a_i + 2\lambda B_{ii} \text{ ₹/MWh} \tag{3.31}$$

$$\lambda^{\text{new}} = \lambda + \Delta\lambda,\ K_i = \frac{\partial P_L}{\partial P_{g_i}} \tag{3.32}$$

From Eq. (3.29),

$$\Delta P_{g_i} = \frac{(1 - K_i)\lambda^{\text{new}} - X_i}{Y_i} \quad (i = 1, 2, ..., \text{NG}) \tag{3.33}$$

Substituting Eqs. (3.26c) and (3.25b) in Eq. (3.28),

$$\sum_{j=1}^{NG}\left(\frac{\partial P_L}{\partial P_{g_j}} - 1\right)\Delta P_{g_j} = -\left(P_D + P_L - \sum_{i=1}^{NG} P_{g_i}\right)$$

Rearranging the above equation, we get

$$\sum_{j=1}^{NG} (1 - K_j)\,\Delta P_{g_j} = P_D^* \tag{3.34}$$

where

$$P_D^* = P_D + P_L - \sum_{i=1}^{NG} P_{g_i} \tag{3.35}$$

Substituting Eq. (3.33) in Eq. (3.34), we get

$$\sum_{j=1}^{NG} (1 - K_j)\left(\frac{(1 - K_j)\lambda^{\text{new}} - X_j}{Y_j}\right) = P_D^*$$

or

$$\sum_{j=1}^{NG} \frac{(1 - K_j)^2}{Y_j}\,\lambda^{\text{new}} - \sum_{j=1}^{NG} \frac{(1 - K_j)X_j}{Y_j} = P_D^*$$

or

$$\lambda^{\text{new}} = \frac{P_D^* + \sum_{j=1}^{NG} \frac{(1 - K_j)X_j}{Y_j}}{\sum_{j=1}^{NG} \frac{(1 - K_j)^2}{Y_j}} \tag{3.36}$$

By iterating Eq. (3.36) and Eq. (3.33), the solution can be obtained. The stepwise procedure is given in Algorithem 3.4.

Algorithm 3.4: Economic Dispatch (Approximate Newton–Raphson Method)

1. Read data, namely a_i, b_i, c_i (cost coefficients); B_{ij}, B_{i0}, B_{00} (*B*-coefficients); (i = 1, 2, ..., NG; j = 1, 2, ..., NG); ε (convergence tolerance); and ITMAX (maximum allowed iterations), etc.
2. Compute the initial values of P_{g_i}(i = 1, 2, ..., NG) and λ by assuming $P_L = 0$. The values of λ and P_{g_i} (i = 1, 2, ..., NG) can be obtained directly using Eqs. (3.10) and (3.9), respectively.
3. Set iteration counter, IT = 1.
4. Compute λ^{new} using Eq. (3.36).
5. Find ΔP_{g_i} (i = 1, 2, ..., NG) using Eq. (3.33) and $P_{g_i}^{\text{new}} = P_{g_i} + \Delta P_{g_i}$ (i = 1, 2, ..., NG).
6. Compute P_L using Eq. (3.14) or Eq. (3.13).
7. Compute $\Delta P = P_D + P_L - \sum_{i=1}^{NG} P_{g_i}$.
8. Check $|\Delta P| \le \varepsilon$, if 'yes' GOTO Step 10,
   or $|\lambda^{\text{new}} - \lambda| \le \varepsilon$, if 'yes' GOTO Step 10,
   or IT > ITMAX , if 'yes' GOTO Step 10 (premature end of procedure).

9. Modify $P_{gi} = P_{gi}^{new}$ (i = 1, 2, ..., NG) and $\lambda = \lambda^{new}$

   IT = IT + 1, GOTO Step 4 and repeat.

10. Compute optimal total cost and transmission loss.

11. Stop.

The limits of generations can be handled by the procedure outlined in Section 3.4.

EXAMPLE 3.6 Determine the economic schedule to meet the demand of 150 MW using the approximate Newton–Raphson method. Use the data of Example 3.3.

Solution Find the initial values (as calculated in Example 3.3),

$$\lambda = 11.81812 \text{ ₹/MWh}; \qquad P_{g1} = 83.5276 \text{ MW}; \qquad P_{g2} = 66.47239 \text{ MW}$$

Calculate incremental losses, using Eq. (3.32),

$$K_1 = 2B_{11}P_{g1} + 2B_{12}P_{g2} = 0.1404663$$

$$K_2 = 2B_{21}P_{g1} + 2B_{22}P_{g2} = 0.23324785$$

Calculate transmission losses as

$$P_L = B_{11}P_{g1}^2 + B_{22}P_{g2}^2 + 2B_{12}P_{g1}P_{g2} = 13.59311 \text{ MW}$$

$$P_D^* = P_D + P_L - (P_{g1} + P_{g2}) = 13.59311 \text{ MW}$$

Find constants as defined in Eqs. (3.30) and (3.31), i.e.

$$X_1 = 2a_1P_{g1} + b_1 = 11.81812 \text{ ₹/MWh}$$

$$X_2 = 2a_2P_{g2} + b_2 = 11.81812 \text{ ₹/MWh}$$

$$Y_1 = 2(a_1 + \lambda B_{11}) = 0.041416 \text{ ₹/MWh}$$

$$Y_2 = 2(a_2 + \lambda B_{22}) = 0.06209 \text{ ₹/MWh}$$

Compute the new value of λ, using Eq. (3.36),

$$\lambda^{new} = \frac{P_D^* + \sum_{j=1}^{2} \dfrac{(1 - K_j)X_j}{Y_j}}{\sum_{j=1}^{2} \dfrac{(1 - K_j)^2}{Y_j}} = 14.81919 \text{ ₹/MWh}$$

Find the change in generations using Eq. (3.33),

$$\Delta P_{g1} = \frac{(1 - K_1)\lambda^{new} - X_1}{Y_1} = 22.20076 \text{ MW}; \quad \Delta P_{g2} = \frac{(1 - K_2)\lambda^{new} - X_2}{Y_2} = -7.151834 \text{ MW}$$

Check the convergence criteria

$$|\lambda^{\text{new}} - \lambda| = |14.81919 - 11.81812| = 3.00106 > 0.001$$

The required convergence is not met, so modify the generation values and update λ.

$$P_{g1} = P_{g1} + \Delta P_{g1} = 83.5276 + 22.20076 = 105.7284 \text{ MW}$$

$$P_{g2} = P_{g2} + \Delta P_{g2} = 66.47239 - 7.151834 = 59.32056 \text{ MW}$$

The above procedure is repeated till convergence is obtained. The generation schedule λ, load mismatch and advancement in λ during each iteration are given in Table 3.7.

TABLE 3.7 Generation schedule during each iteration (Example 3.6)

IT	P_{g1} (MW)	P_{g2} (MW)	λ (₹/MWh)	ΔP (MW)	$\|\lambda^{\text{new}} - \lambda\|$ (₹/MWh)
1	105.7284	59.32056	14.81919	0.6586790	3.001060
2	103.0107	62.88092	14.87678	0.0366457	0.057589
3	103.3456	62.57994	14.86997	0.0003290	0.006812
4	103.3156	62.61134	14.87035	0.0000038	0.000442
5	103.3186	62.60832	14.87041	0.0000172	0.000058

Cost = 2309.771 ₹/h, P_L = 15.92689 MW

3.7 ECONOMIC DISPATCH USING EFFICIENT METHOD

For the economic dispatch problem, the necessary conditions for optimality are given by Eq. (3.16) and Eq. (3.17) as

$$\frac{\partial F_i}{\partial P_{g_i}} = \lambda\left(1 - \frac{\partial P_L}{\partial P_{g_i}}\right) \qquad (i = 1, 2, ..., \text{NG}) \tag{3.37}$$

$$\sum_{i=1}^{\text{NG}} P_{g_i} = P_L + P_D \tag{3.38}$$

The initial values $P_{g_i}^0$ and λ^0 are calculated by assuming that the transmission losses are absent (i.e. $P_L = 0$) and using Eq. (3.10) and Eq. (3.9).

The nonlinear equations in P_{g_i} and λ can be solved iteratively. Let $P_{g_i}^0 (i = 1, 2, ..., \text{NG})$ and λ^0 be approximated solutions to Eq. (3.37). To find the new approximation, let

$$\lambda^{\text{new}} = \lambda^0 + \Delta\lambda$$

$$P_{g_i}^{\text{new}} = P_{g_i}^0 + \Delta P_{g_i} \qquad (i = 1, 2, ..., \text{NG})$$

So, here the aim is to obtain

$$\frac{\partial F_i^{\text{new}}}{\partial P_{g_i}} = \lambda^{\text{new}}\left(1 - \frac{\partial P_L^{\text{new}}}{\partial P_{g_i}}\right) \qquad (i = 1, 2, ..., \text{NG}) \tag{3.39}$$

and

$$\sum_{i=1}^{NG} P_{gi}^{new} = P_L^{new} + P_D \tag{3.40}$$

Taylor's expression to first order of P_L^{new} is

$$P_L^{new} = P_L^0 + \sum_{i=1}^{NG} \frac{\partial P_L^0}{\partial P_{gi}} \Delta P_{gi} \tag{3.41}$$

where P_L^0 is the initial transmission losses.

The modified value of the incremental transmission loss from the initial value can be obtained as

$$\frac{\partial P_L^{new}}{\partial P_{gi}} = \frac{\partial P_L^0}{\partial P_{gi}} + \sum_{j=1}^{NG} \frac{\partial^2 P_L^0}{\partial P_{gi} \partial P_{gj}} \Delta P_{gj}$$

To retain the classical form and at the same time to improve the convergence, it is possible to include only the *i*th item of the summation of the above equation, i.e.

$$\frac{\partial P_L^{new}}{\partial P_{gi}} = \frac{\partial P_L^0}{\partial P_{gi}} + \frac{\partial^2 P_L^0}{\partial P_{gi}^2} (P_{gi}^{new} - P_{gi}^0) \tag{3.42}$$

Similarly, the modified incremental cost from the initial value can be obtained as

$$\frac{\partial F_i^{new}}{\partial P_{gi}} = \frac{\partial F_i^0}{\partial P_{gi}} + \frac{\partial^2 F_i^0}{\partial P_{gi}^2} (P_{gi}^{new} - P_{gi}^0) \tag{3.43}$$

Substituting Eq. (3.42) and Eq. (3.43) into Eq. (3.39),

$$\frac{\partial F_i^0}{\partial P_{gi}} + \frac{\partial^2 F_i^0}{\partial P_{gi}^2} (P_{gi}^{new} - P_{gi}^0) = \lambda^{new} \left[1 - \frac{\partial P_L^0}{\partial P_{gi}} - \frac{\partial^2 P_L^0}{\partial P_{gi}^2} (P_{gi}^{new} - P_{gi}^0) \right]$$

The above equation can be rewritten in terms of P_{gi}^{new} and λ^{new} as

$$\frac{\partial^2 F_i^0}{\partial P_{gi}^2} P_{gi}^{new} + \left(\frac{\partial F_i^0}{\partial P_{gi}} - \frac{\partial^2 F_i^0}{\partial P_{gi}^2} P_{gi}^0 \right) = \lambda^{new} \left(1 - \frac{\partial P_L^0}{\partial P_{gi}} \right) - (\lambda^0 + \Delta\lambda) \frac{\partial^2 P_L^0}{\partial P_{gi}^2} (P_{gi}^{new} - P_{gi}^0)$$

Finally, the following is obtained by rearranging the above equation,

$$\left(\frac{\partial^2 F_i^0}{\partial P_{gi}^2} + \lambda^0 \frac{\partial^2 P_L^0}{\partial P_{gi}^2} \right) P_{gi}^{new} + \left(\frac{\partial F_i^0}{\partial P_{gi}} - \frac{\partial^2 F_i^0}{\partial P_{gi}^2} P_{gi}^0 - \lambda^0 \frac{\partial^2 P_L^0}{\partial P_{gi}^2} P_{gi}^0 \right) = \lambda^{new} \left(1 - \frac{\partial P_L^0}{\partial P_{gi}} \right) \tag{3.44}$$

Here the term $\Delta\lambda \dfrac{\partial^2 P_L^0}{\partial P_{gi}^2} \Delta P_g$ is ignored, being small.

Substituting Eq. (3.41) into Eq. (3.40),

$$\sum_{i=1}^{NG} P_{g_i}^{\text{new}} = P_L^0 + \sum_{i=1}^{NG} \frac{\partial P_L^0}{\partial P_{g_i}} (P_{g_i}^{\text{new}} - P_{g_i}^0) + P_D$$

By separating the $P_{g_i}^{\text{new}}$ terms in the above equation,

$$\sum_{i=1}^{NG} \left(1 - \frac{\partial P_L^0}{\partial P_{g_i}}\right) P_{g_i}^{\text{new}} = P_D + P_L^0 - \sum_{i=1}^{NG} \frac{\partial P_L^0}{\partial P_{g_i}} P_{g_i}^0 \tag{3.45}$$

Let

$$P_{g_i}^* = \left(1 - \frac{\partial P_L^0}{\partial P_{g_i}}\right) P_{g_i}^{\text{new}}$$

Substituting the value of $P_{g_i}^*$ into Eq. (3.45),

$$\sum_{i=1}^{NG} P_{g_i}^* = P_D^* \tag{3.46}$$

where $P_D^* = P_D + P_L^0 - \sum_{i=1}^{NG} \frac{\partial P_L^0}{\partial P_{g_i}} P_{g_i}^0$

Substituting the value of $P_{g_i}^*$ into Eq. (3.44),

$$\frac{\left(\frac{\partial^2 F_i^0}{\partial P_{g_i}^2} + \lambda^0 \frac{\partial^2 P_L^0}{\partial P_{g_i}^2}\right)}{\left(1 - \frac{\partial P_L^0}{\partial P_{g_i}}\right)^2} P_{g_i}^* + \frac{\left(\frac{\partial F_i^0}{\partial P_{g_i}} - \frac{\partial^2 F_i^0}{\partial P_{g_i}^2} P_{g_i}^0 - \lambda^0 \frac{\partial^2 P_L^0}{\partial P_{g_i}^2} P_{g_i}^0\right)}{\left(1 - \frac{\partial P_L^0}{\partial P_{g_i}}\right)} = \lambda^{\text{new}} \tag{3.47}$$

The above Eqs. (3.47) and (3.46) can be rewritten in the simplified form of equal incremental cost, when the transmission loss is not considered, as

$$2a_i^* P_{g_i}^* + b_i^* = \lambda^{\text{new}} \qquad (i = 1, 2, ..., NG) \tag{3.48a}$$

and

$$\sum_{i=1}^{NG} P_{g_i}^* = P_D^* \tag{3.48b}$$

where

$$a_i^* = \frac{1}{2} \frac{\frac{\partial^2 F_i^0}{\partial P_{g_i}^2} + \lambda^0 \frac{\partial^2 P_L^0}{\partial P_{g_i}^2}}{\left(1 - \frac{\partial P_L^0}{\partial P_{g_i}}\right)^2} \tag{3.48c}$$

$$b_i^* = \frac{\left(\dfrac{\partial F_i^0}{\partial P_{gi}} - \dfrac{\partial^2 F_i^0}{\partial P_{gi}^2} P_{gi}^0 - \lambda^0 \dfrac{\partial^2 P_L^0}{\partial P_{gi}^2} P_{gi}^0\right)}{\left(1 - \dfrac{\partial P_L^0}{\partial P_{gi}}\right)} \tag{3.48d}$$

$$P_{gi}^* = \left(1 - \frac{\partial P_L^0}{\partial P_{gi}}\right) P_{gi}^{\text{new}} \tag{3.48e}$$

$$P_D^* = P_D + P_L^0 - \sum_{i=1}^{NG} \frac{\partial P_L^0}{\partial P_{gi}} P_{gi}^0 \tag{3.48f}$$

In view of the defined problem in Eq. (3.12), the problem in Eq. (3.48) is redefined as

$$2a_i^* P_{gi}^* + b_i^* = \lambda^{\text{new}} \qquad (i = 1, 2, ..., \text{NG}) \tag{3.49a}$$

and

$$\sum_{i=1}^{NG} P_{gi}^* = P_D^* \tag{3.49b}$$

where

$$K_i^0 = \frac{\partial P_L^0}{\partial P_{gi}} = \sum_{j=1}^{NG} 2 B_{ij} P_{gj}^0 \tag{3.49c}$$

$$a_i^* = \frac{a_i + \lambda^0 B_{ii}}{(1 - K_i^0)^2} \tag{3.49d}$$

$$b_i^* = \frac{b_i - \lambda^0 B_{ii} P_{gi}^0}{1 - K_i^0} \tag{3.49e}$$

$$P_{gi}^* = (1 - K_i^0) P_{gi}^{\text{new}} \tag{3.49f}$$

$$P_L^0 = \sum_{i=1}^{NG} \sum_{j=1}^{NG} P_{gi}^0 B_{ij} P_{gj}^0 \tag{3.49g}$$

$$P_D^* = P_D + P_L^0 - \sum_{i=1}^{NG} K_i^0 P_{gi}^0 \tag{3.49h}$$

The problem in Eq. (3.49) can be solved using Eqs. (3.9) and (3.10) as explained in Section 3.3. The stepwise procedure to implement the efficient method is outlined below (Algorithm 3.5.).

Algorithm 3.5: Economic Dispatch Using Efficient Method

1. Read data, namely a_i, b_i, c_i (cost coefficients); B_{ij}(B-coefficients) (i = 1, 2, ..., NG; j = 1, 2, ..., NG); ε (convergence tolerance); and ITMAX (maximum allowable iterations), etc.

2. Compute the initial values of $P_{g_i}(i = 1, 2, ..., \text{NG})$ and λ by assuming that $P_L = 0$. The values of λ and $P_{g_i}(i = 1, 2, ..., \text{NG})$ can be obtained directly using Eqs. (3.10) and (3.9), respectively.
3. Set iteration counter IT = 1.
4. Compute K_i^0, a_i^*, b_i^*, P_L^0, and P_D^* using Eqs. (3.49c), (3.49d), (3.49e), (3.49g), and (3.49h), respectively.
5. Compute λ^{new}, $P_{g_i}^*$, and $P_{g_i}^{\text{new}}$ using Eqs. (3.10), (3.49a), and (3.49f), respectively.
6. Check $|\lambda^{\text{new}} - \lambda^0| \le \varepsilon$, if 'yes' then GOTO Step 8.
7. Modify $P_{g_i}^0 = P_{g_i}^{\text{new}}$ $(i = 1, 2, ..., \text{NG})$ and $\lambda^0 = \lambda^{\text{new}}$

   IT = IT + 1 GOTO Step 4 and repeat.
8. Compute the optimal total cost, and the transmission loss.
9. Stop.

The limits of generations can be added by the procedure outlined in Section 3.4.

EXAMPLE 3.7 Determine the economic schedule to meet the demand of 150 MW using the Efficient method. Use the data of Example 3.3.

Solution Initial values as calculated in Example 3.3 are:

$$\lambda^0 = 11.81812 \text{ ₹/MWh}, \qquad P_{g_1}^0 = 83.5276 \text{ MW}, \qquad P_{g_2}^0 = 66.47239 \text{ MW}$$

Find the incremental transmission loss with initial values of generation, i.e.

$$K_1^0 = 2B_{11}P_{g_1}^0 + 2B_{12}P_{g_2}^0 = 0.1404663$$

$$K_2^0 = 2B_{21}P_{g_1}^0 + 2B_{22}P_{g_2}^0 = 0.23324785$$

Calculate transmission losses as

$$P_L^0 = B_{11}(P_{g_1}^0)^2 + B_{22}(P_{g_2}^0)^2 + 2B_{12}P_{g_1}^0 P_{g_2}^0 = 13.59311 \text{ MW}$$

Find the constants defined in Eqs. (3.49), i.e.

$$a_1^* = \frac{a_1 + \lambda^0 B_{11}}{(1 - K_1^0)^2} = 0.028029 \text{ ₹/MW}^2\text{h}$$

$$a_2^* = \frac{a_2 + \lambda^0 B_{22}}{(1 - K_2^0)^2} = 0.232479 \text{ ₹/MW}^2\text{h}$$

$$b_1^* = \frac{b_1 - \lambda^0 B_{11} P_{g_1}^0}{1 - K_1^0} = 9.724716 \text{ ₹/MWh}$$

$$b_2^* = \frac{b_2 - \lambda^0 B_{22} P_{g2}^0}{1 - K_2^0} = 10.02016 \text{ ₹/MWh}$$

$$P_D^* = P_D + P_L^0 - \sum_{i=1}^{2} K_i^0 P_{g_i}^0 = 136.40690 \text{ MW}$$

Find the new value of λ using Eq. (3.10), i.e.

$$\lambda^{\text{new}} = \frac{P_D^* + \sum_{i=1}^{2} \frac{b_i^*}{2a_i^*}}{\sum_{i=1}^{2} \frac{1}{2a_i^*}} = 14.81919 \text{ ₹/MWh}$$

Calculate new generations as

$$P_{g_1}^* = \frac{\lambda^{\text{new}} - b_1^*}{2a_1^*} = 90.84408 \text{ MW}; \qquad P_{g_2}^* = \frac{\lambda^{\text{new}} - b_2^*}{2a_2^*} = 45.5298 \text{ MW}$$

The required new modified generations are

$$P_{g1}^{\text{new}} = \frac{P_{g1}^*}{1 - K_1^0} = 105.7283 \text{ MW}; \qquad P_{g2}^{\text{new}} = \frac{P_{g2}^*}{1 - K_2^0} = 59.32055 \text{ MW}$$

Check $|\lambda^{\text{new}} - \lambda^0| = 3.00107 > 0.001$.

For iteration IT = 2 ·

$$P_{g1}^0 = 105.7283 \text{ MW}, \; P_{g2}^0 = 59.32055 \text{ MW}, \; \lambda^0 = 14.81919 \text{ ₹/MWh}$$

The above procedure is repeated till convergence is obtained. The generation schedule, λ, transmission losses, load mismatch and advancement in λ during each iteration are given in Table 3.8.

TABLE 3.8 Generation schedule during each iteration (Example 3.7)

IT	P_{g1} (MW)	P_{g2} (MW)	λ (₹/MWh)	P_L (MW)	ΔP (MW)	$\lvert\lambda^{\text{new}} - \lambda\rvert$ (₹/MWh)
2	103.0107	62.88093	14.87678	15.92827	0.0366220	0.0575914
3	103.3456	62.57993	14.86997	15.92586	0.0034237	0.0068140
4	103.3157	62.61134	14.87041	15.92699	0.0000038	0.0004434
5	103.3186	62.60833	14.87035	15.92699	0.0000095	0.0000582

Cost = 2309.77 ₹/h

3.8 CLASSICAL METHOD TO CALCULATE LOSS COEFFICIENTS

The simple method to calculate the penalty factors and total transmission losses for generator economic allocation is the '*B*-coefficient' or the 'loss-coefficients' method. This analysis was initiated by Kron [1951, 1952], and popularized by Kirchamayer [1958]: it has had numerous extensions by Happ [1963, 1963, 1964, 1964], and has been formulated using efficient sparse matrix computation technique by Meyer [1971]. Physically, these constants represent the parameters of a fictitious network through which the power flows from the generating buses to a hypothetical load. These coefficients or constants are in fact not constants but depend on the loading conditions as well as the configuration of the power system.

In classical approach the B_{ij}, B_{i0} and B_{00} coefficients can be calculated directly as discussed here. The voltage magnitude and phase angles at each bus of the network power system are obtained by performing the load flow. The variable load or generator currents are separated from the fixed network. All shunt elements, namely, line charging capacitances from bus to ground and transformer neutrals, etc. are connected to dummy bus L as shown in Figure 3.4.

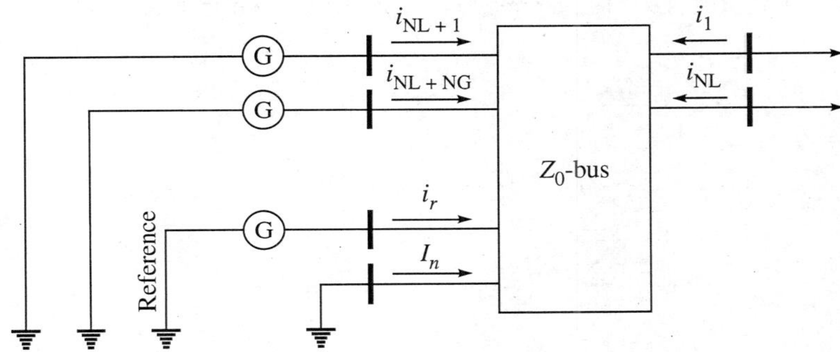

Figure 3.4 Power system with separated shunt elements from load and generators.

Select the reference bus and shunt elements to bus L to be part of the network. Form Z_0-bus using the Z-bus algorithm and add zero row and column in order to obtain (NL+NG+1) × (NL+NG+1) matrix, where NL is the number of loads, NG is the number of generators and NB is the number of buses.

$$E_0 = Z_0 I_0 \tag{3.50}$$

where I_0 is a vector of load and generator currents and

$$Z_0 = \begin{bmatrix} Z & Z_{Li}^T \\ Z_{Li} & Z_{LL} \end{bmatrix}$$

The voltage vector transforms into the new reference Frame 1 (Figure 3.5) by the conjugate transpose of T_1, i.e.

$$E_1 = (T_1^*)^T E_0 = ((T_1^*)^T Z_0 T_1) I_1 = Z_1 I_1 \tag{3.51}$$

where

$$Z_1 = (T_1^*)^T Z_0 T_1 \tag{3.52}$$

$$T_1 = \begin{vmatrix} I & 0 \\ -\lambda & 1 \end{vmatrix}$$

with

$$\lambda = (\lambda_1, \lambda_2, \ldots, \lambda_{NB-1})^T$$

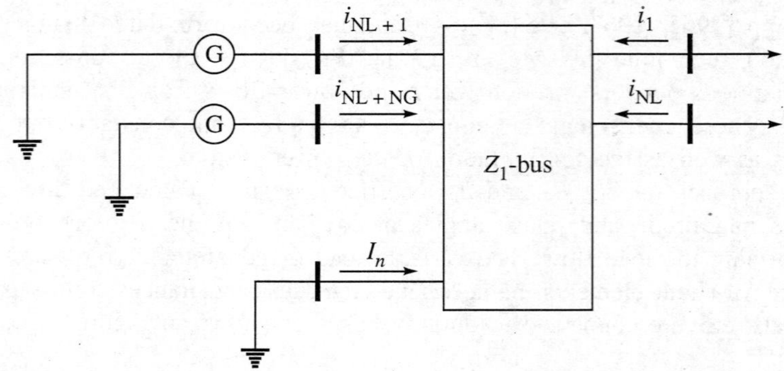

FIGURE 3.5 Transformation of power system to Frame 1.

$$\lambda_k = \frac{Z_{Lk}}{Z_{LL}} \qquad (k = 1, 2, ..., \text{NB} - 1) \tag{3.53}$$

$$t_k = 1 - \lambda_k^* \qquad (k = 1, 2, ..., \text{NB} - 1) \tag{3.54}$$

$$I_L = -\sum_{k=1}^{\text{NB}-1} \lambda_k I_k + I_n \tag{3.55}$$

Now,

$$I_n = -\frac{e_r}{Z_{LL}} \tag{3.56}$$

where I_n is the neutral current.

t_k is complex calculated off nominal turns ratio between bus k and the reference bus. Z_1 is calculated with the following rule [Kirchamayer, 1958]:

$$Z_1(n, k) = Z_0(n, k) - \lambda_k Z_0(n, L); \qquad (n = 1, 2, ...,\text{NB}; k = 1, 2, ..., \text{NB}) \tag{3.56a}$$

$$Z_1(n, L) = Z_0(n, L) - \lambda_k^* Z_0(L, L); \qquad (n = 1, 2, ..., \text{NB}; L = \text{NB} + 1) \tag{3.56b}$$

$$Z_1(L, n) = Z_1(n, L); \qquad (n = 1, 2, ..., \text{NB}; L = \text{NB} + 1) \tag{3.56c}$$

Each of the NL load currents is assumed to vary as a constant complex fraction of the total load.

$$I_k = l_k I_T \qquad (k = 1, 2, ..., \text{NL}) \tag{3.57}$$

$$I_T = \sum_{i=1}^{\text{NL}} I_k \tag{3.58}$$

where

l_k is the complex fraction

I_T is the total load current.

The voltages are transformed according to

$$E_2 = (T_2^*)^T E_1 \tag{3.59}$$

which maintain the generator buses at $e_k - t_k e_r$ and defines an equivalent bus

$$e_l - t_l e_r = \sum_{k=1}^{\text{NL}} l_k^* (e_k - t_k e_r) \tag{3.60}$$

Transmission loss is represented as

$$P_L + jQ_L = (I_1^*)^T E_1 = [(I_2^* T_2^*)^T Z_1 (T_2 I_2)] = (I_2^*)^T Z_2 I_2 \tag{3.61}$$

where

$$Z_2 = (T_2^*)^T Z_1 T_2 \quad \text{and} \quad I_1 = T_2 I_2$$

$$T_2 = \begin{bmatrix} l & 0 \\ 0 & I \end{bmatrix}$$

$$l = [l_1, l_2, \ldots, l_{\text{NL}}]^T$$

To implement on computer, Z_2 can be obtained as (Figure 3.6)

$$Z_2 = \begin{vmatrix} W & a \\ b & Z_1 \end{vmatrix} \tag{3.62}$$

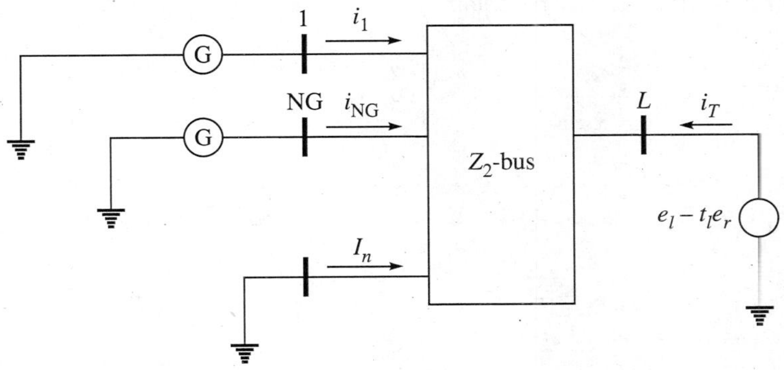

FIGURE 3.6 Transformation of power system to Frame 2.

where

$$W = \sum_{i=1}^{\text{NL}} \sum_{j=1}^{\text{NL}} l_i^* \, Z_1(i, j) \, l_j$$

$$a_j = \sum_{k=1}^{\text{NL}} l_k^* \, Z_1(k, j) \qquad (j = \text{NL} + 1, 2, \ldots, \text{NB} + 1)$$

$$b_i = \sum_{k=1}^{NL} Z_1(i, k)\, l_k \qquad (i = \text{NL} + 1, .., \text{NB} + 1)$$

$Z_2(m, n) = Z_1(i, j) \quad (i = \text{NL} + 1, ..., \text{NB};\ j = \text{NL} + 1, ..., \text{NB} + 1;\ m = i - \text{NL} + 1;\ n = j - \text{NL} + 1)$

The load bus is eliminated which is a further dimension reducing transformation to Frame 3 (Figure 3.7)

$$P_L + jQ_L = (I_2^*)^T E_2 = [(I_3^* T_3^*)^T Z_2\ (T_3 I_3)] = (I_3^*)^T Z_3 I_3 \tag{3.63}$$

where

$$Z_3 = (T_3^*)^T Z_2 T_3 \qquad \text{and} \qquad I_2 = T_3 I_3, \quad E_2 = Z_2\, I_2$$

$$T_3 = \begin{bmatrix} t \\ I \end{bmatrix} \tag{3.64}$$

with

$$t = [t_{\text{NL+1}}, 2, ..., t_{\text{NL+NG+1}}]$$

$$t_k = -\frac{t^*_{\text{NL}+k}}{t^*_L}, \qquad t_{\text{NL+NG+1}} = -\frac{1}{t^*_L} \qquad (k = 1, 2, ..., \text{NG}) \tag{3.65}$$

$$t_L = \sum_{k=1}^{NL} l_k^* t_k$$

To implement on computer, Z_3 can be obtained as (Figure 3.7)

$$Z_3 = [-t \qquad I]\begin{bmatrix} W & a \\ b & Z_1 \end{bmatrix}\begin{bmatrix} -t \\ I \end{bmatrix} \tag{3.66}$$

where

$$Z_3(n, m) = T_n W T_m^* - b_n T_m^* - T_n a_m + Z_1(n, m) \quad (n = 1, 2, ..., \text{NG} + 1;\ m = 1, 2, ..., \text{NG} + 1) \tag{3.67}$$

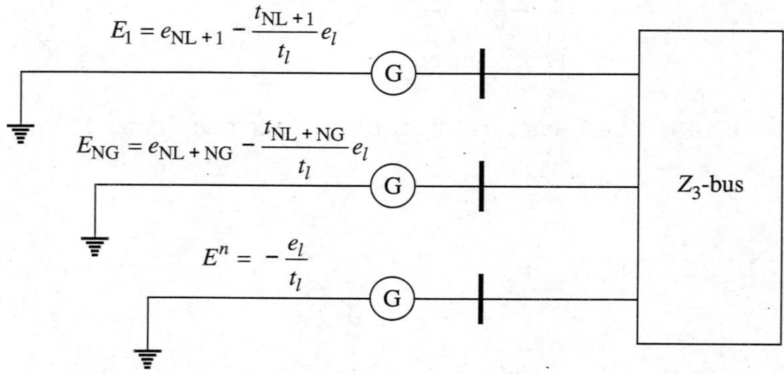

FIGURE 3.7 Transformation of power system to Frame 3.

It is assumed that the real and reactive powers at each generator bus are linearly related as shown in Figure 3.8.

$$e_k^* i_k = P_k - jQ_k \quad (k = \text{NL} + 1, ..., \text{NB}) \tag{3.68}$$

$$i_k = \frac{P_k - jQ_k}{e_k^*} = \alpha_m P_k + \beta_m \quad (m = k - \text{NL}) \tag{3.69}$$

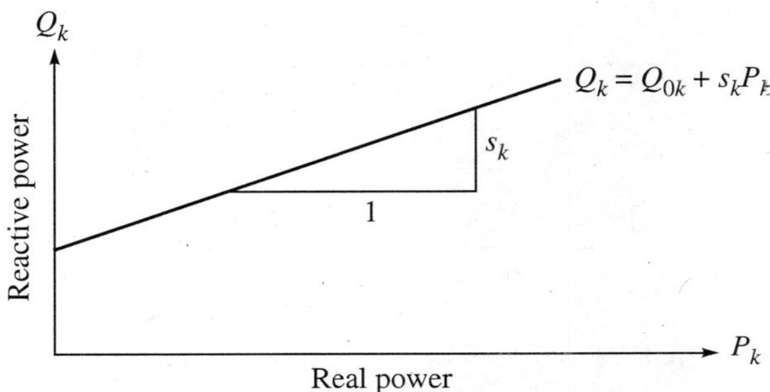

FIGURE 3.8 Linearly related real and reactive powers.

The real power transmission loss is expressed as

$$P_L = \text{Re } [(I_3^*)^T Z_3 I_3] \tag{3.70}$$

where $\quad I_3 = [\alpha_1 P_1 + \beta_1, ..., \alpha_{\text{NG}} P_{\text{NG}} + \beta_{\text{NG}}]^T$

Equating to the loss formula given by Eq. (3.14)

$$B_{ii} = \alpha_i \alpha_i^* \text{ Re } [Z_3 (i, j)] \tag{3.71}$$

$$B_{ij} = \text{Re}\left[\frac{1}{2}[\alpha_i^* \alpha_j Z_3(i, j) + \alpha_i \alpha_j^* Z_3(i, j)]\right] \quad (i = 1, 2, ..., \text{NG}; \; j = 1, 2, ..., \text{NG}) \tag{3.72}$$

$$B_{i0} = \text{Re}[\alpha_i^* I_n Z_3(i, \text{NG} + 1) + \alpha_i I_n^* Z_3(\text{NG} + 1, i)] \tag{3.73}$$

$$B_{00} = I_n I_n^* \text{ Re}[Z_3(\text{NG} + 1, \text{NG} + 1)] \tag{3.74}$$

Algorithm 3.6: Evaluation of *B*-coefficients Using Classical Method

1. Perform load flow to find V_i and $P_i + jQ_i$ at each bus.
2. Build Z_0-bus using the *Z*-bus algorithm by fixing all charging to neutral common point, *L*. Then add zero row and column for the reference bus.
3. Compute λ_k and t_k using Eqs. (3.53) and (3.54), respectively.
   Compute neutral current I_n and I_L current using Eqs. (3.56) and (3.55), respectively.
4. Build Z_1 bus using Eqs. (3.56a)–(3.56c).
5. Compute I_T using Eq. (3.58) and then l_k (k = 1, 2, ..., NL) from Eq. (3.57)).
6. Build Z_2-bus using Eq. (3.62).
7. Compute t_k using Eq. (3.65).
8. Build Z_3-bus following Eq. (3.67).

9. Compute $\alpha_k (k = 1, 2, ..., NG)$ using Eq. (3.69).
10. Build *B*-coefficients following Eqs. (3.71)–(3.74).
11. Stop.

EXAMPLE 3.8 Use the classical method to determine the *B*-coefficients for a 5-bus system shown in Figure 3.9. Bus 5 is taken as the slack bus. The series impedance and line charging of each line is taken from Table 3.9.

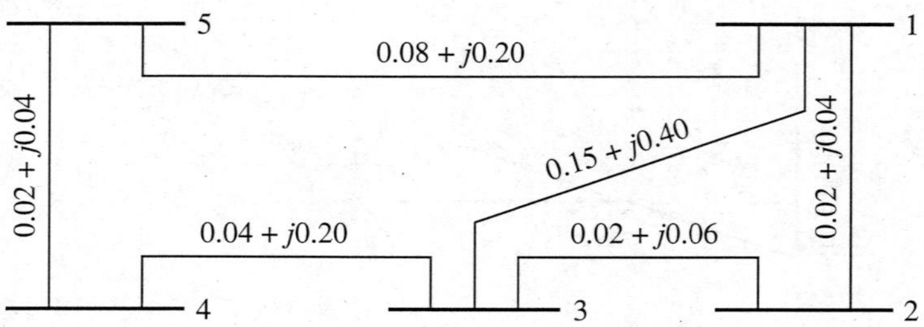

FIGURE 3.9 Power system (Example 3.8).

TABLE 3.9 Power system network data (Example 3.8)

Line	*Impedance* Z	*Line charging* $\omega C/2$
1–2	0.02 + *j*0.04	*j*0.010
1–3	0.15 + *j*0.40	*j*0.025
1–5	0.08 + *j*0.20	*j*0.020
2–3	0.02 + *j*0.06	*j*0.010
3–4	0.04 + *j*0.20	*j*0.020
4–5	0.02 + *j*0.04	*j*0.020

Solution The base case power flow (p.u.100 MVA) is achieved for 1.0×10^{-4} convergence in 10 iterations using the decoupled load flow method. The voltage and power at each bus are given in Table 3.10.

Number of load buses, NL = 3
number of buses, NB = 5
number of generator buses, NG = 2
Reference bus = 5

TABLE 3.10 System data after load flow calculations (Example 3.8)

Bus	V_i	$P_i + jQ_i$
1	0.865666 – *j*0.125030	– 0.500 – *j*0.25000
2	0.858383 – *j*0.133531	– 0.400 – *j*0.15000
3	0.871962 – *j*0.125507	– 0.450 – *j*0.20000
4	0.984324 – *j*0.008090	0.350 + *j*0.15000
5	1.0 + *j*0.0	1.089 + *j*0.55608

The total line charging admittance to ground at each bus is given in Table 3.11.

TABLE 3.11 Line charging admittance to ground (Example 3.8)

Bus	*Admittance,* Y_p	*Impedance,* Z_p
1	$j0.055$	$-j18.1818$
2	$j0.020$	$-j50.0$
3	$j0.055$	$-j18.1818$
4	$j0.04$	$-j25.0$
5	$j0.04$	$-j25.0$

The bus impedance algorithm is applied to obtain the Z_0 for the power system shown in Figure 3.10. All line chargings are fixed to neutral and treated as bus L (common point). The bus 5 is a reference bus. The data to build the Z_0-bus is given in Table 3.12.

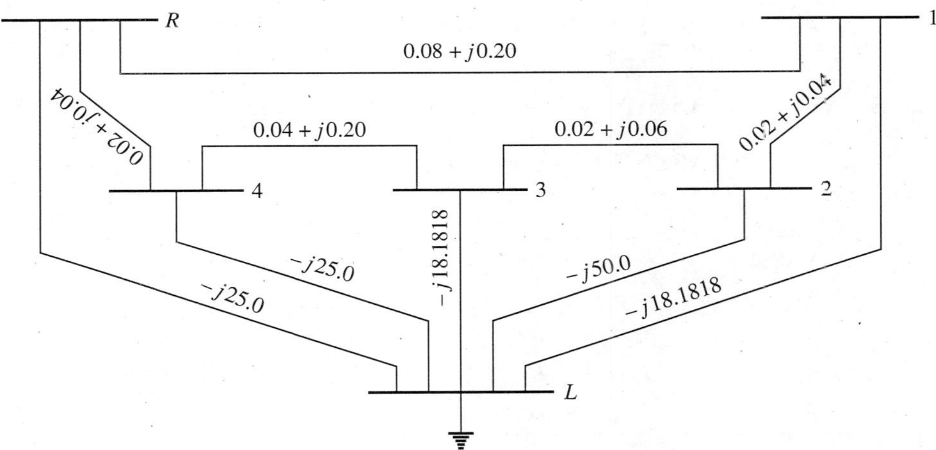

FIGURE 3.10 Power system configuration for Z_0-bus algorithm (Example 3.8).

$$
Z_0 = \begin{bmatrix}
0.044087+j0.123897 & 0.037080+j0.111712 & 0.029675+j0.093206 & 0.008838+j0.015219 & 0.0+j0.0 & 0.024534+j0.070400 \\
0.037080+j0.111712 & 0.046785+j0.134466 & 0.035772+j0.108198 & 0.010502+j0.017624 & 0.0+j0.0 & 0.025537+j0.073760 \\
0.029675+j0.093206 & 0.035772+j0.108198 & 0.041282+j0.130368 & 0.012351+j0.021325 & 0.0+j0.0 & 0.024344+j0.072924 \\
0.008838+j0.015219 & 0.010502+j0.017624 & 0.012351+j0.021325 & 0.017940+j0.037025 & 0.0+j0.0 & 0.009967+j0.018302 \\
0.0+j0.0 & 0.0+j0.0 & 0.0+j0.0 & 0.0+j0.0 & 0.0+j0.0 & 0.0+j0.0 \\
0.024534+j0.070400 & 0.025537+j0.073760 & 0.024344+j0.072924 & 0.009967+j0.018302 & 0.0+j0.0 & 0.017132-j4.713606
\end{bmatrix}
$$

TABLE 3.12 Data for building Z_0-bus (Example 3.8)

Line	*Sending bus*	*Ending bus*	Z_{ser} (p.u.)	*Type of link from bus to bus*
1	*R*	1	$0.08 + j0.20$	1 (new to reference)
2	1	2	$0.02 + j0.04$	2 (new to old)
3	2	3	$0.02 + j0.06$	2 (new to old)
4	3	4	$0.04 + j0.20$	2 (new to old)
5	4	*R*	$0.02 + j0.04$	3 (old to reference)
6	3	1	$0.15 + j0.40$	4 (old to old)
7	1	*L*	$-j18.1818$	2 (new to old)
8	*L*	2	$-j50.0$	4 (old to old)
9	*L*	3	$-j18.1818$	4 (old to old)
10	*L*	4	$-j25.0$	4 (old to old)
11	*L*	*R*	$-j25.0$	3 (old to reference)

From Eqs. (3.53) and (3.54), the values of λ_k and t_k are obtained and are given in Table 3.13.

TABLE 3.13 λ and t values (Example 3.8)

k	λ_k	t_k
1	$-0.014916 + j0.005259$	$1.014916 + j0.005259$
2	$-0.015628 + j0.005475$	$1.015628 + j0.005475$
3	$-0.015452 + j0.005221$	$1.015452 + j0.005221$
4	$-0.003875 + j0.002129$	$1.003875 + j0.002129$
5	$0.0 + j0.0$	$1.0 + j0.0$

From Eq. (3.56) the neutral current, $I_n = -0.000771 - j0.212149$.
The λ_k are used in matrix T_1 to transfer the system to

$$Z_1 = \begin{bmatrix} 0.044823 & 0.037849 & 0.030422 & 0.009083 & 0.0 & 0.049579 \\ +j0.124818 & +j0.112678 & +j0.094166 & +j0.015440 & +j0.0 & +j0.000180 \\ 0.037849 & 0.047588 & 0.036552 & 0.010758 & 0.0 & 0.051609 \\ +j0.112678 & +j0.135479 & +j0.109204 & +j0.017855 & +j0.0 & +j0.000188 \\ 0.030422 & 0.036552 & 0.042039 & 0.012601 & 0.0 & 0.049217 \\ +j0.094166 & +j0.109204 & +j0.131368 & +j0.021556 & +j0.0 & +j0.000179 \\ 0.009083 & 0.010758 & 0.012601 & 0.018018 & 0.0 & 0.020067 \\ +j0.015440 & +j0.017855 & +j0.021556 & +j0.037075 & +j0.0 & +j0.000073 \\ 0.0 & 0.0 & 0.0 & 0.0 & 0.0 & 0.0 \\ +j0.0 & +j0.0 & +j0.0 & +j0.0 & +j0.0 & +j0.0 \\ 0.0 & 0.0 & 0.0 & 0.0 & 0.0 & 0.017132 \\ +j0.0 & +j0.0 & +j0.0 & +j0.0 & +j0.0 & -j4.713606 \end{bmatrix}$$

The load current is defined as the sum of bus injection currents. Using Eq. (3.58),

$$I_T = -1.426628 - j0.903495$$
$$I_1 = -0.524929 - j0.364612$$
$$I_2 = -0.428441 - j0.241396$$
$$I_3 = -0.473258 - j0.297487$$

Using Eq. (3.57), we obtain

$$l_1 = \frac{I_1}{I_T} = 0.378144 - j0.016095$$

$$l_2 = \frac{I_2}{I_T} = 0.290831 + j0.014978$$

$$l_3 = \frac{I_3}{I_T} = 0.331025 + j0.001116$$

The ratios l_i are used in transformation T_2. Using Eq. (3.62), Z_2 is obtained as

$$Z_2 = \begin{bmatrix} 0.038024 + j0.113096 & 0.010778 + j0.018138 & 0.0 + j0.0 & 0.050050 + j0.000152 \\ 0.010691 + j0.018196 & 0.018018 + j0.037075 & 0.0 + j0.0 & 0.020067 + j0.000073 \\ 0.0 + j0.0 & 0.0 + j0.0 & 0.0 + j0.0 & 0.0 + j0.0 \\ 0.0 + j0.0 & 0.0 + j0.0 & 0.0 + j0.0 & 0.017132 + j4.713606 \end{bmatrix}$$

Using Eq. (3.65), we calculate

$$t_4 = 1.015304 + j0.005298$$
$$t_1 = 0.988727 + j0.003063$$
$$t_2 = 0.984900 + j0.005139$$
$$t_3 = 0.984900 + j0.005139$$

In Step 4, the turn-ratios are used to define transformation T_3, to eliminate load. Equation (3.67) is used to build Z_3 as

$$Z_3 = \begin{bmatrix} 0.033962 + j0.111712 & 0.026638 + j0.092190 & -0.002780 + j0.091959 \\ 0.026273 + j0.092294 & 0.036885 + j0.109709 & -0.012408 + j0.109302 \\ 0.026273 + j0.092294 & 0.036885 + j0.109709 & 0.004724 - j4.604303 \end{bmatrix}$$

The currents are calculated as implicit functions of the real power using Eq. (3.69),

$$\alpha_1 = 1.012279 - j0.443717$$
$$\alpha_2 = 1.000000 - j0.510634$$

The B-coefficients are obtained using Eqs. (3.71), (3.72), (3.73) and (3.74) as

$$B_{11} = 0.04149 \text{ MW}^{-1}, \; B_{12} = 0.03277 \text{ MW}^{-1}$$
$$B_{21} = 0.03277 \text{ MW}^{-1}, \; B_{22} = 0.04650 \text{ MW}^{-1}$$
$$B_{01} = 0.00212 \text{ MW}, \quad B_{02} = 0.00255 \text{ MW}, \; B_{00} = 0.00021 \text{ MW}$$

3.9 LOSS COEFFICIENT CALCULATION USING Y_{BUS}

Another method to evaluate the loss coefficients is presented here. The method inverts the remote equivalent independent (REI) matrix [Meyer, 1971] to obtain the loss coefficients.

Exploiting the load flow analysis, we compute the phase angle and voltage magnitude for every bus in the system. The real and reactive loads are converted into lumped elements so that the system is linearized. All autotransformers are converted to Π equivalents such that shunt currents to ground may be combined at adjoining buses.

All the load buses are connected to a common node called L node. The potential at the L node is arbitrarily set to zero. The network will appear as shown in Figure 3.11.

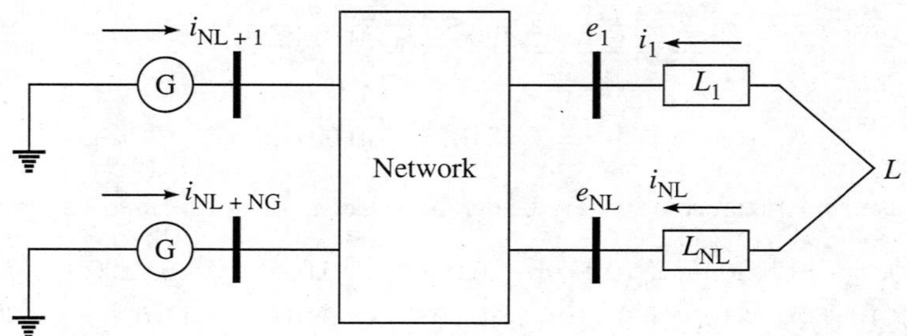

FIGURE 3.11 Lumped loads connected to a common node L, where L_1 to L_{NL} are the lumped loads.

An artificial node R, is defined whose injection current and power is equivalent to the NL load buses, i.e.

$$i_R = \sum_{i=1}^{NL} i_i \tag{3.75}$$

$$S_R = \sum_{i=1}^{NL} S_i = \sum_{i=1}^{NL} e_i i_i^* \tag{3.76}$$

$$e_R = \frac{S_R}{i_R^*} \tag{3.77}$$

Equivalent admittance from R node to 'zero potential' L node is calculated as

$$Y_{RL} = -\frac{S_R^*}{e_R e_R^*} \tag{3.78}$$

The network can be redrawn as in Figure 3.12.

The system nodal equations are

$$I_{BUS} = \begin{bmatrix} Y_{11} & Y_{12} \\ Y_{21} & Y_{22} \end{bmatrix}_{M \times M} E_{BUS} \tag{3.79}$$

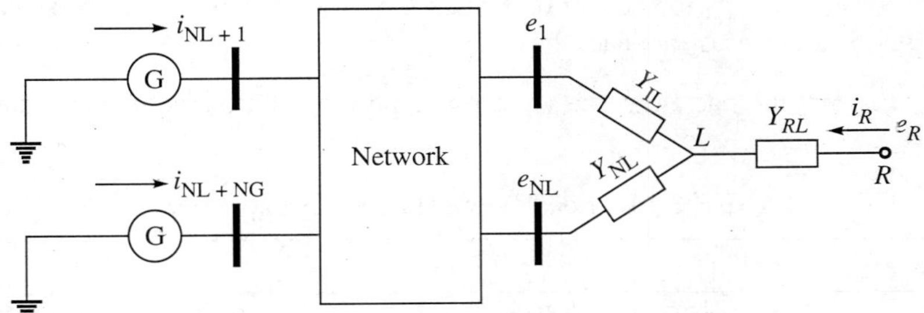

FIGURE 3.12 Remote equivalent independent of a network.

where

M = NL + NG + 2

I_{BUS} = bus injection vector = $[i_1,.., i_{NL}, i_L, i_{NL+1},...., i_{NL+NG}, i_R]^T$

E_{BUS} = bus voltage vector = $[0,.., 0, i_{NL+1},...., i_{NL+NG}, i_R]^T$

Eliminating zero injection currents from Eq. (3.79),

$$I_{BUS} = [Y_{22} - Y_{21}Y_{11}^{-1} Y_{12}] E_{BUS} \tag{3.80}$$

where

Y_{22} is partitioned matrix of size (NG + 1 × NG + 1)

Y_{21} is partitioned matrix of size (NG + 1 × NL + 1)

Y_{12} is partitioned matrix of size (NL + 1 × NG + 1)

Y_{11} is partitioned matrix of size (NL + 1 × NL + 1)

Another admittance matrix Y_3 is considered by eliminating bus R elements from the admittance matrix of Eq. (3.80). The inverse of Y_3 is the required impedance matrix, i.e.

$$Z_3 = Y_3^{-1} \tag{3.81}$$

Transmission loss can be obtained as

$$P_L = \text{Re } [(I_3^*)]^T Z_3 I_3] \tag{3.82}$$

where

$I_3 = [i_1, i_2, ..., i_{NG}]$

$\text{Re}(I_{3i}) = \dfrac{P_i}{|V_i|} \qquad (i = 1, 2,.., \text{NG})$

with

P_i as the power injection at the generator bus

$|V_i|$ as the voltage magnitude of generator bus

The above procedure is demonstrated with Example 3.9.

EXAMPLE 3.9 Use the Y_{BUS} method to determine the *B*-coefficients for a 5-bus system shown in Figure 3.9. Bus 5 is taken as the slack bus.

Solution Performing load flow, the voltage and power at all buses are calculated and given in Table 3.14.

TABLE 3.14 Load flow solution (Example 3.9)

Busbar	$P + jQ$	V	I
1	$-0.50 - j0.25$	$0.865666 - j0.125030$	$-0.524929 + j0.364612$
2	$-0.40 - j0.15$	$0.858383 - j0.133531$	$-0.428441 + j0.241396$
3	$-0.45 - j0.20$	$0.871962 - j0.125507$	$-0.473258 + j0.297487$
4	$-0.15 - j0.10$	$0.984324 - j0.008090$	$-0.151544 + j0.102838$

Load buses 1, 2, 3, and 4 are converted into lumped admittances as below:

$$Y_{1L} = -0.653587 + j0.326793$$
$$Y_{2L} = -0.530046 + j0.198767$$
$$Y_{3L} = -0.579845 + j0.257709$$
$$Y_{4L} = -0.154805 + j0.103204$$

Using Eq. (3.78),

$$Y_{RL} = 1.917881 - j0.895011$$

The network will be represented as shown in Figure 3.13.

The system admittance matrix is obtained using the *Y*-bus algorithm as

$$Y_{BUS} = \begin{bmatrix} 21.93304 - j24.75117 & -10.0 + j20.0 & -0.82192 + j2.19178 & 0.65359 - j0.32679 & 0.0 + j0.0 & -11.76471 + j2.94118 & 0.0 + j0.0 \\ -10.0 + j20.0 & 14.46995 - j34.78123 & -5.0 + j15.0 & 0.53005 - j0.19877 & 0.0 + j0.0 & 0.0 + j0.0 & 0.0 + j0.0 \\ -0.82192 + j2.19178 & -5.0 + j15.0 & 6.20362 - j21.68676 & 0.57985 - j0.25771 & -0.96154 + j4.80769 & 0.0 + j0.0 & 0.0 + j0.0 \\ 0.65359 - j0.32679 & 0.53005 - j0.19877 & 0.57985 - j0.25771 & 0.0 - j0.00854 & 0.15481 - j0.10320 & 0.0 + j0.0 & -1.91788 + j0.89501 \\ 0.0 + j0.0 & 0.0 + j0.0 & -0.96154 + j4.80769 & 0.15481 - j0.10320 & 10.80674 - j24.66449 & -10.0 + j20.0 & 0.0 + j0.0 \\ -17.76471 + j2.94118 & 0.0 + j0.0 & 0.0 + j0.0 & 0.0 + j0.0 & -10.0 + j20.0 & 21.76471 - j22.90118 & 0.0 + j0.0 \\ 0.0 + j0.0 & 0.0 + j0.0 & 0.0 + j0.0 & -1.91788 + j0.89501 & 0.0 + j0.0 & 0.0 + j0.0 & 1.91788 + j0.89501 \end{bmatrix}$$

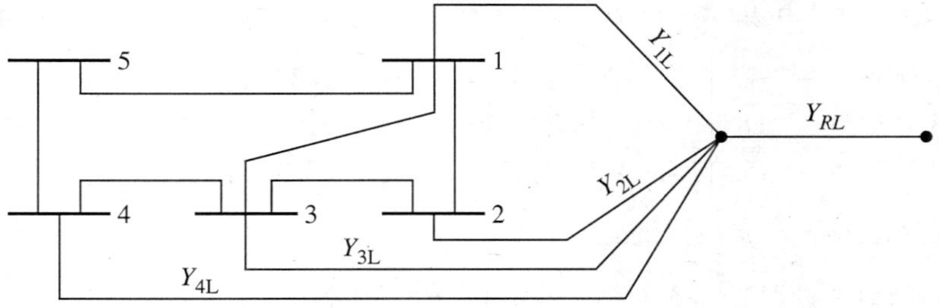

FIGURE 3.13 The power system (Example 3.9).

Partitioning the Y_{BUS} matrix into four matrices, namely, Y_{11}, Y_{12}, Y_{21}, Y_{22} gives

$$Y_{11} = \begin{bmatrix} 21.93304 & -10.0 & -0.82192 & 0.65359 \\ -j24.75117 & +j20.0 & +j2.19178 & -j0.32679 \\ -10.0 & 14.46995 & -5.0 & 0.53005 \\ +j20.0 & -j34.78123 & +j15.0 & -j0.19877 \\ -0.82192 & -5.0 & 6.20362 & 0.57985 \\ +j2.19178 & +j15.0 & -j21.68676 & -j0.25771 \\ 0.65359 & 0.53005 & 0.57985 & 0.0 \\ -j0.32679 & -j0.19877 & -j0.25771 & -j0.00854 \end{bmatrix}_{4\times 4}$$

$$Y_{12} = \begin{bmatrix} 0.0 & -11.76471 & 0.0 \\ +j0.0 & +j2.94118 & +j0.0 \\ 0.0 & 0.0 & 0.0 \\ +j0.0 & +j0.0 & +j0.0 \\ -0.96154 & 0.0 & 0.0 \\ +j4.80769 & +j0.0 & +j0.0 \\ 0.15481 & 0.0 & -1.91788 \\ -j0.10320 & +j0.0 & +j0.89501 \end{bmatrix}_{4\times 3}$$

$$Y_{21} = \begin{bmatrix} 0.0 & 0.0 & -0.96154 & 0.15481 \\ +j0.0 & +j0.0 & +j4.80769 & -j0.10320 \\ -17.76471 & 0.0 & 0.0 & 0.0 \\ +j2.94118 & +j0.0 & +j0.0 & +j0.0 \\ 0.0 & 0.0 & 0.0 & -1.91788 \\ +j0.0 & +j0.0 & +j0.0 & +j0.89501 \end{bmatrix}_{3\times 4}$$

$$Y_{22} = \begin{bmatrix} 10.80674 & -10.0 & 0.0 \\ -j24.66449 & +j20.0 & +j0.0 \\ -10.0 & 21.76471 & 0.0 \\ +j20.0 & -j22.90118 & +j0.0 \\ 0.0 & 0.0 & 1.91788 \\ +j0.0 & +j0.0 & +j0.89501 \end{bmatrix}_{3\times 3}$$

From Eq. (3.80), the admittance matrix is obtained as (by eliminating zero injection current)

$$YR = \begin{bmatrix} 10.968810 - j25.17223 & -8.442224 + j18.791480 & -2.530555 + j6.417044 \\ -8.442224 + j18.791480 & 19.727650 - j24.495150 & -11.288740 + j5.742203 \\ -2.530555 + j6.417044 & -11.288740 + j5.742203 & 13.825880 - j12.01144 \end{bmatrix}$$

Eliminating the reference bus elements

$$Y_3 = \begin{bmatrix} 10.968810 - j25.17223 & -8.442224 + j18.791480 \\ -8.442224 + j18.791480 & 19.727650 - j24.495150 \end{bmatrix}$$

The inverse of Y_3 matrix is

$$Z_3 = \begin{bmatrix} 0.040235 + j0.054666 & 0.034555 + j0.027974 \\ 0.034555 + j0.027974 & 0.046476 + j0.036764 \end{bmatrix}$$

Using Eq. (3.82), B-coefficients are real values of Z matrix, i.e.

$B_{11} = 0.0415240 \qquad B_{12} = 0.0351046$

$B_{21} = 0.0351046 \qquad B_{22} = 0.0464764$

3.10 LOSS COEFFICIENTS USING SENSITIVITY FACTORS

3.10.1 DC Load Flow

The complex power flow from bus i to j is expressed as

$$S_{ij} = V_i \left[\frac{V_i - V_j}{Z_m} \right]^* \tag{3.83}$$

where

Z_m is the impedance of the branch connecting the ith and jth buses

V_i is the voltage of the ith bus

V_j is the voltage of jth bus.

Thus,

$$S_{ij} = \frac{1}{Z_m^*} [|V_i^2| - |V_i||V_j| \angle(\theta_i - \theta_j)]$$

where

$|V_i|$ and θ_i are the magnitude and angle of voltage of the ith bus, respectively

$|V_j|$ and θ_j are the magnitude and angle of voltage of the jth bus, respectively

Therefore,

$$S_{ij} = \frac{1}{Z_m^*} [|V_i^2| - |V_i||V_j| \{\cos(\theta_i - \theta_j) + j \sin(\theta_i - \theta_j)\}] \tag{3.84}$$

The reciprocal of conjugate impedance is

$$\frac{1}{Z_m^*} = \frac{1}{R_m - jX_m} = \frac{R_m + jX_m}{R_m^2 + X_m^2} = \frac{R_m + jX_m}{|Z_m|^2}$$

where

R_m is the resistance of the *m*th line

X_m is the reactance of the *m*th line.

Substituting the above equation into Eq. (3.84),

$$S_{ij} = \frac{1}{|Z_m|^2}[|V_i^2| - |V_i||V_j|(\cos(\theta_i - \theta_j) + j\sin(\theta_i - \theta_j))](R_m + jX_m) \tag{3.85}$$

The real part of Eq. (3.85) is the active power flow from the sending bus *i* to the receiving bus *j*, i.e.

$$P_m = P_{ij} = \frac{1}{|Z_m|^2}[R_m|V_i^2| - |V_i||V_j|(R_m\cos(\theta_i - \theta_j) - X_m\sin(\theta_i - \theta_j))] \tag{3.86}$$

Assume

(i) $|V_i| = |V_j| = 1$

(ii) $X_m >> R_m$

(iii) $(\theta_i - \theta_j)$ is small such that $\cos(\theta_i - \theta_j) = 1$ and $\sin(\theta_i - \theta_j) = \theta_i - \theta_j$

Thus, Eq. (3.86) can be rewritten, in view of the above assumptions, as

$$P_m = P_{ij} = \frac{1}{|Z_m|^2}[R_m - (R_m - X_m(\theta_i - \theta_j))] = \frac{X_m}{|Z_m|^2}(\theta_i - \theta_j) = \frac{X_m}{R_m^2 + X_m^2}(\theta_i - \theta_j)$$

$$= \frac{\theta_i - \theta_j}{X_m} \tag{3.87}$$

Generalizing, we have

$$P_i = \sum_{j=1}^{k} \frac{1}{X_{ij}}(\theta_i - \theta_j) \tag{3.88}$$

where *k* is the number of directly connected branches to *i*.

In the matrix form

$$\theta = [X]\,P$$

$$\Delta\theta = [X]\,\Delta P \tag{3.89}$$

3.10.2 Power Loss in a Line

The active power flow from the sending bus *j* to the receiving bus *i* can be expressed in the similar way as Eq. (3.86), i.e.

$$P_m = P_{ji} = \frac{1}{|Z_m|^2}[R_m|V_j|^2 - |V_i||V_j|(R_m\cos(\theta_i - \theta_j) + X_m\sin(\theta_i - \theta_j))] \tag{3.90}$$

So, power loss of line can be obtained as

$$PL_m = P_{ij} + P_{ji}$$

$$= \frac{1}{|Z_m|^2}[(|V_i|^2 + |V_j|^2)R_m - 2|V_i||V_j|R_m \cos(\theta_i - \theta_j)] \tag{3.91}$$

Equation (3.91) is rewritten as

$$PL_m = \frac{1}{|Z_m|^2}[2R_m - 2R_m \cos(\theta_i - \theta_j)] \qquad (\because |V_i| = |V_j| = 1)$$

$$= \frac{2R_m}{R_m^2 + X_m^2}\left[2\sin^2\left(\frac{\theta_i - \theta_j}{2}\right)\right] \qquad \left(\because 1 - \cos\alpha = 2\sin^2\frac{\alpha}{2}\right)$$

$$= \frac{4R_m}{X_m^2}\left[\sin\left(\frac{\theta_i - \theta_j}{2}\right)\right]^2 \qquad [\because X_m \gg R_m, \text{Section 3.10.2}]$$

$$= \frac{4R_m}{X_m^2}\left(\frac{\theta_i - \theta_j}{2}\right)^2 \qquad [\because (\theta_i - \theta_j)/2 \text{ is small, Section 3.10.2}]$$

$$= R_m\left(\frac{\theta_i - \theta_j}{X_m}\right)^2$$

The above equation can be written as a function of real power generation by substituting Eq. (3.87) into it. Thus,

$$PL_m = R_m P_m^2 \tag{3.92}$$

3.10.3 Generation Shift Distribution (GSD) Factors

Using a DC load flow model, the GSD factor is expressed as

$$A(m, i) = \frac{\partial P_m}{\partial P_{g_i}} = \frac{\partial}{\partial P_{g_i}}\left(\frac{\theta_j - \theta_k}{X_m}\right) = \frac{1}{X_m}\left(\frac{\partial \theta_j}{\partial P_{g_i}} - \frac{\partial \theta_k}{\partial P_{g_i}}\right)$$

From Eq. (3.89), it is concluded that $\frac{\partial \theta_j}{\partial P_{g_i}} = X_{ji}$ and $\frac{\partial \theta_k}{\partial P_{g_i}} = X_{ki}$. Thus,

$$A(m, i) = \frac{X_{ji} - X_{ki}}{X_m} \tag{3.93}$$

where

P_m is the real power flow on transmission line m from sending bus j to receiving bus k.

X_{ji} and X_{ki} are the elements of the X matrix.

X_m is the reactance of line m.

In Eq. (3.93), since all generation changes are compensated by the reference bus, the total system generation is assumed to be unchanged.

$$\sum_{i=1}^{NG} P_{g_i} = \sum_{i=1}^{NLOAD} P_i = \text{constant}$$

where

NG is number of generators

NLOAD is number of loads.

Using GSD factors, the new line flows, after rescheduling generation, can be expressed as

$$P_m = P_m^0 + \sum_{i=1}^{NG} A(m, i)\, \Delta P_{g_i} \qquad (m = 1, 2,.., NL) \tag{3.94}$$

where

P_m^0 is the base case line flow

NL is the number of lines in the system.

By this incremental summation, any change in line flow due to change in generation can easily be calculated.

3.10.4 Generalized Generation Shift Distribution (GGSD) Factor

Since all the generation changes are absorbed by the reference bus, the total system generation remains unchanged. If not, because the initial line flows are known quantities, a new load flow must be executed to re-establish the initial values. That is, the solution accuracy of GSD factors is guaranteed only when the total system generation remains unchanged. A new sensitivity factor, the generalized generation shift distribution factor (GGDF) was developed by Ng [1981] to update the GSD factor. Instead of an incremental form, an integral generation form has been defined as

$$P_m = \sum_{i=1}^{NG} D(m, i)\, P_{g_i} \qquad (m = 1, 2,.., NL) \tag{3.95}$$

where

$D(m, i)$ is the GGDF

NG is number of generation buses

NL is number of lines.

3.10.5 Derivation of GGDF

From Eq. (3.95), the definition of GGDF, if a particular generator k is increased in generation by some amount ΔP_{g_k}, the flow on line m will be

$$P'_m = \sum_{i=1}^{NG} D(m, i)\, P_{g_i} + D(m, k)\Delta P_{g_k} \qquad (m = 1, 2,.., NL) \tag{3.96}$$

The reference generator ($R \neq k$) will adjust the change in ΔP_{g_k} by decreasing its generation by ΔP_{g_k}. The real power flow on line m after this generation shift will be

$$\overline{P}_m = \sum_{i=1}^{NG} D(m,i)\, P_{g_i} + D(m,k)\Delta P_{g_k} - D(m,R)\Delta P_{g_k} \qquad (m = 1, 2,\ldots, NL) \tag{3.97}$$

Subtracting Eq. (3.95) from Eq. (3.97),

$$\overline{P}_m - P_m = [D(m, k) - D(m, R)]\Delta P_{g_k} \qquad (m = 1, 2,\ldots, NL)$$

or

$$\frac{\overline{P}_m - P_m}{\Delta P_{g_k}} = D(m, k) - D(m, R) \qquad (m = 1, 2,\ldots, NL) \tag{3.98}$$

The change in line flow with respect to the generation change is termed the GSD factor. So, the above equation can be rewritten as

$$A(m, k) = D(m, k) - D(m, R) \tag{3.99}$$

$$\overline{P}_m - P_m = A(m, k)\Delta P_{g_k} \tag{3.100}$$

There is a need to calculate $D(m, R)$. By shifting all the generations from all the generators to the reference generator R, i.e. by making

$$-\sum_{\substack{i=1\\ i\neq R}}^{NG} P_{g_i} = \Delta P_{g_R}$$

Equation (3.100) can be rewritten as

$$\overline{P}_m - P_m = -\sum_{\substack{i=1\\ i\neq R}}^{NG} A(m, i)\, P_{g_i} \tag{3.101}$$

From Eq. (3.95),

$$\overline{P}_m = \sum_{\substack{i=1\\ i\neq R}}^{NG} D(m, i)\, \overline{P}_{g_i} + D(m, R)\overline{P}_{g_R} \tag{3.102}$$

where

$\overline{P}_{g_i}$ is final generation from generator i, which is now reduced to zero

$\overline{P}_{g_R}$ is final generation from the reference generator which contains the total system generation.

Equation (3.102) is reduced to

$$\overline{P}_m = D(m, R)\overline{P}_{g_R} = D(m, R) \sum_{i=1} P_{g_i} \tag{3.103}$$

Substituting Eq. (3.103) into Eq. (3.101),

$$D(m, R) \sum_{i=1}^{NG} P_{g_i} = P_m - \sum_{\substack{i=1 \\ i \neq R}}^{NG} A(m, i)\, P_{g_i}$$

or

$$D(m, R) = \frac{P_m - \sum\limits_{\substack{i=1 \\ i \neq R}}^{NG} A(m, i)\, P_{g_i}}{\sum\limits_{i=1}^{NG} P_{g_i}} \tag{3.104}$$

where

P_m is base case flow of *m*th line

P_{g_i} is generation of the *i*th generator.

3.10.6 Evaluation of *B*-Coefficients

The power loss expressed by using *B*-coefficients and bus generation in a quadratic form was first developed by Kusic [1996] as

$$P_L = \sum_{i=1}^{NG} \sum_{j=1}^{NG} B_{ij}\, P_{g_i}\, P_{g_j} \tag{3.105}$$

The incremental loss can be obtained by differentiating with respect to P_{g_i}, i.e.

$$\frac{\partial P_L}{\partial P_{g_i}} = \sum_{j=1}^{NG} 2\, B_{ij}\, P_{g_j} \tag{3.106a}$$

Taking the derivatives of Eq. (3.106a) with respect to P_{g_i},

$$B_{ij} = \frac{1}{2} \left(\frac{\partial^2 P_L}{\partial P_{g_i}\, \partial P_{g_j}} \right) \tag{3.106b}$$

We know from Eq. (3.92),

$$PL_m = R_m\, P_m^2$$

The system power loss is expressed by the summation over all transmission lines as

$$P_L = \sum_{m=1}^{NL} R_m P_m^2 \tag{3.107}$$

Substituting Eq. (3.95) into Eq. (3.107),

$$P_L = \sum_{m=1}^{NL} R_m \left[\sum_{i=1}^{NG} D(m, i) P_{g_i} \right]^2 \tag{3.108}$$

Taking the derivatives of Eq. (3.108) with respect to P_{g_i},

$$\frac{\partial P_L}{\partial P_{g_i}} = \sum_{m=1}^{NL} R_m \sum_{j=1}^{NG} [2D(m,i)\, D(m,j)\, P_{g_j}] \tag{3.109}$$

and

$$\frac{\partial^2 P_L}{\partial P_{g_i}\, \partial P_{g_j}} = \sum_{m=1}^{NL} 2\, R_m\; D(m,i)\, D(m,j) \tag{3.110}$$

Comparing Eq. (3.105) with Eq. (3.110),

$$B_{ij} = \sum_{m=1}^{NL} R_m\; D(m,i)\, D(m,j) \qquad (i = 1, 2, ..., \text{NG};\;\; j = 1, 2, ..., \text{NG}) \tag{3.111}$$

where

R_m is the resistance of line m

$D(m, i)$ are Generalized Generation Shift Distribution Factors.

Algorithm 3.7: *B*-coefficients Using Sensitivity Factors

The stepwise algorithm to calculate *B*-coefficients is outlined below:

1. Form the reactance matrix using Z_{BUS} algorithm.
2. Compute $A(m, i)$ using Eq. (3.93).
3. Compute $D(m, R)$ using Eq. (3.104).
4. Compute $D(m, i)$ using Eq. (3.99).

$$D(m, i) = A(m, i) + D(m, R) \qquad (m = 1, 2, ..., \text{NL};\; i = 1, 2, ..., \text{NG})$$

5. Compute B_{ij} using Eq. (3.111).

EXAMPLE 3.10 Using the GGDF method, determine the *B*-coefficients for a 5-bus system shown in Figure 3.14. Bus 2 is taken as the slack bus. The line data is given in Table 3.15.

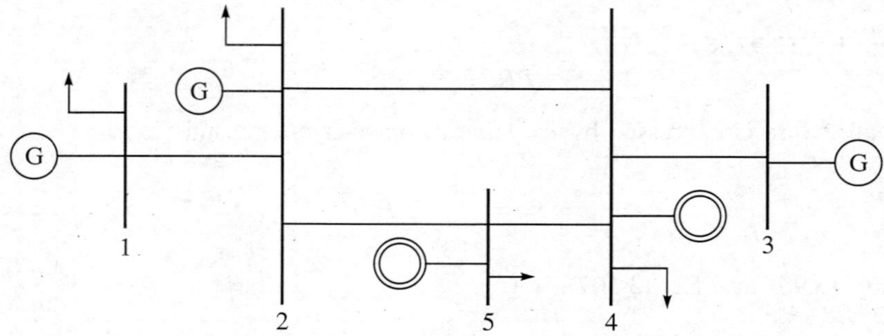

FIGURE 3.14 Power System (Example 3.10).

TABLE 3.15 Line data (Example 3.10)

m	*Sending bus*	*Ending bus*	*Z(series)*
1	1	2	$0.060 + j0.206$
2	2	5	$0.160 + j0.524$
3	5	4	$0.066 + j0.236$
4	2	4	$0.210 + j0.694$
5	4	3	$0.212 + j0.806$

Solution From the base case load flow, the line flows are obtained as given in Table 3.16.

TABLE 3.16 Line flows from base case data (Example 3.10)

m	P_m	$P_m + jG_m$	$PL_m + jQL_m$	$\lvert V_m \rvert$	δ^0
1	58.00	$144.000 + j39.023$	$86.0 + j20.0$	1.015	11.1139
2	12.78	$10.668 + j13.549$	$46.8 + j5.8$	0.980	7.9999
3	–04.77	$34.600 + j10.528$	$0.0 + j0.0$	0.953	15.6829
4	8.00	$0.0 + j10.884$	$36.2 + j3.0$	0.981	5.8130
5	–33.07	$0.0 + j25.343$	$17.4 + j12.0$	0.972	6.3267

To obtain the *Z*-bus, the *Z*-bus algorithm is used. The data for *Z*-bus is described in Table 3.17.

TABLE 3.17 Data for *Z*-bus building (Example 3.10)

m	*SB–EB*	*Z(series)*	*Type of link*
1	1–R	$0.060 + j0.206$	1 (new bus to slack bus)
2	R–5	$0.160 + j0.524$	1 (new bus to slack bus)
3	R–4	$0.210 + j0.694$	1 (new bus to slack bus)
4	4–3	$0.212 + j0.806$	3 (new bus to old bus)
5	5–4	$0.066 + j0.236$	4 (old bus to old bus)

Note: SB—Sending bus node, EB—Ending bus node

The reactance matrix is the imaginary part of *Z*-bus, i.e.

$$X = \begin{bmatrix} j0.206 & 0 & 0 & 0 & 0 \\ 0 & 0 & 0 & 0 & 0 \\ 0 & 0 & j1.1692 & j0.3632 & j0.2500 \\ 0 & 0 & j0.3632 & j0.3632 & j0.2500 \\ 0 & 0 & j0.2500 & j0.2500 & j0.335 \end{bmatrix}$$

The GSD factors are calculated using Eq. (3.93) for generator 1, i.e.

$$A(1, 1) = \frac{X_{11} - X_{21}}{X_1} = 1.0 \qquad (m = 1, i = 1, j = 1 \text{ and } k = 2)$$

For generators 1, 2 and 3, the GSD factors are calculated and are tabulated in Table 3.18.

TABLE 3.18 GSD factors

m	$A(m, 1)$	$A(m, 2)$	$A(m, 3)$
1	1.0	0.0	0.0
2	0.0	0.0	−0.477099
3	0.0	0.0	−0.479661
4	0.0	0.0	−0.523343
5	0.0	0.0	−1.0

Using Eq. (3.104), the values of $D(m, R)$ are calculated as

$$D(1, R) = \frac{58.0 - (1.0 \times 144.0 + 0.0 \times 34.60)}{144.0 + 10.668 + 34.60} = -0.454382$$

$D(2, R) = 0.154742$; $D(3, R) = 0.062484$; $D(4, R) = 0.137940$; $D(5, R) = 0.008084$

Similarly, the GGDF are calculated using Eq. (3.99) and are tabulated in Table 3.19.

TABLE 3.19 GGDF factors (Example 3.10)

m	$D(m, 1)$	$D(m, 2)$	$D(m, 3)$	R_m
1	0.545618	−0.454382	−0.454382	0.060
2	0.154742	0.154742	−0.322358	0.160
3	0.062484	0.062484	−0.411770	0.066
4	0.137940	0.137940	−0.385403	0.210
5	0.008084	0.008084	−0.991916	0.212

Using Eq. (3.111), the B-coefficients are calculated as

$$\begin{aligned} B_{11} = {} & [0.06 \times (0.545618 \times 0.545618) + 0.16 \times (0.154742 \times 0.154742) \\ & + 0.066 \times (0.062484 \times 0.062484) + 0.21 \times (0.13794 \times 0.13794) \\ & + 0.212 \times (0.008084 \times 0.008084)] = 0.02596 \end{aligned}$$

$$\begin{aligned} B_{12} = {} & 0.06 \times (0.545618 \times -0.454382) + 0.16 \times (0.154742 \times 0.154742) \\ & + 0.066 \times (0.062484 \times 0.062484) + 0.21 \times (0.13794 \times 0.13794) \\ & + 0.212 \times (0.008084 \times 0.008084) = -0.006777 \text{ MW}^{-1} \end{aligned}$$

Similarly,

$B_{13} = -0.037441 \text{ MW}^{-1}$,

$B_{21} = -0.006777 \text{ MW}^{-1}$, $B_{22} = 0.020486 \text{ MW}^{-1}$, $B_{23} = -0.010178 \text{ MW}^{-1}$,

$B_{31} = -0.037441 \text{ MW}^{-1}$, $B_{32} = -0.010178 \text{ MW}^{-1}$, $B_{33} = 0.280279 \text{ MW}^{-1}$

3.11 TRANSMISSION LOSS COEFFICIENTS

An exact transmission loss formula has been derived by Dopazo et al. [1967], using bus powers and system parameters. The total system loss is the sum of the bus powers, i.e.

$$P_L + jQ_L = \sum_{i=1}^{NB} S_i = \sum_{i=1}^{NB} V_i I_i^* \tag{3.112}$$

where

NB is the number of buses

P_L is real power loss of the power system

Q_L is reactive power loss of the power system

V_i is the voltage of the ith bus

S_i is the injected power at the ith bus

For the network shown in Figure 3.15,

$$V_i = \sum_{j=1}^{NB} Z_{ij} I_j \tag{3.113}$$

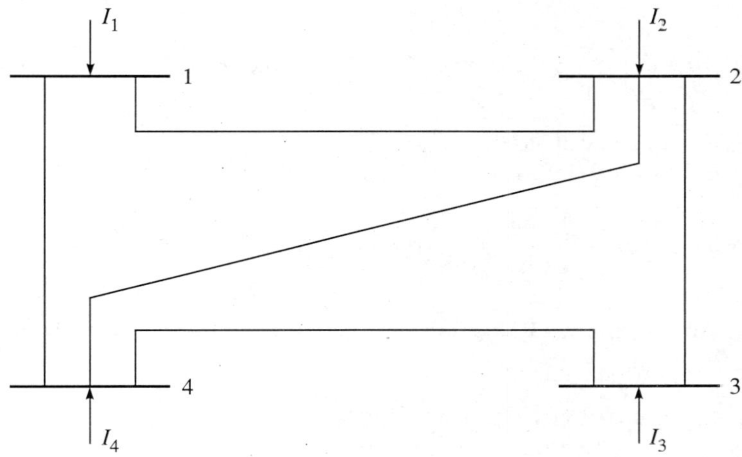

FIGURE 3.15 Power system network.

where Z_{ij} are the clements of impedance matrix, i.e.

$$Z_{ij} = R_{ij} + jX_{ij} \qquad (i = 1, 2,..., \text{NB}; \quad j = 1, 2,..., \text{NB})$$

Substituting Eq. (3.113) into Eq. (3.112),

$$P_L + jQ_L = \sum_{i=1}^{NB} \sum_{j=1}^{NB} Z_{ij}\, I_j\, I_i^* \tag{3.114}$$

Equation (3.114) can be rewritten in rectangular form as

$$P_L + jQ_L = \sum_{i=1}^{NB} \sum_{j=1}^{NB} Z_{ij} \left(|I_j| \cos\theta_j + j|I_j| \sin\theta_j\right)\left(|I_i| \cos\theta_i - j|I_i| \sin\theta_i\right) \tag{3.115}$$

where

$\theta_i = \delta_i - \phi_i$

δ_i is voltage angle at ith bus.

$$\phi_i = \tan^{-1}\frac{Q_i}{P_i}$$

Equation (3.115) is expanded as

$$P_L + jQ_L = \sum_{i=1}^{NB}\sum_{j=1}^{NB} Z_{ij}\,|I_i||I_j|\left(\cos\theta_i\cos\theta_j + \sin\theta_i\sin\theta_j\right) \tag{3.116}$$

$$+\,j\sum_{i=1}^{NB}\sum_{j=1}^{NB} Z_{ij}\,|I_i||I_j|\left(\cos\theta_i\sin\theta_j - \cos\theta_j\sin\theta_i\right)$$

Equating the following

$$\sum_{i=1}^{NB}|I_i|\cos\theta_i\left(\sum_{j=1}^{NB} Z_{ij}\,|I_j|\sin\theta_j\right) = \sum_{j=1}^{NB}|I_j|\cos\theta_j\left(\sum_{i=1}^{NB} Z_{ji}\,|I_i|\sin\theta_i\right) \tag{3.116a}$$

and because Z_{BUS} is symmetric matrix, so $Z_{ij} = Z_{ji}$.

Substituting Eq. (3.116a) into Eq. (3.116) and simplifying

$$P_L + jQ_L = \sum_{i=1}^{NB}\sum_{j=1}^{NB} Z_{ij}\,|I_i||I_j|(\cos\theta_i\cos\theta_j + \sin\theta_i\sin\theta_j) \tag{3.117}$$

Substituting the following in Eq. (3.117) and separating the real and imaginary parts

$$Z_{ij} = R_{ij} + jX_{ij}$$

$$\cos(\theta_i - \theta_j) = \cos\theta_i\cos\theta_j + \sin\theta_i\sin\theta_j$$

$$P_L = \sum_{i=1}^{NB}\sum_{j=1}^{NB} R_{ij}\,|I_i||I_j|\cos(\theta_i - \theta_j) \tag{3.118}$$

$$Q_L = \sum_{i=1}^{NB}\sum_{j=1}^{NB} X_{ij}\,|I_i||I_j|\cos(\theta_i - \theta_j) \tag{3.119}$$

We know

$$S_i = |V_i||I_i| \quad\text{or}\quad |I_i| = \frac{S_i}{|V_i|} \tag{3.120}$$

From Figure 3.16,

$$P_i = S_i\cos\phi_i \tag{3.121}$$

$$Q_i = S_i\sin\phi_i \tag{3.122}$$

From Eqs. (3.120) and (3.121)

$$|I_i| = \frac{P_i}{|V_i|\cos\phi_i} \tag{3.123}$$

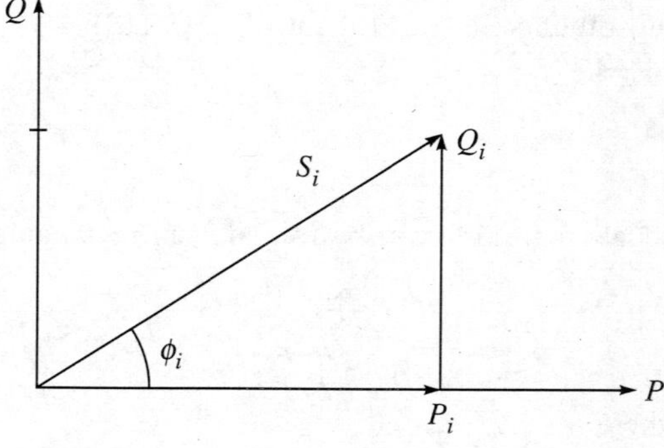

FIGURE 3.16 Power triangle for ith bus.

From Eqs. (3.120) and (3.122),

$$|I_i| = \frac{Q_i}{|V_i| \sin \phi_i} \tag{3.124}$$

Substituting Eq. (3.123) into Eq. (3.118),

$$P_L = \sum_{i=1}^{NB} \sum_{j=1}^{NB} R_{ij} \frac{P_i P_j}{|V_i||V_j|} \frac{\cos(\theta_i - \theta_j)}{\cos \phi_i \cos \phi_j} \tag{3.125}$$

Equation (3.125) can be rewritten in terms of *B*-coefficients as

$$P_L = \sum_{i=1}^{NB} \sum_{j=1}^{NB} P_i B_{ij} P_j \tag{3.126}$$

where

$$B_{ij} = \frac{R_{ij}}{|V_i||V_j|} \frac{\cos(\theta_i - \theta_j)}{\cos \phi_i \cos \phi_j}$$

$$P_i = P_{g_i} - P_{d_i} \qquad (i = 1, 2,.., \text{NB})$$

Similarly, substituting Eq. (3.124) into Eq. (3.119)

$$Q_L = \sum_{i=1}^{NB} \sum_{j=1}^{NB} X_{ij} \frac{Q_i Q_j}{|V_i||V_j|} \frac{\cos(\theta_i - \theta_j)}{\sin \phi_i \sin \phi_j} \tag{3.127}$$

Equation (3.127) can be rewritten in simple form as

$$Q_L = \sum_{i=1}^{NB} \sum_{j=1}^{NB} Q_i C_{ij} Q_j \tag{3.128}$$

where

$$C_{ij} = \frac{X_{ij}}{|V_i||V_j|} \frac{\cos(\theta_i - \theta_j)}{\sin \phi_i \sin \phi_j}$$

$$Q_i = Q_{g_i} - Q_{d_i} \qquad (i = 1, 2,.., \text{NB})$$

Let NG be the number of generating buses and for rest of buses, $P_{g_i} = 0.0$ Equation (3.126) can be rewritten as

$$P_L = \sum_{i=1}^{NB} \sum_{j=1}^{NB} (P_{g_i} - P_{d_i}) B_{ij} (P_{g_j} - P_{d_j})$$

$$= \sum_{i=1}^{NB} \sum_{j=1}^{NB} P_{d_i} B_{ij} P_{d_j} - \sum_{i=1}^{NG} P_{g_i} \left[\sum_{j=1}^{NB} B_{ij} P_{d_j} \right] - \sum_{i=1}^{NB} P_{d_i} \left[\sum_{j=1}^{NG} B_{ij} P_{g_j} \right] + \sum_{i=1}^{NG} \sum_{j=1}^{NG} P_{g_i} B_{ij} P_{g_j}$$

$$= \sum_{i=1}^{NB}\sum_{j=1}^{NB} P_{d_i} B_{ij} P_{d_j} - \sum_{i=1}^{NG}\left[\sum_{j=1}^{NB} (B_{ij} + B_{ji}) P_{d_j}\right] P_{g_i} + \sum_{i=1}^{NG}\sum_{j=1}^{NG} P_{g_i} B_{ij} P_{g_j}$$

The above equation can be written in the form of Kron's loss formula

$$P_L = B_{00} + \sum_{i=1}^{NG} B_{i0} P_{g_i} + \sum_{i=1}^{NG}\sum_{j=1}^{NG} P_{g_i} B_{ij} P_{g_j} \tag{3.129}$$

where

$$B_{00} = \sum_{i=1}^{NB}\sum_{j=1}^{NB} P_{d_i} B_{ij} P_{d_j}$$

$$B_{i0} = -\sum_{j=1}^{NB} (B_{ij} + B_{ji}) P_{d_j}$$

Similarly from Eq. (3.128), we can obtain

$$Q_L = C_{00} + \sum_{i=1}^{NG} C_{i0}\, Q_{g_i} + \sum_{i=1}^{NG}\sum_{j=1}^{NG} Q_{g_i}\, C_{ij}\, Q_{g_j} \tag{3.130}$$

where

$$C_{00} = \sum_{i=1}^{NB}\sum_{j=1}^{NB} Q_{d_i}\, C_{ij}\, Q_{d_j}$$

$$C_{i0} = -\sum_{j=1}^{NB} (C_{ij} + C_{ji})\, Q_{d_j}$$

The system power losses are based on the assumption that (i) the generator bus-voltage magnitudes and angles are constant and (ii) the power factor of each source is constant. However, the use of loss coefficients can account for any change in load demand at the buses while scheduling the generations for the system.

3.12 TRANSMISSION LOSS FORMULA: FUNCTION OF GENERATION AND LOADS

Substituting $\theta_i = \delta_i - \phi_i$ in Eq. (3.125),

$$P_L = \sum_{i=1}^{NB}\sum_{j=1}^{NB} R_{ij} \frac{P_i P_j}{|V_i||V_j|} \frac{\cos[(\delta_i - \delta_j) - (\phi_i - \phi_j)]}{\cos\phi_i \cos\phi_j}$$

$$= \sum_{i=1}^{NB}\sum_{j=1}^{NB} R_{ij} \frac{P_i P_j}{|V_i||V_j|}\left[\frac{\cos(\delta_i - \delta_j)\cos(\phi_i - \phi_j) + \sin(\delta_i - \delta_j)\sin(\phi_i - \phi_j)}{\cos\phi_i \cos\phi_j}\right] \tag{3.131}$$

Let

$$a_{ij} = \frac{R_{ij}}{|V_i||V_j|} \cos(\delta_i - \delta_j); \qquad b_{ij} = \frac{R_{ij}}{|V_i||V_j|} \sin(\delta_i - \delta_j)$$

Equation (3.131) can be rewritten, considering a_{ij} and b_{ij}, as

$$P_L = \sum_{i=1}^{NB} \sum_{j=1}^{NB} P_i P_j \left[a_{ij} \frac{\cos\phi_i \cos\phi_j + \sin\phi_i \sin\phi_j}{\cos\phi_i \cos\phi_j} + b_{ij} \frac{\sin\phi_i \cos\phi_j - \cos\phi_i \sin\phi_j}{\cos\phi_i \cos\phi_j} \right]$$

$$= \sum_{i=1}^{NB} \sum_{j=1}^{NB} P_i P_j [a_{ij}(1 + \tan\phi_i \tan\phi_j) + b_{ij}(\tan\phi_i - \tan\phi_j)]$$

Substituting $\tan\phi_i = \dfrac{Q_i}{P_i}$ in the above equation,

$$P_L = \sum_{i=1}^{NB} \sum_{j=1}^{NB} P_i P_j \left[a_{ij} \left(1 + \frac{Q_i Q_j}{P_i P_j}\right) + b_{ij} \left(\frac{Q_i}{P_i} - \frac{Q_j}{P_j}\right) \right]$$

On simplification,

$$P_L = \sum_{i=1}^{NB} \sum_{j=1}^{NB} [a_{ij}(P_i P_j + Q_i Q_j) + b_{ij}(Q_i P_j - P_i Q_j)] \tag{3.132}$$

Similarly in Eq. (3.127), substituting $\theta_i = \delta_i - \phi_i$ and

$$c_{ij} = \frac{X_{ij}}{|V_i||V_j|} \cos(\delta_i - \delta_j), \qquad d_{ij} = \frac{X_{ij}}{|V_i||V_j|} \sin(\delta_i - \delta_j), \text{ we get}$$

$$Q_L = \sum_{i=1}^{NB} \sum_{j=1}^{NB} Q_i Q_j [c_{ij}(\cot\phi_i \cot\phi_j + 1) + d_{ij}(\cot\phi_j - \cot\phi_i)] \tag{3.133}$$

Substituting $\tan\phi_i = \dfrac{Q_i}{P_i}$ and $\cot\phi_i = \dfrac{1}{\tan\phi_i}$ in Eq. (3.133), we get

$$Q_L = \sum_{i=1}^{NB} \sum_{j=1}^{NB} [c_{ij}(P_i P_j + Q_i Q_j) + d_{ij}(Q_i P_j - P_i Q_j)] \tag{3.134}$$

The above method requires the evaluation of bus impedance matrix.

3.13 ECONOMIC DISPATCH USING EXACT LOSS FORMULA

The economic dispatch problem is defined as to minimize the total operating cost of a power system while meeting the total load plus transmission losses within the generator limits. Mathematically, the problem is defined as

$$\text{Minimize} \qquad F(P_{g_i}) = \sum_{i=1}^{NG} F_i = \sum_{i=1}^{NG} (a_i P_{g_i}^2 + b_i P_{g_i} + c_i) \ \text{₹/h} \tag{3.135a}$$

subject to (i) the energy balance equation

$$\sum_{i=1}^{NG} P_{g_i} = \sum_{i=1}^{NB} P_{d_i} + P_L \tag{3.135b}$$

(ii) the inequality constraints

$$P_{g_i}^{\min} \le P_{g_i} \le P_{g_i}^{\max} \qquad (i = 1, 2,..., \text{NG}) \tag{3.135c}$$

where

a_i, b_i, c_i are the cost coefficients

P_{d_i} is the load demand at the *i*th bus

P_{g_i} is the real power generation (decision variable)

NG is the number of generation buses

P_L is the transmission power loss.

This method uses the fact that under normal operating conditions, the transmission loss is quadratic in the injected bus real powers. The general form of the loss formula using *B*-coefficients is [Eq. (3.126)]

$$P_L = \sum_{i=1}^{NB} \sum_{j=1}^{NB} P_i B_{ij} P_j \text{ MW} \tag{3.136}$$

where

$$B_{ij} = \frac{R_{ij} \cos(\theta_i - \theta_j)}{|V_i||V_j| \cos\phi_i \cos\phi_j}$$

$$\theta_i = \delta_i - \phi_i \qquad (i = 1, 2, ..., \text{NB})$$

$$P_i = P_{g_i} - P_{d_i} \qquad (i = 1, 2, ..., \text{NB})$$

$$\phi_i = \tan^{-1}(Q_i/P_i)$$

P_i and P_j are the real power injections at *i*th and *j*th buses, respectively

Q_i and Q_j are the reactive power injections at *i*th and *j*th buses, respectively

NB is the number of buses in the network

$Z_{ij} = R_{ij} + jX_{ij}$ (elements of impedance matrix).

The above constrained optimization problem is converted into an unconstrained one. Lagrange multiplier method is used in which a function is minimized (or maximized) with conditions in the form of equality constraints. Using Lagrange multipliers, an augmented function is defined as

$$L(P_{g_i}, \lambda_p) = F(P_{g_i}) + \lambda_p \left(\sum_{i=1}^{NB} P_{d_i} + P_L - \sum_{i=1}^{NG} P_{g_i} \right) \tag{3.137}$$

where λ_p is the Lagrangian multiplier.

The necessary conditions for the optimization problem, Eq. (3.137), state that the derivatives with respect to control/decision variables P_{g_i} $(i = 1, 2,..., \text{NG})$ and λ_p, are equal to zero, i.e.

$$\frac{\partial L(P_{g_i}, \lambda_p)}{\partial P_{g_i}} = \frac{\partial F_i}{\partial P_{g_i}} + \lambda_p \left(\frac{\partial P_L}{\partial P_{g_i}} - 1 \right) = 0 \qquad (i = 1, 2, ..., \text{NG}) \tag{3.138a}$$

$$\frac{\partial L(P_{g_i}, \lambda_p)}{\partial \lambda_p} = \sum_{i=1}^{\text{NB}} P_{d_i} + P_L - \sum_{i=1}^{\text{NG}} P_{g_i} = 0 \tag{3.138b}$$

where

$$\frac{\partial P_L}{\partial P_{g_i}} = \sum_{j=1}^{\text{NB}} (B_{ij} + B_{ji}) P_j \qquad (i = 1, 2, ..., \text{NG}) \tag{3.138c}$$

$$\frac{\partial F(P_{g_i})}{\partial P_{g_i}} = 2a_i P_{g_i} + b_i \qquad (i = 1, 2, ..., \text{NG}) \tag{3.138d}$$

The solution of nonlinear Eqs. (3.138a) and (3.138b) can be obtained using the Newton–Raphson method in which change in $P_{g_i} (i = 1, 2, ..., \text{NG})$ and λ_p is obtained by expanding Eqs. (3.138a) and (3.138b) about the initial value using Taylor's expansion, i.e.

$$\frac{\partial^2 L}{\partial P_{g_i}^2} \Delta P_{g_i} + \sum_{\substack{j=1 \\ j \neq i}}^{\text{NG}} \frac{\partial^2 L}{\partial P_{g_i} \partial P_{g_j}} \Delta P_{g_j} + \frac{\partial^2 L}{\partial P_{g_i} \partial \lambda_p} \Delta \lambda_p = -\frac{\partial L}{\partial P_{g_i}} \qquad (i = 1, 2, ..., \text{NG}) \tag{3.139a}$$

or

$$\sum_{j=1}^{\text{NG}} \frac{\partial^2 L}{\partial \lambda_p \partial P_{g_j}} \Delta P_{g_j} + \frac{\partial^2 L}{\partial \lambda_p^2} \Delta \lambda_p = -\frac{\partial L}{\partial \lambda_p} \tag{3.139b}$$

In the matrix form the above equations can be rewritten as

$$\begin{bmatrix} \nabla_{P_g P_g} L & \nabla_{P_g \lambda_p} L \\ (\nabla_{P_g \lambda_p} L)^T & \nabla_{\lambda_p \lambda_p} L \end{bmatrix} \begin{bmatrix} \Delta P_g \\ \Delta \lambda_p \end{bmatrix} = - \begin{bmatrix} \nabla_{P_g} L \\ \nabla_{\lambda_p} L \end{bmatrix} \tag{3.140}$$

Differentiating transmission loss Eq. (3.138c) with respect to P_{g_j},

$$\frac{\partial^2 P_L}{\partial P_{g_i} \partial P_{g_j}} = B_{ij} + B_{ji} \quad (i = 1, 2, ..., \text{NG}; j = 1, 2, ..., \text{NG}) \tag{3.141}$$

All the derivatives for expressions given in Eqs. (3.139a) and (3.139b) are known and available from Eqs. (3.26a–3.26d) and Eq. (3.141). Using Gauss elimination method, Eqs. (3.139a) and (3.139b) are solved to find change in control variables, namely $P_{g_i} (i = 1, 2, ..., \text{NG})$ and λ_p. The control variables are updated. The above procedure is repeated till no further improvement is achieved. The detailed stepwise procedure is outlined in Algorithm 3.8.

Algorithm 3.8: Economic Dispatch Using Exact Loss Formula

1. Read data: NG (number of buses having generators), NB (number of buses), NV (number of *PV* buses). V_1, δ_1 for slack bus, P_{d_i}, $Q_{d_i} (i = 1, 2, ..., \text{NB})$. V_i^s $(i = 2, 3, ..., \text{NV})$ for *PV* buses,

$V_i^{\min}$, $V_i^{\max}$ (i = NV + 1, NV + 2,..., NB) for *PQ* buses. $Q_i^{\min}$, $Q_i^{\max}$ (i = 2, 3,.., NV) for *PV* buses, a_i, b_i, c_i(cost coefficients); i = 1, 2, ..., NG. R_1, R_2 (maximum number of iterations), ε_1, ε_2 (tolerance in convergence), etc.

2. Obtain Y_{BUS} and by inverting it obtain Z_{BUS}.
3. Compute the initial values of P_{g_i} (i = 1, 2,.., NG) and λ_p by assuming that $P_L = 0$. So, the initial values can be obtained directly using Eqs. (3.9) and (3.10).
4. For *PQ* buses, the total reactive power demand is distributed to various generators in such a manner that the power factor at all the generator buses remains the same.

$$Q_{g_i} = P_{g_i} \times \sum_{i=1}^{\text{NB}} Q_{d_i} \Bigg/ \sum_{i=1}^{\text{NB}} P_{d_i}$$

5. Compute cost on initial schedule and consider as previous cost (F^{prev}).
6. Set iteration counter II = 1.
7. Compute $P_i = P_{g_i} - P_{d_i} \qquad (i = 1, 2, ..., \text{NB})$

   $Q_i = Q_{g_i} - Q_{d_i} \qquad (i = 1, 2, ..., \text{NB for } PQ\text{-buses only})$

   Take $P_{g_i} = 0$, $Q_{g_i} = 0$ for non-generating buses.
8. Perform load flow to obtain real and reactive power, P_i, Q_i and voltage magnitude and angles $|V_i|$, δ_i at each bus as explained in Section 2.11.
9. Check at slack bus that $|P_{g_s} - P_{d_s} - P_s| \le \varepsilon_2$, if 'yes' GOTO Step 23.
10. Compute $P_{g_s} = P_s + P_{d_s}$ for slack bus and $Q_{g_i} = Q_i + Q_{d_i}$ for slack and *PV*-buses.
11. Compute loss coefficients B_{ij}, using Eq. (3.136).
12. Assume/set that no generation has been fixed either at lower or at upper limits.
13. Set iteration counter, III = 1.
14. Evaluate Hessian and Jacobian matrix elements using Eq. (3.138) and Eq. (3.141). Deactivate the row and column of Hessian matrix and the row of Jacobian matrix representing the generator whose generation is fixed either at lower limit or at upper limit. This is done so that fixed generators do not participate in generation allocation.
15. Gauss elimination method is used in which triangularization and back substitution processes are performed to find ΔP_{g_i} (i = 1, 2, ..., R) and $\Delta\lambda_p$. Here R is the number of generators which can participate in the allocation.
16. Check $\sqrt{\sum_{i=1}^{R} (\Delta P_{g_i})^2 + (\Delta\lambda_p)^2} \le \varepsilon_1$ or $\sqrt{\sum_{i=1}^{R} \left(\dfrac{\partial L}{\partial P_{g_i}}\right)^2 + \left(\dfrac{\partial L}{\partial \lambda_p}\right)^2} \le \varepsilon_1$

    if 'yes' then GOTO Step 20.
17. Modify $P_{g_i}^{\text{new}} = P_{g_i} + \Delta P_{g_i}$ (i = 1, 2, ..., R) and $\lambda_p^{\text{new}} = \lambda_p + \Delta\lambda_p$.
18. Check III $\ge R_1$,
    if 'yes' then GOTO Step 20 (without convergence),
    else III = III + 1, $P_{g_i} = P_{g_i}^{\text{new}}$ (i = 1, 2, ..., R), $\lambda_p = \lambda_p^{\text{new}}$,
    GOTO Step 14 and repeat.

19. Check the limits of generators. If, no limit is violated further, then GOTO Step 20, else fix the limits as following:

    If $P_{g_i} < P_{g_i}^{\min}$ then $P_{g_i} = P_{g_i}^{\min}$

    If $P_{g_i} > P_{g_i}^{\max}$ then $P_{g_i} = P_{g_i}^{\max}$

    GOTO Step 13.
20. Compute the optimal total cost F, transmission loss P_L, etc.
21. Check cost $|F^{\text{prev}} - F| \le \varepsilon_2$, if 'yes' GOTO Step 23.
22. Check II $\ge R_2$

    if 'yes' then GOTO Step 23 (without convergence)

    else II = II + 1, $F^{\text{prev}} = F$, GOTO Step 7 and repeat.
23. Stop.

EXAMPLE 3.11 A 5-bus system is shown in Figure 3.17. The series impedance and the shunt admittance of each line are given in Table 3.20. The system has three generators. The operating cost characteristics of three generators are given below. Find the economic generation schedule.

$$F_1(P_{g_1}) = 60P_{g_1}^2 + 200\,P_{g_1} + 140 \quad ₹/\text{h}$$

$$F_2(P_{g_2}) = 75P_{g_2}^2 + 150\,P_{g_2} + 120 \quad ₹/\text{h}$$

$$F_3(P_{g_3}) = 70P_{g_3}^2 + 180\,P_{g_3} + 80 \quad ₹/\text{h}$$

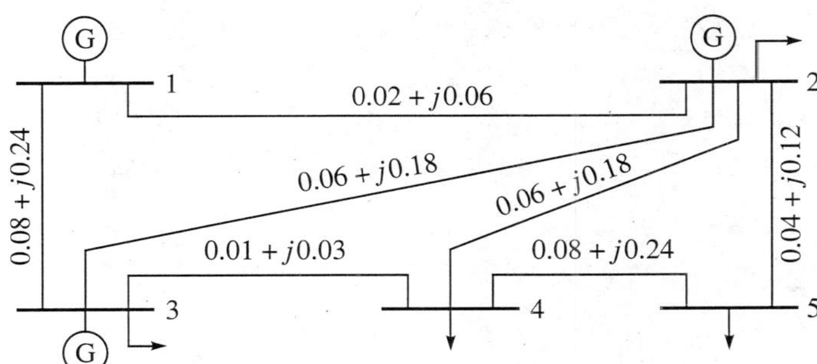

FIGURE 3.17 Power system network (Example 3.11).

TABLE 3.20 Line data (p.u.) (Example 3.11)

Line	*Sending end bus*	*Ending end bus*	Y_{SH}	Z_{SER}
1	1	2	$j0.06$	$0.02 + j0.06$
2	1	3	$j0.05$	$0.08 + j0.24$
3	2	3	$j0.04$	$0.06 + j0.18$
4	2	4	$j0.04$	$0.06 + j0.18$
5	2	5	$j0.03$	$0.04 + j0.12$
6	3	4	$j0.02$	$0.01 + j0.03$
7	4	5	$j0.05$	$0.08 + j0.24$

Solution Y_{BUS} is obtained and its elements are given below:

$$Y_{BUS} = \begin{bmatrix} 6.25 & -5.0 & -1.25 & 0.0 & 0.0 \\ +j18.64 & +j15.0 & +j3.75 & +j0.0 & +j0.0 \\ -5.0 & 10.83333 & -1.66667 & -1.66667 & -2.5 \\ +j15.0 & -j32.33 & +j5.0 & +j5.0 & +j7.5 \\ -1.25 & -1.66667 & 12.91667 & -10.0 & 0.0 \\ +j3.75 & +j5.0 & -j38.640 & +j30.0 & +j0.0 \\ 0.0 & -1.66667 & -10.0 & 12.91667 & -1.25 \\ +j0.0 & +j5.0 & -j30.0 & -j38.640 & +j3.75 \\ 0.0 & -2.5 & 0.0 & -1.25 & 3.75 \\ +j0.0 & +j7.5 & +j0.0 & +j3.75 & -j11.170 \end{bmatrix}$$

By taking the inverse of Y_{BUS}, Z_{BUS} is obtained and its elements are given below:

$$Z_{BUS} = \begin{bmatrix} 0.0126019 & 0.0004494 & -0.0042793 & -0.0052528 & -0.0051958 \\ -j1.686547 & -j1.722815 & -j1.736837 & -j1.739733 & -j1.739577 \\ 0.0004494 & 0.0052821 & -0.0037327 & -0.0038368 & -0.0014543 \\ -j1.722815 & -j1.708342 & -j1.735222 & -j1.735530 & -j1.728464 \\ -0.0042793 & -0.0037327 & 0.0089317 & 0.0045316 & -0.0047161 \\ -j1.736837 & -j1.735222 & -j1.697483 & -j1.710662 & -j1.738149 \\ -0.0052528 & -0.0038368 & 0.0045316 & 0.0090162 & -0.0032723 \\ -j1.739733 & -j1.735530 & -j1.710662 & -j1.697234 & -j1.733854 \\ -0.0051958 & -0.0014543 & -0.0047161 & -0.0032723 & 0.0211987 \\ -j1.739577 & -j1.728464 & -j1.738149 & -j1.733854 & -j1.660936 \end{bmatrix}$$

Total real demand, $\sum_{i=1}^{5} P_{d_i} = 1.65$ p.u. and total reactive demand, $\sum_{i=1}^{5} Q_{d_i} = 0.40$ p.u.

The initial λ_p is obtained as

$$\lambda_p = \left(\sum_{i=1}^{5} P_{d_i} + \sum_{i=1}^{3} \frac{b_i}{2a_i} \right) \Big/ \sum_{i=1}^{3} \frac{1}{2a_i} = 253.0107$$

The initial P_{g_i} $(i = 1, 2, 3)$ and Q_{g_i} $(i = 1, 2, 3)$ are calculated following Step 2 and Step 3 of Algorithm 3.8 and are tabulated in Table 3.21 along with loads at each bus. The load flow solution is obtained using the decoupled load flow method (see Section 2.11). The convergence 0.00001 is achieved in six iterations. The voltage magnitudes, voltage angles, real and reactive powers injected at each bus are obtained and are given in Table 3.22. Loss coefficients are calculated using Eq. (3.126). The values of angles δ, ϕ, and θ are tabulated in Table 3.23. For example, $\phi_1 = \tan^{-1}(Q_1/P_1) = -0.8224847$, $\theta_1 = \delta_1 - \phi_1 = 0.8224847$; and so on for ϕ_2, ϕ_3, ϕ_4 and ϕ_5.

TABLE 3.21 Initial values considered to start procedure (Example 3.11)

Bus	P_g (p.u.)	Q_g (p.u.)	P_d (p.u.)	Q_d (p.u.)	*Type*
1	0.441756	0.107092	0.00	0.00	Slack
2	0.686738	0.166482	0.20	0.10	*PQ*
3	0.521505	0.126426	0.45	0.15	*PQ*
4	0.0	0.0	0.40	0.05	*PQ*
5	0.0	0.0	0.60	0.10	*PQ*

TABLE 3.22 Load flow solution after six iterations (Example 3.11)

Bus	$\lvert V \rvert$ (p.u.)	δ (rad)	P (p.u.)	Q (p.u.)
1	1.06	0.0	0.4600578	–0.4955128
2	1.070723	–0.0208256	0.4867406	0.0664834
3	1.068388	–0.0403014	0.0715068	–0.0235741
4	1.065702	–0.0495604	–0.4000043	–0.0499996
5	1.051906	–0.0728022	–0.6000013	–0.0999998

TABLE 3.23 Values of δ, ϕ, and θ (Example 3.11)

Bus	δ(rad)	ϕ(rad)	θ(rad)
1	0.0	– 0.8224847	0.8224847
2	– 0.0208256	0.1357489	– 0.1565744
3	– 0.0403014	– 0.3184559	0.2781545
4	– 0.0495604	0.1243527	– 0.1739131
5	– 0.0728022	0.1651480	– 0.2379501

The values of B-coefficients are given below:

$$B = \begin{bmatrix} 0.02423 & 0.00033 & -0.00500 & -0.00374 & -0.00339 \\ 0.00033 & 0.00469 & -0.00315 & -0.00342 & -0.00132 \\ -0.00500 & -0.00315 & 0.00868 & 0.00380 & -0.00390 \\ -0.00374 & -0.00342 & 0.00380 & 0.00806 & -0.00298 \\ -0.00339 & -0.00132 & -0.00390 & -0.00298 & 0.01969 \end{bmatrix}$$

Set $R = 3$ initially. Using the Newton–Raphson method, elaborated in Steps 9, 10, 11, and 12 of Algorithm 3.8, the real power generations are obtained in five iterations. The implementation of Newton–Raphson method is outlined next:

Iteration = 1: The elements of Hessian are computed as below:

$$\begin{bmatrix} 132.259200 & 0.165799 & -2.531326 & -0.971042 \\ 0.165783 & 152.374900 & -1.591480 & -0.991264 \\ -2.531279 & -1.591449 & 144.389900 & -1.004788 \\ -0.971042 & -0.991264 & -1.004788 & 0.0 \end{bmatrix} \begin{bmatrix} \Delta P_{g_1} \\ \Delta P_{g_2} \\ \Delta P_{g_3} \\ \Delta \lambda_p \end{bmatrix} = \begin{bmatrix} -7.326804 \\ -2.210387 \\ 1.211333 \\ -0.018300 \end{bmatrix}$$

Triangularizing the above equations, we get

$$\begin{bmatrix} 132.259200 & 0.165799 & -2.531326 & -0.971042 \\ 0.0 & 152.374900 & -1.588307 & -0.990046 \\ 0.0 & 0.0 & 144.3249 & -1.033692 \\ 0.0 & 0.0 & 0.0 & -0.020966 \end{bmatrix} \begin{bmatrix} \Delta P_{g_1} \\ \Delta P_{g_2} \\ \Delta P_{g_3} \\ \Delta \lambda_p \end{bmatrix} = \begin{bmatrix} -7.326804 \\ -2.201203 \\ 1.048162 \\ -0.078888 \end{bmatrix}$$

Change in generation P_{g_i} and incremental cost λ_p are obtained using back substitution,

$\Delta P_{g_1} = -0.027130$ p.u., $\Delta P_{g_2} = 0.010359$ p.u., $\Delta P_{g_3} = 0.034212$ p.u., $\Delta\lambda_p = 3.7628$ ₹/p.u.h

The different convergence criteria can be applied as

$$\left| \sum_{i=1}^{3} P_{g_i} - \sum_{i=1}^{5} P_{d_i} - P_L \right| = 0.4562624\text{E-03} > 0.0001$$

or

$$\sqrt{\sum_{i=1}^{3} \left(\frac{\partial L}{\partial P_{g_i}} \right)^2 + \left(\frac{\partial L}{\partial \lambda_p} \right)^2} = 7.74826000 > 0.0001$$

or

$$\sqrt{\sum_{i=1}^{3} (\Delta P_{g_i})^2 + (\Delta \lambda_p)^2} = 3.76299500 > 0.0001$$

No convergence criteria is satisfied, so to go for next iteration, i.e. 2, the values are updated as

$$P_{g_i} = P_{g_i} + \Delta P_{g_i} \quad (i = 1, 2, ..., \text{NG}) \quad \text{and} \quad \lambda_p = \lambda_p + \Delta\lambda_p$$

$$P_{g_1} = 0.441756 - 0.027130 = 0.414626 \text{ p.u.}$$

$$P_{g_2} = 0.686738 + 0.010359 = 0.697097 \text{ p.u.}$$

$$P_{g_3} = 0.521505 + 0.034212 = 0.555717 \text{ p.u.}$$

$$\lambda_p = 253.0107 + 3.7628 = 256.7735 \text{ ₹/p.u.h}$$

The above procedure is repeated till any one convergence criterion is satisfied. After the fourth iteration, one of the criteria is satisfied.

The real generations are given in Table 3.24 which are used as initial values for the next iteration. The cost and transmission loss after the first overall iteration is

$$\text{Cost} = 696.1357\ ₹/\text{h}, \quad P_L = 0.01829998 \text{ p.u.} \quad \text{and} \quad \lambda_p = 256.9161\ ₹/\text{p.u.h}$$

TABLE 3.24 Initial values for next iterations (Example 3.11)

Bus	P_g (p.u.)	Q_g (p.u.)	P_d (p.u.)	Q_d (p.u.)	*Type*
1	0.412303	0.107092	0.00	0.00	Slack
2	0.697811	0.166482	0.20	0.10	*PQ*
3	0.558186	0.126426	0.45	0.15	*PQ*
4	0.0	0.0	0.40	0.05	*PQ*
5	0.0	0.0	0.60	0.10	*PQ*

The above procedure is repeated following Algorithm 3.8 till the slack bus balance is achieved. After three iterations, the convergence is obtained and the final solution is given in Tables 3.25 and 3.26. The slack bus mismatch is

$$\left| P_{g_1} - P_{d_1} - P_1 \right| = 3.507733\text{E-05} < 0.0001$$

$$\lambda_p = 256.9665\ ₹/\text{p.u.h}, \ \text{Cost} = 695.8972\ ₹/\text{h}, \ P_L = 0.01739662 \text{ p.u.}$$

TABLE 3.25 Final solution (Example 3.11)

Bus	P_g (p.u.)	Q_g (p.u.)	P_d (p.u.)	Q_d (p.u.)	P (p.u.)	Q (p.u.)
1	0.412717	0.107092	0.00	0.00	0.412752	−0.499097
2	0.698472	0.166482	0.20	0.10	0.498471	0.066481
3	0.556208	0.126426	0.45	0.15	0.106206	−0.023574
4	0.0	0.0	0.40	0.05	−0.399999	−0.049999
5	0.0	0.0	0.60	0.10	−0.600000	−0.100001

TABLE 3.26 Voltage magnitude and angle (Example 3.11)

Bus	V (p.u.)	δ (rad)
1	1.060000	0.0
2	1.071436	−0.0191460
3	1.069649	−0.0371636
4	1.066864	−0.0466896
5	1.052804	−0.0706622

3.14 ECONOMIC DISPATCH USING LOSS FORMULA WHICH IS FUNCTION OF REAL AND REACTIVE POWER

The economic dispatch problem is defined as to minimize the total operating cost of a power system while meeting total real load plus real transmission losses within the generator limits of real power. Mathematically, the problem is defined in Eqs. (3.135a)–(3.135c).

This method uses the fact that under normal operating conditions, the transmission loss is quadratic in the injected bus real powers. The general form of the loss formula using Eq. (3.132) is

$$P_L = \sum_{i=1}^{NB} \sum_{j=1}^{NB} [a_{ij}(P_iP_j + Q_iQ_j) + b_{ij}(Q_iP_j - P_iQ_j)] \tag{3.142}$$

where

$$a_{ij} = \frac{R_{ij}}{|V_i||V_j|}\cos(\delta_i - \delta_j);\; b_{ij} = \frac{R_{ij}}{|V_i||V_j|}\sin(\delta_i - \delta_j)$$

$$P_i + jQ_i = (P_{g_i} - P_{d_i}) + j(Q_{g_i} - Q_{d_i}) \qquad (i = 1, 2, ..., \text{NB})$$

P_i and P_j are real power injections at ith and jth buses, respectively

Q_i and Q_j are reactive power injections at ith and jth buses, respectively

P_{d_i} and Q_{d_i} are real and the reactive power load demands at the ith bus, respectively

P_{g_i} and Q_{g_i} are real and the reactive power generations at the ith bus, respectively

NB is the number of buses in the network

$Z_{ij} = R_{ij} + jX_{ij}$ (elements of impedance matrix).

Using the Lagrange multiplier method, the constrained optimization problem given in Eq. (3.135) is converted into unconstrained one and is given in Eq. (3.137).

The necessary conditions for the optimization problem, given by Eq. (3.137), state that the derivatives with respect to control/decision variables (P_{g_i}, Q_{g_i} $(i = 1, 2,.., \text{NG})$ and λ_p), are equal to zero, i.e.

$$\frac{\partial L}{\partial P_{g_i}} = \frac{\partial F_i}{\partial P_{g_i}} + \lambda_p\left(\frac{\partial P_L}{\partial P_{g_i}} - 1\right) = 0 \qquad (i = 1, 2, ..., \text{NG}) \tag{3.143a}$$

$$\frac{\partial L}{\partial Q_{g_i}} = \lambda_p\left(\frac{\partial P_L}{\partial Q_{g_i}}\right) = 0 \qquad (i = 1, 2, ..., \text{NG}) \tag{3.143b}$$

$$\frac{\partial L}{\partial \lambda_p} = \sum_{i=1}^{NB} P_{d_i} + P_L - \sum_{i=1}^{NG} P_{g_i} = 0 \tag{3.143c}$$

where incremental transmission losses are expressed as

$$\frac{\partial P_L}{\partial P_{g_i}} = 2a_{ii}P_i + \sum_{\substack{j=1\\ j\neq i}}^{NB} [(a_{ij} + a_{ji})P_j + (b_{ji} - b_{ij})Q_j] \qquad (i = 1, 2, ..., \text{NG}) \tag{3.144}$$

$$\frac{\partial P_L}{\partial Q_{g_i}} = 2a_{ii}Q_i + \sum_{\substack{j=1 \\ j \neq i}}^{NB} [(a_{ij} + a_{ji})\, Q_j + (b_{ij} - b_{ji})\, P_j] \qquad (i = 1, 2, ..., NG) \qquad (3.145)$$

The solution of nonlinear Eqs. (3.143a–3.143c) can be obtained using the Newton–Raphson method in which change in P_{g_i}, Q_{g_i} (i = 1, 2,.., NG) and λ_p is obtained by expanding Eqs. (3.143a–3.143c) about the initial values using Taylor's expansion.

$$\sum_{j=1}^{NG} \frac{\partial^2 L}{\partial P_{g_i} \partial P_{g_j}} \Delta P_{g_j} + \sum_{j=1}^{NG} \frac{\partial^2 L}{\partial P_{g_i} \partial Q_{g_j}} \Delta Q_{g_j} + \frac{\partial^2 L}{\partial P_{g_i} \partial \lambda_p} \Delta \lambda_p = -\frac{\partial L}{\partial P_{g_i}} \quad (i = 1, 2, ..., NG) \quad (3.146a)$$

$$\sum_{j=1}^{NG} \frac{\partial^2 L}{\partial Q_{g_i} \partial P_{g_j}} \Delta P_{g_j} + \sum_{j=1}^{NG} \frac{\partial^2 L}{\partial Q_{g_i} \partial Q_{g_j}} \Delta Q_{g_j} + \frac{\partial^2 L}{\partial Q_{g_i} \partial \lambda_p} \Delta \lambda_p = -\frac{\partial L}{\partial Q_{g_i}} \quad (i = 1, 2, ..., NG) \quad (3.146b)$$

$$\sum_{j=1}^{NG} \frac{\partial^2 L}{\partial \lambda_p \partial P_{g_j}} \Delta P_{g_j} + \sum_{j=1}^{NG} \frac{\partial^2 L}{\partial \lambda_p \partial Q_{g_j}} \Delta Q_{g_j} + \frac{\partial^2 L}{\partial \lambda_p^2} \Delta \lambda_p = -\frac{\partial L}{\partial \lambda_p} \quad (3.146c)$$

Second-order differential expressions are presented below which are required for expressions given by Eqs. (3.146a–3.146c). In addition to Eq. (3.26a–3.26d), the following expressions are obtained from Eq. (3.143a)

$$\frac{\partial^2 L}{\partial P_{g_i} \partial P_{g_j}} = \lambda_p \frac{\partial^2 P_L}{\partial P_{g_i} \partial P_{g_j}} \qquad (i = 1, 2, ..., NG; \;\; j = 1, 2, ..., NG; \; i \neq j) \qquad (3.147a)$$

From Eq. (3.143b), the following can be obtained:

$$\frac{\partial^2 L}{\partial Q_{g_i} \partial Q_{g_j}} = \lambda_p \frac{\partial^2 P_L}{\partial Q_{g_i} \partial Q_{g_j}} \qquad (i = 1, 2, ..., NG; \;\; j = 1, 2, ..., NG) \qquad (3.147b)$$

$$\frac{\partial^2 L}{\partial \lambda_p \partial Q_{g_i}} = \frac{\partial^2 L}{\partial Q_{g_i} \partial \lambda_p} = \frac{\partial P_L}{\partial Q_{g_i}} \qquad (i = 1, 2, ..., NG) \qquad (3.147c)$$

The second-order differential expressions, required for Eqs. (3.26a–3.26d) and Eqs. (3.147a–3.147c) are presented below. These are obtained from Eqs. (3.144) and (3.145), respectively.

$$\frac{\partial^2 P_L}{\partial P_{g_i} \partial P_{g_j}} = \frac{\partial^2 P_L}{\partial Q_{g_i} \partial Q_{g_j}} = a_{ij} + a_{ji} \qquad (i = 1, 2, ..., NG; \;\; j = 1, 2, ..., NG) \qquad (3.147d)$$

$$\frac{\partial^2 P_L}{\partial P_{g_i} \partial Q_{g_j}} = -\frac{\partial^2 P_L}{\partial Q_{g_i} \partial P_{g_j}} = b_{ji} - b_{ij} \qquad (i = 1, 2, .., NG, \;\; j = 1, 2, .., NG) \qquad (3.147e)$$

The nonlinear equations (3.146a) to (3.146c) are solved using the Newton–Raphson method. To solve this problem, Algorithm 3.8 can be implemented.

EXAMPLE 3.12 Consider the 5-bus system of Example 3.11 and obtain the optimum schedule.

Solution Y_{BUS} elements and Z_{BUS} elements are same as those given in Example 3.11. The initial values and load flow solutions are also same as those of Example 3.11 and are given in Table 3.21 and Table 3.22, respectively. The units are in p.u. system.

The loss coefficients are tabulated in Tables 3.27 and 3.28.

$$a_{ij} = \frac{R_{ij}}{|V_i||V_j|}\cos(\delta_i - \delta_j) \qquad (i = 1, 2, ..., 5; \;\; j = 1, 2, ..., 5)$$

$$b_{ij} = \frac{R_{ij}}{|V_i||V_j|}\sin(\delta_i - \delta_j) \qquad (i = 1, 2, ..., 5; \;\; j = 1, 2, ..., 5)$$

TABLE 3.27 *a*-coefficients for real transmission loss P_L (Example 3.12)

i	a_{i1}	a_{i2}	a_{i3}	a_{i4}	a_{i5}
1	0.0112156	0.0003959	−0.0037756	−0.0046442	−0.0046475
2	0.0003959	0.0046074	−0.0032624	−0.0033610	−0.0012894
3	−0.0037755	−0.0032623	0.0078249	0.0039799	−0.0041942
4	−0.0046442	−0.0033610	0.0039799	0.0079387	−0.0029183
5	−0.0046474	−0.0012894	−0.0041942	−0.0029183	0.0191583

TABLE 3.28 *b*-coefficients for real transmission loss P_L (Example 3.12)

i	b_{i1}	b_{i2}	b_{i3}	b_{i4}	b_{i5}
1	0.0	0.0000082	−0.0001522	−0.0002304	−0.0003389
2	−0.0000082	0.0	−0.0000635	−0.0000966	−0.0000671
3	0.0001522	0.0000635	0.0	0.0000369	−0.0001364
4	0.0002304	0.0000966	−0.0000369	0.0	−0.0000678
5	0.0003389	0.0000671	0.0001364	0.0000678	0.0

Using the Newton–Raphson method elaborated in Algorithm 3.8, the real and reactive power generations are obtained in six iterations

The Hessian matrix elements are obtained as given below:

$$H_{ii} = \frac{\partial^2 L}{\partial P_{g_i}^2} = 2a_i + 2\lambda_p a_{ii} \qquad (i = 1, 2, 3)$$

$$H_{ij} = \frac{\partial^2 L}{\partial P_{g_i}\partial P_{g_j}} = \lambda_p(a_{ij} + a_{ji}) \qquad (i = 1, 2, 3; \;\; j = 1, 2, 3; \;\; \text{and} \;\; i \neq j)$$

$$H_{ik} = \frac{\partial^2 L}{\partial P_{g_i}\partial Q_{g_j}} = \lambda_p(b_{ji} - b_{ij}) \qquad (i = 1, 2, 3; \;\; j = 1, 2, 3; \;\; k = 4, 5, 6)$$

$$H_{ik} = \frac{\partial^2 L}{\partial P_{g_i} \partial \lambda_p} = \left[\sum_{j=1}^{NB} \left((a_{ij} + a_{ji}) P_j + (b_{ji} - b_{ij}) Q_j \right) \right] - 1.0 \qquad (i = 1, 2, 3; \quad k = 7)$$

$$H_{kj} = \frac{\partial^2 L}{\partial Q_{g_i} \partial P_{g_j}} = \lambda_p (b_{ij} - b_{ji}) \qquad (i = 1, 2, 3; \quad j = 1, 2, 3; \quad k = 4, 5, 6)$$

$$H_{kl} = \frac{\partial^2 L}{\partial Q_{g_i} \partial Q_{g_j}} = \lambda_p (a_{ij} + a_{ji}) \qquad (i = 1, 2, 3; \quad j = 1, 2, 3; \quad k = 4, 5, 6; \quad l = 4, 5, 6)$$

$$H_{kl} = \frac{\partial^2 L}{\partial Q_{g_i} \partial \lambda_p} = \sum_{j=1}^{NB} \left((a_{ij} + a_{ji}) Q_j + (b_{ij} - b_{ji}) P_j \right) \qquad (i = 1, 2, 3; \quad k = 4, 5, 6; \quad l = 7)$$

$$H_{ki} = \frac{\partial^2 L}{\partial \lambda_p \partial P_{g_i}} = \frac{\partial^2 L}{\partial P_{g_i} \partial \lambda_p} \qquad (i = 1, 2, 3; \quad k = 7)$$

$$H_{kl} = \frac{\partial^2 L}{\partial \lambda_p \, \partial Q_{g_i}} = \frac{\partial^2 L}{\partial Q_{g_i} \partial \lambda_p} \qquad (i = 1, 2, 3; \quad k = 7; \quad l = 4, 5, 6)$$

$$H_{77} = \frac{\partial^2 L}{\partial \lambda_p^2} = 0.0$$

The Jacobian matrix elements are given below:

$$\frac{\partial L}{\partial P_{g_i}} = 2 a_i P_{g_i} + b_i + \lambda_p \left[\left(\sum_{j=1}^{NB} \left((a_{ij} + a_{ji}) P_j + (b_{ji} - b_{ij}) Q_j \right) \right) - 1.0 \right] \qquad (i = 1, 2, 3)$$

$$\frac{\partial L}{\partial Q_{g_i}} = \lambda_p \left(\sum_{j=1}^{NB} \left((a_{ij} + a_{ji}) Q_j + (b_{ij} - b_{ji}) P_j \right) \right) \qquad (i = 1, 2, 3)$$

$$\frac{\partial L}{\partial \lambda_p} = \sum_{i=1}^{NB} P_{d_i} + \left(\sum_{i=1}^{NB} \sum_{j=1}^{NB} a_{ij} (P_i P_j + Q_i Q_j) + b_{ij} (Q_i P_j - P_i Q_j) \right) - \sum_{i=1}^{NG} P_{g_i}$$

Seven simultaneous equations are solved using Gauss elimination method. The modified generations which are used as initial values for the next iteration are given below:

$P_{g_1} = 0.412302$ p.u., $P_{g_2} = 0.697811$ p.u., $P_{g_3} = 0.558187$ p.u.

$Q_{g_1} = 0.077638$ p.u., $Q_{g_2} = 0.177555$ p.u., $Q_{g_3} = 0.163107$ p.u.

The convergence obtained in six iterations is given below:

$$\left| \sum_{i=1}^{3} P_{g_i} - \sum_{i=1}^{5} P_{d_i} - P_L \right| = 0.00000002 < 0.00001$$

The values of λ_p, P_L, and cost at present obtained schedule are given as

$$\lambda_p = 255.8830 \text{ ₹/p.u.h}; \quad \text{Cost} = 695.35550 \text{ ₹/h}, \quad P_L = 0.01558135 \text{ p.u.}$$

The initial values for the next iteration are given in Table 3.29.

TABLE 3.29 Initial values for next iterations (Example 3.12)

Bus	P_g (p.u.)	Q_g (p.u.)	P_d (p.u.)	Q_d (p.u.)	*Type*
1	0.426539	−0.151419	0.00	0.00	Slack
2	0.691533	−0.116962	0.20	0.10	*PQ*
3	0.547509	−0.062548	0.45	0.15	*PQ*
4	0.0	0.0	0.40	0.05	*PQ*
5	0.0	0.0	0.60	0.10	*PQ*

After four iterations the convergence is obtained to get the final solution as

$$\left| P_{g_1} - P_{d_1} - P_1 \right| = 6.437302\text{E-}06 < 0.00001$$

The final solution is given in Table 3.30. The convergence is obtained in each iteration and is tabulated in Table 3.31.

TABLE 3.30 Final solution (Example 3.12)

Bus	P_g (p.u.)	Q_g (p.u.)	P_d (p.u.)	Q_d (p.u.)	P (p.u.)	Q (p.u.)	$\lvert V \rvert$ (p.u.)	δ rad
1	0.426513	−0.150459	0.00	0.00	0.426959	−0.013799	1.06	0.00
2	0.691320	−0.114766	0.20	0.10	0.491319	−0.214768	1.0503470	−0.0134539
3	0.548050	−0.056980	0.45	0.15	0.098050	−0.206977	1.0423970	−0.0305688
4	0.0	0.0	0.40	0.05	−0.400000	−0.050002	1.0405910	−0.0408579
5	0.0	0.0	0.60	0.10	−0.599999	−0.099999	1.0292560	−0.0667104

TABLE 3.31 Convergence during iterations (Example 3.12)

Iteration	$P_{g_1} - P_{d_1}$	P_1	$\lvert P_{g_1} - P_{d_1} - P_1 \rvert$
1	4.417562E-01	4.600578E-01	−1.830158E-02
2	4.265393E-01	4.273436E-01	−8.042753E-04
3	4.265130E-01	4.269587E-01	−4.456937E-04

The optimum values of λ_p, cost, and transmission loss are given below:

$$\lambda_p = 255.9390 \text{ ₹/p.u.h, Cost} = 695.43360 \text{ ₹/h, } P_L = 0.01612151 \text{ p.u.}$$

3.15 ECONOMIC DISPATCH FOR ACTIVE AND REACTIVE POWER BALANCE

The objective of economic dispatch problem is to minimize the total operating cost of a power system while meeting the total real load plus real transmission losses within the generator limits of real power as well as reactive load plus reactive transmission losses within the generator limits of reactive power. Mathematically, the problem is defined as

Minimize
$$F(P_{g_i}) = \sum_{i=1}^{NG} (a_i P_{g_i}^2 + b_i P_{g_i} + c_i) \text{ ₹/h} \tag{3.148a}$$

subject to
$$\sum_{i=1}^{NG} P_{g_i} = \sum_{i=1}^{NB} P_{d_i} + P_L \tag{3.148b}$$

$$\sum_{i=1}^{NG} Q_{g_i} = \sum_{i=1}^{NB} Q_{d_i} + Q_L \tag{3.148c}$$

$$P_{g_i}^{\min} \le P_{g_i} \le P_{g_i}^{\max} \quad (i = 1, 2, \ldots, NG) \tag{3.148d}$$

$$Q_{g_i}^{\min} \le Q_{g_i} \le Q_{g_i}^{\max} \quad (i = 1, 2, \ldots, NG) \tag{3.148e}$$

where

a_i, b_i, and c_i are cost coefficients

P_{d_i} is the real power load demand at the ith bus

P_{g_i} is the real power generation (decision variable)

Q_{d_i} is the reactive power load demand at the ith bus

Q_{g_i} is the reactive power generation (decision variable)

P_L is transmission real power loss

Q_L is transmission reactive power loss

NG is the number of generation buses

NB is the number of buses in the network.

This method uses the fact that under normal operating conditions, the transmission loss is quadratic in the injected bus real powers. The general form of the loss formula obtained in Eqs. (3.132) and (3.134) is

$$P_L = \sum_{i=1}^{NB} \sum_{j=1}^{NB} [a_{ij}(P_i P_j + Q_i Q_j) + b_{ij}(Q_i P_j - P_i Q_j)] \tag{3.149a}$$

$$Q_L = \sum_{i=1}^{NB} \sum_{j=1}^{NB} [c_{ij}(P_i P_j + Q_i Q_j) + d_{ij}(Q_i P_j - P_i Q_j)] \tag{3.149b}$$

where

$$a_{ij} = \frac{R_{ij}}{|V_i||V_j|} \cos(\delta_i - \delta_j)$$

$$b_{ij} = \frac{R_{ij}}{|V_i||V_j|} \sin(\delta_i - \delta_j)$$

$$c_{ij} = \frac{X_{ij}}{|V_i||V_j|} \cos(\delta_i - \delta_j)$$

$$d_{ij} = \frac{X_{ij}}{|V_i||V_j|} \sin(\delta_i - \delta_j)$$

$P_i + jQ_i = (P_{g_i} - P_{d_i}) + j(Q_{g_i} - Q_{d_i}) \qquad (i = 1, 2, ..., \text{NB})$

P_i and P_j are the real power injections at the ith and jth buses, respectively

Q_i and Q_j are the reactive power injections at the ith and jth buses, respectively

$Z_{ij} = R_{ij} + jX_{ij}$ (elements of impedance matrix).

Using the Lagrange multiplier method, the constrained optimization problem given by Eq. (3.148) is converted into an unconstrained one.

$$L(P_{g_i}, Q_{g_i}, \lambda_p) = F(P_{g_i}) + \lambda_p\left(\sum_{i=1}^{\text{NB}} P_{d_i} + P_L - \sum_{i=1}^{\text{NG}} P_{g_i}\right) + \lambda_q\left(\sum_{i=1}^{\text{NB}} Q_{d_i} + Q_L - \sum_{i=1}^{\text{NG}} Q_{g_i}\right) \tag{3.150}$$

where λ_p is the Lagrangian multiplier.

Necessary conditions for optimization problem stated by Eq. (3.150) are

$$\frac{\partial L}{\partial P_{g_i}} = \frac{\partial F}{\partial P_{g_i}} + \lambda_p\left(\frac{\partial P_L}{\partial P_{g_i}} - 1\right) + \lambda_q \frac{\partial Q_L}{\partial P_{g_i}} = 0 \qquad (i = 1, 2, ..., \text{NG}) \tag{3.151a}$$

$$\frac{\partial L}{\partial Q_{g_i}} = \lambda_p\left(\frac{\partial P_L}{\partial Q_{g_i}}\right) + \lambda_q\left(\frac{\partial Q_L}{\partial Q_{g_i}} - 1\right) = 0 \qquad (i = 1, 2, ..., \text{NG}) \tag{3.151b}$$

$$\frac{\partial L}{\partial \lambda_p} = \sum_{i=1}^{\text{NB}} P_{d_i} + P_L - \sum_{i=1}^{\text{NG}} P_{g_i} = 0 \tag{3.151c}$$

$$\frac{\partial L}{\partial \lambda_q} = \sum_{i=1}^{\text{NB}} Q_{d_i} + Q_L - \sum_{i=1}^{\text{NG}} Q_{g_i} = 0 \tag{3.151d}$$

where the incremental transmission loss expressions are expressed here as

$$\frac{\partial P_L}{\partial P_{g_i}} = 2a_{ii}P_i + \sum_{\substack{j=1 \\ j\neq i}}^{\text{NB}} [(a_{ij} + a_{ji})P_j + (b_{ji} - b_{ij})Q_j] \qquad (i = 1, 2, ..., \text{NG}) \tag{3.152a}$$

$$\frac{\partial P_L}{\partial Q_{g_i}} = 2a_{ii}Q_i + \sum_{\substack{j=1 \\ j\neq i}}^{\text{NB}} [(a_{ij} + a_{ji})Q_j + (b_{ij} - b_{ji})P_j] \qquad (i = 1, 2, ..., \text{NG}) \tag{3.152b}$$

$$\frac{\partial Q_L}{\partial P_{g_i}} = 2c_{ii}P_i + \sum_{\substack{j=1\\ j\neq i}}^{NB} [(c_{ij} + c_{ji})P_j + (d_{ji} - d_{ij})Q_j] \qquad (i = 1, 2, ..., NG) \qquad (3.152c)$$

$$\frac{\partial Q_L}{\partial Q_{g_i}} = 2c_{ii}Q_i + \sum_{\substack{j=1\\ j\neq i}}^{NB} [(c_{ij} + c_{ji})Q_j + (d_{ij} - d_{ji})P_j] \qquad (i = 1, 2, ..., NG) \qquad (3.152d)$$

The solution of nonlinear Eqs. (3.151a) to (3.151d) can be obtained using the Newton–Raphson method in which change in variables, P_{g_i}, Q_{g_i} $(i = 1, 2,.., NG)$, λ_p and λ_q are obtained by expanding Eqs. (3.151a) to (3.151d) about the initial values of the variables using Taylor's expansion. In the matrix form the above equations can be rewritten as

$$\begin{bmatrix} \nabla_{P_gP_g}L & \nabla_{P_gQ_g}L & \nabla_{P_g\lambda_p}L & \nabla_{P_g\lambda_q}L \\ \nabla_{P_gQ_g}L & \nabla_{Q_gQ_g}L & \nabla_{Q_g\lambda_p}L & \nabla_{Q_g\lambda_q}L \\ (\nabla_{P_g\lambda_p}L)^T & (\nabla_{Q_g\lambda_p}L)^T & \nabla_{\lambda_p\lambda_p}L & \nabla_{\lambda_p\lambda_q}L \\ (\nabla_{P_g\lambda_q}L)^T & (\nabla_{Q_g\lambda_q}L)^T & \nabla_{\lambda_q\lambda_p}L & \nabla_{\lambda_q\lambda_q}L \end{bmatrix} \begin{bmatrix} \Delta P_g \\ \Delta Q_g \\ \Delta\lambda_p \\ \Delta\lambda_q \end{bmatrix} = - \begin{bmatrix} \nabla_{P_g}L \\ \nabla_{Q_g}L \\ \nabla_{\lambda_p}L \\ \nabla_{\lambda_q}L \end{bmatrix} \qquad (3.153)$$

Elements of Hessian matrix derived from Eqs. (3.151a) to (3.151d) are as discussed in previous section. Equation (3.153) can be solved on the basis of the detailed Algorithm 3.8.

EXAMPLE 3.13 Consider a 5-bus system of Example 3.11 and obtain the optimum schedule.

Solution Y_{BUS} elements and Z_{BUS} elements are same as those of Example 3.11. The initial values and load flow solutions will also come out same as in Example 3.11 and are given in Table 3.21 and Table 3.22, respectively. The loss coefficients a_{ij} and b_{ij} are tabulated in Tables 3.27 and 3.28. (see Example 3.12). The loss coefficients c_{ij} and d_{ij} are tabulated below in Tables 3.32 and 3.33.

$$c_{ij} = \frac{X_{ij}}{|V_i||V_j|}\cos(\delta_i - \delta_j) \qquad (i = 1, 2, ..., 5;\ j = 1, 2, ..., 5)$$

$$d_{ij} = \frac{X_{ij}}{|V_i||V_j|}\sin(\delta_i - \delta_j) \qquad (i = 1, 2, ..., 5;\ j = 1, 2, ..., 5)$$

TABLE 3.32 *c*-coefficients for reactive transmission loss Q_L (Example 3.13)

i	c_{i1}	c_{i2}	c_{i3}	c_{i4}	c_{i5}
1	–1.501021	–1.517614	–1.532397	–1.538181	–1.555997
2	–1.517614	–1.490116	–1.516583	–1.520337	–1.532566
3	–1.532397	–1.516583	–1.487124	–1.502384	–1.545794
4	–1.538181	–1.520337	–1.502384	–1.494412	–1.546260
5	–1.555997	–1.532566	–1.545794	–1.546260	–1.501063

TABLE 3.33 *d*-coefficients for reactive transmission loss Q_L (Example 3.13)

i	d_{i1}	d_{i2}	d_{i3}	d_{i4}	d_{i5}
1	0.0	−0.0316097	−0.0617913	−0.0762953	−0.1134805
2	0.0316098	0.0	−0.0295405	−0.0436986	−0.0797294
3	0.0617913	0.0295405	0.0	−0.0139109	−0.0502571
4	0.0762953	0.0436986	0.0139109	0.0	−0.0359443
5	0.1134805	0.0797294	0.0502571	0.0359443	0.0

The Hessian matrix elements are computed using equations as given below:

$$H_{ii} = \frac{\partial^2 L}{\partial P_{g_i}^2} = 2a_i + 2\lambda_p a_{ii} + 2\lambda_q c_{ii} \qquad (i = 1, 2, 3)$$

$$H_{ij} = \frac{\partial^2 L}{\partial P_{g_i} \partial P_{g_j}} = \lambda_p(a_{ij} + a_{ji}) + \lambda_q(c_{ij} + c_{ji}) \qquad (i = 1, 2, 3; \ \ j = 1, 2, 3; \ \ \text{and} \ \ i \neq j)$$

$$H_{ik} = \frac{\partial^2 L}{\partial P_{g_i} \partial Q_{g_j}} = \lambda_p(b_{ji} - b_{ij}) + \lambda_q(d_{ji} - d_{ij}) \qquad (i = 1, 2, 3; \ \ j = 1, 2, 3; \ \ k = 4, 5, 6)$$

$$H_{i7} = \frac{\partial^2 L}{\partial P_{g_i} \partial \lambda_p} = \left[\sum_{j=1}^{NB} \left((a_{ij} + a_{ji})\, P_j + (b_{ji} - b_{ij}) Q_j \right) \right] -1.0; \qquad (i = 1, 2, 3)$$

$$H_{i8} = \frac{\partial^2 L}{\partial P_{g_i} \partial \lambda_q} = \sum_{j=1}^{NB} \left((c_{ij} + c_{ji})\, P_j + (d_{ji} - d_{ij}) Q_j \right) \qquad (i = 1, 2, 3)$$

$$H_{i+3j} = \frac{\partial^2 L}{\partial Q_{g_i} \partial P_{g_j}} = \lambda_p(b_{ij} - b_{ji}) + \lambda_q(d_{ij} - d_{ji}) \qquad (i = 1, 2, 3; \ \ j = 1, 2, 3)$$

$$H_{i+3,\,j+3} = \frac{\partial^2 L}{\partial Q_{g_i} \partial Q_{g_j}} = \lambda_p(a_{ij} + a_{ji}) + \lambda_q(c_{ij} + c_{ji}) \qquad (i = 1, 2, 3; \ \ j = 1, 2, 3)$$

$$H_{i+3,\,7} = \frac{\partial^2 L}{\partial Q_{g_i} \partial \lambda_p} = \sum_{j=1}^{NB} \left((a_{ij} + a_{ji})\, Q_j + (b_{ij} - b_{ji}) P_j \right) \qquad (i = 1, 2, 3)$$

$$H_{i+3,\,8} = \frac{\partial^2 L}{\partial Q_{g_i} \partial \lambda_q} = \sum_{j=1}^{NB} \left((c_{ij} + c_{ji})\, Q_j + (d_{ij} - d_{ji}) P_j \right) \qquad (i = 1, 2, 3)$$

$$H_{7i} = \frac{\partial^2 L}{\partial \lambda_p \partial P_{g_i}} = \frac{\partial^2 L}{\partial P_{g_i} \partial \lambda_p} \qquad (i = 1, 2, 3)$$

$$H_{7,i+3} = \frac{\partial^2 L}{\partial \lambda_p \partial Q_{g_i}} = \frac{\partial^2 L}{\partial Q_{g_i} \partial \lambda_p} \qquad (i = 1, 2, 3)$$

$$H_{8i} = \frac{\partial^2 L}{\partial \lambda_q \partial P_{gi}} = \frac{\partial^2 L}{\partial P_{gi} \partial \lambda_q} \qquad (i = 1, 2, 3)$$

$$H_{8,i+3} = \frac{\partial^2 L}{\partial \lambda_q \partial Q_{gi}} = \frac{\partial^2 L}{\partial Q_{gi} \partial \lambda_q} \qquad (i = 1, 2, 3)$$

$$H_{77} = \frac{\partial^2 L}{\partial \lambda_p^2} = H_{88} = \frac{\partial^2 L}{\partial \lambda_q^2} = H_{78} = \frac{\partial^2 L}{\partial \lambda_p \partial \lambda_q} = H_{87} = \frac{\partial^2 L}{\partial \lambda_p \partial \lambda_q} = 0.0$$

The computed Jacobian matrix elements are given below:

$$\frac{\partial L}{\partial P_{gi}} = 2a_i P_{g_i} + b_i + \lambda_p \left[\left(\sum_{j=1}^{NB} \left((a_{ij} + a_{ji}) P_j + (b_{ji} - b_{ij}) Q_j \right) \right) - 1.0 \right]$$

$$+ \lambda_q \left[\sum_{j=1}^{NB} \left((c_{ij} + c_{ji}) P_j + (d_{ji} - d_{ij}) Q_j \right) \right] \qquad (i = 1, 2, 3)$$

$$\frac{\partial L}{\partial Q_{gi}} = \lambda_p \left[\sum_{j=1}^{NB} \left((a_{ij} + a_{ji}) Q_j + (b_{ij} - b_{ji}) P_j \right) \right]$$

$$+ \lambda_q \left[\left(\sum_{j=1}^{NB} \left((c_{ij} + c_{ji}) Q_j + (d_{ij} - d_{ji}) P_j \right) \right) - 1.0 \right] \qquad (i = 1, 2, 3)$$

$$\frac{\partial L}{\partial \lambda_p} = \sum_{i=1}^{NB} P_{d_i} + \left(\sum_{i=1}^{NB} \sum_{j=1}^{NB} a_{ij}(P_i P_j + Q_i Q_j) + b_{ij}(Q_i P_j - P_i Q_j) \right) - \sum_{i=1}^{NG} P_{g_i}$$

$$\frac{\partial L}{\partial \lambda_q} = \sum_{i=1}^{NB} Q_{d_i} + \left(\sum_{i=1}^{NB} \sum_{j=1}^{NB} c_{ij}(P_i P_j + Q_i Q_j) + d_{ij}(Q_i P_j - P_i Q_j) \right) - \sum_{i=1}^{NG} Q_{g_i}$$

The modified generations which are used as initial values for the next iteration are given below:

$$P_{g_1} = 0.4265079 \text{ p.u.}, \quad P_{g_2} = 0.6916468 \text{ p.u.}, \quad P_{g_3} = 0.5479347 \text{ p.u.}$$

$$Q_{g_1} = -0.005769 \text{ p.u.}, \quad Q_{g_2} = 0.2211785 \text{ p.u.}, \quad Q_{g_3} = 0.2322112 \text{ p.u.}$$

The convergence obtained during Newton–Raphson in six interations, is given below:

$$\left| \sum_{i=1}^{3} P_{g_i} - \sum_{i=1}^{5} P_{d_i} - P_L \right| = -0.00000019 < 0.00001$$

The values of λ_p, λ_q, P_L and Q_L for the obtained schedule are given as

$$\lambda_p = 255.9044, \quad \lambda_q = 0.34058850$$

$$P_L = 0.01608950 \text{ p.u.}, \quad Q_L = 0.04761850 \text{ p.u.}$$

In two iterations, the overall convergence value obtained is 4.025251E-03. The final solution achieved is given in Tables 3.34 and 3.35.

TABLE 3.34 Final solution (Example 3.13)

Bus	P_g (p.u.)	Q_g (p.u.)	P_d (p.u.)	Q_d (p.u.)	P (p.u.)	Q (p.u.)
1	0.426485	−0.006340	0.00	0.00	0.430511	−0.660293
2	0.691690	0.220934	0.20	0.10	0.491695	0.120932
3	0.547717	0.232224	0.45	0.15	0.097717	0.082224
4	0.0	0.0	0.40	0.05	−0.400000	−0.049996
5	0.0	0.0	0.60	0.10	−0.600002	−0.100001

TABLE 3.35 Voltage magnitude and angle (Example 3.13)

Bus	$\lvert V \rvert$ (p.u.)	δ (rad)
1	1.060000	0.0
2	1.077662	–0.0216892
3	1.080308	–0.0411964
4	1.076726	–0.0502282
5	1.060467	–0.0729917

λ_p = 255.8771, λ_q = 0.3357378, P_L = 0.01589258 p.u., Q_L = 0.04680709 p.u.,
Cost = 695.4353 ₹/h

3.16 EVALUATION OF INCREMENTAL TRANSMISSION LOSS

The transmission loss can be expressed in terms of *B*-coefficients. Hence penalty factors or incremental losses can be evaluated in terms of *B*-coefficients. The transmission losses can also be expressed in terms of power flow equations. The transmission loss in terms of power injection at various buses is

$$P_L = \sum_{i=1}^{\text{NB}} P_i = \sum_{i=1}^{\text{NG}} P_{g_i} - \sum_{i=1}^{\text{NB}} P_{d_i} \tag{3.154}$$

where

P_{g_i} is the power generated at the *i*th unit

P_{d_i} is the bus power demand at the *i*th bus

P_i is the bus power at the *i*th bus

NB is the total number of buses in the power system network

NG is the number of generating buses supplying real power.

The incremental transmission loss for the *i*th generating unit is

$$\frac{\partial P_L}{\partial P_{g_i}} = \frac{\partial P_L}{\partial P_i} \qquad (i = 1, 2, ..., \text{NG})$$

The real and reactive power injection P_i and Q_i are

$$P_i = \sum_{j=1}^{NB} |V_i||V_j|[G_{ij}\cos(\delta_i - \delta_j) + B_{ij}\sin(\delta_i - \delta_j)] \quad (i = 1, 2, ..., NB) \tag{3.155}$$

$$Q_i = \sum_{j=1}^{NB} |V_i||V_j|[G_{ij}\sin(\delta_i - \delta_j) - B_{ij}\cos(\delta_i - \delta_j)] \quad (i = 1, 2, ..., NB) \tag{3.156}$$

where

$Y_{ij} = G_{ij} + jB_{ij}$ are the elements of the bus admittance matrix

$|V_i|$ is the voltage magnitude at the ith bus.

δ_i is the voltage angle at the ith bus.

For slack or reference bus, angle is zero or fixed one in some cases.
Substituting Eq. (3.155) into Eq. (3.154),

$$P_L = \sum_{i=1}^{NB}\sum_{j=1}^{NB} |V_i||V_j|[G_{ij}\cos(\delta_i - \delta_j) + B_{ij}\sin(\delta_i - \delta_j)]$$

On rearrangement,

$$P_L = \sum_{i=1}^{NB} |V_i|^2 G_{ii} + \sum_{i=1}^{NB}\sum_{\substack{j=1\\ j\neq i}}^{NB} |V_i||V_j|[G_{ij}\cos(\delta_i - \delta_j) + B_{ij}\sin(\delta_i - \delta_j)] \tag{3.157}$$

Equations (3.155) and (3.157) show that the distribution of P_i and P_L depend on the bus voltage magnitude and angle.

$$\frac{\partial P_L}{\partial \delta_i} = \sum_{j=2}^{NB}\left[\frac{\partial P_j}{\partial \delta_i}\times\frac{\partial P_L}{\partial P_j} + \frac{\partial Q_j}{\partial \delta_i}\times\frac{\partial P_L}{\partial Q_j}\right] \tag{3.158}$$

$$\frac{\partial P_L}{\partial |V_i|} = \sum_{j=2}^{NB}\left[\frac{\partial P_j}{\partial |V_i|}\times\frac{\partial P_L}{\partial P_j} + \frac{\partial Q_j}{\partial |V_i|}\times\frac{\partial P_L}{\partial Q_j}\right] \tag{3.159}$$

Equations (3.158) and (3.159) can be written in matrix form as

$$\begin{bmatrix} \dfrac{\partial P}{\partial \delta} & \dfrac{\partial Q}{\partial \delta} \\ \dfrac{\partial P}{\partial |V|} & \dfrac{\partial Q}{\partial |V|} \end{bmatrix} \begin{bmatrix} \dfrac{\partial P_L}{\partial P} \\ \dfrac{\partial P_L}{\partial Q} \end{bmatrix} = \begin{bmatrix} \dfrac{\partial P_L}{\partial \delta} \\ \dfrac{\partial P_L}{\partial |V|} \end{bmatrix} \tag{3.160}$$

or

$$\begin{bmatrix} \dfrac{\partial P}{\partial \delta} & \dfrac{\partial P}{\partial |V|} \\ \dfrac{\partial Q}{\partial \delta} & \dfrac{\partial Q}{\partial |V|} \end{bmatrix}^T \begin{bmatrix} \dfrac{\partial P_L}{\partial P} \\ \dfrac{\partial P_L}{\partial Q} \end{bmatrix} = \begin{bmatrix} \dfrac{\partial P_L}{\partial \delta} \\ \dfrac{\partial P_L}{\partial |V|} \end{bmatrix} \tag{3.161}$$

Expressions for the elements of the preceding matrix can be obtained from Eq. (3.157):

$$\frac{\partial P_L}{\partial \delta_i} = -\sum_{\substack{j=1 \\ j\neq i}}^{NB} [2|V_i||V_j|G_{ij}\sin(\delta_i - \delta_j)] \qquad (i = 1, 2, ..., NB) \tag{3.162a}$$

$$\frac{\partial P_L}{\partial |V_i|} = 2|V_i|G_{ii} + \sum_{\substack{j=1 \\ j\neq i}}^{NB} [2|V_j|G_{ij}\cos(\delta_i - \delta_j)] \qquad (i = MB + 1, MB + 2, ..., NB) \tag{3.162b}$$

where MB is the voltage controlled bus.

From Eq. (3.155), we can obtain

$$\frac{\partial P_i}{\partial \delta_i} = \sum_{\substack{j=1 \\ j\neq i}}^{NB} |V_i||V_j|[-G_{ij}\sin(\delta_i - \delta_j) + B_{ij}\cos(\delta_i - \delta_j)] \qquad (i = 2, 3,..., NB) \tag{3.162c}$$

$$\frac{\partial P_i}{\partial \delta_j} = |V_i||V_j|[G_{ij}\sin(\delta_i - \delta_j) - B_{ij}\cos(\delta_i - \delta_j)]$$

$$(i = 2, 3,..., NB; \quad j = 2, 3,..., NB; \quad i \neq j) \tag{3.162d}$$

$$\frac{\partial P_i}{\partial |V_i|} = 2|V_i|G_{ii} + \sum_{\substack{j=1 \\ j\neq i}}^{NB} |V_j|[G_{ij}\cos(\delta_i - \delta_j) - B_{ij}\sin(\delta_i - \delta_j)]$$

$$(i = MB + 1; \quad MB + 2, ..., NB) \tag{3.162e}$$

$$\frac{\partial P_i}{\partial |V_j|} = |V_i|[G_{ij}\cos(\delta_i - \delta_j) + B_{ij}\sin(\delta_i - \delta_j)]$$

$$(i = MB + 1,..., NB; \quad j = MB + 1, ..., NB; \quad i \neq j) \tag{3.162f}$$

From Eq. (3.155), we can obtain

$$\frac{\partial Q_i}{\partial \delta_i} = \sum_{\substack{j=1 \\ j\neq i}}^{NB} |V_i||V_j|[G_{ij}\cos(\delta_i - \delta_j) + B_{ij}\sin(\delta_i - \delta_j)] \qquad (i = 2, 3, ..., NB) \tag{3.162g}$$

$$\frac{\partial Q_i}{\partial \delta_j} = |V_i||V_j|[-G_{ij}\cos(\delta_i - \delta_j) - B_{ij}\sin(\delta_i - \delta_j)]$$

$$(i = 2, 3,..., NB; \quad j = 2, 3,..., NB; \quad i \neq j) \tag{3.162h}$$

$$\frac{\partial Q_i}{\partial |V_i|} = -2|V_i|B_{ii} + \sum_{\substack{j=1 \\ j\neq i}}^{NB} |V_j|[G_{ij}\sin(\delta_i - \delta_j) - B_{ij}\cos(\delta_i - \delta_j)];$$

$$(i = MB + 1, MB + 2,..., NB) \tag{3.162i}$$

$$\frac{\partial Q_i}{\partial |V_j|} = |V_i|[G_{ij}\sin(\delta_i - \delta_j) - B_{ij}\cos(\delta_i - \delta_j)];$$

$$(i = MB + 1,..., NB; \quad j = MB+1,..., NB; \quad i \neq j) \tag{3.162j}$$

The (2NB + MB – 1) simultaneous equations represented by Eq. (3.160) or Eq. (3.161) can be solved using the Gauss elimination method or Gauss Jordon method or matrix inverse method.

An intrinsic characteristic of any practical electric power system operating in steady state is strong inter-reliance between real power and bus voltage angles and between reactive powers and voltage magnitudes. Equation (3.158) can be simplified if it is considered that for the given voltage, real power, P_i depends on the bus voltage angles. Then the new expression appears as

$$\frac{\partial P_L}{\partial \delta_i} = \sum_{j=2}^{\text{NB}} \left[\frac{\partial P_j}{\partial \delta_i} \times \frac{\partial P_L}{\partial P_j} \right] \tag{3.163}$$

Equation (3.163) can be written in matrix form as

$$\left[\frac{\partial P}{\partial \delta}\right]\left[\frac{\partial P_L}{\partial P}\right] = \left[\frac{\partial P_L}{\partial \delta}\right] \tag{3.164}$$

The elements for the above matrix are already expressed in Eqs. (3.162a), (3.162c), and (3.162d). The (NB – 1) simultaneous equations can be solved by any suitable numerical technique, viz. Gauss elimination, Gauss Jordon, etc.

3.16.1 Alternative Method to Evaluate Incremental Loss

Further, from the Newton–Raphson power flow algorithm the real and reactive bus power mismatches can be written as

$$\begin{bmatrix} \Delta P \\ \Delta Q \end{bmatrix} = \begin{bmatrix} \dfrac{\partial P}{\partial \delta} & \dfrac{\partial P}{\partial |V|} \\[2ex] \dfrac{\partial Q}{\partial \delta} & \dfrac{\partial Q}{\partial |V|} \end{bmatrix} \begin{bmatrix} \Delta \delta \\ \Delta |V| \end{bmatrix} \tag{3.165}$$

The slack bus (reference bus) real power mismatch is

$$\Delta P_1 = \begin{bmatrix} \dfrac{\partial P_1}{\partial \delta} & \dfrac{\partial P_1}{\partial |V|} \end{bmatrix} \begin{bmatrix} \Delta \delta \\ \Delta |V| \end{bmatrix} \tag{3.166}$$

where

$$\delta = [\delta_2, \delta_3, \ldots, \delta_{\text{NB}}]^T$$
$$|V| = [|V_{\text{MB}+1}|, \quad |V_{\text{MB}+2}, \ldots, |V_{\text{NB}}|]^T$$

From Eqs. (3.165) and (3.166),

$$\Delta P_1 = \begin{bmatrix} \dfrac{\partial P_1}{\partial \delta} \\[2ex] \dfrac{\partial P_1}{\partial |V|} \end{bmatrix}^T \begin{bmatrix} \dfrac{\partial P}{\partial \delta} & \dfrac{\partial P}{\partial |V|} \\[2ex] \dfrac{\partial Q}{\partial \delta} & \dfrac{\partial Q}{\partial |V|} \end{bmatrix}^{-1} \begin{bmatrix} \Delta P \\ \Delta Q \end{bmatrix} \tag{3.167}$$

or

$$\Delta P_1 = \left[\begin{bmatrix} \dfrac{\partial P}{\partial \delta} & \dfrac{\partial P}{\partial |V|} \\[2ex] \dfrac{\partial Q}{\partial \delta} & \dfrac{\partial Q}{\partial |V|} \end{bmatrix}^{-1} \right]^T \begin{bmatrix} \dfrac{\partial P_1}{\partial \delta} \\[2ex] \dfrac{\partial P_1}{\partial |V|} \end{bmatrix} \begin{bmatrix} \Delta P \\ \Delta Q \end{bmatrix} \tag{3.168a}$$

The elements of Jacobian matrices are given by Eqs. [(3.162c)–(3.162j)].

Equation (3.168a) can be rewritten as

$$\Delta P_1 = \begin{bmatrix} \alpha \\ \beta \end{bmatrix}^T \begin{bmatrix} \Delta P \\ \Delta Q \end{bmatrix} \tag{3.168b}$$

where

$$\begin{bmatrix} \alpha \\ \beta \end{bmatrix} = \left[\begin{bmatrix} \dfrac{\partial P}{\partial \delta} & \dfrac{\partial P}{\partial |V|} \\ \dfrac{\partial Q}{\partial \delta} & \dfrac{\partial Q}{\partial |V|} \end{bmatrix}^{-1} \right]^T \begin{bmatrix} \dfrac{\partial P_1}{\partial \delta} \\ \dfrac{\partial P_1}{\partial |V|} \end{bmatrix}$$

and

$$\alpha = [\alpha_2, \alpha_3, ..., \alpha_{\text{NB}}]^T$$

$$\beta = [\beta_{\text{MB}+1}, \beta_{\text{MB}+2}, ..., \beta_{\text{NB}}]^T$$

Eq. (3.168b) can be rewritten as

$$\Delta P_1 = \sum_{i=2}^{\text{NB}} \alpha_i \Delta P_i + \sum_{i=\text{MB}+1}^{\text{NB}} \beta_i \Delta Q_i \tag{3.169}$$

From Eq. (3.154),

$$\Delta P_L = \Delta P_1 + \sum_{i=2}^{\text{NB}} \Delta P_i \tag{3.170}$$

To get the change in total transmission loss, substitute Eq. (3.169) into Eq. (3.170)

$$\Delta P_L = \sum_{i=2}^{\text{NB}} (1 + \alpha_i) \Delta P_i + \sum_{i=\text{MB}+1}^{\text{NB}} \beta_i \Delta Q_i \tag{3.171}$$

The incremental transmission loss is defined as the change in transmission loss due to a change in generation i, keeping all other generators constant, i.e.

$$\frac{\partial P_L}{\partial P_{g_i}} = \frac{\partial P_L}{\partial P_i} = 1 + \alpha_i \qquad (i = 2, ..., \text{NG}) \tag{3.172a}$$

For slack bus

$$\frac{\partial P_L}{\partial P_{g_1}} = \frac{\partial P_L}{\partial P_1} = 0.0 \tag{3.172b}$$

3.17 ECONOMIC DISPATCH BASED ON PENALTY FACTORS

The problem of optimal allocation of generation to various plants utilizes a set of penalized incremental cost functions for generating units as described by Eq. (3.16), i.e.

$$\frac{\partial F}{\partial P_{g_i}} = \lambda \left(1 - \frac{\partial P_L}{\partial P_{g_i}} \right) \tag{3.173}$$

Penalty factor is termed by Eq. (3.20) as

$$L_i = \frac{1}{\left(1 - \frac{\partial P_L}{\partial P_{g_i}}\right)} \tag{3.174}$$

From this equation,

$$\frac{1}{L_i} = 1 - \frac{\partial P_L}{\partial P_{g_i}} \tag{3.174a}$$

or

$$\frac{\partial P_L}{\partial P_{g_i}} = 1 - \frac{1}{L_i} \tag{3.174b}$$

Incremental operating cost is defined by Eq. (3.19),

$$\frac{\partial F}{\partial P_{g_i}} = 2a_i P_{g_i} + b_i$$

Substituting the above equation into Eq. (3.173),

$$2a_i P_{g_i} + b_i = \lambda \left(1 - \frac{\partial P_L}{\partial P_{g_i}}\right) \tag{3.175}$$

From Eqs. (3.174a) and (3.175),

$$2a_i P_{g_i} + b_i = \frac{\lambda}{L_i}$$

or

$$P_{g_i} = \left(\frac{\lambda}{L_i} - b_i\right)\frac{1}{2a_i} \qquad (i = 1, 2, ..., \text{NG}) \tag{3.176a}$$

or

$$P_{g_i} = \frac{\lambda}{2a_i L_i} - \frac{b_i}{2a_i} \qquad (i = 1, 2, ..., \text{NG}) \tag{3.176b}$$

The total transmission loss is expressed as a function of initial power loss P_L^0 and the change in total transmission loss ΔP_L, i.e.

$$P_L = P_L^0 + \Delta P_L$$

Since power changes in load buses are zero

$$P_L = P_L^0 + \sum_{i=1}^{\text{NG}} \frac{\partial P_L}{\partial P_{g_i}} (P_{g_i} - P_{g_i}^0) \tag{3.177}$$

Substituting Eqs. (3.174b) and (3.176b) in Eq. (3.177),

$$P_L = P_L^0 + \sum_{i=1}^{\text{NG}} \left(1 - \frac{1}{L_i}\right)\left[\frac{\lambda}{2a_i L_i} - \frac{b_i}{2a_i} - P_{g_i}^0\right]$$

or

$$P_L = P_L^0 + \sum_{i=1}^{NG} \lambda\left(\frac{1}{2a_i L_i} - \frac{1}{2a_i L_i^2}\right) - \sum_{i=1}^{NG}\left[\left(\frac{b_i}{2a_i} + P_{g_i}^0\right) - \frac{1}{L_i}\left(\frac{b_i}{2a_i} + P_{g_i}^0\right)\right] \quad (3.178)$$

The power balance equation is

$$\sum_{i=1}^{NG} P_{g_i} = P_L + \sum_{i=1}^{NB} P_{d_i} \quad (3.179)$$

When $P_L = 0$, then

$$\sum_{i=1}^{NG} P_{g_i} = \sum_{i=1}^{NB} P_{d_i} \quad (3.179a)$$

Substituting Eq. (3.178) and Eq. (3.176a) into Eq. (3.179),

$$\sum_{i=1}^{NG}\left[\frac{\lambda}{2a_i L_i} - \frac{b_i}{2a_i}\right] = \sum_{i=1}^{NB} P_{d_i} + P_L^0 + \sum_{i=1}^{NG} \lambda\left(\frac{1}{2a_i L_i} - \frac{1}{2a_i L_i^2}\right) - \sum_{i=1}^{NG}\left[\left(\frac{b_i}{2a_i} + P_{g_i}^0\right) - \frac{1}{L_i}\left(\frac{b_i}{2a_i} + P_{g_i}^0\right)\right]$$

On simplification and cancelling the equal terms,

$$\sum_{i=1}^{NG} \frac{\lambda}{2a_i L_i^2} = \left[\sum_{i=1}^{NB} P_{d_i} - \sum_{i=1}^{NG} P_{g_i}^0 + P_L^0 + \sum_{i=1}^{NG} \frac{1}{L_i}\left(\frac{b_i}{2a_i} + P_{g_i}^0\right)\right]$$

Considering Eq. (3.179a),

$$\lambda = \frac{P_L^0 + \sum_{i=1}^{NG} \frac{1}{L_i}\left(\frac{b_i}{2a_i} + P_{g_i}^0\right)}{\sum_{i=1}^{NG} \frac{1}{2a_i L_i^2}} \quad (3.180)$$

To obtain the final schedule the detailed algorithm is outlined below:

Algorithm 3.9: Economic Dispatch Based on Penalty Factors

1. Read data a_i, b_i, and c_i (i = 1, 2,..., NG), load on each bus, line data for the power system network.
2. Obtain Y_{BUS} using the Y-bus algorithm.
3. Calculate initial values of P_{g_i}(i = 1, 2,..., NG) and λ by assuming that $P_L = 0$. Then the problem can be stated by Eqs. (3.2a) and (3.2b) and the solution can be obtained directly using the equations already stated as Eq. (3.10) and Eq. (3.9).
4. The total reactive power demand is distributed to various generators in such a manner that the power factor at all the generator buses remains the same.

$$Q_{g_i} = P_{g_i} \times \sum_{i=1}^{MB+1} Q_{d_i} \Big/ \sum_{i=1}^{NB} P_{d_i} \qquad (i = 1, 2, ..., \text{NB (in case of } PQ \text{ buses))}$$

5. Calculate $P_i = P_{g_i} - P_{d_i}$ $\quad (i = 1, 2, ..., \text{NB})$

   $Q_i = Q_{g_i} - Q_{d_i}$ $\quad (i = 1, 2, ..., \text{NB})$

   taking $P_{g_i} = 0$, $Q_{g_i} = 0$ for non-generating buses.
6. Perform load flow to obtain the real and reactive powers, P_i, Q_i and voltage magnitude and angles $|V_i|$, δ_i at each bus.
7. Calculate $\alpha_i(i = 2,.., \text{NG})$ from Eq. (3.168) and then calculate $L_i(i = 1, 2, ..., \text{NG})$ from Eq. (3.174a).
8. Calculate λ from Eq. (3.180) and $P_{g_i}(i = 1, 2, ..., \text{NG})$ from Eq. (3.176a).
9. Calculate the total cost.
10. Stop.

Generator limits can be implemented by fixing the generation to the maximum or minimum limit as per requirement and then such generators are not allowed to participate in generation scheduling.

EXAMPLE 3.14 Consider the 4-bus system of Figure 3.18. The series impedance of each line is given in Table 3.36. The system has three generators. The operating cost characteristics of three generators are given below:

$$F_1 = 50P_{g_1}^2 + 351P_{g_1} + 44.4 \text{ ₹/h}$$

$$F_2 = 50P_{g_2}^2 + 389P_{g_2} + 40.6 \text{ ₹/h}$$

$$F_3 = 60P_{g_3}^2 + 340P_{g_3} + 40.0 \text{ ₹/h}$$

Given the number of buses NB as 4, the number of lines NL as 4, and the number of voltage-controlled buses MB as 2, find the economic generation schedule.

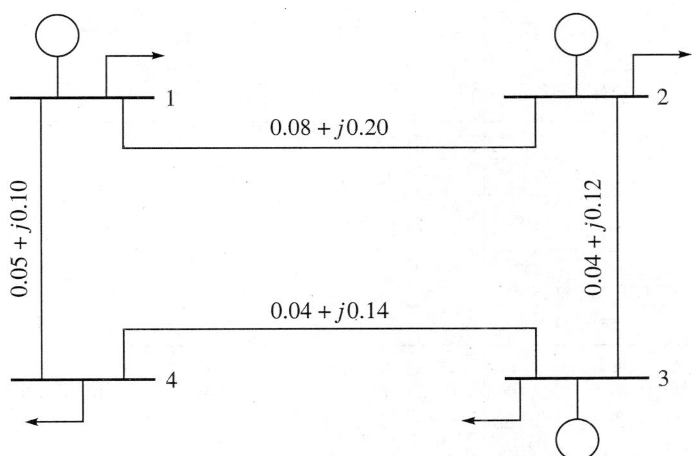

FIGURE 3.18 Power network system.

TABLE 3.36 Line data (Example 3.14)

Line no.	*From bus*	*To bus*	Z_{SER} (p.u.)
1	1	2	0.08 + *j*0.20
2	1	4	0.05 + *j*0.10
3	2	3	0.04 + *j*0.12
4	3	4	0.04 + *j*0.14

Solution Y_{BUS} is given below as

$$Y_{BUS} = \begin{bmatrix} 5.724138 & -1.724138 & 0.0 & -4.0 \\ -j12.31034 & +j4.310345 & +j0.0 & +j8.0 \\ -1.724138 & 4.224138 & -2.5 & 0.0 \\ +j4.310345 & -j11.810340 & +j7.5 & +j0.0 \\ 0.0 & -2.5 & 4.386792 & -1.886792 \\ +j0.0 & +j7.5 & -j14.103770 & +j6.603774 \\ -4.0 & 0.0 & -1.886792 & 5.886792 \\ +j8.0 & +j0.0 & +j6.603774 & -j14.603770 \end{bmatrix}$$

Initial values are obtained and tabulated in Table 3.37.

$$\lambda = \left(\sum_{i=1}^{4} P_{d_i} + \sum_{i=1}^{3} \frac{b_i}{2a_i} \right) \Bigg/ \sum_{i=1}^{3} \frac{b_i}{2a_i} = 608.2354$$

$$P_{g_i} = (\lambda - b_i)/(2 \times a_i) \qquad (i = 1, 2, 3)$$

$$P_{g_1} = 2.572354 \text{ p.u.}; \quad P_{g_2} = 2.192353 \text{ p.u.}; \quad P_{g_3} = 0.235295 \text{ p.u.}$$

TABLE 3.37 Initial values (Example 3.14)

Bus	P_g (p.u.)	P_d (p.u.)	Q_d (p.u.)	*Type*
1	2.572354	1.85	–	Slack
2	2.192353	1.45	–	*PV*
3	2.235295	2.10	–	*PV*
4	0.0	1.60	0.80	*PQ*

Load flow is performed using the decoupled load flow method. The convergence is obtained in four iterations (see Tables 3.38 and 3.39). Bus 1 is taken as the slack bus, buses 2 and 3 are taken as *PV* buses and bus 4 as *PQ* bus

$$P_L^0 = 1.079061\text{E-01 p.u.}$$

TABLE 3.38 Load flow solution (Example 3.14)

Bus	P_g (p.u.)	P_d (p.u.)	Q_d (p.u.)	P (p.u.)	Q (p.u.)
1	2.572354	1.85	–	0.8302581	0.2982449
2	2.192353	1.45	–	0.7423536	−0.3050831
3	2.235295	2.10	–	0.1352946	1.0886370
4	0.0	1.60	0.80	−1.60	−0.7999994

TABLE 3.39 Voltage magnitude and angle (Example 3.14)

Bus	$\lvert V\rvert$ (p.u.)	δ (rad)
1	1.02	0.0
2	1.04	0.03898002
3	1.06	−0.03029456
4	0.94239460	−0.09235717

The expressions of Eq. (3.161) are represented in matrix form as

$$\begin{bmatrix} 13.079150 & -8.0574 & 0.0 & 0.0 \\ -8.438936 & 14.75836 & -6.467162 & 2.290310 \\ 0.0 & -6.700961 & 13.76972 & -6.828105 \\ 0.0 & -1.56199 & 3.849879 & 12.91362 \end{bmatrix} \begin{bmatrix} \alpha_2 \\ \alpha_3 \\ \alpha_4 \\ \beta_4 \end{bmatrix} = \begin{bmatrix} 0.524086 \\ -0.015651 \\ 0.563112 \\ -1.022160 \end{bmatrix}$$

where $\alpha_i = \partial P_L/\partial P_i \quad (i = 2, 3, 4), \qquad \beta_4 = \partial P_L/\partial Q_4$

Using the Gauss elimination method, the above matrix is triangularized as

$$\begin{bmatrix} 13.079150 & -8.0574 & 0.0 & 0.0 \\ 0.0 & 9.559563 & -6.467162 & 2.290310 \\ 0.0 & 0.0 & 9.236439 & -5.222668 \\ 0.0 & 0.0 & 0.0 & 14.86722 \end{bmatrix} \begin{bmatrix} \alpha_2 \\ \alpha_3 \\ \alpha_4 \\ \beta_4 \end{bmatrix} = \begin{bmatrix} 0.524086 \\ 0.322500 \\ 0.789175 \\ -1.208117 \end{bmatrix}$$

Back substitution gives

$$\frac{\partial P_L}{\partial P_1} = 0.0 \text{ (slack bus)}, \quad \frac{\partial P_L}{\partial P_2} = 0.089306, \quad \frac{\partial P_L}{\partial P_3} = 0.079922$$

From Eq. (3.174a), we get

$$L_1 = 1.0\Big/\left(1.0 - \frac{\partial P_L}{\partial P_1}\right) = 1.0, \quad L_2 = 1.0\Big/\left(1.0 - \frac{\partial P_L}{\partial P_2}\right) = 1.098064, \quad L_3 = 1.0\Big/\left(1.0 - \frac{\partial P_L}{\partial P_3}\right) = 1.086865$$

From Eq. (3.180), new incremental cost is computed as

$$\lambda = \frac{P_L^0 + \sum_{i=1}^{3} \frac{1}{L_i}\left(\frac{b_i}{2a_i} + P_{g_i}^0\right)}{\sum_{i=1}^{3} \frac{1}{2a_i L_i^2}} = 646.711900$$

From Eq. (3.176a), generations are computed as

$$P_{g_1} = \left(\frac{\lambda}{L_1} - b_1\right)\frac{1}{2a_1} = 2.957119 \text{ p.u.}$$

$$P_{g_2} = \left(\frac{\lambda}{L_2} - b_2\right)\frac{1}{2a_2} = 1.999563 \text{ p.u.}$$

$$P_{g_3} = \left(\frac{\lambda}{L_3} - b_3\right)\frac{1}{2a_3} = 2.125210 \text{ p.u.}$$

The overall computed cost is

$$F = 3571.481 \quad ₹/\text{h}$$

3.18 OPTIMAL POWER FLOW BASED ON NEWTON METHOD

The optimal power flow is a power flow problem in which certain controllable variables are adjusted to minimize an objective function such as the cost of active power generation or the losses, while satisfying physical and operating limits on various controls, dependent variables and function of variables. The types of controls that an optimal power flow must be able to accommodate are active and reactive power injections, generator voltages, transformer tap ratios and phase-shift angles. In other words, the optimal power problem seeks to find an optimal profile of active and reactive power generations along with voltage magnitudes in such a manner as to minimize the total operating costs of a thermal electric power system, while satisfying network security constraints. For example:

Minimize operating cost of thermal stations

$$F = \sum_{i=1}^{NG} F_i = \sum_{i=1}^{NG} (a_i P_{g_i}^2 + b_i\, P_{g_i} + c_i) \text{ ₹/h} \tag{3.181a}$$

subject to (a) active power balance in the network

$$P_i(V, \delta) - P_{g_i} + P_{d_i} = 0 \qquad (i = 1, 2, ..., \text{NB}) \tag{3.181b}$$

(b) reactive power balance in the network

$$Q_i(V, \delta) - Q_{g_i} + Q_{d_i} = 0 \qquad (i = \text{NV} + 1, \text{NV} + 2,.., \text{NB}) \tag{3.181c}$$

(c) Security-related constraints called soft constraints.
– limits on real power generations

$$P_{g_i}^{\min} \le P_{g_i} \le P_{g_i}^{\max} \qquad (i = 1, 2, ..., \text{NG}) \tag{3.181d}$$

– limits on voltage magnitudes

$$V_i^{\min} \le V_i \le V_i^{\max} \qquad (i = \text{NV} + 1;\quad \text{NV} + 2, ..., \text{NB}) \tag{3.181e}$$

– limits on voltage angles

$$\delta_i^{\min} \le \delta_i \le \delta_i^{\max} \qquad (i = 2, 3, ..., \text{NB}) \tag{3.181f}$$

(d) Functional constraint which is a function of control variables.
– limits on reactive power

$$Q_{g_i}^{\min} \le Q_{g_i} \le Q_{g_i}^{\max} \qquad (i = 1, 2, ..., \text{NG}) \tag{3.181g}$$

– limits on active power flow of line and reactive power flow of line can be applied.

Real power flow equations are

$$P_i(V, \delta) = V_i \sum_{j=1}^{\text{NB}} V_j \left(G_{ij} \cos(\delta_i - \delta_j) + B_{ij} \sin(\delta_i - \delta_j)\right) \tag{3.181h}$$

Reactive power flow equations are

$$Q_i(V, \delta) = V_i \sum_{j=1}^{\text{NB}} V_j \left(G_{ij} \sin(\delta_i - \delta_j) - B_{ij} \cos(\delta_i - \delta_j)\right) \tag{3.181i}$$

where

NG is the number of generator buses.

NB is the number of buses

NV is the number of voltage controlled buses

P_i is the active power injection into bus i

Q_i is the reactive power injection into bus i

P_{d_i} is the active load on bus i

Q_{d_i} is the reactive load on bus i

P_{g_i} is the active generation on bus i

Q_{g_i} is the reactive generation on bus i

V_i is the magnitude of voltage at bus i

δ_i is the voltage phase angle at bus i

$Y_{ij} = G_{ij} + jB_{ij}$ (are the elements of admittance matrix).

The constraint minimization problem can be transformed into an unconstrained one by augmenting the load flow constraints into the objective function. The additional variables are known as Lagrange multiplier functions or incremental cost functions in power system terminology. The cost function becomes:

$$L(P_g, V, \delta) = F(P_{g_i}) + \sum_{i=1}^{\text{NB}} \lambda_{p_i} [P_i(V, \delta) - P_{g_i} + P_{d_i}] + \sum_{i=\text{NV}+1}^{\text{NB}} \lambda_{q_i} [Q_i(V, \delta) - Q_{g_i} + Q_{d_i}] \tag{3.182}$$

The optimization problem is solved, if the following equations of optimality are satisfied.

$$\frac{\partial L}{\partial P_{g_i}} = \frac{\partial F}{\partial P_{g_i}} - \lambda_{p_i} \qquad (i = 1, 2, ..., \text{NG}) \tag{3.183a}$$

$$\frac{\partial L}{\partial \delta_i} = \sum_{j=1}^{\text{NB}} \left[\lambda_{p_j} \frac{\partial P_j}{\partial \delta_i} \right] + \sum_{j=\text{NV}+1}^{\text{NB}} \left[\lambda_{q_j} \frac{\partial Q_j}{\partial \delta_i} \right] \qquad (i = 2, 3, ..., \text{NB}) \tag{3.183b}$$

$$\frac{\partial L}{\partial V_i} = \sum_{j=1}^{\text{NB}} \left[\lambda_{p_j} \frac{\partial P_j}{\partial V_i} \right] + \sum_{j=\text{NV}+1}^{\text{NB}} \left[\lambda_{q_j} \frac{\partial Q_j}{\partial V_i} \right] \qquad (i = \text{NV} + 1, \text{NV} + 2, ..., \text{NB}) \tag{3.183c}$$

$$\frac{\partial L}{\partial \lambda_{p_i}} = P_i(V, \delta) - P_{g_i} + P_{d_i} \qquad (i = 1, 2, ..., \text{NB}) \tag{3.183d}$$

$$\frac{\partial L}{\partial \lambda_{q_i}} = Q_i(V, \delta) - Q_{g_i} + Q_{d_i} \qquad (i = \text{NV} + 1, \text{NV} + 2, ..., \text{NB}) \tag{3.183e}$$

Any small variation in control variables about their initial values is obtained by forming the total differentials.

$$\sum_{j=1}^{\text{NG}} \frac{\partial^2 L}{\partial P_{g_i} \partial P_{g_j}} \Delta P_{g_j} + \sum_{j=2}^{\text{NB}} \frac{\partial^2 L}{\partial P_{g_i} \partial \delta_j} \Delta \delta_j + \sum_{j=\text{NV}+1}^{\text{NB}} \frac{\partial^2 L}{\partial P_{g_i} \partial V_j} \Delta V_j + \sum_{j=1}^{\text{NB}} \frac{\partial^2 L}{\partial P_{g_i} \partial \lambda_{p_j}} \Delta \lambda_{p_j}$$

$$+ \sum_{j=\text{NV}+1}^{\text{NB}} \frac{\partial^2 L}{\partial P_{g_i} \partial \lambda_{q_j}} \Delta \lambda_{q_j} = -\frac{\partial L}{\partial P_{g_i}} \qquad (i = 1, 2, ..., \text{NG}) \tag{3.184a}$$

$$\sum_{j=1}^{\text{NG}} \frac{\partial^2 L}{\partial \delta_i \partial P_{g_j}} \Delta P_{g_j} + \sum_{j=2}^{\text{NB}} \frac{\partial^2 L}{\partial \delta_i \partial \delta_j} \Delta \delta_j + \sum_{j=\text{NV}+1}^{\text{NB}} \frac{\partial^2 L}{\partial \delta_i \partial V_j} \Delta V_j + \sum_{j=1}^{\text{NB}} \frac{\partial^2 L}{\partial \delta_i \partial \lambda_{p_j}} \Delta \lambda_{p_j}$$

$$+ \sum_{j=\text{NV}+1}^{\text{NB}} \frac{\partial^2 L}{\partial \delta_i \partial \lambda_{q_j}} \Delta \lambda_{q_j} = -\frac{\partial L}{\partial \delta_i} \qquad (i = 2, 3, ..., \text{NB}) \tag{3.184b}$$

$$\sum_{j=1}^{\text{NG}} \frac{\partial^2 L}{\partial V_i \partial P_{g_j}} \Delta P_{g_j} + \sum_{j=2}^{\text{NB}} \frac{\partial^2 L}{\partial V_i \partial \delta_j} \Delta \delta_j + \sum_{j=\text{NV}+1}^{\text{NB}} \frac{\partial^2 L}{\partial V_i \partial V_j} \Delta V_j + \sum_{j=1}^{\text{NB}} \frac{\partial^2 L}{\partial V_i \partial \lambda_{p_j}} \Delta \lambda_{p_j}$$

$$+ \sum_{j=\text{NV}+1}^{\text{NB}} \frac{\partial^2 L}{\partial V_i \partial \lambda_{q_j}} \Delta \lambda_{q_j} = -\frac{\partial L}{\partial V_i} \qquad (i = \text{NV} + 1, \text{NV} + 2,..., \text{NB}) \tag{3.184c}$$

$$\sum_{j=1}^{\text{NG}} \frac{\partial^2 L}{\partial \lambda_{p_i} \partial P_{g_j}} \Delta P_{g_j} + \sum_{j=2}^{\text{NB}} \frac{\partial^2 L}{\partial \lambda_{p_i} \partial \delta_j} \Delta \delta_j + \sum_{j=\text{NV}+1}^{\text{NB}} \frac{\partial^2 L}{\partial \lambda_{p_i} \partial V_j} \Delta V_j + \sum_{j=1}^{\text{NB}} \frac{\partial^2 L}{\partial \lambda_{p_i} \partial \lambda_{p_j}} \Delta \lambda_{p_j}$$

$$+ \sum_{j=\text{NV}+1}^{\text{NB}} \frac{\partial^2 L}{\partial \lambda_{p_i} \partial \lambda_{q_j}} \Delta \lambda_{q_j} = -\frac{\partial L}{\partial \lambda_{p_i}} \qquad (i = 1, 2, ..., \text{NB}) \tag{3.184d}$$

$$\sum_{j=1}^{NG} \frac{\partial^2 L}{\partial \lambda_{q_i} \partial P_{g_j}} \Delta P_{g_j} + \sum_{j=2}^{NB} \frac{\partial^2 L}{\partial \lambda_{q_i} \partial \delta_j} \Delta \delta_j + \sum_{j=NV+1}^{NB} \frac{\partial^2 L}{\partial \lambda_{q_i} \partial V_j} \Delta V_j + \sum_{j=1}^{NB} \frac{\partial^2 L}{\partial \lambda_{q_i} \partial \lambda_{p_j}} \Delta \lambda_{p_j}$$

$$+ \sum_{j=NV+1}^{NB} \frac{\partial^2 L}{\partial \lambda_{q_i} \partial \lambda_{q_j}} \Delta \lambda_{q_j} = -\frac{\partial L}{\partial \lambda_{q_i}} \qquad (i = NV + 1, NV + 2, ..., NB) \qquad (3.184e)$$

We now differentiate Eqs. (3.183a–3.183e) with respect to control variables, i.e. P_{g_i}, δ_i, V_i, λ_{p_i} and λ_{q_i}. Second-order partial derivatives required for Eq. (3.183a) are obtained by differentiating Eq. (3.183a) with respect to control variables.

$$\frac{\partial^2 L}{\partial P_{g_i}^2} = 2a_i \qquad (i = 1, 2,..,NG) \qquad (3.185a)$$

$$\frac{\partial^2 L}{\partial P_{g_i} \partial P_{g_j}} = 0 \qquad (i = 1, 2, ..., NG; \quad j = 1, 2, ..., NG; \quad i \neq j) \qquad (3.185b)$$

$$\frac{\partial^2 L}{\partial P_{g_i} \partial \delta_j} = \frac{\partial^2 L}{\partial \delta_j \partial P_{g_i}} = 0 \qquad (i = 1, 2, ..., NG; \quad j = 2, ..., NB) \qquad (3.185c)$$

$$\frac{\partial^2 L}{\partial P_{g_i} \partial V_j} = \frac{\partial^2 L}{\partial V_j \partial P_{g_i}} = 0 \qquad (i = 1, 2, ..., NG; \quad j = NV + 1, NV + 2, ..., NB) \qquad (3.185d)$$

$$\frac{\partial^2 L}{\partial P_{g_i} \partial \lambda_{p_i}} = \frac{\partial^2 L}{\partial \lambda_{p_i} \partial P_{g_i}} = -1 \qquad (i = 1, 2, ..., NG) \qquad (3.185e)$$

$$\frac{\partial^2 L}{\partial P_{g_i} \partial \lambda_{p_j}} = \frac{\partial^2 L}{\partial \lambda_{p_j} \partial P_{g_i}} = 0 \qquad (i = 1, 2, ..., NG; \quad j = 1, 2, ..., NB; \quad i \neq j) \qquad (3.185f)$$

$$\frac{\partial^2 L}{\partial P_{g_i} \partial \lambda_{q_j}} = \frac{\partial^2 L}{\partial \lambda_{q_j} \partial P_{g_i}} = 0 \qquad (i = 1, 2, ..., NG; \quad j = NV + 1, NV + 2, ..., NB) \qquad (3.185g)$$

Second-order partial derivatives required for Eq. (3.183b) are obtained by differentiating Eq. (3.183b) with respect to control variables.

$$\frac{\partial^2 L}{\partial \delta_i \partial \delta_k} = \sum_{j=1}^{NB} \left[\lambda_{p_j} \frac{\partial^2 P_j}{\partial \delta_i \partial \delta_k} \right] + \sum_{j=NV+1}^{NB} \left[\lambda_{q_j} \frac{\partial^2 Q_j}{\partial \delta_i \partial \delta_k} \right]$$

$$(i = 2, 3, ..., NB; \quad k = 2, 3, ..., NB) \qquad (3.186a)$$

$$\frac{\partial^2 L}{\partial \delta_i \partial V_k} = \sum_{j=1}^{NB} \left[\lambda_{p_j} \frac{\partial^2 P_j}{\partial \delta_i \partial V_k} \right] + \sum_{j=NV+1}^{NB} \left[\lambda_{q_j} \frac{\partial^2 Q_j}{\partial \delta_i \partial V_k} \right]$$

$$(i = 2, 3, ..., NB; \quad k = NV + 1, NV + 2, ..., NB) \qquad (3.186b)$$

$$\frac{\partial^2 L}{\partial \delta_i \partial \lambda_{p_j}} = \frac{\partial P_j}{\partial \delta_i} \qquad (i = 2, 3, ..., NB; \quad j = 1, 2, ..., NB) \qquad (3.186c)$$

$$\frac{\partial^2 L}{\partial \delta_i \partial \lambda_{q_j}} = \frac{\partial Q_j}{\partial \delta_i} \quad (i = 2, 3, ..., \text{NB}; \quad j = \text{NV} + 1, \text{NV} + 2, ..., \text{NB}) \tag{3.186d}$$

Second-order partial derivatives required for Eq. (3.183c) are obtained by differentiating Eq. (3.183c) with respect to control variables, i.e.

$$\frac{\partial^2 L}{\partial V_i \partial \delta_k} = \sum_{j=1}^{\text{NB}} \left[\lambda_{p_j} \frac{\partial^2 P_j}{\partial V_i \partial \delta_k} \right] + \sum_{j=\text{NV}+1}^{\text{NB}} \left[\lambda_{q_j} \frac{\partial^2 Q_j}{\partial V_i \partial \delta_k} \right]$$
$$(i = \text{NV} + 1, \text{NV} + 2, ..., \text{NB}; \quad k = 2, 3, ..., \text{NB}) \tag{3.187a}$$

$$\frac{\partial^2 L}{\partial V_i \partial V_k} = \sum_{j=1}^{\text{NB}} \left[\lambda_{p_j} \frac{\partial^2 P_j}{\partial V_i \partial V_k} \right] + \sum_{j=\text{NV}+1}^{\text{NB}} \left[\lambda_{q_j} \frac{\partial^2 Q_j}{\partial V_i \partial V_k} \right]$$
$$(i = \text{NV} + 1, \text{NV} + 2, ..., \text{NB}; \quad k = \text{NV} + 1, \text{NV} + 2, ..., \text{NB}) \tag{3.187b}$$

$$\frac{\partial^2 L}{\partial V_i \partial \lambda_{p_j}} = \frac{\partial P_j}{\partial V_i} \quad (i = \text{NV} + 1, \text{NV} + 2, ..., \text{NB}; \quad j = 1, 2, ..., \text{NB}) \tag{3.187c}$$

$$\frac{\partial^2 L}{\partial V_i \partial \lambda_{q_j}} = \frac{\partial Q_j}{\partial V_i} \quad (i = \text{NV} + 1, \text{NV} + 2, ..., \text{NB}; \quad j = \text{NV} + 1, \text{NV} + 2, ..., \text{NB}) \tag{3.187d}$$

Second-order partial derivatives required for Eq. (3.183d) are obtained by differentiating Eq. (3.183d) with respect to control variables, i.e.

$$\frac{\partial^2 L}{\partial \lambda_{p_i} \partial \delta_j} = \frac{\partial P_i}{\partial \delta_j} \quad (i = 1, 2, ..., \text{NB}; \quad j = 2, 3, ..., \text{NB}) \tag{3.188a}$$

$$\frac{\partial^2 L}{\partial \lambda_{p_i} \partial V_j} = \frac{\partial P_i}{\partial V_j} \quad (i = 1, 2, ..., \text{NB}; \quad j = \text{NV} + 1, \text{NV} + 2, ..., \text{NB}) \tag{3.188b}$$

$$\frac{\partial^2 L}{\partial \lambda_{p_i} \partial \lambda_{p_j}} = 0 \quad (i = 1, 2, ..., \text{NB}; \quad j = 1, 2, ..., \text{NB}) \tag{3.188c}$$

$$\frac{\partial^2 L}{\partial \lambda_{p_i} \partial \lambda_{q_j}} = 0 \quad (\text{i} = 1, 2, ..., \text{NB}; \quad j = \text{NV} + 1, \text{NV} + 2, ..., \text{NB}) \tag{3.188d}$$

Second-order partial derivatives required for Eq. (3.183e) are obtained by differentiating Eq. (3.183e) with respect to control variables, i.e.

$$\frac{\partial^2 L}{\partial \lambda_{q_i} \partial \delta_j} = \frac{\partial Q_i}{\partial \delta_j} \quad (i = \text{NV} + 1, \text{NV} + 2,.., \text{NB}; \quad j = 2, 3, ..., \text{NB}) \tag{3.189a}$$

$$\frac{\partial^2 L}{\partial \lambda_{q_i} \partial V_j} = \frac{\partial Q_i}{\partial V_j} \quad (i = \text{NV} + 1, \text{NV} + 2,.., \text{NB}; \quad j = \text{NV} + 1, \text{NV} + 2, ..., \text{NB}) \tag{3.189b}$$

$$\frac{\partial^2 L}{\partial \lambda_{q_i} \partial \lambda_{p_j}} = 0 \quad (i = \text{NV} + 1, \text{NV} + 2, ..., \text{NB}; \quad j = 1, 2, ..., \text{NB}) \tag{3.189c}$$

$$\frac{\partial^2 L}{\partial \lambda_{q_i} \partial \lambda_{q_j}} = 0 \qquad (i = \text{NV} + 1, \text{NV} + 2,.., \text{NB}; \;\; j = \text{NV} + 1, \text{NV} + 2,.., \text{NB}) \tag{3.189d}$$

Equations (3.184a) to (3.184e) can be rewritten considering second-partial derivative terms having zero values as defined by Eqs. (3.185–3.189).

$$\frac{\partial^2 L}{\partial P_{g_i}^2} \Delta P_{g_i} + \frac{\partial^2 L}{\partial P_{g_i} \partial \lambda_{p_i}} \Delta \lambda_{p_i} = -\frac{\partial L}{\partial P_{g_i}} \qquad (i = 1, 2, ..., \text{NG}) \tag{3.190a}$$

$$\sum_{j=2}^{\text{NB}} \frac{\partial^2 L}{\partial \delta_i \partial \delta_j} \Delta \delta_j + \sum_{j=\text{NV}+1}^{\text{NB}} \frac{\partial^2 L}{\partial \delta_i \partial V_j} \Delta V_j + \sum_{j=1}^{\text{NB}} \frac{\partial^2 L}{\partial \delta_i \partial \lambda_{p_j}} \Delta \lambda_{p_j} + \sum_{j=\text{NV}+1}^{\text{NB}} \frac{\partial^2 L}{\partial \delta_i \partial \lambda_{q_j}} \Delta \lambda_{q_j} = -\frac{\partial L}{\partial \delta_i}$$

$$(i = 2, 3, ..., \text{NB}) \tag{3.190b}$$

$$\sum_{j=2}^{\text{NB}} \frac{\partial^2 L}{\partial V_i \partial \delta_j} \Delta \delta_j + \sum_{j=\text{NV}+1}^{\text{NB}} \frac{\partial^2 L}{\partial V_i \partial V_j} \Delta V_j + \sum_{j=1}^{\text{NB}} \frac{\partial^2 L}{\partial V_i \partial \lambda_{p_j}} \Delta \lambda_{p_j} + \sum_{j=\text{NV}+1}^{\text{NB}} \frac{\partial^2 L}{\partial V_i \partial \lambda_{q_j}} \Delta \lambda_{q_j} = -\frac{\partial L}{\partial V_i}$$

$$(i = \text{NV} + 1, ..., \text{NB}) \tag{3.190c}$$

$$\frac{\partial^2 L}{\partial \lambda_{p_i} \partial P_{g_i}} \Delta P_{g_i} + \sum_{j=2}^{\text{NB}} \frac{\partial^2 L}{\partial \lambda_{p_i} \partial \delta_j} \Delta \delta_j + \sum_{j=\text{NV}+1}^{\text{NB}} \frac{\partial^2 L}{\partial \lambda_{p_i} \partial V_j} \Delta V_j = -\frac{\partial L}{\partial \lambda_{p_i}} \quad (i = 1, 2, ..., \text{NB}) \tag{3.190d}$$

$$\sum_{j=2}^{\text{NB}} \frac{\partial^2 L}{\partial \lambda_{q_i} \partial \delta_j} \Delta \delta_j + \sum_{j=\text{NV}+1}^{\text{NB}} \frac{\partial^2 L}{\partial \lambda_{q_i} \partial V_j} \Delta V_j = -\frac{\partial L}{\partial \lambda_{q_i}} \qquad (i = \text{NV} + 1, \text{NV} + 2, ..., \text{NB}) \tag{3.190e}$$

3.18.1 Limits on Variables

Kuhn–Tucker conditions

The Kuhn–Tucker approach can be adopted and thus inequality constraints are assumed to be inactive during the initial solution process. Any variable can be fixed to its lower or upper limit values. Then, the partial derivative terms with respect to the fixed variables can be deactivated so that these terms do not participate in the evaluation of rest of variables. A variable should be moved to its relevant limit in coordination with the simultaneous correction of all other variables.

$$x_i^{\text{new}} = \begin{cases} x_i^{\max} & \text{if} \;\; x_i^{\text{old}} + \Delta x_i > x_i^{\max} \\ x_i^{\min} & \text{if} \;\; x_i^{\text{old}} + \Delta x_i < x_i^{\min} \\ x_i^{\text{old}} + \Delta x_i & \end{cases} \tag{3.191a}$$

In accordance with the Kuhn–Tucker theorem, the necessary conditions for minimization of L under constraints are

$$\left\{ \begin{aligned} &\frac{\partial L}{\partial x_i} = 0 \;\; \text{if} \;\; x_i^{\min} < x_i < x_i^{\max} \\ &\frac{\partial L}{\partial x_i} \le 0 \;\; \text{if} \;\; x_i = x_i^{\min} \\ &\frac{\partial L}{\partial x_i} \ge 0 \;\; \text{if} \;\; x_i = x_i^{\max} \end{aligned} \right\} \tag{3.191b}$$

Penalty function method

Penalty functions ideally fulfil the requirements for inequality constraints enforcement. The penalty function is described below which is quadratic.

$$\sum_{i=1}^{R} \alpha_i = \sum_{i=1}^{R} \frac{S_i}{2}(y_i - \bar{y}_i) \tag{3.192a}$$

where

$\bar{y}_i$ is the target value

y_i is the current value

S_i is the weighting factor.

The weighting factor is automatically controlled to give the appropriate amount of hardness of enforcement. The first and second derivatives of α_i are:

$$\frac{\partial \alpha_i}{\partial y_i} = S_i (y_i - \bar{y}_i) \tag{3.192b}$$

$$\frac{\partial^2 \alpha_i}{\partial y_i^2} = S_i \tag{3.192c}$$

The proper value of S_i is automatically controlled. For the larger values of S_i, the target limit acts as the rigid/hard limit and for small value of S_i, the target limit acts as the soft limit.

The detailed algorithm for optimal load flow is outlined below.

Algorithm 3.10: Optimal Power Flow Based on Newton Method

1. Read data a_i, b_i, and c_i (i = 1, 2, ..., NG), load on each bus, line data for the power system network.
2. Obtain Y_{BUS} using the Y-bus algorithm.
3. Calculate the initial values of P_{g_i} (i = 1, 2, ..., NG) and λ by assuming that $P_L = 0$. Then the problem can be stated by Eqs. (3.2a) and (3.2b) and the solution can be obtained directly using Eq. (3.10) and Eq. (3.9). Initialize all $\lambda_{p_i} = \lambda$ (i = 1, 2, ..., NB), $\lambda_{q_i} = 0$ (i = NV + 1, NV + 2, ..., NB), V_i = 1 p.u. (i = 2, 3, ..., NB) and $\delta_i = 0$ (i = 2, 3, ..., NB).
4. Calculate the Jacobian and Hessian matrix elements from Eqs. (3.183) to (3.189)

$$[H]\begin{bmatrix} \Delta P_g \\ \Delta\delta \\ \Delta\lambda_p \\ \Delta V \\ \Delta\lambda_q \end{bmatrix} = -\begin{bmatrix} \dfrac{\partial L}{\partial P_g} \\ \dfrac{\partial L}{\partial \delta} \\ \dfrac{\partial L}{\partial \lambda_p} \\ \dfrac{\partial L}{\partial V} \\ \dfrac{\partial L}{\partial \lambda_q} \end{bmatrix}$$

The Gauss elimination method is used to find ΔP_g, $\Delta\delta$, $\Delta\lambda_p$, ΔV, and $\Delta\lambda_q$.

5. Check convergence

$$\left[\sum_{i=1}^{NG}(\Delta P_{g_i})^2+\sum_{i=2}^{NB}(\Delta\delta_i)^2+\sum_{i=1}^{NB}(\Delta\lambda_{p_i})^2+\sum_{i=NV+1}^{NB}(\Delta V_i)^2+\sum_{i=NV+1}^{NB}(\Delta\lambda_q)^2\right]^{\frac{1}{2}}\le\varepsilon$$

and optimality conditions. If condition is not satisfied then GOTO Step 6 else GOTO Step 8.

6. Modify the variables,

$$P_{g_i} = P_{g_i} + \Delta P_{g_i} \qquad (i = 1, 2, ..., \text{NG})$$
$$\delta_i = \delta_i + \Delta\delta_i \qquad (i = 2, 3, ..., \text{NB})$$
$$\lambda_{p_i} = \lambda_{p_i} + \Delta\lambda_{p_i} \qquad (i = 1, 2, ..., \text{NB})$$
$$V_i = V_i + \Delta V_i \qquad (i = \text{NV} + 1, \text{NV} + 2, ..., \text{NB})$$
$$\lambda_{q_i} = \lambda_{q_i} + \Delta\lambda_{q_i} \qquad (i = \text{NV} + 1, \text{NV} + 2, ..., \text{NB})$$

7. Check the limits, if any limit of a variable is violated, then impose or remove power flow equation or a penalty for inequality. Add or remove derivatives for penalty or equation change and GOTO Step 4 to update the solution.
8. Calculate the total cost.
9. Stop.

EXAMPLE 3.15 Consider the 3-bus system of Figure 3.19. The series impedance of each line is given in Table 3.40. The system has three generators. Find the economic generation schedule. The operating cost characteristic of two generators are given below:

$$F_1 = 50\,P_{g_1}^2 + 351\,P_{g_1} + 44.4 \quad ₹/\text{h}$$
$$F_2 = 50\,P_{g_2}^2 + 389\,P_{g_2} + 40.6 \quad ₹/\text{h}$$

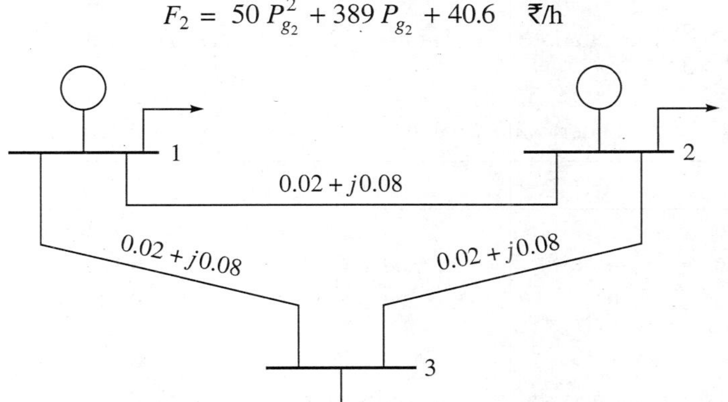

FIGURE 3.19 Power network system.

TABLE 3.40 Line data (Example 3.15)

Line no.	*From bus*	*To bus*	Z_{SER} (p.u.)	Y_{SH} (p.u.)
1	1	2	0.02 + *j*0.08	*j*0.02
2	1	3	0.02 + *j*0.08	*j*0.02
3	2	3	0.02 + *j*0.08	*j*0.02

Solution Number of generators, NG = 2; Number of buses, NB = 3; Number of lines, NL = 3; Number of *PV* buses, NV = 1.

Y-bus is given below:

$$Y_{\text{BUS}} = \begin{bmatrix} 5.882353 - j23.48941 & -2.941177 + j11.76471 & -2.941177 + j11.76471 \\ -2.941177 + j11.76471 & 5.882353 - j23.48941 & -2.941177 + j11.76471 \\ -2.941177 + j11.76471 & -2.941177 + j11.76471 & 5.882353 - j23.48941 \end{bmatrix}$$

Initial values are obtained and tabulated in Table 3.41.

$$\lambda = \left(\sum_{i=1}^{3} P_{d_i} + \sum_{i=1}^{3} \frac{b_i}{2a_i} \right) \Bigg/ \left(\sum_{i=1}^{3} \frac{1}{2a_i} \right) = 291.1111$$

$$P_{g_i} = \frac{\lambda - b_i}{2 \times a_i} \qquad (i = 1, 2)$$

$$P_{g_1} = 0.759259; \quad P_{g_2} = 0.940741$$

TABLE 3.41 Initial values (Example 3.15)

Bus	P_g (p.u.)	P_d (p.u.)	Q_d (p.u.)	*Type*	V (p.u.)	δ (rad)	λ_p	λ_q
1	0.759259	2.0	1.0	Slack	1.04	0	291.1111	0.0
2	0.940741	0.0	0.5	*PV*	1.04	0	291.1111	0.0
3	0.0	1.5	0.6	*PQ*	1.00	0	0.0	0.0

P_g, Q_g, P, and Q are tabulated in Table 3.42 which are obtained with the initial values tabulated in Table 3.41.

TABLE 3.42 Other values calculated with initial values (Example 3.15)

Bus	P_g (p.u.)	Q_g (p.u.)	P_d (p.u.)	Q_d (p.u.)	P (p.u.)	Q (p.u.)
1	0.7592590	0.5461460	0.20	0.10	0.1223528	0.4461460
2	0.9407406	0.4461460	0.00	0.00	0.1223528	0.4461460
3	0.0	0.0	1.50	0.06	−0.2352939	−0.9811764

Cost = 653.9258 ₹/h

With the above-mentioned values, the following equation is solved to initiate the iterative process to implement the Newton–Raphson method.

$$\begin{bmatrix} \nabla_{P_g P_g} L & 0 & \nabla_{P_g \lambda_p} L & 0 & 0 \\ 0 & \nabla_{\delta\delta} L & \nabla_{\delta\lambda_p} L & \nabla_{\delta V} L & \nabla_{\delta\lambda_q} L \\ \nabla_{\lambda_p P_g} L & \nabla_{\lambda_p \delta} L & 0 & \nabla_{\lambda_p V} L & 0 \\ 0 & \nabla_{V\delta} L & \nabla_{V\lambda_p} L & \nabla_{VV} L & \nabla_{V\lambda_q} L \\ 0 & \nabla_{\lambda_q \delta} L & 0 & \nabla_{\lambda_q V} L & 0 \end{bmatrix} \begin{bmatrix} \Delta P_g \\ \Delta\delta \\ \Delta\lambda_p \\ \Delta V \\ \Delta\lambda_q \end{bmatrix} = - \begin{bmatrix} -\nabla_{P_g} L \\ -\nabla_{\delta} L \\ -\nabla_{\lambda_p} L \\ -\nabla_{V} L \\ -\nabla_{\lambda_q} L \end{bmatrix} \qquad \text{(i)}$$

To solve the preceding equation, there is a need to calculate the Jacobian and Hessian matrix elements. These elements are obtained as described below:

Partial derivatives with respect to P_g:

$$\frac{\partial L}{\partial P_{g_1}} = 2.0a_1P_{g_1} + b_1 - \lambda_{P_1} = -0.0000014$$

$$\frac{\partial L}{\partial P_{g_2}} = 2.0a_2P_{g_2} + b_2 - \lambda_{P_2} = -0.0000038$$

Partial derivatives with respect to δ_2:

$$\frac{\partial L}{\partial \delta_2} = \sum_{j=1}^{3} \lambda_{p_j} \frac{\partial P_j}{\partial \delta_2} + \lambda_{q_3} \frac{\partial Q_3}{\partial \delta_2} = -3561.829$$

where the partial derivative of P_1 with respect to δ_2 is

$$\frac{\partial P_1}{\partial \delta_2} = V_1V_2\,(G_{12} \sin\,(\delta_1 - \delta_2) - B_{12} \cos\,(\delta_1 - \delta_2)) = -12.72471$$

Partial derivative of P_2 with respect to δ_2:

$$\frac{\partial P_2}{\partial \delta_2} = V_2 \sum_{\substack{j=1\\ j\neq 2}}^{3} V_j\,(-G_{2j} \sin(\delta_2 - \delta_j) + B_{2j} \cos(\delta_2 - \delta_j)) = 24.960$$

Partial derivative of P_3 with respect to δ_2:

$$\frac{\partial P_3}{\partial \delta_2} = V_3V_2(G_{32} \sin\,(\delta_3 - \delta_2) - B_{32} \cos\,(\delta_3 - \delta_2)) = -12.23529$$

Partial derivative of Q_3 with respect to δ_2:

$$\frac{\partial Q_3}{\partial \delta_2} = V_3V_2(-G_{32} \cos\,(\delta_3 - \delta_2) - B_{32} \sin\,(\delta_3 - \delta_2)) = 3.0588240$$

Partial derivative of L with respect to δ_3:

$$\frac{\partial L}{\partial \delta_3} = \sum_{j=1}^{3} \lambda_{p_j} \frac{\partial P_j}{\partial \delta_3} + \lambda_{q_3} \frac{\partial Q_3}{\partial \delta_3}$$

where the partial derivative of P_1 with respect to δ_3 is

$$\frac{\partial P_1}{\partial \delta_3} = V_1V_3(G_{13} \sin\,(\delta_1 - \delta_3) - B_{13} \cos\,(\delta_1 - \delta_3)) = -12.23529$$

Partial derivative of P_2 with respect to δ_3:

$$\frac{\partial P_2}{\partial \delta_3} = V_2V_3(G_{23} \sin\,(\delta_2 - \delta_3) - B_{23} \cos\,(\delta_2 - \delta_3)) = -12.23529$$

Partial derivative of P_3 with respect to δ_3:

$$\frac{\partial P_3}{\partial \delta_3} = V_3 \sum_{\substack{j=1\\ j\neq 3}}^{3} V_j\,(-G_{3j} \sin\,(\delta_3 - \delta_j) + B_{3j} \cos\,(\delta_3 - \delta_j)) = 24.47059$$

Partial derivative of Q_3 with respect to δ_3:

$$\frac{\partial Q_3}{\partial \delta_3} = V_3 \sum_{\substack{j=1 \\ j\neq 3}}^{3} V_j \left(G_{3j} \cos(\delta_3 - \delta_j) + B_{3j} \sin(\delta_3 - \delta_j)\right) = -6.117647$$

Partial derivative of L with respect to V_3:

$$\frac{\partial L}{\partial V_3} = \sum_{j=1}^{3} \lambda_{pj} \frac{\partial P_j}{\partial V_3} + \lambda_{q3} \frac{\partial Q_3}{\partial V_3} = 1780.915$$

where the partial derivative of P_1 with respect to V_3 is

$$\frac{\partial P_1}{\partial V_3} = V_1(G_{13} \cos(\delta_1 - \delta_3) + B_{13} \sin(\delta_1 - \delta_3)) = -3.058824$$

Partial derivative of P_2 with respect to V_3:

$$\frac{\partial P_2}{\partial V_3} = V_2(G_{23} \cos(\delta_2 - \delta_3) + B_{23} \sin(\delta_2 - \delta_3)) = -3.058824$$

Partial derivative of P_3 with respect to V_3:

$$\frac{\partial P_3}{\partial V_3} = 2V_3 G_{33} + \sum_{\substack{j=1 \\ j\neq 3}}^{3} V_j \left(G_{3j} \cos(\delta_3 - \delta_j) + B_{3j} \sin(\delta_3 - \delta_j)\right) = 5.647059$$

Partial derivative of Q_3 with respect to V_3:

$$\frac{\partial Q_3}{\partial V_3} = -2V_3 B_{33} + \sum_{\substack{j=1 \\ j\neq 3}}^{3} V_j \left(G_{3j} \sin(\delta_3 - \delta_j) - B_{3j} \cos(\delta_3 - \delta_j)\right) = 22.508240$$

Partial derivative of L with respect to λ_{p1}:

$$\frac{\partial L}{\partial \lambda_{p1}} = P_1 - P_{g_1} + P_{d_1} = 0.4369062$$

Partial derivative of L with respect to λ_{p2}:

$$\frac{\partial L}{\partial \lambda_{p2}} = P_2 - P_{g_2} + P_{d_2} = 0.8183877$$

Partial derivative of L with respect to λ_{p3}:

$$\frac{\partial L}{\partial \lambda_{p3}} = P_3 - P_{g_3} + P_{d_3} = -1.264706$$

Partial derivative of L with respect to λ_{q3}:

$$\frac{\partial L}{\partial \lambda_{q3}} = Q_3 - Q_{g_3} + Q_{d_3} = 0.9211764$$

Hessian matrix elements are obtained from Eqs. (3.185–3.189).

Second partial derivatives of L with respect to P_g:

$$\frac{\partial^2 L}{\partial P_{g_1}^2} = 2.0a_1 = 120.0; \quad \frac{\partial^2 L}{\partial P_{g_1}\partial P_{g_2}} = 0.0$$

$$\frac{\partial^2 L}{\partial P_{g_2}\partial P_{g_1}} = 0.0; \quad \frac{\partial^2 L}{\partial P_{g_2}^2} = 2.0a_2 = 150.0$$

Second partial derivatives of L with respect to λ_p:

$$\frac{\partial^2 L}{\partial P_{g_1}\partial \lambda_{p_1}} = -1.0; \quad \frac{\partial^2 L}{\partial P_{g_1}\partial \lambda_{p_2}} = 0.0; \quad \frac{\partial^2 L}{\partial P_{g_1}\partial \lambda_{p_3}} = 0.0$$

$$\frac{\partial^2 L}{\partial P_{g_2}\partial \lambda_{p_1}} = 0.0; \quad \frac{\partial^2 L}{\partial P_{g_2}\partial \lambda_{p_2}} = -1.0; \quad \frac{\partial^2 L}{\partial P_{g_2}\partial \lambda_{p_3}} = 0.0$$

Second partial derivatives of L with respect to δ_2:

$$\frac{\partial^2 L}{\partial \delta_2^2} = \sum_{j=1}^{3} \lambda_{p_j} \frac{\partial^2 P_j}{\partial \delta_2^2} + \lambda_{q_3} \frac{\partial^2 Q_3}{\partial \delta_2^2} = 2742.609$$

where

$$\frac{\partial^2 P_1}{\partial \delta_2^2} = -V_1V_2\big(G_{12}\cos(\delta_1 - \delta_2) + B_{12}\sin(\delta_1 - \delta_2)\big) = 3.181176$$

$$\frac{\partial^2 P_2}{\partial \delta_2^2} = V_2 \sum_{\substack{j=1\\ j\neq 2}}^{3} V_j\big(-G_{2j}\cos(\delta_2 - \delta_j) - B_{2j}\sin(\delta_2 - \delta_j)\big) = 6.240000$$

$$\frac{\partial^2 P_3}{\partial \delta_2^2} = -V_3V_2\big(G_{32}\cos(\delta_3 - \delta_2) + B_{32}\sin(\delta_3 - \delta_2)\big) = 3.058824$$

$$\frac{\partial^2 Q_3}{\partial \delta_2^2} = V_3V_2\big(-G_{32}\sin(\delta_3 - \delta_2) + B_{32}\cos(\delta_3 - \delta_2)\big) = 12.23529$$

Second partial derivatives of L with respect to δ_2 and δ_3:

$$\frac{\partial^2 L}{\partial \delta_2 \partial \delta_3} = \sum_{j=1}^{3} \lambda_{p_j} \frac{\partial^2 P_j}{\partial \delta_2 \partial \delta_3} + \lambda_{q_3} \frac{\partial^2 Q_3}{\partial \delta_2 \partial \delta_3} = -890.4575$$

where

$$\frac{\partial^2 P_1}{\partial \delta_2 \partial \delta_3} = 0.0$$

$$\frac{\partial^2 P_2}{\partial \delta_2 \partial \delta_3} = V_2V_3\big(G_{23}\cos(\delta_2 - \delta_3) + B_{23}\sin(\delta_2 - \delta_3)\big) = -3.058824$$

$$\frac{\partial^2 P_3}{\partial \delta_2 \partial \delta_3} = V_3V_2\big(G_{32}\cos(\delta_3 - \delta_2) + B_{32}\sin(\delta_3 - \delta_2)\big) = -3.058824$$

$$\frac{\partial^2 Q_3}{\partial \delta_2 \partial \delta_3} = V_3V_2\left(G_{32} \sin(\delta_3 - \delta_2) - B_{32} \cos(\delta_3 - \delta_2)\right) = -12.23529$$

Second partial derivatives of L with respect to δ_3 and δ_2:

$$\frac{\partial^2 L}{\partial \delta_3 \partial \delta_2} = \sum_{j=1}^{3} \lambda_{p_j} \frac{\partial^2 P_j}{\partial \delta_3 \partial \delta_2} + \lambda_{q3} \frac{\partial^2 Q_3}{\partial \delta_3 \partial \delta_2} = -890.4575$$

where

$$\frac{\partial^2 P_1}{\partial \delta_3 \partial \delta_2} = 0.0$$

$$\frac{\partial^2 P_2}{\partial \delta_3 \partial \delta_2} = V_3V_2\left(G_{32} \cos(\delta_3 - \delta_2) + B_{32} \sin(\delta_3 - \delta_2)\right) = -3.058824$$

$$\frac{\partial^2 P_3}{\partial \delta_3 \partial \delta_2} = V_3V_2\left(G_{32} \cos(\delta_3 - \delta_2) + B_{32} \sin(\delta_3 - \delta_2)\right) = -3.058824$$

$$\frac{\partial^2 Q_3}{\partial \delta_3 \partial \delta_2} = V_3V_2\left(G_{32} \sin(\delta_3 - \delta_2) - B_{32} \cos(\delta_3 - \delta_2)\right) = -12.23529$$

Second partial derivatives of L with respect to δ_3:

$$\frac{\partial^2 L}{\partial \delta_3^2} = \sum_{j=1}^{3} \lambda_{p_j} \frac{\partial^2 P_j}{\partial \delta_3^2} + \lambda_{q3} \frac{\partial^2 Q_3}{\partial \delta_3^2} = 1780.915$$

where

$$\frac{\partial^2 P_1}{\partial \delta_3^2} = -V_1V_3\left(G_{13} \cos(\delta_1 - \delta_3) + B_{13} \sin(\delta_1 - \delta_3)\right) = 3.058824$$

$$\frac{\partial^2 P_2}{\partial \delta_3^2} = -V_2V_3\left(G_{23} \cos(\delta_2 - \delta_3) + B_{23} \sin(\delta_2 - \delta_3)\right) = 3.058824$$

$$\frac{\partial^2 P_3}{\partial \delta_3^2} = V_3 \sum_{\substack{j=1\\ j\neq 3}}^{3} V_j\left(-G_{3j} \cos(\delta_3 - \delta_j) - B_{3j} \sin(\delta_3 - \delta_j)\right) = 6.117647$$

$$\frac{\partial^2 Q_3}{\partial \delta_3^2} = \sum_{\substack{j=1\\ j\neq 3}}^{3} V_3V_j\left(-G_{3j} \sin(\delta_3 - \delta_j) + B_{3j} \cos(\delta_3 - \delta_j)\right) = 24.47059$$

Second partial derivatives of L with respect to δ_2 and V_3:

$$\frac{\partial^2 L}{\partial \delta_2 \partial V_2} = \sum_{j=1}^{3} \lambda_{p_j} \frac{\partial^2 P_j}{\partial \delta_2 \partial V_3} + \lambda_{q3} \frac{\partial^2 Q_3}{\partial \delta_2 \partial V_3} = -3424.836$$

where

$$\frac{\partial^2 P_1}{\partial \delta_2 \partial V_3} = 0.0$$

$$\frac{\partial^2 P_2}{\partial\delta_2 \partial V_3} = V_2\big(G_{23}\ \sin\,(\delta_2 - \delta_3) + B_{23}\cos\,(\delta_2 - \delta_3)\big) = -11.76471$$

$$\frac{\partial^2 P_3}{\partial\delta_2 \partial V_3} = V_2\big(G_{32}\ \sin\,(\delta_3 - \delta_2) - B_{32}\cos\,(\delta_3 - \delta_2)\big) = 11.76471$$

$$\frac{\partial^2 Q_3}{\partial\delta_2 \partial V_3} = V_2\big(-G_{32}\ \cos\,(\delta_3 - \delta_2) - B_{32}\sin\,(\delta_3 - \delta_2)\big) = -2.941177$$

Second partial derivatives of L with respect to δ_3 and V_3:

$$\frac{\partial^2 L}{\partial\delta_3 \partial V_3} = \sum_{j=1}^{3} \lambda_{p_j} \frac{\partial^2 P_j}{\partial\delta_3 \partial V_3} + \lambda_{q3} \frac{\partial^2 Q_3}{\partial\delta_3 \partial V_3} = -7123.66$$

where

$$\frac{\partial^2 P_1}{\partial\delta_3 \partial V_3} = V_1\big(G_{13}\ \sin\,(\delta_1 - \delta_2) - B_{13}\cos\,(\delta_1 - \delta_3)\big) = -12.23529$$

$$\frac{\partial^2 P_2}{\partial\delta_3 \partial V_3} = V_2\big(G_{23}\ \sin\,(\delta_2 - \delta_3) - B_{23}\cos\,(\delta_2 - \delta_3)\big) = -12.23529$$

$$\frac{\partial^2 P_3}{\partial\delta_3 \partial V_3} = \sum_{\substack{j=1\\ j\neq 3}}^{3} V_j\Big((-G_{3j}\sin\,(\delta_3 - \delta_j) + B_{3j}\cos\,(\delta_3 - \delta_j)\Big) = 24.47059$$

$$\frac{\partial^2 Q_3}{\partial\delta_3 \partial V_3} = \sum_{\substack{j=1\\ j\neq 3}}^{3} V_j\Big(G_{3j}\cos\,(\delta_3 - \delta_j) + B_{3j}\sin\,(\delta_3 - \delta_j)\Big) = 6.117647$$

Second partial derivatives of L with respect to δ and λ_p:

$$\frac{\partial^2 L}{\partial\delta_2 \partial\lambda_{p_1}} = \frac{\partial P_1}{\partial\delta_2} = -12.72471; \quad \frac{\partial^2 L}{\partial\delta_3 \partial\lambda_{p_1}} = \frac{\partial P_1}{\partial\delta_3} = -12.23529$$

$$\frac{\partial^2 L}{\partial\delta_2 \partial\lambda_{p_2}} = \frac{\partial P_2}{\partial\delta_2} = 24.960; \quad \frac{\partial^2 L}{\partial\delta_3 \partial\lambda_{p_2}} = \frac{\partial P_2}{\partial\delta_3} = -12.23529$$

$$\frac{\partial^2 L}{\partial\delta_2 \partial\lambda_{p_3}} = \frac{\partial P_3}{\partial\delta_2} = -12.23529; \quad \frac{\partial^2 L}{\partial\delta_3 \partial\lambda_{p_3}} = \frac{\partial P_3}{\partial\delta_3} = 24.47059$$

Second partial derivatives of L with respect to δ and λ_q:

$$\frac{\partial^2 L}{\partial\delta_2 \partial\lambda_{q_3}} = \frac{\partial Q_3}{\partial\delta_2} = 3.058824; \quad \frac{\partial^2 L}{\partial\delta_3 \partial\lambda_{q_3}} = \frac{\partial Q_3}{\partial\delta_3} = -6.117647$$

Second partial derivatives of L with respect to V_3:

$$\frac{\partial^2 L}{\partial V_3^2} = \sum_{j=1}^{3} \lambda_{p_j} \frac{\partial^2 P_j}{\partial V_3^2} + \lambda_{q3} \frac{\partial^2 Q_3}{\partial V_3^2} = 0.0$$

where

$$\frac{\partial^2 P_1}{\partial V_3^2} = 0.0; \quad \frac{\partial^2 P_2}{\partial V_3^2} = 0.0; \quad \frac{\partial^2 P_3}{\partial V_3^2} = 2.0G_{33} = 11.76471$$

$$\frac{\partial^2 Q_3}{\partial V_3^2} = -2.0B_{33} = 46.97882$$

Second partial derivatives of L with respect to V_3 and λ_p:

$$\frac{\partial^2 L}{\partial V_3 \partial \lambda_{p1}} = \frac{\partial P_1}{\partial V_3} = -3.058824; \quad \frac{\partial^2 L}{\partial V_3 \partial \lambda_{p2}} = \frac{\partial P_2}{\partial V_3} = -3.0588240; \quad \frac{\partial^2 L}{\partial V_3 \partial \lambda_{p3}} = \frac{\partial P_3}{\partial V_3} = 5.647059$$

Second partial derivatives of L with respect to V_3 and λ_q:

$$\frac{\partial^2 L}{\partial V_3 \partial \lambda_{q3}} = \frac{\partial Q_3}{\partial V_3} = 22.50824$$

Using the Gauss elimination method, the change in variables is obtained and the updated values of variables are shown in Tables 3.43 and 3.44.

TABLE 3.43 Updated values of variables (Example 3.15)

Bus	P_g (p.u.)	Q_g (p.u.)	P_d (p.u.)	Q_d (p.u.)	P (p.u.)	Q (p.u.)
1	0.7284207	0.1295118	0.20	0.10	0.5485317	0.0295118
2	0.9690878	−0.0730638	0.00	0.00	0.9946736	−0.0730638
3	0.0	0.0	1.50	0.06	−1.5205370	0.0056424

TABLE 3.44 Updated values of variables (Example 3.15)

Bus	V (p.u.)	δ (rad)	λ_p	λ_q
1	1.04	0.0	287.4105	–
2	1.04	0.01169348	295.3632	–
3	1.025293	–0.05167275	300.2588	–10.70845

Cost = 653.3180 ₹/h

The final solution after four iterations is given in Tables 3.45 and 3.46.

TABLE 3.45 Final solution obtained after four iterations (Example 3.15)

Bus	P_g (p.u.)	Q_g (p.u.)	P_d (p.u.)	Q_d (p.u.)	P (p.u.)	Q (p.u.)
1	0.7767037	0.1521580	0.20	0.10	0.5767038	0.0521580
2	0.9452465	−0.0327489	0.00	0.00	0.9452462	−0.0327489
3	0.0	0.0	1.50	0.06	−1.5000000	−0.0599991

TABLE 3.46 Final solution obtained after four iterations (Example 3.15)

Bus	V (p.u.)	δ (rad)	λ_p	λ_q
1	1.04	0.0	293.2044	–
2	1.04	0.00966759	291.7870	–
3	1.02297	−0.05142289	301.0320	0.58963730

Cost = 660.3356 ₹/h

3.18.2 Decoupled Method for Optimal Power Flow

An important characteristic of any practical power transmission system operating in steady state is that the change in real power from the specified value at a bus is more dependent on the changes in voltage angles at various buses than the change in voltage magnitudes, and the change in reactive power from the specified value at a bus is more dependent on the changes in voltage magnitudes at various buses than the changes in voltage angles. This is due to the fact that transmission lines are mostly reactive, the conductance G's are very small compared to the susceptance, B's. Also, under the normal operating conditions the angle $(\delta_i - \delta_j)$ is small (typically less than 10°). In view of this, the terms, $\frac{\partial P}{\partial V}$ and $\frac{\partial Q}{\partial \delta}$ being small can be ignored.

The terms $\frac{\partial^2 P}{\partial V \partial \delta}$ and $\frac{\partial^2 Q}{\partial \delta \partial V}$ will also be small and can be ignored. Ignoring these terms being small, Eqs. (3.186b), (3.186d), (3.187a), (3.187c), (3.188b), and (3.188a) can also be ignored being small. Equations (3.190a) to (3.190e) become:

$$\frac{\partial^2 L}{\partial P_{g_i}^2}\Delta P_{g_i} + \frac{\partial^2 L}{\partial P_{g_i}\partial \lambda_{p_i}}\Delta \lambda_{p_i} = -\frac{\partial L}{\partial P_{g_i}} \quad (i = 1, 2, \ldots, \text{NG}) \tag{3.193a}$$

$$\sum_{j=2}^{\text{NB}} \frac{\partial^2 L}{\partial \delta_i \partial \delta_j}\Delta \delta_j + \sum_{j=1}^{\text{NB}} \frac{\partial^2 L}{\partial \delta_i \partial \lambda_{p_j}}\Delta \lambda_{p_j} = -\frac{\partial L}{\partial \delta_i} \quad (i = 2, 3, \ldots, \text{NB}) \tag{3.193b}$$

$$\frac{\partial^2 L}{\partial \lambda_{p_i}\partial P_{g_i}}\Delta P_{g_i} + \sum_{j=2}^{\text{NB}} \frac{\partial^2 L}{\partial \lambda_{p_i}\partial \delta_j}\Delta \delta_j = -\frac{\partial L}{\partial \lambda_{p_i}} \quad (i = 1, 2, \ldots, \text{NB}) \tag{3.193c}$$

$$\sum_{j=\text{NV}+1}^{\text{NB}} \frac{\partial^2 L}{\partial V_i \partial V_j}\Delta V_j + \sum_{j=\text{NV}+1}^{\text{NB}} \frac{\partial^2 L}{\partial V_i \partial \lambda_{q_j}}\Delta \lambda_{q_j} = -\frac{\partial L}{\partial V_i} \quad (i = \text{NV} + 1, \text{NV} + 2, \ldots, \text{NB}) \tag{3.193d}$$

$$\sum_{j=2}^{\text{NB}} \frac{\partial^2 L}{\partial \lambda_{q_i}\partial \delta_j}\Delta \delta_j + \sum_{j=\text{NV}+1}^{\text{NB}} \frac{\partial^2 L}{\partial \lambda_{q_i}\partial V_j}\Delta V_j = -\frac{\partial L}{\partial \lambda_{q_i}} \quad (i = \text{NV} + 1, \text{NV} + 2, \ldots, \text{NB}) \tag{3.193e}$$

The above equations can be decoupled and can be solved separately. In the matrix notation these can be represented as

$$\begin{bmatrix} \nabla_{P_g P_g} L & \nabla_{P_g \delta} L & \nabla_{P_g \lambda_p} L \\ \nabla_{\delta P_g} L & \nabla_{\delta\delta} L & \nabla_{\delta \lambda_p} L \\ \nabla_{\lambda_p P_g} L & \nabla_{\lambda_p \delta} L & \nabla_{\lambda_p \lambda_p} L \end{bmatrix} \begin{bmatrix} \Delta P_g \\ \Delta \delta \\ \Delta \lambda_p \end{bmatrix} = - \begin{bmatrix} -\nabla_{P_g} L \\ -\nabla_{\delta} L \\ -\nabla_{\lambda_p} L \end{bmatrix} \tag{3.194a}$$

The size of the Hessian matrix is (NG + 2 × NB – 1) × (NG + 2 × NB – 1) and the size of Jacobian is (NG + 2 × NB – 1) × 1.

$$\begin{bmatrix} \nabla_{VV} L & \nabla_{V\lambda_q} L \\ \nabla_{\lambda_q V} L & \nabla_{\lambda_q \lambda_q} L \end{bmatrix} \begin{bmatrix} \Delta V \\ \Delta \lambda_q \end{bmatrix} = \begin{bmatrix} -\nabla_V L \\ -\nabla_{\lambda_q} L \end{bmatrix} \tag{3.194b}$$

The size of the Hessian matrix is [2 × (NB – NV – 1)] × [2 × (NB – NV – 1)] and the size of Jacobian is [2 × (NB – NV – 1)] × 1. The elements of Hessian and Jacobian matrices has already been described in Eqs. (3.183)–(3.189).

The detailed algorithm is outlined below.

Algorithm 3.11: Decoupled Method for Optimal Power Flow

1. Read data a_i, b_i, and c_i (i = 1, 2, ..., NG), load on each bus, and the line data for the power system network.
2. Obtain Y_{BUS} using the Y-bus algorithm.
3. Calculate the initial values of P_{g_i} (i = 1, 2, ..., NG) and λ by assuming that $P_L = 0$. Then the problem can be stated by Eqs. (3.2a) and (3.2b) and the solution can be obtained directly using Eqs. (3.9) and (3.10).

   Initialize all $\lambda_{p_i} = \lambda$ (i = 1, 2, ..., NB), $\lambda_{q_i} = 0$ (i = NV + 1, NV + 2, ..., NB), V_i = 1 p.u. (i = 2, 3, ..., NB) and $\delta_i = 0$ (i = 2, 3, ..., NB).
4. Calculate the Jacobian and Hessian matrix elements from Eqs. (3.183a) to (3.189) and solve Eqs. (3.194a) and (3.194b).

$$[H^1] \begin{bmatrix} \Delta P_g \\ \Delta \delta \\ \Delta \lambda_p \end{bmatrix} = - \begin{bmatrix} \dfrac{\partial L}{\partial P_g} \\ \dfrac{\partial L}{\partial \delta} \\ \dfrac{\partial L}{\partial \lambda_p} \end{bmatrix} \quad \text{and} \quad [H^2] \begin{bmatrix} \Delta V \\ \Delta \lambda_q \end{bmatrix} = - \begin{bmatrix} \dfrac{\partial L}{\partial V} \\ \dfrac{\partial L}{\partial \lambda_q} \end{bmatrix}$$

   Gauss elimination method is used to solve these simultaneous equations separately to find ΔP_g, $\Delta\delta$, $\Delta\lambda_p$, ΔV, and $\Delta\lambda_q$.
5. Check convergence

$$\left[\sum_{i=1}^{NG} (\Delta P_{g_i})^2 + \sum_{i=2}^{NB} (\Delta \delta_i)^2 + \sum_{i=1}^{NB} (\Delta \lambda_{p_i})^2 \right]^{\frac{1}{2}} \le \varepsilon \quad \text{and} \quad \left[\sum_{i=NV+1}^{NB} (\Delta V_i)^2 + \sum_{i=NV+1}^{NB} (\Delta \lambda_q)^2 \right]^{\frac{1}{2}} \le \varepsilon$$

   and optimality conditions. If convergence condition is not satisfied then GOTO Step 6 else GOTO Step 8.
6. Modify the variables,

$$\begin{aligned} P_{g_i} &= P_{g_i} + \Delta P_{g_i} && (i = 1, 2, ..., NG) \\ \delta_i &= \delta_i + \Delta \delta_i && (i = 2, 3, ..., NB) \\ \lambda_{p_i} &= \lambda_{p_i} + \Delta \lambda_{p_i} && (i = 1, 2, ..., NB) \\ V_i &= V_i + \Delta V_i && (i = NV + 1, NV + 2, ..., NB) \\ \lambda_{q_i} &= \lambda_{q_i} + \Delta \lambda_{q_i} && (i = NV + 1, NV + 2, ..., NB) \end{aligned}$$

7. Check the limits, if any limit of a variable is violated then impose or remove power flow equation or a penalty for inequality. Add or remove derivatives for penalty or equation change and GOTO Step 4 to update the solution.
8. Calculate the total cost.
9. Stop.

EXAMPLE 3.16 Consider the 3-bus system of Figure 3.19 (see Example 3.15). The series impedance of each line is given in Table 3.40. The system has two generators. The operating cost characteristics of two generators are given below.

$$F_1 = 50P_{g_1}^2 + 352P_{g_1} + 44.4 \quad ₹/\text{h}$$

$$F_2 = 50P_{g_2}^2 + 389P_{g_2} + 40.6 \quad ₹/\text{h}$$

Obtain the optimal schedule using the decoupled method.

Solution Number of generators, NG = 2; Number of buses, NB = 3; Number of lines, NL = 3; Number of *PV* buses, NV = 1.

Y-bus and calculations for initial values are given in Example 3.15 and tabulated in Table 3.41. P_g, Q_g, P and Q are tabulated in Table 3.42 which are obtained with the initial values. The updated values after first iteration are tabulated in Tables 3.47 and 3.48.

TABLE 3.47 Updated values of variables after the first iteration (Example 3.16)

Bus	P_g (p.u.)	Q_g (p.u.)	P_d (p.u.)	Q_d (p.u.)	P (p.u.)	Q (p.u.)
1	0.7689440	0.0714597	0.20	0.10	0.4721372	−0.1497818
2	0.9404750	0.0885403	0.00	0.00	0.8499779	−0.2373549
3	0.0	0.0	1.50	0.06	−1.3046550	0.3271568

Final solution is obtained in 11 iterations for convergence 0.0001 and is tabulated in Tables 3.49 and 3.50

TABLE 3.48 Updated values of variables after the first iteration (Example 3.16)

Bus	V (p.u.)	δ (rad)	λ_p (₹/p.u.h)	λ_q (₹/p.u.h)
1	1.04	0.0	292.2733	–
2	1.04	0.00985892	291.0713	–
3	1.040301	−0.04675326	295.4336	−6.128246

Cost = 656.6735 ₹/h

TABLE 3.49 Final solution obtained after 11 iterations (Example 3.16)

Bus	P_g (p.u.)	Q_g (p.u.)	P_d (p.u.)	Q_d (p.u.)	P (p.u.)	Q (p.u.)
1	0.7766963	0.0714597	0.2	0.10	0.5766963	0.0521595
2	0.9452535	0.0885403	0.0	0.00	0.9452534	−0.0327503
3	0.0	0.0	1.5	0.06	−1.5000000	−0.0599991

TABLE 3.50 Final solution obtained after 11 iterations (Example 3.16)

Bus	V (p.u.)	δ (rad)	λ_p (₹/p.u.h)	λ_q (₹/p.u.h)
1	1.04	0.0	293.2036	–
2	1.04	0.00966797	291.7880	–
3	1.02297	–0.05142269	300.8488	0.00000542

Cost = 660.3355 ₹/h

3.19 OPTIMAL POWER FLOW BASED ON GRADIENT METHOD

The objective function to be minimized is the operating cost

$$F = \sum_{i=1}^{NG} F_i(P_{g_i}) \tag{3.195a}$$

where

$$F_i(P_{g_i}) = a_i P_{g_i}^2 + b_i P_{g_i} + c_i$$

If the system real power loss is to be minimized, the objective function is

$$F = P_1(V, \delta)$$

In this case the net injected real powers are fixed, the minimization of the real injected power P_1 at the slack bus is equivalent to minimization of total system loss. This is known as optimal reactive power flow problem, subject to the load flow equations:

(a) Real power balance in the network for each *PV* bus

$$P_i(V, \delta) - P_{g_i} + P_{d_i} = 0 \qquad (i = 2, 3, ..., \text{NV}) \tag{3.195b}$$

(b) Active and reactive power balance in the network for each *PQ* bus

$$P_i(V, \delta) - P_{g_i} + P_{d_i} = 0 \qquad (i = \text{NV} + 1, \text{NV} + 2, ..., \text{NB}) \tag{3.195c}$$

$$Q_i(V, \delta) - Q_{g_i} + Q_{d_i} = 0 \qquad (i = \text{NV} + 1, \text{NV} + 2, ..., \text{NB}) \tag{3.195d}$$

with

$$P_i(V, \delta) = V_i \sum_{j=1}^{NB} V_j \left(G_{ij} \cos(\delta_i - \delta_j) + B_{ij} \sin(\delta_i - \delta_j)\right) \tag{3.195e}$$

and

$$Q_i(V, \delta) = V_i \sum_{j=1}^{NB} V_j \left(G_{ij} \sin(\delta_i - \delta_j) - B_{ij} \cos(\delta_i - \delta_j)\right) \tag{3.195f}$$

where

NG is the number of generator buses
NB is the number of buses
NV is the number of voltage controlled buses
P_i is the active power injection into bus i
Q_i is the reactive power injection into bus i
P_{d_i} is the active load on bus i

Q_{d_i} is the reactive load on bus i
P_{g_i} is the active generation on bus i
Q_{g_i} is the reactive generation on bus i
V_i is the magnitude of voltage at bus i
δ_i is the voltage phase angle at bus i
$Y_{ij} = G_{ij} + jB_{ij}$ are the elements of the admittance matrix.

Equations (3.195b), (3.195c) and (3.195d) can be expressed in vector form as

$$g(x, y) = \begin{bmatrix} \text{Eq. (3.195b) for each } PV \text{ bus} \\ \left.\begin{matrix} \text{Eq. (3.195c)} \\ \text{Eq. (3.195d)} \end{matrix}\right\} \text{for } PQ \text{ bus} \end{bmatrix} \tag{3.196}$$

where the vector of dependent variables is

$$x = \begin{bmatrix} \delta_i \text{ for } PV \text{ bus} \\ \left.\begin{matrix} \delta_i \\ V_i \end{matrix}\right\} \text{for } PQ \text{ bus} \end{bmatrix} \tag{3.197a}$$

and the vector of independent variables is

$$y = \begin{bmatrix} \left.\begin{matrix} V_1 \\ \delta_1 \end{matrix}\right\} \text{for slack bus} \\ \left.\begin{matrix} P_i \\ V_i \end{matrix}\right\} \text{for } PV \text{ bus} \\ \left.\begin{matrix} P_i \\ Q_i \end{matrix}\right\} \text{for } PQ \text{ bus} \end{bmatrix} = \begin{bmatrix} u \\ p \end{bmatrix} \tag{3.197b}$$

In the above formulation, the objective function must include the slack bus power. The vector of independent variables y can be partitioned into two parts—a vector u of control variables which are to be varied to achieve optimum value of objective function and a vector p of fixed or disturbance or uncontrollable parameters.

Control parameters may be voltage magnitudes on PV buses, or P_{g_i} at buses with controllable power. Slack bus voltage and regulating transformer tap-setting may be employed as additional control variables. Q_{g_i} may be used as control variables on buses with reactive power control [Singh, 1993].

The optimization problem can be restated as

$$\min_{u} \quad F(x, u) \tag{3.198a}$$

subject to equality constraints

$$g(x, u, p) = 0 \tag{3.198b}$$

To solve the optimization problem, we define Lagrange functions as

$$L(x, u, p) = F(x, u) + \lambda^T g(x, u, p) \tag{3.198c}$$

where λ is the vector of Lagrange multipliers of the same dimension as $g(x, u, p)$.

The necessary conditions to minimize the unconstrained Lagrangian function are

$$\frac{\partial L}{\partial x} = \frac{\partial F}{\partial x} + \left[\frac{\partial g}{\partial x}\right]^T \lambda = 0 \tag{3.198d}$$

$$\frac{\partial L}{\partial u} = \frac{\partial F}{\partial u} + \left[\frac{\partial g}{\partial u}\right]^T \lambda = 0 \tag{3.198e}$$

$$\frac{\partial L}{\partial \lambda} = g(x, u, p) = 0 \tag{3.198f}$$

Equations (3.198d), (3.198e) and (3.198f) are nonlinear algebraic equations and can only be solved iteratively. A simple yet efficient iteration scheme, that can be employed, is the steepest descent method (also called the gradient method). The basic technique is to adjust the control vector u, so as to move from one feasible solution point (a set of values of x which satisfies Eq. (3.198f) for given u and p; it indeed is the load flow solution) in the direction of steepest descent (negative gradient) to a new feasible solution point with lower value of objective function.

The computational procedure for the gradient method with relevant details is given below:

1. Make an initial guess for u, the control variables.
2. Find a feasible load flow solution from Eq. (2.111a and b) by the Newton–Raphson iterative method. The method improves the solution x as follows:

$$x^{r+1} = x^r + \Delta x$$

where Δx is obtained by solving the set of linear equations given below:

$$\left[\frac{\partial g}{\partial x}(x^r, y)\right] \Delta x = -g(x^r, y)$$

$$\Delta x = -\left[\frac{\partial g}{\partial x}(x^r, y)\right]^{-1} g(x^r, y)$$

The end result of Step 2 is a feasible solution of x and the Jacobian matrix.

3. Solve Eq. (3.198d) for λ

$$\lambda = -\left[\left(\frac{\partial g}{\partial x}\right)\right]^{-1} \frac{\partial F}{\partial x} \tag{3.199a}$$

4. Insert λ from Eq. (3.199a) into Eq. (3.198e) and compute the gradient

$$\nabla L = \frac{\partial F}{\partial u} + \left[\frac{\partial g}{\partial u}\right]^T \lambda \tag{3.199b}$$

5. If ∇L equals zero within the prescribed tolerance, the minimum has been reached. Otherwise:
6. Find a new set of control variables

$$u^{\text{new}} = u^{\text{old}} + \Delta u \qquad \text{where} \qquad \Delta u = -\alpha \nabla L \tag{3.199c}$$

Here Δu is a step in negative direction of the gradient. The step size is adjusted by the positive scalar α.

In the algorithm, the choice of α is very critical. Too small a value of α guarantees the convergence, but slows down the rate of convergence; too high a value causes oscillations around the minimum. Several methods are available for optimum choice of step size.

3.19.1 Inequality Constraints on Control Variables

Though in the earlier discussion, the control variables are assumed to be unconstrained, the permissible values are, in fact, always constrained, i.e.

$$u^{\min} \le u \le u^{\max}$$

e.g.
$$P_{g_1}^{\min} \le P_{g_i} \le P_{g_i}^{\max}$$

These inequality constraints on control variables can be easily handled. If the correction Δu_i in Eq. (3.199c) causes u_i to exceed one of the limits, u_i is set equal to the corresponding limit, i.e.

$$u_i^{\text{new}} = \begin{cases} u_i^{\max} & \text{if} \quad u_i^{\text{old}} + \Delta u_i > u_i^{\max} \\ u_i^{\min} & \text{if} \quad u_i^{\text{old}} + \Delta u_i < u_i^{\min} \\ u_i^{\text{old}} + \Delta u_i & \end{cases} \tag{3.200a}$$

After a control variables reaches any of the limits, its component in the gradient should continue to be computed in later iterations, as the variable may come within limits at some later stage.

In accordance with the Kuhn-Tucker theorem, the necessary conditions for minimization of L under constraint are:

$$\left\{ \begin{aligned} \frac{\partial L}{\partial u_i} &= 0 \quad \text{if} \quad u_i^{\min} < u_i < u_i^{\max} \\ \frac{\partial L}{\partial u_i} &\le 0 \quad \text{if} \quad u_i = u_i^{\min} \\ \frac{\partial L}{\partial u_i} &\ge 0 \quad \text{if} \quad u_i = u_i^{\max} \end{aligned} \right\} \tag{3.200b}$$

Therefore, now, in Step 5 of the computational algorithm, the gradient vector has to satisfy the optimality condition (3.200b).

3.19.2 Inequality Constraints on Dependent Variables

Often, the upper and lower limits on dependent variables are specified as

$$x^{\min} \le x \le x^{\max}$$

e.g.
$$V_i^{\min} \le V_i \le V_i^{\max} \quad \text{on a } PQ \text{ bus}$$

Such inequality constraints can be conveniently handled by the penalty function method. The objective function is augmented by penalties for inequality constraints violations. This forces the solution to lie sufficiently close to the constraint limits, when these limits are violated. The penalty function method is valid in this case, because these constraints are seldom rigid limits in the strict sense, but are in fact, soft limits (e.g. $V \le 1.0$ on a PQ bus really means V should not exceed 1.0 too much and $V = 1.01$ may still be permissible).

The penalty method calls for augmentation of the objective function so that the new objective function becomes

$$\overline{F} = F(x, u) + \sum_j W_j \tag{3.201}$$

where the penalty W_j is introduced for each violated inequality constraint.

A suitable penalty function is defined as

$$W_j = \begin{cases} \gamma_j\,(x_j - X_j^{\max})^2, & \text{for} \quad x_j > x_j^{\max} \\ \gamma_j\,(x_j - x_j^{\min})^2, & \text{for} \quad x_j < x_j^{\min} \end{cases} \tag{3.202}$$

where γ_j is a real positive number which controls the degree of penalty and is called the penalty factor.

The necessary conditions (3.198d) and (3.198e) would now be modified as given below, while the condition (3.198f), i.e. load flow equation, remains unchanged.

$$\frac{\partial L}{\partial x} = \frac{\partial F}{\partial x} + \sum_j \frac{\partial W_j}{\partial x} + \left[\frac{\partial g}{\partial x}\right]^T \lambda = 0 \tag{3.203a}$$

$$\frac{\partial L}{\partial u} = \frac{\partial F}{\partial u} + \sum_j \frac{\partial W_j}{\partial u} + \left[\frac{\partial g}{\partial u}\right]^T \lambda = 0 \tag{3.203b}$$

The vector $\partial W_i/\partial x$ obtained from Eq. (3.202) would contain only one non-zero term corresponding to the dependent variable x_j; while $\partial W_j/\partial u = 0$, as the penalty functions on dependent variables are independent of the control variables.

By choosing a higher value of γ_j, the penalty function can be made steeper so that the solution lies closer to the rigid limits; the convergence, however, will become poorer. A good scheme is to start with a low value of γ_j and to increase it during the optimization process, if the solution exceeds a certain tolerance limit.

EXAMPLE 3.17 Consider the 5-bus system of Figure 3.17 (see Example 3.11). The series impedance of each line is given in Table 3.20. The system has three generators. Find the economic generation schedule. The operating cost characteristics of the three generators are given below:

$$F_1 = 60P_{g_1}^2 + 200P_{g_1} + 140.0 \text{ ₹/h}$$

$$F_2 = 75P_{g_2}^2 + 150P_{g_2} + 120.0 \text{ ₹/h}$$

$$F_3 = 70P_{g_3}^2 + 180P_{g_3} + 80.0 \text{ ₹/h}$$

Solution Considering bus 1 as slack bus, buses 2 and 3 as *PV* buses and 4 and 5 buses are considered as *PQ* buses. For the sample system, the optimization problem is stated as

Minimize
$$F = \sum_{i=1}^{3} (a_i P_{g_i}^2 + b_i P_{g_i} + c_i) \text{ ₹/h} \tag{3.204a}$$

subject to the load flow equations

$$P_i(V, \delta) - P_{g_i} + P_{d_i} = 0 \qquad (i = 2, 3, \ldots, 5) \tag{3.204b}$$

$$Q_i(V, \delta) - Q_{g_i} + Q_{d_i} = 0 \qquad (i = 4, 5) \tag{3.204c}$$

where

$$P_i(V, \delta) = V_i \sum_{i=1}^{5} V_j \left((G_{ij} \cos(\delta_i - \delta_j) + B_{ij} \sin(\delta_i - \delta_j) \right) \tag{3.204d}$$

$$Q_i(V, \delta) = V_i \sum_{i=1}^{5} V_j \left(G_{ij} \sin(\delta_i - \delta_j) - B_{ij} \cos(\delta_i - \delta_j) \right) \tag{3.204e}$$

For this problem, the dependent variables are

$$x = [\delta_2, \delta_3, \delta_4, \delta_5, V_4, V_5]^T$$

The independent variables are

$$y = [u\ p]^T$$

The control variables are

$$u = [P_{g_2}, P_{g_3}]^T$$

The fixed variables are

$$p = [V_1, \delta_1, P_2, V_2, P_3, V_3, P_4, Q_4, P_5, Q_5]^T$$

The Lagrangian function can be written as

$$L(x, u, p) = \sum_{i=1}^{3} \left(a_i P_{g_i}^2 + b_i P_{g_i} + c_i \right) + \sum_{i=2}^{5} \lambda_{p_i} \left(P_i(V, \delta) - P_{g_i} + P_{d_i} \right) + \sum_{i=4}^{5} \lambda_{q_i} \left(Q_i(V, \delta) - Q_{g_i} + Q_{d_i} \right) \tag{3.205}$$

The necessary conditions for minimization are

$$\frac{\partial L}{\partial P_{g_i}} = 2a_i P_{g_i} + b_i + \lambda_{p_i} \qquad (i = 2, 3) \tag{3.206a}$$

$$\frac{\partial L}{\partial \delta_i} = (2a_1 P_{g_1} + b_1) \frac{\partial P_1}{\partial \delta_i} + \sum_{j=2}^{5} \lambda_{p_j} \frac{\partial P_j}{\partial \delta_i} + \sum_{j=4}^{5} \lambda_{q_j} \frac{\partial Q_j}{\partial \delta_i} \qquad (i = 2, 3, 4, 5) \tag{3.206b}$$

$$\frac{\partial L}{\partial V_i} = (2a_1 P_{g_1} + b_1) \frac{\partial P_1}{\partial V_i} + \sum_{j=2}^{5} \lambda_{p_j} \frac{\partial P_j}{\partial V_i} + \sum_{j=4}^{5} \lambda_{q_j} \frac{\partial Q_j}{\partial V_i} \qquad (i = 4, 5) \tag{3.206c}$$

$$\frac{\partial L}{\partial \lambda_{p_i}} = P_i(V, \delta) - P_{g_i} + P_{d_i} = 0 \qquad (i = 2, 3, 4, 5) \tag{3.206d}$$

$$\frac{\partial L}{\partial \lambda_{q_i}} = Q_i(V, \delta) - Q_{g_i} + Q_{d_i} = 0 \qquad (i = 4, 5) \tag{3.206e}$$

Equations (3.206d) and (3.206e) can be solved using the Newton–Raphson method. To implement the Newton–Raphson method, the following equation is solved.

$$\begin{bmatrix} \dfrac{\partial P}{\partial \delta} & \dfrac{\partial P}{\partial V} \\ \dfrac{\partial Q}{\partial \delta} & \dfrac{\partial Q}{\partial V} \end{bmatrix}_{6\times 6} \begin{bmatrix} \Delta\delta \\ \Delta V \end{bmatrix}_{6\times 1} = \begin{bmatrix} \Delta P \\ \Delta Q \end{bmatrix}_{6\times 1} \tag{3.207a}$$

The size of the Jacobian matrix is $(2 \times (5 - 1) - 2) = 6$. δ_i and V_i are calculated using the Gauss elimination method and are updated till no further improvement is achieved.

$$\delta_i^{r+1} = \delta_i^r + \Delta\delta_i \qquad (i = 2, 3, 4, 5)$$

$$V_i^{r+1} = V_i^r + \Delta V_i \qquad (i = 4, 5)$$

Initial data to implement the Newton–Raphson method is given in Table 3.51.

TABLE 3.51 Initial data for N–R load flow (Example 3.17)

Bus	*Type*	P_g (p.u.)	P_d (p.u.)	Q_d (p.u.)	V (p.u.)	δ (rad)	λ_p (₹/p.u.h)	λ_q (₹/p.u.h)
1	Slack	0.45858	0.00	0.00	1.0600	0.0	–	–
2	*PV*	0.68674	0.20	0.10	1.0593	0.0	253.0107	–
3	*PV*	0.52151	0.45	0.15	1.0525	0.0	253.0107	–
4	*PQ*	0.0	0.40	0.05	1.0	0.0	0.0	0.0
5	*PQ*	0.0	0.60	0.10	1.0	0.0	0.0	0.0

After four iterations, the solution obtained is given in Table 3.52.

TABLE 3.52 Solution of N–R load flow method after four iterations (Example 3.17)

Bus	P_g (p.u.)	Q_g (p.u.)	P_d (p.u.)	Q_d (p.u.)	P (p.u.)	Q (p.u.)
1	0.4585800	−0.1633390	0.00	0.00	0.4585800	−0.1633390
2	0.6867383	0.1656168	0.20	0.10	0.4867383	0.0656168
3	0.5215049	0.1267469	0.45	0.15	0.0715049	−0.0232531
4	0.0	0.0	0.40	0.05	−0.4000005	−0.0500000
5	0.0	0.0	0.60	0.10	−0.6000000	−0.0999998

The voltage and the voltage angle at each bus are given below:

$V_1 = 1.0600$ p.u. $\qquad \delta_1 = 0.0$ rad

$V_2 = 1.0593$ p.u. $\qquad \delta_2 = -0.01739534$ rad

$V_3 = 1.0525$ p.u. $\qquad \delta_3 = -0.03591179$ rad

$V_4 = 1.048837$ p.u. $\qquad \delta_4 = -0.04516637$ rad

$V_5 = 1.034887$ p.u. $\qquad \delta_5 = -0.06914239$ rad

Equations (3.206b) and (3.206c) can be rewritten in matrix form as

$$\begin{bmatrix} (2a_1 P_{g_1} + b_1)\dfrac{\partial P_1}{\partial \delta} \\ (2a_1 P_{g_1} + b_1)\dfrac{\partial P_1}{\partial V} \end{bmatrix}_{6\times 1} + \begin{bmatrix} \dfrac{\partial P}{\partial \delta} & \dfrac{\partial Q}{\partial \delta} \\ \dfrac{\partial P}{\partial V} & \dfrac{\partial Q}{\partial V} \end{bmatrix}_{6\times 6} \begin{bmatrix} \lambda_p \\ \lambda_q \end{bmatrix}_{6\times 1} = \begin{bmatrix} 0 \\ 0 \end{bmatrix}$$

$$\begin{bmatrix} \dfrac{\partial P}{\partial \delta} & \dfrac{\partial Q}{\partial V} \\ \dfrac{\partial Q}{\partial \delta} & \dfrac{\partial Q}{\partial V} \end{bmatrix}_{6\times 6}^{T} \begin{bmatrix} \lambda_p \\ \lambda_q \end{bmatrix}_{6\times 1} = -\begin{bmatrix} (2a_1 P_{g_1} + b_1)\dfrac{\partial P_1}{\partial \delta} \\ (2a_1 P_{g_1} + b_1)\dfrac{\partial P_1}{\partial V} \end{bmatrix} \quad (3.207b)$$

The Jacobian matrix is available from the last iteration of the N–R method.

$$J = \begin{bmatrix} 36.307790 & -5.539205 & -5.501603 & -8.069157 & 2.00526 & 3.162241 \\ -5.608016 & 42.88788 & -33.01344 & 0.0 & 11.34501 & 0.0 \\ -5.604438 & -3.21776 & 42.61675 & -4.036656 & -14.60909 & 1.453976 \\ -8.35267 & 0.0 & -4.101710 & 12.10581 & 1.258813 & -4.616217 \\ -1.617749 & -10.23234 & 13.1661 & -1.386275 & 40.53706 & -3.848699 \end{bmatrix}$$

$$[J]\begin{bmatrix} \lambda_{p2} \\ \lambda_{p3} \\ \lambda_{p4} \\ \lambda_{p5} \\ \lambda_{q4} \\ \lambda_{q5} \end{bmatrix} = \begin{bmatrix} 4319.686 \\ 1079.046 \\ 0.0 \\ 0.0 \\ 0.0 \\ 0.0 \end{bmatrix}$$

From the above equation the values of λ_p and λ_q are obtained by implementing the Gauss elimination method.

$$\lambda_{p_2} = 258.0038, \quad \lambda_{p_3} = 261.2159, \quad \lambda_{p_4} = 262.8730, \quad \lambda_{p_5} = 267.5160$$

$$\lambda_{q_4} = 0.1127377, \quad \lambda_{q_5} = 1.167928$$

From Eq. (3.206a), the gradient can be obtained as

$$\frac{\partial L}{\partial P_{g_2}} = 2a_2 P_{g_2} + b_2 + \lambda_{p_2} = -4.993037$$

$$\frac{\partial L}{\partial P_{g_3}} = 2a_3 P_{g_3} + b_3 + \lambda_{p_3} = -8.205190$$

If the norm of gradient is more than the required tolerance, then update the values.

$$\| \nabla_{P_g} \| = 9.604976 > 0.1$$

The total operating cost at this schedule is 704.0132 ₹/h.

After updating the values, the Newton–Raphson method is implemented and the solution is given in Table 3.53 and voltage magnitude and angles at each bus are given below:

$V_1 = 1.060$ p.u. $\qquad \delta_1 = 0.0$ rad

$V_2 = 1.0593$ p.u. $\qquad \delta_2 = -0.01598643$ rad

$V_3 = 1.0525$ p.u. $\qquad \delta_3 = -0.03360153$ rad

$V_4 = 1.048837$ p.u. $\qquad \delta_4 = -0.04303627$ rad

$V_5 = 1.034892$ p.u. $\qquad \delta_5 = -0.06749434$ rad

TABLE 3.53 Solution of N–R load flow method after two iterations (Example 3.17)

Bus	P_g (p.u.)	Q_g (p.u.)	P_d (p.u.)	Q_d (p.u.)	P (p.u.)	Q (p.u.)
1	0.4249496	−0.1529420	0.00	0.00	0.4249496	−0.1529420
2	0.6992280	0.1606308	0.20	0.10	0.4992280	0.0606308
3	0.5420235	0.1194695	0.45	0.15	0.0920235	−0.0305305
4	0.0	0.0	0.40	0.05	−0.3999993	−0.0499971
5	0.0	0.0	0.60	0.10	−0.6000002	−0.0999995

The final solution as obtained after eight iterations is given in Table 3.54.

TABLE 3.54 Solution of N–R load flow method after two iterations (Example 3.17)

Bus	P_g (p.u.)	Q_g (p.u.)	P_d (p.u.)	Q_d (p.u.)	P (p.u.)	Q (p.u.)
1	0.4258551	−0.1532493	0.00	0.00	0.4258551	−0.1532493
2	0.6924789	0.1627552	0.20	0.10	0.4924789	0.0627552
3	0.5478095	0.1174821	0.45	0.15	0.0978096	−0.0325179
4	0.0	0.0	0.40	0.05	−0.3999997	−0.0500030
5	0.0	0.0	0.60	0.10	−0.5999999	−0.0999999

For the above solution, the voltage and voltage magnitude at each bus are given below:

$V_1 = 1.06$ p.u. $\quad \delta_1 = 0.0$ rad

$V_2 = 1.0593$ p.u. $\quad \delta_2 = -0.01610616$ rad

$V_3 = 1.0525$ p.u. $\quad \delta_3 = -0.03333623$ rad

$V_4 = 1.048837$ p.u. $\quad \delta_4 = -0.04284792$ rad

$V_5 = 1.034894$ p.u. $\quad \delta_5 = -0.06751189$ rad

The obtained values of λ_p and λ_q are given below:

$$\lambda_{p_2} = 253.8127, \quad \lambda_{p_3} = 256.7519, \quad \lambda_{p_4} = 258.4248, \quad \lambda_{p_5} = 263.1094$$

$$\lambda_{q_4} = 0.1106465, \quad \lambda_{q_5} = 1.146384$$

The gradient as obtained is given below:

$$\frac{\partial L}{\partial P_{g_2}} = 2a_2P_{g_2} + b_2 + \lambda_{p_2} = 5.915117\text{E-02}$$

$$\frac{\partial L}{\partial P_{g_2}} = 2a_2P_{g_2} + b_2 + \lambda_{p_2} = -5.852580\text{E-02}$$

$$\|\nabla_{P_g}\| = 8.321136\text{E-02} < 0.1$$

The total operating cost at this schedule is = 695.5009 ₹/h

REFERENCES

Books

Arrillaga, J. and C.P. Arnold, *Computer Analysis of Power Systems*, John Wiley & Sons, Singapore, 1990.

Elgerd, O.I., *Electric Energy Systems Theory: An Introduction,* 2nd ed., Tata McGraw-Hill, New Delhi, 1983.

El-Hawary, M.E. and G.S. Christensen, *Optimal Economic Operation of Power Systems*, Academic Press, New York, 1979.

Gross, C.A., *Power System Analysis*, Wiley, New York, 1979.

Kirchmayer, L.K., *Economic Operation of Power Systems*, Wiley Eastern, New Delhi, 1958.

Kothari, D.P. and I.J. Nagrath, *Modern Power System Analysis*, 3rd ed., Tata McGraw-Hill, New Delhi, 2003.

Kusic, G.L., *Computer Aided Power Systems Analysis*, Prentice-Hall of India, New Delhi, 1986.

Mahalanabis, A.K., D.P. Kothari and S.I. Ahson, *Computer Aided Power System Analysis and Control*, Tata McGraw-Hill, New Delhi, 1991.

Nagrath, I.J. and D.P. Kothari, *Power System Engineering*, Tata McGraw-Hill, New Delhi, 1994.

Singh, L.P., *Advanced Power System Analysis and Dynamics*, 2nd ed., Wiley Eastern, New Delhi, 1993.

Stagg, G.W. and A.H. Ei-Abiad, *Computer Methods in Power Systems Analysis*, McGraw-Hill, New Delhi, 1968.

Stevenson, W.D., *Elements of Power System Analysis*, 4th ed., McGraw-Hill, New York, 1982.

Wood, A.J. and B. Wollenberg, *Power Generation, Operation and Control*, John Wiley, New York, 1984.

Papers

Alguacil, N. and A.J. Conejo, Multiperiod optimal power flow using benders decomposition, *IEEE Trans. on Power Systems*, Vol. **PWRS-5,** No. 1, pp. 196–201, February 2000.

Alvarado, F.L., Penalty factors from Newton's method, *IEEE Trans. on Power Apparatus and Systems*, Vol. **PAS-97,** No. 6, pp. 2031–2037, 1978.

Aoki, K. and T. Satoh, New algorithms for classic economic load dispatch, *IEEE Trans. on Power Apparatus and Systems*, Vol. **PAS-103,** No. 6, pp. 1423–1431, 1984.

Burchett, R.C., H.H. Happ, D.R. Vierath and K.A. Wirgau, Developments in optimal power flow, *IEEE Trans. on Power Apparatus and Systems*, Vol. **PAS-101,** No. 5, pp. 406–414, 1982.

Carpentier, J.L., Optimal power flow, *Int. J. Electric Power and Energy Systems*, Vol. **1,** No. 1, pp. 1315, 1979.

Carpentier, J.L., Optimal power flow: uses, methods and developments, *Proceedings of IFAC Conference*, R.J. Brazil, pp. 1125, 1985.

Chen, L., S. Matoba, H. Inabe and T. Okabe, Surrogate constraint method for optimal power flow, *IEEE Trans. on Power Systems*, Vol. **PWRS-13,** No. 3, pp. 1084–1089, 1998.

Chowdhury B.H. and S. Rahman, A review of recent advances in economic dispatch, *IEEE Trans. on Power Systems,* Vol. **PWRS-5,** No. 4, pp. 1248–1259, 1990.

Chun-Lung, Chen and Nanming Chen, Direct search method for solving economic dispatch problem considering transmission capacity constraints, *IEEE Trans. on Power Systems*, Vol. **PWRS-16,** No. 4, pp. 764–769, November 2001.

da Costa, G.R.M., C.E.U. Costa and A.M. de Souza, Comparative studies of optimization methods for the optimal power flow problem, *Electric Power Systems Research*, Vol. **56,** pp. 249–254, 2000.

da Costa, V.M., N. Martins and J.L.R. Pereira, Developments in the Newton-Raphson power flow formulation based on current injections, *IEEE Trans. on Power Systems*, Vol. **PWRS-14(4)**, pp. 1320–1326, 1999.

da Costa, V.M., N. Martins and J.L.R. Pereira, An augmented Newton-Raphson power flow formulation based on current injections, *Int. J. Electric Power and Energy Systems*, Vol. **23**, pp. 305–312, 2001.

Dommel, H.W. and W.F. Tinney, Optimal power flow solutions, *IEEE Trans. on Power Apparatus and Systems,* Vol. **PAS-87,** No. 10, pp. 1866–1876, 1968.

Dopazo J.F., O.A. Klitin, G.W. Stagg and M. Watson, An optimization technique for real and reactive power allocation, *Proceedings of IEEE*, Vol. **55,** No. 11, pp. 1877–1885, 1967.

Early, E.D., R.E. Watson and G.L. Smith, A general transmission loss equation, *AIEE Trans. Part-III: Power Apparatus and Systems*, Vol. **74,** No. 3, pp. 510–520, 1955.

Ekwue A.O. and J.F. Macqueen, Comparison of load flow solution methods, *Electric Power Systems Research*, Vol. **22,** No. 3, pp. 213–222, 1991.

Famideh–Vojdani, A.R. and F.D. Galiana, Economic dispatch with generation constraints, *IEEE Trans. on Automatic Control*, Vol. **25,** No. 2, pp. 213–217, 1980.

Fang, R.S. and A.K. David, Optimal dispatch under transmission contracts, *IEEE Trans. on Power Systems*, Vol. **PWRS-14,** No. 2, pp. 732–737, May 1999.

Happ, H.H., Analysis of networks with complex auto-transformers, *IEEE Trans. on Power Apparatus and Systems*, Vol. **PAS-82,** No. 65, pp. 75–81, 1963.

Happ, H.H., Analysis of networks with complex auto-transformers II: Relations between all open-path and open-path-closed-path impedance Matrices, *IEEE Trans. on Power Apparatus and Systems*, Vol. **PAS-82,** No. 69, pp. 958–965, 1963.

Happ, H.H., Analysis of networks with complex auto-transformers, III: Invariant Reduction for purpose of loss formula, *IEEE Trans. on Power Apparatus and Systems*, Vol. **PAS-83**, pp. 707–714, 1964.

Happ, H.H., Optimal power dispatch, *IEEE Trans. on Power Apparatus and Systems*, Vol. **PAS-93,** No. 3, pp. 820–830, 1974.

Happ, H.H., Optimal power dispatch: A comprehensive survey, *IEEE Trans. on Power Apparatus and Systems*, Vol. **PAS-96,** No. 3, pp. 841–854, 1977.

Happ, H.H., J.F. Hohenstein, L.K. Kirchmayer and G.W. Stagg, Direct calculation of transmission loss formula-II, *IEEE Trans. on Power Apparatus and Systems*, Vol. **PAS-83**, pp. 702–707, 1964.

Hill, E.F. and W.D. Stevenson, A new method of determining loss coefficients, *IEEE Trans. on Power Apparatus and Systems,* Vol. **PAS-87,** No. 7, pp. 1548–1553, July 1968.

Housos, E.C. and G.D. Irisarri, A sparse variable metric optimisation method applied to the solution of power system problems, *IEEE Trans. on Power Apparatus and Systems*, Vol. **PAS-101,** No. 1, pp. 195–202, January 1982.

Huneault, M. and F.D. Galiana, A survey of the optimal power flow literature, *IEEE Trans. on Power Systems*, Vol. **PWRS-6,** No. 2, pp. 762–770, 1991.

IEEE working group, Description and bibliography of major economic security function, Parts II and III, Bibliography (1959–1972 and 1973–1979), *IEEE Trans. on Power Apparatus and Systems*, Vol. **PAS-100,** No. 1, 215–235, 1981.

Kron, G., Tensorial analysis of integrated transmission systems, Part I: The six basic reference frame, *AIEE Trans. Part-III: Power Apparatus and Systems*, Vol. **70**, pp. 1239–1248, 1951.

Kron, G., Tensorial Analysis of integrated transmission systems, Part II: Off-Nominal Turn Ratios, *AIEE Trans. Part-III: Power Apparatus and Systems*, Vol. **71**, pp. 505–512, 1952.

Liang, Z. and J.D. Glover, Improved cost functions for economic dispatch computations, *IEEE Trans. on Power Systems*, Vol. **PWRS-6,** No. 2, 821–829, 1991.

Lin, C.E., S.T. Chen and C.L. Huang, A two-step sensitivity approach for realtime line flow calculation, *Electric Power Systems Research*, Vol. **21,** No. 1, pp. 63–69, 1991.

Meyer, W.S. and V.D. Albertson, Improved loss formula computation by optimally ordered elimination techniques, *IEEE Trans. on Power Apparatus and Systems,* Vol. **PAS-90,** No. 1, pp. 62–69, 1971.

Mohamed-Nor, K. and A.H.A. Rashid, Efficient economic dispatch algorithm for thermal unit commitment, *IEE Proceedings C*, Vol. **138,** No. 3, pp. 213–217, 1991.

Momoh, J.A, R.J. Koessler, M.S. Bond, B. Stott, D. Sun, A. Papalexopoulos and P. Ristaanovic, Challenges to optimal power flow, *IEEE Trans. on Power Systems*, Vol. **PWRS-12,** No. 1, pp. 444–447, February 1997.

Monticelli, A. and W.H.E. Liu, Adaptive movement penalty method for the Newton optimal power flow, *IEEE Trans. on Power Systems*, Vol. **PWRS-7,** No. 1, pp. 334–342, 1992.

Nanda, J., L. Hari, M.L. Kothari and J. Henry, Extremely fast economic load dispatch algorithm through modified coordination equations, *IEE Proceedings C*, Vol. **139,** No. 1, pp. 39–46, 1992.

Ng, W.Y., Generalized generation distribution factors for power system security evaluations, *IEEE Trans. on Power Apparatus and Systems*, Vol. **PAS-100,** No. 3, pp. 1001–1005, 1981.

Palanichamy, C. and K. Srikrishna, Simple algorithm for economic power dispatch, *Electric Power Systems Research*, Vol. **21,** No. 2, pp. 147–153, 1991.

Palanichamy, C. and K. Srikrishna, A method for shortterm generation redispatch, *Electric Power Systems Research*, Vol. **17,** No. 2, pp. 129–138, 1989.

Prasad, C. Durga, A.K. Jana and S.C. Tripathy, Modificaions to Newton–Raphson load flow for ill-conditioned power systems, *Int. J. Electrical Power and Energy Systems*, Vol. **12,** No. 3, pp. 192–196, 1990.

Ramaraj, N., R. Rajaram and K. Parthasarathy, A new analytical approach to optimize a generation schedule, *Electric Power Systems Research*, Vol. **11,** No. 2, pp. 147–152, 1986.

Sasson, A.M. and H.M. Merrill, Some applications of optimization techniques to power system problems, *Proceedings of IEEE*, Vol. **62,** No. 7, pp. 959–972, 1974.

Shoults, R.R, W.M. Grady and S. Helmick, An efficient method for computing loss formula coefficients based upon the method of least squares, *IEEE Trans. on Power Apparatus and Systems*, Vol. **PAS-98,** No. 6, pp. 2144–2152, 1979.

Sun, D.I., B. Ashley, B. Brewer, A. Hughes and W.F. Tinney, Optimal power flow by Newton approach, *IEEE Trans. on Power Apparatus and Systems*, Vol. **PAS-103,** No. 10, pp. 2864–2880, 1984.

Talukdar, S.N. and F.F. Wu, Computer-aided dispatch for electric power systems, *Proceedings of IEEE*, Vol. **69,** No. 10, pp. 1212–1231, 1981.

Tao Guo, Mark I Henwood and Mariekevan Oaijen, An algorithm for combined heat power economic dispatch, *IEEE Trans. on Power Systems*, Vol. **PWRS-11,** No. 4, pp. 1778–1784, November 1996.

Wang, C. and S.M. Shahidehopur, Optimal generation scheduling with ramping costs, *IEEE Trans. on Power Systems*, Vol. **PWRS-10,** No. 1, pp. 60–67, February 1995.

Whei-Min Lin, Fu-Sheng and Ming-Tong Tsay, An improved Tabu search for economic dispatch with multiple minima, *IEEE Trans. on Power Systems*, Vol. **PWRS-17,** No. 1, pp. 108–112, February 2002.

Wong, K.P. and K. Doan, A recursive economic dispatch algorithm for assessing the costs of thermal generator schedules, *IEEE Trans. on Power Systems*, Vol. **PWRS-7,** No. 2, pp. 577–583, 1992.

Yung-Chung Chang, Wei-Tzen Yang and Chun Chang Liu, A new method for calculating loss coefficients, *IEEE Trans. on Power Systems*, Vol. **PWRS-9,** No. 3, pp. 1665–1671, August 1994.

CHAPTER

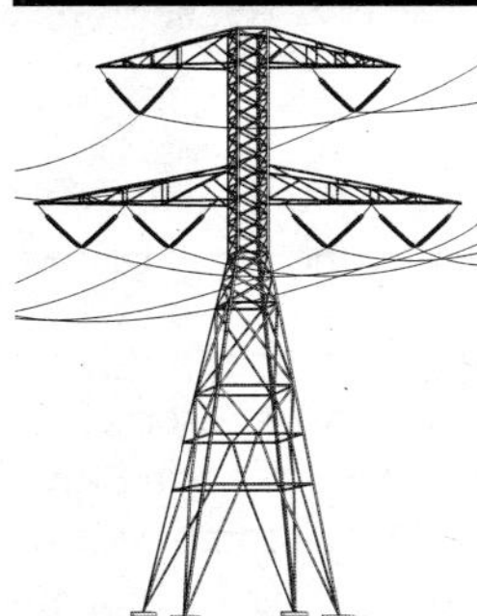

4

Optimal Hydrothermal Scheduling

4.1 INTRODUCTION

In the present set-up of large systems with hydro and thermal power stations, the integrated operation of these power stations is inevitable and the economic aspect of such an operation cannot therefore be overlooked. The underlying idea of integrated operation is for optimum utilization of all energy sources in the most economical manner, so that an uninterrupted supply can be made available to the consumers. The operating cost of thermal plants is very high, though their capital cost is low. On the other hand, the operating cost of hydroelectric plants is low but their capital cost is high. So, it has become economical as well as convenient to have both thermal and hydro plants in the same grid. The hydroelectric plant can be started quickly and it has higher reliability and greater speed of response. Hence hydro plants can take up fluctuating loads. In contrast to hydro plants, the starting of thermal plants is slow and their speed of response is slow as well. Normally the thermal plant is preferred as a base load plant whereas the hydroelectric plant is run as a peak load plant.

In a hydrothermal integration, it is essential to use the total quantity of water available from the hydro system to the fullest extent. In a hydro system, fixed charges continue regardless of the amount of power generated, as there is no fuel cost associated with hydropower. Therefore, minimum overall cost is achieved by maximum exploitation of hydro resources, i.e. water available over a period of time. Most hydroelectric plants are multipurpose. In such cases, it is necessary to meet certain obligations other than power generation. These may include a maximum fore-bay elevation, not to be exceeded because of danger of flooding, and a minimum plant discharge and spillage to meet irrigational and navigational commitments. Other distinctions among hydro power systems are the number of hydro stations, their locations and operating characteristics. The problem is different in cases when the hydro plants are located on the same stream or on different streams. The operation of a downstream plant depends on the immediate upstream plant. But downstream plant influences the immediate upstream plant by its effect on tail water elevation and effective head.

4.1.1 Classification of Hydro Plants

Classification on the basis of type

Hydro power plants, on the basis of their type, are classified into pumped storage plants and conventional plants. The conventional plants are further classified into storage plants and run-of-river plants.

Pumped storage plants. A pumped storage plant is associated with upper and lower reservoirs. During light load periods, water is pumped from the lower to the upper reservoirs using the available energy from other sources as surplus energy. During peak load the water stored in the upper reservoir is released to generate power to save fuel costs of thermal plants. The pumped storage plant is operated until the added pumping cost exceeds the savings in thermal costs due to the peak sharing operation.

Conventional plants. Conventional plants are classified into run-of-river plants and storage plants.

Run-of-river plants. Run-of-river plants have little storage capacity, and use water as it becomes available. The water not utilized is spilled.

Storage plants. Storage plants are associated with reservoirs which have significant storage capacity. During periods of low power requirements, water can be stored and then released when the demand is high.

Classification on the basis of location

On the basis of their location, hydro plants are classified into three different categories:

(a) Hydro plants on different streams
(b) Hydro plants on the same stream (serial or cascaded plants)
(c) Multi-chain hydro plants.

Hydro plants on different streams. The hydro plants are located on different streams and are independent of each other as shown in Figure 4.1.

Hydro plants on the same stream (serial or cascaded plants). When hydro plants are located on the same stream, the downstream plant depends on the immediate upstream plant. The arrangement is shown in Figure 4.2. The downstream plant influences the immediate upstream plant by its effect on tail water elevation and effective head.

Multi-chain hydro plants. These hydro plants are located on different streams as well as on the same stream. The arrangement is shown in Figure 4.3.

Further, in view of the increase in the demand for electricity for purposes such as industrial, agricultural, commercial and domestic together with the high cost of fuel as well as its limited reserve, considerable attention is being given to hydrothermal coordination problem. Besides the insignificant incremental cost involved in hydro generation, the problem of minimizing the operational cost of a hydrothermal system can be reduced essentially to that of minimizing the fuel cost for thermal plants under the constraints of the water available for hydro generation in a given period of time. The cyclic nature of water flows and load demand, as well as the validity of model assumptions, suggest splitting the problem into long and short periods.

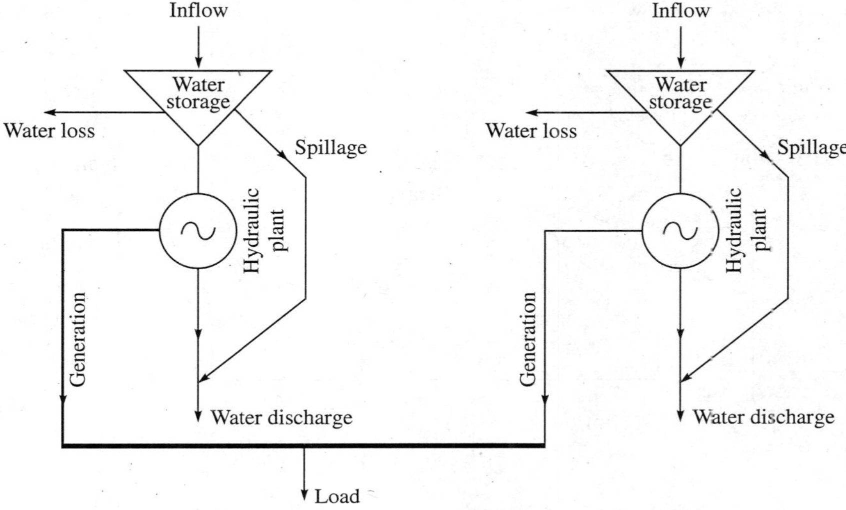

FIGURE 4.1 Hydro plants on different streams.

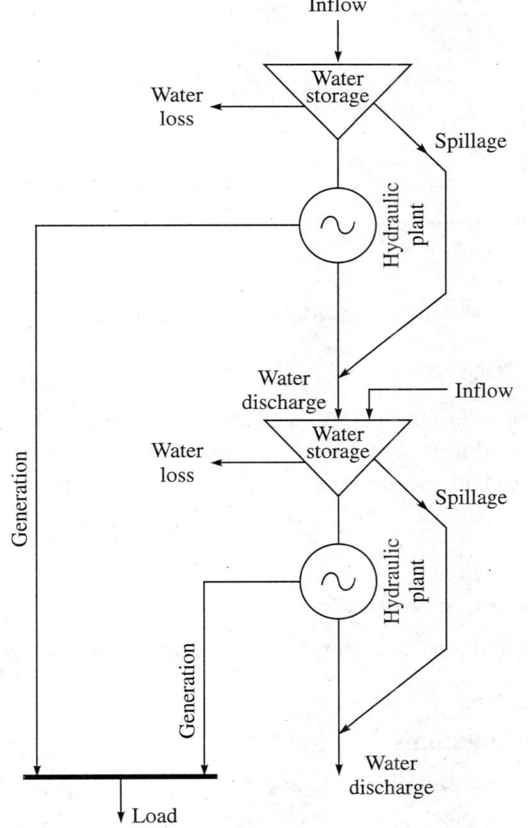

FIGURE 4.2 Hydro plants on the same stream.

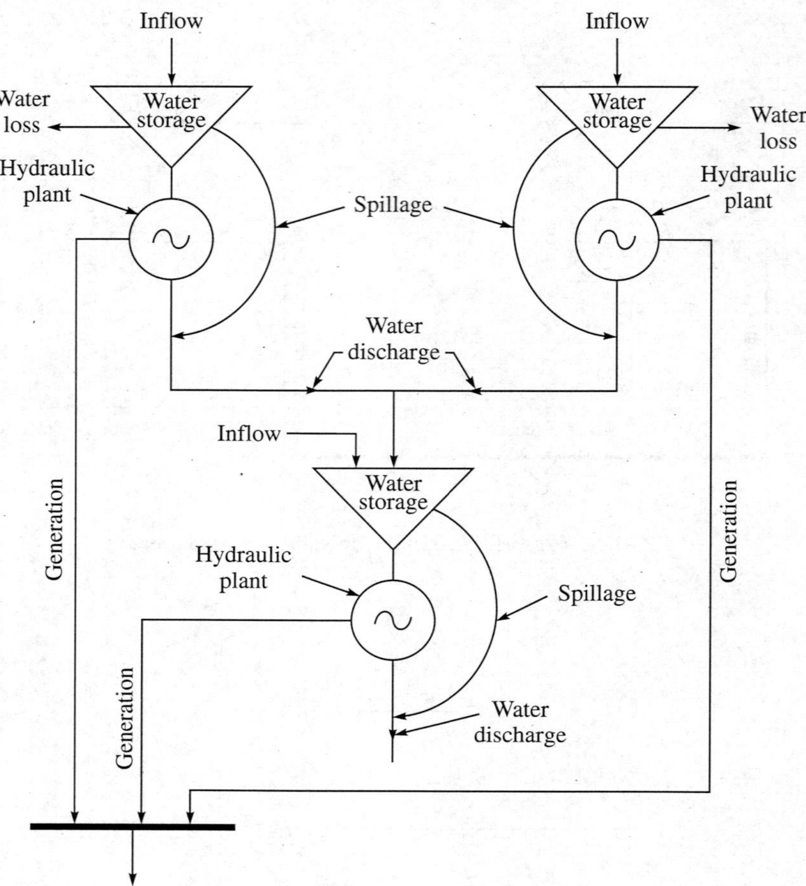

FIGURE 4.3 Multi-chain hydro plants.

4.1.2 Long-Range Problem

The long-range problem considers the yearly cyclic nature of reservoir water inflows and seasonal load demand and correspondingly a scheduling period of one year is used. The solution of the long-range problem considers the dynamics of head variations through the water flow continuity equation. The long-term hydrothermal scheduling problem can be classified in three groups:

Multi-storage hydroelectric systems

The hydro plants are located on different streams. The hydro plants have their independent reservoirs.

Cascaded hydroelectric systems

The hydro plants are located on same streams and the downstream plant depends on the immediate upstream plant. The upstream plant inputs discharge to the immediate downstream plant.

Multi-chain hydroelectric systems

The hydro plants are located on different streams as well as follow the same stream.

4.1.3 Short-Range Problem

Here, the load demand on the power system exhibits cyclic variation over a day or a week and correspondingly the scheduling interval for short-range is either a day or a week. As the scheduling interval of short-range problem is small, the solution of the short-range problem can assume the head to be fairly constant. The amount of water to be utilized for the short-range scheduling problem is known from the solution of the long-range scheduling problem. The short-term hydrothermal scheduling problem is classified into two groups:

Fixed-head hydrothermal scheduling

The head of reservoir can be assumed fixed if the hydro plants have reservoirs of large capacity.

Variable-head hydrothermal scheduling

The head of reservoir is variable if the hydro plants are having reservoirs of small capacity.

The short-range or long-range hydrothermal scheduling problem can also contain pumped storage plants. During light load period, water is pumped from the lower to upper reservoirs using the available energy from other sources as surplus energy. During peak load, the water stored in the upper reservoir is released to generate power to save fuel costs of thermal plants.

The main objective of hydrothermal operation is to minimize the total system operating cost, represented by the fuel cost required for the system's thermal generation, over the optimization interval. Each hydro plant is constrained by the amount of water available for draw-down in the scheduling period. A prediction of the future power system demand and water supply is assumed to be known for the optimization interval.

Several researchers have tried to investigate the complex problem of hydrothermal coordination at various levels. A description of the computational experience and potential economics had been achieved for a utility system using coordination equations which were later on extended for a system with multiple chains of plants, neglecting variations in head. In succession, head variations were taken into account and transmission losses were included into coordination equations. Variational calculus was employed in the initial stage to achieve the optimal solution. In continuation, many approaches have been used to solve the hydrothermal scheduling problem. These include dynamic programming [Singh and Agarwal, 1972], Pontryagin's maximum principle (continuous as well as discrete versions) [Dhillon and Morsztyn, 1971; Nagrath, Dayal, and Kothari, 1973], local variational method [Rao, Prabhu, and Agarwal, 1975], functional analysis technique [El-Hawary and Christensen, 1972], non-linear programming [Agarwal and Nagrath, 1972], principle of progressive optimality [Bijwe and Nanda, 1977], decomposition and coordination method [Bonaert, Fl-Abiad, Koivo, 1972; Soares, Lyra, and Tavares, 1980] and network flow-based methods [Carvalho and Soares, 1987]. Basic requirements of proposed algorithms for implementing optimum operation strategies in integrated electric power systems are speed, efficiency, and reliable convergence.

4.2 HYDRO PLANT PERFORMANCE MODELS

A number of proposed models can be found in the literature [El-Hawary and Christensen, 1979], due to diversity of the plant types and their characteristics. A basic physically-based relationship between the active power generated, the rate of water discharge, and the effective head is given by

$$P = 0.0085qh\ \eta(q, h) \tag{4.1}$$

where

P is the active power generation (MW)
q is the rate of water discharge (m^3/s)
h is the effective head (m)
η is function of q and h.

Here, some widely used models are presented which are used in economic dispatch studies.

4.2.1 Glimn–Kirchmayer Model

The mathematical model which defines the rate of water discharge is

$$q = K\ \psi(h)\phi(P) \tag{4.2}$$

where

K is the constant of proportionality
h is the effective head (m)
P is the real power generation (MW).

The functions ψ and ϕ are defined as quadratic functions, i.e.

$$\psi(h) = \alpha h^2 + \beta h + \gamma, \quad \text{where } \alpha, \beta, \text{ and } \gamma \text{ are coefficients.} \tag{4.3}$$

$$\phi(P) = xP^2 + yP + z, \quad \text{where } x, y, \text{ and } z \text{ are coefficients.} \tag{4.4}$$

4.2.2 Hildebrand's Model

In this model, the rate of water discharge is defined by

$$q = \sum_{i=0}^{L} \sum_{j=0}^{K} C_{ij}\ P^i\ h^j \tag{4.5}$$

where

C_{ij} are coefficients
L is the maximum exponent (usually taken as 2)
K is the maximum exponent (usually taken as 2)
h is the effective head (m)
P is the real power generation (MW)

4.2.3 Hamilton–Lamonts's Model

The rate of water discharge is defined by

$$q = \psi(h)\ \frac{\phi(P)}{h} \tag{4.6}$$

where

h is the effective head (m)
P is the real power generation (MW).

The functions ψ and ϕ are defined by

$$\psi(h) = \alpha h^2 + \beta h + \gamma, \quad \text{where } \alpha, \beta, \text{ and } \gamma \text{ are coefficients.} \tag{4.7}$$

and

$$\phi(P) = xP^3 + yP + z, \quad \text{where } x, y, \text{ and } z \text{ are coefficients.} \tag{4.8}$$

4.2.4 Arvanitidis–Rosing Model

The real power is defined by

$$P = qh[\beta - e^{-\alpha A (h - h_0)}] \tag{4.9}$$

where

P is the real power generation (MW)
h is the effective head (m)
q is the rate of water discharge (m^3/s)
A is the area of reservoir (m^2)
h_0 is the minimum head (m)
α and β are coefficients.

4.3 SHORT-RANGE FIXED-HEAD HYDROTHERMAL SCHEDULING

The basic problem considered involves the short-range optimal economic operation of an electric power system that includes both hydro and thermal generation resources. The object is to minimize the total system operating cost, represented by the fuel cost required for the system's thermal generation, over the optimization interval. Each hydro plant is constrained by the amount of water available for draw-down in the interval. A prediction of the system's future power demand and water supply is assumed to be available for the optimization interval.

The output of each hydro unit varies with effective head and the rate of water discharged through the turbines. For large capacity reservoirs, it is practical to assume that the effective head is constant over the optimization interval. Here the short-range fixed-head hydrothermal problem is defined.

Consider an electric power system network having N thermal generating plants and M hydro plants, where $M + N$ is the total number of generating plants. The basic problem is to find the active power generation of each plant in the system as a function of time over a finite time period from 0 to T.

4.3.1 Thermal Model

The objective function to be minimized is the total system operating cost, represented by the fuel cost of thermal generation, over the optimization interval, i.e.

$$J_1 = \int_0^T \sum_{i=1}^{N} F_i(P_i(t))\, dt \tag{4.10}$$

where $F_i(P_i(t))$ is the fuel cost of the ith thermal unit which is a function of the level of active power generation at the ith unit and is approximated by

$$F_i(P_i(t)) = a_i P_i^2(t) + b_i P_i(t) + c_i \quad ₹/\text{h} \qquad (i = 1, 2, ..., N) \tag{4.11}$$

where a_i, b_i, and c_i are the cost coefficients of the ith generating unit.

4.3.2 Hydro Model

In a hydro system, there is no fuel cost incurred in the operation of hydro units. The input-output characteristic of a hydro generator is expressed by the variation of water discharge $q(t)$ as a function of power output $P(t)$ and net head h. According to Glimn-Kirchmayer model, discharge is [El-Hawary and Tsang, 1986; El Hawary and Ravindranath, 1988]:

$$q_j(t) = K\psi(h_j)\ \phi(P_{j+N}(t)) \quad \text{m}^3/\text{h} \qquad (j = 1, 2, ..., M) \tag{4.12}$$

where ϕ and ψ are two independent functions.

For a large capacity reservoir, it is practical to assume that the effective head is constant over the optimization interval. In the case of constant head, $\psi(h_j)$ becomes constant, then Eq. (4.12) is rewritten as

$$q_j(t) = K'\ \phi(P_{j+N}(t)) \qquad (j = 1, 2, ..., M) \tag{4.13}$$

where K' is a new constant formed by the multiplication of K and $\psi(h_j)$.

Each hydro plant is constrained by the amount of water available for the optimization interval, i.e.

$$\int_0^T q_j(t)\, dt = V_j \qquad (j = 1, 2, ..., M) \tag{4.14}$$

where V_j is the predefined volume of water available in cubic metres and hydro performance q_j is represented by the conventional quadratic model, i.e.

$$q_j(t) = x_j\, P_{j+N}^2(t) + y_j\, P_{j+N}(t) + z_j \quad \text{m}^3/\text{h} \qquad (j = 1, 2, ..., M) \tag{4.15}$$

where x_j, y_j, and z_j are the discharge coefficients of the jth hydro plant.

4.3.3 Equality and Inequality Constraints

(i) Load demand equality constraint:

$$\sum_{i=1}^{N+M} P_i(t) = P_D(t) + P_L(t) \tag{4.16}$$

where

$P_D(t)$ is the load demand during the sub-interval

$P_L(t)$ is the transmission loss during the sub-interval.

(ii) Limits are imposed as

$$P_i^{\min} \le P_i(t) \le P_i^{\max} \qquad (i = 1, 2, ..., N + M) \tag{4.17}$$

where

$P_i^{\min}$ is the lower limit of the ith generator output

$P_i^{\max}$ is the upper limit of the ith generator output.

4.3.4 Transmission Losses

A common approach to model transmission losses in the system is to use Kron's approximated loss formula in terms of B-coefficients.

$$P_L(t) = \sum_{i=1}^{N+M} \sum_{j=1}^{N+M} P_i(t)\, B_{ij}\, P_j(t) + \sum_{i=1}^{N+M} B_{i0} P_i(t) + B_{00} \quad \text{MW} \tag{4.18}$$

where B_{00}, B_{i0}, and B_{ij} are B-coefficients.

The fixed-head hydrothermal problem can be defined considering the operating cost over the optimization interval to meet the load demand in each interval. Each hydro plant is constrained by the amount of water available for draw-down in the interval. The problem is defined as

Minimize
$$J = \int_0^T \sum_{i=1}^{N} F_i(P_i(t))\, dt \tag{4.19a}$$

subject to
$$\sum_{i=1}^{N+M} P_i(t) = P_D(t) + P_L(t) \tag{4.19b}$$

$$\int_0^T q_j(t)\, dt = V_j \qquad (j = 1, 2, ..., M) \tag{4.19c}$$

$$P_i^{\min} \le P_i(t) \le P_i^{\max} \qquad (i = 1, 2, ..., N + M) \tag{4.19d}$$

The simplest way to arrive at the required equations is to use the calculus of variations. Each constraint equation is associated with an unknown multiplier function known as a Lagrange multiplier. The above objective as augmented by the constraints is given as

$$L(P_i(t), \lambda(t), (v_j)) = \int_0^T \left[\sum_{i=1}^{N} F_i(P_i(t)) + \sum_{j=1}^{M} v_j q_j(t) + \lambda(t) \left[P_D(t) + P_L(t) - \sum_{i=1}^{N+M} P_i(t) \right] \right] dt - \sum_{j=1}^{M} v_j V_j \tag{4.20}$$

where

$\lambda(t)$ is the incremental cost of power delivered in the system

v_j are water conversion factors.

The optimality conditions are described by taking the partial derivatives of augmented objective function with respect to the decision variables, $P_i(t)$ $(i = 1, 2, ..., N + M)$, $\lambda(t)$, v_j $(j = 1, 2, ..., M)$

$$\frac{\partial F_i(P_i(t))}{\partial P_i(t)} + \lambda(t) \left[\frac{\partial P_L(t)}{\partial P_i(t)} - 1 \right] = 0 \qquad (i = 1, 2, ..., N) \tag{4.21a}$$

$$v_j \frac{\partial q_j(t)}{\partial P_{j+N}(t)} + \lambda(t) \left[\frac{\partial P_L(t)}{\partial P_{j+N}(t)} - 1 \right] = 0 \qquad (j = 1, 2, ..., M) \tag{4.21b}$$

$$\int_0^T q_j(t)\,dt = V_j \qquad (j = 1, 2, ..., M) \tag{4.21c}$$

$$\sum_{i=1}^{N+M} P_i(t) = P_D(t) + P_L(t) \tag{4.21d}$$

4.3.5 Discrete Form of Short-Range Fixed-Head Hydrothermal Scheduling Problem

The problem stated in Eq. (4.19) can be redefined in discrete form as

$$\text{Minimize} \qquad J = \sum_{k=1}^{T} \sum_{i=1}^{N} t_k\, F_i(P_{ik}) \tag{4.22a}$$

$$\text{subject to} \qquad \sum_{i=1}^{N+M} P_{ik} = P_{Dk} + P_{Lk} \qquad (k = 1, 2, ..., T) \tag{4.22b}$$

$$\sum_{k=1}^{T} t_k\, q_{jk} = V_j \qquad (j = 1, 2, ..., M) \tag{4.22c}$$

$$P_i^{\min} \le P_{ik} \le P_i^{\max} \qquad (i = 1, 2, ..., N; \quad k = 1, 2, ..., T) \tag{4.22d}$$

where

$F_i(P_{ik})$ is the cost function of thermal units in the interval k and is defined by

$$F_i(P_{ik}) = a_i P_{ik}^2 + b_i P_{ik} + c_i \text{ ₹/h}$$

with a_i, b_i, and c_i as the cost coefficients.

P_{ik} is the output of thermal and hydro units during the kth interval.

P_{Lk} are the transmission losses during the kth interval given by

$$P_{Lk} = \sum_{i=1}^{N+M} \sum_{j=1}^{N+M} P_{ik}\, B_{ij}\, P_{jk} + \sum_{i=1}^{N+M} B_{i0}\, P_{ik} + B_{00} \text{ MW}$$

with

B_{ij}, B_{i0}, and B_{00} as the B-coefficients

P_{Dk} is the demand during the kth interval

$P_i^{\min}$ is the lower limit of ith generator output

$P_i^{\max}$ is the upper limit of ith generator output

q_{jk} is the rate of discharge from the jth hydro unit in interval k and is defined by

$$q_{jk} = x_j\, P^2_{j+N,k} + y_j\, P_{j+N,k} + z_j \text{ m}^3\text{/h}$$

(with x_j, y_j, and z_j as the discharge coefficients of the jth unit)

V_j is the pre-specified volume of water available for unit j for whole of the period
N is the number of thermal units
M is the number of hydro units
T is the overall period for scheduling.

The above objective as augmented by the constraints is given as

$$L(P_{ik}, \lambda_k, v_j) = \sum_{k=1}^{T}\left[\sum_{i=1}^{N} t_k F_i(P_{ik}) + \sum_{j=1}^{M} v_j t_k q_{jk} + \lambda_k\left[P_{Dk} + P_{Lk} - \sum_{i=1}^{M+N} P_{ik}\right]\right] - \sum_{j=1}^{M} v_j V_j \quad (4.23)$$

where

λ_k is the incremental cost of power delivered in the system during the kth interval
v_j are the water conversion factor.

The optimality conditions are described by taking the partial derivatives of augmented objective function with respect to the decision variables, P_{ik} $(i = 1, 2, ..., N + M)$, λ_k, v_j $(j = 1, 2, ..., M)$

$$t_k \frac{\partial F_i}{\partial P_{ik}} + \lambda_k\left[\frac{\partial P_{Lk}}{\partial P_{ik}} - 1\right] = 0 \qquad (i = 1, 2, ..., N;\quad k = 1, 2, ..., T) \quad (4.24a)$$

$$v_j t_k \frac{\partial q_{jk}}{\partial P_{mk}} + \lambda_k\left[\frac{\partial P_{Lk}}{\partial P_{mk}} - 1\right] = 0 \qquad (j = 1, 2, ..., M;\quad m = N + j;\quad k = 1, 2, ..., T) \quad (4.24b)$$

$$\sum_{k=1}^{T} t_k\, q_{jk} = V_j \qquad (j = 1, 2, ..., M) \quad (4.24c)$$

$$\sum_{i=1}^{N+M} P_{ik} = P_{Dk} + P_{Lk} \qquad (k = 1, 2, ..., T) \quad (4.24d)$$

Inherently, these equations are nonlinear. The Newton–Raphson method has been applied to find P_{ik} $(i = 1, 2, ..., N + M)$, λ_k for all the T intervals. v_j, the water conversion factor is modified to satisfy the available water constraint.

The small changes in P_{ik} and λ_k from their respective initial values are obtained from Eqs. (4.24a), (4.24b), and (4.24d).

$$\left(t_k \frac{\partial^2 F_i}{\partial P_{ik}^2} + \lambda_k \frac{\partial^2 P_{Lk}}{\partial P_{ik}^2}\right)\Delta P_{ik} + \lambda_k \sum_{\substack{j=1\\ j\neq i}}^{N+M} \frac{\partial^2 P_{Lk}}{\partial P_{ik}\partial P_{jk}}\Delta P_{jk} + \left(\frac{\partial P_{Lk}}{\partial P_{ik}} - 1\right)\Delta\lambda_k$$

$$= -\left[t_k \frac{\partial F_i}{\partial P_{ik}} + \lambda_k\left(\frac{\partial P_{Lk}}{\partial P_{ik}} - 1\right)\right] \qquad (i = 1, 2, ..., N) \quad (4.25a)$$

$$\left(v_j t_k \frac{\partial^2 q_{jk}}{\partial P_{mk}^2} + \lambda_k \frac{\partial^2 P_{Lk}}{\partial P_{mk}^2}\right)\Delta P_{mk} + \lambda_k \sum_{\substack{l=1\\ l\neq m}}^{N+M} \frac{\partial^2 P_{Lk}}{\partial P_{mk}\partial P_{lk}}\Delta P_{lk} + \left(\frac{\partial P_{Lk}}{\partial P_{mk}} - 1\right)\Delta\lambda_k$$

$$= -\left[v_j t_k \frac{\partial q_{jk}}{\partial P_{mk}} + \lambda_k \left(\frac{\partial P_{Lk}}{\partial P_{mk}} - 1 \right) \right] \qquad (j = 1, 2, ..., M; \quad m = j + N) \qquad (4.25b)$$

$$\sum_{j=1}^{N+M} \left(\frac{\partial P_{Lk}}{\partial P_{jk}} - 1 \right) \Delta P_{jk} = -\left(P_{Dk} + P_{Lk} - \sum_{i=1}^{N+M} P_{ik} \right) \qquad (4.25c)$$

Substituting the values of the partial derivatives with initial values of control variables, in Eqs. (4.25a), (4.25b), and (4.25c).

$$(2t_k a_i + \lambda_k^0 B_{ii})\, \Delta P_{ik} + \lambda_k^0 \sum_{\substack{j=1 \\ j \neq i}}^{N+M} 2B_{ij} \Delta P_{jk} + \left(\sum_{j=1}^{N+M} 2B_{ij} P_{jk}^0 + B_{i0} - 1 \right) \Delta\lambda_k$$

$$= -\left[t_k (2a_i P_{ik}^0 + b_i) + \lambda_k^0 \left(\sum_{j=1}^{N+M} 2B_{ij} P_{jk}^0 + B_{i0} - 1 \right) \right] \qquad (i = 1, 2, ..., N) \qquad (4.26a)$$

$$(2v_j^0 t_k x_j + \lambda_k^0 B_{mm})\, \Delta P_{mk} + \sum_{\substack{l=1 \\ l \neq m}}^{N+M} 2B_{ml} \Delta P_{jl} + \left(\sum_{l=1}^{N+M} 2B_{ml} P_{lk}^0 + B_{m0} - 1 \right) \Delta\lambda_k$$

$$= -\left[v_j^0 t_k (2x_j P_{mk}^0 + y_j) + \lambda_k^0 \left(\sum_{l=1}^{N+M} 2B_{ml} P_{lk}^0 + B_{m0} - 1 \right) \right] \qquad (j = 1, 2, ..., M; \quad m = j + N) \quad (4.26b)$$

$$\sum_{j=1}^{N+M} \left(\sum_{l=1}^{N+M} 2B_{jl} P_{lk}^0 + B_{j0} - 1 \right) \Delta P_{jk} = -\left(P_{Dk} + P_{Lk}^0 - \sum_{i=1}^{N+M} P_{ik}^0 \right) \qquad (4.26c)$$

The above equation can be written in matrix form as

$$\begin{bmatrix} \nabla_{P_k P_k} L & \nabla_{P_k \lambda_k} L \\ (\nabla_{P_k \lambda_k} L)^T & \nabla_{\lambda_k \lambda_k} L \end{bmatrix} \begin{bmatrix} \Delta P_k \\ \Delta\lambda_k \end{bmatrix} = \begin{bmatrix} -\nabla_{P_k} L \\ -\nabla_{\lambda_k} L \end{bmatrix} \qquad (4.27)$$

The size of Hessian matrix is $S \times S$
where

$S = (N + M + 1)$
N is the number of thermal units
M is the number of hydro units.

Suppose there are three thermal units and two hydro units. The scheduling is required for 24 intervals, then the size of the Hessian matrix will be 6×6 ($(3 + 2 + 1) = 6$).

4.3.6 Initial Guess

It is assumed that the hydro generator requires more water for a longer time period or higher load demand [El-Hawary, 1986]. Thus the whole available amount of water is allocated in each time interval, considering time interval and load demand.

$$V_{jk} = \frac{V_j\,(P_{Dk}\,t_k)}{\left(\sum_{k=1}^{T} P_{Dk}t_k\right)} \qquad (j = 1, 2, ..., M;\quad k = 1, 2, ..., T) \qquad (4.28)$$

where

V_j is the available water for whole period
V_{jk} is the water allocated during the kth sub-interval
P_{Dk} is the demand during the kth sub-interval
t_k is the duration of the kth sub-interval

The volume of each sub-interval is represented as

$$t_k\, q_{jk} = V_{jk} \qquad (j = 1, 2, ..., M;\quad k = 1, 2, ..., T) \qquad (4.29)$$

or

$$t_k(x_jP^2_{mk} + y_jP_{mk} + z_j) = V_{jk} \qquad (j = 1, 2, ..., M;\quad m = j + N)$$

$$x_jP^2_{mk} + y_j\,P_{mk} + z_j = V_{jk}/t_k \qquad (j = 1, 2, ..., M;\quad m = j + N)$$

$$x_jP^2_{mk} + y_jP_{mk} + (z_j - q_{jk}) = 0 \qquad (j = 1, 2, ..., M;\quad m = j + N) \qquad (4.30)$$

The roots of a quadratic equation can be obtained as

$$P^0_{mk} = \frac{-y_j + \sqrt{y_j^2 - 4x_j(z_j - q_{jk})}}{2x_j} \qquad (j = 1, 2, ..., M;\quad k = 1, 2, ..., T) \qquad (4.31)$$

The rest of the demand during the interval is allocated to thermal generators.

$$PD_k = P_{Dk} - \sum_{j=1}^{M} P_{(j+N)k} \qquad (k = 1, 2, ..., T)$$

Further, it is assumed that there are no transmission losses. So, from Eqs. (4.24a), (4.24b), and (4.24d), we get

$$\lambda^0_k = \left(PD_k + \sum_{i=1}^{N} \frac{b_i}{2a_i}\right) \div \left(\sum_{i=1}^{N} \frac{1}{2a_i}\right) \qquad (k = 1, 2, ..., T) \qquad (4.32)$$

Thermal power generations are obtained as

$$2a_iP^0_{ik} + b_i = \lambda^0_k \qquad (i = 1, 2, ..., N;\quad k = 1, 2, ..., T)$$

$$P^0_{ik} = (\lambda^0_k - b_i) \div (2a_i) \qquad (i = 1, 2, ..., N;\quad k = 1, 2, ..., T) \qquad (4.33)$$

The water conversion factor can be obtained as

$$v_j^0\,(2x_jP^0_{mk} + y_j) = \lambda^0_k \qquad (j = 1, 2, ..., M)$$

or

$$v_j^0 = \frac{\lambda^0_k}{2x_jP^0_{mk} + y_j} \qquad (j = 1, 2, ..., M; \quad m = N + j) \tag{4.34}$$

4.3.7 Alternative Method for Initial Guess

The power demand is equally distributed among thermal and hydro units during each interval.

$$P^0_{ik} = \frac{P_{Dk}}{N+M} \qquad (i = 1, 2, ..., N + M; \quad k = 1, 2, ..., T) \tag{4.35}$$

Further, it is assumed that there are no transmission losses.

$$\lambda^0_k = 2a_iP^0_{ik} + b_i \qquad (k = 1, 2, ..., T) \tag{4.36}$$

Any thermal unit can be considered as the ith unit.

The water conversion factor can be obtained as

$$v^0_j(2x_jP^0_{mk} + y_j) = \lambda^0_k \qquad (j = 1, 2, ..., M)$$

$$v^0_j = \frac{\lambda^0_k}{2x_jP^0_{mk} + y_j} \qquad (j = 1, 2, ..., M; \quad m = N + j) \tag{4.37}$$

To implement the above procedure to solve the fixed-head hydrothermal scheduling problem, the stepwise procedure is outlined below. This procedure is known as iterative procedure.

Algorithm 4.1: Classical Method for Short-Range Fixed-Head Hydrothermal Scheduling

1. Read the number of thermal units N, the number of hydro units M, the number of sub-intervals T, cost coefficients, a_i, b_i, c_i (i = 1, 2, ..., N), B-coefficients, B_{ij}(i = 1, 2, ..., N + M; j = 1, 2, ..., N + M), discharge coefficients, x_i, y_i, z_i (i = 1, 2, ..., M), demand P_{Dk}(k = 1, 2, ..., T), and pre-specified available water V_j (j = 1, 2, ..., M).
2. Calculate the initial guess values of P^0_{ik} (i = 1, 2, ..., N + M), λ^0_k and v^0_j (j = 1, 2, ..., M).
3. Consider v_j^0 (j = 1, 2, ..., M) as calculated in Step 2.
4. Start the iteration counter, r = 1.
5. Start hourly count, k = 1.
6. Consider P^0_{ik} (i = 1, 2, ..., N + M) and λ_k^0.
7. Calculate ΔP_{ik} (i = 1, 2, ..., N + M) and $\Delta\lambda_k$, using the Newton–Raphson method. Gauss Elimination method is used to solve the following equations.

$$\begin{bmatrix} \nabla_{P_kP_k}L & \nabla_{P_k\lambda_k}L \\ (\nabla_{P_k\lambda_k}L)^T & \nabla_{\lambda_k\lambda_k}L \end{bmatrix}\begin{bmatrix} \Delta P_k \\ \Delta\lambda_k \end{bmatrix} = \begin{bmatrix} -\nabla_{P_k}L \\ -\nabla_{\lambda_k}L \end{bmatrix}$$

8. Check the convergence

$$\text{if } \left|\sum_{i=1}^{M+N} \Delta P_{ik} + \Delta\lambda_k\right| \le \varepsilon \quad \text{or} \quad \left|\sum_{i=1}^{M+N} \frac{\partial L}{\partial P_{ik}} + \frac{\partial L}{\partial \lambda_k}\right| \le \varepsilon \text{ then GOTO Step 11.}$$

9. Calculate the new values of P_{ik} (i = 1, 2, ..., $N + M$) and λ_k as

$$P_{ik}^{\text{new}} = P_{ik}^{\ 0} + \Delta P_{ik} \quad \text{and} \quad \lambda_k^{\text{new}} = \lambda_k^{\ 0} + \Delta\lambda_k$$

10. Set limits correspondingly as

$$P_{ik}^{\text{new}} = \begin{cases} P_i^{\max} & ; \text{if } P_{ik}^{\text{new}} \geq P_i^{\max} \\ P_i^{\min} & ; \text{if } P_{ik}^{\text{new}} \leq P_i^{\min} \\ P_{ik}^{\text{new}} & ; \text{otherwise} \end{cases}$$

Disallow generator to participate, whose limits have been set either to lower or upper limit, in the scheduling by deleting that row and column.

11. Set $P_{ik}^{\ 0} = P_{ik}^{\text{new}}$ (i = 1, 2, ..., $N + M$) and $\lambda_k^{\ 0} = \lambda_k^{\text{new}}$ GOTO Step 7 and repeat.
12. If $k \geq T$, then GOTO Step 13, else $k = k + t_k$, GOTO Step 6 and repeat.
13. Calculate water withdrawals V_j (j = 1,.., M).
14. If ($|V_j - V_j^s| \leq \varepsilon$) or if ($r \geq R$) then GOTO Step 15,

else
$$v_j^{\text{new}} = v_j^{\ 0} + (V_j - V_j^s)/V_j^s,$$
$$v_j^{\ 0} = v_j^{\text{new}} \quad (i = 1, 2, ..., M)$$
$$r = r + 1; \text{ GOTO Step 5 and repeat.}$$

15. Calculate the optimal cost and loss and stop.

EXAMPLE 4.1 A hydrothermal system is considered which consists of one thermal and one hydro generating station as shown in Figure 4.4. The operating cost of thermal station is given by

$$F_{1k} = 0.01P_{1k}^2 + 0.1P_{1k} + 100.0 \quad ₹/\text{h}$$

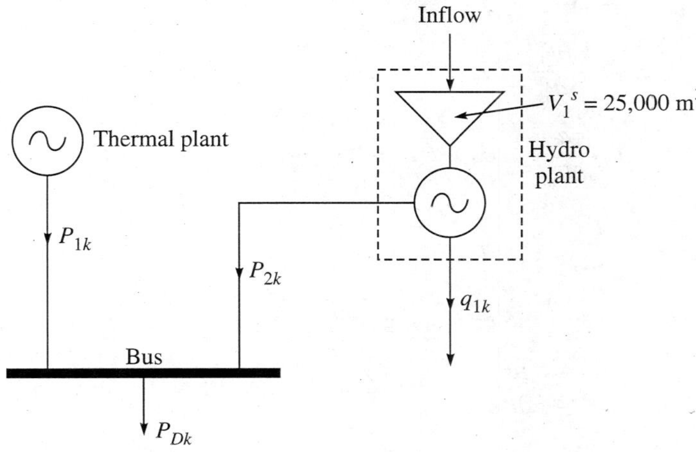

FIGURE 4.4 Hydrothermal system.

The rate of discharge of hydro-generating station is given by

$$q_{1k} = 0.05P_{2k}^2 + 20.0P_{2k} + 140.0 \quad \text{m}^3/\text{h}$$

Minimum and maximum limits of the generations are

$$P_{1k}^{\min} = 50.0 \text{ MW}, \quad P_{1k}^{\max} = 205.0 \text{ MW}$$
$$P_{2k}^{\min} = 10.0 \text{ MW}, \quad P_{2k}^{\max} = 200.0 \text{ MW}$$

Assume that the transmission losses are zero and the reservoir is large. The water available in the reservoir is

$$V_1^s = 25{,}000.0 \text{ m}^3$$

The demand curve is given in Figure 4.5. Find the economic schedule for two hours.

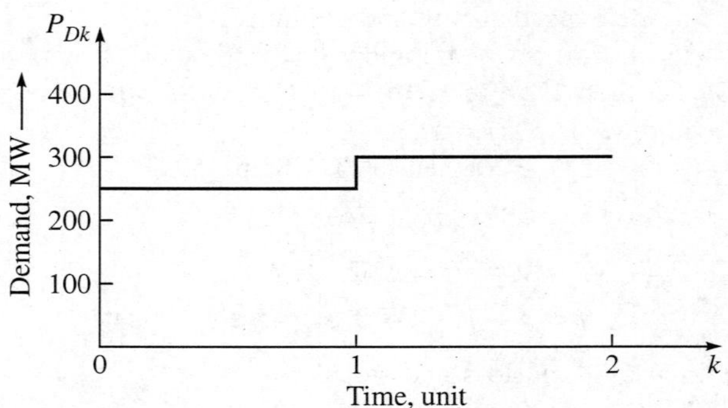

FIGURE 4.5 Demand curve.

Solution When $P_{Lk} = 0.0$, the incremental transmission losses will become zero. Substituting this effect in Eqs. (4.24a), (4.24b) and (4.24d), with number of thermal generators, $N = 1$ and number of hydro generators, $M = 1$, we get

$$t_k \frac{\partial F_i}{\partial P_{ik}} = \lambda_k \quad (i = 1) \tag{i}$$

$$v_j t_k \frac{\partial q_{jk}}{\partial P_{mk}} = \lambda_k \quad (j = 1; \quad m = 1 + j) \tag{ii}$$

$$\sum_{i=1}^{1+1} P_{ik} = P_{Dk} \tag{iii}$$

From the characteristics of the generating units, we get

$$\frac{\partial F_1}{\partial P_{1k}} = 0.02P_{1k} + 0.1 \text{ ₹/MWh}$$

$$\frac{\partial q_{1k}}{\partial P_{2k}} = 0.10P_{2k} + 20.3 \text{ m}^3\text{/MWh}$$

Substituting the above equations into (i) and (ii), respectively, we get

$$t_k(0.02P_{1k} + 0.1) = \lambda_k \quad \text{or} \quad P_{1k} = \left(50\frac{\lambda_k}{t_k} - 5\right) \tag{iv}$$

and

$$v_1 t_k(0.10P_{2k} + 20.0) = \lambda_k \quad \text{or} \quad P_{2k} = \left(10\frac{\lambda_k}{v_1 t_k} - 200\right) \qquad \text{(v)}$$

Substituting Eqs. (iv) and (v) into Eq. (iii), we get

$$\left(50\frac{\lambda_k}{t_k} - 5\right) + \left(10\frac{\lambda_k}{v_1 t_k} - 200\right) = P_{Dk}$$

or

$$\frac{\lambda_k}{t_k}\left(50 + \frac{10}{v_1}\right) = P_{Dk} + 205$$

or

$$\frac{\lambda_k}{t_k} = \frac{P_{Dk} + 205}{50 + 10/v_1} \qquad \text{(vi)}$$

Computation of initial values

Power generations, P_{ik} are computed using Eq. (4.35) for $k = 1$

$$P_{11} = 250/2 = 125 \text{ MW} \quad \text{and} \quad P_{21} = 250/2 = 125 \text{ MW}$$

Incremental cost λ_k is computed using Eq. (4.36) for $k = 1$

$$\lambda_1/t_1 = 2 \times 0.01 \times 125 + 0.1 = 2.6 \text{ ₹/MWh}$$

Power generations, P_{ik} are computed using Eq. (4.35) for $k = 2$

$$P_{12} = 300/2 = 150 \text{ MW and } P_{22} = 300/2 = 150 \text{ MW}$$

Incremental cost λ_k is computed using Eq. (4.36) for $k = 2$

$$\lambda_2/t_2 = 2 \times 0.01 \times 150 + 0.1 = 3.1 \text{ ₹/MWh}$$

From Eq. (4.37), water conversion factor, v_1 is calculated

$$v_1 = \frac{\lambda_1}{2x_1 P_{21} + y_1} = \frac{2.6}{2 \times 0.05 \times 125 + 20} = 0.08 \text{ ₹/m}^3$$

Set iteration counter IT = 1

Compute λ_k, for $k = 1$ from Eq. (vi), i.e.

$$\frac{\lambda_k}{t_k} = \frac{250 + 205}{50 + 10/0.08} = 2.6 \text{ ₹/MWh}$$

Compute P_{1k}, for $k = 1$ from Eq.(iv), i.e.

$$P_{11} = (50 \times 2.6 - 5) = 125 \text{ MW}$$

Compute P_{2k}, for $k = 1$ from Eq. (v), i.e.

$$P_{21} = (10 \times (2.6/0.08) - 200) = 125 \text{ MW}$$

Compute λ_k, for k = 2 from Eq. (vi), i.e.

$$\frac{\lambda_k}{t_k} = \frac{300 + 205}{50 + 10/0.08} = 2.885714 \text{ ₹/MWh}$$

Compute P_{1k}, for k = 2 from Eq. (iv), i.e.

$$P_{21} = (50 \times 2.885714 - 5) = 139.2857 \text{ MW}$$

Compute P_{2k}, for k = 2 from Eq. (v), i.e.

$$P_{22} = (10 \times 2.885714/0.08 - 200) = 160.7143 \text{ MW}$$

Total volume utilized is computed from Eq. (4.24c), i.e.

$$V = 6[(0.05 \times 125.00 + 20)125.00 + 140 + (0.05 \times 160.7143 + 20)160.7143 + 140]$$
$$= 48401.94 \text{ m}^3$$

Now,

$$|\ 48401.94 - 25000.00\ | \le 0.01$$

Convergence criteria is not satisfied, so update v_1

$$v_1 = 0.08 + 0.0955 \times 23401.94/25000.00 = 0.1693954 \text{ ₹/m}^3$$

Increment the iteration counter, IT = 2

$$P_{11} = 203.6515 \text{ MW}, \quad P_{21} = 46.3485 \text{ MW}, \quad \lambda_1 = 4.173030 \text{ ₹/MWh}$$
$$P_{12} = 226.5803 \text{ MW}, \quad P_{22} = 73.41977 \text{ MW}, \quad \lambda_2 = 4.631605 \text{ ₹/MWh}$$
$$V_1 = 18313.79 \text{ m}^3 \quad \text{and} \quad |\ 18313.79 - 25000.00\ | \ge 0.01$$

Convergence criteria is not satisfied, so update v_1.

$$v_1 = 0.1438541 \text{ ₹/m}^3$$

Solution is obtained in the seventh iteration and is given below:

$$P_{11} = 184.1549 \text{ MW}, \quad P_{21} = 65.84509 \text{ MW}, \quad \lambda_1 = 3.783098 \text{ ₹/MWh}$$
$$P_{12} = 204.9412 \text{ MW}, \quad P_{22} = 95.05884 \text{ MW}, \quad \lambda_2 = 4.198823 \text{ ₹/MWh}$$
$$V_1 = 25000.0019 \text{ m}^3 \quad \text{and} \quad |\ 25000.0019 - 25000.00\ | \le 0.01$$

The convergence is achieved and the total operating cost = ₹ 5988.293.

Sample System 4.1

A hydrothermal system is considered which consists of three thermal and one hydro generating station. The operating cost of the thermal station is given by

$$F_{1k} = 0.01P_{1k}^2 + 0.1P_{1k} + 100.0 \text{ ₹/h}$$
$$F_{2k} = 0.02P_{2k}^2 + 0.1P_{2k} + 120.0 \text{ ₹/h}$$
$$F_{3k} = 0.01P_{3k}^2 + 0.2P_{3k} + 150.0 \text{ ₹/h}$$

The rate of discharge of hydro generating station is given by

$$q_{1k} = 0.06P_{4k}^2 + 20.0P_{4k} + 140.0 \quad \text{m}^3/\text{h}$$

Minimum and maximum limits of the generations are:

$$P_{1k}^{\min} = 50.0 \text{ MW}, \quad P_{1k}^{\max} = 200.0 \text{ MW}$$

$$P_{2k}^{\min} = 40.0 \text{ MW}, \quad P_{2k}^{\max} = 170.0 \text{ MW}$$

$$P_{3k}^{\min} = 30.0 \text{ MW}, \quad P_{3k}^{\max} = 215.0 \text{ MW}$$

$$P_{4k}^{\min} = 10.0 \text{ MW}, \quad P_{4k}^{\max} = 100.0 \text{ MW}$$

Assume that the reservoir is large and water available in the reservoir is

$$V_1 = 25000.0 \text{ m}^3$$

The B-coefficients of the power system network are given by

$$B = \begin{bmatrix} 0.00050 & 0.00005 & 0.00020 & 0.00003 \\ 0.00005 & 0.00004 & 0.00018 & -0.00011 \\ 0.00020 & 0.00018 & 0.00050 & -0.00012 \\ 0.00003 & -0.00011 & -0.00012 & 0.00023 \end{bmatrix} \text{MW}^{-1}$$

Using the iterative method, the generation schedule of this hydrothermal problem for 24 hours is obtained and is given below, when hourly demand is given. Table 4.1 represents the operating cost of thermal units, incremental cost λ_k, and discharge of hydro units during 24 hours. Thermal and hydro generations to meet the hourly demand along with transmission losses are given in Table 4.2.

4.4 NEWTON–RAPHSON METHOD FOR SHORT-RANGE FIXED-HEAD HYDROTHERMAL SCHEDULING

The optimality conditions to solve the fixed-head hydrothermal scheduling are given in Eqs. (4.24a–4.24d), which are obtained by taking the partial derivatives of augmented objective function defined by Eq. (4.23), with respect to the decision variables, $P_{ik}(i = 1, 2, ..., N + M)$, λ_k, $v_j(j = 1, 2, ..., M)$. Inherently, these equations are nonlinear. The Newton–Raphson method has been applied to find $P_{ik}(i = 1, 2, ..., N + M, k = 1, 2, ..., T)$; λ_k $(k = 1, 2, ..., T)$ and v_j $(j = 1, 2, ..., M)$.

Assume that P_{ik}^0, λ_k^0 and v_j^0 are known and can be considered as initial values. Small change in these variables can be obtained from the following equations to update the values of control variables. Thus,

$$P_{ik}^{\text{new}} = P_{ik}^0 + \Delta P_{ik} \qquad (i = 1, 2, ..., N + M; \quad k = 1, 2, ..., T)$$

$$\lambda_k^{\text{new}} = \lambda_k^0 + \Delta\lambda_k \qquad (k = 1, 2, ..., T)$$

$$v_j^{\text{new}} = v_j^0 + \Delta v_j \qquad (j = 1, 2, ..., M)$$

$$\left(t_k \frac{\partial^2 F_i}{\partial P_{ik}^2} + \lambda_k^0 \frac{\partial^2 P_{Lk}}{\partial P_{ik}^2}\right)\Delta P_{ik} + \lambda_k^0 \sum_{\substack{j=1 \\ j\neq i}}^{N+M} \frac{\partial^2 P_{Lk}}{\partial P_{ik}\partial P_{jk}} \Delta P_{jk} + \left(\frac{\partial P_{Lk}}{\partial P_{ik}} - 1\right)\Delta\lambda_k$$

TABLE 4.1 Hourly cost, λ_k, and discharge for required demand

k	P_{Dk} (MW)	F_k (₹/h)	λ_k (₹/MWh)	q_{1k} (Mm^3/h)
1	175.0	512.9634	2.4006	346.0000
2	190.0	539.1459	2.5141	346.0000
3	220.0	598.9467	2.7451	346.0000
4	280.0	750.3193	3.2228	346.0000
5	320.0	876.8282	3.5534	346.0000
6	360.0	993.5868	3.8938	515.9093
7	390.0	1064.7850	4.1557	787.7614
8	410.0	1114.4540	4.3335	981.0660
9	440.0	1192.2800	4.6051	1289.7430
10	475.0	1288.1440	4.9295	1679.2540
11	525.0	1434.6410	5.4070	2293.1200
12	550.0	1512.1370	5.6522	2626.4530
13	565.0	1560.0060	5.8014	2835.1800
14	540.0	1480.7970	5.5536	2490.9580
15	500.0	1359.9820	5.1661	1977.5480
16	450.0	1219.1110	4.6970	1397.7510
17	425.0	1152.8670	4.4686	1132.5560
18	400.0	1089.3990	4.2443	883.1864
19	375.0	1028.6910	4.0240	649.1595
20	340.0	948.3157	3.7223	346.3888
21	300.0	810.8776	3.3869	346.0000
22	250.0	669.1344	2.9813	346.0000
23	200.0	557.9610	2.5905	346.0000
24	180.0	521.4222	2.4383	346.0000

$$= -\left[t_k \frac{\partial F_i}{\partial P_{ik}} + \lambda_k^0 \left(\frac{\partial P_{Lk}}{\partial P_{ik}} - 1\right)\right] \qquad (i = 1, 2, ..., N; \quad k = 1, 2, ..., T) \tag{4.38a}$$

$$\left(v_j^0 t_k \frac{\partial^2 q_{jk}}{\partial P_{mk}^2} + \lambda_k^0 \frac{\partial^2 P_{Lk}}{\partial P_{mk}^2}\right) \Delta P_{mk} + \lambda_k^0 \sum_{\substack{l=1 \\ l \neq m}}^{N+M} \frac{\partial^2 P_{Lk}}{\partial P_{mk} \partial P_{lk}} \Delta P_{lk} + \left(\frac{\partial P_{Lk}}{\partial P_{mk}} - 1\right) \Delta \lambda_k + \frac{\partial q_{jk}}{\partial P_{mk}} \Delta v_j$$

$$= -\left[v_j^0 t_k \frac{\partial q_{jk}}{\partial P_{mk}} + \lambda_k^0 \left(\frac{\partial P_{Lk}}{\partial P_{mk}} - 1\right)\right] \quad (j = 1, 2, ..., M; \quad m = j + N; \quad k = 1, 2, ..., T) \tag{4.38b}$$

$$\sum_{j=1}^{N+M} \left(\frac{\partial P_{Lk}}{\partial P_{jk}} - 1\right) \Delta P_{jk} = -\left(P_{Dk} + P_{Lk}^0 - \sum_{i=1}^{N+M} P_{ik}^0\right) \qquad (k = 1, 2, ..., T) \tag{4.38c}$$

TABLE 4.2 Thermal and hydro generations, transmission losses, and demand error

k	P_{Dk} (MW)	P_{1k} (MW)	P_{2k} (MW)	P_{3k} (MW)	P_{4k} (MW)	P_{Lk} (MW)
1	175.0	69.919	37.834	64.614	10.000	7.3674
2	190.0	76.333	41.599	70.916	10.000	8.8492
3	220.0	89.288	49.376	83.594	10.000	12.2573
4	280.0	115.705	66.008	109.218	10.000	20.9314
5	320.0	133.697	77.994	126.473	10.000	28.1641
6	360.0	148.008	88.176	140.467	17.841	34.4939
7	390.0	155.757	94.120	148.438	29.735	38.0511
8	410.0	160.876	98.149	153.731	37.773	40.5294
9	440.0	168.482	104.294	161.637	49.990	44.4045
10	475.0	177.242	111.622	170.809	64.487	49.1608
11	525.0	189.535	122.393	183.808	85.649	56.3841
12	550.0	195.579	127.914	190.258	96.428	60.1799
13	565.0	199.172	131.272	194.112	102.958	62.5151
14	540.0	193.170	125.695	187.682	92.100	58.6471
15	500.0	183.422	116.962	177.324	75.002	52.7103
16	450.0	170.998	106.371	164.264	54.105	45.7378
17	425.0	164.690	101.206	157.689	43.857	42.4434
18	400.0	158.321	96.128	151.086	33.743	39.2796
19	375.0	151.893	91.133	144.457	23.764	36.2483
20	340.0	142.798	84.279	135.133	10.018	32.2306
21	300.0	124.663	71.905	117.830	10.000	24.3982
22	250.0	102.411	57.502	96.363	10.000	16.2771
23	200.0	80.630	44.154	75.132	10.000	9.9186
24	180.0	72.052	39.081	66.712	10.000	7.8451

Total cost = ₹ 24,276.79, v_1 = 0.1822 ₹/m^3, V_1 = 25,000.03 m^3

$$\sum_{k=1}^{T}\left(\frac{\partial q_{jk}}{\partial P_{mk}}-1\right)\Delta P_{mk} = -\left(\sum_{k=1}^{T} q_{jk}^{0} - V_j\right) \qquad (j = 1, 2, \ldots, M; \quad m = j + N) \qquad (4.38d)$$

Substituting the values of derivatives into the above equations, the derivatives are evaluated about initial conditions

$$(2t_k a_i + \lambda_k^0 B_{ii})\Delta P_{ik} + \lambda_k^0 \sum_{\substack{j=1\\ j\neq i}}^{N+M} 2B_{ij}\Delta P_{jk} + \left(\sum_{j=1}^{N+M} 2B_{ij}P_{jk}^0 + B_{i0} - 1\right)\Delta\lambda_k$$

$$= -\left[t_k\,(2a_i P_{ik}^0 + b_i) + \lambda_k^0\left(\sum_{j=1}^{N+M} 2B_{ij}P_{jk}^0 + B_{i0} - 1\right)\right] \qquad (i = 1, 2, \ldots, N; \quad k = 1, 2, \ldots, T) \qquad (4.39a)$$

$$(2v_j^0 t_k x_j + \lambda_k^0 B_{mm})\Delta P_{mk} + \lambda_k \sum_{\substack{l=1\\ l\neq m}}^{N+M} 2B_{ml}\,\Delta P_{lk} + \left(\sum_{l=1}^{N+M} 2B_{ml}P_{lk}^0 + B_{m0} - 1\right)\Delta\lambda_k + t_k(2x_j P_{mk}^0 + y_j)\,\Delta v_j$$

$$= -\left[v_j^0 t_k(2x_j P_{mk}^0 + y_j) + \lambda_k^0\left(\sum_{l=1}^{N+M} 2B_{ml}P_{lk}^0 + B_{m0} - 1\right)\right]$$

$$(j = 1, 2, ..., M; \quad m = j + N; \quad k = 1, 2, ..., T) \tag{4.39b}$$

$$\sum_{j=1}^{N+M}\left(\sum_{l=1}^{N+M} 2B_{jl}P_{lk}^0 + B_{j0} - 1\right)\Delta P_{jk} = -\left(P_{Dk} + P_{Lk}^0 - \sum_{i=1}^{N+M} P_{ik}^0\right) \quad (k = 1, 2, ..., T) \tag{4.39c}$$

$$\sum_{k=1}^{T} t_k(2x_j P_{mk}^0 + y_j)\,\Delta P_{mk} = -\left(\sum_{k=1}^{T} t_k q_{jk}^0 - V_j\right) \quad (j = 1, 2, ..., M; \quad m = j + N) \tag{4.39d}$$

The above Eqs. (4.39a) to (4.39d), can be written in matrix form as

$$[H][\Delta z] = [-J] \tag{4.40}$$

where Z is a vector defined below:

$$Z = [Z_1, Z_2, ..., Z_T, \quad v_1, v_2, ..., v_M]^T$$

$$Z_k = [P_{1k}, ..., P_{Nk}, P_{(N+1)k}, ..., P_{(N+M)k}, \lambda_k]^T \qquad (k = 1, 2, ..., T)$$

The size of vector Z and Jacobian matrix, J is $S \times 1$ and the size of the Hessian matrix, H is $S \times S$; where

$$S = [(N + M + 1) \times T + M];$$

N is the number of thermal units; M is the number of hydro units; T is the number of time intervals.

Suppose there are three thermal units and two hydro units. The scheduling is required for 24 intervals, then the size of the Hessian matrix will be 146 × 146 [(3 + 2 + 1)24 + 2 = 146)].

To implement the Newton–Raphson method to solve the fixed-head hydrothermal scheduling problem, the detailed stepwise procedure is outlined below.

Algorithm 4.2: Newton–Raphson Method for Short-Range Fixed-Head Hydrothermal Scheduling

1. Read the number of thermal units N, the number of hydro units M, the number of sub-intervals T, cost coefficients, a_i, b_i, c_i (i = 1, 2, ..., N), B–coefficients, B_{ij} (i = 1, 2, ..., $N + M$; j = 1, 2, ..., $N + M$), discharge coefficients, x_i, y_i, z_i (i = 1, 2, ..., M), demand P_{Dk} (k = 1, 2, ..., T), and pre-specified available water v_j (j = 1, 2, ..., M)
2. Calculate the initial guess for P_{ik}^0(i = 1, 2, ..., $N + M$), λ_k^0 and v_j^0(j = 1, 2, ..., M) as discussed in Sectior 4.3.6 or 4.3.7 for T intervals.
3. Start iterations, $r = 1$
4. Calculate ΔP_{ik} (i = 1, 2, ..., $N + M$); (k = 1, 2, ..., T); $\Delta\lambda_k$ (k = 1, 2, ..., T) and v_j (j = 1, 2, ..., M) by solving Eq. (4.42) using the gauss elimination method. Hessian and Jacobian matrix elements are obtained from Eqs. (4.24a), (4.24b), (4.24c), and (4.24d).

5. Check the convergence

$$\text{if } \left| \sum_{k=1}^{T} \left[\sum_{i=1}^{M+N} \Delta P_{ik} + \Delta\lambda_k \right] + \sum_{j=1}^{M} \Delta v_j \right| \le \varepsilon \quad \text{or} \quad \left| \sum_{k=1}^{T} \left[\sum_{i=1}^{M+N} \frac{\partial L}{\partial P_{ik}} + \frac{\partial L}{\partial \lambda_k} \right] + \sum_{j=1}^{M} \frac{\partial L}{\partial v_j} \right| \le \varepsilon$$

then GOTO Step 10.

6. Calculate the new values of P_{ik}^{new} $(i = 1, 2, ..., N + M;\ \ k = 1, 2, ..., T)$; λ_k^{new} $(k = 1, 2, ..., T)$ and v_j^{new} $(j = 1, 2, ..., M)$ by adding the changes to the previous values as

$$P_{ik}^{\text{new}} = P_{ik}^{0} + \Delta P_{ik} \qquad (i = 1, 2, ..., N + M;\ \ k = 1, 2, ..., T)$$
$$\lambda_k^{\text{new}} = \lambda_k^{0} + \Delta\lambda_k \qquad (k = 1, 2, ..., T)$$
$$v_j^{\text{new}} = v_j^{0} + \Delta v_j \qquad (j = 1, 2, ..., M)$$

7. Set limits correspondingly as:

$$P_{ik}^{\text{new}} = \begin{cases} P_i^{\max} & ;\text{if } P_{ik}^{\text{new}} \ge P_i^{\max} \\ P_i^{\min} & ;\text{if } P_{ik}^{\text{new}} \le P_i^{\min} \\ P_{ik}^{\text{new}} & ;\text{otherwise} \end{cases}$$

Disallow to participate the generation, whose limits have been set either to lower or to upper limit, in the scheduling by deleting that row and column.

8. If $r \ge R$, then GOTO Step 9,

else $\quad r = r + 1,$

$$P_{ik}^{0} = P_{ik}^{\text{new}} \qquad (i = 1, 2, ..., N + M;\ \ k = 1, 2, ..., T)$$
$$\lambda_k^{0} = \lambda_k^{\text{new}} \qquad (k = 1, 2, ..., T)$$
$$v_j^{0} = v_j^{\text{new}} \qquad (j = 1, 2, ..., M)$$

GOTO Step 4 and repeat.

9. Calculate the optimal total cost and transmission loss for T intervals.
10. Stop.

Sample System 4.2

A hydrothermal system has been considered, which consists of one thermal and one hydro generating station as shown in Figure 4.6. The operating cost of the thermal station is given by

$$F_{1k} = 0.001991P_{1k}^2 + 9.606P_{1k} + 373.7 \text{ ₹/h}$$

The rate of discharge of hydro generating station is given by

$$q_{1k} = 2.19427 \times 10^{-5}P_{2k}^2 + 2.5709 \times 10^{-4}P_{2k} + 1.742333 \text{ Mm}^3\text{/h}$$

Assume that the reservoirs are large and water available in the reservoirs is $V_1 = 72.4797$ Mm^3. The B-coefficients of the power system network are given by

$$B = \begin{bmatrix} 0.00005 & 0.00001 \\ 0.00001 & 0.00150 \end{bmatrix} \text{MW}^{-1}$$

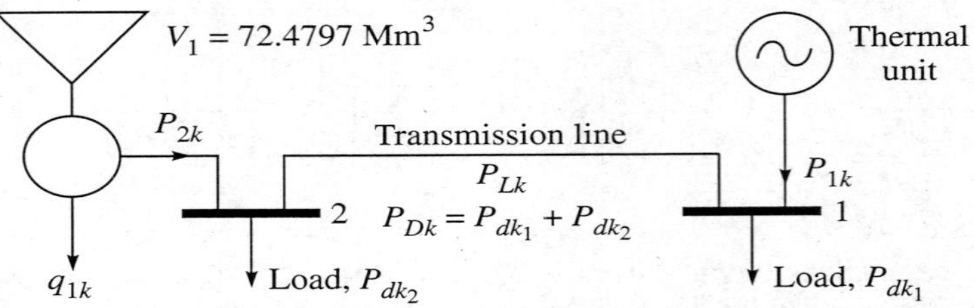

FIGURE 4.6 Hydrothermal system.

Using the Newton–Raphson method, the generation schedule of the hydrothermal problem for 24 hours was obtained and is given below, when hourly demand was taken. Table 4.3 represents the operating cost of thermal units, incremental cost, λ_k, and discharge of hydro units during 24 hours. Thermal and hydro generations to meet the hourly demand along with transmission losses are given in Table 4.4. The unsatisfied demand during each sub-interval is given in Table 4.4.

4.5 APPROXIMATE NEWTON–RAPHSON METHOD FOR SHORT-RANGE FIXED-HEAD HYDROTHERMAL SCHEDULING

Suppose the initial values of control variables P_{ik}, λ_k, and v_j are known. The updated values of control variables in the next iteration are

$$P_{ik}^{\text{new}} = P_{ik}^{0} + \Delta P_{ik} \qquad (i = 1, 2, ..., N + M; \quad k = 1, 2, ..., T)$$

$$\lambda_k^{\text{new}} = \lambda_k^{\text{new}} + \Delta\lambda_k \qquad (k = 1, 2, ..., T)$$

$$v_j^{\text{new}} = v_j^{0} + \Delta v_j \qquad (j = 1, 2, ..., M)$$

Any small change in control values from their previous values can be obtained from Eqs. (4.39a–4.39d), by substituting the values of derivatives into Eq. (4.39a). The derivatives are evaluated about initial values.

$$(2t_k a_i + 2\lambda_k^0 B_{ii})\,\Delta P_{ik} + \lambda_k^0 \sum_{\substack{j=1 \\ j\neq i}}^{N+M} 2B_{ij}\,\Delta P_{jk} + (K_{ik}^0 - 1)\,\Delta\lambda_k$$

$$= -\,[t_k(2a_i P_{ik}^0 + b_i) + \lambda_k^0\,(K_{ik}^0 - 1)] \qquad (i = 1, 2, ..., N) \qquad (4.41)$$

where

$$K_{ik}^0 = \sum_{j=1}^{N+M} 2B_{ij} P_{jk}^0 + B_{i0}$$

To simplify calculations, neglect all the terms with B_{ij}, $i \neq j$. So neglecting the term $\sum_{\substack{j=1 \\ j\neq i}}^{N+M} 2B_{ij}\,\Delta P_{jk}$, Eq. (4.41) becomes

$$(2t_k a_i + 2\lambda^0{}_k\, B_{ii})\,\Delta P_{ik} = (1 - K_{ik}^0)\lambda_k^{\text{new}} - t_k(2a_i P_{ik}^0 + b_i) \qquad (i = 1, 2, ..., N)$$

TABLE 4.3 Hourly cost, λ, and discharge for required demand

k	P_{Dk} (MW)	F_k (₹/h)	λ_k (₹/MWh)	q_{1k} (Mm^3/h)
1	455.0	2707.811	10.83128	2.896065
2	425.0	2412.680	10.68426	2.868330
3	415.0	2315.171	10.63554	2.859194
4	407.0	2237.477	10.59665	2.851924
5	400.0	2169.721	10.56270	2.845591
6	420.0	2363.871	10.65988	2.863755
7	487.0	3026.957	10.98948	2.926196
8	604.0	4232.544	11.58041	3.041320
9	665.0	4885.654	11.89655	3.104540
10	675.0	4994.354	11.94892	3.115120
11	695.0	5213.143	12.05413	3.136465
12	705.0	5323.235	12.10697	3.147231
13	580.0	3980.234	11.45755	3.017057
14	605.0	4243.114	11.58554	3.042338
15	616.0	4359.680	11.64215	3.053579
16	653.0	4755.824	11.83391	3.091925
17	721.0	5500.355	12.19183	3.164587
18	740.0	5712.244	12.29314	3.185407
19	700.0	5268.131	12.08053	3.141840
20	678.0	5027.054	11.96466	3.118306
21	630.0	4508.835	11.71447	3.067991
22	585.0	4032.583	11.48308	3.022083
23	540.0	3565.490	11.25469	2.977363
24	503.0	3188.220	11.06912	2.941476

Total operating cost = ₹ 96,024.39.

V_1 = 72.47969 Mm^3

v_1 = 995.0728 ₹/Mm^3

or

$$\Delta P_{ik} = \frac{(1 - K_{ik}^0)\lambda_k^{\text{new}} - t_k\,(2a_i P_{ik}^0 + b_i)}{2t_k a_i + 2\lambda_k^0 B_{ii}} \qquad (i = 1, 2, ..., N)$$

$$= \frac{(1 - K_{ik}^0)\,\lambda_k^{\text{new}} - \alpha_{ik}}{X_{ik}} \qquad (i = 1, 2, ..., N) \tag{4.42}$$

where

$$\alpha_{ik} = t_k(2a_i P_{ik}^0 + b_i) \tag{4.42a}$$

$$X_{ik} = 2(t_k a_i + \lambda_k^0 B_{ii}) \tag{4.42b}$$

TABLE 4.4 Thermal and hydro generations, transmission losses, and demand error

k	P_{Dk} (MW)	P_{1k} (MW)	P_{2k} (MW)	P_{Lk} (MW)	*Error* (MW)
1	455.0	231.8438	235.2348	12.07865	0.000003
2	425.0	203.6639	232.4628	11.12668	−0.000030
3	415.0	194.2866	231.5423	10.82885	−0.000009
4	407.0	186.7905	230.8071	10.59757	0.000007
5	400.0	180.2357	230.1646	10.40029	−0.000012
6	420.0	198.9742	232.0023	10.97655	−0.000012
7	487.0	261.9826	238.2088	13.19139	−0.000014
8	604.0	372.8918	249.2372	18.12907	−0.000021
9	665.0	431.1694	255.0860	21.25539	−0.000042
10	675.0	440.7532	256.0515	21.80463	0.000004
11	695.0	459.9464	257.9880	22.93442	0.000031
12	705.0	469.5559	258.9591	23.51503	0.000011
13	580.0	350.0487	246.9549	17.00364	−0.000023
14	605.0	373.8447	249.3326	18.17723	−0.000011
15	616.0	384.3315	250.3824	18.71383	−0.000036
16	653.0	419.6801	253.9300	20.61002	−0.000027
17	721.0	484.9490	260.5168	24.46588	0.000000
18	740.0	503.2571	262.3731	25.63015	0.000025
19	700.0	464.7501	258.4733	23.22341	0.000023
20	678.0	443.6300	256.3415	21.97144	0.000002
21	630.0	397.6930	251.7217	19.41472	−0.000015
22	585.0	354.8037	247.4295	17.23327	0.000051
23	540.0	312.0835	243.1742	15.25767	−0.000020
24	503.0	277.0832	239.7025	13.78570	−0.000064

Subsituting the values of derviatives into Eq. (4.39b) and rearranging the terms, we have

$$(2v_j^0 t_k x_j + \lambda_k^0 B_{mm})\Delta P_{mk} + \lambda_k^0 \sum_{\substack{l=1 \\ l\neq m}}^{N+M} 2B_{ml}\, \Delta P_{lk}$$

$$= -[t_k(2x_j P_{mk}^0 + y_j)(v_j^0 + \Delta v_j) + (\lambda_k^0 + \Delta\lambda_k)(K_{mk}^0 - 1)] \qquad (j = 1, 2, ..., M; \quad m = j + N)$$

where

$$K_{mk}^0 = \sum_{l=1}^{N+M} 2B_{ml} P_{lk}^0 + B_{m0}$$

Neglecting the term $\sum_{\substack{l=1\\l\neq m}}^{N+M} 2B_{jl}\Delta P_{lk}$ for simplification, we have

$$(2\nu_j^0 t_k x_j + \lambda_k^0 B_{mm})\Delta P_{mk} = -[t_k(2x_j P_{mk}^0 + y_j)\nu_j^{\text{new}} + \lambda_k^{\text{new}}(K_{mk}^0 - 1)] \quad (j = 1, 2, ..., M; \quad m = j + N)$$

or

$$\Delta P_{mk} = \frac{(1 - K_{mk}^0)\lambda_k^{\text{new}} - t_k(2x_j P_{mk}^0 + y_j)\nu_j^{\text{new}}}{2\nu_j^0 t_k x_j + 2\lambda_k^0 B_{mm}} \quad (j = 1, 2, ..., M; \quad m = j + N)$$

$$= \frac{(1 - K_{mk}^0)\lambda_k^{\text{new}} - \beta_{jk}\nu_j^{\text{new}}}{Y_{jk}} \quad (j = 1, 2, ..., M; \quad m = j + N) \tag{4.43}$$

where

$$\beta_{jk} = t_k(2x_j P_{mk}^0 + y_j) \tag{4.43a}$$

$$Y_{jk} = 2(\nu_j^0 t_k x_j + \lambda_k^0 B_{mm}) \tag{4.43b}$$

Substiuting the values of derivatives into Eq. (4.39c), we get

$$\sum_{j=1}^{N+M} (1 - K_{jk}^0)\Delta P_{jk} = \left(P_{Dk} + P_{Lk}^0 - \sum_{i=1}^{N+M} P_{ik}^0\right) \quad (k = 1, 2, ..., T)$$

Substituting Eq. (4.42) and Eq. (4.43) into the above equation, we have

$$\sum_{i=1}^{N}(1 - K_{ik}^0)\left(\frac{\lambda_k^{\text{new}}(1 - K_{ik}^0) - \alpha_{ik}}{X_{ik}}\right) + \sum_{j=1}^{M}(1 - K_{mk}^0)\left(\frac{\lambda_k^{\text{new}}(1 - K_{mk}^0) - \beta_{jk}\nu_j^{\text{new}}}{Y_{jk}}\right)$$

$$= \left(P_{Dk} + P_{Lk}^0 - \sum_{i=1}^{N+M} P_{ik}^0\right) \quad (k = 1, 2, ..., T; \quad m = j + N)$$

or

$$\left(\sum_{i=1}^{N}\frac{(1 - K_{ik}^0)^2}{X_{ik}} + \sum_{j=1}^{M}\frac{(1 - K_{mk}^0)^2}{Y_{jk}}\right)\lambda_k^{\text{new}} - \sum_{j=1}^{M}\left(\frac{(1 - K_{mk}^0)\beta_{jk}}{Y_{jk}}\right)\nu_j^{\text{new}}$$

$$= \left(P_{Dk} + P_{Lk}^0 - \sum_{i=1}^{N+M} P_{ik}^0\right) + \sum_{i=1}^{N}\left(\frac{(1 - K_{ik}^0)\alpha_{ik}}{X_{ik}}\right) \quad (k = 1, 2, ..., T; \quad m = j + N)$$

or

$$C_k\lambda_k^{\text{new}} - \sum_{j=1}^{M} D_{jk}\nu_j^{\text{new}} = F_k \tag{4.44}$$

where

$$C_k = \sum_{i=1}^{N}\frac{(1 - K_{ik}^0)^2}{X_{ik}} + \sum_{j=1}^{M}\frac{(1 - K_{mk}^0)^2}{Y_{jk}} \tag{4.44a}$$

$$D_{jk} = \frac{(1 - K_{mk}^0)\beta_{jk}}{Y_{jk}} \tag{4.44b}$$

$$F_k = \left(P_{Dk} + P_{Lk}^0 - \sum_{i=1}^{N+M} P_{ik}^0 \right) + \sum_{i=1}^{N} \left(\frac{(1 - K_{mk}^0)\alpha_{ik}}{X_{ik}} \right) \tag{4.44c}$$

Substituting Eq. (4.42) into Eq. (4.39d)

$$\sum_{k=1}^{T} \beta_{jk} \left(\frac{(1 - K_{mk}^0)\,\lambda_k^{\text{new}} - \beta_{jk} v_j^{\text{new}}}{Y_{jk}} \right) = -\left(\sum_{k=1}^{T} t_k\, q_{jk}^0 - V_j \right) \qquad (j = 1, 2, ..., M; \quad m = j + N)$$

or

$$\sum_{k=1}^{T} D_{jk} \lambda_k^{\text{new}} - H_j v_j^{\text{new}} = O_j \qquad (j = 1, 2, ..., M; \quad m = j + N) \tag{4.45}$$

where

$$H_j = \sum_{k=1}^{T} \frac{\beta_{jk}\beta_{jk}}{Y_{jk}} \tag{4.45a}$$

$$O_j = V_j - \sum_{k=1}^{T} t_k\, q_{jk}^0 \tag{4.45b}$$

From Eq. (4.44), we find the value of λ_k^{new} as

$$\lambda_k^{\text{new}} = \frac{F_k}{C_k} + \sum_{j=1}^{M} \frac{D_{jk}}{C_k} v_j^{\text{new}} \tag{4.46}$$

Substituting the value of λ_k^{new} into Eq. (4.45), we get

$$\sum_{k=1}^{T} D_{jk} \left(\frac{F_k}{C_k} + \sum_{l=1}^{M} \frac{D_{lk}}{C_k} v_l^{\text{new}} \right) - H_j v_j^{\text{new}} = O_j \qquad (j = 1, 2, ..., M; \quad m = j + N)$$

or

$$\sum_{l=1}^{M} \left(\sum_{k=1}^{T} \frac{D_{jk} D_{lk}}{C_k} v_l^{\text{new}} \right) - H_j v_j^{\text{new}} = O_j - \sum_{k=1}^{T} \frac{D_{jk} F_k}{C_k} \qquad (j = 1, 2, ..., M; \quad m = j + N) \tag{4.47}$$

Equation (4.47) can be written in matrix form as

$$[Q_{jl}]_{M\times M}\, [v_j]_{M\times l} = [R_j]_{M\times l}$$

where

$$Q_{jj} = \left(\sum_{k=1}^{T} \frac{D_{jk} D_{jk}}{C_k} \right) - H_j \qquad (j = 1, 2, ..., M) \tag{4.48}$$

$$Q_{jl} = \left(\sum_{k=1}^{T} \frac{D_{jk} D_{lk}}{C_k} \right) \qquad (j = 1, 2, ..., M; \quad l = 1, 2, ..., M; \quad j \neq l) \tag{4.49}$$

$$R_j = O_j - \sum_{k=1}^{T} \frac{D_{jk} F_k}{C_k} \qquad (j = 1, 2, ..., M) \tag{4.50}$$

Here only the matrix of $M \times M$ is to be solved to calculate v_j^{new}. λ_k^{new} can be computed from Eq. (4.46). When the values of v_j^{new} and λ_k^{new} are known then ΔP_{ik} (i = 1, 2, ..., N) and ΔP_{mk} (j = 1, 2, ..., M; $m = j + N$), can be computed from Eqs. (4.42) and (4.43), respectively. The detailed algorithm is elaborated below:

Algorithm 4.3: Approximate Newton–Raphson Method for Short-Range Fixed-Head Hydro-thermal Scheduling

1. Read the number of thermal units N, the number of hydro units M, the number of sub-intervals T, cost coefficients, a_i, b_i, c_i (i = 1, 2, ..., N), B–coefficients, B_{ij} (i = 1, 2, ..., $N + M$; j =1, 2, ..., $N + M$), discharge coefficients, x_i, y_i, z_i (i = 1, 2, ..., M), demand P_{Dk} (k = 1, 2, ..., T) and pre-specified available water V_j (j = 1, 2, ..., M).
2. Calculate the initial guess values of P_{ik}^0 (i = 1, 2, ..., $N + M$); λ_k^0 and v_j^0 (j = 1, 2, ..., M)
3. Start the iteration counter, r = 1.
4. Compute the variables, K_{ik}^0, α_{ik} from Eq. (4.42a), X_{ik} from Eq. (4.42b), β_{jk} from Eq. (4.43a), Y_{jk} from Eq. (4.43b), C_k from Eq. (4.44a), D_{jk} from Eq. (4.44b), F_k from Eq. (4.44c), H_j from Eq. (4.45a) and O_j from Eq. (4.45b).
5. Compute v_j^{new} by solving the following simultaneous equations using the Gauss elimination method.
$$[Q_{jl}]_{M \times M} \, [v_j]_{M \times l} = [R_j]_{M \times l}$$
6. Check the convergence
$$\text{if } | v_j^{new} - v_j^0 | \le \varepsilon \text{ then GOTO Step 12.}$$
7. Compute λ_k^{new} from Eq. (4.46).
8. Calculate ΔP_{ik} (i = 1, 2, ..., $N + M$), using Eqs. (4.42) and (4.43).
9. Calculate the new values of P_{ik}^{new}
$$P_{ik}^{new} = P_{ik}^0 + \Delta P_{ik} \qquad (i = 1, 2, ..., N + M; \quad k = 1, 2, ..., T)$$
10. Set limits correspondingly as
$$P_{ik}^{new} = \begin{cases} P_i^{max} & ; \text{if } P_{ik}^{new} \ge P_i^{max} \\ P_i^{min} & ; \text{if } P_{ik}^{new} \le P_i^{min} \\ P_{ik}^{new} & ; \text{otherwise} \end{cases}$$
Disallow the generators to participate, whose limit have been set either to lower or upper limits by setting the relating coefficients to zero.
11. If ($r \ge$ IT) then GOTO Step 12,
else
$$r = r + 1,$$
$$P_{ik}^0 = P_{ik}^{new} \qquad (i = 1, 2, ..., N + M; \quad k = 1, 2, ..., T)$$
$$\lambda_k^0 = \lambda_k^{new} \qquad (k = 1, 2, ..., T)$$
$$v_j^0 = v_j^{new} \qquad (j = 1, 2, ..., N)$$
GOTO Step 4 and repeat.

12. Calculate the optimal cost and loss, etc..
13. Stop.

EXAMPLE 4.2 A hydrothermal system is given which consists of one thermal and one hydro generating station as shown in Figure 4.7. The operating cost of the thermal station is given by

$$F_{1k} = 0.00184P_{1k}^2 + 9.2P_{1k} + 575 \text{ ₹/h}$$

The rate of discharge of hydro generating station is given by

$$q_{1k} = 4.97P_{2k} + 330 \text{ m}^3\text{/h}$$

The B-coefficients for transmission losses are

$$B_{11} = B_{12} = B_{21} = 0.0,\ B_{22} = 0.00008 \text{ MW}^{-1},\quad B_{10} = B_{20} = B_{00} = 0.0$$

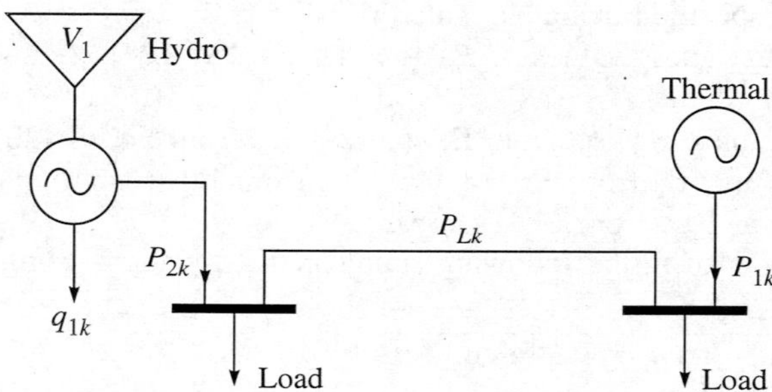

FIGURE 4.7 Hydrothermal system.

Assume that the reservoir is large. The water available in the reservoir is $V_1^s = 100{,}000.0$ m^3 The demand curve is given in Figure 4.8. Find the optimal generation schedule.

Solution The initial values are computed as:
Power generations P_{ik}^0 are computed using Eq. (4.35) for $k = 1$

$$P_{11}^0 = 1200/2 = 600 \text{ MW} \quad \text{and} \quad P_{21}^0 = 1200/2 = 600 \text{ MW}$$

Incremental cost λ_k^0 is computed using Eq. (4.36) for $k = 1$

$$\lambda_1 = 2 \times 0.00184 \times 600 + 9.2 = 11.408 \text{ ₹/MWh}$$

Power generations P_{ik}^0 are computed using Eq. (4.35) for $k = 2$

$$P_{12}^0 = 1500/2 = 750 \text{ MW} \quad \text{and} \quad P_{22}^0 = 1500/2 = 750 \text{ MW}$$

Incremental cost λ_k^0 is computed using Eq. (4.36) for $k = 2$

$$\lambda_2^0 = 2 \times 0.00184 \times 750 + 9.2 = 11.96 \text{ ₹/MWh}$$

From Eq. (4.37), water conversion factor v_1 is calculated as

$$v_1^0 = \frac{\lambda_1}{2x_1P_{21} + y_1} = \frac{11.408}{4.97} = 2.295372 \text{ ₹/m}^3$$

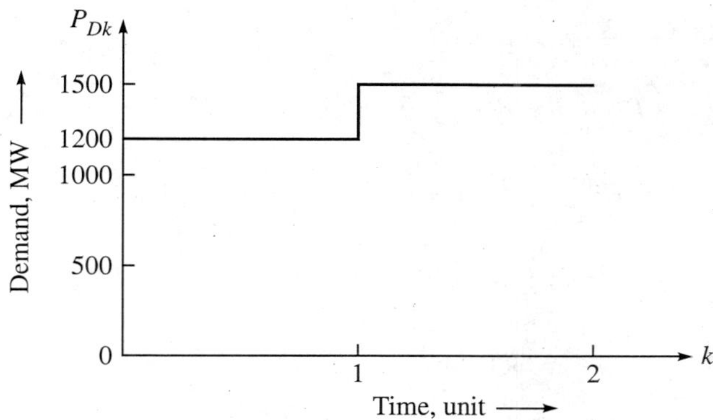

FIGURE 4.8 Demand curve.

Set the iteration counter IT = 1.

K_{ik}, α_{ik}, X_{ik}, β_{jk}, Y_{jk}, C_k, D_{jk}, P_{Lk} and F_k are computed from Eqs. (4.41), (4.42a), (4.42b), (4.43a), (4.43b), (4.44a), (4.44b), and (4.44c), respectively, for $k = 1$

$$K_{11}^0 = 2B_{11}P_{11}^0 + 2B_{12}\,P_{21}^0 = 0.0$$
$$K_{21}^0 = 2B_{21}P_{11}^0 + 2B_{22}\,P_{21}^0 = 0.096$$
$$\alpha_{11} = t_1(2a_1P_{11}^0 + b_1) = 136.896$$
$$X_{11} = 2(t_1a_1 + \lambda_1^0\,B_{11}) = 0.04416$$
$$\beta_{11} = t_1(2x_1P_{21}^0 + y_1) = 59.64;$$
$$Y_{11} = 2\,(v_1^0t_1x_1 + \lambda_1^0\,B_{22}) = 0.00182528$$
$$C_1 = \frac{(1-K_{11}^0)^2}{X_{11}} + \frac{(1-K_{21}^0)^2}{Y_{11}} = 470.3658$$
$$D_{11} = \frac{(1-K_{21}^0)^2\,\beta_{11}}{Y_{11}} = 29537.69$$
$$P_{L1} = B_{00} + (B_{10}\,P_{11}^0 + B_{20}\,P_{21}^0) + (B_{11}\,(P_{11}^0)^2 + B_{22}\,(P_{21}^0)^2 + 2B_{12}\,P_{11}^0\,P_{12}^0) = 28.8\text{ MW}$$
$$F_1 = P_{D1} + P_{L1}^0 - (P_{11}^0 + P_{21}^0) + \frac{(1-K_{11}^0)^2\,\alpha_{11}}{X_{11}} = 3128.80$$

K_{ik}, α_{ik}, X_{ik}, β_{jk}, Y_{jk}, C_k, D_{jk}, P_{Lk} and F_k are computed from Eqs. (4.41), (4.42a), (4.42b), (4.43a), (4.43b), (4.44a), (4.44b) and (4.44c), respectively, for $k = 2$

$$K_{12}^0 = 2B_{11}^0\,P_{12}^0 + 2B_{12}\,P_{22}^0 = 0.0; \qquad K_{22}^0 = 2B_{21}\,P_{12}^0 + 2B_{22}\,P_{22}^0 = 0.12$$
$$\alpha_{12}^0 = t_2(2a_1P_{12}^0 + b_1) = 143.52; \qquad X_{12} = 2(t_2a_1 + \lambda_2^0\,B_{11}) = 0.04416$$
$$\beta_{12} = t_2(2x_1P_{22}^0 + y_1) = 59.64$$
$$Y_{12} = 2(v_1^0t_2x_1 + \lambda_2^0\,B_{22}) = 0.0019136$$
$$C_2 = \frac{(1-K_{12}^0)^2}{X_{12}} + \frac{(1-K_{22}^0)^2}{Y_{12}} = 427.3272$$

$$D_{12} = \frac{(1 - K_{22}^0)^2 \beta_{12}}{Y_{12}} = 27426.42$$

$$P_{L2} = B_{00} + (B_{10} P_{12}^0 + B_{20} P_{22}^0) + (B_{11}(P_{12}^0)^2 + B_{22}(P_{22}^0)^2 + 2B_{12} P_{12}^0 P_{22}^0) = 45.0 \text{ MW}$$

$$F_2 = P_{D2} + P_{L2}^0 - (P_{12}^0 + P_{22}^0) + \frac{(1 - K_{12}^0)^2 \alpha_{12}}{X_{12}} = 3295.0$$

H_j and O_j are computed from Eqs. (4.45a) and (4.45b), respectively

$$H_1 = \frac{(\beta_{11})^2}{Y_{11}} + \frac{(\beta_{12})^2}{Y_{12}} = 3807467.0$$

$$O_1 = V_1^s - [t_1(x_1 P_{21}^2 + y_1 P_{21} + z_1) + t_2(x_1 P_{22}^2 + y_1 P_{22} + z_1)] = 11566$$

Q_{ij} and R_j are computed from Eqs. (4.49) and (4.50), respectively, as

$$Q_{11} = \frac{(D_{11})^2}{C_1} + \frac{(D_{12})^2}{C_2} - H_1 = -0.1923168 \times 10^6$$

$$R_1 = O_1 - \left[\frac{(D_{11}F_1)}{C_1} + \frac{(D_{12}F_2)}{C_2}\right] = -0.3963915$$

Equation (4.48) becomes

$$-0.1923168 \times 10^6 \; v_1^{\text{new}} = -0.3963915 \quad \text{or} \quad v_1^{\text{new}} = 2.061139 \text{ ₹/m}^3$$

$$| v_1^{\text{new}} - v_1^0 | = | 2.061139 - 2.295372 | = 0.234233 \geq 0.0001$$

λ_k is computed from Eq. (4.46) as

$$\lambda_1^{\text{new}} = \frac{F_1}{C_1} + \left(\frac{D_{11}}{C_1}\right) v_1^{\text{new}} = 136.0858 \text{ ₹/MWh}$$

$$\lambda_2^{\text{new}} = \frac{F_2}{C_2} + \left(\frac{D_{12}}{C_2}\right) v_1^{\text{new}} = 139.9973 \text{ ₹/MWh}$$

Compute the new generations from Eqs. (4.42) and (4.43) as

$$P_{11}^{\text{new}} = P_{11}^0 + \frac{(1 - K_{11}^0)\lambda_1^{\text{new}} - \alpha_{11}}{X_{11}} = 581.6522 \text{ MW}$$

$$P_{21}^{\text{new}} = P_{21}^0 + \frac{(1 - K_{21}^0)\lambda_1^{\text{new}} - \beta_{11}}{Y_{11}} = 652.1569 \text{ MW}$$

$$P_{12}^{\text{new}} = P_{12}^0 + \frac{(1 - K_{12}^0)\lambda_2^{\text{new}} - \alpha_{12}}{X_{12}} = 670.2289 \text{ MW}$$

$$P_{22}^{\text{new}} = P_{22}^0 + \frac{(1 - K_{22}^0)\lambda_2^{\text{new}} - \beta_{12}}{Y_{12}} = 891.7831 \text{ MW}$$

For the second iterations, the previous new values will become the old values and the above described procedure is repeated till the convergence is achieved.

$$v_1^0 = 2.061139 \text{ ₹/m}^3$$

$$P_{11}^0 = 581.6522 \text{ MW}, \quad P_{21}^0 = 652.1569 \text{ MW}, \quad \lambda_1^0 = 136.0858 \text{ ₹/MWh}$$

$$P_{12}^0 = 670.2289 \text{ MW}, \quad P_{22}^0 = 891.7831 \text{ MW}, \quad \lambda_2^0 = 139.9973 \text{ ₹/MWh}$$

$$V_1 = 100000.60 \text{ m}^3$$

Increment the iteration counter, IT = 2

$$v_1^{new} = 2.028335 \text{ ₹/m}^3$$

$$|v_1^{new} - v_1^0| = |\ 2.028335 - 2.061139\ | = 0.032804 \geq 0.0001$$

$$P_{11}^{new} = 567.3913 \text{ MW}, \quad P_{21}^{new} = 668.3195 \text{ MW}, \quad \lambda_1^{new} = 135.4560 \text{ ₹/MWh}$$

$$P_{12}^{new} = 685.7040 \text{ MW}, \quad P_{22}^{new} = 875.6098 \text{ MW} \quad \lambda_2^{new} = 140.6807 \text{ ₹/MWh}$$

$$V_1 = 99999.94 \text{ m}^3$$

Increment the iteration counter, IT = 3

$$v_1^{new} = 2.028377 \text{ ₹/m}^3$$

$$|\ v_1^{new} - v_1^0\ | = |\ 2.028377 - 2.028335\ | = 0.000042 \leq 0.0001$$

$$P_{11}^{new} = 567.4131 \text{ MW}, \quad P_{21}^{new} = 668.3188 \text{ MW}, \quad \lambda_1^{new} = 135.4570 \text{ ₹/MWh}$$

$$P_{12}^{new} = 685.7244 \text{ MW}, \quad P_{22}^{new} = 875.6116 \text{ MW}, \quad \lambda_2^{new} = 140.6816 \text{ ₹/MWh}$$

$$V_1 = 100000.00 \text{ m}^3$$

The convergence is achieved.
The total operating cost = ₹ 169,637.60

Sample System 4.3

A hydrothermal system is given which consists of one thermal and two hydro generating stations. The operating cost of the thermal station is given by

$$F_{1k} = 0.01P_{1k}^2 + 3.0P_{1k} + 15.0 \text{ ₹/h}$$

The rate of discharge of hydro-generating station is given by

$$q_{1k} = 1.41584 \times 10^{-6}\ P_{2k}^2 + 8.49504 \times 10^{-4}\ P_{2k} + 5.66336 \times 10^{-3} \text{ Mm}^3\text{/h}$$

$$q_{2k} = 2.83168 \times 10^{-6}\ P_{3k}^2 + 1.699008 \times 10^{-3}\ P_{3k} + 1.1132672 \times 10^{-2} \text{ Mm}^3\text{/h}$$

Assume that the reservoirs are large and water available in the reservoirs is

$$V_1 = 0.707920 \text{ Mm}^3 \quad \text{and} \quad V_2 = 0.991088 \text{ Mm}^3$$

The B-coefficients of the power system network are given by

$$B = \begin{bmatrix} 0.0 & 0.0 & 0.0 \\ 0.0 & 0.001 & 0.0 \\ 0.0 & 0.0 & 0.0005 \end{bmatrix} \text{MW}^{-1}$$

Using the approximate Newton–Raphson method, the generation schedule of the hydro-thermal problem for 24 hours is obtained below, where the hourly demand is given. Table 4.5 represents the operating cost of thermal units, incremental cost, λ_k, and discharge of hydro units during 24 hours. Thermal and hydro generations to meet the hourly demand along with transmission losses are given in Table 4.6. The unsatisfied demand during each sub-interval is given in Table 4.6.

TABLE 4.5 Hourly cost, λ, and discharge for required demand

k	P_{Dk} (MW)	F_k (₹/h)	λ_k (₹/MWh)	q_{1k} (Mm^3/h)	q_{2k} (Mm^3/h)
1	30.0	18.88393	3.025783	0.023420	0.026740
2	33.0	21.30824	3.041765	0.024339	0.028918
3	35.0	22.93648	3.052452	0.024954	0.030377
4	38.0	25.39685	3.068531	0.025877	0.032573
5	40.0	27.04923	3.079282	0.026494	0.034044
6	45.0	31.22307	3.106272	0.028040	0.037743
7	50.0	35.45866	3.133425	0.029592	0.041472
8	59.0	43.24131	3.182713	0.032403	0.048264
9	61.0	44.99889	3.193738	0.033030	0.049787
10	58.0	42.36639	3.177210	0.032090	0.047504
11	56.0	40.62426	3.166224	0.031464	0.045989
12	57.0	41.49409	3.171714	0.031777	0.046746
13	60.0	44.11881	3.188222	0.032716	0.049025
14	61.0	44.99889	3.193738	0.033030	0.049787
15	65.0	48.54493	3.215868	0.034288	0.052848
16	68.0	51.23193	3.232535	0.035233	0.055157
17	71.0	53.94256	3.249262	0.036181	0.057478
18	62.0	45.88152	3.199260	0.033344	0.050550
19	55.0	39.75698	3.160741	0.031151	0.045233
20	50.0	35.45866	3.133425	0.029592	0.041472
21	43.0	29.54618	3.095457	0.027421	0.036259
22	33.0	21.30824	3.041765	0.024339	0.028918
23	31.0	19.68965	3.031104	0.023726	0.027465
24	30.0	18.88393	3.025783	0.023420	0.026740

Total cost = ₹ 848.3436

v_1 = 3201.953 ₹/Mm^3, v_2 = 1713.925 ₹/Mm^3

V_1 = 0.7079206 Mm^3, V_2 = 0.9910884 Mm^{3*}

4.6 SHORT-RANGE VARIABLE-HEAD HYDROTHERMAL SCHEDULING—CLASSICAL METHOD

The basic problem considered involves the short-range optimal economic operation of an electric power system that includes both hydro and thermal generation resources. The object is to minimize

TABLE 4.6 Thermal and hydro generations, transmission losses and demand error

k	P_{Dk} (MW)	P_{1k} (MW)	P_{2k} (MW)	P_{3k} (MW)	P_{Lk} (MW)	*Error* (MW)
1	30.0	1.289103	20.22103	8.938724	0.448840	−0.000012
2	33.0	2.088211	21.23325	10.181180	0.502679	0.000036
3	35.0	2.622568	21.90774	11.010260	0.540562	−0.000007
4	38.0	3.426480	22.91892	12.254970	0.600369	−0.000003
5	40.0	3.964031	23.59267	13.085510	0.642229	0.000018
6	45.0	5.313577	25.27580	15.164500	0.753847	−0.000036
7	50.0	6.671205	26.95706	17.247110	0.875414	0.000041
8	59.0	9.135574	29.97866	21.005090	1.119327	0.000011
9	61.0	9.686846	30.64929	21.841800	1.177911	−0.000022
10	58.0	8.860439	29.64323	20.586940	1.090632	0.000025
11	56.0	8.311170	28.97214	19.751120	1.034438	0.000011
12	57.0	8.585651	29.30774	20.168980	1.062338	−0.000033
13	60.0	9.411045	30.31401	21.423360	1.148419	0.000009
14	61.0	9.686846	30.64929	21.841800	1.177911	−0.000022
15	65.0	10.79333	31.98961	23.516880	1.299857	0.000037
16	68.0	11.62671	32.99407	24.774750	1.395503	−0.000020
17	71.0	12.46309	33.99782	26.033890	1.494734	−0.000071
18	62.0	9.962971	30.98449	22.260350	1.207800	−0.000002
19	55.0	8.037014	28.63647	19.333420	1.006938	0.000033
20	50.0	6.671205	26.95706	17.247110	0.875414	0.000041
21	43.0	4.772795	24.60278	14.332470	0.708007	−0.000036
22	33.0	2.088211	21.23325	10.181180	0.502679	0.000036
23	31.0	1.555154	20.55850	9.352715	0.466389	0.000019
24	30.0	1.289103	20.22103	8.938724	0.448840	−0.000012

the total system operating cost, represented by the fuel cost required for the system's thermal generation, over the optimization interval. Each hydro plant is constrained by the amount of water available for draw-down in the interval. A prediction of the system's future power demand and water supply is assumed to be available for the optimization interval.

The output of each hydro unit varies with effective head and the rate of water discharged through the turbines. For large capacity reservoirs, it is practical to assume that the effective head is constant over the optimization interval. If the system reservoirs are of small capacities then the variations in effective heads cannot be ignored. Here the short-range, variable-head hydrothermal problem is defined:

Consider an electric power system network having N thermal power generating plants and M hydro power generation plants, where $M + N$ is the total number of generating plants. The basic problem is to find the active power generation of each plant in the system as a function of time over a finite time period from 0 to T.

4.6.1 Thermal Model

The objective function to be minimized is the total system operating cost represented by the fuel cost of thermal generation, over the optimization interval, i.e.

$$J_1 = \int_0^T \sum_{i=1}^{N} F_i(P_i(t))\,dt \tag{4.51}$$

where $F_i(P_i(t))$ is the fuel cost of the ith thermal unit which is a function of the level of power generation at the ith unit and is approximated by

$$F_i(P_i(t)) = a_i P_i^2(t) + b_i P_i(t) + c_i \quad ₹/\text{h} \qquad (i = 1, 2, ..., N) \tag{4.52}$$

with a_i, b_i and c_i as the cost coefficients of the ith generating unit.

4.6.2 Hydro Model

In a hydro system, there is no fuel cost incurred in the operation of hydro units. The input-output characteristic of a hydro generator is expressed by the variation of water discharge $q(t)$ as a function of power output $P_j(t)$ and net head h. According to Glimn-Kirchmayer model, discharge is [Wood and Wollenberg, 1996]:

$$q_j(t) = K_j \psi(h_j(t)\ \phi(P_{j+N}(t)) \quad \text{m}^3/\text{h} \qquad (j = 1, 2, ..., M) \tag{4.53}$$

where ψ and ϕ are two independent functions of head and hydro generations and K_j is a constant of proportionality. These functions are represented by conventional quadratic equations,

$$\phi(P_{mk}) = x_j P_{mk}^2 + y_j P_{mk} + z_j \qquad (m = j + N) \tag{4.54}$$

$$\psi(h_{jk}) = \alpha_j h_{jk}^2 + \beta_j h_{jk} + \gamma_j \tag{4.55}$$

where

x_j, y_j, and z_j are the discharge coefficients of the mth hydro plant.

α_j, β_j, and γ_j are the discharge coefficients of the jth hydro plant.

Each hydro plant is constrained by the amount of water available for the optimization interval.

$$\int_0^T q_j(t)\,dt = V_j \qquad (j = 1, 2, ..., M) \tag{4.56}$$

where V_j is the predefined volume of water available in cubic metres.

4.6.3 Reservoir Dynamics

Reservoir dynamics are obtained by assuming that the jth hydro plant's reservoir has vertical sides and is of small capacity. So, the net head is given by

$$h_j(t) = h_j(0) + \frac{1}{S_j}\int_0^T (I_j(t) - q_j(t))\,dt \tag{4.57}$$

where

$h_j(0)$ is the initial hydraulic head of the jth hydro unit

$I_j(t)$ is the natural inflow of water to the jth hydro unit

$q_j(t)$ is the rate of water discharge at the jth hydro unit

S_j is the surface area of the reservoir.

4.6.4 Equality and Inequality Constraints

(i) Load demand equality constraint:

$$\sum_{i=1}^{N+M} P_i(t) = P_D(t) + P_L(t) \tag{4.58}$$

where

$P_D(t)$ is the load demand during the sub-interval.

$P_L(t)$ is the transmission loss during the sub-interval.

(ii) Limits are imposed as

$$P_i^{\min} \le P_i(t) \le P_i^{\max} \qquad (i = 1, 2 ..., N + M) \tag{4.59}$$

where

$P_i^{\min}$ is the lower limit of the ith generator output

$P_i^{\max}$ is the upper limit of the ith generator output

4.6.5 Transmission Losses

A common approach to model transmission losses in the system is to use the Kron's approximated loss formula in terms of B-coefficients.

$$P_L(t) = B_{00} + \sum_{i=1}^{N+M} B_{i0} P_i(t) + \sum_{i=1}^{N+M} \sum_{j=1}^{N+M} P_i(t)\, B_{ij}\, P_j(t) \text{ MW} \tag{4.60}$$

where B_{ij}, B_{i0} and B_{00} are B-coefficients.

The variable-head hydrothermal problem is framed considering the operating cost over the optimization interval to meet the load demand in each interval. Each hydro plant is constrained by the amount of water available for draw-down in the interval. The problem is defined as

$$\text{Minimize} \qquad J = \int_0^T \sum_{i=1}^{N} F_i(P_i(t))\, dt \tag{4.61a}$$

subject to

$$\sum_{i=1}^{N+M} P_i(t) = P_D(t) + P_L(t) \tag{4.61b}$$

$$\int_0^T q_j(t)\,dt = V_j \qquad (j = 1, 2, ..., M) \tag{4.61c}$$

$$P_i^{\min} \le P_i\,(t) \le P_i^{\max} \qquad (i = 1, 2, ..., N + M) \tag{4.61d}$$

4.6.6 Discrete Form of Short-Range Variable-Head Hydrothermal Scheduling Problem

The problem stated in Eq. (4.61) can be redefined in discrete form as

Minimize
$$J = \sum_{k=1}^{T} \sum_{i=1}^{N} t_k\, F_i(P_{ik}) \tag{4.62a}$$

subject to
$$\sum_{i=1}^{N+M} P_{ik} = P_{Dk} + P_{Lk} \qquad (k = 1, 2, ..., T) \tag{4.62b}$$

$$\sum_{k=1}^{T} t_k\, q_{jk} = V_j \qquad (j = 1, 2, ..., M) \tag{4.62c}$$

$$P_i^{\min} \le P_{ik} \le P_i^{\max} \quad (i = 1, 2\ , ..., N + M; \quad k = 1, 2, ..., T) \tag{4.62d}$$

$$h_{jk+1} = h_{jk} + \frac{t_k}{S_j}(I_{jk} - q_{jk}) \quad (j = 1, 2\ , ..., M; \quad k = 1, 2, ..., T) \tag{4.62e}$$

where

$F_i(P_{ik})$ is the cost function of thermal units in interval k and is defined by

$$F_i(P_{ik}) = a_i P_{ik}^2 + b_i P_{ik} + c_i \text{ ₹/h}$$

with a_i, b_i and c_i as the cost coefficients.

P_{ik} is the output of thermal and hydro units during the kth interval.

P_{Lk} is the transmission losses during the kth interval and is given by

$$P_{Lk} = \sum_{i=1}^{N+M} \sum_{j=1}^{N+M} P_{ik}\, B_{ij}\, P_{jk} + \sum_{i=1}^{N+M} B_{i0}\, P_{ik} + B_{00} \ \text{MW}$$

with B_{ij}, B_{i0}, B_{00} as B-coefficients.

P_{Dk} is the demand during the kth interval.

$P_i^{\min}$ is the lower limit of generations during the kth interval.

$P_i^{\max}$ is the upper limit of generations during the kth interval.

q_{jk} is the rate of discharge from the jth hydro unit in interval k and is defined by

$$q_{jk} = K_j \phi(P_{mk})\ \psi(h_{jk}) \qquad (m = j + N)$$

with

$$\phi(P_{mk}) = x_j P_{mk}^2 + y_j P_{mk} + z_j$$

$$\psi(h_{jk}) = \alpha_j h_{jk}^2 + \beta_j h_{jk} + \gamma_j$$

x_j, y_j and z_j as the discharge coefficients of the *m*th hydro plant.
a_j, β_j and γ_j as the discharge coefficients of the *j*th hydro plant.
K_j as constant of proportionality.
V_j is the pre-specified volume of water available for unit *j* for whole of the period.
h_{jk} is the head of the *j*th hydro unit during the *k*th sub-interval.
I_{jk} is the natural inflow in *j*th hydro plant during the *k*th sub-interval.
N is the number of thermal units.
M is the number of hydro units.
T is the overall period for scheduling.

The above objective as augmented by the constraints is given as

$$L(P_{ik}, \lambda_k, v_j) = \sum_{k=1}^{T} \left[\sum_{i=1}^{N} t_k F_i(P_{ik}) + \sum_{j=1}^{M} v_{jk} t_k q_{jk} + \lambda_k \left[P_{Dk} + P_{Lk} - \sum_{i=1}^{N+M} P_{ik} \right] \right] \quad (4.63)$$

where

λ_k is the incremental cost of power delivered in the system during the *k*th interval.
v_{jk} are the water conversion factors.

The optimality conditions are described by taking the partial derivatives of augmented objective function with respect to the decision variables, $P_{ik}(i = 1, 2, ..., N + M)$, λ_k, $v_{jk}(j = 1, 2, ..., M)$. Thus,

$$t_k \frac{\partial F_i(P_{ik})}{\partial P_{ik}} + \lambda_k \left[\frac{\partial P_{Lk}}{\partial P_{ik}} - 1 \right] = 0 \qquad (i = 1, 2, ..., N; \quad k = 1, 2, ..., T) \quad (4.64a)$$

$$v_{jk} t_k \frac{\partial q_{jk}}{\partial P_{mk}} + \lambda_k \left[\frac{\partial P_{Lk}}{\partial P_{mk}} - 1 \right] = 0 \qquad (j = 1, 2, ..., M; \quad m = N + j; \quad k = 1, 2, ..., T) \quad (4.64b)$$

$$\sum_{k=1}^{T} t_k q_{jk} = V_j \qquad (j = 1, 2, ..., M) \quad (4.64c)$$

$$\sum_{i=1}^{N+M} P_{ik} = P_{Dk} + P_{Lk} \qquad (k = 1, 2, ..., T) \quad (4.64d)$$

$$h_{jk+1} = h_{jk} + \frac{t_k}{S_j} (I_{jk} - q_{jk}) \quad (j = 1, 2, ..., M; \quad k = 1, 2, ..., T) \quad (4.64e)$$

The time-varying water conversion factor, v_{jk}, is defined by

$$v_{jk} = v_{0j} e^{(M_{jk} t_k)} \qquad (j = 1, 2, ..., M; \quad k = 1, 2, ..., T) \quad (4.64f)$$

where

$$M_{jk} = \frac{1}{S_j} \frac{\partial q_{jk}}{\partial h_{jk}}$$

v_{0j} is the water conversion factor
S_j is the surface area of the reservoir.

The function M_{jk} depends on the rate of change of effective head at the hydro plant with the rate of water discharge. In many situations, this rate is a straight line, and therefore, M_{jk} can be assumed constant, M_j. So, the problem becomes like the fixed-head hydrothermal scheduling problem. v_{0j} is varied to meet the water availability demand. The approximate Newton–Raphson method has been applied to find P_{ik} $(i = 1, 2, ..., N + M)$, λ_k for all the T intervals and then the reservoir dynamics is obtained for each interval.

4.6.7 Approximate Newton–Raphson Method for Hydrothermal Generations

Here, it is assumed that v_{0j} is known then from Eq. (4.64a), small deviations from initial values of $P_{ik}(i = 1, 2, ..., N + M)$, λ_k are given by

$$\left(t_k \frac{\partial^2 F_i}{\partial P_{ik}^2} + \lambda_k^0 \frac{\partial^2 P_{Lk}}{\partial P_{ik}^2}\right)\Delta P_{ik} + \lambda_k^0 \sum_{\substack{j=1\\ j\neq i}}^{N+M} \frac{\partial^2 P_{Lk}}{\partial P_{ik}\partial P_{jk}}\Delta P_{jk} + \left(\frac{\partial P_{Lk}}{\partial P_{ik}} - 1\right)\Delta\lambda_k$$

$$= -\left[t_k \frac{\partial F_i}{\partial P_{ik}} + \lambda_k^0\left(\frac{\partial P_{Lk}}{\partial P_{ik}} - 1\right)\right] \quad (i = 1, 2, ..., N) \tag{4.65a}$$

$$\left(v_{jk} t_k \frac{\partial^2 q_{jk}}{\partial P_{mk}^2} + \lambda_k^0 \frac{\partial^2 P_{Lk}}{\partial P_{mk}^2}\right)\Delta P_{mk} + \lambda_k^0 \sum_{\substack{l=1\\ l\neq m}}^{N+M} \frac{\partial^2 P_{Lk}}{\partial P_{mk}\partial P_{lk}}\Delta P_{lk} + \left(\frac{\partial P_{Lk}}{\partial P_{mk}} - 1\right)\Delta\lambda_k$$

$$= -\left[v_{jk} t_k \frac{\partial q_{jk}}{\partial P_{mk}} + \lambda_k^0\left(\frac{\partial P_{Lk}}{\partial P_{mk}} - 1\right)\right] \quad (j = 1, 2, ..., M; \quad m = j + N) \tag{4.65b}$$

$$\sum_{j=1}^{N+M}\left(\frac{\partial P_{Lk}}{\partial P_{jk}} - 1\right)\Delta P_{jk} = -\left(P_{Dk} + P_{Lk}^0 - \sum_{i=1}^{N+M} P_{ik}\right) \tag{4.65c}$$

Equation (4.65a) can be simplified by rearranging the terms, and neglecting the term $\sum_{\substack{j=1\\ j\neq i}}^{N+M} \frac{\partial^2 P_{Lk}}{\partial P_{ik}\partial P_{jk}}\Delta P_{jk}$ and taking $\lambda_k^{\text{new}} = \lambda_k^0 + \Delta\lambda_k$.

$$\left(t_k \frac{\partial^2 F_i}{\partial P_{ik}^2} + \lambda_k^0 \frac{\partial^2 P_{Lk}}{\partial P_{ik}^2}\right)\Delta P_{ik} = -\left[t_k \frac{\partial F_i(P_{ik})}{\partial P_{ik}} + \lambda_k^{\text{new}}\left(\frac{\partial P_{Lk}}{\partial P_{ik}} - 1\right)\right] \quad (i = 1, 2, ..., N) \tag{4.66}$$

or

$$\Delta P_{ik} = \frac{(1 - K_{ik})\lambda_k^{\text{new}} - A_{ik}}{X_{ik}} \quad (i = 1, 2, ..., N) \tag{4.67}$$

where

$$A_{ik} = t_k \frac{\partial F_{ik}}{\partial P_{ik}} \tag{4.67a}$$

$$X_{ik} = t_k \frac{\partial^2 F_{ik}}{\partial P_{ik}^2} + \lambda_k^0 \frac{\partial^2 P_{Lk}}{\partial P_{ik}^2} \tag{4.67b}$$

$$K_{ik} = \frac{\partial P_L}{\partial P_{ik}} \tag{4.67c}$$

Equation (4.65b) can be simplified by rearranging the terms and neglecting the term $\sum_{\substack{l=1 \\ l \neq m}}^{N+M} \frac{\partial^2 P_{Lk}}{\partial P_{mk} \partial P_{lk}} \Delta P_{lk}$ the above equation becomes

$$\left(v_{jk} t_k \frac{\partial^2 q_{jk}}{\partial P_{mk}^2} + \lambda_k^0 \frac{\partial^2 P_{Lk}}{\partial P_{mk}^2} \right) \Delta P_{mk} = -\left[v_j t_k \frac{\partial q_{jk}}{\partial P_{mk}} + \lambda_k^{\text{new}} \left(\frac{\partial P_{Lk}}{\partial P_{mk}} - 1 \right) \right] \quad (j = 1, 2, \ldots, M; \quad m = j + N) \tag{4.68}$$

or

$$\Delta P_{mk} = \frac{(1 - K_{mk}) \lambda_k^{\text{new}} - \beta_{jk}}{Y_{jk}} \qquad (j = 1, 2, \ldots, M; \quad m = j + N) \tag{4.69}$$

where

$$\beta_{jk} = t_k v_{jk} \frac{\partial q_{jk}}{\partial P_{mk}} \tag{4.69a}$$

$$Y_{jk} = t_k v_{jk} \frac{\partial^2 q_{jk}}{\partial P_{mk}^2} + \lambda_k^0 \frac{\partial^2 P_{Lk}}{\partial P_{mk}^2} \tag{4.69b}$$

Substituting Eqs. (4.67) and (4.69) into Eq. (4.65c), we get

$$\sum_{i=1}^{N} (1 - K_{ik}) \left(\frac{\lambda_k^{\text{new}} (1 - K_{ik}) - A_{ik}}{X_{ik}} \right) + \sum_{j=1}^{M} (1 - K_{mk}) \left(\frac{\lambda_k^{\text{new}} (1 - K_{mk}) - \beta_{jk}}{Y_{jk}} \right) = \left(P_{Dk} + P_{Lk}^0 - \sum_{i=1}^{N+M} P_{ik}^0 \right)$$

$$\left(\sum_{i=1}^{N} \frac{(1 - K_{ik})^2}{X_{ik}} + \sum_{j=1}^{M} \frac{(1 - K_{mk})^2}{Y_{jk}} \right) \lambda_k^{\text{new}} = \left(P_{Dk} + P_{Lk}^0 - \sum_{i=1}^{N+M} P_{ik}^0 \right) + \sum_{i=1}^{N} \frac{(1 - K_{ik}) A_{ik}}{X_{ik}} + \sum_{j=1}^{M} \frac{(1 - K_{mk}) \beta_{jk}}{Y_{jk}}$$

$$\lambda_k^{\text{new}} = D_k / C_k \tag{4.70}$$

where

$$C_k = \sum_{i=1}^{N} \frac{(1 - K_{ik})^2}{X_{ik}} + \sum_{j=1}^{M} \frac{(1 - K_{mk})^2}{Y_{ik}} \tag{4.70a}$$

$$D_k = \left(P_{Dk} + P_{Lk}^0 - \sum_{i=1}^{N+M} P_{ik}^0 \right) + \sum_{i=1}^{N} \frac{(1-K_{ik})A_{ik}}{X_{ik}} + \sum_{j=1}^{M} \frac{(1-K_{mk})\beta_{jk}}{Y_{jk}} \tag{4.70b}$$

Using the derivatives, the variables A_{ik}, X_{ik}, K_{ik}, β_{jk} and Y_{jk} are rewritten as

$$A_{ik} = t_k(2a_i P_{ik}^0 + b_i) \tag{4.71a}$$

$$X_{ik} = 2(t_k a_i + \lambda_k^0 B_{ii}) \tag{4.71b}$$

$$K_{ik} = \sum_{j=1}^{N+M} 2B_{ij} P_{jk}^0 + B_{i0} \tag{4.71c}$$

$$\beta_{jk} = t_k v_{jk}\ K_j\ \psi(h_{jk})\ (2x_j P_{mk}^0 + y_j) \qquad (m = j + N) \tag{4.71d}$$

$$Y_{jk} = 2(t_k v_{jk}\ K_j \psi\ (h_{jk})\ x_j + \lambda_k^0\ \beta_{mm}) \qquad (m = j + N) \tag{4.71e}$$

4.6.8 Initial Guess

The power demand is equally distributed among thermal and hydro units during each interval.

$$P_{ik}^0 = \frac{P_{Dk}}{N+M} \qquad (i = 1, 2, ..., N + M; \qquad k = 1, 2, ..., T) \tag{4.72a}$$

Further, it is assumed that there are no transmission losses.

$$\lambda_k^0 = 2a_i P_{ik}^0 + b_i \qquad (k = 1, 2, ..., T) \tag{4.72b}$$

Any thermal unit can be considered as the ith unit.
The water conversion factor can be obtained as

$$v_{j0}^0\ K\psi(h_{jk})\ (2x_j P_{mk}^0 + y_j) = \lambda_k^0 \qquad (j = 1, 2, ..., M)$$

$$v_{0j}^0 = \frac{\lambda_k^0}{K_j\ \psi(h_{jk})\,(2x_j P_{mk}^0 + y_j)} \qquad (j = 1, 2, ..., M; \quad m = N + j) \tag{4.72c}$$

To implement the above procedure to solve the short-range variable-head hydrothermal scheduling problem, the stepwise procedure is outlined below. This procedure is known as iterative procedure.

Algorithm 4.4: Short-Range Variable-Head, Hydrothermal Scheduling Problem

1. Read the number of thermal units N, the number of hydro units M, the number of sub-intervals T, cost coefficients, a_i, b_i, $c_i (i = 1, 2, ..., N)$, B-coefficients, $B_{ij} (i = 1, 2, ..., N + M;$ $j = 1, 2, ..., N + M)$, discharge coefficients, x_i, y_i, z_i $(i = 1, 2, ..., M)$, discharge coefficients, α_i, β_i, γ_i $(i = 1, 2, ..., M)$, surface area of reservoir, S_j $(j = 1, 2, ..., M)$, natural water inflows, I_{jk} $(j = 1, 2, ..., M; k = 1, 2, ..., T)$, demand P_{Dk} $(k = 1, 2, ..., T)$ and pre-specified available water V_j $(j = 1, 2, ..., M)$.
2. Calculate the initial guess values of $P_{ik}^0 (i = 1, 2, ..., N + M)$ λ_k^0 and v_{0j}^0 $(j = 1, 2, ..., M;$ $k = 1, 2, ..., T)$
3. Start the iteration counter, $r = 1$.
4. Start hourly count, k.
5. Compute v_{jk} using Eq. (4.64f).

6. Compute A_{ik} from Eq. (4.71a), X_{ik} from Eq. (4.71b), K_{ik} from Eq. (4.71c), β_{jk} from Eq. (4.71d), Y_{jk} from Eq. (4.71e), C_k from Eq. (4.70a), and D_k from Eq. (4.70b).
7. Calculate $\Delta\lambda_k^{\text{new}}$, ΔP_{ik} $(i = 1, 2, \ldots, N + M)$, using Eqs. (4.70), Eq. (4.67) and Eq. (4.69), respectively.
8. Check the convergence

$$\text{if } \left| \sum_{i=1}^{M} \Delta P_i + \Delta\lambda \right| \le \varepsilon \quad \text{or} \quad \left| \lambda_k^{\text{new}} - \lambda^0 \right| \le \varepsilon \text{ then GOTO Step 11.}$$

9. Calculate new values of P_{ik}^{new} as

$$P_{ik}^{\text{new}} = P_{ik}^{0} + \Delta P_{ik} \qquad (i = 1, 2, ..., N + M)$$

10. Set limits correspondingly as

$$P_{ik}^{\text{new}} = \begin{cases} P_i^{\max} & ;\text{if } P_{ik}^{\text{new}} \ge P_i^{\max} \\ P_i^{\min} & ;\text{if } P_{ik}^{\text{new}} \le P_i^{\min} \\ P_{ik}^{\text{new}} & ;\text{otherwise} \end{cases}$$

Disallow to participate the generation, whose limits has been set either to lower or upper limits, in the scheduling by setting the corresponding variables equal to zero.

11. Compute head variation $h_{jk+1}(j = 1, 2, ..., M)$ from Eq. (4.64e).
12. If $k \ge T$, then GOTO Step 13,

else
$$k = k + t_k,$$
$$P_{ik}^{0} = P_{ik}^{\text{new}} \qquad (i = 1, 2, ..., N + M)$$
$$\lambda_k^0 = \lambda_k^{\text{new}} \text{ and GOTO Step 5 and repeat.}$$

13. Calculate water withdrawals V_i $(i = 1, 2, ..., M)$.
14. If $((\,|\, V_i - V_i^s \,|, \le \varepsilon_2)$ or if $(r \ge R))$
then GOTO Step 15,

else $v_j^{\text{new}} = v_j^{\text{new}} + \dfrac{\Delta v_j}{V_j^s}$, $r = r + 1$; GOTO Step 5 and repeat.

15. Calculate the optimal cost and loss, etc.
16. Stop.

Sample System 4.4

A hydrothermal system is given which consists of two thermal and two hydro generating station. The operating cost of the thermal station is given by

$$F_{1k} = 0.0025P_{1k}^2 + 3.20P_{1k} + 25.0 \text{ ₹/h}$$
$$F_{2k} = 0.0008P_{2k}^2 + 3.40P_{2k} + 30.0 \text{ ₹/h}$$

The variations of rate of discharge of hydro generating station are given by bi-quadratic function of effective head and active power:

$$\phi(P_{3k}) = 0.000216P_{3k}^2 + 0.306P_{3k} + 0.198 \text{ Mft}^3\text{/h}$$

$$\phi(P_{4k}) = 0.000360P_{4k}^2 + 0.612P_{4k} + 0.936 \text{ Mft}^3\text{/h}$$
$$\Psi(h_{1k}) = 0.00001h_{1k}^2 - 0.0030h_{1k} + 0.90 \text{ ft}$$
$$\Psi(h_{2k}) = 0.00002h_{2k}^2 - 0.0025h_{2k} + 0.95 \text{ ft}$$

The reservoirs have small capacity and vertical sides. The water available, surface area and initial height of head are given in Table 4.7.

TABLE 4.7 Reservoir data

Unit no. j	*Volume of water, V_j* (Mft^3)	*Surface area, S_j* (Mft^2)	*Constant, K_j*	*Initial height, h_{j0}* (ft)
1	2850.00	1000.00	1.0	300.0
2	2450.00	400.00	1.0	250.0

The *B*-coefficients of the power system network are given by

$$B = \begin{bmatrix} 0.000140 & 0.000010 & 0.000015 & 0.000015 \\ 0.000010 & 0.000060 & 0.000010 & 0.000013 \\ 0.000015 & 0.000010 & 0.000068 & 0.000065 \\ 0.000015 & 0.000013 & 0.000065 & 0.000070 \end{bmatrix} \text{MW}^{-1}$$

The generation schedule of hydrothermal problem for 24 hours is obtained and results are given below, when hourly demand is known. Table 4.8 represents the operating cost of thermal units, transmission loss and incremental cost, λ_k during 24 hours. Thermal and hydro generations to meet the hourly demand are given in Table 4.9. Variations in head and water conversion factors are presented in Table 4.10.

$$\text{Total operating cost} = ₹\ 69{,}801.08$$
$$v_{01} = 10.45437\ ₹/\text{Mft}^3$$
$$v_{02} = 3.99630\ ₹/\text{Mft}^3$$
$$V_1 = 2850.001 \text{ Mft}^3$$
$$V_2 = 2450.0 \text{ Mft}^3$$

4.7 APPROXIMATE NEWTON–RAPHSON METHOD FOR SHORT-RANGE VARIABLE-HEAD HYDROTHERMAL SCHEDULING

Short-range, variable-head hydrothermal scheduling problem is stated by Eqs. (4.62a–4.62e) in discrete form. The problem stated by Eq. (4.62) is defined as unconstrainted problem and is given as

$$L(P_{ik}, \lambda_k, v_{jk}) = \sum_{k=1}^{T}\left[\sum_{i=1}^{N} t_k F_i + \lambda_k\left[P_{Dk} + P_{Lk} - \sum_{i=1}^{N+M} P_{ik}\right]\right]$$
$$+ \sum_{k=1}^{T}\left[\sum_{j=1}^{M} v_{jk}\left(h_{jk} - h_{jk-1} - \frac{t_k}{S_j} I_{jk} + \frac{t_k}{S_j} q_{jk}\right)\right] \tag{4.73}$$

TABLE 4.8 Operating cost, transmission loss, and λ_k

k	P_{Dk} (MW)	F_k (₹/h)	P_{Lk} (MW)	λ_k (₹/MWh)
1	800.0	1958.288	22.31927	4.212352
2	750.0	1824.227	19.54996	4.153136
3	700.0	1692.264	16.97533	4.094674
4	700.0	1691.458	16.97506	4.094349
5	700.0	1690.652	16.97485	4.094025
6	750.0	1820.731	19.54867	4.151735
7	800.0	1952.392	22.31691	4.209999
8	1000.0	2500.537	35.37994	4.450924
9	1330.0	3472.733	64.20802	4.873655
10	1350.0	3530.056	66.25757	4.898669
11	1450.0	3839.658	77.05112	5.032588
12	1500.0	3994.371	82.78933	5.099574
13	1300.0	3360.388	61.19382	4.826056
14	1350.0	3509.680	66.25397	4.890889
15	1350.0	3504.844	66.25372	4.889043
16	1370.0	3561.997	68.34032	4.914008
17	1450.0	3808.275	77.04802	5.020662
18	1570.0	4189.647	91.21260	5.185397
19	1430.0	3732.734	74.81849	4.988676
20	1350.0	3477.560	66.25736	4.878622
21	1270.0	3228.490	58.26852	4.770926
22	1150.0	2867.947	47.33380	4.614373
23	1000.0	2435.483	35.38715	4.425542
24	900.0	2156.670	28.45605	4.303146

TABLE 4.9 Thermal and hydro generations

k	P_{1k} (MW)	P_{2k} (MW)	P_{3k} (MW)	P_{4k} (MW)
1	152.6337	367.2675	273.1757	29.24221
2	144.3205	341.7363	263.0292	20.46395
3	136.0556	316.3388	252.9392	11.64142
4	136.0034	316.1842	253.1028	11.68409
5	135.9514	316.0297	253.2667	11.72709
6	144.0968	341.0726	263.6768	20.70216
7	152.2604	366.1595	274.1201	29.77642
8	185.4327	467.9445	315.8770	66.12560
9	241.3728	639.0975	386.2605	127.47750
10	244.5627	648.8575	390.5567	132.28080
11	261.6290	700.9557	412.2740	152.19260
12	270.0417	726.6292	423.1396	162.97880

(*Contd.*)

TABLE 4.9 *(Contd.)*

k	P_{1k} (MW)	P_{2k} (MW)	P_{3k} (MW)	P_{4k} (MW)
13	235.0588	619.8972	379.5179	126.71970
14	243.4070	645.4152	390.2769	137.15470
15	243.1324	644.5975	390.2168	138.30700
16	246.3064	654.3096	394.4836	143.24090
17	259.8796	695.7527	411.8091	159.60690
18	280.5305	758.7271	438.0397	183.91530
19	255.7227	683.1213	407.1636	158.81140
20	241.5828	639.9796	389.6908	145.00480
21	227.5740	597.1954	372.3810	131.11820
22	206.9143	534.0148	346.7274	109.67720
23	181.4929	456.1472	315.1064	82.64043
24	164.7030	404.6500	294.3029	64.80006

TABLE 4.10 Variation of head and v_{jk}

k	h_{1k} (ft)	h_{2k} (ft)	v_{1k} (₹/Mft^3)	v_{2k} (₹/Mft^3)
1	300.0000	250.0000	10.45467	3.997735
2	299.9101	249.9246	10.45458	3.997320
3	299.8241	249.8711	10.45449	3.996908
4	299.7419	249.8392	10.45449	3.996910
5	299.6596	249.8071	10.45449	3.996912
6	299.5773	249.7750	10.45458	3.997330
7	299.4911	249.7209	10.45467	3.997758
8	299.4010	249.6443	10.45504	3.999516
9	299.2946	249.4754	10.45571	4.002642
10	299.1594	249.1423	10.45576	4.002884
11	299.0224	248.7965	10.45599	4.003932
12	298.8761	248.3955	10.45611	4.004498
13	298.7252	247.9647	10.45569	4.002551
14	298.5930	247.6361	10.45581	4.003087
15	298.4564	247.2793	10.45582	4.003134
16	298.3199	246.9199	10.45588	4.003381
17	298.1815	246.5474	10.45606	4.004236
18	298.0359	246.1299	10.45635	4.005533
19	297.8788	245.6441	10.45605	4.004155
20	297.7353	245.2307	10.45589	4.003408
21	297.5993	244.8563	10.45573	4.002673
22	297.4706	244.5206	10.45550	4.001570
23	297.3526	244.2429	10.45520	4.000222
24	297.2473	244.0361	10.45502	3.999356

where

λ_k is the incremental cost of power delivered in the system during the kth interval.

v_{jk} are water conversion factors.

The optimality conditions are described by taking the partial derivatives of augmented objective function with respect to the decision variables, P_{ik} (i = 1, 2, ..., N + M), λ_k, v_{jk} (j = 1, 2, ..., M); i.e.

$$t_k \frac{\partial F_i}{\partial P_{ik}} + \lambda_k \left[\frac{\partial P_{Lk}}{\partial P_{ik}} - 1\right] = 0 \qquad (i = 1, 2, ..., N; \quad k = 1, 2, ..., T) \tag{4.74a}$$

$$v_{jk} \frac{t_k \partial q_{jk}}{S_j \partial P_{mk}} + \lambda_k \left[\frac{\partial P_{Lk}}{\partial P_{mk}} - 1\right] = 0 \qquad (j = 1, 2, ..., M; \quad m = N + j; \quad k = 1, 2, ..., T) \tag{4.74b}$$

$$P_{Dk} + P_{Lk} - \sum_{i=1}^{N+M} P_{ik} = 0 \qquad (k = 1, 2, ..., T) \tag{4.74c}$$

$$h_{jk} - h_{jk-1} - \frac{t_k}{S_j} I_{jk} + \frac{t_k}{S_j} q_{jk} = 0 \qquad (j = 1, 2, ..., M; \quad k = 1, 2, ..., T) \tag{4.74d}$$

$$v_{jk}\left(1 + \frac{t_k}{S_j}\frac{\partial q_{jk}}{\partial h_{jk}}\right) - v_{jk+1}\left(1 - \frac{t_{k+1}}{S_j}\frac{\partial q_{jk+1}}{\partial h_{jk}}\right) = 0 \qquad (j = 1, 2, ..., M; \quad k = 1, 2, ..., T) \tag{4.74e}$$

Suppose the initial values of control variables P_{ik}, λ_k, and v_j are known. The updated values of control variables in the next iteration are

$$P_{ik}^{\text{new}} = P_{ik}^0 + \Delta P_{ik} \qquad (i = 1, 2, ..., N + M; \quad k = 1, 2, ..., T)$$

$$\lambda_k^{\text{new}} = \lambda_k^0 + \Delta\lambda_k \qquad (k = 1, 2, ..., T)$$

$$v_{jk}^{\text{new}} = v_{jk}^0 + \Delta v_{jk} \qquad (j = 1, 2, ..., M; \quad k = 1, 2, ..., T)$$

Any small change in control variables from their previous values can be obtained as given below:

$$\left(t_k \frac{\partial^2 F_i}{\partial P_{ik}^2} + \lambda_k^0 \frac{\partial^2 P_{Lk}}{\partial P_{ik}^2}\right)\Delta P_{ik} + \lambda_k^0 \sum_{\substack{j=1\\ j\neq i}}^{N+M} \frac{\partial^2 P_{Lk}}{\partial P_{ik}\partial P_{jk}}\Delta P_{jk} + \left(\frac{\partial P_{Lk}}{\partial P_{ik}} - 1\right)\Delta\lambda_k$$

$$= -\left[t_k \frac{\partial F_i}{\partial P_{ik}} + \lambda_k^0 \left(\frac{\partial P_{Lk}}{\partial P_{ik}} - 1\right)\right] \qquad (i = 1, 2, ..., N) \tag{4.75a}$$

$$\left(v_{jk}^0 \frac{t_k}{S_j}\frac{\partial^2 q_{jk}}{\partial P_{mk}^2} + \lambda_k^0 \frac{\partial^2 P_{Lk}}{\partial P_{mk}^2}\right)\Delta P_{mk} + \lambda_k^0 \sum_{\substack{l=1\\ l\neq m}}^{N+M} \frac{\partial^2 P_{Lk}}{\partial P_{mk}\partial P_{lk}}\Delta P_{lk} + \left(\frac{\partial P_{Lk}}{\partial P_{mk}} - 1\right)\Delta\lambda_k$$

$$+\left(\frac{t_k}{S_j}\frac{\partial Q_{jk}}{\partial P_{mk}}\right)\Delta v_{jk} + \left(v_{jk}^0\frac{t_k}{S_j}\frac{\partial^2 q_{jk}}{\partial P_{mk}\partial h_{jk}}\right)\Delta h_{jk} = -\left[v_{jk}^0\frac{t_k}{S_j}\frac{\partial q_{jk}}{\partial P_{mk}} + \lambda_k^0\left(\frac{\partial P_{Lk}}{\partial P_{mk}} - 1\right)\right]$$

$$(j = 1, 2, ..., M; \quad m = j + N) \tag{4.75b}$$

$$\sum_{j=1}^{N+M}\left(\frac{\partial P_{Lk}}{\partial P_{jk}} - 1\right)\Delta P_{jk} = -\left(P_{Dk} + P_{Lk}^0 - \sum_{i=1}^{N+M} P_{ik}^0\right) \tag{4.75c}$$

$$\left(\frac{t_k}{S_j}\frac{\partial q_{jk}}{\partial P_{mk}}\right)\Delta P_{mk} + \left(1 + \frac{t_k}{S_j}\frac{\partial q_{jk}}{\partial h_{jk}}\right)\Delta h_{jk} - \left(1 - \frac{t_{k+1}}{S_j}\frac{\partial q_{jk+1}}{\partial h_{jk}}\right)\Delta h_{jk}$$

$$= -\left(h_{jk} - h_{jk-1} - \frac{t_k}{S_j}I_{jk} + \frac{t_k}{S_j}q_{jk}\right) \qquad (j = 1, 2, ..., M; \quad m = j + N) \tag{4.75d}$$

$$\left(1 + \frac{t_k}{S_j}\frac{\partial q_{jk}}{\partial h_{jk}}\right)\Delta v_{jk} - \left(1 - \frac{t_{k+1}}{S_j}\frac{\partial q_{jk+1}}{\partial h_{jk}}\right)\Delta v_{jk+1} + \left(v_{jk}^0\frac{t_k}{S_j}\frac{\partial^2 q_{jk}}{\partial h_{jk}\partial P_{mk}}\right)\Delta P_{mk}$$

$$+\left(v_{jk+1}^0\frac{t_{k+1}}{S_j}\frac{\partial^2 q_{jk+1}}{\partial h_{jk}\partial P_{mk+1}}\right)\Delta P_{mk+1} + \left(v_{jk}^0\frac{t_k}{S_j}\frac{\partial^2 q_{jk}}{\partial h_{jk}\partial h_{jk}}\right)\Delta h_{jk} + \left(v_{jk+1}^0\frac{t_{k+1}}{S_j}\frac{\partial^2 q_{jk}}{\partial h_{jk}\partial h_{jk+1}}\right)\Delta h_{jk+1}$$

$$-\left[v_{jk}^0\left(1 + \frac{t_k}{S_j}\frac{\partial q_{jk}}{\partial h_{jk}}\right) - v_{jk+1}^0\left(1 - \frac{t_{k+1}}{S_j}\frac{\partial q_{jk+1}}{\partial h_{jk}}\right)\right] \tag{4.75e}$$

In Eq. (4.75a), the term $\sum_{\substack{j=1\\j\neq i}}^{N+M}\frac{\partial^2 P_{Lk}}{\partial P_{ik}\partial P_{jk}}\Delta P_{jk}$ is neglected. Thus, we get

$$\left(t_k\frac{\partial^2 F_i}{\partial P_{ik}^2} + \lambda_k^0\frac{\partial^2 P_{Lk}}{\partial P_{ik}^2}\right)\Delta P_{ik} = -\left[t_k\frac{\partial F_{ik}}{\partial P_{ik}} + \lambda_k^{\text{new}}\left(\frac{\partial P_{Lk}}{\partial P_{ik}} - 1\right)\right] \qquad (i = 1, 2, ..., N)$$

$$\Delta P_{ik} = \frac{(1 - K_{ik})\,\lambda_k^{\text{new}} - A_{ik}}{\beta_{ik}} \qquad (i = 1, 2, ..., N) \tag{4.76}$$

where

$$A_{ik} = t_k\,\frac{\partial F_{ik}}{\partial P_{ik}} \tag{4.76a}$$

$$\beta_{ik} = t_k\,\frac{\partial F_{ik}^2}{\partial P_{ik}^2} + \lambda_k^0\,\frac{\partial^2 P_{Lk}}{\partial P_{ik}^2} \tag{4.76b}$$

$$K_{ik} = \frac{\partial P_L}{\partial P_{ik}} \tag{4.76c}$$

In Eq. (4.75b), the term $\sum_{\substack{l=1\\l\neq m}}^{N+M} \frac{\partial^2 P_{Lk}}{\partial P_{mk}\partial P_{lk}} \Delta P_{lk}$ is neglected. Thus, we get

$$\left(\nu_{jk}^0 \frac{t_k}{S_j}\frac{\partial^2 q_{jk}}{\partial P_{mk}^2} + \lambda_k^0 \frac{\partial^2 P_{Lk}}{\partial P_{mk}^2}\right)\Delta P_{mk} + \left(\nu_{jk}^0 \frac{t_k}{S_j}\frac{\partial^2 q_{jk}}{\partial P_{mk}\partial h_{jk}}\right)\Delta h_{jk} = -\left[\nu_{jk}^{\text{new}} \frac{t_k}{S_j}\frac{\partial q_{jk}}{\partial P_{mk}} + \lambda_k^{\text{new}}\left(\frac{\partial P_{Lk}}{\partial P_{mk}} - 1\right)\right]$$

$$(j = 1, 2, ..., M; \quad m = j + N)$$

$$\Delta P_{mk} = \frac{(1 - K_{mk})\lambda_k^{\text{new}} - D_{jk}\nu_{jk}^{\text{new}} - E_{jk}\Delta h_{jk}}{C_{jk}} \qquad (j = 1, 2, ..., M; \quad m = j + N) \tag{4.77}$$

where

$$D_{ik} = \frac{t_k}{S_j}\frac{\partial q_{jk}}{\partial P_{mk}} \tag{4.77a}$$

$$C_{jk} = \nu_{jk}\frac{t_k}{S_j}\frac{\partial^2 q_{jk}}{\partial P_{mk}^2} + \lambda_k^0 \frac{\partial^2 P_{Lk}}{\partial P_{mk}^2} \tag{4.77b}$$

$$E_{ik} = \nu_{jk}\frac{t_k}{S_j}\frac{\partial^2 q_{jk}}{\partial P_{mk}\partial h_{jk}} \tag{4.77c}$$

Substituting Eqs. (4.76) and (4.77) into Eq. (4.75c), we have

$$\sum_{i=1}^{N} (1 - K_{ik})\left(\frac{(1 - K_{ik})\lambda_k^{\text{new}} - A_{ik}}{\beta_{ik}}\right) + \sum_{j=1}^{M} (1 - K_{mk})\left(\frac{(1 - K_{mk})\lambda_k^{\text{new}} - D_{jk}\nu_{jk}^{\text{new}} - E_{jk}\Delta h_{jk}}{C_{jk}}\right)$$

$$= P_{Dk} + P_{Lk}^0 - \sum_{i=1}^{N+M} P_{ik}^0$$

or

$$\left[\sum_{i=1}^{N} \frac{(1 - K_{ik})^2}{\beta_{ik}} + \sum_{j=1}^{M} \frac{(1 - K_{mk})^2}{C_{jk}}\right]\lambda_k^{\text{new}} - \sum_{j=1}^{M} \frac{(1 - K_{mk})D_{jk}}{C_{jk}}\nu_{jk}^{\text{new}} - \sum_{j=1}^{M} \frac{(1 - K_{mk})E_{jk}}{C_{jk}}\Delta h_{jk}$$

$$= P_{Dk} + P_{Lk}^0 - \sum_{i=1}^{N+M} P_{ik}^0 + \sum_{i=1}^{N} \frac{(1 - K_{ik})A_{ik}}{\beta_{ik}}$$

or

$$F_k^1 \lambda_k^{\text{new}} = \sum_{j=1}^{M} G_{jk}^1 \nu_{jk}^{\text{new}} + \sum_{j=1}^{M} H_{jk}^1 \Delta h_{jk} + J_k^1 \tag{4.78}$$

where

$$F_k^1 = \sum_{i=1}^{N} \frac{(1-K_{ik})^2}{\beta_{ik}} + \sum_{j=1}^{M} \frac{(1-K_{mk})^2}{C_{jk}} \tag{4.78a}$$

$$G_{jk}^1 = \frac{(1-K_{mk})\,D_{jk}}{C_{jk}} \tag{4.78b}$$

$$H^1{}_{jk} = \frac{(1-K_{mk})\,E_{jk}}{C_{jk}} \tag{4.78c}$$

$$J_k^1 = P_{Dk} + P_{Lk}^0 - \sum_{i=1}^{N+M} P_{ik}^0 + \sum_{i=1}^{N} \frac{(1-K_{ik})\,A_{ik}}{\beta_{ik}} \tag{4.78d}$$

Substituting Eq. (4.77) into Eq. (4.75d)

$$\left(\frac{t_k}{S_j}\frac{\partial q_{jk}}{\partial P_{mk}}\right)\left(\frac{(1-K_{mk})\,\lambda_k^{\text{new}} - D_{jk}\,v_{jk}^{\text{new}} - E_{jk}\Delta h_{jk}}{C_{jk}}\right) + \left(\frac{t_k}{S_j}\frac{\partial q_{jk}}{\partial h_{jk}} + \frac{t_{k+1}}{S_j}\frac{\partial q_{jk+1}}{\partial h_{jk}}\right)\Delta h_{jk}$$

$$= -\left(h_{jk} - h_{jk-1} - \frac{t_k}{S_j} I_{jk} + \frac{t_k}{S_j} q_k\right) \qquad (j = 1, 2, ..., M; \quad m = j + N)$$

or

$$\frac{D_{jk}(1-K_{mk})}{C_{jk}}\,\lambda_k^{\text{new}} - \frac{D_{jk}D_{jk}}{C_{jk}}\,v_{jk}^{\text{new}} + \left(\frac{t_k}{S_j}\frac{\partial q_{jk}}{\partial h_{jk}} + \frac{t_{k+1}}{S_j}\frac{\partial q_{jk+1}}{\partial h_{jk}} - \frac{D_{jk}E_{jk}}{C_{jk}}\right)\Delta h_{jk}$$

$$= -\left(h_{jk} - h_{jk-1} - \frac{t_k}{S_j} I_{jk} + \frac{t_k}{S_j} q_{jk}\right) \qquad (j = 1, 2, ..., M; \quad m = j + N).$$

or

$$F_{jk}^2\lambda_k^{\text{new}} - G_{jk}^2\, v_{jk}^{\text{new}} + H_{jk}^2\,\Delta h_{jk} = L_{jk}^2 \qquad (j = 1, 2, ..., M) \tag{4.79}$$

where

$$F_{jk}^2 = \frac{D_{jk}\,(1-K_{mk})}{C_{jk}} \tag{4.79a}$$

$$G_{jk}^2 = \frac{D_{jk}D_{jk}}{C_{jk}} \tag{4.79b}$$

$$H_{jk}^2 = \frac{t_k}{S_j}\frac{\partial q_{jk}}{\partial h_{jk}} + \frac{t_{k+1}}{S_j}\frac{\partial q_{jk+1}}{\partial h_{jk}} - \frac{D_{jk}E_{jk}}{C_{jk}} \tag{4.79c}$$

$$L_{jk}^2 = -\left(h_{jk} - h_{jk-1} - \frac{t_k}{S_j} I_{jk} + \frac{t_k}{S_j} q_{jk}\right) \tag{4.79d}$$

Solving Eq. (4.75e)

$$\left(1+\frac{t_k}{S_j}\frac{\partial q_{jk}}{\partial h_{jk}}\right)v_{jk}^{\text{new}} - \left(1-\frac{t_{k+1}}{S_j}\frac{\partial q_{jk+1}}{\partial h_{jk}}\right)v_{jk+1}^{\text{new}} + \left(v_{jk}^0\frac{t_k}{S_j}\frac{\partial^2 q_{jk}}{\partial h_{jk}\partial P_{mk}}\right)\Delta P_{mk}$$

$$+\left(v_{jk+1}^0\frac{t_{k+1}}{S_j}\frac{\partial^2 q_{jk+1}}{\partial h_{jk}\partial P_{mk+1}}\right)\Delta P_{mk+1} + \left(v_{jk}^0\frac{t_k}{S_j}\frac{\partial^2 q_{jk}}{\partial h_{jk}\partial h_{jk}}\right)\Delta h_{jk}$$

$$+\left(v_{jk+1}^0\frac{t_{k+1}}{S_j}\frac{\partial^2 q_{jk}}{\partial h_{jk}\partial h_{jk+1}}\right)\Delta h_{jk+1} = 0$$

Substituting Eq. (4.77) into the above equation, we have

$$\left(1+\frac{t_k}{S_j}\frac{\partial q_{jk}}{\partial h_{jk}}\right)v_{jk}^{\text{new}} - \left(1-\frac{t_{k+1}}{S_j}\frac{\partial q_{jk+1}}{\partial h_{jk}}\right)v_{jk+1}^{\text{new}}$$

$$+\left(v_{jk}^0\frac{t_k}{S_j}\frac{\partial^2 q_{jk}}{\partial h_{jk}\partial P_{mk}}\right)\left(\frac{(1-K_{mk})\lambda_k^{\text{new}} - D_{jk}v_{jk}^{\text{new}} - E_{jk}\Delta h_{jk}}{C_{jk}}\right)$$

$$+\left(v_{jk+1}^0\frac{t_{k+1}}{S_j}\frac{\partial^2 q_{jk+1}}{\partial h_{jk}\partial P_{mk+1}}\right)\left(\frac{(1-K_{mk+1})\lambda_{k+1}^{\text{new}} - D_{jk+1}v_{jk+1}^{\text{new}} - E_{jk+1}\Delta h_{jk-1}}{C_{jk+1}}\right)$$

$$+\left(v_{jk}^0\frac{t_k}{S_j}\frac{\partial^2 q_{jk}}{\partial h_{jk}\partial h_{jk}}\right)\Delta h_{jk} + \left(v_{jk+1}^0\frac{t_{k+1}}{S_j}\frac{\partial^2 q_{jk}}{\partial h_{jk}\partial h_{jk+1}}\right)\Delta h_{jk+1} = 0$$

or

$$\left(v_{jk}^0\frac{t_k}{S_j}\frac{\partial^2 q_{jk}}{\partial h_{jk}\partial P_{mk}}\frac{(1-K_{km})}{C_{jk}}\right)\lambda_k^{\text{new}} + \left(1+\frac{t_k}{S_j}\frac{\partial q_{jk}}{\partial h_{jk}} - v_{jk}^0\frac{t_k}{S_j}\frac{\partial^2 q_{jk}}{\partial h_{jk}\partial P_{mk}}\frac{D_{jk}}{C_{jk}}\right)v_{jk}^{\text{new}}$$

$$+v_{jk}^0\frac{t_k}{S_j}\left(\frac{\partial^2 q_{jk}}{\partial h_{jk}\partial h_{jk}} - \frac{\partial^2 q_{jk}}{\partial h_{jk}\partial P_{mk}}\frac{E_{jk}}{C_{jk}}\right)\Delta h_{jk} = -\left(v_{jk+1}^0\frac{t_{k+1}}{S_j}\frac{\partial^2 q_{jk+1}}{\partial h_{jk}\partial P_{mk+1}}\frac{(1-K_{mk+1})}{C_{jk+1}}\right)\lambda_{k+1}^{\text{new}}$$

$$+\left(1-\frac{t_{k+1}}{S_j}\frac{\partial q_{jk+1}}{\partial h_{jk}} + v_{jk+1}^0\frac{t_{k+1}}{S_j}\frac{\partial^2 q_{jk+1}}{\partial h_{jk}\partial P_{mk+1}}\frac{D_{k+1}}{C_{k+1}}\right)v_{jk+1}^{\text{new}}$$

$$-v_{jk+1}^0\frac{t_{k+1}}{S_j}\left(\frac{\partial^2 q_{jk+1}}{\partial h_{jk}\partial h_{jk+1}} - \frac{\partial^2 q_{jk+1}}{\partial h_{jk}\partial P_{mk+1}}\frac{E_{jk+1}}{C_{jk+1}}\right)\Delta h_{jk+1}$$

or

$$F_{jk}^3 \lambda_k^{\text{new}} + (1 + G_{jk}^3) v_{jk}^{\text{new}} + H_{jk}^3 \Delta h_{jk} = -F_{jk+1}^3 \lambda_{k+1}^{\text{new}} + (1 - G_{jk+1}^3)\; v_{jk+1}^{\text{new}} - H_{jk+1}^3\; \Delta h_{jk+1} \tag{4.80}$$

where

$$F_{jk}^3 = v_{jk}^0 \frac{t_k}{S_j} \frac{\partial^2 q_{jk}}{\partial h_{jk} \partial P_{mk}} \frac{(1 - K_{km})}{C_{jk}} \tag{4.80a}$$

$$G_{jk}^3 = \frac{t_k}{S_j} \frac{\partial q_{jk}}{\partial h_{jk}} - v_{jk}^0 \frac{t_k}{S_j} \frac{\partial^2 q_{jk}}{\partial h_{jk} \partial P_{mk}} \frac{D_{jk}}{C_{jk}} \tag{4.80b}$$

$$H_{jk}^3 = v_{jk}^0 \frac{t_k}{S_j} \left(\frac{\partial^2 q_{jk}}{\partial h_{jk} \partial h_{jk}} - \frac{\partial^2 q_{jk}}{\partial h_{jk} \partial P_{mk}} \frac{E_{jk}}{C_{jk}} \right) \tag{4.80c}$$

From Eq. (4.78)

$$\lambda_k^{\text{new}} = \sum_{j=1}^{M} \frac{G_{jk}^1}{F_k^1} v_{jk}^{\text{new}} + \sum_{j=1}^{M} \frac{H_{jk}^1}{F_k^1} \Delta h_{jk} + \frac{J_k^1}{F_k^1} \tag{4.81}$$

Substituting into Eq. (4.79),

$$F_{jk}^2 \left(\sum_{j=1}^{M} \frac{G_{jk}^1}{F_k^1} v_{jk}^{\text{new}} + \sum_{j=1}^{M} \frac{H_{jk}^1}{F_k^1} \Delta h_{jk} + \frac{J_k^1}{F_k^1} \right) - G_{jk}^2 v_{jk}^{\text{new}} + H_{jk}^2 \Delta h_{jk} = L_{jk}^2$$

or

$$\left[-G_{jk}^2 v_{jk}^{\text{new}} + F_{jk}^2 \sum_{j=1}^{M} \frac{G_{jk}^1}{F_k^1} v_{jk}^{\text{new}} \right] + \left[H_{jk}^2 \Delta h_{jk} + F_{jk}^2 \sum_{j=1}^{M} \frac{H_{jk}^1}{F_k^1} \Delta h_{jk} \right] = \left[L_{jk}^2 - F_{jk}^2 \frac{J_k^1}{F_k^1} \right]$$

The above equation can be rewritten in matrix form as

$$R_k v_k^{\text{new}} + Q_k \; \Delta h_k = S_k \tag{4.82}$$

where

$$R_k = -\begin{bmatrix} G_{1k}^2 & & \\ & \cdot & \\ & & G_{mk}^2 \end{bmatrix} + \frac{1}{F_k^1} \begin{bmatrix} F_{1k}^2 \\ \cdot \\ F_{mk}^2 \end{bmatrix} \begin{bmatrix} G_{1k}^1 \\ \cdot \\ G_{mk}^1 \end{bmatrix}^T \tag{4.82a}$$

$$Q_k = \begin{bmatrix} H_{1k}^2 & & \\ & \cdot & \\ & & H_{mk}^2 \end{bmatrix} + \frac{1}{F_k^1} \begin{bmatrix} F_{1k}^2 \\ \cdot \\ F_{mk}^2 \end{bmatrix} \begin{bmatrix} H_{1k}^1 \\ \cdot \\ H_{mk}^1 \end{bmatrix}^T \tag{4.82b}$$

$$S_k = \begin{bmatrix} L_{1k}^2 \\ \cdot \\ L_{mk}^2 \end{bmatrix} - \frac{1}{F_k^1} \begin{bmatrix} F_{1k}^2 \; J_k^1 \\ \cdot \\ F_{mk}^2 \; J_k^1 \end{bmatrix} \tag{4.82c}$$

Substituting Eq. (4.81) into Eq. (4.80),

$$F_{jk}^{3}\left(\sum_{j=1}^{M}\frac{G_{jk}^{1}}{F_{k}^{1}}v_{jk}^{\text{new}}+\sum_{j=1}^{M}\frac{H_{jk}^{1}}{F_{k}^{1}}\Delta h_{jk}+\frac{J_{k}^{1}}{F_{k}^{1}}\right)+(1+G_{jk}^{3})\,v_{jk}^{\text{new}}+H_{jk}^{3}\Delta h_{jk}$$

$$=-F_{jk+1}^{3}\left(\sum_{j=1}^{M}\frac{G_{jk+1}^{1}}{F_{k+1}^{1}}v_{jk+1}^{\text{new}}+\sum_{j=1}^{M}\frac{H_{jk+1}^{1}}{F_{k+1}^{1}}\Delta h_{jk+1}+\frac{J_{k+1}^{1}}{F_{k+1}^{1}}\right)+(1-G_{jk+1}^{3})\,v_{jk+1}^{\text{new}}-H_{jk+1}^{3}\,\Delta h_{jk+1}$$

or

$$\left((1+G_{jk}^{3})\,v_{jk}^{\text{new}}+F_{jk}^{3}\sum_{j=1}^{M}\frac{G_{jk}^{1}}{F_{k}^{1}}v_{jk}^{\text{new}}\right)+\left(H_{jk}^{3}\Delta h_{jk}+\sum_{j=1}^{M}\frac{H_{jk}^{1}}{F_{k}^{1}}\Delta h_{jk}\right)+\left(F_{jk}^{3}\frac{J_{k}^{1}}{F_{k}^{1}}\right)$$

$$=\left((1-G_{jk+1}^{3})\,v_{jk+1}^{\text{new}}-F_{jk+1}^{3}\sum_{j=1}^{M}\frac{G_{jk+1}^{1}}{F_{k+1}^{1}}v_{jk+1}^{\text{new}}\right)-\left(H_{jk+1}^{3}\Delta h_{jk+1}+\sum_{j=1}^{M}F_{jk+1}^{3}\frac{H_{jk+1}^{1}}{F_{k+1}^{1}}\Delta h_{jk+1}\right)$$

$$-\left(F_{jk+1}^{3}\frac{J_{k+1}^{1}}{F_{k+1}^{1}}\right)$$

The above equation can be rewritten in matrix form as

$$(1-T_{k+1})\,v_{k+1}^{\text{new}}-U_{k1}\,\Delta h_{k+1}=(1+T_{k})v_{k}^{\text{new}}+U_{k}\Delta h_{k}+S_{k} \tag{4.83}$$

where

$$T_{k}=\begin{bmatrix}G_{1k}^{3} & & \\ & \cdot & \\ & & G_{mk}^{3}\end{bmatrix}+\frac{1}{F_{k}^{1}}\begin{bmatrix}F_{1k}^{3}\\ \cdot\\ F_{mk}^{3}\end{bmatrix}\begin{bmatrix}G_{1k}^{1}\\ \cdot\\ G_{mk}^{1}\end{bmatrix}^{T} \tag{4.83a}$$

$$U_{k}=\begin{bmatrix}H_{1k}^{3} & & \\ & \cdot & \\ & & H_{mk}^{3}\end{bmatrix}+\frac{1}{F_{k}^{1}}\begin{bmatrix}F_{1k}^{3}\\ \cdot\\ F_{mk}^{3}\end{bmatrix}\begin{bmatrix}H_{1k}^{1}\\ \cdot\\ H_{mk}^{1}\end{bmatrix}^{T} \tag{4.83b}$$

$$S_{k}=-\frac{1}{F_{k+1}^{1}}\begin{bmatrix}F_{1k+1}^{2} & & J_{k+1}\\ & \cdot & \\ F_{mk+1}^{2} & & J_{k+1}\end{bmatrix}-\frac{1}{F_{k}^{1}}\begin{bmatrix}F_{1k}^{2} & & J_{k}^{1}\\ & \cdot & \\ F_{mk}^{2} & & J_{k}^{1}\end{bmatrix} \tag{4.83c}$$

From the solutions of Eqs. (4.82) and (4.83), the values of ΔP_{ik}, $\Delta\lambda_k$ can be obtained by back substitutions of Eq. (4.81), Eq. (4.76) and (4.77), respectively. This is repeated till ΔP_{ik}, tends to zero.

To implement the above procedure to solve the short-range variable-head hydrothermal scheduling problem, the stepwise procedure is outlined below. This procedure is known as iterative procedure.

Algorithm 4.5: Approximate Newton–Raphson Method for Short-Range Variable-Head Hydro-thermal Problem

1. Read the number of thermal units N, the number of hydro units M, the number of sub-intervals T, cost coefficients, a_i, bi, c_i $(i = 1, 2, ..., N)$ B-coefficients, B_{ij} $(i = 1, 2, ..., N + M; j = 1, 2, ..., N + M)$, discharge coefficients, x_i, y_i, z_i $(i = 1, 2, ..., M)$, discharge coefficients, α_i, β_i, γ_i $(i = 1, 2, ..., M)$, surface area of reservoir, S_j $(j = 1, 2, ..., M)$, natural water inflows, I_{jk} $(j = 1, 2, ..., M; k = 1, 2, ..., T)$, demand P_{Dk} $(k = 1, 2, ..., T)$ and pre-specified available water V_j^s $(j = 1, 2, ..., M)$.
2. Perform the short-range fixed-head hydrothermal Algorithm 4.3 to calculate the initial guess values of P^0_{ik} $(i = 1, 2, ..., N + M)$ λ^0_k $(k = 1, 2, ..., T)$. Then compute ν^0_{jk} and h_{jk} for $j = 1, 2, ..., M$; $T = 1, 2, ..., T$ using the following equations.
$$\nu^0_{jk} \frac{t_k}{S_j} \frac{\partial q_{jk}}{\partial P_{mk}} + \lambda^0_k \left[\frac{\partial P_{Lk}}{\partial P_{mk}} - 1 \right] = 0$$
3. Start the iteration counter, $r = 1$.
4. Set the largest value variable for convergence, Big = 0.
5. Start hourly count, $k = 1$.
6. Compute Δh_{jk} $(j = 1, 2, ..., M)$ from Eqs. (4.82) and (4.83).
7. Compute ν_{jk}^{new} using Eq. (4.79).
8. Compute λ_k^{new} using Eq. (4.81).
9. Compute ΔP_{ik} $(i = 1, 2, ..., N + M)$ using Eqs. (4.76) and (4.77), respectively.
10. Check the convergence
$$\text{if } \left| \sum_{i=1}^{N+M} \Delta P_{ik} \right| \geq \text{Big} \quad \text{then} \quad \text{Big} = \left| \sum_{i=1}^{N+M} \Delta P_{ik} \right| \quad \text{and GOTO Step 12.}$$
11. Calculate the new values of P_{ik}^{new} $(i = 1, 2, ..., N + M)$ as
$$P_{ik}^{\text{new}} = P_{ik}^0 + \Delta P_{ik}$$
12. Set limits correspondingly as
$$P_{ik}^{\text{new}} = \begin{cases} P_i^{\max} & ;\text{if } P_{ik}^{\text{new}} \geq P_i^{\max} \\ P_i^{\min} & ;\text{if } P_{ik}^{\text{new}} \leq P_i^{\min} \\ P_{ik}^{\text{new}} & ;\text{otherwise} \end{cases}$$
Disallow to participate the generation, whose limits has been set either to lower or upper limits, in the scheduling by setting the corresponding variables equal to zero.
13. If $k \geq T$, then GOTO Step 14,

    else $k = k + t_k$,
    $P^0_{ik} = P_{ik}^{\text{new}}$ $(i = 1, 2, ..., N + M)$, $\lambda^0_k = \lambda_k^{\text{new}}$
    $\nu^0_{jk} = \nu_{jk}^{\text{new}}$ $(j = 1, 2, ..., M)$, and GOTO Step 6 and repeat.

14. Calculate water withdrawals V_i ($i = 1, 2, ..., M$).
15. If ((Big $\leq \varepsilon$) or ($r \geq R$))

    then GOTO Step 16,

    else $r = r + 1$; GOTO Step 4 and repeat.

16. Calculate the optimal cost and loss etc.
17. Stop.

4.8 HYDRO PLANT MODELLING FOR LONG-TERM OPERATION

Storage plants are associated with reservoirs that have significant storage capacity. During periods of low power requirements, water can be stored and then released when the demand is high. The run-of-river plants have little storage capacity, and use water as it becomes available. Water not utilized is spilled.

Modelling of storage plants, for a long-term study depends on water head variation. For hydro plants in which the water head variation is small, the power generated by the plants can be considered as a constant times the discharge, i.e.

$$P_k = hQ_k \tag{4.84}$$

where

h is a constant (MWh/m^3)

Q_k is the discharge through the turbine during the kth interval (m^3).

But for power systems in which the water head varies by a considerable amount, the head does not remain constant. The average hydro generation during any sub-interval depends on the water discharge through the turbine and on the average head, which is also a function of the storage. The average hydro generation during the kth sub-interval is given by [Kothari and Nagrath, 2003; Nagrath and Kothari, 1994].

$$P_j^k = 9.81 \times 10^{-3}\ \overline{h}_j^k\ (Q_j^k - \mu_j) \text{ MW} \tag{4.85}$$

where

$Q_j^k - \mu_j$ is the effective discharge (m^3/s)

$\overline{h}_j^k$ is the average head in the kth interval

The average head is defined by

$$\overline{h}_j^k = h_j^0 + \frac{\Delta T\,(X_j^{k+1} + X_j^k)}{2A} \tag{4.86}$$

where

A is the area of cross-section of the reservoir at the given storage

h_j^0 is the basic water head corresponding to dead storage

x_j^k is water storage of jth reservoir at the end of the kth interval

ΔT is time length.

Equation (4.86) can be rewritten as

$$\bar{h}_j^k = h_j^0 [1 + 0.5g_j(X_j^{k+1} + X_j^k)] \tag{4.87}$$

where $g_j = \Delta T/Ah_j^0$ is tabulated for various storage values.

Substituting Eq. (4.87) into Eq. (4.85), we get

$$P_j^k = h_j[1+ 0.5g_j(X_j^{k+1} + X_j^k)] (Q_j^k - \mu_j) \tag{4.88}$$

where $h_j = 9.81 \times 10^{-3} h_j^0$.

For run-of-rivers, the storage capacity is small. So with no storage, Eq. (4.88) becomes

$$P_j^k = h_j(Q_j^k - \mu_j) \tag{4.89}$$

The source of energy in a hydro system is the water inflows which can be stored in reservoirs located along the rivers. The inflow of water into a reservoir depends upon the amount of water released from the upstream reservoirs, tributary inflows into the river section between itself and the neighbouring upstream reservoirs and the losses. This value depends upon the amount of rainfall and the geographical connection of any of the rivers associated with hydro schemes.

In a hydro system, the geographical areas with different amounts of rainfall are defined as hydrological areas. There may be several rivers within each hydrological area having their own tributaries and reservoirs located along them. This situation is modelled by using different attributes for each reservoir. Each reservoir in the model belongs to a particular hydrological area and can have any number of upstream reservoirs located at defined distances from it. This produces the water which reaches downstream reservoirs after some time delay. The reservoir storage equations following different locations and arrangements are discussed below:

4.8.1 Hydro Plants on Different Water Streams

We assume here that all hydro plants are on different water streams as shown in Figure 4.9. The reservoir inflows include the tributary inflows and some of it may be lost on its way due to irrigation schemes, drainage, etc. The complete water inflow model for the reservoir is therefore

$$I_j^k = J_j^k - L_j^k \quad (j = 1, 2, ..., M) \tag{4.90}$$

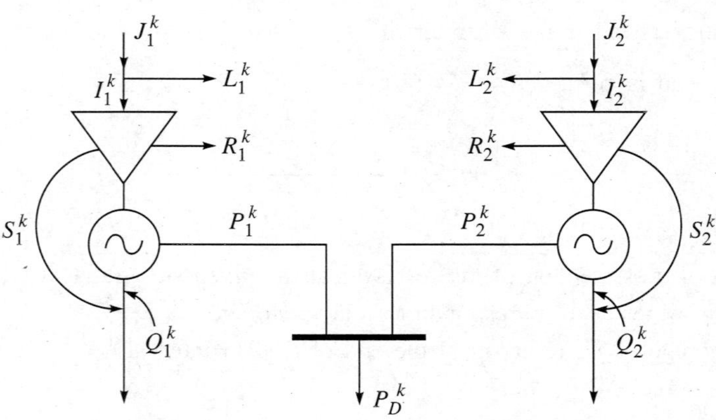

FIGURE 4.9 Hydro plants on different water streams.

where

J_j^k is water inflow into the reservoir during the kth sub-interval

L_j^k is water losses in incoming flow into the reservoir during the kth sub-interval.

The water outflows from the reservoir include the water released for energy production, water spilled due to overflow and water losses due to irrigation schemes, evaporation and other causes. The outflow model is

$$O_j^k = Q_j^k + S_j^k + R_j^k \qquad (j = 1, 2, ..., M) \tag{4.91}$$

where

M is the number of hydro plants

Q_j^k is water discharge from the jth reservoir during the kth sub-interval for production of energy

S_j^k is spillage from the jth reservoir during the kth sub-interval

R_j^k is water losses at the jth reservoir during the kth sub-interval.

The spillage occurs only when the reservoir storage limit is exceeded. The storage of water in the jth reservoir at the beginning of the kth hour is given by

$$X_j^k = X_j^{k-1} + I_j^k - O_j^{k-1} \qquad (j = 1, 2, ..., M) \tag{4.92}$$

where

X_j^k is water storage in the jth reservoir during the kth sub-interval.

By substituting Eqs. (4.90) and (4.91) into Eq. (4.92), the storage at the end of the kth sub-interval can be obtained as

$$X_j^{k+1} = X_j^k + J_j^k - Q_j^k - S_j^k - (L_j^k + R_j^k) \qquad (j = 1, 2, ..., M) \tag{4.93}$$

Substituting Eq. (4.93) into Eq. (4.88), we get

$$P_j^k = h_j[1 + 0.5g_j \{2X_j^k + J_j^k - Q_j^k - S_j^k - (L_j^k + R_j^k)\}] (Q_j^k - \mu_j) \qquad (j = 1, 2, ..., M) \tag{4.94}$$

where

h_j is the basic head of the jth hydro plant

g_j is the water head correction factor to account for variation in head with storage of the jth hydro plant.

μ_j is the non-effective water discharge of the jth hydro plant.

The total volume of water available at the end of the kth sub-interval can be obtained from initial storage by adding Eq. (4.93) for $k = 1, 2, ..., T$. Thus,

$$X_j^{T+1} - X_j^1 + \sum_{k=1}^{T} [Q_j^k + S_j^k + (L_j^k + R_j^k) - J_j^k] = 0 \tag{4.95}$$

4.8.2 Hydro Plants on the Same Water Stream

We now assume that all M, hydro plants are on the same water stream, i.e. in series as shown in Figure 4.10.

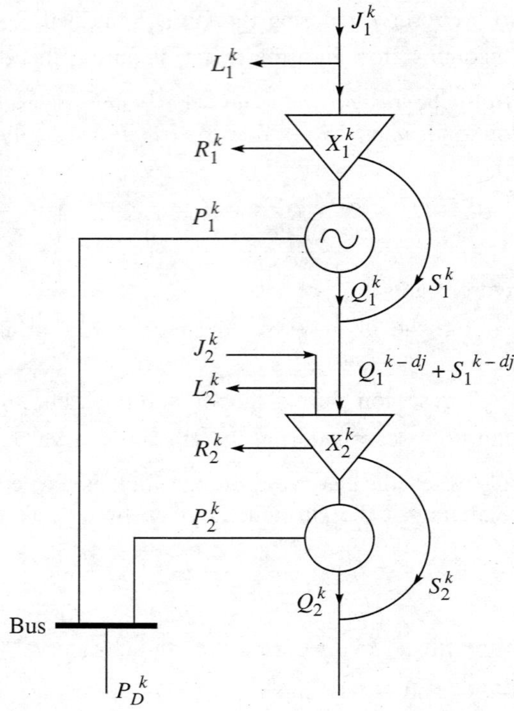

FIGURE 4.10 Hydro plants on the same water stream.

The reservoir inflows include the tributary inflows, and the water released from the upstream reservoir. Water released from the upstream reservoir will reach the reservoir after a time delay and some of it may be lost on its way due to irrigation schemes, drainage, etc. The complete water inflow model for the reservoir is therefore given by

$$I_j^k = J_j^k + Q_{j-1}^{k-dj} + S_{j-1}^{k-dj} - L_j^k \qquad (j = 1, 2, ..., M) \tag{4.96}$$

where

J_j^k is the water inflow into the reservoir during the kth sub-interval

L_j^k is water losses in incoming flow into the reservoir during the kth sub-interval

Q_{j-1}^{k-dj} is the water discharge from the $(j - 1)$th upstream reservoir flowing into the reservoir during the kth sub-interval after a delay time of d_j.

S_{j-1}^{k-dj} is the spillage from the $(j - 1)$th upstream reservoir during the kth sub-interval after a delay time of d_j.

The water outflows from the reservoir include the water released for energy production, water spilled due to overflow and water losses due to irrigation schemes, evaporation and other causes. The outflow model is therefore given by

$$O_j^{k-1} = Q_j^{k-1} + S_j^{k-1} + R_j^{k-1} \qquad (j = 1, 2, ..., M) \tag{4.97}$$

where

M is the number of hydro plants

Q_j^k is water discharge from the jth reservoir during the kth sub-interval for production of energy

S_j^k is spillage from the jth reservoir during the kth sub-interval

R_j^k is water losses at the jth reservoir during the kth sub-interval.

The storage of water in the jth reservoir at beginning of the kth hour is given by

$$X_j^k = X_j^{k-1} + I_j^k - O_j^{k-1} \qquad (j = 1, 2, ..., M) \tag{4.98}$$

where X_j^k is water storage in the jth reservoir during the kth sub-interval.

By substituting Eqs. (4.96) and (4.97) into Eq. (4.98), the storage at the end of the kth sub-interval can be obtained as

$$X_j^{k+1} = X_j^k + J_j^k - Q_j^k + Q_{j-1}^{k-dj} - S_j^k + S_{j-1}^{k-dj} - (L_j^k + R_j^k) \qquad (j = 1, 2, ..., M) \tag{4.99}$$

Total volume of water available at the end of the kth sub-interval can be obtained from initial storage by adding Eq. (4.99) for $k = 1, 2, ..., T$. Thus,

$$X_j^{T+1} - X_j^1 + \sum_{k=1}^{T} [Q_j^k + S_j^k + (L_j^k + R_j^k) - J_j^k] - \sum_{k=1}^{T-d_j} (Q_{j-1}^k + S_{j-1}^k) = 0 \tag{4.100}$$

4.8.3 Multi-Chain Hydro Plants

The hydro plants may be on the same water stream or on different streams as shown in Figure 4.11. The system consists of M rivers. The rivers may or may not be independent of each other, each with one or several reservoirs and power plants in series and interconnected to neighbouring systems. The reservoir inflows include the tributary inflows, and water released from the upstream reservoirs. Water from the upstream reservoirs will reach the reservoir after a time delay. Some of it may be lost on its way due to irrigation schemes, drainage, etc. The complete water inflow model for reservoir is therefore given by

$$I_j^k = J_j^k + \sum_{u=1}^{M_u} [Q_u^{k-d_u} + S_u^{k-d_u}] - L_j^k \qquad (i = 1, 2, ..., M) \tag{4.101}$$

where

J_{ij}^k is the water inflow into the jth reservoir during the kth sub-interval

L_{ij}^k is the water losses in the incoming flow into the jth reservoir during the kth sub-interval

$Q_u^{k-d_u}$ is the water discharge from the uth immediate upstream reservoir during the kth sub-interval after a delay time of d_u.

$S_u^{k-d_u}$ is spillage from the uth immediate upstream reservoir during the kth sub-interval after a delay time of d_u.

M_u is number of immediate upstream reservoirs.

The water outflows from a reservoir include the water released for energy production, water spilled due to overflow and water losses due to irrigation schemes, evaporation and other causes. The outflow model is

$$O_j^{k-1} = Q_j^{k-1} + S_j^{k-1} + R_j^{k-1} \qquad (j = 1, 2, ..., M) \tag{4.102}$$

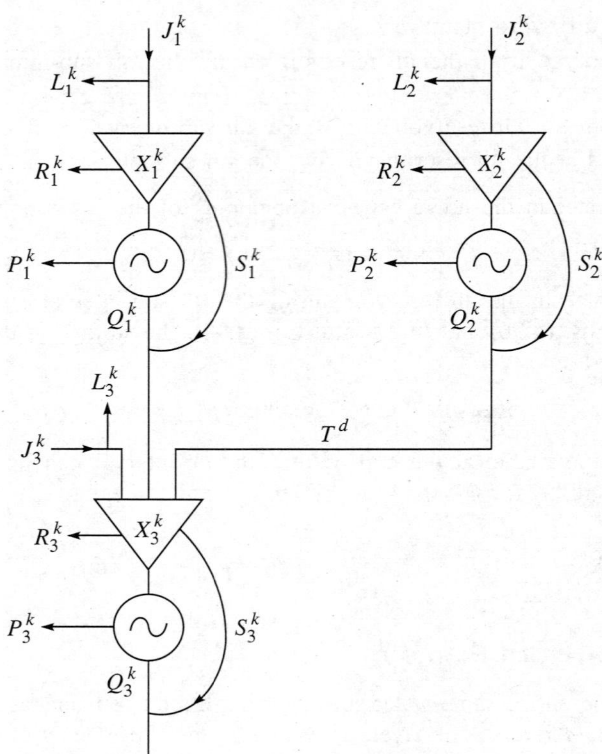

FIGURE 4.11 Multi-chain hydro plants.

where

M is the number of hydro plants

Q_j^k is the water discharge from the jth reservoir during the kth sub-interval.

S_j^k is the spillage from the jth reservoir during the kth sub-interval

R_j^k is the water losses at the jth reservoir during the kth sub-interval.

The storage of water in the jth reservoir at the beginning of the kth hour is given by

$$X_j^k = X_j^{k-1} + I_j^k - O_j^{k-1} \quad (j = 1, 2, ..., M) \tag{4.103}$$

where X_j^k is water storage in the jth reservoir during the kth sub-interval.

By substituting Eqs. (4.101) and (4.102) into Eq. (4.103), the storage at the end of the kth sub-interval can be obtained as

$$X_j^{k+1} = X_j^k + J_j^k - Q_j^k - S_j^k + Z_j^k - (L_j^k + R_j^k) \quad (j = 1, 2, ..., M) \tag{4.104}$$

$$Z_j^k = \sum_{u=1}^{M_u} (Q_u^{k-d_u} + S_u^{k-d_u})$$

Total volume of water available at the end of the kth sub-interval can be obtained from the initial storage by adding Eq. (4.104) for $k = 1, 2, ..., T$.

$$X_j^{T+1} - X_j^1 + \sum_{k=1}^{T} [Q_j^k + S_j^k + (L_j^k + R_j^k) - J_j^k + Z_j^k] = 0 \tag{4.105}$$

4.8.4 Pumped Storage Plants

Figure 4.12 shows the system consisting of pumped storage plants. The model for pumping units considers reversible turbines only. The unit is assumed to operate at full rate, pumping water from the lowest reservoir to the upper reservoir. It is represented by a similar model to that of turbine generator.

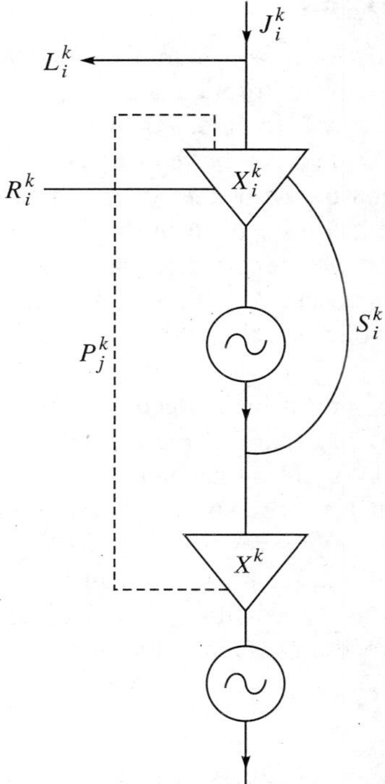

FIGURE 4.12 Pumped storage hydro system.

The power generated during generation mode is

$$P_j^k = \frac{h_j^p}{\eta_g} [1 + 0.5g_j(X_j^k + X_j^{k-1})] (Q^k{}_j - \mu_j) \tag{4.106}$$

where

h_j^p is the basic head of the pumped hydro plant

η_g is the efficiency of pumped hydro plant in generation mode

The power utilized during the pumping mode is

$$P_j^k = \frac{h_j^p}{\eta_p}\,[1 + 0.5g_j\,(X_j^k + X_j^{k-1})]\,(Q_j^k - \mu_j) \tag{4.107}$$

where η_p is the efficiency of pumped hydro plant in pumping mode.

The water storage at the end of the kth sub-interval can be obtained as

$$X_j^{k+1} = X_j^k + J_j^k - Q_j^k - S_j^k + Z_j^k - (L_j^k + R_j^k) \qquad (j = 1, 2, ..., M) \tag{4.108}$$

where Q_j^k is negative during the pumping mode and positive during the generation mode.

4.9 LONG-RANGE GENERATION SCHEDULING OF HYDROTHERMAL SYSTEMS

A modern power system may consist of a large number of thermal and conventional hydro power plants connected to various load centres through a lossy transmission network. Since there is insignificant incremental cost involved in hydro generation, the problem of minimizing the operational cost of a hydrothermal system can be reduced essentially to that of minimizing the fuel cost for thermal plants under the constraints of the water available for hydro generation in a given period of time. Considerable work has been done in the area of hydrothermal optimal scheduling and a number of study results have been reported [Agarwal and Nagrath, 1972; Rao, Prabhu, and Aggarwal, 1975; Mohan, Kuppusamy, and Abdullah, 1992] on this problem, with the assumption, that the water inflows to the reservoirs and the load demands are known with complete certainty. Nevertheless, this is not true.

The availability of limited amount of hydroelectric energy, in the form of stored water in the system reservoirs makes the optimal operation complex, because of the link between an operating decision in a given stage and the future consequences of this decision in subsequent stages. Further, it is impossible to have perfect forecasts of the future inflow sequence as well as the load variation during a given period. Therefore, for long-term storage regulation, it becomes necessary to account for the random nature of the load and river inflows.

A hydrothermal system is considered with N thermal and M hydro plants. The problem is visualized as a T stage decision process by subdividing the planning period into T sub-intervals.

4.9.1 Fuel Cost

The aim is to optimize the running cost of thermal stations with full utilization of water available during the optimization period. The objective function, which is fuel cost of the thermal plant, is assumed to be approximated by a quadratic function of generator power output and is given as

$$F = \sum_{k=1}^{T}\left[\sum_{i=1}^{N} a_i (P_i^k)^2 + b_i P_i^k + c_i\right] \tag{4.109}$$

where

a_i, b_i, and c_i are cost coefficients

P_i^k is the thermal power generation during the kth sub-interval

4.9.2 Water Storage Equation

By assuming that all hydro plants are on different water streams, the storage at the end of *k*th sub-interval can be obtained, from Eq. (4.110) given below.

$$X_j^{k+1} = X_j^k + J_j^k - Q_j^k - S_j^k \qquad (j = 1, 2, ..., M) \tag{4.110}$$

where

X_j^k is water storage for the *j*th turbine during the *k*th sub-interval

J_j^k is water inflow into the reservoir for the *j*th turbine during the *k*th sub-interval

Q_j^k is water discharge through the *j*th turbine during the *k*th sub-interval

S_j^k is spillage from the *j*th turbine during the *k*th sub-interval.

4.9.3 Hydro Generation

The average hydro generation during any sub-interval depends on the water discharge through the turbine and on the average head, which is also a function of the storage. The average hydro generation during the *k*th sub-interval is given by [5,8]

$$P_{j+N}^k = h_j\,[1 + 0.5g_j(2X_j^k + J_j^k - Q_j^k - S_j^k)]\,(Q_j^k - u_j) \qquad (j = 1, 2, ..., M) \tag{4.111}$$

where

h_j is the basic head of the *j*th hydro plant

g_j is the water head correction factor to account for variation in head with storage of the *j*th hydro plant

u_j is the non-effective water discharge of the *j*th hydro plant

X_j^k is the water storage for the *j*th turbine during the *k*th sub-interval.

J_j^k is the water inflow into reservoir for the *j*th turbine during the *k*th sub-interval

Q_j^k is the water discharge through the *j*th turbine during the *k*th sub-interval

S_j^k is the spillage from the *j*th turbine during the *k*th sub-interval.

4.9.4 Power Balance Equation

The power balance equation for the *k*th sub-interval is

$$\sum_{i=1}^{M+N} P_i^k - P_D^k - P_L^k = 0 \tag{4.112}$$

where

P_D^k is the load demand during the *k*th sub-interval

P_L^k is transmission losses during the *k*th sub-interval.

The transmission power loss expressed through the well known loss formula expression is given by [Kothari and Nagrath, 2003; Nagrath and Kothari, 1994]

$$P_L^k = B_{00} + \sum_{i=1}^{M+N} B_{i0}P_i^k + \sum_{i=1}^{M+N}\sum_{j=1}^{M+N} P_i^k B_{ij} P_j^k \tag{4.113}$$

where B_{ij}, B_{i0}, and B_{00} are *B*-coefficients.

Equality and inequality constraints

(a) Fore-bay limit of reservoir

$$X_j^{\min} \le X_j^k \le X_j^{\max} \qquad (j = 1, 2, ..., M) \tag{4.114}$$

(b) Water discharge limit

$$Q_j^{\min} \le Q_j^k \le Q_j^{\max} \qquad (j = 1, 2, ..., M) \tag{4.115}$$

(c) Output of thermal plants

$$P_i^{\min} \le P_i^k \le P_i^{\max} \qquad (i = 1, 2, ..., N) \tag{4.116}$$

(d) Total volume of water available at the end of kth sub-interval can be obtained from initial storage as follows.

$$X_j^{T+1} - X_j^1 - \sum_{k=1}^{T} J_j^k + \sum_{k=1}^{T} Q_j^k + \sum_{k=1}^{T} S_j^k = 0 \quad (j = 1, 2, ..., M) \tag{4.117}$$

Aggregating the above equations, the hydrothermal multiobjective optimization problem is defined below:

Minimize

$$F = \sum_{k=1}^{T} \left[\sum_{i=1}^{N} a_i (P_i^k)^2 + b_i P_i^k + c_i \right] \tag{4.118a}$$

subject to:

(i) Load demand constraint

$$\sum_{i=1}^{M+N} P_i^k - P_D^k - P_L^k = 0 \tag{4.118b}$$

(ii) Storage continuity constraint

$$X_j^{k+1} = X_j^k + J_j^k - Q_j^k - S_j^k \qquad (j = 1, 2, ..., M) \tag{4.118c}$$

(iii) Total volume of water available constraint

$$X_j^{T-1} - X_j^1 - \sum_{k=1}^{T} J_j^k + \sum_{k=1}^{T} Q_j^k + \sum_{k=1}^{T} S_j^k = 0 \quad (j = 1, 2, ..., M) \tag{4.118d}$$

(iv) Hydro generation equation

$$P_{j+N}^k = h_j[1 + 0.5g_j(2X_j^k + J_j^k - Q_j^k - S_j^k)]\,(Q_j^k - u_j) \qquad (j = 1, 2, ..., M) \tag{4.118e}$$

$$X_j^{\min} \le X_j^k \le X_j^{\max} \qquad (j = 1, 2, ..., M) \tag{4.118f}$$

$$Q_j^{\min} \le Q_j^k \le Q_j^{\min} \qquad (j = 1, 2, ..., M) \tag{4.118g}$$

$$P_i^{\min} \le P_i^k \le P_i^{\max} \qquad (i = 1, 2, ..., N) \tag{4.118h}$$

4.9.5 Optimal Control Strategy

The scalarized optimization problem is further solved by forming the Lagrangian function, which is obtained by augmenting the objective function with various equality constraints expressed in terms of the expected values, through dual variables as follows:

$$L = \sum_{k=1}^{T}\left[\sum_{i=1}^{N}(a_i(P_i^k)^2 + b_iP_i^k + c_i) + \lambda_1^k\left\{P_D^k + P_L^k - \sum_{i=1}^{N+M} P_i^k\right\}\right. \tag{4.119}$$

$$+\sum_{j=1}^{M}\lambda_{2j}^k\left\{P_{j+N}^k - h_j(1+0.5g_j(2X_j^k + J_j^k - Q_j^k - S_j^k))(Q_j^k - u_j)\right\}$$

$$\left. + \lambda_{3j}^k(X_j^{k+1} - X_j^k - J_j^k + Q_j^k + S_j^k)\right]$$

$$+\sum_{j=1}^{M}\lambda_{4j}\left(X_j^{T+1} - X_j^1 - \sum_{k=1}^{T}J_j^k + \sum_{k=1}^{T}Q_j^k + \sum_{k=1}^{T}S_j^k\right)$$

The control variables are the water discharges through the turbines of hydro plants, during each sub-interval. The reservoir storage and the hydro power generation at the end of the sub-interval are obtained from Eqs. (4.110) and (4.111), respectively. Irrespective of the hydro generations, the thermal generations satisfy the power transfer constraint to achieve the minimal fuel cost.

The dual variables λ_1^k, λ_{2j}^k and λ_{3j}^k are obtained by equating the partial derivatives (of the Lagrangian function with respect to the dependent variables) to zero, i.e.

$$\frac{\partial L}{\partial P_i^k} = (2a_iP_i^k + b_i) + \lambda_1^k\left[\frac{\partial P_L^k}{\partial P_i^k} - 1\right] = 0 \qquad (i = 1, 2, ..., N) \tag{4.120}$$

$$\frac{\partial L}{\partial P_{j+N}^k} = \lambda_{2j}^k + \lambda_1^k\left[\frac{\partial P_L^k}{\partial P_{j+N}^k} - 1\right] = 0 \qquad (j = 1, 2, ..., M) \tag{4.121}$$

$$\left.\frac{\partial L}{\partial X_j^k}\right|_{k\neq 1} = \lambda_{3j}^{k-1} - \lambda_{3j}^k - \lambda_{2j}^k\, h_jg_j\,(Q_j^k - u_j) \qquad (j = 1, 2, ..., M) \tag{4.122}$$

$\lambda_{3j}^1 = 0$, since the equations associated with this set of dual variables are redundant.

$$\frac{\partial L}{\partial Q_j^k} = \lambda_{3j}^k + \lambda_{4j} - \lambda_{2j}^k h_j[1 + 0.5g_j(2X_j^k + J_j^k - S_j^k - 2Q_j^k + u_j)] \qquad (j = 1, 2, ..., M) \tag{4.123}$$

The dual variables λ_{4j} are to be adjusted to maximize the Lagrangian function under the constraints of other optimality conditions. The corresponding gradient vector is

$$\frac{\partial L}{\partial \lambda_{4j}} = X_j^{T+1} - X_j^1 - \sum_{k=1}^{T}J_j^k + \sum_{k=1}^{T}Q_j^k + \sum_{k=1}^{T}S_j^k \qquad (j = 1, 2, ..., M) \tag{4.124}$$

The upper and lower limits on the control variables are taken care by making these variables equal to the respective bounded values, whenever such limits are violated. For the dependent variables, these limits can be considered by augmenting the cost function through the Powell's Penalty functions [Agarwal and Nagrath, 1972].

Algorithm 4.6: Long-Range Generation Scheduling of Hydrothermal System

The steps of the algorithm are given below:

1. Assume a set of λ_{4j} for (j = 1, 2, ..., M).
2. Set k = 1, $\lambda_{3j}^k = 0$ (j = 1, 2, ..., M).
3. Assume/modify Q_j^k (j = 1, 2, ..., M)..
4. Calculate hydro generations P_l^k (j = 1, 2, ..., L; $l = j + N$) from Eqs. (4.111).
5. Calculate thermal generations P_i^k (i = 1, 2, ..., N) and λ_1^k by solving the following set of equations.

$$\frac{\partial L}{\partial P_i^k} = (2a_i P_i^k + b_i) + \lambda_1{}^k \left[\frac{\partial P_L^k}{\partial P_i^k} - 1\right] = 0 \qquad (i = 1, 2, ..., N)$$

$$\sum_{i=1}^{M+N} P_i^k - P_D^k - P_L^k = 0$$

The above set of equations is solved by an efficient method which is a less time consuming algorithm (see Algorithm 3.5).

6. λ_{2j}^k and λ_{3j}^k for (j = 1, 2, ..., L) are solved using Eqs. (4.121) and (4.122), respectively.

$$\lambda_{2j}^k = \lambda_1{}^k \left[1 - \frac{\partial P_L^k}{\partial P_{j+N}^k}\right] = 0 \qquad (j = 1, 2, ..., M)$$

$$\lambda_{3j}^k = \lambda_{3j}^{k-1} - \lambda_{2j}^k - h_j g_j (Q_j^k - u_j) \qquad (j = 1, 2, ..., M)$$

7. Calculate the gradient vector from Eq. (4.123). If the optimality conditions are achieved within the prescribed accuracy then GOTO Step 8, else adjust the water discharges using the conjugate gradient method [Agarwal and Nagrath, 1972] and GOTO Step 4.
8. Calculate the values of storages from Eqs. (4.110) for all the hydro generators.

$$X_j^{k+1} = X_j^k + J_j^k - Q_j^k - S_j^k \qquad (j = 1, 2, ..., M)$$

9. If ($k \geq T$) then GOTO Step 10,

   else set $k = k + 1$ and GOTO Step 3.

10. Check the convergence using the gradient vector given by Eq. (4.124). If convergence is not achieved then adjust λ_{4j} using the steepest ascent method.

$$\frac{\partial L}{\partial \lambda_{4j}} = X_j^{T+1} - X_j^1 - \sum_{k=1}^{T} J_j^k + \sum_{k=1}^{T} Q_j^k + \sum_{k=1}^{T} S_j^k \qquad (j = 1, 2, ..., M)$$

$$\lambda_{4j} = \lambda_{4j} + s\left(\frac{\partial L}{\partial \lambda_{4j}}\right) \qquad (j = 1, 2, ..., M), \quad \text{where } s \text{ is step size.}$$

and GOTO Step 2 else Stop.

To implement the conjugate gradient method, the stepwise procedure is outlined below:

Algorithm 4.6.1: Conjugate Gradient Method to Compute Optimal Discharge

1. Start with the initial point Q^1

   where $Q^1 = [Q_{1k} \quad Q_{2k} \cdots Q_{mk}]^T$
2. Set the direction $S^1 = -\nabla L_Q = -\nabla L_Q^1$

   where $$\nabla L_Q = \left[\frac{\partial L}{\partial Q_{1k}} \quad \frac{\partial L}{\partial Q_{2k}} \quad \cdots \quad \frac{\partial L}{\partial Q_{mk}}\right]^T$$
3. Find Q^2 according to the relation

   $$Q^2 = Q^1 + \mu^* S^1$$

   where μ^* is the optimal step length in the direction S^1. It is evaluated by the direct root method explained in Algorithm 4.6.2.
4. Set, $i = 2$.
5. Evaluate ∇L_Q^i and S^i

   $$S^i = -\nabla L_Q^i + \frac{|\nabla L_Q^i|^2}{|\nabla L_Q^{i-1}|^2} S^{i-1}$$
6. Compute the optimum step length μ^* in the direction S^i and compute the new point

   $$Q^{i+1} = Q^i + \mu^* S^i$$
7. Check the convergence.
   If $|\nabla L_Q^i| \le \varepsilon$ or $|Q^{i+1} - Q^i| \le \varepsilon$
   then GOTO Step 8
   else $i = i + 1$ and GOTO Step 5 and repeat.
8. Stop, solution is optimal.

4.9.6 Direct Root Method

In this method, the derivative of L is linearized rather than the function. This method is also known as the Regula falsi technique in numerical method and is used to find the root of an equation.

Let

$$\mu = A \text{ and its slope } L'_A = \left.\frac{dL_Q}{d\mu}\right|_{\mu=A} \quad \text{be known}$$

$$\mu = B \text{ and its slope } L'_B = \left.\frac{dL_Q}{d\mu}\right|_{\mu=B} \quad \text{be known}$$

Assume that slope $L'(\mu)$ varies linearly, i.e.

$$L'_{\mu} = a + b_{\mu} = 0 \quad \text{or} \quad \mu = -\frac{a}{b} \tag{4.125}$$

Form two equations to find a and b points from Eq. (4.125), i.e.

$$L'_A = a + bA \tag{4.126}$$

$$L'_B = a + bB \tag{4.127}$$

Subtracting Eq. (4.127) from Eq. (4.126),

$$b = \frac{L'_A - L'_B}{A - B} \tag{4.128}$$

Substituting the value of b into Eq. (4.126) and solving for a, we get

$$a = \frac{L'_B A - L'_A B}{A - B} \tag{4.129}$$

Dividing Eq. (4.129) by Eq. (4.128), we get

$$\mu^* = \frac{L'_B A - L'_A B}{L'_A - L'_B} \tag{4.130}$$

To implement the direct root method, the stepwise procedure is outlined here

Algorithm 4.6.2: Direct Root Method to Compute Optimal Step

1. Choose the initial value t and set $A = 0$,

   compute $L'_A = \left.\dfrac{dL_Q}{d\mu}\right|_{\mu=A}$

2. Search B by setting $B = t$, such that

   $$L'_B = \left.\frac{dL_Q}{d\mu}\right|_{\mu=B} > 0 \text{ if yes GOTO Step 3,}$$

   otherwise update t ($t = 2 \times t$) and GOTO Step 2 and repeat.
3. Compute μ^* from Eq. (4.130).
4. If $L'(\mu^*) \leq \varepsilon$ then stop.
5. If $L'(\mu^*) < 0$ then set $A = \mu^*$, $L'_A = L'(\mu^*)$ GOTO Step 3 and repeat.
6. If $L'(\mu^*) > 0$ then set $B = \mu^*$, $L'_A = L'(\mu^*)$ GOTO Step 3 and repeat.

EXAMPLE 4.3 Consider a fundamental hydrothermal system shown in Figure 4.13. The objective is to find the optimal generation schedule for a typical day, wherein the load varies in three steps of eight hours each as 7 MW, 10 MW, and 5 MW, respectively. There is no water inflow into the reservoir of the hydro plant. The spillage also does not occur. The initial water storage in the reservoir is 100 m^3/s and the final water storage should be 60 m^3/s, i.e. the total water available for hydro generation during the day is 40 m^3/s.

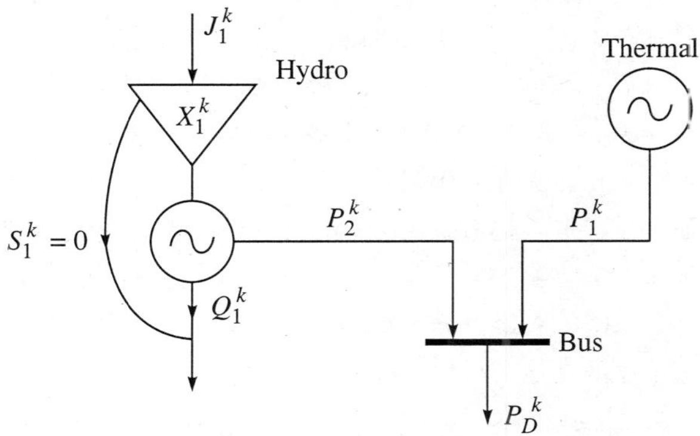

FIGURE 4.13 Hypothetical system.

Basic head is 20 m. Water head correction factor is given to be 0.005. Assume for simplicity that the reservoir is rectangular so that the head correction factor does not change with water storage. Let the non-effective water discharge be assumed as 2 m^3/s. Incremental fuel cost of the thermal plant is

$$\frac{\partial F}{\partial P_1} = 1.0P_1 + 25.0 \text{ ₹/h}$$

Further, the transmission losses are also neglected. (The problem is a hypothetical problem.)

Solution Assume $\lambda_{41} = 8.0$ and

$$Q_1^1 = 10 \text{ m}^3\text{/s}, \quad Q_1^2 = 15 \text{ m}^3\text{/s}, \quad Q_1^3 = 15 \text{ m}^3\text{/s}$$

It is given that, $h_1 = 9.81 \times 10^{-3} \times 20 = 0.1962$ and the head correction factor, $g_1 = 0.005$

$$J_1^1 = J_1^2 = J_1^3 = 0 \text{ m}^3\text{/s}$$

$$S_1^1 = S_1^2 = S_1^3 = 0 \text{ m}^3\text{/s}$$

$$X_1^1 = 100 \text{ m}^3\text{/s}, X_1^4 = 60 \text{ m}^3\text{/s}$$

From Eq. (4.110)

$$X_1^2 = X_1^1 + J_1^1 - Q_1^1 - S_1^1 = 100 - 10 = 90 \text{ m}^3\text{/s}$$

$$X_1^3 = X_1^2 + J_1^2 - Q_1^2 - S_1^2 = 90 - 15 = 75 \text{ m}^3\text{/s}$$

The values of hydro generations in the sub-intervals can be obtained using Eq. (4.111), i.e.

$$\begin{aligned} P_2^1 &= h_1[1 + 0.5g_1(X_1^2 + X_1^1)]\ (Q_1^1 - \mu_1) \\ &= 0.1962\ [1+ 0.5 \times 0.005\ (90.0 + 100.0)]\ (10.0 - 2.0) \\ &= 2.31516 \text{ MW} \end{aligned}$$

$$P_2^2 = h_1[1 + 0.5g_1(X_1^3 + X_1^2)]\ (Q_1^2 - \mu_1) = 3.60272 \text{ MW}$$

$$P_2^3 = h_1[1 + 0.5g_1(X_1^4 + X_1^3)]\ (Q_1^3 - \mu_1) = 3.41143 \text{ MW}$$

The thermal generations in the three intervals are obtained from Eq. (4.118), i.e

$$P_1^1 = P_D^1 - P_2^1 = 7 - 2.31516 = 4.68484 \text{ MW}$$
$$P_1^2 = P_D^2 - P_2^2 = 10 - 3.60272 = 6.39728 \text{ MW}$$
$$P_1^3 = P_D^3 - P_2^3 = 5 - 3.41143 = 1.58857 \text{ MW}$$

λ_1^k can be computed from Eq. (4.120), i.e

$$\lambda_1^1 = 1.0P_1^1 + 25.0 = 1.0 \times 4.68484 + 25 = 29.68484 \text{ ₹/h}$$
$$\lambda_1^2 = 1.0P_1^2 + 25.0 = 1.0 \times 6.39728 + 25 = 31.39728 \text{ ₹/h}$$
$$\lambda_1^3 = 1.0P_1^3 + 25.0 = 1.0 \times 1.58857 + 25 = 26.58857 \text{ ₹/h}$$

λ_{21}^k can be computed from Eq. (4.121), i.e.

$$\lambda_{21}^1 = \lambda_1^1 = 29.68484$$
$$\lambda_{21}^2 = \lambda_1^2 = 31.39728$$
$$\lambda_{21}^3 = \lambda_1^3 = 26.58857$$

λ_{31}^k can be computed from Eq. (4.122), i.e.

$$\lambda_{31}^1 = 0.0$$
$$\lambda_{31}^2 = \lambda_{31}^1 - \lambda_{21}^2 h_1 g_1 (Q_1^2 - u_1)$$
$$= 0 - 31.39728 \times 0.1962 \times 0.005 \times (15-2) = -0.40041$$
$$\lambda_{31}^3 = \lambda^2_{31} - \lambda^3_{21} h_1 g_1 (Q_1^3 - u_1)$$
$$= 0.40041 - 26.58857 \times 0.1962 \times 0.005 \times (15-2) = -0.73949$$

Using Eq. (4.123), the gradient vector is obtained as

$$\frac{\partial L}{\partial Q_1^1} = \lambda_{31}^1 + \lambda_{41} - \lambda_{21}^1 h_1 [1 + 0.5g_1(2X_1^1 - 2Q_1^1 + u_1)] = 0$$

Solving the above equation, Q_1^1 can be obtained as

$$Q_1^1 = 33.45237$$

$$\frac{\partial L}{\partial Q_1^2} = \lambda_{31}^2 + \lambda_{41} - \lambda_{21}^2 h_1 [1 + 0.5g_1(2X_1^2 - 2Q_2^1 + u_1)] = 0$$

Solving the above equation, Q_1^2 can be obtained as

$$Q_1^2 = 45.49931$$

$$\frac{\partial L}{\partial Q_1^3} = \lambda_{31}^3 + \lambda_{41} - \lambda_{21}^3 h_1 [1 + 0.5g_1(2X_1^3 - 2Q_1^3 + u_1)] = 0$$

Solving the above equation, Q_1^3 can be obtained as

$$Q_1^3 = 1.81165$$

Using Eq. (4.124), gradient with respect to λ_{41} is obtained as

$$\frac{\partial L}{\partial \lambda_{41}} = X_1^4 - X_1^1 + (Q_1^1 + Q_1^2 + Q_1^3) = 40.7633$$

The value of λ_{41} can be updated using the steepest descent method, i.e.

$$\lambda_{41} = \lambda_{41} + s\left(\frac{\partial L}{\partial \lambda_{41}}\right)$$

$$= 7.5 + s \times 40.7633$$

Sample System Study 4.5

A long-range hydrothermal problem with four thermal and three hydro electric plant is considered here. The incremental fuel costs of the thermal plants considered here are given in Table 4.11.

TABLE 4.11 Incremental cost equation (₹/MWh) coefficients

Plant no.	a_i	b_i
1	0.10	2.5
2	0.12	2.6
3	0.16	2.8
4	0.16	2.8

The average transmission loss coefficients are as follows (MW^{-1}):

$$B_{11} = B_{22} = 0.0005,\ B_{33} = B_{44} = 0.0008$$

$$B_{55} = B_{66} = 0.0007,\ B_{77} = 0.0009$$

$$B_{ij} = 0.0\ (i = 1, 2, ..., 7;\ \ j = 1, 2, ..., 7;\ \ i = j).$$

The data for the three hydro plants are presented in Table 4.12. The water inflows and the standard deviations for the 12 sub-intervals are given in Table 4.13.

TABLE 4.12 Hydro plant data [Rao, Prabhu, Aggarwal, 1975]

Parameter	*Hydro plants*		
	1	2	3
Basic head, h(m)	0.98	0.50	0.75
Water head correction factor, g	0.004	0.002	0.004
Non-effective water discharge, u	0.0	0.0	0.0
Initial storage X^1 (m^3/s-month)	100.0	150.0	100.0
Final storage X^{T+1} (m^3/s-month)	100.0	150.0	150.0
Maximum allowable discharge (m^3/s)	73.0	75.0	80.0
Minimum allowable discharge (m^3/s)	3.334	4.353	3.543
Maximum hydro generation (MW)	89.03	57.61	75.0
Minimum hydro generation (MW)	0.0	0.0	0.0

TABLE 4.13 Demand, water inflows, and assumed discharge

Sub Interval. k	*Demand* (MW)	*Water inflows* (m^3/s)			*Assumed discharge* (m^3/s)		
		J_1	J_2	J_3	Q_1	Q_2	Q_3
1	200.0	0.8	23.0	0.8	10.8	33.0	10.8
2	210.0	10.0	17.0	10.0	20.0	27.0	20.0
3	205.0	11.0	12.0	11.0	21.0	22.0	21.0
4	180.0	11.9	10.0	11.9	21.9	20.0	21.9
5	195.0	17.0	20.0	17.0	27.0	30.0	27.0
6	200.0	18.5	40.0	18.5	28.5	50.0	28.5
7	220.0	27.6	53.0	27.6	37.6	63.0	37.6
8	204.0	43.8	63.0	43.8	33.8	53.0	33.8
9	189.0	56.4	62.0	56.4	46.4	52.0	46.4
10	199.0	40.3	48.0	40.0	30.3	38.0	30.3
11	207.0	30.1	55.0	30.1	20.1	45.0	20.1
12	198.0	46.3	48.0	46.3	36.0	38.0	36.3

The overall incremental cost, transmission losses, and λ_1^k are shown in Table 4.14. The optimal generation schedule of hydro and thermal units is given in Table 4.15. The discharge and water storage are given in Table 4.16. The values of λ_{2j}^k and λ_{3j}^k are given in Table 4.17. The number of iterations for the discussed algorithm depends on the initial guess of discharge and dual variable λ_{4j}.

Total Incremental cost = ₹ 6800.129

$$\lambda_{41} = 11.87719 \text{ ₹/m}^3$$

$$\lambda_{42} = 5.47397 \text{ ₹/m}^3$$

$$\lambda_{43} = 9.18910 \text{ ₹/m}^3$$

TABLE 4.14 Incremental cost, transmission loss and λ_1^k

k	*Incremental cost* (₹)	P_L^k (MW)	λ_1^k (₹/MWh)
1	914.4983	4.817092	8.595355
2	888.8978	4.876114	8.478350
3	844.6933	4.625056	8.273413
4	746.6210	3.722375	7.803908
5	707.3789	3.970779	7.609521
6	638.8799	4.012251	7.260437
7	575.6646	4.952200	6.925622
8	478.8612	4.359816	6.386069
9	329.1064	4.298597	5.466470
10	281.6418	5.722983	5.146189
11	218.5313	7.046297	4.691734
12	175.3546	7.344503	4.357534

TABLE 4.15 Generation schedule

Sub-interval	*Thermal generations*				*Hydro generations*		
k	P_1 (MW)	P_2 (MW)	P_3 (MW)	P_4 (MW)	P_5 (MW)	P_6 (MW)	P_7 (MW)
1	56.1384	46.6299	33.3599	33.3599	8.2891	20.0544	6.9853
2	55.1204	45.7616	32.7218	32.7218	17.9360	15.9589	14.6556
3	53.3309	44.2362	31.6001	31.6001	20.0194	12.7912	16.0470
4	49.2053	40.7228	29.0141	29.0141	13.7595	9.5839	12.4227
5	47.4881	39.2618	27.9377	27.4149	22.1493	15.7179	18.4783
6	44.3878	36.6261	25.9943	25.9943	23.8280	27.4329	19.7489
7	41.3970	34.0862	24.1196	23.1196	36.5886	35.4898	29.1515
8	36.5334	29.9610	21.0709	21.0709	38.9559	30.9251	29.8425
9	28.1316	22.8503	15.8044	15.8044	45.4372	28.0715	37.1992
10	25.1735	20.3514	13.9502	13.9502	62.9309	25.3447	43.0221
11	20.9442	16.7829	11.2992	11.2992	72.6481	34.4400	46.6329
12	17.8091	14.1408	9.3340	9.3340	74.7947	27.0234	52.9086

TABLE 4.16 Discharge and water storage

Sub-interval	*Discharge* (m^3/s)			*Water storage* (m^3/s)		
k	Q_1	Q_2	Q_3	X_1	X_2	X_3
1	6.0876	31.0451	6.7093	100.0000	150.0000	100.0000
2	13.3380	25.0161	14.2864	94.7124	141.9549	94.0907
3	15.0494	20.3105	15.8547	91.3744	133.9388	89.8043
4	10.3823	15.3851	12.3714	87.3250	125.6283	84.9496
5	16.6672	25.4535	18.4552	88.8428	120.2431	84.4782
6	17.9059	44.7964	19.8061	89.1756	114.7896	83.0229
7	27.4656	58.4413	29.3720	89.7696	109.9932	81.7169
8	28.5974	50.6359	29.5099	89.9041	104.5519	79.9449
9	31.5372	44.8802	34.9319	105.1067	116.9150	94.2350
10	42.3654	39.7140	39.1527	129.9695	134.0358	115.7031
11	50.3937	53.5579	43.1389	127.9041	142.3217	116.8504
12	53.9303	41.7753	50.1166	107.6103	143.7638	103.8115

TABLE 4.17 λ_{2j}^k and λ_{3j}^k

k	λ_{21} (₹/MWh)	λ_{22} (₹/MWh)	λ_{23} (₹/MWh)	λ_{31} (₹/m^3)	λ_{32} (₹/m^3)	λ_{33} (₹/m^3)
1	8.495608	8.354030	8.487282	0.0	0.0	0.0
2	8.265455	8.288922	8.254690	−0.432158	−0.207356	−0.353790
3	8.041533	8.125255	8.034438	−0.906556	−0.372384	−0.735941

(Contd.)

TABLE 4.17 (*Contd.*)

k	λ_{21} (₹/MWh)	λ_{22} (₹/MWh)	λ_{23} (₹/MWh)	λ_{31} (₹/m^3)	λ_{32} (₹/m^3)	λ_{33} (₹/m^3)
4	7.653579	7.699199	7.629407	−1.218046	−0.490838	−1.019100
5	7.373557	7.442072	7.356421	−1.699800	−0.680265	−1.426394
6	7.018235	6.981592	7.002343	−2.192419	−0.993015	−1.842461
7	6.570863	6.581517	6.562216	−2.899872	−1.377647	−2.420697
8	6.037784	6.109583	6.043032	−3.576717	−1.687012	−2.955685
9	5.118736	5.251637	5.100442	−4.209526	−1.922706	−3.490190
10	4.692792	4.963589	4.747669	−4.988869	−2.119831	−4.047841
11	4.214550	4.465517	4.297914	−5.821426	−2.358994	−4.604064
12	3.901246	4.192677	3.942543	−6.646176	−2.534145	−5.196824

REFERENCES

Books

Christensen, G.S. and S.A. Soliman, *Optimal Long-Term Operation of Electric Power Systems*, Plenum Press, New York, 1988.

El-Hawary, M.E. and G.S. Christensen, *Optimal Economic Operation of Power Systems*, Academic Press, New York, 1979.

Kothari, D.P. and I.J. Nagrath, *Modern Power System Analysis*, 3rd ed., Tata McGraw-Hill, New Delhi, 2003.

Mahalanabis, A.K., D.P. Kothari and S.I. Ahson, *Computer-Aided Power System Analysis and Control*, Tata McGraw-Hill, New Delhi, 1991.

Nagrath, I.J. and D.P. Kothari, *Power System Engineering*, Tata McGraw-Hill, New Delhi, 1994.

Rao, S.S., *Optimization, Theory and Applications*, 2nd ed., Wiley Eastern, New Delhi, 1987.

Singh, L.P., *Advanced Power System Analysis and Dynamics*, 2nd ed., Wiley Eastern, New Delhi, 1993.

Wood, A.J. and B. Wollenberg, *Power Generation, Operation and Control*, John Wiley, New York, 1984.

Papers

Agarwal, S.K. and I.J. Nagrath, Optimal scheduling of hydrothermal systems, *IEE Proceedings-C*, Vol. **119(2)**, pp. 169–173, 1972.

Arvanitidis, N.V. and J. Rosing, Optimal operation of multireservoir systems using a composite representation, *IEEE Trans. on Power Apparatus and Systems*, Vol. **PAS-89(2)**, pp. 327–335, 1970.

Bijwe, P.R. and J. Nanda, Optimum scheduling in hydrothermal system using progressive optimality algorithm, *IEEE PES Summer Meeting*, Mexico City, Paper No. A-77-600-0, 1977.

Bonaert, A.P., A.H. El-Abiad and A.J. Koivo, Optimal scheduling of hydrothermal power systems, *IEEE Trans. on Power Apparatus and Systems*, Vol. **PAS-91(1)**, pp. 263–270, 1972.

Calderon, L.R. and F.D. Galiana, Continuous solution simulation in the short-term hydrothermal co-ordination problem, *IEEE Trans. on Power Systems*, Vol. **PWRS-2(3)**, pp. 737–743, 1987.

Carvalho, M.F. and S. Soares, An efficient hydrothermal scheduling algorithm, *IEEE Trans. on Power Systems,* Vol. **PWRS-2(3)**, pp. 537–542, 1987.

Dey, S.C. and L.P. Singh, Optimal load scheduling of hydrothermal system including transmission losses using dynamic programming, *Journal Institution of Engineers (India)*, Vol. **EL68**, pp. 206–213, 1988.

Dillon, T.S. and K. Morsztyn, Mathematical solution of the problem of optimal control of integrated power systems, *International Journal of Control*, Vol. **13**, pp. 831–851, 1971.

Duran, H., C. Puech, J. Diaz and G. Sanchez, Optimal operation of multireservoir systems using an aggregation-decomposition approach, *IEEE Trans. on Power Apparatus and Systems*, Vol. **PAS-104(8)**, pp. 2086–2092, 1985.

Edwin, K.W. and R.D. Machate, Influence of inaccurate input data on the optimal short-term operation of power generation systems, *IFAC Symposium on Automatic Control in Power, Generation, Distribution and Protection*, Pretoria, South Africa, 1980.

El-Hawary, E.L. and K.M. Ravindranath, Optimal operation of variable head hydrothermal systems using the Glimn-Kirchmayer model and the Newton–Raphson method, *Electric Power Systems Research*, Vol. **14(1)**, pp. 11–22, 1988.

El-Hawary, M.E. and D.H. Tsang, The hydrothermal optimal load flow, a practical formulation and solution techniques using Newton's approach, *IEEE Trans. on Power Systems*, Vol. **PWRS-1(3)**, pp. 157–167, 1986.

El-Hawary, M.E. and G.S. Christensen, Application of functional analysis to optimization of electric power system, *International Journal of Control,* Vol. **16(6)**, pp. 1063–1072, 1972.

El-Hawary, M.E. and J.K. Landrigan, Optimum operation of fixed-head hydrothermal electric power systems: Powell's hybrid method versus Newton–Raphson method, *IEEE Trans. on Power Apparatus and Systems*, Vol. **PAS-101(3)**, pp. 547–554, 1982.

El-Hawary, M.E. and K.M. Ravindranath, A general overview of multiple objective optimal power flow in hydrothermal electric power systems, *Electric Machines and Power Systems*, Vol. **PWRS-19(3)**, pp. 313–327, 1991.

Heredia, F.J. and N. Nabona, Optimum short-term hydrothermal scheduling with spinning reserve through network flows, *IEEE Trans. on Power Systems*, Vol. **PWRS-10(3)**, pp. 1642–1651, August 1995.

Kothari, D.P., Optimal hydrothermal scheduling: A review, *Journal of Scientific and Industrial Research*, Vol. **47**, pp. 98–101, 1988.

Mohan, M.R., K. Kuppusamy and M. Abdullah, Optimal short-term hydrothermal scheduling using decomposition approach and linear programming method, *Int. J. Electrical Power and Energy Systems*, Vol. **14(1)**, pp. 39–44, 1992.

Nagrath, I.J., G. Dayal and D.P. Kothari, Application of the discrete maximum principle to the optimum scheduling of multi-reservoir systems, *Journal Institution of Engineers (India)*, Vol. **EL-53**, pp. 101–105, 1973.

Parti, S.C., *Stochastic optimal power generation scheduling*, Ph.D. (Thesis), TIET, Patiala, 1987.

Pereira, M.V.F. and L.M.V.G. Pinto, Application of decomposition techniques to the mid- and short-term scheduling of hydrothermal systems, *IEEE Trans. on Power Apparatus and Systems*, Vol. **PAS-102(11)**, pp. 3611–3618, 1983.

Prakasa Rao, K.S., S.S. Prabhu and R.P. Aggarwal, A two level approach for optimal scheduling in hydrothermal power systems using the method of local variations, *IEEE conference*, Paper C-75 078 1, Winter Power Meeting, 1975.

Rashid, A.H.A. and K.M. Nor, An efficient method for optimal scheduling of fixed-head hydro and thermal plants, *IEEE Trans. on Power Systems*, Vol. **PWRS-6(2)**, pp. 632–636, 1991.

Rashid, A.H.A. and K.M. Nor, An algorithm for the optimal scheduling of variable head hydro and thermal plants, *IEEE Trans. on Power Systems*, Vol. **PWRS-8(3)**, pp. 1242–1249, 1993.

Sachdeva, S.S., Bibliography on optimal reservoir drawdown for the hydroelectric thermal power system operation, *IEEE Trans. on Power Apparatus and Systems*, Vol. **PAS-101(6)**, pp. 1487–1496, 1982.

Shaw, J.J., R.F. Gendron and D.P. Bertsekas, Optimal scheduling of large hydrothermal power systems, *IEEE Trans. on Power Apparatus and Systems*, Vol. **PAS-104 (2)**, pp. 286–293, 1985.

Singh, L.P. and R.P. Aggarwal, Economic load scheduling of hydrothermal stations using dynamic programming, *Journal Institution of Engineers (India)*, Vol. **EL-52(4)**, pp. 175–180, 1972.

Soares, S., C. Lyra and H. Tavares, Optimal generation scheduling of hydrothermal power systems, *IEEE Trans. on Power Apparatus and Systems*, Vol. **PAS-99(3)**, pp. 1107–1118, 1980.

Vemuri, S. and E.F. Hill, Sensitivity analysis of optimum operation of hydrothermal plants, *IEEE Trans. on Power Apparatus and Systems*, Vol. **PAS-96(2)**, pp. 688–696, 1977.

Wu, Y.C., Efficient two-level interior point method for optimal pumped hydrostorage scheduling exploiting the non-sparse matrix structure, *IEE Proceedings: Generation, Transmission and Distribution*, Vol. **148(1)**, pp. 41–47, January 2001.

Yan, H., P.B. Luh, X. Guan and P.M. Rogan, Scheduling of hydrothermal power systems, *IEEE Trans. on Power Systems*, Vol. **PWRS-8(3)**, pp. 1358–1365, 1993.

Yang, P.C., H.T. Yang and C.L. Huang, Scheduling short-term hydrothermal generation using evolutionary programming techniques, *IEE Proceedings: Generation, Transmission and Distribution*, Vol. **143(4)**, pp. 371–376, July 1996.

Yu, Z., F.T. Sparrow and D. Nderitu, Long-term hydrothermal scheduling using composite thermal and composite hydro representations, *IEE Proceedings: Generation, Transmission and Distribution*, Vol. **145(2)**, pp. 210–216, March 1998.

Zaghlool, M.F. and F.C. Trutt, Efficient methods for optimal scheduling of fixed head hydrothermal power systems, *IEEE Trans. on Power Systems*, Vol. **PWRS-3(1)**, pp. 24–30, 1988.

CHAPTER 5

Multiobjective Generation Scheduling

5.1 INTRODUCTION

Environmental pollution is on the increase due to industrial advancement. Though technology has made economic development possible, it at the same time, produces enormous quantities of harmful products and wastes. Also, the existing energy production processes are not ecologically clean. For instance, thermal power plants pollute air, soil, and water. The combustion of fossil fuels gives rise to particulate material and gaseous pollutants apart from the discharge of heat to water courses. The particulate material does not cause a serious problem in air contamination but the three principal gaseous pollutants, oxides of carbon (CO_x), oxides of sulphur (SO_x) and oxides of nitrogen (NO_x) cause detrimental effects on human beings. The usual control practice is to reduce offensive emissions through post-combustion cleaning systems such as electrostatic precipitators, stack gas scrubbers, or switching permanently to fuels with low emission potentials. Post-combustion removal systems require time for engineering design, construction, and testing before they can be brought online. Hence, the obvious alternative is to go for fuels having low emission potentials, i.e. from coal to oil; oil is however extremely expensive and supplies are uncertain. Thus, there is a sheer need for optimum operating strategy, which can ensure minimum pollution level at minimum operating cost.

There are several literature reviews in the area of minimizing pollution level, but still much progress is possible in the area of economic-emission load dispatch. Gent and Lamont [1971] have published a paper on minimum-emission dispatch wherein a computer program has been developed for online steam unit dispatch resulting in minimization of NO_x emission employing the Newton–Raphson convergence for curve fitting. They have described an ingenious application of the incremental loading technique for minimization of emission of nitrogen oxides by a fossil steam boiler. Sullivan [1972] also attempted only the minimum pollution dispatch by applying Kuhn-Tucker conditions. Delson [1974] has advocated a method for achieving controlled emission dispatch employing constant penalty factors, to schedule the generators to meet the given environ-mental restrictions at minimum operating cost. Finnigan and Fouad [1974] have implemented two nonlinear programming solution procedures for economic dispatch treating the system pollutants as an additional constraint to the problem.

Zahavi and Eisenberg [1975] suggested an interactive search method based on golden section search technique to solve the economic emission load dispatch problem. Cadogan and Eisenberg [1977] treated the economic environmental dispatch problem as a multi-strategy problem and suggested a nonlinear programming technique, for the implementation of the dispatch strategies. Kothari et al. [1977] exhibited a computer-oriented technique for thermal power generation scheduling which resulted in minimum NO_x emission. Effluent dispatching is the operation of a generation system so that power plants with higher emission rates generate less than they would under least cost dispatching. An example of joint evaluation of fuel switching and effluent dispatching was hinted by Tsuji [1981]. Brodsky and Hahn [1986] presented a model to show that emissions and production costs were a function of the organizational form. They illustrated that the production costs and emission could be reduced with an increased coordination among the utilities, i.e. when inter-utility pooling arrangements are brought into a single framework.

Nanda et al. [1987] suggested a new computational approach employing an improved complex box method to find minimum emission dispatch. In an another attempt Nanda et al. [1988] expressed a goal programming technique to solve the economic emission load dispatch problem for thermal generating units running with natural gas and fuel oil. Yokoyama et al. [1988] have expertised an efficient algorithm to obtain the optimal power flow in power system operation and planning phases by solving a multiobjective optimization problem. Minimum interference in the environment was one of the objectives in deciding the optimal system operation. Heslin and Hobbs [1989] have stated a model for evaluating the cost and employment impacts of effluent dispatching and fuel switching as means for reducing emissions from power plants. They have produced trade-off curves, which show the costs and coal field job losses associated with reduction in SO_2 emissions. Kermanshahi et al. [1990] propounded a decision-making methodology to determine the optimal generation dispatch and environmental marginal cost for power system operation with multiple conflicting objectives. Palanichamy and Srikrishna [1991] have enumerated an algorithm for successful operation of the system subject to economical and environmental constraints. A price penalty factor has been defined which has blended the emission costs with the normal fuel costs. The familiar quadratic form of objective function has been used and the positive root of such a function gives the optimal dispatch directly.

In the context of increasing public awareness of the environmental situation and the plea for clean air, the US congress has passed acid rain legislation [Hobbs, 1993]. In an effort to alleviate acid rain, Title IV of the 1990 U.S. clean act amendments mandated a restriction of over 50% in electric utility SO_2 emissions by the year 2000. Hobbs has further summarized a method for including, underutilization constraints in probabilistic production costing models and real-time dispatch. Andrews [1993] introduced a simple system-level emissions model suitable for assisting regulators to design efficient integrated resource planning rules. Specifically, regulations should not only focus on marginal capacity choices, but also compare complete resource portfolios.

5.2 MULTIOBJECTIVE OPTIMIZATION—STATE-OF-THE-ART

In recent years there has been an increase in research on multiobjective optimization methods. Decisions with multiobjectives are quite prevalent in government, military, industry, and other organizations. Researchers from a wide variety of disciplines such as mathematics, management, science, economics, engineering and others have contributed to the solution methods for multi-objective optimization problems. The situation is formulated as a multiobjective optimization problem (also called multi-performance, multi-criterion or vector optimization) in which the engineer's goal is to maximize or minimize not a single objective function but several objective functions simultaneously. The purpose of multiobjective problems in the mathematical programming framework is to optimize the different objective functions, G in number, subject to a set of system constraints. For example,

$$\text{Maximize} \qquad f(x) = [f_1(x), \ldots, f_i(x), \ldots, f_G(x)]^T \tag{5.1a}$$

$$\text{subject to} \qquad x \in X \tag{5.1b}$$

where x is an n-dimensional vector of decision variables, X is the decision space, and $f(x)$ is a vector of G real-valued functions.

In order to make the problem non-trivial, it is assumed that the objectives are in conflict and incommensurable. Owing to the conflicting nature of objectives, an optimal solution that simultaneously maximizes/minimizes all the criteria is usually not attainable. Instead, there are several solutions, called efficient solutions, that have the property, that no improvement in any objective is possible without sacrificing one or more of the other objectives.

Definition

A solution $x^e \in X$ is said to be efficient if for any $x \in X$ satisfying $f_k(x) > f_k(x^e)$, $f_j(x) < f_j(x^e)$ for at least one other index $j \neq k$.

The set of all efficient solutions in the continuous case is known as the *efficient frontier*. An efficient solution is also known as a non-dominated solution, non-inferior solution or pareto optimal solution.

Many approaches and methods have been proposed in recent years to solve multiobjective optimization problems (MOP). These methods are broadly grouped under two major titles—non-interactive and interactive. In the non-interactive method, a global preference function of the objectives is identified and optimized with respect to the constraints. On the other hand, in the interactive method, the decision maker (DM) identifies a local preference function or trade-off among objectives and the solution process proceeds gradually towards the globally satisfactory solution. Wallenius [1975] made a comparative evaluation of some interactive methods by using such measures of performance with ease of use, ease of understanding, and DM's confidence in the solution, etc. He revealed that a simple trial and error type unstructured approach competed successfully with more sophisticated methods. Since the DM is involved in the entire solution process, interactive methods have been better accepted in practice. The primary objective of the MOP solution methods is to find the best compromise solution.

Definition

The best compromise solution is an efficient solution that maximizes the DM's preference function.

It may be obvious that trade-offs among these objectives are difficult because of their different nature. This implies that the objectives are 'non-commensurable'. In other words, it is difficult to treat respective objectives under the identical criterion. Principally, methodologies for solving multiobjective problems differ in two major ways:

(i) The procedure used to generate non-inferior solutions.
(ii) The ways and means used to interact with the DM and the type of information made available to the DM such as trade-offs.

The weighting and constraint methods identify the non-inferior set, within which the best compromise solution lies [Haimes, 1977]. In almost all decision-making problems, there are several criteria for judging the possible alternatives. The main concern of DM is to fulfil the conflicting goals while satisfying the constraints of the system. There are two different approaches to solve such problems.

(i) One approach assumes that there exists a utility function for the particular problem. Such a function is used to obtain the best alternative.
(ii) The other approach makes no assumption regarding the existence of any utility function, but provides the DM with a set of simple but effective tools to obtain the best alternative.

Here, a few methods for generating efficient points are reviewed. These are the min-max optimum, the weighting method, the ε-constraint method, and the weighted min-max method.

5.2.1 Weighting Method

The weighting method [Haimes, 1977], also known as the parametric approach, has been the most common method used for solving multiobjective problems until recently. Multiobjective problem (5.1) is converted in this method into scalar optimization as given below:

$$\text{Minimize} \quad \sum_{i=1}^{G} w_i \, f_i(x) \tag{5.2a}$$

$$\text{subject to} \quad x \in X \tag{5.2b}$$

$$\sum_{i=1}^{G} w_i = 1, \quad w_i \geq 0 \qquad (i = 1, 2, ..., G) \tag{5.2c}$$

where w_i are the weighting coefficients.

The approach yields meaningful results to the decision maker only when solved many times for different values of w_i $(i = 1, 2, ..., G)$. Though very little is usually known about the values of weighting coefficients, the DM still chooses them, presumably on the basis of his intuition. The weighting coefficients do not reflect proportionally the relative importance of the objectives but are only factors which, when varied, locate points in the non-inferior set.

5.2.2 Min-Max Optimum

The idea of stating the min-max optimum and applying it to multi-criterion optimization problems was taken from game theory, which deals with solving conflicting situations [Osyczka and Davies, 1984]. The min-max optimum compares relative deviations from the separately attainable minima. Consider the ith objective function for which the relative deviation can be calculated from

$$Z_i'(x) = \frac{|f_i(x) - f_i^0|}{|f_i^0|} \tag{5.3}$$

or from

$$Z_i''(x) = \frac{|f_i(x) - f_i^0|}{|f_i(x)|} \tag{5.4}$$

where f_i^0 denotes the minimum value of the ith function.

It is clear from Eqs. (5.3) and (5.4), that for every $i \in I$ and for every $x \in X$, $f_i(x)$ should not be equal to zero. Let, $Z(x) = [Z_1(x),\ldots, Z_i(x),\ldots, Z_G(x)]^T$ be a vector of the relative increments which are defined in Euclidian space. The components of the vector $Z(x)$ will be evaluated from the formula

$$i \in I \; Z_i(x) = \max \left[Z_i'(x),\, Z_i''(x)\right] \tag{5.5}$$

The min-max optimum is defined as follows:

1. $$V_1(x^*) = \min_{x \in X} \max_{i \in I} \left[Z_i(x)\right] \tag{5.6}$$

   and then $I_1 = \{i_1\}$, where i_1 is the index for which the value is maximal.
   If there is a set of solutions $x \subset X$, which satisfies step 1, then:

2. $$V_2(x^*) = \min_{x \in X} \max_{\substack{i \in I \\ i \notin I_1}} \left[Z_i(x)\right] \tag{5.7}$$

. and then $I_2 = \{i_1, i_2\}$, where i_2 is the index for which the value of $Z_i(x)$ in this step is maximal.
. If there is a set of solutions $x_{r-1} \subset X$ which satisfies the step $r - 1$, then:

r. $$V_r\,(x^*) = \min_{x \in X} \max_{\substack{i \in I \\ i \notin I_{r-1}}} \left[Z_i(x)\right] \tag{5.8}$$

   and then $I_r = \{i_{r-1}, i_r\}$, where i_r is the index for which the value of $Z_i(x)$ in the rth step is maximal.
   If there is a set of solutions $x_{k-1} \subset X$ which satisfies the step $(k - 1)$, then:

k. $$V_k(x^*) = \min_{x \in X_{k-1}} [Z_i(x)] \text{ for } i \in I \text{ and } i \notin I_{k-1} \tag{5.9}$$

   where $V_1(x),\ldots,V_k(x)$ is the set of optimal values of fractional deviations ordered non-increasingly.

   This optimum can be described as follows:

   Knowing the extremes of the objective functions, which can be obtained by solving the optimization problems for each criterion separately, the desirable solution is the one that gives the smallest values of the relative increments of all the objective functions. The point $x^* \in X$ which satisfies Eq. (5.7) may be called the best compromised solution considering all the criteria simultaneously and on equal terms of importance.

5.2.3 ε-Constraint Method [Haimes, 1977]

In this method, one of the objective functions constitutes the primary objective function and all other objectives act as constraints. In other words, the ε-constraint approach replaces the $(G - 1)$ objective functions by $(G - 1)$ constraints as given below:

Minimize $\quad f_i(x) \quad$ (5.10a)

subject to $\quad f_j(x) \le \varepsilon_j \quad (j = 1, 2, \ldots, G;\ \ j \ne i) \quad$ (5.10b)

$\quad x \in X$

where $\varepsilon_j (j = 1, 2, \ldots, G;\ \ j \ne i)$ are maximum tolerable levels.

In order to obtain a reasonable choice of n_j it is useful to assume $\varepsilon_j = f_j + \varepsilon_j$ and $\varepsilon_j > 0$, where f_j is obtained by minimizing each objective function separately as

$$f_j = \min f_j(x) \qquad (j = 1, 2, ..., G) \tag{5.11a}$$

subject to

$$x \in X \tag{5.11b}$$

The levels of satisfactory ε_j are varied parametrically to evaluate the impact on the single objective function $f_i(x)$. Of course, the ith objective $f_i(x)$ can be replaced by the jth objective, $f_j(x)$ and the solution procedure is repeated. The ε-constraint approach facilitates the generation of non-inferior solutions as well as the trade-off functions. The major weakness of the constraint method is computational efficiency where there are several objectives. The number of solutions of the scalarized problem required to identify completely, or even approximate the non-inferior set, increases exponentially with the number of objectives [Tapia and Murtagh, 1991].

5.2.4 Weighted Min-Max Method [Charalambous, 1989]

In this method, the following minimization problem is solved:

$$\min_x \left[\max_{1 \le i \le G} w_i f_i(x) \right] \tag{5.12a}$$

subject to

$$w_i \ge 0 \qquad (i = 1, 2, ..., G) \tag{5.12b}$$

$$\sum_{i=1}^{G} w_i = 1 \tag{5.12c}$$

$$x \in X \tag{5.12d}$$

A systematic variation of the w_is will generate the non-inferior solution. The solution obtained in this way is a local efficient point. This method overcomes the convexity limitations of the non-negative convex combination method.

5.2.5 Utility Function Method [Rao, 1987]

In this method, a utility function $U_i(f_i)$ is defined for each objective function depending on the importance of f_i compared to the other objective functions. Then a total or overall utility function U is defined. A common one is the additive utility function method for which Eq. (5.1) has the form

$$f(x) = \sum_{i=1}^{G} U_i(f_i(x)) \tag{5.13}$$

where $U_i(f_i(x))$ are the utility functions of the criteria.

The utility function method requires that $U_i(f_i(x))$ should be known prior to solving the problem. Since the evaluation of $U_i(f_i(x))$ for even a simple linear programming is difficult, the application of this method is rather limited.

5.2.6 Global Criterion Method [Osyczka and Davies, 1984]

In this method an optimal solution is a vector of decision variables that minimizes some global criterion. A function which describes this global criterion is a measure of how close the decision maker can get to the ideal vector f_0. The most common form of this function is

$$f(x) = \sum_{i=1}^{G} \left[\frac{f_i^0 - f_i(x)}{f_i^0} \right]^p \tag{5.14}$$

The solution obtained after minimizing Eq. (5.14) differs according to the value of p chosen. Thus the problem is to determine which p would result in a solution that is the most satisfactory for the decision maker.

Another possible measure of 'closeness to the ideal solution' is a family of L_p-metric defined as follows:

$$L_p(f) = \left[\sum_{i=1}^{G} |f_i^0 - f_i(x)|^p \right]^{\frac{1}{p}} ; \qquad 1 \le p \le \infty \tag{5.15}$$

For example,

$$L_1(f) = \sum_{i=1}^{G} |f_i^0 - f_i(x)| \tag{5.16}$$

$$L_2(f) = \left[\sum_{i=1}^{G} |f_i^0 - f_i(x)|^2 \right]^{\frac{1}{2}} \tag{5.17}$$

$$L(f) = \min |f_i^0 - f_i(x)| \tag{5.18}$$

Instead of deviation in an absolute sense it is recommended to use relative deviations in Eq. (5.15). The relevant L_p-metrics are

$$L_p(f) = \left[\sum_{i=1}^{G} \left| \frac{f_i^0 - f_i(x)}{f_i^0} \right|^p \right]^{\frac{1}{p}} ; \qquad 1 \le p \le \infty \tag{5.19}$$

There are several excellent literature reviews on multiobjective problems. Cohon and Marks [1975] classified the multiobjective approaches into generating techniques, which rely on the prior articulation of preferences and techniques that are faster interactive to the definitions of preferences. The methods in various classes are reviewed and evaluated by them in terms of the hypothesized criteria. In another survey paper, Brayton et al. [1981] have described multiobjective constrained optimization techniques which are appropriate to the design of integrated circuits. Shin and Ravindran [1991] conducted an excellent survey of the interactive methods developed for solving continuous multiobjective optimization problems and other applications. Most of the methods fall into the following categories—feasible region, reduction methods, feasible direction methods, criterion weight space methods, trade-off cutting plane methods, Lagrange multiplier methods, visual interactive methods, branch and bound methods, relaxation methods, sequential methods, and scalarizing function methods. The methods of each category have been reviewed based on the nature of preference assessments, functional assumptions, and relationship between methods.

Saber and Ravindran [1993] outlined a literature review of Nonlinear Goal Programming (NLGP) methods and applications. The methodological articles are categorized into four major approaches, namely (i) simplex method, (ii) direct search, (iii) gradient search, and (iv) interactive approaches to help practitioners in selecting the proper NLGP method to use. In order to identify areas where NLGP methods can be used, the application articles are categorized into nine major areas including engineering design. The number of publications in the field of MOP problems is too many; a few representative ones are discussed here.

A computational procedure for calculating a set of the so-called non-inferior vectors to an unconstrained optimization problem has been described by Vemuri [1974]. Haimes and Hall [1974] developed the Surrogate Worth tradeoff method for solving non-commensurable multi-objective functions. The trade-off and surrogate worth functions have been constructed in the functional space, where an interaction with the decision maker takes place and only these functions have been transformed into the decision space. In another attempt, Haimes et al. [1981] integrated two existing methodologies into a single objective dynamic programming method for capacity expansion and surrogate worth trade-off method for optimizing multiple objectives into a unified scheme.

A multiobjective optimization problem is generally solved for finding the set of all non-inferior solutions to the problem. A new methodology termed the 'envelop approach' has been presented by Li and Haimes [1987] for generating the set of non-inferior solutions. The relationship between envelop approach and multiobjective optimization also has been explored. A man-machine interactive approach to solve multiple objective linear programming problems was elaborated by Quaddus and Holzman [1986]. An interactive paired comparison method has been developed by Malakooti [1988] using the heuristic approaches for finding the local adjacent discrete points to approximate the gradient and for identifying discrete points for the one-dimensional search.

A new method for generating efficient points for the multi-criterion optimization by defining a shifted minimax function has been brought out by Charalambous [1989] and has been applied to design 1-D filters. Rarig and Haimes [1983] have developed the risk/dispersion index method to find solutions to the multiobjective problem that minimizes sensitivity and maximizes overall utility, using the information provided by dispersion index. The index is interpreted as a first-order approximation to the standard deviation in the optimal solutions of the nonlinear programs. Kaunas and Haimes [1985] have applied Risk/Dispersion method to solve the groundwater contamination problem.

In general, a large-scale system typified by an electric power system also possesses multiple objectives to be achieved, namely economics operation, reliability, security, and environment. Yokoyama et al. [1988] have solved the power flow optimization problem with multiobjective optimization formulation and the ε-constraint technique has been applied to obtain a set of non-inferiority solutions. Kothari et al. [1988] have applied linear and nonlinear goal programming algorithms to solve the economic-emission load dispatch problem. Wadhwa and Jain [1990] have formulated the optimal load flow problem as multiobjective programming problem. They minimized both the cost of generation and transmission loss simultaneously through priority goal programming.

Types of conflict among planning objectives have been illustrated and categorized by Merrill et al. [1993]. The purpose of their paper was to document new discoveries and observations in dealing with conflicting objectives and risk. Risk is defined as the hazard to which a utility is exposed because of uncertainty. Differentiating uncertainty and risk, they concluded that uncertainty cannot be eliminated but risk can certainly be managed. Hwang et al. [1993] developed a technique for order preference by similarity to ideal solution for multiple-objective decision making in which the chosen solution should be as close to positive ideal solution as possible and as far away from negative ideal solution as possible. To resolve the conflict existing between two distance objective functions, membership functions and satisfactory/preference levels have been employed.

Since multiple criterion problems tend to correlate with particular application fields, it seems unlikely that a single multiple criteria decision-making technique can solve all such problems. In fact, a variety of multiple criterion decision-making methods with well-developed theories are used for different purposes.

5.3 FUZZY SET THEORY IN POWER SYSTEMS

Advancements in computer technology associated with intellectual activities resulted in new fields of inquiry such as systems science, information science, decision analysis, or artificial intelligence. Approaches in computer technology have been steadily extending the capabilities for coping with systems of an increasingly broad range, including systems that were previously intractable by virtue of their nature and complexity. One way of simplifying a very complex system is to allow some degree of uncertainty in its description. It is realized that there are several fundamentally different types of uncertainties and that each of them plays a distinct role in the simplification of problem [Hwang, Lai, and Liu, 1993]. A mathematical formulation within which these various types of uncertainties can be properly characterized and investigated is now available in terms of the theory of fuzzy sets and fuzzy measures. In fact, one of Lofti Zadeh's contributions to system modelling is the representation of vague, incomplete knowledge or qualitative information that does not have a random nature and therefore cannot be represented by a probabilistic approach. Fuzzy set theory provided a basis for the interpretation of membership function as possibility distributions, which is a very useful concept in many practical applications.

In some papers, fuzzy set theory has been applied for power system operation. Su and Hsu [1991] implemented fuzzy dynamic programming for the unit commitment of a power system. Basically, forecasted hourly load has been taken into account by using fuzzy set notations, making this approach superior to the conventional dynamic programming method which assumes that hourly loads are exactly known and there exists no error in the forecasted loads. To reach an optimal commitment schedule under the fuzzy environment, security constraints, generation cost, and load demand have been expressed in fuzzy set notation. A fuzzy model for power system operation was presented by Miranda and Saraiva [1992]. Uncertainties in loads and generations have been modelled as fuzzy numbers. System behaviour and known injections have been dealt with the D.C. fuzzy power flow model. System optimal operation while uncertain has been calculated with linear programm-ing procedures where the problem nature and structure allowed some efficient techniques.

Application of fuzzy set theory within the field of decision making have, for the most part, consisted of extensions or fuzzifications of the classical theories of decision making. Decision-making under conditions of risk and uncertainty has been modelled by probabilistic decision theories and by game theories. Fuzzy decision theories attempt to deal with the vagueness or fuzziness inherent in subjective or imprecise determinations of preferences, constraints, and goals. The fuzzy programming approach to multiobjective linear programming problem was developed by Leberling [1981] and Hannan [1981]. Sakawa et al. [1987] have presented a fuzzy satisficing method for solving linear multiobjective programming problems by assuming that the decision maker has fuzzy goals for each of the objective functions. Through the interaction with the decision maker, the fuzzy goals of the decision maker were quantified by eliciting the corresponding membership functions, including nonlinear functions.

An interactive sequential goal programming has been introduced by Sasaki et al. [1988] in which the method of displaced ideal with fuzzy set theory was taken up for solving multiobjective decision making problems interactively. The method of displaced ideal was utilized to determine the best compromised solution among many non-dominated solutions to the decision making and to consider the fuzziness of the judgment by the decision maker. Tapia and Murtagh [1991] put up a methodology for solving a decision-making problem involving a multiplicity of objectives as well as selection criteria for the best compromised solution. The concept of fuzzy sets has been utilized in such a way that the decision maker can vary interactively the fuzzy membership values that are functions of the decision maker's input information known as preference criteria.

Niimura and Yokoyama [1991] employed fuzzy set theory to decision making in optimal generation dispatch of thermal generating units. Membership functions have been introduced to measure generation-load balance, fuel cost, and time to stay in a zone. Approximate reasoning, using such indices yielded information of a compatibility of output zones. Tomsovic [1992] modelled multiple objectives and soft constraints using fuzzy sets for the reactive power/voltage control problem. Piece-wise linear convex membership function for the fuzzy sets was defined. The fuzzy optimization problem was reformulated as standard linear programming problem.

5.3.1 Basics of Fuzzy Set Theory

Zadeh introduced a theory whose objects, i.e. fuzzy sets are sets with boundaries that are not precise. The membership in a fuzzy set is not a matter of affirmation or denial, but rather a matter of a degree. When A is a fuzzy set and x is a relevant object, the proposition "x is a member of A" is not necessarily either true or false, but it may be true only to some degree, the degree to which x is actually a member of A.

There are three basic methods by which sets can be defined within a given universal set X:

1. A set is defined by naming all its members. This method can be used for finite sets. Set A, whose members are a_1, a_2, ..., a_n, is written as

$$A = \{a_1, a_2, \ldots, a_n\}$$

2. A set is defined by a property satisfied by its members, i.e.

$$A = \{x \mid P(x)\}$$

where the symbol | denotes the phrase *such that*, and $P(x)$ designates a proposition of the form x *has the property* P. In other words, A is defined as the set of all elements of X for which the rule $P(x)$ is true. The property P is such that for any given $x \in X$, the condition or rule or proposition $P(x)$ is either true or false.

3. A set is defined by a function, called *characteristic function*, that declares which elements of X are members of the set and which are not. Set A is defined by its characteristic function or discrimination function $\mu_A(x)$, as follows:

$$\mu_A(x) = \begin{cases} 1 & \text{for } x \in A \\ 0 & \text{for } x \notin A \end{cases}$$

The characteristic function maps elements of X to elements of the set $\{0, 1\}$, which is expressed by

$$\mu_A: X \to \{0, 1\}$$

The characteristic function assigns a value either 1 or 0 to each individual in the universal set, thereby discriminating between members and non-members of the crisp set under consideration. This function can be generalized such that values assigned to the elements of the universal set fall within a specified range and indicate the membership grade of these elements in the set. Larger values represent higher degree of set membership. Such a function is called a *membership function*, and the set defined by it is a fuzzy set.

Let X be the universe of objects with elements x, where A is called a fuzzy set of X. Membership of x in classical set 'A' can be viewed as a characteristic function μ_A from X to (0,1) such that

$$\mu_A(x) = \begin{cases} 1 & \text{for } x \in A \\ 0 & \text{for } x \notin A \end{cases} \tag{5.20}$$

For a fuzzy set A of universe X, their grade of membership of x in A is defined as

$$\mu_A(x) = [1, 0] \tag{5.21}$$

where μ_A is called the *membership function*.

The value of μ_A can be anywhere from 0 to 1, and this range is what makes it different from the crisp set. The closer the value of $\mu_A(x)$ is to 1.0, the more x belongs to 1. Thus a fuzzy set has no sharp boundary. Each of the crisp subsets of X can be shown to have a one-to-one correspondence with the characteristic function. Because the membership functions are extensions of characteristic functions, fuzzy sets are extensions of crisp sets.

Fuzzy set elements are ordered pairs indicating the value of a set element and the grade of membership, i.e.

$$A = \{(x, \mu_A(x)) \mid x \in X\} \tag{5.22}$$

Let us take an example of a group of tall people whose heights are given in feet as shown in Table 5.1, i.e.

$$X = \{5.2, 5.3, 5.4, 5.5, 5.6, 5.7, 5.8, 5.9, 6.0, 6.1, 6.2\}$$

TABLE 5.1 Example of a fuzzy set

Height *x, in feet*	*Tall* μ_A	*Not Tall* $\bar{\mu}_A$	*Union* $\mu_A \cup \bar{\mu}_A$	*Intersection* $\mu_A \cap \bar{\mu}_A$
5.2	0	1	1	0
5.3	0	1	1	0
5.4	0	1	1	0
5.5	0.14	0.86	0.86	0.14
5.6	0.29	0.71	0.71	0.29
5.7	0.43	0.57	0.57	0.43
5.8	0.57	0.43	0.57	0.43
5.9	0.71	0.29	0.71	0.29
6.0	0.86	0.14	0.81	0.14
6.1	1	0	1	0
6.2	1	0	1	0

A person having height more than or equal to 6.1 feet can be considered as *tall* person. On the other hand, a person having height less than or equal to 5.4 can be considered as *not tall* person. The membership function is defined by Eq. (5.20) as

$$\mu_A(x) = \begin{cases} 1 & ; \quad x \geq 6.1 \\ \dfrac{x - 5.4}{6.1 - 5.4} & ; \quad 5.4 < x < 6.1 \\ 0 & ; \quad x \leq 5.4 \end{cases} \tag{5.23}$$

Fuzzy set is defined as

$$A = \{(x, \mu_A(x)) \mid x \in X\} \tag{5.24}$$

The fuzzy sets can be combined pairwise with minimum, maximum, and order reversal. In the following, the most common set operators are defined. For two fuzzy sets A and B, their union operation is defined as

$$\mu_A \cup \mu_B(x) = \max[\mu_A(x), \mu_B(x)] \tag{5.25}$$

For two fuzzy sets A and B, their intersection operation is defined as

$$\mu_A \cap \mu_B(x) = \min[\mu_A(x), \mu_B(x)] \tag{5.26}$$

For a fuzzy set A, the complement operation is defined as:

$$\overline{\mu}_A(x) = 1 - \mu_A(x) \tag{5.27}$$

For example:

$$\begin{aligned} A &= (1.0 \quad 0.7 \quad 0.4 \quad 0.5) \\ B &= (0.9 \quad 0.3 \quad 0.0 \quad 0.8) \\ A \cup B &= (1.0 \quad 0.7 \quad 0.4 \quad 0.8) \\ A \cap B &= (0.9 \quad 0.3 \quad 0.0 \quad 0.5) \\ \overline{A} &= (0.0 \quad 0.3 \quad 0.6 \quad 0.4) \end{aligned}$$

In Table 5.1, all the three operations are defined to give $\overline{\mu}_A$, $\mu_A \cup \overline{\mu}_A$, and $\mu_A \cap \overline{\mu}_A$.

$$\begin{aligned} &A \text{ is properly fuzzy iff } A \cap \overline{A} \neq \phi \\ &\qquad\qquad\qquad\quad \text{iff } A \cup \overline{A} \neq X \end{aligned} \tag{5.28}$$

In addition, to perform certain mathematical operations, a crisp set (non-fuzzy) may be required. An α-cut can be used to create a family of crisp sets from a given fuzzy set:

$$\mu_A(x) = \begin{cases} 1 & ; \text{if } \mu_A(x) >= \alpha \\ 0 & ; \text{otherwise} \end{cases} \tag{5.29}$$

5.4 THE SURROGATE WORTH TRADE-OFF APPROACH FOR MULTIOBJECTIVE THERMAL POWER DISPATCH PROBLEM

In the past, it was normal to formulate optimization models concerning minimizing (or maximizing) a single scalar-valued objective function. The optimization models and the analysts' perception of a problem become more realistic if many objectives are considered. The power system can also operate most efficiently when optimized with respect to several objectives or criteria under many constraints. Obviously, trade-offs among these objectives are impossible because of their different nature. So, it is stated that objectives are non-commensurable.

Generally, the multiobjective problems are solved to find non-inferior (pareto-optimal, non-dominated) solutions. Qualitatively, a non-inferior solution of a multiobjective problem is one where any improvement of one objective function can be achieved only at the expense of another. The most widely used methods of generating such non-inferior solutions are the ε-constraint and weighted minimax methods [Osyczka and Davies, 1984]. Methodologies for solving multiobjective problems differ in two major ways:

(i) The procedure used to generate non-inferior solutions
(ii) The ways used to interact with the decision makers (DMs) and the type of information made available to the DM such as trade-offs.

In almost all decision-making problems, there are several criteria for judging the possible alternatives. The main concern of the decision maker is to fulfil the conflicting goals while satisfying the constraints of the system. Further, there are two approaches to solve such problems:

(i) One approach assumes that there exists a utility function for the particular problem. Such a function is used to obtain the best alternative.
(ii) The other approach makes no assumptions regarding the existence of utility function, but provides the DM with a set of simple but effective tools to obtain the best alternative. The SWT method provides the facility to interact with the DM.

Apart from heat, power utilities using fossil fuels as a primary energy source, produce particulates and gaseous pollutants. The particulates and the gaseous pollutants such as carbon dioxide (CO_2), oxides of sulphur (SO_x) and oxides of nitrogen (NO_x) cause detrimental effects on human beings. Pollution control agencies restrict the amount of emissions of pollutants depending upon their relative harmfulness to human beings. Therefore, a priority structure can be formed for the multiobjective problem. Here, a classical economic dispatch as a multiobjective optimization problem is formulated. Four objectives are considered to minimize, namely operating cost and impacts on environment of NO_x, SO_2 and CO_2 emissions. The formulated multiobjective problem adopts a ε-constraint form, which allows explicit trade-offs between objective levels for each non-inferior solution [Li and Haimes, 1987]. The SWT method is used to find the best alternative among the non-inferior solutions.

5.4.1 Multiobjective Problem Formulation

In the multiobjective problem formulation, four important noncommensurable objectives in an electrical thermal power system have been considered. These are economy and environmental impacts because of NO_x, SO_2, and CO_2 gaseous pollutants.

Economy objective

The fuel cost of a thermal unit is regarded as an essential criterion for economic feasibility. The fuel cost curve is assumed to be approximated by a quadratic function of generator power output P_{g_i} as

$$F_1 = \sum_{i=1}^{NG} (a_i P_{g_i}^2 + b_i P_{g_i} + c_i) \text{ ₹/h} \tag{5.30}$$

where a_i, b_i, and c_i are cost coefficients and NG is the number of generators.

Environmental objectives

The emission curves can be directly related to the cost curve through the emission rate per Mkcal, which is a constant factor for a given type of fuel. Therefore, the amount of NO_x emission is given as a quadratic function of generator output P_{g_i}, i.e.

$$F_2 = \sum_{i=1}^{NG} (d_{1i} P_{g_i}^2 + e_{1i} P_{g_i} + f_{1i}) \text{ kg/h} \tag{5.31}$$

where d_{1i}, e_{1i}, and f_{1i} are NO_x emission coefficients [Zahavi and Eisenberg, 1975].

Similarly, the amount of SO_2 emission is given as a quadratic function of generator output P_{g_i}, i.e.

$$F_3 = \sum_{i=1}^{NG} (d_{2i} P_{g_i}^2 + e_{2i} P_{g_i} + f_{2i}) \text{ kg/h} \tag{5.32}$$

where d_{2i}, e_{2i}, and f_{2i} are SO_2 emission coefficients [Zahavi and Eisenberg, 1975].

The amount of CO_2 emission is also represented as a quadratic function of generator output P_{g_i}, i.e.

$$F_4 = \sum_{i=1}^{NG} (d_{3i} P_{g_i}^2 + e_{3i} P_{g_i} + f_{3i}) \text{ kg/h} \tag{5.33}$$

where d_{3i}, e_{3i}, and f_{3i} are CO_2 emission coefficients [Wong et al., 1995].

Constraints

To ensure a real power balance, an equality constraint is imposed, i.e.

$$\sum_{i=1}^{NG} P_{g_i} - (P_D + P_L) = 0 \tag{5.34}$$

where

P_D is the power demand

P_L is the transmission losses, which are approximated in terms of B-coefficients as

$$P_L = B_{00} + \sum_{i=1}^{NG} B_{i0} P_{g_i} + \sum_{i=1}^{NG} \sum_{j=1}^{NG} P_{g_i} B_{ij} P_{g_j} \text{ MW} \tag{5.35}$$

The inequality constraints imposed on generator output are

$$P_{g_i}^{\min} \le P_{g_i} \le P_{g_i}^{\max} \qquad (i = 1, 2, ..., \text{NG}) \tag{5.36}$$

where $P_{g_i}^{\min}$ is the lower limit, and $P_{g_i}^{\max}$ is the upper limit of generator output.

Aggregating Eqs. (5.30) to (5.36), the multiobjective optimization problem is defined as

$$\begin{aligned} &\text{Minimize} && [F_1(P_g), F_2(P_g), F_3(P_g), F_4(P_g)]^T \\ &\text{subject to} && \sum_{i=1}^{\text{NG}} P_{g_i} - (P_D + P_L) = 0 \\ & && P_{g_i}^{\min} \le P_{g_i} \le P_{g_i}^{\max} \qquad (i = 1, 2, ..., \text{NG}) \end{aligned} \tag{5.37}$$

where $F_1(P_g)$, $F_2(P_g)$, $F_3(P_g)$, and $F_4(P_g)$ are the objective functions to be minimized over the set of admissible decision vector P_g.

5.4.2 The ε-Constraint Method

To generate non-inferior solutions to the multiobjective problem, the ε-constraint method is used [Haimes and Hall, 1974; Cohon and Marks, 1975]. For the ε-constraint method, one of the objective functions constitutes the primary objective function and all other objectives act as constraints. To be more specific, this procedure is implemented by replacing three objectives in the problem of Eq. (5.37) with three constraints.

$$\begin{aligned} &\text{Minimize} && [F_1(P_g)] \\ &\text{subject to} && F_j(P_g) \le \varepsilon_j \qquad (j = 2, 3, 4) \\ & && \sum_{i=1}^{\text{NG}} P_{g_i} - (P_D + P_L) = 0 \\ & && P_{g_i}^{\min} \le P_{g_i} \le P_{g_i}^{\max} \qquad (i = 1, 2, ..., \text{NG}) \end{aligned} \tag{5.38}$$

where ε_j is maximum tolerable objective level. The value of ε_j is chosen for which the objective constraints in the problem of Eq. (5.38) are binding at the optimal solution. The level ε_j is varied parametrically to evaluate the impact on the single objective function $F_1(P_g)$.

To find the solution, the constrained optimization problem of Eq. (5.38) is converted into an unconstrained optimization problem. The generalized Lagrangian L is formed as

$$L = F_1 + \sum_{j=2}^{4} \lambda_{1j} (F_j - \varepsilon_j) - \mu \left(\sum_{i=1}^{\text{NG}} P_{g_i} - (P_L + P_D) \right) \tag{5.39}$$

where λ_{1j} and μ are Lagrange multipliers. The subscript $1j$ denotes that λ is a Lagrange multiplier associated with the jth constraint, where the prime objective function is F_1.

Necessary conditions to obtain the solution are

$$\frac{\partial L}{\partial P_{g_i}} = \frac{\partial F_1}{\partial P_{g_i}} + \sum_{j=2}^{4} \lambda_{1j} \frac{\partial F_j}{\partial P_{g_i}} - \mu \left(1 - \frac{\partial P_L}{\partial P_{g_i}} \right) \qquad (i = 1, 2, ..., \text{NG}) \tag{5.39a}$$

$$\frac{\partial L}{\partial \lambda_{1j}} = F_j - \varepsilon_j \qquad (j = 2, 3, 4) \tag{5.39b}$$

$$\frac{\partial L}{\partial \mu} = P_D + P_L - \sum_{j=1}^{NG} P_{g_i} \tag{5.39c}$$

where

$$\frac{\partial F_1}{\partial P_{g_i}} = 2a_i P_{g_i} + b_i$$

$$\frac{\partial F_{j+1}}{\partial P_{g_i}} = 2d_{ji} P_{g_i} + e_{ji} \qquad (j = 1, 2, 3)$$

$$\frac{\partial P_L}{\partial P_{g_i}} = 2B_{ii} P_{g_i} + \sum_{\substack{j=1 \\ j \neq i}}^{NG} 2B_{ij} P_{g_j} + B_{i0}$$

Newton–Raphson method can be applied to solve the above equations. To implement the Newton–Raphson method, the following equation is solved iteratively till the required convergence criteria is satisfied.

$$\begin{bmatrix} \nabla_{P_g P_g} L & \nabla_{P_g \lambda} L & \nabla_{P_g \mu} L \\ (\nabla_{P_g \lambda} L)^T & \nabla_{\lambda\lambda} L & \nabla_{\lambda\mu} L \\ (\nabla_{P_g \mu} L)^T & \nabla_{\mu\lambda} L & \nabla_{\mu\mu} L \end{bmatrix} \begin{bmatrix} \Delta P_g \\ \Delta \lambda \\ \Delta \mu \end{bmatrix} = \begin{bmatrix} -\nabla_{P_g} L \\ -\nabla_{\lambda} L \\ -\nabla_{\mu} L \end{bmatrix} \tag{5.40}$$

Hessian matrix elements can be obtained from Eqs. (5.39a) to (5.39c) by differentiating with respect to control variables one by one.

$$\frac{\partial^2 L}{\partial P_{g_i}^2} = \frac{\partial^2 F_1}{\partial P_{g_i}^2} + \sum_{j=2}^{M} \lambda_{1j} \frac{\partial^2 F_j}{\partial P_{g_i}^2} + \mu \frac{\partial^2 P_L}{\partial P_{g_i}^2} \qquad (i = 1, 2, ..., NG) \tag{5.40a}$$

$$\frac{\partial^2 L}{\partial P_{g_i} \partial P_{g_j}} = \mu \frac{\partial^2 P_L}{\partial P_{g_i} \partial P_{g_j}} \qquad (i = 1, 2, ..., NG; \;\; j \neq i; \;\; j = 1, 2, ..., NG) \tag{5.40b}$$

$$\frac{\partial^2 L}{\partial P_{g_i} \partial \lambda_{1j}} = \frac{\partial^2 L}{\partial \lambda_{1j} \partial P_{g_i}} = \frac{\partial F_j}{\partial P_{g_i}} \qquad (j = 2, 3, 4; \;\; i = 1, 2, ..., NG) \tag{5.40c}$$

$$\frac{\partial^2 L}{\partial P_{g_i} \partial \mu} = \frac{\partial^2 L}{\partial \mu \partial P_{g_i}} = \frac{\partial P_L}{\partial P_{g_i}} - 1 \qquad (i = 1, 2, ..., NG) \tag{5.40d}$$

$$\frac{\partial^2 L}{\partial \lambda_{1j}^2} = \frac{\partial^2 L}{\partial \mu^2} = \frac{\partial^2 L}{\partial \mu \partial \lambda_{1j}} = \frac{\partial^2 L}{\partial \lambda_{1j} \partial \mu} = 0 \qquad (j = 2, 3, 4) \tag{5.40e}$$

The optimal solution must satisfy the Kuhn-Tucker conditions and the main condition is

$$\lambda_{1j}\,(F_j(P_g) - \varepsilon_j) = 0.0;\quad \lambda_{1j} > 0.0 \qquad (j = 2, 3, 4) \tag{5.41}$$

The Lagrange multipliers related to the objectives, as constraints may be zero or non-zero. The set of non-zero Lagrange multipliers corresponds to the non-inferior set of solutions. The set of non-zero Lagrange multipliers represents the set of trade-off ratios between the principal objective and each of the constraining objectives, respectively. The system given by Eq. (5.39) is solved using the Newton–Raphson method for R values of ε_j (j = 2 or 3 or 4) while other ε_k ($k \neq j$) are fixed at some level. Only those values of $\lambda^r_{1j} > 0.0$ which correspond to active constraints $F_j^r(P_g) = \varepsilon_j^r$ ($r = 1, 2, ..., R$) are considered, since they belong to the non-inferior solution [Haimes and Hall, 1974].

In the problem, the initial value of ε_j has been taken such that $\varepsilon_j > F_j^{min}$ and $\varepsilon_j < F_j^{max}$. Since objectives are of conflicting nature, the value of one objective will be maximum when the value of another objective is minimum and vice versa.

Algorithm 5.1: Non-Inferior Solution by ε-Constraint Method

1. Read data, namely cost coefficients, emission coefficients and B-coefficients, ε (convergence tolerance) and ITMAX (maximum allowed iterations), M (number of objectives), NG (number of generators) and K_1 to K_j (minimum number of inferior solutions required for each objective as constraint), etc.
2. Set objective index, $j = 1$
3. If ($j \geq M$), then GOTO Step 20.
   else increment the objective index, $j = j + 1$
4. Fix ε_j = 2, 3, 4 such that $F_j^{min} < \varepsilon_j < F_j^{max}$
5. Set iteration for non-inferior solutions, $k = 1$
6. Increment count of non-inferior solutions, $k = k + 1$
7. If ($k \geq K_j$) GOTO Step 3.
8. Compute initial values of P_{g_i} (i = 1, 2, ..., NG) and λ.
9. Assume that no generator has been fixed either at lower limit or at upper limit.
10. Set iteration counter IT=1
11. Compute Hessian and Jacobian matrix elements using Eqs. (5.39a) to (5.39c) and Eqs. (5.40a) to (5.40e), respectively.
    Deactivate row and column of Hessian matrix and row of Jacobian matrix representing the generator whose generation is fixed either at lower limit or at upper limit. This is done so that fixed generators cannot participate in allocation.
12. Gauss elimination method is employed in which triangularization and back-substitution processes are performed to find ΔP_{g_i} ($i = 1, 2, ..., R$), $\Delta\lambda_{1j}$, $\Delta\mu$. Here R is the number of generators which can participate in allocation.
13. Check either $\sqrt{\sum_{i=1}^{R} (\Delta P_{g_i})^2 + (\Delta\lambda_{1j})^2 + (\Delta\mu)^2} \leq \varepsilon$.
    if convergence condition is 'satisfied' then GOTO Step 16.
14. Modify control variables,

$$P_{g_i}^{new} = P_{g_i} + \Delta P_{g_i} \qquad (i = 1, 2, ..., R)$$

$$\lambda_{1j}^{\text{new}} = \lambda_{1j} + \Delta\lambda_{1j}; \quad \mu^{\text{new}} = \mu + \Delta\mu$$

15. IT = IT + 1, $P_{g_i} = P_{g_i}^{\text{new}}$ $(i = 1, 2, ..., R)$, $\lambda_{1j} = \lambda_{1j}^{\text{new}}$ and $\mu = \mu^{\text{new}}$ and GOTO Step 11 and repeat.
16. Check the limits of generators and fix up as following:

$$\text{If } P_{g_i} < P_{g_i}^{\min} \quad \text{then} \quad P_{g_i} = P_{g_i}^{\min}$$

$$\text{If } P_{g_i} > P_{g_i}^{\max} \quad \text{then} \quad P_{g_i} = P_{g_i}^{\max}$$

If no more violations then GOTO Step 18.
17. GOTO Step 10.
18. Check the condition is satisfied

$$\lambda_{1j}(F_j(P) - \varepsilon_j) = 0.0; \; \lambda_{1j} > 0.0$$

If 'yes' then GOTO Step 19.
else modify ε_j and GOTO Step 6.
19. Record it as non-inferior solution, compute values of all objectives and transmission loss and modify ε_j for the next non-inferior solution and GOTO Step 6.
20. Stop.

5.4.3 The Surrogate Worth Trade-Off (SWT) Function

The SWT function assigns a scalar value (on an ordinal scale) to any given non-inferior solution. One way to specify non-inferior solution is trade-off functions. Moreover, there is close relationship between the SWT function W_{1j} and the partial derivatives of the utility functions. In multiobjective analysis, it is assumed implicitly that the DM maximizes his utility which is a monotonic decreasing function of the objective functions. Given a decision vector P_g and the associated consequences F_i, the utility is given by :

$$U = U[F_1(P_g), F_2(P_g), F_3(P_g), F_4(P_g)] \tag{5.42}$$

By linearizing utility function for a small change in F_1 (the prime objective), the following can be obtained [Haimes, 1977].

$$\Delta U = \sum_{j=1}^{4} \frac{\partial U}{\partial F_j} \Delta F_j \tag{5.43}$$

However for non-inferior points, $F_1 = F_1(F_2, F_3, F_4)$ or

$$\Delta F_1 = \sum_{j=2}^{4} \frac{\partial F_1}{\partial F_j} \Delta F_j \tag{5.44}$$

λ_{1j} indicates the marginal benefit of the objective function F_1 due to an additional unit of ε_j. So it can be defined as

$$\lambda_{1j} = -\frac{\partial F_1}{\partial \varepsilon_j} \qquad (j = 2, 3, 4) \tag{5.45}$$

In the non-inferior set, only those $\lambda_{1j} > 0$ corresponding to $F_j = \varepsilon_j$ are considered. Thus, for $F_j = \varepsilon_j$, Eq. (5.45) can be written as

$$\lambda_{1j} = -\frac{\partial F_1}{\partial F_j} \qquad (j = 2, 3, 4) \tag{5.46}$$

By substituting Eq. (5.46) into Eq. (5.44)

$$\Delta F_1 = -\sum_{j=2}^{4} \lambda_{1j}\Delta F_j \tag{5.47}$$

Eliminating ΔF_1 from Eq. (5.43) using Eq. (5.47)

$$\Delta U = \sum_{j=2}^{4}\left(\frac{\partial U}{\partial F_j} - \frac{\partial U}{\partial F_1}\lambda_{1j}\right)\Delta F_j = \sum_{j=2}^{4} \Delta U_{1j}\Delta F_j \tag{5.48}$$

where

$$\Delta U_{1j} = \frac{\partial U}{\partial F_j} - \frac{\partial U}{\partial F_1}\lambda_{1j} \tag{5.49}$$

Let

$$\delta_i = \frac{\partial U}{\partial F_i}$$

$$\Delta U_{1j} = \delta_j - \delta_1\lambda_{1j} \tag{5.50}$$

The SWT function W_{1j} is a monotonic function of ΔU_{1j}, with the property that $W_{1j} = 0 \leftrightarrow \Delta U_{1j}$ and is written as

$$W_{1j} = h_j\, \Delta U_{1j} \qquad (j = 2, 3, 4) \tag{5.51}$$

where h_j is some monotonic increasing function of its argument, with a range of –20 to 20 and with property that $h_j(0) = 0$. If δ_j is considered constant or varies only slightly with $F_j(j = 2, 3, 4)$, then it is possible to assume that W_{1j} depends only on λ_{1j}.

5.4.4 Utility Function

It can be assumed that there exists a very general utility function that for a given DM can predict his behaviour and interest. Let the DM's utility function be defined for each objective function depending on the importance of the objective function F_i to other objective functions. So, the total or overall utility function is defined as [Haimes, 1977]:

$$\text{Maximize} \qquad U = \sum_{k=1}^{4} w_k\, F_k(P_g) \tag{5.52a}$$

The solution vector P_g is then bound by maximizing the total utility subject to the technology constraints as defined below:

$$\sum_{i=1}^{NG} P_{g_i} - (P_D + P_L) = 0.0 \tag{5.52b}$$

$$P_{g_i}^{\min} \geq P_{g_i} \geq P_{g_i}^{\max} \qquad (i = 1, 2, ..., NG) \tag{5.52c}$$

Further define

$$w_j = \frac{\alpha_j}{\sum_{k=1}^{4} \alpha_k} \qquad (j = 1, 2, 3, 4) \tag{5.52d}$$

such that

$$\sum_{k=1}^{4} w_k = 1.0 \ (w_k \geq 0.0) \tag{5.52e}$$

where α_k is the scalar weights assigned to the kth objective. The DM assigns the weight α_k on the attribute between (0, 100).

The generalized Lagrangian L is formed as

$$L = \sum_{k=1}^{4} w_k F_k + \lambda \left((P_L + P_D) - \sum_{i=1}^{NG} P_{g_i} \right) \tag{5.53}$$

where λ is the Lagrange multiplier.

Necessary conditions to obtain the solution are

$$\frac{\partial L}{\partial P_{g_i}} = \sum_{k=1}^{4} w_k \frac{\partial F_k}{\partial P_{g_i}} + \lambda \left(\frac{\partial P_L}{\partial P_{g_i}} - 1 \right) = 0 \qquad (i = 1, 2, ..., NG) \tag{5.54a}$$

$$\frac{\partial L}{\partial \lambda} = P_D + P_L - \sum_{i=1}^{NG} P_{g_i} = 0 \tag{5.54b}$$

Newton–Raphson method can be applied to solve the above equations. To implement the Newton–Raphson method, the following equation is solved iteratively. Here, the Hessian matrix must be negative definite to ensure that the function is maximum.

$$\begin{bmatrix} \nabla_{P_g P_g} L & \nabla_{P_g \lambda} L \\ (\nabla_{P_g \lambda} L)^T & \nabla_{\lambda\lambda} L \end{bmatrix} \begin{bmatrix} \Delta P_g \\ \Delta \lambda \end{bmatrix} = \begin{bmatrix} -\nabla_{P_g} L \\ -\nabla_{\lambda} L \end{bmatrix} \tag{5.55}$$

Algorithm 5.2: The SWT Algorithm

The SWT analysis is most useful when objectives are in conflict. The trade-off analysis can then be conducted between the cost and each of the other objectives. A stepwise procedure corresponding to an algorithm outlined by Haimes and Hall [1974], is given below:

1. Find the minimum and maximum values for each of the multiple objectives separately $F_j^{\min}$, $F_j^{\max}$ (j = 2, 3, 4). This is carried out by performing economic and minimum NO_x, SO_2, and CO_2 emission dispatch separately.

   Minimize $\qquad [F_j(P_g)]$

subject to
$$(P_D + P_L) - \sum_{i=1}^{NG} P_{g_i} = 0$$

$$P_{g_i}^{\min} \le P_{g_i} \le P_{g_i}^{\max} \qquad (i = 1, 2, ..., NG)$$

2. **Generation of Trade-off function:** The system Eq. (5.39) is solved using the Newton–Raphson method (Algorithm 5.1). The solution procedure is repeated for K_1 values of ε_2, say, $\varepsilon_2^1, ..., \varepsilon_2^{K_1}$, where ε_3 and ε_4 are held at some fixed level, say, ε_3^0 and ε_4^0. Only those values of $\lambda^r{}_{12} > 0$ which correspond to the active constraint $F_2^r = \varepsilon_2^r$ $(r = 1, 2, ..., K_1)$ are considered since they belong to non-inferior solution. Similarly the trade-off function λ_{13} is generated, where Eq. (5.39) is solved for K_2 different values of ε_3^r $(r = 1, 2, ..., K_2)$, with fixed level of ε_2^0 and ε_4^0. λ_{14} is generated by varying ε_4^r $(r = 1, 2, ..., K_3)$ while ε_2^0 and ε_3^0 are held at fixed level as explained in Algorithm 5.1. Regression analysis is performed to yield the trade-off functions $\lambda_{12}[F_2]$, $\lambda_{13}[F_3]$, and $\lambda_{14}[F_4]$.
3. **Generation of SWT function:** The SWT functions W_{12}, W_{13}, and W_{14} are obtained by solving Eqs. (5.49) and (5.51) at the time when the trade-off functions are generated.
4. Find functions $W_{1j}(\lambda_{1j})$ $(j = 2, 3, 4)$ by regression analysis or by interpolation. Then the values $\lambda_{1j} = \lambda^*_{1j}$ is chosen where $W_{1j}(\lambda_{1j}) = 0.0$ $(j = 2, 3, 4)$.
5. The optimal set of decision vectors is found by solving the following problem:

Minimize
$$\left[F_1 + \sum_{j=2}^{4} \lambda^*_{1j} F_j\right] \tag{5.56a}$$

subject to
$$(P_D + P_L) - \sum_{i=1}^{NG} P_{g_i} = 0 \tag{5.56b}$$

$$P_{g_i}^{\min} \ge P_{g_i} \ge P_{g_i}^{\max} \qquad (i = 1, 2, ..., NG) \tag{5.56c}$$

The generalized Lagrangian L is formed as

$$L = \left(F_1 + \sum_{j=2}^{4} \lambda^*_{1j} F_j\right) + \mu\left((P_L + P_D) - \sum_{i=1}^{NG} P_{g_i}\right) \tag{5.57}$$

where μ is Lagrange multiplier.
Necessary conditions to obtain the solution are:

$$\frac{\partial L}{\partial P_{g_i}} = \frac{\partial F_1}{\partial P_{g_i}} + \sum_{j=2}^{4} \lambda^*_{1j} \frac{\partial F_j}{\partial P_{g_i}} + \mu\left(\frac{\partial P_L}{\partial P_{g_i}} - 1\right) = 0 \qquad (i = 1, 2, ..., NG) \tag{5.58a}$$

$$\frac{\partial L}{\partial \lambda} = P_D + P_L - \sum_{j=1}^{NG} P_{g_i} = 0 \tag{5.58b}$$

Newton–Raphson method has been applied in which the following equation is solved iteratively. Here the Hessian matrix must be positive definite to ensure that the function is minimum.

$$\begin{bmatrix} \nabla_{P_g P_g} L & \nabla_{P_g \mu} L \\ (\nabla_{P_g \mu} L)^T & \nabla_{\mu\mu} L \end{bmatrix} \begin{bmatrix} \nabla_{P_g} \\ \nabla_{\mu} \end{bmatrix} = \begin{bmatrix} -\nabla_{P_g} L \\ -\nabla_{\mu} L \end{bmatrix} \tag{5.59}$$

6. Stop.

5.4.5 Test System and Results

A six-generator system [Lamont and Obessis, 1995; Wong et al., 1955] is considered. The fuel cost, NO_x emission, SO_2 emission and CO_2 emission equations are given in Tables 5.2 to 5.5. Transmission loss coefficients are given in Table 5.6. The power demand is considered to be 1800 MW.

TABLE 5.2 Fuel cost (₹/h) equations

$F_{11} = 0.002035P_1^2 + 8.43205P_1 + 85.6348$
$F_{12} = 0.003866P_2^2 + 6.41031P_2 + 303.7780$
$F_{13} = 0.002182P_3^2 + 7.42890P_3 + 847.1484$
$F_{14} = 0.001345P_4^2 + 8.30154P_4 + 274.2241$
$F_{15} = 0.002182P_5^2 + 7.42890P_5 + 847.1484$
$F_{16} = 0.005963P_6^2 + 6.91559P_6 + 202.0258$

TABLE 5.3 NO_x emission (kg/h) equations

$F_{21} = 0.006323P_1^2 - 0.38128P_1 + 80.9019$
$F_{22} = 0.006483P_2^2 - 0.79027P_2 + 28.8249$
$F_{23} = 0.003174P_3^2 - 1.36061P_3 + 324.1775$
$F_{24} = 0.006732P_4^2 - 2.39928P_4 + 610.2535$
$F_{25} = 0.003174P_5^2 - 1.36061P_5 + 324.1775$
$F_{26} = 0.006181P_6^2 - 0.39077P_6 + 50.3808$

TABLE 5.4 SO_2 emission (kg/h) equations

$F_{31} = 0.001206P_1^2 + 5.05928P_1 + 51.3778$
$F_{32} = 0.002320P_2^2 + 3.84624P_2 + 182.2605$
$F_{33} = 0.001284P_3^2 + 4.45647P_3 + 508.5207$
$F_{34} = 0.000813P_4^2 + 4.97641P_4 + 165.3433$
$F_{35} = 0.001284P_5^2 + 4.45647P_5 + 508.5207$
$F_{36} = 0.003578P_6^2 + 4.14938P_6 + 121.2133$

TABLE 5.5 CO_2 emission (kg/h) equations

$$F_{41} = 0.265110P_1^2 - 61.01945P_1 + 5080.148$$
$$F_{42} = 0.140053P_2^2 - 29.95221P_2 + 3824.770$$
$$F_{43} = 0.105929P_3^2 - 9.552794P_3 + 1342.851$$
$$F_{44} = 0.106409P_4^2 - 12.73642P_4 + 1819.625$$
$$F_{45} = 0.105929P_5^2 - 9.552794P_5 + 1342.851$$
$$F_{46} = 0.403144P_6^2 - 121.9812P_6 + 11381.070$$

TABLE 5.6 Loss coefficients

0.000200	0.000010	0.000015	0.000005	0.000000	−0.000030
0.000010	0.000300	−0.000020	0.000001	0.000012	0.000010
0.000015	−0.000020	0.000100	−0.000010	0.000010	0.000008
0.000005	0.000001	−0.000010	0.000150	0.000006	0.000050
0.000000	0.000012	0.000010	0.000006	0.000250	0.000020
−0.000030	0.000010	0.000008	0.000050	0.000020	0.000210

To obtain the solution, seven different cases are considered as below:

Case I Minimum cost dispatch

Case II Minimum NO_x emission dispatch

Case III Minimum SO_2 emission dispatch

Case IV Minimum CO_2 emission dispatch

Case V Utility function value when weights 25, 25, 25, and 25 are assigned to cost, NO_x emission, SO_2 emission, and CO_2 emission, respectively.

Case VI Utility function value when weights 40, 30, 20, and 10 are assigned to cost, NO_x emission, SO_2 emission, and CO_2 emission, respectively.

Case VII Utility function value when weights 60, 20, 10, and 10 are assigned to cost, NO_x emission, SO_2 emission, and CO_2 emission, respectively.

The results obtained for case I, II, III, and IV are shown in Table 5.10 and the corresponding generation schedule is given in Table 5.11. The conflicting objectives F_1 and F_2, trade-off function λ_{12}, and SWT function W_{12}, corresponding to each non-inferior solution are shown in Table 5.7. The values of objective functions F_3 and F_4 are fixed at 11289.0 kg/h and 62502.0 kg/h, respectively. W_{12}^1, W_{12}^2 and W_{12}^3 are the SWT functions for case V, VI and VII, respectively. Similarly, the conflicting objective F_1 and F_3, trade-off function λ_{13}, and SWT function W_{13}, corresponding to each non-inferior solution are shown in Table 5.8. The values of objective functions F_2 and F_4 are fixed at 2265.0 kg/h and 60899.9 kg/h, respectively. Table 5.9 shows the non-inferior solution for F_1, F_4, λ_{14}, and W_{14} when F_2 and F_3 are fixed at 2265.0 kg/h and 11223.0 kg/h, respectively.

TABLE 5.7 Non-inferior solutions when F_3 and F_4 are fixed

Sr. no.	F_1 (₹/h)	F_2 (kg/h)	λ_{12} (₹/kg)	W_{12}^1	W_{12}^2	W_{12}^3
1	18835.3716	2142.50	0.004335	−5.63	−2.68	−2.67
2	18835.3577	2145.50	0.004665	−5.73	−2.74	−2.72
3	18835.3435	2148.50	0.004968	−5.83	−2.80	−2.77
4	18835.3284	2151.50	0.005249	−5.93	−2.86	−2.83
5	18835.3118	2154.50	0.005512	−6.04	−2.93	−2.89
6	18835.2952	2157.50	0.005756	−6.14	−3.00	−2.95
7	18835.2776	2160.50	0.005985	−6.26	−3.07	−3.01
8	18835.2600	2162.50	0.006202	−6.37	−3.15	−3.08
9	18835.2403	2166.50	0.006403	−6.49	−3.22	−3.15
10	18835.2210	2169.50	0.006595	−6.61	−3.31	−3.22
11	18835.1812	2172.50	0.006778	−6.74	−3.39	−3.29
12	18835.1591	2175.50	0.006952	−6.87	−3.48	−3.37
13	18835.1377	2178.50	0.007116	−7.00	−3.57	−3.45
14	18835.1159	2181.50	0.007273	−7.15	−3.67	−3.54
15	18835.0933	2184.50	0.007425	−7.29	−3.77	−3.63

TABLE 5.8 Non-inferior solutions when F_2 and F_4 are fixed

Sr. no.	F_1 (₹/h)	F_3 (kg/h)	λ_{13} (₹/kg)	W_{13}^1	W_{13}^2	W_{13}^3
1	18721.9870	11223.02	1414.0000	15.82	7.36	7.56
2	18722.0529	11223.06	1346.0540	15.08	7.02	7.21
3	18722.1011	11223.09	702.8373	7.87	3.66	3.76
4	18722.1344	11223.11	309.2964	3.56	1.61	1.65
5	18722.1497	11223.12	284.0960	3.17	1.47	1.52
6	18722.1815	11223.14	61.1189	0.66	0.31	0.32
7	18722.1973	11223.15	25.0154	0.26	0.12	0.13
8	18722.2061	11223.16	2.3095	0.08	0.04	0.04

By regression analysis, the SWT functions are shown as:

Case V

$$W_{12} = -2.28937 - 685.559\lambda_{12}$$
$$W_{13} = 0.0304876 + 0.01116\lambda_{13}$$
$$W_{14} = 0.315173 + 2.53935\lambda_{14}$$

with 0.998, 0.999, and 0.999 standard deviations, respectively. The values of λ_{12}^*, λ_{13}^*, and λ_{14}^* are – 0.00334, – 2.7318, and – 0.12412, respectively as discussed in Step 4 of the Algorithm 5.2. The optimal decision vector P_g^* is obtained by solving the problem of Eq. (5.52) as discussed in Step 5 of the algorithm. The obtained optimal values of cost, NO_x emission, SO_2 emission, CO_2 emission, transmission losses, and utility function are shown in Table 5.9 and the corresponding generation schedule is given in Table 5.10.

TABLE 5.9 Non-inferior solutions when F_2 and F_3 are fixed

Sr. no.	F_1 (₹/h)	F_4 (kg/h)	λ_{14} (₹/kg)	W_{14}^1	W_{14}^2	W_{14}^3
1	18721.3741	60100.0	5.774911	15.0	19.67	20.07
2	18721.3813	60110.0	2.708273	7.1	9.33	9.52
3	18721.3963	60130.0	2.483800	6.6	8.67	8.85
4	18721.4262	60170.0	2.079267	5.6	7.46	7.62
5	18721.4708	60230.0	1.368308	3.9	5.12	5.23
6	18721.4782	60240.0	1.356113	3.8	5.11	5.21
7	18721.5223	60300.0	0.451489	1.4	1.82	1.86

TABLE 5.10 Comparison of results

Case no.	F_1 (₹/h)	F_2 (kg/h)	F_3 (kg/h)	F_4 (kg/h)	P_L (MW)	*Utility*
I	18721.39	2282.964	11222.99	60482.22	130.1478	–
II	18950.87	2070.127	11356.50	66939.14	148.2133	–
III	18721.49	2277.155	11222.94	60620.73	130.0320	–
IV	18790.84	2361.644	11266.52	58066.35	141.3971	–
V	18772.61	2339.327	11255.35	58112.39	139.3240	22619.919
VI	18848.86	2424.725	11301.76	58348.42	138.8742	16362.155
VII	18837.92	2412.389	11295.13	58262.36	146.3526	18740.978

Case VI

$$W_{12} = -0.335545 - 473.126\lambda_{12}$$
$$W_{13} = -0.001487 + 0.005224\lambda_{13}$$
$$W_{14} = 0.465315 + 0.332834\lambda_{14}$$

with 0.969, 0.999, and 0.999 as standard deviations, respectively. The values of λ_{12}^*, λ_{13}^*, and λ_{14}^* are –0.00071, 0.2848, and –0.1398, respectively. The optimal values of cost, NO_x emission, SO_2 emission, CO_2 emission, transmission losses, and utility function are shown in Table 5.9 and the corresponding generation schedule is shown in Table 5.10.

Case VII

$$W_{12} = -0.655027 - 410.04\lambda_{12}$$
$$W_{13} = -0.0003438 + 0.00534722\lambda_{13}$$
$$W_{14} = 0.48 + 3.39042\lambda_{14}$$

with 0.961, 0.999, and 0.999 as standard deviations, respectively. The values of λ_{12}^*, λ_{13}^*, and λ_{14}^* are – 0.0015974, 0.0643, and –0.14157, respectively. The optimal values of cost, NO_x emission, SO_2 emission, CO_2 emission, transmission losses, and utility function are shown in Table 5.9 and the corresponding generation schedule is shown in Table 5.11.

TABLE 5.11 Generation schedules

Case no.	P_1 (MW)	P_2 (MW)	P_3 (MW)	P_4 (MW)	P_5 (MW)	P_6 (MW)
I	251.6940	303.7786	503.4812	372.3225	301.4699	197.4014
II	195.4008	215.4831	536.1999	329.1267	479.4874	192.5153
III	250.9675	302.7487	507.4563	369.7902	302.6239	196.4453
IV	249.4354	334.0286	393.4468	383.3058	345.6173	235.5632
V	250.0781	327.7353	409.1765	380.8321	339.5446	231.9571
VI	247.1096	351.6184	355.5683	386.7245	363.8378	242.5683
VII	247.5022	348.3443	361.8217	386.1796	360.9468	241.5578

It may be noted that each step for reduction of emission becomes increasingly expensive. The solution procedure is sensitive to the utility function. The obtained results from the test system show the effectiveness and flexibility of the proposed algorithm for multiobjective optimization problems of any number of objectives. For different quality of coal, the cost characteristics and emission characteristics will change but the method remains the same. The method is also applicable if the exact loss formula is used.

5.5 MULTIOBJECTIVE THERMAL POWER DISPATCH PROBLEM—WEIGHTING METHOD

In the multiobjective problem formulation, four important non-commensurable objectives in an electrical thermal power system are considered. These are economy and environmental impacts because of NO_x, SO_2, and CO_2 emissions. The multiobjective optimization problem is defined as

$$\text{Minimize} \quad F_1 = \sum_{i=1}^{NG} (a_i P_{g_i}^2 + b_i P_{g_i} + c_i) \text{ ₹/h} \tag{5.60a}$$

$$\text{Minimize} \quad F_2 = \sum_{i=1}^{NG} (d_{1i} P_{g_i}^2 + e_{1i} P_{g_i} + f_{1i}) \text{ kg/h} \tag{5.60b}$$

$$\text{Minimize} \quad F_3 = \sum_{i=1}^{NG} (d_{2i} P_{g_i}^2 + e_{2i} P_{g_i} + f_{2i}) \text{ kg/h} \tag{5.60c}$$

$$\text{Minimize} \quad F_4 = \sum_{i=1}^{NG} (d_{3i} P_{g_i}^2 + e_{3i} P_{g_i} + f_{3i}) \text{ kg/h} \tag{5.60d}$$

subject to
$$\sum_{i=1}^{NG} P_{g_i} - (P_D + P_L) = 0 \tag{5.60e}$$

$$P_{g_i}^{\min} \le P_{g_i} \le P_{g_i}^{\max} \qquad (i = 1, 2, ..., \text{NG}) \tag{5.60f}$$

where

NG is the number of generators
a_i, b_i, and c_i are cost coefficients
d_{1i}, e_{1i}, and f_{1i} are NO_x emission coefficients
d_{2i}, e_{2i}, and f_{2i} are SO_2 emission coefficients
d_{3i}, e_{3i}, and f_{3i} are CO_2 emission coefficients [Wong et al., 1995]
P_D is the power demand

P_L is the transmission losses, which are approximated in terms of *B*-coefficients as

$$P_L = B_{00} + \sum_{i=1}^{NG} B_{i0} P_{g_i} + \sum_{i=1}^{NG} \sum_{j=1}^{NG} P_{g_i} B_{ij} P_{g_j} \quad \text{MW}$$

where $P_i^{\min}$ is the lower limit of generator output and $P_i^{\max}$ is the upper limit of generator output.

$F_1(P_g)$, $F_2(P_g)$, $F_3(P_g)$, and $F_4(P_g)$ are the objective functions to be minimized over the set of admissible decision vector *P*. To generate the non-inferior solution to the multiobjective problem, the weighting method is applied. In this method, the problem is converted into a scalar optimization as given below:

Minimize
$$\sum_{k=1}^{M} w_k F_k(P_{g_i}) \tag{5.61a}$$

subject to
$$\sum_{i=1}^{NG} P_{g_i} - (P_D + P_L) = 0 \tag{5.61b}$$

$$P_{g_i}^{\min} \le P_{g_i} \le P_{g_i}^{\max} \qquad (i = 1, 2, ..., \text{NG}) \tag{5.61c}$$

$$\sum_{k=1}^{M} w_k = 1 \quad (w_k \ge 0) \tag{5.61d}$$

where

M is number of objectives
w_k are levels of normalized weights.

This approach yields meaningful result to the decision maker when solved many times for different values of $w_k(k = 1, 2, ..., M)$. Weighting factors w_k are determined based on the relative importance of various objectives, which may vary from place to place and utility to utility. The constraint scalar optimization problem is converted into unconstraint scalar optimization problem. Each constraint equation is associated with an unknown multiplier function known as Lagrange multiplier. The augmented objective function is

$$L = \sum_{k=1}^{M} w_k F_k + \lambda \left(P_D + P_L - \sum_{i=1}^{NG} P_{g_i} \right) \tag{5.62}$$

The optimality conditions are described by taking the partial derivatives of the augmented objective function with respect to the decision variables.

$$\frac{\partial L}{\partial P_{g_i}} = \sum_{k=1}^{M} w_k \frac{\partial F_k}{\partial P_{g_i}} + \lambda \left(\frac{\partial P_L}{\partial P_{g_i}} - 1 \right) = 0 \quad (i = 1, 2, ..., NG) \tag{5.63a}$$

$$\frac{\partial L}{\partial \lambda} = P_D + P_L - \sum_{i=1}^{NG} P_{g_i} = 0 \tag{5.63b}$$

where

$$\frac{\partial F_1}{\partial P_{g_i}} = 2a_i P_{g_i} + b_i$$

$$\frac{\partial F_{k+1}}{\partial P_{g_i}} = 2d_{ki} P_{g_i} + e_{ki} \quad (k = 1, 2, 3)$$

$$\frac{\partial P_L}{\partial P_{g_i}} = B_{i0} + \sum_{j=1}^{NG} 2B_{ij} P_{g_j}$$

Inherently, these equations are nonlinear. The Newton–Raphson method has been applied to find the solution. To implement the Newton–Raphson method, the following equation is solved iteratively.

$$\begin{bmatrix} \nabla_{P_g P_g} L & \nabla_{P_g \lambda} L \\ (\nabla_{P_g \lambda} L)^T & \nabla_{\lambda\lambda} L \end{bmatrix} \begin{bmatrix} \Delta P_g \\ \Delta \lambda \end{bmatrix} = \begin{bmatrix} -\nabla_{P_g} L \\ -\nabla_{\lambda} L \end{bmatrix} \tag{5.64}$$

Hessian matrix elements can be obtained from Eqs. (5.63a) and (5.63b) by differentiating these equations with respect to P_{g_i}.

$$\frac{\partial^2 L}{\partial P_{g_i}^2} = \sum_{k=1}^{M} w_k \frac{\partial^2 F_k}{\partial P_{g_i}^2} + \lambda \frac{\partial^2 P_L}{\partial P_{g_i}^2} \quad (i = 1, 2, ..., NG) \tag{5.65a}$$

$$\frac{\partial^2 L}{\partial P_{g_i} \partial P_{g_j}} = \lambda \frac{\partial^2 P_L}{\partial P_{g_i} \partial P_{g_j}} \quad (i = 1, 2, ..., NG;\ j = 1, 2, ..., NG;\ i \neq j) \tag{5.65b}$$

$$\frac{\partial^2 L}{\partial P_{g_i} \partial \lambda} = \frac{\partial^2 L}{\partial \lambda \partial P_{g_i}} = \frac{\partial P_L}{\partial P_{g_i}} - 1 \quad (i = 1, 2, ..., NG) \tag{5.65c}$$

$$\frac{\partial^2 L}{\partial \lambda^2} = 0 \tag{5.65d}$$

By varying weights, Eq. (5.64) is solved using the Newton–Raphson method to generate non-inferior solutions as explained in Algorithm 5.3.

Algorithm 5.3: Non-Inferior Solution by Weighting Method

1. Read data, namely cost coefficients, emission coefficients and B-coefficients, demand, ε (convergence tolerance) and ITMAX (maximum allowed iterations), M (number of objectives), NG (number of generators), and K (number of inferior solutions), etc.
2. Set iteration for non-inferior solutions, $k = 1$
3. Increment count of non-inferior solutions, $k = k + 1$
4. If ($k \geq K$) GOTO Step 17.
5. Feed or generate weights, w_i (i = 1, 2, ..., M)
6. Compute initial values of P_{g_i} (i = 1, 2, ..., NG) and λ by presuming that $P_L = 0$. The values of λ and P_{g_i} (i = 1, 2, ..., NG) can be computed directly using Eqs. (3.10) and (3.9), respectively.
7. Assume that no generator has been fixed either at lower limit or at upper limit.
8. Set iteration counter IT = 1.
9. Compute Hessian and Jacobian matrix elements using Eqs. (5.65a) and (5.65d) and Eqs. (5.63a) to (5.63b), respectively.
   Deactivate row and column of Hessian matrix and row of Jacobian matrix representing the generator whose generation is fixed either at lower limit or at upper limit. This is done so that fixed generators cannot participate in allocation.
10. Gauss elimination method is employed in which triangularization and back-substitution processes are performed to find ΔP_{g_i} (i = 1, 2, ..., R) and $\Delta\lambda$. Here R is the number of generators which can participate in allocation.
11. Check either $\sqrt{\sum_{i=1}^{R} (\Delta P_{g_i})^2 + (\Delta\lambda)^2} \leq \varepsilon$ or $\sqrt{\sum_{i=1}^{R} \left(\frac{\partial L}{\partial P_{g_i}}\right)^2 + \left(\frac{\partial L}{\partial \lambda}\right)^2} \leq \varepsilon$

    if convergence condition is 'yes' then GOTO Step 14.
    Check IT > ITMAX, if condition is 'yes', GOTO Step 14. (It means the procedure proceeds without obtaining required convergence.)
12. Modify control variables,

$$P_{g_i}^{\text{new}} = P_{g_i} + \Delta P_{g_i}; \qquad (i = 1, 2, ..., R) \quad \lambda^{\text{new}} = \lambda + \Delta\lambda$$

13. IT = IT + 1, $P_{g_i} = P_{g_i}^{\text{new}}$, $\lambda = \lambda^{\text{new}}$ and GOTO Step 9 and repeat.
14. Check the limits of generators and fix up as following:

$$\text{If } P_{g_i} < P_{g_i}^{\min} \text{ then } P_{g_i} = P_{g_i}^{\min}$$

$$\text{If } P_{g_i} > P_{g_i}^{\max} \text{ then } P_{g_i} = P_{g_i}^{\max}$$

    If no more violations then GOTO Step 16.

15. GOTO Step 8.
16. Record as non-inferior solution and compute F_k (k = 1, 2, ..., M) and transmission loss and GOTO Step 3 for another non-inferior solution.
17. Stop.

5.5.1 Decision Making

Considering the imprecise nature of the decision maker's judgment, it is natural to assume that the decision maker may have fuzzy or imprecise goals for each objective functions. The fuzzy sets are defined by equations called membership functions. These functions represent the degree of membership in some fuzzy sets using values from 0 to 1. The membership value 0, indicates incompatibility with the sets, while 1 means full compatibility. By taking account of the minimum and maximum values of each objective function together with the rate of increase of membership satisfaction, the decision maker must detect membership function $\mu(F_i)$ in a subjective manner. Here it is assumed that $\mu(F_i)$ is a strictly monotonic decreasing and continuous function defined as

$$\mu(F_i) = \begin{cases} 1 & ; F_i \le F_i^{\min} \\ \dfrac{F_i^{\max} - F_i}{F_i^{\max} - F_i^{\min}} & ; F_i^{\min} < F_i < F_i^{\max} \\ 0 & ; F_i \ge F_i^{\max} \end{cases} \quad (i = 1, 2,., M) \tag{5.66}$$

The value of membership function suggests how far (in the scale from 0 to 1) a non-inferior (non-dominated) solution has satisfied the F_i objective. The sum of membership function values $\mu(F_i)$ (i = 1, 2, ..., M) for all the objectives can be computed in order to measure the *accomplishment* of each solution in satisfying the objectives. The accomplishment of each non-dominated solution can be rated with respect to all the K non-dominated solutions by normalizing its accomplishment over the sum of the accomplishments of K non-dominated solutions as follows:

$$\mu_D^k = \frac{\left[\sum_{i=1}^{M} \mu(F_i^k)\right]}{\left[\sum_{k=1}^{K}\sum_{i=1}^{M} \mu(F_i^k)\right]} \tag{5.67}$$

The function μ_D in (5.67) can be treated as a membership function for non-dominated solutions, in a fuzzy set and represented as fuzzy cardinal priority ranking of the non-dominated solutions. The solution that attains the maximum membership μ_D^k, in the fuzzy set so obtained can be chosen as the 'best' solution or the one having the highest cardinal priority ranking.

$$\text{Max}\{\mu_D^k : k = 1, 2, ..., K\} \tag{5.68}$$

5.5.2 Sample System Study

A six-generator system [Lamont and Obessis, 1995; Wong et al., 1995] is considered. The fuel cost, NO_x emission, SO_2 emission, and CO_2 emission equations are given in Tables 5.2 to 5.5. Transmission loss coefficients are given in Table 5.6. The power demand is considered 1800 MW. Only two objectives are considered—operating cost F_1, and NO_x emission, F_2. For various combinations of weights, the non-inferior solution is obtained and is given in Table 5.12 and corresponding to these solutions generation schedules are given in Table 5.13.

TABLE 5.12 Non-inferior solution for two objectives

Sr. no.	w_1	w_2	F_1 (₹/h)	F_2 (kg/h)	P_L (MW)	λ (₹/MWh)
1	1.0	0.0	18721.39	2282.9640	130.1478	10.670020
2	0.9	0.1	18723.45	2243.7440	130.6770	9.851715
3	0.8	0.2	18729.63	2208.7390	131.4510	9.028288
4	0.7	0.3	18739.99	2177.6680	132.4773	8.200272
5	0.6	0.4	18754.58	2150.3640	133.7666	7.368015
6	**0.5**	**0.5**	**18773.86**	**2126.8640**	**135.3374**	**6.531866**
7	0.4	0.6	18797.98	2107.1810	137.2075	5.691980
8	0.3	0.7	18827.27	2091.4390	139.4004	4.848495
9	0.2	0.8	18862.17	2079.8310	141.9434	4.001504
10	0.1	0.9	18903.18	2072.6170	144.8686	3.151061
11	0.0	1.0	18950.87	2070.1270	148.2133	2.297187

TABLE 5.13 Generation schedules corresponding to non-inferior solutions given in Table 5.12.

Sr. no.	P_1 (MW)	P_2 (MW)	P_3 (MW)	P_4 (MW)	P_5 (MW)	P_6 (MW)
1	251.6940	303.7787	503.4812	372.3225	301.4699	197.4014
2	245.0595	295.9456	508.8663	365.6243	317.2672	197.9141
3	238.6702	287.8311	513.6334	359.8299	333.3319	198.1545
4	232.5323	279.4695	517.8625	354.7182	349.7373	198.1575
5	226.6340	270.8847	521.6100	350.1326	365.5513	197.9437
6	**220.9682**	**262.0978**	**524.9345**	**345.9697**	**383.8309**	**197.5291**
7	215.5119	253.1213	527.8673	342.1431	401.6404	196.9189
8	210.2457	243.9654	530.4387	338.5901	420.0428	196.1152
9	205.1505	234.6374	532.6724	335.2613	439.1036	195.1167
10	200.2081	225.1423	534.5878	332.1176	458.8931	193.9191
11	195.4008	215.4831	536.1999	329.1267	479.4874	192.5153

To decide the best solution, minimum and maximum values of objective functions are required. Minimum values of objectives are obtained by giving full weightage to one of the objectives and neglecting others. When the given weightage value is 1.0, it means that full weightage is given to the objective and when the weightage is zero the objective is neglected. Owing to the conflicting nature of objectives, F_2 will have the maximum value when F_1 has the minimum value and vice versa. The minimum and maximum values are obtained and are given below:

$$F_1^{\min} = 18721.39 \text{ ₹/h}, \quad F_1^{\max} = 18950.87 \text{ ₹/h}$$

$$F_2^{\min} = 2070.127 \text{ kg/h}, \quad F_2^{\max} = 2282.964 \text{ kg/h}$$

Using Eq. (5.66), the membership functions of F_1 and F_2 objectives corresponding to each non-inferior solution are obtained and are given in Table 5.14. The membership functions also follow the conflicting nature. Using Eq. (5.67), the normalized membership function μ_D of each non-inferior is obtained and is shown in Table 5.14. The non-inferior solution attains the maximum normalized membership function and is the best solution. From Table 5.14, solution number 6, having weights, $w_1 = w_2 = 0.5$ shows the maximum value of μ_D, i.e. 0.104843, so this solution is considered as the best solution. With the increase in the number of objectives, the number of weight combinations increases.

TABLE 5.14 Decision making

Sr. no.	w_1	w_2	F_1	F_2	$\mu(F_1)$	$\mu(F_2)$	μ_D
1	1.0	0.0	18721.39	2282.964	1.000000	0.000000	0.069674
2	0.9	0.1	18723.45	2243.744	0.991012	0.184273	0.081887
3	0.8	0.2	18729.63	2208.739	0.964092	0.348740	0.091470
4	0.7	0.3	18739.99	2177.668	0.918932	0.494726	0.098495
5	0.6	0.4	18754.58	2150.364	0.855371	0.623012	0.103005
6	**0.5**	**0.5**	**18773.86**	**2126.864**	**0.771342**	**0.733424**	**0.104843**
7	0.4	0.6	18797.98	2107.181	0.666264	0.825905	0.103965
8	0.3	0.7	18827.27	2091.439	0.538598	0.899865	0.100223
9	0.2	0.8	18862.17	2079.831	0.386513	0.954404	0.093427
10	0.1	0.9	18903.18	2072.617	0.207832	0.988299	0.083339
11	0.0	1.0	18950.87	2070.127	0.000000	1.000000	0.069674

Three objectives

Here, three objectives are considered—operating cost F_1, NO_x emission F_2, and SO_2 emission, F_3. For various combinations of weights, the non-inferior solution is obtained and is given in Table 5.15 and corresponding to these solutions generation schedules are given in Table 5.16.

TABLE 5.15 Non-inferior solutions of three objectives

Sr. no.	w_1	w_2	w_3	F_1 (₹/h)	F_2 (kg/h)	F_3 (kg/h)	P_L (MW)	λ (₹/MWh)
1	1.00	0.00	0.00	18721.39	2282.964	11223.00	130.1478	10.6700
2	0.85	0.15	0.00	18726.03	2225.736	11225.24	131.0331	9.4406
3	0.70	0.30	0.00	18739.99	2177.668	11233.08	132.4773	8.2003
4	0.55	0.45	0.00	18763.64	2138.142	11246.68	134.5159	6.9504
5	0.40	0.60	0.00	18797.98	2107.181	11266.67	137.2075	5.6920
6	0.25	0.75	0.00	18843.99	2085.104	11293.62	140.6262	4.4254
7	0.10	0.90	0.00	18903.18	2072.617	11328.41	144.8686	3.1511
8	0.85	0.00	0.15	18721.39	2282.401	11222.99	130.1364	10.0281
9	0.70	0.15	0.15	18726.73	2221.922	11225.62	131.1027	8.7982
10	**0.55**	**0.30**	**0.15**	**18742.52**	**2171.707**	**11234.50**	**132.7029**	**7.5568**
11	0.40	0.45	0.15	18769.67	2131.215	11250.18	134.9872	6.3054
12	0.25	0.60	0.15	18809.00	2100.428	11273.12	138.0252	5.0451
13	0.10	0.75	0.15	18861.97	2079.882	11304.17	141.9136	3.7761
14	0.70	0.00	0.30	18721.39	2281.762	11222.97	130.1236	9.3862
15	0.55	0.15	0.30	18727.59	2217.645	11226.08	131.1855	8.1557
16	0.40	0.30	0.30	18745.78	2165.171	11236.37	132.9750	6.9131
17	0.25	0.45	0.30	18777.07	2123.760	11254.47	135.5561	5.6601

(Contd.)

TABLE 5.15 (*Contd.*)

Sr. no.	w_1	w_2	w_3	F_1 (₹/h)	F_2 (kg/h)	F_3 (kg/h)	P_L (MW)	λ (₹/MWh)
18	0.10	0.60	0.30	18822.58	2093.524	11281.05	139.0195	4.3975
19	0.55	0.00	0.45	18721.40	2281.033	11222.96	130.1090	8.7442
20	0.40	0.15	0.45	18728.66	2212.817	11226.66	131.2854	7.5131
21	0.25	0.30	0.45	18749.83	2157.928	11238.68	133.3056	6.2692
22	0.10	0.45	0.45	18786.26	2115.766	11259.82	136.2530	5.0143
23	0.40	0.00	0.60	18721.41	2280.192	11222.96	130.0922	8.1023
24	0.25	0.15	0.60	18730.01	2207.326	11227.40	131.4076	6.8705
25	0.10	0.30	0.60	18754.92	2149.876	11241.61	133.7138	5.6250
26	0.25	0.00	0.75	18721.43	2279.210	11222.95	130.0726	7.4603
27	0.10	0.15	0.75	18731.73	2201.028	11228.36	131.5599	6.2277
28	0.10	0.00	0.90	18721.46	2278.052	11222.94	130.0497	6.8184

TABLE 5.16 Generation schedules corresponding to non-inferior solutions given in Table 5.15

Sr. no.	P_1 (MW)	P_2 (MW)	P_3 (MW)	P_4 (MW)	P_5 (MW)	P_6 (MW)
1	251.6940	303.7787	503.4812	372.3225	301.4699	197.4014
2	241.8334	291.9211	511.3216	362.6299	325.2613	198.0658
3	232.5323	279.4695	517.8625	354.7182	349.7373	198.1575
4	223.7737	266.5157	523.3234	348.0045	375.1287	197.7612
5	215.5119	253.1213	527.8673	342.1431	401.6404	196.9189
6	207.6779	239.3226	531.5964	336.9003	429.4866	195.6405
7	200.2081	225.1423	534.5878	332.1176	458.8931	193.9191
8	251.6260	303.6805	503.8584	372.0807	301.5803	197.3105
9	241.1435	291.0351	512.1447	361.8672	326.9285	197.9837
10	**231.2934**	**277.7245**	**518.9722**	**353.6080**	**353.0754**	**198.0158**
11	222.0635	263.8584	524.6151	346.6488	380.2867	197.5071
12	213.3777	249.5018	529.2388	340.5885	408.8187	196.4959
13	205.1582	234.6954	532.9645	335.1746	438.9290	194.9904
14	251.5483	303.5689	504.2879	371.8058	301.7057	197.2070
15	240.3607	290.0240	513.0734	361.0122	328.8295	197.8857
16	229.9002	275.7305	520.2182	352.3786	356.8910	197.8438
17	220.1391	260.8155	526.0402	345.1485	386.2122	197.1936
18	210.9826	245.3522	530.7303	338.8723	417.1050	195.9740
19	251.4588	303.4408	504.7813	371.4904	301.8495	197.0881
20	239.4649	288.8593	514.1296	360.0470	331.0173	197.7674
21	228.3142	273.4288	521.6181	351.0030	361.2996	197.6301
22	217.9575	257.2978	527.6193	343.4781	393.0923	196.8021
23	251.3544	303.2922	505.3543	371.1249	302.0162	196.9503
24	238.4298	287.5031	515.3414	358.9485	333.5624	197.6223
25	226.4926	270.7429	523.2018	349.4525	366.4525	197.3605
26	251.2311	303.1179	506.0276	370.6964	302.2114	196.7883
27	237.2204	285.9045	516.7459	357.6869	336.5604	197.4417
28	251.0834	302.9104	506.8299	370.1870	302.4434	196.5956

The minimum and maximum values are obtained and are given below:

$$F_1^{min} = 18721.39 \text{ ₹/h}, \quad F_1^{max} = 18903.18 \text{ ₹/h}$$
$$F_2^{min} = 2072.617 \text{ kg/h}, \quad F_2^{max} = 2282.964 \text{ kg/h}$$
$$F_3^{min} = 11222.94 \text{ kg/h}, \quad F_3^{max} = 11328.41 \text{ kg/h}$$

Using Eq. (5.66), the membership functions of F_1, F_2, and F_3 objectives corresponding to each non-inferior solution are obtained and are given in Table 5.17. The membership functions also follow the conflicting nature. Using Eq. (5.67), the normalized membership function μ_D of each non-inferior is obtained and is shown in Table 5.17. The non-inferior solution that attains the maximum normalized membership function is the best solution. From Table 5.17, solution number 10, having weights, $w_1 = 0.55$, $w_2 = 0.30$, and $w_3 = 0.15$ shows the maximum value of μ_D, i.e. 0.0490096, so this solution is considered the best solution.

TABLE 5.17 Decision making

Sr. no.	w_1	w_2	w_3	$\mu(F_1)$	$\mu(F_2)$	$\mu(F_3)$	μ_D
1	1.00	0.00	0.00	1.000000	0.000000	0.999463	0.034811
2	0.85	0.15	0.00	0.974504	0.272064	0.978158	0.038732
3	0.70	0.30	0.00	0.897663	0.500584	0.903901	0.040080
4	0.55	0.45	0.00	0.767607	0.688492	0.774867	0.038841
5	0.40	0.60	0.00	0.578705	0.835683	0.585336	0.034815
6	0.25	0.75	0.00	0.325565	0.940638	0.329833	0.027787
7	0.10	0.90	0.00	0.000000	1.000000	0.000000	0.017410
8	0.85	0.00	0.15	0.999989	0.002678	0.999565	0.034859
9	0.70	0.15	0.15	0.970637	0.290197	0.974593	0.038919
10	**0.55**	**0.30**	**0.15**	**0.883793**	**0.528922**	**0.890346**	**0.040096**
11	0.40	0.45	0.15	0.734397	0.721422	0.741729	0.038259
12	0.25	0.60	0.15	0.518045	0.867784	0.524273	0.033255
13	0.10	0.75	0.15	0.226656	0.965462	0.229790	0.024755
14	0.70	0.00	0.30	0.999978	0.005713	0.999667	0.034913
15	0.55	0.15	0.30	0.965909	0.310530	0.970195	0.039114
16	0.40	0.30	0.30	0.865829	0.559996	0.872679	0.040017
17	0.25	0.45	0.30	0.693731	0.756864	0.701055	0.037460
18	0.10	0.60	0.30	0.443374	0.900607	0.449015	0.031216
19	0.55	0.00	0.45	0.999936	0.009180	0.999759	0.034974
20	0.40	0.15	0.45	0.960021	0.333480	0.964686	0.039315
21	0.25	0.30	0.45	0.843578	0.594427	0.850717	0.039846
22	0.10	0.45	0.45	0.643159	0.794870	0.650371	0.036359
23	0.40	0.00	0.60	0.999871	0.013179	0.999852	0.035044
24	0.25	0.15	0.60	0.952608	0.359587	0.957686	0.039518
25	0.10	0.30	0.60	0.815547	0.632706	0.822968	0.039542
26	0.25	0.00	0.75	0.999774	0.017844	0.999944	0.035126
27	0.10	0.15	0.75	0.943110	0.389529	0.948640	0.039717
28	0.10	0.00	0.90	0.999602	0.023351	1.000000	0.035219

Four objectives

Here, four objectives are considered—operating cost F_1, NO_x emission F_2, SO_2 emission F_3, and CO_2 emission, F_4. For various combinations of weights, the non-inferior solution is obtained and is given in Table 5.18 and corresponding to these solutions the generation schedules are given in Table 5.19.

TABLE 5.18 Non-inferior solutions of three objectives

Sr. no.	w_1	w_2	w_3	w_4	F_1 (₹/h)	F_2 (kg/h)	F_3 (kg/h)	F_4 (kg/h)	P_L (MW)	λ (₹/MWh)
1	1.0	0.0	0.0	0.0	18721.39	2282.964	11223.00	60482.22	130.1478	10.67002
2	**0.7**	**0.3**	**0.0**	**0.0**	**18739.99**	**2177.668**	**11233.08**	**61111.65**	**132.4773**	**8.20027**
3	0.4	0.6	0.0	0.0	18797.98	2107.181	11266.67	62781.85	137.2075	5.69198
4	0.1	0.9	0.0	0.0	18903.18	2072.617	11328.41	65658.73	144.8686	3.15106
5	0.7	0.0	0.3	0.0	18721.39	2281.762	11222.97	60509.79	130.1236	9.38615
6	0.4	0.3	0.3	0.0	18745.78	2165.171	11236.37	61316.44	132.9750	6.91310
7	0.1	0.6	0.3	0.0	18822.58	2093.524	11281.05	63493.86	139.0195	4.39752
8	0.4	0.0	0.6	0.0	18721.41	2280.192	11222.96	60546.66	130.0922	8.10228
9	0.1	0.3	0.6	0.0	18754.92	2149.876	11241.61	61623.00	133.7138	5.62504
10	0.1	0.0	0.9	0.0	18721.46	2278.052	11222.94	60598.46	130.0497	6.81839
11	0.7	0.0	0.0	0.3	18781.96	2352.074	11261.10	58076.32	140.3862	31.45539
12	0.4	0.3	0.0	0.3	18781.96	2337.578	11260.98	58083.58	140.5167	28.96449
13	0.1	0.6	0.0	0.3	18782.47	2323.575	11261.17	58104.17	140.6782	26.46876
14	0.4	0.0	0.3	0.3	18783.31	2353.416	11261.93	58073.43	140.5443	30.16703
15	0.1	0.3	0.3	0.3	18783.31	2338.709	11261.80	58080.92	140.6769	27.67565
16	0.1	0.0	0.6	0.3	18784.72	2354.807	11262.78	58070.99	140.7075	28.87851
17	0.4	0.0	0.0	0.6	18788.10	2358.696	11264.85	58067.24	141.0891	52.20033
18	0.1	0.3	0.0	0.6	18788.02	2350.910	11264.74	58069.23	141.1536	49.70889
19	0.1	0.0	0.3	0.6	18788.88	2359.477	11265.33	58066.79	141.1785	50.91130
20	0.1	0.0	0.0	0.9	18790.37	2361.139	11266.23	58066.36	141.3446	72.94322

The minimum and maximum values are obtained and are given below:

$$F_1^{\min} = 18721.39 \text{ ₹/h}, \quad F_1^{\max} = 18903.18 \text{ ₹/h}$$
$$F_2^{\min} = 2072.617 \text{ kg/h}, \quad F_2^{\max} = 2361.139 \text{ kg/h}$$
$$F_3^{\min} = 11222.94 \text{ kg/h}, \quad F_3^{\max} = 11328.41 \text{ kg/h}$$
$$F_4^{\min} = 58066.36 \text{ kg/h}, \quad F_4^{\max} = 65658.73 \text{ kg/h}$$

Using Eq. (5.66), the membership functions of F_1, F_2, and F_3 objectives corresponding to each non-inferior solutions are obtained and are given in Table 5.20. The membership functions also follow the conflicting nature. Using Eq. (5.67), the normalized membership function μ_D of each non-inferior is obtained and is shown in Table 5.20. The non-inferior solution that attains the maximum normalized membership function is the best solution. From Table 5.20, solution

TABLE 5.19 Generation schedules corresponding to non-inferior solutions given in Table 5.15

Sr. no.	P_1 (MW)	P_2 (MW)	P_3 (MW)	P_4 (MW)	P_5 (MW)	P_6 (MW)
1	251.6940	303.7787	503.4812	372.3225	301.4699	197.4014
2	**232.5323**	**279.4695**	**517.8625**	**354.7182**	**349.7373**	**198.1575**
3	215.5119	253.1213	527.8673	342.1431	401.6404	196.9189
4	200.2081	225.1423	534.5878	332.1176	458.8931	193.9191
5	251.5483	303.5689	504.2879	371.8058	301.7057	197.2070
6	229.9002	275.7305	520.2182	352.3786	356.8910	197.8438
7	210.9826	245.3522	530.7303	338.8723	417.1050	195.9740
8	251.3544	303.2922	505.3543	371.1249	302.0162	196.9503
9	226.4926	270.7429	523.2018	349.4525	366.4525	197.3605
10	251.0834	302.9104	506.8299	370.1870	302.4434	196.5956
11	249.8242	331.1674	400.6613	382.5285	342.2422	233.9626
12	248.5857	327.9310	401.9333	380.2640	347.6911	234.1114
13	247.3434	324.6653	403.1582	378.0903	353.1761	234.2450
14	249.7580	331.5945	399.5382	382.6179	342.8188	234.2169
15	248.5098	328.3045	400.8224	380.3266	348.3470	234.3665
16	249.6899	332.0354	398.3879	382.7086	343.4109	234.4748
17	249.5542	333.1542	395.6131	383.0761	344.5988	235.0924
18	248.9153	331.4234	396.2828	381.8715	347.4895	235.1711
19	249.5169	333.3977	394.9947	383.1243	344.9178	235.2269
20	249.4557	333.8794	393.8141	383.2671	345.4443	235.4839

TABLE 5.20 Decision making

Sr. no.	w_1	w_2	w_3	w_4	$\mu(F_1)$	$\mu(F_2)$	$\mu(F_3)$	$\mu(F_4)$	μ_D
1	1.0	0.0	0.0	0.0	1.000000	0.270951	0.999463	0.681805	0.059818
2	**0.7**	**0.3**	**0.0**	**0.0**	**0.897663**	**0.635901**	**0.903901**	**0.598901**	**0.061523**
3	0.4	0.6	0.0	0.0	0.578705	0.880205	0.585336	0.378918	0.049099
4	0.1	0.9	0.0	0.0	0.000000	1.000000	0.000000	0.000000	0.020262
5	0.7	0.0	0.3	0.0	0.999978	0.275116	0.999667	0.678174	0.059833
6	0.4	0.3	0.3	0.0	0.865829	0.679215	0.872679	0.571928	0.060577
7	0.1	0.6	0.3	0.0	0.443374	0.927538	0.449015	0.285139	0.042653
8	0.4	0.0	0.6	0.0	0.999871	0.280559	0.999852	0.673317	0.059846
9	0.1	0.3	0.6	0.0	0.815547	0.732225	0.822968	0.531551	0.058807
10	0.1	0.0	0.9	0.0	0.999602	0.287975	1.000000	0.666494	0.059856
11	0.7	0.0	0.0	0.3	0.666785	0.031418	0.638167	0.998689	0.047313
12	0.4	0.3	0.0	0.3	0.666839	0.081662	0.639343	0.997733	0.048337
13	0.1	0.6	0.0	0.3	0.663981	0.130195	0.637519	0.995021	0.049171
14	0.4	0.0	0.3	0.3	0.659361	0.026769	0.630353	0.999069	0.046918
15	0.1	0.3	0.3	0.3	0.659382	0.077742	0.631519	0.998083	0.047955
16	0.1	0.0	0.6	0.3	0.651646	0.021946	0.622251	0.999391	0.046506
17	0.4	0.0	0.0	0.6	0.633027	0.008469	0.602631	0.999885	0.045469
18	0.1	0.3	0.0	0.6	0.633468	0.035454	0.603678	0.999622	0.046040
19	0.1	0.0	0.3	0.6	0.628730	0.005762	0.598113	0.999944	0.045236
20	0.1	0.0	0.0	0.9	0.620553	0.000000	0.589521	1.000000	0.044781

number 2, having weights, $w_1 = 0.7$, $w_2 = 0.30$, $w_3 = 0.0$ and $w_4 = 0.0$ shows the maximum value of μ_D, i.e. 0.061523, so this solution is considered the best solution.

5.6 MULTIOBJECTIVE DISPATCH FOR ACTIVE AND REACTIVE POWER BALANCE

The multiobjective dispatch problem is defined so as to minimize the number of objectives, namely, the total operating cost, emission, etc. of a power system while meeting the total real load plus real transmission losses within the generator limits of real power as well as reactive load plus reactive transmission losses within the generator limits of reactive power. Mathematically, the problem is defined as follows:

$$\text{Minimize} \quad F_1 = \sum_{i=1}^{NG} (a_i P_{g_i}^2 + b_i P_{g_i} + c_i) \text{ ₹/h} \tag{5.69a}$$

$$\text{Minimize} \quad F_2 = \sum_{i=1}^{NG} (d_{1i} P_{g_i}^2 + e_{1i} P_{g_i} + f_{1i}) \text{ kg/h} \tag{5.69b}$$

$$\text{Minimize} \quad F_3 = \sum_{i=1}^{NG} (d_{2i} P_{g_i}^2 + e_{2i} P_{g_i} + f_{2i}) \text{ kg/h} \tag{5.69c}$$

$$\text{Minimize} \quad F_4 = \sum_{i=1}^{NG} (d_{3i} P_{g_i}^2 + e_{3i} P_{g_i} + f_{3i}) \text{ kg/h} \tag{5.69d}$$

subject to

$$\sum_{i=1}^{NG} P_{g_i} = \sum_{i=1}^{NB} P_{d_i} + P_L \tag{5.69e}$$

$$\sum_{i=1}^{NG} Q_{g_i} = \sum_{i=1}^{NB} Q_{d_i} + Q_L \tag{5.69f}$$

$$P_{g_i}^{\min} \le P_{g_i} \le P_{g_i}^{\max} \quad (i = 1, 2, ..., NG) \tag{5.69g}$$

$$Q_{g_i}^{\min} \le Q_{g_i} \le Q_{g_i}^{\max} \quad (i = 1, 2, ..., NG) \tag{5.69h}$$

where

a_i, b_i, and c_i are cost coefficients
d_{1i}, e_{1i}, and f_{1i} are NO_x emission coefficients
d_{2i}, e_{2i}, and f_{2i} are SO_2 emission coefficients
d_{3i}, e_{3i}, and f_{3i} are CO_2 emission coefficients
P_{d_i} is the real power load demand at the ith bus
P_{g_i} is the real power generation (decision variable)
Q_{d_i} is the reactive power load demand at the ith bus

Q_{g_i} is the reactive power generation (decision variable)
NG is the number of generation buses
NB is the number of buses in the network
P_L is the transmission real power loss
Q_L is the transmission reactive power loss.

This method uses the fact that under normal operating condition, the transmission loss is quadratic in the injected bus real powers. The general form of the loss formula using Eqs. (3.132) and (3.134) is

$$P_L = \sum_{i=1}^{NB} \sum_{j=1}^{NB} [a_{ij}(P_i P_j + Q_i Q_j) + b_{ij}(Q_i P_j - P_i Q_j)] \tag{5.70a}$$

$$Q_L = \sum_{i=1}^{NB} \sum_{j=1}^{NB} [c_{ij}(P_i P_j + Q_i Q_j) + h_{ij}(Q_i P_j - P_i Q_j)] \tag{5.70b}$$

where

$$a_{ij} = \frac{R_{ij}}{|V_i||V_j|} \cos(\delta_i - \delta_j)$$

$$b_{ij} = \frac{R_{ij}}{|V_i||V_j|} \sin(\delta_i - \delta_j)$$

$$c_{ij} = \frac{X_{ij}}{|V_i||V_j|} \cos(\delta_i - \delta_j)$$

$$h_{ij} = \frac{X_{ij}}{|V_i||V_j|} \sin(\delta_i - \delta_j)$$

$$P_i = P_{g_i} - P_{d_i} \qquad (i = 1, 2, ..., \text{NB})$$

$$Q_i = Q_{g_i} - Q_{d_i} \qquad (i = 1, 2, ..., \text{NB})$$

P_i and P_j are the real power injections at the *i*th and the *j*th buses, respectively
Q_i and Q_j are the reactive power injections at the *i*th and *j*th buses, respectively
$Z_{ij} = R_{ij} + jX_{ij}$: elements of impedance matrix.

To generate the non-inferior solution to the multiobjective problem, the weighting method is applied. In this method, the problem is converted into a scalar optimization as given below:

Minimize $$\sum_{k=1}^{M} w_k F_k(P_{g_i}) \tag{5.71a}$$

subject to
$$\sum_{i=1}^{\mathrm{NG}} P_{g_i} = \sum_{i=1}^{\mathrm{NB}} P_{d_i} + P_L \tag{5.71b}$$

$$\sum_{i=1}^{\mathrm{NG}} Q_{g_i} = \sum_{i=1}^{\mathrm{NB}} Q_{d_i} + Q_L \tag{5.71c}$$

$$P_{g_i}^{\min} \le P_{g_i} \le P_{g_i}^{\max} \quad (i = 1, 2, ..., \mathrm{NG}) \tag{5.71d}$$

$$Q_{g_i}^{\min} \le Q_{g_i} \le Q_{g_i}^{\max} \quad (i = 1, 2, ..., \mathrm{NG}) \tag{5.71e}$$

$$\sum_{k=1}^{M} w_k = 1 \qquad (w_k \ge 0) \tag{5.71f}$$

where

M is the number of objectives

w_k are levels of normalized weights.

This approach yields meaningful result to the decision maker when solved many times for different values of w_k ($k = 1, 2, ..., M$). Weighting factors w_k are determined based on the relative importance of various objectives, which may vary from place to place and utility to utility.

Problem defined by Eq. (5.71) can be redefined in simplified form as

Minimize
$$F_T(P_{g_i}) = \sum_{i=1}^{\mathrm{NG}} (A_i P_{g_i}^2 + B_i P_{g_i} + C_i) \tag{5.72a}$$

subject to Eqs. (5.71b) to (5.71e), and 5.72(b)

$$\sum_{k=1}^{M} w_k = 1 \ (w_k \ge 0) \tag{5.72b}$$

where
$$A_i = w_1 a_i + \sum_{k=2}^{M} w_k d_{(k-1)i}$$

$$B_i = w_1 b_i + \sum_{k=2}^{M} w_k e_{(k-1)i}$$

$$C_i = w_1 c_i + \sum_{k=2}^{M} w_k f_{(k-1)i}$$

Using the Lagrange multiplier method, the constrained optimization problem given by Eq. (5.72) is converted into an the unconstrained optimization problem.

$$L = F_T + \lambda_p \left(\sum_{i=1}^{\mathrm{NB}} P_{d_i} + P_L - \sum_{i=1}^{\mathrm{NG}} P_{g_i} \right) + \lambda_q \left(\sum_{i=1}^{\mathrm{NB}} Q_{d_i} + Q_L - \sum_{i=1}^{\mathrm{NG}} Q_{g_i} \right) \tag{5.73}$$

where λ_p and λ_q are Lagrangian multipliers.

Necessary conditions for optimization problem stated by Eq. (5.73) are:

$$\frac{\partial L}{\partial P_{g_i}} = \frac{\partial F_T}{\partial P_{g_i}} + \lambda_p\left(\frac{\partial P_L}{\partial P_{g_i}} - 1\right) + \lambda_q \frac{\partial Q_L}{\partial P_{g_i}} = 0 \qquad (i = 1, 2, ..., \text{NG}) \tag{5.74a}$$

$$\frac{\partial L}{\partial Q_{g_i}} = \lambda_p\left(\frac{\partial P_L}{\partial Q_{g_i}}\right) + \lambda_q\left(\frac{\partial Q_L}{\partial Q_{g_i}} - 1\right) = 0 \qquad (i = 1, 2, ..., \text{NG}) \tag{5.74b}$$

$$\frac{\partial L}{\partial \lambda_p} = \sum_{i=1}^{\text{NB}} P_{d_i} + P_L - \sum_{i=1}^{\text{NG}} P_{g_i} = 0 \tag{5.74c}$$

$$\frac{\partial L}{\partial \lambda_q} = \sum_{i=1}^{\text{NB}} Q_{d_i} + Q_L - \sum_{i=1}^{\text{NG}} Q_{g_i} = 0 \tag{5.74d}$$

where incremental transmission loss expressions are expressed here:

$$\frac{\partial P_L}{\partial P_{g_i}} = 2a_{ii}P_i + \sum_{j=1}^{\text{NB}} [(a_{ij} + a_{ji})P_j + (b_{ji} - b_{ij})Q_j] \qquad (i = 1, 2, ..., \text{NG}) \tag{5.75a}$$

$$\frac{\partial P_L}{\partial Q_{g_i}} = 2a_{ii}Q_i + \sum_{j=1}^{\text{NB}} [(a_{ij} + a_{ji})Q_j + (b_{ij} - b_{ji})P_j] \qquad (i = 1, 2, ..., \text{NG}) \tag{5.75b}$$

$$\frac{\partial Q_L}{\partial P_{g_i}} = 2c_{ii}P_i + \sum_{j=1}^{\text{NB}} [(c_{ij} + c_{ji})P_j + (h_{ji} - h_{ij})Q_j] \qquad (i = 1, 2, ..., \text{NG}) \tag{5.75c}$$

$$\frac{\partial Q_L}{\partial Q_{g_i}} = 2c_{ii}Q_i + \sum_{j=1}^{\text{NB}} [(c_{ij} + c_{ji})Q_j + (h_{ij} - h_{ji})P_j] \qquad (i = 1, 2, ..., \text{NG}) \tag{5.75d}$$

Incremental fuel cost is given by

$$\frac{\partial F_T}{\partial P_{g_i}} = 2A_iP_{g_i} + B_i \qquad (i = 1, 2, ..., \text{NG}) \tag{5.75e}$$

The solution of nonlinear Eqs. (5.74a) to (5.74d) can be obtained using the Newton–Raphson method in which change in variables, P_{g_i}, $Q_{g_i}(i = 1, 2, ..., \text{NG})$, λ_p and λ_q are obtained by expanding Eqs. (5.74a) to (5.74d) about the initial values of the variables using Taylor's expansion. In the matrix form the above equations can be rewritten as

$$\begin{bmatrix} \nabla_{P_gP_g}L & \nabla_{P_gQ_g}L & \nabla_{P_g\lambda_p}L & \nabla_{P_g\lambda_q}L \\ \nabla_{Q_gP_g}L & \nabla_{Q_gQ_g}L & \nabla_{Q_g\lambda_p}L & \nabla_{Q_g\lambda_q}L \\ (\nabla_{P_g\lambda_p}L)^T & (\nabla_{Q_g\lambda_p}L)^T & \nabla_{\lambda_p\lambda_p}L & \nabla_{\lambda_p\lambda_q}L \\ (\nabla_{P_g\lambda_q}L)^T & (\nabla_{Q_g\lambda_q}L)^T & \nabla_{\lambda_q\lambda_p}L & \nabla_{\lambda_q\lambda_q}L \end{bmatrix} \begin{bmatrix} \Delta P_g \\ \Delta Q_g \\ \Delta\lambda_p \\ \Delta\lambda_q \end{bmatrix} = -\begin{bmatrix} \nabla_{P_g}L \\ \nabla_{Q_g}L \\ \nabla_{\lambda_p}L \\ \nabla_{\lambda_q}L \end{bmatrix} \tag{5.76}$$

Elements of Hessian matrix as derived from Eq. (5.74a) to (5.74d) are given below:

$$\frac{\partial^2 L}{\partial P_{g_i}^2} = 2A_i + 2\lambda_p a_{ii} + 2\lambda_q c_{ii} \qquad (i = 1, 2, ..., \text{NG}) \tag{5.77a}$$

$$\frac{\partial^2 L}{\partial P_{g_i} \partial P_{g_j}} = \lambda_p(a_{ij} + a_{ji}) + \lambda_q(c_{ij} + c_{ji}) \qquad (i = 1, 2, ..., \text{NG}; \quad j = 1, 2, ..., \text{NG}; \quad i \neq j) \tag{5.77b}$$

$$\frac{\partial^2 L}{\partial P_{g_i} \partial Q_{g_j}} = \lambda_p(b_{ji} - b_{ij}) + \lambda_q(d_{ji} - d_{ij}) \qquad (i = 1, 2, ..., \text{NG}; \quad j = 1, 2, ..., \text{NG}) \tag{5.77c}$$

$$\frac{\partial^2 L}{\partial \lambda_p \partial P_{g_i}} = \frac{\partial^2 L}{\partial P_{g_i} \partial \lambda_p} = \frac{\partial P_L}{\partial P_{g_i}} - 1 \qquad (i = 1, 2, ..., \text{NG}) \tag{5.77d}$$

$$\frac{\partial^2 L}{\partial \lambda_q \partial P_{g_i}} = \frac{\partial^2 L}{\partial P_{g_i} \partial \lambda_q} = \frac{\partial Q_L}{\partial P_{g_i}} \qquad (i = 1, 2, ..., \text{NG}) \tag{5.77e}$$

$$\frac{\partial^2 L}{\partial Q_{g_i} \partial Q_{g_j}} = \lambda_p(a_{ij} + a_{ji}) \quad (i = 1, 2, ..., \text{NG}; \quad j = 1, 2, ..., \text{NG}) \tag{5.77f}$$

$$\frac{\partial^2 L}{\partial Q_{g_i} \partial P_{g_j}} = \lambda_p(b_{ij} - b_{ji}) + \lambda_q(d_{ij} - d_{ji}) \qquad (i = 1, 2, ..., \text{NG}; \quad j = 1, 2, ..., \text{NG}) \tag{5.77g}$$

$$\frac{\partial^2 L}{\partial Q_{g_i} \partial \lambda_p} = \frac{\partial^2 L}{\partial \lambda_p \partial Q_{g_i}} = \frac{\partial P_L}{\partial Q_{g_i}} \qquad (i = 1, 2, ..., \text{NG}) \tag{5.77h}$$

$$\frac{\partial^2 L}{\partial \lambda_q \partial Q_{g_i}} = \frac{\partial^2 L}{\partial Q_{g_i} \partial \lambda_q} = \frac{\partial Q_L}{\partial Q_{g_i}} - 1 \qquad (i = 1, 2, ..., \text{NG}) \tag{5.77i}$$

$$\frac{\partial^2 L}{\partial \lambda_p^2} = \frac{\partial^2 L}{\partial \lambda_q^2} = \frac{\partial^2 L}{\partial \lambda_p \partial \lambda_q} = \frac{\partial^2 L}{\partial \lambda_q \partial \lambda_p} = 0 \tag{5.77j}$$

Equation (5.76) can be solved to obtain the non-inferior solutions and a detailed algorithm is outlined here.

Algorithm 5.4: Non-Inferior Solution for Multiobjective Dispatch for Active and Reactive Power Balance

1. Read data: NG (number of buses having generators), NB (number of buses), NV (number of PV buses). V_1, δ_1 for slack bus, P_{d_i}, Q_{d_i} $(i = 1, 2, ..., \text{NB})$. V_i^s $(i = 2, 3,.., \text{NV})$ for PV buses, $V_i^{\min}$, $V_i^{\max}$ $(i = \text{NV} + 1, \text{NV} + 2, ..., \text{NB})$ for PQ buses. $Q_i^{\min}$, $Q_i^{\max}$ $(i = 2, 3,.., \text{NV})$ for PV buses, cost coefficients, emission coefficients, R_1, R_2 (maximum number of iterations), ε_1, ε_2 (tolerance in convergence), K the number of non-inferior solutions, etc.
   Obtain Y_{bus} and by inverting it obtain Z_{bus}.
2. Set iteration for non-inferior solutions, $k = 1$.
3. Increment count of non-inferior solutions, $k = k + 1$.
4. If $(k \geq K)$ GOTO Step 24.

5. Feed or generate the weights, $w_i(i = 1, 2, ..., M)$.
6. Assume initial values of real power, $P_{g_i}(i = 1, 2, ..., \text{NG})$, reactive power Q_{g_i} for *PQ*-buses and λ_p, and compute initial cost (F^{prev}).
5. Set iteration counter II = 1.
8. Calculate $P_i = P_{g_i} - P_{d_i}(i = 1, 2, ..., \text{NB})$ and $Q_i = Q_{g_i} - Q_{d_i}(i = 1, 2, ..., \text{NB})$ for *PQ*-buses. Take $P_{g_i} = 0$, $Q_{g_i} = 0$ for non-generating buses.
9. Perform load flow to obtain real and reactive power, P_i, Q_i and voltage magnitude and angles $|V_i|$, δ_i at each bus.
10. Check at slack bus that $|P_{g_s} - P_{d_s} - P_s)| \le \varepsilon_1$, if 'yes' GOTO Step 24.
11. Compute $P_{g_s} = P_s + P_{d_s}$ for slack bus and $Q_{g_i} = Q_i + Q_{d_i}$ for slack and *PV*-buses.
12. Calculate loss coefficients a_{ij}, b_{ij}, c_{ij}, and d_{ij} using Eqs. (5.70c) to (5.70f), respectively.
13. Assume/set that no generation has been fixed either at lower or at upper limits.
14. Set iteration counter, III = 1.
15. Calculate Hessian and Jacobian matrix elements using Eqs. (5.74) and (5.77). Size of Hessian matrix is [(2NG + 2) × (2NG + 2)] and size of Jacobian is [(2NG + 2) × 1]. Deactivate row and column of Hessian matrix and row of Jacobian matrix representing the generator whose generation is fixed either at lower limit or at upper limit. This is done so that fixed generators cannot participate in allocation.
16. Using Gauss elimination method, find ΔP_{g_i}, ΔQ_{g_i} $(i = 1, 2, ..., R)$, λ_p and λ_q. Here R is the number of generators which can participate in allocation.
17. Check $\sqrt{\sum_{i=1}^{R}[(\Delta P_{g_i})^2 + (\Delta Q_{g_i})^2] + (\Delta\lambda_p)^2 + (\Delta\lambda_q)^2} \le \varepsilon_2$

    or $\sqrt{\sum_{i=1}^{R}\left[\left(\frac{\partial L}{\partial P_{g_i}}\right)^2 + \left(\frac{\partial L}{\partial Q_{g_i}}\right)^2\right] + \left(\frac{\partial L}{\partial \lambda_p}\right)^2 + \left(\frac{\partial L}{\partial \lambda_q}\right)^2} \le \varepsilon_2$

    If 'yes' then GOTO Step 19.
18. Modify $P_{g_i}^{\text{new}} = P_{g_i} + \Delta P_{g_i} \quad (i = 1, 2, ..., R)$

    $Q_{g_i}^{\text{new}} = Q_{g_i} + \Delta Q_{g_i} \quad (i = 1, 2, ..., R)$

    $\lambda_p^{\text{new}} = \lambda_p + \Delta\lambda_p$

    $\lambda_q^{\text{new}} = \lambda_q + \Delta\lambda_q$
19. If (III $\ge R_1$) then GOTO Step 20 (without convergence),

    else III = III + 1, $P_{g_i} = P_{g_i}^{\text{new}}$, $Q_{g_i} = Q_{g_i}^{\text{new}} \quad (i = 1, 2, ..., R)$

    $\lambda_p = \lambda_p^{\text{new}}$, $\lambda_q = \lambda_q^{\text{new}}$, GOTO Step 15 and repeat.
20. Check the limits of generators if no limit is violated further then GOTO Step 21, else fix the limits as following:

    If $P_{g_i} < P_{g_i}^{\min}$ then $P_{g_i} = P_{g_i}^{\min}$

If $P_{g_i} > P_{g_i}^{max}$ then $P_{g_i} = P_{g_i}^{max}$ and GOTO Step 15.

Reactive power limits are considered in the load flow as *PV* buses.

21. Compute optimal total cost *F*, transmission loss P_L, etc.
22. If ($|F_T^{prev} - F_T| \le \varepsilon_1$), then GOTO Step 24.
23. If (II $\ge R_2$) then GOTO Step 24 (without convergence).

$$\text{else II} = \text{II} + 1,\ F^{prev} = F, \text{ GOTO Step 8 and repeat}$$

24. Record as non-inferior solution and compute F_k (k = 1, 2, ..., *M*) and transmission loss and GOTO Step 3 for another non-inferior solution.
25. Stop.

5.6.1 Sample System Study

A six-generator system is considered. The fuel cost, NO_x emission, SO_2 emission, and CO_2 emission equations are given in Tables 5.21 to 5.24. Line data consisting of line charging and line impedance of a power system is given in Table 5.25. The scheduled generation, load and specified voltage on various types of buses are given in Table 5.26.

Number of generators, NG = 5, Number of buses, NB = 11

Number of lines, NL = 17, Number of *PQ* buses, NPQ = 6

TABLE 5.21 Fuel cost (₹/h) equations

$F_{11} = 20.35P_1^2 + 843.205P_1 + 85.6348$
$F_{12} = 38.66P_2^2 + 641.031P_2 + 303.7780$
$F_{13} = 21.82P_3^2 + 742.890P_3 + 847.1484$
$F_{14} = 13.45P_4^2 + 830.154P_4 + 274.2241$
$F_{15} = 59.63P_5^2 + 691.559P_5 + 202.0258$

TABLE 5.22 NO_x emission (kg/h) equations

$F_{21} = 63.23P_1^2 - 38.128P_1 + 80.9019$
$F_{22} = 64.83P_2^2 - 79.027P_2 + 28.8249$
$F_{23} = 31.74P_3^2 - 136.061P_3 + 324.1775$
$F_{24} = 67.32P_4^2 - 239.928P_4 + 610.2535$
$F_{25} = 61.81P_5^2 - 39.077P_5 + 50.3808$

TABLE 5.23 SO_2 emission (kg/h) equations

$F_{31} = 12.06P_1^2 + 505.928P_1 + 51.3778$
$F_{32} = 23.20P_2^2 + 384.624P_2 + 182.2605$
$F_{33} = 12.84P_3^2 + 445.647P_3 + 508.5207$
$F_{34} = 8.13P_4^2 + 497.641P_4 + 165.3433$
$F_{35} = 35.78P_5^2 + 414.938P_5 + 121.2133$

TABLE 5.24 CO_2 emission (ton/h) equations

$$F_{41} = 26.5110P_1^2 - 61.019450P_1 + 50.80148$$
$$F_{42} = 14.0053P_2^2 - 29.952210P_2 + 38.24770$$
$$F_{43} = 10.5929P_3^2 - 9.552794P_3 + 13.42851$$
$$F_{44} = 10.6409P_4^2 - 12.736420P_4 + 18.19625$$
$$F_{45} = 40.3144P_5^2 - 121.98120P_5 + 113.81070$$

TABLE 5.25 Line data

Line No	*Link p–q*	*Line charging* Y_{pq}	*Impedance* Z_{pq}
1	1–9	$j0.030$	$0.15 + j0.50$
2	1–11	$j0.010$	$0.05 + j0.16$
3	2–3	$j0.030$	$0.15 + j0.50$
4	2–7	$j0.020$	$0.10 + j0.28$
5	2–10	$j0.010$	$0.05 + j0.16$
6	3–4	$j0.015$	$0.08 + j0.24$
7	4–6	$j0.020$	$0.10 + j0.28$
8	4–8	$j0.020$	$0.10 + j0.28$
9	4–9	$j0.030$	$0.15 + j0.50$
10	5–6	$j0.025$	$0.12 + j0.36$
11	5–9	$j0.010$	$0.05 + j0.16$
12	7–8	$j0.010$	$0.05 + j0.16$
13	7–10	$j0.015$	$0.08 + j0.24$
14	8–9	$j0.025$	$0.12 + j0.36$
15	8–10	$j0.015$	$0.08 + j0.24$
16	8–11	$j0.020$	$0.10 + j0.28$
17	10–11	$j0.025$	$0.12 + j0.36$

TABLE 5.26 Generation, load, and voltage at buses

Bus no.	*Generation*		*Load*		*Voltage*		*Type of bus*
	P_{g_i} (p.u.)	Q_{g_i} (p.u.)	P_{d_i} (p.u.)	Q_{d_i} (p.u.)	V_i (p.u.)	δ_i (rad)	
1	–	–	0.25	0.05	1.07	0.0	Slack
2	0.6625	–	0.25	0.05	1.088	–	*PV*
3	0.6625	–	0.25	0.05	1.095	–	*PV*
4	0.4778	–	0.25	0.05	1.062	–	*PV*
5	0.4778	–	0.25	0.05	1.046	–	*PV*
6	0.0	0.0	0.10	0.02	–	–	*PQ*
7	0.0	0.0	0.40	0.10	–	–	*PQ*
8	0.0	0.0	0.90	0.45	–	–	*PQ*
9	0.0	0.0	0.70	0.35	–	–	*PQ*
10	0.0	0.0	0.25	0.05	–	–	*PQ*
11	0.0	0.0	0.25	0.05	–	–	*PQ*

TABLE 5.27 Power generated, power demand, and power injected at buses for'best' solution

Bus no.	P_{g_i} (p.u.)	Q_{g_i} (p.u.)	P_{d_i} (p.u.)	Q_{d_i} (p.u.)	P_i (p.u.)	Q_i (p.u.)
1	0.379955	0.181061	0.25	0.05	0.132610	0.293544
2	1.234042	0.399560	0.25	0.05	0.934047	0.365405
3	0.796340	−0.076570	0.25	0.05	0.546340	−0.255580
4	0.730913	0.340662	0.25	0.05	0.480917	0.212154
5	0.828159	0.261883	0.25	0.05	0.578162	0.063813
6	0.0	0.0	0.10	0.02	−0.100000	−0.020000
7	0.0	0.0	0.40	0.10	−0.400001	−0.100000
8	0.0	0.0	0.90	0.45	−0.900015	−0.450001
9	0.0	0.0	0.70	0.35	−0.700002	−0.349999
10	0.0	0.0	0.25	0.05	−0.249999	−0.049998
11	0.0	0.0	0.25	0.05	−0.249994	−0.050000

Reactive power limit of *PV*-bus is given below:

$$-0.1 \le Q_2 \le 0.5$$
$$-0.1 \le Q_3 \le 0.5$$
$$-0.1 \le Q_4 \le 0.5$$
$$-0.1 \le Q_5 \le 0.5$$

Four objectives are considered—operating cost F_1, NO_x emission F_2, SO_2 emission F_3 and CO_2 emission F_4. For various combinations of weights, the non-inferior solution is obtained and is given in Table 5.28. The minimum and maximum values are obtained and are given below:

$$F_1^{\min} = 4687.4430 \text{ ₹/h} \qquad F_1^{\max} = 5075.7230 \text{ ₹/h}$$
$$F_2^{\min} = 708.3177 \text{ kg/h} \qquad F_2^{\max} = 1075.3240 \text{ kg/h}$$
$$F_3^{\min} = 2813.2380 \text{ kg/h} \qquad F_3^{\max} = 3044.8280 \text{ kg/h}$$
$$F_4^{\min} = 87.1483 \text{ ton/h} \qquad F_4^{\max} = 228.6567 \text{ ton/h}$$

Using Eq. (5.67), the normalized membership function μ_D of each non-inferior is obtained and is shown in Table 5.29. The non-inferior solution that attains the maximum normalized membership function is the best solution. From Table 5.29, solution number 17, having weights, $w_1 = 0.5$, $w_2 = 0.25$, $w_3 = 0.0$ and $w_4 = 0.25$ shows the maximum value of μ_D, i.e. 0.032639, so this solution is considered as the best solution. Corresponding to the 'best' solution, the power generated and power injected at each bus is given in Table 5.27 and the voltage at each bus is given in Table 5.30.

5.7 MULTIOBJECTIVE SHORT-RANGE FIXED-HEAD HYDROTHERMAL SCHEDULING—APPROXIMATE NEWTON–RAPHSON METHOD

The basic problem considered involves short-range optimal economic operation of an electric power system that includes both hydro and thermal generation resources. The multiobjective problem minimizes the number of objectives, namely the total system operating cost, minimal

TABLE 5.28 Non-inferior solutions of three objectives

Sr. no.	w_1	w_2	w_3	w_4	F_1 (₹/h)	F_2 (kg/h)	F_3 (kg/h)	F_4 (ton/h)
1	1.00	0.00	0.00	0.00	4695.180	1043.0820	2817.868	117.6251
2	0.75	0.25	0.00	0.00	4710.326	891.9321	2826.716	125.7260
3	0.50	0.50	0.00	0.00	4769.232	798.0036	2861.836	141.7259
4	0.25	0.75	0.00	0.00	4877.221	734.4222	2926.303	172.6481
5	0.00	1.00	0.00	0.00	5075.723	708.3177	3044.828	228.6567
6	0.75	0.00	0.25	0.00	4695.207	1042.8190	2817.884	117.6244
7	0.50	0.25	0.25	0.00	4712.087	881.8958	2827.758	126.4017
8	0.25	0.50	0.25	0.00	4788.665	780.7662	2873.433	147.3127
9	0.00	0.75	0.25	0.00	4929.639	720.7835	2957.603	187.5849
10	0.50	0.00	0.50	0.00	4695.241	1042.4900	2817.904	117.6234
11	0.25	0.25	0.50	0.00	4719.115	868.6505	2831.949	127.7508
12	0.00	0.50	0.50	0.00	4816.497	761.6649	2890.046	155.3030
13	0.25	0.00	0.75	0.00	4695.286	1042.0680	2817.930	117.6222
14	0.00	0.25	0.75	0.00	4729.467	849.8852	2838.119	130.4602
15	0.00	0.00	1.00	0.00	4695.347	1041.5070	2817.966	117.6205
16	0.75	0.00	0.00	0.25	4687.443	1046.3850	2813.238	114.4185
17	**0.50**	**0.25**	**0.00**	**0.25**	**4719.066**	**872.7900**	**2831.938**	**121.6086**
18	0.25	0.50	0.00	0.25	4809.232	767.2090	2885.723	145.7614
19	0.00	0.75	0.00	0.25	4992.908	713.4568	2995.393	195.7333
20	0.50	0.00	0.25	0.25	4688.990	1048.7480	2814.173	113.3293
21	0.25	0.25	0.25	0.25	4728.251	854.5417	2837.411	123.2027
22	0.00	0.50	0.25	0.25	4848.224	747.2574	2909.006	155.2110
23	0.25	0.00	0.50	0.25	4689.761	1048.6700	2814.637	112.5303
24	0.00	0.25	0.50	0.25	4742.296	832.4686	2845.786	125.7116
25	0.00	0.00	0.75	0.25	4690.277	1048.0580	2814.947	111.5577
26	0.50	0.00	0.00	0.50	4692.470	1049.7440	2816.269	107.7859
27	0.25	0.25	0.00	0.50	4744.718	840.5290	2847.264	118.0445
28	0.00	0.50	0.00	0.50	4905.849	737.4598	2943.462	157.6548
29	0.25	0.00	0.25	0.50	4694.153	1049.5460	2817.281	106.0032
30	0.00	0.25	0.25	0.50	4769.112	816.8209	2861.834	120.3674
31	0.00	0.00	0.50	0.50	4697.462	1049.3250	2819.266	103.4786
32	0.25	0.00	0.00	0.75	4709.341	1051.8970	2826.394	98.0417
33	0.00	0.25	0.00	0.75	4830.345	810.7568	2898.509	116.3684
34	0.00	0.00	0.25	0.75	4725.573	1053.6540	2836.124	93.8266
35	0.00	0.00	0.00	1.00	4820.885	1075.3240	2893.228	87.1483

TABLE 5.29 Membership functions of each non-inferior solution

Sr. no.	w_1	w_2	w_3	w_4	$\mu(F_1)$	$\mu(F_2)$	$\mu(F_3)$	$\mu(F_4)$	μ_D
1	1.00	0.00	0.00	0.00	0.980074	0.087853	0.980009	0.784629	0.029386
2	0.75	0.25	0.00	0.00	0.941066	0.499697	0.941802	0.727382	0.032264
3	0.50	0.50	0.00	0.00	0.789356	0.755629	0.790157	0.614316	0.030599
4	0.25	0.75	0.00	0.00	0.511234	0.928872	0.511786	0.395797	0.024356
5	0.00	1.00	0.00	0.00	0.000000	1.000000	0.000000	0.000000	0.010374
6	0.75	0.00	0.25	0.00	0.980005	0.088569	0.979942	0.784534	0.029392
7	0.50	0.25	0.25	0.00	0.936530	0.527044	0.937305	0.722607	0.032404
8	0.25	0.50	0.25	0.00	0.739306	0.802596	0.740081	0.574835	0.029638
9	0.00	0.75	0.25	0.00	0.376233	0.966034	0.376638	0.290243	0.020844
10	0.50	0.00	0.50	0.00	0.979916	0.089464	0.979853	0.784641	0.029400
11	0.25	0.25	0.50	0.00	0.918429	0.563134	0.919205	0.713074	0.032304
12	0.00	0.50	0.50	0.00	0.667627	0.854642	0.668346	0.518370	0.028104
13	0.25	0.00	0.75	0.00	0.979800	0.090613	0.979741	0.784649	0.029409
14	0.00	0.25	0.75	0.00	0.891768	0.614264	0.892565	0.693927	0.032083
15	0.00	0.00	1.00	0.00	0.979643	0.092144	0.979586	0.784661	0.029422
16	0.75	0.00	0.00	0.25	1.000000	0.078852	1.000000	0.807289	0.029942
17	**0.50**	**0.25**	**0.00**	**0.25**	**0.918555**	**0.551854**	**0.919254**	**0.756479**	**0.032639**
18	0.25	0.50	0.00	0.25	0.686337	0.839536	0.687011	0.585798	0.029034
19	0.00	0.75	0.00	0.25	0.213288	0.985997	0.213459	0.232660	0.017070
20	0.50	0.00	0.25	0.25	0.996016	0.072414	0.995964	0.814986	0.029872
21	0.25	0.25	0.25	0.25	0.894900	0.601577	0.895624	0.745214	0.032548
22	0.00	0.50	0.25	0.25	0.585916	0.893899	0.586475	0.519020	0.026821
23	0.25	0.00	0.50	0.25	0.994029	0.072625	0.993962	0.820633	0.029891
24	0.00	0.25	0.50	0.25	0.858727	0.661720	0.859459	0.727434	0.032237
25	0.00	0.00	0.75	0.25	0.992700	0.074293	0.992621	0.827505	0.029952
26	0.50	0.00	0.00	0.50	0.987052	0.069699	0.986912	0.854160	0.030063
27	0.25	0.25	0.00	0.50	0.852491	0.639758	0.853079	0.781666	0.032440
28	0.00	0.50	0.00	0.50	0.437505	0.920595	0.437698	0.501750	0.023836
29	0.25	0.00	0.25	0.50	0.982718	0.070240	0.982544	0.866758	0.030109
30	0.00	0.25	0.25	0.50	0.789664	0.704356	0.790162	0.765250	0.031636
31	0.00	0.00	0.50	0.50	0.974196	0.070841	0.973971	0.884598	0.030123
32	0.25	0.00	0.00	0.75	0.943601	0.063832	0.943193	0.923019	0.029812
33	0.00	0.25	0.00	0.75	0.631962	0.720879	0.631804	0.793510	0.028822
34	0.00	0.00	0.25	0.75	0.901797	0.059046	0.901181	0.952806	0.029202
35	0.00	0.00	0.00	1.00	0.656326	0.000000	0.654606	1.000000	0.023974

TABLE 5.30 Voltage at buses for 'best' solution

Bus no.	V_i (p.u.)	δ_i (rad)
1	1.070	0.0
2	1.088	0.1242089
3	1.062	0.2124252
4	1.062	0.1255512
5	1.046	0.1061312
6	1.053988	0.1015301
7	1.014140	0.0176067
8	0.993863	−0.0009264
9	1.002498	0.0247629
10	1.034458	0.0349411
11	1.034815	−0.0117819

emissions levels, etc. required for the system's thermal generation, over the optimization interval. Each hydro plant is constrained by the amount of water available for draw-down in the interval. A prediction of the system's future power demand and water supply is assumed to be available for the optimization interval.

Minimize
$$F_1 = \sum_{k=1}^{T}\left(\sum_{i=1}^{N} t_k(a_i P_{ik}^2 + b_i P_{ik} + c_i)\right) \text{ ₹} \tag{5.78a}$$

Minimize
$$F_2 = \sum_{k=1}^{T}\left(\sum_{i=1}^{N} t_k(d_{1i} P_{ik}^2 + e_{1i} P_{ik} + f_{1i})\right) \text{ kg} \tag{5.78b}$$

Minimize
$$F_3 = \sum_{k=1}^{T}\left(\sum_{i=1}^{N} t_k(d_{2i} P_{ik}^2 + e_{2i} P_{ik} + f_{2i})\right) \text{ kg} \tag{5.78c}$$

subject to
$$\sum_{i=1}^{N+M} P_{ik} = P_{Dk} + P_{Lk} \tag{5.78d}$$

$$\sum_{k=1}^{T} t_k\, q_{jk} = V_j \qquad (j = 1, 2, ..., M) \tag{5.78e}$$

$$P_{ik}^{\min} \le P_{ik} \le P_{ik}^{\max} \qquad (i = 1, 2, ..., N + M) \tag{5.78f}$$

where

F_1 is cost function of thermal units for the whole planning period, T
F_2 is NO_x emission function of thermal units for the whole planning period, T
F_3 is SO_2 emission function of thermal units for the whole planning period, T
a_i, b_i, and c_i are cost coefficients

d_{1i}, e_{1i}, and f_{1i} are NO_x emission coefficients
d_{2i}, e_{2i}, and f_{2i} are SO_2 emission coefficients
P_{ik} is the output of thermal and hydro units during the kth interval

P_{Lk} is transmission losses during the kth interval and is given by

$$P_{Lk} = \sum_{i=1}^{N+M} \sum_{j=1}^{N+M} P_{ik} B_{ij} P_{jk} + \sum_{i=1}^{N+M} B_{i0} P_{ik} + B_{00}$$

where
B_{ij}, B_{i0}, and B_{00} are B-coefficients
P_{Dk} is the demand during the kth interval
$P_{ik}^{\min}$ is the lower limit of generations during the kth interval
$P_{ik}^{\max}$ is the upper limit of generations during the kth interval

q_{jk} is rate of discharge from the jth hydro unit in interval k and is defined by

$$q_j(t) = x_j P_{j+N,k}^2 + y_j P_{j+N,k} + z_i \quad \text{M}^3\text{/h}$$

where
x_j, y_j, and z_j are discharge coefficients of the jth unit
V_j is the pre-specified volume of water available for unit j for whole of the period
N is the number of thermal units
M is the number of hydro units
T is the overall period for scheduling.

To generate the non-inferior solution to the multiobjective problem, the weighting method is applied. In this method the problem is converted into a scalar optimization as given below:

$$\text{Minimize} \quad \sum_{k=1}^{3} w_k F_k \tag{5.79a}$$

$$\text{subject to} \quad \sum_{i=1}^{N+M} P_{ik} = P_{Dk} + P_{Lk} \tag{5.79b}$$

$$\sum_{k=1}^{T} t_k q_{jk} = V_j \qquad (j = 1, 2, ..., M) \tag{5.79c}$$

$$P_{ik}^{\min} \le P_{ik} \le P_{ik}^{\max} \qquad (i = 1, 2,...., N + M) \tag{5.79d}$$

$$\sum_{k=1}^{3} w_k = 1.0, \quad w_k \ge 0 \tag{5.79e}$$

This approach yields meaningful result to the decision maker when solved many times for different values of w_k (k = 1, 2, 3). Weighting factors w_k are determined based on the relative importance of various objectives, which may vary from place to place and utility to utility. The problem defined by Eqs. (5.79) can be redefined in simplified form as

$$\text{Minimize} \quad \sum_{k=1}^{T}\left(\sum_{i=1}^{N} t_k F_T\right) \tag{5.80a}$$

subject to

$$\sum_{i=1}^{N+M} P_{ik} = P_{Dk} + P_{Lk} \tag{5.80b}$$

$$\sum_{k=1}^{T} t_k\, q_{jk} = V_j \quad (j = 1, 2, ..., M) \tag{5.80c}$$

$$P_{ik}^{\min} \le P_{ik} \le P_{ik}^{\max} \quad (i = 1, 2, ..., N + M) \tag{5.80d}$$

where

$$F_T = \alpha_i P_{ik}^2 + \beta_i P_{ik} + \gamma_i$$
$$\alpha_i = w_1 a_i + w_2 d_{1i} + w_3 d_{2i}$$
$$\beta_i = w_1 b_i + w_2 e_{1i} + w_3 e_{2i}$$
$$\gamma_i = w_1 c_i + w_2 f_{1i} + w_3 f_{2i}$$

The above objective function as augmented by the constraints is given as

$$L(P_{ik}, \lambda_k, v_j) = \sum_{k=1}^{T}\left[\sum_{i=1}^{N} t_k F_T + \sum_{j=1}^{M} v_j t_k q_{jk} + \lambda_k\left[P_{Dk} + P_{Lk} - \sum_{i=1}^{N+M} P_{ik}\right]\right] - \sum_{j=1}^{M} v_j V_j \tag{5.81}$$

where

λ_k is incremental cost of power delivered in the system during the kth interval

v_j are water conversion factors.

The optimality conditions are described by taking the partial derivatives of augmented objective function with respect to the decision variables, $P_{ik}(i = 1, 2, ..., N + M; \lambda_k)$; $v_j(j = 1, 2, ..., M)$.

$$t_k \frac{\partial F_T}{\partial P_{ik}} + \lambda_k\left[\frac{\partial P_{Lk}}{\partial P_{ik}} - 1\right] = 0 \quad (i = 1, 2, ..., N;\ k = 1, 2, ..., T) \tag{5.82a}$$

$$v_j t_k \frac{\partial q_{jk}}{\partial P_{mk}} + \lambda_k\left[\frac{\partial P_{Lk}}{\partial P_{mk}} - 1\right] = 0 \quad (j = 1, 2, ..., M;\ m = N + j;\quad k = 1, 2, ..., T) \tag{5.82b}$$

$$\sum_{k=1}^{T} t_k\, q_{jk} = V_j \quad (j = 1, 2, ..., M) \tag{5.82c}$$

$$\sum_{i=1}^{N+M} P_{ik} = P_{Dk} + P_{Lk} \quad (k = 1, 2, ..., T) \tag{5.82d}$$

Inherently, these equations are nonlinear. The Newton–Raphson method has been applied to find $P_{ik}(i = 1, 2,..., N + M)$, λ_k for all the T intervals. v_j water conversion factor is modified to satisfy the available water constraint. Suppose the initial values of control variables P_{ik}, λ_k and v_j are known. The updated values of control variables in the next iteration are

$$P_{ik}^{\text{new}} = P_{ik}^{0} + \Delta P_{ik} \qquad (i = 1, 2, ..., N + M; \quad k = 1, 2, ..., T)$$

$$\lambda_{k}^{\text{new}} = \lambda_{k}^{0} + \Delta \lambda_{k} \qquad (k = 1, 2, ..., T)$$

$$v_{j}^{\text{new}} = v_{j}^{0} + \Delta v_{j} \qquad (j = 1, 2, ..., M)$$

Any small change in control values from their previous values can be obtained as given below:

$$\left(t_k \frac{\partial^2 F_T}{\partial P_{ik}^2} + \lambda_k^0 \frac{\partial^2 P_{Lk}}{\partial P_{ik}^2} \right) \Delta P_{ik} + \lambda_k^0 \sum_{\substack{j=1 \\ j \neq i}}^{N+M} \frac{\partial^2 P_{Lk}}{\partial P_{ik} \partial P_{jk}} \Delta P_{jk} + \left(\frac{\partial P_{Lk}}{\partial P_{ik}} - 1 \right) \Delta \lambda_k$$

$$= - \left[t_k \frac{\partial F_T}{\partial P_{ik}} + \lambda_k^0 \left(\frac{\partial P_{Lk}}{\partial P_{ik}} - 1 \right) \right] \qquad (i = 1, 2,..., N; \quad k = 1, 2, ..., T) \qquad (5.83a)$$

$$\left(v_j^0 t_k \frac{\partial^2 q_{jk}}{\partial P_{mk}^2} + \lambda_k^0 \frac{\partial^2 P_{Lk}}{\partial P_{mk}^2} \right) \Delta P_{mk} + \lambda_k^0 \sum_{\substack{l=1 \\ l \neq m}}^{N+M} \frac{\partial^2 P_{Lk}}{\partial P_{mk} \partial P_{lk}} \Delta P_{lk} + \left(\frac{\partial P_{Lk}}{\partial P_{mk}} - 1 \right) \Delta \lambda_k + \frac{\partial q_{jk}}{\partial P_{mk}} \Delta v_j$$

$$= - \left[v_j^0 t_k \frac{\partial q_{jk}}{\partial P_{mk}} + \lambda_k^0 \left(\frac{\partial P_{Lk}}{\partial P_{mk}} - 1 \right) \right] \qquad (j = 1, 2, ..., M; \quad m = j + N; \quad k = 1, 2, ..., T) \qquad (5.83b)$$

$$\sum_{j=1}^{N+M} \left(\frac{\partial P_{Lk}}{\partial P_{ik}} - 1 \right) \Delta P_{jk} = - \left(P_{Dk} + P_{Lk}^0 - \sum_{i=1}^{N+M} P_{ik}^0 \right) \qquad (k = 1, 2,..., T) \qquad (5.83c)$$

$$\sum_{k=1}^{T} \left(t_k \frac{\partial q_{jk}}{\partial P_{mk}} - 1 \right) \Delta P_{mk} = - \left(\sum_{k=1}^{T} t_k q_{ik}^0 - V_j \right) \qquad (j = 1, 2, ..., M; \quad m = j + N) \qquad (5.83d)$$

Substituting the values of derivatives defined in Eqs. (4.42a), the derivatives are evaluated about initial values.

$$(2 t_k \alpha_i + 2 \lambda_k^0 B_{ii}) \Delta P_{ik} + \lambda_k^0 \sum_{\substack{j=1 \\ j \neq i}}^{N+M} 2 B_{ij} \Delta P_{jk} + (K_{ik}^0 - 1) \Delta \lambda_k$$

$$= - \left[t_k (2 \alpha_i P_{ik}^0 + \beta_i) + \lambda_k^0 (K_{ik}^0 - 1) \right] \qquad (i = 1, 2, ..., N) \qquad (5.83e)$$

where

$$K_{ik}^0 = \sum_{j=1}^{N+M} 2 B_{ij} P_{jk}^0 + B_{i0}$$

Equation (5.83e) can be simplified by rearranging the terms as

$$(2t_k\alpha_i + 2\lambda_k^0 B_{ii})\,\Delta P_{ik} + \lambda_k^0 \sum_{\substack{j=1\\ j\neq i}}^{N+M} 2B_{ij}\Delta P_{jk} = (1 - K_{ik}^0)(\lambda_k^0 + \Delta\lambda_k) - t_k(2\alpha_i P_{ik}^0 + \beta_i) \qquad (i = 1, 2, ..., N)$$

To simplify calculations, neglect all the terms with B_{ij} $(i \neq j)$. So, neglecting the term $\sum_{\substack{j=1\\ j\neq i}}^{N+M} 2B_{ij}\Delta P_{jk}$, the above equation becomes

$$(2t_k\alpha_i + 2\lambda_k^0 B_{ii})\,\Delta P_{ik} = (1 - K_{ik}^0)\,\lambda_k^{\text{new}} - t_k(2\alpha_i P_{ik}^0 + \beta_i) \qquad (i = 1, 2, ..., N)$$

or

$$\Delta P_{ik} = \frac{(1 - K_{ik}^0)\,\lambda_k^{\text{new}} - t_k(2\alpha_i\, P_{ik}^0 + \beta_i)}{2t_k\alpha_i + 2\lambda_k^0 B_{ii}} \qquad (i = 1, 2, ..., N)$$

$$\Delta P_{ik} = \frac{(1 - K_{ik}^0)\,\lambda_k^{\text{new}} - A_{ik}}{X_{ik}} \qquad (i = 1, 2, ..., N) \tag{5.84}$$

where

$$A_{ik} = t_k(2\alpha_i P_{ik}^0 + \beta_i) \tag{5.84a}$$

$$X_{ik} = 2(t_k\alpha_i + \lambda_k^0 B_{ii}) \tag{5.84b}$$

Substituting the values of derivatives in Eq. (5.83b),

$$(2v_j^0 t_k x_j + 2\lambda_k^0 B_{mm})\,\Delta P_{mk} + \lambda_k^0 \sum_{\substack{l=1\\ l\neq m}}^{N+M} 2B_{ml}\Delta P_{lk} + (K_{mk}^0 - 1)\,\Delta\lambda_k + t_k(2x_j P_{mk}^0 + y_j)\,\Delta v_j^0$$

$$= -\,[v_j^0 t_k(2x_j P_{mk}^0 + y_j) + \lambda_k^0\;(K_{mk}^0 - 1)] \qquad (j = 1, 2, ..., M;\quad m = j + N;\quad k = 1, 2, ..., T)$$

where

$$K_{mk}^0 = \sum_{l=1}^{N+M} 2B_{ml}P_{lk}^0 + B_{m0}$$

The above equation can be simplified by rearranging the terms, i.e.

$$(2v_j^0 t_k x_j + \lambda_k^0 B_{mm})\Delta P_{mk} + \lambda_k^0 \sum_{\substack{l=1\\ l\neq m}}^{N+M} 2B_{ml}\Delta P_{lk}$$

$$= -[t_k(2x_j P_{mk}^0 + y_j)(v_j^0 + \Delta v_j) + (\lambda_k^0 + \Delta\lambda_k)\,(K_{mk}^0 - 1)] \qquad (j = 1, 2, ..., M;\quad m = j + N)$$

Neglecting the term $\sum_{\substack{l=1\\ l\neq m}}^{N+M} 2B_{ml}\Delta P_{lk}$ for simplification, we have

$$(2v_j^0 t_k x_j + 2\lambda_k^0 B_{mm})\,\Delta P_{mk} = -\left[t_k(2x_j P_{mk}^0 + y_j)v_j^{\text{new}} + \lambda_k^{\text{new}}\,(K_{mk}^0 - 1)\right] \quad (j = 1, 2, ..., M;\quad m = j + N)$$

or

$$\Delta P_{mk} = \frac{(1-K^0_{mk})\,\lambda^{\text{new}}_k - t_k\,(2x_j P^0_{mk} + y_j)\,\nu^{\text{new}}_j}{2\nu^0_j\,t_k x_j + 2\lambda^0_k B_{mm}} \qquad (j = 1, 2, ..., M; \quad m = j + N)$$

$$= \frac{(1-K^0_{mk})\,\lambda^{\text{new}}_k - E_{jk}\nu^{\text{new}}_j}{Y_{jk}} \qquad (j = 1, 2, ..., M; \quad m = j + N) \qquad (5.85)$$

where

$$E_{jk} = t_k(2x_j P^0_{mk} + y_i) \qquad (5.85a)$$

$$Y_{jk} = 2(\nu^0_j\, t_k x_j + \lambda^0_k\, B_{mm}) \qquad (5.85b)$$

Substituting the values of derivatives in Eq. (5.83c)

$$\sum_{j=1}^{N+M}\left(\sum_{l=1}^{N+M} 2B_{jl}P^0_{lk} - 1\right)\Delta P_{jk} = -\left(P_{Dk} + P^0_{Lk} - \sum_{i=1}^{N+M} P^0_{ik}\right) \qquad (k = 1, 2, ..., T)$$

or

$$\sum_{j=1}^{N+M}(1-K^0_{jk})\Delta P_{jk} = P_{Dk} + P^0_{Lk} - \sum_{i=1}^{N+M} P^0_{ik} \qquad (k = 1, 2, ..., T)$$

Substituting Eq. (5.84) and Eq. (5.85) into the above equation

$$\sum_{i=1}^{N}(1-K^0_{ik})\left(\frac{\lambda^{\text{new}}_k(1-K^0_{ik}) - A_{ik}}{X_{ik}}\right) + \sum_{j=1}^{M}(1-K^0_{mk})\left(\frac{\lambda^{\text{new}}_k(1-K^0_{mk}) - E_{jk}\,\nu^{\text{new}}_j}{Y_{jk}}\right)$$

$$= \left(P_{Dk} + P^0_{Lk} - \sum_{i=1}^{N+M} P^0_{ik}\right) \qquad (k = 1, 2, ..., T)$$

or

$$\left(\sum_{i=1}^{N}\frac{(1-K^0_{ik})^2}{X_{ik}} + \sum_{j=1}^{M}\frac{(1-K^0_{mk})^2}{Y_{jk}}\right)\lambda^{\text{new}}_k - \sum_{j=1}^{M}\left(\frac{(1-K^0_{mk})E_{jk}}{Y_{jk}}\right)\nu^{\text{new}}_j$$

$$= \left(P_{Dk} + P^0_{Lk} - \sum_{i=1}^{N+M} P^0_{ik}\right) + \sum_{i=1}^{N}\left(\frac{(1-K^0_{mk})A_{ik}}{X_{ik}}\right) \qquad (k = 1, 2, ..., T)$$

or

$$C_k\lambda^{\text{new}}_k - \sum_{j=1}^{M} D_{jk}\,\nu^{\text{new}}_j = F_k \qquad (5.86)$$

where

$$C_k = \sum_{i=1}^{N}\frac{(1-K^0_{ik})^2}{X_{ik}} + \sum_{j=1}^{M}\frac{(1-K^0_{mk})^2}{Y_{jk}} \qquad (5.86a)$$

$$D_{jk} = \frac{(1 - K_{mk}^0) E_{jk}}{Y_{jk}} \tag{5.86b}$$

$$F_k = \left(P_{Dk} + P_{Lk}^0 - \sum_{i=1}^{N+M} P_{ik}^0 \right) + \sum_{i=1}^{N} \left(\frac{(1 - K_{ik}^0) A_{ik}}{X_{ik}} \right) \tag{5.86c}$$

Substituting the values of derivatives in Eq. (5.83d)

$$\sum_{k=1}^{T} t_k (2x_j P_{mk}^0 + y_j) \Delta P_{mk} = - \left(\sum_{k=1}^{T} t_k q_{ik}^0 - V_j) \right) \qquad (j = 1, 2, ..., M; \quad m = j + N)$$

Substituting Eq. (5.85) into the above equation

$$\sum_{k=1}^{T} E_{jk} \left(\frac{(1 - K_{mk}^0) \lambda_k^{\text{new}} - E_{jk} v_j^{\text{new}}}{Y_{jk}} \right) = - \left(\sum_{k=1}^{T} t_k q_{jk}^0 - V_j \right) \qquad (j = 1, 2, ..., M; \quad m = j + N)$$

or

$$\sum_{k=1}^{T} D_{jk} \lambda_k^{\text{new}} - H_j v_j^{\text{new}} = O_j \qquad (j = 1, 2, ..., M; \quad m = j + N) \tag{5.87}$$

where

$$H_j = \sum_{k=1}^{T} \frac{E_{jk} E_{jk}}{Y_{jk}} \tag{5.87a}$$

$$O_j = V_j - \sum_{k=1}^{T} t_k q_{ik}^0 \tag{5.87b}$$

From Eq. (5.86), we find the value of λ_k^{new} as

$$\lambda_k^{\text{new}} = \frac{F_k}{C_k} + \sum_{j=1}^{M} \frac{D_{jk}}{C_k} v_j^{\text{new}} \tag{5.88}$$

Substituting the value of λ_k^{new} into Eq. (5.87),

$$\sum_{k=1}^{T} D_{jk} \left(\frac{F_k}{C_k} + \sum_{j=1}^{M} \frac{D_{jk}}{C_k} v_j^{\text{new}} \right) - H_j v_j^{\text{new}} = O_j \qquad (j = 1, 2, ..., M; \quad m = j + N)$$

or

$$\sum_{j=1}^{M} \left(\sum_{k=1}^{T} \frac{D_{jk} D_{jk}}{C_k} v_j^{\text{new}} \right) - H_j v_j^{\text{new}} = O_j - \sum_{k=1}^{T} \frac{D_{jk} F_k}{C_k} \qquad (j = 1, 2, ..., M; \quad m = j + N) \tag{5.89}$$

Equation (5.89) can be written in matrix form as

$$[Q_{jl}]_{M \times M} \, [v_j]_{M \times l} = [R_j]_{M \times l}$$

where

$$Q_{jj} = \left(\sum_{k=1}^{T} \frac{D_{jk} D_{jk}}{C_k} \right) - H_j \qquad (j = 1, 2, ..., M) \tag{5.90}$$

$$Q_{jl} = \left(\sum_{k=1}^{T} \frac{D_{jk} D_{lk}}{C_k} \right) \qquad (j = 1, 2, ..., M; \quad l = 1, 2, ..., M; \quad j \neq l) \tag{5.91}$$

$$R_j = O_j - \sum_{k=1}^{T} \frac{D_{jk} F_k}{C_k} \qquad (j = 1, 2, ..., M) \tag{5.92}$$

Here only matrix of $M \times M$ is to be solved to calculate v_j^{new} whereas λ_k^{new} can be computed from Eq. (4.47). When the values of v_j^{new} and λ_k^{new} are known then ΔP_{ik} $(i = 1, 2, ..., N)$ and ΔP_{mk} $(j = 1, 2, ..., M;\ m = j + N)$, can be computed from Eqs. (4.43) and (4.44), respectively. The detailed algorithm is elaborated below:

Algorithm 5.5: Generation of Non-Inferior Solution for Multiobjective Hydrothermal Scheduling Using the Approximate Newton–Raphson Method

1. Read the number of thermal units N, the number of hydro units M, the number of sub-intervals T, cost coefficients, emission coefficients, B-coefficients, discharge coefficients, demand $P_{Dk}(k = 1, 2, ..., T)$ and pre-specified available water V_j $(j = 1, 2, ..., M)$, etc.
2. Set iteration for non-inferior solutions, $k = 1$.
3. Increment count of non-inferior solutions, $k = k + 1$.
4. If $(k \geq K)$ GOTO Step 17.
5. Feed or generate the weights, $w_i(i = 1, 2, ..., M)$.
6. Calculate the initial guess values of P_{ik}^0 $(i = 1, 2, ..., N + M)$, λ_k^0 and v_j^0 $(j = 1, 2, ..., M)$.
7. Start the iteration counter, $r = 1$.
8. Compute the variables, K_{ik}^0, A_{ik} from Eq. (5.84a), X_{ik} from Eq. (5.84b), E_{jk} from Eq. (5.85a), Y_{jk} from Eq. (5.85b), C_k from Eq. (5.86a), D_{jk} from Eq. (5.86b), F_k from Eq. (5.86c), H_j from Eq. (5.87a) and O_j from Eq. (5.87b).
9. Compute v_j^{new} by solving the following simultaneous equations using the Gauss elimination method.
$$[Q_{jl}]_{M\times M}\ [v_j]_{M\times 1} = [R_j]_{M\times 1}$$
10. Check the convergence
if $|\ v_j^{\text{new}} - v_j^0\ | \leq \varepsilon$ then GOTO Step 16.
11. Compute λ_k^{new} from Eq. (5.88).
12. Calculate ΔP_{ik} $(i = 1, 2,..., N + M)$, using Eqs. (5.84) and (5.85).
13. Calculate the new values of P_{ik}^{new} $(i = 1, 2,..., N + M)$
$$P_{ik}^{\text{new}} = P_{ik}^0 + \Delta P_{ik} \qquad (i = 1, 2, ..., N + M; \quad k = 1, 2, ..., T)$$
14. Set limits correspondingly as:
$$P_{ik}^{\text{new}} = \begin{cases} P_{ik}^{\max} & ; \text{if } P_{ik}^{\text{new}} \geq P_{ik}^{\max} \\ P_{ik}^{\min} & ; \text{if } P_{ik}^{\text{new}} \geq P_{ik}^{\min} \\ P_{ik}^{\text{new}} & ; \text{otherwise} \end{cases}$$

Disallow to participate the generation, whose limits have been set either to lower or upper limits by setting the relating coefficients zero.

15. If ($r \geq$ IT), then GOTO Step 5,

$$\text{else} \quad r = r + 1,$$
$$P_{ik}^0 = P_{ik}^{\text{new}} \quad (i = 1, 2,..., N + M; \quad k = 1, 2, ..., T)$$
$$\lambda_k^0 = \lambda_k^{\text{new}} \quad (k = 1, 2,..., T)$$
$$v_j^0 = v_j^{\text{new}} \quad (j = 1, 2, ..., N) \text{ and GOTO Step 12 and repeat.}$$

16. Record it as non-inferior solution and calculate the objective values and loss etc. and GOTO Step 3.
17. Stop.

5.7.1 Sample System

A hydro-thermal system is given which consists of two thermal and two hydro generating stations as shown in Figure 5.7. The fuel cost, NO_x emission, SO_2 emission and CO_2 emission equations are given below. Transmission loss coefficients are given in Table 5.31.

TABLE 5.31 *B*-coefficients (MW^{-1})

0.000140	0.000010	0.000015	0.000015
0.000010	0.000060	0.000010	0.000013
0.000015	0.000010	0.000068	0.000065
0.000015	0.000013	0.000065	0.000070

Operating cost of thermal station is given by

$$F_{11} = 0.0025P_{1k}^2 + 3.20P_{1k} + 25.0 \text{ ₹/h}$$
$$F_{12} = 0.0008P_{2k}^2 + 3.40P_{2k} + 30.0 \text{ ₹/h}$$

NO_x emission of thermal station is given by

$$F_{21} = 0.006483P_{1k}^2 - 0.79027P_{1k} + 28.82488 \text{ kg/h}$$
$$F_{22} = 0.006483P_{2k}^2 - 0.79027P_{2k} + 28.82488 \text{ kg/h}$$

SO_2 emission of thermal station is given by

$$F_{31} = 0.00232P_{1k}^2 + 3.84632P_{1k} + 182.2605 \text{ kg/h}$$
$$F_{32} = 0.00232P_{2k}^2 + 3.84632P_{2k} + 182.2605 \text{ kg/h}$$

CO_2 emission of thermal station is given by

$$F_{41} = 0.084025P_{1k}^2 - 2.944584P_{1k} + 137.7043 \text{ kg/h}$$
$$F_{42} = 0.084025P_{2k}^2 - 2.944584P_{2k} + 137.7043 \text{ kg/h}$$

Rate of discharge of hydro generating station is given by:

$$q_{1k} = 6.1160 \times 10^{-6}P_{3k}^2 + 0.00866494P_{3k} + 0.05606727 \text{ Mm}^3\text{/h}$$
$$q_{2k} = 1.0194 \times 10^{-5}P_{4k}^2 + 0.01732988P_{4k} + 0.02650452 \text{ Mm}^3\text{/h}$$

Volume of water available in reservoir is given by

$$V_1 = 71.0 \text{ Mm}^3$$

$$V_2 = 60.0 \text{ Mm}^3$$

Only two objectives are considered—operating cost F_1, and NO_x emission F_2. For various combinations of weights, the non-inferior solution is obtained and is given in Table 5.32.

TABLE 5.32 Non-inferior solutions

Sr. no.	w_1	w_2	F_1	F_2	v_1	v_2
1	100	0	52874.40	28478.67	333.7511	201.0085
2	90	10	53061.94	24446.85	330.3708	200.9783
3	80	20	53409.74	22414.03	325.1166	199.8769
4	70	30	53797.45	21231.71	318.8058	197.6761
5	60	40	54154.64	20553.30	312.5692	195.1165
6	50	50	54425.04	20216.39	307.5677	192.6615
7	40	60	54652.34	20028.74	302.7455	190.0693
8	30	70	54861.16	19915.32	297.6498	187.2390
9	20	80	55032.20	19857.73	292.8095	184.4094
10	10	90	55187.26	19829.91	287.8277	181.4550
11	0	100	55311.99	19823.11	283.0973	178.5556

To decide the best solution, minimum and maximum values of objective functions are required. Minimum values of objectives are obtained by giving full weightage to one of the objectives and neglecting others. When the given weightage value is 1.0, it means that full weightage is given to the objective and when the weightage is zero the objective is neglected. Owing to the conflicting nature of objectives, F_2 will have maximum value when F_1 is having minimum value or vice versa. The minimum and maximum values are obtained and are given below:

$$F_1^{\min} = 52874.40 \text{ ₹} \qquad F_1^{\max} = 55311.99 \text{ ₹}$$

$$F_2^{\min} = 19823.11 \text{ kg} \qquad F_2^{\max} = 28478.67 \text{ kg}$$

Using Eq. (5.57), the membership functions of F_1 and F_2 objectives corresponding to each non-inferior solutions are obtained and are given in Table 5.33. The membership functions also follow the conflicting nature. Using Eq. (5.58), the normalized membership function μ_D of each non-inferior is obtained and is shown in Table 5.33. The non-inferior solution that attains the maximum normalized membership function is the best solution. From Table 5.33, solution number 3, having weights, $w_1 = 80$ $w_2 = 20$ shows the maximum value of μ_D, i.e. 0.108743, so this solution is considered the best solution. Corresponding to the preferred solution shown in Table 5.33 at serial number 3, the generation schedules are given in Tables 5.34 and 5.35. The solution is obtained in 6 iterations and the obtained convergence to meet the volume of water utilized is 0.30517580E–04. Cost, NO_x emission, and λ_k, discharge during 24 intervals is shown in Table 5.33. Generation schedule, transmission loss and mismatch in demand during each interval is given in Table 5.35.

TABLE 5.33 Decision making

Sr. no.	w_1	w_2	F_1(₹)	F_2(kg)	$\mu(F_1)$	$\mu(F_2)$	μ_D
1	100	0	52874.40	28478.67	1.000000	0.000000	0.073423
2	90	10	53061.94	24446.85	0.923065	0.465806	0.101975
3	**80**	**20**	**53409.74**	**22414.03**	**0.780383**	**0.700664**	**0.108743**
4	70	30	53797.45	21231.71	0.621328	0.837261	0.107094
5	60	40	54154.64	20553.30	0.474792	0.915639	0.102090
6	50	50	54425.04	20216.39	0.363866	0.954563	0.096803
7	40	60	54652.34	20028.74	0.270617	0.976242	0.091548
8	30	70	54861.16	19915.32	0.184950	0.989347	0.086221
9	20	80	55032.20	19857.73	0.114781	0.996000	0.081557
10	10	90	55187.26	19829.91	0.051170	0.999214	0.077122
11	0	100	55311.99	19823.11	0.0	1.0	0.073423

TABLE 5.34 Schedule corresponding to preferred solution

Interval	P_{Dk} (MW)	F_{1k} (₹/h)	F_{2k} (kg/h)	λ_k	q_{1k} (Mm^3/h)	q_{2k} (Mm^3/h)
1	400	1097.6650	117.6295	3.328275	1.083662	0.026505
2	300	864.8992	53.7901	3.146616	0.700009	0.026505
3	250	751.0435	32.4935	3.057208	0.516236	0.026505
4	250	751.0435	32.4935	3.057208	0.516236	0.026505
5	250	751.0435	32.4935	3.057208	0.516236	0.026505
6	300	864.8992	53.7901	3.146616	0.700009	0.026505
7	450	1216.6080	160.3328	3.420560	1.283657	0.026505
8	900	2137.8950	698.5376	4.136229	2.800091	1.859663
9	1230	2849.1700	1333.3430	4.683991	3.990101	3.716961
10	1250	2894.3720	1379.3890	4.718668	4.066558	3.836010
11	1350	3124.1470	1623.3200	4.894739	4.456507	4.442635
12	1400	3241.4240	1754.0160	4.984497	4.656306	4.753106
13	1200	2781.8270	1265.9510	4.632308	3.876354	3.539791
14	1250	2894.3720	1379.3890	4.718668	4.066558	3.836010
15	1250	2894.3720	1379.3890	4.718668	4.066558	3.836010
16	1270	2939.8240	1426.3420	4.753520	4.143528	3.955816
17	1350	3124.1470	1623.3200	4.894739	4.456507	4.442635
18	1470	3408.3510	1947.0370	5.112148	4.941507	5.195874
19	1330	3077.6850	1572.6860	4.859160	4.377492	4.319787
20	1250	2894.3720	1379.3890	4.718668	4.066558	3.836010
21	1170	2715.0360	1200.5590	4.581015	3.763741	3.364311
22	1050	2453.2790	958.5485	4.379663	3.324467	2.679101
23	900	2137.8950	698.5376	4.136229	2.800091	1.859663
24	600	1544.3720	311.2523	3.674913	1.831046	0.341105

TABLE 5.35 Generation schedule corresponding to preferred solution

Interval	PD_k (MW)	PL_k (MW)	P_{1k} (MW)	P_{2k} (MW)	P_{3k} (MW)	P_{4k} (MW)	ΔPD_k
1	400	5.90149	120.0833	175.7740	110.04420	0.0	0.000004
2	300	3.43064	97.6003	135.0508	70.77953	0.0	0.000006
3	250	2.46342	86.3903	114.8204	51.25274	0.0	–0.000015
4	250	2.46342	86.3903	114.8204	51.25274	0.0	–0.000015
5	250	2.46342	86.3903	114.8204	51.25274	0.0	–0.000015
6	300	3.43064	97.6003	135.0508	70.77953	0.0	0.000006
7	450	7.40881	131.3568	196.2686	129.78350	0.0	0.000027
8	900	29.32106	214.1518	348.7260	266.53480	99.90861	–0.000168
9	1230	56.27325	273.4134	459.7775	361.67940	191.40330	–0.000427
10	1250	58.22341	277.0599	466.6632	367.50480	196.99530	0.000145
11	1350	68.53664	295.3904	501.3708	396.73770	225.03800	–0.000221
12	1400	74.04928	304.6173	518.9008	411.42020	239.11110	–0.000038
13	1200	53.41735	267.9555	449.4828	352.95390	183.02520	0.000111
14	1250	58.22341	277.0599	466.6632	367.50480	196.99530	0.000145
15	1250	58.22341	277.0599	466.6632	367.50480	196.99530	0.000145
16	1270	60.21088	280.7130	473.5676	373.33750	202.59280	0.000061
17	1350	68.53664	295.3904	501.3708	396.73770	225.03800	–0.000221
18	1470	82.17287	317.6052	543.6446	432.05110	258.87190	0.000023
19	1330	66.39845	291.7112	494.3918	390.87710	219.41850	–0.000114
20	1250	58.22341	277.0599	466.6632	367.50480	196.99530	0.000145
21	1170	50.64431	262.5120	439.2290	344.24390	174.65950	–0.000095
22	1050	40.36893	240.8798	398.6149	309.55670	141.31750	0.000038
23	900	29.32106	214.1518	348.7260	266.53480	99.90861	–0.000168
24	600	12.96472	161.6948	251.7323	181.57380	17.96381	0.000064

With the increase in the number of objectives, the number of weight combinations increases.

Four objectives

Here, four objectives are considered—operating cost F_1, NO_x emission F_2, SO_2 emission F_3, and CO_2 emission F_4. For various combinations of weights, the non-inferior solution is obtained and is given in Table 5.36. Water conversion values are given in Table 5.37 corresponding to non-inferior solutions depicted in Table 5.36. The minimum and maximum values are obtained and are given below:

$$F_1^{\min} = 52874.40 \text{ ₹}, \quad F_1^{\max} = 55311.99 \text{ ₹}$$
$$F_2^{\min} = 19823.11 \text{ kg}, \quad F_2^{\max} = 28478.67 \text{ kg}$$
$$F_3^{\min} = 72928.52 \text{ kg}, \quad F_3^{\max} = 74744.49 \text{ kg}$$
$$F_4^{\min} = 348655.80 \text{ kg}, \quad F_4^{\max} = 457290.50 \text{ kg}$$

Using Eq. (5.57), the membership functions of F_1, F_2, F_3, and F_4 objectives corresponding to each non-inferior solutions are obtained and are given in Table 5.38. The membership functions also follow the conflicting nature. Using Eq. (5.58), the normalized membership function μ_D of each non-inferior is obtained and is shown in Table 5.38. The non-inferior solution attains the maximum normalized membership function as the best solution. From Table 5.38, solution number 7, having weights, $w_1 = 50$, $w_2 = 50$, $w_3 = 25$, and $w_4 = 0$ shows the maximum value of μ_D, i.e. 0.033349; so this solution is considered the best solution.

TABLE 5.36 Non-inferior solutions

Sr. no.	w_1	w_2	w_3	w_4	F_1	F_2	F_3	F_4
1	100	0	0	0	52874.40	28478.67	74744.49	457290.50
2	75	25	0	0	53601.90	21749.15	73270.84	371722.20
3	50	50	0	0	54425.04	20216.39	73369.79	353000.70
4	25	75	0	0	54948.63	19882.05	73535.41	349171.60
5	0	100	0	0	55311.99	19823.11	73678.74	348698.30
6	75	0	25	0	53007.07	25201.09	73745.94	415118.50
7	**50**	**25**	**25**	**0**	**53894.93**	**21080.80**	**73129.66**	**363233.00**
8	25	50	25	0	54669.08	20036.30	73352.74	350750.40
9	0	75	25	0	55155.26	19843.80	73555.35	348735.20
10	50	0	50	0	53308.69	23307.60	73230.12	390863.30
11	25	25	50	0	54209.27	20683.10	73077.13	358237.10
12	0	50	50	0	54921.27	19957.13	73372.78	349809.80
13	25	0	75	0	53695.83	22229.83	72992.69	377156.50
14	0	25	75	0	54535.83	20485.06	73089.48	355817.50
15	0	0	100	0	54127.01	21665.29	72928.52	370083.30
16	75	0	0	25	54985.22	19872.40	73522.26	349029.40
17	50	25	0	25	55081.74	19848.21	73560.48	348798.80
18	25	50	0	25	55173.25	19833.17	73599.98	348683.00
19	0	75	0	25	55260.10	19825.39	73640.16	348658.20
20	50	0	25	25	55034.61	19861.57	73526.64	348903.50
21	25	25	25	25	55129.13	19841.97	73565.75	348731.10
22	0	50	25	25	55218.71	19830.63	73605.85	348662.20
23	25	0	50	25	55084.26	19854.02	73532.29	348820.50
24	0	25	50	25	55176.74	19838.62	73572.16	348701.00
25	0	0	75	25	55133.29	19850.01	73538.19	348781.40
26	50	0	0	50	55151.05	19836.62	73586.62	348702.00
27	25	25	0	50	55201.48	19830.42	73609.26	348665.50
28	0	50	0	50	55250.47	19826.40	73632.11	348656.40
29	25	0	25	50	55177.20	19834.51	73589.87	348681.70
30	0	25	25	50	55227.04	19829.39	73612.67	348658.90
31	0	0	50	50	55203.41	19833.24	73593.43	348672.30
32	25	0	0	75	55212.55	19829.54	73612.97	348661.20
33	0	25	0	75	55246.64	19826.82	73628.89	348655.90
34	0	0	25	75	55230.31	19828.93	73615.38	348658.00
35	0	0	0	100	55244.59	19827.06	73627.18	348655.80

TABLE 5.37 Water conversion values for non-inferior solutions

Sr. no.	w_1	w_2	w_3	w_4	v_1	v_2
1	100	0	0	0	333.7611	201.0085
2	75	25	0	0	322.0613	198.8873
3	50	50	0	0	307.5677	192.6615
4	25	75	0	0	295.2536	185.8432
5	0	100	0	0	283.0973	178.5556
6	75	0	25	0	361.7179	218.6732
7	**50**	**25**	**25**	**0**	**345.7772**	**213.7634**
8	25	50	25	0	329.3097	206.2584
9	0	75	25	0	315.7425	198.6967
10	50	0	50	0	387.2522	234.7726
11	25	25	50	0	368.2651	227.8276
12	0	50	50	0	350.2087	219.3220
13	25	0	75	0	410.8825	249.6600
14	0	25	75	0	389.7021	241.1986
15	0	0	100	0	432.9778	263.5066
16	75	0	0	25	1304.5800	820.7986
17	50	25	0	25	1292.0840	813.4811
18	25	50	0	25	1279.3640	805.9796
19	0	75	0	25	1266.4530	798.3171
20	50	0	25	25	1325.1520	833.7022
21	25	25	25	25	1312.4480	826.2583
22	0	50	25	25	1299.5410	818.6426
23	25	0	50	25	1345.5700	846.5084
24	0	25	50	25	1332.6700	838.9446
25	0	0	75	25	1365.9270	859.2541
26	50	0	0	50	2276.0590	1433.5390
27	25	25	0	50	2263.2130	1425.9410
28	0	50	0	50	2250.2630	1418.2560
29	25	0	25	50	2296.3020	1446.2420
30	0	25	25	50	2283.3520	1438.5810
31	0	0	50	50	2316.4640	1458.8940
32	25	0	0	75	3247.0400	2045.8880
33	0	25	0	75	3234.0750	2038.1940
34	0	0	25	75	3267.1640	2058.5190
35	0	0	0	100	4217.8880	2658.1320

TABLE 5.38 Decision making

Sr. no.	w_1	w_2	w_3	w_4	$\mu(F_1)$	$\mu(F_2)$	$\mu(F_3)$	$\mu(F_4)$	μ_D
1	100	0	0	0	1.000000	0.000000	0.000000	0.000000	0.010451
2	75	25	0	0	0.701549	0.777479	0.811494	0.787670	0.032169
3	50	50	0	0	0.363866	0.954563	0.757008	0.960005	0.031723
4	25	75	0	0	0.149065	0.993191	0.665808	0.995252	0.029297
5	0	100	0	0	0.000000	1.000000	0.586877	0.999609	0.027031
6	75	0	25	0	0.945576	0.378668	0.549874	0.388200	0.023643
7	**50**	**25**	**25**	**0**	**0.581339**	**0.854695**	**0.889242**	**0.865814**	**0.033349**
8	25	50	25	0	0.263748	0.975369	0.766395	0.980719	0.031208
9	0	75	25	0	0.064299	0.997610	0.654824	0.999270	0.028384
10	50	0	50	0	0.821838	0.597427	0.833921	0.611473	0.029938
11	25	25	50	0	0.452382	0.900643	0.918170	0.911802	0.033265
12	0	50	50	0	0.160292	0.984516	0.755360	0.989378	0.030198
13	25	0	75	0	0.663016	0.721945	0.964667	0.737647	0.032264
14	0	25	75	0	0.318415	0.923523	0.911364	0.934075	0.032265
15	0	0	100	0	0.486130	0.787168	1.000000	0.802756	0.032147
16	75	0	0	25	0.134056	0.994305	0.673048	0.996561	0.029241
17	50	25	0	25	0.094460	0.997100	0.651998	0.998684	0.028658
18	25	50	0	25	0.056919	0.998838	0.630251	0.999749	0.028068
19	0	75	0	25	0.021289	0.999737	0.608125	0.999978	0.027476
20	50	0	25	25	0.113795	0.995557	0.670635	0.997720	0.029029
21	25	25	25	25	0.075018	0.997821	0.649098	0.999307	0.028439
22	0	50	25	25	0.038268	0.999131	0.627016	0.999941	0.027845
23	25	0	50	25	0.093424	0.996429	0.667524	0.998484	0.028801
24	0	25	50	25	0.055487	0.998209	0.645571	0.999584	0.028205
25	0	0	75	25	0.073310	0.996893	0.664276	0.998844	0.028565
26	25	25	0	50	0.045335	0.999156	0.625140	0.999911	0.027899
27	0	50	0	50	0.025239	0.999620	0.612556	0.999995	0.027563
28	50	0	0	50	0.066026	0.998439	0.637607	0.999575	0.028234
29	25	0	25	50	0.055296	0.998683	0.635818	0.999762	0.028108
30	0	25	25	50	0.034853	0.999275	0.623260	0.999972	0.027771
31	0	0	50	50	0.044546	0.998830	0.633856	0.999848	0.027978
32	25	0	0	75	0.040797	0.999257	0.623096	0.999950	0.027831
33	0	25	0	75	0.026810	0.999572	0.614329	0.999999	0.027597
34	0	0	25	75	0.033508	0.999328	0.621771	0.999980	0.027742
35	0	0	0	100	0.027653	0.999544	0.615271	1.000000	0.027616

REFERENCES

Books

Christensen, G.S. and S.A. Soliman, *Optimal Long-Term Operation of Electric Power Systems*, Plenum Press, New York, 1988.

El-Hawary, M.E. and G.S. Christensen, *Optimal Economic Operation of Power Systems*, Academic Press, New York, 1979.

Haimes, Y.Y., *Hierarchical Analysis of Water Resource System: Modelling and Optimization of Large Scale Systems*, McGrawHill, New York, 1977.

Klir, G.J. and B.Yuan, *Fuzzy Sets and Fuzzy Logic: Theory and Applications*, Prentice-Hall India, New Delhi, 1997.

Klir, G.J. and T.A. Folger, *Fuzzy Sets, Uncertainty and Information*, Prentice-Hall India, 1993.

Kosko, B., *Neural Networks and Fuzzy Systems: Dynamical Systems Approach to Machine Intelligence*, Prentice-Hall of India, New Delhi, 1994.

Kothari, D.P. and I.J. Nagrath, *Modern Power System Analysis*, 3rd ed., Tata McGraw-Hill, New Delhi, 2003.

Mahalanabis, A.K., D.P. Kothari and S.I. Ahson, *Computer Aided Power System Analysis and Control*, Tata McGraw-Hill, New Delhi, 1991.

Nagrath, I.J. and D.P. Kothari, *Power System Engineering*, Tata McGraw-Hill, New Delhi, 1994.

Osyczka, A. and B.J. Davies, *Multicriterion Optimization in Engineering with FORTRAN Programs*, Ellis Horwood, 1984.

Rao, S.S., *Optimization, Theory and Applications*, 2nd ed., Wiley Eastern, New Delhi, 1987.

Wood, A.J. and B. Wollenberg, *Power Generation, Operation and Control*, John Wiley, New York, 1984.

Papers

Andrews, C.J., A simple model for assessing emissions reduction options, *IEEE Trans. on Power Systems*, Vol. **8(4)**, pp. 1471–1477, 1993.

Brar, Y.S., Jaspreet S. Dhillon and D.P. Kothari, Multiobjective load dispatch by fuzzy logic based weightage pattern, *Electric Power Systems Research*, Vol. **63(2)**, pp. 149–160, 2002.

Brayton, R.K., G.D. Hachtel and A.L. Sangiovanni Vincentelli, A survey of optimization techniques for integrated-circuit design, *IEEE Proceedings*, Vol. **69(10)**, pp. 1334–1361, 1981.

Brodsky, S.F.J. and R.W. Hahn, Assessing the influence of power pools on emission constrained economic dispatch, *IEEE Trans. on Power Systems*, Vol. **107(1)**, pp. 57–62, 1986.

Cadogan, J.B. and L. Eisenberg, Sulphur oxide emissions management for electric power systems, *IEEE Trans. on Power Apparatus and Systems*, Vol. **96(2)**, pp. 393–401, 1977.

Chang, C.S. and W. Fu, Stochastic multiobjective generation dispatch of combined heat and power systems, *IEE Proceedings—Generation, Transmission and Distribution*, Vol. **145(5)**, pp. 583–591, September 1998.

Chankong, V., Y.Y. Haimes and D.M. Gemperline, A multiobjective dynamic programming method for capacity expansion, *IEEE Trans. on Automatic Control*, Vol. **26(5),** pp. 1195–1207, 1981.

Charalambous, C., A new approach to multicriterion optimization problem and its application to the design of 1D digital filters, *IEEE Trans. on Circuits and Systems*, Vol. **36(6)**, pp. 773–784, 1989.

Cohon, J.L. and D.H. Marks, A review and evaluation of multiobjective programming techniques, *Water Resources Research*, Vol. **11(2)**, pp. 208–220, 1975.

Crousillat, E.O., P. Dorfner, P. Alvarado and H.M. Merrill, Conflicting objectives and risk in power system planning, *IEEE Trans. on Power Systems*, Vol. **8(3)**, pp. 887–893, 1993.

David, A.K. and Z. Rongda, An expert system with fuzzy sets for optimal planning, *IEEE Trans. on Power Systems*, Vol. **6(1)**, pp. 59–65, 1991.

Delson, J.K., Controlled emission dispatch, *IEEE Trans. on Power Apparatus and Systems*, Vol. **93(5)**, pp. 1359–1366, 1974.

Dhillon, J.S. and D.P. Kothari, The surrogate worth trade-off approach for multiobjective thermal power dispatch problem, *Electric Power System Research*, Vol. **56(2)**, pp. 103–110, 2000.

Finnigan, O.E. and A.A. Fouad, Economic dispatch with pollution constraints, *IEEE Winter Power Society Meeting*, Paper no. C-74, 1558, N.Y., 1974.

Gent, M.R. and J.W. Lamont, Minimumemission dispatch, *IEEE Trans. on Power Apparatus and Systems*, Vol. **90(6)**, pp. 2650–2660, 1971.

Haimes, Y.Y. and W.A. Hall, Multiobjectives in water resource systems analysis: The Surrogate worth trade-off method, *Water Resources Research*, Vol. **10(4)**, pp. 615–624, 1974.

Hannan, E.L., Linear programming with multiple fuzzy goals, *Fuzzy Sets and Systems*, Vol. **6(3)**, pp. 235–248, 1981.

Heslin, J.S. and B.F. Hobbs, A multiobjective production costing model for analyzing emissions dispatching and fuel switching, *IEEE Trans. on Power Systems*, Vol. **4(3)**, pp. 836–842, 1989.

Hobbs, B.F., Emissions dispatch under the underutilization provision of the 1990 U.S. clean air act amendments: Models and analysis, *IEEE Trans. on Power Systems*, Vol. **8(1)**, pp. 177–183, 1993.

Hota, P.K., R. Chakrabarti and P.K. Chattopadhyay, A simulated annealing-based goal-attainment method for economic-emission load dispatch with nonsmooth fuel cost and emission level functions, *Electric Machines and Power Systems*, Vol. **28(11)**, pp. 1037–1051, 2000.

Hwang, C.L., Y.J. Lai and T.Y. Liu, A new approach for multiple objective decision making, *Computers Operations Research*, Vol. **20(8)**, pp. 889–899, 1993.

Kaunas, J.R. and Y.Y. Haimes, Risk management of groundwater contamination in a multi-objective framework, *Water Resources Research*, Vol. **21(11)**, pp. 1721–1730, 1985.

Kermanshahi, B.S., Y. Wu, K. Yasuda and R. Yokoyama, Environmental marginal cost evaluation by non-inferiority surface, *IEEE Trans. on Power Systems*, Vol. **5(4)**, pp. 151–1159, 1990.

Kothari, D.P., S.K. Maheshwari and K.G. Sharma, Minimization of air pollution due to thermal plants, *Journal Institution of Engineers (India)*, Vol. **EN-57(2)**, pp. 65–68, 1977.

Lamont, J.W. and E.V. Obessis, Emission dispatch models and algorithms for 1990's, *IEEE Trans. on Power Systems*, Vol. **10(2)**, pp. 941–947, 1995.

Leberling, H., On finding compromise solution in multicriteria problems using the fuzzy min-operator, *Fuzzy Sets and Systems*, Vol. **6(2)**, pp. 105–118, 1981.

Li, D. and Y.Y. Haimes, The envelope approach for multiobjective optimization problems, *IEEE Trans. on Systems, Man and Cybernetics*, Vol. **17(6)**, pp. 1026–1037, 1987.

Malakooti, B., A decision support system and a heuristic interactive approach for solving discrete multiple criteria problems, *IEEE Trans. on Systems, Man and Cybernetics*, Vol. **18(2)**, pp. 273–284, 1988.

Miranda, V. and J.T. Saraiva, Fuzzy modelling of power system optimal load flow, *IEEE Trans. on Power Systems*, Vol. **7(2)**, pp. 843–849, 1992.

Nanda, J., D.P. Kothari and K.S. Lingamurthy, A new approach to economic and minimum emission dispatch, *Journal Indian Institute of Science*, Vol. **67**, pp. 249–256, 1987.

Nanda, J., D.P. Kothari and K.S. Lingamurthy, Economic emission load dispatch through goal programming techniques, *IEEE Trans. on Energy Conversion*, Vol. **3(1)**, pp. 26–32, 1988.

Nangia, U., N.K. Jain and C.L. Wadhwa, Multiobjective optimal load flow based on ideal distance minimization in 3D space, *Int. J. Electrical Power and Energy Systems*, Vol. **23(8)**, pp. 847–855, 2001.

Nangia, U., N.K. Jain and C.L. Wadhwa, Surrogate worth trade-off technique for multiobjective optimal power flows, *IEE Proceedings: Generation, Transmission and Distribution*, Vol. **144(6)**, pp. 547–553, November 1997.

Niimura, T. and R. Yokoyama, An approximate reasoning approach for optimal dynamic dispatch of thermal generating units including auxiliary control, *IEEE Trans. on Power Systems*, Vol. **6(2)**, pp. 651–657, 1991.

Palanichamy, C. and K. Srikrishna, Economic thermal power dispatch with emission constraint, *Journal Institution of Engineers* (*India*), Vol. **EL-72**, pp. 11–18, 1991.

Quaddus, M.A. and A.G. Holzman, IMOLP: An interactive method for multiple objective linear programs, *IEEE Trans. on Systems, Man and Cybernetics*, Vol. **16(3)**, pp. 462–468, 1986.

Rarig, H.M. and Y.Y. Haimes, Risk/Dispersion index method, *IEEE Trans. on Systems, Man and Cybernetics*, Vol. **13(3)**, pp. 317–328, 1983.

Saber, H.M. and A. Ravindran, Nonlinear goal programming theory and practice: A survey, *Computers Operations Research*, Vol. **20(3)**, pp. 275–291, 1993.

Sakawa, M., H. Yano and T. Yumine, An interactive fuzzy satisficing method for multiobjective linear-programming problems and its application, *IEEE Trans. on Systems, Man and Cybernetics*, Vol. **17(4)**, pp. 654–661, 1987.

Sasaki, M., M. Gen and K. Ida, Interactive sequential fuzzy goal programming, *Computers Industrial Engineering*, Vol. **19(1-4)**, pp. 567–571, 1990.

Shin, W.S. and A. Ravindran, Interactive multiple objective optimization: Survey I continuous case, *Computers Operations Research*, Vol. **18(1)**, pp. 97–114, 1991.

Su, C.C. and Y.Y. Hsu, Fuzzy dynamic programming: An application to unit commitment, *IEEE Trans. on Power Systems*, Vol. **6(3)**, pp. 1231–1237, 1991.

Sullivan, R.L., Minimum pollution dispatching, *IEEE Summer Power Society Meeting*, Paper No. C-72-4687, July 914, 1972, San Francisco.

Tapia, C.G. and B.A. Murtagh, Interactive fuzzy programming with preference criteria in multiobjective decision making, *Computers Operations Research*, Vol. **18(3)**, pp. 307–316, 1991.

Tomsovic, K., A fuzzy linear programming approach to the reactive power/voltage control problem, *IEEE Trans. on Power Systems*, Vol. **7(1)**, pp. 287–293, 1992.

Tsuji, A., Optimal fuel mix and load dispatching under environmental constraints, *IEEE Trans. on Power Apparatus and Systems*, Vol. **100(5)**, pp. 2357–2364, 1981.

Vemuri, V., Multiple objective optimization in water resource systems, *Water Resources Research*, Vol. **10(1)**, pp. 44–48, 1974.

Wadhwa, C.L. and N.K. Jain, Multiple objective optimal load flow: A new perspective, *IEE Proceedings, Part C*, Vol. **137(1)**, pp. 13–18, 1990.

Wallenius, J., Comparative evaluation of some interactive approaches to multicriterion optimization, *Management Science*, Vol. **21(14)**, pp. 1387–1396, 1975.

Wong, K.P., B. Fan, C.S. Chang and A.C. Liew, Multiobjective generation dispatch using bi-criterion global optimization, Vol. **10(4)**, pp. 1813–1819, 1995.

Yokoyama, R., S.H. Bae, T. Morita and H. Sasaki, Multiobjective optimal generation dispatch based on probability security criteria, *IEEE Trans. on Power Systems*, Vol. **3(1)**, pp. 317–324, 1988.

Zahavi, J. and L. Eisenberg, Economic-environmental power dispatch, *IEEE Trans. on Systems, Man and Cybernetics*, Vol. **5(5)**, pp. 485–489, 1975.

Zhu, J.Z. and M.R. Irving, Combined active and reactive dispatch with multiple objectives using an analytic hierarchical process, *IEE Proceedings—Generation, Transmission and Distribution*, Vol. **143(4)**, pp. 344–352, July 1996.

CHAPTER 6

Stochastic Multiobjective Generation Scheduling

6.1 INTRODUCTION

Optimal economic dispatch in electric power systems has gained increasing importance as the cost associated with generation and transmission of electric energy keeps on increasing. The procedure involves the allocation of total generation requirements among the available generating units in the system in such a manner that the constraints imposed on different system variables are adequately satisfied and the achieved overall cost associated with it is a minimum.

Despite extensive research focussing on thermal power dispatch problem, much of the effort todate has involved the development of deterministic models applicable to steady-state conditions. Most of these attempts assume the system data to be deterministic. It means that all input information is known with complete certainty and the optimal plans of dispatch are always realized exactly. In practice, there are several inaccuracies and uncertainties in the input information (Figure 6.1), which lead to deviations from optimal operation.

The operating cost functions representing the performance characteristics of thermal plants are computed by calculating the overall thermodynamic performance of a unit consisting of boiler, turbine, condenser, heat cycle, and associated plant auxiliaries. Such cost functions are inaccurate in most cases. The inaccuracies may be viewed due to the following reasons [Kirchmayer, 1958].

- Inaccuracies in the process of measuring the basic data used for computation of thermodynamic performance of the unit
- Deviations from the computed thermodynamic performance of the unit because of errors encountered in operation due to operating at other than standard pressure and temperature
- Effect of time on equipment conditions which influences some of its operating characteristics, notably its efficiency
- Inaccuracies resulting from inability to hold generation at exact desired output
- Fuel cost variations
- Load forecasting errors
- Inaccuracies introduced by various types of transmission loss equations.

Further, because of great difficulty in determining the dependency of maintenance costs from the power output, the additional costs for maintenance, supplies, and water are very inaccurate.

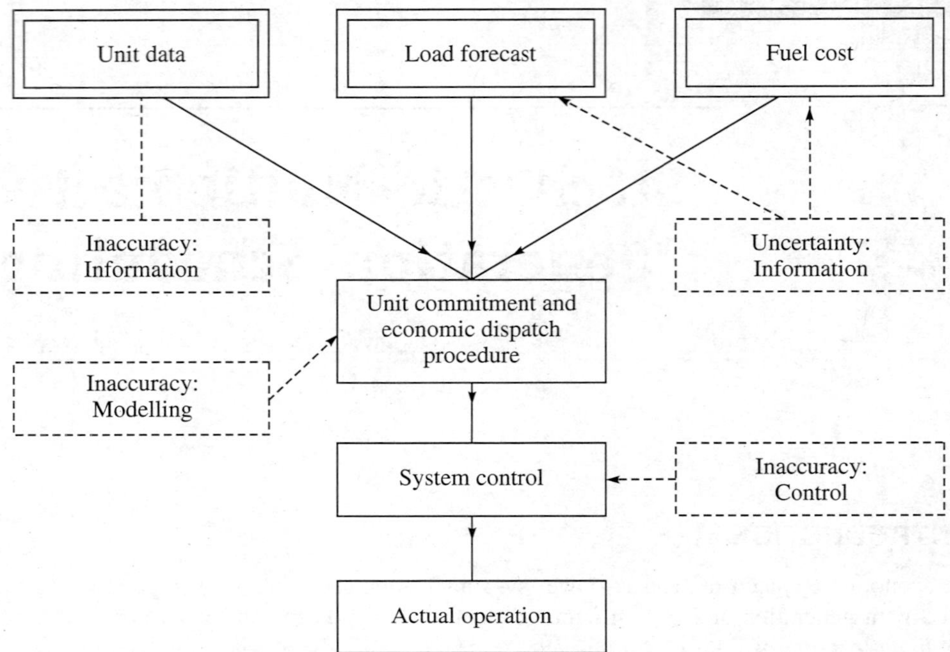

FIGURE 6.1 Optimal power system operation: inaccuracies and uncertainties.

If all these factors are taken together, these will cause inaccuracies of great magnitude in the steady-state operation. The effect of inaccuracies is in an increase in the overall cost. Viviani and Heydt [1981] have outlined the computational details of the stochastic optimal energy dispatch problem. The stochastic optimal energy dispatch algorithm employed the multivariate Gram-Charlier series to statistically model the probability density function of the control vector. The applicability of the series has been limited by the high computational requirements of calculating high order statistical moments. The method obviates some of the difficulty through a polynomial transformation of the variates to be modelled in order to enhance normality. The main aim of the method was to produce a tool which would be useful from an operational standpoint but fails to consider stochastic cost function.

Yakin [1985] has articulated an approach to the optimal generation scheduling of a power system by treating the electricity demand at a node as a random variable with a known probabi-lity distribution. Particularly, a two-stage stochastic programming with recourse model has been developed for stochastic economic dispatch. An equivalent problem to this two-stage model has been defined. The penalties for discrepancies in the generation have been included in the objective function of the equivalent problem. The major difficulty during the implementation is to draw out the exact values of these penalties.

El-Hawary and Mbamalu [1988] have investigated the perturbations in the system thermal fuel cost and the system equality constraints as stochastic and normally distributed with zero mean and a given variance. In an another attempt, El-Hawary and Mbamalu [1989] introduced a method in which the system power demand was assumed random with zero mean and unit variance. In the third attempt, El-Hawary and Mbamalu (1991) considered the perturbations in system power demand as random and normally distributed with zero mean and some variance. They observed that optimality conditions in terms of the active power generations were biased by parameters obtained from the variances of active power generations. But these methods do not provide trade-off between economy and risk measures due to uncertainties in system production cost and randomness of demand.

Parti [1987] has expounded an economic dispatch of thermal generation while incorporating the randomness in system production cost and system load through generator outputs, which were treated as random variables. He appended the traditional objective function of economic dispatch with a penalty term accounting for the possible deviations proportional to the expectations of the square of unsatisfied load because of randomness of generator power. This approach suppresses the true character of the problem by considering only the monetary aspects and fails to express the non-commensurability of the conflicting objectives.

Besides electric energy, power plants also produce sizeable quantities of solid wastes, sludge, and pollutants that affect air and water quality. The pollutants affecting air quality are of the greatest interest. They include particulates, NO_x, CO_x, SO_x and other sundry oxides of sulphur that can travel over considerable distances, and have long-term effects both in space and time. The traditional means for controlling emissions, such as precipitators and scrubbers, are hardware intensive, relatively inflexible and limit the ratio of emissions to energy produced in each plant, but not the total emission produced in a region. In contrast, dispatching requires little hardware and is flexible and effective at the regional level. Here, the objective function used in dispatching can be changed in a few moments.

In general, a large-scale system as typified by an electric power system, possesses multiple objectives to be achieved, namely economic operation, reliability, security and minimal impact on environment. It may be obvious that trade-offs among these objectives are difficult because of their different nature. This implies that objectives are *non-commensurable.*

6.2 MULTIOBJECTIVE STOCHASTIC OPTIMAL THERMAL POWER DISPATCH—ε-CONSTRAINT METHOD

Extensive studies, associated with the optimal power dispatch, have been centred on making it more efficient in algorithm and applicable to online with deterministic data. In actual practice, it is a misleading assumption that data is known with complete certainty. In spite of this, it is also true that the optimal dispatch is to optimize just one specific objective, or single performance index. Now the trend is to formulate multiobjective optimization problem with due consideration of uncertainties for a more realistic approach. The multiobjective stochastic optimization problem is described in the subsequent sections.

6.2.1 Stochastic Problem Formulation

The objective function to be minimized is the total operating cost for thermal generating units in the system. The operating cost curve is assumed to be approximated by a quadratic function of generator active power output as

$$F_1 = \sum_{i=1}^{\mathrm{NG}} (a_i P_i^2 + b_i P_i + c_i) \tag{6.1}$$

where

NG is the total number of generators

a_i, b_i, and c_i are cost coefficients

P_i is the active power generation of the ith generator.

A stochastic model of function F_1, is formulated by considering the otherwise deterministic a_i, b_i, and c_i as random variables. Any possible deviation of operating cost coefficients and load demand from their respective expected values are manipulated through the randomness of generator power P_i [Parti, et al., 1983]. A specific way of reducing a stochastic model to its deterministic equivalent is to take its expected value [Sen Gupta, 1972; Fredric Soloman, 1987]. Assuming that the random variables are normally distributed and statistically independent, the expected value of operating cost becomes:

$$\overline{F}_1 = \sum_{i=1}^{NG} (\overline{a}_i \overline{P}_i^2 + \overline{b}_i \overline{P}_i + \overline{c}_i + \overline{a}_i \text{ var}(P_i)) \tag{6.2}$$

where

$\overline{P}_i$ is expected power generation of the ith generator

$\overline{a}_i$, $\overline{b}_i$, and $\overline{c}_i$ are expected cost coefficients

The variance of power P_i is given as

$$\text{var}(P_i) = C_{P_i}^2 \, \overline{P}_i^2 \tag{6.3}$$

where C_{P_i} is the coefficient of variation of random variable P_i.

Therefore, the expected operating cost as given by Eq. (6.2) is modified as

$$\overline{F}_1 = \sum_{i=1}^{NG} [(1 + C_{P_i}^2)\, \overline{a}_i \overline{P}_i^2 + \overline{b}_i \overline{P}_i + \overline{c}_i] \tag{6.4}$$

The load demand constraint is

$$\sum_{i=1}^{NG} \overline{P}_i = \overline{P}_D + \overline{P}_L \tag{6.5}$$

where

$\overline{P}_D$ is the expected power demand

$\overline{P}_L$ is the expected transmission loss.

The expected limits on the power generation imposed are

$$\overline{P}_i^{\min} \le \overline{P}_i \le \overline{P}_i^{\max} \qquad (i = 1, 2, ..., \text{NG}) \tag{6.6}$$

where

$\overline{P}_i^{\min}$ is the expected lower limit of generator power output

$\overline{P}_i^{\max}$ is the expected upper limit of generator power output.

The transmission line losses are expressed in terms of B-coefficients as

$$P_L = \sum_{i=1}^{NG} \sum_{j=1}^{NG} P_i B_{ij} P_j + \sum_{i=1}^{NG} P_i B_{i0} + B_{00} \tag{6.7}$$

where B_{ij}, B_{i0}, and B_{00} are B-coefficients.

With P_is as independent random variables, the expected transmission loss can be written as

$$\overline{P}_L = \sum_{i=1}^{NG} \sum_{j=1}^{NG} \overline{P}_i \overline{B}_{ij} \overline{P}_j + \sum_{i=1}^{NG} \overline{B}_{ii} \text{ var}(P_i) + \sum_{i=1}^{NG} \overline{B}_{i0} \overline{P}_i + \overline{B}_{00} \tag{6.8a}$$

$$\overline{P}_L = \sum_{i=1}^{NG} (1 + C_{P_i}^2) \overline{B}_{ii} \overline{P}_i^2 + \sum_{i=1}^{NG} \sum_{\substack{j=1 \\ j \neq i}}^{NG} \overline{P}_i \overline{B}_{ij} \overline{P}_j + \sum_{i=1}^{NG} \overline{B}_{i0} \overline{P}_i + \overline{B}_{00} \tag{6.8b}$$

where $\overline{B}_{ij}$, $\overline{B}_{i0}$, and $\overline{B}_{00}$ are expected B-coefficients.

The variance of transmission loss has been neglected because in the deterministic case, the B-matrix represents only the approximate transmission loss which is normally not more than 5 percent of the total power transferred from generators to the using substations [Parti et. al., 1983].

Since generator outputs P_is are treated as random variables, the expected deviations are proportional to the expectation of the square of the unsatisfied load demand. These expected deviations are given as

$$E\left[\left(\overline{P}_D + \overline{P}_L - \sum_{i=1}^{NG} P_i\right)^2\right] \tag{6.9}$$

Using Eq. (6.5), the above equation can be rewritten as

$$E\left[\left(\sum_{i=1}^{NG} \overline{P}_i - \sum_{i=1}^{NG} P_i\right)^2\right]$$

This on simplification reduces to

$$\sum_{i=1}^{NG} \text{var}(P_i) \tag{6.10}$$

Equation (6.10) is equated to $\overline{F}_2$, the new objective function, i.e.

$$\overline{F}_2 = \sum_{i=1}^{NG} \text{var}(P_i) \tag{6.11a}$$

Substituting Eq. (6.3) into Eq. 6.11(a), we get

$$\overline{F}_2 = \sum_{i=1}^{NG} C_{P_i}^2 \overline{P}_i^2 \tag{6.11b}$$

A multiobjective optimization problem can be formulated considering (a) the expected operating cost and (b) the risk associated with possible deviations of the random variables from their expected values while satisfying the expected load demand constraints and expected generation limits. The multiple objective optimization problem is defined as

Minimize $\quad [\overline{F}_1, \overline{F}_2]^T \quad$ (6.12a)

subject to $\quad \sum_{i=1}^{NG} \overline{P}_i = \overline{P}_D + \overline{P}_L \quad$ (6.12b)

$$\overline{P}_i^{\min} \le \overline{P}_i \le \overline{P}_i^{\max} \qquad (i = 1, 2, ..., NG) \tag{6.12c}$$

6.2.2 Algorithm

To generate a non-inferior solution to the multiobjective optimization problem, the ε-constraint method is used. In this method, one specific objective function, which is arbitrarily chosen or preferably corresponding to the most important objective, is taken as the scalar performance index to be minimized. The multiple objective optimization problem as identified by Eq. (6.12) with ε-constraint approach expressed as

Minimize $\quad \overline{F}_1 \quad$ (6.13a)

subject to $\quad \overline{F}_2 \le \varepsilon_2 \quad$ (6.13b)

$$\sum_{i=1}^{NG} \overline{P}_i = \overline{P}_D + \overline{P}_L \tag{6.13c}$$

$$\overline{P}_i^{\min} \le \overline{P}_i \le \overline{P}_i^{\max} \qquad (i = 1, 2, ..., NG) \tag{6.13d}$$

where ε_2 is interpreted as the maximum tolerable objective level. The values of ε are chosen for which the objective constraints in Eq. (6.13) are binding at the optimal solution. As a constraint is varied parametrically, a set of non-inferior solutions (with their corresponding trade-offs) is generated.

The well-known method of Lagrange multipliers is quite popular in the power system planning studies. The Lagrangian L formed for the system is given by Eq. (6.14)

$$L = \overline{F}_1 + \lambda_{12}(\overline{F}_2 - \varepsilon_2) + \mu\left(\overline{P}_L + \overline{P}_D - \sum_{i=1}^{NG} \overline{P}_i\right) \tag{6.14}$$

where λ_{12} and μ are Lagrange multipliers.

The necessary conditions to obtain solution are given as

$$\frac{\partial L}{\partial \overline{P}_i} = \frac{\partial \overline{F}_1}{\partial \overline{P}_i} + \lambda_{12}\frac{\partial \overline{F}_2}{\partial \overline{P}_i} + \mu\left[\frac{\partial \overline{P}_L}{\partial \overline{P}_i} - 1\right] = 0 \tag{6.15a}$$

$$\frac{\partial L}{\partial \lambda_{12}} = \overline{F}_2 - \varepsilon_2 = 0 \tag{6.15b}$$

$$\frac{\partial L}{\partial \mu} = \overline{P}_L + \overline{P}_D - \sum_{i=1}^{NG} \overline{P}_i = 0 \tag{6.15c}$$

The optimal solution to Eq. (6.14) must satisfy the Kuhn Tucker conditions. The main condition is

$$\lambda_{12}(\overline{F}_2 - \varepsilon_2) = 0; \qquad \lambda_{12} \geq 0 \tag{6.15d}$$

The value of λ_{12} corresponding to the binding constraints indicates the marginal benefit of the objective function due to an additional unit of ε. The Lagrange multipliers related to the objectives, as constraints may be zero or nonzero. The set of nonzero Lagrange multipliers corresponds to the non-inferior set of solutions. The set of nonzero Lagrange multipliers represents the set of trade-off ratios between the principal objective and each of the constraining objectives, respectively. The system given by Eq. (6.14) is solved using the Newton–Raphson method for R values of ε_2. Only those values of $\lambda^r_{12} > 0.0$ which correspond to active constraints $F_2^r(P) = \varepsilon_2^r$, $r = 1, 2, ..., R$ are considered, since they belong to the non-inferior solution [Haimes and Hall, 1974].

In the problem, the initial value of ε_2 is taken such that $\varepsilon_2 > F_2^{\min}$ and $\varepsilon_2 < F_2^{\max}$. Since objectives are of conflicting nature, the value of one objective will be maximum, when the value of another objective is minimum and vice versa.

To implement the Newton–Raphson method, the following equation is solved iteratively till no further improvement in decision variables is achieved.

$$\begin{bmatrix} \nabla_{PP}L & \nabla_{P\lambda_{12}}L & \nabla_{P\mu}L \\ (\nabla_{P\lambda_{12}}L)^T & \nabla_{\lambda_{12}\lambda_{12}}L & \nabla_{\lambda_{12}\mu}L \\ (\nabla_{P\mu}L)^T & \nabla_{\mu\lambda_{12}}L & \nabla_{\mu\mu}L \end{bmatrix} \begin{bmatrix} \Delta P_g \\ \Delta\lambda_{12} \\ \Delta\mu \end{bmatrix} = \begin{bmatrix} -\nabla_{P_g}L \\ -\nabla_{\lambda_{12}}L \\ -\nabla_{\mu}L \end{bmatrix} \tag{6.16}$$

The Newton–Raphson method shows very effective results when the initial guess is in the domain of solution. Utilization of factorized matrix is another aspect of the aforementioned method.

Algorithm 6.1: Non-Inferior Solution by the ε-Constraint Method

1. Read data, namely cost coefficients, emission coefficients and B-coefficients, Err (convergence tolerance) and ITMAX (maximum allowed iterations), NG (number of generators) and K (minimum number of non-inferior solutions required for the objective as constraint), etc.
2. Fix ε_2 such that $\overline{F}_2^{\min} < \varepsilon_2 < \overline{F}_2^{\max}$.
3. Set iteration for non-inferior solutions, $k = 1$.
4. Increment count of non-inferior solutions, $k = k + 1$.
5. If $(k \geq K)$ GOTO Step 18.

6. Compute the initial values of $\overline{P}_i$ $(i = 1, 2, ..., \text{NG})$ and μ.
7. Assume that no generator has been fixed either at lower limit or at upper limit.
8. Set iteration counter, IT = 1.
9. Compute Hessian and Jacobian matrix elements using Eqs. (6.15a) to (6.15c), respectively. Deactivate row and column of Hessian matrix and row of Jacobian matrix representing the generator whose generation is fixed either at lower limit or at upper limit. This is done so that fixed generators cannot participate in allocation.
10. Gauss elimination method is employed in which triangularization and back substitution processes are performed to find, $\Delta\overline{P}_i$ $(i = 1, 2, ..., R)$, $\Delta\lambda_{12}$, $\Delta\mu$. Here R is the number of generators that can participate in allocation.
11. If $\left(\sqrt{\sum_{i=1}^{R} (\Delta\overline{P}_i)^2 + (\Delta\lambda_{12})^2 + (\Delta\mu)^2}\right) \le \text{err}$ or $\left(\sqrt{\sum_{i=1}^{R} \left(\frac{\partial L}{\partial \overline{P}_i}\right)^2 + \left(\frac{\partial L}{\partial \lambda_{12}}\right)^2 + \left(\frac{\partial L}{\partial \mu}\right)^2}\right) \le \text{err}$

    then GOTO Step 14.
12. Modify control variables,

$$\overline{P}_i^{\text{new}} = \overline{P}_i + \Delta\overline{P}_i \qquad (i = 1, 2, ..., R)$$

$$\lambda_{12}^{\text{new}} = \lambda_{12} + \Delta\lambda_{12}$$

$$\mu^{\text{new}} = \mu + \Delta\mu$$

13. Update iteration counter, IT = IT + 1,
    Assign new values to old variables to continue the process.

$$\overline{P}_i = \overline{P}_i^{\text{new}} \qquad (i = 1, 2, ..., R)$$

$$\lambda_{12} = \lambda_{12}^{\text{new}} \quad \text{and} \quad \mu = \mu^{\text{new}}$$

    GOTO Step 9 and repeat.
14. Check the limits of generators and fix up as following:

    If $\overline{P}_i < \overline{P}_i^{\min}$ then $\overline{P}_i = \overline{P}_i^{\min}$

    If $\overline{P}_i > \overline{P}_i^{\max}$ then $\overline{P}_i = \overline{P}_i^{\max}$

    If no more violations of limits are there then GOTO Step 16.
15. GOTO Step 8.
16. Check the condition is satisfied

$$\lambda_{12}(\overline{F}_2(\overline{P}) - \varepsilon_2) = 0.0; \qquad \lambda_{12} > 0.0,$$

    If 'yes' then GOTO Step 17.

    else modify ε_2 and GOTO Step 5.
17. Record it as non-inferior solutions, compute values of all objectives and transmission loss and modify ε_2 for the next non-inferior solution and GOTO Step 4
18. Stop.

6.2.3 Application of the Method

Two sample systems are taken up to illustrate the method to evaluate the possible economic significance with respect to risk.

Case 1: In this case, a three generator sample system is selected. The expected generator characteristics are given in Table 6.1. The expected *B*-coefficients of transmission loss formula are presented in Table 6.2. In addition, the following coefficients of variation of random variables are assumed.

$$C_{P_i} = 0.1 \qquad (i = 1, 2, 3)$$

TABLE 6.1 Expected generation characteristics

Plant *i*	a_i (\$/MW²h)	b_i (\$/MWh)	c_i (\$/h)	P_i^{max} (MW)	P_i^{min} (MW)
1	0.010	2.00	10.0	200.0	10.0
2	0.012	1.50	10.0	200.0	10.0
3	0.004	1.80	20.0	200.0	10.0

TABLE 6.2 Expected *B*-coefficients

i	*j*	B_{ij}	*i*	*j*	B_{ij}
1	1	0.0002725	1	2	–0.000351
2	2	0.0003090	1	3	–0.003679
3	3	0.0032295	2	3	–0.000565

Using this data, the resulting expected generation schedules with expected cost and risk are given in Table 6.4 for various values of ε. The deterministic results are shown in Table 6.3 ($C_{P_i} = 0.0$; $i = 1, 2, 3$).

TABLE 6.3 Deterministic results

Sr. no.	P_D (MW)	F_1 (\$/h)	P_1 (MW)	P_2 (MW)	P_3 (MW)
1	140.0	361.263	46.146	54.781	45.182
2	180.0	484.010	64.417	68.727	56.251
3	220.0	621.695	83.441	83.200	66.471

The percentage deviation in the cost for different schedules corresponding to risk is shown in Figure 6.2 for different expected demands. In representation, ε_2 is interpreted as risk, which is proportional to the expected power demand ($\varepsilon_2 = \alpha P_D$, where α is a risk factor). The curve indicates an increase in the percentage deviation of cost of operation for different levels of risk. The operator has a choice to select the risk factor from the curve.

Case 2: A large system consisting of eight generators is selected for this case. The data for the loss formula coefficients as well as the expected incremental production cost coefficients of a

TABLE 6.4 Expected non-inferior generation schedules

Sr. no.	$\overline{F}_1$ (\$/h)	$\overline{F}_2$ (MW^2)	$\overline{P}_L$ (MW)	$\overline{P}_1$ (MW)	$\overline{P}_2$ (MW)	$\overline{P}_3$ (MW)
	$\overline{P}_D$ = 140 MW					
1	362.2143	71.400	5.76481	47.982	53.900	43.960
2	362.2040	71.428	5.79242	47.835	53.975	44.060
3	362.1948	71.456	5.81940	47.692	54.048	44.158
4	362.1868	71.484	5.84578	47.554	54.118	44.252
5	362.1797	71.512	5.87163	47.419	54.187	44.345
6	362.1736	71.540	5.89695	47.287	54.253	44.436
7	362.1684	71.568	5.92182	47.159	54.318	44.524
8	362.1641	71.596	5.94625	47.034	54.381	44.611
9	362.1604	71.624	5.97028	46.911	54.443	44.696
10	362.1575	71.652	5.99390	46.791	54.503	44.780
	$\overline{P}_D$ = 180 MW					
11	485.8152	119.880	8.35676	67.572	67.980	52.921
12	485.7668	119.916	8.44331	67.297	68.054	53.210
13	485.7271	119.952	8.52199	67.051	68.120	53.470
14	485.6942	119.988	8.59468	66.827	68.178	53.709
15	485.6663	120.024	8.66279	66.619	68.232	53.932
16	485.6427	120.060	8.72711	66.425	68.282	54.141
17	485.6222	120.096	8.78810	66.238	68.335	54.337
18	485.6051	120.132	8.84665	66.066	68.377	54.526
19	485.5905	120.168	8.90289	65.902	68.418	54.706
20	485.5782	120.204	8.95712	65.745	68.456	54.879
	$\overline{P}_D$ = 220 MW					
21	624.3509	182.600	11.78805	85.728	83.587	62.642
22	624.2985	182.644	11.90598	85.524	83.556	62.996
23	624.2559	182.688	12.01322	85.342	83.526	63.316
24	624.2204	182.732	12.11263	85.175	83.499	63.610
25	624.1908	182.776	12.20563	85.021	83.474	63.884
26	624.1658	182.820	12.29350	84.877	83.450	64.141
27	624.1448	182.864	12.37712	84.741	83.427	64.384
28	624.1272	182.908	12.45700	84.612	83.405	64.616
29	624.1125	182.952	12.53376	84.489	83.384	64.837
30	624.0999	182.996	12.60826	84.368	83.367	65.050

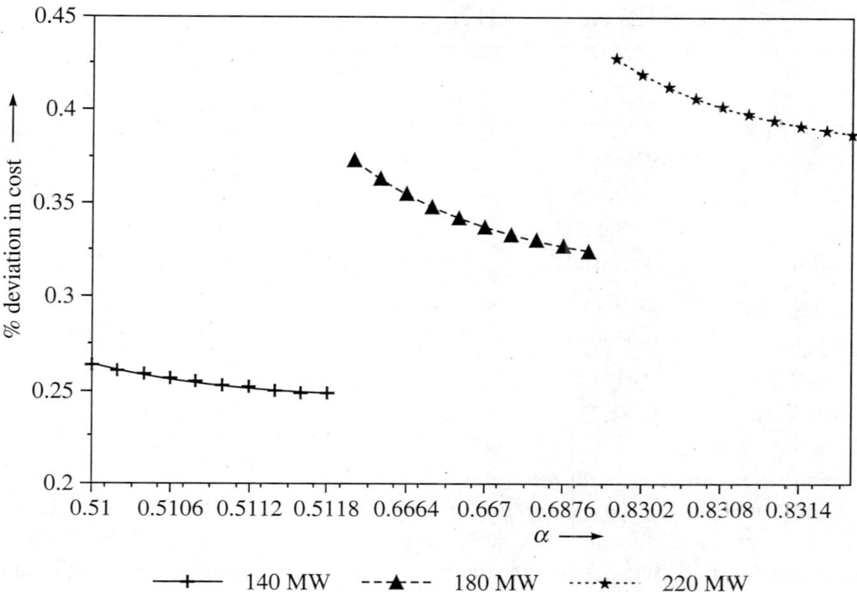

FIGURE 6.2 Percentage deviation in cost vs. α.

system are given in Tables 6.5 and 6.6 respectively. The values of coefficient of variation are given below:

$$C_{P_i} = 0.1 \qquad (i = 1, 2, ..., 8).$$

TABLE 6.5 Transmission loss formula coefficients ($B_{ij} \times 10^2$)

i	j	B_{ij}	i	j	B_{ij}	i	j	B_{ij}
1	1	0.07863	2	3	0.04624	5	6	0.01224
2	2	0.06098	2	4	0.01246	5	7	0.00721
3	3	0.09163	2	5	−0.01218	5	8	0.00378
4	4	0.02646	2	6	−0.01810	6	7	0.02166
5	5	0.02311	2	7	−0.01750	6	8	0.01682
6	6	0.03723	2	8	−0.01754	7	8	0.05768
7	7	0.06285	3	4	0.01242			
8	8	0.12010	3	5	−0.01198			
1	2	−0.00999	3	6	−0.02204			
1	3	−0.01402	3	7	−0.02530			
1	4	−0.00695	3	8	−0.02841			
1	5	−0.01136	4	5	0.00179			
1	6	−0.02076	4	6	−0.00707			
1	7	−0.02892	4	7	−0.00876			
1	8	−0.03292	4	8	−0.00992			

TABLE 6.6 Expected generation characteristics

Plant i	$\overline{a}_i$ ($/MW^{2}h)	$\overline{b}_i$ ($/MWh)	$\overline{P}_i^{max}$ (MW)	$\overline{P}_i^{min}$ (MW)
1	0.004100	1.280	200.0	50.0
2	0.002200	0.795	210.0	210.0
3	0.000950	1.809	200.0	10.0
4	0.002145	0.657	400.0	150.0
5	0.001110	0.889	310.0	310.0
6	0.006000	0.300	200.0	100.0
7	0.010400	0.635	100.0	50.0
8	0.006350	0.572	150.0	50.0

The deterministic generation schedules obtained are given in Table 6.7 for various demands. Generators 2 and 5 remain at their maximum loads of 210 MW and 310 MW, respectively, and are not included. The expected cost and risk with expected transmission losses for various expected demands are presented in Table 6.8 which are in the non-inferior set. The corresponding expected generation schedules are given in Table 6.9 for various values of ε.

TABLE 6.7 Deterministic results

$\overline{P}_D$ (MW)	$\overline{F}_1$ ($/h)	$\overline{P}_1$ (MW)	$\overline{P}_3$ (MW)	$\overline{P}_4$ (MW)	$\overline{P}_6$ (MW)	$\overline{P}_7$ (MW)	$\overline{P}_8$ (MW)
1240	1731.691	137.199	61.047	272.756	145.826	65.254	101.165
1320	1909.131	153.874	84.645	292.947	157.232	71.308	109.476
1400	2097.048	170.938	107.708	313.463	169.036	77.534	117.916

TABLE 6.8(a) Expected cost, risk and transmission loss when $\overline{P}_D$ = 1240 MW

Sr. no.	$\overline{F}_1$ ($/h)	$\overline{F}_2$ (MW^2)	$\overline{P}_L$ (MW)
1	1751.636	2554.00	59.87011
2	1751.404	2555.44	59.89693
3	1751.175	2556.88	59.92381
4	1750.951	2558.32	59.95073
5	1750.731	2559.76	59.97770
6	1750.514	2561.20	60.00471
7	1750.302	2562.64	60.03177
8	1750.093	2564.08	60.05887
9	1749.887	2565.52	60.08601
10	1749.685	2566.96	60.11319

TABLE 6.8(b) Expected cost, risk and transmission loss when $\overline{P}_D$ = 1320 MW

Sr. no.	$\overline{F}_1$ (\$/h)	$\overline{F}_2$ (MW^2)	$\overline{P}_L$ (MW)
1	1993.124	2682.000	64.35024
2	1990.328	2683.600	64.30595
3	1987.785	2685.200	64.27084
4	1985.448	2686.800	64.24320
5	1983.286	2688.400	64.22176
6	1981.271	2690.000	64.20558
7	1975.903	2694.822	64.17964
8	1974.342	2696.400	64.18031

TABLE 6.8(c) Expected cost, risk and transmission loss when $\overline{P}_D$ = 1400 MW

Sr. no.	$\overline{F}_1$ (\$/h)	$\overline{F}_2$ (MW^2)	$\overline{P}_L$ (MW)
1	2195.333	2950.80	70.79334
2	2191.831	2952.56	70.72950
3	2188.678	2954.32	70.67876
4	2185.806	2956.08	70.63847
5	2183.164	2957.84	70.60670
6	2180.714	2959.60	70.58204
7	2178.429	2961.36	70.56348
8	2174.271	2964.88	70.54130
9	2172.365	2966.64	70.53646

TABLE 6.9(a) Expected generation schedules corresponding to the results given in Table 6.8(a)

Sr. no.	$\overline{P}_1$ (MW)	$\overline{P}_3$ (MW)	$\overline{P}_4$ (MW)	$\overline{P}_6$ (MW)	$\overline{P}_7$ (MW)	$\overline{P}_8$ (MW)
1	145.128	91.628	218.912	145.615	74.500	105.194
2	145.035	91.354	219.479	145.624	74.368	105.144
3	144.943	91.079	220.044	145.633	74.237	105.094
4	144.852	90.806	220.604	145.641	74.109	105.045
5	144.762	90.534	221.162	145.649	73.981	104.996
6	144.672	90.262	221.717	145.656	73.856	104.948
7	144.583	89.991	222.268	145.664	73.732	104.900
8	144.494	89.721	222.816	145.671	73.609	104.852
9	144.407	89.452	223.362	145.678	73.489	104.805
10	144.320	89.183	223.904	145.684	73.369	104.758

TABLE 6.9(b) Expected generation schedules corresponding to the results given in Table 6.8(b)

Sr. no.	$\overline{P_1}$ (MW)	$\overline{P_3}$ (MW)	$\overline{P_4}$ (MW)	$\overline{P_6}$ (MW)	$\overline{P_7}$ (MW)	$\overline{P_8}$ (MW)
1	175.189	132.842	166.741	154.820	114.548	121.527
2	174.948	132.543	168.513	155.099	113.140	121.378
3	174.713	132.248	170.190	155.348	111.853	121.232
4	174.483	131.955	171.787	155.572	110.667	121.089
5	174.258	131.665	173.318	155.775	109.566	120.949
6	174.037	131.376	174.792	155.961	108.537	120.812
7	173.395	130.503	178.961	156.425	105.798	120.401
8	173.186	130.234	180.233	156.565	104.982	120.283

TABLE 6.9(c) Expected generation schedules corresponding to the results given in Table 6.8(c)

Sr. no.	$\overline{P_1}$ (MW)	$\overline{P_3}$ (MW)	$\overline{P_4}$ (MW)	$\overline{P_6}$ (MW)	$\overline{P_7}$ (MW)	$\overline{P_8}$ (MW)
1	194.355	148.276	181.225	169.693	127.185	131.552
2	194.111	148.103	183.135	169.944	125.534	131.393
3	193.873	147.927	184.929	170.163	124.039	131.237
4	193.639	147.748	186.629	170.357	122.668	131.085
5	193.410	147.566	188.252	170.529	121.400	130.935
6	193.184	147.383	189.809	170.683	120.219	130.787
7	192.962	147.197	191.310	170.822	119.113	130.642
8	192.526	146.820	194.167	171.062	117.088	130.357
9	192.312	146.629	195.534	171.167	116.155	130.218

The percentage deviation in the cost for different schedules, corresponding to risk, is shown in Figures 6.3 and 6.4 for different expected demands where risk ε is proportional to expected power demand. The curve indicates an increase in the percentage deviation of cost of operation for different levels of risk. The operator (DM) has a choice to take risk for minimum expected cost, and for this minimum risk he has to pay more.

6.3 MULTIOBJECTIVE STOCHASTIC OPTIMAL THERMAL POWER DISPATCH—THE SURROGATE WORTH TRADE-OFF METHOD

In this section, the effect of uncertain system parameters is incorporated, explicitly in the optimal multiobjective power dispatch. Multiobjective problem is stated by considering (i) the expected operating cost, (ii) the expected minimum NO_x emission, (iii) the expected transmission loss, and (iv) the expected deviations because of the unsatisfied demand. The surrogate worth trade-off technique is discussed to find the compromised solution.

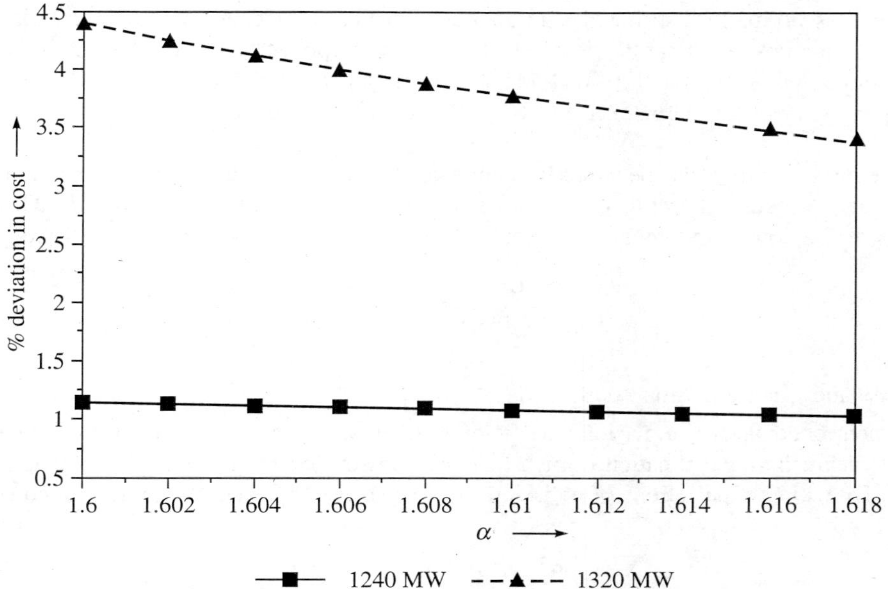

FIGURE 6.3 Percentage deviation in cost vs. α.

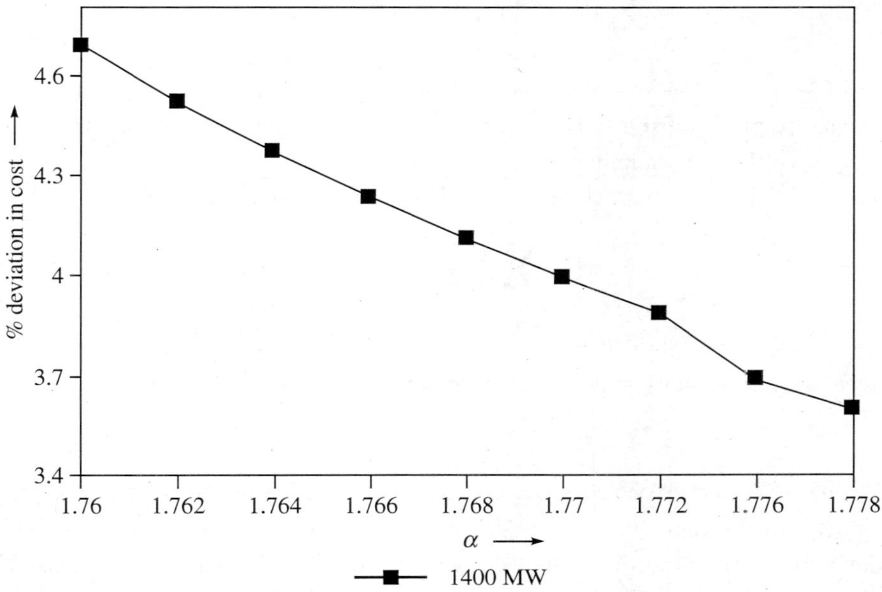

FIGURE 6.4 Percentage deviation in cost vs. α.

6.3.1 Multiobjective Optimization Problem Formulation

The multiobjective optimization problem is viewed as a stochastic multiobjective optimization problem by considering the system power demand, cost coefficients, NO_x emission coefficients and

B-coefficients as normally distributed, and also as statistically independent random variables. The random generator power output P_i [Parti et al., 1983] manipulates any possible deviations in the above-mentioned parameters and in load demand from their expected values. The stochastic model of operating cost has been defined in Eq. (6.2). The stochastic NO_x emission model is described as under:

The emission curve can be directly related to the cost curve through the emission rate per Mkcal, which is a constant factor for a given type of fuel. Therefore, the amount of NO_x emission is given as a function of the generator output P_i which is quadratic [Nanda et al., 1988]. i.e.

$$F_2 = \sum_{i=1}^{NG} (d_i P_i^2 + e_i P_i + f_i) \tag{6.17}$$

where d_i, e_i, and f_i are emission coefficients.

As mentioned above, emission coefficients are realized as independent random variables because of measuring or estimation error. Moreover, power generation level is random, since load is random. By taking expectations of emission Eq. (6.17), the expected NO_x emission comes out as

$$\overline{F}_2 = \sum_{i=1}^{NG} [\overline{d}_i \overline{P}_i^2 + \overline{e}_i \overline{P}_i + \overline{f}_i + \overline{d}_i \text{ var } (P_i)] \tag{6.18a}$$

$$\overline{F}_2 = \sum_{i=1}^{NG} [(1 + C_{P_i}^2)\, \overline{d}_i \overline{P}_i^2 + \overline{e}_i \overline{P}_i + \overline{f}_i] \tag{6.18b}$$

where $\overline{d}_i$, $\overline{e}_i$, and $\overline{f}_i$ are expected emission coefficients.

The stochastic model of transmission losses has been defined in Eq. (6.8a) and is considered as another objective $\overline{F}_3$ to be minimized.

To ensure a real power balance, an equality constraint is stated as

$$\sum_{i=1}^{NG} \overline{P}_i - \overline{P}_D = 0 \tag{6.19}$$

where $\overline{P}_D$ is the expected load demand.

The inequality constraints imposed on generator output are

$$\overline{P}_i^{\min} \le \overline{P}_i \le \overline{P}_i^{\max} \qquad (i = 1, 2, ..., \text{NG}) \tag{6.20}$$

where $\overline{P}_i^{\min}$ and $\overline{P}_i^{\max}$ are expected lower and upper limits of generator outputs, respectively.

Since generator outputs P_i are normally distributed independent random variables, so the expected deviations are proportional to the expectation of the square of the unsatisfied load demand. These expected deviations are considered as another objective to be minimized and the objective is given as

$$\overline{F}_4 = E\left[\left(\overline{P}_D - \sum_{i=1}^{NG} \overline{P}_i\right)^2\right] \tag{6.21}$$

This on simplification reduces to

$$\overline{F}_4 = \sum_{i=1}^{NG} \text{var}(P_i)$$

or

$$\overline{F}_4 = \sum_{i=1}^{NG} C_{P_i}^2 \overline{P}_i^2 \tag{6.22}$$

Aggregating Eqs. (6.4), (6.18), (6.8), and (6.22), the deterministic equivalent of stochastic multi-objective optimization problem is defined as

Minimize $$[\overline{F}_1, \overline{F}_2, \overline{F}_3, \overline{F}_4]^T \tag{6.23a}$$

subject to $$\sum_{i=1}^{NG} \overline{P}_i - \overline{P}_D = 0 \tag{6.23b}$$

$$\overline{P}_i^{\min} \le \overline{P}_i \le \overline{P}_i^{\max} \qquad (i = 1, 2, ..., NG) \tag{6.23c}$$

where $\overline{F}_1$ and $\overline{F}_3$ are expected cost and transmission losses respectively (see Section 6.2.1). $\overline{F}_1, \overline{F}_2, \overline{F}_3$, and $\overline{F}_4$ are the expected values of objective functions to be minimized over the set of admissible decision vector $\overline{P}_i$.

6.3.2 Solution Procedure

To generate non-inferior solutions to a multiobjective optimization problem, the ε-constraint method is utilized [Haimes, 1977]. The ε-constraint approach replaces three objective functions to constraints as given below.

Minimize $$\overline{F}_1 \tag{6.24a}$$

subject to $$\overline{F}_j \le \varepsilon_j \qquad (j = 2, 3, ..., 4) \tag{6.24b}$$

$$\sum_{i=1}^{NG} \overline{P}_i - \overline{P}_D = 0 \tag{6.24c}$$

$$\overline{P}_i^{\min} \le \overline{P}_i \le \overline{P}_i^{\max} \qquad (i = 1, 2, ..., NG) \tag{6.24d}$$

where ε_j is the maximum tolerable objective level for the jth objective.

Generation of non-inferior solutions

Form the generalized Lagrangian L to the system respresented by Eqs. (6.24)

$$L = \overline{F}_1 + \sum_{j=2}^{4} \lambda_{1j}(\overline{F}_j - \varepsilon_j) + \mu\left(\overline{P}_D - \sum_{i=1}^{NG} \overline{P}_i\right) \tag{6.25}$$

where λ_{1j} $(j = 2, 3, 4)$ and μ are generalized Lagrangian multipliers. The subscript $1j$ denotes that λ is the Lagrange multiplier associated with the jth constraint, where the prime objective function is $\overline{F}_1$.

The necessary conditions to obtain solution are given as

$$\frac{\partial L}{\partial \overline{P}_i} = \frac{\partial \overline{F}_1}{\partial \overline{P}_i} + \sum_{j=2}^{4} \lambda_{1j} \frac{\partial \overline{F}_j}{\partial \overline{P}_i} - \mu = 0 \qquad (i = 1, 2, ..., \text{NG}) \tag{6.26a}$$

$$\frac{\partial L}{\partial \lambda_{1j}} = \overline{F}_j - \varepsilon_j = 0 \qquad (j = 2, 3, 4) \tag{6.26b}$$

$$\frac{\partial L}{\partial \mu} = \overline{P}_D - \sum_{i=1}^{N} \overline{P}_i = 0 \tag{6.26c}$$

where

$$\frac{\partial \overline{F}_1}{\partial \overline{P}_i} = 2(1 + C_{P_i}^2)\overline{a}_i \overline{P}_i + \overline{b}_i \tag{6.26d}$$

$$\frac{\partial \overline{F}_2}{\partial \overline{P}_i} = 2(1 + C_{P_i}^2)\overline{d}_i \overline{P}_i + \overline{e}_i \tag{6.26e}$$

$$\frac{\partial \overline{F}_3}{\partial \overline{P}_i} = 2(1 + C_{P_i}^2)\overline{B}_{ii} \overline{P}_i + \sum_{\substack{j=1 \\ j \neq i}}^{\text{NG}} 2B_{ij}\overline{P}_j + B_{i0} \tag{6.26f}$$

$$\frac{\partial \overline{F}_3}{\partial \overline{P}_i} = 2C_{P_i}^2 \overline{P}_i \tag{6.26g}$$

The Newton–Raphson method can be applied to solve the above nonlinear equations. To implement the Newton–Raphson method the following equation is solved iteratively.

$$\begin{bmatrix} \nabla_{PP}L & \nabla_{P\lambda}L & \nabla_{P\mu}L \\ (\nabla_{P\lambda}L)^T & \nabla_{\lambda\lambda}L & \nabla_{\lambda\mu}L \\ (\nabla_{P\mu}L)^T & \nabla_{\mu\lambda}L & \nabla_{\mu\mu}L \end{bmatrix} \begin{bmatrix} \Delta P \\ \Delta\lambda \\ \Delta\mu \end{bmatrix} = \begin{bmatrix} -\nabla_P L \\ -\nabla_\lambda L \\ -\nabla_\mu L \end{bmatrix} \tag{6.27}$$

Algorithm 6.2: Non-Inferior Solutions by the ε-Constraint Method

1. Read data, namely cost coefficients, emission coefficients and *B*-coefficients, Err (Convergence tolerance) and ITMAX (maximum allowed iterations), *M* (number of objectives), NG (number of generators) and *K* (minimum number of non-inferior solutions required for the objective as constraint), etc.
2. Set objective index $j = 1$.
3. If $(j \geq M)$ then GOTO Step 20.
   else increment the objective index, $j = j + 1$.
4. Fix ε_m such that $\overline{F}_m^{\min} < \varepsilon_m < \overline{F}_m^{\max}$ $(m = 2, 3, 4;\ m \neq j)$.
5. Set iteration for non-inferior solutions, $k = 0$.
6. Increment the count of non-inferior solutions, $k = k + 1$.

7. If $(k \geq K)$ GOTO Step 3.
8. Compute initial values of $\overline{P}_i$ $(i = 1, 2, ..., \text{NG})$ and μ.
9. Assume that no generator has been fixed either at lower limit or at upper limit.
10. Set iteration counter, IT = 1.
11. Compute Hessian and Jacobian matrix elements using Eqs. (6.26) and (6.27), respectively. Deactivate row and column of Hessian matrix and row of Jacobian matrix representing the generator whose generation is fixed either at lower limit or at upper limit. This is done so that fixed generators cannot participate in allocation.
12. Gauss elimination method is employed in which triangularization and back substitution processes are performed to find $\Delta\overline{P}_i$ $(i = 1, 2, ..., R)$, $\Delta\lambda_{1j}$ $(j = 2, 3, 4)$, and $\Delta\mu$. Here R is the number of generators which can participate in allocation.
13. Check either $\left(\sqrt{\sum_{i=1}^{R} (\Delta\overline{P}_i)^2 + \sum_{j=2}^{4} (\Delta\lambda_{1j})^2 + (\Delta\mu)^2}\right) \leq \text{err.}$

    if convergence condition is 'yes' then GOTO Step 16.
14. Modify control variables,

$$\overline{P}_i^{\text{new}} = \overline{P}_i + \Delta\overline{P}_i \qquad (i = 1, 2, ..., R)$$
$$\lambda_{1j}^{\text{new}} = \lambda_{1j} + \Delta\lambda_{1j} \qquad (j = 2, 3, 4)$$
$$\mu^{\text{new}} = \mu + \Delta\mu$$

15. Update the iteration counter, IT = IT + 1.
    Assign new values to old value variables

$$\overline{P}_i = \overline{P}_i^{\text{new}} \quad (i = 1, 2, ..., R)$$
$$\lambda_{1j} = \lambda_{1j}^{\text{new}} \quad \text{and} \quad \mu = \mu^{\text{new}}$$

    GOTO Step 11 and repeat.
16. Check the limits of generators and fix up as following

$$\text{If } \overline{P}_i < \overline{P}_i^{\min} \text{ then } \overline{P}_i = \overline{P}_i^{\min}$$
$$\text{If } \overline{P}_i > \overline{P}_i^{\max} \text{ then } \overline{P}_i = \overline{P}_i^{\max}$$

    If no more violations then GOTO Step 18.
17. GOTO Step 10.
18. Check the condition is satisfied

$$\lambda_{12}[\overline{F}_2(\overline{P}) - \varepsilon_2] = 0.0; \quad \lambda_{1j} > 0.0$$

    If 'yes' then GOTO Step 19.

    else modify ε_j and GOTO Step 7.
19. Record it as non-inferior solutions, compute values of all objectives and transmission loss, and modify ε_j for the next non-inferior solution and GOTO Step 6.
20. Stop.

6.3.3 Surrogate Worth Trade-Off Algorithm

The Surrogate Worth Trade-off (SWT) analysis is most useful when objectives are in conflict. The trade-off analysis can then be conducted between the cost and each of the other objectives. A stepwise procedure corresponding to an algorithm outlined by Haimes and Hall [1974], for the problem is given below.

1. Find the minimum and maximum values for each of the multiple objectives separately, i.e. $\overline{F}_j^{\min}$ and $\overline{F}_j^{\max}$ $(j = 1, 2, 3, 4)$. This is carried out by performing economic and minimum emission dispatch separately.
2. **Generation of trade-off function:** Optimal solution to problem of Eq. (6.25) must satisfy Kuhn Tucker conditions. The main condition is

$$\lambda_{1j}[\overline{F}_j - \varepsilon_j] = 0; \quad \lambda_{1j} \geq 0; \qquad (j = 2, 3, 4) \tag{6.28}$$

   The system given by Eq. (6.25) is solved for K values of ε_2, say, ε_2^1, ..., ε_2^K, where ε_3^0 and ε_4^0 are held at some level ε_j^0. Set initial values of ε_j such that $\varepsilon_j > \overline{F}_j^{\min}$ and $\varepsilon_j < \overline{F}_j^{\max}$. Only those values of $\lambda_{12}^k > 0$, which correspond to the active constraint $\overline{F}_2^k = \varepsilon_2^k$ $(k = 1, 2, ..., K)$ are considered since they belong to the non-inferior solution. Similarly the trade-off function λ_{13} is generated, where Eq. (6.25) is solved for K' different values of ε_3^k $(k = 1, 2, ..., K')$, with fixed level ε_2^0 and ε_4^0. Similarly, λ_{14} is generated. Regression analysis is performed to yield the trade-off functions $\lambda_{12}[\overline{F}_2]$, $\lambda_{13}[\overline{F}_3]$ and $\lambda_{14}[\overline{F}_4]$.
3. **Generation of SWT function:** SWT function assigns a scalar value (on an ordinal scale) to any given non-inferior solution. One way of specifying non-inferior solution is by trade-off functions. Moreover, there is close relationship between the SWT function W_{1j} and the partial derivatives of the utility functions. In multiobjective analysis, it is assumed implicitly that the DM maximizes his utility which is a monotonic decreasing function of the objective functions. Given a decision vector $\overline{P}$ and the associated consequences $\overline{F}_i$, the utility is given by

$$U = U[\overline{F}_1, \overline{F}_2, \overline{F}_3, \overline{F}_4] \tag{6.29}$$

   By linearizing the utility function for a small change in $\overline{F}_1$, the following can be obtained [Haimes, 1977].

$$\Delta U_{1j} = \left(\frac{\partial U}{\partial \overline{F}_j} - \frac{\partial U}{\partial \overline{F}_1} \lambda_{1j} \right) \tag{6.30}$$

   The SWT function W_{1j} is a monotonic function of U_{1j}, with the property that $W_{1j} = 0$ <-> $U_{1j} = 0$ and is written as

$$W_{1j} = h_j \, U_{1j} \qquad (j = 2, 3, 4) \tag{6.31}$$

   where h_j is some monotonic increasing function of its argument, with a range of –10 to +10 and with the property that $h_j(0) = 0$.
4. Find functions $W_{1j}(\lambda_{1j})$ for $(j = 2, 3, 4)$ by regression analysis or by interpolation. Then the values $\lambda_{1j} = \lambda_{1j}^*$ is chosen where $W_{1j}(\lambda_{1j}) = 0$ $(j = 2, 3, 4)$.

5. The optimal set of decision vector is found by solving the following problem.

Minimize $$\left[\overline{F}_1 + \sum_{j=2}^{4} \lambda_{1j}^{*} \overline{F}_j\right] \tag{6.32a}$$

subject to $$\sum_{i=1}^{NG} \overline{P}_i - \overline{P}_D = 0 \tag{6.32b}$$

$$\overline{P}_i^{\min} \le \overline{P}_i \le \overline{P}_i^{\max} \qquad (i = 1, 2, ..., NG) \tag{6.32c}$$

Form the generalized Lagrangian L to the system of Eq. (6.32)

$$L(\overline{P}_i, \mu) = \overline{F}_1 + \sum_{j=2}^{4} \lambda_{1j}^{*} \overline{F}_j + \mu\left(\overline{P}_D - \sum_{i=1}^{NG} \overline{P}_i\right) \tag{6.33}$$

where μ is generalized Lagrangian multipliers.

The first-order derivative necessary optimal conditions to obtain solution are given as

$$\frac{\partial L}{\partial \overline{P}_i} = \frac{\partial \overline{F}_1}{\partial \overline{P}_i} + \sum_{j=1}^{4} \lambda_{1j} \frac{\partial \overline{F}_2}{\partial \overline{P}_i} - \mu = 0 \qquad (i = 1, 2, ..., NG) \tag{6.34a}$$

$$\frac{\partial L}{\partial \mu} = \overline{P}_D - \sum_{i=1}^{N} \overline{P}_i = 0 \tag{6.34b}$$

The Newton–Raphson method can be applied to solve the above equations. To implement the Newton–Raphson method, the following equation is solved iteratively.

$$\begin{bmatrix} \nabla_{PP} L & \nabla_{P\mu} L \\ (\nabla_{P\mu} L)^T & \nabla_{\mu\mu} L \end{bmatrix} \begin{bmatrix} \Delta P \\ \Delta \mu \end{bmatrix} = \begin{bmatrix} -\nabla_P L \\ -\nabla_\mu L \end{bmatrix} \tag{6.35}$$

Hessian matrix elements can be obtained from above equations by differentiating with respect to to control variables, one by one.

$$\frac{\partial^2 L}{\partial \overline{P}_i^2} = \frac{\partial^2 \overline{F}_1}{\partial \overline{P}_i^2} + \sum_{j=2}^{4} \lambda_{1j} \frac{\partial^2 \overline{F}_j}{\partial \overline{P}_i^2} \qquad (i = 1, 2, ..., NG) \tag{6.36a}$$

$$\frac{\partial^2 L}{\partial \overline{P}_i \partial \overline{P}_j} = \lambda_{13} \frac{\partial^2 \overline{F}_3}{\partial \overline{P}_i \partial \overline{P}_j} \qquad (i = 1, 2, ..., NG;\ \ j \ne i;\ \ j = 1, 2, ..., NG) \tag{6.36b}$$

$$\frac{\partial^2 L}{\partial \overline{P}_i \partial \mu} = \frac{\partial^2 L}{\partial \mu \partial \overline{P}_i} = -1 \qquad (i = 1, 2, ..., NG) \tag{6.36c}$$

$$\frac{\partial^2 L}{\partial \mu^2} = 0 \tag{6.36d}$$

Utility function

Here, it is assumed that there exists a very general utility function for a given DM that can predict his behaviour and interest. Let the DM's utility function be defined for each objective function depending on the importance of the objective function $\overline{F}_i$ to the other objective functions. So, total or overall utility function is defined as [Osyczka and Davies, 1984].

Maximize
$$U = \sum_{i=1}^{4} k_i' \overline{F}_i \tag{6.37}$$

The solution vector $\overline{P}$ is then found by maximizing the total utility subjected to the technology constants as defined below.

$$\sum_{i=1}^{NG} \overline{P}_i - \overline{P}_D = 0 \tag{6.38}$$

$$\overline{P}_i^{\min} \le \overline{P}_i \le \overline{P}_i^{\max} \qquad (i = 1, 2, ..., \text{NG}) \tag{6.39}$$

Further define

$$k_i' = \frac{w_i'}{\sum_{i=1}^{4} w_i'} \tag{6.40a}$$

such that

$$\sum_{i=1}^{4} k_i' = 1; \quad k_i' \ge 0 \tag{6.40b}$$

The DM gives the weight w_i on the attribute between (0, 99). The solution of the above optimization problem can be obtained as explained in Section 5.4.4.

6.3.4 Sample System Study

The method is applied to a six-generator sample system to demonstrate its applicability. The expected fuel cost characteristics ($/h) undertaken for study are as:

$$\overline{F}_{11} = 0.005\,\overline{P}_1^2 + 2.0\,\overline{P}_1 + 100.0$$

$$\overline{F}_{12} = 0.010\,\overline{P}_2^2 + 2.0\,\overline{P}_2 + 200.0$$

$$\overline{F}_{13} = 0.020\,\overline{P}_3^2 + 2.0\,\overline{P}_3 + 300.0$$

$$\overline{F}_{14} = 0.003\,\overline{P}_4^2 + 1.95\,\overline{P}_4 + 80.0$$

$$\overline{F}_{15} = 0.015\,\overline{P}_5^2 + 1.45\,\overline{P}_5 + 100.0$$

$$\overline{F}_{16} = 0.010\,\overline{P}_6^2 + 0.95\,\overline{P}_6 + 120.0$$

The expected NO_x emission (kg/h) characteristics are:

$$\overline{F}_{21} = 0.0006572\,\overline{P}_1^2 - 0.05497\,\overline{P}_1 + 4.111$$
$$\overline{F}_{22} = 0.0005916\,\overline{P}_2^2 - 0.05880\,\overline{P}_2 + 2.593$$
$$\overline{F}_{23} = 0.0004906\,\overline{P}_3^2 - 0.05014\,\overline{P}_3 + 4.268$$
$$\overline{F}_{24} = 0.0003780\,\overline{P}_4^2 - 0.03150\,\overline{P}_4 + 5.526$$
$$\overline{F}_{25} = 0.0004906\,\overline{P}_5^2 - 0.05014\,\overline{P}_5 + 4.268$$
$$\overline{F}_{26} = 0.0005713\,\overline{P}_6^2 - 0.05548\,\overline{P}_6 + 6.132$$

The expected *B*-coefficients are given in Table 6.10.

TABLE 6.10 Expected transmission loss coefficients

0.000200	0.000010	0.000015	0.000005	0.000000	0.000030
0.000010	0.000300	−0.000020	0.000001	0.000012	0.000010
0.000015	−0.000020	0.000100	0.000010	0.000010	0.000008
0.000005	0.000001	0.000010	0.000150	0.000006	0.000050
0.000000	0.000012	0.000010	0.000006	0.000250	0.000020
0.000030	0.000010	0.000008	0.000050	0.000020	0.000210

Table 6.11 shows the conflicting objectives, trade-off functions, utility and SWT function at each non-inferior set. The decision vector P_i corresponding to each non-inferior set is shown in

TABLE 6.11 Expected cost, emission, risk and transmission loss along with utility and SWT function, when demand is 200 MW

Sr. no.	$\overline{F}_1$ (\$/h)	$\overline{F}_2$ (kg/h)	$\overline{F}_3$ (MW)	$\overline{F}_4$ (MW^2)	λ_{12} (\$/kg)	U	W_{12}
1	1306.917	20.6380	1.9145	74.2663	371.5998	542.1307	–5
2	1305.927	20.6434	1.9096	73.9515	395.2971	541.6717	–6
3	1304.951	20.6559	1.9071	73.7545	429.8842	541.2439	–6
4	1303.951	20.6776	1.9076	73.7125	470.5927	540.8398	–6
5	1302.847	20.7181	1.9140	73.9955	500.6307	540.4645	–7
6	1301.824	20.7492	1.9178	74.1388	507.7724	540.0906	–7
7	1300.818	20.7684	1.9187	74.0979	500.6307	539.6841	–7
8	1299.788	20.7950	1.9219	74.1986	473.6248	539.2983	–6
9	1298.892	20.8104	1.9215	74.1169	415.0005	538.9263	–5
10	1298.068	20.8202	1.9207	73.9351	328.1359	538.5624	–4
11	1297.315	20.8361	1.9221	73.8146	221.8559	538.2407	–3
12	1296.801	20.8649	1.9263	73.8709	99.3539	538.0527	–1

(*Contd.*)

TABLE 6.11 (*Contd.*)

Sr. no.	$\overline{F_1}$ (\$/h)	$\overline{F_2}$ (kg/h)	$\overline{F_3}$ (MW)	$\overline{F_4}$ (MW^2)	λ_{13} (\$/MW)	U	W_{13}
1	1306.095	20.9202	2.2559	85.5850	669.5771	544.1902	–9
2	1305.819	20.9517	2.2899	86.1714	599.0720	544.2104	–9
3	1305.374	21.0180	2.3305	87.2824	519.2102	544.2756	–8
4	1304.760	21.1077	2.3764	88.7415	428.1452	544.3491	–7
5	1303.964	21.2135	2.4269	90.4506	325.0275	544.4036	–6
6	1303.231	21.3202	2.4790	92.1787	207.9039	544.4880	–3
7	1302.859	21.4210	2.5315	93.8470	76.1176	544.7037	–0
	$\overline{F_1}$ (\$/h)	$\overline{F_2}$ (kg/h)	$\overline{F_3}$ (MW)	$\overline{F_4}$ (MW^2)	λ_{14} (\$/MW^2)	U	W_{14}
1	1287.161	21.6603	2.4001	88.7065	27.0517	537.4181	90
2	1286.748	21.6600	2.4000	88.8003	21.7904	537.2714	50
3	1286.421	21.6600	2.4000	88.9000	18.0675	537.1605	3
4	1286.156	21.6600	2.4000	89.0000	15.2434	537.0745	2
5	1285.938	21.6600	2.4000	89.1000	12.9794	537.0071	1
6	1285.757	21.6600	2.4000	89.2000	11.0854	536.9547	1
7	1285.607	21.6600	2.4000	89.3000	9.4471	536.9147	0
8	1285.485	21.6600	2.4000	89.4000	7.9907	536.8859	0
9	1285.387	21.6600	2.4000	89.5000	6.6679	536.8669	0

Table 6.12. In this case, the coefficients of variations of cost, emission and B-coefficients are assumed 1%. The scalar weights are 40.0, 20.0, 20.0 and 20.0 for cost, emission, and power loss and risk objectives in sequence.

By regression analysis, the trade-off functions for 200 MW are represented as

$$\lambda_{12}(\overline{F_2}) = 5717.199 + 284.376\,\overline{F_2} - 26.306\,\overline{F_2}^2$$

$$\lambda_{13}(\overline{F_3}) = -316.958 + 2719.367\,\overline{F_3} - 1012.223\,\overline{F_3}^2$$

$$\lambda_{14}(\overline{F_4}) = 1550.06 - 17.228\,\overline{F_4}$$

with 0.5167, 0.9998, and 0.9636 being standard deviations respectively.

The Surrogate worth function as straight line is shown as

$$W_{12}(\lambda_{12}) = 0.3583 - 0.01427\lambda_{12}$$

$$W_{13}(\lambda_{13}) = 0.1603 - 0.01526\lambda_{13}$$

$$W_{14}(\lambda_{14}) = -3.8614 + 0.4278\lambda_{14}$$

with 0.9802, 0.9701, and 0.9661 being standard deviations respectively. The values of λ_{12}^*, λ_{13}^*, and λ_{14}^* are 25.10, 10.50, and 9.026, respectively as discussed in Step 3 of the algorithm. The

TABLE 6.12 Generation schedules for demand 200 MW corresponding to non-inferior solutions

Sr. no.	$\overline{P_1}$ (MW)	$\overline{P_2}$ (MW)	$\overline{P_3}$ (MW)	$\overline{P_4}$ (MW)	$\overline{P_5}$ (MW)	$\overline{P_6}$ (MW)
1	18.202	28.223	35.765	33.441	29.102	55.267
2	18.599	28.117	35.050	35.165	28.432	54.637
3	18.684	28.080	34.243	37.103	27.990	53.900
4	18.458	28.073	33.313	39.318	27.781	53.057
5	17.918	28.049	32.223	41.989	27.569	52.252
6	17.803	28.291	30.952	44.467	27.387	51.099
7	18.316	28.726	29.443	46.652	27.210	49.654
8	19.149	29.088	27.750	48.885	26.953	48.174
9	20.870	29.488	25.949	50.623	26.512	46.558
10	23.672	29.855	24.006	51.751	25.690	45.027
11	27.294	29.855	21.944	52.352	24.977	43.579
12	31.205	29.465	19.976	52.523	24.633	42.198
1	45.895	33.262	26.579	4.368	27.924	61.971
2	47.782	34.575	24.333	3.313	29.414	60.583
3	49.836	34.931	22.167	2.569	30.559	59.939
4	51.900	34.735	19.927	2.101	31.503	59.834
5	53.734	34.199	17.548	1.958	32.270	60.292
6	55.165	33.794	14.994	2.090	33.023	60.934
7	56.232	33.862	12.276	2.332	33.882	61.416
1	44.092	12.349	6.573	43.918	31.185	61.884
2	42.698	12.427	6.933	45.111	30.436	62.394
3	41.316	12.597	7.122	46.274	29.966	62.724
4	39.997	12.812	7.264	47.337	29.620	62.969
5	38.735	13.059	7.388	48.308	29.346	63.163
6	37.516	13.335	7.506	49.202	29.118	63.322
7	36.332	13.638	7.623	50.029	28.924	63.454
8	35.174	13.966	7.744	50.799	28.752	63.566
9	34.035	14.320	7.870	51.517	28.596	63.663

optimal decision vector $\overline{P}$ is shown in Table 6.14. Corresponding to the non-inferior set, Tables 6.13 and 6.14 depict the values of various objective functions at the expected minimum cost, expected minimum emission and maximum expected utility schedules. It may be noted that each step involved in reduction of emission of NO_x becomes increasingly expensive.

In the multiobjective framework it is realized that expected cost and risk are conflicting objectives and are subject to mutual interface. The solution set of the formulated problems is non-inferior due to contradictions among objectives taken and has been obtained through the ε-constrained

TABLE 6.13 Comparison of results

$\overline{P}_D$ (MW)	$\overline{F}_1$ ($/h)	$\overline{F}_2$ (kg/h)	$\overline{F}_3$ (MW)	$\overline{F}_4$ (MW^2)
Minimum cost dispatch				
200.0	1284.365	21.8230	2.4749	93.8997
400.0	1788.286	24.8757	8.7164	355.1199
600.0	2386.325	37.0950	19.1952	813.4999
Minimum emission dispatch				
200.0	1318.079	20.2491	1.7949	68.6702
400.0	1878.364	20.6196	6.8843	268.3033
600.0	2577.228	27.7313	15.3691	605.5116
Maximum utility approach				
200.0	1309.791	20.3447	1.7755	66.9215
400.0	1856.690	20.9385	7.2144	268.1749
600.0	2575.991	28.2854	15.3633	600.4974

TABLE 6.14 Generation schedules corresponding to the results given in Table 6.13

$\overline{P}_D$ (MW)	$\overline{P}_1$ (MW)	$\overline{P}_2$ (MW)	$\overline{P}_3$ (MW)	$\overline{P}_4$ (MW)	$\overline{P}_5$ (MW)	$\overline{P}_6$ (MW)
Minimum cost dispatch						
200.0	28.431	14.216	7.108	55.719	27.810	66.716
400.0	75.490	37.745	18.873	134.150	43.497	90.245
600.0	122.549	61.275	30.637	212.582	59.183	113.775
Minimum emission dispatch						
200.0	30.409	37.017	35.812	21.824	35.812	39.125
400.0	56.052	65.505	70.164	66.409	70.164	71.704
600.0	81.696	93.992	104.517	110.994	104.517	104.284
Maximum Utility						
200.0	33.216	32.507	30.186	33.650	33.264	37.177
400.0	66.707	69.890	56.570	65.791	68.520	72.522
600.0	98.381	96.977	103.874	104.148	98.156	98.463

technique. The novel formulation as a multiobjective optimization problem has made it possible to quantitatively grasp trade-off relations among conflicting objectives.

The trade-off approach is effective only up to two objectives; as the number of objectives increases, the selection of the best solution becomes cumbersome. An interactive method SWT has been applied to identify the best compromised solution for multiobjective power dispatch problem, when conflicting objectives are more than two. The major characteristics and advantages of the SWT method are that the surrogate worth functions, which relate the decision maker's preferences to the non-inferior solutions through the trade-off functions, are constructed in the functional space and only then are transformed into the decision space.

6.4 MULTIOBJECTIVE STOCHASTIC OPTIMAL THERMAL POWER DISPATCH—WEIGHTING METHOD

The economic dispatch problem was defined so as to determine the allocation of electricity demand among the committed generating units to minimize the operating costs subject to physical and technological constraints. Most of the existing formulations of the economic dispatch are solved as static deterministic optimization problems. Actually, there are many inaccuracies and uncertainties in the input information which lead to deviations from optimal operation and cause an increase in the cost over the optimal value [Edwin and Machate, 1980]. As a result of the rise in production costs due to uncertain factors, the electric energy system has been represented as a network characterized by random variables and investigated by numerous researchers at various levels [Dopazo et al., 1975; Parti et al., 1983; and El-Hawary and Mbamalu, 1991]. Although, these approaches have been successful in applications involving stochastic economic dispatch, but all the methods do not provide trade-off between economy and risk measures due to uncertainties in system production cost and random nature of demand. Typically, such conflicts exist because no such feasible solution has been found which would minimize them all.

The pollution minimization problem has attracted a lot of attention due to the public demand for clean air. Thermal power stations are major causes of atmospheric pollution, because of high concentration of pollutants they cause. Since optimum economic dispatch is not environmentally the best solution, many organizations in their fight against air pollution have come up with a new method, the so-called minimum emission dispatch (MED). MED is used to minimize the total stack emission (NO_x) for the entire system, although this may be controlled either through post-combustion cleaning systems (electrostatic precipitators, stack gas scrubbers) or automatically (controlling unit loading). MED may be obtained by introducing a different set of generator representations into the economic dispatch problem. The MED generator equation for each unit is a function of stack emission (NO_x) versus megawatt output, instead of input ($/h) versus megawatt output for the economic dispatch.

6.4.1 Stochastic Multiobjective Optimization Problem Formulation

The objective function to be minimized is the total operating cost for thermal generating units in the system and a quadratic operating cost curve is assumed.

$$F_1 = \sum_{i=1}^{NG} (a_i P_i^2 + b_i P_i + c_i) \tag{6.41}$$

where

a_i, b_i, and c_i are the cost coefficients

NG is the total number of generating units.

A stochastic model of function F_1, is formulated by considering cost coefficients and load demand as random variables. By taking expectation, the stochastic model can be converted into its deterministic equivalent. The random variables are assumed to be normally distributed and statistically dependent on each other. The expected value of operating cost function may be obtained through expanding the function using Taylor's series, about mean [Rao, 1987]. By taking the expectation of the expanded form, the expected operating cost function is represented by

$$\overline{F_1} = \sum_{i=1}^{NG} [(\overline{a}_i \overline{P}_i^2 + \overline{b}_i \overline{P}_i + \overline{c}_i + \overline{a}_i \operatorname{var}(P_i) + \operatorname{cov}(b_i, P_i) + 2\overline{P}_i \operatorname{cov}(a_i, P_i)] \tag{6.42}$$

where

$\overline{a}_i$, $\overline{b}_i$, and $\overline{c}_i$ are expected cost coefficients

$\overline{P}_i$ is the expected value of the *i*th generator output.

In this study, variance and covariance are replaced by coefficient of variation (CV) and correlation coefficient (CC), respectively. In general, variance and covariance are defined as

$$\text{var}(X) = C_X^2\, \overline{X}^2 \tag{6.43}$$

$$\text{cov}(X,Y) = R_{XY}\, C_X\, C_Y\, \overline{X}\,\overline{Y} \tag{6.44}$$

where

C_X and C_Y are the CV of random variables X and Y, respectively.

R_{XY} is the CC of random variables X and Y.

The value of CC is positive or negative depending upon the sign of the covariance and its value lies between –1.0 and 1.0.

Using Eqs. (6.43) and (6.44), Eq. (6.42) can be rewritten in the simplified form as

$$\overline{F}_1 = \sum_{i=1}^{NG} [(1 + C_{P_i}^2 + 2R_{a_iP_i} C_{a_i} C_{P_i})\, \overline{a}_i \overline{P}_i^2 + (1 + R_{b_iP_i} C_{b_i} C_{P_i})\overline{b}_i \overline{P}_i + \overline{c}_i] \tag{6.45}$$

For the fixed network configuration and random load demand, the equality constraint in the classical dispatch problem is represented by the expected power balance equation stated as

$$\sum_{i=1}^{NG} \overline{P}_i = \overline{P}_D + \overline{P}_L \tag{6.46}$$

where $\overline{P}_D$ and $\overline{P}_L$ are the expected load demand and the expected transmission loss, respectively.

Expected transmission loss

The transmission power loss expressed through the simplified well known loss formula expression as a quadratic function of power generations is given by [Kusic, 1986]:

$$P_L = \sum_{i=1}^{NG} \sum_{j=1}^{NG} P_i B_{ij} P_j \tag{6.47}$$

Power generations P_i are dependent random variables. B_{ij} are also considered as inaccurate B-coefficients. The expected transmission losses using Taylor's series are represented as

$$\overline{P}_L = \sum_{i=1}^{NG} \sum_{j=1}^{NG} \overline{P}_i \overline{B}_{ij} \overline{P}_j + \sum_{i=1}^{NG} \overline{B}_{ii}\, \text{var}(P_i) + \sum_{i=1}^{NG-1} \sum_{j=i+1}^{NG} 2\overline{B}_{ij}\, \text{cov}(P_i, P_j) + \sum_{i=1}^{NG} 2\overline{P}_i\, \text{cov}(P_i, B_{ii})$$

$$+ \sum_{i=1}^{NG} \sum_{\substack{j=1 \\ j \neq i}}^{NG} 2\overline{P}_j\, \text{cov}\,(P_i, B_{ij}) \tag{6.48}$$

On simplification, the above equation can be rewritten as

$$\overline{P}_L = \sum_{i=1}^{NG} [1 + C_{P_i}^2 + 2R_{P_iB_{ii}} C_{P_i} C_{B_{ii}}] \overline{B}_{ii} \overline{P}_i^{\,2} + \sum_{i=1}^{NG} \sum_{\substack{j=1 \\ j \neq i}}^{NG} [1 + R_{P_i,P_j} C_{P_i} C_{P_j} + 2R_{P_iB_{ij}} C_{P_i} C_{B_{ij}}] \overline{P}_i \overline{B}_{ij} \overline{P}_j \qquad (6.49)$$

where

$\overline{B}_{ij}$ are expected B-coefficients

$R_{P_iB_{ij}}$ are the correlation coefficients of random variables P_i and B_{ij}

$C_{B_{ij}}$ are the coefficients of variation of uncertain parameter B_{ij}

Expected deviations

Since generator outputs P_is are treated as random variables, the expected deviations are proportional to the expectation of the square of the unsatisfied load demand. These expected deviations are considered as the second objective to be minimized. The second objective function $\overline{F}_2$ is represented as

$$\overline{F}_2 = E\left[\left(\overline{P}_D + \overline{P}_L - \sum_{i=1}^{NG} P_i\right)^2\right] \qquad (6.50)$$

which on simplification reduces to

$$\overline{F}_2 = \sum_{i=1}^{NG} \text{var}(P_i) + \sum_{i=1}^{NG-1} \sum_{j=i+1}^{NG} 2\,\text{cov}(P_i, P_j) \qquad (6.51a)$$

$$\overline{F}_2 = \sum_{i=1}^{NG} C_{P_i}^2 \overline{P}_i^{\,2} + \sum_{i=1}^{NG} \sum_{\substack{j=1 \\ j \neq i}}^{NG} R_{P_iP_j} C_{P_i} C_{P_j} \overline{P}_i \overline{P}_j \qquad (6.51b)$$

The deterministic equivalent of multiobjective stochastic optimization problem is formulated by taking (a) the expected operating cost, and (b) the expected risk associated with the possible deviation of the random variables from their expected values. These two different objectives are to be minimized, while satisfying the expected equality and inequality constraints. Mathematically:

Minimize $\quad [\overline{F}_1, \overline{F}_2]^T \qquad (6.52a)$

subject to

$$\sum_{i=1}^{NG} \overline{P}_i = \overline{P}_D + \overline{P}_L \qquad (6.52b)$$

$$\overline{P}_i^{\min} \le \overline{P}_i \le \overline{P}_i^{\max} \qquad (i = 1, 2, ...; \text{NG}) \qquad (6.52c)$$

where $\overline{F}_1$ and $\overline{F}_2$ are the expected values of objective functions to be minimized over the set of admissible decision variables $\overline{P}_i$.

6.4.2 Solution Approach

To generate the non-inferior solution of multiobjective optimization problem, the weighting method is used. In this method the problem is converted into a scalar optimization problem as

$$\text{Minimize} \qquad (w_1\overline{F}_1 + w_2\overline{F}_2) \tag{6.53a}$$

$$\text{subject to} \qquad \sum_{i=1}^{NG} \overline{P}_i = \overline{P}_D + \overline{P}_L \tag{6.53b}$$

$$\overline{P}_i^{\min} \le \overline{P}_i \le \overline{P}_i^{\max} \qquad (i = 1, 2, ..., NG) \tag{6.53c}$$

$$\sum_{k=1}^{2} w_k = 1, \; w_k \ge 0$$

where w_k are the levels of the weighting coefficients. This approach yields meaningful result to the decision maker when solved many times for different values of w_k, $k = 1, 2$. The values of weighting coefficients vary from 0 to 1.

To solve the scalar optimization problem (6.53), the Lagrangian function is defined as

$$L(\overline{P}_i, \lambda) = w_1\overline{F}_1 + w_2\overline{F}_2 + \lambda\left(\overline{P}_D + \overline{P}_L - \sum_{i=1}^{NG} \overline{P}_i\right) \tag{6.54}$$

where λ is Lagrangian multiplier.

The necessary conditions to minimize the unconstrained Lagrangian function are:

$$\frac{\partial L}{\partial \overline{P}_i} = w_1\frac{\partial \overline{F}_1}{\partial \overline{P}_i} + w_2\frac{\partial \overline{F}_2}{\partial \overline{P}_i} + \lambda\left(\frac{\partial \overline{P}_L}{\partial \overline{P}_i} - 1\right) = 0 \qquad (i = 1, 2, ..., NG) \tag{6.55a}$$

$$\frac{\partial L}{\partial \lambda} = \overline{P}_D + \overline{P}_L - \sum_{i=1}^{NG} \overline{P}_i = 0 \tag{6.55b}$$

The scalar optimization problem is solved using the Newton–Raphson algorithm. The size of the formulated Hessian matrix is the same as that for the deterministic problem, because single objective function is solved in both cases with the same number of constraints. To implement the Newton–Raphson method, the following equation is solved iteratively till no further improvement in decision variables is achieved.

$$\begin{bmatrix} \nabla_{PP}L & \nabla_{P\lambda}L \\ (\nabla_{P\lambda}L)^T & \nabla_{\lambda\lambda}L \end{bmatrix} \begin{bmatrix} \Delta P \\ \Delta\lambda \end{bmatrix} = \begin{bmatrix} -\nabla_P L \\ -\nabla_\lambda L \end{bmatrix} \tag{6.56}$$

Hessian matrix elements can be obtained from the above equations by differentiating with respect to control variables, one by one.

$$\frac{\partial^2 L}{\partial \overline{P}_i^2} = w_1\frac{\partial^2 \overline{F}_1}{\partial \overline{P}_i^2} + w_2\frac{\partial^2 \overline{F}_2}{\partial \overline{P}_i^2} + \lambda\left(\frac{\partial^2 \overline{P}_L}{\partial \overline{P}_i^2} - 1\right) = 0 \qquad (i = 1, 2, ..., NG) \tag{6.57a}$$

$$\frac{\partial^2 L}{\partial \overline{P}_i \partial \overline{P}_j} = w_2 \frac{\partial^2 \overline{F}_2}{\partial \overline{P}_i \partial \overline{P}_j} + \lambda \frac{\partial^2 \overline{P}_L}{\partial \overline{P}_i \partial \overline{P}_j} \qquad (i = 1, 2, ..., \text{NG}; \quad j = 1, 2, ..., \text{NG}; \quad i \neq j) \tag{6.57b}$$

$$\frac{\partial^2 L}{\partial \overline{P}_i \partial \lambda} = \frac{\partial \overline{P}_L}{\partial \overline{P}_i} - 1 \qquad (i = 1, 2, ..., \text{NG}) \tag{6.57c}$$

$$\frac{\partial^2 L}{\partial \lambda^2} = 0 \tag{6.57d}$$

Algorithm 6.3: Non-Inferior Solutions by the Weighting Method

1. Read data, namely cost coefficients, emission coefficients and B-coefficients, demand, Err (convergence tolerance) and ITMAX (maximum allowed iterations), M (number of objectives), NG (number of generators) and K (number of inferior solutions), etc.
2. Set iteration for non-inferior solutions, $k = 1$.
3. Increment count of non-inferior solutions, $k = k + 1$.
4. If $(k \geq K)$ GOTO Step 17.
5. Feed or generate weights, w_i [$i = 1, 2, ..., M$ (here $M = 2$)]
6. Compute the initial values of $\overline{P}_i$ $(i = 1, 2, ..., \text{NG})$ and λ by presuming that $\overline{P}_L = 0$. The values of λ and $\overline{P}_i$ $(i = 1, 2, ..., \text{NG})$ can be computed directly using Eqs. (3.10) and (3.9), respectively.
7. Assume that no generator has been fixed either at lower limit or at upper limit.
8. Set iteration counter, IT = 1.
9. Compute Hessian and Jacobian matrix elements using Eqs. (6.55a) and (6.55b) and Eqs. (6.57a) to Eq. (6.57d), respectively. Deactivate row and column of Hessian matrix and row of Jacobian matrix representing the generator whose generation is fixed either at lower limit or at upper limit. This is done so that fixed generators cannot participate in allocation.
10. Gauss elimination method is employed in which triangularization and backsubstitution processes are performed to find $\Delta\overline{P}_i$ $(i = 1, 2, ..., R)$ and $\Delta\lambda$. Here R is the number of generators that can participate in allocation.
11. Check either $\sqrt{\sum_{i=1}^{R} (\Delta\overline{P}_i)^2 + (\Delta\lambda)^2} \leq \varepsilon$ or $\sqrt{\sum_{i=1}^{R} \left(\frac{\partial L}{\partial \overline{P}_i}\right)^2 + \left(\frac{\partial L}{\partial \lambda}\right)^2} \leq \varepsilon$

    if convergence condition is 'satisfied' then GOTO Step 14.
    Check IT > ITMAX, if condition is 'yes', GOTO Step 14 (It means the procedure proceeds without obtaining the required convergence).
12. Modify control variables,

$$\overline{P}_i^{\text{new}} = \overline{P}_i + \Delta\overline{P}_i \qquad (i = 1, 2, ..., R)$$
$$\lambda^{\text{new}} = \lambda + \Delta\lambda$$

13. Update iteration counter, IT = IT + 1,

Update the old values with new values.

$$\overline{P}_i = \overline{P}_i^{\text{new}} \qquad (i = 1, 2, ..., R),\ \lambda = \lambda^{\text{new}} \text{ and GOTO Step 9 and repeat.}$$

14. Check the limits of generators and fix up as following:

    If $\overline{P}_i < \overline{P}_i^{\min}$ then $\overline{P}_i = \overline{P}_i^{\min}$

    If $\overline{P}_i > \overline{P}_i^{\max}$ then $\overline{P}_i = \overline{P}_i^{\max}$

    If no more violations then GOTO Step 16.
15. GOTO Step 8.
16. Record as non-inferior solution. Compute $\overline{F}_k (k = 1, 2, ..., M)$ and transmission loss and GOTO Step 3 for another non-inferior solution.
17. Stop.

6.4.3 Decision Making

Considering the imprecise nature of the DM's judgement, it is natural to assume that the DM may have fuzzy or imprecise goals for each objective function. The fuzzy sets are defined by equations called membership functions. These functions represent the degree of membership in certain fuzzy sets using values from 0 to 1 [Klir and Folger, 1993]. The membership value 0, indicates incompatibility with the sets, while 1 denotes full compatibility. By taking account of the minimum and maximum values of each objective function together with the rate of increase of membership satisfaction, the DM must determine membership function $\mu(\overline{F}_i)$ in a subjective manner. Here it is assumed that $\mu(\overline{F}_i)$ is a strictly monotonic decreasing and continuous function defined as

$$\mu(\overline{F}_i) = \begin{cases} 1 & ; \overline{F}_i \le \overline{F}_i^{\min} \\ \dfrac{\overline{F}_i^{\max} - \overline{F}_i}{\overline{F}_i^{\max} - \overline{F}_i^{\min}} & ; \overline{F}_i^{\min} < \overline{F}_i < \overline{F}_i^{\max} \\ 0 & ; \overline{F}_i \ge \overline{F}_i^{\max} \end{cases} \tag{6.58}$$

The value of membership function indicates how much (in the scale from 0 to 1) a non-inferior (non-dominated) solution has satisfied the $\overline{F}_i$ objective. The sum of the membership function values $(\mu(\overline{F}_i))$ $(i = 1, 2, ..., M)$ for all the objectives can be computed in order to measure the accomplishment of each solution in satisfying the objectives. The *accomplishment* of each non-dominated solution can be rated with respect to all the K non-dominated solutions by normalizing its accomplishment over the sum of the accomplishments of the K non-dominated solutions as follows [Tapia and Murtagh, 1991]:

$$\mu_D^k = \frac{\sum_{i=1}^{M} \mu(\overline{F}_i)^k}{\sum_{k=1}^{K} \sum_{i=1}^{M} \mu(\overline{F}_i)^k} \tag{6.59}$$

where M is the total number of objectives.

The function μ_D^k in Eq. (6.59) can be treated as a membership function for non-dominated solutions in a fuzzy set and represented as fuzzy cardinal priority ranking of the non-dominated solutions. The solution that attains the maximum membership μ_D^k in the fuzzy set so obtained can be chosen as the best solution or that having the highest cardinal priority ranking.

$$\text{Max } \{\mu_D^k : k = 1, 2, ..., K\} \tag{6.60}$$

6.4.4 Results and Discussion

The proposed method is applied to a three-generator sample system in order to demonstrate its applicability. Expected generator characteristics are given in Table 6.16. Expected B-coefficients of transmission loss formula are presented in Table 6.15. In addition, the following values of CVs and CCs of random variables are assumed.

$$C_{a_i} = 0.1,\ C_{b_i} = 0.1,\ C_{P_i} = 0.1,\ C_{B_{ii}} = 0.1 \qquad (i = 1, 2, 3)$$

$$R_{a_iP_i} = 1.0,\ R_{b_iP_i} = 1.0,\ R_{P_i,B_{ii}} = 1.0 \qquad (i = 1, 2, 3)$$

$$R_{P_iP_j} = 1.0,\ R_{P_i,B_{ij}} = 1.0, \qquad (i = 1, 2, 3;\ \ j = 1, 2, 3;\ \ i \neq j)$$

TABLE 6.15 Expected B-coefficients

0.0000166	−0.0000102	−0.0000136
−0.0000102	0.0000219	0.0000025
−0.0000136	0.0000025	0.0000168

TABLE 6.16 Expected generation characteristics

Generator *no.*	$\bar{a}_i$ (\$/MW^{2}h)	$\bar{b}_i$ (\$/MWh)	$\bar{c}_i$ (\$/h)	$\bar{P}_i^{\max}$ (MW)	$\bar{P}_i^{\min}$ (MW)
1	0.012	6.515	135.33	105.0	30.0
2	0.005	5.627	261.19	225.0	50.0
3	0.006	5.506	264.63	250.0	70.0

The parameter statistics is known from past history. Otherwise Monte Carlo simulation technique is a useful tool in simulating the parameter statistics. Analysis of variation in expected cost seems necessary because of the existence of covariance of two random variables in the problem formulation. In the study, covariance of bivariate random variables is considered positive (increasing) or negative (decreasing) pairwise. Covariance of one pair of random variables is considered at a time, whereas the rest of the random variables are considered independent (uncorrelated). All correlation coefficients are varied from –1.0 to 1.0 in steps of 0.5. Assuming both the weights w_1 and w_2 as 0.5, the percentage relative changes in $\overline{F_1}$ from deterministic $\overline{F_1}$ with respect to $R_{a_iP_i}$, $R_{b_iP_i}$, $R_{P_iB_{ij}}$ and $R_{P_iP_j}$ $(i \neq j)$ are shown in Figure 6.5 for the expected demand of 350 MW. It can be observed that (i) there is increase in the percentage relative cost as $R_{P_iP_j}$ is changed from positive value to negative value, and (ii) there is decrease in the percentage relative cost as CC of rest of the random variables are changed accordingly.

Minimum and maximum values of $\overline{F}_1$ and $\overline{F}_2$

Minimum values of the objectives $\overline{F}_1$ and $\overline{F}_2$ are obtained by setting weights w_1 to 1.0 and w_2 to 0.0 and vice versa [Dhillon et al., 1993]. Owing to the conflicting nature of the objectives, $\overline{F}_2$ will have maximum value, when $\overline{F}_1$ is minimum. The respective minimum and maximum values of objectives thus obtained are given below when $\overline{P}_D$ is 350 MW.

$$\overline{F}_1^{\min} = 3013.519 \text{ (\$/h)}, \qquad \overline{F}_1^{\max} = 3063.917 \text{ (\$/h)}$$

$$\overline{F}_2^{\min} = 1239.593 \text{ (MW}^2\text{)}, \qquad \overline{F}_2^{\max} = 1267.589 \text{ (MW}^2\text{)}$$

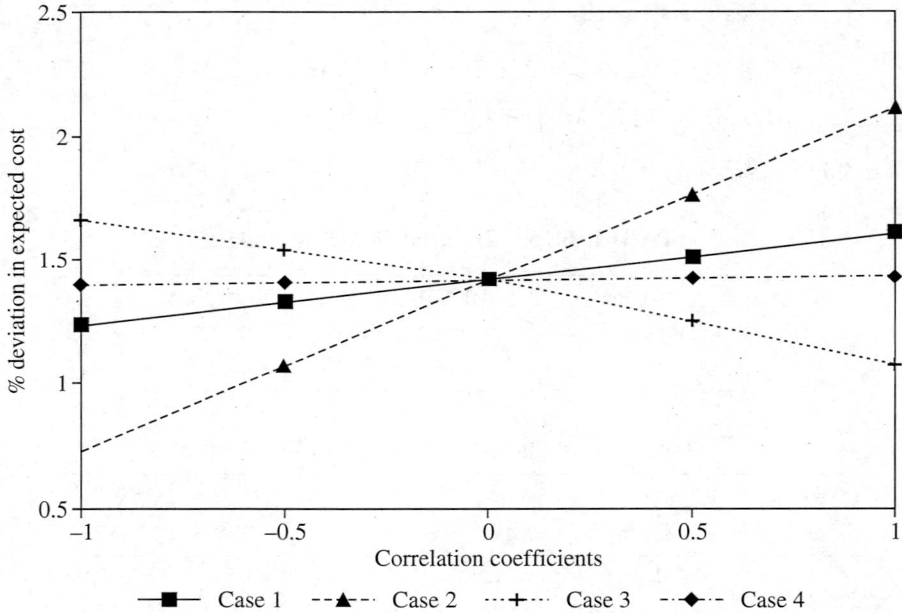

FIGURE 6.5 Percentage deviation in expected fuel cost with respect to correlation coefficients.

Over and above, an experienced decision maker can adopt any other suitable method to select the minimum and maximum values of the objectives within which he is expecting the compromised solution, like minimizing and maximizing each objective function separately subject to the required constraints.

Determination of optimal preferred solution

Four different cases are considered to realize the effect of covariance of the random variables to each other (pairwise).

Case 1: In this case all random variables are considered dependent on each other pairwise. By varying weights w_1 (from 1.0 to 0.0) and w_2 (from 0.0 to 1.0) with a regular decrement and increment of 0.1 respectively, $\overline{F}_1$ and $\overline{F}_2$ as computed for non-inferior schedules are shown in

Table 6.17, when $\overline{P}_D$ is 350 MW. Corresponding to the above schedules, the expected generations and expected transmission losses are shown in Table 6.18. Table 6.17 also shows the membership functions $\mu(\overline{F}_1)^k$ and $\mu(\overline{F}_2)^k$ of objectives $\overline{F}_1$ and $\overline{F}_2$, respectively, along with the normalized membership function μ_D^k. Percentage relative deviations of $\overline{F}_1$ and $\overline{F}_2$ from their respective deterministic values with respect to weights are shown in Figure 6.6. Best solutions so obtained from generated non-inferior solutions are given in Table 6.19 for expected demands of 350, 400, 450, and 500 MW respectively.

TABLE 6.17 Expected cost and risk with their membership functions (expected demand is 350 MW)

Sr. no.	w_1	w_2	$\overline{F}_1$ (\$/h)	$\overline{F}_2$ (MW^2)	$\mu(\overline{F}_1)$	$\mu(\overline{F}_2)$	μ_D^k
1	1.0	0.0	3013.519	1267.589	1.00000	0.00000	0.07702
2	0.9	0.1	3013.612	1265.897	0.99815	0.06044	0.08153
3	0.8	0.2	3013.963	1263.941	0.99119	0.13030	0.08638
4	0.7	0.3	3014.734	1261.658	0.97589	0.21185	0.09148
5	0.6	0.4	3016.204	1258.961	0.94672	0.30818	0.09665
6	0.5	0.5	3018.876	1255.734	0.89371	0.42345	0.10145
7	0.4	0.6	3023.724	1251.824	0.79751	0.56312	0.10479
8	0.3	0.7	3032.785	1247.033	0.61773	0.73425	0.10413
9	0.2	0.8	3050.861	1241.166	0.25906	0.94381	0.09264
10	0.1	0.9	3056.911	1239.891	0.13902	0.98936	0.08691
11	0.0	1.0	3063.917	1239.593	0.00000	1.00000	0.07702

TABLE 6.18 Expected generation schedules and transmission losses (expected demand is 350 MW)

Sr. no.	w_1	w_2	$\overline{P}_1$ (MW)	$\overline{P}_2$ (MW)	$\overline{P}_3$ (MW)	$\overline{P}_L$ (MW)
1	1.0	0.0	60.869	148.662	146.501	6.0322
2	0.9	0.1	62.952	146.829	146.013	5.7944
3	0.8	0.2	65.424	144.658	145.438	5.5195
4	0.7	0.3	68.407	142.042	144.750	5.1982
5	0.6	0.4	72.079	138.828	143.911	4.8184
6	0.5	0.5	76.714	134.781	142.868	4.3634
7	0.4	0.6	82.756	129.523	141.532	3.8113
8	0.3	0.7	90.975	122.400	139.759	3.1335
9	0.2	0.8	102.849	112.164	137.289	2.3018
10	0.1	0.9	105.000	105.215	141.906	2.1209
11	0.0	1.0	105.000	94.313	152.766	2.0785

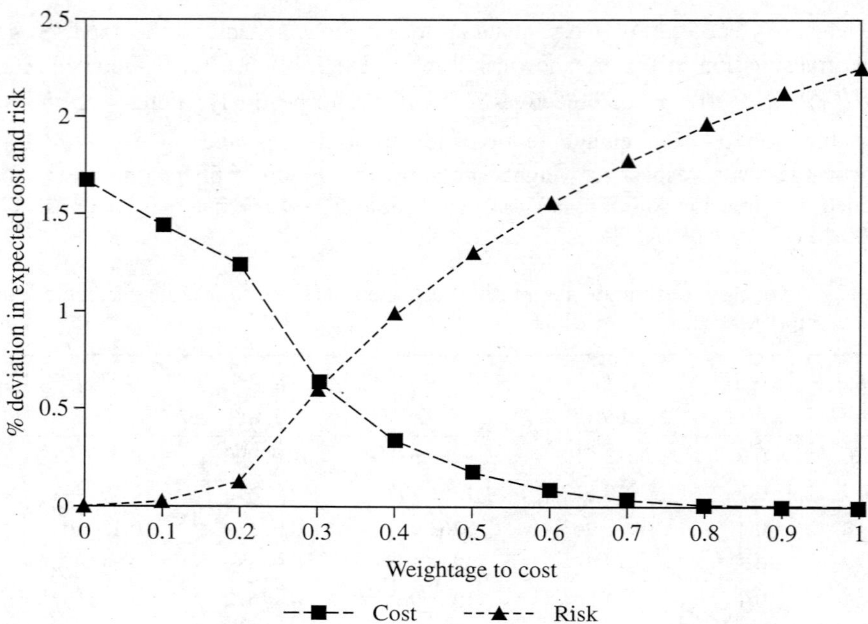

FIGURE 6.6 Percentage deviations in expected cost and risk with respect to weight, w_1.

TABLE 6.19 Comparison of results of four cases

$\overline{P}_D$ (MW)	$\overline{F}_1$ ($/h)	$\overline{F}_2$ (MW^2)	$\overline{P}_L$ (MW)	$\overline{P}_1$ (MW)	$\overline{P}_2$ (MW)	$\overline{P}_3$ (MW)
Case I						
350.0	3023.724	1251.824	3.8113	82.756	129.523	141.532
400.0	3412.855	1641.559	5.1616	93.337	150.823	161.001
450.0	3816.886	2084.501	6.5633	105.000	169.608	181.956
500.0	4232.638	2595.893	9.4991	105.000	187.694	216.805
Case II						
350.0	3022.618	1251.642	3.7855	82.160	129.996	141.630
400.0	3418.477	1635.635	4.4299	99.791	145.026	159.613
450.0	3815.815	2083.319	6.4339	105.000	169.769	181.665
500.0	4230.941	2593.945	9.3079	105.000	187.889	216.419
Case III						
350.0	2996.655	1251.488	3.7637	82.396	129.705	141.663
400.0	3380.846	1641.037	5.0971	92.917	151.043	161.138
450.0	3779.968	2083.282	6.4298	105.000	169.582	181.848
500.0	4189.928	2593.896	9.3031	105.000	187.639	216.664
Case IV						
350.0	2996.306	436.721	3.9232	82.219	135.354	136.351
400.0	3379.355	577.308	5.4529	90.929	157.121	157.403
450.0	3776.131	739.849	7.3192	99.080	179.538	178.702
500.0	4186.835	929.535	9.7847	105.000	203.656	201.128

Case 2: The results for this case are computed with the consideration that B-coefficients and power generations are independent of each other ($R_{P_iB_{ij}} = 0$; $i = 1, 2, 3, j = 1, 2, 3$), and rest of the random variables are dependent on each other. With demands 350, 400, 450 and 500 MW, the obtained 'best' solutions are given in Table 6.19.

Case 3: Cost coefficients and power generations are considered independent to each other ($R_{a_iP_i} = 0$, $R_{b_iP_i} = 0$; $i = 1, 2, 3$) along with the consideration that power generations and B-coefficients are also independent ($R_{P_iB_{ij}} = 0$; $i = 1, 2, 3, j = 1, 2, 3$), whereas expected power generations are dependent on each other. 'Best' solutions for this case are given in Table 6.19 for the same demands as in the case 2.

Case 4: All random variables are independent of each other. In this case ($R_{a_iP_i} = 0$, $R_{b_iP_i} = 0$, $R_{P_iB_{ij}} = 0$; $i = 1, 2, 3, j = 1, 2, 3$) and $R_{P_iP_j} = 0$; $i = 1, 2, 3, j = 1, 2, 3, i \neq j$) are taken. The obtained 'best' solutions are given along with the expected schedules and transmission losses in Table 6.19 for various demands under consideration. It can be concluded from Figure 6.6 that a decrease in the weightage to expected operating cost leads to a decrease in the expected risk and an increase in the expected operating cost. Comparison of four cases clearly shows an increase in expected cost as the correlation (dependence) of bivariate random variables comes into existence (Cases 1, 2, and 3) compared to independent random variables (case 4). Expected risk is always in conflict with the expected operating cost in all the cases. Expected risk is lower in the case 4 where random variables are independent of each other. Case 2 gives very small decrease in expected operating cost as well as in expected risk compared to the Case 1. This is because of fixed network configuration. Case 3 gives significant decrease in the expected cost compared to the Cases 1 and 2. However, there is insignificant decrease in expected risk compared to the ones obtained in Cases 1 and 2, respectively. This is attributed to the fact that the objective function $\overline{F_1}$ is a separable function and $R_{a_iP_i}$, $R_{b_iP_i}$ and $R_{P_iB_{ij}}$ have no effect on the objective $\overline{F_2}$. Considering, the results of all four cases, Table 6.19 shows an increase in the expected operating cost as the correlation (dependence) of each random variables increases.

6.5 STOCHASTIC ECONOMIC-EMISSION LOAD DISPATCH

The economic-emission load dispatch (EELD) problem is a multiple non-commensurable objective problem that minimizes both cost and emission together. In this section, a stochastic EELD problem is formulated with the consideration of the uncertainties in the system production cost and random nature of load demand. In addition, risk is considered as another conflicting objective to be minimized because of random load and uncertain system production cost. Validity of the method has been demonstrated by analyzing a sample system consisting of six generators.

6.5.1 Stochastic Economic-Emission Problem Formulation

In this section, the multiobjectives with the equality and inequality constraints pertaining to the power system optimization problem are described. The important non-commensurable objectives taken into account here are:

- Economic operation
- Minimal impacts on environment
- Expected deviations due to unsatisfied load

The stochastic economic emission formulation is adopted by considering fuel cost coefficients, emission coefficients and load demand as random variables. The stochastic models are converted to their deterministic equivalents by taking their expected values, with the assumption that all the random variables are normally distributed and statistically dependent on each other. Expected fuel cost and expected deviations are described in Section 6.3.1 and are given in Eqs. (6.42) and (6.51), respectively.

Expected NO$_x$ emission

The emission curve can be directly related to the cost curve through the emission rate per Mkcal, which is a constant factor for a given type of fuel. Therefore, the amount of NO$_x$ emission is given as a function of generator output P_i, which is quadratic, i.e.

$$F_2 = \sum_{i=1}^{NG} (d_i P_i^2 + e_i P_i + f_i) \tag{6.61}$$

where d_i, e_i, and f_i are emission coefficients.

A stochastic NO$_x$ emission model is formulated by considering emission coefficients and load demand at random. Using Taylor's series and taking the expectation, the expected NO$_x$ emission is obtained as

$$\overline{F}_2 = \sum_{i=1}^{NG} (\overline{d}_i \overline{P}_i^2 + \overline{e}_i \overline{P}_i + \overline{f}_i + \overline{d}_i \text{ var } (P_i) + \text{cov } (e_i, P_i) + 2\overline{P}_i \text{ cov } (d_i, P_i)) \tag{6.62}$$

where $\overline{d}_i$, $\overline{e}_i$, and $\overline{f}_i$ are expected emission coefficients.

Rewriting the above equation as

$$\overline{F}_2 = \sum_{i=1}^{NG} [(1 + C_{P_i}^2 + 2R_{d_iP_i} C_{d_i} C_{P_i}) \overline{d}_i \overline{P}_i^2 + (1 + R_{e_iP_i} C_{e_i} C_{P_i}) \overline{e}_i \overline{P}_i + \overline{f}_i] \tag{6.63}$$

where C_{d_i} and C_{e_i} are the coefficients of variation of random variables d_i and e_i respectively. $R_{d_iP_i}$ is the correlation coefficient of random variables d_i and P_i. $R_{e_iP_i}$ is the correlation coefficient of the random variables e_i and P_i.

The expected deviations due to variance in power mismatch are considered as the third objective F_3, to be minimized and are given by Eqs. (6.51a) and (6.51b).

Equality and inequality constraints

When the network configuration is fixed and the load demand is random, then the equality constraint is imposed to ensure real power balance. This is stated as

$$\overline{P}_D + \overline{P}_L - \sum_{i=1}^{NG} \overline{P}_i = 0 \tag{6.64}$$

The expected power generation is bounded as inequality constraint

$$\overline{P}_i^{\min} \le \overline{P}_i \le \overline{P}_i^{\max} \qquad (i = 1, 2, ..., NG) \tag{6.65}$$

where $\overline{P}_i^{\min}$ and $\overline{P}_i^{\max}$ are expected lower and upper limits, respectively, of generator output.

Expected transmission loss

The transmission power losses expressed through the simplified well known loss formula expression as a quadratic function of the power generation are given by

$$P_L = \sum_{i=1}^{NG} \sum_{j=1}^{NG} P_i B_{ij} P_j \tag{6.66}$$

The power generations P_i are random variables dependent on each other. B_{ij} are also considered as inaccurate B-coefficients. The expected transmission loss using Taylor series is represented as

$$\overline{P}_L = \sum_{i=1}^{NG} \sum_{j=1}^{NG} \overline{P}_i \overline{B}_{ij} \overline{P}_j + \sum_{i=1}^{NG} \overline{B}_{ii} \operatorname{var}(P_i) + \sum_{i=1}^{NG-1} \sum_{j=i+1}^{NG} 2\overline{B}_{ij} \operatorname{cov}(P_i, P_j) \tag{6.67}$$

This, on simplification, can be written as

$$\overline{P}_L = \sum_{i=1}^{NG} (1 + C_{P_i}^2) \overline{B}_{ii} \overline{P}_i^2 + \sum_{i=1}^{NG} \sum_{\substack{j=1 \\ j \neq i}}^{NG} (1 + R_{P_i P_j} C_{P_i} C_{P_j}) \overline{P}_i \overline{B}_{ij} \overline{P}_j \tag{6.68}$$

The deterministic equivalent of stochastic economic emission problem is defined by considering three objectives: (i) expected fuel cost, (ii) expected NO_x emission, and (iii) expected deviations due to unsatisfied load demand subject to expected equality.

Minimize $\quad [\overline{F}_1, \overline{F}_2, \overline{F}_3]^T \quad$ (6.69a)

subject to

$$\sum_{i=1}^{NG} \overline{P}_i = \overline{P}_D + \overline{P}_L \tag{6.69b}$$

$$\overline{P}_i^{\min} \le \overline{P}_i \le \overline{P}_i^{\max} \qquad (i = 1, 2, ..., NG) \tag{6.69c}$$

where $\overline{F}_1$, $\overline{F}_2$, and $\overline{F}_3$ are the expected values of the objective functions to be minimized over the set of admissible decision variable $\overline{P}_i$.

6.5.2 Solution Approach

The weighting method is used to generate the non-inferior solutions of the multiobjective optimization problem. In this method the problem is converted into a scalar optimization problem and is given below as

Minimize

$$\sum_{k=1}^{3} w_k \overline{F}_k \tag{6.70a}$$

subject to

$$\sum_{i=1}^{NG} \overline{P}_i = \overline{P}_D + \overline{P}_L \tag{6.70b}$$

$$\overline{P}_i^{\min} \le \overline{P}_i \le \overline{P}_i^{\max} \qquad (i = 1, 2, ..., NG) \tag{6.70c}$$

$$\sum_{k=1}^{3} w_k = 1, \; w_k \ge 0 \tag{6.70d}$$

where w_k are the levels of the weighting coefficients. This approach yields meaningful result to the DM when solved many times for different values of $w_k(k = 1, 2, 3)$. The values of the weighting coefficients vary from 0 to 1.

To solve the scalar optimization problem defind by Eqs. (6.70), the Lagrangian function is defined as

$$L(\overline{P}_i, \lambda) = \sum_{k=1}^{3} w_k \overline{F}_k + \lambda \left(\overline{P}_D + \overline{P}_L - \sum_{i=1}^{NG} \overline{P}_i \right) \tag{6.71}$$

where λ is Lagrangian multiplier.

The necessary conditions to minimize the unconstrained Lagrangian function are:

$$\frac{\partial L}{\partial \overline{P}_i} = \sum_{k=1}^{3} w_k \frac{\partial \overline{F}_k}{\partial \overline{P}_i} + \lambda \left[\frac{\partial \overline{P}_L}{\partial \overline{P}_i} - 1 \right] = 0 \qquad (i = 1, 2, ..., \text{NG}) \tag{6.72a}$$

$$\frac{\partial L}{\partial \lambda} = \overline{P}_D + \overline{P}_L - \sum_{i=1}^{NG} \overline{P}_i = 0 \tag{6.72b}$$

where

$$\frac{\partial \overline{F}_1}{\partial \overline{P}_i} = (1 + C_{P_i}^2 + 2R_{a_iP_i}\, C_{a_i} C_{P_i})\, \overline{a}_i \overline{P}_i + (1 + R_{b_iP_i}\, C_{b_i} C_{P_i})\, \overline{b}_i \tag{6.72c}$$

$$\frac{\partial \overline{F}_2}{\partial \overline{P}_i} = (1 + C_{P_i}^2 + 2R_{d_iP_i}\, C_{d_i} C_{P_i})\, \overline{d}_i \overline{P}_i + (1 + R_{e_iP_i}\, C_{e_i} C_{P_i})\, \overline{e}_i \tag{6.72d}$$

$$\frac{\partial \overline{F}_3}{\partial \overline{P}_i} = 2C_{P_i}^2 \overline{P}_i + \sum_{\substack{j=1 \\ j \neq i}}^{NG} R_{P_iP_j} C_{P_i} C_{P_j} \overline{P}_j \tag{6.72e}$$

$$\frac{\partial \overline{P}_L}{\partial \overline{P}_i} = 2(1 + C_{P_i}^2)\, \overline{B}_{ii} \overline{P}_i + \sum_{\substack{j=1 \\ j \neq i}}^{NG} [1 + R_{P_i,P_j} C_{P_i} C_{P_j}]\, \overline{B}_{ij} \overline{P}_j \tag{6.72f}$$

The scalar optimization problem is solved using the Newton–Raphson algorithm. The size of the formulated Hessian matrix is the same as that for the deterministic problem, because single objective function is solved in both cases with the same number of constraints. To implement the Newton–Raphson method, the following equation is solved iteratively till no further improvement in decision variables is achieved.

$$\begin{bmatrix} \nabla_{PP} L & \nabla_{P\lambda} L \\ (\nabla_{P\lambda} L)^T & \nabla_{\lambda\lambda} L \end{bmatrix} \begin{bmatrix} \Delta P \\ \Delta \lambda \end{bmatrix} = \begin{bmatrix} -\nabla_P L \\ -\nabla_\lambda L \end{bmatrix} \tag{6.73}$$

Hessian matrix elements can be obtained from the above equations by differentiating with respect to control variables, one by one.

$$\frac{\partial^2 L}{\partial \overline{P}_i^2} = \sum_{k=1}^{3} w_k \frac{\partial^2 \overline{F}_k}{\partial \overline{P}_i^2} + \lambda \left(\frac{\partial^2 \overline{P}_L}{\partial \overline{P}_i^2} - 1 \right) \qquad (i = 1, 2, ..., \text{NG}) \tag{6.74a}$$

$$\frac{\partial^2 L}{\partial \overline{P}_i \partial \overline{P}_j} = w_3 \frac{\partial^2 \overline{F}_3}{\partial \overline{P}_i \partial \overline{P}_j} + \lambda \frac{\partial^2 \overline{P}_L}{\partial \overline{P}_i \partial \overline{P}_j} \quad (i = 1, 2, ..., \text{NG}; \; j = 1, 2, ..., \text{NG}; \; i \neq j) \tag{6.74b}$$

$$\frac{\partial^2 L}{\partial \overline{P}_i \partial \lambda} = \frac{\partial \overline{P}_L}{\partial \overline{P}_i} - 1 \quad (i = 1, 2, ..., \text{NG}) \tag{6.74c}$$

$$\frac{\partial^2 L}{\partial \lambda^2} = 0 \tag{6.74d}$$

where

$$\frac{\partial^2 \overline{F}_1}{\partial \overline{P}_i^2} = 2(1 + C_{P_i}^2 + 2R_{a_i P_i} C_{a_i} C_{P_i})\, \overline{a}_i \tag{6.74e}$$

$$\frac{\partial^2 \overline{F}_2}{\partial \overline{P}_i^2} = 2(1 + C_{P_i}^2 + 2R_{d_i P_i} C_{d_i} C_{P_i})\, \overline{d}_i \tag{6.74f}$$

$$\frac{\partial^2 \overline{F}_3}{\partial \overline{P}_i^2} = 2C_{P_i}^2 \tag{6.74g}$$

$$\frac{\partial^2 \overline{F}_3}{\partial \overline{P}_i \partial \overline{P}_j} = R_{P_i P_j} C_{P_i} C_{P_j} \quad (i \neq j) \tag{6.74h}$$

$$\frac{\partial^2 \overline{P}_L}{\partial \overline{P}_i^2} = 2(1 + C_{P_i}^2)\, \overline{B}_{ii} \tag{6.74i}$$

$$\frac{\partial^2 \overline{P}_L}{\partial \overline{P}_i \partial \overline{P}_j} = 2(1 + R_{P_i P_j} C_{P_i} C_{P_j})\, \overline{B}_{ij} \quad (i \neq j) \tag{6.74j}$$

Algorithm 6.3 can be used to generate the non-inferior solutions. Use Eqs. (6.72) and (6.74) to obtain the Hessian and Jacobian matrix elements.

6.5.3 Test System and Results

The validity of the proposed method is illustrated on a six-generator sample system. The fuel cost and NO_x emission equations are given in Tables 6.20 and 6.21 and the average expected transmission loss coefficients are given in Table 6.22 [Hill and Stevenson, 1968]. The operating limits of the generators are given in Table 6.23.

TABLE 6.20 Expected fuel cost (₹/h) equations

$$\overline{F}_{11} = 0.15247\, \overline{P}_1^2 + 38.53973\, \overline{P}_1 + 756.79886$$
$$\overline{F}_{12} = 0.10587\, \overline{P}_2^2 + 46.15916\, \overline{P}_2 + 451.32513$$
$$\overline{F}_{13} = 0.02803\, \overline{P}_3^2 + 40.39655\, \overline{P}_3 + 1049.99770$$
$$\overline{F}_{14} = 0.03546\, \overline{P}_4^2 + 38.30553\, \overline{P}_4 + 1243.53110$$
$$\overline{F}_{15} = 0.02111\, \overline{P}_5^2 + 36.32782\, \overline{P}_5 + 1658.56960$$
$$\overline{F}_{16} = 0.01799\, \overline{P}_6^2 + 38.27041\, \overline{P}_6 + 1356.65920$$

TABLE 6.21 Expected NO_x emission (kg/h) equations

$$\overline{F}_{21} = 0.00419\,\overline{P}_1^2 + 0.32767\,\overline{P}_1 + 13.85932$$
$$\overline{F}_{22} = 0.00419\,\overline{P}_2^2 + 0.32767\,\overline{P}_2 + 13.85932$$
$$\overline{F}_{23} = 0.00683\,\overline{P}_3^2 + 0.54551\,\overline{P}_3 + 40.2669$$
$$\overline{F}_{24} = 0.00683\,\overline{P}_4^2 + 0.54551\,\overline{P}_4 + 40.2669$$
$$\overline{F}_{25} = 0.00461\,\overline{P}_5^2 + 0.51116\,\overline{P}_5 + 42.89553$$
$$\overline{F}_{26} = 0.00461\,\overline{P}_6^2 + 0.51116\,\overline{P}_6 + 42.89553$$

TABLE 6.22 Expected loss coefficients

0.002022	−0.000286	−0.000534	−0.000565	−0.000454	0.000103
−0.000286	0.003243	0.000016	−0.000307	−0.000422	−0.000147
−0.000533	0.000016	0.002085	0.000831	0.000023	−0.000270
−0.000565	−0.000307	0.000831	0.001129	0.000113	−0.000295
−0.000454	−0.000422	0.000023	0.000113	0.000460	−0.000153
0.000103	−0.000147	−0.000270	−0.000295	−0.000153	0.000898

TABLE 6.23 Operating limits

Generator no.	*Lower limit* $\overline{P}_i^{min}$ (MW)	*Upper limit* $\overline{P}_i^{max}$ (MW)
1	10	125
2	10	150
3	35	225
4	35	210
5	130	325
6	125	315

In addition, the following values of the coefficients of variation and correlation coefficients are assumed in the study:

$$C_{a_i} = 0.1,\; C_{b_i} = 0.1,\; C_{P_i} = 0.1; \qquad (i = 1, 2, ..., 6)$$
$$R_{P_iP_j} = 1.0 \qquad (i = 1, 2, ..., 6;\; j = 1, 2, ..., 6;\; i \neq j)$$

Owing to the existence of variances and covariances of the random variables in the formulation of the problem, an analysis of the variations in objectives $\overline{F}_1$ and $\overline{F}_2$ seems necessary. In the study, the covariance of bivariate random variables is considered positive or negative. Since the covariance is represented by correlation coefficients, the correlation coefficients are varied from −1.0 to 1.0 in steps of 0.5. One pair of random variables is considered at a time, while the rest of random variables are considered independent of each other (uncorrelated). By taking the weights w_1, w_2, and w_3 as 0.4, 0.3, and 0.3, respectively, the percentage relative deviations in $\overline{F}_1$ and $\overline{F}_2$

from their deterministic values with respect to $R_{a_iP_i}$, $R_{b_iP_i}$, $R_{d_iP_i}$, $R_{e_iP_i}$, and $R_{P_iP_j}$ $(i \neq j)$ are shown, one by one, in Figures 6.7 and 6.8 respectively, for an expected demand of 700 MW. It is observed that (i) there is an increase in the percentage relative deviation in $\overline{F_1}$ and $\overline{F_2}$ as the value of $R_{P_iP_j}$ $(i \neq j)$ is changed from a positive value to a negative value, (ii) there is a decrease in the

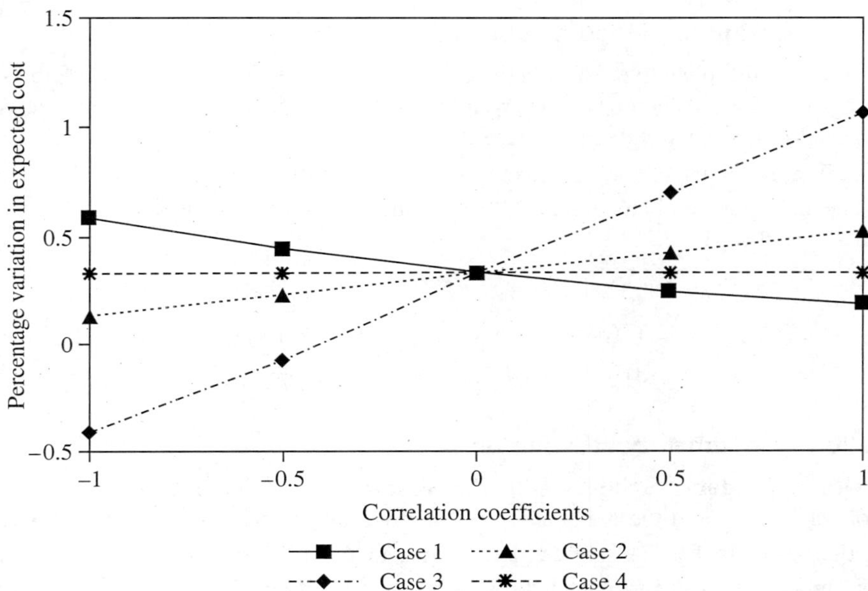

FIGURE 6.7 Percentage deviation in expected cost with respect to correlation coefficients.

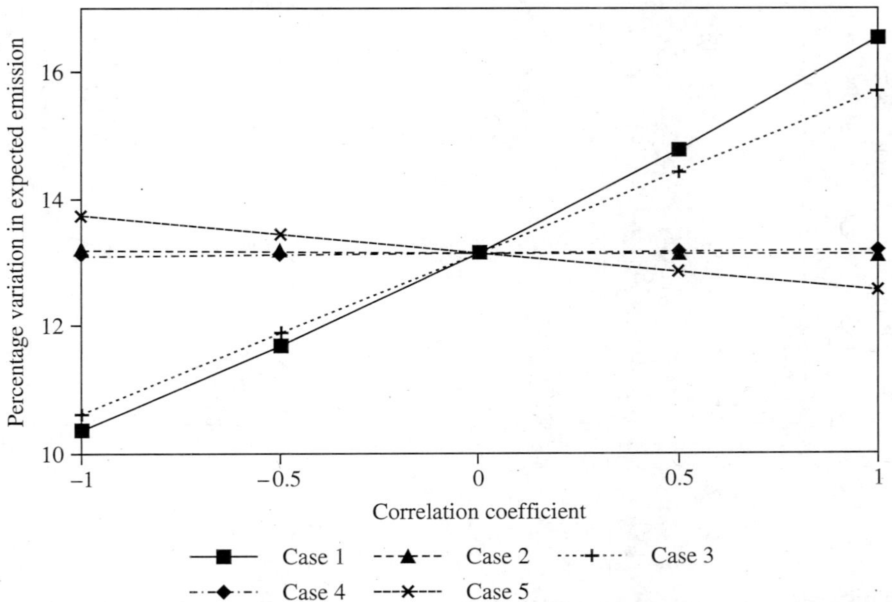

FIGURE 6.8 Percentage variation in expected emission with respect to correlation coefficients.

percentage relative deviation in $\overline{F}_1$ and a small change in the percentage deviation in $\overline{F}_2$ as the values of $R_{a_iP_i}$ and $R_{b_iP_i}$ are changed from positive to negative values, and (iii) there is a decrease in the percentage deviation in $\overline{F}_2$ and a small change in the percentage deviation in $\overline{F}_1$ as the values of $R_{d_iP_i}$ and $R_{e_iP_i}$ are changed in the above manner.

Minimum and maximum values of objectives

Minimum values of the objectives are obtained by giving full weightage to one of the objectives and neglecting the others. When the given weight value is 1.0, it means full weightage is given to the objective and when the weightage is zero the objective is neglected.

The conflicting nature of the objectives, $\overline{F}_2$ and $\overline{F}_3$, will have maximum values when $\overline{F}_1$ is minimum. The minimum and maximum values of the objectives obtained in this way are given below when P_D is 700 MW.

$$\overline{F}_1^{\min} = 38932.88 \text{ ₹/h} \qquad \overline{F}_1^{\max} = 39964.13 \text{ ₹/h}$$

$$\overline{F}_2^{\min} = 491.320 \text{ kg/h} \qquad \overline{F}_2^{\max} = 582.766 \text{ kg/h}$$

$$\overline{F}_3^{\min} = 5291.854 \text{ MW}^2 \qquad \overline{F}_3^{\max} = 5613.872 \text{ MW}^2$$

Determination of optimal 'best' solution

(a) Case with dependent variables: In this case all the random variables are considered dependent on each other and the weights w_1, w_2, and w_3 are varied in the range 0.0 to 1.0 in such a way that their sum is 1.0. The percentage deviation in objectives $\overline{F}_1$, $\overline{F}_2$, and $\overline{F}_3$ from their expected minimum values are shown in Figures 6.9, 6.10, and 6.11, respectively, with respect to

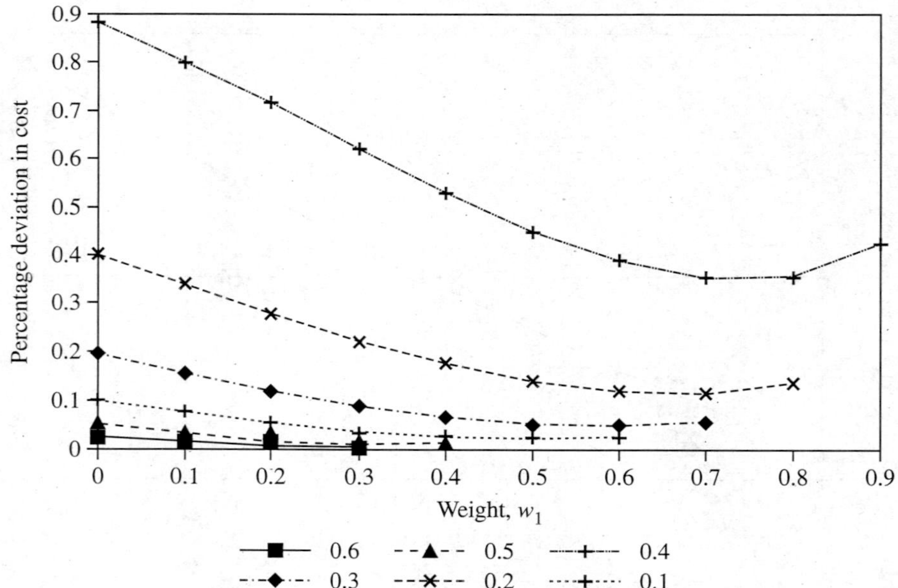

FIGURE 6.9 Percentage deviation in expected cost with respect to weights.

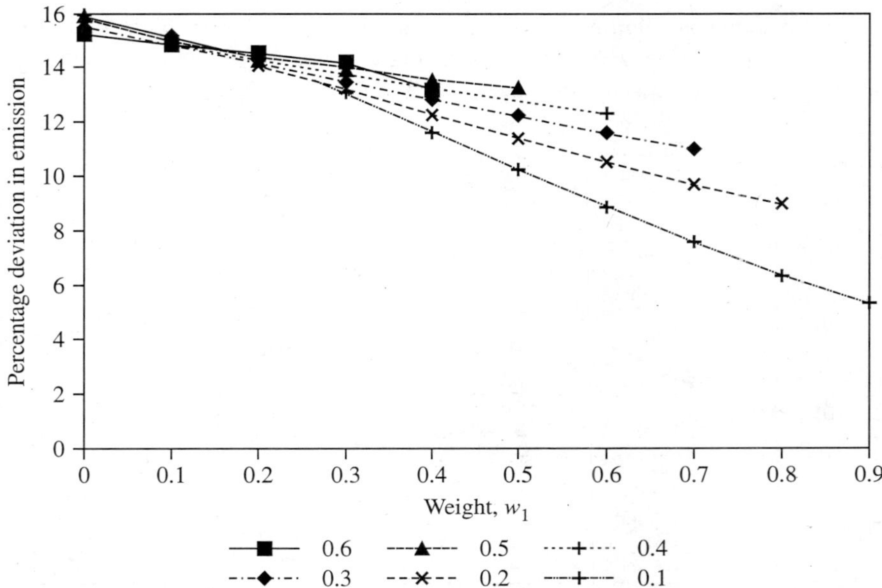

FIGURE 6.10 Percentage deviation in expected emission with respect to weights.

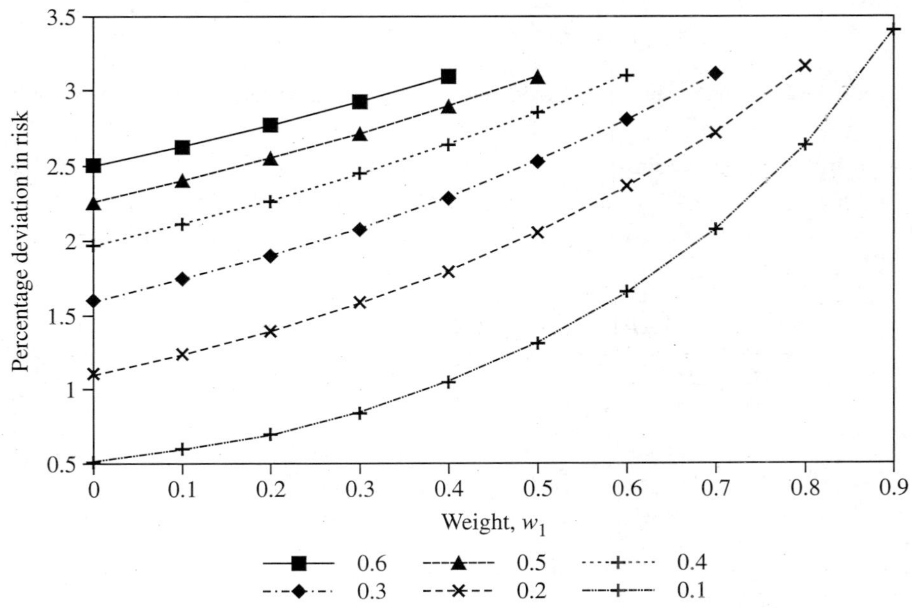

FIGURE 6.11 Percentage deviation in risk with respect to weights.

the various combinations of weights when the expected demand is 700 MW. The conflicting nature of the objectives can be observed from these diagrams. The normalized membership function μ_D^k, obtained from membership functions $\mu(\overline{F}_1)$, $\mu(\overline{F}_2)$, and $\mu(\overline{F}_3)$ of objective functions $\overline{F}_1$, $\overline{F}_2$, and

$\overline{F}_3$ respectively, is shown in Figure 6.12 for each non-inferior solution. The 'best' solutions are given in Table 6.24 and corresponding to the schedules obtained, the expected power generations are given in Table 6.25 for expected demands of 500, 700, and 900 MW.

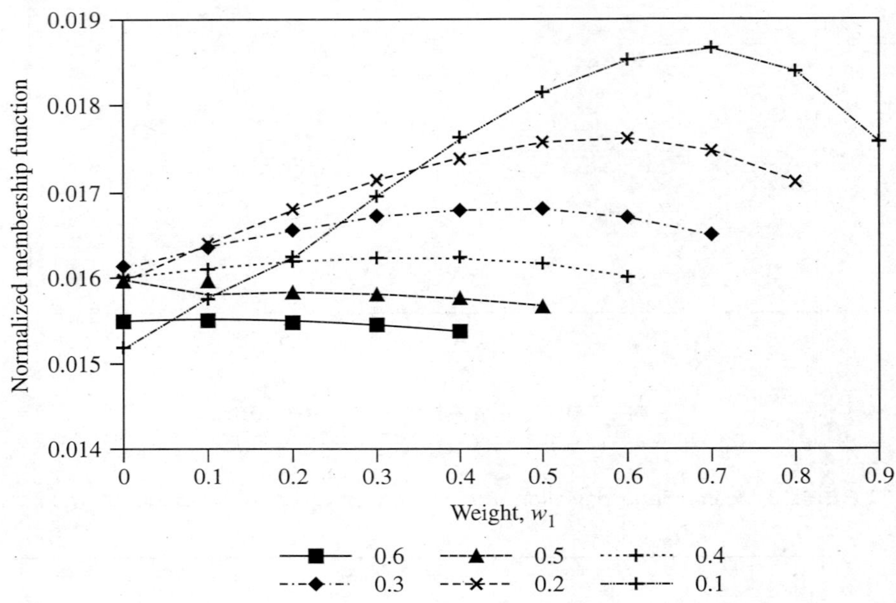

FIGURE 6.12 Variation in normalized membership function with respect to weights.

(b) Case with independent variables: In this case all the random variables are considered independent of each other $R_{a_iP_i} = R_{b_iP_i} = R_{d_iP_i} = R_{e_iP_i} = R_{P_iP_j} = 0.0$ ($i = 1, 2, ..., 6; j = 1, 2, ..., 6$). The 'best' solutions so obtained are given in Tables 6.24 and 6.25 for expected demands of 500, 700, and 900 MW respectively.

TABLE 6.24 Best optimal solutions

Demand (MW)	$\overline{F}_1$ (₹/h)	$\overline{F}_2$ (kg/h)	$\overline{F}_3$ (MW^2)	$\overline{P}_L$ (MW)
Case 1				
500.0	28550.15	312.513	2674.567	17.162
700.0	39070.74	528.447	5401.182	34.927
900.0	50807.24	864.060	9110.655	54.498
Case 2				
500.0	28476.63	287.483	558.758	20.799
700.0	39010.74	493.977	1114.285	39.669
900.0	50854.86	800.629	1861.073	65.032

TABLE 6.25 Expected optimal generation schedules

Demand (MW)	$\overline{P}_1$ (MW)	$\overline{P}_2$ (MW)	$\overline{P}_3$ (MW)	$\overline{P}_4$ (MW)	$\overline{P}_5$ (MW)	$\overline{P}_6$ (MW)
Case 1						
500.0	59.873	39.651	35.000	72.397	185.241	125.000
700.0	85.924	60.963	53.909	107.124	250.503	176.504
900.0	122.004	86.523	59.947	140.959	325.000	220.063
Case 2						
500.0	59.671	41.410	51.870	83.265	157.831	126.752
700.0	89.098	65.475	69.713	116.079	223.596	175.709
900.0	124.610	92.965	87.093	152.743	286.843	220.777

The conventional economic thermal power dispatch method allocates generation schedules to the individual thermal generating units based upon deterministic cost function or NO_x emission function, ignoring inaccuracies and uncertainties during estimation. Furthermore, the load demand is considered constant, but in practice it is random. Such generation schedules result in the lowest cost or lowest emission, but the cost or emission always associated with such a schedule has a relatively large variance and covariance, which can be interpreted as risk measure.

Conventional dispatch schedules, therefore, render the solution non-optimal to a power system analyst who would like to avoid risks. The feasibility of quantitative representation of inaccuracies and uncertainties of the input data and power demand for the power dispatch problem in terms of probability and statistics, has been investigated in this chapter. The economic dispatch problem is formulated as the multiobjective optimization problem, considering expected cost and risk measure as two conflicting objectives to be minimized. The solution set of such a formulated problem is non-inferior due to the conflicting nature of objectives. Fuzzy set theory has been applied to get the efficient optimal dispatch from the non-inferior set.

Through this study, a theoretical basis and methodological grounding for optimal economical dispatch problem in a unified multiobjective framework are established. The proposed formulation enables the DM to consider the inaccuracies and uncertainties in the economic dispatch procedure. It allows explicit trade-off between the operating cost of units and the risk levels, and provides the efficient optimal solution from the non-inferior solutions.

6.6 MULTIOBJECTIVE OPTIMAL THERMAL POWER DISPATCH—RISK/DISPERSION METHOD

Until recently, most optimization models have been formulated in terms of minimizing (or maximizing) a single scalar-valued objective function. Nevertheless, the order of the day shows a definite trend towards formulating such problems in terms of multiple objectives. The consideration of many objectives in the planning process accomplishes three major improvements in the problem solving. First, multiobjective programming promotes a more appropriate role for the decision-making process. Second, a wider range of alternatives is usually identified when a multiobjective methodology is employed. Third, models as well as the analysts' perception of a problem become more realistic if many objectives are considered.

Since mathematical models are an idealization of actual system models, the various system responses are bound to deviate. The magnitude of such deviations can be best evaluated by sensitivity analysis because:

(a) The stability of the optimal solution may be critically dependent on changes in the model parameters.
(b) Some parameters may be controllable and, therefore, it is important to know what effects may result from changing their values.
(c) Other parameters may be estimated more accurately if the solution is critically dependent on one of them.

The multiobjective formulation is amenable to sensitivity analysis by specifying a sensitivity index, that is, a function which indicates the relative size of the perturbations in the solution due to variations in the parameters. The specified index may be included as an additional objective to be minimized in a multiobjective system. Then the problem is solved to find the preferred solution [Osyczka and Davies, 1984]. A second way of applying sensitivity analysis to a multiobjective problem is to solve the system without treating the sensitivity index as an objective and then to evaluate the sensitivity after a preferred solution of the original system is found [Rarig and Haimes, 1983; Kaunas and Haimes, 1985].

A classical economic dispatch problem is formulated as a multiobjective optimization problem considering two non-commensurable objectives to be minimized, namely the operating cost and the impacts on the environment. The formulated problem adopts an ε-constraint form, which allows explicit trade-offs between objective levels for each non-inferior solution [Rarig and Haimes, 1983]. The effects of random variations in the model parameters of the optimal solution to nonlinear programming lead to a sensitivity measure called *dispersion* (Ω). The index is interpreted as a first-order approximation to the standard deviation in the optimal solution of the nonlinear programs. A sensitivity trade-off $\nabla_z\Omega$ (Appendix D) is considered that gives an explicit representation of the trade-offs between the sensitivity and the objective levels. Results are obtained for two sample systems having three and six generators, respectively.

6.6.1 Multiobjective Optimization Problem Formulation

In the multiobjective optimization problem formulation, two important non-commensurable objectives in an electrical thermal power system are considered. These are economy and environmental impacts. The multiobjective optimization problem is defined as

$$\text{Minimize} \qquad F_1(P) = \sum_{i=1}^{N} (a_i P_i^2 + b_i P_i + c_i) \tag{6.75a}$$

$$\text{Minimize} \qquad F_2(P) = \sum_{i=1}^{N} (d_i P_i^2 + e_i P_i + f_i) \tag{6.75b}$$

$$\text{subject to} \qquad \sum_{i=1}^{N} P_i - (P_D + P_L) = 0 \tag{6.75c}$$

$$P_i^{\min} \le P_i \le P_i^{\max} \qquad (i = 1, 2, \ldots, N) \tag{6.75d}$$

where

a_i, b_i, and c_i are cost coefficients of the *i*th generating unit

d_i, e_i, and f_i are emission coefficients

P_D is the power demand to be met

P_L is the transmission losses, which are approximated in terms of *B*-coefficients as

$$P_L = \sum_{i=1}^{N} \sum_{j=1}^{N} P_i B_{ij} P_j$$

$P_i^{\min}$ is the lower operating generation limit

$P_i^{\max}$ is the upper operating generation limit

N is the number of generators in the power system

$F_1(P)$, $F_2(P)$ are the objective functions to be minimized over the set of admissible decision vector *P*.

6.6.2 The ε-Constraint Method

To generate non-inferior solutions to the multiobjective optimization problem, the ε-constraint method is used [Haimes, 1977]. For the ε-constraint method, one of the objective functions constitutes the primary objective function and all other objects act as constraints. To be more specific, this procedure is implemented by replacing one objective in the problem as defined by Eq. (4.5) with one constraint.

$$\text{Minimize} \quad [F_1(P)] \tag{6.76a}$$

subject to

$$F_j(P) \le \varepsilon_j \; (j = 2) \tag{6.76b}$$

$$\sum_{i=1}^{N} P_i - (P_D + P_L) = 0 \tag{6.76c}$$

$$P_i^{\min} \le P_i \le P_i^{\max} \qquad (i = 1, 2, ..., N) \tag{6.76d}$$

where ε_j is the maximum tolerable objective level. The value of ε_j is chosen for which the objective constraints in problem defined by Eqs. (6.76) are binding at the optimal solution. The level of ε_j is varied parametrically to evaluate the impact on the single objective function $F_1(P)$. To find the solution, the constrained problem of Eqs. (6.76) is converted into an unconstrained problem. The generalized Lagrangian *L* is formed as

$$L = F_1(P) + \lambda_{12}(F_2(P) - \varepsilon_2) - \mu\left(\sum_{i=1}^{N} P_i - P_L - P_D\right) \tag{6.77}$$

where λ_{12} and μ are the Lagrange multipliers. The subscript 12 denotes that λ is Lagrange multiplier associated with the second constraint, where the prime objective function is $F_1(P)$.

The necessary conditions for the optimal solution of Eq. (6.77) are

$$\frac{\partial L}{\partial P_i} = \frac{\partial F_1(P)}{\partial P_i} + \lambda_{12}\frac{\partial F_2(P)}{\partial P_i} - \mu\left(1 - \frac{\partial P_L}{\partial P_i}\right) = 0 \qquad (i = 1, 2, ..., N) \tag{6.78a}$$

$$\frac{\partial L}{\partial \lambda_{12}} = F_2(P) - \varepsilon_2 = 0 \tag{6.78b}$$

$$\frac{\partial L}{\partial \mu} = P_D + P_L - \sum_{i=1}^{N} P_i = 0 \tag{6.78c}$$

The optimal solution must satisfy the above Kuhn-Tucker conditions besides the following main condition:

$$\lambda_{12}(F_2(P) - \varepsilon_2) = 0.0; \quad \lambda_{12} > 0.0 \tag{6.78d}$$

The Lagrange multipliers related to the objectives, as constraints may be zero or non-zero. The set of non-zero Lagrange multipliers correspond to the non-inferior set of solutions. The set of non-zero Lagrange multipliers represents the set of trade-off ratios between the principle objective and each of the constraining objectives respectively. The system given by Eq. (6.77) is solved for R values of ε_2 using the Newton–Raphson method. Only those values of $\lambda_{12}^r > 0.0$ which correspond to the active constraints $F_2^r(P) = \varepsilon_2^r$, $r = 1, 2, ..., R$ are considered, since they belong to the non-inferior solution [Haimes, 1977].

In the problem, the initial value of ε_2 is taken such that $\varepsilon_2 > F_2^{\min}$ and $\varepsilon_2 < F_2^{\max}$. Since cost and emission are of conflicting nature, the value of objective F_2 will be maximum, when the value of F_1 objective is minimum and vice versa. So, minimum and maximum values of F_1 and F_2 are obtained by performing economic dispatch and minimum emission dispatch separately.

Similarly, for more than two objectives, the trade-off functions λ_{ij} $(i \neq j)$, can be generated with respect to the ith prime objective, by varying ε_j while other ε_k $(k = j)$ are fixed at some level. The ith prime objective can be replaced by the jth objective and the solution procedure is repeated for more information. Algorithm 6.2 can be implemented for the generation of non-inferior solution.

6.6.3 Parameter Sensitivity

In practice, the values of the objective and the constraints of the problem defined by Eqs. (6.76) will actually depend on the decision variable P_i. Usually each equation is formulated in terms of the decision variables and other parameters. Therefore, the values of the objective and constraint functions depend not only on the values of the decision variables but also on the values of coefficients a_i, b_i, c_i, d_i, e_i, f_i and B_{ij} $(i = 1, 2, ..., N;\ j = 1, 2, ..., N)$. The values for these coefficients are approximated by estimation, statistical averaging, past experience, and are therefore prone to error.

In the light of the above consideration, the multiobjective optimization problem as given by the problem defined by Eqs. (6.76) is reformulated by assuming that, in addition to the decision variables $P_i (i = 1, 2, ..., N)$, the objective function is dependent on a set of cost parameters $x_i (i = 1, 2, ..., 3N)$ and the constrained objective along with the problem constraints is dependent on a technology parameter set $y_i (i = 1, 2, ..., m)$.

Minimize $\quad [F_1(P, x)] \tag{6.79a}$

subject to $\quad F_j(P, y) \le \varepsilon_j \quad (j = 2) \tag{6.79b}$

$$q(P, y) = u \tag{6.79c}$$

where

$$x = [a_1,..,a_N,\quad b_1,..,b_N,\quad c_1,..,c_N]^T$$

$$y = [d_1,..,d_N,\quad e_1,..,e_N,\quad f_1,..,f_N,\quad B_{11},\quad B_{12},\ ...,B_{NN}]^T$$

$$q(P,\ y) = \sum_{i=1}^{N} P_i - P_L \tag{6.79d}$$

$$u = P_D \quad \text{and} \quad m = 3N + (N(N + 1))/2$$

6.6.4 Risk Index and Sensitivity Trade-Offs

Suppose the problem defined by Eqs. (6.79) admits a local solution P^* and satisfies the sufficiency conditions for a strict local minimum. Then, for a sufficiently small neighbourhood about z, the optimal P^* to problem (6.79) is the same [Rarig and Haimes, 1983] as the solution to:

$$\text{Minimize} \qquad [F(P,\ x)] \tag{6.80a}$$

$$\text{subject to} \qquad g(P,\ y) = z \tag{6.80b}$$

where the set of binding constraints set $F_2(P,\ y)$ and $q(P,\ y)$ by vector equation $g(P,\ y) = z$ ($z = [\varepsilon_2|u]^T$). The dependence of the optimal solution F^* to the problem of Eqs. (6.80) on the parameter sets x, y, z can be expressed by writing

$$F^* = F^*(x,\ y,\ z) = F^*(w,\ z) \tag{6.81}$$

where $w = [\ x \mid y\ ]^T$.

The parameter sets x and y are the approximated values of the actual unknown parameters. Keeping this in view, x and y can be treated as random variables. Therefore, w_i are considered random variables with means and variances as

$$\overline{w}_i = E(w_i) \tag{6.82a}$$

$$\sigma_i^2 = \text{var}(w_i) \qquad \text{where } i = 1,\ 2,\ ...,\ 3N + m. \tag{6.82b}$$

If it is assumed that z is a deterministic variable, then the optimal solution F^* to the problem defined by Eqs. (6.80) can be considered as a function of the random variable w_i, i.e. $F^* = F^*(w)$.

By ignoring higher terms, Taylor series expansion about the mean is

$$F^* = F^*(\overline{w}_i) + \sum_{i=1}^{L} \left[\frac{\partial F^*}{\partial w_i}\right](w_i - \overline{w}_i) \tag{6.83}$$

where $L = 3N + m$.

Taking the expected value and variance of the above expression, a first-order approximation to the mean values and variances of random variable F^* is

$$E(F^*) = F^*(w)$$

$$\text{var}(F^*) = \sum_{i=1}^{L} \left(\frac{\partial F^*}{\partial w_i}\right)^2 \sigma_i^2 \tag{6.84}$$

where the partial derivatives are evaluated at the point $w_i = \overline{w}_i$.

A new scalar sensitivity variable to be the standard deviation of F^* is defined. This can be interpreted as a measure of dispersion of the optimal solution F^* due to uncertainties in the cost and technology parameters.

By definition:

$$\Omega = \sqrt{\sum_{i=1}^{L} \left(\frac{\partial F^*}{\partial w_i}\right)^2 \sigma_i^2} \tag{6.85}$$

Since ε_j corresponds to a particular value of z_j, the trade-off between the sensitivity and objective levels, $\partial\Omega/\partial\varepsilon_j$ is obtained by differentiating (6.85) with respect to z_j, i.e.

$$\frac{\partial\Omega}{\partial z_j} = \left[\sum_{i=1}^{L} \sigma_i^2 \left(\frac{\partial F^*}{\partial w_i}\right)\left(\frac{\partial^2 F^*}{\partial z_j \partial w_i}\right)\right]\frac{1}{\Omega} \tag{6.86a}$$

where

$$\nabla_w F^* = [\lambda^* \nabla_y g \mid \nabla_x F] \tag{6.86b}$$

$$\nabla_{zw} F^* = \begin{bmatrix} \sum_{i=1}^{k} \lambda \nabla_{py} g_i & (\nabla_y g)^T \\ \nabla_{Px} F^* & 0 \end{bmatrix} \begin{bmatrix} \nabla_{PP} L & (\nabla_P g)^T \\ \nabla_P g & 0 \end{bmatrix}^{-1} \begin{bmatrix} 0 \\ I \end{bmatrix} \tag{6.86c}$$

where k is number of binding constraints.

For simplicity, the sensitivity trade-off ratio can be evaluated as

$$\frac{\partial F_1}{\partial \Omega} = \frac{\dfrac{\partial F_1}{\partial \varepsilon_j}}{\dfrac{\partial \Omega}{\partial \varepsilon_j}} \quad \text{where} \quad \frac{\partial \Omega}{\partial \varepsilon_j} \neq 0 \tag{6.87}$$

As the number of objectives increases, the trade-offs of sensitivity with respect to objective levels, ε_j^s, are determined. These trade-offs are the components of the vector z_j, which are calculated from Eq. (6.86a). The calculated trade-offs are submitted to the decision maker to select the preferred solution.

Derivatives of $\nabla_w F^*$ and $\nabla_{zw} F^*$

In order to calculate $\nabla_w F^*$ and $\nabla_{zw} F^*$, consider the following optimization problem:

Minimize $F(P, x)$ (6.88a)

subject to $g(P, y) = z$ (6.88b)

In order to satisfy the second-order sufficiency conditions, the following system of equations, the so-called Kuhn-Tucker necessary conditions are to be solved:

$$\nabla_P F(P, x) + \lambda \nabla_P g(P, y) = 0 \tag{6.89a}$$

$$g(P, y) = z \tag{6.89b}$$

If the second-order sufficiency conditions are satisfied then the optimal solution P^* and the corresponding Lagrange Multiplier λ, satisfying the optimization problem as defined by Eqs. (6.88), are implicitly defined as function of y, x, and z in Eqs. (6.89a) and (6.89b). So P^* and λ^* can be written as

$$P^* = P^*(y, x, z) \tag{6.90}$$

$$\lambda^* = \lambda^*(y, x, z) \tag{6.91}$$

Using these functional relationships, $\nabla_w F^*$ and $\nabla_{zw} F^*$ can be calculated in a straightforward manner. This is done by using the chain rule to successively differentiate the system of Eqs. (6.89a) and (6.89b) with respect to the vector variables y, x, and z. Since $w = [y \mid x]^T$, to obtain the desired quantities, $\nabla_w F^*$ and $\nabla_{zw} F^*$, it is required to calculate $\nabla_y F^*$, $\nabla_x F^*$, $\nabla_{zy} F^*$, and $\nabla_{zx} F^*$.

Calculations for $\nabla_y F^*$ and $\nabla_{zy} F^*$

The optimal objective value F^* to the problem of the system of Eqs. (6.88) can be written as $F^* = F(P^*)$. Differentiating this expression using the chain rule with respect to y gives

$$\nabla_y F^* = \nabla_P F \; \nabla_y P^* \tag{6.92}$$

Next differentiating the expression $g(P^*, y) = z$ yields

$$\nabla_P g \; \nabla_y P^* + \nabla_y g = 0 \tag{6.93}$$

Multiply Eq. (6.89a) by $\nabla_y P^*$ to obtain

$$\nabla_P F \; \nabla_y P^* + \lambda^* \nabla_P g \; \nabla_y P^* = 0 \tag{6.94}$$

Multiply Eq. (6.93) by λ^* to obtain

$$\lambda^* \; \nabla_P F \; \nabla_y P^* + \lambda^* \; \nabla_y g = 0 \tag{6.95}$$

After solving Eqs. (6.94) and (6.95), the following can be obtained

$$\nabla_P F \; \nabla_y P^* = \lambda^* \; \nabla_y g \tag{6.96}$$

Equating Eqs. (6.96) and Eq. (6.92), gives

$$\nabla_y F^* = \lambda^* \nabla_y g \tag{6.97}$$

Finally, transposing Eq. (6.97) and differentiating with respect to z by successive application of the chain rule yields

$$\nabla_{zy} F^* = (\nabla_y g)^T \, (\nabla_z \lambda^*)^T + \left(\sum_{i=1}^{k} \lambda_i^* \nabla_{Py} g_i \right) \nabla_z P^* \tag{6.98}$$

Calculations for $\nabla_z F^*$ and $\nabla_{zx} F^*$

Assume that the optimal solution to problem in Eq. (6.88) depends only on the parameter sets x and z, whereas all other parameters are kept fixed. Then, the optimal objective value can be written as

$$F^* = F(P^*, x)$$

Using the chain rule, differentiate the optimal objective function with respect to x, to obtain

$$\nabla_x F^* = \nabla_P F \; \nabla_x P^* + \nabla_x F \tag{6.99}$$

The Kuhn-Tucker necessary conditions for an optimal solution to the problem defined by Eqs. (6.88) are

$$\nabla_P F + \lambda \nabla_P g = 0 \tag{6.100}$$

$$g(P) = 0 \tag{6.101}$$

In these equations, P^* and λ^* are defined implicitly as functions of x and z. Multiplying Eq. (6.100) by $\nabla_x P^*$, and differentiating Eq. (6.101) with respect to x yields two new set of equations

$$\nabla_P F \; \nabla_x P^* + \lambda^* \nabla_P g \; \nabla_x P^* = 0 \tag{6.102}$$

$$\nabla_P g \; \nabla_x P^* = 0 \tag{6.103}$$

Substituting the result of Eq. (6.103) into the Eq. (6.102), we get

$$\nabla_P F \; \nabla_x P^* = 0$$

Substituting this result into Eq. (6.99), gives

$$\nabla_x F^* = \nabla_x F \tag{6.104}$$

Differentiating the above result with respect to z, using chain rule yields

$$\nabla_{zx} F^* + \nabla_P (\nabla_x F)^T \; \nabla_z P^* = \nabla_{Px} F \; \nabla_z P^* \tag{6.105}$$

Combining Eqs. (6.97), (6.98), (6.104) and (6.105), the following expressions for $\nabla_w F^*$ and $\nabla_{zw} F^*$ are obtained

$$\nabla_w F^* = [\nabla_y F^* \mid \nabla_x F^*] = [\lambda^* \; \nabla_y g \mid \nabla_x F^*] \tag{6.106}$$

and

$$\nabla_{zw} F^* = \begin{bmatrix} \nabla_z (\nabla_y F^*)^T \\ \nabla_z (\nabla_x F^*)^T \end{bmatrix}$$

$$\nabla_{zw} F^* = \begin{bmatrix} \sum_{i=1}^{k} \lambda_i \nabla_{Py} g_i \; \nabla_z P^* + (\nabla_y g)^T \, (\nabla_z \lambda^*)^T \\ \nabla_{Px} F \, \nabla_z P^* \end{bmatrix}$$

$$\nabla_{zw} F^* = \begin{bmatrix} \sum_{i=1}^{k} \lambda_i \nabla_{Py} g_i & (\nabla_y g)^T \\ \nabla_{Px} F & 0 \end{bmatrix} \begin{bmatrix} \nabla_z P^* \\ (\nabla_z \lambda^*)^T \end{bmatrix} \tag{6.107}$$

The implicit function theorem assures that in some neighbourhood of zero, there exists a unique solution $\lambda^* = \lambda^*(z)$, $P^* = P^*(z)$ to the system Eqs. in (6.100) and (6.101). Moreover, there exists a neighbourhood of zero such that for any z in neighbourhood the function $P^*(z)$ and $\lambda^*(z)$ satisfy the second order sufficiency conditions for a strict local minimum.

So $\nabla_z P^*$ and $\nabla_z \lambda^*$ can be expressed as

$$\begin{bmatrix} \nabla_z P^* \\ \nabla_z \lambda^* \end{bmatrix} = \begin{bmatrix} \nabla_{PP} L & (\nabla_P g)^T \\ \nabla_P g & 0 \end{bmatrix} \begin{bmatrix} 0 \\ I \end{bmatrix} \tag{6.108}$$

Putting Eq. (6.106) in Eq. (6.107), the required expression is obtained.

6.6.5 Test System and Results

Results are obtained for two sample systems to illustrate the applicability of the method.

Case 1: The method is applied to a three-generator sample system including transmission loss. Expected generator characteristics are given below:

Fuel characteristics ($/h) are:

$$F_{11} = 0.010P_1^2 + 2.0P_1 + 10.0$$
$$F_{12} = 0.012P_2^2 + 1.5P_2 + 20.0$$
$$F_{13} = 0.004P_3^2 + 1.8P_3 + 20.0$$

NO_x emission characteristics (kg/h) are:

$$F_{21} = 0.0006570P_1^2 - 0.05497P_1 + 4.111$$
$$F_{22} = 0.0005916P_2^2 - 0.05880P_2 + 2.593$$
$$F_{23} = 0.0004906P_3^2 - 0.05014P_3 + 4.268$$

The B-coefficients of transmission losses (MW^{-1}) are:

$$B_{11} = 0.0002725 \quad B_{22} = 0.0003090 \quad B_{33} = 0.0032295$$
$$B_{12} = B_{21} = -0.0000351$$
$$B_{13} = B_{31} = -0.0003679$$
$$B_{23} = B_{32} = -0.0000565$$

Table 6.26 displays the results obtained for particular values of the standard deviations $\sigma_i = C_{w_i} \times \overline{w_i}$ (i = 1, 2, ..., 24) at various objective levels. The values of coefficients of variation for cost and emission parameters are assumed 0.1%. Coefficients of variations for transmission loss parameters are assumed to be zero, because of small variations in these parameters [Parti et al., 1983]. This table depicts the mean values F_1 and F_2, dispersion Ω of F_1 about the mean value, the sensitivity trade-off ratio of the function F_1 at the dispersion and trade-off λ_{12} along with non-inferior optimal generation schedules. Ω conveys the significant information to decision maker, i.e. the larger the value of Ω, the greater the possibility that the actual solution deviates significantly from the nominal solution. In the case of 150 MW, there is dispersion of 74.641 about the mean cost 390.852 $/h and sensitivity trade-off ratio is 0.0708. About the mean cost 390.705 $/h, there is dispersion of 35.041 and the ratio is 0.0643. The decision maker will prefer the schedule of cost 390.705 $/h which has comparatively small dispersion. The decision maker will prefer the next schedule, corresponding to the cost 390.603 $/h (5th set) rather than the cost 389.10 $/h (6th set) because of a small dispersion. It is apparent that each step in the reduction of emission of NO_x becomes increasingly expensive with more power losses. Table 6.27 gives the non-inferior schedules corresponding to the results in Table 6.26.

From the results of minimum cost and minimum emission shown in Table 6.28, it is observed that there is an increase in the expected fuel cost by 1.161 $/h and the reduction in the expected emission is 0.039 kg/h for 150 MW. For 200 and 250 MW, the expected costs increase by 2.022 and 3.251 $/h and the expected emissions reduce at the rate of 0.08 and 1.141 kg/h respectively. It is observed from the preferred expected solutions given in Table 6.29 that for a reduction in 1 kg of NO_x per hour, the costs are increased by 1 $/h for 150 MW, 1.77 $/h for 200 MW, and 3.02 $/h for 250 MW, respectively.

TABLE 6.26 Results

Demand (MW)	*Sr. no.*	F_1 (\$/h)	F_2 (kg/h)	λ_{12} (\$/kg)	Ω	$\partial F_1/\partial\Omega$
150.0	1	390.852	7.148	320.677	74.641	0.0708
	2	390.705	7.155	142.071	35.041	0.0643
	3	390.656	7.160	92.093	23.060	0.0588
	4	390.623	7.165	54.196	13.901	0.0465
	5	390.603	7.170	24.080	6.752	0.0285
	6	389.110	7.192	189.161	48.066	0.2269
200.0	1	551.639	8.138	345.844	163.159	0.0335
	2	551.306	8.160	199.028	104.528	0.0927
	3	551.225	8.165	187.008	98.854	0.0994
	4	551.104	8.170	85.359	44.513	0.0569
	5	551.061	8.175	63.112	33.432	0.0521
	6	551.026	8.180	43.707	23.763	0.0448
	7	550.998	8.185	26.561	15.381	0.0352
	8	550.997	8.190	11.238	8.685	0.0244
	9	550.134	8.330	336.504	167.771	0.5855
250.0	1	735.869	10.503	133.873	114.751	0.0540
	2	735.414	10.525	17.343	20.103	0.0412
	3	735.413	10.526	11.025	15.741	0.0339
	4	735.402	10.528	4.082	12.474	0.0304
	5	735.319	10.529	53.768	51.964	0.0898
	6	735.193	10.539	65.702	63.461	0.1090
	7	734.878	10.555	119.829	115.424	0.1591

TABLE 6.27 Non-inferior schedules corresponding to the results given in Table 6.26

Demand (MW)	*Sr. no.*	*Generation schedules* (MW)		
		P_1	P_2	P_3
150.0	1	48.219	58.024	51.688
	2	49.566	57.447	50.560
	3	49.824	57.641	49.932
	4	50.062	57.816	49.364
	5	50.285	57.974	48.841
	6	50.358	59.224	46.390
200.0	1	68.645	77.694	66.869
	2	72.379	74.110	66.372
	3	72.605	74.246	65.810
	4	71.807	76.238	64.103
	5	72.075	76.212	63.724

(Contd.)

TABLE 6.27 (*Contd.*)

Demand (MW)	*Sr. no.*	*Generation schedules* (MW)		
		P_1	P_2	P_3
	6	72.337	76.175	63.371
	7	72.591	76.135	63.037
	8	72.836	76.094	62.718
	9	70.576	84.195	53.529
250.0	1	92.289	99.351	75.475
	2	96.169	94.176	77.240
	3	96.077	94.456	76.952
	4	96.025	94.723	76.616
	5	96.803	92.978	78.025
	6	97.251	92.792	77.532
	7	97.804	92.894	76.339

TABLE 6.28 Minimum economic and emission dispatch

Demand (MW)	*Generation output* (MW)			*Power loss* (MW)	*Fuel cost* ($/h)	NO_x *output* (kg/h)
	P_1	P_2	P_3			
Economic dispatch						
150.0	50.64	58.21	48.03	6.8	390.592	7.179
200.0	73.84	75.90	61.46	11.12	550.938	8.212
250.0	98.14	94.36	73.63	16.13	735.209	10.592
Minimum emission dispatch						
150.0	47.46	55.67	56.12	9.25	391.753	7.140
200.0	67.31	76.45	71.11	14.87	552.960	8.132
250.0	88.62	98.32	83.91	20.85	738.460	10.451

TABLE 6.29 Comparison of results

Demand (MW)	*Dispatch*	F_1 ($/h)	F_2 (kg/h)	P_L (MW)
150.0	Minimum cost	390.592	7.179	6.80
	Minimum emission	391.753	7.140	9.25
	Combined	390.603	7.170	7.10
200.0	Minimum cost	550.938	8.212	11.12
	Minimum emission	552.938	8.132	14.87
	Combined	550.977	8.190	11.64
250.0	Minimum cost	735.209	10.592	16.13
	Minimum emission	738.460	10.451	20.85
	Combined	735.402	10.528	17.48

Case 2: A large system is considered to illustrate the method and to evaluate the compromised solution from the combined economic-emission dispatch. The expected generator characteristics are given below:

Fuel characteristics ($/h) are:

$$F_{11} = 0.005P_1^2 + 2.00P_1 + 100.0$$
$$F_{12} = 0.010P_2^2 + 2.00P_2 + 200.0$$
$$F_{13} = 0.020P_3^2 + 2.00P_3 + 300.0$$
$$F_{14} = 0.003P_4^2 + 1.95P_4 + 80.0$$
$$F_{15} = 0.015P_5^2 + 1.45P_5 + 100.0$$
$$F_{16} = 0.010P_6^2 + 0.95P_6 + 120.0$$

NO_x emission characteristics (kg/h) are:

$$F_{21} = 0.0006572P_1^2 - 0.05497P_1 + 4.111$$
$$F_{22} = 0.0005916P_2^2 - 0.05880P_2 + 2.593$$
$$F_{23} = 0.0004906P_3^2 - 0.05014P_3 + 4.268$$
$$F_{24} = 0.0003780P_4^2 - 0.03150P_4 + 5.526$$
$$F_{25} = 0.0004906P_5^2 - 0.05014P_5 + 4.268$$
$$F_{26} = 0.0005173P_6^2 - 0.05548P_6 + 6.132$$

The *B*-coefficients are given in Table 6.30. The values of coefficients of variation for cost, emission, and *B*-coefficients are assumed to be 0.1%.

TABLE 6.30 *B*-coefficients $\times$ 10^2 (MW^{-1})

0.0200	0.0010	0.0015	0.0005	0.0000	0.0030
0.0010	0.0300	−0.0020	0.0001	0.0012	0.0010
0.0015	−0.0020	0.0100	0.0010	0.0010	0.0008
0.0005	0.0001	0.0010	0.0150	0.0006	0.0050
0.0000	0.0012	0.0010	0.0006	0.0250	0.0020
0.0030	0.0010	0.0008	0.0050	0.0020	0.0210

Table 6.31 displays the mean cost, emission and losses obtained from minimum cost and minimum emission dispatch procedures. The combined dispatch of economic emission is also presented in Table 6.31. The generation schedules are shown in Table 6.32, corresponding to depicted results in Table 6.31.

This section incorporates the sensitivity measure into multiobjective optimization, which generates non-inferior optimal solution with respect to the objective level utility and the sensitivity index. The index provides useful information about the distribution of the optimal solution in the presence of variations in the model parameters defining the problem. The decision maker is able to analyze the sensitivity information conveniently, since sensitivity index is a scalar-valued quantity, regardless of the number of objectives. The most important characteristic of sensitivity index is that a sensitivity trade-off, is calculated at each non-inferior point. This allows the decision maker to know the trade-offs between the objective levels and parameter sensitivity.

TABLE 6.31 Results for six-generator system

Dispatch	*Demand*	*Cost*	*Emission*
Minimum cost	200.0	1290.071	21.665
	400.0	1813.156	24.782
	600.0	2452.050	37.672
Minimum emission	200.0	1322.448	20.222
	400.0	1901.441	20.757
	600.0	2643.043	28.601
Combined dispatch	200.0	1291.411	21.100
	400.0	1818.828	23.036
	600.0	2462.224	33.749

TABLE 6.32 Generation schedules corresponding to the results given in Table 6.31

Demand	P_1	P_2	P_3	P_4	P_5	P_6
Minimum cost						
200.0	29.886	15.560	8.193	55.416	28.222	65.186
400.0	78.397	40.818	21.732	133.535	45.451	88.959
600.0	127.878	66.914	36.060	212.450	63.449	113.216
Minimum emission						
200.0	30.638	37.356	35.994	22.133	36.194	39.514
400.0	57.015	66.398	71.747	68.100	71.222	72.634
600.0	84.025	95.916	108.806	114.795	106.622	105.969
Combined dispatch						
200.0	31.863	21.725	12.158	45.445	30.215	60.856
400.0	63.349	51.969	31.384	120.118	52.952	88.463
600.0	112.823	79.026	48.674	187.398	74.179	116.628

The proposed method suffers from the drawback of computational efficiency of the ε-constraint method when there are several objectives. The number of solutions of the problem in the non-inferior set increases exponentially with the number of objectives. There exists additional difficulties of evaluation of derivatives before calculating the sensitivity trade-off. For large systems, there is computation burden for which the only recourse is to calculate the derivatives numerically.

The results summarized in this chapter reinforce the following premises upon which the study was based.

- Risk assessment must be considered as an integral part of the multiobjective decision making process.
- Uncertainty in model parameters can be quantified in terms of sensitivity, where trade-offs among cost, sensitivity of model parameters can be evaluated within a multiobjective risk-based decision-making process.

- The Risk/Dispersion index method can be modified to incorporate some specific attributes associated with the power dispatch problem.

6.7 STOCHASTIC MULTIOBJECTIVE SHORT-TERM HYDROTHERMAL SCHEDULING

The scheduling problem requires an appropriate objective function such as minimum generation cost or transmission losses or minimum pollution level, etc. In essence, the hydrothermal scheduling problem can be visualized as a multiobjective one. A simple and robust solution methodology for a class of multiobjective short-range fixed-head hydrothermal problems is undertaken which makes use of fuzzy set theory. First of all, a short-range fixed-head hydrothermal scheduling problem is formulated in a unified multiobjective framework considering stochastic cost, NO_x emission, SO_2 emission, and CO_2 emission curves for thermal power generation units and uncertainty in system load demand. The expected values of thermal fuel costs, NO_x emissions, SO_2 emission and CO_2 emission, over whole of the planning period are the four conflicting objectives to be minimized. Further, the expectation of the square of the unsatisfied load because of possible variance of generator outputs over whole of the planning period, is incorporated as another objective to be minimized. Basically, the solution procedure for the multiobjective problem is based on the generation of non-inferior solutions. The weighting technique is used to generate non-inferior solutions, which allows explicit trade-off between objective levels. Exploiting fuzzy set theory [18], a cardinal priority ranking of the non-inferior solutions is defined that maximizes the satisfaction of all the objectives and is utilized to find the best compromising solution from the non-inferior solution set. The practical viability of the multiobjective hydrothermal scheduling problem has been demonstrted on three sample systems.

6.7.1 Stochastic Multiobjective Optimization Problem Formulation

Consider an electric power system network having N thermal generating plants and M hydro plants, where $M + N$ is the total number of generating plants. The basic problem is to find the active power generation of each plant in the system as a function of time over a finite time period from 0 to T.

Stochastic thermal model

The objective function to be minimized is the total system operating cost, represented by the fuel cost of thermal generation, over the optimization interval.

$$J_1 = \int_0^T \left(\sum_{i=1}^{N} (a_i P_i^2 + b_i P_i + c_i) \right) dt \tag{6.109}$$

where

a_i, b_i, and c_i are the cost coefficients of the ith generating unit
N is the number of thermal units
T is the total planning period

A stochastic model of function J_1 is formulated by considering errors in coefficients of input-output characteristics and load demand during each subinterval as random variables. Any possible deviation in coefficients of input-output characteristics and load demand from their expected values are manipulated through the randomness of random variable, P_i. A stochastic model is reduced to its deterministic equivalent by taking its expected value. Presuming that random variables are normally distributed and statistically independent, the expected value of operating cost comes out to be

$$\overline{J}_1 = \int_0^T \left(\sum_{i=1}^{N} (\overline{a}_i \overline{P}_i^2 + \overline{b}_i \overline{P}_i + \overline{c}_i + \overline{a}_i \, \mathrm{var}(P_i)) \right) dt \tag{6.110}$$

where

$\overline{a}_i$, $\overline{b}_i$, and $\overline{c}_i$, are expected cost coefficients of the ith thermal unit

$\overline{P}_i$ is the expected power generation by the ith thermal unit.

The variance of power P_i is given as

$$\mathrm{var}(P_i) = C_{P_i}^2 \overline{P}_i^2 \qquad (i = 1, 2, ..., N) \tag{6.111}$$

where C_{P_i} is the coefficient of variation of random variable P_i

The zero value of the coefficient of variation implies no randomness, in other words, there is complete certainty about the value of the random variable. Equation (6.110) can be rewritten as

$$\overline{J}_1 = \int_0^T \left(\sum_{i=1}^{N} \left((1 + C_{P_i}^2) \overline{a}_i \overline{P}_i^2 + \overline{b}_i \overline{P}_i + \overline{c}_i \right) \right) dt \tag{6.112}$$

Stochastic emission models

Thermal power stations are a major cause of atmospheric pollution because of high concentration of pollutants caused by them. The emission curves for a thermal plant can be directly related to the cost curve through the emission rate per MJ (1 Btu = 1055.06 J), which is a constant factor for a given type or grade of fuel, thus yielding quadratic NO_x, SO_2, and CO_2 emission curves in terms of active power generation. The aim is to optimize the NO_x, SO_2, and CO_2 emissions of thermal plant with full utilization of water available during the optimization period.

NO_x emission objective can be defined as

$$J_2 = \int_0^T \left(\sum_{i=1}^{N} (d_{1i} P_i^2 + e_{1i} P_i + f_{1i}) \right) dt \tag{6.113}$$

where d_{1i}, e_{1i}, and f_{1i} are NO_x emission coefficients of the ith thermal unit.

SO_2 emission objective can be stated as

$$J_3 = \int_0^T \left(\sum_{i=1}^{N} (d_{2i} P_i^2 + e_{2i} P_i + f_{2i}) \right) dt \tag{6.114}$$

where d_{2i}, e_{2i}, and f_{2i} are SO_2 emission coefficients of the ith thermal unit.

CO_2 emission objective can be described as

$$J_4 = \int_0^T \left(\sum_{i=1}^{N} (d_{3i}P_i^2 + e_{3i}P_i + f_{3i}) \right) dt \tag{6.115}$$

where d_{3i}, e_{3i}, and f_{3i}, are CO_2 emission coefficients of the ith thermal unit.

Development of stochastic models of functions J_2, J_3, and J_4 are adopted after much cogitation that thermal generations and load demand during each subinterval are random variables. Any possible deviation of NO_x, SO_2, and CO_2 emission coefficients and load demand from their expected values is managed through the random power generation P_i. Presuming that random variables are normally distributed and statistically independent, the expected value of NO_x emission is estimated as

$$\bar{J}_2 = \int_0^T \left(\sum_{i=1}^{N} (\bar{d}_{1i}\bar{P}_i^2 + \bar{e}_{1i}\bar{P}_i + \bar{f}_{1i} + \bar{d}_{1i}\ \mathrm{var}(P_i)) \right) dt \tag{6.116a}$$

where $\bar{d}_{1i}$, $\bar{e}_{1i}$, and $\bar{f}_{1i}$ are expected NO_x emission coefficients of the ith thermal unit.

Substituting Eq. (6.111) into Eq. (6.116a),

$$\bar{J}_2 = \int_0^T \left(\sum_{i=1}^{N} \left[(1 + C_{P_i}^2)\, \bar{d}_{1i}\bar{P}_i^2 + \bar{e}_{1i}\bar{P}_i + \bar{f}_{1i}\right] \right) dt \tag{6.116b}$$

The expected value of SO_2 emission becomes

$$\bar{J}_3 = \int_0^T \left(\sum_{i=1}^{N} (\bar{d}_{2i}\bar{P}_i^2 + \bar{e}_{2i}\bar{P}_i + \bar{f}_{2i} + \bar{d}_{2i}\ \mathrm{var}(P_i)) \right) dt \tag{6.117a}$$

where $\bar{d}_{2i}$, $\bar{e}_{2i}$, and $\bar{f}_{2i}$, are expected SO_2 emission coefficients of the ith thermal unit.

Equation (6.117a) can be rewritten as

$$\bar{J}_3 = \int_0^T \left(\sum_{i=1}^{N} \left[(1 + C_{P_i}^2)\, \bar{d}_{2i}\bar{P}_i^2 + \bar{e}_{2i}\bar{P}_i + \bar{f}_{2i}\right] \right) dt \tag{6.117b}$$

The expected value of CO_2 emission becomes

$$\bar{J}_4 = \int_0^T \left(\sum_{i=1}^{N} (\bar{d}_{3i}\bar{P}_i^2 + \bar{e}_{3i}\bar{P}_i + \bar{f}_{3i} + \bar{d}_{3i}\ \mathrm{var}(P_i)) \right) dt \tag{6.118a}$$

where $\bar{d}_{3i}$, $\bar{e}_{3i}$, and $\bar{f}_{3i}$, are expected CO_2 emission coefficients of the ith thermal unit.

Eq. (6.118a) can be rewritten as

$$\bar{J}_4 = \int_0^T \left(\sum_{i=1}^{N} \left[(1 + C_{P_i}^2)\, \bar{d}_{3i}\bar{P}_i^2 + \bar{e}_{3i}\bar{P}_i + \bar{f}_{3i}\right] \right) dt \tag{6.118b}$$

Stochastic hydro model

In a short-range hydrothermal scheduling problem, there is insignificant fuel cost incurred in the operation of hydro units [Rashid and Nor, 1991]. The input-output characteristic of a hydro generator is expressed by the variation of water discharge $q(t)$ as a function of power output P_j and net head h. According to Glimn–Kirchmayer model, the discharge is

$$q_j = K\ \phi(h)\ \tau(P_j) \qquad (j = 1, 2, ..., M) \tag{6.119}$$

where

ϕ, τ are functions of head and hydro generations, respectively
K is a constant
M is total number of hydro units.

For a large capacity reservoir, it is practical to assume that the effective head is constant over the optimization interval. In the case of constant head, $\phi(h)$ becomes constant, Eq. (6.119) is rewritten as

$$q_j = K'\ \tau(P_j) \qquad (j = 1, 2,..., M) \tag{6.120}$$

where K' becomes the new constant and is formed by the multiplication of K and $\phi(h)$. Each hydro plant is constrained by the amount of water available for the optimization interval, i.e.

$$\int_0^T q_j\, dt = R_j \qquad (j = 1, 2, ..., M) \tag{6.121}$$

where R_j is the predefined volume of water available for the jth hydro plant.

The performance of hydro q_j is represented by

$$q_j = x_j P_j^2 + y_j P_j + z_j \qquad (j = 1, 2,..., M) \tag{6.122}$$

where x_j, y_j, and z_j are the discharge coefficients of the jth hydro plant.

Since the thermal generations and load demand are random, the hydro generations also become random in view of the load demand constraint as given by Eq. (6.124). A stochastic model of function q_j is developed by deeming hydro generation and discharge coefficients during each subinterval as random variables. Any possible deviation in discharge and load demand from the expected value is manipulated through the randomness of hydro generator P_j. Since the random variables are assumed normally distributed and independent, the expected value of discharge becomes

$$\overline{q}_j = \overline{x}_j \overline{P}_j^2 + \overline{y}_j \overline{P}_j + \overline{z}_j + \overline{x}_j \operatorname{var}(P_j) \qquad (j = 1, 2, ..., M) \tag{6.123a}$$

where $\overline{x}_j$, $\overline{y}_j$, and $\overline{z}_j$ are the expected discharge coefficients of the jth hydro plant.

Eq. (6.123a) can be rewritten as

$$\overline{q}_j = (1 + C_{P_j}^2)\overline{x}_j \overline{P}_j^2 + \overline{y}_j \overline{P}_j + \overline{z}_j \qquad (j = 1, 2, ..., M) \tag{6.123b}$$

Equality and inequality constraints

(i) The expected load demand equality constraint is

$$\sum_{i=1}^{M+N} \overline{P_i} = \overline{P_D} + \overline{P_L} \tag{6.124}$$

where

$\overline{P_D}$ is the expected load demand during the interval

$\overline{P_L}$ is the expected transmission loss during the interval

(ii) The expected limits are imposed as

$$\overline{P_i}^{\min} \le \overline{P_i} \le \overline{P_i}^{\max} \qquad (i = 1, 2, ..., N + M) \tag{6.125}$$

where

$\overline{P_i}^{\min}$ is the expected lower limit of the *i*th generator output during the interval.

$\overline{P_i}^{\max}$ is the expected upper limit of the *i*th generator output during the interval.

Expected transmission losses

A common approach to model transmission losses in the system is to use the Kron's approximated loss formula through *B*-coefficients

$$P_L = \sum_{i=1}^{M+N} \sum_{j=1}^{M+N} P_i B_{ij} P_j \tag{6.126}$$

The evaluation of *B*-coefficients is very sensitive to operating conditions so *B*-coefficients will also become random because generation levels are random. With normally distributed random variables, the expected transmission losses are

$$\overline{P_L} = \sum_{i=1}^{M+N} \sum_{j=1}^{M+N} \overline{P_i}\,\overline{B_{ij}}\,\overline{P_j} + \sum_{i=1}^{M+N} \overline{B_{ii}}\ \mathrm{var}(P_i) + \sum_{i=1}^{M+N} \sum_{\substack{j=1 \\ j\neq i}}^{M+N} \overline{B_{ij}}\ \mathrm{cov}(P_i, P_j) \tag{6.127}$$

Covariance of power P_i and P_j is given by

$$\mathrm{cov}(P_i, P_j) = R_{P_iP_j} C_{P_i} C_{P_j} \overline{P_i}\,\overline{P_j} \qquad (i = 1,.., M + N;\ \ j = 1,.., M + N;\ \ i \neq j) \tag{6.128}$$

where $R_{P_iP_j}$ are the correlation coefficients of P_i and P_j, random variables.

Substituting Eqs. (6.111) and (6.128) into Eq. (6.127),

$$\overline{P_L} = \sum_{i=1}^{M+N} (1 + C_{P_i}^2)\,\overline{B_{ii}}\,\overline{P_i}^2 + \sum_{i=1}^{N+M} \sum_{\substack{j=1 \\ j\neq i}}^{N+M} (1 + R_{P_iP_j} C_{P_i} C_{P_j})\,\overline{P_i}\,\overline{B_{ij}}\,\overline{P_j}$$

or

$$\overline{P_L} = \sum_{i=1}^{N+M} \sum_{j=1}^{N+M} \overline{P_i}\,T_{ij}\,\overline{P_j} \tag{6.129}$$

where

$$T_{ii} = (1 + C_{P_i}^2)\,\overline{B}_{ii}$$

$$T_{ij} = (1 + R_{P_iP_j} C_{P_i} C_{P_j})\,\overline{B}_{ij} \qquad (i \neq j)$$

Expected deviations

Generator outputs P_i are treated as random variables, and the stochastic model is converted into its deterministic equivalent by taking its expected value. So, the solution will provide only the expected values of power generations. By virtue of the above consideration, there will be mismatch in load demand. The variance of a random variable quantifies the degree of uncertainty associated with the mean value of the random variable. The active power loss, the system fuel cost and emission curves are quadratic functions of decision variable P_i, and their variances quantify the degree of uncertainty associated with their expected values. So, the expected mismatch can be estimated through minimization of the squared error of the unsatisfied power demand, i.e.

$$E = \left[\left(\overline{P}_D + \overline{P}_L - \sum_{i=1}^{M+N} P_i\right)^2\right] \qquad (6.130)$$

where P_i is the actual power generation required to meet the load, which is considered random.

Using Eq. (6.124), Eq. (6.130) can be rewritten as

$$E = \left[\left(\sum_{i=1}^{M+N} P_i - \sum_{i=1}^{M+N} \overline{P}_i\right)^2\right] \qquad (6.131)$$

This on simplification reduces to

$$\overline{J}_5 = \int_0^T \left(\sum_{i=1}^{M+N} \operatorname{var}(P_i) + \sum_{i=1}^{M+N-1} \sum_{j=i+1}^{M+N} 2\operatorname{cov}(P_i, P_j)\right) dt \qquad (6.132)$$

Substituting Eqs. (6.111) and (6.128) into Eq. (6.132),

$$\overline{J}_5 = \int_0^T \left(\sum_{i=M}^{M+N} C_{P_i}^2 \overline{P}_i^2 + \sum_{i=1}^{M+N-1} \sum_{j=i+1}^{M+N} 2\,R_{P_iP_j} C_{P_i} C_{P_j} \overline{P}_i \overline{P}_j\right) dt$$

or

$$\overline{J}_5 = \int_0^T \left(\sum_{i=1}^{M+N} \sum_{j=1}^{M+N} S_{ij} \overline{P}_i \overline{P}_j\right) dt \qquad (6.133)$$

where

$$S_{ii} = C_{P_i}^2$$

$$S_{ij} = R_{P_iP_j} C_{P_i} C_{P_j} \qquad (i \neq j)$$

Multiobjective optimization formulation

Multiobjective optimization problem is framed considering: (a) the expected operating cost, (b) the expected NO_x emission, (c) the expected SO_2 emission, (d) the expected CO_2 emission of thermal units, and (e) the expected risk associated with possible deviation of random variables P_i from their respective expected values over the optimization interval to meet the expected load demand in each interval. Each hydro plant is constrained by expected amount of water available for draw-down in the interval. Mathematically, the multiobjective optimization problem is defined as

$$\text{Minimize} \qquad [\bar{J}_1, \bar{J}_2, \bar{J}_3, \bar{J}_4, \bar{J}_5]^T \tag{6.134a}$$

$$\text{subject to} \qquad \sum_{i=1}^{M+N} \bar{P}_i = \bar{P}_D + \bar{P}_L \tag{6.134b}$$

$$\int_0^M \bar{q}_j \, dt = \bar{R}_j \qquad (j = 1, 2, ..., M) \tag{6.134c}$$

$$\bar{P}_i^{\min} \le \bar{P}_i \le \bar{P}_i^{\max} \qquad (i = 1, 2, ..., M + N) \tag{6.134d}$$

The above multiobjective optimization problem can be redefined in discrete form as

$$\text{Minimize} \qquad \bar{J}_1 = \sum_{k=1}^{T} t_k \left(\sum_{i=1}^{N} \left((1 + C_{P_i}^2)\, \bar{a}_i \bar{P}_i^2 + \bar{b}_i \bar{P}_i + \bar{c}_i \right) \right) \tag{6.135a}$$

$$\text{Minimize} \qquad \bar{J}_2 = \sum_{k=1}^{T} t_k \left(\sum_{i=1}^{N} \left[(1 + C_{P_i}^2)\, \bar{d}_{1i} \bar{P}_i^2 + \bar{e}_{1i} \bar{P}_i + \bar{f}_{1i} \right] \right) \tag{6.135b}$$

$$\text{Minimize} \qquad \bar{J}_3 = \sum_{k=1}^{T} t_k \left(\sum_{i=1}^{N} \left[(1 + C_{P_i}^2)\, \bar{d}_{2i} \bar{P}_i^2 + \bar{e}_{2i} \bar{P}_i + \bar{f}_{2i} \right] \right) \tag{6.135c}$$

$$\text{Minimize} \qquad \bar{J}_4 = \sum_{k=1}^{T} t_k \left(\sum_{i=1}^{N} \left[(1 + C_{P_i}^2)\, \bar{d}_{3i} \bar{P}_i^2 + \bar{e}_{3i} \bar{P}_i + \bar{f}_{3i} \right] \right) \tag{6.135d}$$

$$\text{Minimize} \qquad \bar{J}_5 = \sum_{k=1}^{T} t_k \left(\sum_{i=M}^{M+N} \sum_{j=1}^{M+N} S_{ij} \bar{P}_i \bar{P}_j \right) \tag{6.135e}$$

$$\text{subject to} \qquad \sum_{i=1}^{M+N} \bar{P}_{ik} = \bar{P}_{Dk} + \bar{P}_{Lk} \tag{6.135f}$$

$$\sum_{k=1}^{T} t_k \bar{q}_{jk} = \bar{R}_j \qquad (j = 1, 2, ..., M) \tag{6.135g}$$

$$\bar{P}_{ik}^{\min} \le \bar{P}_{ik} \le \bar{P}_{ik}^{\max} \qquad (i = 1, 2, ..., M + N) \tag{6.135h}$$

6.7.2 Solution Procedure

To generate the non-inferior solutions to the multiobjective optimization problem, the weighting method is applied. In this method, the problem is converted into a scalar optimization as given below:

$$\text{Minimize} \quad \sum_{k=1}^{5} w_k \bar{J}_k \tag{6.136a}$$

$$\text{subject to} \quad \sum_{i=1}^{M+N} \bar{P}_{ik} = \bar{P}_{Dk} + \bar{P}_{Lk} \tag{6.136b}$$

$$\sum_{k=1}^{T} t_k \bar{q}_{jk} = \bar{R}_j \qquad (j = 1, 2, ..., M) \tag{6.136c}$$

$$\bar{P}_{ik}^{\min} \leq \bar{P}_{ik} \leq \bar{P}_{ik}^{\max} \qquad (i = 1, 2, ..., M + N) \tag{6.136d}$$

$$\sum_{k=1}^{5} w_k = 1 \qquad (w_k \geq 0) \tag{6.136e}$$

where w_k are levels of the weighting coefficients.

This approach yields meaningful result to the decision maker when solved many times for different values of w_k (k = 1, 2, ..., 5). Weighting factors w_k are determined based on the relative importance of various objectives, which may vary from place to place and utility to utility.

The simplest way to arrive at the required equations is to use the calculus of variations. Each constraint equation is associated with an unknown multiplier function known as the Lagrange multiplier. The augmented objective function is

$$L = \sum_{k=1}^{5} w_k \bar{J}_k + \sum_{k=1}^{T} \left[\sum_{j=N+1}^{M} \nu_j \bar{q}_{jk} + \lambda_k \left(\bar{P}_{Dk} + \bar{P}_{Lk} - \sum_{i=1}^{M} \bar{P}_{ik} \right) \right] - \sum_{j=N+1}^{M} \nu_j \bar{R}_j \tag{6.137}$$

where

ν_j is water conversion factor of the jth plant

λ_k is incremental cost of power delivered in the system.

The optimality conditions are described by taking the partial derivatives of augmented objective function with respect to the decision variables.

$$\sum_{k=1}^{5} w_k \frac{\partial \bar{J}_k}{\partial \bar{P}_{ik}} + \lambda_k \left(\frac{\partial \bar{P}_{Lk}}{\partial \bar{P}_{ik}} - 1 \right) = 0 \qquad (i = 1, 2, ..., N\ ) \tag{6.138a}$$

$$\nu_j \frac{\partial \bar{q}_{jk}}{\partial \bar{P}_{mk}} + w_5 \frac{\partial \bar{J}_5}{\partial \bar{P}_{mk}} + \lambda_k \left(\frac{\partial \bar{P}_{Lk}}{\partial \bar{P}_{mk}} - 1 \right) = 0 \qquad (j = 1, 2, ..., M; \quad m = j + N) \tag{6.138b}$$

$$\sum_{k=1}^{T} t_k \bar{q}_{jk} - \bar{R}_j = 0 \qquad (j = 1, 2, ..., M) \tag{6.138c}$$

$$\overline{P}_{Dk} + \overline{P}_{Lk} - \sum_{i=1}^{M} \overline{P}_{ik} = 0 \tag{6.138d}$$

Inherently, these equations are nonlinear. The Newton–Raphson method has been applied to find $\overline{P}_{ik}$ $(i = 1, 2, ..., N + M)$, λ_k for all the T intervals. v_j water conversion factor is modified to satisfy the available water constraint. Suppose the initial values of control variables $\overline{P}_{ik}$, λ_k, and v_j are known. The updated values of control variables in the next iteration are

$$\begin{aligned} \overline{P}_{ik}^{\text{new}} &= \overline{P}_{ik}^{0} + \Delta\overline{P}_{ik} \qquad (i = 1, 2, ..., N + M; \quad k = 1, 2, ..., T) \\ \lambda_k^{\text{new}} &= \lambda_k^0 + \Delta\lambda_k \qquad (k = 1, 2, ..., T) \\ v_j^{\text{new}} &= v_j^0 + \Delta v_j \qquad (j = 1, 2, ..., M) \end{aligned}$$

Any small change in control values from their previous values can be obtained as given below:

$$\left(\sum_{j=1}^{5} w_j \frac{\partial^2 \overline{J}_j}{\partial \overline{P}_{ik}^2} + \lambda_k^0 \frac{\partial^2 \overline{P}_{Lk}}{\partial \overline{P}_{ik}^2}\right)\Delta\overline{P}_{ik} + \sum_{\substack{j=1\\ j\neq i}}^{N+M}\left(\frac{\partial^2 \overline{J}_5}{\partial \overline{P}_{ik}\partial \overline{P}_{jk}} + \lambda_k^0 \frac{\partial^2 \overline{P}_{Lk}}{\partial \overline{P}_{ik}\partial \overline{P}_{jk}}\right)\Delta\overline{P}_{jk} + \left(\frac{\partial \overline{P}_{Lk}}{\partial \overline{P}_{ik}} - 1\right)\Delta\lambda_k$$

$$= -\left[\sum_{j=1}^{5} w_j \frac{\partial \overline{J}_j}{\partial \overline{P}_{ik}} + \lambda_k^0 \left(\frac{\partial \overline{P}_{Lk}}{\partial \overline{P}_{ik}} - 1\right)\right] \qquad (i = 1, 2,..., N; \quad k = 1, 2,..., T) \tag{6.139a}$$

$$\left(w_5 \frac{\partial^2 \overline{J}_5}{\partial \overline{P}_{mk}^2} + v_j^0 t_k \frac{\partial^2 \overline{q}_{jk}}{\partial \overline{P}_{mk}^2} + \lambda_k^0 \frac{\partial^2 \overline{P}_{Lk}}{\partial \overline{P}_{mk}^2}\right)\Delta\overline{P}_{mk} + \sum_{\substack{l=1\\ l\neq m}}^{N+M}\left(w_5 \frac{\partial^2 \overline{J}_5}{\partial \overline{P}_{mk}\partial \overline{P}_{lk}} + \lambda_k^0 \frac{\partial^2 \overline{P}_{Lk}}{\partial \overline{P}_{mk}\partial \overline{P}_{lk}}\right)\Delta\overline{P}_{lk}$$

$$+ \left(\frac{\partial \overline{P}_{Lk}}{\partial \overline{P}_{mk}} - 1\right)\Delta\lambda_k + \frac{\partial \overline{q}_{jk}}{\partial \overline{P}_{mk}}\Delta v_j = -\left[w_5 \frac{\partial \overline{J}_5}{\partial \overline{P}_{mk}} + v_j^0 t_k \frac{\partial \overline{q}_{jk}}{\partial \overline{P}_{mk}} + \lambda_k^0 \left(\frac{\partial \overline{P}_{Lk}}{\partial \overline{P}_{mk}} - 1\right)\right]$$

$$(j = 1, 2, ..., M; \quad m = j + N; \quad k = 1, 2, ..., T) \tag{6.139b}$$

$$\sum_{j=1}^{N+M}\left(\frac{\partial \overline{P}_{Lk}}{\partial \overline{P}_{jk}} - 1\right)\Delta\overline{P}_{jk} = -\left(\overline{P}_{Dk} + \overline{P}_{Lk}^0 - \sum_{i=1}^{N+M} \overline{P}_{ik}^0\right) \qquad (k = 1, 2, ..., T) \tag{6.139c}$$

$$\sum_{k=1}^{T}\left(t_k \frac{\partial \overline{q}_{jk}}{\partial \overline{P}_{mk}} - 1\right)\Delta\overline{P}_{mk} = -\left(\sum_{k=1}^{T} t_k \overline{q}_{jk}^0 - V_j\right) \qquad (j = 1, 2,..., M; \quad m = j + N) \tag{6.139d}$$

Substituting the values of derivatives which are evaluated about initial values in Eq. (6.139a)

$$(2t_k\alpha_i + 2t_k w_5 S_{ii} + 2\lambda_k^0 T_{ii})\,\Delta\overline{P}_{ik} + \sum_{\substack{j=1\\ j\neq i}}^{N+M} (2t_k w_5 S_{ij} + 2\lambda_k^0 T_{ij})\,\Delta\overline{P}_{jk} + (K_{ik}^0 - 1)\,\Delta\lambda_k$$

$$= -\left[t_k \left(2\alpha_i P_{ik}^0 + \beta_i + w_5 \sum_{j=1}^{M+N} 2S_{ij}\overline{P}_{jk}^0\right) + \lambda_k^0 (K_{ik}^0 - 1)\right] \qquad (i = 1, 2, ..., N) \tag{6.139e}$$

where

$$K_{ik}^0 = \sum_{j=1}^{N+M} 2T_{ij}\overline{P}_{ik}^0$$

$$\alpha_i = w_1(1+C_{P_i}^2)\,\overline{a}_i + \sum_{j=2}^{5} w_j(1+C_{P_i}^2)\,\overline{d}_{ji}$$

$$\beta_i = w_1\overline{b}_i + \sum_{j=2}^{5} w_j\overline{e}_{ji}$$

Equation (6.139e) can be simplified by rearranging the terms, i.e.

$$(2t_k\alpha_i + 2t_kw_5S_{ii} + 2\lambda_k^0\,T_{ii})\,\Delta\overline{P}_{ik} + \sum_{\substack{j=1\\ j\neq i}}^{N+M}(2t_kw_5S_{ij} + 2\lambda_k^0T_{ij})\,\Delta\overline{P}_{jk}$$

$$= -\left[t_k\left(2\alpha_iP_{ik}^0 + \beta_i + w_5\sum_{j=1}^{M+N}2S_{ij}\overline{P}_{jk}^0\right) + (\lambda_k^0 + \Delta\lambda_k)(K_{ik}^0 - 1)\right] \qquad (i = 1, 2, ..., N)$$

To simplify calculations, neglect all the terms with S_{ij} and T_{ij}, $i \neq j$. So, neglecting the term, $\sum_{\substack{j=1\\ j\neq i}}^{N+M}(2t_kw_5S_{ij} + 2T_{ij})\,\Delta P_{jk}$, the above equation becomes

$$(2t_k\alpha_i + 2t_kw_5S_{ii} + 2\lambda_k^0\,T_{ii})\,\Delta\overline{P}_{ik} = -\left[t_k\left(2\alpha_iP_{ik}^0 + \beta_i + w_5\sum_{j=1}^{M+N}2S_{ij}\overline{P}_{jk}^0\right) + \lambda_k^{\text{new}}\,(K_{ik}^0 - 1)\right]$$

$$(i = 1, 2, ..., N)$$

or

$$\Delta\overline{P}_{ik} = \frac{(1-K_{ik}^0)\,\lambda_k^{\text{new}} - t_k(2\alpha_iP_{ik}^0 + \beta_i + L_{ik})}{2t_k\alpha_i + 2t_kw_5S_{ii} + 2\lambda_k^0T_{ii}} \qquad (i = 1, 2, ..., N)$$

or

$$\Delta\overline{P}_{ik} = \frac{(1-K_{ik}^0)\,\lambda_k^{\text{new}} - A_{ik}}{X_{ik}} \qquad (i = 1, 2, ..., N) \tag{6.140}$$

where

$$A_{ik} = t_k(2\alpha_iP_{ik}^0 + \beta_i + L_{ik}) \tag{6.140a}$$

$$X_{ik} = 2(t_k\alpha_i + t_kw_5S_{ii} + \lambda_k^0T_{ii}) \tag{6.140b}$$

$$L_{ik} = t_kw_5\sum_{j=1}^{N+M}2S_{ij}\overline{P}_{jk}^0$$

Substituting the values of derivatives in Eq. (6.139b),

$$(2t_k w_5 S_{mm} + 2v_j^0 t_k C_{P_j}^2 \bar{x}_j + 2\lambda_k^0 T_{mm})\Delta\bar{P}_{mk}$$

$$+ \sum_{\substack{l=1 \\ l\neq m}}^{N+M} (2t_k w_5 S_{ij} + 2T_{ml})\,\Delta\bar{P}_{lk} + (K_{mk}^0 - 1)\,\Delta\lambda_k + t_k(2\gamma_j \bar{P}_{mk}^0 + y_j)\Delta v_j$$

$$= -\left[t_k w_5 \sum_{l=1}^{M+N} S_{ml}\bar{P}_{lk} + v_j^0 t_k (2\gamma_j \bar{P}_{mk}^0 + y_j) + \lambda_k^0 (K_{mk}^0 - 1)\right] \quad (j = 1, 2, ..., M;\ m = j + N;\ k = 1, 2, ..., T)$$

where

$$\gamma_j = (1 + C_{P_j}^2)\, x_j$$

$$K_{mk}^0 = \sum_{l=1}^{M+N} 2T_{ml} P_{lk}^0$$

The above equation can be simplified by rearranging the terms, i.e.

$$(2t_k w_5 S_{mm} + 2v_j^0 t_k C_{P_j}^2 \bar{x}_j + 2\lambda_k^0 T_{mm})\,\Delta\bar{P}_{mk} + \sum_{\substack{l=1 \\ l\neq m}}^{N+M} (2t_k w_5 S_{ml} + 2T_{ml})\,\Delta\bar{P}_{lk}$$

$$= -\left[t_k w_5 \sum_{l=1}^{M+N} S_{ml}\bar{P}_{lk} + (v_j^0 + \Delta v_j)\, t_k\,(2\gamma_j \bar{P}_{mk}^0 + y_j) + (\lambda_k^0 + \Delta\lambda_k^0)\,(K_{mk}^0 - 1)\right]$$

$$(j = 1, 2, ..., M; \quad m = j + N; \quad k = 1, 2, ..., T)$$

Neglecting the term $\sum_{\substack{l=1 \\ l\neq m}}^{N+M} 2(t_k w_5 S_{ml} + T_{ml})\,\Delta\bar{P}_{lk}$ for simplification,

$$(2t_k w_5 S_{mm} + 2v_j^0 t_k C_{P_j}^2 \bar{x}_j + 2\lambda_k^0 T_{mm})\Delta\bar{P}_{mk}$$

$$= -\left[t_k w_5 \sum_{l=1}^{M+N} S_{ml}\bar{P}_{lk} + v_j^{\text{new}} t_k (2\gamma_j \bar{P}_{mk}^0 + y_j) + \lambda_k^{\text{new}} (K_{mk}^0 - 1)\right]$$

$$(j = 1, 2, ..., M; \quad m = j + N; \quad k = 1, 2, ..., T)$$

or

$$\Delta P_{mk} = \frac{(1 - K_{mk}^0)\,\lambda_k^{\text{new}} - t_k\,(2\gamma_j \bar{P}_{mk}^0 + y_j)\, v_j^{\text{new}} - L_{mk}}{2t_k w_5 S_{mm} + 2v_j^0 t_k C_{P_j}^2 \bar{x}_j + 2\lambda_k^0 T_{mm}} \qquad (j = 1, 2, ..., M; \quad m = j + N)$$

or

$$\Delta P_{mk} = \frac{(1 - K_{mk}^0)\,\lambda_k^{\text{new}} - E_{jk} v_j^{\text{new}} - L_{mk}}{Y_{jk}} \qquad (j = 1, 2, ..., M; \quad m = j + N) \tag{6.141}$$

where

$$E_{jk} = t_k\,(2\gamma_j \bar{P}_{mk}^0 + y_i) \tag{6.141a}$$

$$Y_{jk} = 2(t_k w_5 S_{mm} + v_j^0 t_k C_{P_j}^2 \bar{x}_j + \lambda_k^0 T_{mm}) \tag{6.141b}$$

$$L_{mk} = t_k w_5 \sum_{j=1}^{N+M} 2S_{mj}\overline{P}_{jk}^0 \tag{6.141c}$$

Substituting the values of derivatives in Eq. (6.139c),

$$\sum_{j=1}^{N+M}\left(\sum_{l=1}^{N+M} 2T_{jl}\overline{P}_{lk}^0 - 1\right)\Delta\overline{P}_{jk} = -\left(\overline{P}_{Dk} + \overline{P}_{Lk}^0 - \sum_{i=1}^{N+M}\overline{P}_{ik}^0\right) \qquad (k = 1, 2, ..., T)$$

or

$$\sum_{j=1}^{N+M}(1 - K_{jk}^0)\,\Delta\overline{P}_{jk} = \overline{P}_{Dk} + \overline{P}_{Lk}^0 - \sum_{i=1}^{N+M}\overline{P}_{ik}^0 \qquad (k = 1, 2, ..., T)$$

Substituting Eqs. (6.140) and (6.141) into the above equation,

$$\sum_{i=1}^{N}(1 - K_{ik}^0)\left(\frac{\lambda_k^{\text{new}}(1 - K_{ik}^0) - A_{ik}}{X_{ik}}\right) + \sum_{j=1}^{M}(1 - K_{mk}^0)\left(\frac{\lambda_k^{\text{new}}(1 - K_{mk}^0) - E_{jk}\nu_j^{\text{new}} - L_{mk}}{Y_{jk}}\right)$$

$$= \left(\overline{P}_{Dk} + \overline{P}_{Lk}^0 - \sum_{i=1}^{N+M}\overline{P}_{ik}^0\right) \qquad (k = 1, 2, ..., T; \quad m = j + N)$$

or

$$\left(\sum_{i=1}^{N}\frac{(1 - K_{ik}^0)^2}{X_{ik}} + \sum_{j=1}^{M}\frac{(1 - K_{mk}^0)^2}{Y_{jk}}\right)\lambda_k^{\text{new}} - \sum_{j=1}^{M}\left(\frac{(1 - K_{mk}^0)E_{jk}}{Y_{jk}}\right)\nu_j^{\text{new}}$$

$$= \left(P_{Dk} + P_{Lk}^0 - \sum_{i=1}^{N+M}\overline{P}_{ik}^0\right) + \sum_{i=1}^{N}\left(\frac{(1 - K_{ik}^0)A_{ik}}{X_{ik}}\right) + \sum_{j=1}^{M}\left(\frac{(1 - K_{mk}^0)L_{mk}}{Y_{jk}}\right) \qquad (k = 1, 2, ..., T; \quad m = j + N)$$

or

$$C_k\,\lambda_k^{\text{new}} - \sum_{j=1}^{M} D_{jk}\nu_j^{\text{new}} = F_k \tag{6.142}$$

where

$$C_k = \sum_{i=1}^{N}\frac{(1 - K_{ik}^0)^2}{X_{ik}} + \sum_{j=1}^{M}\frac{(1 - K_{mk}^0)^2}{Y_{jk}} \tag{6.142a}$$

$$D_{jk} = \frac{(1 - K_{mk}^0)\,E_{jk}}{Y_{jk}} \tag{6.142b}$$

$$F_k = \left(\overline{P}_{Dk} + \overline{P}_{Lk}^0 - \sum_{i=1}^{N+M}\overline{P}_{ik}^0\right) + \sum_{i=1}^{N}\left(\frac{(1 - K_{ik}^0)\,A_{ik}}{X_{ik}}\right) + \sum_{j=1}^{M}\left(\frac{(1 - K_{mk}^0)\,L_{mk}}{Y_{jk}}\right) \tag{6.142c}$$

Substituting the values of derivatives in Eq. (6.139d),

$$\left(\sum_{k=1}^{T} t_k (2\bar{x}_j \, \bar{P}_{mk}^0 + \bar{y}_j)\, \Delta \bar{P}_{mk}\right) = -\left(\sum_{k=1}^{T} t_k \bar{q}_{jk}^0 - V_j\right) \qquad (j = 1, 2, ..., M; \quad m = j + N)$$

Substituting Eq. (6.141) into the above equation,

$$\sum_{k=1}^{T} E_{jk} \left(\frac{(1 - K_{mk}^0)\, \lambda_k^{\text{new}} - E_{jk} v_j^{\text{new}} - L_{mk}}{Y_{jk}}\right) = -\left(\sum_{k=1}^{T} t_k \, \bar{q}_{jk}^0 - V_j\right) \qquad (j = 1, 2, ..., M; \quad m = j + N)$$

or

$$\sum_{k=1}^{T} D_{jk} \lambda_k^{\text{new}} - H_j v_j^{\text{new}} = O_j \qquad (j = 1, 2, ..., M; \quad m = j + N) \tag{6.143}$$

where

$$H_j = \sum_{k=1}^{T} \frac{E_{jk} E_{jk}}{Y_{jk}} \tag{6.143a}$$

$$O_j = V_j - \sum_{k=1}^{T} t_k \, \bar{q}_{jk}^0 + \sum_{k=1}^{T} \frac{E_{jk} L_{mk}}{Y_{jk}} \tag{6.143b}$$

From Eq. (6.142), we find the value of λ_k^{new}, i.e.

$$\lambda_k^{\text{new}} = \frac{F_k}{C_k} + \sum_{j=1}^{M} \frac{D_{jk}}{C_k} v_j^{\text{new}} \tag{6.144}$$

Substituting the value of λ_k^{new} into Eq. (6.143),

$$\sum_{k=1}^{T} D_{jk} \left(\frac{F_k}{C_k} + \sum_{j=1}^{M} \frac{D_{jk}}{C_k} v_j^{\text{new}}\right) - H_j v_j^{\text{new}} = O_j \qquad (j = 1, 2, ..., M; \quad m = j + N)$$

$$\sum_{j=1}^{M} \left(\sum_{k=1}^{T} \frac{D_{jk} D_{jk}}{C_k} v_j^{\text{new}}\right) - H_j v_j^{\text{new}} = O_j - \sum_{k=1}^{T} \frac{D_{jk} F_k}{C_k} \qquad (j = 1, 2, ..., M; \quad m = j + N) \tag{6.145}$$

Equation (6.145) can be written in matrix form, i.e.

$$[Q_{jl}]_{M \times M} \, [v_j]_{M \times l} = [R_j]_{M \times l}$$

where

$$Q_{jj} = \left(\sum_{k=1}^{T} \frac{D_{jk} D_{jk}}{C_k}\right) - H_j \qquad (j = 1, 2, ..., M) \tag{6.145a}$$

$$Q_{jl} = \sum_{k=1}^{T} \frac{D_{jk} D_{lk}}{C_k} \qquad (j = 1, 2, ..., M; \quad l = 1, 2, ..., M; \quad j \neq l) \tag{6.145b}$$

$$R_j = O_j - \sum_{k=1}^{T} \frac{D_{jk} F_k}{C_k} \qquad (j = 1, 2, ..., M) \tag{6.145c}$$

Here only matrix of $M \times M$ is to be solved to calculate v_j^{new}, and λ_k^{new} can be computed from Eq. (6.144). When the values of v_j^{new} and λ_k^{new} are known then $\Delta\overline{P}_{ik}$ $(i = 1, 2, ..., N)$ and $\Delta\overline{P}_{mk}$ $(j = 1, 2, ..., M;\ m = j + N)$ can be computed from Eqs. (6.140) and (6.141), respectively. The detailed algorithm is elaborated below:

Algorithm 6.4: Generation of Non-Inferior Solution for Multiobjective Hydrothermal Scheduling Using Approximate Newton–Raphson Method

1. Read the number of thermal units N, the number of hydro units M, the number of sub-intervals T, expected cost coefficients, expected emission coefficients, expected B-coefficients, expected discharge coefficients, expected demand and pre-specified available water, coefficient of variance and correlation coefficients, etc.
2. Set iteration for non-inferior solutions, $k = 1$.
3. Increment count of non-inferior solutions, $k = k + 1$.
4. If $(k \geq K)$ GOTO Step 17.
5. Feed or generate the weights, $w_i (i = 1, 2, ..., M)$.
6. Calculate the initial guess values of $\overline{P}_{ik}^0$ $(i = 1, 2, ..., N + M)$, λ_k^0 and v_j^0 $(j = 1, 2, ..., M)$.
7. Start the iteration counter, $r = 1$.
8. Compute the variables, K_{ik}^0, A_{ik} from Eq. (6.140a), X_{ik} from Eq. (6.140b), E_{jk} from Eq. (6.141a), Y_{jk} from Eq. (6.141b), C_k from Eq. (6.142a), D_{jk} from Eq. (6.142b), F_k from Eq. (6.142c), H_j from Eq. (6.143a), and O_j from Eq. (6.143b).
9. Compute v_j^{new} by solving the following simultaneous equations using the Gauss elimination method
$$[Q_{jl}]_{M \times M}\, [v_j]_{M \times l} = [R_j]_{M \times l}$$
10. Check the convergence
if $|\, v_j^{\text{new}} - v_j^0 \,| \leq \text{Err}$ then GOTO Step 16.
11. Compute λ_k^{new} from Eq. (6.144).
$$\lambda_k^{\text{new}} = \frac{F_k}{C_k} + \sum_{j=1}^{M} \frac{D_{jk}}{C_k} v_j^{\text{new}}$$
12. Calculate ΔP_{ik} $(i = 1, 2,..\ N + M)$ using Eqs. (6.140) and (6.141).
13. Calculate the new values of $\overline{P}_{ik}^{\text{new}}$:
$$\overline{P}_{ik}^{\text{new}} = \overline{P}_{ik}^0 + \Delta\overline{P}_{ik} \qquad (i = 1, 2, ..., N + M; \quad k = 1, 2, ..., T)$$
14. Set limits correspondingly as
$$P_{ik}^{\text{new}} = \begin{cases} P_{ik}^{\max} & ;\text{if } P_{ik}^{\text{new}} \geq P_{ik}^{\max} \\ P_{ik}^{\min} & ;\text{if } P_{ik}^{\text{new}} \leq P_{ik}^{\min} \\ P_{ik}^{\text{new}} & ;\text{otherwise} \end{cases}$$

Disallow to participate the generation, whose limits have been set either to lower or upper limits by setting the relating coefficients to zero.

15. If $r \geq$ IT, then GOTO Step 5,

else $r = r + 1$,

$$\overline{P}_{ik}^{0} = \overline{P}_{ik}^{new} \qquad (i = 1, 2, ..., N + M; \quad k = 1, 2, ..., T)$$

$$\lambda_k^0 = \lambda_k^{new} \qquad (k = 1, 2, ..., T)$$

$$v_j^0 = v_j^{new} \qquad (j = 1, 2, ..., N)$$

GOTO Step 12 and repeat.

16. Record it as non-inferior solution and calculate the objective values and loss and GOTO Step 3.
17. Stop.

6.7.3 Decision Making

Considering the imprecise nature of the decision maker's judgment, it is natural to assume that the decision maker may have fuzzy or imprecise goals for each objective function. The fuzzy sets are defined by equations called the membership functions. These functions represent the degree of membership in some fuzzy sets using values from 0 to 1 [Klir and Folger, 1993]. The membership value 0, indicates incompatibility with the sets, while 1 means full compatibility. By taking account of the minimum and maximum values of each objective function together with the rate of increase of membership satisfaction, the decision maker must detect membership function $\mu(\overline{J}_i)$ in a subjective manner. Here it is assumed that $\mu(\overline{J}_i)$ is a strictly monotonic decreasing and continuous function defined as

$$\mu(\overline{J}_i) = \begin{cases} 1 & ; \overline{J}_i \leq \overline{J}_i^{min} \\ \dfrac{\overline{J}_i^{max} - \overline{J}_i}{\overline{J}_i^{max} - \overline{J}_i^{min}} & ; \overline{J}_i^{min} < \overline{J}_i < \overline{J}_i^{max} \\ 0 & ; \overline{J}_i \geq \overline{J}_i^{max} \end{cases} \tag{6.146}$$

where

$\mu(\overline{J}_i)$ is membership function of objective, $\overline{J}_i$

$\overline{J}_i^{min}$, $\overline{J}_i^{max}$ are minimum and maximum values of the ith objective, respectively.

The value of membership function suggests how far (in the scale from 0 to 1) a non-inferior (non-dominated) solution has satisfied the $\overline{J}_i$ objective. The sum of membership function values $\mu(\overline{J}_i)$ $(i = 1, 2, ..., 5)$ for all the objectives can be computed in order to measure the accomplishment of each solution in satisfying the objectives. The accomplishment of each non-dominated solution can be rated with respect to all the K non-dominated solutions by normalizing its accomplishment over the sum of the accomplishments of K non-dominated solutions as follows [Tapia and Murtagh, 1991]:

$$\mu_D^k = \frac{\sum_{i=1}^{5} \mu(\overline{J}_i^k)}{\sum_{k=1}^{K} \sum_{i=1}^{5} \mu(\overline{J}_i^k)} \tag{6.147}$$

where μ_D^k is cardinal priority of the kth non-inferior solution.

The function μ_D in Eq. (6.147) can be treated as a membership function for non-dominated solutions in a fuzzy set and represented as fuzzy cardinal priority ranking of the non-dominated solutions. The solution that attains the maximum membership μ_D^k, in the fuzzy set so obtained can be chosen as the best solution or the one having the highest cardinal priority ranking, i.e.

$$\text{Max } \{\mu_D^k : k = 1, 2, ..., K\} \tag{6.148}$$

6.7.4 Test Systems and Results

Test system 1

A short-range hydrothermal load scheduling problem of 24 hours duration has been undertaken. The entire optimization period has been divided into 24 intervals and each interval is of one hour. The system under study consists of one thermal plant and one hydro plant [Rashid and Nor, 1991]. The expected cost, NO_x emission, SO_2 emission, CO_2 emission of thermal plant and discharge characteristics of hydro plant are given below:

Expected cost characteristic of thermal plant ($/h):

$$\overline{J}_{11}(t) = 0.001991\,\overline{P}_1^2 + 9.606\,\overline{P}_1(t) + 373.7$$

Expected NO_x emission characteristic of thermal plant (kg/h):

$$\overline{J}_{21}(t) = 0.006483\,\overline{P}_1^2 - 0.79027\,\overline{P}_1(t) + 28.82488$$

Expected SO_2 emission characteristic of thermal plant (kg/h):

$$\overline{J}_{31}(t) = 0.00232\,\overline{P}_1^2 + 3.84632\,\overline{P}_1(t) + 182.2605$$

Expected CO_2 emission characteristic of thermal plant (kg/h):

$$\overline{J}_{41}(t) = 0.084025\,\overline{P}_1^2 - 2.9445484\,\overline{P}_1(t) + 137.7043$$

Expected hydro plant characteristics (Mm^3/h):

$$\overline{q}_2(t) = 2.19427 \times 10^{-5}\ \overline{P}_2^2(t) - 2.5709 \times 10^{-4}\ \overline{P}_2(t) + 1.742333$$

$$\overline{R}_2 = 72.4797 \text{ Mm}^3$$

Expected B-coefficients (MW^{-1}):

$$\overline{B}_{11} = 0.00005,\ \overline{B}_{12} = \overline{B}_{21} = 0.00001,\ \overline{B}_{22} = 0.00015$$

If the coefficient of variation is zero and random variables are uncorrelated to each other, then the problem is considered deterministic. The random variables are uncorrelated only if the correlation coefficient is zero.

Effect of variance and covariance

Owing to the existence of the variances and covariance of random variables in the formulation of the problem, an analysis of variations in objectives $\overline{J}_i$ seems necessary. Actually, probability properties are known from past history or can be estimated via the Monto Carlo simulation techniques [Sen Gupta, 1972]. There is a need to use exact values of coefficients of variation and correlation coefficients as and when required. But, here in the study assumed values of these coefficients are chosen. Since variance is represented by the coefficient of variation, the coefficients of variation are varied from 0% to 10% in steps of 1%. By giving full weightage to one objective and neglecting others, the effect of variance is obtained on all objectives one by one. The percentage relative deviations in total expected cost, NO_x emission, SO_2 emission and CO_2 emission from their respective deterministic values with respect to the coefficients of variation are shown in Figure 6.13. It is observed from this figure that the relative percentage deviation of all the objectives increases as the variance increases. The effect of variance is more on the operating cost as compared to NO_x, SO_2, and CO_2 emissions, respectively.

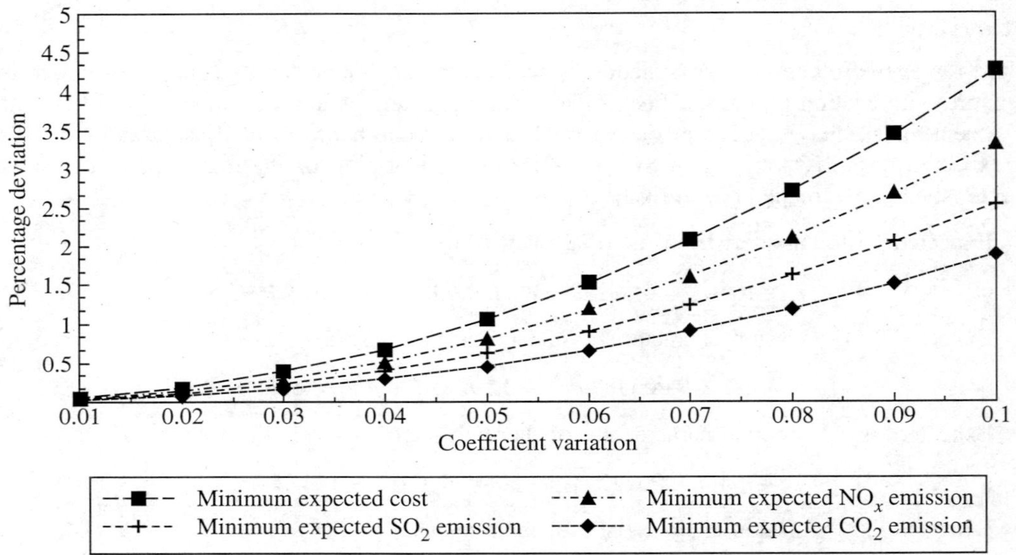

FIGURE 6.13 Percentage deviation in expected minimum cost, NO_x emission, SO_2 emission, and CO_2 emission with respect to coefficients of variation, respectively.

By setting all weights equal to 0.20, the effect of variance on all objectives is observed simultaneously. The percentage relative deviations in total expected cost, NO_x emission, SO_2 emission and CO_2 emission from their respective deterministic values, with respect to coefficients of variation are shown in Figure 6.13. It is observed from Figure 6.14 that the relative percentage deviations of all the objectives increase as the variance increases. The effect on cost becomes smaller compared to NO_x, SO_2 and CO_2 emissions when equal importance is given to all objectives. Further, the effect of variance on water conversion factor is also considerable.

The covariance of bivariate random variables can be considered positive or negative. The covariance is represented by correlation coefficients. The correlation coefficients are varied from –1.0 to 1.0 in steps of 0.2. The percentage relative deviations in total expected cost, NO_x emission, SO_2 emission and CO_2 emission from their respective deterministic values with respect to correlation coefficient ($R_{P_iP_j}$ $(i \neq j)$ are shown in Figure 6.15. The weights w_1, w_2, w_3, w_4, and w_5 are taken as 0.25, 0.25, 0.25, 0.25, and 0.0 respectively. It is examined that (i) there is an increase in the percentage relative deviations in total expected cost $\overline{J}_1$ as the value of $R_{P_iP_j}$ $(i \neq j)$ is changed from a negative value to a positive value, (ii) there is a decrease in the percentage relative deviations in total expected NO_x emission $\overline{J}_2$ as the value of $R_{P_iP_j}$ $(i \neq j)$ is changed from a negative value to a positive value, and (iii) there is a very small effect on SO_2 emission $\overline{J}_3$, and

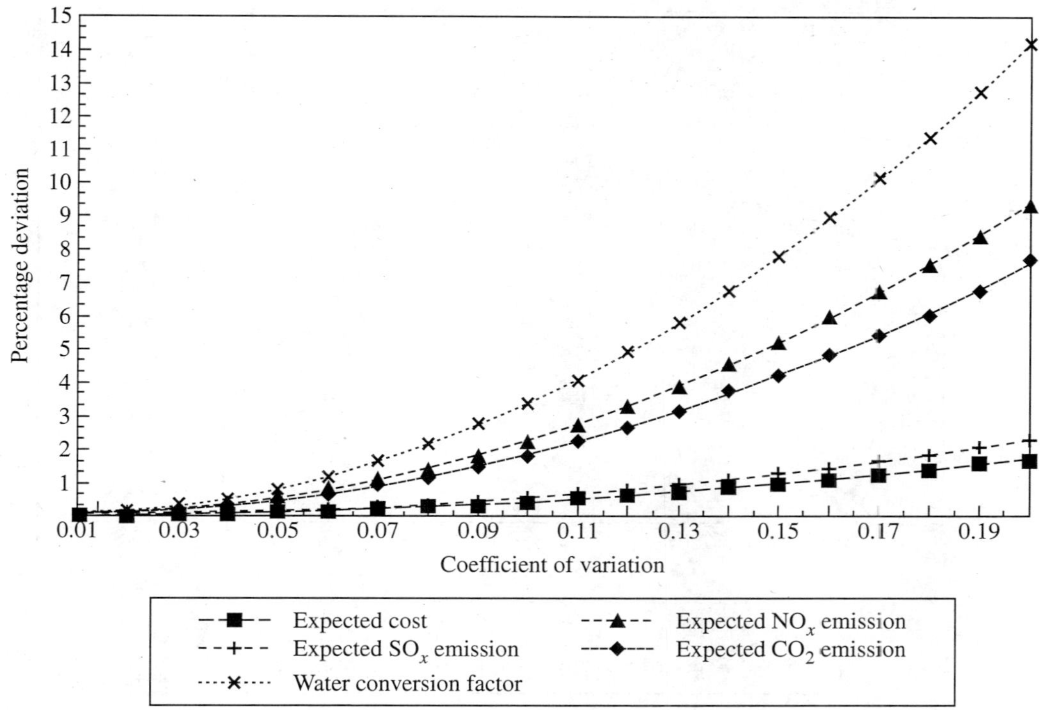

FIGURE 6.14 Percentage deviation in expected minimum cost, NO$_x$ emission, SO$_2$ emission, CO$_2$ emission, and water conversion factor with respect to coefficients of variation, respectively when all weights are set equal to 0.2.

CO_2 emission $\bar{J}_4$ objectives. From this study, it has been observed that the existence of random variables gives a significant effect to each objective either considered individually or in the multiobjective framework.

Minimum and maximum values of objectives

To compute the membership function, $\mu(\bar{J}_i)$ of $\bar{J}_i$ objective, there is a need to find the minimum and maximum values of that objective. Minimum objective values are obtained by giving full weightage to one objective and neglecting others. When the assigned weight value is 1.0, it means that full weightage is given to the objective and when the assigned weightage is zero, the objective is neglected. Maximum objective values are obtained by exploiting their conflicting nature. If an objective $\bar{J}_i$ is in conflict with another objective $\bar{J}_j$ then the $\bar{J}_j$ objective will have the maximum value corresponding to the minimum value of $\bar{J}_i$ objective or vice versa.

Owing to the conflicting nature of objectives, $\bar{J}_2$, $\bar{J}_4$, and $\bar{J}_5$ will have maximum values when $\bar{J}_1$ is minimum. The objective $\bar{J}_3$ will have maximum value when $\bar{J}_2$ is minimum. The obtained minimum and maximum objective values are given below for uncorrelated random variables having 10% variance.

$$\bar{J}_1^{\min} = 96450.77 \ \$ \qquad \bar{J}_1^{\max} = 97475.09 \ \$$$

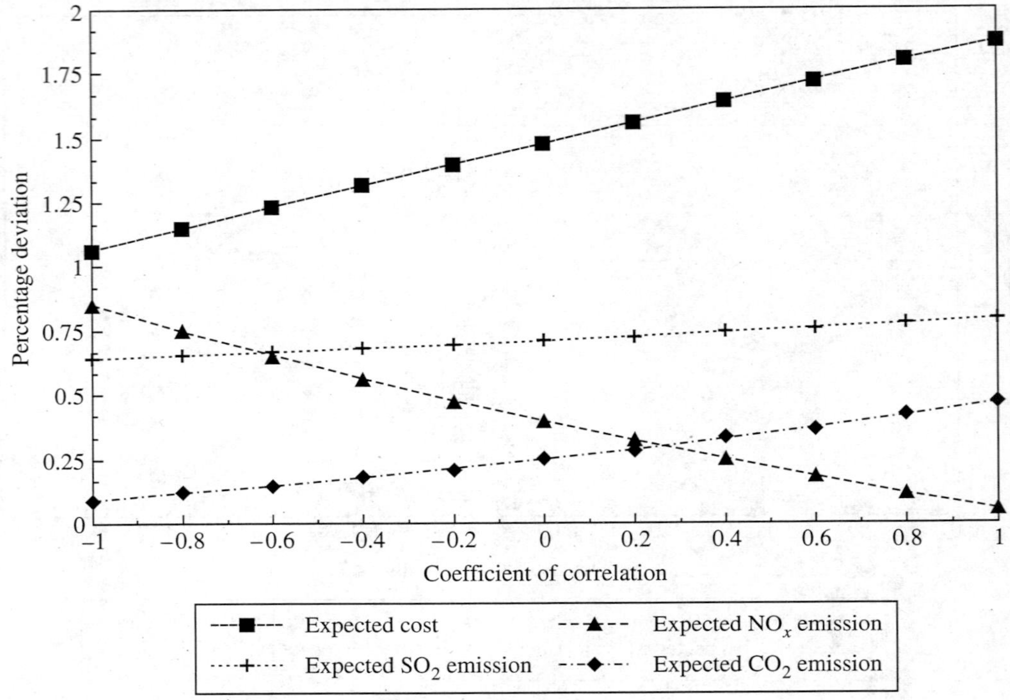

FIGURE 6.15 Percentage deviation in expected cost, NO_x emission, SO_2 emission, and CO_2 emission with respect to coefficients of correlation, respectively.

$$\bar{J}_2^{\min} = 14591.82 \text{ kg} \qquad \bar{J}_2^{\max} = 15169.05 \text{ kg}$$
$$\bar{J}_3^{\min} = 44337.36 \text{ kg} \qquad \bar{J}_3^{\max} = 44665.53 \text{ kg}$$
$$\bar{J}_4^{\min} = 245810.70 \text{ kg} \qquad \bar{J}_4^{\max} = 252472.20 \text{ kg}$$
$$\bar{J}_5^{\min} = 46044.05 \text{ MW}^2 \qquad \bar{J}_5^{\max} = 46805.98 \text{ MW}^2$$

Determination of the optimal or 'best' solution

First, the optimal or best solution is found for only two objectives $\bar{J}_1$ and $\bar{J}_2$. The weights w_1 and w_2, respectively, are varied in the range 0.0 to 1.0 so that their sum is 1.0. The weights, w_3, w_4, and w_5 are taken zero to neglect the other objectives, $\bar{J}_3$, $\bar{J}_4$, and $\bar{J}_5$, respectively. The random variables are assumed independent of each other with 10% variation. Total expected cost $\bar{J}_1$, and NO_x emission $\bar{J}_2$, are tabulated in Table 6.33, for eleven possible weight combinations which corresponds to non-inferior solutions. The membership functions $\mu(\bar{J}_1)$ and $\mu(\bar{J}_2)$ of $\bar{J}_1$ and $\bar{J}_2$ objectives, respectively, are also presented in Table 6.33. The expected cost $\bar{J}_1$ rises and the expected NO_x emission $\bar{J}_2$ declines, when w_1 weightage to expected cost is reducing. Similar trend has been explored in the membership functions of $\bar{J}_1$ and $\bar{J}_2$ objectives. It can be concluded from Figure 6.16 that if two objectives are in conflict then their membership functions are also in

TABLE 6.33 Total expected cost and NO_x emissions correspond to non-inferior set, when only two objectives are considered ($w_3 = w_4 = w_5 = 0$)

Sr. no.	w_1	w_2	$\bar{J}_1$ ($)	$\bar{J}_2$ (kg)	$\mu(\bar{J}_1)$	$\mu(\bar{J}_2)$	μ_D
1	1.0	0.0	96028.66	14839.02	1.0000	0.0000	0.07057
2	0.9	0.1	96032.16	14774.06	0.9966	0.1125	0.07827
3	0.8	0.2	96044.09	14707.11	0.9851	0.2284	0.08564
4	0.7	0.3	96067.08	14638.60	0.9628	0.3471	0.09244
5	0.6	0.4	96104.64	14569.26	0.9265	0.4671	0.09835
6	0.5	0.5	96161.45	14500.20	0.8716	0.5867	0.10292
7	**0.4**	**0.6**	**96243.90**	**14433.13**	**0.7919**	**0.7029**	**0.10549**
8	0.3	0.7	96360.71	14370.68	0.6789	0.8110	0.10515
9	0.2	0.8	96523.80	14316.77	0.5213	0.9044	0.10061
10	0.1	0.9	96750.05	14277.39	0.3026	0.9726	0.08999
11	0.0	1.0	97063.07	14261.55	0.0000	1.0000	0.07057

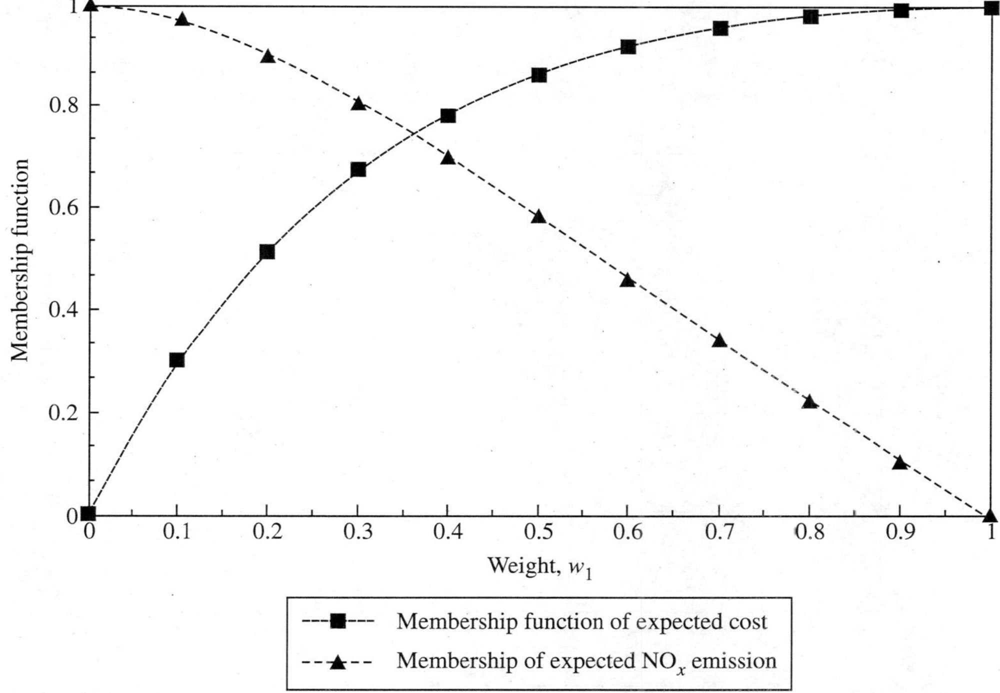

FIGURE 6.16 Variation of membership functions of expected cost and NO_x emission with respect to weight, w_1, where $w_2 = 1.0 - w_1$ and $w_3 = w_4 = w_5 = 0$.

conflict or vice-versa. Figures 6.17 and 6.18 show that the expected cost is in conflict with the expected SO_2 emission and risk level, respectively.

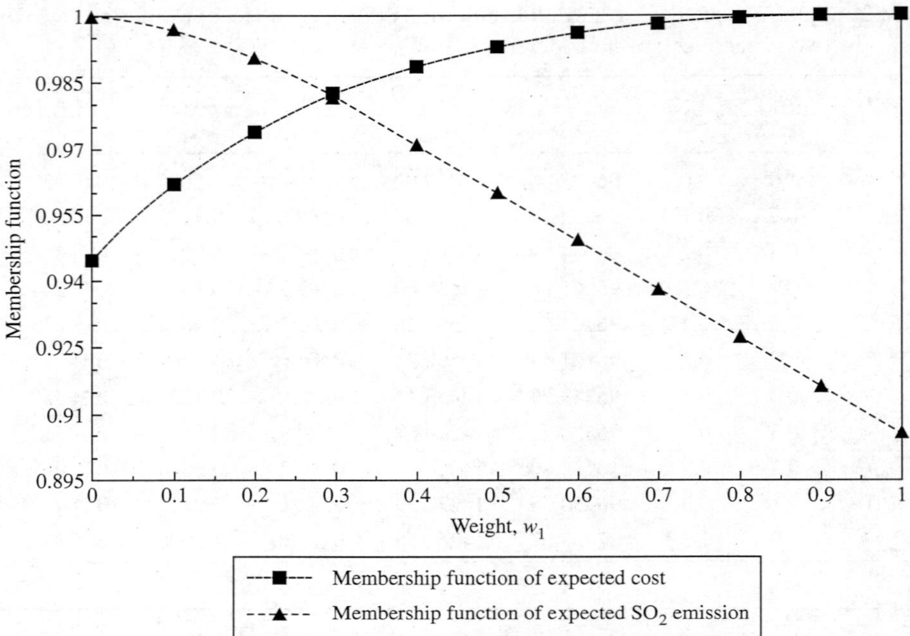

FIGURE 6.17 Variation of membership functions of expected cost and SO_2 emission with respect to weight, w_1, where $w_3 = 1 - w_1$ and $w_2 = w_4 = w_5 = 0$.

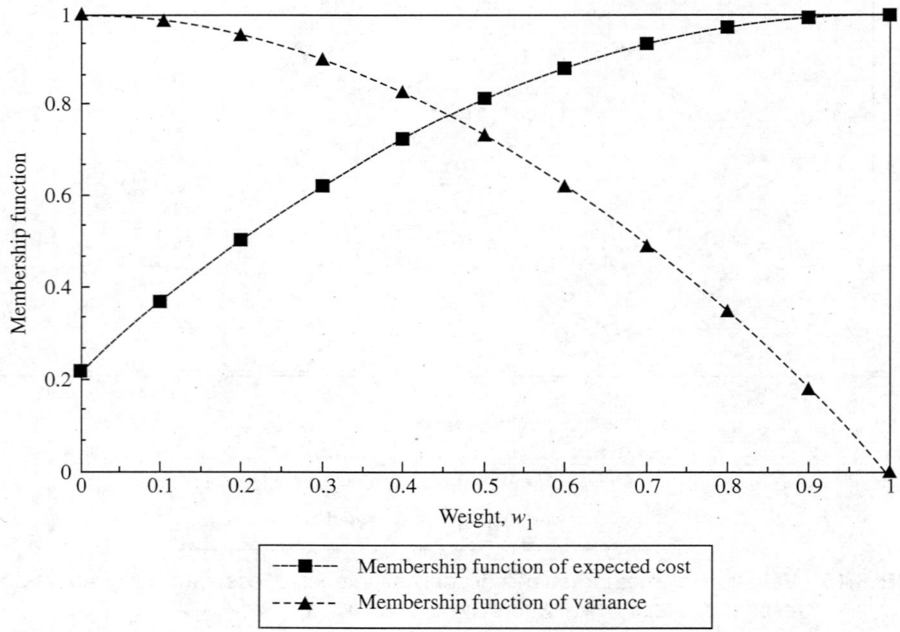

FIGURE 6.18 Variation of membership functions of expected cost and variance in power generation with respect to weight, w_1, where $w_5 = 1 - w_1$ and $w_2 = w_3 = w_4 = 0$.

The non-inferior solution that attains the maximum membership μ_D^k, is distinguished as the best solution among the non-inferior solutions. The weight combination, presented at serial number 7 in Table 6.33 gives the maximum value of μ_D^k, i.e. 0.105049 and therefore provides the best or preferred weight combination.

The non-inferior solutions for 126 different simulated weight combinations are generated considering all the objectives simultaneously. Non-inferior solution that acquires the maximum membership μ_D^k, is chosen as the best solution and is furnished in Table 6.34. The best solutions are secured for distinct values of coefficients of variation and correlation coefficients in diverse situations and are conferred in Table 6.34.

TABLE 6.34 Best expected optimal schedules from non-inferior set

	C_{P_i}	$R_{P_iP_j}$	*Cost* ($)	NO_x *emission* (kg)	SO_2 *emission* (kg)	CO_2 *emission* (kg)	*Risk* (MW^2)	μ_D^k
1	0.01	0.0	96386.43	14360.12	44156.03	242124.3	460.4161	0.0081656
2	0.05	0.0	96480.82	14442.96	44214.58	243249.0	11516.66	0.0081787
3	0.10	0.0	96814.30	14686.73	44410.25	246600.1	46125.73	0.0082168
4	0.10	1.0	96735.22	14726.10	44386.10	247042.8	88802.83	0.0091561
5	0.10	–1.0	96802.52	14687.49	44405.48	246601.1	3455.506	0.0082648

The weight combinations and the water conversion factor ν_2, corresponding to best schedules are depicted in Table 6.36. For case one, the achieved expected generation schedules of 24 hours have been furnished in Table 6.35. For each sub-interval, expected transmission loss $\overline{P_L}$, incremental cost λ, and expected discharge $\overline{q}_2$, are exhibited in Table 6.35. The attained equality constraint ΔP_D, during each sub-interval, is also provided in Table 6.35 which shows the accuracy of the obtained solutions.

Conventional economic short-term fixed-head hydrothermal power dispatch method allocates generation schedule to the individual generating units based upon deterministic cost function and load demand, ignoring inaccuracies and uncertainties. Such generation schedules result in the lowest expected total cost, but this cost is also associated with a relatively large variance that can be interpreted as risk measure. Moreover, in power system operation planning, there exist multiple objectives to be attained, which conflict with each other and are subject to a mutual interface. It means that any one objective can be improved only at the expense of other objectives. In the multiobjective framework, the analysis of hydrothermal short-range fixed-head is undertaken with explicit recognition of uncertainties in production cost, NO_x, SO_2 and CO_2 emissions and load demand.

6.8 STOCHASTIC MULTIOBJECTIVE LONG-TERM HYDROTHERMAL SCHEDULING

A modern power system may consist of several thermal, conventional hydro power plants connected to various load centres through a lossy transmission network. With the insignificant incremental cost involved in hydro generation, the problem of minimizing the operational cost of a hydrothermal system can be reduced essentially to that of minimizing the fuel cost for thermal

TABLE 6.35 Expected schedule of each sub-interval for case one given in Table 6.34 and Table 6.36

k	$\overline{P}_{Dk}$ (MW)	λ	$\overline{P}_{Lk}$ (MW)	$\overline{P}_{1k}$ (MW)	$\overline{P}_{2k}$ (MW)	ΔP_{Dk} (MW)	$\overline{q}_{2k}$ (Mm^3/h)
1	455.0	9.90335	10.81529	259.7265	206.0888	0.000013	2.621406
2	425.0	9.41218	9.57364	237.7585	196.8151	–0.000001	2.541795
3	415.0	9.24936	9.17863	230.4520	193.7266	–0.000021	2.516119
4	407.0	9.11942	8.86938	224.6126	191.2567	0.000061	2.495888
5	400.0	9.00595	8.60373	219.5074	189.0963	–0.000033	2.478411
6	420.0	9.33072	9.37495	234.1042	195.2707	0.000047	2.528904
7	487.0	10.43174	12.23392	283.2397	215.9942	0.000034	2.710609
8	604.0	12.40413	18.26157	369.9315	252.3301	–0.000023	3.074704
9	665.0	13.45848	21.93765	415.5875	271.3502	–0.000034	3.288394
10	675.0	13.63308	22.57575	423.1026	274.4732	–0.000053	3.324999
11	695.0	13.98378	23.88217	438.1586	280.7235	0.000032	3.399545
12	705.0	14.15988	24.55056	445.6997	283.8509	–0.000040	3.437487
13	580.0	11.99425	16.91630	352.0551	244.8612	–0.000019	2.995131
14	605.0	12.42126	18.31886	370.6774	252.6415	0.000011	3.078074
15	616.0	12.61010	18.95552	378.8879	256.0677	–0.000048	3.115442
16	653.0	13.24963	21.18517	406.5808	267.6043	–0.000013	3.245054
17	721.0	14.44270	25.64107	457.7835	288.8576	0.000021	3.499124
18	740.0	14.78024	26.96996	472.1622	294.8078	–0.000010	3.573808
19	700.0	14.07176	24.21510	441.9281	282.2870	0.000013	3.418460
20	678.0	13.68556	22.76915	425.3588	275.4104	–0.000019	3.336067
21	630.0	12.85130	19.78314	389.3524	260.4306	0.000120	3.163771
22	585.0	12.07942	17.19190	355.7754	246.4165	–0.000059	3.011500
23	540.0	11.31724	14.79958	322.3689	232.4306	0.000024	2.868128
24	503.0	10.69769	12.97995	295.0278	220.9521	0.000093	2.756874

where $\Delta P_D = (\overline{P}_1 + \overline{P}_2) - \overline{P}_L - \overline{P}_D$

TABLE 6.36 Weight combination and water conversion factor corresponding to the schedule given in Table 6.34

	$(w_1, w_2, w_3, w_4, w_5)$	v_2 (\$/Mm^3)
1	(0.4,0.1,0.2,0.2,0.1)	1050.432
2	(0.5,0.2,0.1,0.1,0.1)	1140.219
3	(0.4,0.1,0.3,0.1,0.1)	1122.390
4	(0.6,0.1,0.1,0.1,0.1)	1197.168
5	(0.5,0.1,0.2,0.1,0.1)	1195.220

plants under the constraints of the water available for hydro generation in a planned period. Mostly, hydrothermal optimal scheduling is achieved, with the assumption, that the water inflows to the reservoirs and the load demands are known with complete certainty. However, this is not true.

The availability of limited amount of hydroelectric energy, as stored water in the system reservoirs, makes the optimal operation complex, because it creates a link between an operating decision in a given stage and the future consequences of this decision. Further, it is impossible to have perfect forecasts of the future inflow sequence and the load variation during a given period. Therefore, for long-term storage regulation, it becomes necessary to account for the random nature of the load and river inflow and so a stochastic representation of these must be used.

Most of the algorithms incorporate uncertainties in the system load demand and water inflows, but choose a deterministic cost function for thermal generating units. A major source of uncertainty in optimal dispatch is that associated with cost coefficients [Dhillon et al., 1993]. However with the increasing concern recently given to the environmental considerations [Dhillon et al., 1993; Dhillon et al., 1994], a revised generation scheduling for the hydrothermal power system is required that meets the constraints of available water at hydro plants and load demand for power while accounting for both cost and NO_x emission.

Fuzzy sets were first introduced in solving power system long-range decision making problems. Fuzzy decision making theories attempt to deal with the vagueness or fuzziness inherent in subjective or imprecise determinations of preferences, constraints, and goals. Tapia and Murtagh [1991] put up a methodology for solving a decision making problem involving a multiplicity of objectives and selection criteria for the best compromised solution.

The intent of this section is to provide a technique that allows scheduling of long-range hydrothermal system probabilistically considering stochastic cost and NO_x emission curves for thermal power generation units and uncertainty in load demand and reservoir water inflows. However, there is a growing trend towards formulating a multiobjective optimization problem [El-Hawary and Ravindranath, 1991], so, the approach is developed by formulating hydrothermal scheduling as multiobjective optimization problem. The expected fuel costs and NO_x emission over whole of the planning period are considered as two conflicting objectives. The formulation also incorporates any possible deviations in generations over whole of the planning period as the third objective to be minimized. The weighted minimax technique [Klir and Folger, 1993] is used to generate the non-inferior set by converting the problem into a scalar optimization problem. To reduce the complexity of the problem, interval-wise decomposition is carried out. Each sub-problem is separately solved by using the conjugate gradient method to obtain the optimal discharge [Parti, 1987]. In each subinterval, thermal generations are calculated by a simplified technique, which reduces the economic dispatch problem into an equivalent lossless problem. The method is less time consuming. A numerical example of a power system consisting of three-hydro and four-thermal plants is solved and the results are presented.

6.8.1 Stochastic Multiobjective Optimization Problem Formulation

In this section, the multiobjectives with equality and inequality constraints concerning the power system optimization problem are described. The important objectives are considered here, i.e.

1. Economic operations
2. Minimal impacts on environment
3. Expected deviations due to unsatisfied loads.

The stochastic formulation is adopted by considering fuel cost coefficients, NO_x emission coefficients, load demand and water inflows into reservoirs as random variables. Water inflows into reservoirs of various hydro plants are assumed to be statistically correlated during the same subintervals but independent at different subintervals. The stochastic models are converted to their deterministic equivalents by taking their expected values, with the assumption that all the random variables are normally distributed. A hydrothermal system is considered with N thermal and L hydro plants. The problem is visualized as an M stage decision process by subdividing the planning period into M subintervals.

Expected fuel cost

The aim is to optimize the running cost of thermal stations with full use of water available during the optimization period. The objective function, which is fuel cost of the thermal plant, is assumed to be approximated by a quadratic function of generator power output and is given as

$$F_1 = \sum_{m=1}^{M}\left[\sum_{i=1}^{N}(a_i(P_i^m)^2 + b_iP_i^m + c_i)\right]\$ \tag{6.149}$$

where a_i, b_i, and c_i are cost coefficients. P_i^m is the thermal power generation during the mth subinterval.

A stochastic model of function F_1 during the mth subinterval is formulated by considering cost coefficients and load demand, during the mth subinterval as random variables. The expected value of fuel cost function may be obtained through expanding the function using Taylor's series about the mean. The obtained expected fuel cost during the mth subinterval is represented by

$$\overline{F}_1^m = \sum_{i=1}^{N}[\overline{a}_i(\overline{P}_i^m)^2 + \overline{b}_i\overline{P}_i^m + \overline{c}_i + \overline{a}_i \text{ var }(P_i^m) + 2\overline{P}_i^m \text{ cov }(a_i, P_i^m) + \text{cov }(b_i, P_i^m)] \text{ \$/h} \tag{6.150}$$

where

$\overline{P}_i^m$ is the expected value of thermal generator output during the mth subinterval.

$\overline{a}_i$, $\overline{b}_i$, and $\overline{c}_i$ are the expected cost coefficients.

Equation (6.150) can be rewritten as

$$\overline{F}_1^m = \sum_{i=1}^{N}[A_i^m(\overline{P}_i^m)^2 + B_i^m\overline{P}_i^m + \overline{c}_i] \text{ \$/h} \tag{6.151}$$

where

$$A_i^m = [1.0 + (C(P_i^m))^2 + 2R(a_i, P_i^m)\,C(a_i)\,C(P_i^m)]\,\overline{a}_i$$

$$B_i^m = [1.0 + R(b_i, P_i^m)\,C(b_i)\,C(P_i^m)\,\overline{b}_i$$

$C(P_i^m)$, $C(a_i)$, and $C(b_i)$ are the coefficients of variation of random variables P_i^m, a_i, and b_i, respectively. $R(a_i, P_i^m)$ is the correlation coefficient of random variables a_i and P_i^m, and $R(b_i, P_i^m)$ is the correlation coefficient of random variables b_i and P_i^m.

Expected NO_x emission

Only thermal power stations are major causes of atmospheric pollution because of high concentration of pollutants caused by them. The NO_x emission curve for thermal power plants can be directly related to the cost curve through the emission rate per MBtu, which is a constant factor for a given type of fuel. The aim is to optimize the NO_x emission of thermal stations with full use of water available during the optimization period. The amount of NO_x emission is given as a function of the generator output P_i^m, which is quadratic.

$$F_2 = \sum_{m=1}^{M} \left[\sum_{i=1}^{N} (d_i (P_i^m)^2 + e_i P_i^m + f_i) \right] \text{kg} \tag{6.152}$$

where d_i, e_i, and f_i are emission coefficients.

A stochastic model is formulated by considering emission coefficients and load demand as random. Using Taylor's series and taking expectations, the expected NO_x emission for the mth subinterval is obtained as

$$\overline{F}_2^m = \sum_{i=1}^{N} \left[\overline{d}_i (\overline{P}_i^m)^2 + \overline{e}_i \overline{P}_i^m + \overline{f}_i + \overline{d}_i \text{ var}(P_i^m) + 2\overline{P}_i^m \text{ cov}(d_i, P_i^m) + \text{cov}(e_i, P_i^m) \right] \text{kg/h} \tag{6.153}$$

where $\overline{d}_i$, $\overline{e}_i$, and $\overline{f}_i$ are the expected emission coefficients.

Rewriting the above equation as

$$\overline{F}_2^m = \sum_{i=1}^{N} \left[(D_i^m (\overline{P}_i^m)^2 + E_i^m \overline{P}_i^m + \overline{f}_i \right] \text{kg/h} \tag{6.154}$$

where

$D_i^m = [1.0 + (C(P_i^m)^2 + 2R(d_i, P_i^m)\, C(d_i)\, C(P_i^m)]\, \overline{d}_i$

$E_i^m = [1.0 + R(e_i, P_i^m)\, C(e_i)\, C(P_i^m)]\, \overline{e}_i$

$C(d_i)$ and $C(e_i)$ are the coefficients of variation of random variables, d_i and e_i, respectively.

$R(d_i, P_i^m)$ is the correlation coefficient of random variables d_i and P_i^m.

$R(e_i, P_i^m)$ is the correlation coefficient of random variables e_i and P_i^m.

Expected deviations

Since generator outputs P_i^m of hydro and thermal plants are treated as random variables, the expected deviations are proportional to the expectation of the square of unsatisfied load demand during the mth subinterval. These expected deviations on the whole of the planning period are considered as an objective to be minimized. The expected deviations during the mth subinterval are represented as

$$\overline{F}_3^m = E\left[\left(\overline{P}_D^m + \overline{P}_L^m - \sum_{i=1}^{T} P_i^m \right)^2 \right] \text{MW}^2 \tag{6.155}$$

where

$\overline{P}_D^m$ is the expected load demand during the mth subinterval

$\overline{P}_L^m$ are the expected transmission losses during the mth subinterval

T is the total number of hydro and thermal plants.

This on simplification reduces to:

$$\overline{F}_3^m = \sum_{i=1}^{T} \text{var}\,(P_i^m) + \sum_{i=1}^{T} \sum_{\substack{j=1 \\ j \neq i}}^{T} \text{cov}\,(P_i^m, P_j^m) \quad \text{MW}^2 \tag{6.156}$$

It is assumed here that thermal and hydro generations are independent of each other during the different subintervals.

Expected transmission losses

The power transmission losses expressed through the well-known loss formula are given by

$$P_L^m = \sum_{i=1}^{T} \sum_{j=1}^{T} P_i^m\, B_{ij}\, P_j^m \quad \text{MW} \tag{6.157}$$

where B_{ij}s are the B-coefficients.

The power generations' P_i^m during the mth subinterval are dependent random variables. The expected transmission losses during the mth subinterval are obtained using the Taylor's series with the assumption that thermal and hydro generations are independent of each other during the different subintervals.

$$\overline{P}_L^m = \sum_{i=1}^{T} \sum_{j=1}^{T} \overline{P}_i^m \overline{B}_{ij} \overline{P}_j^m + \sum_{i=1}^{T} \overline{B}_{ii}\, \text{var}\,(P_i^m) + \sum_{i=1}^{T-1} \sum_{j=i+1}^{T} 2\,\overline{B}_{ij}\, \text{cov}\,(P_i^m, P_j^m) \quad \text{MW} \tag{6.158}$$

Expected water storage equation

If all hydro plants are on different water streams, the storage at the end of the mth subinterval can be obtained from the equation,

$$X_j^{m+1} = X_j^m + J_j^m - Q_j^m - S_j^m \qquad (j = 1, 2, ..., L) \tag{6.159}$$

where

X_j^m is water storage at the jth reservoir during the mth subinterval
J_j^m is water inflow into the jth reservoir during the mth subinterval
S_j^m is spillage from the jth reservoir during the mth subinterval
Q_j^m is water discharge through the jth turbine during the mth subinterval.

Spillage occurs only when the reservoir storage limit is exceeded. With the assumptions that water inflows and water storage during the mth subinterval are uncorrelated to each other, the expected value of Eq. (6.159) is given as

$$\overline{X}_j^{m+1} = \overline{X}_j^m + \overline{J}_j^m - Q_j^m - S_j^m \qquad (j = 1, 2,.., L) \tag{6.160}$$

where

$\overline{X}_j^m$ is the expected water storage at the jth reservoir during the mth interval
$\overline{J}_j^m$ is the expected water inflow into the jth reservoir during the mth interval.

The corresponding variances are:

$$\text{var}(X_j^{m+1}) = E[(X_j^{m+1} - \overline{X}_j^{m+1})^2] \qquad (j = 1, 2,.., L)$$

On simplification, the previous equation can be rewritten as

$$\mathrm{var}(X_j^{m+1}) = \mathrm{var}(X_j^m) + \mathrm{var}(J_j^m) \qquad (j = 1, 2, ..., L) \qquad (6.161)$$

Similarly, covariance is

$$\mathrm{cov}(X_j^{m+1}, X_k^{m+1}) = E[(X_j^{m+1} - \overline{X}_j^{m+1})(X_k^{m+1} - \overline{X}_k^{m+1})] \qquad (j = 1, 2, ..., L; \quad k = 1, 2, ..., L; \quad j \neq k)$$

After simplification, covariance is

$$\mathrm{cov}(X_j^{m+1}, X_k^{m+1}) = \mathrm{cov}(X_j^m, X_k^m) + \mathrm{cov}(J_j^m, J_k^m) \quad (j = 1, 2, ..., L; \quad k = 1, 2, ..., L; \quad j \neq k) \quad (6.162)$$

Expected hydro generation

The average hydro generation during any subinterval depends on the water discharge through the turbine and on the average head, which is also a function of the storage. The average hydro generation during the mth subinterval is given by

$$P_{j+N}^m = h_j [1 + 0.5g_j(2X_j^m + J_j^m - Q_j^m - S_j^m)] (Q_j^m - \mu_j) \qquad (j = 1, 2, ..., L) \qquad (6.163)$$

where

h_j is the basic head of the jth hydro plant

g_j is the water head correction factor to account for variation in head with storage of the jth hydro plant

μ_j is the non-effective water discharge of the jth hydro plant.

Since water inflows and water storage are random variables, so hydro generations will also be random. Expected hydro generation of the mth subinterval can be written as

$$\overline{P}_{j+N}^m = h_j [1 + 0.5g_j(2\overline{X}_j^m + \overline{J}_j^m - Q_j^m - S_j^m)](Q_j^m - \mu_j) \qquad (j = 1, 2, ..., L) \qquad (6.164)$$

Variance of hydro power is given as follows:

$$\mathrm{var}(P_{j+N}^m) = E[(P_{j+N}^m - \overline{P}_{j+N}^m)^2] \qquad (j = 1, 2, ..., L)$$

On simplification, the above equation can be written as

$$\mathrm{var}(P_{j+N}^m) = h_j^2 (Q_j^m - \mu_j)^2 g_j^2 [\mathrm{var}(X_j^m) + 0.25\, \mathrm{var}(J_j^m)] \qquad (j = 1, 2, ..., L) \qquad (6.165)$$

Covariance of hydro generation is given as below

$$\mathrm{cov}(P_{j+N}^m, P_{k+N}^m) = h_j h_k (Q_j^m - \mu_j)(Q_k^m - \mu_k) g_j g_k [\mathrm{cov}(X_j^m, X_k^m) + 0.25\, \mathrm{cov}(J_j^m, J_k^m)]$$

$$(j = 1, 2, ..., L; \quad k = 1,2, ..., L; \quad j \neq k) \qquad (6.166)$$

Aggregating the above equations, the hydrothermal multiobjective optimization problem is defined below:

$$\text{Minimize} \qquad \left[\sum_{m=1}^{M} \overline{F}_1^m, \sum_{m=1}^{M} \overline{F}_2^m, \sum_{m=1}^{M} \overline{F}_3^m \right]^T \qquad (6.167a)$$

subject to: (a) expected load demand constraint for the mth subinterval

$$\sum_{i=1}^{T} \overline{P}_i^m - \overline{P}_D^m - \overline{P}_L^m = 0 \tag{6.167b}$$

(b) expected storage continuity constraint

$$\overline{X}_j^{m+1} = \overline{X}_j^m + \overline{J}_j^m - Q_j^m - S_j^m \qquad (j = 1, 2, ..., L) \tag{6.167c}$$

(c) total expected volume of water available constraint

$$\overline{X}_j^{m+1} - \overline{X}_j^1 - \sum_{m=1}^{M} \overline{J}_j^m + \sum_{m=1}^{M} Q_j^m + \sum_{m=1}^{M} S_j^m = 0 \qquad (j = 1, 2, ..., L) \tag{6.167d}$$

(d) expected hydro generation equation

$$\overline{P}_{j+N}^m = h_j [1 + 0.5 g_j (2\overline{X}_j^m + \overline{J}_j^m - Q_j^m - S_j^m)](Q_j^m - \mu_j) \qquad (j = 1, 2, ..., L) \tag{6.167e}$$

(e) expected output of thermal plants

$$\overline{P}_i^{\min} \le \overline{P}_i^m \le \overline{P}_i^{\max} \qquad (i = 1, 2, ..., N) \tag{6.167f}$$

(f) water discharge limit

$$Q_j^{\min} \le Q_j^m \le Q_j^{\max} \qquad (j = 1, 2, ..., L) \tag{6.167g}$$

(g) expected forebay limit of reservoir

$$\overline{X}_j^{\min} \le \overline{X}_j^m \le \overline{X}_j^{\max} \qquad (j = 1, 2, ..., L) \tag{6.167h}$$

6.8.2 Optimal Control Strategy

The scalarized form of the optimization problem is obtained by using the weighted minimax method and is given below:

Minimize
$$\sum_{m=1}^{M} \left[\sum_{k=1}^{3} (w_k \overline{F}_k^m) \right] \tag{6.168a}$$

subject to
$$\sum_{k=1}^{3} w_k = 1 \qquad (w_k \ge 0) \tag{6.168b}$$

where w_k is the level of the weighting coefficients. All other equality and inequality constraints are same as mentioned in the problem of the system of Eqs. (6.167).

This approach yields meaningful result to the decision maker when solved often for different values of $w_k (k = 1, 2, 3)$. The scalarized optimization problem is solved by forming the Lagrangian function, which is obtained by augmenting the objective function with various equality constraints expressed in terms of expected values, through dual variables as follows:

$$L = \sum_{m=1}^{M}\left[\sum_{k=1}^{3} w_k \bar{F}_k^m + \lambda_1^m\left(\bar{P}_D^m + \bar{P}_L^m - \sum_{i=1}^{T}\bar{P}_i^m\right)\right.$$
$$+ \sum_{j=1}^{L}\lambda_{2j}^m\,[\bar{P}_{j+N}^m - h_j\,(1 + 0.5g_j\,(2\bar{X}_j^m + \bar{J}_j^m - Q_j^m - S_j^m))\,(Q_j^m - \mu_j)]$$
$$+ \lambda_{3j}^m\,(\bar{X}_j^{m+1} - \bar{X}_j^m - \bar{J}_j^m + Q_j^m + S_j^m)$$
$$\left.+ \sum_{j=1}^{L}\lambda_{4j}\left(\bar{X}_j^{m+1} - X_j^1 - \sum_{m=1}^{M}\bar{J}_j^m + \sum_{m=1}^{M}Q_j^m + \sum_{m=1}^{M}S_j^m\right)\right] \quad (6.169)$$

During each subinterval, the control variables are the water discharges through the turbines of hydro plants. The reservoir storage and the hydro power generation at the end of the subinterval are obtained from Eqs. (6.160) and (6.163), respectively. Irrespective of the hydro generations the thermal generations satisfy the power transfer constraint to achieve the minimal fuel cost. The dual variables λ_1^m, λ_{2j}^m, and λ_{3j}^m are obtained by equating the partial derivatives of the Lagrangian function with respect to the dependent variables to zero.

$$\frac{\partial L}{\partial \bar{P}_i^m} = \sum_{k=1}^{3} w_k\,\frac{\partial \bar{F}_k^m}{\partial \bar{P}_i^m} + \lambda_1^m\left(\frac{\partial \bar{P}_L^m}{\partial \bar{P}_i^m} - 1\right) = 0 \qquad (i = 1, 2, ..., N) \quad (6.170)$$

$$\frac{\partial L}{\partial \bar{P}_{j+N}^m} = \lambda_{2j}^m + \lambda_i^m\left(\frac{\partial \bar{P}_L^m}{\partial \bar{P}_{j+N}^m} - 1\right) = 0 \qquad (j = 1, 2, ..., L) \quad (6.171)$$

$$\left(\frac{\partial L}{\partial \bar{X}_j^m}\right)_{m\neq 1} = \lambda_{3j}^{m-1} - \lambda_{3j}^m + \lambda_{2j}^m\,h_j\,g_j\,(Q_j^m - \mu_j) = 0 \qquad (j = 1, 2, ..., L) \quad (6.172)$$

$\lambda_{3j}^1 = 0$, since the equations associated with this set of dual variables are redundant.

$$\frac{\partial L}{\partial Q_j^m} = \lambda_{3j}^m + \lambda_{4j} - \lambda_{2j}^m\,h_j\,[1 + 0.5g_j\,(2\bar{x}_j^m + \bar{J}_j^m - S_j^m - 2Q_j^m + \mu_j)]$$
$$+ w_3\left[\frac{\partial}{\partial Q_j^m}(\text{var}\,(P_{j+N}^m)) + \sum_{k=1}^{L} 2\frac{\partial}{\partial Q_j^m}(\text{cov}\,(P_{j+N}^m, P_{k+N}^m))\right] + \lambda_1^m\left(\frac{\partial \bar{P}_L^m}{\partial Q_j^m}\right)$$
$$(j = 1, 2, ..., L) \quad (6.173)$$

where

$$\frac{\partial}{\partial Q_j^m}[\text{var}\,(P_{j+N}^m)] = 2h_j^2 g_j^2\,(Q_j^m - \mu_j)[\text{var}(X_j^m) + 0.25\,\text{var}(J_j^m)] \qquad (j = 1, 2, ..., L) \quad (6.174)$$

$$\frac{\partial}{\partial Q_j^m}[\text{cov}\,(P_{j+N}^m, P_{k+N}^m)] = h_j h_k g_j g_k (Q_k^m - \mu_k)[\text{cov}(X_j^m, X_k^m) + 0.25\,\text{cov}(J_j^m, J_k^m)]$$
$$(j = 1, 2, ..., L;\ j \neq k) \quad (6.175)$$

$$\frac{\partial \bar{P}_L^m}{\partial Q_j^m} = B_{j+N,j+N}\,\frac{\partial}{\partial Q_j^m}[\text{var}(P_{j+N}^m)] + \sum_{k=1}^{L} 2B_{j+N,k+N}\,\frac{\partial}{\partial Q_j^m}[\text{cov}\,(P_{j+N}^m, P_{k+N}^m)]$$
$$(j = 1, 2, ..., L) \quad (6.176)$$

The dual variables λ_{4j} are to be adjusted to maximize the Lagrangian function under the constraints of other optimality conditions. The corresponding gradient vector is

$$\frac{\partial L}{\partial \lambda_{4j}} = \overline{X}_j^{m+1} - \overline{X}_j^1 - \sum_{m=1}^{M} \overline{J}_j^m + \sum_{m=1}^{M} Q_j^m + \sum_{m=1}^{M} S_j^m \qquad (j = 1, 2, ..., L) \tag{6.177}$$

The upper and lower limits on the control variables are taken care of by making these variables equal to the respective bounded values whenever such limits are violated. For the dependent variables, these limits can be considered by augmenting the cost function through the Powell's Penalty function.

Algorithm 6.5: Stochastic Multiobjective Long-Term Hydrothermal Scheduling

To generate the non-inferior solutions, the steps of the algorithm are given below:

1. Set weights w_k $(k = 1, 2, ..., K)$ at desired values.
2. Assume a set of λ_{4j} for $j = 1, 2, ..., L$.
3. Set $m = 1$, $\lambda_{3j}^m = 0$, $\text{var}(X_j^m) = 0$, $\text{cov}(X_j^m, X_k^m) = 0$ $(j = 1, 2, ..., L; k = 1, 2,.. L; j \neq k)$
4. Assume Q_j^m $(j = 1, 2, ..., L)$.
5. Calculate hydro generations P_l^m, $\text{var}(P_l^m)$, $\text{cov}(P_l^m, P_n^m)$ $(j = 1, 2, ..., L; k = 1, 2, ..., L; l = j + N; n = k + N; j \neq k)$ from Eqs. (6.164), (6.165) and (6.166), respectively.
6. Calculate thermal generations P_i^m $(i = 1, 2, ..., N)$ and λ_1^m by solving the following set of equations.

$$\sum_{k=1}^{3} w_k \frac{\partial \overline{F}_k^m}{\partial \overline{P}_i^m} + \lambda_1^m \left(\frac{\partial \overline{P}_L^m}{\partial \overline{P}_i^m} - 1 \right) = 0 \qquad (i = 1, 2, ..., N)$$

$$\overline{P}_D^m + \overline{P}_L^m - \sum_{i=1}^{T} \overline{P}_i^m = 0$$

   The above set of equations is solved by a simplified method [Osyczka and Davies, 1984] which is a less time-consuming algorithm. The method reduces the economic dispatch problem into an equivalent lossless problem.
7. λ_{2j}^m and λ_{3j}^m for $j = 1, 2, ..., L$ are solved using Eqs. (6.171) and (6.172), respectively.
8. Calculate the gradient vector from Eq. (6.173).
   If the optimality conditions are achieved within the prescribed accuracy, then GOTO Step 9, else adjust the expected water discharges using the conjugate gradient method, and GOTO Step 5.
9. Calculate the expected values, variances and covariances of storage from Eqs. (6.160), (6.161) and (6.162), respectively for all the hydro generators.
10. If $(m \geq M)$ then GOTO Step 11,

    else set $m = m + 1$ and GOTO Step 4.
11. Check the convergence using the gradient vector given by Eq. (6.177). If convergence is not achieved, then adjust λ_{4j} using the steepest ascent method.

$$\lambda_{4j} = \lambda_{4j} + \alpha \frac{\partial L}{\partial \lambda_{4j}} \quad (j = 1, 2, ..., L) \text{ and GOTO Step 3.}$$

12. Calculate the overall cost, emission and risk objectives. Calculate the membership function $\mu(F_k)$ with $k = 1, 2, ..., K$, from Eq. (6.149).
13. Check if the schedule with new weight combinations is required.
    If 'Yes' then modify the weights and GOTO Step 2 and repeat
    else stop.

Efficient algorithm to solve the thermal problem during the sub-interval

A simple and efficient economic dispatch algorithm is used. The algorithm reduces the economic dispatch into an equivalent lossless problem from which solutions can be easily obtained. The first step of the algorithm is to calculate an initial guess. It is obtained by solving the following lossless problem for the *m*th subinterval. Time dependence is suppressed here for simplicity.

$$2\alpha_i \overline{P_i} + \beta_i = \lambda_1 \tag{6.178}$$

$$\sum_{i=1}^{N} \overline{P_i} = \overline{P_D} \tag{6.179}$$

where

$$\alpha_i = (w_1 A_i + w_2 D_i + w_3 (C(P_i^m))^2)$$

$$\beta_i = (w_1 B_i + w_2 E_i)$$

From Eq. (6.178)

$$\overline{P_i} = \frac{\lambda_1 - \beta_i}{2\alpha_i} \tag{6.180}$$

After summing Eq. (6.180) over *i*, and on simplification,

$$\lambda_1 = \frac{\overline{P_D} + \sum_{i=1}^{N} \frac{\beta_i}{2\alpha_i}}{\sum_{i=1}^{N} \frac{1}{2\alpha_i}} \tag{6.181}$$

Equations (6.180) and (6.181) determine the analytical solution for the lossless case. The availability of an analytical solution to Eq. (6.181) significantly increases the efficiency of the algorithm.

Consider now the lossy case. The equation to be solved has the form:

$$\sum_{k=1}^{3} w_k \overline{F_k}^{\text{new}} = \lambda_1 \left(1 - \frac{\partial \overline{P_L}}{\partial \overline{P_i}}\right) \qquad (i = 1, 2, ..., N) \tag{6.182}$$

$$\sum_{i=1}^{N} \overline{P_i}^{\text{new}} = \overline{P_D} + \overline{P_L}^{\text{new}} \tag{6.183}$$

These are nonlinear equations in $\overline{P_i}$s and λ_1 and can only be solved iteratively. Let λ_1^{old} and $\overline{P_i}^{\text{old}}$ $(i = 1, 2, ..., N)$ be approximated solutions to Eqs. (6.182) and (6.183). Here the aim is to find new approximations

$$\lambda_1^{\text{new}} = \lambda_1^{\text{old}} + \delta\lambda_1$$

$$\overline{P_i}^{\text{new}} = \overline{P_i}^{\text{old}} + \delta\overline{P_i}$$

The expressions λ_1^{new} and $\overline{P}_i^{\text{new}}$ $(i = 1, 2, ..., N)$ depend on the transmission loss formula used. By Taylor's expression to first order

$$\overline{P}_L^{\text{new}} = \overline{P}_L^{\text{old}} + \sum_{i=1}^{N} \frac{\partial \overline{P}_L^{\text{old}}}{\partial \overline{P}_i} (\overline{P}_i^{\text{new}} - \overline{P}_i^{\text{old}})$$

$$\frac{\partial \overline{P}_L^{\text{new}}}{\partial \overline{P}_i} = \frac{\partial \overline{P}_L^{\text{old}}}{\partial \overline{P}_i} + \sum_{i=1}^{N} \frac{\partial^2 \overline{P}_L^{\text{old}}}{\partial \overline{P}_i \partial \overline{P}_i} (\overline{P}_i^{\text{new}} - \overline{P}_i^{\text{old}})$$

Similarly, the objective functions can be approximated to first order.

$$\frac{\partial \overline{F}_k^{\text{new}}}{\partial \overline{P}_i} = \frac{\partial \overline{F}_k^{\text{old}}}{\partial \overline{P}_i} + \sum_{i=1}^{N} \frac{\partial^2 \overline{F}_k^{\text{old}}}{\partial \overline{P}_i^2} (\overline{P}_i^{\text{new}} - \overline{P}_i^{\text{old}})$$

Putting these values in Eqs. (6.182) and (6.183), the lossless problem can be defined as

$$2\alpha_i^* \overline{P}_i^* + \beta_i^* = \lambda_1^{\text{new}} \qquad (i = 1, 2, ..., N) \tag{6.184}$$

$$\sum_{i=1}^{N} \overline{P}_i^* = \overline{P}_D^* \tag{6.185}$$

where

$$\alpha_i^* = \frac{1}{2}\left(\sum_{k=1}^{3} w_k \frac{\partial^2 \overline{F}_k}{\partial \overline{P}_I^2} + \lambda_1^{\text{old}} \frac{\partial^2 \overline{P}_L^{\text{old}}}{\partial \overline{P}_i^2}\right)\left(\frac{1}{\phi^2}\right)$$

$$\beta_i^* = \left[\sum_{k=1}^{3} w_k \left(\frac{\partial \overline{F}_k^{\text{old}}}{\partial \overline{P}_i} - \frac{\partial^2 \overline{F}_k^{\text{old}}}{\partial \overline{P}_i^2} \overline{P}_i^{\text{old}}\right) - \lambda_1^{\text{old}} \frac{\partial^2 \overline{P}_L^{\text{old}}}{\partial \overline{P}_i^2} \overline{P}_i^{\text{old}}\right]\left(\frac{1}{\phi}\right)$$

$$P_D^* = \overline{P}_L^{\text{old}} + \overline{P}_D - \sum_{i=1}^{N} \frac{\partial \overline{P}_L^{\text{old}}}{\partial \overline{P}_i} \overline{P}_i^{\text{old}}$$

$$P_i^* = \overline{P}_i^{\text{new}} \phi$$

$$\phi = 1 - \frac{\partial \overline{P}_L^{\text{old}}}{\partial \overline{P}_i}$$

To retain the classical form and at the same time to improve convergence, it is possible to include only the *i*th term of summation, i.e.

$$\lambda_1^{\text{old}} \left[\frac{\partial^2 \overline{P}_L^{\text{old}}}{\partial \overline{P}_i^2} (P_i^{\text{new}} - P_i^{\text{old}})\right]$$

6.8.3 Sample System Study

A long range hydrothermal problem with four thermal and three hydro electric plants is considered here. The expected incremental fuel cost and NO_x emission equation of thermal plants considered here are given below:

Expected incremental cost equations (\$/MW-h) are:

$$\partial \overline{F}_{11} / \partial \overline{P}_1^{\,m} = 0.1\, \overline{P}_1^{\,m} + 2.5$$

$$\partial \overline{F}_{12} / \partial \overline{P}_2^{\,m} = 0.12\, \overline{P}_2^{\,m} + 2.6$$

$$\partial \overline{F}_{13} / \partial \overline{P}_3^{\,m} = 0.16\, \overline{P}_3^{\,m} + 2.8$$

$$\partial \overline{F}_{14} / \partial \overline{P}_4^{\,m} = 0.16\, \overline{P}_4^{\,m} + 2.8$$

Expected NO_x emission equations (kg/h) are:

$$\overline{F}_{21}(\overline{P}_1^{\,m}) = 0.00767\,(\overline{P}_1^{\,m})^2 + 0.80507\, \overline{P}_1^{\,m} + 363.7048$$

$$\overline{F}_{22}(\overline{P}_2^{\,m}) = 0.00767\,(\overline{P}_2^{\,m})^2 + 0.80507\, \overline{P}_2^{\,m} + 363.7048$$

$$\overline{F}_{23}(\overline{P}_3^{\,m}) = 0.01378\,(\overline{P}_3^{\,m})^2 - 1.24885\, \overline{P}_3^{\,m} + 137.3701$$

$$\overline{F}_{24}(\overline{P}_4^{\,m}) = 0.01378\,(\overline{P}_4^{\,m})^2 - 1.24885\, \overline{P}_4^{\,m} + 137.3701$$

The average transmission loss coefficients are as follows (MW^{-1}):

$$\overline{B}_{11} = \overline{B}_{22} = 0.0005,\ \overline{B}_{33} = \overline{B}_{44} = 0.0008,\ \overline{B}_{55} = \overline{B}_{66} = 0.0007$$

$$\overline{B}_{77} = 0.0009,\ \overline{B}_{ij} = 0.0 \qquad (i = 1, 2, ..., T;\ \ j = 1, 2, ..., T;\ \ i \neq j)$$

The following values of coefficients of variations and of correlation coefficients are assumed in the study.

$$C(a_i) = C(b_i) = C(P_i^m) = C(d_i) = C(e_i) = 0.1 \qquad (i = 1, 2, ..., N;\ \ m = 1, 2, ..., M).$$

$$R(a_i, P_i^m) = R(b_i, P_i^m) = R(e_i, P_i^m) = R(d_i, P_i^m) = R(P_i^m, P_j^m) = 1.0$$

$$(i = 1, 2, ..., N;\ \ j = 1, 2, ..., N;\ \ i \neq j;\ \ m = 1, 2, ..., M)$$

Actually, probability properties are known from past history or can be estimated via the Monto Carlo simulation techniques. Data for the three hydro plants are presented in Table 6.37. The expected water inflows and the standard deviations for 12 subintervals are given in Table 6.38. The correlation coefficients between the water inflows for the same subinterval are assumed to be 0.5.

The overall expected incremental cost, emission and risk associated are shown in Table 6.39 for different weight combinations. The normalized membership function values are also presented in Table 6.39 that correspond to non-inferior solutions. Each non-inferior solution is optimal. As the partial derivatives of Lagrangian with respect to control variables are achieved equivalent to zero within the prescribed accuracy for each subinterval, the available water is used in full for the whole of planning period by adjusting λ_{4j} so that Eq. (6.177) becomes equivalent to zero within the prescribed accuracy.

TABLE 6.37 Hydro plants data

	Hydro plants		
	1	2	3
h—basic head (metres)	0.98	0.50	0.75
ε—water head correction factor	0.004	0.002	0.004
f—non-effective water discharge	0.0	0.0	0.0
X^1—initial storage (cu.m/sec-month)	100.0	150.0	100.0
X^{m+1}—final storage (cu.m/sec-month)	100.0	150.0	150.0
Maximum allowable discharge (cu.m/sec)	73.0	75.0	80.0
Minimum allowable discharge (cu.m/sec)	3.334	4.353	3.543
Maximum hydro generation (MW)	89.03	57.61	75.0
Minimum hydro generation (MW)	0.0	0.0	0.0

TABLE 6.38 Expected demand, water inflows, standard deviations and assumed discharge

Sub int.	*Demand* (MW)	*Water inflows* (cu.m/sec)			*Standard deviation* (cu.m/sec)			*Assumed discharge* (cu.m/sec)		
		$\bar{J}_1$	$\bar{J}_2$	$\bar{J}_3$	$\sigma(J_1)$	$\sigma(J_2)$	$\sigma(J_3)$	Q_1	Q_2	Q_3
1	200.0	0.8	23.0	0.8	0.08	2.30	0.08	10.8	33.0	10.8
2	210.0	10.0	17.0	10.0	1.00	1.70	1.00	20.0	27.0	20.0
3	205.0	11.0	12.0	11.0	1.10	1.20	1.10	21.0	22.0	21.0
4	180.0	11.9	10.0	11.9	1.19	1.00	1.19	21.9	20.0	21.9
5	195.0	17.0	20.0	17.0	1.70	2.00	1.70	27.0	30.0	27.0
6	200.0	18.5	40.0	18.5	1.85	4.00	1.85	28.5	50.0	28.5
7	220.0	27.6	53.0	27.6	2.76	5.30	2.76	37.6	63.0	37.6
8	204.0	43.8	63.0	43.8	4.38	6.30	4.38	33.8	53.0	33.8
9	189.0	56.4	62.0	56.4	5.64	6.20	5.64	46.4	52.0	46.4
10	199.0	40.3	48.0	40.0	4.03	4.80	4.03	30.3	38.0	30.3
11	207.0	30.1	55.0	30.1	3.01	5.50	3.01	20.1	45.0	20.1
12	198.0	46.3	48.0	46.3	4.63	4.80	4.63	36.0	38.0	36.3

The deterministic incremental cost is given in Table 6.39. The optimal expected schedules for two weight combinations (0.8, 0.0, 0.2) and (0.6, 0.2, 0.2) are shown in Tables 6.40 and 6.41, respectively.

The expected discharges for the corresponding weight combinations are shown in Table 6.42. It can be observed from Figure 6.19 that the relative percentage deviation in total expected incremental cost with respect to the value of the total incremental cost obtained from deterministic schedule is considerable, about 2%. The deviation of total expected incremental cost from deterministic total incremental cost varies as the weight combinations are changed. The variation in weight combination also gives the change in the expected generation schedule of each subinterval which can be observed by comparing the schedules depicted in Tables 6.40 and 6.42, respectively.

TABLE 6.39 Variation in the overall expected incremental cost, NO_x emission, risk, and transmission losses

Sr. no.	w_1	w_2	w_3	$\overline{F_1}$	$\overline{F_2}$	$\overline{F_3}$	μ_D^k
		Deterministic		6800.129	–	–	–
1	1.0	0.0	0.0	6954.131	12511.94	1877.853	0.0834
2	0.8	0.0	0.2	6954.910	12500.69	1875.123	0.1104
3	0.6	0.0	0.4	6964.954	12484.13	1872.207	0.1308
4	0.4	0.0	0.6	6974.381	12457.82	1870.487	0.1466
5	**0.2**	**0.0**	**0.8**	**7001.399**	**12406.45**	**1867.762**	**0.1621**
6	0.7	0.1	0.2	6967.032	12456.69	1874.578	0.1218
7	0.6	0.2	0.2	6974.552	12402.58	1872.525	0.1558
8	0.5	0.3	0.2	7026.460	12336.67	1877.205	0.0887

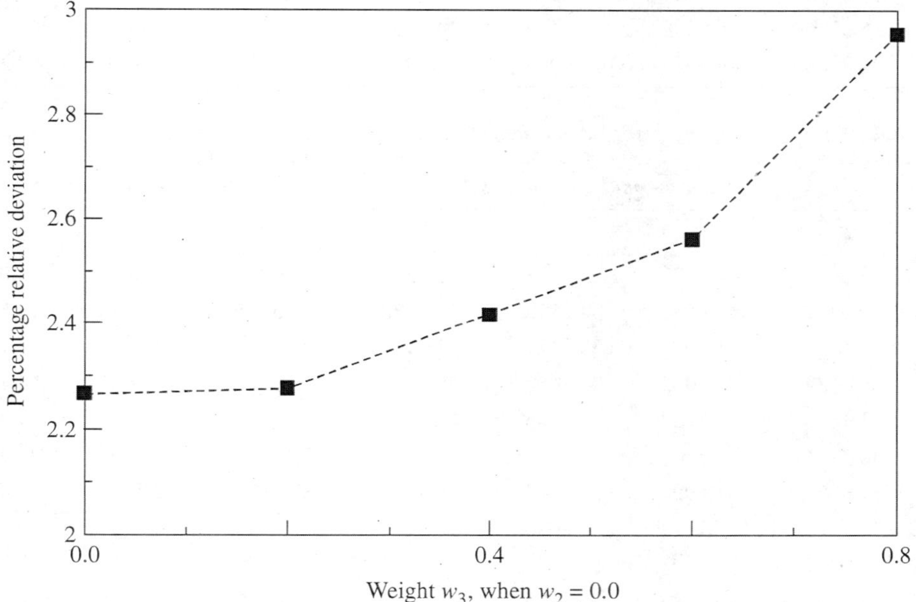

FIGURE 6.19 Percentage relative deviation of expected cost from deterministic value of cost with respect to weights.

The expected discharge of each subinterval is also changed as the weight combination is changed as shown in Table 6.43. The overall expected incremental cost and expected NO_x emission are of conflicting nature. It can be observed from Figure 6.20 that the membership function of cost decreases whereas the membership function of NO_x emission increases. The weight w_3 is fixed to 0.2 and the weights w_1 and w_2 are varied in the range of 0.0 to 0.1 in such a way that the sum of weights is 1.0. The overall expected cost is also in conflict with the variance of generation mismatch for whole of the planning period and is shown in Figure 6.21. The membership function

TABLE 6.40 Expected generation schedule when weights are $w_1 = 0.8$, $w_2 = 0.0$, and $w_3 = 0.2$

Sub int.	Expected thermal generations				Expected hydro generations		
	$\overline{P_1}$	$\overline{P_2}$	$\overline{P_3}$	$\overline{P_4}$	$\overline{P_5}$	$\overline{P_6}$	$\overline{P_7}$
1	55.1656	46.2359	33.4042	33.4042	8.9242	20.1417	7.53448
2	53.4951	44.7988	32.3410	32.3410	19.7942	16.0753	15.9887
3	51.9912	43.5057	31.3838	31.3838	21.4292	12.8346	17.0834
4	47.0929	39.2989	28.2657	28.2657	16.4695	10.0243	14.2158
5	45.7565	38.1525	27.4149	27.4149	24.2367	16.1531	19.8258
6	42.8291	35.6431	25.5411	25.5411	25.6652	27.8963	20.8856
7	40.2646	33.4469	23.9181	23.9181	37.8828	35.7316	29.8465
8	36.5718	30.2881	21.5664	21.5664	38.2481	30.7843	29.3271
9	27.2141	22.3022	15.6054	15.6054	46.6932	28.3645	37.5962
10	29.4934	24.2449	17.0576	17.0576	54.1909	24.2649	37.6414
11	25.3906	20.7490	14.4436	14.4436	63.6741	33.4735	40.8310
12	20.4548	16.5504	11.2983	11.2983	69.9625	25.6950	49.3836

TABLE 6.41 Expected generation schedule when weights are $w_1 = 0.6$, $w_2 = 0.2$, and $w_3 = 0.2$

Sub int.	Expected thermal generations				Expected hydro generations		
	$\overline{P_1}$	$\overline{P_2}$	$\overline{P_3}$	$\overline{P_4}$	$\overline{P_5}$	$\overline{P_6}$	$\overline{P_7}$
1	52.2826	44.2199	35.1298	35.1298	9.6734	20.3345	8.0045
2	50.9489	43.0630	34.2878	34.2878	20.0005	16.1455	16.1094
3	49.4997	41.8065	33.3729	33.3729	21.5401	12.8879	17.1451
4	44.4189	37.4065	30.1650	30.1650	16.8690	10.1635	14.4446
5	43.1654	36.3222	29.3735	29.3735	24.4937	16.2737	19.9634
6	40.2784	33.8265	27.5503	27.5503	25.8547	27.9998	20.9763
7	37.8151	31.6991	25.9946	25.9946	37.9125	35.7750	29.8340
8	34.3518	28.7112	23.8070	23.8070	37.8655	30.7354	29.0764
9	24.8012	20.4902	17.7731	17.7731	46.6259	28.4318	37.4912
10	27.9561	23.2028	19.7665	19.7665	52.8726	23.7285	36.5602
11	21.5810	17.7245	15.7382	15.7382	66.5194	33.6934	42.3091
12	20.1380	16.4862	14.8263	14.8263	65.5869	25.4539	46.8206

of cost decreases but the membership function of risk increases when weights w_1 and w_3 are varied and weight w_2 is set to zero.

The minimum and maximum values of the functions cost, emission and risk are selected from the non-inferior solutions. It also depends on the choice of the operator or decision maker, i.e. the range in which he is seeking the solution.

The operator can select the best operating point corresponding to the weight combinations when the value of the normalized membership function is maximum among the non-inferior

TABLE 6.42 Expected discharge for different combination of weights

Sub int.	$w_1 = 0.8,\ w_2 = 0.0,\ w_3 = 0.2$			$w_1 = 0.6,\ w_2 = 0.2,\ w_3 = 0.2$		
	Q_1	Q_2	Q_3	Q_1	Q_2	Q_3
1	6.5585	31.1836	7.2424	7.1148	31.4895	7.6992
2	14.7710	25.2078	15.6412	14.9532	25.3325	15.7839
3	16.2284	20.3914	17.0027	16.3512	20.4918	17.0971
4	12.5828	16.1121	14.3291	12.9281	16.3529	14.5961
5	18.5799	26.2228	20.1538	18.8530	26.4567	20.3598
6	19.7593	45.7307	21.4299	20.0018	45.9855	21.6066
7	29.2961	59.1508	30.9220	29.4813	59.3567	31.0446
8	29.0027	50.7031	29.8922	28.8706	50.7470	29.7642
9	33.5294	45.6405	36.4336	33.6552	45.8687	36.4761
10	37.5595	38.2404	35.1948	36.7977	37.4751	34.2723
11	44.8651	52.2372	38.3460	47.1684	52.6787	39.8671
12	50.6312	39.7570	46.9679	47.7347	39.4613	44.6337

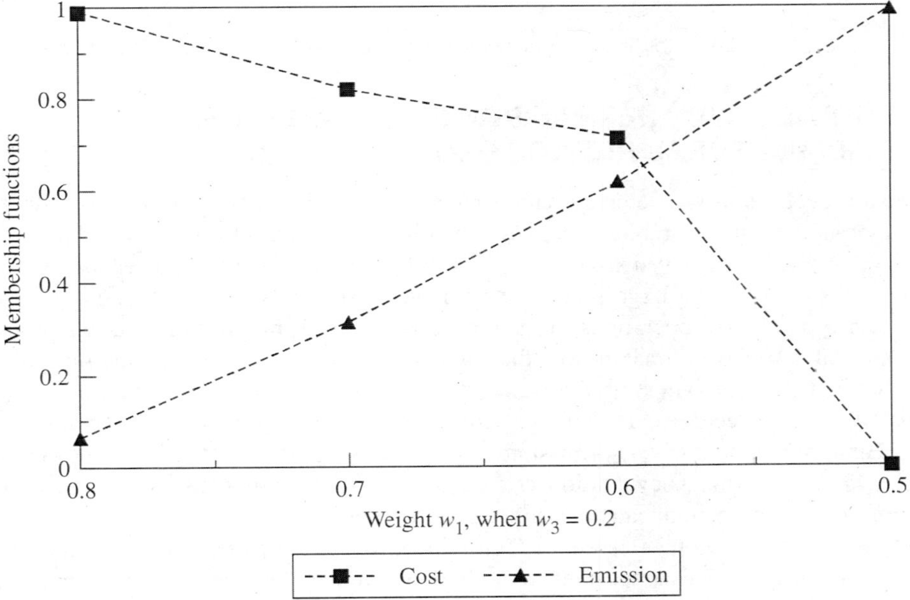

FIGURE 6.20 Conflicting nature of objective, expected cost, and expected NO_x emission.

solutions. The weight combination presented at serial number 5 in the Table 6.39 gives the maximum value of μ_D^k. The number of iterations for the discussed algorithm depends on the initial guess of discharge and dual variable λ_{4j}. An initial guess of discharge can be made by using the local variation method.

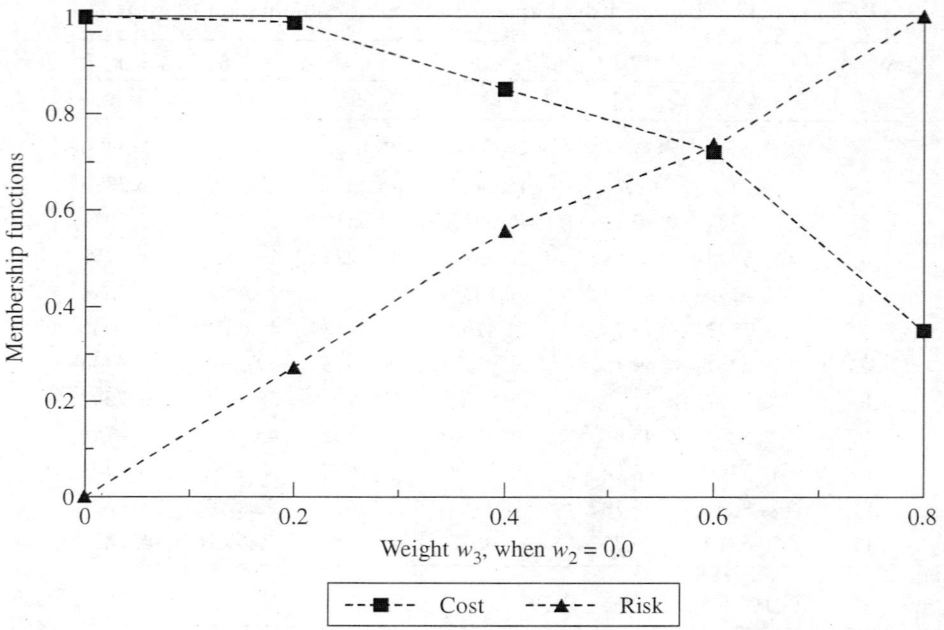

FIGURE 6.21 Conflicting nature of objectives, expected cost, and risk.

6.9 MULTIOBJECTIVE THERMAL POWER DISPATCH USING ARTIFICIAL NEURAL NETWORK (ANN)

Many complex, real-world problems are characterized as decision making problems with multiple, conflicting and non-commensurable objectives. The objectives are said to be in conflict, if the level of one objective can be improved only at the expense of others. Thermal power dispatch problem is also one of them, because a large-scale electric power system possesses multiple objectives to be achieved, namely economic operations, reliability, security, and minimal impacts on environment. The main purpose of the optimal power dispatch problem has so far been mainly confined to minimize the total generation cost of a power system. However, in order to meet the environmental regulations enforced in recent years, emission control has become one of the important operational objectives. Here, the amount of NO_x emission, which is in proportion to the active power output of a generator, is selected as an evaluation criteria and the minimum emission is sought within a small region around an economically feasible operating point.

System security is another essential factor in power system operation and also in system planning. To be specific, it is important to limit line flows within the prescribed upper bounds. Then a security index as a function of overloads will be minimized by some preventive control actions. The use of a sensitivity method to calculate flows in system security and contingency analysis still remains popular, even though fast load flow algorithms have been developed. The reasons are its linearity, the speedy computation, its simplicity and its accuracy. These inherent properties make it widely acceptable for power system applications.

Decision making is the process of choosing some possible course of action among the various alternatives. In analyzing such problems, because of the multiplicity of objectives and the variety of alternatives, it is often desirable to obtain complete knowledge of the decision maker's global preference structure explicitly represented by a prescriptive decision model, such as a multi-attribute utility function based on the decision maker's prior articulated preference. But some studies indicate that the existing multi-attribute utility theory and methods have some inherent disadvantages. Further, in a complex phenomenon, the determination of a functional structure of formal representation for general preferences is usually difficult. So the assessment of a decision maker underlying multi-attribute value function using the existing methods is not an easy task. The recent advances in ANN's make it possible to develop and apply the new methodology and technology to extract decision rules from available information in the setting of decision making with multiple objectives.

An ANN consists of many highly interconnected, simple and similar processing elements, called neurons operating in parallel to perform useful computation tasks such as recognizing pre-programmed or learned patterns. A neural network can be used to classify patterns by selecting the output which represents an unknown input pattern, in the event where an exact input-output relationship is not easily defined. The approach can also be used to recognize patterns of behaviours in decision making [Wang and Malakooti, 1992], where exact functional relationships are not easily defined. A multilayer ANN with an error backward propagation learning rule is the most popular adaptive neural network model based on supervised learning strategy [Dhillon et al., 1991].

Actually, there are several inaccuracies and uncertainties in the input information which lead to deviations from the optimal value. For a more realistic approach, the electric energy system has been represented as a network characterized by random variables and investigated by numerous researchers at various levels [Parti et al., 1983; El-Hawary and Mbamalu, 1989]. In this section, a classical stochastic thermal power dispatch as a multiobjective problem is formulated with explicit recognition of inaccuracies and uncertainties in the system data. The stochastic behaviour of random variables is known in advance from past history. Variance due to mismatch in generation is incorporated as another objective to be minimized. Some authors [Tapia and Murtagh, 1991; David and Rongda, 1991] suggest formulating a multiobjective decision problem as a fuzzy programming problem wherein the aspiration levels for the objectives are expressed by means of fuzzy set membership functions. So, multiobjectives are reformulated as fuzzy sets, keeping in mind that higher the membership the greater the satisfaction with a solution. The fuzzy sets more accurately represent the operational constraints of the power system. Therefore, transmission line flow constraints are included in the formulation by using the fuzzy set theory. A three-layered neural network is trained to achieve the desired level of objectives using the backpropagation algorithm, when input to the network is known in advance. The network is trained for a sample system consisting of three generators. The range of objective level is provided for the decision maker by performing minimum cost and emission algorithms. The emission will be maximum at the schedule, when cost is minimum and vice versa, because the objectives are in conflict. Minimum cost generation schedules are considered as inputs for various patterns for different load demands.

6.9.1 Stochastic Economic-Emission Problem Formulation

In this section, the multiobjectives with the equality and inequality constraints pertaining to the power system optimization problem are described. The important non-commensurable objectives, taken into account are economic operation, minimal impacts on environment, and expected deviations due to unsatisfied load. The stochastic multiobjective formulation is adopted by considering fuel cost coefficients, NO_x emission coefficients, and load demand as random variables. The stochastic models are converted to their deterministic equivalents by taking their expected values, with the assumption that all the random variables are normally distributed and statistically dependent on each other [Mahalanabis, et al., 1991]. The deterministic equivalent of the stochastic economic emission load dispatch problem of Eqs. (6.69) is described in Section. 6.4.1. Inclusion of security constraint in the optimization problem of Eqs. (6.69) is considered for the better reliable solution strategy.

Transmission line security

The active power flow in transmission line m connecting bus 1 to bus j must satisfy the following constraint:

$$\overline{P}_{T_m}^{\min} \le \overline{P}_{T_m} \le \overline{P}_{T_m}^{\max} \qquad (m = 1, 2, ..., \text{NL}) \tag{6.186}$$

where $\overline{P}_{T_m}^{\min}$ and $\overline{P}_{T_m}^{\max}$ are minimum and maximum limits of active power flow in line m. NL is the total number of lines.

The linearized formula based on the generalized generation distribution factors (GGDF) (see Chapter 3) relating the bus incremental power injection with the network power flow is used to have linear security constraints. Therefore, the expected active power flow is expressed as with the assumption that the network configuration is fixed under normal conditions.

$$\overline{P}_{T_m} = \sum_{i=1}^{\text{NG}} D_{l-j,\,i}\,\overline{P}_i \tag{6.187}$$

and

$$D_{l\text{-}j,i} = A_{l\text{-}j,i} + D_{l\text{-}j,R} \tag{6.187a}$$

where

$D_{l\text{-}j,R}$ = GGDF for line l-j, due to the slack generator R

$$= \frac{P_{T_m} - \sum\limits_{\substack{i=1\\ i\neq R}}^{\text{NG}} A_{l-j,\,i}\overline{P}_i}{\sum\limits_{i=1}^{\text{NG}} \overline{P}_i} \tag{6.187b}$$

$A_{l\text{-}j,i}$ = generation shift distribution factors for line l-j, due to shift of generation on generator i

$$= \frac{X_{l-i} - X_{j-i}}{X_{l-j}} \tag{6.187c}$$

where X_{l-i}, X_{j-i} are elements of the bus reactance matrix, and X_{l-j} is the reactance of line l–j.

The stochastic economic emission problem is extended to consider three objectives: (i) expected fuel cost, (ii) expected NO_x emission, and (iii) variance of generation mismatch to meet the expected load demand within (a) the expected generation limits and (b) the transmission line flow bounds. The multiobjective optimization problem is defined as

Minimize $$[\overline{F}_1, \overline{F}_2, \overline{F}_3]^T \quad (6.188a)$$

subject to $$\overline{P}_D + \overline{P}_L - \sum_{i=1}^{NG} \overline{P}_i = 0 \quad (6.188b)$$

$$\overline{P}_i^{\min} \le \overline{P}_i \le \overline{P}_i^{\max} \qquad (i = 1, 2, ..., NG) \quad (6.188c)$$

$$\overline{P}_{T_m}^{\min} \le \overline{P}_{T_m} \le \overline{P}_{T_m}^{\max} \qquad (m = 1, 2, ..., NL) \quad (6.188d)$$

where $\overline{F}_1$, $\overline{F}_2$, and $\overline{F}_3$ are the expected values of the three objective functions, namely the operating cost, emission and deviations, respectively, to be minimized over the set of admissible decision variable $\overline{P}_i$ and are defined by Eqs. (6.45), (6.63), and (6.68) respectively.

$$\overline{F}_1 = \sum_{i=1}^{NG} \left[(1 + C_{P_i}^2 + 2R_{a_iP_i}C_{a_i}C_{P_i})\overline{a}_i\overline{P}_i^2 + (1 + R_{b_iP_i}C_{b_i}C_{P_i})\overline{b}_i\overline{P}_i + \overline{c}_i\right] \quad (6.188e)$$

$$\overline{F}_2 = \sum_{i=1}^{NG} \left[(1 + C_{P_i}^2 + 2R_{d_iP_i}C_{d_i}C_{P_i})\overline{d}_i\overline{P}_i^2 + (1 + R_{e_iP_i}C_{e_i}C_{P_i})\overline{e}_i\overline{P}_i + \overline{f}_i\right] \quad (6.188f)$$

$$\overline{F}_3 = \sum_{i=1}^{NG} C_{P_i}^2\overline{P}_i^2 + \sum_{i=1}^{NG}\sum_{\substack{j=1\\ j\ne i}}^{NG} R_{P_iP_j}C_{P_i}C_{P_j}\overline{P}_i\overline{P}_j \quad (6.188g)$$

$$\overline{P}_L = \sum_{i=1}^{NG} (1 + C_{P_i}^2)\,\overline{B}_{ii}\overline{P}_i^2 + \sum_{i=1}^{NG}\sum_{\substack{j=1\\ j\ne i}}^{NG} (1 + R_{P_iP_j}C_{P_i}C_{P_j})\overline{P}_i\overline{B}_{ij}\overline{P}_j \quad (6.188h)$$

where

$\overline{a}_i$, $\overline{b}_i$, and $\overline{c}_i$ are the expected cost coefficients

$\overline{d}_i$, $\overline{e}_i$, and $\overline{f}_i$ are the expected NO_x emission coefficients

$\overline{B}_{ij}$ are the expected *B*-coefficients

C_x is the coefficient of variation of random variable x (where x may be a_i, b_i, d_i, e_i, P_i)

$R_{x,y}$ is the correlation coefficient of random variables x and y (where x and y may be a_i, b_i, d_i, e_i, P_i).

6.9.2 Membership Functions

In an approximate reasoning, logical decision making can be defined by fuzzy sets using the operating conditions. The fuzzy sets are defined by equations called the *membership functions*. These functions represent the degree of membership in some fuzzy sets using the values from 0 to 1. The membership value 0 indicates incompatibility with the sets, while 1 means full compatibility. When neither is true, a value between 0 and 1 is taken.

Objective functions $\mu(\overline{F}_k)$

By taking account of the calculated individual minimum and maximum of each objective function together with the rate of increase of membership of satisfaction (Figure 6.22), the decision maker must determine his membership function $\mu(\overline{F}_k)$, in a subjective manner. Here, it is assumed that $\mu(\overline{F}_k)$ is a strictly monotonic decreasing and continuous function with respect to $\overline{F}_k$ and is given below:

$$\mu(\overline{F}_k) = \begin{cases} 1 & ; \overline{F}_k \le \overline{F}_k^{\min} \\ \left(\dfrac{\overline{F}_k^{\max} - \overline{F}_k}{\overline{F}_k^{\max} - \overline{F}_k^{\min}}\right) & ; \overline{F}_k^{\min} < \overline{F}_k < \overline{F}_k^{\max} \\ 0 & ; \overline{F}_k \ge \overline{F}_k^{\max} \end{cases} \tag{6.189}$$

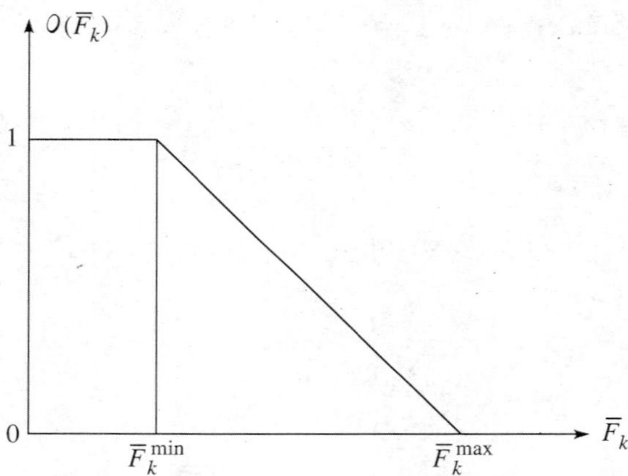

FIGURE 6.22 Membership function of the objectives.

Transmission line flow $\mu(\overline{P}_{T_m})$

The fuzzy sets more accurately represent the operational constraints of the power system. Fuzzy membership may have a variety of shapes but for simplicity, here the line flow constraints are represented by the triangular membership function (Figure 6.23). Mathematically, the membership function is defined as

$$\mu(\overline{P}_{T_m}) = \begin{cases} 0 & ; \overline{P}_{T_m} \le \overline{P}_{T_m}^{\min} \\ \left(\dfrac{\overline{P}_{T_m} - \overline{P}_{T_m}^{\min}}{\overline{P}_{T_m}^{c} - \overline{P}_{T_m}^{\min}}\right) & ; \overline{P}_{T_m}^{\min} > \overline{P}_{T_m} > \overline{P}_{T_m}^{c} \\ \left(\dfrac{\overline{P}_{T_m}^{\max} - \overline{P}_{T_m}}{\overline{P}_{T_m}^{\max} - \overline{P}_{T_m}^{c}}\right) & ; \overline{P}_{T_m}^{c} > \overline{P}_{T_m} > \overline{P}_{T_m}^{\max} \\ 0 & ; \overline{P}_{T_M} \ge \overline{P}_{T_m}^{\max} \end{cases} \tag{6.190}$$

where $\overline{P}_{T_m}^{c}$ is the mean of the lower and upper limits of line flows.

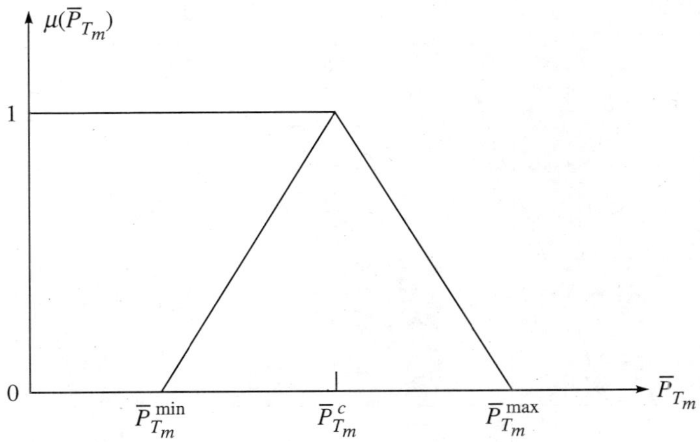

FIGURE 6.23 Membership function of line flow.

Generator limits $\mu(\overline{P}_i)$

The fuzzy sets more accurately represent the operational inequality constraints of the power system. Fuzzy membership may have a variety of shapes but for simplicity, here the generator limits are represented by a rectangular membership function (Figure 6.24). Mathematically, the membership function is defined as

$$\mu(\overline{P}_i) = \begin{cases} 0 & ; \overline{P}_i \le \overline{P}_i^{\min} \\ \left(\dfrac{\overline{P}_i - \overline{P}_i^{\min}}{\Delta\overline{P}_i}\right) & ; \overline{P}_i^{\min} > \overline{P}_i > \overline{P}_i^{\min} + \Delta\overline{P}_i \\ 1 & ; \overline{P}_i^{\min} + \Delta\overline{P}_i > \overline{P}_i > \overline{P}_i^{\max} - \Delta\overline{P}_i \\ \left(\dfrac{\overline{P}_i^{\max} - \overline{P}_i}{\Delta\overline{P}_i}\right) & ; \overline{P}_i^{\max} - \Delta\overline{P}_i > \overline{P}_i > \overline{P}_i^{\max} \\ 0 & ; \overline{P}_i \ge \overline{P}_i^{\max} \end{cases} \tag{6.191}$$

where $\Delta\overline{P}_i$ is the range in which the membership function varies linearly, as decided by the decision maker.

6.9.3 Performance Index

The objective functions are reformulated as fuzzy sets and each line flow constraint defines a fuzzy region of acceptability. Keeping in mind that higher the membership value the better is the solution, the line flow constraints are viewed as objectives that maximize the membership functions for each line. The multiobjective optimization problem of Eqs. (6.188) can be rewritten as

$$\text{Maximize} \quad [\mu(\overline{F}_1), \mu(\overline{F}_2), \mu(\overline{F}_3), \mu(\overline{P}_{T_m}); \; (m = 1, 2, ..., \text{NL}, \; \mu(\overline{P}_i) \; (i = 1, 2, ..., N)]^T \tag{6.192a}$$

subject to

$$\overline{P}_D + \overline{P}_L - \sum_{i=1}^{\text{NG}} \overline{P}_i = 0 \tag{6.192b}$$

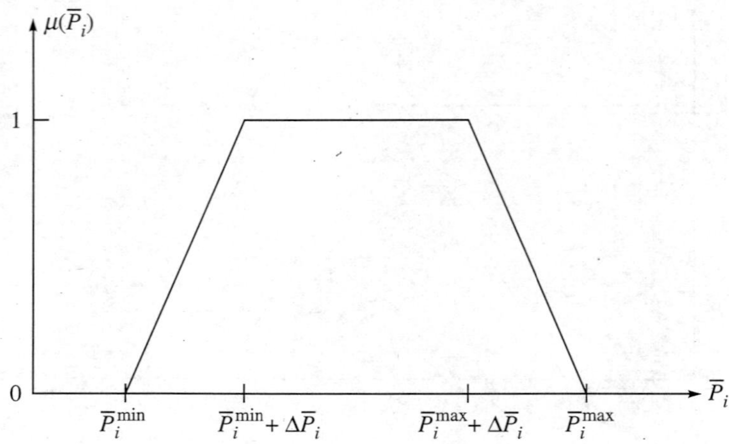

FIGURE 6.24 Membership function of generator limits.

A multiobjective optimization problem can be changed to a scalar optimization problem by defining a global function which minimizes the deviations from the so-called reference (target) objective [Osyczka and Davies, 1984]. Any reasonable or desirable point in space of the objective is chosen by the decision maker and may be considered as the reference objective. The scalar optimization problem obtained from the above stated multiobjective problem is given below:

Minimize
$$E = \frac{1}{2}\sum_{j=1}^{NL+NG+3}(\mu_j(.) - \mu_j^r)^2 \tag{6.193a}$$

subject to
$$\overline{P}_D + \overline{P}_L - \sum_{i=1}^{NG}\overline{P}_i = 0 \tag{6.193b}$$

To solve such optimization problems, the original constrained problem is transformed into an auxiliary unconstrained problem, whose minimum is the same as the minimum of the original problem [Sasson, 1969]. By applying the Zangwill's method [Sasson, 1969], the constrained problem of Eqs. (6.193) is transformed into an unconstrained form and is given below:

$$E = \frac{1}{2}\sum_{j=1}^{NL+NG+3}(\mu_j(.) - \mu_j^r)^2 + \frac{1}{R}\left(\overline{P}_D + \overline{P}_L - \sum_{i=1}^{NG}\overline{P}_i\right)^2 \tag{6.194}$$

where μ_j^r is the target membership function for the *j*th function and *R* is a constant.

6.9.4 Structure of ANN

The design of ANNs is based on the neural structure of the human brain. The network is a massively interconnected dynamic system with interacting parts based on neurobiological models. It consists of a large network of processors called neurons which are arranged in layers and connected to each other by means of information channels called *interconnections.* The input neurons, forming the input layer, receive data from the outside world. The output neurons, forming the output layer, send information to the user. The hidden neurons form the hidden layers in between, and store the information obtained through training.

A schematic of a three-layered feedforward neural network model is shown in Figure 6.25. There is a connection strength, synapses or weight associated with each connection. Each neuron can have multiple inputs while there can be only one output. The inputs to a neuron could be from external stimuli or could be from the output of the other neurons. In the simplest case each neuron produces its output by computing the inner product of its input signal(s) and associated weights which is passed through a nonlinear function as shown in Figure 6.26. One commonly used nonlinear monotonic function is the sigmoidal one, which can be defined as follows:

$$f(x) = \frac{1.0}{1.0 + \exp(-x)} \tag{6.195}$$

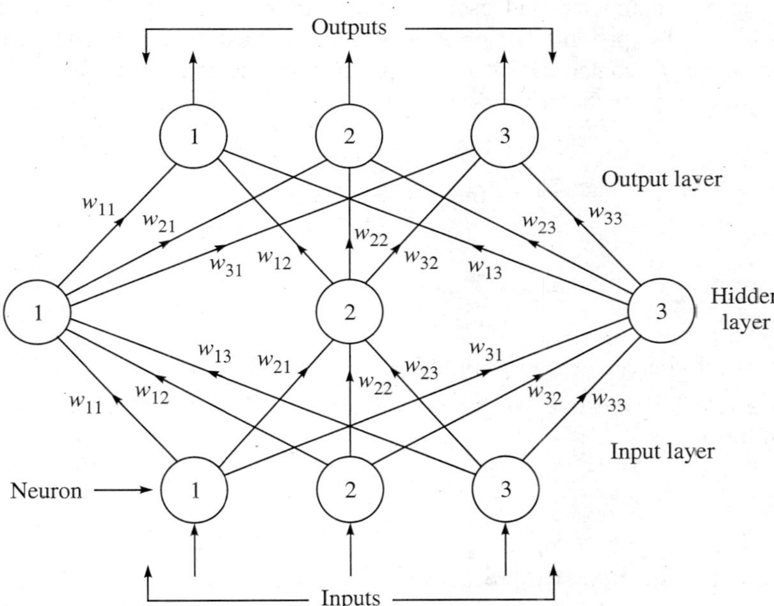

FIGURE 6.25 Schematic of feedforward three-layered neural network.

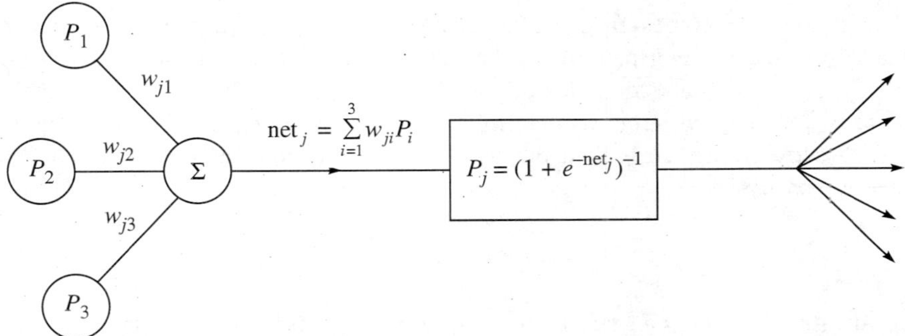

FIGURE 6.26 Mathematical model of neuron.

A crucial property of these networks is their ability to improve their performance by learning new information. This is accomplished by modifying the interconnection strengths among neurons according to some prescribed rules. Further, the network architecture permits high computational rate to be obtained through the massively parallel distributed processing. It is also due to this kind of processing that such networks have a great degree of robustness or fault tolerance to local damages. Various learning algorithms have been recently introduced, the most famous being the backpropagation algorithm.

6.9.5 Backpropagation Algorithm

Backpropagation learning algorithm finds the values of all of the weights that minimize the function using a method of gradient descent. That is, after each pattern has been presented, the error on that pattern is computed and each weight is moved down the error gradient towards its minimum value for that pattern. The error is actually defined by Eq. (6.194), where μ_j^r is the target objective for the jth component of the output pattern and $\mu_j(.)$ is the jth component of the actual objective produced by the network representation with the input pattern. The network is specified as

$$\overline{P}_j = f_j(\text{net}_j) = \frac{1}{1+e^{-\text{net}_j}} \tag{6.196}$$

$$\text{net}_j = \sum_{k=1}^{N} w_{jk}\overline{P}_k \tag{6.197}$$

where w_{jk} is the weight on connection from unit j at layer $(l–1)$ to unit k at layer l to be adjusted.

To obtain a rule for adjusting weights, the gradient of E with respect to w_{jk} is used and is represented as follows:

$$\frac{\partial E}{\partial w_{jk}} = \delta_j \overline{P}_k \tag{6.198}$$

The determination of k is a recursive process that starts with the output unit, and is given by

$$\delta_j = \frac{\partial E}{\partial \overline{P}_j}(1-\overline{P}_j)\overline{P}_j \tag{6.199}$$

The backpropagation rule comes from its assignment of deltas to hidden units that receive no direct feedback from the training patterns in the outside world. These deltas actually influence the modification of weights to connections leading to the hidden units. The delta term for hidden units, for which there is no specified target, is determined recursively in terms of delta terms of the units to which it is directly connected to and the weights of these connections. A unit in an arbitrary hidden layer is given by

$$\delta_j = \overline{P}_j(1-\overline{P}_j)\sum_k \delta_k w_{kj} \tag{6.200}$$

The rule for adjusting weights can be obtained by using Eq. (6.198) and is given by

$$w_{jk}^{m+1} = \eta\delta_j\overline{P}_k + \alpha w_{jk}^{m} \tag{6.201}$$

where

η is the learning rate parameter

α is the momentum constant to determine the effect of previous weight changes.

6.9.6 Sample System Study

The applicability of the method is demonstrated on a sample three-generator power system, whose expected cost and emission characteristics are given in Tables 6.43 and 6.44, respectively. Expected B-coefficients for transmission loss are depicted in Table 6.45. Table 6.46 shows the expected GGDFs (Section 3.10.5). In addition, the values of the CVs and CCs are taken as 0.1 and 1.0, respectively, for all random variables.

TABLE 6.43 Expected cost coefficients and generator limits

Plant no.	$\overline{a}_i$ ($/MW^{2}h)	$\overline{b}_i$ ($/MWh)	$\overline{c}_i$ ($/h)	$\overline{P}_i^{\max}$ (MW)	$\overline{P}_i^{\min}$ (MW)
1	0.006	5.506	264.634	225.0	40.0
2	0.016	5.268	154.298	240.0	20.0
3	0.005	5.627	261.186	114.0	20.0

TABLE 6.44 Expected NO_x emission coefficients

Plant no.	$\overline{d}_i$ (kg/MW^{2}h)	$\overline{e}_i$ (kg/MWh)	$\overline{f}_i$ (kg/h)
1	0.00626	−0.56699	62.36725
2	0.00626	−0.56699	62.36725
3	0.00348	0.36551	165.12520

TABLE 6.45 Expected B-coefficients $\times 10^2$

0.027251	−0.003506	−0.036788
−0.003506	0.030896	−0.005653
−0.036788	−0.005653	0.32295

TABLE 6.46 Expected GGDF

Line no.	*Line l–k*	$D_{l\text{-}k,1}$	$D_{l\text{-}k,2}$	$D_{l\text{-}k,3}$
1	1–2	0.5456	−0.4544	−0.4544
2	2–5	0.1548	0.1548	−0.3225
3	5–4	0.0621	0.0621	−0.4154
4	2–4	0.1378	0.1378	−0.3850
5	4–3	0.0081	0.0081	−0.9919

Input for the network

The input for the network is obtained by performing the minimum expected cost thermal power dispatch. Minimum expected NO_x emission dispatch problem can also be used to get the input for the network. The schedule obtained from the expected economic dispatch and minimum expected emission dispatch are given in Table 6.47 for different expected demands.

TABLE 6.47 Expected thermal load dispatch

Demand (MW)	*Generation schedule* (MW)			*Cost*	*Emission*
$\overline{P}_D$	$\overline{P}_1$	$\overline{P}_2$	$\overline{P}_3$	$\overline{F}_1$ (\$/h)	$\overline{F}_2$ (kg/h)
Minimum cost dispatch					
150.0	84.4225	40.3451	28.1497	1599.194	288.092
200.0	114.6325	52.6536	38.0150	1948.139	315.879
250.0	145.4129	65.3569	47.6014	2319.782	358.735
Minimum emission dispatch					
150.0	66.4080	66.0222	25.1528	1615.615	279.379
200.0	90.0697	89.0362	25.1528	1981.723	302.269
250.0	108.7398	106.2855	42.7061	2361.683	338.105

Range of objective levels for decision maker

$\overline{F}_1^{\min}$ and $\overline{F}_2^{\min}$ are obtained by solving the expected economic dispatch and minimum emission dispatch separately. As economy, environmental impacts and risk are mutually conflicting objectives, therefore F_2 will have maximum value at the schedule, when $\overline{F}_1$ is minimum. Further, the range of objective levels may be decided by the experienced decision maker from the expected solution trend.

Neural network design

A three-layered ANN is formed with three neurons in input, hidden, and output layers. The network is trained with backpropagation algorithm to achieve the target values given in Table 6.49, whereas the input given is obtained from the minimum cost dispatch and is depicted in Table 6.48. The values of n and α are chosen as 0.000005 and 0.5 respectively. The value of R is taken as 500. The strength of connections as weights is given in Table 6.50.

The choice of the number of neurons is a difficult task. But good results are achieved if the number of neurons is equal to the number of patterns. The number of iterations depend on the initial weights and the value of n. The choice of membership function is again a crucial point because the objective functions are in conflict. Choice becomes easy if the solution trend is known a priori.

A theoretical basis and methodology for optimal dispatch problem in a unified multi-objective framework is established. The chapter also investigates the feasibility of quantitative representation of inaccuracies and uncertainties of the input data and power demand for the power dispatch problem in terms of probability and statistics. An artificial neural network model is established to capture the optimal generation dispatch for power system operations with multiple conflicting objectives.

TABLE 6.48 Real power line flows

Line	*Minimum cost dispatch*			*Minimum emission dispatch*		
	Power demand (MW)			*Power demand* (MW)		
l–k	150	200	250	150	200	250
1–2	14.9368	21.3436	28.0090	−2.8563	−2.7454	−8.3733
2–5	10.2357	13.6360	17.2757	14.0502	19.6138	19.5132
5–4	−3.9453	−5.3916	−6.6848	−0.0841	0.6739	−4.3870
2–4	6.3553	8.4162	10.7175	10.5488	14.9969	13.1886
4–3	−26.9110	−36.3521	−45.5085	−23.4983	−45.5085	−40.6184

TABLE 6.49 Target membership functions

Membership function of	*Expected power demand*		
	150 (MW)	200 (MW)	250 (MW)
F_1	0.85	0.9	0.89
F_2	0.5	0.46	0.46
F_3	0.7	0.62	0.44
PT_1	0.36	0.45	0.4
PT_2	0.83	0.7	0.12
PT_3	0.54	0.7	0.1
PT_4	0.83	0.7	0.2
PT_5	0.83	0.8	0.2
P_1	1.0	1.0	1.0
P_2	1.0	1.0	1.0
P_3	1.0	1.0	1.0

TABLE 6.50 Weights

j	w_{j1}	w_{j2}	w_{j3}
Layer 1			
1	72.17046	−32.42737	−175.75860
2	12.15827	−90.55445	85.64217
3	14.84998	−107.18450	99.38823
Layer 2			
1	2.62108	−40.59315	44.89007
2	1.22125	−42.65099	45.73229
3	0.73992	−59.48106	63.13109

The synthesized neural networks are very robust and possess generalization capability of examples. The main advantage of the neural networks is that they are independent of functional form and insensitive to parameter perturbation. Although training requires considerable computation, determining the optimal alternative can be executed on personal computers. This feature also facilitates the applications of the new technology to me decisions.

REFERENCES

Books

Arrillaga, J. and C.P. Arnold, *Computer Analysis of Power Systems*, John Wiley & Sons, Singapore 1990.

Christensen, G.S. and S.A. Soliman, *Optimal Long-Term Operation of Electric Power Systems*, Plenum Press, New York, 1988.

Elgerd, O.I., Electric energy systems theory: An introduction, Tata McGraw-Hill, 2nd ed., 1983.

El-Hawary, M.E. and G.S. Christensen, *Optimal Economic Operation of Power Systems*, Academic Press, New York, 1979.

Frederick, Soloman, *Probability and Stochastic Processes*, Prentice Hall, New Jersey, 1987.

Haimes, Y.Y., Hierarchical *Analysis of Water Resource System: Modeling and Optimization of Large Scale Systems*, McGraw-Hill, New York, 1977.

Kirchmayer, L.K., *Economic Operation of Power Systems*, Wiley Eastern, New Delhi, 1958.

Klir, G.J. and B. Yuan, *Fuzzy Sets and Fuzzy Logic: Theory and Applications*, Prentice-Hall India, 1997.

Klir, G.J. and T.A. Folger, *Fuzzy Sets, Uncertainty and Information*, Prentice-Hall India, New Delhi, 1993.

Kosko, B., *Neural Networks and Fuzzy Systems: Dynamical Systems Approach to Machine Intelligence*, Prentice-Hall India, New Delhi, 1994.

Kothari, D.P. and I.J. Nagrath, *Modern Power System Analysis*, 3rd ed., Tata McGraw-Hill, New Delhi, 2003.

Kusic, G.L., *Computer Aided Power Systems Analysis*, Prentice-Hall of India, New Delhi, 1986.

Mahalanabis, A.K., D.P. Kothari and S.I. Ahson, *Computer Aided Power System Analysis and Control*, Tata McGraw-Hill, New Delhi, 1991.

Nagrath, I.J. and D.P. Kothari, *Power System Engineering*, Tata McGraw-Hill, New Delhi, 1994.

Osyczka, A. and B.J. Davies, *Multicriterion Optimization in Engineering with FORTRAN Programs*, Ellis Horwood, 1984.

Papoulis, A., *Probability, Random Variables and Stochastic Processes*, Tata McGraw-Hill, New Delhi, 1991.

Rao, S.S., *Optimization, Theory and Applications*, 2nd ed., Wiley Eastern, New Delhi, 1987.

Sen Gupta, J.K., *Stochastic Programming*, North Holland, 1972.

Stagg, G.W. and A.H. Ei-Abiad, *Computer Methods in Power Systems Analysis*, McGraw-Hill, New Delhi, 1968.

Stevenson, W.D., *Elements of Power System Analysis*, 4th ed., McGraw-Hill, New York, 1982.

Wood, A.J. and B. Wollenberg, *Power Generation, Operation and Control*, John Wiley, New York, 1984.

Zurada, J.M., *Introduction to Artificial Neural Network*, Jaico Publishing House, Delhi, 1997.

Papers

Blaszczynski, G.M., Sensitivity study of the economic dispatch, *Proceedings of PICA Conference*, New Orleans, LA, U.S.A., 1975.

Brar, Y.S., J.S. Dhillon and D.P. Kothari, Genetic-fuzzy logic based weightage pattern search for multi-objective load dispatch problem, *Asian Journal of Information Technology*, Vol. **2(4)**, pp. 365–373, October-December 2003.

Brar, Y.S., Jaspreet S. Dhillon and D.P. Kothari, Multiobjective load dispatch by fuzzy logic based weightage pattern, *Electric Power Systems Research*, Vol. **63(2)**, pp. 149–160, 2002.

Chowdhury, N. and R. Billinton, Risk constrained economic load dispatch in interconnected generating systems, *IEEE Trans. on Power Systems*, Vol. **5(4)**, pp. 1239–1247, 1990.

Cohon, J.L. and D.H. Marks, A review and evaluation of multi-objective programming techniques, *Water Resources Research*, Vol. **11(2)**, pp. 208–220, 1975.

David, A.K. and Z. Rongda, An expert system with fuzzy sets for optimal planning, *IEEE Trans. on Power Systems*, Vol. **6(1)**, pp. 59–65, 1991.

Dhillon, J.S. and D.P. Kothari, The surrogate worth trade-off approach for multi-objective thermal power dispatch problem, *Electric Power System Research*, Vol. **56(2)**, pp. 103–110, 2000.

Dhillon, J.S., S.C. Parti and D.P. Kothari, Multi-objective optimal thermal power dispatch, *Int. Journal of Electric Power and Energy Systems*, Vol. **16(6)**, pp. 383–390, 1994.

Dhillon, J.S., S.C. Parti and D.P. Kothari, Stochastic economic emission load dispatch, *Electric Power System Research*, Vol. **26(3)**, pp. 179–186, 1993.

Dhillon, J.S., S.C. Parti and D.P. Kothari, Fuzzy decision making in multi-objective long-term scheduling of hydrothermal system, *Int. Journal of Electric Power and Energy Systems*, Vol. **23(1)**, pp. 19–29, 2001.

Dhillon, J.S., S.C. Parti and D.P. Kothari, Multi-objective decision making in stochastic economic dispatch, *Electric Machines and Power Systems*, Vol. **23(3)**, pp. 289–301, 1995.

Dillon, T.S., R.W. Martin and D. Sjelvgren, Stochastic optimization and modelling of large hydrothermal systems for long-term regulation, *Int. Journal of Electrical Power and Energy Systems*, Vol. **2(1)**, pp. 220, 1980.

Dillon, T.S., S. Sestito and S. Leung, Short-term load forecasting using an adaptive neural network, *Int. J. Electrical Power and Energy Systems*, Vol. **13(4)**, pp. 186–192, 1991.

Dopazo, J.F., O.A. Klitin and A.M. Sasson, Stochastic load flows, *IEEE Trans. on Power Apparatus and Systems*, Vol. **94(2)**, pp. 299–309, 1975.

Edwin, K.W. and R.D. Machate, Influençe of inaccurate input data on the optimal short-term operation of power generation systems, IFAC Symposium on Automatic Control in Power, Generation, Distribution and Protection, Pretoria, South Africa, 1980.

El-Hawary, M.E. and G.A.N. Mbamalu, Stochastic optimal load flow using Newton–Raphson iterative technique, *Electric Machines and Power Systems*, Vol. **15(6)**, pp. 371–380, 1988.

El-Hawary, M.E. and G.A.N. Mbamalu, Stochastic optimal load flow using a combined quasi-Newton and conjugate gradient technique, *Int. Journal of Electrical Power and Energy Systems*, Vol. **11(2)**, pp. 85–93, 1989.

El-Hawary, M.E. and G.A.N. Mbamalu, A comparison of probablistic perturbation and deterministic based optimal power flow solutions, *IEEE Trans. on Power Systems*, Vol. **6(3)**, pp. 1099–1105, 1991.

El-Hawary, M.E. and K.M. Ravindranath, A general overview of multiple objective optimal power flow in hydrothermal electric power systems, *Electric Machines and Power Systems*, Vol. **19(3)**, pp. 313–327, 1991.

Glimn, A.F., L.K. Kirchmayer, G.W. Stagg and V.R. Peterson, Accuracy considerations in economic dispatch of power systems, *AIEE Trans. Part-III: Power Apparatus and Systems,* Vol. **75(3)**, pp. 1125–1137, 1956.

Haimes, Y.Y. and W.A. Hall, Multi-objectives in water resource systems analysis: The Surrogate worth trade-off method, *Water Resources Research*, Vol. **10(4)**, pp. 615–624, 1974.

Hannan, E.L., Linear programming with multiple fuzzy goals, *Fuzzy Sets and Systems*, Vol. **6(3)**, pp. 235–248, 1981.

Hill, E.F. and W.D. Stevenson, A new method of determining loss coefficients, *IEEE Trans. on Power Apparatus and Systems*, Vol. **87(7)**, pp. 1548–1553, 1968.

Hsu, Y.Y. and C.C. Yang, Design of artificial neural networks for short-term load forecasting. Part II: Multilayer feed forward networks for peak load and valley load forecasting, *IEE Proceeding, Part C*, Vol. **138(5)**, pp. 414–418, 1991.

Kaunas, J.R. and Y.Y. Haimes, Risk management of groundwater contamination in a multi-objective framework, *Water Resources Research*, Vol. **21(11)**, pp. 1721–1730, 1985.

Kothari, D.P. and I.J. Nagrath, Optimal stochastic scheduling of hydrothermal systems using discrete maximum principle, *Journal Institution of Engineers (India)*, Vol. **61**, pp. 22–26, 1980.

Leberling, H., On finding compromise solution in multicriteria problems using the fuzzy min-operator, *Fuzzy Sets and Systems*, Vol. **6(2)**, pp. 105–118, 1981.

Lee, K.Y., Y.T. Cha and J.H. Park, Short-term load forecasting using an artificial neural network, *IEEE Trans. on Power Systems*, Vol. **7(1)**, pp. 124–132, 1992.

Leite da Silva, A.M., R.N. Allan, S.M. Soares and V.L. Arienti, Probablistic load flow considering network outages, *IEE Proceedings, Part C*, Vol. **132(3)**, pp. 139–145, 1985.

Mazumdar, M. and C.K. Yin, Variance of power generating system production costs, *IEEE Trans. on Power Systems*, Vol. **4(2)**, pp. 662–667, 1989.

Meliopoulos, A.P. X. Chao, G.J. Cokkinides and R. Monsalvatge, Transmission loss evaluation based on probabilistic power flow, *IEEE Trans. on Power Systems*, Vol. **6(1)**, pp. 364–371, 1991.

Miranda, V. and J.T. Saraiva, Fuzzy modelling of power system optimal load flow, *IEEE Trans. on Power Systems*, Vol. **7(2)**, pp. 843–849, 1992.

Mo, B., J. Hegge and I. Wangensteen, Stochastic generation expansion planning by means of stochastic dynamic programming, *IEEE Trans. on Power Systems*, Vol. **6(2)**, pp. 662–668, 1991.

Nanda, J., D.P. Kothari and K.S. Lingamurthy, A new approach to economic and minimum emission dispatch, *Journal Indian Institute of Science*, Vol. **67**, pp. 249–256, 1987.

Neto, T.A.A., M.V.F. Pereira and J. Kelman, A risk-constrained stochastic dynamic programming approach to the operation plannng of hydrothermal systems, *IEEE Trans. on Power Apparatus and Systems*, Vol. **104(2)**, pp. 273–279, 1985.

Ouyang, Z. and S.M. Shahidehpour, A hybrid artificial neural network-dynamic programming approach to unit commitment, *IEEE Trans. on Power Systems*, Vol. **7(1)**, pp. 236–242, 1992.

Park, D.C., M.A. El-Sharkawi, R.J. Marks II, An adaptively trained neural network, *IEEE Trans. on Neural Networks*, Vol. **2(3)**, pp. 334–345, 1991.

Parti, S.C., *Stochastic optimal power generation scheduling*, Ph.D. (Thesis), TIET, Patiala, 1987.

Parti, S.C., D.P. Kothari and P.V. Gupta, Economic thermal power dispatch, *Journal Institution of Engineers (India)*, Vol. **63(EL-2)**, pp. 126–132, 1983.

Pereira, M.V.F., Optimal stochastic operations scheduling of large hydroelectric systems, *Int. J. Electrical Power and Energy Systems*, Vol. **11(3)**, pp. 161–169, 1989.

Rarig, H.M. and Y.Y. Haimes, Risk/Dispersion index method, *IEEE Trans. on Systems, Man and Cybernetics*, Vol. **13(3)**, pp. 317–328, 1983.

Rashid, A.H.A. and K.M. Nor, An efficient method for optimal scheduling of fixed-head hydro and thermal plants, *IEEE Trans. on Power Systems*, Vol. **6(2)**, pp. 632–636, 1991.

Rau, N.S. and C. Necsulescu, Probability distributions of incremental cost of production and production cost, *IEEE Trans. on Power Apparatus and Systems*, Vol. **104(12)**, pp. 3493–3499, 1985.

Sasson, A.M., Combined use of the Powell and Fletcher-Powell nonlinear programming methods for optimal load flows, *IEEE Trans. on Power Apparatus and Systems*, Vol. **88(10)**, pp. 1530–1535, 1969.

Sasson, A.M., Non-linear programming solutions for load-flow, minimum loss and economic dispatching problems, *IEEE Trans. on Power Apparatus and Systems*, Vol. **88(4)**, pp. 399–409, 1969.

Sherkat, V.R., R. Campo, K. Moslehi and E.O. Lo, Stochastic long-term hydrothermal optimization for a multireservoir system, *IEEE Trans. on Power Apparatus and Systems*, Vol. **104(8)**, pp. 2040–2050, 1985.

Su, C.C. and Y.Y. Hsu, Fuzzy dynamic programming: An application to unit commitment, *IEEE Trans. on Power Systems*, Vol. **6(3)**, pp. 1231–1237, 1991.

Tapia, C.G. and B.A. Murtagh, Interactive fuzzy programming with preference criteria in multi-objective decision-making, *Computers Operations Research*, Vol. **18(3)**, pp. 307–316, 1991.

Tsuji, A., Optimal fuel mix and load dispatching under environmental constraints, *IEEE Trans. on Power Apparatus and Systems*, Vol. **100(5)**, pp. 2357–2364, 1981.

Vemuri, V., Multiple objective optimization in water resource systems, *Water Resources Research*, Vol. **10(1)**, pp. 44–48, 1974.

Viviani, G.L. and G.T. Heydt, Stochastic optimal energy dispatch, *IEEE Trans. on Power Apparatus and Systems*, Vol. **100(7)**, pp. 3221–3228, 1981.

Wang, J. and B. Malakooti, A feedforward neural network for multiple criteria decision making, *Computers Operations Research*, Vol. **19(2)**, pp. 151–167, 1992.

Yakin, M.Z., Stochastic economic dispatch in electrical power systems, *Engineering Optimization*, Vol. **8**, pp. 119–135, 1985.

Zimmermann, H.J., Fuzzy programming and linear programming with several objective functions, *Fuzzy Sets and Systems*, Vol. **1(1)**, pp. 44–55, 1978.

CHAPTER 7

Evolutionary Programming for Generation Scheduling

7.1 INTRODUCTION

A global optimization technique known as genetic algorithm (GA) has emerged as a candidate due to its flexibility and efficiency for many optimization applications. Genetic algorithm is a stochastic searching algorithm. It combines an artificial, i.e. the Darwinian Survival of the Fittest principle with genetic operation, abstracted from nature to form a robust mechanism that is very effective at finding optimal solutions to complex-real world problems. Evolutionary computing is an adaptive search technique based on the principles of genetics and natural selection. They operate on string structures. The string is a combination of binary digits representing a coding of the control parameters for a given problem. Many such string structures are considered simultaneously, with the most fit of these structures receiving exponentially increasing opportunities to pass on genetically important material to successive generation of string structures. In this way, genetic algorithms search for many points in the search space at once, and yet continually narrow the focus of the search to the areas of the observed best performance. The basic elements of genetic algorithms are reproduction, crossover, and mutation.

The first step is the coding of control variables as strings in binary numbers. In reproduction, the individuals are selected based on their fitness values relative to those of the population. In the crossover operation, two individual strings are selected at random from the mating pool and a crossover site is selected at random along the string length. The binary digits are interchanged between two strings at the crossover site. In mutation, an occasional random alteration of a binary digit is done. The above procedure to implement genetic algorithms is outlined below:

Algorithm 7.1: Genetic Algorithm

1. Code the problem variables into binary strings.
2. Randomly generate initial population strings. Tossing of a coin can be used.
3. Evaluate fitness values of population members.
4. Is solution available among the population?
   If 'yes' then GOTO Step 9.
5. Select highly fit strings as parents and produce offsprings according to their fitness.

6. Create new strings by mating current offspring. Apply crossover and mutation operators to introduce variations and form new strings.
7. New strings replace existing one.
8. GOTO Step 4 and repeat.
9. Stop.

Genetic algorithms differ from more traditional optimization techniques as:

- Genetic algorithms use objective function information to guide the search, not derivative or other auxiliary information. Evaluation of a given function uses the parameters, encoded in the string structures.
- Genetic algorithms use a coding of the parameters used to calculate the objective function in guiding the search, not the parameters themselves.
- Genetic algorithms search through many points in the solution space at one time, not a single point.
- Genetic algorithms use probabilistic rules, not deterministic rules, in moving from one set of solution (a population) to the next.

Genetic algorithms are computerized search and optimization algorithms based on the principles of natural genetics and natural selection. Although genetic algorithms were first presented systematically by Professor John Holland of the University of Michigan, the basic ideas of analysis and design based on the concept of biological evolution can be found in the work of Goldberg [1989]. Philosophically, genetic algorithms are based on the Darwin's theory of survival of the fittest.

7.1.1 Coding

Implementation of a problem in a genetic algorithm starts from the parameter encoding. The encoding must be carefully designed to utilize the genetic algorithm's ability to efficiently transfer information between chromosome strings and objective function of the problem. Binary coded strings having 1s and 0s are used. The equivalent decimal integer of binary string y is obtained as

$$y_j = \sum_{i=1}^{l} 2^{i-1} b_{ij} \quad (j = 1, 2, ..., L) \tag{7.1}$$

where

y_j is the decimal-coded value of the binary string

b_{ij} is the ith binary digit of the jth string

l is the length of the string

L is the number of strings or population size.

The continuous variable x_j can be obtained to represent a point in the search space according to a fixed mapping rule, i.e.

$$x_j = x^{\min} + \frac{x^{\max} - x^{\min}}{2^l - 1} y_j \qquad (j = 1, 2, ..., L) \tag{7.2}$$

where

$x^{\min}$ is the minimum value of variable x_j

$x^{\max}$ is the maximum value of variable x_j
y_j is the binary-coded value of the string
l is the length of the string
L is the number of strings or population size.

The number of binary digits needed to represent a continuous variation in accuracy of Δx can be computed from the relation

$$2^l \geq \frac{x^{\max} - x^{\min}}{\Delta x} + 1 \tag{7.3}$$

or

$$1 \geq \log_2 \left(\frac{x^{\max} - x^{\min}}{\Delta x} + 1 \right) \tag{7.4}$$

Evaluation of a chromosome represented by a string is accomplished by decoding the binary chromosome string by an alternative method in which the fractional part is considered.

$$y_j = \sum_{i=1}^{l} 2^{-i} b_{ij} \qquad (j = 1, 2, ..., L) \tag{7.5}$$

where

b_{ij} is the ith binary digit of the jth string
l is the length of the string
L is the number of strings or population size.

The continuous variable x_j can be obtained to represent a point in the search space according to a fixed mapping rule, i.e.

$$x_j = x_j^{\min} + (x_j^{\max} - x_j^{\min})\, y_j \qquad (j = 1, 2, ..., L) \tag{7.6}$$

where

$x_j^{\min}$ is the minimum value of variable x_j
$x_j^{\max}$ is the maximum value of variable x_j
y_j is the binary-coded value of the string
L is the number of strings or population size.

EXAMPLE 7.1 Find the value of x represented by 110011001101, a string of 12 binary digits*. The value of x lies between 2.5 to 10.0.

Solution The decimal value of 110011001101 is

$$= 1 \times 2^0 + 1 \times 2^1 + 0 \times 2^2 + 0 \times 2^3 + 1 \times 2^4 + 1 \times 2^5 + 0 \times 2^6 + 0 \times 2^7 + 1 \times 2^8 + 1 \times 2^9 + 0 \times 2^{10} + 1 \times 2^{11}$$

$$= 1 + 2 + 16 + 32 + 256 + 512 + 2048 = 2867$$

$$x = 2.5 + \frac{10.0 - 2.5}{2^{12} - 1} \times 2867 = 7.75092$$

*Though, normally, the rightmost bit of a binary string is the least significant bit (LSB), in the chapter the leftmost bit is taken as the LSB. It is easy to store the number this way being a combination of bits in an array.

EXAMPLE 7.2 Find the value of x_1 and x_2 represented by 110011001101, a string of 12 binary digits. The value of x_1 lies between 0 to 5.0, and x_2 between 7.5 and 2.0. Each variable is represented by a 6-bit string.

Solution

$$\text{12-bit string} = 110011001101$$

$$\text{6-bit string for } x_1 = 110011 \text{ and 6-bit string for } x_2 = 001101$$

$$\text{Decimal value of } 110011 = 1 \times 2^0 + 1 \times 2^1 + 0 \times 2^2 + 0 \times 2^3 + 1 \times 2^4 + 1 \times 2^5$$

$$= 1 + 2 + 16 + 32 = 51$$

$$\text{Decimal value of } 001101 = 0 \times 2^0 + 0 \times 2^1 + 1 \times 2^2 + 1 \times 2^3 + 0 \times 2^4 + 1 \times 2^5$$

$$= 4 + 8 + 32 = 44$$

$$x_1 = 0.0 + \frac{5.0 - 0.0}{2^6 - 1} \times 51 = 4.04762$$

$$x_2 = 2.0 + \frac{7.5 - 2.0}{2^6 - 1} \times 44 = 5.84127$$

EXAMPLE 7.3 Find the value of x represented by 11010110, a string of 8 binary digits. The value of x lies between 2.5 to 10.0

Solution The fractional decimal value of 11010110 is

$$= 1 \times 2^{-1} + 1 \times 2^{-2} + 0 \times 2^{-3} + 1 \times 2^{-4} + 0 \times 2^{-5} + 1 \times 2^{-6} + 1 \times 2^{-7} + 0 \times 2^{-8}$$

$$= 1\times\frac{1}{2} + 1\times\frac{1}{4} + 0\times\frac{1}{8} + 1\times\frac{1}{16} + 0\times\frac{1}{32} + 1\times\frac{1}{64} + 1\times\frac{1}{128} + 0\times\frac{1}{256}$$

$$= 0.5 + 0.25 + 0.0625 + 0.015625 + 0.0078125$$

$$= 0.8359375$$

$$x = 2.5 + 0.8359375 \times (10.0 - 2.5) = 8.76953125$$

7.2 FITNESS FUNCTION

Genetic algorithms mimic the survival-of-the fittest principle of nature to make a search process. Therefore, genetic algorithms are naturally suitable for solving maximization problems. Minimization problems are usually converted into maximization problems using some suitable transformation. Fitness function $f(x)$ is derived from the objective function and used in successive genetic operations. The fitness function for maximization problems can be used the same way as the objective function $F(x)$, i.e.

$$f(x) = F(x) \tag{7.7}$$

The fitness function for the minimization problem can be obtained from the objective function using the following relation

$$f(x) = \frac{1}{1 + F(x)} \tag{7.8}$$

This transformation does not change the location of minimum. But it only converts a minimization problem to an equivalent maximization problem. The fitness value of the string is termed the *string's fitness*. In many case, the fitness value corresponds to the number of offspring that an individual can expect to produce in next generation. A commonly used transformation is of proportional fitness assignment, i.e.

$$f(x) = \frac{F(x_i)}{\sum_{i=1}^{L} F(x_i)} \tag{7.9}$$

where

L is the population size

x_i is the phenotype value of the individual

$F(x)$ is the objective function corresponding to the individual.

The fitness assignment assures that each individual has a probability of reproducing according to its relative fitness but it fails to account for negative objective function, i.e.

$$f(x) = aF(x) + b \tag{7.10}$$

where

a is a positively scaling factor—positive for the maximization problem and negative for the minimization problem

b is offset to ensure that the resulting fitness values are non-negative.

A further method of transferring the objective function values to fitness measures is the power law scaling [Goldberg, 1989]

$$f(x) = (F(x))^k \tag{7.11}$$

where k is problem dependant and may be changed during the execution of genetic algorithm to shrink or stretch the range of fitness as required.

Another transfer function that can be used is

$$f(x) = e^{-(aF(x))^b} \tag{7.12}$$

where a and b are constants and their value is problem dependant.

In order to emphasize the best chromosome and speed up the convergence of the evolutionary process, fitness is normalized into the range between 0 and 1. The fitness function of the ith chromosome is

$$f_i(x) = \frac{1}{1 + k\left(\frac{F_i(x)}{F^{\min}} - 1\right)} \tag{7.13}$$

where

$F_i(x)$ is the solution corresponding to the ith chromosome

$F^{\min}$ is the solution of the highest ranking chromosome

k is the scaling constant.

At the start of genetic algorithm runs, it is common to have a few extraordinary individuals in a population of mediocre colleagues. During the selection process, extraordinary individuals would take over a significant proportion of finite population in a single generation, and this is undesirable. This is a leading cause of premature convergence. As the genetic algorithm searching process bases on the fitness information of the population, the art lies in how distinguishable the potential solutions can be selected and recombined. Regulation of the number of copies is especially important in small population genetic algorithms. One of the useful scaling procedures is sigma truncation. This method improves linear scaling both to deal with negative evaluation values and to incorporate problem dependant information into mapping itself. The new fitness is calculated according to

$$f_i^{\mathrm{s}} = f_i + (f_{\mathrm{avg}} - c \times \sigma) \tag{7.14}$$

where

f_i is the raw fitness of the individual

f_i^{s} is the scaled fitness of the individual string

f_{avg} is the average fitness of the population

c is the number of expected copies desired for the best population member

σ is the standard deviation of the population.

7.3 GENETIC ALGORITHM OPERATORS

7.3.1 Reproduction

The first genetic algorithm operator is reproduction. The reproduction genetic algorithm operator selects good strings in a population and forms a mating pool. So, sometime the operator is also named as the selection operator. The commonly used reproduction operator is the proportionate reproduction operator where a string is selected for the mating pool with a probability proportional to its fitness. Therefore, the probability for selecting the ith string is

$$p_i = \frac{f_i}{\sum_{j=1}^{L} f_j} \tag{7.15}$$

where

L is the population size

f_i is the fitness of the ith population.

One way to implement this selection scheme is to imagine a roulette-wheel with its circumference marked for each string proportionate to the string's fitness (see Figure 7.1). The roulette-wheel is spun L times, each time the pointer of the roulette-wheel selects the string. As the circumference of the wheel is marked according to a string's fitness, the roulette-wheel mechanism is expected to make f_i/f_{av} copies of the ith string in the mating pool. The average fitness of the population is obtained as

$$f_{\mathrm{av}} = \left(\sum_{i=1}^{L} f_i \right) \times \frac{1}{L} \tag{7.16}$$

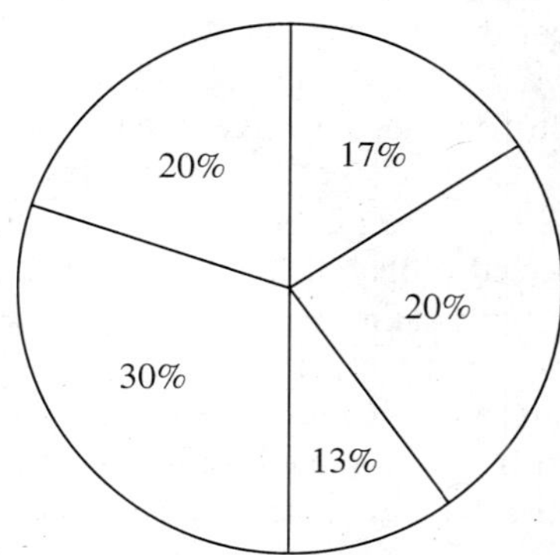

FIGURE 7.1 Roulette wheel selection.

where

L is the population size

f_i is the fitness of the ith population.

Using the fitness value f_i of all strings, the probability of selecting a string p_i can be calculated. Thereafter, the cumulative probability of each string being copied can be calculated by adding the individual probabilities from top of the list. Thus, the bottom-most string in the population should have a cumulative probability equal to 1. In order to choose L strings, L random numbers between 0 and 1 are generated at random. A string that represents the chosen random number in the cumulative probability range for the string is copied to the mating pool. No new strings are formed in the reproduction phase.

To implement the roulette wheel selection, a step-wise procedure is outlined below.

Algorithm 7.2: Roulette Wheel Selection

1. Input the fitness values of all individuals, f_i (i = 1, 2, ..., L), population size, L.
2. Initialize the population counter, $i = 0$.
3. Increment the population counter, $i = i + 1$.
4. Initialize the selection counter, $j = 0$ and initialize the cumulative sum, $S = 0$.
5. Generate a random number X.
6. Increment the selection counter, $j = j + 1$.
7. Obtain the cumulative sum of fitness,

$$S = S + \frac{f_i}{\sum_{k=1}^{L} f_i}$$

8. Check, if ($X \geq S$) then GOTO Step 6 and repeat.
9. Select the individual, $\text{SEL}_i = j$.
10. Check, if ($i < L$) then GOTO Step 3 and repeat.
11. Stop.

The basic roulette wheel selection method is stochastic sampling with replacement (SSR). The segment size and selection probability remain same throughout the selection phase and individuals are selected according to the procedure outlined above.

Stochastic sampling with partial replacement (SSPR) extends upon SSR by resizing an individual's segment if it is selected. Each time an individual is selected, the size of its segment is reduced by 1.0. If the segment size becomes negative, then it is set to zero.

Remainder sampling methods involve two distinct phases. In the integral phase, the individuals are selected deterministically according to the integer part of their expected trials. The remaining individuals are then selected probabilistically from the fractional part of the individuals expected values. Remainder stochastic sampling with replacement (RSSR) uses roulette wheel selection phase, individual's fraction parts remain unchanged and compete for selection between spin. Remainder stochastic sampling without replacement (RSSWR) sets the fractional part of an individual's expected value to zero if it is sampled during the fractional phase. To implement the stochastic remainder roulette wheel selection, a step-wise procedure is outlined below.

Algorithm 7.3: Stochastic Remainder Roulette Wheel Selection

1. Input the fitness values of all individuals, $f_i(i = 1, 2, ..., L)$, population size, L.
2. Initialize the population counter, $i = 0$ and initialize the selection counter, $j = 0$.
3. Increment the selection counter, $j = j + 1$.
4. Find $y = f_j \div \left[\left(\sum_{i=1}^{L} f_i\right) \div L\right]$.
5. Separate the integer part Y real number, I = integer (Y).
6. Separate the fractional part of Y, $F_j = Y - \text{I}$.
7. If ($I < 0$) then GOTO Step 12.
8. Increment the population counter, $k = k + 1$.
9. Decrease the integer part to zero, $I = I - 1$.
10. $\text{SEL}_k = j$.
11. GOTO Step 7 and repeat.
12. Check, if ($j < L$) GOTO Step 3 and repeat.
13. Reset the selection counter, $j = 0$.
14. If ($k > L$) GOTO Step 19.
15. Increment the selection counter, $j = j + 1$.
16. If ($j > L$) then set $j = 1$.
17. If ($F_j > 0.0$) then $\{W = i\text{flip}(F_j)\}$
    If ($W = 1$) then $\{K = k + 1, \text{SEL}_k = j, F_j = F_j - 1\}$.
18. GOTO Step 14 and repeat.
19. Stop.

7.3.2 Competition and Selection

Each individual x_i in the combined population has to compete with some other individuals to have a chance to be copied to the next generation. The score for each trial vector after stochastic competition is given by

$$w_i = \sum_{n=1}^{L} w_n \tag{7.17}$$

$$w_n = \begin{cases} 1 & \text{if } u_1 < \dfrac{f_t}{f_i + f_t} \\ 0 & \text{otherwise} \end{cases} \tag{7.18}$$

where

L is the population size or the number of competitors

f_t is the fitness value of the randomly selected competitor in the combined population

f_i is the fitness value of x_i

u_1 and u_2 are randomly selected from a uniform distribution set $u(0,1)$

$$t = \text{int}\ (2 \times L \times u_2 + 1)\ \text{int}\ (x_i).$$

After competing, the trial $2L$ solutions, including the parents and the offspring, are ranked in descending order of the score obtained in Eq. (7.17). The first L trial solutions survive and are copied along with their objective functions into the survivor set as the individuals of the next generation.

7.3.3 Crossover Operator

The basic operator for producing new chromosome in the genetic algorithm is that of crossover. In the crossover operator, information is exchanged among strings of the mating pool to create new strings. In other words, crossover produces new individuals that have some parts of both parent's genetic materials. It is expected from the crossover operator that good substrings from parent strings will be combined to form a better child offspring. There are three forms of crossover: (i) one point crossover, (ii) multipoint crossover, and (iii) uniform crossover.

One point crossover

Two individual strings are selected at random from the mating pool. Next, a crossover site is selected randomly along the string length and binary digits (alleles) are swapped (exchanged) between the two strings at the crossover site.

Parent 1: $x_1 = \{010\ 1101011\}$

Parent 2: $x_2 = \{100\ 0011100\}$

Suppose site 3 is selected at random. It means starting from the 4th bit and onwards, bits of strings will be swapped to produce offspring which are given below:

Offspring 1: $x_1 = \{010\ 0011100\}$

Offspring 2: $x_2 = \{100\ 1101011\}$

Multipoint crossover

For multipoint crossover, from m crossover positions along the string length, l are chosen at random with no duplicates and sorted into ascending order.

$$k_i \in \{1, 2, ..., l - 1\}$$

where

k_i is the ith crossover point

l is the length of the chromosome.

The bits between successive crossover points are exchanged alternatively between two parents to produce two new offspring.

Parent 1: $x_1 = \{000\ 000\ 000\ 000\}$

Parent 2: $x_2 = \{111\ 111\ 111\ 111\}$

Suppose $k_i \in \{3, 6, 9\}$ is selected at random. It means that bits 4th, 5th, and 6th of parent strings are exchanged, bits 7th, 8th, and 9th of parent strings are not exchanged and bits 10th, 11th, and 12th of parent strings are exchanged to produce offspring.

Offspring 1: $x_1 = \{000\ 111\ 000\ 111\}$

Offspring 2: $x_2 = \{111\ 000\ 111\ 000\}$

Uniform crossover

Single and multipoint crossover define cross points as places within length of string where a chromosome can be split. Uniform crossover generalizes this scheme to make every locus a potential crossover point. A crossover mask having same length as the chromosome structures is created at random and the parity of the bits in the mask indicates which parent will supply the offspring with which bits. The '1' in the random mask means bits swapping and the '0' means bit replicating.

Parent 1: $x_1 = \{1011000111\}$

Parent 2: $x_2 = \{0001111000\}$

mask: $= \{0011001100\}$

Offspring 1: $x_1 = \{1001001011\}$

Offspring 2: $x_2 = \{0011110100\}$

It is intuitively obvious from this construction that good strings from parent strings can be combined to form a better child string, if an appropriate site is chosen. With a random site, the children strings produced may or may not have a combination of good substrings from parent strings, depending on whether or not the crossing site falls in the appropriate place. If good strings are created by crossover, there will be more copies of them in the next mating pool generated by the reproduction operator. But if good strings are not created by crossover, they will not survive too long, because reproduction will select against those strings in subsequent generations. So it is clear that the effect of crossover may be detrimental or beneficial. Thus in order to preserve some of good strings, those that are already present in the mating pool are used in crossover. Hence, it can be concluded that the crossover operator has three distinct sub-steps, namely:

- Slice each of the parent strings in substrings
- Exchange a pair of corresponding substrings of parents
- Merge the two respective substrings to form offspring

7.3.4 Mutation

Mutation is the important operator, because newly created individuals have no new inheritance information and the number of alleles is constantly decreasing. This process results in the contraction of the population to one point, which is only wished at the end of the convergence process, after the population works in a very promising part of the search space. Diversity is necessary to search a big part of the search space. It is one goal of the learning algorithm to search always in regions not viewed before. Therefore, it is necessary to enlarge the information contained in the population. One way to achieve this goal is mutation. Mutation operator changes 1 to 0 and vice versa with a small mutation probability p_m. The bit-wise mutation is performed bit-by-bit by flipping the coin with required probability.

Child A: 1 1 1 1 0 1 0

↓

New child A: 1 1 0 1 0 1 0

Suppose there are x_i individuals and x_i is mutated and assigned to x_{i+m} in accordance with the equation

$$x_{i+m,j} = x_{ij} + N\left(0,\ p_m(x_j^{\max} - x_j^{\min})\frac{f_i}{f^{\max}}\right) \qquad (j = 1, 2, ..., l) \tag{7.19}$$

where

x_{ij} is the jth element of the ith individual

$N(\mu, \sigma^2)$ is a Gaussian random variable with mean μ and variance σ^2

f_i is the fitness value of the ith individual

$f^{\max}$ is the maximum fitness value of old generation

$x_j^{\max}$ is the maximum limit of the jth element

$x_j^{\min}$ is the minimum limit of the jth element

p_m is the mutation scale in the range between 0 and 1.

In general, mutation probability is fixed throughout the whole search processing. However, a small fixed mutation probability can only result in a premature convergence, while the search with a large fixed mutation probability will not converge. An adaptive scale is given to change the mutation probability to solve the problem as follows.

$$p_m(k+1) = \begin{cases} p_m(k) = p_m^{\text{step}} & ;\text{if } f^{\min}(k) \text{ unchanged} \\ p_m(k) & ;\text{if } f^{\min}(k) \text{ decreased} \\ p_m^{\text{final}} & ;\text{if } p_m(k) - p_m^{\text{step}} < p_m^{\text{final}} \end{cases} \tag{7.20}$$

$$p_m(0) = p_m^{\text{init}}$$

where

k is the generation number

p_m^{init}, p_m^{final}, and p_m^{step} are fixed numbers having values around 1, 0.005 and 0.001 to 0.01 respectively.

7.4 RANDOM NUMBER GENERATION

The important part to implement the genetic algorithm is random number generation. The random numbers are stored in an array whose index (location) is randomly selected. The random number array can be reshuffled when all the random numbers stored in an array are utilized. The random number generation is performed as given below.

$$R_i = R_j^{\text{new}} \quad (j = 1, 2, ..., 54) \tag{7.21a}$$

where i = MOD ($j \times 21$, 55); MOD means remainder from division of two numbers.

To initiate the process of random number generation, the following values are assumed.

$$R_1^{prev} = 0.234, \quad R_1^{new} = 1.0 \times 10^{-9}, \qquad R_{55} = R_1^{prev}$$

These values are updated to continue the process of random number generation.

$$R_{j+1}^{prev} = R_j \tag{7.21b}$$

$$R_{j+1}^{new} = X_j^{prev} - X_j^{new} \tag{7.21c}$$

Positiveness of the random number is checked. Negative numbers are changed to positive.

$$R_{j+1}^{new} = \begin{cases} R_{j+1}^{new} & ; R_{j+1}^{new} > 0.0 \\ R_{j+1}^{new} + 1 ; & R_{j+1}^{new} < 0.0 \end{cases} \tag{7.21d}$$

Pseudorandom numbers

Random numbers are not numbers generated by a random process. These are generated by a completely deterministic arithmetical process. The resulting set of numbers having various statistical properties which are called randomness. A typical arithmetic process is

$$S_{n+1} = (M \times S_n) \text{ MOD } N \tag{7.21e}$$

where

S_0 is seed number which is used to start the process to generate random numbers.
M and N are specific numbers. N should be large to generate large numbers.

Some generators include additive numbers

$$S_{n+1} = (M \times S_n + A) \text{ MOD } N \tag{7.21c}$$

where

S_0 is seed number which is used to start the process to generate random numbers.
M and N are specific numbers. N should be large to generate large numbers.
A is additive number

The sequence S_n must satisfy certain statistical test and must be evenly distributed over the interval (0, N). A successful sequence is said to consist pseudorandom numbers. A simple example produces well scrambled arrangement of the integers from 0 to $(2^{16} - 1)$ or 65535.

$$S_{n+1} = (25173 \times S_n + 13849) \text{ MOD } 2^{16}$$

Uniform random numbers

The sequence R_n must satisfy certain statistical test and must be uniformly distributed over the interval (0, 1). A successful sequence is said to consist uniform random numbers. Uniform random numbers generated by normalizing the pseudorandom numbers.

$$R_n = S_n/N \tag{7.21d}$$

where

S_0 is seed number which is used to start the process to generate random numbers.
$S_n = (M \times S_{n-1} + A) \text{ MOD } N$

with

M and N are specific numbers. N should be large to generate large numbers.
A is additive number.

Normally distributed random numbers

Normally distributed random numbers with mean zero and standard deviation one is represented by $N(0, 1)$ and can be generated with the following relation

$$R_i = \sqrt{-2\ln(1 - R_{j-1})} \times \cos(2\pi R_j) \quad \text{or} \quad R_i = \sqrt{-2\ln(1 - R_{j-1})} \times \sin(2\pi R_j) \tag{7.22a}$$

where

$$i = \text{MOD}\,(j \times 21, 55)$$
$$R_j = S_j \div N$$
$$S_{j+1} = (M \times S_j + A)\ \text{MOD}\ N$$

The random number generator will produce the maximum cycle length N of pseudorandom numbers with any initial value of S under either of these conditions.

1. N is a power of 10
   A ends in (unit digits) 1, 3, 7, or 9
   $M - 1$ is multiple of 20
2. N is a power of 12
   A is odd.
   $M - 1$ is multiple of 4

For example, the numbers may be taken as: $N = 10000$, $A = 4857$, and $M = 8601$.

Cauchy random numbers

Cauchy random numbers with scaling parameter $t(= 1)$ can be generated with following relation

$$R_n = -t\tan(\pi R_j) \tag{7.22b}$$

where

$R_j = S_j/N$ and $S_j = (M \times S_{j-1} + A)$ MOD N

with

S_0 is seed number which is used to start the process to generate random numbers.
M and N are specific numbers. N should be large to generate large numbers.
A is additive number.

Reshuffling of random numbers

The generated random numbers can be reshuffled to get more numbers as given below and are ensured to be positive number.

$$R_i = \begin{cases} R_i & ; R_i > 0.0 \\ R_i + 1 & ; R_i < 0.0 \end{cases} \tag{7.23}$$

where

$R_i = R_i - R_{i+31}$ $\quad (i = 1, 2, ..., 24)$

$R_i = R_i - R_{i-24}$ $\quad (i = 25, 26,.., 55)$

Random integer numbers

A random integer number can be created between two integer numbers, I^{min} and I^{max} with the following relation

$$I = I^{min} + \text{TRUNC}\ [(I^{max} - I^{min} + 1) \times X)] \tag{7.24}$$

where

X is a random number.

I^{min} is the minimum value of integer number.

I^{max} is the maximum value of integer number.

Flipping the coin

Flipping of coin is a powerful tool to decide for an action, whether to take place or not to take place. The coin is flipped with a probability to decide for an action. To flip a coin with a probability, p the following relation can be used.

$$\text{flip} = \begin{cases} 1\ ;\ p = 1.0 \text{ or } R \le p \\ 0\ ;\ \text{otherwise} \end{cases} \tag{7.25}$$

where

R is any random number.

p is the probability to flip a coin.

7.5 ECONOMIC DISPATCH PROBLEM

From the unit commitment table of a given plant, the fuel cost curve of the plant can be determined in the form of a polynomial of suitable degree by the method of least squares fit. If the transmission losses are neglected, the total system load can be optimally divided among the various generating plants using equal incremental cost criteria of Eq. (3.7). It is, however, unrealistic to neglect transmission losses particularly when long distance transmission of power is involved. A modern electric utility serves over a vast area of relatively low load density. The transmission losses may vary from 5 to 15% of total load. Therefore, it is essential to account for losses while developing an economic load dispatch policy.

The economic dispatch problem is defined so as to minimize the total operating cost of a power system while meeting the total load plus transmission losses within generator limits. Mathematically, the problem is defined as

Minimize
$$F(P_i) = \sum_{i=1}^{NG} (a_i P_i^2 + b_i P_i + c_i) \text{ ₹/h} \tag{7.26a}$$

subject to (i) the energy balance equation

$$\sum_{i=1}^{NG} P_i = P_D + P_L \tag{7.26b}$$

(ii) the inequality constraints

$$P_i^{min} \le P_i \le P_i^{max} \qquad (i = 1, 2, ..., NG) \tag{7.26c}$$

where

a_i, b_i, and c_i are cost coefficients

P_D is load demand

P_i is real power generation and will act as decision variable
P_L is power transmission loss
NG is the number of generation buses.

One of the most important, simple but approximate methods of expressing transmission loss as a function of generator powers is through B-coefficients. This method uses the fact that under normal operating condition, the transmission loss is quadratic in the injected bus real powers. The general form of the loss formula (derived later in this section) using B-coefficients is

$$P_L = \sum_{i=1}^{NG} \sum_{j=1}^{NG} P_i B_{ij} P_j \text{ MW} \tag{7.27}$$

where

P_i, P_j are real power injections at the ith, jth buses
B_{ij} are loss coefficients which are constant under certain assumed conditions
NG is number of generation buses.

The above loss formula is known as the George's formula. The above constrained optimization problem is converted into an unconstrained one. Lagrange multiplier method is used in which a function is minimized (or maximized) subject to side conditions in the form of equality constraints. Using Lagrange multipliers, an augmented function is defined as

$$L(P_i, \lambda) = F(P_i) + \lambda \left(P_D + P_L - \sum_{i=1}^{NG} P_i \right) \tag{7.28}$$

where λ is the Lagrangian multiplier.

Necessary conditions for the optimization problem are

$$\frac{\partial L(P_i, \lambda)}{\partial P_i} = \frac{\partial F(P_i)}{\partial P_i} + \lambda \left(\frac{\partial P_L}{\partial P_i} - 1 \right) = 0 \qquad (i = 1, 2, ..., NG)$$

Rearranging the above equation

$$\frac{\partial F(P_i)}{\partial P_i} = \lambda \left(1 - \frac{\partial P_L}{\partial P_i} \right) \qquad (i = 1, 2, ..., NG) \tag{7.29}$$

where

$\frac{\partial F(P_i)}{\partial P_i}$ is the incremental cost of the ith generator (₹/MWh).

$\frac{\partial P_L}{\partial P_i}$ represent the incremental transmission losses.

Equation (7.29) is known as *the exact coordination equation* and

$$\frac{\partial L(P_i, \lambda)}{\partial \lambda} = P_D + P_L - \sum_{i=1}^{NG} P_i = 0 \tag{7.30}$$

By differentiating the transmission loss equation, Eq. (7.27), with respect to P_i, the incremental transmission loss can be obtained as

$$\frac{\partial P_L}{\partial P_i} = \sum_{j=1}^{NG} 2B_{ij}P_j \qquad (i = 1, 2, ..., NG) \tag{7.31}$$

and by differentiating the cost function of Eq. (7.26a) with respect to, P_i, the incremental cost can be obtained as

$$\frac{\partial F(P_i)}{\partial P_i} = 2a_iP_i + b_i \qquad (i = 1, 2, ..., NG) \tag{7.32}$$

To find the solution, substitute Eqs. (7.31) and (7.32) into Eq. (7.29) to obtain

$$2a_iP_i + b_i = \lambda\left(1 - \sum_{j=1}^{NG} 2B_{ij}P_j\right) \qquad (i = 1, 2, ..., NG)$$

Rearranging the above equation to get P_i, i.e.

$$2a_iP_i + b_i = \lambda\left(1 - 2B_{ii}P_i - \sum_{\substack{j=1\\ j\neq i}}^{NG} 2B_{ij}P_j\right) \qquad (i = 1, 2, ..., NG)$$

or

$$2(a_i + \lambda B_{ii})P_i + \lambda \sum_{\substack{j=1\\ j\neq i}}^{NG} 2B_{ij}P_j = \lambda - b_i \qquad (i = 1, 2, ..., NG) \tag{7.33}$$

The above linear equations can be solved using the Gauss elimination method to obtain the value of P_i if λ is known. Here λ is obtained using genetic algorithms.

7.6 GENETIC ALGORITHM SOLUTION METHODOLOGY

The detailed solution methodology includes: the encoding and decoding techniques, constrained generation output calculation, the fitness function, parent selection, and parameter selection.

7.6.1 Encoding and Decoding

Decoding a binary string into an unsigned integer can play very important roles in genetic algorithm implementation. The inequality power limit constraint is performed in such a way that the individual string is normalized over the unit's operating region. The inequality constraints are handled in the manner, which efficiently reduces the searching space, and thus enhances the performance of the system. Binary coded strings having 1s and 0s are used. The equivalent decimal integer of binary string λ is obtained as

$$y^j = \sum_{i=1}^{l} 2^{i-1} b_i^j \qquad (j = 1, 2, ..., L) \tag{7.34}$$

where

b_i^j is the *i*th binary digit of the *j*th string
l is the length of the string
L is the number of strings or population size.

The continuous variable λ can be obtained to represent a point in the search space according to a fixed mapping rule, i.e.

$$\lambda^j = \lambda^{\min} + \frac{\lambda^{\max} - \lambda^{\min}}{2^l - 1} y^j \qquad (j = 1, 2, ..., L) \tag{7.35}$$

where

$\lambda^{\min}$ is the minimum value of variable, λ
$\lambda^{\max}$ is the maximum value of variable, λ
y^j is the binary coded value of the string
l is the length of the string
L is the number of strings or population size.

The number of binary digits needed to represent a continuous variation in accuracy of $\Delta\lambda$ can be computed from the relation

$$l \geq \log_2 \left(\frac{\lambda^{\max} - \lambda^{\min}}{\Delta\lambda} + 1 \right) \tag{7.36}$$

7.6.2 Calculation for Generation and Transmission Losses

When the incremental cost λ^j is known for whole population, then the generation can be obtained from Eq. (7.33), i.e.

$$2(a_i + \lambda^j B_{ii})P_j^i + \lambda^j \sum_{\substack{k=1 \\ k \neq i}}^{\text{NG}} 2B_{ik} P_k^i = \lambda^j - b_i \qquad (i = 1, 2, ..., \text{NG}; \;\; j = 1, 2, ..., L) \tag{7.37}$$

The above equation can be rewritten as

$$\sum_{k=1}^{\text{NG}} A_{ik}^j P_k^j = C_i^j \qquad (i = 1, 2, ..., \text{NG}; \;\; j = 1, 2, ..., L) \tag{7.38}$$

where

$$A_{ii}^j = 2(a_i + \lambda^j B_{ii})$$

$$A_{ik}^j = 2\lambda^j B_{ik} \qquad (i \neq k)$$

$$C_i^j = \lambda^j - b_i$$

Transmission loss for whole population can be obtained as

$$P_L^j = \sum_{i=1}^{\text{NG}} \sum_{k=1}^{\text{NG}} P_i^j B_{ik} P_k^j \qquad (j = 1, 2, ..., L) \tag{7.39}$$

7.6.3 Fitness Function and Parent Selection

Implementation of a problem in a genetic algorithm is realized within the fitness function. Since the proposed approach uses the equal incremental cost criterion as its basis, the constraint Eq. (7.30) can be rewritten as

$$\varepsilon^j = \left| P_D + P_L^j - \sum_{i=1}^{NG} P_i^j \right| \tag{7.40}$$

Then the converging rule is when ε decreases to within a specific tolerance.

In order to emphasize the 'best' chromosomes and speed up convergence of the iteration procedure, fitness is normalized into range between 0 and 1. The fitness function adopted is

$$f^j = 1 \div \left(1 + \alpha \frac{\varepsilon^j}{P_D}\right) \qquad (j = 1, 2, \ldots, L) \tag{7.41}$$

where α is the scaling constant.

When the fitness of each chromosome is calculated, the "stochastic remainder roulette wheel selection" technique is used to select the best parents according to their fitness.

Algorithm 7.4: Economic Dispatch Using Genetic Algorithm

The step-wise procedure is outlined below:

1. Read data, namely cost coefficients, a_i, b_i, c_i, B-coefficients, B_{ij} (i = 1, 2, ..., NG; j = 1, 2, ..., NG), convergence tolerance, error, step size α, and maximum allowed iterations, ITMAX, l length of string, L population size, p_c probability of crossover, p_m probability of mutation, S seed number, $\lambda^{\min}$ and $\lambda^{\max}$, etc.
2. Generate an array of random numbers. Generate the population λ^j (j = 1, 2, ..., L) by flipping the coin. The bit is set according to the coin flip as

$$b_{ij} = \begin{cases} 1 & \text{if } p = 1 \text{ or random } 0 \le p \\ 0 & \text{otherwise} \end{cases}$$

   where p is the probability (0.5).

3. Set generation counter, k = 0, BIG =1.0, $f^{\max}$ = 0.0 and $f^{\min}$ = 1.0.
4. Increment the generation counter, $k = k + 1$ and set population counter, j = 0.
5. Increment the population counter, $j = j + 1$.
6. Decode the string using Eq. (7.34) and Eq. (7.35).
7. Using Gauss elimination method, find P_i^j (i = 1, 2, ..., NG) from Eq. (7.38).
8. Calculate transmission loss from Eq. (7.39).
9. Using Eq. (7.40), find ε^j and check if (ε^j < BIG), then set BIG = ε^j.
10. Find fitness from Eq. (7.41).
    If ($f^j > f^{\max}$) then set $f^{\max} = f^j$ and if ($f^j < f^{\min}$) then set $f^{\min} = f^j$.
11. If ($j < L$) then GOTO Step 5 and repeat.
12. If (BIG $\le$ error) then GOTO Step 18.
13. Find population with maximum fitness and average fitness of the population.

14. Select the parents for crossover using stochastic remainder roulette wheel selection using Algorithm 7.3.
15. Perform single point crossover for the selected parents.
16. Perform the mutation.
17. If ($k <$ ITMAX) then GOTO Step 4 and repeat.
18. Stop.

EXAMPLE 7.4 Find the generation schedule of a three-generator power system to meet a demand of 300 MW. The cost characteristics of generators are given as below.

$$F_1 = 0.00525P_1^2 + 8.663P_1 + 328.13 \text{ ₹/h}$$
$$F_2 = 0.00609P_2^2 + 10.040P_2 + 136.91 \text{ ₹/h}$$
$$F_3 = 0.00592P_3^2 + 9.760P_3 + 59.16 \text{ ₹/h}$$

The cost characteristics are valid for the following minimum and maximum limits of power generation.

$$P_1^{\min} = 50 \text{ MW}, \qquad P_1^{\max} = 250 \text{ MW}$$
$$P_2^{\min} = 5 \text{ MW}, \qquad P_2^{\max} = 150 \text{ MW}$$
$$P_3^{\min} = 15 \text{ MW}, \qquad P_3^{\max} = 100 \text{ MW}$$

The transmission loss coefficients are given as

$$B = \begin{bmatrix} 0.000136 & 0.0000175 & 0.000184 \\ 0.0000175 & 0.000154 & 0.000283 \\ 0.000184 & 0.000283 & 0.000161 \end{bmatrix} \text{MW}^{-1}$$

Solution To implement the genetic algorithm, incremental cost λ is considered as variable to be searched. Assume the following:

$$\text{Length of string, } l = 16 \text{ bits}$$
$$\text{Population size, } L = 20$$
$$\text{Crossover probability, } p_c = 0.8$$
$$\text{Mutation probability, } p_m = 0.01$$

The minimum and maximum values of incremental cost are assumed as

$$\lambda^{\min} = 10 \text{ ₹/MWh}, \qquad \lambda^{\max} = 12.5 \text{ ₹/MWh}$$

Generate an array of random numbers using Eq. (7.21) and reshuffle three times using Eq. (7.23).

Each bit of the individual of population is created randomly by flipping a coin with probability 0.5.

Equation (7.25) is used to flip the coin with probability of 0.5 for 16 times (length of string). For the first bit, coin is flipped as given below with $p = 0.5$ and random number $R = 0.0862$.

$$\text{flip} = \begin{cases} 1 & ; p = 1.0 \text{ or } 0.0862 \le 0.5 \\ 0 & ; \text{otherwise} \end{cases}$$

So flip = 1 and the first bit has a 1 value.

Similarly, the second bit is obtained by flipping the coin where $p = 0.5$ and $R = 0.8747$.

$$\text{flip} = \begin{cases} 1 \; ; p = 1.0 \text{ or } 0.8747 \geq 0.5 \\ 0 \; ; \text{otherwise} \end{cases}$$

so flip = 0 and the second bit has a 0 value.

The above process is repeated for the rest of 14 bits and the string so obtained is given below.

The first member of population is 1001001111010000.

Decode the value of string using Eq. (7.34)

$$y^1 = 1 \times 2^0 + 0 \times 2^1 + 0 \times 2^2 + 1 \times 2^3 + 0 \times 2^4 + 0 \times 2^5 + 1 \times 2^6 + 1 \times 2^7 + 1 \times 2^8 + 1 \times 2^9 + 0 \times 2^{10} + 1 \times 2^{11} + 0 \times 2^{12} + 0 \times 2^{13} + 0 \times 2^{14} + 0 \times 2^{15}$$

$$= 1 + 8 + 64 + 128 + 256 + 512 + 2048 = 3017$$

Using Eq. (7.35), find the value of λ

$$\lambda^1 = \lambda^{\min} + \frac{\lambda^{\max} - \lambda^{\min}}{2^l - 1} y^1$$

or

$$\lambda^1 = 10.0 + \frac{12.5 - 10.0}{2^{16} - 1} 3017 = 10.11509 \text{ ₹/MWh}$$

Solve the simultaneous equations given by Eq. (7.38), i.e.

$$\sum_{k=1}^{3} A_{ik}^j P_k^j = C_i^j \qquad (i = 1, 2, 3; \;\; j = 1)$$

where

$$A_{11}^1 = 2(a_1 + \lambda^1 B_{11}) = 2(0.00525 + 10.11509 \times 0.000136) = 0.0133$$
$$A_{22}^1 = 2(a_2 + \lambda^1 B_{22}) = 2(0.00609 + 10.11509 \times 0.000154) = 0.0153$$
$$A_{33}^1 = 2(a_3 + \lambda^1 B_{33}) = 2(0.00592 + 10.11509 \times 0.000161) = 0.0151$$
$$A_{12}^1 = A_{21}^1 = 2\lambda^1 B_{12} = 2 \times 10.11509 \times 0.0000175 = 0.0004$$
$$A_{13}^1 = A_{31}^1 = 2\lambda^1 B_{13} = 2 \times 10.11509 \times 0.000184 = 0.0037$$
$$A_{23}^1 = A_{32}^1 = 2\lambda^1 B_{23} = 2 \times 10.11509 \times 0.000283 = 0.0057$$
$$C_1^1 = \lambda^1 - b_1 = 10.11509 - 8.663 = 1.4521$$
$$C_2^1 = \lambda^1 - b_2 = 10.11509 - 10.040 = 0.0751$$
$$C_3^1 = \lambda^1 - b_3 = 10.11509 - 9.760 = 0.3551$$

The above equations can be written in matrix form

$$\begin{bmatrix} 0.0133 & 0.0004 & 0.0037 \\ 0.0004 & 0.0153 & 0.0057 \\ 0.0037 & 0.0057 & 0.0151 \end{bmatrix} \begin{bmatrix} P_1^1 \\ P_2^1 \\ P_3^1 \end{bmatrix} = \begin{bmatrix} 1.4521 \\ 0.0751 \\ 0.3551 \end{bmatrix}$$

Triangularizing the previous equations

$$\begin{bmatrix} 0.0133 & 0.0004 & 0.0037 \\ 0.0 & 0.0153 & 0.0056 \\ 0.0 & 0.0 & 0.0120 \end{bmatrix} \begin{bmatrix} P_1^1 \\ P_2^1 \\ P_3^1 \end{bmatrix} = \begin{bmatrix} 1.4521 \\ 0.0363 \\ -0.0662 \end{bmatrix}$$

With back substitution, generation can be obtained as

$$P_1^1 = 111.0146 \text{ MW}, \quad P_2^1 = 2.390850 \text{ MW}, \quad P_3^1 = -5.52260 \text{ MW}$$

The P_2^1 and P_3^1 violate the lower limits, so adjust these at lower limits.

$$P_1^1 = 111.0146 \text{ MW}, \quad P_2^1 = 5.0 \text{ MW}, \quad P_3^1 = 15.0 \text{ MW}$$

Cost and transmission loss are given below.

$$\text{Total cost,} \quad F = \sum_{i=1}^{3} (a_i (P_i^1)^2 + b_i P_i^1 + c_i) = 1748.706 \text{ ₹/h}$$

$$P_L{}^1 = \sum_{i=1}^{3} \sum_{k=1}^{3} P_i^1 B_{ik} P_k^1 = 2.39085 \text{ MW}$$

From Eq. (7.40)

$$\varepsilon^1 = \left| P_D + P_L^1 - \sum_{i=1}^{\text{NG}} P_i^1 \right| = 171.3763$$

Fitness value can be obtained from Eq. (7.41) as

$$f^1 = 1 \div \left(1 + \alpha \frac{\varepsilon^1}{P_D}\right)$$

$$= 1 \div \left(1 + 1.0 \times \frac{171.3763}{300.0}\right) = 0.63643$$

Similarly, the rest of the members are generated and are given in Table 7.1.

Generate an integer random number I between 1 and 20 (population size) and where X is a random number, i.e.

$$I = 1 + \text{TRUNC}\ ((20 - 1 + 1) \times X) = 12$$

From Table 7.1, the selected string is 9 corresponding to population at serial 12. So, the 9th string is selected for crossover. Further to find the string for crossover from the remaining 19 strings, generate integer number I between 1 and 19, and where X is a random number.

$$I = 1 + \text{TRUNC}\ ((19 - 1 + 1) \times X) = 5$$

From Table 7.1, the selected string is 4, corresponding to population at serial 5. So, the 4th string is selected for crossover. To continue, find the string for crossover from the remaining 18 strings.

TABLE 7.1 Population of generation *l*

Population	*Binary string, y*	λ	*f*	*Selected strings*
1	1001001111010000	10.11509	0.63643	1
2	1101101111110010	10.77985	0.75558	2
3	0001100001000110	10.95796	0.79507	3
4	0001010010000000	10.01129	0.62785	3
5	0100101110110000	10.13497	0.63873	4
6	0101000011111110	11.24063	0.87907	5
7	1001101111101100	10.54540	0.70893	6
8	0111100011101001	11.47578	0.96134	6
9	0101111100010101	11.65019	0.97036	7
10	1110000001001100	10.48856	0.69839	8
11	0000010001110101	11.70047	0.95173	8
12	1011101111010000	10.11585	0.63650	9
13	0011001110011100	10.56443	0.71252	9
14	0111111011010010	10.73724	0.74669	10
15	1101000100111100	10.59125	0.71763	11
16	0000101001011000	10.25696	0.65825	11
17	1011110110100111	12.24357	0.79198	12
18	1111101110111111	12.47925	0.75247	13
19	1101101111100010	10.70172	0.73943	14
20	0101110001000101	11.58427	0.99607	15

Maximum fitness = 00.99607, Average fitness = 0.76875, Minimum fitness = 0.62785

Best solution so far gives λ = 11.58427 ₹/MWh, with fitness value 0.99607.

To perform single point crossover, a crossover site is required. The crossover site is selected by flipping the coin with the crossover probability, $p_c = 0.8$. If the flipping of coin gives a 1 value then generate any integer number between 1 and 15, and where X is a random number.

$$I = 1 + \text{TRUNC}\ ((15 - 1 + 1) \times X) = 13$$

The bits between the crossover points are exchanged alternatively between two parents to produce two new offspring. Crossover has been done between the 9th and 4th strings at the 13th crossover site and strings 1 and 2 are obtained for next generation.

9th string (parent 1) = 0101111100010101 = 11.65019

4th string (parent 2) = 0001010010000000 = 10.01129

1st string (child 1) = 0101111100010000 = 10.087660

2nd string (child 2) = 0001010010000101 = 11.573820

Similarly, mutation is carried out by flipping the coin with the mutation probability, p_m. Child string given in Table 7.2 is after the crossover and mutation operation. The 10th bit of 9th population, the 1st and 2nd bits of 13th population, the 12th bit of 15th population, the 14th bit of

16th population and the 6th bit of 18th population are muted. The best solution so far gives λ = 11.57382 ₹/MWh, and has fitness value as 0.99971.

TABLE 7.2 Population of generation 1

A	B	Parent string	C	Child string	λ	Fitness	D
1	9	0101111100010101	13	0101111100010000	10.08766	0.63415	1
2	4	0001010010000000	13	0001010010000101	11.57382	0.99971	2
3	8	0111100011101001	13	0111100011101010	10.85077	0.77081	2
4	14	0111111011010010	13	0111111011010001	11.36225	0.92002	3
5	5	0100101110110000	11	0100101110111100	10.60372	0.72003	4
6	13	0011001110011100	11	0011001110010000	10.09567	0.63481	4
7	15	1101000100111100	11	1101000100110000	10.12249	0.63705	5
8	1	1001001111010000	11	1001001111011100	10.58385	0.71621	6
9	10	1110000001001100	10	1110000000000110	10.93778	0.79029	7
10	3	0001100001000110	10	0001100001001100	10.48920	0.69851	8
11	11	0000010001110101	6	0000011100010101	11.64919	0.97074	9
12	9	0101111100010101	6	0101110001110101	11.70146	0.95137	10
13	6	0101000011111110	6	1001000011111110	11.24060	0.87905	11
14	6	0101000011111110	6	0101000011111110	11.24063	0.87907	11
15	3	0001100001000110	2	0001101111111100	10.62348	0.72387	12
16	7	1001101111101100	2	1001100001000010	10.64549	0.72819	12
17	2	1101101111110010	11	1101101111110101	11.71736	0.94565	13
18	11	0000010001110101	11	0000000001110010	10.76173	0.96133	14
19	8	0111100011101001	2	0111101111010000	10.11589	0.96133	14
20	12	1011101111010000	2	1011100011101001	11.47574	0.96133	14

A = Population count, *B* = String selected for mating, *C* = Crossover site selected, *D* = preselected population, Maximum fitness = 0.99971, Average fitness = 0.79746, Minimum fitness = 0.63415

After 2nd generation, the best solution is obtained and population is given in Table 7.3. Maximum fitness = 0.99972, Average fitness = 0.83547, Minimum fitness = 0.62860

Best solution gives λ = 11.57385 ₹/MWh with fitness value 0.99972. To find the final schedule, the simultaneous equations given by Eq. (7.38) are solved.

$$\sum_{k=1}^{3} A_{ik}^{j} P_k^j = C_i^j \qquad (i = 1, 2, 3; \quad j = 1)$$

where

$$A_{11}^1 = 2(a_1 + \lambda^1 B_{11}) = 2(0.00525 + 11.57385 \times 0.000136) = 0.0136$$

$$A_{22}^1 = 2(a_2 + \lambda^1 B_{22}) = 2(0.00609 + 11.57385 \times 0.000154) = 0.0157$$

$$A_{33}^1 = 2(a_3 + \lambda^1 B_{33}) = 2(0.00592 + 11.57385 \times 0.000161) = 0.0156$$

$$A_{12}^1 = A_{21}^1 + 2\lambda^1 B_{12} = 2 \times 11.57385 \times 0.0000175 = 0.0004$$

TABLE 7.3 Population at generation 2

A	*B*	*Parent string*	*C*	*Child string*	λ	*Fitness*
1	11	0000011100010101	5	0000011011010001	11.36110	0.91962
2	4	0111111011010001	5	0111111100010101	11.65034	0.97030
3	3	0111100011101010	6	0111100011111110	11.24140	0.87931
4	14	0101000011111110	6	0101000011101010	10.85000	0.77065
5	12	0101110001110101	9	0101110000010000	10.08034	0.63354
6	6	0011001110010000	9	0011001111110101	11.71679	0.94585
7	7	1101000100110000	13	1001010010000101	11.57385	0.99972
8	2	0001010010000101	13	0101000100110000	10.12245	0.63705
9	5	0100101110111100	8	0100101111111010	10.93576	0.78982
10	14	0101000011111110	8	0101000010111100	10.59609	0.71856
11	9	1110000000000110	8	1110000011011100	10.57645	0.71480
12	8	1001001111011100	8	1001001100000110	10.94518	0.79199
13	12	0101110001110101	7	0101110011010001	11.35966	0.91911
14	4	0111111011010001	7	0111111001110101	11.70405	0.95043
15	11	0000011100010101	7	0000011010000101	11.57595	0.99943
16	2	0001010010000101	7	0001010100010101	11.64706	0.97155
17	10	0001100001001100	12	0001100001000000	10.02045	0.62860
18	1	0101111100010000	12	0101111100011100	10.55642	0.71100
19	13	1001000011111110	5	1001000011111110	11.24060	0.87905
20	13	1001000011111110	5	1001000011111110	11.24060	0.97905

A = Population count, *B* = string selected for mating, *C* = crossover site selected.

$$A_{13}^1 = A_{31}^1 = 2\lambda^1 B_{13} = 2 \times 11.57385 \times 0.000184 = 0.0043$$
$$A_{23}^1 = A_{23}^1 = 2\lambda^1 B_{23} = 2 \times 11.57385 \times 0.000283 = 0.0066$$
$$C_1^1 = \lambda^1 - b_1 = 11.57385 - 8.663 = 2.9109$$
$$C_2^1 = \lambda^1 - b_2 = 11.57385 - 10.040 = 1.5339$$
$$C_3^1 = \lambda^1 - b_3 = 11.57385 - 9.760 = 1.8139$$

The above equations can be written in matrix form as

$$\begin{bmatrix} 0.0136 & 0.0004 & 0.0043 \\ 0.0004 & 0.0157 & 0.0066 \\ 0.0043 & 0.0066 & 0.0156 \end{bmatrix} \begin{bmatrix} P_1^1 \\ P_2^1 \\ P_3^1 \end{bmatrix} = \begin{bmatrix} 2.9109 \\ 1.5339 \\ 1.8139 \end{bmatrix}$$

Triangularizing the above equations,

$$\begin{bmatrix} 0.0133 & 0.0004 & 0.0043 \\ 0.0 & 0.0157 & 0.0064 \\ 0.0 & 0.0 & 0.0116 \end{bmatrix} \begin{bmatrix} P_1^1 \\ P_2^1 \\ P_3^1 \end{bmatrix} = \begin{bmatrix} 2.9109 \\ 1.4475 \\ 0.3144 \end{bmatrix}$$

With back substitution, generation can be obtained as

$$P_1^1 = 202.4288 \text{ MW}, \quad P_2^2 = 80.94910 \text{ MW}, \quad P_3^1 = 27.06991 \text{ MW}$$

No limit of generation is violated.
Cost and transmission loss are given below.

$$\text{Total cost,} \quad F = \sum_{i=1}^{3} \left(a_i (P_i^1)^2 + b_i P_i^1 + c_i\right) = 3614.148 \text{ ₹/h}$$

$$P_L^1 = \sum_{i=1}^{3} \sum_{k=1}^{3} P_i^1 B_{ik} P_k^1 \quad = \quad 10.53036 \text{ MW}$$

From Eq. (7.40),

$$\varepsilon^1 = \left| P_D + P_L^1 - \sum_{i=1}^{NG} P_i^1 \right| = 0.08252$$

More iterations for generation or size of chromosome length can improve the value of ε and hence the fitness value of population.

7.7 GENETIC ALGORITHM SOLUTION BASED ON REAL POWER SEARCH

In Section 7.6, the incremental cost is searched using genetic algorithms. The genetic algorithm can be implemented by searching the generation of power plants P_i within the generator limits. The economic dispatch problem is defined in Section 7.5 by Eq. (7.26) and transmission losses are defined by Eq. (7.27). Necessary conditions for optimization problem are given by Eqs. (7.29) and (7.30).

7.7.1 Encoding and Decoding

Decoding a binary string into an unsigned integer can play very important roles in genetic algorithm implementation. The inequality power limit constraint is performed in such a way that the individual string is converted into the unit's operating region. The inequality constraints are handled in the manner which efficiently reduces the searching space, and thus enhances the performance of the system. Binary coded strings having 1s and 0s are used. The equivalent decimal integer of binary string λ is obtained as

$$y_i^j = \sum_{k=1}^{l_i} 2^{k-1} b_{ik}^j \qquad (i = 1, 2, ..., \text{NG}; \;\; j = 1, 2, ..., L) \tag{7.42}$$

where

b_{ik}^j is the kth binary digit of the jth string and ith substring
l_i is the length of string of the ith substring
L is the number of strings or population size.

The continuous variable P_i can be obtained to represent a point in the search space according to a fixed mapping rule, i.e.

$$P_i^j = P_i^{\min} + \frac{P_i^{\max} - P_i^{\min}}{2^l - 1} y_i^j \qquad (i = 1, 2, ..., \text{NG}; \;\; j = 1, 2, ..., L) \tag{7.43}$$

where

$P_i^{\min}$ is the minimum value of generation of the ith plant
$P_i^{\max}$ is the maximum value of generation of the ith plant
y_i^j is the binary coded value of the ith substring
L is the number of strings or population size.

The number of binary digits needed to represent a continuous variation in accuracy of Δx_j can be computed from the relation

$$l_i \geq \log_2 \left(\frac{P_i^{\max} - P_i^{\min}}{\Delta P_i} + 1 \right) \qquad (i = 1, 2, ..., \text{NG}) \tag{7.44}$$

7.7.2 Fitness Function and Parent Selection

Implementation of a problem in a genetic algorithm is realized within the fitness function. Since the proposed approach uses the equal incremental cost criterion as its basis, the constraint equation (7.40) can be rewritten as

$$\varepsilon^j = \left| P_D + P_L^j - \sum_{i=1}^{\text{NG}} P_i^j \right| \tag{7.45}$$

Then the converging rule is when ε decreases to within a specific tolerance.

In order to emphasize the ‘best’ chromosomes and speed up convergence of the iteration procedure, fitness is normalized into the range between 0 and 1. The fitness function adopted is

$$f^j = 1 \div \left(1 + \alpha \frac{\varepsilon^j}{P_D} \right) \qquad (j = 1, 2, ..., L) \tag{7.46}$$

where α is the scaling constant.

When the fitness of each chromosome is calculated, the “stochastic remainder roulette wheel selection” technique is used to select the best parents according to their fitness.

Algorithm 7.5: Economic Dispatch Using Genetic Algorithm

The step-wise procedure is outlined below.

1. Read data, namely cost coefficients, a_i, b_i, and c_i; B-coefficients, B_{ij}(i = 1, 2, ..., NG; j = 1, 2, ..., NG). Convergence tolerance, error, step size, α and maximum allowed iterations, ITMAX, length of string l, population size L, probability of crossover p_c, probability of mutation p_m, seed number S, $P_i^{\min}$ and $P_i^{\max}$, etc.

2. Generate an array of random numbers. Generate the population $P_i^j (j = 1, 2, ..., L)$ by flipping the coin. The bit is set according to the coin flip as

$$b_{ij} = \begin{cases} 1 & \text{if } p = 1 \text{ or random } 0 \le p \\ 0 & \text{otherwise} \end{cases}$$

where p is the probability and is taken as 0.5.
3. Set generation counter, $j = 0$, BIG = 1.0, $f^{max} = 0.0$, and $f^{min} = 1.0$.
4. Increment the generation counter, $k = k + 1$ and set population counter, $j = 0$.
5. Increment the population counter, $j = j + 1$.
6. Decode the string using Eq. (7.42) and Eq. (7.43) to find P_i^j ($i = 1, 2, ...,$ NG).
7. Calculate the transmission loss from Eq. (7.39).
8. Using Eq. (7.45), find ε^j and check if ($\varepsilon^j <$ BIG) then set BIG = ε^j.
9. Find fitness from Eq. (7.46).
If $(f^j > f^{max})$, then set $f^{max} = f^j$ and if $(f^j < f^{min})$ then set $f^{min} = f^j$.
10. If $(j < L)$ then GOTO Step 5 and repeat.
11. If (BIG $\le$ error) then GOTO Step 18.
12. Find population with maximum fitness and average fitness of the population.
13. Select the parents for crossover using stochastic remainder roulette wheel selection using Algorithm 7.3.
14. Perform single point crossover for the selected parents.
15. Perform the mutation.
16. If ($k <$ ITMAX) then GOTO Step 4 and repeat.
17. Stop.

EXAMPLE 7.5 Find the generation schedule of a three-generator power system to meet a demand of 300 MW. The cost characteristics of generators are given as below.

$$F_1 = 0.00525P_1^2 + 8.663P_1 + 328.13 \text{ ₹/h}$$
$$F_2 = 0.00609P_2^2 + 10.040P_2 + 136.91 \text{ ₹/h}$$
$$F_3 = 0.00592P_3^2 + 9.760P_3 + 59.16 \text{ ₹/h}$$

The cost characteristics are valid for the following minimum and maximum limits of power generation.

$$P_1^{min} = 50 \text{ MW}, \quad P_1^{max} = 250 \text{ MW}$$
$$P_2^{min} = 5 \text{ MW}, \quad P_2^{max} = 150 \text{ MW}$$
$$P_3^{min} = 15 \text{ MW}, \quad P_3^{max} = 100 \text{ MW}$$

The transmission loss coefficients are given as

$$B = \begin{bmatrix} 0.000136 & 0.0000175 & 0.000184 \\ 0.0000175 & 0.000154 & 0.000283 \\ 0.000184 & 0.000283 & 0.000161 \end{bmatrix} \text{MW}^{-1}$$

Solution To implement the genetic algorithm, real power generation of generators is considered as variable to be searched. Assume the following:

Lenght of string, $l = 48$ bits where $l_i = 16$ bits $\quad (i = 1, 2, 3)$

Population size, $L = 20$

Crossover probability, $p_c = 0.8$

Mutation probability, $p_m = 0.01$

Generate an array of random numbers using Eq. (7.21) and reshuffle three times using Eq. (7.23).

Each bit of the individual of population is created randomly by flipping a coin with probability 0.5. Equation (7.25) is used to flip the coin with probability of 0.5 for 48 times (length of string). The whole population of 20 strings is generated and is given in Table 7.4.

TABLE 7.4 Initial population strings

Population	*String of population*		
1	0000010010100101	0100101000111011	1001011001100001
2	0011000011110111	1010111000011001	1110000010000100
3	0110011111000110	0101010111001001	1111101011101111
4	0111001100101100	0000000001011110	1010011000110010
5	1000011011000001	1110010111010100	1010111110100111
6	1100100101111000	1011100111101000	1101101110100110
7	0110100001111011	1011110101010101	1100100110010000
8	0011011011111000	0101110100100011	0010101101110011
9	0111100011100110	1100110010001001	1010000110110000
10	0010011000111001	1100011110010110	0100001110010110
11	0101000101100111	1111011100001000	0111111001101110
12	1111010100001011	0011110011001000	1100111000010011
13	1100001111111000	0110010101110010	0011101110010111
14	0101100111000111	0000100010111001	0100101001100100
15	1111010000010000	1110001000100010	0100101101011111
16	1010010110010010	1111011111011101	0010010111100100
17	1100101110111000	0100110011011000	1001111100100100
18	1110011110100101	0010000011011100	0001011000101010
19	1010111100001010	1101100111111100	0010101001101100
20	0011101111111110	1110010100010110	1110110000001111

The first 48 bit string is 0000010010100101 0100101000111011 1001011001100001. This string represents three substrings, each of 16 bits. These substrings are decoded to get P_i.

The 16 bit substring 1 is 0000010010100101.

Decode the value of this substring using Eq. (7.42)

$$y_1^1 = 0 \times 2^0 + 0 \times 2^1 + 0 \times 2^2 + 0 \times 2^3 + 0 \times 2^4 + 1 \times 2^5 + 0 \times 2^6 + 0 \times 2^7 + 1 \times 2^8 + 0 \times 2^9 + 1 \times 2^{10} + 0 \times 2^{11} + 0 \times 2^{12} + 1 \times 2^{13} + 0 \times 2^{14} + 1 \times 2^{15}$$

$$= 32 + 256 + 1024 + 8192 + 32768 = 42272$$

Using Eq. (7.43), find the value of P_1^1.

$$P_1^1 = P_1^{\min} + \frac{P_1^{\max} - P_1^{\min}}{2^l - 1} y_1^1 \quad \text{or} \quad P_1^1 = 50.0 + \frac{250.0 - 50.0}{2^{16} - 1} \times 42272 = 179.00587 \text{ MW}$$

The 16 bit substring 2 is 0100101000111011. Using Eq. (7.42), we get

$$y_2^1 = 0 \times 2^0 + 1 \times 2^1 + 0 \times 2^2 + 0 \times 2^3 + 1 \times 2^4 + 0 \times 2^5 + 1 \times 2^6 + 0 \times 2^7 + 0 \times 2^8 + 0 \times 2^9 + 1 \times 2^{10} + 1 \times 2^{11} + 1 \times 2^{12} + 0 \times 2^{13} + 1 \times 2^{14} + 1 \times 2^{15}$$

$$= 2 + 16 + 64 + 1024 + 2048 + 4096 + 16384 + 32768 = 56402$$

Using Eq. (7.43), find the value of P_2^1

$$P_2^1 = P_2^{\min} + \frac{P_2^{\max} - P_2^{\min}}{2^l - 1} y_2^1 \quad \text{or} \quad P_2^1 = 5.0 + \frac{150.0 - 5.0}{2^{16} - 1} \times 56402 = 129.7927 \text{ MW}$$

The 16 bit substring 3 is 1001011001100001. Using Eq. (7.42), we get

$$y_3^1 = 1 \times 2^0 + 0 \times 2^1 + 0 \times 2^2 + 1 \times 2^3 + 0 \times 2^4 + 1 \times 2^5 + 1 \times 2^6 + 0 \times 2^7 + 0 \times 2^8 + 1 \times 2^9 + 1 \times 2^{10} + 0 \times 2^{11} + 0 \times 2^{12} + 0 \times 2^{13} + 0 \times 2^{14} + 1 \times 2^{15}$$

$$= 1 + 8 + 32 + 64 + 512 + 1024 + 32768 = 34409$$

Using Eq. (7.43), find the value of P_3^1.

$$P_3^1 = P_3^{\min} + \frac{P_3^{\max} - P_3^{\min}}{2^l - 1} y_3^1 \quad \text{or} \quad P_3^1 = 15.0 + \frac{100.0 - 15.0}{2^{16} - 1} \times 34409 = 59.6291 \text{ MW}$$

Cost and transmission loss are obtained from the real power generation schedule.

$$\text{Total cost, } F = \sum_{i=1}^{3} \left(a_i (P_i^1)^2 + b_i P_i^1 + c_i \right) = 4251.895 \text{ ₹/h}$$

$$P_L^1 = \sum_{i=1}^{3} \sum_{k=1}^{3} P_i^1 B_{ik} P_k^1 = 16.64633 \text{ MW}$$

From Eq. (7.40),

$$\varepsilon^1 = \left| P_D + P_L^1 - \sum_{i=1}^{NG} P_i^1 \right| = 51.781310$$

Fitness value can be obtained from Eq. (7.41),

$$f^1 = 1 \div \left(1 + \alpha \frac{\varepsilon^1}{P_D} \right)$$

or

$$f^1 = 1 \div \left(1 + 1.0 \times \frac{51.78131}{300.0} \right) = 0.8528$$

The above procedure is repeated to decode the strings given in Table 7.4, to obtain real power generation for whole population. The real power generation for whole population is given in Table 7.5. Table 7.5 also shows the fitness value of each population member.

TABLE 7.5 Power generation schedule from initial population

Population	P_1(MW)	P_2(MW)	P_3(MW)	f
1	179.0059	129.7927	59.6291	0.85280
2	236.7582	91.3539	25.9663	0.88054
3	128.0468	88.6391	97.1362	0.99673
4	91.2543	74.1026	40.3658	0.75037
5	152.6413	29.7253	91.3541	0.88822
6	73.8865	18.3749	48.8197	0.64938
7	223.5073	101.7087	18.1789	0.90466
8	74.5487	116.4289	83.6745	0.88907
9	130.5615	87.2430	19.4889	0.81399
10	172.1820	64.9758	50.1154	0.92815
11	230.1114	14.5914	54.3437	0.95593
12	213.0365	15.8946	81.5564	0.98671
13	74.8142	49.5477	92.6498	0.76768
14	227.8164	93.9625	27.7237	0.89176
15	56.3935	43.6733	98.2815	0.73442
16	107.5357	111.4484	28.1621	0.83385
17	73.3005	20.4038	27.2763	0.62365
18	179.6132	38.4273	43.0259	0.86240
19	113.2486	41.0271	33.0389	0.71913
20	149.8917	64.2767	94.7601	0.98290

Maximum fitness = 0.99673, Average fitness = 0.84562, Minimum fitness = 0.62365

By applying the stochastic remainder roulette wheel selection procedure, the 11th and 6th strings are selected from population for crossover. Crossover site is selected at the 9th location to perform single point crossover.

Parent 11 : 010100010 11001111110111000010000111111001101110
Parent 6 : 110010010 11110001011100111101000110110110100110

Crossover site is 9

Child 1 : 010100010 11110001011100111101000110110110100110
Child 2 : 110010010 11001111110111000010000111111001101110

The 42nd bit of child 2 is muted. After mutation the strings become:

Child 1 : 010100010 11110001011100111101000110110110100110
Child 2 : 110010010 11001111110111000010000111111000101110

The crossover and mutation operation is repeated for whole population and new generation is obtained which is given in Table 7.6. The best solution is one which has maximum fitness value. The best solution is obtained at 25th generation. Table 7.7 gives the population corresponding to the

25th generation. The real power generation schedule representing maximum fitness value is given below and is the required schedule.

$$P_1 = 236.3401 \text{ MW}, \quad P_2 = 29.19211 \text{ MW}, \quad P_3 = 47.74708 \text{ MW}$$

TABLE 7.6 Power generation schedule from first generation of population

Population	P_1(MW)	P_2(MW)	P_3(MW)	f
1	73.8590	18.3749	48.81971	0.64934
2	230.1389	14.5914	53.67964	0.95421
3	179.0059	129.7927	24.90272	0.93062
4	221.9448	101.7087	59.62906	0.82119
5	153.0472	29.7253	91.35409	0.88920
6	127.6410	88.6391	91.82361	0.98041
7	230.5631	91.3584	47.21660	0.84967
8	135.9754	87.2430	19.48898	0.82555
9	179.0059	64.0886	50.13619	0.94432
10	160.4631	130.6799	59.62906	0.89445
11	236.7582	91.3539	79.09209	0.77811
12	74.5487	116.4289	30.54864	0.77999
13	235.9709	101.6777	18.17899	0.87427
14	223.5073	14.6224	49.03113	0.92420
15	128.0468	29.7319	91.35409	0.83206
16	165.1415	88.6325	97.13618	0.83206
17	172.1820	15.8924	81.55642	0.90040
18	213.0365	64.9780	50.11543	0.87846
19	74.5487	43.9278	40.36576	0.95345
20	91.2543	146.6391	83.67445	0.37461

Maximum fitness = 0.98112, Average fitness = 0.86578, Minimum fitness = 0.64934

Cost and transmission loss are obtained for the best solution as

$$\text{Total cost, } F = \sum_{i=1}^{3} \left(a_i (P_i^1)^2 + b_i P_i^1 + c_i \right) = 3642.648 \text{ ₹/h}$$

$$P_L^1 = \sum_{i=1}^{3} \sum_{k=1}^{3} P_i^1 B_{ik} P_k^1 = 13.27789 \text{ MW}$$

From Eq. (7.40),

$$\varepsilon^1 = \left| P_D + P_L^1 - \sum_{i=1}^{NG} P_i^1 \right| = 0.001410$$

The convergence achieved is 0.001. Genetic algorithm is a global search technique. By increasing the string length, accuracy can be improved.

TABLE 7.7 Real power generation schedules from the 25th generation of population

Population	P_1(MW)	P_2(MW)	P_3(MW)	f
1	230.17550	11.06683	51.04669	0.93724
2	234.77760	29.12131	47.10895	0.99317
3	236.58430	29.20760	48.39040	0.99732
4	236.34010	29.19211	47.74708	1.00000
5	236.58430	29.19432	50.51232	0.99120
6	230.17550	29.20539	48.39040	0.98336
7	229.79710	29.15228	49.71984	0.98587
8	230.13890	29.33372	49.72374	0.98746
9	236.34010	29.19211	47.76783	0.99993
10	234.77760	29.12131	51.04669	0.99533
11	230.17550	29.19211	50.42931	0.98916
12	229.79710	29.19211	51.04669	0.98980
13	230.09000	28.13008	48.39040	0.97981
14	242.59020	29.26291	48.22438	0.97970
15	228.36880	65.46258	49.71855	0.91225
16	230.17550	29.33372	48.39040	0.98376
17	230.17550	29.15671	49.71984	0.98701
18	230.13890	29.19211	50.42931	0.98905
19	229.79710	29.33372	50.42931	0.98846
20	230.17550	29.19432	49.72374	0.98713

Maximum fitness = 1.00000, Average fitness = 0.98285, Minimum fitness = 0.91225.

7.8 ECONOMIC DISPATCH WITH VALVE POINT LOADING

Economic dispatch (ED) is considered one of the key functions in electric power system operation. The economic dispatch problem is commonly formulated as an optimization problem, with the aim of minimizing the total generation cost of the power system but still satisfying specified constraints. The input–output characteristics (or cost functions) of a generator are approximated using quadratic or piecewise quadratic functions, under the assumption that the incremental cost curves of the units are monotonically increasing piecewise-linear functions. However, real input–output characteristics display higher-order nonlinearities and discontinuities due to valve-point loading in fossil fuel burning plants. The valve-point loading effect has been modelled in as a recurring rectified sinusoidal function, such as the one shown in Figure 7.2.

The generating units with multi-valve steam turbines exhibit a greater variation in the fuel cost functions. The valve-point effects introduce ripples in the heat-rate curves. Mathematically operating cost is defined as:

$$F(P_i) = \sum_{i=1}^{NG} (a_i P_i^2 + b_i P_i + c_i + |d_i \times \sin\{e_i \times (P_i^{\min} - P_i)\}|) \tag{7.47}$$

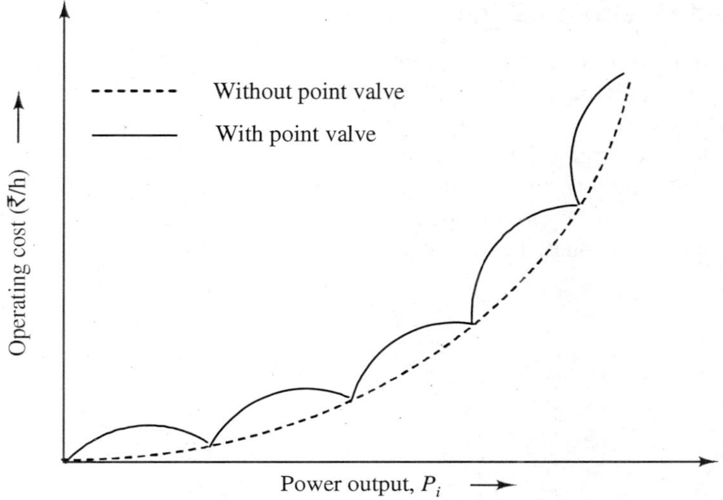

FIGURE 7.2 Operating cost characteristics with valve point loading.

where

a_i, b_i, c_i, d_i, e_i are cost coefficients of the ith unit.

Mathematically, economic dispatch problem considering valve point loading is defined as: Minimize operating cost

$$F(P_i) = \sum_{i=1}^{NG} (a_i P_i^2 + b_i P_i + c_i + |d_i \times \sin\{e_i \times (P_i^{\min} - P_i)\}|) \tag{7.48}$$

Subject to: (i) the energy balance equation is given by Eq. (7.26b) and
(ii) the inequality constraints are given by Eq. (7.26c).

7.9 ECONOMIC DISPATCH WITH RAMP RATE LIMITS AND PROHIBITED OPERATING ZONES

The unit generation output is usually assumed to be adjusted smoothly and instantaneously for convenience in solving the economic dispatch problem. Practically, the operating range of all committed units is restricted by their ramp rate limits for forcing the units operation continually between two adjacent specific operation periods. In addition, the prohibited operating zones in the input–output curve of generator are due to steam valve operation or vibration in a shaft bearing. Because it is difficult to determine the prohibited zone by actual performance testing or operating records. The best economy is achieved by avoiding operation in such areas that are in actual operation. Hence, the two constraints of generator operation must be taken into account to achieve true economic operation.

7.9.1 Prohibited Operating Zone

The prohibited operating zones in the input–output performance curve for a thermal unit can be due to vibrations in a shaft bearing caused by a steam value or can be due to faults in the machines themselves or the associated auxiliary equipment, such as boilers, feed pumps, etc. Further, it is difficult to determine the shape of the input–output curve in the neighbourhood of a prohibited zone by actual performance testing or from operational records. In actual operation, the best economy is achieved by avoiding operation in these areas. Cost functions that take into account prohibited operating zones, can be represented as in Figure 7.3.

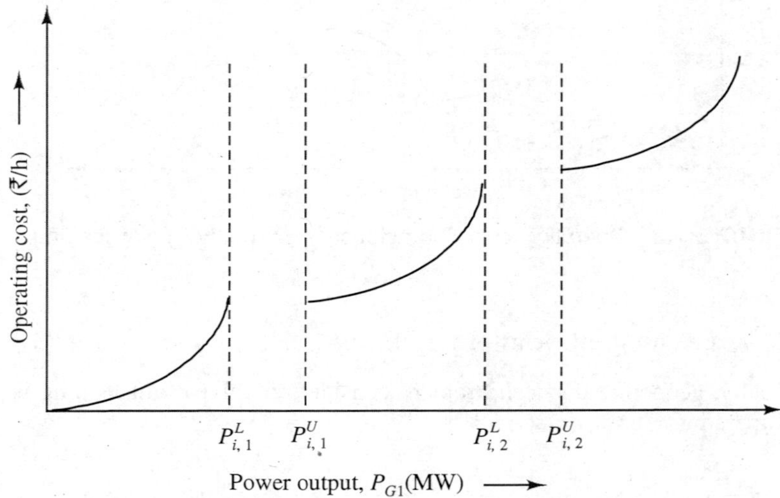

FIGURE 7.3 Operating cost characteristics with prohibited operating zones.

The input–output performance curve for a typical thermal unit with many valve points are given in Figure 7.2. These valve points generate many prohibited zones. In practical operation, adjusting the generation output of a unit must avoid unit operation in the prohibited zones. The feasible operating zones of the unit can be described. For units with prohibited operating zones (POZ), additional constraints on the unit operating range are:

$$P_i^{\min} \leq P_i \leq P_{i,1}^L \quad (i = 1, 2, \ldots, \text{NG}) \tag{7.49a}$$

$$P_{i,j-1}^U \leq P_i \leq P_{i,j}^L \quad (j = 2, 3, \ldots, Nz_i) \quad (i = 1, 2, \ldots, \text{NG}) \tag{7.49b}$$

$$P_{i,Nz_i}^U \leq P_i \leq P_i^{\max} \quad (i = 1, 2, \ldots, \text{NG}) \tag{7.49c}$$

where

Nz_i is the number of prohibited zones of ith generator.

$P_{i,j}^L$ is lower bound of jth prohibited zone of ith generator

$P_{i,j-1}^U$ is upper bound of jth prohibited zone of ith generator

7.9.2 Ramp Rate Limit

Practically, the operating range of all online units is restricted by their ramp rate limits for forcing the units' operation continually between two adjacent specific operation periods. Figure 7.4 shows three possible situations in which a unit is on-line from time interval (t–1) to t. Figure 7.4(a) shows the unit operating in steady-state conditions, Figure 7.4(b) shows the unit increasing its power generation whereas Figure 7.4(c) shows the unit decreasing the power generation output.

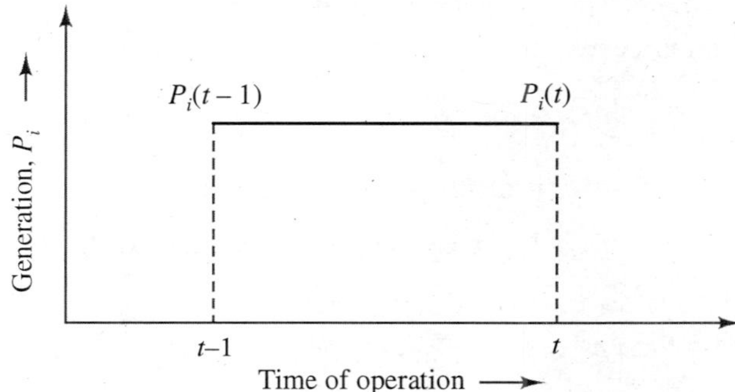

FIGURE 7.4(a) Steady state operation of an on-line of ith generator.

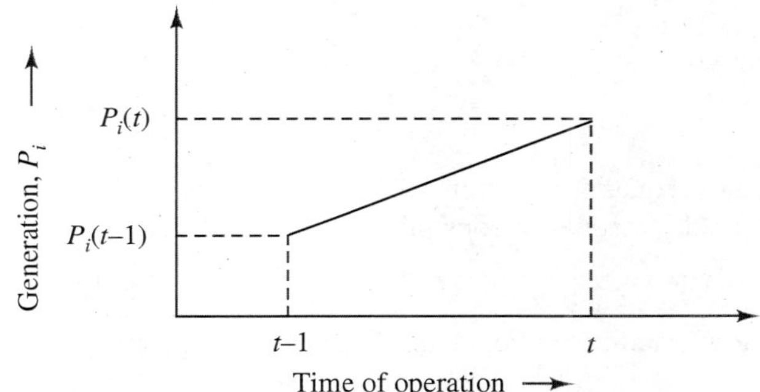

FIGURE 7.4(b) Increasing the power of an on-line of ith generator.

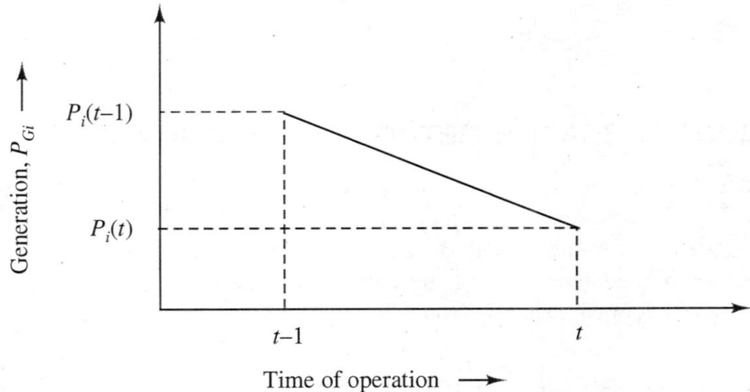

FIGURE 7.4(c) Decreasing the power of an on-line of ith generator.

Inequality constraints due to ramp rate limits that can be imposed are described below:

(i) In case generation increases

$$P_i - P_i^O \leq \text{UR}_i \tag{7.50a}$$

where
UR_i up ramp limit of ith generator (MW/h).
P_i^O is previous output power of ith generating unit.

(ii) In case generation decreases

$$P_i^O - P_i \leq \text{DR}_i \tag{7.50b}$$

where
DR_i is down the ramp limit of ith generator (MW/h).

Generator Ramp Rate inequality constraint is undertaken while solving economic dispatch problem is given below:

$$\max\,(P_i^{\min},\, P_i^O - \text{DR}_i) \leq P_i \leq \min\,(P_i^{\max},\, P_i^O + \text{UR}_i) \quad (i = 1, 2, ..., \text{NG}) \tag{7.51}$$

Mathematically, the economic dispatch problem considering ramp rate limits and POZ is defined as:

$$\text{Minimize operating cost} \qquad F(P_i) = \sum_{i=1}^{\text{NG}} (a_i P_i^2 + b_i P_i + c_i) \tag{7.52a}$$

Subject to:

(i) The energy balance equation is given by Eq. (7.26b) and

(ii) Generator Ramp Rate inequality constraint is undertaken as:

$$\max\,(P_i^{\min},\, P_i^O - \text{DR}_i) \leq P_i \leq \min\,(P_i^{\max},\, P_i^O + \text{UR}_i) \quad (i = 1, 2, ..., \text{NG}) \tag{7.52b}$$

(iii) For units with POZ, additional constraints on the unit operating range are:

$$P_i^{\min} \leq P_i \leq P_{i,1}^L \quad (i = 1, 2, \ldots, \text{NG}) \tag{7.52c}$$

$$P_{i,j-1}^U \leq P_i \leq P_{i,j}^L \quad (j = 2, 3, \ldots, Nz_i)\ (i = 1, 2, \ldots, \text{NG}) \tag{7.52d}$$

$$P_{i,Nz_i}^U \leq P_i \leq P_i^{\max} \quad (i = 1, 2, \ldots, \text{NG}) \tag{7.52e}$$

7.10 EVOLUTIONARY SEARCH METHOD FOR ECONOMIC DISPATCH

Let NG committed generating units deliver the power output subject to their respective energy balance constraints, Eq. (7.26b), and the capacity constraints, Eq. (7.26c). A dependent unit P_d is selected arbitrarily from the committed units to meet the equality constraints. The power output of the dependent unit is computed by rewriting the energy balance Eq. (7.26b).

Transmission loss is represented by Kron's loss formula:

$$P_L = B_{00} + \sum_{i=1}^{\text{NG}} B_{io} P_i + \sum_{i=1}^{\text{NG}} \sum_{j=1}^{\text{NG}} P_i B_{ij} P_j \tag{7.53}$$

Equation (7.26b) can be rewritten considering transmission losses.

$$P_d + \sum_{\substack{i=1 \\ i \neq d}}^{NG} P_i = \left[\sum_{\substack{i=1 \\ i \neq d}}^{NG} \sum_{\substack{j=1 \\ j \neq d}}^{NG} P_i B_{ij} P_j + \sum_{\substack{j=1 \\ j \neq d}}^{NG} P_j (B_{jd} + B_{dj}) P_d + B_{dd} P_d^2 + \sum_{\substack{i=1 \\ i \neq d}}^{NG} B_{io} P_i + B_{do} P_d + B_{oo} \right] + P_D \tag{7.54}$$

Above equation can be simplified as:

$$XP_d^2 + YP_d + Z = 0 \tag{7.55}$$

where $X = B_{dd}$

$$Y = \sum_{\substack{j=1 \\ j \neq d}}^{NG} (B_{jd} + B_{dj}) P_j + B_{do} - 1$$

$$Z = P_D + B_{oo} + \sum_{\substack{i=1 \\ i \neq d}}^{NG} \sum_{\substack{j=1 \\ j \neq d}}^{NG} P_i B_{ij} P_j + \sum_{\substack{i=1 \\ i \neq d}}^{NG} B_{io} P_i - \sum_{\substack{i=1 \\ i \neq d}}^{NG} P_i$$

The positive roots of the equation are obtained as:

$$P_d = \frac{-Y \pm \sqrt{Y^2 - 4XZ}}{2X}, \quad \text{where } Y^2 - 4XZ \geq 0 \tag{7.56}$$

The positive values of roots, which lie within operating limits, are considered only, then the fuel cost is computed. In case, transmission losses are neglected, then coefficients of Eq. (7.55) become

$$X = 0,\ Y = -1 \quad \text{and} \quad Z = P_D - \sum_{\substack{i=1 \\ i \neq d}}^{NG} P_i$$

The value of dependent generator, $P_d = Z$.

7.10.1 Evolutionary Search Optimization Method

In exploratory move, the current point is perturbed in all the possible directions along each variable at a time and the best point is recorded. The current point is changed to the best point at the end of each variable perturbation. If the point found at the end of all variable perturbations is different than the original point, the exploratory move is a success; otherwise, the exploratory move is a failure. In any case, the best point is considered to be the outcome of the exploratory move.

Evolutionary method is proposed to search the optimal power generation. In this method, $2^{(NG-1)}$ feasible solutions are generated for (NG – 1) committed generating units. A (NG – 1) dimensional hypercube of side Δ is formed around the point, P_i^C which represents power

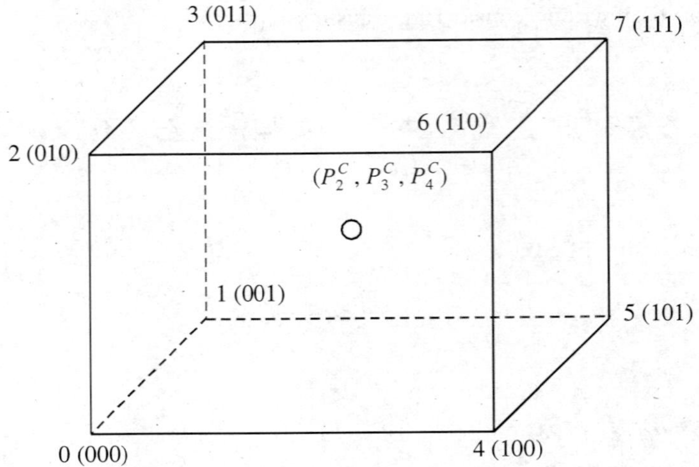

FIGURE 7.5 Three dimensional hypercube representing corners in decimal.

generation pattern of committed units from the current point in the hyperspace. The better feasible solution is obtained from objective function of the problem among the corners of the hypercube, which represent power generation pattern. Another hypercube is formed around the better point, to continue the iterative process. All the corners of the hypercube represented in binary (NG – 1) bits equivalent code, generated around the current set of generating pattern of units, are explored for the desired solution simultaneously. Table 7.8 shows the generating pattern for 4-generating units where 3 bits code is considered to represent the corners of the 3-dimensional hypercube (Figure 7.5) because one generator is taken as slack generator. Serial numbers of hypercube corners in decimal are converted into their binary equivalent code. The deviation from the current centre point is obtained by replacing 0s with –Δ and 1s with +Δ in code associated with hypercube corners.

TABLE 7.8 Power generation vector at hypercube corners for 4-generator power system

Hyper-cube corner	*Possible combinations of 3-bits* $C_2\ C_1\ C_0$	*Distance of hypercube corners from centre point* $P_2^c\ P_3^c\ P_4^c$	*Possible power generation pattern of units at the hypercube corners*		
0	0 0 0	$-\gamma_2 - \gamma_3 - \gamma_4$	$P_2^c - \gamma_2$	$P_3^c - \gamma_3$	$P_4^c - \gamma_4$
1	0 0 1	$-\gamma_2 - \gamma_3 + \gamma_4$	$P_2^c - \gamma_2$	$P_3^c - \gamma_3$	$P_4^c + \gamma_4$
2	0 1 0	$-\gamma_2 + \gamma_3 - \gamma_4$	$P_2^c - \gamma_2$	$P_3^c + \gamma_3$	$P_4^c - \gamma_4$
3	0 1 1	$-\gamma_2 + \gamma_3 + \gamma_4$	$P_2^c - \gamma_2$	$P_3^c + \gamma_3$	$P_4^c + \gamma_4$
4	1 0 0	$+\gamma_2 - \gamma_3 - \gamma_4$	$P_2^c + \gamma_2$	$P_3^c - \gamma_3$	$P_4^c - \gamma_4$
5	1 0 1	$+\gamma_2 - \gamma_3 + \gamma_4$	$P_2^c + \gamma_2$	$P_3^c - \gamma_3$	$P_4^c + \gamma_4$
6	1 1 0	$+\gamma_2 + \gamma_3 - \gamma_4$	$P_2^c + \gamma_2$	$P_3^c + \gamma_3$	$P_4^c - \gamma_4$
7	1 1 1	$+\gamma_2 + \gamma_3 + \gamma_4$	$P_2^c + \gamma_2$	$P_3^c + \gamma_3$	$P_4^c + \gamma_4$

As the number of committed generators is increased, the number of hypercube corners increases exponentially. The process of exploring the better solution from all corners of the hypercube becomes time consuming, which needs some efficient search technique that should explore all the corners of the hypercube with minimum number of function evaluations and comparisons.

$$P_i^j = P_i^c + \gamma_i^j \qquad (i = 1, 2, \ldots, \text{NG} - 1; j = 1, 2, \ldots, 2^{\text{NG} - 1}) \tag{7.57}$$

where

P_i^j is the decision variable giving generation.

P_i^c is the initial value of generations.

γ_i is the distance of sides of the hypercube from the point around which hypercube is generated.

A matrix has been generated from possible combinations of binary bits. '0' bit is replaced by $-\gamma$ and '1' bit is replaced by $+\gamma$. As an illustration, the generation of weight combinations for three objectives has been shown in Table 7.8.

To initiate the iterative process, initial generator values are set at the mean value of minimum and maximum operating limits of that generator as given below:

$$P_i^0 = \frac{P_i^{\max} + P_i^{\min}}{2} \qquad (i = 1, 2, \ldots, \text{NG}) \tag{7.58}$$

The unmet load is distributed among all the committed generating units presuming transmission loss is neglected.

$$P_i^c = P_i^0 + \frac{\left(P_D - \sum_{j=1}^{\text{NG}} P_j^0\right)}{\left(\sum_{j=1}^{\text{NG}} P_j^0\right)} \times P_i^0 \quad (i = 1, 2, \ldots, \text{NG}; i \neq d) \tag{7.59}$$

Initial distance of sides of hypercube from the central point is obtained as:

$$\gamma_i = \frac{P_i^{\max} - P_i^{\min}}{\delta} \qquad (i = 1, 2, \ldots, \text{NG}; i \neq d) \tag{7.60}$$

Algorithm 7.6: Economic Dispatch Using Evolutionary Search Method

To implement the heuristic evolutionary search, the stepwise algorithm is outlined as below:

1. Input the data.
2. Compute initial centre point, P_i^c $(i = 1, 2, \ldots, \text{NG}; i \neq d)$ using Eq. (7.59).
3. Compute the slack generation, P_d^c using Eq. (7.56) and operating cost, F^{Prev} using either Eq. (7.47) or Eq. (7.52a).
4. Initialize iteration counter, $r = 0$.

   DO

   4.1. Increment iteration counter, $r = r + 1$.

4.2. Initialize iteration counter, $k = 0$.
DO
4.3.1 Generate binary (NG – 1) bits combination number of kth hypercube corner. Generate γ_i^k and P_i^k (i = 1, 2, …, NG; $i \neq d$) using Eq. (7.57). Compute P_d^k using Eq. (7.56).
4.3.2 Compute objective, F_k from either Eq. (7.47) or Eq. (7.52a).
4.3.3 Increment iteration counter, $k = k + 1$.
WHILE ($k < 2^{NG-1}$)

5. Find minimum function value, $F^{\min}$ = Min $\{F^k(k = 1, 2, ..., 2^{(NG-1)} + 1)\}$
6. Choose weight combination P_i^c having $F^{\min}$ as a centre of hypercube.
7. IF ($F^{\min} \le F^{\text{Prev}}$) THEN $F^{\text{Prev}} = F^{\min}$ and $P_i^{\text{Prev}} = P_i^c$ (i = 1, 2, …, NG; $i \neq d$)
   ELSE $\gamma_i = \gamma_i/2$ (i = 1, 2, …, NG; $i \neq d$)
   WHILE ($r \le R$)
8. STOP.

EXAMPLE 7.6 Find the generation schedule of a four-generator system to meet a demand of 660 MW. The cost characteristics are given below:

$$F_1 = 0.0015\, P_1^2 + 1.8\, P_1 + 40.0 \text{ (₹/h)}$$
$$F_2 = 0.0030\, P_2^2 + 1.8\, P_2 + 60.0 \text{ (₹/h)}$$
$$F_3 = 0.0012\, P_3^2 + 2.1\, P_3 + 100.0 \text{ (₹/h)}$$
$$F_4 = 0.0010\, P_4^2 + 2.0\, P_4 + 120.0 \text{ (₹/h)}$$

The cost characteristics are valid for the following minimum and maximum limits of power generation.

$$P_1^{\min} = 50 \text{ MW}; \qquad P_1^{\max} = 300 \text{ MW}$$
$$P_2^{\min} = 20 \text{ MW}; \qquad P_2^{\max} = 125 \text{ MW}$$
$$P_3^{\min} = 30 \text{ MW}; \qquad P_3^{\max} = 175 \text{ MW}$$
$$P_4^{\min} = 40 \text{ MW}; \qquad P_4^{\max} = 250 \text{ MW}$$

The transmission loss coefficients are given as:

$$B = \begin{bmatrix} 0.000140 & 0.000010 & 0.000015 & 0.000015 \\ 0.000010 & 0.000060 & 0.000010 & 0.000013 \\ 0.000015 & 0.000010 & 0.000068 & 0.000065 \\ 0.000015 & 0.000013 & 0.000065 & 0.000070 \end{bmatrix} \text{MW}^{-1}$$

Solution To implement the evolutionary search, real power generation of committed generators is considered as variable to be searched. Assume that the first generator is dependent (slack) generator. Initial generator values are obtained using Eq. (7.58).

$$P_1^0 = \frac{P_1^{\max} + P_1^{\min}}{2.0} = 175.0 \text{ MW}, \qquad P_2^0 = \frac{P_2^{\max} + P_2^{\min}}{2.0} = 72.5 \text{ MW}$$

$$P_3^0 = \frac{P_3^{\max} + P_3^{\min}}{2.0} = 102.5 \text{ MW}, \qquad P_4^0 = \frac{P_4^{\max} + P_4^{\min}}{2.0} = 145.5 \text{ MW}$$

Unmet load is distributed among all the committed generating units using Eq. (7.59).

$$P_2^c = P_2^0 + \frac{P_D - (P_1 + P_2 + P_3 + P_4)}{(P_1 + P_2 + P_3 + P_4)} \times P_2^0$$

$$= 72.5 + \frac{(660.0 - 495.0)}{495.0} \times 72.5 = 96.6666 \text{ MW}$$

$$P_3^c = 102.5 + \frac{(660.0 - 495.0)}{495.0} \times 102.5 = 136.6667 \text{ MW}$$

$$P_4^c = 145.0 + \frac{(660.0 - 495.0)}{495.0} \times 145.0 = 193.3333 \text{ MW}$$

Slack generation has been found so that Eq. (7.26) is satisfied.

$$X = B_{dd} = 0.00014 \text{ MW}^{-1}$$

$$Y = \sum_{\substack{j=1 \\ j\neq 1}}^{4} (B_{jd} + B_{dj})\, P_j - 1 = -\,0.9881667$$

$$Z = P_D + \sum_{\substack{i=1 \\ i\neq 1}}^{4} \sum_{\substack{j=1 \\ j\neq 1}}^{4} P_i B_{ij} P_j + \sum_{\substack{i=1 \\ i\neq 1}}^{4} B_{i0} P_i - \sum_{\substack{i=1 \\ i\neq 1}}^{4} P_i = 241.966 \text{ MW}$$

$$0.00014\, P_1^2 - 0.9881667\, P_1 + 241.966 = 0$$

The positive roots of the equation are obtained as:

$$P_1 = \frac{-Y \pm \sqrt{Y^2 - 4XZ}}{2X} = 254.0037 \text{ MW}$$

Cost and transmission loss are obtained from the real power generation schedule.

$$\text{Total Cost, } F^{\text{Prev}} = \sum_{i=1}^{4} (a_i P_i^2 + b_i P_i + c_i) = 1809.475 \text{ ₹/h}$$

$$\text{Transmission loss, } P_L = \sum_{i=1}^{4} \sum_{j=1}^{4} P_i B_{ij} P_j = 20.67044 \text{ MW}$$

Initial distance of the sides of hypercube from the central point is obtained from Eq. (7.60):

$$\gamma_2 = \frac{P_2^{\max} - P_2^{\min}}{2.0} = 52.5 \text{ MW}$$

$$\gamma_3 = \frac{P_3^{\max} - P_3^{\min}}{2.0} = 72.50 \text{ MW}$$

$$\gamma_4 = \frac{P_4^{\max} - P_4^{\min}}{2.0} = 105.0 \text{ MW}$$

Generation schedule at each corner of the hypercube is given in Table 7.9.

TABLE 7.9 Power generation vector at hypercube corners for 4-generators.

Hyper cube corners	*Possible Combinations of 3-bits*			*Distance of hypercube corners from centre points* (96.6666, 136.6667, 193.3333)			*Possible power generations pattern at the hypercube corners*		
	C_2	C_1	C_0	γ_2	γ_3	γ_4	P_2^k	P_3^k	P_4^k
0	0	0	0	–52.51	–72.51	–105.0	44.1666	64.1667	88.3333
1	0	0	1	–52.51	–72.51	105.0	44.1666	64.1667	298.3333
2	0	1	0	–52.51	72.51	–105.0	44.1666	209.1667	88.3333
3	0	1	1	–52.51	72.51	105.0	44.1666	209.1667	298.3333
4	0	0	0	52.51	–72.51	–105.0	149.1667	64.1667	88.3333
5	0	0	1	52.51	–72.51	105.0	149.1667	64.1667	298.3333
6	0	1	0	52.51	72.51	–105.0	149.1667	209.1667	88.3333
7	0	1	1	52.51	72.51	105.0	149.1667	209.1667	298.3333

Cost and transmission loss are obtained from the real power generation schedule of each hypercube corner points and is given in Table 7.10. Corner of hypercube having smaller operating cost value is selected as central point and process is repeated till no improvement in cost is obtained or the size of hypercube becomes small. The best schedule obtained is given in Table 7.11.

TABLE 7.10 New generation pattern, cost and transmission loss corresponding to corners of hypercube

k	P_1^k (MW)	P_2^k (MW)	P_3^k (MW)	P_4^k (MW)	F (₹/h)	P_L (MW)
0	503.3964	44.16666	64.16667	88.33333	2015.738	40.06299
1	327.2324	44.16666	64.16667	250.00	1857.183	25.56577
2	381.2039	44.16666	175.00	88.33333	1898.213	28.70388
3	212.7762	44.16666	175.00	250.00	1823.010	21.94287
4	412.1448	125.00	64.16667	88.33333	1912.691	29.64480
5	**240.4454**	**125.00**	**64.16667**	**250.00**	**1813.589**	**19.61209**
6	293.0365	125.00	175.00	88.33333	1836.866	21.36979
7	128.6095	125.00	175.00	250.00	1814.932	18.60945

TABLE 7.11 Best generation schedule

P_1 (MW)	P_2 (MW)	P_3 (MW)	P_4 (MW)	F (₹/h)	P_L (MW)
201.6681	118.8477	141.1979	217.1224	1802.135	18.83613

7.11 EVOLUTIONARY PROGRAMMING FOR ECONOMIC DISPATCH–I

Evolutionary programming is a probabilistic, global search technique. It starts with a population of randomly generated candidate solutions and evolves towards better solutions over a number of generations or iterations. The main stages of this technique include initialization, mutation, and competition and selection. Mutation is often implemented by adding a random number or a vector from a certain distribution (e.g., a Gaussian distribution in the case of classical EP (CEP)) to a parent. The degree of variation of the Gaussian mutation is controlled by its standard deviation, which is also known as a 'strategy parameter' in evolutionary search. In the self-adaptation scheme of evolutionary programming, this parameter is not prefixed; rather, it is evolved along with the objective variables. Implementation of evolutionary programming for economic dispatch problem using scaled cost based adaption of strategy parameter is as given below. The evolutionary programming can be implemented by searching the generation of power plants, P_i within generator limits. The economic dispatch problem is defined by Eq. (7.26) or Eq. (7.48) or Eq. (7.52) and transmission losses are defined by Eq. (7.27).

Initialization

The initial population comprises combinations of only the candidate dispatch solutions, which satisfy all the constraints and are feasible solutions of economic dispatch. It consists of P_i^j (i = 1, 2, ..., NG; j = 1, 2, ..., L) trial parent individuals. The elements of a parent are the combinations of power outputs of the generating units, which are chosen randomly by a random number ranging over $[P_i^{\min}, P_i^{\max}]$.

$$P_i^j = P_i^{\min} + \text{rand()}\,(P_i^{\max} - P_i^{\min}) \quad (i = 1, 2, \ldots, \text{NG};\ i \neq d;\ j = 1, 2, \ldots, L) \tag{7.61}$$

where rand() is uniform random number ranging over [0, 1].

The elements of parent/offspring P_i^j may violate constraint Eq. (7.26c). This violation is corrected by fixing them either at lower or upper limits as described below:

$$P_i^j = \begin{cases} P_i^{\min} & ;\ P_i^j < P_i^{\min} \\ P_i^{\max} & ;\ P_i^j > P_i^{\max} \\ P_i^j & ;\ P_i^{\min} \le P_i^j \le P_i^{\max} \end{cases} \quad (i = 1, 2, \ldots, \text{NG};\ i \neq d;\ j = 1, 2, \ldots, L) \tag{7.62}$$

Evaluation

In order to satisfy the power balance constraint, a generator is arbitrarily selected as a dependent generator d. $P_L = 0$ means transmission losses are neglected. $P_L \neq 0$ means transmission losses are considered. Output of dependent or slack generator is obtained from Eq. (7.55) and is given below:

$$P_d^j = \begin{cases} Z^j & ; P_L = 0 \\ W^j & ; P_L \neq 0 \end{cases} \quad (j = 1, 2, \ldots, L) \tag{7.63}$$

where $W^j = (-Y^j \pm \sqrt{(Y^j)^2 - 4XZ^j})/(2X)$ when $(Y_j)^2 - 4XZ^j \geq 0$

with $X = \begin{cases} 0 & ; P_L = 0 \\ B_{dd} & ; P_L \neq 0 \end{cases}$,

$$Y^j = \begin{cases} -1 & ; P_L = 0 \\ \sum\limits_{\substack{k=1 \\ k \neq d}}^{NG} (B_{kd} + B_{dk}) P_k^j + B_{do} - 1 & ; P_L \neq 0 \end{cases}$$

$$Z^j = \begin{cases} P_D - \sum\limits_{\substack{k=1 \\ k \neq d}}^{NG} P_k^j & ; P_L = 0 \\ P_D + B_{oo} + \sum\limits_{\substack{k=1 \\ k \neq d}}^{NG} \sum\limits_{\substack{l=1 \\ l \neq d}}^{NG} P_k^j B_{kl} P_l^j + \sum\limits_{\substack{k=1 \\ k \neq d}}^{NG} B_{ko} P_k^j - \sum\limits_{\substack{k=1 \\ k \neq d}}^{NG} P_k^j & ; P_L \neq 0 \end{cases}$$

Similarly, if the output of the dependent generator violates its limits, it is fixed using Eq. (7.62). After limiting the value of the dependent generator as above, a penalty term is introduced in the objective function Eq. (7.47) to penalize its fitness value. When so introduced, Eq. (7.47) is changed to the following generalized form:

$$f^j = F(P_i^j) + \varphi^j \quad (j = 1, 2, \ldots, L) \tag{7.64}$$

where

$$\text{Penalty factor is given by } \varphi^j = \begin{cases} (P_d^j - P_d^{\min})^2 & ; P_d^j < P_d^{\min} \\ (P_d^{\max} - P_d^j)^2 & ; P_d^j > P_d^{\max} \\ 0 & ; P_d^{\min} \leq P_d^j \leq P_d^{\max} \end{cases} \tag{7.65}$$

The standard deviation σ_i^j is given by the following mathematical expression

$$\sigma_i^j = \beta \frac{f^j}{f^{\min}} (P_i^{\max} - P_i^{\min}) \quad (i = 1, 2, \ldots, NG;\ i \neq d;\ j = 1, 2, \ldots, L) \tag{7.66}$$

where

β is the scaling factor which has to be tuned during the process of search for the optimum around the initial points,

f^j the fitness value of the jth individual, and

$f^{\min}$ is the minimum fitness among the M parents.

Mutation

The three Evolutionary Programming (EP) techniques undertaken using different kinds of mutations like: Gaussian, Cauchy and combined Gaussian–Cauchy mutation. The conventional EP employing the Gaussian mutation operator is known as Classical Evolutionary Programming (CEP). The EP employing Cauchy mutation operator is known as Fast Evolutionary Programming (FEP) as it converges faster than CEP. The EP that uses the better of the Gaussian and Cauchy mutation operators is known as Improved Fast Evolutionary Programming (IFEP) as it has higher convergence rate compared to CEP and FEP when used alone.

Classical evolutionary programming

Each vector element is updated by adding a normally distributed random number with mean zero and standard deviation σ_j denoted as $N(0, \sigma_j^2)$. To generate offspring, the following equation is used.

$$P_i^{j+L} = P_i^j + \sigma_i^j \, N(0, 1) \qquad (i = 1, 2, \ldots, \text{NG};\ i \neq d;\ j = 1, 2, \ldots, L) \tag{7.67}$$

where

$N(0, 1)$ is a normally distributed random variable with zero mean and unit variance.

Fast evolutionary programming

Using Cauchy mutation, an offspring is created by

$$P_i^{j+L} = P_i^j + \sigma_i^j \, C_i^j(0, 1) \quad (i = 1, 2, \ldots, \text{NG};\ i \neq d;\ j = 1, 2, \ldots, L) \tag{7.68}$$

where

$C_i^j(0, 1)$ is a Cauchy random variable with scale parameter $t = 1$ centred at zero that is generated anew for each value of j.

Mean of Gaussian and Cauchy mutations

In this method, mean of two Gaussian and Cauchy mutations is taken to generate offspring from each parent are generated as

$$P_i^{j+L} = P_i^j + \frac{1}{2}\, \sigma_i^j \,(N(0, 1) + C_i^j(0, 1)) \quad (i = 1, 2, \ldots, \text{NG};\ i \neq d;\ j = 1, 2, \ldots, L) \tag{7.69}$$

Better of Gaussian and Cauchy mutations

In this method, the better of two offspring from each parent are generated by Gaussian mutation and the other by Cauchy mutation is taken as

$$P_i^{j+L} = P_i^j + \sigma_i^j \quad N(0, 1) \quad (i = 1, 2, \ldots, \text{NG};\ i \neq d;\ j = 1, 2, \ldots, L) \tag{7.70}$$

$$P_i^{j+L} = P_i^j + \sigma_i^j \quad (0, 1) \quad (i = 1, 2, \ldots, \text{NG};\ i \neq d;\ j = 1, 2, \ldots, L) \tag{7.71}$$

Mutation results in creation of I offspring individuals. The parent individuals are candidate dispatch solutions, which satisfy all constraints. However, after mutation, the elements of offspring P_i^{j+L} may violate constraint Eq. (7.26c). This violation is corrected by using Eq. (7.62).

Evaluation of offspring

In order to satisfy the power balance constraint, a generator is arbitrarily selected as a dependent generator *d*. Its output is obtained by solving Eq. (7.56). In case the output of the dependent generator violates its limits, it is fixed either at lower or upper limits using Eq. (7.64). After limiting the value of the dependent generator as above, a penalty term is introduced in the objective function to penalize its fitness value using Eq. (7.65). The values of the objective function of both offspring are evaluated, compared, and the better individual is chosen as offspring for competition and selection. The initial population and their offspring created by mutation form a combined population of $2L$ individuals.

Competition and selection

The $2L$ individuals compete with each other for selection. A weight value w^r is assigned to each individual as follows:

$$W^j = \sum_{k=1}^{R} w^k \quad (j = 1, 2, \ldots, 2L) \tag{7.72}$$

where

$$w^k = \begin{cases} 1 & ; \text{if} \quad u_1 > \dfrac{f^j}{f^r + f^j} \\ 0 & ; \text{else} \end{cases} \tag{7.73}$$

with f^r is the fitness of the rth competitor randomly selected from $2L$ individuals and u_1 is a uniform random number ranging over [0, 1]. $r = [2L \times u_2 + 1]$ and u_2 are uniform random numbers ranging over [0, 1].

While computing the weight for each individual, it is ensured that each individual is selected only once from the combined population. Even though relative fitness values are used during the process of mutation, competition and selection, it leads to slow convergence. This weight assignment is found to yield proper selection and good convergence. When all the $2L$ individuals obtain their weights, they are ranked in descending order and the first L individuals are selected as parents along with their fitness values for next generation.

Algorithm 7.7: Economic Dispatch Using Evolutionary Programming

The stepwise procedure is outlined below:

1. Read data, viz. Cost Coefficients, a_i, b_i, c_i (i = 1, 2, .., NG). Convergence tolerance error, scaling factor, β and maximum allowed iterations, ITMAX, population size, L, seed number, S, $P_i^{\min}$ and $P_i^{\max}$ (i = 1, 2, ..., NG) etc.
2. Generate an array of (NG × L) size of uniform random numbers.
3. Set population counter, $j = 0$
   DO
4. Increment population counter, $j = j + 1$
5. Set generation counter, $i = 0$
   DO

6. Increment the generation counter, $i = i + 1$
7. IF ($i \neq d$) THEN Generate the population P_i^j
   WHILE (i < NG)
8. Compute P_d^j using Eq. (7.63), Check limits using Eq. (7.62). Compute penalty factor, φ^j and f^j using Eq. (7.65) and Eq. (7.64), respectively.
   WHILE ($j < L$)
9. Set iteration counter, IT = 0
   DO
10. Increment iteration counter, IT = IT + 1
11. Set population counter, $j = 1$, $f^{\min} = f^j$ and $j^{\min} = j$
    DO
12. Increment population counter, $j = j + 1$
13. IF ($f^j < f^{\min}$) THEN set $f^{\min} = f^j$, $j^{\min} = j$ and $P_i^{\min} = P_i^j$ (i = 1, 2, …, NG)
    WHILE ($j < L$)
14. IF (IT = 1) THEN $f^{\text{Prev}} = f^{\min}$ and $P_{\text{i}}^{\text{Prev}} = P_i^{\min}$ (i = 1, 2, …, NG)
    ELSE IF ($f^{\min} < f^{\text{Prev}}$) THEN $f^{\text{Prev}} = f^{\min}$ and $P_{\text{i}}^{\text{Prev}} = P_i^{\min}$ (i = 1, 2, …, NG)
15. Set population counter, $j = 0$
    DO
16. Increment population counter, $j = j + 1$
17. Set generation counter, $i = 0$
    DO
18. Increment the generation counter, $i = i + 1$
19. Compute standard deviation, σ_i^j using Eq. (7.66)
    WHILE (i < NG)
    WHILE ($j < L$)
20. Generate array of (NG × L) size of normalized random numbers or Cauchy random numbers or arrays of both normalized and Cauchy random numbers
21. Set population counter, $j = 0$
    DO
22. Increment population counter, $j = j + 1$
23. Set generation counter, $i = 0$
    DO
24. Increment the generation counter, $i = i + 1$
25. IF ($i \neq d$) THEN
    1. Generate the population P_i^{j+L} using Eq. (7.67) or Eq. (7.68) or Eq. (7.69) or Eq. (7.70) and Eq. (7.71)
    2. Check the limits if violated set as per Eq. (7.62)
       WHILE (i < NG)
26. Compute P_d^{j+L} using Eq. (7.63), Check limits using Eq. (7.62). Compute penalty factor, φ^{j+L} and f^{j+L} using Eq. (7.65) and Eq. (7.64), respectively.
    WHILE ($j < L$)
27. Set population counter, $j = 0$
    DO
28. Increment population counter, $j = j + 1$
29. Generate array of (6 × L) size of uniform numbers
30. Initialize $m = 0$ and SUM = 0.0

31. Set population counter, $k = 0$ and compute $R = [2L \times u_2 + 1]$
    DO
32. Increment population counter, $k = k + 1$
33. Select u_1 as $(m + 1)$th element and u_2 as $(m + 2)$th element of array of uniform random numbers
34. Compute $r = [2L \times u_2 + 1]$ and compute w^k using Eq. (7.73).
35. Update SUM = SUM + w^k and $m = m + 2$
    WHILE ($k < $ R) GOTO 39
36. Set W^j = SUM
    WHILE ($j < (2 \times L)$)
37. Set population counter, $j = 0$
    DO
38. Increment population counter, $j = j + 1$
39. Set counter, $k = j$
    DO
40. Increment counter, $k = k + 1$
41. IF ($W^j < W^k$) THEN $f^k \leftrightarrow f^j$, $W^k \leftrightarrow W^j$ and $P_i^k \leftrightarrow P_i^j$
    (i = 1, 2, ..., NG)
    WHILE ($k < (2 \times \mathrm{L})$)
    WHILE ($j < (2 \times \mathrm{L} - 1)$)
    WHILE (IT < IT MAX)
42. STOP

EXAMPLE 7.7 Find the generation schedule of a five-generator system to meet a demand of 730 MW. The cost characteristics are given below:

$$F_1 = 0.0015 P_1^2 + 1.8\, P_1 + 40.0 + \left|200.0 \times \sin\{0.035 \times (50.0 - P_1)\}\right| \; (₹/\text{h})$$

$$F_2 = 0.0030\, P_2^2 + 1.8\, P_2 + 60.0 + \left|140.0 \times \sin\{0.040 \times (20.0 - P_2)\}\right| \; (₹/\text{h})$$

$$F_3 = 0.0012\, P_3^2 + 2.1\, P_3 + 100.0 + \left|160.0 \times \sin\{0.038 \times (30.0 - P_3)\}\right| \; (₹/\text{h})$$

$$F_4 = 0.0080\, P_4^2 + 2.0\, P_4 + 25.0 + \left|100.0 \times \sin\{0.042 \times (10.0 - P_4)\}\right| \; (₹/\text{h})$$

$$F_5 = 0.0010\, P_5^2 + 2.0\, P_5 + 120.0 + \left|180.0 \times \sin\{0.037 \times (40.0 - P_5)\}\right| \; (₹/\text{h})$$

The cost characteristics are valid for the following minimum and maximum limits of power generation.

$$P_1^{\min} = 50 \text{ MW} \quad P_1^{\min} = 300 \text{ MW}$$

$$P_2^{\min} = 20 \text{ MW} \quad P_2^{\max} = 125 \text{ MW}$$

$$P_3^{\min} = 30 \text{ MW} \quad P_3^{\max} = 175 \text{ MW}$$

$$P_4^{\min} = 10 \text{ MW} \quad P_4^{\max} = 75 \text{ MW}$$

$$P_5^{\min} = 40 \text{ MW} \quad P_5^{\max} = 250 \text{ MW}$$

The transmission loss is neglected.

Solution Each real power generation of committed generator is searched. First generator is considered as dependent (slack) generator. Let there be ten members in the population. Initially seed number is taken as 305. An array of uniform random numbers is generated and stored. Initial generation output of generators P_i^j (i = 2, 3, 4, 5; j = 1, 2, ..., 10) are obtained using Eq. (7.58) and using uniform random numbers.

$$P_2^1 = P_2^{\min} + \text{rand}()\,(P_2^{\max} - P_2^{\min}) = 20.0 + 0.42193 \times (125.0 - 20) = 64.3026 \text{ MW}$$

$$P_3^1 = P_3^{\min} + \text{rand}()\,(P_3^{\max} - P_3^{\min}) = 30.0 + 0.66462 \times (175.0 - 30.0) = 126.3696 \text{ MW}$$

$$P_4^1 = P_4^{\min} + \text{rand}()\,(P_4^{\max} - P_4^{\min}) = 10.0 + 0.45834 \times (75.0 - 10.0) = 39.7920 \text{ MW}$$

$$P_5^1 = P_5^{\min} + \text{rand}()\,(P_5^{\max} - P_5^{\min}) = 40.0 + 0.78525 \times (250.0 - 40.0) = 204.9031 \text{ MW}$$

Compute the generation by slack (dependent) generator using Eq. (7.63)

$$P_1^1 = P_D - \sum_{i=2}^{5} P_i^1 = 294.6328 \text{ MW}$$

All the generations are within the limits so it is a feasible solution given in Table 7.13 along with operating cost. Second member of population is generated as given below.

$$P_2^2 = P_2^{\min} + \text{rand}()\,(P_2^{\max} - P_2^{\min}) = 20.0 + 0.77921 \times (125.0 - 20) = 101.8168 \text{ MW}$$

$$P_3^2 = P_3^{\min} + \text{rand}()\,(P_3^{\max} - P_3^{\min}) = 30.0 + 0.49275 \times (175.0 - 30.0) = 101.4492 \text{ MW}$$

$$P_4^2 = P_4^{\min} + \text{rand}()\,(P_4^{\max} - P_4^{\min}) = 10.0 + 0.47453 \times (75.0 - 10.0) = 40.8443 \text{ MW}$$

$$P_5^2 = P_5^{\min} + \text{rand}()\,(P_5^{\max} - P_5^{\min}) = 40.0 + 0.03545 \times (250.0 - 40.0) = 47.4440 \text{ MW}$$

Generation by slack (dependent) generator is obtained using Eq. (7.63) for second member of population.

$$P_1^2 = P_D - \sum_{i=2}^{5} P_i^2 = 438.4456 > P_1^{\max}$$

So this solution is not feasible. There is the need to find the feasible solution by updating the generation P_i^j (i = 1, 2, ..., NG; $i \neq d$) by the following relations until it becomes feasible solution.

$$\text{If } P_d^j > P_d^{\max} \text{ Then } P_i^j = P_i^j + w \times \text{rand}()\,(P_i^{\max} - P_i^{\min}) \quad (i = 1, 2, \ldots, \text{NG}; i \neq d)$$

$$\text{If } P_d^j < P_d^{\min} \text{ Then } P_i^j = P_i^j - w \times \text{rand}()\,(P_i^{\max} - P_i^{\min}) \quad (i = 1, 2, \ldots, \text{NG}; i \neq d)$$

New generation are obtained

$$P_2^2 = P_2^2 + 0.2 \times \text{rand}()\,(P_2^{\max} - P_2^{\min}) = 101.8168 + 0.2 \times 0.06922 \times (125.0 - 20) = 103.2704 \text{ MW}$$

$$P_3^2 = P_3^2 + 0.2 \times \text{rand}()\,(P_3^{\max} - P_3^{\min}) = 101.4492 + 0.2 \times 0.57184 \times (175.0 - 30.0) = 118.0327 \text{ MW}$$

$$P_4^2 = P_4^2 + 0.2 \times \text{rand}()\,(P_4^{\max} - P_4^{\min}) = 40.8443 + 0.2 \times 0.81691 \times (75.0 - 10.0) = 51.4642 \text{ MW}$$

$$P_5^2 = P_5^2 + 0.2 \times \text{rand}()\,(P_5^{\max} - P_5^{\min}) = 47.4440 + 0.2 \times 0.240579 \times (250.0 - 40.0) = 57.5483 \text{ MW}$$

$$P_1^2 = P_D - \sum_{i=2}^{5} P_i^2 = 399.6843 > P_1^{\max}$$

It is repeated till the feasible solution is obtained as given in Table 7.12. Fitness function is computed for each initial feasible member of population using Eq. (7.64) and is given in Table 7.13. The fourth member of population gives the best feasible solution.

TABLE 7.12 Feasible solution of 2nd member of population

S.No.	P_1^2(MW)	P_2^2(MW)	P_3^2(MW)	P_4^{2*}(MW)	P_5^2(MW)
1	438.4456	101.8168	101.4492	40.8443	47.4440
2	399.6843	103.2704	118.0327	51.4642	57.5483
3	357.7761	113.2121	135.6025	51.7173	71.6920
4	296.5495	125.0000	158.7305	57.1025	92.6175

TABLE 7.13 Initial feasible population

Population	*Cost, F*(₹/h)	P_1^j(MW)	P_2^j(MW)	P_3^j(MW)	P_4^j(MW)	P_5^j(MW)
1	2458.408	294.6328	64.3026	126.3696	39.7920	204.9031
2	2661.784	296.5495	125.0000	158.7305	57.1025	92.6175
3	2350.085	226.0638	87.8925	128.8514	55.8396	231.3528
4	**2233.816**	**238.3305**	**103.0034**	**120.9787**	**70.1695**	**197.5180**
5	2382.427	266.2323	110.3057	112.3916	42.7657	198.3047
6	2624.228	297.6123	125.0000	145.5710	73.4031	88.4136
7	2501.074	293.2327	115.9073	74.8884	23.7921	222.1796
8	2645.562	265.2730	125.0000	131.5770	49.5311	158.6189
9	2626.593	270.7329	121.3881	65.0996	46.8745	225.9049
10	2533.099	275.3370	116.1050	110.6062	53.2241	174.7277

Standard deviation for the initial feasible population is computed from Eq. (7.66) and is given in Table 7.14.

$$\sigma_2^1 = \beta \frac{f^1}{f^{\min}} (P_2^{\max} - P_2^{\min}) = 0.01 \times \frac{2458.408}{2233.816} (125.0 - 20.0) = 1.15557$$

TABLE 7.14 Standard deviations

Population j	σ_2^j	σ_3^j	σ_4^j	σ_5^j
1	1.15557	1.59579	0.71535	2.31114
2	1.25117	1.72780	0.77453	2.50233
3	1.10465	1.52547	0.68383	2.20930
4	1.05000	1.45000	0.65000	2.10000
5	1.11985	1.54647	0.69324	2.23971
6	1.23351	1.70342	0.76360	2.46702
7	1.17562	1.62348	0.72777	2.35125
8	1.24354	1.71727	0.76981	2.48708
9	1.23462	1.70496	0.76429	2.46925
10	1.19068	1.64427	0.73709	2.38135

An array of 2 × (20 × 4) normalized random numbers with zero mean and unit variance are generated to compute offspring using Eq. (7.69). The computed offspring is given in Table 7.15.

$$P_i^{j+10} = P_i^j + 0.05\,\sigma_i^j\,(N(0,1) + C_i^j(0,1)) \quad (i = 2, \ldots, 5;\ j = 1, 2, \ldots, 10)$$

P_1^{j+10} and f^{j+10} (j = 1, 2, ..., 10) using Eqs. (7.63) and (7.64) respectively.

TABLE 7.15 Offspring

Population j	*Cost, F*(₹/h)	P_1(MW)	P_2(MW)	P_3(MW)	P_4^*(MW)	P_5(MW)
11	25678.630	300.0000	62.8824	126.2517	39.9026	196.1465
12	2667.231	294.3535	125.0000	160.2477	56.7485	93.6502
13	2390.794	223.0001	87.4052	133.1511	55.7009	230.7427
14	2225.991	235.2505	102.9855	123.7573	70.0640	197.9426
15	2377.367	256.4420	110.6051	119.4663	42.9416	200.5450
16	2628.276	296.8299	124.0896	147.2411	73.3034	88.5361
17	2471.170	277.3073	115.4839	92.8929	25.0941	219.2219
18	2646.617	269.4036	124.5352	131.3928	48.0062	156.6622
19	2625.248	270.2908	121.5766	65.2656	47.2986	225.5684
20	2513.762	268.6528	115.9064	111.2729	56.5467	177.6213

Find the number of competitors from the combined population.

$$I = 1 + (20 - 1 + 1) \times 0.09843 = 2.96876 \approx 3$$

Find the kth member of combined population who will compete with the first member of combined population.

$$k = [20 \times 0.49363 + 1] = 9$$

Find the weight after competition with 10th member of population with the first member.

$$w^1 = \begin{cases} 1 & ; \text{if} \quad u_1 > \dfrac{f^1}{f^9 + f^1} \\ 0 & ; \text{else} \end{cases}$$

Table 7.16 lists weight value assigned to first member of population with three competitors. Weight assigned to each member of combined population after competing with R number of competitors is shown in Table 7.17. Ten members are selected for the next generation who have the largest assigned weights after completion and are given in Table 7.18.

TABLE 7.16 Weight value assigned to the first member of population with three competitors

r	u_2	K	f^1	f^r	u_1	w^r
1	0.49363	9	2458.408	2626.593	0.41267	1
2	0.10325	4	2458.408	2233.816	0.19331	0
3	0.79261	17	2458.408	2471.170	0.81669	1
					Total	**2**

TABLE 7.17 Weight value assigned to each member of population with r competitors

Population j	W^j	*Cost, F*(₹/h)	P_1(MW)	P_2(MW)	P_3(MW)	P_4(MW)	P_5(MW)
1	2	2458.4080	294.6328	64.3026	126.3696	39.7920	204.9031
2	7	2661.7840	296.5495	125.0000	158.7305	57.1025	92.6175
3	0	2350.0850	226.0638	87.8925	128.8514	55.8396	231.3528
4	8	2233.8160	238.3305	103.0034	120.9787	70.1695	197.5180
5	5	2382.4270	266.2323	110.3057	112.3916	42.7657	198.3047
6	5	2624.2280	297.6123	125.0000	145.5710	73.4031	88.4136
7	9	2501.0740	293.2327	115.9073	74.8884	23.7921	222.1796
8	5	2645.5620	265.2730	125.0000	131.5770	49.5311	158.6189
9	5	2626.5930	270.7329	121.3881	65.0996	46.8745	225.9049
10	2	2533.0990	275.3370	116.1050	110.6062	53.2241	174.7277
11	0	25678.630	300.0000	62.8824	126.2517	39.9026	196.1465
12	7	2667.2310	294.3535	125.0000	160.2477	56.7485	93.6502
13	1	2390.7940	223.0001	87.4052	133.1511	55.7009	230.7427
14	4	2225.9910	235.2505	102.9855	123.7573	70.0640	197.9426
15	2	2377.3670	256.4420	110.6051	119.4663	42.9416	200.5450
16	8	2628.2760	296.8299	124.0896	147.2411	73.3034	88.5361
17	8	2471.1700	277.3073	115.4839	92.8929	25.0941	219.2219
18	2	2646.6170	269.4036	124.5352	131.3928	48.0062	156.6622
19	10	2625.2480	270.2908	121.5766	65.2656	47.2986	225.5684
20	2	2513.7620	268.6528	115.9064	111.2729	56.5467	177.6213

TABLE 7.18 Second generation of population.

Population j	W^j	*Cost, F*(₹/h)	P_1(MW)	P_2(MW)	P_3(MW)	P_4^*(MW)	P_5(MW)
1	10	2625.2480	270.2908	121.5766	65.2656	47.2986	225.5684
2	9	2501.0740	293.2327	115.9073	74.8884	23.7921	222.1796
3	8	2628.2760	296.8299	124.0896	147.2411	73.3034	88.5361
4	8	2471.1700	277.3073	115.4839	92.8929	25.0941	219.2219
5	8	2233.8160	238.3305	103.0034	120.9787	70.1695	197.5180
6	7	2667.2310	294.3535	125.0000	160.2477	56.7485	93.6502
7	7	2661.7840	296.5495	125.0000	158.7305	57.1025	92.6175
8	5	2624.2280	297.6123	125.0000	145.5710	73.4031	88.4136
9	5	2382.4270	266.2323	110.3057	112.3916	42.7657	198.3047
10	5	2645.5620	265.2730	125.0000	131.5770	49.5311	158.6189

Best solutions after 50 iterations with different population size are given in Table 7.19.

TABLE 7.19 Best solutions

Population size	β	*Cost, F*(₹/h)	P_1(MW)	P_2(MW)	P_3(MW)	P_4^*(MW)	P_5(MW)
10	0.01	2210.420	232.9946	103.7412	124.1196	70.3564	198.7881
20	0.01	2043.468	229.1989	101.730	114.2774	74.0693	210.7243
30	0.07	2040.679	232.7736	99.3956	112.2255	75.0000	210.6054

7.12 EVOLUTIONARY PROGRAMMING FOR ECONOMIC DISPATCH–II

Evolutionary programming is a probabilistic, global search technique that starts with a population of randomly generated candidate solutions and evolves towards better solutions over a number of generations or iterations. The main stages of this technique include initialization, mutation, and competition and selection. Implementation of evolutionary programming for economic dispatch problem using self-adaption based on empirical learning rate of strategy parameter is as given below. The evolutionary programming can be implemented by searching the generation of power plants, P_i within generator limits. The economic dispatch problem is defined by Eq. (7.26) or Eq. (7.48) or Eq. (7.52) and transmission losses are defined by Eq. (7.27).

Initialization

The initial population comprises combinations of only the candidate economic dispatch solutions, which satisfy all the constraints. It consists of P_i^j (i = 1, 2, ..., NG; j = 1, 2, ..., L) trial parent individuals. The elements of a parent are the combinations of power outputs of the generating units randomly chosen by a random number ranging over $[P_j^{\min}, P_j^{\max}]$, using Eq. (7.61). The elements of offspring P_i^j may violate operating constraint. This violation is corrected by using Eq. (7.62).

Evaluation

In order to satisfy the power balance constraint, a generator is arbitrarily selected as a dependent generator d and is obtained by Eq. (7.63). If the output of the dependent generator violates its operating limits, it is fixed using Eq. (7.62). After limiting the value of the dependent generator as above, a penalty term is introduced in the objective function to penalize its fitness value. So, fitness value is defined by Eq. (7.64). The standard deviation σ_i^j is obtained from Eq. (7.66).

Mutation

The evolutionary programming (EP) technique considers different kinds of mutation viz. Gaussian, Cauchy and combined Gaussian–Cauchy mutations. To create offspring, the following scheme is adopted for self-adaption of strategy parameter based on empirical learning rates τ_1 and τ_2.

$$\sigma_i^{j+L} = \sigma_i^j \exp\{\tau_1 N(0, 1) + \tau_2 N^j(0, 1)\} \tag{7.74}$$

where

$$\tau_1 = (\sqrt{(2 \times (\text{NG}-1))})^{-1} \tag{7.75}$$

and

$$\tau_2 = (\sqrt{(2\sqrt{(\text{NG}-1)})})^{-1} \tag{7.76}$$

Classical evolutionary programming

Offspring is evaluated by adding a normally distributed random number with mean zero and standard deviation σ_i^{j+L} denoted as $N(0, \sigma_j^2)$. This results in

$$P_i^{j+L} = P_i^j + \sigma_i^{j+L} N(0, 1) \quad (i = 1, 2, \ldots, \text{NG}; i \neq d; j = 1, 2, \ldots, L) \tag{7.77}$$

Fast evolutionary programming

Using Cauchy mutation, an offspring is created by using the following equation.

$$P_i^{j+L} = P_i^j + \sigma_i^j C_i^{j+L}(0, 1) \quad (i = 1, 2, \ldots, \text{NG}; i \neq d; j = 1, 2, \ldots, L) \tag{7.78}$$

where $C_i^j(0, 1)$ is a Cauchy random variable with scale parameter $t = 1$ centred at zero that is generated anew for each value of j.

Mean of Gaussian and Cauchy mutation

In this method, mean of two Gaussian and Cauchy mutations is taken to generate offspring from each parent are generated as

$$P_i^{j+L} = P_i^j + \frac{1}{2}\sigma_i^{j+L}(N(0, 1) + C_i^j(0, 1)) \quad (i = 1, 2, \ldots, \text{NG}; i \neq d; j = 1, 2, \ldots, L) \tag{7.79}$$

Better of Gaussian and Cauchy mutation

In this method, the better of two offspring from each parent are generated by Gaussian mutation and the other by Cauchy mutation is taken as

$$P_i^{j+L} = P_i^j + \sigma_i^{j+L} N(0, 1) \quad (i = 1, 2, \ldots, \text{NG}; i \neq d; j = 1, 2, \ldots, L) \tag{7.80a}$$

$$P_i^{j+L} = P_i^j + \sigma_i^{j+L} C(0, 1) \quad (i = 1, 2, \ldots, \text{NG}; i \neq d; j = 1, 2, \ldots, L) \tag{7.80b}$$

Mutation results in creation of L offspring individuals. The parent individuals are candidate dispatch solutions, which satisfy all constraints. However, after mutation, the elements of offspring P_i^{j+L} may violate operating constraint. This violation is corrected by using Eq. (7.62).

Evaluation of offspring

In order to satisfy the power balance constraint, a generator is arbitrarily selected as a dependent generator d. Its output is obtained from Eq. (7.63). If, the output of the dependent generator violates its operating limits, its value is fixed by Eq. (7.62). After limiting the value of the dependent generator as above, a penalty term is introduced in the objective function to penalize its fitness value given by Eq. (7.64). The values of the objective function of both offspring are evaluated, compared, and the better individual is chosen as offspring for competition and selection. The initial population and their offspring created by mutation form a combined population of $2L$ individuals.

Competition and selection

The $2L$ individuals compete with each other for selection. A weight value w^r is assigned to each individual. The weights are computed from Eq. (7.71) as explained in Section 7.11.8. While computing the weight for each individual, it is ensured that each individual is selected only once from the combined population. Even though relative fitness values are used during the process of mutation, competition and selection, it leads to slow convergence. This weight assignment is found to yield proper selection and good convergence. When all the $2L$ individuals obtain their weights, they are ranked in descending order and the first L individuals are selected as parents along with their fitness values for the next generation.

Algorithm 7.8: Economic Dispatch Using Evolutionary Programming

The stepwise procedure is outlined below:

1. Read data, viz. Cost Coefficients, a_i, b_i, c_i (i = 1, 2, ..., NG). Convergence tolerance error, scaling factor, β and the maximum allowed iterations, ITMAX, population size, L, seed number, S, $P_i^{\min}$ and $P_i^{\max}$ (i = 1, 2, ..., NG), etc.
2. Generate an array of (NG × L) size of uniform random numbers.
3. Set population counter, $j = 0$
   DO
4. Increment population counter, $j = j + 1$
5. Set generation counter, $i = 0$
   DO
6. Increment the generation counter, $i = i + 1$
7. IF ($i \neq d$) THEN Generate the population P_i^j
   WHILE ($i \leq$ NG)
8. Compute P_d^j using Eq. (7.63), check limits. Compute penalty factor, φ^j and f^j using Eq. (7.65) and Eq. (7.64) respectively.
   WHILE ($j \leq L$)
9. Set iteration counter, IT = 0
   DO
10. Increment iteration counter, IT = IT + 1

11. Set population counter, $j = 1$, $f^{\min} = f^j$ and $j^{\min} = j$
DO
12. Increment population counter, $j = j + 1$
13. IF ($f^j < f^{\min}$) THEN set $f^{\min}$ and f^j, $j^{\min} = j$ and $P_i^{\min} = P_i^j$ ($i = 1, 2, \ldots,$ NG)
WHILE ($j \le L$)
14. IF (IT = 1) THEN $f^{\text{Prev}} = f^{\min}$ and $P_i^{\text{Prev}} = P_i^{\min}$ ($i = 1, 2, \ldots,$ NG)

ELSEIF ($f^{\min} < f^{\text{Prev}}$) THEN $f^{\text{Prev}} = f^{\min}$ and $P_i^{\text{Prev}} = P_i^{\min}$ ($i = 1, 2, \ldots,$ NG)
15. Set population counter, $j = 0$
DO
16. Increment population counter, $j = j + 1$
17. Set generation counter, $i = 0$
DO
18. Increment the generation counter, $i = i + 1$
19. Compute standard deviation, σ_i^j using Eq. (7.66)
WHILE ($i \le$ NG)
WHILE ($j \le$ L)
20. Generate array of (NG × L) size of normalized random numbers or Cauchy random numbers or arrays of both normalized and Cauchy random numbers
21. Set population counter, $j = 0$
DO
22. Increment population counter, $j = j + 1$
23. Set generation counter, $i = 0$
DO
24. Increment the generation counter, $i = i + 1$
25. IF ($i \ne d$) THEN
1. Generate the population P_i^{j+L} using Eq. (7.77) or Eq. (7.78) or Eq. (7.79) or (7.80)
2. Check the limits if violated set as per Eq. (7.62)
WHILE ($i \le$ NG)
26. Compute P_d^{j+L} using Eq. (7.63), check limits. Compute penalty factor, φ^{j+L} and f^{j+L} using Eq. (7.65) and Eq. (7.64), respectively.
WHILE ($j \le L$)
27. Set population counter, $j = 0$
DO
28. Increment population counter, $j = j + 1$
29. Generate array of (2 × L) size of uniform numbers
30. Initialize $m = 0$ and SUM = 0.0
31. Set population counter, $k = 0$
DO
32. Increment population counter, $k = k + 1$
33. Select u_1 as $(m + 1)$th element and u_2 as $(m + 2)$th element of array of uniform random numbers
34. Compute $r = [2M \times u_2 + 1]$ and compute w^k using Eq. (7.73).
35. Update SUM = SUM + w^k, $m = m + 2$
WHILE ($k \le L$)

36. Set W^j = SUM
    WHILE ($j \le (2 \times L)$)
37. Set population counter, $j = 0$
    DO
38. Increment population counter, $j = j + 1$
39. Set counter, $k = j$
    DO
40. Increment counter, $k = k + 1$
41. IF ($W^j < W^k$) THEN $f^k \leftrightarrow f^j$, $W^k \leftrightarrow W^j$ and $P_i^k \leftrightarrow P_i^j$ (i = 1, 2, ..., NG)
    WHILE ($k \le (2 \times$ L))
    WHILE ($j \le (2 \times$ L $- 1)$)
    WHILE (IT $\le$ ITMAX)
42. STOP

EXAMPLE 7.8 Consider a five-generator system given in Example 7.7 and obtain the optimum schedule.

Solution Each real power generation is searched. First generator is considered as dependent (slack) generator. Let there be ten members in the population. Initially seed number is taken as 305. An array of uniform random numbers is generated. Initial feasible member of population is given in Table 7.20. Fourth member of population gives the best feasible solution. Standard deviation for the initial feasible population is computed using Eq. (7.66) and is given in Table 7.21.

TABLE 7.20 Initial feasible population

Population j	*Cost, F*(₹/h)	P_1^j(MW)	P_2^j(MW)	P_3^j(MW)	P_4^j(MW)	P_5^j(MW)
1	2601.5760	264.1686	94.3391	150.5532	55.8799	165.0593
2	2439.1290	270.5951	98.5831	122.0463	50.6965	188.0789
3	2270.8650	249.1604	92.0834	118.7740	66.3138	203.6684
4	2428.8070	291.7780	94.4020	105.6862	52.1894	185.9445
5	2456.6880	281.1795	90.3467	115.5920	59.3012	183.5805
6	2554.9870	289.1436	86.0397	130.9660	42.1751	181.6757
7	2544.6430	290.9409	88.8931	128.0085	48.6814	173.4762
8	2587.7260	264.6671	107.1970	149.2984	58.9488	149.8887
9	2426.8330	299.5311	90.6927	115.3235	52.0035	172.4492
10	2338.5280	260.2577	89.5289	118.8632	55.2777	206.0726

The Gaussian EP technique considered differ in the kind of mutation used as described below:

$$\tau_1 = \left(\sqrt{(2 \times (5-1))}\right)^{-1} = 0.353553, \quad \tau_2 = \left(\sqrt{(2\sqrt{(5-1)})}\right)^{-1} = 0.5$$

$$\sigma_2^{1+10} = \sigma_2^1 \exp\{\tau_1 N(0, 1) + \tau_2 N^1(0, 1)\} \quad (i = 2, \ldots, 5;\ j = 1, 2, \ldots, 10)$$

TABLE 7.21 Standard Deviations

Population j	σ_2^j	σ_3^j	σ_4^j	σ_5^j
1	1.20291	1.66117	0.74466	2.40583
2	1.12780	1.55744	0.69816	2.25560
3	1.05000	1.45000	0.65000	2.10000
4	1.12303	1.55085	0.69521	2.24606
5	1.13592	1.56865	0.70319	2.27184
6	1.18137	1.63142	0.73133	2.36274
7	1.17659	1.62481	0.72836	2.35318
8	1.19651	1.65232	0.74070	2.39302
9	1.12212	1.54959	0.69464	2.24423
10	1.08129	1.49320	0.66937	2.16257

Standard deviations for offspring are given in Table 7.22. It is performed on each vector element by adding a normally distributed random number with mean zero and standard deviation σ_j denoted as $N(0, \sigma_j^2)$. This results in

$$P_i^{j+10} = P_i^j + \sigma_i^{j+10}\ N(0, 1) \quad (i = 2, \ldots, 5;\ j = 1, 2, \ldots, 10)$$

P_1^{j+10} and f^{j+10} (j = 1, 2, ..., 10) using Eqs. (7.63) and (7.65) respectively.

Weight assigned to each member of combined population after competing with 10 competitors is shown in Table 7.23. Ten members are selected for the next generation who have the largest assigned weights after compitition. Best solutions after 100 iterations with different population sizes are given in Table 7.24.

TABLE 7.22 Standard deviations

Population j	σ_2^j	σ_3^j	σ_4^j	σ_5^j
11	1.40844	1.51371	0.55460	3.10759
12	1.53049	0.89747	1.20480	1.64824
13	0.74132	2.31694	1.10470	6.68148
14	4.70499	2.54395	1.55102	1.70811
15	2.74435	2.63679	0.90265	10.68512
16	0.36354	1.81494	0.29266	6.14791
17	1.10704	3.33786	1.08357	4.74449
18	1.35031	3.17550	0.46059	3.04714
19	1.19446	1.16431	0.54056	3.45810
20	0.35509	0.35748	0.60985	0.67325

TABLE 7.23 Weight value assigned to each member of population with *r* competitors

Population j	W^j	*Cost, F*(₹/h)	P_1(MW)	P_2(MW)	P_3(MW)	P_4(MW)	P_5(MW)
1	9	2576.795	282.5451	88.3497	131.5831	49.1424	178.3796
2	7	2327.643	262.1355	91.6183	114.9008	58.5778	202.7676
3	7	2430.468	296.6466	90.9700	114.7827	51.7903	175.8104
4	6	2442.456	269.0768	99.8923	121.2765	52.3061	187.4483
5	6	2601.576	264.1686	94.3391	150.5532	55.8799	165.0593
6	6	2428.807	291.7780	94.4020	105.6862	52.1894	185.9445
7	6	2338.528	260.2577	89.5289	118.8632	55.2777	206.0726
8	5	2439.129	270.5951	98.5831	122.0463	50.6965	188.0789
9	5	2472.073	280.1102	105.2477	106.7795	53.8100	184.0526
10	5	2544.643	290.9409	88.8931	128.0085	48.6814	173.4762
11	5	2270.865	249.1604	92.0834	118.7740	66.3138	203.6684
12	5	2582.332	260.0977	107.3654	153.0753	58.4572	151.0043
13	5	2589.350	261.2397	95.1047	150.6171	55.6795	167.3590
14	5	2338.630	260.2377	89.3640	118.4712	56.2391	205.6881
15	4	2456.688	281.1795	90.3467	115.5920	59.3012	183.5805
16	4	2587.726	264.6671	107.1970	149.2984	58.9488	149.8887
17	4	2426.833	299.5311	90.6927	115.3235	52.0035	172.4492
18	4	2554.987	289.1436	86.0397	130.9660	42.1751	181.6757
19	3	2193.085	234.3106	91.3318	120.2100	67.1348	217.0127
20	2	2527.708	277.9920	85.1997	131.4370	41.6526	193.7188

TABLE 7.24 Best solutions

Population j	*Cost, F*(₹/h)	β	P_1(MW)	P_2(MW)	P_3(MW)	P_4(MW)	P_5(MW)
10	2215.592	0.01	234.2455	107.87640	117.9686	75.0	194.9096
20	2119.965	0.04	229.2625	95.0717	111.4175	75.0	219.2483
30	2065.454	0.03	227.1188	101.64380	116.7434	75.0	209.4939
40	2030.673	0.02	229.8030	101.5736	113.7999	75.0	209.8235

7.13 PARTICLE SWARM OPTIMIZATION FOR ECONOMIC DISPATCH

In 1995, Kennedy and Eberhart [1995] first introduced the Particle Swarm Optimization (PSO) method, motivated by social behaviour of organisms such as fish schooling and bird flocking. PSO, as an optimization tool, provides a population-based search procedure in which individuals called particles change their positions (states) with time. In a PSO system, particles fly around in a multidimensional search space. During flight, each particle adjusts its position according to its own experience, and the experience of neighbouring particles, making use of the best position encountered by itself and its neighbours. The swarm direction of a particle is defined by the set of particles neighbouring the particle and its history experience.

Let P and v denote a particle's coordinates (position) and its corresponding flight speed (velocity) in a search space, respectively. Therefore, the jth particle is represented as $P_i = [P_{i1}, P_{i2}, ..., P_{iNG}]$ in the NP-dimensional space. The best previous position of each particle is recorded and represented as $\text{Pb}_i = [\text{Pb}_{i1}, \text{Pb}_{i2}, ..., \text{Pb}_{iNG}]$. The index of the best particle among all the particles in the group is represented by the $[G_1, G_2, ..., G_{NG}]$. The rate of the velocity for particle is represented as $v_i = [v_{i1}, v_{i2}, ..., v_{iNP}]$. The modified velocity and position of each particle can be calculated using the current velocity and the distance from Pb_{ij} to G_j as shown in the following formulas:

$$v_{ij}^{r+1} = w \times v_{ij}^r + C_1 \times R_1 \times (\text{Pb}_{ij}^r - P_{ij}^r) + C_2 \times R_2 \times (G_j^r - P_{ij}^r) \quad (i = 1, 2, \ldots, \text{NP};\ j = 1, 2, \ldots, \text{NG}) \tag{7.81}$$

$$P_{ij}^{r+1} = P_{ij}^r + v_{ij}^{r+1} \quad (i = 1, 2, \ldots, \text{NP};\ j = 1, 2, \ldots, \text{NG}) \tag{7.82}$$

where

NP is the number of particles in a group.
NG is the number of members in a particle.
R is the pointer of iterations (generations).
w is the inertia weight factor.
C_1 and C_2 are the acceleration constants.
R_1 and R_2 are uniform random values in the range [0, 1].
v_{ij}^r is the velocity of jth member of ith particle at rth iteration, $V_j^{\min} \le v_{ij}^r \le V_j^{\max}$
P_{ij}^r is the current position of jth member of ith particle at rth iteration.

In the above procedures, the parameter $V_j^{\min}$ determined the resolution, or fitness, with which regions are to be searched between the present position and the target position. If $V_j^{\max}$ is too high, particles might fly past good solutions. If $V_j^{\max}$ is too small, particles may not explore sufficiently beyond local solutions. In many experiences with PSO, $V_j^{\max}$ was often set at 10–20% of the dynamic range of the variable on each dimension.

The constants c_1 and c_2 represent the weighting of the stochastic acceleration terms that pull each particle toward the Pb_{ij}^r, G_j^r positions. Low values allow particles to roam far from the target regions before being tugged back. On the other hand, high values result in abrupt movement toward, or past, target regions. Hence, the acceleration constants C_1, and C_2 were often set to be 2.0 according to past experiences.

Suitable selection of inertia weight w in Eq. (7.83) provides a balance between global and local explorations, thus requiring less iteration on average to find a sufficiently optimal solution. As originally developed, w often decreases linearly from about 0.9 to 0.4 during a run. In general, the inertia weight w is set according to the following equation:

$$w = w^{\max} - \frac{w^{\max} - w^{\min}}{\text{IT}^{\text{MAX}}} \times \text{IT} \tag{7.83}$$

where

IT^{MAX} is the maximum number of iterations (generations), and
IT is the current number of iterations.

Implementation

The main objective of economic dispatch is to obtain the amount of real power to be generated by each committed generator, while achieving a minimum generation cost within the constraints. The particle swarm optimization can be implemented by searching the generation of power plants, P_i within generator limits. The economic dispatch problem is defined by Eq. (7.26) or Eq. (7.48) or Eq. (7.52) and transmission losses are defined by Eq. (7.27). This section provides the solution methodology to the economic dispatch problems through PSO.

Representation of the swarm

Since the decision variables of the economic dispatch problems are real power generations, they are used to form the swarm. The set of real power output (P) of all the generators is represented as the position of the particle in the swarm. For a system with NG generators, the particle position is represented as a vector of length NG. If there are NP particles in the swarm, the complete swarm is represented as a matrix as below:

$$\text{Swarm} = \begin{bmatrix} P_{12} & P_{12} & .. & .. & P_{1\text{NG}} \\ P_{21} & P_{22} & .. & .. & P_{2\text{NG}} \\ . & . & P_{ij} & .. & . \\ . & . & .. & .. & . \\ P_{\text{NP}1} & P_{\text{NP}2} & .. & .. & P_{\text{NPNG}} \end{bmatrix}$$

where P_{ij} is the jth position component of particle i and it represents the real power generation of generator j of the possible solution i.

Initialization of the swarm

Each element of the swarm matrix is initialized randomly within the real power operating limits based on Eq. (7.26c). The velocities of the particles are initialized randomly according to the following inequality:

$$V_j^{\min} \le V_{ij} \le V_j^{\max} \quad (i = 1,\ 2,\ \ldots,\ \text{NP};\ j = 1,\ 2,\ \ldots,\ \text{NG}) \tag{7.84}$$

This velocity-initialization scheme always guarantees to produce new particles satisfying real power operating limit constraints. The maximum velocity limit in the jth dimension is computed as:

$$V_j^{\max} = \frac{P_j^{\max} - P_j^{\min}}{\alpha} \tag{7.85}$$

where α is the chosen number of intervals in the jth dimension.

Evaluation of objective function

In order to satisfy the power balance constraint, a generator is arbitrarily selected as a dependent generator d and is obtained by Eq. (7.63). If the output of the dependent generator violates its operating limits, it is fixed using Eq. (7.62). After limiting the value of the dependent generator as above, a penalty term is introduced in the objective function to penalize its fitness value. So, function value is defined by Eq. (7.64).

Initialization of the best positions

In the strategy of PSO, the particle's best position, P_{ij}^{best} and global best position G_j^{best} are the key factors. The position with minimum objective function value is the particle's best position. The best position out of all the P_{ij}^{best} is taken as G_j^{best}.

Movement of the particles

The particles in the swarm are accelerated to new positions by adding new velocities to their present positions. The new velocities are calculated using Eq. (7.80) and the positions of the particles are updated using Eq. (7.81).

$$v_{ij}^{new} = w \times v_{ij} + c_1 \times \text{rand}() \times (P_{ij}^{best} - P_{ij}) + c_2 \times \text{rand}() \times (G_j^{best} - P_{ij}) \tag{7.86}$$

$$P_{ij}^{new} = P_{ij} + v_{ij}^{new} \quad (i = 1, 2, \ldots, \text{NP}; \; j = 1, 2, \ldots, \text{NG}) \tag{7.87}$$

If any P_{ij} violates the real power operating limit constraints, it is clamped at the boundary value using Eq. (7.62).

Updating the best and the worst positions

The particles are evaluated in the new positions by objective function values. Then P_{ij}^{best} of particle j are updated. The best position out of all the new P_{ij}^{best} is taken as G_j^{best}. An objective value at G_j^{best} is saved as f_{best}.

Stopping criterion

There are various criteria available to stop a stochastic optimization algorithm. Some examples are tolerance, number of function evaluations and the maximum number of iterations. In this section, the maximum number of iterations is chosen as the stopping criterion. If the stopping criterion is not satisfied, the above procedure is repeated with incremented t value. Otherwise, G_j^{best} is the optimum generation schedule and f_{best} is the minimum generation cost of the economic dispatch problem.

Algorithm 7.9: Economic Dispatch using Particle Swarm Algorithm

The search procedures of the proposed method were as shown below:

1. Read data; viz. Cost Coefficients, a_i, b_i, c_i (i = 1, 2, ..., NG). Convergence tolerance error, scaling factor, β and maximum allowed iterations, ITMAX, population size, L, seed number, S, P_i^{min} and P_i^{max} (i = 1, 2, ..., NG), etc. Initialize randomly the individuals of the population according to the limit of each unit including individual dimensions, searching points, and velocities. These initial individuals must be feasible candidate solutions that satisfy the practical operation constraints.
2. Generate an array of (NP × NG) size of uniform random numbers.
3. Set particle (bird) counter, $i = 0$

   DO

4. Increment particle counter, $i = i + 1$
5. Set generation counter, $j = 0$
   DO
6. Increment the generation counter, $j = j + 1$
7. IF ($j \neq d$) THEN generate the position of particle P_{ij} and velocity of particle v_{ij}
8. Initialize $P_{ij}^{\text{best}} = P_{ij}$
   WHILE (j < NG)
9. Compute P_{id} using Eq. (7.63), check limits and adjust using Eq. (7.62). Compute penalty factor, φ_i and f_i using Eq. (7.65) and Eq. (7.64) respectively.
10. Initialize $f_i^{\text{best}} = f_i$
    WHILE (i < NP)
11. Set particle counter, $i = 1$, $f^{\text{global}} = f_i^{\text{best}}$
    DO
12. Increment particle counter, $i = i + 1$
13. IF ($f_i^{\text{best}} < f^{\text{global}}$) THEN set $f^{\text{global}} = f_i^{\text{best}}$ and $G_j^{\text{best}} = P_{ij}^{\text{best}}$ (j = 1, 2, ..., NG)
    WHILE (i < NP)
14. Set iteration counter, IT = 0
    DO
15. Increment iteration counter, IT = IT + 1
16. Compute $w = w^{\max} - \dfrac{w^{\max} - w^{\min}}{\text{IT}^{\max}} \times \text{IT}$
17. Generate an array of (2NP × NG) size of uniform random numbers
18. Set particle (bird) counter, $i = 0$
    DO
19. Increment particle counter, $i = i + 1$
20. Set generation counter, $j = 0$
    DO
21. Increment the generation counter, $j = j + 1$
22. IF ($j \neq d$) THEN generate the velocity of particle
23. Compute v_{ij}^{new} using Eq. (7.86)
24. IF ($v_{ij}^{\text{new}} > V_j^{\max}$) THEN update ($v_{ij}^{\text{new}} = V_j^{\max}$)
25. IF ($v_{ij}^{\text{new}} < V_j^{\min}$) THEN update ($v_{ij}^{\text{new}} = V_j^{\min}$)
26. Compute $P_{ij}^{\text{new}} = P_{ij} + v_{ij}^{\text{new}}$
    WHILE (j < NG)
27. Compute P_{id}^{new} using Eq. (7.63), check limits and adjust using Eq. (7.62). Compute penalty factor, φ_i and f_i^{new} using Eq. (7.65) and Eq. (7.64) respectively.
28. IF ($f_i^{\text{new}} < f_i^{\text{best}}$) THEN set $f_i^{\text{best}} = f_i^{\text{new}}$ and $P_{ij}^{\text{best}} = P_{ij}^{\text{new}}$ (j = 1, 2, ..., NG)
    WHILE (i < NP)
29. Set particle counter, $i = 1$, $f^{\min} = f_i^{\text{best}}$
    DO
30. Increment particle counter, $j = j + 1$
31. IF ($f_i^{\text{best}} < f^{\min}$) THEN set $f^{\min} = f_i^{\text{best}}$ and $P_j^{\min} = P_{ij}^{\text{best}}$ (j = 1, 2, ..., NG)
    WHILE (i < NP)

32. IF $|f_{\min} - f^{\text{global}}| \le$ error THEN GOTO 39
33. Set particle (bird) counter, $i = 0$
    DO
34. Increment particle counter, $i = i + 1$
35. Set generation counter, $j = 0$
    DO
36. Increment the generation counter, $j = j + 1$
37. Set $v_{ij} = v_{ij}^{\text{new}}$
    WHILE (j < NG)
38. IF ($f^{\min} < f^{\text{global}}$) THEN set $f^{\text{global}} = f^{\min}$
    WHILE (IT < IT^{MAX})
39. STOP

EXAMPLE 7.9 Find the generation schedule of a three-generator system to meet a demand of 850 MW. The cost characteristics are given below:

$$F_1 = 0.0016\, P_1^2 + 7.92\, P_1 + 561.0 + |\,300.0 \times \sin\{0.032 \times (100.0 - P_1)\}|\ (₹/\text{h})$$

$$F_2 = 0.0048\, P_2^2 + 7.92\, P_2 + 78.0 + |\,150.0 \times \sin\{0.063 \times (50.0 - P_2)\}|\ (₹/\text{h})$$

$$F_3 = 0.0019\, P_3^2 + 7.85\, P_3 + 310.0 + |\,200.0 \times \sin\{0.042 \times (100.0 - P_3)\}|\ (₹/\text{h})$$

The cost characteristics are valid for the following minimum and maximum limits of power generation.

$$P_1^{\min} = 100 \text{ MW} \qquad P_1^{\max} = 600 \text{ MW}$$

$$P_2^{\min} = 50 \text{ MW} \qquad P_2^{\max} = 200 \text{ MW}$$

$$P_3^{\min} = 100 \text{ MW} \qquad P_3^{\max} = 400 \text{ MW}$$

The transmission loss is neglected.

Solution Each real power generation is searched. NG is taken as three (03). First generator is considered as slack (dependent) generator. Power demand P_D is given as 850.0 MW. The number of particles is assumed as 10. Inertia weights $w^{\min}$ and $w^{\max}$ are taken as 0.4 and 0.9, respectively. Acceleration constants c_1 and c_2 are set to 2.0.

$$-0.5\, P_j^{\min} \le V_{ij} \le +0.5\, P_j^{\max} \quad (i = 1, 2, \ldots, 10;\ j = 2, \ldots, 5)$$

$$V_1^{\min} = -50.0 \quad V_1^{\max} = 300.0$$

$$V_2^{\min} = -25.0 \quad V_2^{\max} = 100.0$$

$$V_3^{\min} = -50.0 \quad V_2^{\max} = 200.0$$

Randomly generated velocities of particles are given in Table 7.25.

$$V_{12} = V_2^{\min} + \text{rand}()\,(V_2^{\max} - V_2^{\min}) = 11.7058$$

$$V_{13} = V_3^{\min} + \text{rand}()\,(V_3^{\max} - V_3^{\min}) = 173.5719$$

TABLE 7.25 Initial velocities of swarm particles

k	V_2^k	V_3^k
1	11.7058	173.5719
2	18.7115	36.7349
3	94.8622	185.5265
4	73.17744	19.10394
5	64.0033	159.0559
6	–9.7003	22.52941
7	92.7420	116.6663
8	92.8973	5.206111
9	65.5699	162.0610
10	–18.0511	103.4914

Initial position of particles is evaluated

$$P_2^1 = P_2^{\min} + \text{rand}()\,(P_2^{\max} - P_2^{\min}) = 113.2894 \text{ MW}$$

$$P_3^1 = P_3^{\min} + \text{rand}()\,(P_3^{\max} - P_3^{\min}) = 299.3854 \text{ MW}$$

TABLE 7.26 Initial positions of particles in swarm

i	f_i	P_{i1}	P_{i2}	P_{i3}
1	8765.712	437.3252	113.2894	299.3854
2	8615.356	391.3047	167.7879	290.9074
3	8571.593	483.7287	123.9130	242.3583
4	8736.489	360.5306	195.4909	293.9785
5	8630.727	320.8552	155.7837	373.3611
6	8717.307	541.8311	146.7960	161.3728
7	8616.495	413.9023	86.9878	349.1099
8	8618.146	463.5411	135.2327	251.2262
9	8394.182	399.2423	103.3758	347.3819
10	8938.913	549.7467	178.6239	121.6294

Initially, the initial positions of particles in swarm (Table 7.26) are considered as best positions of particles P_{ij}^{best}.

The best position out of all the particles P_{ij}^{best} is taken as G_j^{best}. Global best solution is selected from the initial positions of particles.

$$f^{\text{best}} = 8394.182\ (₹/\text{h}),\ G_j^{\text{best}} = 399.2423\ (\text{MW}),\ G_2^{\text{best}} = 103.3758\ (\text{MW}),\ G_3^{\text{best}} = 347.3819\ (\text{MW})$$

The particles in the swarm are accelerated to new positions by adding new velocities to their present positions. The new velocities are calculated using Eq. (7.85) and are shown in Table 7.27.

$$V_{ij}^{\text{new}} = w \times v_{ij} + c_1 \times R_1 \times (P_{ij}^{\text{best}} - P_{ij}) + c_2 \times R_2 \times (G_j^{\text{best}} - P_{ij})\ (i = 1, 2, \ldots, 10;\ j = 2, 3)$$

TABLE 7.27 New velocities of particles in swarm

i	V_{i2}^{new}	V_{i3}^{new}
1	0.232869	200.000000
2	10.131950	136.003900
3	74.445290	200.000000
4	–22.504640	25.914420
5	–25.000000	117.785900
6	–25.000000	198.504000
7	99.634800	91.295990
8	35.904940	128.821100
9	52.455930	129.648800
10	–25.000000	135.271500

The new positions of the particles are updated using Eq. (7.86) and are given in Table 7.28.

$$P_{ij}^{\text{new}} = P_{ij} + v_{ij}^{\text{new}} \quad (i = 1, 2, \ldots, 10;\ j = 2, 3)$$

P_{i1}^{new} is computed as given by Eq. (7.63) and then f_i^{new} is computed using Eq. (7.64) and also depicted in Table 7.28.

TABLE 7.28 New positions of swarm population

i	f_i^{new}	P_{i1}^{new}	P_{i2}^{new}	P_{i3}^{new}
1	8851.248	237.0923	113.5223	499.3854
2	8887.665	245.1688	177.9199	426.9113
3	8592.429	209.2834	198.3583	442.3583
4	8687.032	357.1208	172.9863	319.8929
5	8789.813	228.0693	130.7837	491.1470
6	8791.115	368.3271	121.7960	359.8769
7	8788.688	222.9716	186.6226	440.4059
8	8520.342	298.8151	171.1377	380.0473
9	8525.531	217.1376	155.8317	477.0307
10	8597.358	439.4753	153.6239	256.9009

First particle of swarm is compared and the best position is selected.

If $f_1^{\text{best}} < f_1^{\text{new}}$ then f_1^{best} and best position of particle P_{1j}^{best} (j = 1, 2, 3) are selected otherwise f_1^{new} and new position of particle P_{1j}^{new} (j = 1, 2, 3) are selected as given in Table 7.29.

G_j^{best} Global best solution is selected from the best positions of particles.

f^{best} = 8394.182 (₹/h); G_j^{best} = 399.2423 (MW); G_2^{best} = 103.3758 (MW); G_3^{best} = 347.3819 (MW)

Repeating the abovementioned procedure, solution is obtained after 100 iterations and is given in Table 7.30. Results are obtained by considering 20 particles as well as 40 particles in the swarm and results are shown in Table 7.30.

TABLE 7.29 Best positions of swarm population

i	f_i^{best}	P_{i1}^{best}	P_{i2}^{best}	P_{i3}^{best}
1	8765.712	437.3252	113.2894	299.3854
2	8615.356	391.3047	167.7879	290.9074
3	8571.593	483.7287	123.9130	242.3583
4	8687.032	357.1208	172.9863	319.8929
5	8630.727	320.8552	155.7837	373.3611
6	8717.307	541.8311	146.7960	161.3728
7	8616.495	413.9023	86.98782	349.1099
8	8520.342	298.8151	171.1377	380.0473
9	8394.182	399.2423	103.3758	347.3819
10	8597.358	439.4753	153.6239	256.9009

TABLE 7.30 Results achieved with different population sizes

Particles in Swarm	*Iterations*	f(₹/h)	P_1(MW)	P_2(MW)	P_3(MW)
10	100	8233.667	299.4729	149.7358	400.7914
20	88	8233.665	299.4711	149.7349	400.7939
40	61	8233.824	299.7189	149.7635	400.5177

7.14 ANTI-PREDATORY PARTICLE SWARM OPTIMIZATION

A new variation in the classical Particle Swarm Optimization (PSO) by splitting the cognitive component of the classical PSO into two different components is proposed by Selvakumar and Thanushkodi [2007]. The first component is called good experience component. That is, the bird has a memory about its previously visited best position. This component is exactly the same as the cognitive component of the basic PSO. The second component is given the name bad experience component. The bad experience component helps the particle to remember its previously visited worst position. To calculate the new velocity, the bad experience of the particle is also taken into consideration. This gives the new model of the PSO. The new velocity update equation proposed is given by

$$v_{ij}^{r+1} = w^r v_{ij}^r + C_1R_1(\text{Pb}_{ij}^r - P_{ij}^r) + C_2R_2(P_{ij}^r - \text{Pw}_{ij}^r) + C_3R_3(G_j^r - P_{ij}^r)$$
$$(j = 1, 2, \ldots, \text{NG};\ i = 1, 2, \ldots, \text{NP}) \quad (7.88)$$

$$P_{ij}^{r+1} = P_{ij}^r + v_{ij}^r \quad (j = 1, 2, \ldots, \text{NG};\ i = 1, 2, \ldots, \text{NP}) \quad (7.89)$$

where

NP is the number of particles in a group;
NG is the number of members in a particle;
r is the pointer of iterations (generations);
w is the inertia weight factor;

v_{ij}^r is velocity of particle i at iteration r, $V_j^{min} \le v_{ij}^r \le V_j^{max}$
P_{ij}^r is the current position of ith particle at rth iteration.
C_1 is the acceleration coefficient, which accelerates the particle toward its best position
C_2 is the acceleration coefficient, which accelerates the particle away from its worst position
C_3 is the acceleration coefficient, which accelerates the particle towards the global best position
G_i^r is the global best position of particle i until iteration r
Pb_{ij}^r is dimension j of the own best position of particle i until iteration r
Pw_{ij}^r is dimension j of the own worst position of particle i until iteration r

R_1, R_2 and R_3 are three separately generated uniformly distributed random numbers in the range [0, 1].

The positions are updated using Eq. (7.89). The inclusion of the worst experience component in the behaviour of the particle gives additional exploration capacity to the swarm. By using the bad experience component, the bird (particle) can bypass its previous worst position and always try to occupy a better position.

The central idea of the classical PSO relies on the foraging activity of the swarm. Both the cognitive and social behaviours are the constituents of the foraging activity. This paper introduces a new PSO variant with an included anti-predatory activity. The anti-predatory property, which is natural among birds, helps the swarm to escape from the predators. In the original PSO, food locations are modelled as the best result points of the optimization problem. This paper models the predators as the worst result points and the new model is named as anti-predatory particle swarm optimization (APSO). The APSO is developed by splitting both the cognitive and social behaviours of the classical PSO. The cognitive behaviour is divided into good experience and bad experience components. The former refers to a bird's memory about previously visited best position and the latter refers to a bird's memory about the previously visited worst position (location of predator). Similarly, the social behaviour of the classical PSO is split into global good experience and global bad experience components. To calculate the velocity of a particle, the proposed APSO considers the particle's bad experience along with good experience. Then, the velocity-update equation for APSO model is given by

$$v_{ij}^{r+1} = w^r v_{ij}^r + C_1 R_1 (\mathrm{Pb}_{ij}^r - P_{ij}^r) + C_2 R_2 (P_{ij}^r - \mathrm{Pw}_{ij}^r) + C_3 R_3 (\mathrm{Gb}_j^r - P_{ij}^{r-1}) + C_4 R_4 (P_{ij}^r - \mathrm{Gw}_j^r)$$
$$(j = 1, 2, \ldots, \mathrm{NG};\ i = 1, 2, \ldots, \mathrm{NP}) \quad (7.90)$$

$$P_{ij}^{r+1} = P_{ij}^r + v_{ij}^{r+1} \quad (j = 1, 2, \ldots, \mathrm{NG};\ i = 1, 2, \ldots, \mathrm{NP}) \quad (7.91)$$

where

NP is the number of particles in a group;
NG is the number of members in a particle;
r is the pointer of iterations (generations);
w is the inertia weight factor;
v_{ij}^r is velocity of particle i at iteration r, $V_j^{min} \le v_{ij}^r \le V_j^{max}$
P_{ij}^r is the current position of ith particle at rth iteration.
Gb_j^r is the global best position of member j until iteration r
Gw_j^r is the global worst position of member j until iteration r
Pb_{ij}^r is dimension j of the own best position of particle i until iteration r
Pw_{ij}^r is dimension j of the own worst position of particle i until iteration r

C_1 is the acceleration coefficient, which accelerates the particle towards its best position
C_2 is the acceleration coefficient, which accelerates the particle away from its worst position
C_3 is the acceleration coefficient, which accelerates the particle towards the global best position
C_4 is the acceleration coefficient, which accelerates the particle away from the global worst position
R_1, R_2, R_3 and R_4 are three separately generated uniformly distributed random numbers in the range [0, 1].

The positions are updated using Eq. (7.90). The inclusion of the bad experience components in the behaviour of the particle gives additional exploration capacity to the swarm. By using the bad experiences, the particle (bird) always by-passes its previous worst positions (predator positions) and try to occupy a better position.

7.15 DIFFERENTIAL EVOLUTION FOR ECONOMIC DISPATCH

Differential Evolution is a population-based stochastic function minimizer (or maximizer) relating to evolutionary computation, whose simple yet powerful and straightforward features make it very attractive for numerical optimization. Differential evolution uses a rather greedy and less stochastic approach to problem solving than do evolutionary algorithms. Differential evolution combines simple arithmetic operators with the classical operators of recombination, mutation, and selection to evolve from a randomly generated starting population to a final solution. Differential evolution differs from conventional genetic algorithms in its use of perturbing vectors, which are the difference between two randomly chosen parameter vectors, a concept borrowed from the operators of Nelder and Mead's simplex optimization technique. The differential evolution algorithm was first introduced by Storn and Price in 1995 and was successfully applied in the optimization of some well-known nonlinear, non-differentiable, and non-convex functions by Storn [1997].

The different variants of differential evolution are classified using the following notation: DE/α/β/δ, where α indicates the method for selecting the parent chromosome that will form the base of the mutated vector, β indicates the number of difference vectors used to perturb the base chromosome, and δ indicates the recombination mechanism used to create the offspring population. The bin acronym indicates that the recombination is controlled by a series of independent binomial experiments. The fundamental idea behind differential evolution is a scheme whereby it generates the trial parameter vectors. In each step, the differential evolution mutates vectors by adding weighted, random vector differentials to them. If the cost of the trial vector is better than that of the target, the trial vector replaces the target vector in the next generation. The variant implemented here was the *DE/rand/1/bin*, which involved the following steps and procedures. The differential evolution algorithm can be implemented by searching the generation of power plants, P_i within generator limits. The economic dispatch problem is defined by Eq. (7.26) or Eq. (7.48) or Eq. (7.52) and transmission losses are defined by Eq. (7.27). This section provides the solution methodology to the economic dispatch problems through differential evolution.

Parameter setup

The user must choose the key parameters that control the differential evolution, i.e. population size (L), boundary constraints of optimization variables (NG), mutation factor (f_m), crossover rate (CR), and the stopping criterion of maximum number of iterations (generations) $t_{\max}$.

The set of real power output (P) of all generators is represented as the population. For a system with NG generators, the population is represented as a vector of length NG. If there are L members in the population, the complete population is represented as a matrix as below:

$$\text{Population} = \begin{bmatrix} P_{11}^t & P_{12}^t & \cdots & \cdots & P_{1\,\text{NG}}^t \\ P_{21}^t & P_{22}^t & \cdots & \cdots & P_{2\,\text{NG}}^t \\ \vdots & \vdots & P_{ij}^t & \cdots & \vdots \\ \vdots & \vdots & \cdots & \cdots & \vdots \\ P_{L1}^t & P_{L2}^t & \cdots & \cdots & P_{L\,\text{NG}}^t \end{bmatrix}$$

where P_{ij}^t is the jth element of NG set of committed generators giving ith individual of a population. In other words it represents the real power generation of generator j of the possible solution i. Further, $P_i^t = [P_{i1}^t, P_{i2}^t, ..., P_{i\text{NG}}^t]^T$ stands for the position of the ith individual of a population of real valued NG-dimensional vectors.

Initialization of an Individual Population

Set generation $t = 0$. Initialize a population P_{ij}^t ($j = 1, 2, \ldots$, NG; $i = 1, 2, ..., L$) individuals (real valued NG-dimensional solution vectors) with random values generated according to a uniform probability distribution in the NG-dimensional problem space. Initialize the entire solution vector population within the given upper and lower limits of the search space. Let d be dependent generator.

$$P_{ij}^t = P_i^{\min} + \text{rand}()\,(P_i^{\max} - P_i^{\min}) \quad (j = 1, 2, \ldots, \text{NG}; j \neq d; i = 1, 2, \ldots, L) \tag{7.92}$$

The vector population may violate constraint Eq. (7.26c). This violation is corrected by fixing them either at lower or at upper limit.

Evaluation of the Individual Population

The goal is to minimize the operating cost function. In order to satisfy the power balance constraint, a generator is arbitrarily selected as a dependent generator d and is obtained by Eq. (7.63). If the output of the dependent generator violates its operating limits, it is fixed using Eq. (7.62). After limiting the value of the dependent generator as above, a penalty term is introduced in the objective function to penalize its fitness value. So, the function value is defined by Eq. (7.64).

Mutation operation (differential operation)

Mutation is an operation that adds a vector differential to a population vector of individuals according to the following equation:

$$Z_{ij}^t = P_{R_1 j}^t + f_m (P_{R_2 j}^t - P_{R_3 j}^t) \quad (j = 1, 2, \ldots, \text{NG}; j \neq d;\; i = 1, 2, \ldots, L) \tag{7.93}$$

where

$i = 1, 2, ..., L$ is the individual's population index;

$j = 1, 2, ...,$ NG is the position in NG of the dimensional individual;

t is the time (generation);

$P_i^t = [P_{i1}^t, P_{i2}^t, ..., P_{i\mathrm{NG}}^t]^T$ stands for the position of the ith individual of a population of real valued NG-dimensional vectors;

$Z_i^t = [Z_{i1}^t, Z_{i2}^t, ..., Z_{i\mathrm{NG}}^t]^T$ stands for the position of the ith individual of a mutant vector

R_1, R_2 and R_3 are mutually different integers that are also different from the running index, i, randomly selected with uniform distribution from the set $\{1, 2, ..., i-1, i+1, ..., L\}$.

$f_m(t)$ is the mutation factor and $f_m(t) > 0$ is a real parameter, which controls the amplification of the difference between two individuals with indexes R_2 and R_3 so as to avoid search stagnation and is usually a constant value taken from the range [0.4, 1].

The mutation operation using the difference between two randomly selected individuals may cause the mutant individual to escape from the search domain. If an optimized variable for the mutant individual is outside of the domain search, then this variable is replaced by its lower bound or its upper bound so that each individual can be restricted to remain within the search domain.

Recombination operation

Following the mutation operation, recombination is applied to the population. Recombination is employed to generate a trial vector by replacing certain parameters of the target vector by the corresponding parameters of a randomly generated donor vector.

For each vector, Z_i^{t+1}, an index $R_5(i) \in \{1, 2, ..., \mathrm{NG}\}$ is randomly chosen using a uniform distribution, and a trial vector, $U_i^{t+1} = [U_{i1}^{t+1},\ U_{i2}^{t+1}, ..., U_{i\mathrm{NG}}^{t+1}]^T$

$$U_{ij}^{t+1} = \begin{cases} Z_{ij}^t & \text{if } (R_4(j) \le \mathrm{CR}) \text{ or } (j = R_5(i)) \\ P_{ij}^t & \text{if } (R_4(j) > \mathrm{CR}) \text{ or } (j \ne R_5(i)) \end{cases} \quad (j = 1, 2, \ldots, \mathrm{NG};\ j \ne d;\ i = 1, 2, \ldots, L) \tag{7.94}$$

where

$R_4(j)$ is the jth evaluation of a uniform random number generation with [0, 1]

CR is the crossover or recombination rate in the range [0, 1].

Usually, the performance of a DE algorithm depends on three variables: the population size, the mutation factor $f_m(t)$ and the CR.

Selection operation

Selection is the procedure whereby better offspring are produced. To decide whether the vector U_i^{t+1} should be a member of the population comprising the next generation, it is compared with the corresponding vector P_i^t. Thus, if f denotes the cost function under minimization, then

$$P_{ij}^{t+1} = \begin{cases} U_{ij}^{t+1} & (j = 1, 2, ..., \mathrm{NG}) \quad ; \text{if } f(U_i^{t+1}) < f(P_i^t) \\ P_{ij}^t & (j = 1, 2, ..., \mathrm{NG}) \quad ; \text{otherwise} \end{cases} \quad (i = 1, 2, \ldots, L) \tag{7.95}$$

In this case, the cost of each trial vector U_i^{t+1} is compared with that of its parent target vector P_i^t. If the cost f, of the target vector P_i^t, is lower than that of the trial vector, the target is allowed to advance to the next generation. Otherwise, a trial vector replaces the target vector in the next generation.

Verification of the stopping criterion

Set the generation number for $t = t + 1$. Repeat mutation, recombination and selection operation until a stopping criterion is met, usually a maximum number of iterations (generations), $t_{\max}$. The stopping criterion depends on the type of problem.

Various differential (mutation) strategies

There are several variations of differential evolution algorithm strategies that can be employed for optimization as mentioned by Sum-Im et al. [2009]. Ten variations, which are defined as the following mutation strategies.

***Differential Strategy* 1:** In this mutation strategy, the mutant vector can be generated according to the following equation from the randomly chosen base vector

$$Z_{ij}^t = P_{R_1 j}^t + f_m(P_{R_2 j}^t - P_{R_3 j}^t) \quad (j = 1, 2, \ldots, \text{NG}; j \neq d; i = 1, 2, \ldots, L) \tag{7.96}$$

where

t is the time (generation);

$P_i^t = [P_{i1}^t, P_{i2}^t, \ldots, P_{i\text{NG}}^t]^T$ stands for the position of the ith individual of a population of real valued NG-dimensional vectors;

$Z_i^t = [Z_{i1}^t, Z_{i2}^t, \ldots, Z_{i\text{NG}}^t]^T$ stands for the position of the ith individual of a mutant vector

R_1, R_2 and R_3 are mutually different integers that are also different from the running index, i, randomly selected with uniform distribution from the set $\{1, 2, \ldots, i-1, i+1, \ldots, L\}$.

f_m is the mutation factor and $f_m > 0$ is a real parameter.

***Differential Strategy* 2:** In this mutation strategy, the mutant vector can be generated according to the following equation from the best performing vector of the current generation by considering it as base vector

$$Z_{ij}^t = P_{Bj}^t + f_m(P_{R_1 j}^t - P_{R_2 j}^t) \quad (j = 1, 2, \ldots, \text{NG}; j \neq d; i = 1, 2, \ldots, L) \tag{7.97}$$

where

P_{Bj}^t is the best performing vector of the current generation.

R_1 and R_2 are mutually different integers that are also different from the running index, i, randomly selected with uniform distribution from the set $\{1, 2, \ldots, i-1, i+1, \ldots, L\}$.

***Differential Strategy* 3:** In this mutation strategy, the perturbation is applied at a location between the best performing vector and a randomly selected population vector according to the following equation

$$Z_{ij}^t = P_{ij}^t + f_B(P_{Bj}^t - P_{ij}^t) + f_m(P_{R_1 j}^t - P_{R_2 j}^t) \quad (j = 1, 2, \ldots, \text{NG}; j \neq d; i = 1, 2, \ldots, L) \tag{7.98}$$

where

f_B is applied to control the greediness of the scheme, which usually it is set equally to f_mF to reduce the number of control variables.

***Differential Strategy* 4:** Two difference vectors are used as a perturbation in this mutation strategy

$$Z_{ij}^t = P_{Bj}^t + f_m(P_{R_1 j}^t + P_{R_2 j}^t - P_{R_3 j}^t - P_{R_4 j}^t) \quad (j = 1, 2, \ldots, \text{NG}; j \neq d; i = 1, 2, \ldots, L) \tag{7.99}$$

***Differential Strategy* 5:** This mutation strategy replaces the best performing vector by a randomly selected vector

$$Z_{ij}^t = P_{R_5 j}^t + f_m (P_{R_1 j}^t + P_{R_2 j}^t - P_{R_3 j}^t - P_{R_4 j}^t) \quad (j = 1, 2, \ldots, \text{NG}; j \neq d; i = 1, 2, \ldots, L) \tag{7.100}$$

***Differential Strategy* 6:** In this mutation strategy, the mutant vector can be generated according to the following equation

$$Z_{ij}^t = P_{Bj}^t + f_m (P_{Bj}^t - P_{ij}^t) \quad (j = 1, 2, \ldots, \text{NG}; j \neq d; i = 1, 2, \ldots, L) \tag{7.101}$$

***Differential Strategy* 7:** The mutant vector can be generated according to the following equation for this mutation strategy

$$Z_{ij}^t = P_{Bj}^t + f_m (P_{Bj}^t - P_{ij}^t - P_{R_1 j}^t - P_{R_2 j}^t) \quad (j = 1, 2, \ldots, \text{NG}; j \neq d; i = 1, 2, \ldots, L) \tag{7.102}$$

***Differential Strategy* 8:** This mutation strategy generates the mutant vector according to the following equation

$$Z_{ij}^t = P_{Bj}^t + f_B (P_{Bj}^t - P_{ij}^t) + f_m (P_{R_1 j}^t - P_{R_2 j}^t) \quad (j = 1, 2, \ldots, \text{NG}; j \neq d; i = 1, 2, \ldots, L) \tag{7.103}$$

***Differential Strategy* 9:** This mutation strategy generates the mutant vector according to the following equation

$$Z_{ij}^t = P_{Bj}^t + f_m (P_{Bj}^t + P_{ij}^t - P_{R_1 j}^t - P_{R_2 j}^t) \quad (j = 1, 2, \ldots, \text{NG}; j \neq d; i = 1, 2, \ldots, L) \tag{7.104}$$

***Differential Strategy* 10:** This mutation strategy considers the previous generation's best performing vector in order to create the mutant vector

$$Z_{ij}^t = P_{Bj}^t + f_m (P_{Bj}^t - P_{Bj}^{t-1}) \quad (j = 1, 2, \ldots, \text{NG}; j \neq d; i = 1, 2, \ldots, L) \tag{7.105}$$

Algorithm 7.10: Economic Dispatch Using Differential Evolution Algorithm

The search procedures of the proposed differential evolution method have been outlined below:

1. Read data, viz. Cost Coefficients, a_i, b_i, c_i (i = 1, 2, ..., NG). Convergence tolerance error, population size (L), boundary constraints of optimization variables (NG), mutation factor (f_m), crossover rate (CR), and the stopping criterion of maximum number of iterations (generations) $t_{\max}$, seed number, S, $P_i^{\min}$ and $P_i^{\max}$ (i = 1, 2, ..., NG), etc. Initialize randomly the individuals of the population according to the limit of each unit including individual dimensions and searching points. These initial individuals must be feasible candidate solutions that satisfy the practical operation constraints.
2. Generate an array of (NG × L) size of uniform random numbers.
3. Set population counter, $i = 0$,
   DO
4. Increment population counter, $i = i + 1$
5. Set generation counter, $j = 0$
   DO

6. Increment the generation counter, $j = j + 1$
7. IF $(j \neq d)$ THEN generate the individual P_{ij}^{0}
8. WHILE $(j < \text{NG})$
9. Compute P_{id}^{0} using Eq. (7.63), check limits and adjust using Eq. (7.62). Compute penalty factor, φ_i and $f_i(P_{ij}^{0})$ using Eq. (7.65) and Eq. (7.64) respectively.
   WHILE $(i < L)$
10. Set population counter, $i = 1$, $f_i^{\text{best}} = f_i$
    DO
11. Increment population counter, $i = i + 1$
12. IF $(f_i^{\text{best}} < f_i)$ THEN $f_i^{\text{best}} = f_i$
    WHILE $(i < L)$
13. Set iteration counter, $t = 0$
    DO
14. Increment iteration counter, $t = t + 1$
15. Set population counter, $i = 0$
    DO
16. Increment particle counter, $i = i + 1$
17. Set generation counter, $j = 0$
    DO
18. Increment the electric generation counter, $j = j + 1$
19. Generate an array of uniform random numbers and generate three different integer random numbers within the range of 1 to L.
20. IF $(j \neq d)$ Compute Z_{ij}^{t} using Eq. (7.92)
    WHILE $(j < \text{NG})$
21. Compute P_{id}^{t} using Eq. (7.63), check limits and adjust using Eq. (7.62). Compute penalty factor, φ_i and $f_i(Z_{ij}^{t+1})$ using Eq. (7.65) and Eq. (7.64) respectively.
    WHILE $(i < L)$
22. Set population counter, $i = 0$ and a counter, $k = 0$
23. Generate an array R_4 of random numbers, R and of size, $2L$
    DO
24. Increment particle counter, $i = i + 1$
25. Set generation counter, $j = 0$
    DO
26. Increment the generation counter, $j = j + 1$
    IF $(j \neq d)$ THEN
27. Increment counter, $k = k + 1$. Generate a random integer $R_5()$ within the range from 1 to L.
28. IF $((R_4(k) \leq \text{CR})$ or $(j = \text{R}_5(\text{i})))$ THEN $U_{ij}^{t+1} = Z_{ij}^{t+1}$
29. IF $((R_4(k) < \text{CR})$ or $(j \neq \text{R}_5(\text{i})))$ THEN $U_{ij}^{t+1} = P_{ij}^{i+1}$
    ENDIF
30. Compute U_{id}^{t+1} using Eq. (7.63), check limits and adjust using Eq. (7.62). Compute penalty factor, φ_i and $f_i(Z_{ij}^{t+1})$ using Eq. (7.65) and Eq. (7.64) respectively.
    WHILE $(j < \text{NG})$
    WHILE $(i < L)$
31. Set population counter, $i = 0$.
    DO

32. Increment particle counter, $i = i + 1$
33. IF $(f_i(U_{ij}^{t+1}) < f_i(P_{ij}^t))$
    THEN $P_{ij}^{t+1} = U_{ij}^{t+1}$ $(j = 1, 2, ..., NG)$ ELSE $P_{ij}^{t+1} = P_{ij}^t$ $(j = 1, 2, ..., NG)$
    WHILE $(i < L)$
34. Set population counter, $i = 1$,
    DO
35. Increment population counter, $i = i + 1$
36. IF $(f_i^{\min} < f_i)$ THEN $f_i^{\min} = f_i$
    WHILE $(i < L)$
    IF abs$(f_i^{\min} \le f_i^{\text{best}}) \le$ err THEN GOTO 37
    IF $(f_i^{\min} \le f_i^{\text{best}})$ THEN $(f_i^{\text{best}} = f_i^{\min})$
    WHILE $(t < t_{\max})$
37. STOP

EXAMPLE 7.10 Consider a five-generator system given in Example 7.7 and obtain the optimum schedule.

Solution Each real power generation is searched. First generator is considered as dependent (slack) generator. Let there be 20 members in the population. Initially seed number is taken as 305. An array of uniform random numbers is generated. Initial feasible member of population is given in Table 7.31. Fourth member of population gives the best feasible solution. First generator is considered as slack (dependent) generator. Power demand P_D is given as 730 MW. Mutation is performed by taking f_m as 0.48. The results obtained after mutation in first generation are given in Table 7.32. Recombination operation is performed by setting CR to 0.88. Results obtained after performing recombination and selection are given in Tables 7.33 and 7.34, respectively.

TABLE 7.31 Initial population generated randomly for differential evolution

Members (k)	*Operating cost* (₹/h)	*Power generation in MW* P_1	P_2	P_3	P_4	P_5
1	2350.976	254.9027	70.31945	121.1615	75.0	208.6164
2	2577.610	278.8007	70.13274	104.2399	26.82669	250.0
3	2609.252	263.9504	125.0	175.0	64.52831	101.5212
4	2453.134	247.4525	84.4595	175.0	75.0	148.0881
5	2580.667	271.0399	53.52147	111.8689	43.56982	250.0
6	2604.347	291.0753	113.4982	162.3449	75.0	88.08161
7	2620.666	278.1912	125.0	148.5518	70.70692	107.5501
8	2518.897	279.8595	95.88496	170.4772	37.33062	145.9477
9	2588.394	262.9962	53.11413	175.0	44.22339	194.6662
10	2659.616	279.8397	73.89268	147.4371	42.14243	186.6881
11	2593.818	273.1825	115.0334	86.18538	75.0	180.5987
12	2418.499	293.8149	63.83953	108.9140	47.96465	215.4669
13	2470.572	291.6728	78.64346	110.8254	67.15330	181.7050

(*Contd.*)

TABLE 7.31 Initial population generated randomly for differential evolution (*Contd.*)

Members (k)	*Operating cost* (₹/h)	*Power generation in MW*				
		P_1	P_2	P_3	P_4	P_5
14	2474.945	269.8487	69.08971	114.5516	54.25695	222.2530
15	2345.693	251.0819	100.6990	134.6187	19.30112	224.2993
16	2509.153	286.5265	120.1132	109.1192	33.53798	180.7031
17	2525.489	293.5890	35.4327	124.7191	39.71367	236.5455
18	2639.507	266.6966	110.8340	58.63712	45.79224	248.0400
19	2620.968	248.8822	122.3131	157.7612	40.41310	160.6305
20	2449.533	217.3338	120.6668	133.9313	23.42717	234.6410

TABLE 7.32 Population after mutation during first generation

Members (k)	*Operating cost* (₹/h)	*Power generation in MW*				
		P_1	P_2	P_3	P_4	P_5
1	2452.150	257.3656	85.40227	122.6713	35.09802	229.4628
2	2537.863	261.6107	93.06307	92.97926	37.98719	244.3597
3	2519.435	255.3188	108.1873	175.0000	41.02346	150.4704
4	2354.507	231.5450	88.25891	160.5783	52.03096	197.5869
5	2527.427	244.3915	72.17113	127.7501	35.68726	250.0
6	2633.117	263.7841	86.82455	151.2073	51.33307	176.8510
7	2386.573	268.0099	86.49856	112.8487	41.10189	221.5410
8	2545.699	286.5857	77.23427	130.7360	41.77176	193.6723
9	2453.040	280.9715	71.23512	119.4758	42.99594	215.3217
10	2314.845	249.8368	96.34423	140.9543	33.15910	209.7055
11	2449.443	282.8268	103.9217	111.6336	64.86626	166.7516
12	2390.060	272.3832	102.2732	114.5833	47.37717	193.3831
13	2604.794	275.3941	80.03883	162.7063	26.34295	185.5178
14	2504.885	257.2645	91.04937	150.0263	37.47815	194.1817
15	2406.782	244.4821	119.1597	145.5233	20.46693	200.3678
16	2437.281	265.7909	89.24206	130.0252	10.0	234.9419
17	2338.146	263.2820	91.33356	118.5612	42.89262	213.9306
18	2435.991	268.2382	97.51484	99.01176	15.23520	250.0
19	2659.194	267.4305	125.0	134.3247	45.36584	157.8789
20	2298.024	229.8077	103.1088	137.9232	29.48337	229.6770

TABLE 7.33 Population after performing combination

Members (*k*)	*Operating cost* (₹/h)	*Power generation in MW*				
		P_1	P_2	P_3	P_4	P_5
1	2350.976	254.9027	70.31945	121.1615	75.0	208.6164
2	2577.610	278.8007	70.13274	104.2399	26.82669	250.0
3	2609.252	263.9504	125.0	175.0	64.52831	101.5212
4	2453.134	247.4525	84.45950	175.0	75.0	148.0881
5	2580.667	271.0399	53.52147	111.8689	43.56982	250.0
6	2604.347	291.0753	113.4982	162.3449	75.0	88.08161
7	2620.666	278.1912	125.0000	148.5518	70.70692	107.5501
8	2518.897	279.8595	95.88496	170.4772	37.83062	145.9477
9	2588.394	262.9962	53.11413	175.0	44.22339	194.6662
10	2659.616	279.8397	73.89268	147.4371	42.14243	186.6881
11	2377.953	247.7343	115.0334	111.6336	75.0	180.5987
12	2418.499	293.8149	63.83953	108.9140	47.96465	215.4669
13	2470.572	291.6728	78.64346	110.8254	67.15330	181.7050
14	2474.945	269.8487	69.08971	114.5516	54.25695	222.2530
15	2345.693	251.0819	100.6990	134.6187	19.30112	224.2993
16	2509.153	286.5265	120.1132	109.1192	33.53798	180.7031
17	2525.489	293.5890	35.43270	124.7191	39.71367	236.5455
18	2639.507	266.6966	110.8340	58.63712	45.79224	248.0400
19	2620.968	248.8822	122.3131	157.7612	40.41310	160.6305
20	2449.533	217.3338	120.6668	133.9313	23.42717	234.6410

TABLE 7.34 Population after selection

Members (*k*)	*Operating cost* (₹/h)	*Power generation in MW*				
		P_1	P_2	P_3	P_4	P_5
1	2350.976	254.9027	70.31945	121.1615	75.0	208.6164
2	2577.610	278.8007	70.13274	104.2399	26.82669	250.0
3	2609.252	263.9504	125.0	175.0	64.52831	101.5212
4	2453.134	247.4525	84.4595	175.0	75.0	148.0881
5	2580.667	271.0399	53.52147	111.8689	43.56982	250.0
6	2604.347	291.0753	113.4982	162.3449	75.0	88.08161
7	2620.666	278.1912	125.0	148.5518	70.70692	107.5501
8	2518.897	279.8595	95.88496	170.4772	37.83062	145.9477
9	2588.394	262.9962	53.11413	175.0	44.22339	194.6662
10	2659.616	279.8397	73.89268	147.4371	42.14243	186.6881

(*Contd.*)

TABLE 7.34 Population after selection (*Contd.*)

Members (*k*)	*Operating cost* (₹/h)	*Power generation in MW* P_1	P_2	P_3	P_4	P_5
11	2377.953	247.7343	115.0334	111.6336	75.0	180.5987
12	2418.499	293.8149	63.83953	108.9140	47.96465	215.4669
13	2470.572	291.6728	78.64346	110.8254	67.15330	181.7050
14	2474.945	269.8487	69.08971	114.5516	54.25695	222.2530
15	2345.693	251.0819	100.6990	134.6187	19.30112	224.2993
16	2509.153	286.5265	120.1132	109.1192	33.53798	180.7031
17	2525.489	293.5890	35.4327	124.7191	39.71367	236.5455
18	2639.507	266.6966	110.8340	58.63712	45.79224	248.0400
19	2620.968	248.8822	122.3131	157.7612	40.41310	160.6305
20	2449.533	217.3338	120.6668	133.9313	23.42717	234.6410

Best result obtained after 50 generations is given in Table 7.35.

TABLE 7.35 Best individual from a population of 20 members after 50 generations

Operating cost (₹/h)	*Power generation in MW* P_1	P_2	P_3	P_4	P_5
2140.97	225.3845	113.020	109.4146	73.11176	209.0692

Differential evolution algorithm has a number of significant advantages which are summarized below:

- Differential evolution algorithm has the ability to find the true global minimum regardless of the initial parameter values;
- Differential evolution algorithm is fast and simple with regard to application;
- Differential evolution algorithm requires few control parameters;
- Differential evolution algorithm has parallel processing nature and fast convergence;
- Differential evolution algorithm is capable of providing multiple solutions in a single run;
- The method is effective on integer, discrete and mixed parameter optimization;
- Differential evolution algorithm has the ability to find the optimal solution for a nonlinear constrained optimization problem with penalty functions.

Although differential evolution algorithm has many advantages explained above, there are also a number of disadvantages of differential evolution algorithm that are summarized below:

- Differential evolution algorithm does not always give an exact global optimum due to premature convergence;
- Differential evolution algorithm may require tremendously high-computation time because of a large number of fitness evaluations.
- In differential evolution algorithm, there exist many trials vector generation strategies out of which a few may be suitable for solving a particular problem.

- Moreover, three crucial control parameters involved in differential evolution algorithm, i.e. population size, scaling factor, and crossover rate, may significantly influence the optimization performance.

7.16 REAL CODED GENETIC ALGORITHM

A global optimization technique known as Genetic Algorithm (GA) has emerged as a candidate due to its flexibility and efficiency for many optimization applications. Genetic algorithm is a stochastic searching algorithm. It combines an artificial, i.e. the Darwinian Survival of the Fittest, principle with genetic operation, abstracted from nature to form a robust mechanism that is very effective at finding optimal solutions to complex-real world problems. Evolutionary computing is an adaptive search technique based on the principles of genetics and natural selection. They operate on string structures. Many such string structures are considered simultaneously, with the most fit of these structures receiving exponentially increasing opportunities to pass on genetically important material to successive generation of string structures. In this way, genetic algorithms search for many points in the search space at once, and yet continually narrow the focus of the search to the areas of the observed best performance. The basic elements of real coded genetic algorithms are selection, crossover and mutation. In reproduction, the individuals are selected based on their fitness values relative to those of the population. In the crossover operation, two individual strings are selected at random from the mating pool and a crossover is performed using mathematical relations. In mutation, an occasional random alteration of a string is done. In this chapter, a real coded genetic algorithm with genetic operators including arithmetic-average-bound-blend crossover and wavelet mutation are applied to solve the economic dispatch problem. The arithmetic-average-bound-blend crossover operator combines the arithmetic crossover, average crossover, bound and blend crossover. The arithmetic crossover operation produces some children with their parent's features; average crossover manipulates the genes of the selected parents and the minimum and maximum possible values of the genes and bound crossover is capable of moving the offspring near the domain boundary. Therefore, the offspring spreads over the domain so that a higher chance of reaching the global optimum can be obtained. The wavelet mutation operation based on wavelet theory is a powerful tool for finetuning of the genes to search the solution space locally. This property of wavelet mutation operation enhances the searching performance and provides a faster convergence than conventional real coded genetic algorithm. The above procedure to implement real coded genetic algorithm is outlined below:

Algorithm 7.11: Real coded genetic algorithm

1. Randomly generate initial population strings. Tossing of a coin can be used.
2. Evaluate fitness values of population members.
3. Is solution available among the population?
   If '*yes*' then GOTO Step 9.
4. Select highly fit strings as parents using stochastic remainder roulette-wheel selection and produce off-spring according to their fitness.
5. Create new strings by mating current off-spring. Apply crossover and mutation operators to introduce variations and form new strings.
6. New strings replace existing ones by applying competition and selection.
7. GOTO Step 3 and repeat.
8. Stop

Initialization

The initial population comprises combinations of only the candidate of economic dispatch solutions, which satisfy all the constraints and are feasible solution of economic dispatch. It consists of P_{ik} (i = 1, 2, ..., NG; j = 1, 2, ..., L) trial parent individuals. The elements of a parent are the combinations of power outputs of the generating units, which are chosen randomly by a random number ranging over $[P_i^{\min}, P_i^{\max}]$. Let d is a dependent generator.

$$P_{ij}^t = P_i^{\min} + \text{rand}()(P_i^{\max} - P_i^{\min}) \quad (j = 1, 2, \ldots, \text{NG};\ i = 1, 2, \ldots, L) \tag{7.106}$$

The vector population may violate constraint Eq. (7.26c). This violation is corrected by fixing them either at lower or at upper limit.

Feasible Solution

A feasible solution can naturally be defined as an element that satisfies all the equalities and inequalities constraints of optimization problem. It is a set of values for decision variables that satisfy all constraints in a mathematical programming problem. In case of $|\Delta P_D| \le 0.01$, stop otherwise,

$$P_{ij}^0 = \begin{cases} P_{ij}^0 + w_f\ \text{rand}()(P_j^{\max} - P_j^{\min})\left|\Delta P_{Di}^t\right| ;\ \text{if}\ \Delta P_{Di}^t < 0 \\ P_{ij}^0 - w_f\ \text{rand}()(P_j^{\max} - P_j^{\min})\left|\Delta P_{Di}^t\right| ;\ \text{if}\ \Delta P_{Di}^t > 0 \end{cases} \quad (j = 1, 2, \ldots, \text{NG};\ i = 1, 2, \ldots, L) \tag{7.107}$$

where

ω_f is the adjusting parameter
$P_i^{\max}$ is the maximum generation output of the ith generator
$P_i^{\min}$ is the minimum generation output of the ith generator
rand() is the uniform random number within the range of [0, 1]
$\Delta P_{Di}^t = \sum_{j=1}^{\text{NG}} P_{ij}^t - P_D - P_L$ is the equality constraint

Evaluation of the individual population

The role of the evolution function is to represent the requirements to adapt the solution. It forms the basis of selection and thereby it facilitates improvements. More accurately, it defines improvement means from the problem solving perspective and it represents the task to solve in evolutionary context. Technically, it is the function or procedure that assigns the quality measure to individuals. Typically this function is composed from a quality measure in the phenotype space. The evolution function is commonly called as the fitness function in evolutionary computation. The goal is to minimize the operating cost function. In order to satisfy the power balance constraint, exterior penalty function method is used. The minimization problem is converted to maximization problem as follows:

$$f^t = \frac{1}{1 + F(P_{ij}^t, r_k)/\delta} \tag{7.108}$$

where

P_{ij}^t is a vector of variables

$F(P_{ij}^t, r_k)$ is the objective function of the problem to be minimized and is given as:

$$F(P_{ij}^t, r_k) = \left[\sum_{j=1}^{NG} \left(a_j (P_{ij}^t)^2 + b_j P_{ij}^t + c_j + \left|d_j \sin(e_j (P_j^{\min} - P_{ij}^t))\right|\right)\right] + r_k \left(\sum_{j=1}^{NG} P_{ij}^t - P_D - P_L\right)^2 \tag{7.109}$$

with r_k as the penalty factor.

Reproduction

The basic roulette-wheel selection method is Stochastic Sampling with Replacement (SSR). The segment size and selection probability remain same throughout the selection phase and individuals are selected accordingly. Stochastic sampling with partial replacement (SSPR) extends upon SSR by resizing an individual's segment if it is selected. Each time an individual is selected, the size of its segment is reduced by 1.0. If the segment size becomes negative, then it is set to zero. Remainder sampling methods involve two distinct phases. In the integral phase, the individuals are selected deterministically according to the integer part of their expected trials. The remaining individuals are then selected probabilistically from fractional part of the individual's expected values. In the study, the stochastic remainder roulette-wheel selection is applied whose step-wise procedure is outlined in Algorithm 7.3.

Crossover operators

The selection of chromosomes for the arithmetic-average-bound-blend crossover is based on the roulette-wheel mechanism. The arithmetic-average-bound-blend is a combinatorial operator composed of the arithmetic crossover, average crossover and bound and blend crossover. Suppose two vectors are selected as chromosomes in the generation '*t*' of the real coded genetic algorithm execution. Each chromosome has *m* genes, which are real numbers (the generation outputs of schedulable units). The arithmetic-average-bound-blend operator [Amjady and Nasiri-Rad, 2009] creates ten children from the parents; *v* and *u* as follows:

(i) Arithmetic crossover

$$P_{1j}^{t+1} = w_a P_{vj}^t + (1 - w_a) P_{uj}^t \quad (j = 1, 2, ..., NG) \tag{7.110}$$

$$P_{2j}^{t+1} = (1 - w_a) P_{vj}^t + w_a P_{uj}^t \quad (j = 1, 2, ..., NG) \tag{7.111}$$

$$P_{3j}^{t+1} = \text{Min}\,\{P_{vj}^t, P_{uj}^t\} \quad (j = 1, 2, ..., NG) \tag{7.112}$$

$$P_{4j}^{t+1} = \text{Max}\,\{P_{vj}^t, P_{uj}^t\} \quad (j = 1, 2, ..., NG) \tag{7.113}$$

(ii) Average crossover

$$P_{5j}^{t+1} = \frac{1}{2}(P_{vj}^t + P_{uj}^t) \quad (j = 1, 2, ..., NG) \tag{7.114}$$

$$P_{6j}^{t+1} = \frac{1}{2}[w_b (P_{vj}^t + P_{uj}^t) + (1 - w_b)(P_i^L + P_i^U)] \quad (j = 1, 2, ..., NG) \tag{7.115}$$

(iii) Bound crossover

$$P_{7j}^{t+1} = w_c \text{ Min}\{P_{vj}^t, P_{uj}^t\} + (1 - w_c) P_i^L \quad (j = 1, 2, ..., \text{NG}) \tag{7.116}$$

$$P_{8j}^{t+1} = w_c \text{ Min}\{P_{vj}^t, P_{uj}^t\} + (1 - w_c) P_j^U \quad (j = 1, 2, ..., \text{NG}) \tag{7.117}$$

(iv) Blend crossover

$$P_{9j}^{t+1} = w_d P_{vj}^t + (1 - w_d) P_{uj}^t \quad (j = 1, 2, ..., \text{NG}) \tag{7.118}$$

$$P_{10j}^{t+1} = (1 - w_d) P_{vj}^t + w_d P_{uj}^t \quad (j = 1, 2, ..., \text{NG}) \tag{7.119}$$

where

$$\overline{P}_i^{\min} = \text{Min }\{\overline{P}_{vj}^t, \overline{P}_{uj}^t\} \quad (j = 1, 2, ..., \text{NG})$$

$$\overline{P}_i^{\max} = \text{Max }\{\overline{P}_{vj}^t, \overline{P}_{uj}^t\} \quad (j = 1, 2, ..., \text{NG})$$

w_a, w_b and w_c are constant weights. The values are adjusted such that $0 < w_a, w_b, w_c < 1$

w_d is also constant weight such that $1 \le w_d < 2$.

Among these ten children, two children having the highest fitness values are selected as the offspring chromosomes of the crossover operation. These two offspring chromosomes are added to the previous population including the parents. So, after the execution of the crossover operator, the previous population will be enlarged, which is considered for the mutation operator.

Mutation operator

The role of mutation in genetic algorithm is to restore the lost or unexpected genetic material into a population to prevent the premature convergence of genetic algorithm to sub-optimal solutions; it ensures that the probability of reaching any point in the search space is never zero. Non-uniform and wavelet mutation operators give good results.

***Non-uniform mutation*:** Non-uniform mutation operator is applied in a generation t, and $G_{\max}$ is the maximum number of generations then

$$P_{ij}^t = \begin{cases} P_{ij}^t + \Delta(t, y)\,(P_j^{\max} - P_j^{\min})\,; & \text{if rand}() > p_m \\ P_{ij}^t - \Delta(t, y)\,(P_j^{\max} - P_j^{\min})\,; & \text{if rand}() < p_m \end{cases} \quad (j = 1, 2, \ldots, \text{NG};\ i = 1, 2, \ldots, L) \tag{7.120}$$

and

$$\Delta(t, y) = y\left(1 - \text{rand}()^{\left(1 - \frac{t}{G_{\max}}\right)^b}\right) \tag{7.121}$$

where

y is scaling parameter

r is a random number from the interval [0, 1]

b is a parameter chosen by the user, which determines the degree of dependency on the number of iterations.

***Wavelet mutation*:** The wavelet mutation, based on wavelet theory, is used to realize the mutation operation. Morlet wavelet is used which is as shown in Figure 7.6 as the mother wavelet function, which can be mathematically represented as follows with the control of magnitude

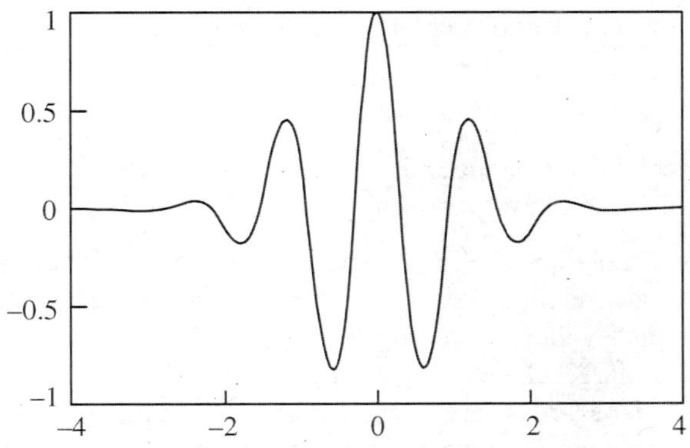

FIGURE 7.6 Morlet.

$$\psi_d(x) = \frac{1}{\sqrt{d}}\left(e^{\frac{-x^2}{2}} \cos 5x\right) \tag{7.122}$$

where

d is the dilation parameter of the wavelet.

$\psi_d(x)$ is the amplitude-scaled version of Morlet wavelet.

The amplitude of $\psi_d(x)$ will be scaled down as the dilation parameter d increases. Every gene of the chromosomes will have a chance to mutate, governed by the probability of the mutation P_m. For each gene of the chromosome, a random number in the range of [0, 1] is generated. If the random number is less than P_m, that gene is selected for the mutation, otherwise it is not selected. In this work, p_m is set at 0.2. The new gene, $\overline{P}_{gik}$ after mutation will be as follows:

$$\overline{P}_{ik} = \begin{cases} \overline{P}_{ik} + \Delta(\phi, t, T)\,(P_i^{\max} - P_{ik}) \text{ if } \Delta \geq p_m \\ \overline{P}_{ik} + \Delta(\phi, t, T)\,(P_{ik} - P_i^{\min}) \text{ if } \Delta < p_m \end{cases} \quad (j = 1, 2, \ldots, M;\ k = 1, 2, \ldots, L) \tag{7.123}$$

where

T is the maximum number of generations of the real coded genetic algorithm.

t is the current generation number.

By using the Morlet wavelet as the mother wavelet, Eq. (7.117) can be rewritten as

$$\Delta(\phi) = \psi_d(\phi) = \frac{1}{\sqrt{d}} e^{\frac{-\phi^2}{2}} \cos 5\phi \tag{7.124}$$

where

ϕ is randomly generated in the range of [– 4, 4] because this wavelet has [– 4, 4] as its effective support.

The dilation parameter d is set to vary with the value of t/T, giving the adaptive search capability to the proposed real coded genetic algorithm.

$$d = \exp\left(\ln g\left(1 - \left(1 - \frac{t}{T}\right)^{\xi}\right)\right) \tag{7.125}$$

where

ξ is the shape parameter of the monotonic increasing function of d
g is the upper limit of the dilation parameter.

The dilation parameter d is a function of t and L, and so Δ is really a function of ϕ, t and L. In the initial generations of the real coded genetic algorithm when t is small compared with T, the dilation parameter d takes small values according to Eq. (7.125). Thus, Δ takes large values based on Eq. (7.124), which makes large changes in the mutating genes. On the other hand, when the genetic algorithm proceeds, d becomes larger and so smaller changes in the mutating genes are made by Δ. In this way, the mutation operator performs a wider search in the solution space at the early stages of the evolution, and at the later stages the search is restricted around the local area of the parameter, resembling a hill climbing operator. Although the wavelet mutation is a non-uniform mutation, it has considerable effects on many generations along the evolution path of the genetic algorithm population. Besides, if the genetic algorithm still traps in a local minimum, it can escape from it by crossover, which has global search characteristic. As another important characteristic, it is seen that Δ takes its values from the Morlet wavelet which has diverse outputs in the range [–1, +1], as shown in Figure 7.6. This characteristic gives more diversity to the mutation operation in searching the solution space and so the chance of finding the optimum solution of the problem increases, which is another advantage of the proposed genetic algorithm. The chromosome owning mutating gene(s) is compared with its parent and the one owning better fitness value is selected and returned to the population.

Competition and selection

Each individual x_i in the combined population has to compete with some other individuals to have a chance to be copied to the next generation. The score for each trial vector after stochastic competition is given by Eqs. (7.72) and (7.73). After competing, the trial $2L$ solutions, including the parents and the offspring, are ranked in descending order of the score obtained in Eq. (7.71). The first L trial solutions survive and are copied along with their objective functions into survivor set as the individuals of the next generation.

EXAMPLE 7.11 Consider a three-generator system given in Example 7.9 and obtain the optimum schedule.

Solution Each real power generation is searched. NG is taken as three (03). First generator is considered as slack (dependant) generator. Power demand P_D is given as 850.0 MW. An array of uniform random numbers is generated to create a population of 20 (L) individuals. Initial feasible member of population is given in Table 7.36.

TABLE 7.36 Initial population generated randomly for real coded genetic algorithm

Individuals	$F(P_{ij}^t, r_k)$	*Power generation in MW*		
		P_1	P_2	P_3
1	8375.901	400.9980	122.7505	326.2533
2	8738.391	366.9366	174.5035	308.5517
3	8566.975	487.1730	123.4398	239.3777
4	8574.248	249.9929	200.0000	400.0000
5	8788.064	368.8599	123.2169	357.9328
6	8601.929	337.0457	199.9164	313.0372
7	8573.049	324.4564	126.5294	399.0168
8	8678.772	513.0711	115.2185	221.7128
9	8590.530	438.1700	96.0726	315.7571
10	8712.348	516.1082	195.3383	138.5527
11	8756.271	451.1765	197.8181	201.0098
12	8705.093	342.7336	176.8437	330.4321
13	8564.678	496.3549	125.6301	228.0126
14	8625.774	283.9576	181.0475	384.9878
15	8715.869	349.0694	167.6029	333.3227
16	8456.449	299.1887	165.1663	385.6456
17	8424.901	405.2048	107.8117	336.9868
18	8464.617	479.9298	51.6311	318.4401
19	8509.189	515.4018	85.4524	249.1465
20	8494.376	316.2368	133.9742	399.7923

Constant weights w_a, w_b and w_c are set to 0.3, 0.4, 0.6, respectively to perform the crossover. Constant weight is set to 1.1. Ten members are generated when crossover operators are applied to u (6th) and v (14th) parents and are given in Table 7.37. Best two individuals are selected to constitute the next generation.

TABLE 7.37 Members after crossover operator's implementation

New Individuals	$F(P_{ij}^t, r_k)$	*Power generation in MW*		
		P_1	P_2	P_3
1	8541.914	299.8841	181.0475	349.0125
2	8549.380	321.1193	199.9164	309.4156
3	13120.890	283.9576	190.4819	196.2849
4	14461.520	337.0457	246.0034	371.8711
5	8583.397	310.5017	186.7081	367.0002
6	8481.383	294.0113	301.5971	331.0249
7	57066.270	242.7935	185.7647	255.9142
8	42783.560	356.2226	195.1992	328.0491
9	8561.348	297.2296	313.0372	236.5355
10	8531.451	323.7737	384.9878	357.1284

Wavelet mutation is performed by taking p_m as 0.25. ξ, the shape parameter of the monotonic increasing function of d is taken as 0.5 and g, the upper limit of the dilation parameter is assumed as 10000. After applying the wavelet mutation operator, best 20 individuals are selected as new generation after arranging the fitness in descending orders. The best result obtained after 50 generations is given in Table 7.38.

TABLE 7.38 Best result after 50 generations

$F(P_{ij}^t, r_k)$ ₹/h	*Power generation in MW* P_1	P_2	P_3
8233.225	398.975	198.1741	249.5032

The conclusion may be drawn from the above study that Evolutionary Algorithms (EAs) inspired by the natural evolution of species, have been successfully applied to solve numerous optimization problems in diverse fields. However, when implementing the EAs, users not only need to determine the appropriate encoding schemes and evolutionary operators, but also need to choose the suitable parameter settings to ensure the success of the algorithm, which may lead to demanding computational costs due to the time-consuming trial-and-error parameter and operator tuning process. To overcome such inconvenience, researchers have actively investigated the adaptation of parameters and operators in EAs. In Differential Evolution (DE), there exist many trial vector generation strategies out of which a few may be suitable for solving a particular problem. Moreover, three crucial control parameters involved in DE, i.e. population size, scaling factor, and crossover rate, may significantly influence the optimization performance of the DE. Therefore, to successfully solve a specific optimization problem at hand, it is generally required to perform a time-consuming trial-and-error search for the most appropriate strategy and to tune its associated parameter values. However, such a trial-and-error searching process requires high computational costs. Moreover, as evolution proceeds, the population of differential evolution may move through different regions in the search space, within which certain strategies associated with specific parameter settings may be more effective than others. Therefore, it is desirable to adaptively determine an appropriate strategy and its associated parameter values at different stages of evolution/search process. As far as particle swarm optimization is concerned, tuning of its associated parameter values is also very much required.

REFERENCES

Books

Arora, J.S., *Introduction to Optimum Design*, 2nd ed. Elsevier Inc., 2004.

Belegundu, A.D. and T.R. Chandrupatla, *Optimization Concepts and Applications in Engineering*, Pearson Education Inc., Delhi, 2002.

Bergen, A.R. and Vijay Vittal, *Power System Analysis*, Pearson Education Inc., Delhi, 2004.

Deb, K., *Multiobjective Optimization using Evolutionary Algorithm*, John Wiley & Sons Ltd, New York, 2001.

Osyczka, A. and B.J. Davies, *Multicriterion Optimization in Engineering with FORTRAN Programs*, Ellis Horwood Ltd., 1984.

Papoulis, A., *Probability, Random Variables and Stochastic Processes*, McGraw Hill, New Delhi, 1991.

Rao, S.S., *Optimization, Theory and Applications*, 2nd ed., Wiley Eastern Limited, New Delhi, 1987.

Song, Yong-Hua, *Modern Optimization Techniques in Power Systems*, Kluwer Academic Publishers, 1999.

Papers

Abido, M.A., A novel multiobjective evolutionary algorithm for environmental/economic power dispatch, *Electric Power Systems Research*, Vol. **65**, pp. 71–81, 2003.

Abido, M.A., Multiobjective evolutionary algorithms for Electric power dispatch problem, *IEEE Transactions on Evolutionary Computation*, Vol. **10(3)**, pp. 315–329, 2006.

Abido, M.A., Optimal design of power system stabilizers using particle swarm optimization, *IEEE Transactions on Energy Conversion*, Vol. **17(3)**, pp. 406–413, 2002.

Amjady, N. and H. Nasiri-Rad, Economic dispatch using an efficient real-coded genetic algorithm, *IET Generation, Transmission, Distribution*, Vol. **3(3)**, pp. 266–278, 2009.

Bergh, F. van den and A.P. Engelbrecht, A cooperative approach to particle swarm optimization, *IEEE Transactions on Evolutionary Computation*, Vol. **8(3)**, pp. 225–239, 2004.

Brar, Y.S., J.S. Dhillon, and D.P. Kothari, Multiobjective load dispatch by fuzzy logic searching weightage pattern, *Electric Power Systems Research*, Vol. **63(2)**, pp. 149–160, 2002.

Chen, Po-Hung and Hong-Chan Chang, Large-scale economic dispatch by genetic algorithm, *IEEE Transactions on Power Systems*, Vol. **10(4)**, pp. 1919–1926, 1995.

Chiang, C.L., Improved genetic algorithm for power economic dispatch of units with valve-point effects and multiple fuels, *IEEE Transactions on Power Systems*, Vol. **20(4)**, pp. 1690–1699, 2005.

Chiou, Ji-Pyng, Variable scaling hybrid differential evolution for large-scale economic dispatch problems, *Electric Power Systems Research*, Vol. **77(3-4)**, pp. 212–218, 2007.

Coelho, L.S. and V.C. Mariani, Combining of chaotic differential evolution and quadratic programming for economic dispatch optimization with valve-point effect, *IEEE Transactions on Power Systems*, Vol. **21(2)**, pp. 989–996, 2006 (Erratum: Vol. **21(3)**, pp. 1465, 2006).

Coelho, L.S. and V.C. Mariani, Improved differential evolution algorithms for handling economic dispatch optimization with generator constraints, *Energy Conversion and Management*, Vol. **48(6)**, pp. 1831–1639, 2007.

Esmin, A.A.A., G. Lambert-Torres and A.C. Zambroni de Souza, A hybrid particle swarm optimization applied to loss power minimization, *IEEE Transactions on Power Systems*, Vol. **20(2)**, pp. 859–866, 2005.

Farag, A., S. Al-Baiyat, and T.C. Cheng, Economic load dispatch multi-objective optimization procedure using linear programming techniques, *IEEE Transactions on Power Systems*, Vol. **10(2)**, pp. 731–738, 1995.

Gaing, Zwe-Lee, Particle Swarm Optimization to Solve the Economic Dispatch Considering the Generator Constraints, *IEEE Transactions on Power Systems*, Vol. **18(3)**, pp. 1187–1195, 2003.

He, Dakuo, Fuli Wang and Zhizhong Mao, A hybrid genetic algorithm approach based on differential evolution for economic dispatch with valve-point effect, *International Journal of Electric Power and Energy Systems*, Vol. **30**, pp. 31–38, 2008

Hota, P.K., R. Chakrabarti, and P.K. Chattopadhyay, A simulated annealing-based goal attainment method for economic emission load dispatch with non-smooth fuel cost and emission level functions, *Electric Machines and Power Systems*, Vol. **28(11)**, pp. 1037–1051, 2000.

Jayabarathi, T., K. Jayaprakash, D.N. Jeyakumar, and T. Raghunathan, Evolutionary Programming Techniques for Different Kinds of Economic Dispatch Problems, *Electric Power Systems Research*, Vol. **73(2)**, pp. 169–176, August 2005.

Kennedy, J. and R. Eberhart, Particle swarm optimization, Proceedings of the *IEEE International Conference on Neural Networks*, pp. 1942–1948, 1995.

Lee, F.N., and A.M. Breipohl, Reserve Constrained Economic Dispatch with Prohibited Operating Zones, *IEEE Transactions on Power Systems*, Vol. **8(1)**, pp. 246–254, 1993.

Lee, K.Y., A. Sode-Yome, and J.H. Park, Adaptive Hopfield neural networks for economic load dispatch, *IEEE Transactions on Power Systems*, Vol. **13(2)**, pp. 519–526, 1998.

Liang, Z.X. and J.D. Glover, A zoom feature for a dynamic programming solution to economic dispatch including transmission losses, *IEEE Transactions on Power Systems*, Vol. **7(2)**, pp. 544–550, 1992.

Lin, W.M., F.S. Cheng and M.T. Tsay, An improved tabu search for economic dispatch with multiple minima, *IEEE Transactions on Power Systems*, Vol. **17(1)**, pp. 108–112, 2002.

Lin, C.E. and G.L. Viviani, Hierarchical economic dispatch for piecewise quadratic cost functions, *IEEE Transactions on Power Apparatus and Systems*, Vol. **103(6)**, pp. 1170–1175, 1984.

Mendes, R., J. Kennedy and J. Neves, The fully informed particle swarm: simpler, maybe better, *IEEE Transactions on Evolutionary Computation*, Vol. **8(3)**, pp. 204–210, 2004.

Naka, S., T. Genji, T. Yura and Y. Fukuyama, A hybrid particle swarm optimization for distribution state estimation, *IEEE Transactions on Power Systems*, Vol. **18(1)**, pp. 60–68, 2003.

Naresh, R., J. Dubey, and J. Sharma, Two-phase neural networks based modeling framework of constrained economic load dispatch, *IEE Proceedings–Generation, Transmission and Distribution*, Vol. **151(3)**, pp. 373–378, 2004.

Park, J.B., K.S. Lee, J.R. Shin, and K.Y. Lee, A particle swarm optimization for economic dispatch with nonsmooth cost functions, *IEEE Transactions on Power Systems*, Vol. **20(1)**, pp. 34–42, 2005.

Pereira-Neto, A., C. Unsihuay, and O.R. Saavedra, Efficient evolutionary strategy optimization procedure to solve the nonconvex economic dispatch problem with generator constraints, *IEE Proceedings–Generation, Transmission and Distribution*, Vol. **153(5)**, pp. 653–660, 2005.

Ratnaweera, A., S.K. Halgamuge and H.C. Watson, Self-organizing hierarchical particle swarm optimizer with time-varying acceleration coefficients, *IEEE Transactions on Evolutionary Computation*, Vol. **8(3)**, pp. 240–255, 2004.

Roa-Sepulveda, C.A., M. Herrera, B. Pavez-Lazo, U.G. Knight and A.H. Coonick, Economic Dispatch Using Fuzzy Decision Trees, *Electric Power Systems Research*, Vol. **66(2)**, pp. 115–122, 2003.

Salomon, Ralf, Evolutionary algorithms and gradient search: Similarities and differences, *IEEE Transactions on Evolutionary Computation*, Vol. **2(2)**, pp. 45–55, 1998

Selvakumar, A.I. and K. Thanushkodi, A new particle swarm optimization solution to nonconvex economic dispatch problems, *IEEE Transactions on Power Systems*, Vol. **22(1)**, pp. 42–51, 2007.

Selvakumar, A.I. and K. Thanushkodi, Anti-predatory particle swarm optimization: Solution to nonconvex economic dispatch problems, *Electric Power Systems Research*, Vol. **78(1)**, pp. 2–10, 2008.

Sinha, N., R. Chakrabarti, and P.K. Chattopadhyay, Evolutionary Programming Techniques for Economic Load Dispatch, *IEEE Transactions on Evolutionary Computation*, Vol. **7(1)**, pp. 83–94, 2003.

Song, Y.H. and C.S.V. Chou, Advanced engineered conditioning genetic approach to power economic dispatch, *IEE Proceedings–Generation, Transmission and Distribution*, Vol. **144(3)**, pp. 285–292, 1997.

Storn, R., Differntial evolution—a simple and efficient heuristic for global optimization over continuous space, *Journal of Global Optimization*, Vol. **11(4)**, pp. 341–359, 1997

Sum-Im, T., G.A. Taylor, M.R. Irving and Y.H. Song, Differential evolution algorithm for static multistage transmission expansion planning, *IET Generation, Transmission, Distribution,* Vol. **3(4)**, pp. 365–384, 2009

Tapia, C.G. and B.A. Murtagh, Interactive fuzzy programming with preference criterion in multi-objective decision making, *Computers and Operations Research*, Vol. **18(3)**, pp. 307–316, 1991.

Varadarajan, M. and K.S. Swarup, Solving multi-objective optimal power flow using differential evolution, *IET Generation, Transmission, Distribution*, Vol. **2(5)**, pp. 720–730, 2008.

Victorie, T.A.A. and A.E. Jeyakumar, Hybrid PSO-SQP for economic dispatch with valve-point effect, *Electric Power Systems Research*, Vol. **71(1)**, pp. 51–59, 2004.

Walter, D.C., and G.B. Sheble, Genetic algorithm solution of economic dispatch with valve point loading, *IEEE Transactions on Power Systems*, Vol. **8(3)**, pp. 1325–1332, 1993.

Wong, K.P. and Y.W. Wong, Genetic and genetic/simulated-annealing approaches to economic dispatch, *IEE Proceedings–Generation, Transmission and Distribution,* Vol. **141(5)**, pp. 507–513, 1994.

Wong, K.P. and Y.W. Wong, Thermal Generator Scheduling Using Hybrid Genetic/Simulated Annealing Approach, *IEE Proceedings–Generation, Transmission and Distribution*, Vol. **142(4)**, pp. 372–380, 1995.

Yang, H.T. and P.C. Yang, C.L. Huang, Evolutionary programming based economic dispatch for units with non-smooth fuel cost functions, *IEEE Transactions on Power Systems*, Vol. **11(1)**, pp. 112–118, 1996.

Yao, X., Yong Liu and Guangming Liu, Evolutionary programming mode faster, *IEEE Transactions on Evolutionary Computation*, Vol. **3(2)**, pp. 82–102, 1999.

Zhao, B., C.X. Guo and Y.J. Cao, A multiagent-based particle swarm optimization approach for optimal reactive power dispatch, *IEEE Transactions on Power Systems*, Vol. **20(2)**, pp. 1070–1078, 2005.

CHAPTER 8

Multiobjective Generation Scheduling: Weight Pattern Search

8.1 INTRODUCTION

Economic load dispatch is a generic term that describes a class of problems in which a specific objective function is optimized while satisfying constraints dictated by operational and physical particulars of the electric network. In addition, the increasing public awareness of the environmental protection and the passage of the clean Air Act Amendments of 1990 have forced the utilities to modify their design or operational strategies to reduce pollution and atmospheric emission of the thermal plants. Apart from heat, power utilities using fossil fuels as primary energy source, produce particulates and gaseous pollutants. The particulates and the gaseous pollutants such as CO, CO_2, SO_x and NO_x cause detrimental effect on human beings.

Society demands adequate and secure electricity not only at the cheapest possible price, but also at minimum levels of pollution. This chapter solves multiobjective thermal power dispatch problem having four objectives, viz. the economic index and impact on environment due to NO_x, SO_2 and CO_2 gaseous pollutants emission taken as individual objectives. Initially the multi-objective optimization problem is converted into single objective optimization problem using weighting method to generate the non-inferior solutions, which are to be presented before the decision-maker to select the best one. Fuzzy decision-making theories attempt to deal with the vagueness or fuzziness inherent in subjective or imprecise determinations of preferences, constraints and goals. So, fuzzy methodology has been applied for solving a decision-making problem involving a multiplicity of objectives and selection criteria for suitable so called 'best' compromised solution. In the weighting method, the weighting pattern is generally simulated by giving suitable variation to generate the non-inferior solution surface, which is a time-consuming process. Moreover, the simulation procedure may skip the weight combination corresponding to actual 'best' solution that is to be selected by decision-maker, thus leads decision-maker to an alternative 'best' solution. To overcome the above said problem, the weight combinations are searched in the non-inferior domain in place of simulating them. Three methods evolutionary optimization method and simplex search method have been discussed in this chapter to search the optimal weightage assigned to the participating objectives.

The main purpose of the optimal power dispatch problem has so far been confined to minimize the total generation cost of a power system. However, in order to meet environment regulations enforced in recent years, emission control has become one of the important operational objectives. System security is another essential factor in power system operation and in system planning. To be specific, it is very important to maintain good voltage profiles and to limit line flows within the prescribed upper bounds. In security analysis, a series of anticipated contingencies are assumed to predict possible overloadings or excessive voltage deviations. Then, a security index as a function of overloads is minimized by some preventive control actions.

8.2 ECONOMIC EMISSION LOAD DISPATCH

Real-world problems involve simultaneous optimization of several objective functions. Objective functions are often competing and of conflicting nature. So, multiobjective optimization problem with such conflicting objective functions gives rise to a set of optimal solutions, instead of one optimal solution. The reason for the optimality of many solutions is that no one solution can be considered better than any other solution with respect to all other objective functions. These optimal solutions are known as non-inferior solutions. To resolve the conflicts between objective functions, decision-maker plays a vital role by providing a tool to select one suitable solution.

8.2.1 Multiobjective Optimization Problem Formulation

A multiobjective thermal power dispatch problem (MOTPDP) is formulated with four important non-commensurable objective functions in an electrical thermal power system. The objectives are economy and environmental impact because of NO_x, SO_2 and CO_2 gaseous pollutants. Thus multiobjective optimization problem minimizes the four competing objective functions while satisfying equality and inequality power system constraints.

$$\text{Minimize operating cost: } F_1 = \sum_{i=1}^{NG} (a_i P_{Gi}^2 + b_i P_{Gi} + c_i) \quad ₹/\text{h} \tag{8.1}$$

$$\text{Minimize NO}_x \text{ Emission: } F_2 = \sum_{i=1}^{NG} (d_{1i} P_{Gi}^2 + e_{1i} P_{Gi} + f_{1i}) \text{ kg/h} \tag{8.2}$$

$$\text{Minimize SO}_2 \text{ Emission: } F_3 = \sum_{i=1}^{NG} (d_{2i} P_{Gi}^2 + e_{2i} P_{Gi} + f_{2i}) \text{ kg/h} \tag{8.3}$$

$$\text{Minimize CO}_2 \text{ Emission: } F_4 = \sum_{i=1}^{NG} (d_{3i} P_{Gi}^2 + e_{3i} P_{Gi} + f_{3i}) \text{ ton/h} \tag{8.4}$$

Constraints: (i) to ensure a real and reactive power balance

$$\sum_{i=1}^{NB} P_{Di} - \sum_{i=1}^{NG} P_{Gi} + P_L = 0 \tag{8.5}$$

$$\sum_{i=1}^{NB} Q_{Di} - \sum_{i=1}^{NG} Q_{Gi} + Q_L = 0 \tag{8.6}$$

(ii) Inequality constraints imposed on generator output are:

$$P_{Gi}^{\min} \le P_{Gi} \le P_{Gi}^{\max} \quad (i = 1, 2, ..., \text{NG}) \tag{8.7}$$

$$Q_{Gi}^{\min} \le Q_{Gi} \le Q_{Gi}^{\max} \quad (i = 1, 2, ..., \text{NG}) \tag{8.8}$$

where

a_i, b_i and c_i are cost coefficients.
d_{1i}, e_{1i} and f_{1i} are NO_x emission coefficients.
d_{2i}, e_{2i} and f_{2i} are SO_2 emission coefficients.
d_{3i}, e_{3i} and f_{3i} are CO_2 emission coefficients.
NG is the number of generators.
NB is the total number of buses.
P_{Di} and Q_{Di} are real power and reactive power demands, respectively at the ith bus.
P_{Gi} and Q_{Gi} are real power and reactive power generation, respectively at the ith bus.
P_L and Q_L are real and reactive transmission losses, respectively.
$P_{Gi}^{\min}$ and $P_{Gi}^{\max}$ are lower and upper limits of active power generation by ith generator, respectively.
$Q_{Gi}^{\min}$ and $Q_{Gi}^{\max}$ are lower and upper limits of reactive power generation by ith generator, respectively.

The real power loss, P_L and reactive power loss, Q_L on transmission lines of power system is given by:

$$P_L = \sum_{i=1}^{\text{NB}}\sum_{j=1}^{\text{NB}} [\alpha_{ij}(P_iP_j + Q_iQ_j) + \beta_{ij}(Q_iP_j - P_iQ_j)] \tag{8.9}$$

$$Q_L = \sum_{i=1}^{\text{NB}}\sum_{j=1}^{\text{NB}} [\gamma_{ij}(P_iP_j + Q_iQ_j) + \eta_{ij}(Q_iP_j - P_iQ_j)] \tag{8.10}$$

with

$$P_i + jQ_i = (P_{Gi} - P_{Di}) + j(Q_{Gi} - Q_{Di})$$

$$\alpha_{ij} = \frac{R_{ij}}{V_iV_j}\cos(\delta_i - \delta_j) \text{ and } \beta_{ij} = \frac{R_{ij}}{V_iV_j}\sin(\delta_i - \delta_j)$$

$$\gamma_{ij} = \frac{X_{ij}}{V_iV_j}\cos(\delta_i - \delta_j) \text{ and } \eta_{ij} = \frac{X_{ij}}{V_iV_j}\sin(\delta_i - \delta_j)$$

δ_i and δ_j are voltage angles at ith and jth buses, respectively.
V_i and V_j are voltage magnitudes at ith and jth buses, respectively.
R_{ij} and X_{ij} are the real and reactive component of impedance bus matrix.

To generate the non-inferior solution, the constrained multiobjective thermal power dispatch problem is converted into a constrained scalar optimization problem using weighting method and is stated below:

Minimize $$\sum_{k=1}^{L} w_k F_k \tag{8.11a}$$

Subject to: $$\sum_{i=1}^{NB} P_{Di} - \sum_{i=1}^{NG} P_{Gi} + P_L = 0 \tag{8.11b}$$

$$\sum_{i=1}^{NB} Q_{Di} - \sum_{i=1}^{NG} Q_{Gi} + Q_L = 0 \tag{8.11c}$$

$$P_{Gi}^{\min} \le P_{Gi} \le P_{Gi}^{\max} \quad (i = 1, 2, ..., NG) \tag{8.11d}$$

$$Q_{Gi}^{\min} \le Q_{Gi} \le Q_{Gi}^{\max} \quad (i = 1, 2, ..., NG) \tag{8.11e}$$

$$\sum_{k=1}^{L} w_k = 1.0, \; w_k \ge 0.0 \tag{8.11f}$$

where

w_k are the levels of the weighting coefficients assigned to objective function.
L is the number of objectives.

This approach yields meaningful results to the decision-maker when the above stated optimization problem is solved many times for different value of w_k (k = 1, 2, …, L). The values of the weighting coefficients vary from 0 to 1 while satisfying Eq. (8.11f). To solve the constrained scalar optimization problem, the Lagrangian function is defined as:

$$L_f(P_{Gi}, Q_{Gi}, \lambda_P, \lambda_Q) = \sum_{j=1}^{L} w_j F_j + \lambda_P \left(\sum_{i=1}^{NB} P_{Di} + P_L - \sum_{i=1}^{NG} P_{Gi} \right) + \lambda_Q \left(\sum_{i=1}^{NB} Q_{Di} + Q_L - \sum_{i=1}^{NG} Q_{Gi} \right) \tag{8.12}$$

where

λ_P and λ_Q are the Lagrangian multipliers for active and reactive powers, respectively.

The necessary conditions to minimize the Lagrangian function are

$$\frac{\partial L}{\partial P_{Gi}} = \sum_{j=1}^{L} w_j \frac{\partial F_j}{\partial P_{Gi}} + \lambda_P \left(\frac{\partial P_L}{\partial P_{Gi}} - 1 \right) + \lambda_Q \left(\frac{\partial Q_L}{\partial P_{Gi}} \right) = 0 \quad (i = 1, 2, ..., NG) \tag{8.13a}$$

$$\frac{\partial L}{\partial Q_{Gi}} = \lambda_P \left(\frac{\partial P_L}{\partial Q_{Gi}} \right) + \lambda_Q \left(\frac{\partial Q_L}{\partial Q_{Gi}} - 1 \right) = 0 \quad (i = 1, 2, ..., NG) \tag{8.13b}$$

$$\frac{\partial L}{\partial \lambda_P} = \sum_{i=1}^{NB} P_{Di} + P_L - \sum_{i=1}^{NG} P_{Gi} = 0 \tag{8.13c}$$

$$\frac{\partial L}{\partial \lambda_Q} = \sum_{i=1}^{NB} Q_{Di} + Q_L - \sum_{i=1}^{NG} Q_{Gi} = 0 \qquad (8.13d)$$

The Newton-Raphson method has been applied to obtain the solution for the weight combinations generated during search moves to search the weights through conventional and non-conventional search techniques.

8.2.2 Operating Limits

Inequality constraints imposed to Problem (8.11) are taken into consideration by fixing the generators to their respective upper or lower limits whenever these are violated. To fix up the limits of active and reactive power generations by the committed generating units, the following strategy has been adopted.

Let

$$\Delta h_p = \sum_{k=1}^{N_{p1}} h_{pk}^{\max} - \sum_{k=1}^{N_{p2}} h_{pk}^{\min} \qquad (8.14a)$$

Let

$$\Delta h_q = \sum_{k=1}^{N_{q1}} h_{qk}^{\max} - \sum_{k=1}^{N_{p2}} h_{qk}^{\min} \qquad (8.14b)$$

where

$h_{pi}^{\max} = P_{Gi} - P_{Gi}^{\max}$ $(i = 1, 2, \ldots, N_{p1})$ upper bound violations for active power generation.
$h_{pi}^{\min} = P_{Gi}^{\min} - P_{Gi}$ $(i = 1, 2, \ldots, N_{p2})$ lower bound violations for active power generation.
$h_{qi}^{\max} = Q_{Gi} - Q_{Gi}^{\max}$ $(i = 1, 2, \ldots, N_{q1})$ upper bound violations for reactive power generation.
$h_{qi}^{\min} = Q_{Gi}^{\min} - Q_{Gi}$ $(i = 1, 2, \ldots, N_{q2})$ lower bound violations for reactive power generation.

The power generations from the units which are violating limits are fixed as below:

1. IF $\Delta h_p > 0$ fix all N_{p1} upper bound violation to the upper limits, i.e. $P_{Gi}^{\max}$
2. IF $\Delta h_p < 0$ fix all N_{p2} lower bound violation to the lower limits, i.e. $P_{Gi}^{\min}$
3. On other side, IF $\Delta h_p = 0$ fix N_{p1} upper and N_{p2} lower bound violations to their respective upper, $P_{Gi}^{\max}$ and lower, $P_{Gi}^{\min}$ limits.
4. IF $\Delta h_q > 0$ fix all N_{q1} upper bound violation to the upper limits, i.e. $Q_{Gi}^{\max}$
5. IF $\Delta h_q < 0$ fix all N_{q2} lower bound violation to the lower limits, i.e. $Q_{Gi}^{\min}$
6. On other side IF $\Delta h_q = 0$ fix both N_{q1} upper and N_{q2} lower bound violations to their respective upper, $Q_{Gi}^{\max}$ and lower, $Q_{Gi}^{\min}$ limits.

Determine the new demand which is original demand minus the sum of fixed generation levels.

$$P_D^{\text{New}} = P_D - \sum_{i=1}^{N_{p1} - N_{p2}} P_{Gi}$$

$$Q_D^{\text{New}} = Q_D - \sum_{i=1}^{N_{q1} - N_{q2}} Q_{Gi}$$

where

P_D^{New} and Q_D^{New} the real and reactive power demand, respectively which are required to meet after scheduling generation.

The new demand is allocated to the rest of committed generators by using Newton Raphson method, by eliminating row and column of the Hessian and Jacobian matrices evaluated corresponding to fixed generator.

8.2.3 Membership Functions of Objectives

Considering the imprecise nature of the decision-maker's judgement, it is natural to assume that the decision-maker may have fuzzy or imprecise goals for each objective function. The fuzzy sets are defined by equations called membership functions. These functions represent the degree of membership in certain fuzzy set using values from 0 to 1. The membership value 0 indicates incompatibility with the set, while 1 means full compatibility. By taking account of the minimum and maximum values of each objective function together with the rate of increase of membership satisfaction, the decision-maker must determine the membership function $\mu(F_i)$ $(i = 1, 2, \ldots, L)$ in a subjective manner by considering the rate of increase of membership satisfaction from two types of membership functions which are given below:

It is assumed that $\mu(F_i)$ $(i = 1, 2, \ldots, L)$ is a strictly monotonic linear decreasing and is defined as:

$$\mu(F_i) = \begin{cases} 1 & ; F_i \le F_i^{min} \\ \dfrac{F_i^{max} - F_i}{F_i^{max} - F_i^{min}} & ; F_i^{min} < F_i < F_i^{max} \\ 0 & ; F_i \ge F_i^{max} \end{cases} \tag{8.15}$$

where

F_i^{min} and F_i^{max} are respectively the minimum and maximum values of ith objective function, in which the solution is expected.

The membership function can also be assumed that $\mu(F_i)$ $(i = 1, 2, \ldots, L)$ is a strictly monotonic exponentially decreasing and is defined as:

$$\mu(F_i) = \begin{cases} 1 & ; F_i \le F_i^{min} \\ U_i\left[1.0 - \exp\left\{-F_{Gi}\left(\dfrac{F_i - F_i^{max}}{F_i^{min} - F_i^{max}}\right)\right\}\right] & ; F_i^{min} < F_i < F_i^{max} \\ 0 & ; F_i \ge F_i^{max} \end{cases} \tag{8.16}$$

where

F_{Gi} is a shape parameter.

U_i is the degree of membership function.

The value of the membership function indicates how much (in scale from 0 to 1) the solution is satisfying the F_i objective. The maximum satisfaction of membership function for any weight combination is obtained by taking intersection of the membership functions of participating objectives and is given below.

$$\mu^{\min} = \text{Min}\ [\mu(F_j)\quad (j = 1, 2, ..., L)] \tag{8.17}$$

8.2.4 Weightage Pattern Search: Evolutionary Search Method

Normally, to solve the multiobjective optimization problem by weighting method, vary the assigned weights to objectives in a systematic manner, in order to generate the non-inferior solutions for optimization problem of Eq. (8.13). Evolutionary search optimization method is applied here to search the assigned weightage to the participating objectives to generate non-inferior solutions that are further provided to decision-maker to select the suitable one. Evolutionary search optimization technique searches for the optimal weight pattern in the non-inferior domain. In this method, a hypercube of weight combination is formed around an initial search point. $(2^{L-1} + 1)$ weight combinations are simulated at 2^{L-1} corner points of $(L-1)$-dimensional hypercube centred on initial point w_i^c $(i = 2, 3, \ldots, L)$. To decide the initial search point, give 50% of the total weightage to main objective and equally divide rest of 50% weightage among all other objectives. $(2^{L-1} + 1)$ non-inferior solutions are generated and membership functions are obtained from Eq. (8.15) when it is assumed that membership functions are monotonic linear decreasing function or from Eq. (8.16) when it is assumed that membership functions are monotonic exponentially decreasing function. To continue with the iterative process, form another hypercube around the relatively better point as compared to the previous one. Repeat this process until the solution criteria for the 'best' compromised solution is met. The 'best' compromised solution is one, which provides the maximum satisfaction level from the participating goals or objectives and is stated below:

$$\text{Max}\{\text{Min}\ [\mu(F_j)^k\ (j = 1, 2, ..., L)]\quad (k = 1, 2, ..., 2^{L-1} + 1)\} \tag{8.18}$$

Generate the weights using the following equations.

$$w_i^j = w_i^c + \gamma_i^j \quad (i = 2, 3, ..., L;\ j = 1, 2, ..., 2^{L-1}) \tag{8.19}$$

$$w_1^j = 1 - \sum_{i=2}^{L} w_i^j \quad (j = 1, 2, ..., 2^{L-1}) \tag{8.20}$$

where

γ is the distance of the corners of the hypercube from the point around which hypercube is generated.

The γ matrix is generated from possible combinations of binary bits. '0' bit is replaced by $-\gamma$ and '1' bit is replaced by $+\gamma$.

Algorithm 8.1: Evolutionary search method to search weight pattern

Evolutionary search method searches the normalized weights, w_i $(i = 1, 2, \ldots, L)$ assigned to participating objectives and are used to generate the non-inferior solutions. Stepwise procedure to implement the evolutionary search method to search the normalized weights is outlined below:

1. Input the data.
2. Find the minimum and maximum values of objectives $F_i^{\min}$ and $F_i^{\max}$ $(i = 1, 2,\ldots, L)$.
3. Set the initial centre w_i^c of the hypercube; $i = 2, 3,\ldots, L$.
4. Set initial maximum value of membership function $\mu^P = 0$.
5. Initialize iteration counter, $r = 0$.
6. Increment the iteration counter, $r = r + 1$.
7. Generate '2^{L-1}' number of weight combinations at the edges of hypercube using Eqs. (8.19–8.20).
8. Initialize iteration counter, $k = 0$.
9. Increment the iteration counter, $k = k + 1$.
10. Generate non-inferior solution for kth weight combination by solving Eqs. (8.14a)–(8.14d) using Newton Raphson method.
11. Find membership function of objectives, $\mu(F_i)^k$ $(i = 1, 2, \ldots, L)$ from Eq. (8.15) or Eq. (8.16).
12. Find the intersection of membership functions,

    $$\mu_k^{\min} = \min \{\mu(\text{Fi})^k \ (i = 1, 2, \ldots, L)\}$$
13. IF $(k \le 2^{L-1} + 1)$, THEN GOTO Step 9.
14. Find maximum satisfied membership function,

    $$\mu^0 = \max \{\mu_k^{\min} \ (k = 1, 2, \ldots, (2^{L-1} + 1)\}$$
15. Choose weight combination w_i^{c0} having maximum satisfied membership function μ^0 as a centre of hypercube.
16. IF $((|\mu^0 - \mu^P| \le e)$, OR $(r > IT^{\max}))$ THEN GOTO Step 18.
17. IF $(\mu^0 > \mu^P)$ THEN $\mu^P = \mu^0$ and $w_i^c = w_i^{c0}$ $(i = 2, 3, \ldots, L)$
    ELSE make $\gamma = \gamma/2$ and GOTO Step 6.
18. STOP.

8.2.5 Weightage Pattern Search: Genetic Algorithm

Genetic algorithm is employed to search for preferred weightage pattern based on fuzzy decision-making methodology to achieve the solution of a multiobjective thermal power dispatch problem, which simultaneously minimizes conflicting objectives while allocating the electricity demand among committed generating units subject to physical and technological constraints.

Encoding of parameters

A binary coding scheme is elaborated here. A finite length of binary numbers, which is concatenated by sub-strings, is first used to represent the scalar weights assigned to objectives. Binary coding has the advantage to be the well-developed mathematical basis and to reduce the optimization space. To obtain the higher precision of the objective weights, define longer binary strings. A decoding method is given by the following equation:

$$y_i^j = \sum_{k=0}^{l_i - 1} \text{bit}_k 2^k \qquad (j = 1, 2, \ldots, P_S; \;\; i = 1, 2, \ldots, L) \tag{8.21}$$

where

l_i is length of sub-string.

bit_k is the binary coefficients.

P_S is the population size.

In case, the parameter belongs to $\alpha^{\min}$ and $\alpha^{\max}$, the decode values of parameter are computed from the mathematical relation given below:

$$\alpha_i^j = \alpha^{\min} + \frac{y_i^j(\alpha^{\max} - \alpha^{\min})}{2^{x_i-1}} \quad (j = 1, 2, ..., P_s;\ \ i = 1, 2, ..., L) \tag{8.22}$$

where

$\alpha^{\max}$ is the maximum value of α, i.e. 100

$\alpha^{\min}$ is the minimum value of α, i.e. 0.

Normalize, the scalar weights using the following mathematical relation:

$$w_i = \frac{\alpha_i}{\sum_{j=1}^{L} \alpha_j}; \quad w_i \geq 0\ (i = 1, 2, ..., L) \tag{8.23}$$

Evaluation of fitness

The fuzzy sets are defined by equations called membership functions as represented either by monotonically linear decreasing function given by Eq. (8.15) or by monotonically exponentially decreasing function represented by Eq. (8.16). The value of the membership function indicates how much a solution is satisfying the objective in the scale from 0 to 1. Equation (8.17) gives the overall satisfaction of participating objectives or goals. It implies that all objectives must have achieved equal or higher satisfaction than $\mu^{\min}$ in terms of membership function of non-inferior solution generated from each string of weight combination in a population. Therefore, the maximum value of $\mu^{\min}$ gives the 'best' solution. The aggregate membership function value selected from Eq. (8.17) is considered as the fitness value of the string of scalar weights.

Algorithm 8.2: Genetic algorithm to search the weight pattern

A stepwise procedure to implement the GA to search for the scalar weights is outlined below:

1. Input the data such as the number of objectives L, population size P_S, chromosome length L_s, length of sub-string l_i, maximum number of generations $R^{\max}$, crossover probability p_c, mutation probability p_m, etc.
2. Find the minimum and maximum values of the objectives, $F_i^{\min}$ and $F_i^{\max}$ $(i = 1, 2, \ldots, L)$.
3. Generate the initial population of weight set randomly by flipping the coin. The bit is set according to the coin flip as:

$$\text{bit}_{ij} = \begin{cases} 1 & \text{if } p = 1 \text{ or random( )} \leq p \\ 0 & \text{otherwise} \end{cases}$$

where p is the probability and is taken as 0.5.

Decode each individual string of population using Eq. (8.22) and normalize the weight set using Eq. (8.23). Find the fitness value corresponding to each string using Eq. (8.17). Initialize generation counter, $R = 0$.

4. Select the string having maximum fitness, f^{pre}.
5. Increment the generation counter, $R = R + 1$.
6. If $(R > R^{\max})$ THEN GOTO Step 16.
7. Select the most fit strings from old population by stochastic remainder roulette-wheel selection (refer Section 7.3.1), for crossover.
8. Select the crossover site by flipping the coin with crossover probability. Perform single point crossover and mutation to get the new population.
9. Decode the string using Eq. (8.22) and normalize the weight set w_i $(i = 1, 2, ..., L)$ for new generated strings using Eq. (8.23).
10. Apply the Newton Raphson method to solve Eqs. (8.14a) to (8.14d) for each weight set of population.
11. Find the fitness value of each weight set of population using Eq. (8.17).
12. Select the string having maximum fitness in the new population, $f^{\max}$.
13. If $(f^{\max} - f^{\text{pre}}) \leq \varepsilon$ THEN GOTO Step 16.
14. Update the value of population fitness.

$$f^{\text{pre}} = \begin{cases} f^{\text{pre}} & ; f^{\text{pre}} > f^{\max} \\ f^{\max} & ; f^{\max} > f^{\text{pre}} \end{cases}$$

15. GOTO Step 5 and repeat.
16. STOP.

8.2.6 Test Systems and Results

Multiobjective thermal power dispatch problem is solved using weighting method where weights are searched by conventional as well as non-conventional search methods. The results are obtained on a sample system; 11-bus, 17-lines comprising five-generator system. Tables 5.2 to 5.5 give operating cost and gaseous pollutants emission chatrateristics. All the committed generating units allowed running between minimum and maximum limits of their power generation. Table 8.1 gives the minimum and maximum limits of active and reactive power generation by generators. Single line diagram of the power system is given in Figure 8.1 This system consists of 11-buses, 17 lines, and 5 generators. Table 8.2 shows the load on the power system. Table 8.3 depicts the line data of the power system.

TABLE 8.1 Generation limits

Generator	*Real power*		*Reactive power*	
	$P_{Gi}^{\min}$(p.u.)	$P_{Gi}^{\max}$(p.u.)	$Q_{Gi}^{\min}$(p.u.)	$Q_{Gi}^{\max}$(p.u.)
1	0.1	2.0	–2.0	2.0
2	0.5	2.0	–2.0	2.0
3	0.5	2.0	–2.0	2.0
4	0.5	2.0	–2.0	2.0
5	0.1	2.0	–2.0	2.0

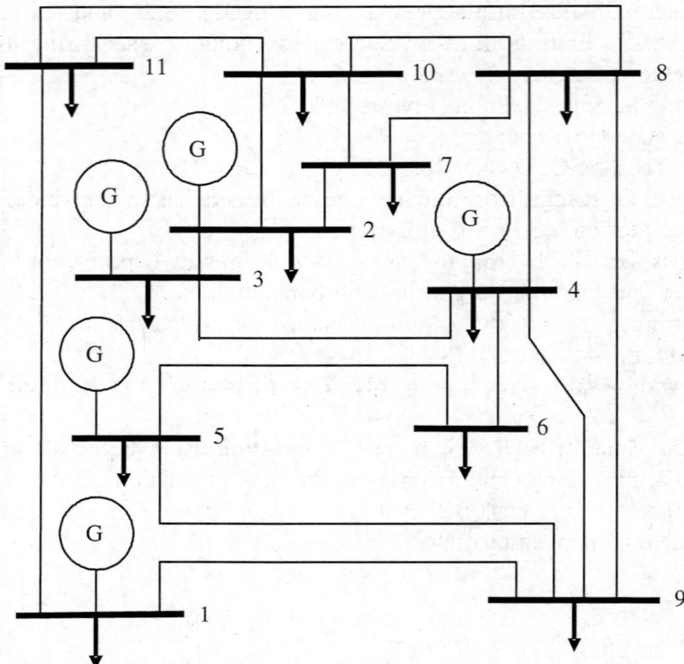

FIGURE 8.1 Single line diagram of 11-bus system.

TABLE 8.2 Load data for 11-bus system

Bus No.	*Real load* (p.u.)	*Reactive load* (p.u.)
1	0.50	0.10
2	0.50	0.10
3	0.50	0.10
4	0.50	0.10
5	0.50	0.10
6	0.10	0.02
7	0.40	0.10
8	0.90	0.45
9	0.70	0.35
10	0.25	0.05
11	0.25	0.05

TABLE 8.3 Line data for 11-bus system

Line No.	*Bus*		*Resistance* (p.u.)	*Reactance* (p.u.)	*Line charging* (p.u.)
	From	*To*			
1	1	9	0.15	0.50	0.030
2	1	11	0.05	0.16	0.100
3	2	3	0.15	0.50	0.030
4	2	7	0.10	0.28	0.020
5	2	10	0.05	0.16	0.010
6	3	4	0.08	0.24	0.015
7	4	6	0.10	0.28	0.020
8	4	8	0.10	0.28	0.020
9	4	9	0.15	0.50	0.030
10	5	6	0.12	0.36	0.025
11	5	9	0.05	0.16	0.010
12	7	8	0.05	0.16	0.010
13	7	10	0.08	0.24	0.015
14	8	9	0.12	0.36	0.025
15	8	10	0.08	0.24	0.015
16	8	11	0.10	0.28	0.020
17	10	11	0.12	0.36	0.025

Minimum value of the objectives, F_i $(i = 1, 2, \ldots, L)$ are obtained by giving full weightage to one of the objectives and neglecting the others. When the given normalized weight value assigned to objective is 1.0, it means that full weightage is given to that objective and when the normalized weightage assigned to objective is zero, the objective is neglected. Owing to the conflicting nature of the objectives, one objective will have maximum value when another objective has minimum value and vice versa. The values of objectives when one objective is given full weightage and rest objectives are ignored are obtained and are given in Table 8.4.

TABLE 8.4 Minimum and maximum values of objectives

Weight	F_1	F_2	F_3	F_4
$w_1 = 1.0$	**5704.104**	1014.770	3676.790	10.919
$w_2 = 1.0$	6066.595	**700.027**	7320.859	20.557
$w_3 = 1.0$	5726.510	1127.250	**3436.576**	11.086
$w_4 = 1.0$	5798.329	1016.747	4051.038	**8.566**

Evolutionary search optimization method

Evolutionary optimization method has been implemented to search the normalized weight pattern to solve the multiobjective thermal power dispatch problem and is examined on a sample system consisting of 11-bus, 17-lines, and 5-generators. The problem is analyzed under three different cases, and results are obtained by applying Algorithm 8.1.

Case 1: Two objectives, operating cost and NO_x emission are minimized of multiobjective thermal power dispatch problem, which have weightages, w_1 and w_2, respectively. Obtain the minimum and maximum values of these objectives from Table 8.4 and are given below.

F_1^{min} = 5704.104 ₹/h, F_1^{max} = 6066.595 ₹/h

F_2^{min} = 700.0267 kg/h, F_2^{max} = 1014.770 kg/h

Construct hypercube around weight, w_2 only. The initial value of hypercube's centre point, select w_2^c as 0.5. The one binary bit can be represented in 2^1 (two) possible different combinations. These two binary bit combinations are the corners of hypercube away from the centre of hypercube. The distance, γ of hypercube corners from its centre point w_2^c is generated from binary digits given in Table 8.5. Figure 8.2 shows the generation of weights at the hypercube corners, graphically. The values of operating cost and NO_x emission obtained during iterations of search are tabulated in Table 8.6. The obtained membership functions, $\mu(F_i)^k$ corresponding to objectives is also given in Table 8.6 alongwith minimum membership function value from two objectives, μ_k^{min}. The injected power and voltage profile at the buses corresponding to 'best' weight combinations are also depicted in Table 8.7. 'Best' weights: w_1 = 0.46, w_2 = 0.54 are obtained in the 5th iteration.

TABLE 8.5 Generation of weights at hypercube corners for two objectives

Hypercube corners	*Possible combinations of one Binary bit b_0*	*Distance of hypercube corners from its centre point w_2^c*	*Possible generated weights at hypercube corners*
1	0	$-\gamma$	$w_2^c - \gamma$
2	1	$+\gamma$	$w_2^c + \gamma$

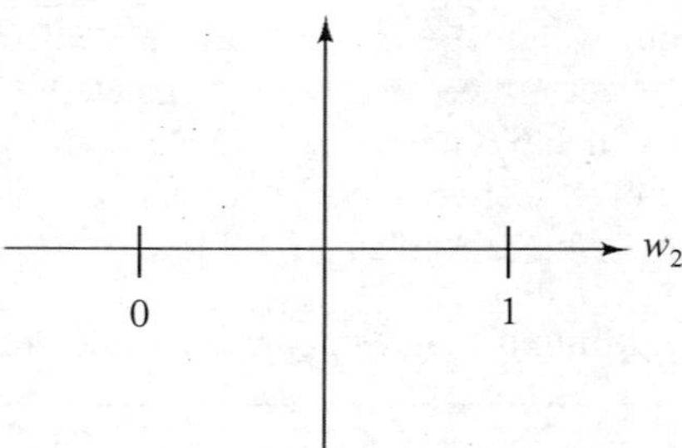

FIGURE 8.2 Generation of weights for two objectives.

TABLE 8.6 Weight search with evolutionary method for two objectives

k	*Weights*		*Objectives*		*Membership function*		μ_k^{min}
	w_1^k	w_2^k	F_1^k(₹/h)	F_2^k(kg/h)	$\mu(F_1)^k$	$\mu(F_2)^k$	
1	0.49	0.51	5780.925	781.8776	0.78808	0.73994	0.73994
2	0.48	0.52	5784.007	778.9422	0.77957	0.74927	0.74927
3	0.47	0.53	5787.163	776.0557	0.77087	0.75844	0.75844
4	0.46	0.54	5790.393	773.2181	0.76196	0.76746	0.76196
5	0.46	0.54	5790.393	773.2181	0.76196	0.76746	0.76196

TABLE 8.7 'Best' power injection and voltage profile with evolutionary method for two objectives (Case I)

Bus	P_{Di} (p.u.)	Q_{Di} (p.u.)	P_i (p.u.)	Q_i (p.u.)	V_i (p.u.)	δ_i (rad)
1	0.50	0.10	0.03597	0.12637	1.07	0.00
2	0.50	0.10	0.72155	0.34094	1.088	0.13471
3	0.50	0.10	0.86804	–0.13703	1.095	0.30421
4	0.50	0.10	0.83787	0.28839	1.062	0.20468
5	0.50	0.10	0.29467	0.20874	1.046	0.09392
6	0.10	0.02	–0.10	–0.02	1.05191	0.14083
7	0.40	0.10	–0.40	–0.10	1.01315	0.03690
8	0.90	0.45	–0.90	–0.45	0.99132	0.02580
9	0.70	0.35	–0.70	–0.35	1.00130	0.03662
10	0.25	0.05	–0.25	–0.05	1.03360	0.05126
11	0.25	0.05	–0.25	–0.05	1.03358	–0.00111

Case 2: In this case, operating cost, NO_x emission and SO_2 emission are considered as objectives to be minimized simultaneously of multiobjective thermal power dispatch problem having weightages w_1, w_2 and w_3, respectively. The minimum and maximum values of these objectives obtained from Table 8.4 are given below:

$$F_1^{\min} = 5704.104 \text{ ₹/h}, \qquad F_1^{\max} = 6066.595 \text{ ₹/h}$$
$$F_2^{\min} = 700.0267 \text{ kg/h}, \qquad F_2^{\max} = 1127.25 \text{ kg/h}$$
$$F_3^{\min} = 3436.576 \text{ kg/h}, \qquad F_3^{\max} = 7320.859 \text{ kg/h}$$

Square is formed around weights w_2 and w_3. The initial centre point of square, (w_2^c, w_3^c) is (0.25, 0.25) taken in this case. If two binary bits are combined then 2^2 (four) different possible combinations can be obtained. These four binary bit combinations are the corners of hypercube away from the centre point of hypercube. The distance γ of hypercube corners from its centre point, (w_2^c and w_3^c) is generated from binary digits as given in Table 8.8. Figure 8.3 shows the pictorial representation of generation of weights at the hypercube corners.

TABLE 8.8 Generation of weights at hypercube corners for three objectives

Hypercube corners	*Possible combinations of two binary bits* b_0, b_1	*Distance of hypercube corners from centre point* (w_2^c, w_3^c)	*Possible generated weights at the hypercube corners*
1	0, 0	$-\gamma, -\gamma$	$w_2^c - \gamma, w_3^c - \gamma$
2	0, 1	$-\gamma, +\gamma$	$w_2^c - \gamma, w_3^c + \gamma$
3	1, 0	$+\gamma, -\gamma$	$w_2^c - \gamma, w_3^c - \gamma$
4	1, 1	$+\gamma, +\gamma$	$w_2^c + \gamma, w_3^c + \gamma$

Attained values of operating cost, NO_x emission and SO_2 emission, and their corresponding weights are given in Table 8.9. The membership functions, $\mu(F_i)^k$ and obtained minimum membership function values among three objectives, $\mu_k^{\min}$ are also shown in Table 8.10. The 'best' weights: $w_1 = 0.50$, $w_2 = 0.48$ and $w_3 = 0.02$ are obtained after 23 iterations. The injected power and voltage profile at the buses corresponding to 'best' weight combinations are depicted in Table 8.11.

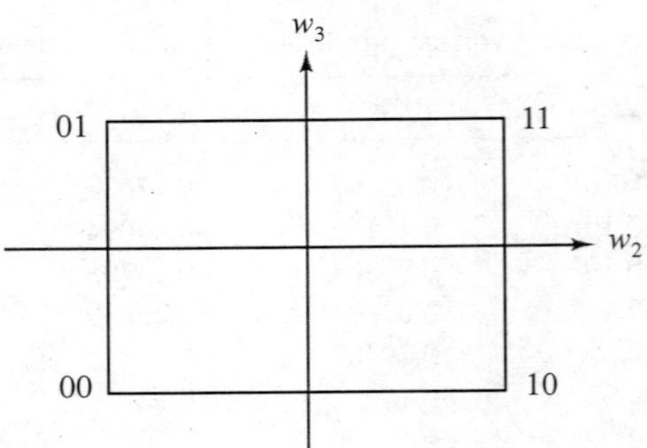

FIGURE 8.3 Generation of weights for three objectives.

TABLE 8.9 Weight search with evolutionary method for three objectives

k	*Weights*			*Objectives*		
	w_1^k	w_2^k	w_3^c	F_1^k(₹/h)	F_2^k(kg/h)	F_3^k(kg/h)
1	0.50	0.26	0.24	5721.610	1020.748	3463.672
2	0.50	0.27	0.23	5722.188	1016.058	3467.747
3	0.50	0.28	0.22	5722.784	1011.232	3472.337
4	0.50	0.29	0.21	5723.396	1006.253	3477.524
5	0.50	0.30	0.20	5724.025	1001.102	3483.407
6	0.50	0.31	0.19	5724.672	995.757	3490.108
7	0.50	0.32	0.18	5725.337	990.194	3497.774
8	0.50	0.33	0.17	5726.022	984.386	3506.589
9	0.50	0.34	0.16	5726.729	978.300	3516.777
10	0.50	0.35	0.15	5727.463	971.900	3528.622
11	0.50	0.36	0.14	5728.227	965.144	3542.482
12	0.50	0.37	0.13	5729.028	957.980	3558.815
13	0.50	0.38	0.12	5729.876	950.353	3578.212
14	0.50	0.39	0.11	5730.789	942.193	3601.451
15	0.50	0.40	0.10	5731.783	933.421	3629.560
16	0.50	0.41	0.09	5732.892	923.941	3663.935
17	0.50	0.42	0.08	5734.159	913.642	3706.469
18	0.50	0.43	0.07	5735.651	902.392	3759.939
19	0.50	0.44	0.06	5737.470	890.034	3828.155
20	0.50	0.45	0.05	5739.773	876.393	3916.904
21	0.50	0.46	0.04	5742.825	861.273	4035.024
22	0.50	0.47	0.03	5747.031	844.487	4196.591
23	0.50	0.48	0.02	5753.141	825.921	4425.211
24	0.50	0.48	0.02	5753.141	825.921	4425.211

TABLE 8.10 Weight search with evolutionary method for three objectives

k	*Weights*			*Membership function*			
	w_1^k	w_2^k	w_3^k	$\mu(F_1)^k$	$\mu(F_2)^k$	$\mu(F_3)^k$	μ_k^{min}
1	0.50	0.26	0.24	0.95171	0.24929	0.99302	0.24929
2	0.50	0.27	0.23	0.95011	0.26027	0.99199	0.26027
3	0.50	0.28	0.22	0.94847	0.27156	0.99079	0.27156
4	0.50	0.29	0.21	0.94678	0.28322	0.98946	0.28322
5	0.50	0.30	0.20	0.94505	0.29528	0.98794	0.29528
6	0.50	0.31	0.19	0.94326	0.30779	0.98622	0.30779
7	0.50	0.32	0.18	0.94143	0.32081	0.98425	0.32081
8	0.50	0.33	0.17	0.93954	0.33440	0.98198	0.33440
9	0.50	0.34	0.16	0.93759	0.34865	0.97935	0.34865
10	0.50	0.35	0.15	0.93556	0.36363	0.97630	0.36363
11	0.50	0.36	0.14	0.93345	0.37944	0.97274	0.37944
12	0.50	0.37	0.13	0.93124	0.39621	0.96853	0.39621
13	0.50	0.38	0.12	0.92890	0.41406	0.96354	0.41406
14	0.50	0.39	0.11	0.92639	0.43316	0.95755	0.43316
15	0.50	0.40	0.10	0.92364	0.45370	0.95032	0.45370
16	0.50	0.41	0.09	0.92059	0.47589	0.94147	0.47589
17	0.50	0.42	0.08	0.91709	0.49999	0.93051	0.49999
18	0.50	0.43	0.07	0.91297	0.52633	0.91675	0.52633
19	0.50	0.44	0.06	0.90796	0.55525	0.89919	0.55525
20	0.50	0.45	0.05	0.90160	0.58718	0.87634	0.58718
21	0.50	0.46	0.04	0.89318	0.62257	0.84593	0.62257
22	0.50	0.47	0.03	0.88158	0.66186	0.80434	0.66186
23	0.50	0.48	0.02	0.86472	0.70532	0.74548	0.70532
24	0.50	0.48	0.02	0.86472	0.70532	0.74548	0.70532

Case 3: All four objectives of the multiobjective thermal power dispatch problem stated in Eqs. (8.11a) to (8.11e) are undertaken for minimization simultaneously that are operating cost, NO_x emission, SO_2 emission and CO_2 emission which have weightages w_1, w_2, w_3 and w_4, respectively. The obtained minimum and maximum values of these objectives from Table 8.4 are given below:

F_1^{min} = 5704.104 ₹/h, F_1^{max} = 6066.595 ₹/h
F_2^{min} = 700.027 kg/h, F_2^{max} = 1127.250 kg/h
F_3^{min} = 3436.859 kg/h, F_3^{max} = 7320.859 kg/h
F_4^{min} = 8.566 ton/h, F_4^{max} = 20.557 ton/h

Considering NO_x, SO_2 and CO_2 gaseous emission pollutant's objective represent the binary bit. The three binary bits can be represented in 2^3 (eight) possible different combinations. These eight binary bit combinations are the corners of hypercube away from the point around which hypercube is generated. The distance γ of hypercube corners from its centre (w_2^c, w_3^c, w_4^c) is generated from binary digits given in Table 8.12. Figure 8.4 shows the pictorial representation of generation of weights at the hypercube corners. The values of operating cost and emission due to pollutants NO_x, SO_2, and CO_2 for each iteration along with their corresponding weightage are tabulated in Table 8.13. The obtained minimum membership function value among four objectives, μ_k^{min} are given in Table 8.14. The 'best' weights: w_1 = 0.38, w_2 = 0.307, w_3 = 0.007, w_4 = 0.306

TABLE 8.11 'Best' power injection and voltage profile with evolutionary method for three objectives (Case 2)

Bus	P_{Di} (p.u.)	Q_{Di} (p.u.)	P_i (p.u.)	Q_i (p.u.)	V_i (p.u.)	δ_i (rad)
1	0.50	0.10	0.11974	0.12320	1.070	0.00
2	0.50	0.10	0.84939	0.31368	1.088	0.13126
3	0.50	0.10	0.93137	–0.13127	1.095	0.27440
4	0.50	0.10	0.44067	0.27646	1.062	0.14755
5	0.50	0.10	0.40374	0.18332	1.046	0.08248
6	0.10	0.02	–0.10	–0.02	1.05322	0.10363
7	0.40	0.10	–0.40	–0.10	1.01372	0.02444
8	0.90	0.45	–0.90	–0.45	0.99319	0.00609
9	0.70	0.35	–0.70	–0.35	1.00254	0.01824
10	0.25	0.05	–0.25	–0.05	1.03410	0.04133
11	0.25	0.05	–0.25	–0.05	1.03446	–0.00852

TABLE 8.12 Generation of hypercube corners for four objectives

Hypercube corners	*Possible combinations of three binary bit* b_2, b_1, b_0	*Distance of hypercube corners from centre* (w_2^c, w_3^c, w_4^c)	*Possible generated weights at the hypercube corners*
1	0, 0, 0	$-\gamma, -\gamma, -\gamma$	$w_2^c - \gamma, w_3^c - \gamma, w_4^c - \gamma$
2	0, 0, 1	$-\gamma, -\gamma, +\gamma$	$w_2^c - \gamma, w_3^c - \gamma, w_4^c + \gamma$
3	0, 1, 0	$-\gamma, +\gamma, -\gamma$	$w_2^c - \gamma, w_3^c + \gamma, w_4^c - \gamma$
4	0, 1, 1	$-\gamma, +\gamma, +\gamma$	$w_2^c - \gamma, w_3^c + \gamma, w_4^c + \gamma$
5	1, 0, 0	$+\gamma, -\gamma, -\gamma$	$w_2^c + \gamma, w_3^c - \gamma, w_4^c - \gamma$
6	1, 0, 1	$+\gamma, -\gamma, +\gamma$	$w_2^c + \gamma, w_3^c - \gamma, w_4^c + \gamma$
7	1, 1, 0	$+\gamma, +\gamma, -\gamma$	$w_2^c + \gamma, w_3^c + \gamma, w_4^c - \gamma$
8	1, 1, 1	$+\gamma, +\gamma, +\gamma$	$w_2^c + \gamma, w_3^c + \gamma, w_4^c + \gamma$

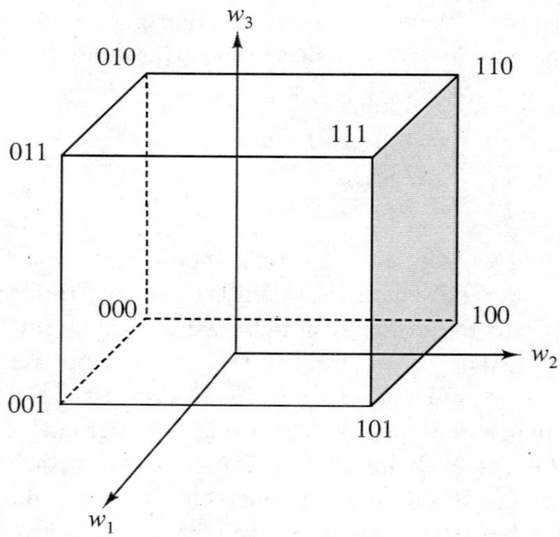

FIGURE 8.4 Generation of weights for four objectives.

TABLE 8.13 Weight search with evolutionary method for four objectives

k	*Weights*				*Objectives*			
	w_1^k	w_2^k	w_3^k	w_4^k	F_1^k(₹/h)	F_2^k(kg/h)	F_3^k(kg/h)	F_4^k(ton/h)
1	0.490	0.177	0.157	0.176	5716.712	1029.023	3466.950	11.105
2	0.480	0.187	0.147	0.186	5717.334	1021.857	3473.540	11.110
3	0.470	0.197	0.137	0.196	5718.043	1014.276	3481.457	11.118
4	0.460	0.207	0.127	0.206	5718.831	1006.217	3491.036	11.127
5	0.450	0.217	0.117	0.216	5719.707	997.604	3502.738	11.138
6	0.440	0.227	0.107	0.226	5720.683	988.342	3517.186	11.152
7	0.430	0.237	0.097	0.236	5721.779	978.317	3535.255	11.168
8	0.420	0.247	0.087	0.246	5723.018	967.382	3558.181	11.189
9	0.410	0.257	0.077	0.256	5724.441	955.352	3587.776	11.214
10	0.400	0.267	0.067	0.266	5726.106	941.991	3626.756	11.245
11	0.390	0.277	0.057	0.276	5728.115	926.990	3679.359	11.286
12	0.380	0.287	0.047	0.286	5730.639	909.947	3752.455	11.342
13	0.370	0.297	0.037	0.296	5733.990	890.345	3857.843	11.422
14	0.360	0.307	0.027	0.306	5738.777	867.532	4016.976	11.546
15	0.350	0.317	0.017	0.316	5746.289	840.799	4272.526	11.751
16	0.380	0.307	0.007	0.306	5749.820	823.721	4608.708	11.886
17	0.380	0.307	0.007	0.306	5749.820	823.721	4608.709	11.886

TABLE 8.14 Weight search with evolutionary method for four objectives

k	*Weights*				*Membership functions*				
	w_1^k	w_2^k	w_3^k	w_4^k	$\mu(F_1)^k$	$\mu(F_2)^k$	$\mu(F_3)^k$	$\mu(F_4)^k$	μ_k^{min}
1	0.490	0.177	0.157	0.176	0.96522	0.22992	0.99218	0.78828	0.22992
2	0.480	0.187	0.147	0.186	096350	0.24670	0.99048	0.78782	0.24670
3	0.470	0.197	0.137	0.196	0.96155	0.26444	0.98845	0.78722	0.26444
4	0.460	0.207	0.127	0.206	0.95937	0.28330	0.98598	0.78647	0.28330
5	0.450	0.217	0.117	0.216	0.95696	0.30346	0.98297	0.78553	0.30346
6	0.440	0.227	0.107	0.226	0.95427	0.32514	0.97925	0.78439	0.32514
7	0.430	0.237	0.097	0.236	0.95124	0.34861	0.97460	0.78300	0.34861
8	0.420	0.247	0.087	0.246	0.94782	0.37420	0.96869	0.78130	0.37420
9	0.410	0.257	0.077	0.256	0.94390	0.40236	0.96107	0.77921	0.40236
10	0.400	0.267	0.067	0.266	0.93931	0.43364	0.95104	0.77657	0.43364
11	0.390	0.277	0.057	0.276	0.93376	0.46875	0.93750	0.77314	0.46875
12	0.380	0.287	0.047	0.286	0.92680	0.50864	0.91868	0.76848	0.50864
13	0.370	0.297	0.037	0.296	0.91755	0.55452	0.89155	0.76179	0.55452
14	0.360	0.307	0.027	0.306	0.90435	0.60792	0.85058	0.75152	0.60792
15	0.350	0.317	0.017	0.316	0.88363	0.67049	0.78479	0.73435	0.67049
16	0.380	0.307	0.007	0.306	0.87389	0.71047	0.69824	0.72317	0.69824
17	0.380	0.307	0.007	0.306	0.87388	0.71047	0.69824	0.72317	0.69824

are obtained after 16 iterations. The injected power and voltage profile at the buses corresponding to 'best' weight combinations are depicted in Table 8.15.

TABLE 8.15 'Best' power injection and voltage profile with evolutionary method for Case 3

Bus	P_{Di}(p.u.)	Q_{Di}(p.u.)	P_i(p.u.)	Q_i(p.u.)	V_i(p.u.)	δ_i(rad)
1	0.50	0.10	0.10567	0.12342	1.070	0.00
2	0.50	0.10	0.86806	0.30306	1.088	0.13206
3	0.50	0.10	0.82541	–0.12362	1.095	0.26265
4	0.50	0.10	0.52637	0.25503	1.062	0.15303
5	0.50	0.10	0.41205	0.18548	1.046	0.08777
6	0.10	0.02	–0.10	–0.02	1.05322	0.10902
7	0.40	0.10	–0.40	–0.10	1.01370	0.02617
8	0.90	0.45	–0.90	–0.45	0.99306	0.00858
9	0.70	0.35	–0.70	–0.35	1.00241	0.02214
10	0.25	0.05	–0.25	–0.05	1.03408	0.04277
11	0.25	0.05	–0.25	–0.05	1.03439	–0.00754

Genetic Algorithm

The genetic algorithm (GA) initially searches the scalar weightage of objectives in the range of 0 to 100, and then normalizes these weights in the range of 0 to 1 using Eq. (8.23). Consider all the four objectives, i.e. operating cost, NO_x emission, SO_2 emission, and CO_2 emission for minimization of multiobjective thermal power dispatch problem defined by Eqs. (3.11a)–(3.11e), which have scalar weightages α_1, α_2, α_3 and α_4, respectively. Scalar weights α_1, α_2, α_3 and α_4 are searched between 0 and 100 by GA and necessary input data are assumed. Size of population, P_S is 10, total length of string, l is 40, length of each sub-string, x_i is 10, crossover probability, p_c is 0.75, mutation probability, p_m is 0.10 and maximum number of generations, R^{max} is 50. At 26th generation of population, the 'best' compromised solution has been obtained and is shown in

TABLE 8.16 Population of scalar weights at 26th generation for four objectives

S. No.	*Scalar weights*				*Objectives*			
	α_1	α_2	α_3	α_4	F_1(₹/h)	F_2(kg/h)	F_3(₹/h)	F_4(ton/h)
1.	70.5767	72.0430	06.1584	54.9365	5746.677	855.544	4054.231	11.755
2.	18.9638	74.8778	04.3011	82.8935	5830.953	778.296	4500.394	13.910
3.	94.0372	73.2160	28.9345	86.0215	5729.693	963.978	3540.422	11.340
4.	70.9677	72.0430	29.5210	46.7253	5737.511	956.363	3536.982	11.556
5.	92.5709	73.1183	98.7292	76.7351	5726.653	1026.454	3451.107	11.340
6.	98.5337	62.0723	02.4438	13.7830	5735.109	854.084	4309.181	11.539
7.	**55.0342**	**61.8768**	**02.4438**	**81.4272**	**5760.272**	**816.197**	**4490.975**	**12.093**
8.	95.4057	73.2160	04.0078	17.6931	5739.720	851.367	4226.688	11.636
9.	32.6491	17.3021	27.8592	38.2209	5721.066	1043.318	3455.521	11.195
10.	06.5494	74.8778	05.9629	42.2287	5877.405	782.031	4318.524	15.201

TABLE 8.17 Population of scalar weights at 26th generation for four objectives

S. No.	*Scalar weights*				*Membership functions*				*Fitness*
	α_1	α_2	α_3	α_4	$\mu(F_1)$	$\mu(F_2)$	$\mu(F_3)$	$\mu(F_4)$	$(\mu^{\min})$
1.	70.5767	72.0430	06.1584	54.9365	0.88256	0.63598	0.84099	0.73403	0.63598
2.	18.9638	74.8778	04.3011	82.8935	0.65006	0.81680	0.72612	0.55433	0.55433
3.	94.0372	73.2160	28.9345	86.0215	0.92941	0.38217	0.97327	0.76865	0.38217
4.	70.9677	72.0430	29.5210	46.7253	0.90784	0.40000	0.97415	0.75064	0.39999
5.	92.5709	73.1183	98.7292	76.7351	0.93780	0.23593	0.99626	0.76868	0.23593
6.	98.5337	62.0723	02.4438	13.7830	0.91447	0.63940	0.77535	0.75207	0.63940
7.	**55.0342**	**61.8768**	**02.4438**	**81.4272**	**0.84505**	**0.72808**	**0.72855**	**0.71588**	**0.71588**
8.	95.4057	73.2160	04.0078	17.6931	0.90175	0.64576	0.79659	0.74397	0.64576
9.	32.6491	17.3021	27.8592	38.2209	0.95321	0.19646	0.99770	0.78076	0.19646
10.	06.5494	74.8778	05.9629	42.2287	0.52192	0.80805	0.77295	0.44667	0.44667

TABLE 8.18 'Best' power injection and voltage profile at buses with genetic-based weightage pattern search

Bus	P_{Di}(p.u.)	Q_{Di}(p.u.)	d_i(rad)	P_i(p.u.)	Q_i(p.u.)	V_i(p.u.)	δ_i(rad)
1	0.50	0.10	0.00	0.12204	0.12343	1.07	0.00
2	0.50	0.10	0.12456	0.79934	0.32005	1.088	0.12600
3	0.50	0.10	0.32611	0.96111	–0.13452	1.095	0.28033
4	0.50	0.10	0.16062	0.46994	0.28368	1.062	0.15248
5	0.50	0.10	0.07367	0.39538	0.18557	1.046	0.08315
6	0.10	0.02	0.10716	–0.10	–0.02	1.05313	0.10670
7	0.40	0.10	0.02199	–0.40	–0.10	1.01385	0.02204
8	0.90	0.45	0.00665	–0.90	–0.45	0.99309	0.00566
9	0.70	0.35	0.01614	–0.70	–0.35	1.00244	0.01923
10	0.25	0.05	0.03828	–0.25	–0.05	1.03422	0.03857
11	0.25	0.05	–0.00904	–0.25	–0.05	1.03448	–0.00924

Tables 8.21 and 8.22 at serial number 07 with maximum fitness value, 0.71588. The values of scalar weights, α_1, α_2, α_3 and α_4 are obtained as 55.0342, 61.8768, 2.4438 and 81.4272, respectively. The value of cost is 5760.272 ₹/h, the value of NO_x emission is 816.197 kg/h, the value of SO_2 emission is 4490.975 kg/h and the value of CO_2 emission is 12.093 ton/h corresponding to 'best' solution. Table 8.23 gives the power injection at buses and voltage profile for the 'best' weight set searched by GA.

The effect of population size on the performance of GA in terms of fitness achieved and time taken to get 'most' fit generation has been studied here. The effect of population size on the achieved fitness is shown in Fig. 8.5, when two, three and four number of objectives are minimized simultaneously. Fix sub-string length to 10. Similarly, Fig. 8.6 gives variation in time to achieve the best population with respect to population size. For less population size, there is less

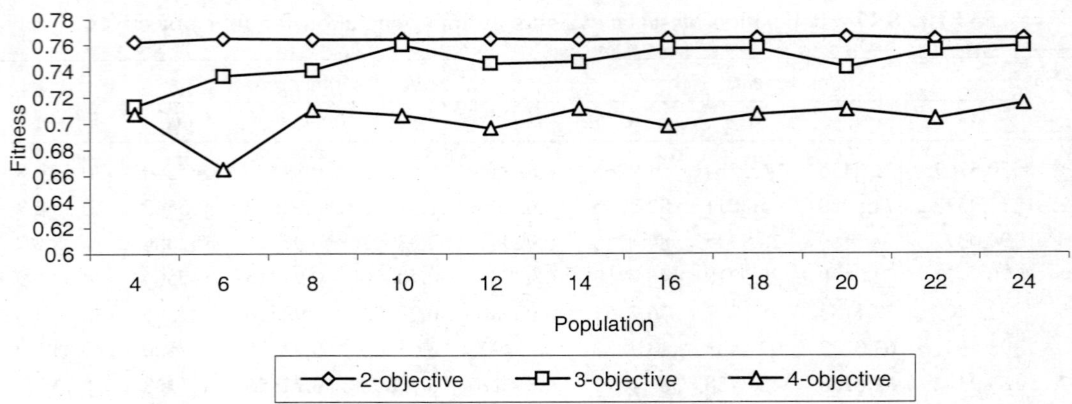

FIGURE 8.5 Variation in fitness with population for different number of objectives undertaken.

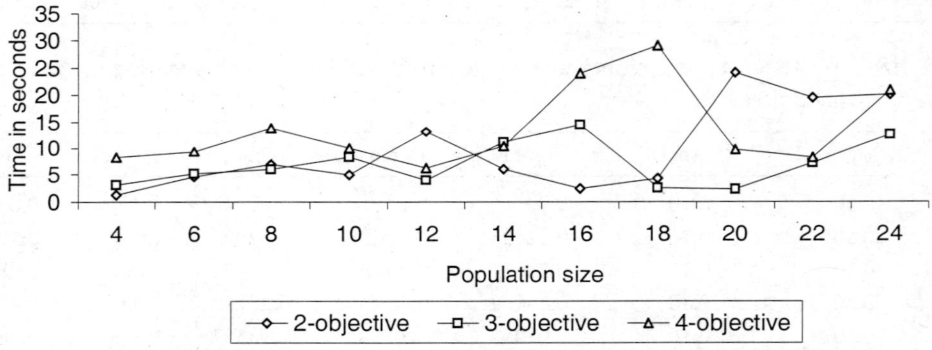

FIGURE 8.6 Variation in time with population for different number of objectives undertaken.

time burden on decision-maker. The solution procedure takes less time if the population size is taken either 4 or 16 for two objectives. For three objectives, the population size, which gives result in less time, is taken either 4 or 18 or 20 and for four objectives, the population size is taken either 4 or 12, which gives the result in less time.

8.2.7 Comparison of Results

The results obtained by searching weight pattern to solve multiobjective thermal power dispatch problem have been compared with weight simulation technique [Dhillon et al., 1993]. In weight simulation technique, multiobjective optimization problem has been solved to generate the non-inferior solutions using weighting method. The weights are varied from 0 to 1 in steps of 0.1 such that their sum remains 1.0. The non-inferior solution that attains the maximum cardinal priority ranking as defined by Eq. (8.17) has been chosen the 'best' compromised solution by the decision-maker [Dhillon et al., 1993]. The comparison of methods has been carried out based on achieved satisfaction of participating objectives and time required to achieve that satisfaction. The weights shown in all comparison tables are normalized weights assigned to objectives. Tables 8.19 to 8.21

TABLE 8.19 Comparison of results: Two objectives

Description	*Linear membership function*			*Exponential membership function*		
	Simulation	*Evolutionary*	*Genetic*	*Simulation*	*Evolutionary*	*Genetic*
P_{G1} (p.u.)	0.54636	0.53597	0.53696	0.54636	0.53597	0.53671
P_{G2} (p.u.)	1.27049	1.22155	1.22613	1.27049	1.22155	1.22498
P_{G3} (p.u.)	1.32163	1.36804	1.36354	1.32163	1.36804	1.36467
P_{G4} (p.u.)	1.28782	1.33787	1.33325	1.28782	1.33787	1.33442
P_{G5} (p.u.)	0.82380	0.79467	0.79742	0.82380	0.79467	0.79673
Q_{G1} (p.u.)	0.22519	0.22637	0.22625	0.22519	0.22637	0.22628
Q_{G2} (p.u.)	0.42680	0.44094	0.43955	0.42680	0.44094	0.43990
Q_{G3} (p.u.)	–0.03114	–0.03703	–0.03646	–0.03114	–0.03703	–0.03661
Q_{G4} (p.u.)	0.37304	0.38839	0.38688	0.37304	0.38839	0.38726
Q_{G5} (p.u.)	0.30267	0.30874	0.30815	0.30267	0.30874	0.30830
Cost (₹/h)	5777.917	5790.393	5789.174	5777.917	5790.393	5789.479
NO_x (kg/h)	784.863	773.218	774.275	784.863	773.218	774.009
w_1	0.50	0.46	0.4637	0.50	0.46	0.4628
w_2	0.50	0.54	0.5363	0.50	0.54	0.5372
$\mu^{\min}$	**0.73046**	**0.76196**	**0.76440**	**0.90915**	**0.92303**	**0.92410**
Time in seconds	**0.16**	**0.27**	**5.11**	**0.16**	**0.22**	**1.15**

TABLE 8.20 Comparison of results: Three objectives

Description	*Linear membership function*			*Exponential membership function*		
	Simulation	*Evolutionary*	*Genetic*	*Simulation*	*Evolutionary*	*Genetic*
P_{G1} (p.u.)	0.72062	0.61974	0.62697	0.70514	0.61974	0.61731
P_{G2} (p.u.)	1.39946	1.34939	1.19088	1.28649	1.34939	1.21185
P_{G3} (p.u.)	1.67333	1.43137	1.73850	1.82271	1.43137	1.66153
P_{G4} (p.u.)	0.47892	0.94067	0.89157	0.54019	0.94067	0.94415
P_{G5} (p.u.)	0.98700	0.90374	0.83018	0.92257	0.90374	0.83572
Q_{G1} (p.u.)	0.22591	0.22320	0.22353	0.22417	0.22320	0.22358
Q_{G2} (p.u.)	0.41851	0.41368	0.46777	0.45025	0.41368	0.45709
Q_{G3} (p.u.)	–0.02786	–0.03127	–0.06133	–0.04765	–0.03127	–0.05552
Q_{G4} (p.u.)	0.41469	0.37646	0.46038	0.46285	0.37646	0.44043
Q_{G5} (p.u.)	0.26974	0.28323	0.28791	0.27234	0.28323	0.28915
Cost (Rs/h)	5749.342	5753.141	5791.194	5773.872	5753.141	5784.649
NO_x (kg/h)	899.648	825.921	802.465	868.766	825.921	800.085
SO_2 (kg/h)	3701.830	4425.211	4348.820	3785.066	4425.211	4451.137
w_1	0.40	0.50	0.3163	0.30	0.50	0.3534
w_2	0.50	0.48	0.6434	0.60	0.48	0.6147
w_3	0.10	0.02	0.0404	0.10	0.02	0.0319
$\mu^{\min}$	**0.53275**	**0.70531**	**0.75975**	**0.84218**	**0.89729**	**0.91475**
Time in Seconds	**0.93**	**2.09**	**8.7**	**0.88**	**1.70**	**9.00**

TABLE 8.21 Comparison of results: Four objectives

Description	*Linear membership function*			*Exponential membership function*		
	Simulation	*Evolutionary*	*Genetic*	*Simulation*	*Evolutionary*	*Genetic*
P_{G1} (p.u.)	0.63615	0.60567	0.62204	0.62906	0.60567	0.61260
P_{G2} (p.u.)	1.67584	1.36806	1.29934	1.56240	1.36806	1.33470
P_{G3} (p.u.)	1.05328	1.32541	1.46111	1.08825	1.32541	1.39625
P_{G4} (p.u.)	0.78786	1.02637	0.96994	0.91492	1.02637	0.99831
P_{G5} (p.u.)	1.06425	0.91205	0.89538	1.02312	0.91205	0.90106
Q_{G1} (p.u.)	0.22741	0.22342	0.22343	0.22525	0.22342	0.22338
Q_{G2} (p.u.)	0.35983	0.40306	0.42005	0.36569	0.40306	0.41200
Q_{G3} (p.u.)	0.01149	–0.02362	0.03452	0.00314	–0.02362	–0.02954
Q_{G4} (p.u.)	0.29134	0.35503	0.38368	0.30018	0.35503	0.37006
Q_{G5} (p.u.)	0.27446	0.28548	0.28557	0.27736	0.28548	0.28559
Cost (₹/h)	5712.839	5749.820	5760.27	5722.244	5749.820	5754.961
NO_x (kg/h)	918.587	823.721	816.197	881.438	823.721	819.369
SO_2 (kg/h)	4110.839	4608.709	4490.97	4354.365	4608.709	4549.258
CO_2 (ton/h)	10.964	11.886	12.093	11.078	11.886	12.005
w_1	0.50	0.380	0.2741	0.30	0.380	0.3521
w_2	0.10	0.307	0.3082	0.10	0.307	0.3358
w_3	0.00	0.007	0.0122	0.00	0.007	0.0106
w_4	0.40	0.306	0.4056	0.60	0.306	0.3015
μ^{min}	**0.48842**	**0.69824**	**0.71588**	**0.82318**	**0.89382**	**0.90150**
Time in Seconds	**3.73**	**2.64**	**6.92**	**3.62**	**2.36**	**8.63**

show the comparison of 'best' compromised solution achieved by weight simulation method, evolutionary weight search method, and GA based weight search method. Linear and exponential membership functions of participating objectives are considered to decide the best solution on 11-bus system, when the multiobjective optimization problem is solved considering two, three and four objectives, respectively.

Table 8.19 depicts the comparison of results when the membership function of participating objectives is monotonic linear decreasing and exponential decreasing and two objectives are considered simultaneously. Tables 8.20 and 8.21 depict the comparison of results when three and four objectives are considered, simultaneously. In weight simulation technique, generally the weights are varied by step length of 0.1. But it is possible that best weight combination may lie in between chosen step length of 0.1. Therefore, this technique has low satisfaction as compared to other methods discussed because the decision-maker has selected the 'best' solution only among the solutions presented to him/her by weight simulation technique. Moreover, as the number of participating objectives increases the number of non-inferior solutions required in this technique increases exponentially. The required number of non-inferior solutions, when the weights are varied in steps of 0.1, 0.01, and 0.001, respectively are given in Table 8.27. If normalized weights are varied from 0 to 1 by the variation of 0.1, then there is need of 11 non-inferior solutions in case of two objectives, 66 non-inferior solutions in case of three objectives, 286 non-inferior solutions in case of four objectives, to interact with the decision-maker to obtain his/her reaction/response to the solution.

The number of non-inferior solutions required in weight search by evolutionary optimization method and weight search by GA for two objectives, three objectives and four objectives are shown in Table 8.22. It is observed from Table 8.22, the method undertaken to search weights pattern take less time as compared to weight simulation method when weights are simulated by a step length less than or equal to 0.01.

TABLE 8.22 Comparison of the number of non-inferior solutions

	Linear membership function			*Exponential membership function*		
	Two objectives	*Three objectives*	*Four objectives*	*Two objectives*	*Three objectives*	*Four objectives*
Simulation						
*Step 0.1	11	66	286	11	66	286
*Step 0.01	101	5151	176851	101	5151	176851
*Step 0.001	1001	501501	167668501	1001	501501	167668501
Evolutionary	15	120	153	15	120	153
Genetic	190	320	270	50	380	360

* The weights are simulated by varying the weights from 0 to 1 in steps of 0.1 or in steps of 0.01 or in steps of 0.001.

The GA provides high-satisfied results as compared to evolutionary search method. But this technique requires more convergence time as compared to other methods due its random nature. Exponential membership function for decision-making gives better membership for the best solution as compared to linear membership function. Genetic algorithm gives better results but takes large time as compared to simulation and evolutionary methods. As the number of objectives increases, the execution time decreases in evolutionary search method.

8.3 MULTIOBJECTIVE SECURE LOAD DISPATCH

In recent years, requirements that are more stringent have been imposed on electric utilities. This tendency has brought about sheer necessity of attaining system planning as well as system operation of higher level and of greater sophistication. Personnel in-charge of system operation are required to determine the optimal system state with satisfying many kinds of operational constraints. In general, a large-scale system, typified by an electric power system, possesses multiple objectives to be achieved. In a power system, economic operation, security, and minimal impact on environments are typical objectives to be satisfied. These objectives are in conflict with each other. So it is difficult to find a single solution, which is optimal for all objectives. In this context, a suitable compromised solution should be determined. The multiobjective optimization method is an ideal solution technique for such kind of problem.

The main purpose of the optimal power dispatch problem has so far been confined to minimize the total generation cost of a power system. However, in order to meet environment regulations enforced in recent years, emission control has become one of important operational objectives. System security is another essential factor in power system operation and in system planning. To be specific, it is very important to maintain good voltage profiles and to limit line flows within prescribed upper bounds. In security analysis, a series of anticipated contingencies are assumed to predict possible overloadings or excessive voltage deviations. Then, minimize a security index as a function of overloads by some preventive control actions.

The intent of this section is to solve multiobjective thermal power dispatch problem having six objectives, viz. the economic index and impact on environment due to NO_x, SO_2, CO_2 gaseous pollutant emissions, overloading of the transmission lines due to active power flow as well as reactive power flow on transmission lines are undertaken as individual objectives. The transmission line power flows are obtained with the help of generalized Z-bus distribution factors (GZBDF). Initially, the multiobjective optimization problem is converted into scalar objective optimization problem using a weighting method. In simple interactive method, the weights are generally simulated with suitable variation to generate non-inferior solutions so that trade-off relationship among participating objectives can be investigated. An enormous amount of computing time is required in the determination of complete non-inferior solution surface. To overcome the above said problem, the weight combinations are searched in the non-inferior domain. So, evolutionary optimization method, simplex search method and GA are again tested in this chapter to search for optimal weight pattern in the non-inferior domain. Fuzzy decision-making theories attempt to deal with the vagueness or fuzziness inherent in subjective or imprecise determinations of preferences, constraints, and goals. Being the imprecise nature of the decision-maker's judgement, assume that the decision maker has fuzzy goals for each of the objective functions. The fuzzy goals are quantified by defining their corresponding membership functions. The 'best' compromised solution is one, which provides the maximum satisfaction level from the membership functions of the participating goals/objectives.

8.3.1 Power Flow on Transmission Lines

Based on the Z-bus concept, the *Z-bus distribution* (ZBD) factor can be expressed as:

$$H(m, i) = \partial S_m / \partial S_i = \left|(Z_{pi} - Z_{qi})/Z_m\right|^* \tag{8.24}$$

where

S_i is the complex power injection of the ith bus.
S_m is the complex line flow on the mth line.
Z_{pi} and Z_{qi} are the elements of the Z-bus matrix.
Z_m is the line impedance of the mth line.
* denotes the complex conjugate.
$H(m, i)$ is the ZBD factor of the ith generator and the mth line.

Based on Eq. (8.24) the complex power flow on transmission lines is expressed as:

$$S_m = S_m^o + \sum_{i=1}^{NB} H(m, i)\, \Delta S_i \qquad (m = 1, 2, \ldots, NL) \tag{8.25}$$

where

S_m^o is a base case of the complex power flow on the mth line.
NL is number of lines in the system.
NB is number of buses.
ΔS_i is the small change in complex power injection at the ith bus.

Since, reference bus compensates any change in generation on rest of buses of power system, so the total system generation remains unchanged.

$$\sum_{i=1}^{NG} \Delta S_{Gi} = \sum_{i=1}^{NB} \Delta S_i = \text{constant} \tag{8.26}$$

where

NG is number of generators,

S_{Gi} is the power generated by ith generator.

Z-bus distribution expresses the new complex power flow on lines after rescheduling the generation.

$$S_m = S_m^o + \sum_{i=1}^{NG} H(m, i)\, \Delta S_{Gi} \qquad (m = 1, 2, \ldots, \text{NL}) \tag{8.27}$$

where

ΔS_{Gi} is the small change in complex power generated by ith generator.

A sensitivity factor, generalized Z-bus distribution factors has been obtained to update the Z-bus distribution factor. Instead of an incremental form, an integral generation form is defined as:

$$S_m = \sum_{i=1}^{NG} D(m, i)\, S_{Gi} \qquad (m = 1, 2, \ldots, \text{NL}) \tag{8.28}$$

where

$D(m, i) = D(m, R) + H(m, i)$

$$D(m, R) = \frac{S_m^o - \sum\limits_{\substack{i=1,\\ i \neq R}}^{NG} H(m, i) S_{Gi}}{\sum\limits_{i=1}^{NG} S_{Gi}}$$

with R denotes the reference bus.

$D(m, i)$ is the GZBDF for ith generator and mth line.

Equation (8.28) can be rewritten as:

$$P_{Tm} + jQ_{Tm} = \sum_{i=1}^{NG} [D_P(m, i) + jD_Q(m, i)]\,(P_{Gi} + jQ_{Gi}) \qquad (m = 1, 2, \ldots, \text{NL}) \tag{8.29}$$

where

P_{Tm} is real part of complex power flow on mth transmission line.

Q_{Tm} is reactive part of complex power flow on mth transmission line.

$D_P(m, i)$ is real part of GZBDF.

$D_Q(m, i)$ is reactive part of GZBDF.

Separating real and imaginary part of Eq. (8.29), real and reactive power flow on transmission lines is obtained as:

$$P_{Tm} = \sum_{i=1}^{NG} [D_P(m, i) P_{Gi} - D_Q(m, i)Q_{Gi}] \qquad (m = 1, 2, ..., NL) \tag{8.30}$$

$$Q_{Tm} = \sum_{i=1}^{NG} [D_P(m, i) Q_{Gi} + D_Q(m, i)P_{Gi}] \qquad (m = 1, 2, ..., NL) \tag{8.31}$$

8.3.2 Generalized *Z*-Bus Distribution Factors (GZBDF)

Considering Eq. (8.28), if a particular *k*th generator is increased in generation by some amount ΔS_{Gk}, the flow on *m*th line will be

$$S_m = \sum_{i=1}^{NG} D(m, i)S_{Gi} + D(m, k)\Delta S_{Gk} \qquad (m = 1, 2, ..., NL) \tag{8.32}$$

The reference generator ($R \neq k$) will adjust the change of ΔS_{Gk} by decreasing owns generation by $\Delta S_{Gk.}$ The complex power flow on line *m*, $\overline{S}_m$ after this generation shift will be

$$\overline{S}_m = \sum_{i=1}^{NG} D(m, i)S_{Gi} + D(m, k)\Delta S_{Gk} - D(m, R)\Delta S_{Gk} \qquad (m = 1, 2, ..., NL) \tag{8.33}$$

Subtract Eq. (8.28) from Eq. (8.33) to obtain:

$$\frac{\overline{S}_m - S_m}{\Delta S_{Gk}} = D(m, k) - D(m, R) \qquad (m = 1, 2, ..., NL) \tag{8.34}$$

The change in line flow with respect to generation change is termed as ZBD factor. So, above equation can be rewritten as:

$$\overline{S}_m - S_m = H(m, k)\, \Delta S_{Gk} \tag{8.35}$$

$\because H(m, k) = D(m, k) - D(m, R)$

There is need to calculate *D*(*m*, *R*) by shifting all the generations from all generators to the reference generator *R*.

i.e.

$$-\sum_{\substack{i=1 \\ i\neq R}}^{NG} S_{Gi} = \Delta S_{Gk} \tag{8.36}$$

From the definition of generalized distribution factors [Ng, 1981], the transmission line flows and generations in the base case load flow must also satisfy Eq. (8.28), which now represent a group of *m* additional equations for calculating all the GZBDF.

Assuming all the Z-bus distribution factors are known, the solution procedure of these (NG $\times$ NB) equations is simple. By shifting all the generations from all the generators to the reference generator R, from Eqs. (8.35) and (8.36), the new equation is obtained [Ng, 1981]:

$$\overline{S}_m - S_m = -\sum_{\substack{i=1 \\ i \neq R}}^{NG} H(m, i)\, S_{Gi} \qquad (m = 1, 2, \ldots, NL) \tag{8.37}$$

From Eq. (8.28)

$$\overline{S}_m = \sum_{\substack{i=1 \\ i \neq R}}^{NG} D(m, i)\overline{S}_{Gi} + D(m, R)\overline{S}_{GR} \qquad (m = 1, 2, \ldots, NL) \tag{8.38}$$

where

$\overline{S}_{Gi}$ is the final generation from ith generator, which is now reduced to zero,
$\overline{S}_{GR}$ is the final generation from the reference generator.

$$\overline{S}_m = D(m, R)\overline{S}_{GR} = D(m, R)\sum_{i=1}^{NG} S_{Gi} \tag{8.39}$$

Substitute Eq. (8.39) into Eq. (8.37) to get

$$D(m, R)\sum_{i=1}^{NG} S_{Gi} = S_m - \sum_{\substack{i=1 \\ i \neq R}}^{NG} H(m, i)S_{Gi}$$

Rearrange above equation to get

$$D(m, R) = \left(S_m - \sum_{\substack{i=1 \\ i \neq R}}^{NG} H(m, i)S_{Gi} \right) \Bigg/ \left(\sum_{i=1}^{NG} S_{Gi} \right) \tag{8.40}$$

and

$$D(m, i) = H(m, i) + D(m, R) \tag{8.41}$$

8.3.3 Multiobjective Optimization Problem Formulation

The multiobjective optimization thermal power dispatch problem (MOTPDP) is defined as to minimize the number of objectives like total operating cost, gaseous pollutants emission level, overloading of transmission lines, etc. of a power system. At the same time the constraints are met like total real power load plus real power transmission losses within generator limits of real power as well as reactive power load plus reactive power transmission losses within generator limits of reactive power. In the multiobjective optimization problem, formulation six non-commensurable objectives in an electrical thermal power system are considered. These are economy, environmental impact because of NO_x, SO_2 and CO_2 pollutants, overloading of transmission lines due to active as well as reactive power flow on transmission lines. Mathematically, multiobjective optimization thermal power dispatch problem is defined as:

$$\text{Minimize operating cost } F_1 = \sum_{i=1}^{NG} (a_i P_{Gi}^2 + b_i P_{Gi} + c_i) \text{ ₹/h} \tag{8.42}$$

$$\text{Minimize NO}_x \text{ emission } F_2 = \sum_{i=1}^{NG} (d_{1i} P_{Gi}^2 + e_{1i} P_{Gi} + f_{1i}) \text{ kg/h} \tag{8.43}$$

$$\text{Minimize SO}_2 \text{ emission } F_3 = \sum_{i=1}^{NG} (d_{2i} P_{Gi}^2 + e_{2i} P_{Gi} + f_{2i}) \text{ kg/h} \tag{8.44}$$

$$\text{Minimize CO}_2 \text{ emission } F_4 = \sum_{i=1}^{NG} (d_{3i} P_{Gi}^2 + e_{3i} P_{Gi} + f_{3i}) \text{ ton/h} \tag{8.45}$$

A mathematical formulation of security constrained mutiobjective optimization thermal power dispatch problem would require a very large number of constraints to consider. However, for typical systems the large proportion of lines has a rather small possibility to become overloaded. The mutiobjective optimization thermal power dispatch problem should consider only the small proportion of lines in violation, or near violation of their respective security limits, which are identified as the critical lines. Here only the critical lines that are binding in the optimal solution are considered. Experienced decision-maker detects the critical lines. An improvement in the security can be obtained by minimizing the following objective functions.

$$\text{Minimize} \qquad F_5 = \sum_{m=1}^{OL} \left(|P_{Tm}| / P_{Tm}^{\max} \right) \tag{8.46}$$

$$\text{Minimize} \qquad F_6 = \sum_{m=1}^{OL} \left(|Q_{Tm}| / Q_{Tm}^{\max} \right) \tag{8.47}$$

Subject to: (i) To ensure real and reactive power balance:

$$\sum_{i=1}^{NB} P_{Di} - \sum_{i=1}^{NG} P_{Gi} + P_L = 0 \tag{8.48}$$

$$\sum_{i=1}^{NB} Q_{Di} - \sum_{i=1}^{NG} Q_{Gi} + Q_L = 0 \tag{8.49}$$

(ii) Inequality constraints imposed on generator output are:

$$P_{Gi}^{\min} \le P_{Gi} \le P_{Gi}^{\max} \qquad (i = 1, 2, \ldots, \text{NG}) \tag{8.50}$$

$$Q_{Gi}^{\min} \le Q_{Gi} \le Q_{Gi}^{\max} \qquad (i = 1, 2, ..., \text{NG}) \tag{8.51}$$

where

a_i, b_i and c_i are cost coefficients.
d_{1i}, e_{1i} and f_{1i} are NO_X emission coefficients.
d_{2i}, e_{2i} and f_{2i} are SO_2 emission coefficients.
d_{3i}, e_{3i} and f_{3i} are CO_2 emission coefficients.
NB is total number of buses.
NG is number of generators.
OL is the set of over loaded lines.
P_{Di} and Q_{Di} are real power and reactive power demands, respectively at the ith bus.
P_{Gi} and Q_{Gi} are real power and reactive power generation, respectively at the ith bus.
P_L and Q_L are real and reactive power transmission losses, respectively.
$P_{Tm}^{\max}$ is the active power rating of the mth line.
$P_{Gi}^{\min}$ is lower limit of active power generation at the ith bus.
$P_{Gi}^{\max}$ is upper limit of active power generation at the ith bus.
$Q_{Gi}^{\min}$ is lower limit of reactive power generation at the ith bus.
$Q_{Gi}^{\max}$ is upper limit of reactive power generation at the ith bus.
$Q_{Tm}^{\max}$ is the reactive power rating of the mth line.

The real and reactive power transmission losses, P_L and Q_L are obtained from Eqs. (8.9) and (8.10), respectively.

8.3.4 Solution Procedure

To generate the non-inferior solutions, the constrained MOTPDP is converted into a constrained scalar optimization problem employing weighting method and is given below:

$$\text{Minimize} \qquad \sum_{k=1}^{L} w_k F_k \tag{8.52a}$$

Subject to: Eqs. (4.26)–(4.29)

$$\sum_{k=1}^{L} w_k = 1.0 \quad ; w_k \ge 0.0 \tag{8.52b}$$

where w_k are the levels of the assigned normalized weightage in the range of [0, 1] in such a way that their sum remains 1.0.

The number of non-inferior solutions are obtained when Eq. (8.52) is solved many times for different values of w_k (k = 1, 2, …, L). To solve the scalar optimization problem, the Lagrangian function is defined as:

$$L_f(P_{Gi}, Q_{Gi}, \lambda_P, \lambda_Q) = \sum_{j=1}^{L} w_j F_j + \lambda_P \left(\sum_{i=1}^{\text{NB}} P_{Di} + P_L - \sum_{i=1}^{\text{NG}} P_{Gi} \right) + \lambda_Q \left(\sum_{i=1}^{\text{NB}} Q_{Di} + Q_L - \sum_{i=1}^{\text{NG}} Q_{Gi} \right) \tag{8.53}$$

where

λ_P and λ_Q are the Lagrangian multipliers for active and reactive power, respectively.

The necessary conditions to minimize the Lagrangian function are

$$\frac{\partial L}{\partial P_{Gi}} = \sum_{j=1}^{L} w_j \frac{\partial F_j}{\partial P_{Gi}} + \lambda_P \left(\frac{\partial P_L}{\partial P_{Gi}} - 1 \right) + \lambda_Q \left(\frac{\partial Q_L}{\partial P_{Gi}} \right) = 0 \qquad (i = 1, 2, \ldots, \text{NG}) \qquad (8.54a)$$

$$\frac{\partial L}{\partial Q_{Gi}} = \sum_{j=1}^{L} w_j \frac{\partial F_j}{\partial Q_{Gi}} + \lambda_P \left(\frac{\partial P_L}{\partial Q_{Gi}} \right) + \lambda_Q \left(\frac{\partial Q_L}{\partial Q_{Gi}} - 1 \right) = 0 \qquad (i = 1, 2, \ldots, \text{NG}) \qquad (8.54b)$$

$$\frac{\partial L}{\partial \lambda_P} = \sum_{i=1}^{\text{NB}} P_{Di} + P_L - \sum_{i=1}^{\text{NG}} P_{Gi} = 0 \qquad (8.54c)$$

$$\frac{\partial L}{\partial \lambda_Q} = \sum_{i=1}^{\text{NB}} Q_{Di} + Q_L - \sum_{i=1}^{\text{NG}} Q_{Gi} = 0 \qquad (8.54d)$$

The Newton-Raphson algorithm has been applied to obtain the solution of Eqs. (8.54a)–(8.54d), for the weight combinations generated during search moves. The weights assigned to the participating objectives are searched by evolutionary optimization method and genetic algorithm. Algorithms (8.1) and (8.2) elaborate to perform the search the weights, respectively. For the decision making of best solution fuzzy set theory has been exploited. Two types of membership functions of participating objectives are explored that are monotonic linear decreasing function and monotonic exponential decreasing function as defined by Eqs. (8.15) and (8.16), respectively. The non-inferior solution that attains the maximum satisfaction of participating objectives as defined by Eq. (8.17) has been designated the 'best' achieved solution. The violated limits are fixed to their upper or lower limits and rescheduling has been done to meet the remaining load as explained in Section 8.2.2.

8.3.5 Sample Study System

A 25-bus, 35-lines sample system comprising 5-generators is selected to show the results. Tables 5.2 to 5.5 give the operating cost, NO_x, SO_2, and CO_2 gaseous emission equations of committed units. Tables 8.23 and 8.24 give line data and load data for testing system, respectively. Minimum values of the cost and emission objectives F_i, (i = 1, 2, …, L-2) are obtained by giving full weightage to one of the objectives and neglecting the others. When the given weight value is 1.0, it means that full weightage is given to the objective and when the weightage is zero, the objective is neglected. Owing to the conflicting nature of the objectives, one objective will have maximum value when the other objective has minimum value, and vice versa. Minimum and maximum values of rest of objectives are estimated by the experienced decision-maker and depend upon the type of problem. For the security of transmission system, the maximum limit of power flow on transmission lines are taken as 10% below than the maximum permissible limit of power flow on lines.

The minimum and maximum values of objectives undertaken in the study are given below:

$F_1^{\min}$ = 7504.825 ₹/h, $F_1^{\max}$ = 7930.314 ₹/h

$F_2^{\min}$ = 770.6549 kg/h, $F_2^{\max}$ = 1374.439 kg/h

TABLE 8.23 Line data for 25-bus system

Line No.	*Bus*		*Resistance*	*Reactance*	*Line charging*
	From	*To*	(p.u.)	(p.u.)	(p.u.)
1	1	3	0.0720	0.2876	0.0179
2	1	16	0.0290	0.1379	0.0337
3	1	17	0.1012	0.2794	0.0148
4	1	19	0.1487	0.3897	0.0224
5	1	23	0.1085	0.2245	0.0573
6	1	25	0.0753	0.3593	0.0873
7	2	6	0.0617	0.2935	0.0186
8	2	7	0.0511	0.2442	0.0155
9	2	8	0.0579	0.2763	0.0175
10	3	13	0.0564	0.1487	0.0085
11	3	14	0.1183	0.3573	0.0185
12	4	19	0.0196	0.0514	0.0113
13	4	20	0.0382	0.1007	0.0220
14	4	21	0.0970	0.2547	0.0558
15	5	10	0.0497	0.2372	0.0577
16	5	17	0.0144	0.1269	0.1335
17	5	19	0.0929	0.2442	0.0140
18	6	13	0.0263	0.0691	0.0040
19	7	8	0.0529	0.1465	0.0078
20	7	12	0.0364	0.1736	0.0110
21	8	9	0.0387	0.1847	0.0118
22	8	17	0.0497	0.2372	0.0572
23	9	10	0.0973	0.2691	0.0085
24	10	11	0.0898	0.2359	0.0135
25	11	17	0.1068	0.2807	0.0161
26	12	17	0.0460	0.2196	0.0135
27	14	15	0.0281	0.0764	0.0044
28	15	16	0.0256	0.0673	0.0148
29	17	18	0.0806	0.2119	0.0122
30	18	19	0.0872	0.2294	0.0132
31	20	21	0.0615	0.1613	0.0354
32	21	22	0.0414	0.1087	0.0238
33	22	23	0.2250	0.3559	0.0169
34	22	24	0.0970	0.2595	0.0567
35	24	25	0.0472	0.1458	0.0317

$F_3^{\min} = 4584.652$ kg/h, $F_3^{\max} = 9966.985$ kg/h

$F_4^{\min} = 10.76423$ ton/h, $F_4^{\max} = 20.24658$ ton/h

$F_5^{\min} = 1.31258$, $F_5^{\max} = 20.24658$

$F_6^{\min} = 19.76106$, $F_6^{\max} = 54.9048$

Figure 8.7 shows the single line diagram of 25-bus power system.

TABLE 8.24 System load data

Bus number	*Real load* (p.u.)	*Reactive load* (p.u.)	*Bus number*	*Real load* (p.u.)	*Reactive load* (p.u.)
01	2.00	0.65	14	0.20	0.07
02	0.10	0.03	15	0.30	0.10
03	0.50	0.17	16	0.30	0.10
04	0.30	0.10	17	0.60	0.20
05	0.25	0.08	18	0.15	0.05
06	0.15	0.05	19	0.15	0.05
07	0.15	0.05	20	0.25	0.08
08	0.25	0.00	21	0.20	0.07
09	0.15	0.05	22	0.20	0.07
10	0.15	0.05	23	0.15	0.05
11	0.05	0.00	24	0.15	0.05
12	0.10	0.00	25	0.25	0.08
13	0.25	0.08			

8.3.6 Results and Discussion

To solve the multiobjective optimization thermal power dispatch problem defined by Eqs. (8.42)–(8.51) the weighting method is used to convert this problem into a single objective constrained problem. The single objective constrained problem is then converted into unconstrained problem using Lagrangian function defined by Eq. (8.53). The Newton Raphson method has been applied to solve the necessary conditions of the problem of Eq. (8.54). The weights assigned to the participating objectives are searched by evolutionary optimization method defined in Section 8.2.4, and by GA described in Section 8.2.5 to generate non-inferior solutions for the problem.

The fuzzy set theory has been exploited to decide the best solution. The real and reactive power flows on transmission lines are obtained using GZBDF from Eqs. (8.30) and (8.31), respectively. The obtained real values, $D_P(m, i)$ and reactive values, $D_Q(m, i)$ of GZBDF from base case flow using Eq. (8.41) are given in Tables 8.25 and 8.26, respectively. For initial discussion monotonic linear decreasing membership function of the participating objectives has been considered. At the end of the chapter the results obtained with the consideration of monotonic exponential decreasing membership function of participating objectives are also compared and are given in Table 8.31.

Evolutionary search optimization method

The evolutionary optimization method is applied to search the optimal normalized weights assigned to participating objectives of multiobjective optimization thermal power dispatch problem. The weights assigned to NO_x emission, SO_2 emission, CO_2 emission, overloading of lines due to active power flow on transmission lines and overloading of transmission lines due to reactive power flow on transmission lines are considered as representing binary bits corresponding to generation of hypercube corners, γ distance away from the centre of hypercube. The five binary bits can be represented in 2^5 (thirty two) possible different combinations. These thirty-two binary

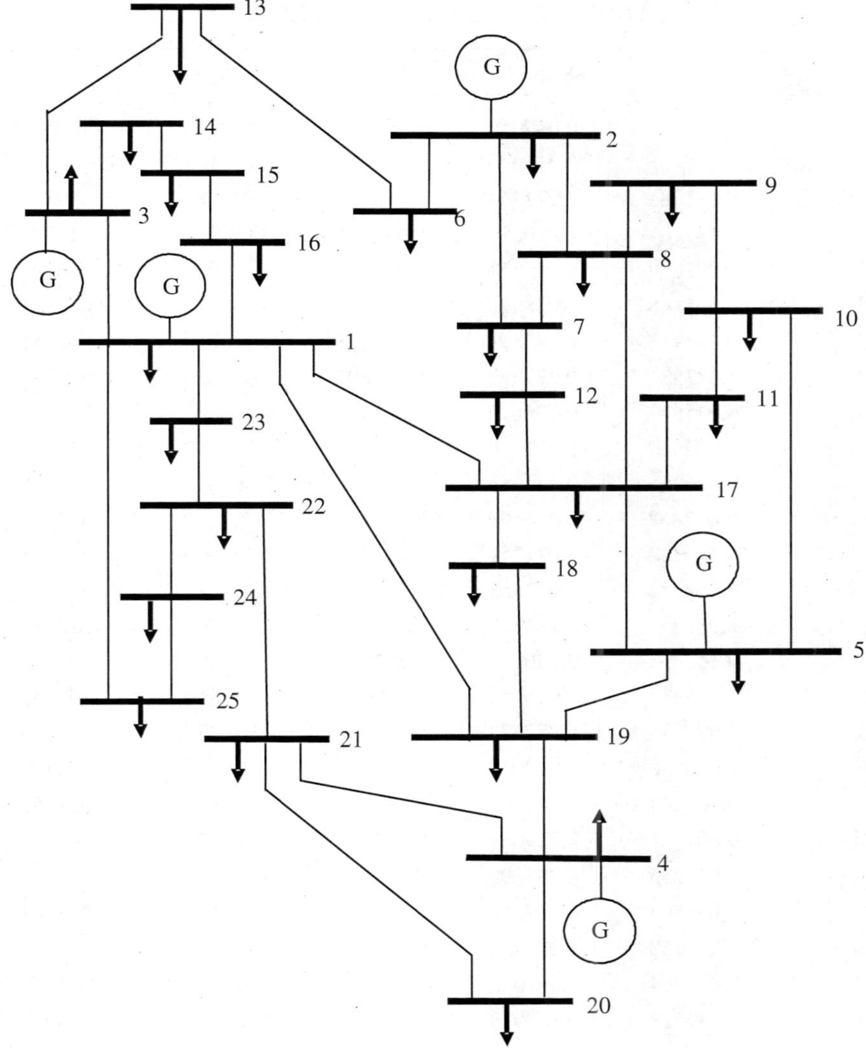

FIGURE 8.7 Single line diagram of 25-bus power system.

bit combinations are used to find out the corners of hypercube away from the point around which hypercube is generated. The best weights searched by this method w_i (1, 2,..., L) are 0.66, 0.16, 0.01, 0.05, 0.07 and 0.05, respectively and are obtained in 11 iterations. The best-achieved satisfaction of participating objectives is 0.72551. The best power injection obtained is given in Table 8.27. The power flow on transmission lines is depicted in Table 8.29.

Genetic algorithm

GA has been applied to search the scalar weightage pattern assigned to participating objectives. The scalar weights α_1, α_2, α_3, α_4, α_5 and α_6 are searched between 0 and 100 by GA which are

TABLE 8.25 Real values of GZBDF

m	$D_P(m, 1)$	$D_P(m, 2)$	$D_P(m, 3)$	$D_P(m, 4)$	$D_P(m, 5)$
1	0.215862	–0.049050	–0.362486	0.161860	0.116001
2	1.027406	–0.011093	–0.154407	0.083197	0.063253
3	0.317781	–0.077344	0.209695	0.046782	–0.151910
4	0.181104	0.020884	0.138220	–0.184037	–0.081592
5	0.082108	0.035351	0.070678	–0.085008	–0.000639
6	0.075448	0.038920	0.066943	–0.650014	0.0087757
7	–0.094923	0.285899	–0.027123	–0.010866	0.0550755
8	–0.027584	0.279340	0.059601	–0.067850	–0.0962289
9	–0.025694	0.296672	0.065915	–0.703854	–0.1061876
10	0.102417	–0.281036	0.275059	0.019887	–0.0465343
11	–0.086954	0.031644	0.168250	–0.061066	–0.0404378
12	0.017868	–0.093100	–0.014454	0.729409	–0.1504808
13	–0.033282	–0.021051	–0.017641	0.111899	0.0501068
14	–0.042182	–0.012853	–0.026600	0.110398	0.0436235
15	–0.015921	–0.114207	–0.043489	0.031497	0.1212477
16	–0.169713	–0.230437	–0.187822	–0.013935	0.3495574
17	–0.135832	0.028744	–0.090077	–0.334810	0.2300555
18	–0.099994	0.282734	–0.274037	–0.016875	0.0494096
19	–0.002538	0.009010	0.023834	–0.019253	–0.0388320
20	–0.003404	0.184400	0.027698	–0.057857	–0.0652552
21	0.009028	0.129669	0.043423	–0.023922	–0.0763546
22	0.061255	0.243190	0.024494	–0.090222	–0.0890431
23	0.004803	0.127156	0.039533	–0.028134	–0.0794813
24	0.027758	0.000251	–0.020057	–0.012544	0.0310961
25	–0.034364	–0.005060	–0.026515	–0.018980	0.0263347
26	–0.039528	0.180948	0.022606	–0.063442	0.0698096
27	–0.093630	0.024017	0.162620	–0.069296	–0.0487989
28	–0.096731	0.019943	0.160013	–0.073736	–0.0533766
29	–0.035195	0.071852	–0.004998	–0.187614	0.0280352
30	–0.040942	0.066418	–0.010841	–0.192513	0.0231012
31	–0.044990	0.008001	–0.030147	0.105331	0.0384564
32	–0.105385	–0.001027	–0.076831	0.205000	0.0624141
33	–0.073241	–0.021808	–0.059987	0.096398	0.0130585
34	–0.049267	–0.000546	–0.035853	0.094512	0.0292014
35	–0.061556	–0.017719	–0.050152	0.081824	0.0124194

TABLE 8.26 Reactive values of GZBDF

m	$D_Q(m, 1)$	$D_Q(m, 2)$	$D_Q(m, 3)$	$D_Q(m, 4)$	$D_Q(m, 5)$
1	–0.002182	–0.009343	–0.015476	–0.008207	–0.014824
2	–0.008905	–0.003580	0.000340	–0.008284	–0.009895
3	–0.013982	–0.023079	–0.015055	–0.019915	–0.028873
4	–0.008081	–0.010263	–0.007886	–0.009935	–0.000662
5	–0.014779	–0.006302	–0.012094	–0.007383	0.001559
6	0.001227	–0.000347	0.001444	–0.014336	–0.000918
7	0.003093	0.006349	0.004457	0.012504	0.020430
8	0.013663	0.011508	0.012774	0.009285	0.006227
9	0.013782	0.015341	0.013950	0.008446	0.004414
10	–0.003192	–0.007270	–0.005417	–0.012083	–0.020324
11	0.012636	0.009334	0.003245	0.014513	0.016250
12	–0.050289	–0.052999	–0.051200	–0.035992	–0.044385
13	0.008425	0.009694	0.008804	0.002559	0.563566
14	–0.002131	–0.001077	–0.001880	–0.006877	–0.005043
15	0.005573	0.006597	0.006080	0.004353	–0.004348
16	0.008707	0.005436	0.007513	0.024219	0.049647
17	0.064896	0.070116	0.066020	0.051632	0.040675
18	0.001907	0.005776	0.003789	0.110055	0.019153
19	0.001721	–0.004846	0.000209	0.001373	0.001685
20	0.005055	0.010715	0.006379	0.000930	–0.001809
21	–0.001122	–0.008962	–0.003408	–0.001832	–0.117488
22	–0.005239	0.000651	0.004047	–0.010385	–0.012934
23	–0.001761	–0.009008	0.003921	–0.002480	–0.001390
24	0.006024	0.001299	0.004635	0.000432	–0.001274
25	0.002796	–0.001400	0.001497	0.001145	–0.003749
26	0.002118	0.008451	0.003583	–0.002036	–0.004305
27	0.007297	0.354703	–0.001894	0.008543	0.010268
28	0.003608	–0.000551	–0.005516	0.004333	0.006035
29	0.009438	0.009032	0.009161	0.008279	0.019962
30	0.006034	0.005815	0.005762	0.005214	0.016899
31	–0.000525	0.000372	–0.000358	–0.004297	0.003345
32	0.002777	0.003595	0.002661	–0.002536	–0.003554
33	0.002826	–0.003912	0.000737	–0.018217	–0.011602
34	–0.018582	–0.012415	–0.017173	–0.001332	–0.011601
35	0.018714	–0.014369	–0.017967	–0.001715	–0.013580

TABLE 8.27 'Best' power injection and voltage profile at buses: Evolutionary method

Bus	P_{Di} (p.u.)	Q_{Di} (p.u.)	P_i (p.u.)	Q_i (p.u.)	V_i (p.u.)	δ_i (rad)
1	2.00	0.65	–0.38257	1.53560	1.020	0.00
2	0.10	0.03	1.42814	–0.00727	0.969	0.55719
3	0.50	0.17	1.45818	–1.50528	0.995	0.38178
4	0.30d	0.10	0.86968	–0.64779	0.960	0.09264
5	0.25	0.08	0.89399	1.72288	0.960	0.22261
6	0.15	0.05	–0.15	–0.05	0.96271	0.41922
7	0.15	0.05	–0.15	–0.05	0.93913	0.33664
8	0.25	0.00	–0.25	0.00	0.93892	0.29884
9	0.15	0.05	–0.15	–0.05	0.93599	0.24040
10	0.15	0.05	–0.15	–0.05	0.94816	0.19580
11	0.05	0.00	–0.05	0.00	0.95261	0.17229
12	0.10	0.00	–0.10	0.00	0.94306	0.24789
13	0.25	0.08	–0.25	–0.08	0.96474	0.39580
14	0.20	0.07	–0.20	–0.07	0.93100	0.07681
15	0.30	0.10	–0.30	–0.10	0.93644	0.02386
16	0.30	0.10	–0.30	–0.10	0.95642	–0.00160
17	0.60	0.20	–0.60	–0.20	0.95596	0.16311
18	0.15	0.05	–0.15	–0.05	0.94753	0.11558
19	0.15	0.05	–0.15	–0.05	0.96011	0.09959
20	0.25	0.08	–0.25	–0.08	0.94717	0.04038
21	0.20	0.07	–0.20	–0.07	0.95047	–0.00131
22	0.20	0.07	–0.20	–0.07	0.95671	–0.04386
23	0.15	0.05	–0.15	–0.05	0.98906	–0.04005
24	0.15	0.05	–0.15	–0.05	0.97009	–0.09209
25	0.25	0.08	–0.25	–0.08	0.98155	–0.09308

then converted into normalized weights w_1, w_2, w_3, w_4, w_5 and w_6 in the range of 0 to 1 using Eq. (8.23). The fitness of each string of weight combinations searched by GA is obtained from satisfaction of participating objectives for that string using Eq. (8.17). The necessary input data is assumed, viz. size of population, P_S is 10, total length of string, l is 60, length of each sub-string, x_i is 10, crossover probability, p_c is 0.75, mutation probability, p_m is 0.10 and maximum number of generations, $R^{\max}$ is 40. During 14th generation of population, the 'best' compromised solution has been obtained. The fitness of all the weight strings of 'best' population, i.e., of 14th generation is shown in Figure 8.8. The string number 5 has the highest achieved fitness/satisfaction, i.e., 0.73481. The 'best' scalar weight α_i (i = 1, 2, …, L) searched by this string are 96.4809, 29.9120, 2.2483, 87.0968, 85.2395 and 27.0772, respectively. The 'best' power injection at buses of the system is given in Table 8.28. The active and reactive power flow on transmission lines are depicted in Table 8.29. The effect of population size on the performance of GA has been depicted in Figure 8.9. The lower size of population has fewer burdens per iteration on decision-maker as compared to high population size.

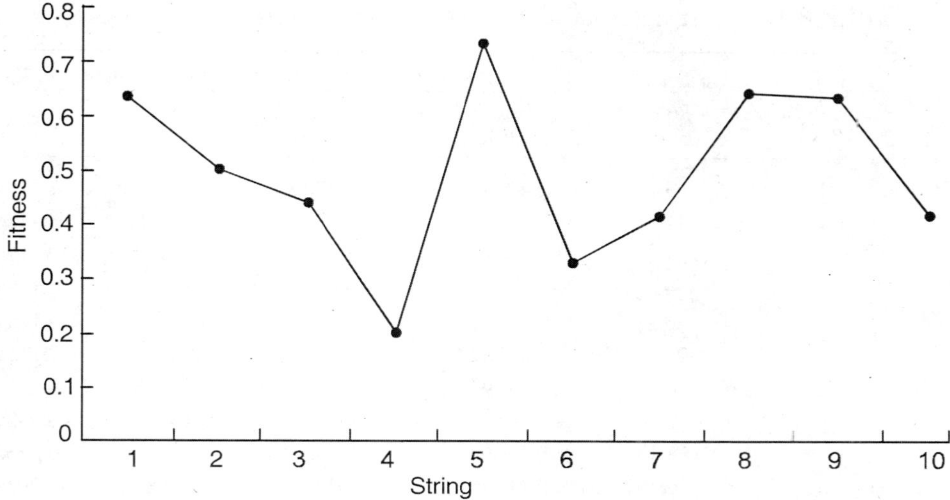

FIGURE 8.8 Fitness of all strings of 'best' population.

TABLE 8.28 'Best' power injection and voltage profile at buses: Genetic algorithm

Bus	P_{Di} (p.u.)	Q_{Di} (p.u.)	P_i (p.u.)	Q_i (p.u.)	V_i (p.u.)	δ_i (rad)
1	2.00	0.65	–0.41280	1.50132	1.020	0.00
2	0.10	0.03	1.42284	0.01893	0.969	0.33219
3	0.50	0.17	1.50957	–1.46524	0.995	0.24098
4	0.30	0.10	0.84458	–0.67204	0.960	0.08505
5	0.25	0.08	0.92228	1.64264	0.960	0.15512
6	0.15	0.05	–0.15	–0.05	0.96248	0.24198
7	0.15	0.05	–0.15	–0.05	0.95340	0.18168
8	0.25	0.00	–0.25	0.00	0.95385	0.15765
9	0.15	0.05	–0.15	–0.05	0.94713	0.12081
10	0.15	0.05	–0.15	–0.05	0.95508	0.10813
11	0.05	0.00	–0.05	0.00	0.95977	0.08995
12	0.10	0.00	–0.10	0.00	0.95672	0.12848
13	0.25	0.08	–0.25	–0.08	0.96475	0.23039
14	0.20	0.07	–0.20	–0.07	0.94533	0.01761
15	0.30	0.10	–0.30	–0.10	0.94889	–0.01695
16	0.30	0.10	–0.30	–0.10	0.96623	–0.02717
17	0.60	0.20	–0.60	–0.20	0.96314	0.08683
18	0.15	0.05	–0.15	–0.05	0.95254	0.06618
19	0.15	0.05	–0.15	–0.05	0.96159	0.07879
20	0.25	0.08	–0.25	–0.08	0.94727	0.03346
21	0.20	0.07	–0.20	–0.07	0.95065	–0.00715
22	0.20	0.07	–0.20	–0.07	0.95685	–0.04821
23	0.15	0.05	–0.15	–0.05	0.93918	–0.04160
24	0.15	0.05	–0.15	–0.05	0.97030	–0.09485
25	0.25	0.08	–0.25	–0.08	0.98171	–0.09499

TABLE 8.29 'Best' real and reactive power flows on transmission lines

Line No.	*Sending Bus*	*Ending Bus*	*Evolutionary* P_{Tm} (p.u.)	*Evolutionary* Q_{Tm} (p.u.)	*Genetic* P_{Tm} (p.u.)	*Genetic* Q_{Tm} (p.u.)
1	1	3	0.44338	1.15170	0.57092	1.31593
2	1	16	0.26733	0.44120	0.25718	0.42286
3	1	17	0.48471	0.01870	0.46590	0.01665
4	1	19	0.16424	0.09683	0.16109	0.10625
5	1	23	0.11351	0.09998	0.11992	0.10313
6	1	25	0.07952	0.11219	0.08691	0.11526
7	2	6	–0.18294	0.36117	–0.16856	0.34769
8	2	7	–0.16598	–0.04499	–0.14478	–0.03255
9	2	8	–0.16467	–0.03585	–0.14297	–0.02350
10	3	13	0.21396	–0.30183	0.19887	–0.29897
11	3	14	–0.15604	–0.18917	–0.14257	–0.17616
12	4	19	0.20759	–0.81381	0.19391	–0.82755
13	4	20	0.02192	0.09764	0.02885	0.08931
14	4	21	0.01091	0.00379	0.01613	–0.00517
15	5	10	0.11243	0.23833	0.10856	0.22145
16	5	17	–0.03115	0.62931	–0.02904	0.58770
17	5	19	–0.16467	0.65942	–0.15410	0.65175
18	6	13	–0.20337	0.30823	–0.18859	0.30275
19	7	8	–0.06097	–0.07662	–0.05358	–0.07007
20	7	12	–0.14816	–0.13792	–0.13451	–0.12518
21	8	9	–0.08142	–0.15958	–0.07215	–0.14847
22	8	17	–0.22227	–0.26730	–0.20498	–0.25260
23	9	10	–0.09285	–0.18603	–0.08446	–0.17424
24	10	11	–0.02201	0.04964	–0.02042	0.04682
25	11	17	–0.03730	0.01115	–0.03691	0.00887
26	12	17	–0.16259	–0.17971	–0.14989	–0.16638
27	14	15	–0.22631	–0.37990	–0.21328	–0.36408
28	15	16	–0.23897	–0.43542	–0.22676	–0.41908
29	17	18	–0.13958	0.13197	–0.13101	0.13826
30	18	19	–0.15338	0.09494	–0.14590	0.10166
31	20	21	–0.00426	–0.00633	0.00009	–0.01388
32	21	22	–0.04320	–0.04368	-0.03867	–0.05724
33	22	23	–0.08170	–0.14708	–0.08429	–0.15161
34	22	24	0.00298	–0.10457	0.00568	–0.11193
35	24	25	–0.03120	–0.17247	0.03258	0.17751

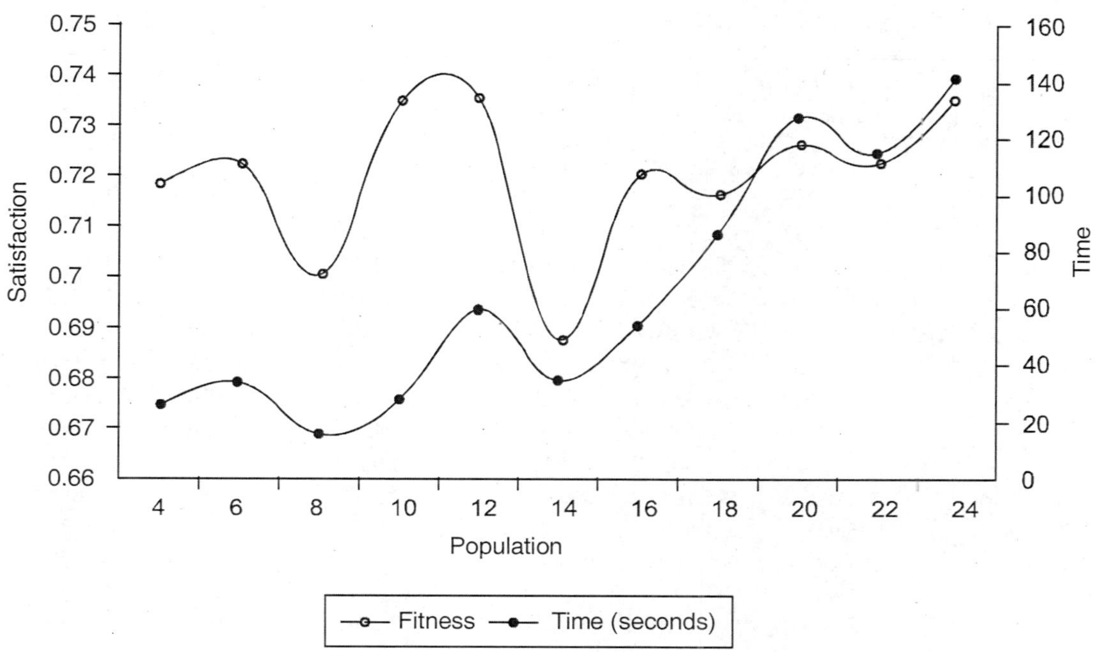

FIGURE 8.9 Performance of GA with different population sizes.

The weights shown in the comparison Table 8.30 are normalized weights assigned to participating objectives. The weight simulation technique is proved as having least satisfied results as compared to all weight pattern search method. This technique leads high burden on the decision-maker. As in this technique large numbers of non-inferior solutions are generated to be presented before the decision-maker, therefore the technique needs large time, to find best-compromised solution. The weight pattern search by evolutionary optimization method has shown faster convergence as compared to all other discussed methods. The method has achieved high satisfaction of participating objectives as compared to weight simulation technique, but this technique has attained lower satisfaction as compared to achieved satisfaction of weight pattern by GA. The results of multiobjective thermal power dispatch problem obtained by searching the weights assigned to participating objectives using GA have attained the high satisfaction of participating objectives as compared to all other methods. The GA also has taken less convergence time as compared to weight simulation technique. This is because for simulation technique, the required number of non-inferior solutions increases exponentially with increase in the number of participating objectives.

8.4 STOCHASTIC MULTIOBJECTIVE OPTIMAL GENERATION ALLOCATION

A stochastic multiobjective active and reactive power dispatch problem is formulated with explicit recognition of uncertainties in the system production cost coefficients, gaseous pollutant emission coefficients, and system's active and reactive power load demand, which are treated as random variables. The expected values of thermal fuel cost and minimal impacts on environments are the

TABLE 8.30 Comparison of methods with linear and exponential membership function of objectives.

	Linear membership function			*Exponential membership function*		
	Simulation	*Evolutionary*	*Genetic*	*Simulation*	*Evolutionary*	*Genetic*
P_{G1} (p.u.)	1.67603	1.61743	1.58720	1.67603	1.61743	1.57956
P_{G2} (p.u.)	1.53266	1.52814	1.52284	1.53266	1.52814	1.52272
P_{G3} (p.u.)	1.82588	1.95818	2.00957	1.82588	1.95818	2.06070
P_{G4} (p.u.)	1.23004	1.16968	1.14458	1.23004	1.16968	1.08467
P_{G5} (p.u.)	1.11618	1.14400	1.17228	1.11618	1.14400	1.19882
Q_{G1} (p.u.)	2.24976	2.18560	2.15132	2.24976	2.18560	2.08161
Q_{G2} (p.u.)	–0.09352	0.02273	0.04893	–0.09352	0.02273	0.11530
Q_{G3} (p.u.)	–1.44050	–1.33528	–1.29524	–1.44050	–1.33528	–1.23032
Q_{G4} (p.u.)	–0.45139	–0.54779	–0.57204	–0.45139	–0.54779	–0.54838
Q_{G5} (p.u.)	1.96594	1.80287	1.72264	1.96594	1.80288	1.62144
Cost (Rs/h)	7573.846	7596.708	7609.707	7573.346	7596.708	7615.767
NO_x (kg/h)	936.928	931.839	930.772	936.928	931.839	937.403
SO_2(kg/h)	6207.486	6062.069	6005.955	6207.486	6062.069	5862.637
CO_2 (ton/h)	12.606	12.687	12.664	12.606	12.687	12.682
Active power flow overloading	2.2660	2.2594	2.2849	2.2660	2.2594	2.3140
Reactive power flow overloading	31.034	25.194	26.221	31.034	25.194	22.898
w_1	0.90	0.66	0.2941	0.90	0.66	0.2084
w_2	0.10	0.16	0.0912	0.10	0.16	0.0790
w_3	0.00	0.01	0.0069	0.00	0.01	0.0076
w_4	0.00	0.05	0.2655	0.00	0.05	0.3077
w_5	0.00	0.07	0.2598	0.00	0.07	0.2329
w_6	0.00	0.05	0.0825	0.00	0.05	0.1643
m^{min}	**0.67924**	**0.72551**	**0.73481**	**0.88421**	**0.90687**	**0.90789**
Time in seconds	**53.83**	**8.84**	**26.91**	**54.54**	**8.90**	**38.39**

conflicting objectives, which need to be minimized simultaneously. Further the expectations of the square of the unsatisfied active and reactive load, because of the possible variance of generator outputs, are incorporated as other objectives to be minimized simultaneously. Basically, the solution procedure for the multiobjective problem is based on the generation of non-inferior solutions. The weighting technique being the simplest tool is used to generate non-inferior solutions, which allow explicit trade-off between conflicting objective levels. The weighting patterns are systematically searched using different optimization methods such as Evolutionary optimization method and Genetic Algorithm. The weightage patterns are used to generate the different non-inferior solutions. The 'best' compromised solution has been decided for the optimal weight pattern by exploiting fuzzy set theory. Being the imprecise nature of the decision-maker's judgement, it is assumed that the decision-maker has fuzzy goals for each of the objective functions. The fuzzy goals are quantified by defining their corresponding membership functions.

The weightage pattern is said to be optimal if it attains maximum satisfaction level from the membership functions of the participating objectives.

8.4.1 Stochastic Multiobjective Optimization Problem Formulation

Real world problems involve simultaneous optimization of several objective functions. So multiobjective optimization problem with such conflicting objective functions gives rise to a set of optimal solutions instead of one optimal solution. The reason for the optimality of many solutions is that no one solution can be considered to be better than any other solution with respect to all other objective functions. Such optimal solutions are known as non-inferior solutions. To resolve the conflicts between competing objective functions, decision-maker plays a vital role and provides a tool to select one compromising solution. Moreover for realistic problems every single datum used to achieve solutions may be uncertain and inaccurate. So inaccuracies and uncertainties are accounted for explicitly. A stochastic multiobjective thermal power dispatch problem is formulated with four important non-commensurable objective functions for an electrical thermal power system network. The multiobjectives taken into account are

i. expected operating cost,
ii. expected minimal impact on environment,
iii. variance of active power mismatch due to unsatisfied active load demand,
iv. variance of reactive power mismatch due to unsatisfied reactive power load demand.

The stochastic economic emission formulation is adopted, considering operating cost coefficients, emission coefficients, and active and reactive power demands as random variables. The stochastic models have been converted into their deterministic equivalents by taking their expected values, with the assumption that all the random variables are normally distributed and statistically dependent on each other. Expected operating cost and expected NO_x gaseous pollutants are described in Section 6.4.1, Chapter 6 and Section 6.5.1, Chaper 6 and are given by Eqs. (6.42) and (6.63), respectively.

Expected deviations

Generator outputs, P_{Gi} and Q_{Gi} are treated as random variables. Stochastic models are converted into their deterministic equivalents by taking their expected values. So the solutions will provide only the expected values of active and reactive power generations. By virtue of the above consideration, there will be a mismatch in active and reactive power demands and generations. So, the expected mismatch can be estimated and minimized by optimizing the following equations: (Section 6.4.1, Chapter 6)

$$\overline{F}_3 = \text{var}(P_{Gi}) + \sum_{i=1}^{NG} \sum_{\substack{j=1 \\ j \neq i}}^{NG} \text{cov}(P_{Gi}, P_{Gj}) \tag{8.55}$$

$$\overline{F}_4 = \sum_{i=1}^{NG} \text{var}(Q_{Gi}) + \sum_{i=1}^{NG} \sum_{\substack{j=1 \\ j \neq i}}^{NG} \text{cov}(Q_{Gi}, Q_{Gj}) \tag{8.56}$$

The above equations are rearranged using Eqs. (6.43) and (6.44) and are rewritten below:

$$\overline{F}_3 = \sum_{i=1}^{NG} \sum_{j=1}^{NG} S_{ij}\overline{P}_{Gi}\overline{P}_{Gj} \tag{8.57}$$

where

$$S_{ij} = \begin{cases} C_{P_{Gi}}^2 & ; \ i = j \\ R_{P_{Gi}P_{Gj}} C_{P_{Gi}} C_{P_{Gj}} & ; \ i \neq j \end{cases}$$

with

$C_{P_{Gi}}$ is the coefficient of variation of generated active power P_{Gi} at ith bus.
$R_{P_iP_j}$ is the correlation coefficient between active power injections at ith and jth buses.

$$\overline{F}_4 = \sum_{i=1}^{NG} \sum_{j=1}^{NG} T_{ij}\overline{Q}_{Gi}\overline{Q}_{Gj} \tag{8.58}$$

where

$$T_{ij} = \begin{cases} C_{Q_{Gi}}^2 & ; \ i = j \\ R_{Q_{Gi}Q_{Gj}} C_{Q_{Gi}} C_{Q_{Gj}} & ; \ i \neq j \end{cases}$$

with

$C_{Q_{Gi}}$ is the coefficient of variation of generated reactive power Q_{Gi} at ith bus.
$R_{Q_iQ_j}$ is the correlation coefficient between reactive power injections at ith and jth buses.

Equality and inequality constraints

Active and reactive power generations must meet their respective load demands plus losses, therefore equality constraints are imposed to ensure active and reactive power balance. This is stated as:

$$\sum_{i=1}^{NB} \overline{P}_{Di} - \sum_{i=1}^{NG} \overline{P}_{Gi} + \overline{P}_L = 0 \tag{8.59}$$

$$\sum_{i=1}^{NB} \overline{Q}_{Di} - \sum_{i=1}^{NG} \overline{Q}_{Gi} + \overline{Q}_L = 0 \tag{8.60}$$

where

NB is the number of buses in the power system network.
NG is the number of generators feeding generation to the power system network.
$\overline{P}_{Gi}$ is expected active power generation by ith generator.
$\overline{Q}_{Gi}$ is expected reactive power generation by ith generator.
$\overline{P}_{Di}$ is expected active power demand at ith bus.

$\bar{Q}_{Di}$ is expected reactive power demand at ith bus.

$\bar{P}_L$ is expected active power transmission loss.

$\bar{Q}_L$ is expected reactive power transmission loss.

The expected active and expected reactive power generations are bounded as inequality constraints given below.

$$\bar{P}_{Gi}^{\min} \le \bar{P}_{Gi} \le \bar{P}_{Gi}^{\max} \qquad (i = 1, 2, \ldots, \text{NG}) \tag{8.61}$$

$$\bar{Q}_{Gi}^{\min} \le \bar{Q}_{Gi} \le \bar{Q}_{Gi}^{\max} \qquad (i = 1, 2, \ldots, \text{NG}) \tag{8.62}$$

where

$\bar{P}_{Gi}^{\min}$ and $\bar{P}_{Gi}^{\max}$ are expected lower and upper limits of active power generation of ith generator, respectively.

$\bar{Q}_{Gi}^{\min}$ and $\bar{Q}_{Gi}^{\max}$ are expected lower and upper limits of reactive power generation of ith generator, respectively.

Expected transmission losses

Since perturbations in system voltages and their angles depend on the random system load demand, hence the loss coefficients become random. The expected active power transmission loss, using Taylor's series expansion is represented as:

$$\bar{P}_L = \sum_{i=1}^{\text{NB}} \sum_{j=1}^{\text{NB}} [\bar{\alpha}_{ij}(\bar{P}_i\bar{P}_j + \bar{Q}_i\bar{Q}_j) + \bar{\beta}_{ij}(\bar{Q}_i\bar{P}_j - \bar{P}_i\bar{Q}_j)] + \sum_{i=1}^{\text{NB}} \bar{\alpha}_{ii}[\text{var}(P_i) + \text{var}(Q_i)]$$
$$+ \sum_{i=1}^{\text{NB}} \sum_{\substack{j=1 \\ j\ne i}}^{\text{NB}} \bar{\alpha}_{ij}[\text{cov}(P_i, P_j) + \text{cov}(Q_i, Q_j)] + \sum_{i=1}^{\text{NB}} \sum_{\substack{j=1 \\ j\ne i}}^{\text{NB}} [(\bar{\beta}_{ji} - \bar{\beta}_{ij}) \text{cov}(P_i, Q_j)] \tag{8.63}$$

Substitute the expressions for variance and covariance to obtain:

$$\bar{P}_L = \sum_{i=1}^{\text{NB}} \sum_{j=1}^{\text{NB}} [\bar{\alpha}_{ij}(\bar{P}_i\bar{P}_j + \bar{Q}_i\bar{Q}_j) + \bar{\beta}_{ij}(\bar{Q}_i\bar{P}_j - \bar{P}_i\bar{Q}_j)] + \sum_{i=1}^{\text{NB}} \bar{\alpha}_{ii}[C_{P_i}^2\bar{P}_i^2 + C_{Q_i}^2\bar{Q}_i^2]$$
$$+ \sum_{i=1}^{\text{NB}} \sum_{\substack{j=1 \\ j\ne i}}^{\text{NB}} \bar{\alpha}_{ij}[R_{P_iP_j}C_{P_i}C_{P_j}\bar{P}_i\bar{P}_j + R_{Q_iQ_j}C_{Q_i}C_{Q_j}\bar{Q}_i\bar{Q}_j]$$
$$+ \sum_{i=1}^{\text{NB}} \sum_{\substack{j=1 \\ j\ne i}}^{\text{NB}} [(\bar{\beta}_{ji} - \bar{\beta}_{ij}) R_{P_iQ_j}C_{P_i}C_{Q_j}\bar{P}_i\bar{Q}_j] \tag{8.64}$$

where

$$\bar{\alpha}_{ij} = \frac{\tau_{ij} R_{ij}}{|\bar{V}_i||\bar{V}_j|} \cos(\bar{\delta}_i - \bar{\delta}_j)$$

$$\bar{\beta}_{ij} = \frac{\tau_{ij} R_{ij}}{|\bar{V}_i||\bar{V}_j|} \sin(\bar{\delta}_i - \bar{\delta}_j)$$

$$\bar{P}_i = \bar{P}_{Gi} - \bar{P}_{Di}$$

$$\bar{Q}_i = \bar{Q}_{Gi} - \bar{Q}_{Di}$$

$$\tau_{ij} = \begin{cases} 1.0 + 3.0 \dfrac{\text{var}(V_i)}{|\bar{V}_i|^2} & ;\text{ if } i = j \\ 1.0 + \dfrac{\text{var}(V_i)}{|\bar{V}_i|^2} + \dfrac{\text{var}(V_j)}{|\bar{V}_j|^2} + \dfrac{\text{cov}(V_i, V_j)}{|\bar{V}_i||\bar{V}_j|} - \dfrac{\text{var}(\delta_i) + \text{var}(\delta_j)}{2.0} + \text{cov}(\delta_i, \delta_j) & ;\text{ if } i \neq j \end{cases}$$

with

$|\bar{V}_i|$ and $|\bar{V}_j|$ are the voltage magnitudes at the *i*th and *j*th buses respectively.

$\bar{\delta}_i$ and $\bar{\delta}_j$ are the voltage phase angles at the *i*th and *j*th buses respectively.

R_{ij} is the real component of the element of bus impedance matrix (z_{ij}).

$\bar{P}_i$ and $\bar{P}_j$ are the expected active powers injected at *i*th and *j*th buses respectively.

$\bar{Q}_i$ and $\bar{Q}_j$ are the expected reactive powers injected at *i*th and *j*th buses respectively.

C_{P_i} and C_{P_j} are the coefficients of variation of active power injections at *i*th and *j*th buses respectively.

C_{Q_i} and C_{Q_j} are the coefficients of variation of reactive power injections at *i*th and *j*th buses respectively.

$R_{P_iP_j}$ is the correlation coefficient between active power injections at *i*th and *j*th buses.

$R_{Q_iQ_j}$ is the correlation coefficient between reactive power injections at *i*th and *j*th buses.

Similarly expected reactive power loss on transmission lines of power system is given by the following equation:

$$\bar{Q}_L = \sum_{i=1}^{NB}\sum_{j=1}^{NB} [\bar{\gamma}_{ij}(\bar{P}_i\bar{P}_j + \bar{Q}_i\bar{Q}_j) + \bar{\eta}_{ij}(\bar{Q}_i\bar{P}_j - \bar{P}_i\bar{Q}_j)] + \sum_{i=1}^{NB} [\bar{\gamma}_{ii}(\text{var}(P_i) + \text{var}(Q_i))]$$
$$+ \sum_{i=1}^{NB}\sum_{\substack{j=1 \\ j\neq i}}^{NB} \bar{\gamma}_{ij}[\text{cov}(P_i, P_j) + \text{cov}(Q_i, Q_j)] + \sum_{i=1}^{NB}\sum_{\substack{j=1 \\ j\neq i}}^{NB} [(\bar{\eta}_{ji} - \bar{\eta}_{ij})\,\text{cov}(P_i, Q_j)] \tag{8.65}$$

Substitute the expressions for variance and covariance to obtain:

$$\bar{Q}_L = \sum_{i=1}^{NB}\sum_{j=1}^{NB} [\bar{\gamma}_{ij}(\bar{P}_i\bar{P}_j + \bar{Q}_i\bar{Q}_j) + \bar{\eta}_{ij}(\bar{Q}_i\bar{P}_j - \bar{P}_i\bar{Q}_j)] + \sum_{i=1}^{NB} \bar{\gamma}_{ii}[C_{P_i}^2\bar{P}_i^2 + C_{Q_i}^2\bar{Q}_i^2]$$
$$+ \sum_{i=1}^{NB}\sum_{\substack{j=1\\ j\neq i}}^{NB} \bar{\gamma}_{ij}[R_{P_iP_j}C_{P_i}C_{P_j}\bar{P}_i\bar{P}_j + R_{Q_iQ_j}C_{Q_i}C_{Q_j}\bar{Q}_i\bar{Q}_j] + \sum_{i=1}^{NB}\sum_{\substack{j=1\\ j\neq i}}^{NB} [(\bar{\eta}_{ji} - \bar{\eta}_{ij})R_{P_iQ_j}C_{P_i}C_{Q_j}\bar{P}_i\bar{Q}_j] \tag{8.66}$$

where

$$\bar{\gamma}_{ij} = \frac{\tau_{ij}X_{ij}}{|\bar{V}_i||\bar{V}_j|}\cos(\bar{\delta}_i - \bar{\delta}_j)$$

$$\bar{\eta}_{ij} = \frac{\tau_{ij}X_{ij}}{|\bar{V}_i||\bar{V}_j|}\cos(\bar{\delta}_i - \bar{\delta}_j)$$

with

X_{ij} is the reactive component of the element of bus impedance matrix (z_{ij}).

Aggregating the above equations, the deterministic equivalent of stochastic multiobjective optimization problem is defined as below:

$$\text{Minimize} \quad [\bar{F}_1(\bar{P}_{Gi}),\ \bar{F}_2(\bar{P}_{Gi}),\ \bar{F}_3(\bar{P}_{Gi}),\ \bar{F}_4(\bar{Q}_{Gi})]^T \tag{8.67a}$$

$$\text{Subject to:} \quad \sum_{i=1}^{NB} \bar{P}_{Di} - \sum_{i=1}^{NG} \bar{P}_{Gi} + \bar{P}_L = 0 \tag{8.67b}$$

$$\sum_{i=1}^{NB} \bar{Q}_{Di} - \sum_{i=1}^{NG} \bar{Q}_{Gi} + \bar{Q}_L = 0 \tag{8.67c}$$

$$\bar{P}_{Gi}^{\min} \le \bar{P}_{Gi} \le \bar{P}_{Gi}^{\max} \quad (i = 1, 2, \ldots, NG) \tag{8.67d}$$

$$\bar{Q}_{Gi}^{\min} \le \bar{Q}_{Gi} \le \bar{Q}_{Gi}^{\max} \quad (i = 1, 2, \ldots, NG) \tag{8.67e}$$

where, $\bar{F}_1(\bar{P}_{Gi}), \bar{F}_2(\bar{P}_{Gi}), \bar{F}_3(\bar{P}_{Gi})$ and $\bar{F}_4(\bar{P}_{Gi})$ are the objective functions to be minimized over the set of admissible decision variables $\bar{P}_{Gi}$ and $\bar{Q}_{Gi}$ ($i = 1, 2, \ldots, NG$).

8.4.2 Solution Approach

To generate the non-inferior solutions of the multiobjective optimization problem, the weighting method is applied. In this method, the multiobjective problem is converted into a scalar optimization problem.

$$\text{Minimize} \quad \sum_{k=1}^{L} w_k \overline{F}_k \tag{8.68a}$$

$$\text{Subject to:} \quad \sum_{i=1}^{NB} \overline{P}_{Di} - \sum_{i=1}^{NG} \overline{P}_{Gi} + \overline{P}_L = 0 \tag{8.68b}$$

$$\sum_{i=1}^{NB} \overline{Q}_{Di} - \sum_{i=1}^{NG} \overline{Q}_{Gi} + \overline{Q}_L = 0 \tag{8.68c}$$

$$\overline{P}_{Gi}^{\min} \le \overline{P}_{Gi} \le \overline{P}_{Gi}^{\max} \quad (i = 1, 2, \ldots, \text{NG}) \tag{8.68d}$$

$$\overline{Q}_{Gi}^{\min} \le \overline{Q}_{Gi} \le \overline{Q}_{Gi}^{\max} \quad (i = 1, 2, \ldots, \text{NG}) \tag{8.68e}$$

$$\sum_{k=1}^{L} w_k = 1.0, \; w_k \ge 0.0 \quad (k = 1, 2, \ldots, L) \tag{8.68f}$$

where α_k are the levels of the assigned scalar weightages and are generated in the range of [0, 100] by giving suitable variations. w_k are the normalized weights. This approach yields meaningful results when solved many times for different values of weights. To solve the scalar optimization problem, the Lagrangian function is defined as:

$$L(\overline{P}_{Gi}, \overline{Q}_{Gi}, \lambda_P, \lambda_Q) = \sum_{j=1}^{L} w_j \overline{F}_j + \lambda_P \left(\sum_{i=1}^{NB} \overline{P}_{Di} + \overline{P}_L - \sum_{i=1}^{NG} \overline{P}_{Gi} \right) + \lambda_Q \left(\sum_{i=1}^{NB} \overline{Q}_{Di} + \overline{Q}_L - \sum_{i=1}^{NG} \overline{Q}_{Gi} \right) \tag{8.69}$$

where

λ_P and λ_Q are the Lagrangian multipliers.

The necessary conditions to minimize the unconstrained Lagrangian function are:

$$\frac{\partial L}{\partial \overline{P}_{Gi}} = \sum_{j=1}^{L-1} w_j \frac{\partial \overline{F}_j}{\partial \overline{P}_{Gi}} + \lambda_P \left(\frac{\partial \overline{P}_L}{\partial \overline{P}_{Gi}} - 1 \right) + \lambda_Q \left(\frac{\partial \overline{Q}_L}{\partial \overline{P}_{Gi}} \right) = 0 \quad (i = 1, 2, \ldots, \text{NG}) \tag{8.70a}$$

$$\frac{\partial L}{\partial \overline{Q}_{Gi}} = w_j \frac{\partial \overline{F}_j}{\partial \overline{Q}_{Gi}} + \lambda_P \left(\frac{\partial \overline{P}_L}{\partial \overline{Q}_{Gi}} \right) + \lambda_Q \left(\frac{\partial \overline{Q}_L}{\partial \overline{Q}_{Gi}} \right) = 0 \quad (i = 1, 2, \ldots, \text{NG}) \tag{8.70b}$$

$$\frac{\partial L}{\partial \lambda_P} = \sum_{i=1}^{NB} \overline{P}_{Di} + \overline{P}_L - \sum_{i=1}^{NG} \overline{P}_{Gi} = 0 \tag{8.70c}$$

$$\frac{\partial L}{\partial \lambda_Q} = \sum_{i=1}^{NB} \overline{Q}_{Di} + \overline{Q}_L - \sum_{i=1}^{NG} \overline{Q}_{Gi} = 0 \tag{8.70d}$$

The Newton-Raphson algorithm has been applied to solve the above non-linear equations, for each searched scalar weight combination. Weights are searched using different optimization techniques such as Evolutionary optimization method and GA. Inequality constraints imposed to problem of Eq. (8.68) are taken into consideration by fixing the generation of generators to their respective upper or lower limits whenever these are violated. To fix up the limits of active and reactive power generations by the committed generating units, the strategy adopted has been discussed in Section 8.2.2.

The decision making is based on fuzzy satisfaction of various objectives. The solution is expected within minimum $\overline{F}_i^{\min}$ and maximum $\overline{F}_i^{\max}$ values of ith objective function. The value of the membership function indicates how much a solution is satisfying the $\overline{F}_i$ objective. By taking account of the minimum and maximum values of each objective function together with the rate of change of membership satisfaction, the membership function $\mu(\overline{F}_i)$ $(i = 1, 2, \ldots, L)$ is defined as follows:

$$\mu(\overline{F}_i) = \begin{cases} 1 & ; \overline{F}_i \le \overline{F}_i^{\min} \\ \dfrac{\overline{F}_i^{\max} - \overline{F}_i}{\overline{F}_i^{\max} - \overline{F}_i^{\min}} & ; \overline{F}_i^{\min} \le \overline{F}_i \le \overline{F}_i^{\max} \\ 0 & ; \overline{F}_i \ge \overline{F}_i^{\max} \end{cases} \tag{8.71}$$

Decision making has been done based on the min–max technique and overall fitness function f is selected as follows:

$$f = [\text{Min}\{\mu(\overline{F}_j) \ \ (j = 1, 2, \ldots, L\}] \tag{8.72}$$

Normally multiobjective optimization problem is solved by weighting method in which the assigned weights to objectives are varied in a systematic manner, in order to generate the non-inferior solutions for optimization problem of Eq. (8.70). Two search methods: Evolutionary optimization and Genetic algorithm are applied to search the assigned weightage to the participating objectives to generate non-inferior solutions that are further provided to decision-maker to select the best compromising one. Evolutionary optimization technique to search the optimal weight pattern is explained in Section 8.2.4 and stepwise procedure is elaborated in Algorithm 8.1. Genetic algorithm to search the optimal weight pattern is explained in Section 8.2.5 and stepwise procedure is elaborated in Algorithm 8.2

8.4.3 Test System and Results

The results of the evolutionary search method and Genetic algorithm to search the optimal weight pattern have been obtained on a test power system of IEEE; 11-bus and 17-lines. The expected operating cost, expected gaseous emission pollutant coefficients and the expected operating generation limits of five committed generating units are given in Tables 8.31 and 8.32. The load data and line data given in Tables 8.33 and 8.3. Normalized weights have been assigned to different objectives to form the Lagrangian function of the multiobjective problem and to solve it using Newton-Raphson method to optimize the power generation schedules. The problem is analyzed under four different cases considering perturbations in variance and covariance of random variables. The following values of coefficients of variation and co-variation of various parameters and variables of the system have been assumed for simulation of their randomness. Practically there is a need to take the exact values of these parameters as and when required.

TABLE 8.31 Expected operating cost and gaseous pollutant emission coefficients of generators for 11-bus power system

Generator No.	$\bar{a}_i$ ₹/(MW^{2}h)	$\bar{b}_i$ ₹/(MWh)	$\bar{c}_i$ ₹/h	$\bar{d}_i$ kg/(MW^{2}h)	$\bar{e}_i$ kg/MWh	$\bar{f}_i$ kg/h
1	0.0015	1.80	040.0	0.0120	–0.555	30.0
2	0.0030	1.80	060.0	0.0150	–0.555	50.0
3	0.0012	2.10	100.0	0.0105	–1.355	60.0
4	0.0080	2.00	025.0	0.0180	–0.805	80.0
5	0.0010	2.00	120.0	0.0080	–0.600	45.0

TABLE 8.32 Generation limits of generators for 11-bus power system.

Generator No.	$\bar{P}_{Gi}^{\min}$ (p.u.)	$\bar{P}_{Gi}^{\max}$ (p.u.)	$\bar{Q}_{Gi}^{\min}$ (p.u.)	$\bar{Q}_{Gi}^{\max}$ (p.u.)
1	0.5	3.0	–0.25	1.5
2	0.2	1.3	–0.45	1.0
3	0.3	1.8	–0.30	1.0
4	0.1	1.0	–0.50	1.0
5	0.4	2.5	–0.50	0.5

TABLE 8.33 11-bus electric power system: load data

Bus no.	*Bus type*	P_{di}(MW)	Q_{di}(MVAR)	$\|V_i\|$ (p.u.)	δ_i(rad)
1	Slack	20.0	05.0	1.070	0.0
2	PV	20.0	05.0	1.088	0.0
3	PV	30.0	10.0	1.095	0.0
4	PV	25.0	05.0	1.062	0.0
5	PV	25.0	05.0	1.046	0.0
6	PQ	10.0	02.0	1.0	0.0
7	PQ	40.0	10.0	1.0	0.0
8	PQ	90.0	45.0	1.0	0.0
9	PQ	70.0	35.0	1.0	0.0
10	PQ	25.0	05.0	1.0	0.0
11	PQ	25.0	05.0	1.0	0.0
Total Load		380.0	132.0		

Case I: All the coefficients of variation and correlation coefficients pairwise are taken as zero. It means all the variables and parameters are considered as deterministic and independent of each other.

$$C_{a_i} = C_{b_i} = C_{d_i} = C_{e_i} = 0;\ C_{P_{Gi}} = C_{Q_{Gi}} = 0 \qquad (i = 1, 2, \ldots, \text{NG})$$

$$C_{P_i} = C_{Q_i} = 0,\ C_{V_i} = C_{\delta_i} = 0 \qquad (i = 1, 2, \ldots, \text{NB})$$

$$R_{a_iP_{Gi}} = R_{b_iP_{Gi}} = R_{d_iP_{Gi}} = R_{e_iP_{Gi}} = 0 \qquad (i = 1, 2, \ldots, \text{NG})$$

$$R_{P_{Gi}P_{Gj}} = R_{Q_{Gi}Q_{Gj}} = R_{P_{Gi}Q_{Gj}} = R_{Q_{Gi}P_{Gj}} = 0 \qquad (i = 1, 2, \ldots, \text{NG})$$

$$R_{P_iP_j} = R_{Q_iQ_j} = R_{P_iQ_j} = R_{Q_iP_j} = 0 \qquad (i = 1, 2, \ldots, \text{NB})$$

$$R_{V_iV_j} = R_{\delta_i\delta_j} = 0 \qquad (i = 1, 2, \ldots, \text{NB})$$

Case II: Operating cost and gaseous emission coefficients of generators and power generations have been considered as random but independent and uncorrelated pair-wise. Bus voltages have been treated as deterministic such as:

$$C_{a_i} = C_{b_i} = C_{d_i} = C_{e_i} = 0.10; C_{P_{Gi}} = C_{Q_{Gi}} = 0.05 \quad (i = 1, 2, \ldots, \text{NG})$$

$$C_{P_i} = C_{Q_i} = 0.05, C_{V_i} = C_{\delta_i} = 0 \quad (i = 1, 2, \ldots, \text{NB})$$

$$R_{a_iP_{Gi}} = R_{b_iP_{Gi}} = R_{d_iP_{Gi}} = R_{e_iP_{Gi}} = 0 \quad (i = 1, 2, \ldots, \text{NG})$$

$$R_{P_{Gi}P_{Gj}} = R_{Q_{Gi}Q_{Gj}} = R_{P_{Gi}Q_{Gj}} = R_{Q_{Gi}P_{Gj}} = 0 \quad (i = 1, 2, \ldots, \text{NG})$$

$$R_{P_iP_j} = R_{Q_iQ_j} = R_{P_iQ_j} = R_{Q_iP_j} = 0 \quad (i = 1, 2, \ldots, \text{NB})$$

$$R_{V_iV_j} = R_{\delta_i\delta_j} = 0 \quad (i = 1, 2, \ldots, \text{NB})$$

Case III: Characteristics coefficients of generators have been considered as random and dependent on active power generations. Power generation and voltage magnitudes are taken as random but are independent and uncorrelated pair-wise to each other such that

$$C_{a_i} = C_{b_i} = C_{d_i} = C_{e_i} = 0.10; \; C_{P_{Gi}} = C_{Q_{Gi}} = 0.05 \quad (i = 1, 2, \ldots, \text{NG})$$

$$C_{P_i} = C_{Q_i} = 0.05; C_{V_i} = C_{\delta_i} = 0.02 \quad (i = 1, 2, \ldots, \text{NB})$$

$$R_{a_iP_{Gi}} = R_{b_iP_{Gi}} = R_{d_iP_{Gi}} = R_{e_iP_{Gi}} = 1.0 \quad (i = 1, 2, \ldots, \text{NG})$$

$$R_{P_{Gi}P_{Gj}} = R_{Q_{Gi}Q_{Gj}} = R_{P_{Gi}Q_{Gj}} = R_{Q_{Gi}P_{Gj}} = 0 \quad (i = 1, 2, \ldots, \text{NG})$$

$$R_{P_iP_j} = R_{Q_iQ_j} = R_{P_iQ_j} = R_{Q_iP_j} = 0 \quad (i = 1, 2, \ldots, \text{NB})$$

$$R_{V_iV_j} = R_{\delta_i\delta_j} = 0 \quad (i = 1, 2, \ldots, \text{NB})$$

Case IV: Characteristics coefficients of generators have been treated as random and also correlated to active power generations. All other variables are correlated pair-wise to each other such that

$$C_{a_i} = C_{b_i} = C_{d_i} = C_{e_i} = 0.10; C_{P_{Gi}} = C_{Q_{Gi}} = 0.05 \quad (i = 1, 2, \ldots, \text{NG})$$

$$C_{P_i} = C_{Q_i} = 0.05; C_{V_i} = C_{\delta_i} = 0.02 \quad (i = 1, 2, \ldots, \text{NB})$$

$$R_{a_iP_{Gi}} = R_{b_iP_{Gi}} = R_{d_iP_{Gi}} = R_{e_iP_{Gi}} = 1.0 \quad (i = 1, 2, \ldots, \text{NG})$$

$$R_{P_{Gi}P_{Gj}} = R_{Q_{Gi}Q_{Gj}} = R_{P_{Gi}Q_{Gj}} = R_{Q_{Gi}P_{Gj}} = 0.5 \quad (i = 1, 2, \ldots, \text{NG})$$

$$R_{P_iP_j} = R_{Q_iQ_j} = R_{P_iQ_j} = R_{Q_iP_j} = 0.5 \quad (i = 1, 2, \ldots, \text{NB})$$

$$R_{V_iV_j} = R_{\delta_i\delta_j} = 0.5 \quad (i = 1, 2, \ldots, \text{NB})$$

The minimum and maximum values of objectives are given below in Tables 8.34 for the power test system undertaken for study.

TABLE 8.34 The minimum and maximum values of objectives

Case No.	$\overline{F}_1^{\min}$ ₹/h	$\overline{F}_1^{\max}$ ₹/h	$\overline{F}_2^{\min}$ kg/h	$\overline{F}_2^{\max}$ kg/h	$\overline{F}_3^{\min}$ (MW)^2	$\overline{F}_3^{\max}$ (MW)^2	$\overline{F}_4^{\min}$ (MVAR)^2	$\overline{F}_4^{\max}$ (MVAR)^2
I	1153.797	1180.055	310.0250	400.3659	0.0	1.0	0.0	1.0
II	1154.074	1182.177	311.0057	401.1603	76.5709	91.2409	4.1360	6.3353
III	1158.527	1186.791	313.1308	403.5071	76.5940	91.2273	4.2152	6.4368
IV	1158.492	1184.905	313.1192	411.9046	228.1772	240.6192	5.7871	16.0451

Evolutionary search optimization method

Evolutionary optimization method as discussed in Section 8.2.4 has been implemented to search the normalized weight pattern to solve the stochastic multiobjective thermal power dispatch problem and is examined on a sample system consisting of 11-bus, 17-lines and 5-generators. All the four objectives that are operating cost, gaseous pollutant emission, variance of active power generation and variance of reactive power generation are optimized simultaneously and results are obtained by applying Algorithm 8.1.

As there are a total of (N = 4) weights and (N = 4–1) of them are to be generated according to Eq. (8.19) and remaining weight is generated using Eq. (8.20). The three binary bits can be represented in 2^3 (eight) possible different combinations as shown in Table 8.35. The eight binary bit combinations are the corners of hypercube away from the point around which hypercube is generated. The distance Δ of hypercube corners from its centre (w_2^c, w_3^c, w_4^c) is generated from binary digits as given in Table 8.35. Table 8.36 shows the results in terms of optimized weight combinations, objective function values and their membership function values corresponding to 'best' solutions. Optimized active power (MW) and reactive power generation (MVAR) schedules are shown in Table 8.37 for all the cases considered.

Initially, the equal weights have been allocated to all the objectives and they forms the base point. The starting value of distance Δ of hypercube corners from the base point has been selected as 0.14 for Case I, 0.16 for Case II, 0.14 for Case III and 0.21 for Case IV. Table 8.36 shows that for all the cases under study, the highest weightage has been assigned to either the cost or the active and/or reactive power variance and least weightage is given to gaseous emission. Level of attainment of fuzzy satisfaction of objectives is very much non-uniform for Case IV.

Genetic algorithm

This section presents the results obtained for stochastic multiobjective thermal power dispatch problem by searching the weightage assigned to objectives by GA as discussed in Section 8.2.5.

TABLE 8.35 Generation of hypercube corners for four objectives

Hyper-cube corners	*Possible combinations of three binary bits* b_2	b_1	b_0	*Distance of hypercube corners from centre*			*Possible generated weights at the hypercube corners*		
1	0	0	0	$-\Delta$	$-\Delta$	$-\Delta$	$w_2^c -\Delta$	$w_3^c -\Delta$	$w_4^c -\Delta$
2	0	0	1	$-\Delta$	$-\Delta$	$+\Delta$	$w_2^c -\Delta$	$w_3^c -\Delta$	$w_4^c +\Delta$
3	0	1	0	$-\Delta$	$+\Delta$	$-\Delta$	$w_2^c -\Delta$	$w_3^c +\Delta$	$w_4^c -\Delta$
4	0	1	1	$-\Delta$	$+\Delta$	$+\Delta$	$w_2^c -\Delta$	$w_3^c +\Delta$	$w_4^c +\Delta$
5	1	0	0	$+\Delta$	$-\Delta$	$-\Delta$	$w_2^c +\Delta$	$w_3^c -\Delta$	$w_4^c -\Delta$
6	1	0	1	$+\Delta$	$-\Delta$	$+\Delta$	$w_2^c +\Delta$	$w_3^c -\Delta$	$w_4^c +\Delta$
7	1	1	0	$+\Delta$	$+\Delta$	$-\Delta$	$w_2^c +\Delta$	$w_3^c +\Delta$	$w_4^c -\Delta$
8	1	1	1	$+\Delta$	$+\Delta$	$+\Delta$	$w_2^c +\Delta$	$w_3^c +\Delta$	$w_4^c +\Delta$

TABLE 8.36 'Best' solutions using weight search by evolutionary method

w_i, F_i, $\mu(\overline{F}_i)$	*Case I*	*Case II*	*Case III*	*Case IV*
w_1	0.39	0.41	0.39	0.04
w_2	0.11	0.09	0.11	0.04
w_3	0.11	0.41	0.39	0.46
w_4	0.39	0.09	0.11	0.46
$\overline{F}_1$, ₹/h	1159.880	1161.163	1166.665	1176.409
$\overline{F}_2$, kg/h	332.6545	335.2182	333.5607	333.1676
$\overline{F}_3$, $(MW)^2$	0.0	81.4505	81.4214	228.9275
$\overline{F}_4$, $(MVAR)^2$	0.0	4.7457	4.8309	12.8973
$\mu(\overline{F}_1)$	0.7684	0.7478	0.7121	0.3217
$\mu(\overline{F}_2)$	0.7495	0.7314	0.7739	0.7971
$\mu(\overline{F}_3)$	1.0	0.6674	0.6701	0.9397
$\mu(\overline{F}_4)$	1.0	0.7228	0.7228	0.3069

TABLE 8.37 Power generation schedules by evolutionary method

Case No.	P_{G1}	P_{G2}	P_{G3}	P_{G4}	P_{G5}	P_{Q1}	P_{Q2}	P_{Q3}	P_{Q4}	P_{Q5}
I	93.383	69.250	91.702	31.985	104.47	21.001	26.212	03.088	21.875	22.293
II	93.021	75.080	85.817	39.551	96.755	21.020	23.214	08.071	19.361	21.853
III	91.094	73.377	88.175	40.355	97.380	21.159	23.136	08.920	19.652	21.991
IV	83.262	76.429	84.309	56.900	89.082	19.922	17.377	11.068	20.360	23.602

The GA initially searches the scalar weightage of objectives in the range of 0 to 100, and then these weights are normalized. The problem is analyzed to minimize four objectives simultaneously like operating cost, gaseous pollutant emission, variance of active power and variance of reactive power to which scalar weightages α_1, α_2, α_3 and α_4 respectively are assigned. Scalar weights α_1, α_2, α_3 and α_4 are searched between 0 and 100 by GA and necessary input data are assumed. Algorithm has been implemented on three test systems to find the best weight combination set and hence to find the optimum generation schedules. Size of population P_S has been taken as 20 for all the test systems. Maximum number of generations IT^{max} is taken as 20. Total length of chromosome L_s is 96; length of each sub-string, l_i is 24. Crossover probability p_c equal to 0.8 gave satisfactory results for all the cases.

With the above presumed input data along with assumed mutation probability p_m of 0.015 for Cases I and IV and 0.01 for Cases II and III, the optimized weight sets have been obtained utilizing GA for the stochastic multiobjective thermal power dispatch problem. The 'best' solutions in terms of objective values and membership function values are presented in Table 8.38. 'Best' solution has been found in 9th generation for Case I, in 16th generation for Case II and III and 18th for Case IV. Table 8.39 depicts the optimized power generation schedules (MW, MVAR).

TABLE 8.38 'Best' solutions by weight search by genetic algorithm

w_i, F_i, $\mu(\overline{F}_i)$	*Case I*	*Case II*	*Case III*	*Case IV*
w_1	0.346749	0.340429	0.340429	0.103507
w_2	0.101792	0.067792	0.067792	0.000847
w_3	0.333777	0.472056	0.472056	0.695297
w_4	0.217682	0.119723	0.119723	0.200349
$\overline{F}_1$, ₹/h	1160.126	1161.943	1166.400	1171.419
$\overline{F}_2$, kg/h	331.7934	336.0529	338.2448	358.4079
$\overline{F}_3$, (MW)^2	0.0	80.6552	80.6964	228.5499
$\overline{F}_4$, (MVAR)^2	0.0	4.5961	4.6849	12.3196
$\mu(\overline{F}_1)$	0.7590	0.7200	0.7215	0.5106
$\mu(\overline{F}_2)$	0.7590	0.7222	0.7221	0.5415
$\mu(\overline{F}_3)$	1.0	0.7216	0.7197	0.9700
$\mu(\overline{F}_4)$	1.0	0.7908	0.7886	0.3632

TABLE 8.39 Power generation schedules by genetic algorithm

Case No.	P_{G1}	P_{G2}	P_{G3}	P_{G4}	P_{G5}	P_{Q1}	P_{Q2}	P_{Q3}	P_{Q4}	P_{Q5}
I	92.886	68.979	92.185	32.285	104.47	20.973	26.257	03.061	21.947	22.288
II	92.595	76.539	84.237	41.997	94.712	20.812	22.251	09.440	18.898	21.539
III	92.609	76.492	84.298	41.923	94.802	20.996	22.469	09.549	19.116	21.717
IV	91.393	88.121	69.771	53.684	86.293	20.358	20.975	07.479	18.838	22.309

Comparison of Results

Optimized solutions of stochastic multiobjective thermal power dispatch problem by different methods show that weight combinations and hence the power generation schedules are different for different situations represented by presuming the values of coefficients of variation of random parameters and variables. Hence, there is a need to consider these variations while performing the generation scheduling to meet the load demand. The results of selected non-inferior solutions in terms of 'best' weight combinations, objective function values, membership function values and generation schedules for different simulated cases obtained by three different optimization methods have already been discussed in the previous section.

Comparison of results in terms of active and reactive power losses, incremental cost, fitness function, and execution time, for all the cases and with all the three methods of optimization, has been given in Table 8.40. From Table 8.40 it can be observed that genetic algorithm has obtained that highest overall fitness f. Depending upon the results obtained for the test power system, GA in general has given best results in terms of overall fitness function than EV method. While coming to execution time, the trend is reverse, i.e. GA takes the highest time 8–11 sec depending upon the population size (20) and the number of generations (20) chosen. Time taken by EV method is least 0.11–0.17 sec for 11-bus system.

TABLE 8.40 Comparison of results

Quantity	*Method*	*Case I*	*Case II*	*Case III*	*Case IV*
Active power loss,	EV	10.789700	10.223300	10.380970	09.982959
$\overline{P}_L$, MW	GA	10.807120	10.080080	10.124060	09.262609
Reactive power loss,	EV	–37.531510	–38.481560	–37.141520	–39.669820
$\overline{Q}_L$, MVAR	GA	–37.475330	–39.059540	–38.152030	–42.040430
λ(Rs/MW)	EV	1.037405	1.243301	1.213126	0.708840
	GA	0.926955	1.081108	1.087012	1.094109
Fitness value, f	EV	0.749510	0.667375	0.670112	0.306864
	GA	0.758951	0.719998	0.719654	0.363181
Time	EV	0.11	0.27	0.17	0.16
seconds	GA	8.24	8.85	7.97	9.40

Generally economic thermal power dispatch method allocates active and reactive power generation to the individual generating units based upon deterministic operating cost function while meeting the active and reactive power demand, ignoring various inaccuracies and uncertainties encountered in system data. Such generation schedules result in the lowest expected total cost, but this cost is always associated with a relatively large variance that can be interpreted as a risk measure. The analysis provides the facility to consider the inaccuracies and uncertainties in the thermal power dispatch procedure as well as allow explicit trade-off among the operating cost of units, the pollutant emission of units and risk levels. The power system operator so-called decision-maker is provided with the most efficient optimal solution from the non-inferior solution surface.

8.5 FUZZY MULTIOBJECTIVE SECURE GENERATION SCHEDULING

Solution methodology of a unified stochastic multiobjective optimization problem with explicit recognition of uncertainty in the system production cost data and nature of load demand is discussed. Operating cost and gaseous pollutant emission are various objectives which are minimized simultaneously so that system active and reactive load may be supplied entirely, securely and most economically. All the objectives have been defined as fuzzy objectives including transmission line flows and system equality constraints of active and reactive power generations. Specific technique is put forth to convert the stochastic model into its deterministic equivalent. So, variations in the parameters are modelled mathematically using variance and covariance. The objectives and goals are graded by membership functions, exploiting fuzzy set theory. Genetic algorithm is applied to solve the optimization problem, in which active and reactive power generations of individual generators are searched simultaneously within their respective minimum and maximum capacity limits. Min–max technique has been applied to select the best optimal solution.

8.5.1 Fuzzy Multiobjective Optimization Problem Formulation

Society demands adequate and secure electricity not only at the cheapest possible price, but also at minimum levels of pollution. A stochastic multi objective generation dispatch problem to meet the active and reactive power demands with explicit recognition of uncertainties in the system production cost coefficients, emission curve coefficients and system's active as well as reactive power load demands, which are treated as random variables. The multiple objectives taken into account are:

i. expected operating cost
ii. expected gaseous pollutant emission
iii. variance of active and reactive power generation
iv. security of transmission lines due to expected active power flow and
v. voltage profile improvement for service quality.

Exploiting fuzzy set theory equality constraints of active and reactive power balance are converted into objectives. All the random variables are assumed to be normally distributed and statistically dependent on each other.

Expected operating cost

The fuel cost curve is approximated by a quadratic function of the generator's active power output P_{Gi}. The expected value of operating cost function $\overline{F}_1$ is obtained by expanding the function about the mean, using Taylor's expansion series by considering the cost coefficients and active power demands and hence active power generations as random. The expected value of operating cost function $\overline{F}_1$ is given by Eq. (6.42) (Section 6.4.1, Chapter 6). The fuzzy sets are defined by equations called membership functions. These functions represent the degree of membership in certain fuzzy sets using values from 0 to 1. The value of the membership function indicates up to which degree a solution is satisfying the F_i objective. For the expected operating cost, a fuzzy objective function $\mu(\overline{F}_1)$, is assumed, which is exponentially decreasing and continuous and defined as follows:

$$\mu(\overline{F}_1) = \begin{cases} 1 & ; \overline{F}_1 \le \overline{F}_1^{\min} \\ \exp\left[-\left(\dfrac{\overline{F}_1 - \overline{F}_1^{\min}}{\overline{F}_1^{\max} - \overline{F}_1^{\min}}\right)\right] & ; \overline{F}_1 > \overline{F}_1^{\min} \end{cases} \tag{8.73}$$

where

$\overline{F}_1^{\min}$ is the minimum value of operating cost

$\overline{F}_1^{\max}$ is the maximum value of operating cost.

Expected gaseous pollutant emission

A stochastic model of F_2 is formulated by considering the emission coefficients as random variables. The expected value of gaseous pollutant emission function $\overline{F}_2$ is given by Eq. (6.63) (Section 6.4.1, Chapter 6). The expected value of gaseous emission is also fuzzified in the form of membership function $\mu(\overline{F}_2)$ and is given below:

$$\mu(\overline{F}_2) = \begin{cases} 1 & ; \overline{F}_2 \le \overline{F}_2^{\min} \\ \exp\left[-\left(\dfrac{\overline{F}_2 - \overline{F}_2^{\min}}{\overline{F}_2^{\max} - \overline{F}_2^{\min}}\right)\right] & ; \overline{F}_2 > \overline{F}_2^{\min} \end{cases} \tag{8.74}$$

The solution is expected within minimum value $\overline{F}_2^{\min}$ and maximum value $\overline{F}_2^{\max}$ of expected polluting gaseous emission.

Expected deviations

The deviations in active and reactive power generations from their respective deterministic values are considered as objectives to be minimized and are represented by Eq. (8.55) and Eq. (8.56) respectively. Fuzzy membership functions of expected active power deviation $\mu(\overline{F}_3)$ and expected reactive power deviation $\mu(\overline{F}_4)$ are defined as follows:

$$\mu(\overline{F}_j) = \begin{cases} 1 & ; \overline{F}_j \le \overline{F}_j^{\min} \\ \exp\left[-\left(\dfrac{\overline{F}_j - \overline{F}_j^{\min}}{\overline{F}_j^{\max} - \overline{F}_j^{\min}}\right)\right] & ; \overline{F}_j > \overline{F}_j^{\min} \end{cases} \quad (j = 3 \text{ and } 4) \tag{8.75}$$

where

$\overline{F}_j^{\min}$ is minimum attainable values of jth objective function.

$\overline{F}_j^{\max}$ is maximum attainable values of jth objective function.

Expected power flow on transmission lines

Active power flow on mth transmission line, connected from bus j to k, is given by:

$$P_{Tm} = g_m \left|V_j\right|^2 - \left|V_j\right|\left|V_k\right| [g_m \cos(\delta_j - \delta_k) + b_m \sin(\delta_j - \delta_k)] \tag{8.76}$$

where

$|V_j|$ and $|V_k|$ are voltage magnitudes at buses j and k respectively.

δ_j and δ_k are voltage phase angles at buses j and k respectively.

Expected value of active power flow on mth transmission line connected from bus j to k, is given by:

$$\overline{P}_{Tm} = g_m\left[\left|\overline{V}_j\right|^2 + \text{var}\left(\left|V_j\right|\right)\right] - \left|\overline{V}_j\right|\left|\overline{V}_k\right| \, [g_m \cos(\overline{\delta}_j - \overline{\delta}_k) + b_m \sin(\overline{\delta}_j - \overline{\delta}_k)]\,\lambda_{jk} \tag{8.77}$$

where

$|\overline{V}_j|$ and $|\overline{V}_k|$ are the expected voltage magnitudes at buses j and k respectively.

$\overline{\delta}_j$ and $\overline{\delta}_k$ are expected values of voltage phase angles at buses j and k respectively.

$$\text{var}\left(\left|V_j\right|\right) = C_{Vj}^2\left|\overline{V}_j\right|^2$$

$$\lambda_{jk} = \left[1.0 - 0.5 \times [\text{var}(\delta_j) + \text{var}(\delta_k)] + \text{cov}(\delta_j\delta_k) + \frac{\text{cov}\left(\left|V_j\right|, \left|V_k\right|\right)}{\left|\overline{V}_j\right|\left|\overline{V}_k\right|}\right]$$

or

$$\lambda_{jk} = \left[1.0 - 0.5 \times (C_{\delta j}^2\overline{\delta}_j^2 + C_{\delta k}^2\overline{\delta}_k^2) + R_{\delta j\delta k}C_{\delta j}C_{\delta k}\overline{\delta}_j\overline{\delta}_k + R_{VjVk}C_{Vj}C_{Vk}\right]$$

For the line flows the membership function similar to that defined by Eq. (8.91) is defined:

$$\mu(\overline{P}_{Tm}) = \begin{cases} 1 & ; \left|\overline{P}_{Tm}\right| \le \sigma_1 P_{TR_m} \\ \exp\left[-\left(\dfrac{\left|\overline{P}_{Tm}\right| - \sigma_1 P_{TR_m}}{\sigma_2 P_{TR_m} - \sigma_1 P_{TR_m}}\right)\right] & ; \left|\overline{P}_{Tm}\right| > \sigma_1 P_{TR_m} \end{cases} \quad (m = 1, 2, \ldots, \text{NL}) \tag{8.78}$$

where

NL is the number of transmission line of the power system network

P_{TR_m} is the active power flow rating of mth transmission line.

σ_1 and σ_2 are the factors to decide the lower and upper tolerable limits of line flows.

From security point of view, line flows should remain within their limits, on all lines.

Voltage profile optimization

Bus voltage is one of the most important security and service quality indices. A good voltage profile is also important to lower the transmission losses. It is incorporated in the problem as another objective to be optimized in the form of membership function of bus voltages as follows:

$$\mu\left(\left|\overline{V}_j\right|\right) = \exp\left[-\left(\frac{abs\left(\left|\overline{V}_j\right| - \left|V_{av}\right|\right)}{\left|V_j^{\max}\right| - \left|V_j^{\min}\right|}\right)\right] \quad (j = 2, 3, \ldots, \text{NB}) \tag{8.79}$$

where

$$|V_{av}| = \left(\left|V_j^{\min}\right| + \left|V_j^{\max}\right|\right)/2$$

with

NB is the number of buses of power system.

$\left|V_j^{\min}\right|$ and $\left|V_j^{\max}\right|$ are the minimum and maximum limits of bus voltage magnitudes and have been selected as equal to 0.90 (p.u.) and 1.10 (p.u.) respectively.

The objective function forces the voltage magnitude to be set around the middle of voltage profile.

Multiobjective optimization problem formulation

The multiobjective optimization problem is formulated to schedule active and reactive power generations of all committed thermal generators. The scheduling multiobjective optimization problem considers all the objectives simultaneously in the form of fuzzy membership functions for optimization and is defined mathematically as:

$$\text{Maximize } \left[\mu(\overline{F}_1), \mu(\overline{F}_2), \mu(\overline{F}_3), \mu(\overline{F}_4)\right]^T \tag{8.80a}$$

Subjected to inequality and equality constraints, such as:

$$\overline{P}_{Gi}^{\min} \le \overline{P}_{Gi} \le \overline{P}_{Gi}^{\max} \quad (i = 1, 2, \ldots, \text{NG}) \tag{8.80b}$$

$$\overline{Q}_{Gi}^{\min} \le \overline{Q}_{Gi} \le \overline{Q}_{Gi}^{\max} \quad (i = 1, 2, \ldots, \text{NG}) \tag{8.80c}$$

$$\sum_{i=1}^{\text{NB}} \overline{P}_{Di} - \sum_{i=1}^{\text{NG}} \overline{P}_{Gi} + \overline{P}_L = 0 \tag{8.80d}$$

$$\sum_{i=1}^{\text{NB}} \overline{Q}_{Di} - \sum_{i=1}^{\text{NG}} \overline{Q}_{Gi} + \overline{Q}_L = 0 \tag{8.80e}$$

$$\left|\overline{P}_{Tm}\right| \le P_{TR_m} \quad (m = 1, 2, \ldots, \text{NL}) \tag{8.80f}$$

$$\left|V_j^{\min}\right| \le \left|\overline{V}_j\right| \le \left|V_j^{\max}\right| \quad (j = 1, 2, \ldots, \text{NB}) \tag{8.80g}$$

Inequality constraints represented by Eqs. (8.80b) and (8.80c) are taken care of while searching the active and reactive power generations within minimum and maximum limits of generators. Equality constraints defined by Eqs. (8.80d) and (8.80e) are considered as additional objectives $\overline{F}_5$ and $\overline{F}_6$ to be minimized.

$$\overline{F}_5 = \left|\sum_{i=1}^{\text{NB}} \overline{P}_{Di} - \sum_{i=1}^{\text{NG}} \overline{P}_{Gi} + \overline{P}_L\right| \tag{8.81}$$

$$\overline{F}_6 = \left|\sum_{i=1}^{\text{NB}} \overline{Q}_{Di} - \sum_{i=1}^{\text{NG}} \overline{Q}_{Gi} + \overline{Q}_L\right| \tag{8.82}$$

Assuming that objectives are of precise nature, exponential membership functions are defined below to define the objectives:

$$\mu(\bar{F}_j) = \begin{cases} 1 & ; \bar{F}_j \le \bar{F}_j^{\min} \\ \exp\left[-\left(\dfrac{\bar{F}_j - \bar{F}_j^{\min}}{\bar{F}_j^{\max} - \bar{F}_j^{\min}}\right)\right] & ; \bar{F}_j > \bar{F}_j^{\min} \end{cases} \quad (j = 5 \text{ and } 6) \tag{8.83}$$

Higher the satisfaction level better is the convergence. $\bar{F}_5^{\min}$ and $\bar{F}_6^{\min}$ are the convergence values to be achieved and $\bar{F}_5^{\max}$ and $\bar{F}_6^{\max}$ give the ranges of feasibility of solution. The inequality constraints defined by Eqs. (8.80f) and (8.80g) are considered in the form of overall membership function of line flows and bus voltages respectively and are defined below:

$$\mu(\bar{F}_7) = \min\left[\mu(\bar{P}_{Tm}) \;\; (m = 1, 2, ..., \text{NL})\right] \tag{8.84}$$

$$\mu(\bar{F}_8) = \min\left[\mu(|V_j|) \;\; (j = 2, 3, ..., \text{NB})\right] \tag{8.85}$$

So the problem is redefined to search the active and reactive power generation schedules as:

$$\text{Maximize } [\mu(\bar{F}_1), \mu(\bar{F}_2), \mu(\bar{F}_3), \mu(\bar{F}_4), \mu(\bar{F}_5), \mu(\bar{F}_6), \mu(\bar{F}_7), \mu(\bar{F}_8)]^T \tag{8.86a}$$

$$\text{Subject to } \bar{P}_{Gi}^{\min} \le \bar{P}_{Gi} \le \bar{P}_{Gi}^{\max} \;\; (i = 1, 2, \ldots, \text{NG}) \tag{8.86b}$$

$$\bar{Q}_{Gi}^{\min} \le \bar{Q}_{Gi} \le \bar{Q}_{Gi}^{\max} \;\; (i = 1, 2, \ldots, \text{NG}) \tag{8.86c}$$

Genetic Algorithm is applied to generate the set of non-inferior solutions and min–max technique is applied to select the optimum solution. Convergence of the method is achieved based on the satisfaction level of $\mu(\bar{F}_5)$ and $\mu(\bar{F}_6)$ objectives.

8.5.2 Generation Search with Genetic Algorithm

Normally multiobjective optimization problems are solved by weighting method in which the problem is converted to a single objective problem by linear combination of different objectives as a weighted sum and these assigned weights to the objectives are varied in a systematic manner, in order to generate the non-inferior solutions for the optimization problem. Unfortunately, this requires multiple runs as many times as the number of desired non-inferior solutions. Furthermore, this method cannot be used to find non-inferior solutions in problems having a non-convex Pareto-optimal front. The multiobjective optimization problem mainly handles multiple objectives simultaneously as competing objectives. To consider the multiple objectives simultaneously fuzzy theory proves to be very useful. A fuzzy based multiobjective thermal power generation dispatch problem has been formulated and thermal power generation schedules have been searched to generate the non-inferior. Genetic Algorithm has been applied. Fitness function is evaluated by utilizing min–max technique.

A global optimization technique known as genetic algorithm has emerged as a candidate for many optimization applications due to its flexibility and efficiency. It is a stochastic searching algorithm. It combines an artificial Darwinian survival of the fittest principle with genetic operation, abstracted from nature to form a robust mechanism that is very effective at finding optimal solutions to complex real world problems. GA operates on string structures. A finite length of binary numbers, which is concatenated by sub-strings, is used to represent the coding of the search parameters (active and reactive power generations). Many such strings are considered simultaneously, with the most fit of these structures receiving exponentially increasing opportunities to pass on genetically important material to successive generation of string structures. In this way, genetic algorithms search for many points in the search space at once, and yet continually narrow the focus of the search to the areas of the observed best performance. A binary string is decoded into an unsigned integer for the GA implementation. The higher precision of the generations can be obtained by defining longer binary strings. A decoding method is formulated in following equation.

$$y_i^j = \sum_{k=1}^{l_i} \text{bit}_k 2^{k-1} \quad (i = 1, 2, \ldots, \text{NG}; j = 1, 2, \ldots, P_S) \tag{8.87}$$

where

bit_k is kth binary coefficient.
l_i is the length of ith sub-string,
P_S is the population size.

In case the parameter belongs to $P_{Gi}^{\min}$ and $P_{Gi}^{\min}$ the decoded value of parameter can be computed.

$$P_{Gi}^j = P_{Gi}^{\min} + \frac{y_i^j \times (P_{Gi}^{\max} - P_{Gi}^{\min})}{(2^{l_i} - 1)} \quad (i = 1, 2, \ldots, \text{NG}; j = 1, 2, \ldots, P_S) \tag{8.88}$$

Similarly if the parameter belongs to $Q_{Gi}^{\min}$ and $Q_{Gi}^{\min}$ the decoded value of parameter would be:

$$Q_{Gi}^j = Q_{Gi}^{\min} + \frac{y_i^j \times (Q_{Gi}^{\max} - Q_{Gi}^{\min})}{(2^{l_i} - 1)} \quad (i = 1, 2, \ldots, \text{NG}; j = 1, 2, \ldots, P_S) \tag{8.89}$$

Algorithm 8.3: Evaluation of objective functions and fitness

Stepwise procedure to evaluate objective functions related to optimization problem and to find out the fitness function corresponding to a non-inferior solution set is outlined below:

1. Perform decoupled load flow.
2. Compute expected active power flow $\overline{P}_{Tm}$ on lines using Eq. (8.77). Compute expected active and reactive power transmission losses using Eq. (8.64) and Eq. (8.66), respectively.
3. Evaluate objective function values; $\overline{F}_1, \overline{F}_2, \overline{F}_3$ and $\overline{F}_4$ using Eqs. (6.42), (6.63), (8.55), and (8.56) and active and reactive power mismatch; $\overline{F}_5$ and $\overline{F}_6$ using Eq. (8.81) and Eq. (8.82), respectively.
4. Evaluate membership functions; $\mu(\overline{F}_i)$ $(i = 1, 2)$ by using Eqs. (8.73) and (8.74), $\mu(\overline{F}_i)$ $(i = 3, 4)$ from Eq. (8.75) and $\mu(\overline{F}_i)$ $(i = 5, 6)$ from Eq. (8.83), respectively.

5. Compute membership functions of line flows $\mu(\overline{P}_{Tm})$ (m = 1, 2, ..., NL), using Eqs. (8.78) and membership functions of bus voltages; $\mu\left(\left|V_j\right|\right)$ (j = 1, 2, ..., M), by using Eqs. (8.79). Then compute $\mu(\overline{F}_7)$ and $\mu(\overline{F}_8)$ by utilizing Eqs. (8.84) and (8.85), respectively.
6. Select fitness function 'f' as given below:

$$f = \text{Min}\{\ \mu(\overline{F}_j)\ ;\ j = 1, 2, ..., 8\} \tag{8.90}$$

Algorithm 8.4: Genetic algorithm to search for active and reactive power generations

To implement the genetic algorithm to search for the active and reactive power generation schedule, a stepwise procedure is outlined below:

1. Input the data; such as number of generators N_G, population size P_S, chromosome length L_s, length of each sub-string l_i, maximum number of generations IT$^{\max}$, crossover probability p_c, mutation probability p_m, and minimum and maximum limits of active and reactive power generations.
2. Find the minimum and maximum values of the objectives and equality constraints, $F_i^{\min}$ and $F_i^{\max}$; i = 1, 2, ..., 6.
3. Generate the initial population of generation schedules randomly. Decode each individual string of population using Eq. (8.87) and map into generation limits using Eqs. (8.88) and (8.89).
4. Evaluate fitness value corresponding to each string by executing Section 8.5.2 and by using Eq. (8.90).
5. Initialize generation counter.
6. Select the string having maximum fitness f^{pre}.
7. Increment the generation counter K = K+1.
8. If (K > IT$^{\max}$) then GO TO 17.
9. Select the most-fit strings from old population by stochastic remainder roulette-wheel selection, for crossover.
10. Select the crossover site by flipping the coin with crossover probability. Perform single point crossover and mutation to get the new population.
11. Decode the string using Eq. (8.87) and find newly generated strings using Eqs. (8.88) and (8.89).
12. Evaluate fitness value corresponding to each string by executing Algorithm 8.3 and by using Eq. (8.90).
13. Select the string having maximum fitness $f^{\max}$ in the new population.
14. If ($f^{\max} - f^{\text{pre}}$) ≤ ε, then GO TO 17.
15. Update the value of population fitness.

$$f^{\text{pre}} = \begin{cases} f^{\text{pre}} & ;\ f^{\text{pre}} > f^{\max} \\ f^{\max} & ;\ f^{\max} > f^{\text{pre}} \end{cases}$$

16. GO TO 7 and repeat.
17. Stop.

8.5.3 Results and Discussions

The validity of the proposed method is demonstrated on standard test system of IEEE 11-bus, 17 lines powered by 5-generators. The operating cost and gaseous emission coefficients and operating limits of generators are given in Tables 8.31 and 8.32. The load data and line data of power system is given in Tables 8.33 and 8.3, respectively. The non-inferior solutions for different generation combinations are generated and searched using GA, considering all the objectives simultaneously. The preferred solutions are selected using min–max technique. To study the effect of random parameters, four cases of coefficients of variance and correlation coefficients of random parameters and variables have been undertaken as given in Section 8.4.5.

The minimum and maximum values of different objectives are given Tables 8.41 and 6.4 those depict the limits of objectives for 11-bus and 25-bus power systems, respectively. To perform the search by GA, individual active and reactive power generations, except the slack bus generation are coded in the form of binary sub-strings of size 40. As total number of generators are 5. Therefore excluding the slack bus generator, total variables to be encoded are 8[2 × (5 – 1)]. Hence total length of binary string becomes 320(8 × 40). Size of population *S*, has been taken as 20 and maximum number of generations, IT^{max} as 30. To search the optimal active and reactive power generation schedules genetic algorithm has been applied on two test power system networks, of which one power system network comprises of 11-bus, 17-lines fed by 5-generators. 'Best' compromised solutions have been obtained by the Genetic Algorithm discussed in Section 8.5.2 for active and reactive power generation optimization. The crossover probability, p_c, is taken equal to 0.6, 0.65, 0.6, 0.8 and 0.8, respectively for all the five cases considered to present the variation and mutation probability, p_m, is taken as 0.01 for Cases I and II, and 0.015 for rest of the cases. The 'best' generation (p.u.) schedules of active and reactive powers of thermal stations searched out by utilizing GA are shown in Table 8.42.

TABLE 8.41 The minimum and maximum values of objectives

Objective	*Case I*	*Case II*	*Case III*	*Case IV*
$\bar{F}_1^{min}$ (₹/h)	1154.0	1154.0	1158.0	1158.0
$\bar{F}_1^{max}$ (₹/h)	1180.0	1182.0	1187.0	1187.0
$\bar{F}_2^{min}$ (kg/h)	310.0	311.0	313.0	313.0
$\bar{F}_2^{max}$ (kg/h)	400.0	401.0	404.0	412.0
$\bar{F}_3^{min}$ (p.u.)^2	0.0	0.0077	0.0077	0.023
$\bar{F}_3^{max}$ (p.u.)^2	0.1	0.0092	0.0092	0.025
$\bar{F}_4^{min}$ (p.u.)^2	0.0	0.00042	0.00042	0.0006
$\bar{F}_4^{max}$ (p.u.)^2	0.1	0.00065	0.00065	0.0019
$\bar{F}_5^{min}$ (p.u.)	0.0	0.0001	0.0001	0.0001
$\bar{F}_5^{max}$ (p.u.)	0.00005	0.004	0.1	0.1
$\bar{F}_6^{min}$ (p.u.)	0.0	0.001	0.001	0.001
$\bar{F}_6^{max}$ (p.u.)	0.00005	0.04	0.1	0.1

Best solutions have been found at generation 23 for Case I, at generation 29 for Case II, at generation 25 for Case III, at generation 29 for Case IV, at generation 28 for Case V. Execution times taken is 4.17, 4.23, 4.40, 4.33 and 4.23 seconds, respectively for Cases I to V. Active power transmission loss P_L, and reactive power transmission loss Q_L, are also given in Table 8.42. These p.u. values are referred to a base value of 100 MVA. Generation schedules are different for

TABLE 8.42 'Best' power generation schedules and losses by combined search using genetic algorithm

Generation	*Case I*	*Case II*	*Case III*	*Case IV*
P_{G1}	1.070631	1.005548	0.827471	1.043623
P_{G2}	0.638584	0.650492	0.688122	0.644990
P_{G3}	0.893257	0.723039	0.874986	0.901869
P_{G4}	0.364916	0.567141	0.437067	0.387073
P_{G5}	0.956633	0.956767	1.081215	0.931299
ΣP_{Gi}	3.924021	3.902987	3.908862	3.908855
P_L	0.124015	0.103284	0.109470	0.109450
Q_{G1}	0.386221	0.177634	0.269716	0.149452
Q_{G2}	0.035168	0.064977	0.265749	0.186487
Q_{G3}	0.039444	0.289441	0.056541	0.117210
Q_{G4}	0.537524	0.103234	0.218553	0.082257
Q_{G5}	0.036942	0.284500	0.143256	0.403199
Σ_{Gi}	1.035299	0.919786	0.953816	0.938605
Q_L	–0.284692	–0.412270	–0.387174	–0.395473

different assumed set of coefficients of variation and correlation coefficients of random parameters and variables being termed as Cases I to IV. Hence it justifies the need to consider the randomness of parameters while performing generation scheduling.

Corresponding to 'best' solutions, the objective function values of operating cost $\overline{F}_1$, gaseous pollutant emission $\overline{F}_2$, variance of active power $\overline{F}_3$, variance of reactive power $\overline{F}_4$, the active power mismatch $\overline{F}_5$ and reactive power mismatch $\overline{F}_6$, are depicted in Table 8.43. Consideration of correlation coefficients of active power injections and reactive power injections at the buses have resulted into decrease in active power mismatch $\overline{F}_5$, and reactive power mismatch $\overline{F}_6$, as is observed by comparing their values for Cases III and IV.

TABLE 8.43 Objective function values at 'best' generation schedule

Case No.	$\overline{F}_1$ ₹/h	$\overline{F}_2$ kg/h	$\overline{F}_3$ (p.u.)^2	$\overline{F}_4$ (p.u.)^2	$\overline{F}_5$ (p.u.)	$\overline{F}_6$ (p.u.)
I	1163.359	342.0086	0.0	0.0	0.000006	0.000009
II	1168.900	343.8119	0.007985	0.000528	0.000296	0.012055
III	1170.405	326.7590	0.008210	0.000537	0.000608	0.020989
IV	1165.780	340.0697	0.023269	0.001401	0.000595	0.014077

Membership function values of objectives $\mu(\overline{F}_1)$, $\mu(\overline{F}_2)$, $\mu(\overline{F}_3)$, $\mu(\overline{F}_4)$, equality constraints, $\mu(\overline{F}_5)$, $\mu(\overline{F}_6)$, and inequality constraints of active power line flow on transmission lines $\mu(\overline{F}_7)$, and bus voltage magnitude limits $\mu(\overline{F}_8)$, corresponding to 'best' non-inferior solutions are given in Table 8.44. Active power line flows on transmission lines and line ratings are shown in Table 8.45. Bus voltage magnitudes and phase angles are given in Tables 8.46 and 8.47. Combined search of active and reactive power generation compels to set the bus voltage magnitudes at the middle of the profile so that the line flows remain within limits.

TABLE 8.44 Membership function values corresponding to 'best' non-inferior solution

Case No.	$\mu(\bar{F}_1)$	$\mu(\bar{F}_2)$	$\mu(\bar{F}_3)$	$\mu(\bar{F}_4)$	$\mu(\bar{F}_5)$	$\mu(\bar{F}_6)$	$\mu(\bar{F}_7)$	$\mu(\bar{F}_8)$
I	0.697693	0.700717	1.0	1.0	0.878281	0.828818	0.776276	0.748467
II	0.587342	0.694491	0.826815	0.625618	0.950939	0.753164	0.793403	0.555047
III	0.651968	0.859678	0.711918	0.600925	0.994929	0.817166	0.725700	0.672103
IV	0.764693	0.760764	0.874359	0.539828	0.995058	0.876259	0.786638	0.585094
V	0.717809	0.641717	0.724613	0.538660	0.972277	0.661719	0.710472	0.522324

TABLE 8.45 Transmission line flows and line flow ratings

Line No.	*Case I*	*Case II*	*Case III*	*Case IV*	*Rating*
1	0.220659	0.195409	0.117384	0.219009	0.350
2	0.649972	0.610139	0.511155	0.625320	0.750
3	–0.212438	–0.186957	–0.230847	–0.224783	0.300
4	0.296231	0.290139	0.318584	0.304648	0.350
5	0.354786	0.347305	0.401913	0.365984	0.450
6	0.373500	0.229753	0.337156	0.370437	0.420
7	–0.019472	–0.006804	–0.034450	–0.005491	0.040
8	0.404859	0.438642	0.445608	0.401835	0.500
9	0.092479	0.110720	0.105955	0.102313	0.150
10	0.122515	0.108034	0.137624	0.107724	0.150
11	0.584112	0.598726	0.694755	0.574358	0.750
12	–0.003147	–0.013013	0.009521	0.004153	0.018
13	–0.109825	–0.105299	–0.100642	–0.108549	0.150
14	–0.164267	–0.174883	–0.183698	–0.164322	0.200
15	–0.107653	–0.096351	–0.105809	–0.110594	0.150
16	–0.247243	–0.223663	–0.175479	–0.236063	0.300
17	–0.121555	–0.112472	–0.066889	–0.113200	0.200

TABLE 8.46 Bus voltage magnitude profile corresponding to 'best' non-inferior solution

Bus No.	*Case I*	*Case II*	*Case III*	*Case IV*
1	1.070000	1.070000	1.070000	1.070000
2	0.993651	1.038792	1.050700	1.047664
3	1.057946	1.117741	1.079469	1.093607
4	1.053975	1.075888	1.055832	1.062992
5	1.020346	1.092498	1.057283	1.107196
6	1.037969	1.082829	1.055546	1.082062
7	0.951239	0.990727	0.991901	0.994917
8	0.949884	0.984945	0.978916	0.986111
9	0.978237	1.026655	1.003526	1.032399
10	0.970389	1.008086	1.011241	1.012790
11	1.006781	1.025336	1.025040	1.026578

TABLE 8.47 Bus voltage angle profile corresponding to 'best' non-inferior solution

Bus No.	*Case I*	*Case II*	*Case III*	*Case IV*
1	0.000000	0.000000	0.000000	0.000000
2	–0.066619	–0.072864	–0.035205	–0.075658
3	0.026727	–0.005570	0.069683	0.020225
4	–0.060192	–0.043322	–0.000833	–0.054210
5	0.002469	–0.010394	0.057459	–0.030902
6	–0.049043	–0.043768	0.009184	–0.058923
7	–0.148536	–0.143707	–0.109475	–0.148032
8	–0.147479	–0.139528	–0.106923	–0.145920
9	–0.085753	–0.083336	–0.039357	–0.095810
10	–0.123278	–0.121246	–0.088890	–0.125036
11	–0.085309	–0.083076	–0.067354	–0.085636

A stochastic fuzzy multiobjective thermal active and reactive power generation scheduling problem has been undertaken and is solved by utilizing Genetic Algorithm. Due to uncertainties and inaccuracies in the measurement and forecasting of various parameters of the power system network, like system production data and load demand power generation is subjected to changes. To consider these variations, random nature has been considered by the coefficients of variation and correlation coefficients of parameters such as operating cost coefficients, gaseous pollutant emission coefficients and variables such as active and reactive powers, bus voltage magnitudes and voltage phase angles. The expected values of all the objective functions are obtained by expanding a function about the mean of the random variable by employing Taylor series expansion. Objectives taken into account for minimization are operating cost, gaseous pollutant emission, variance of active power generation and variance of reactive power generation. Equality constraints of active power balance and reactive power balance are also considered in terms of objectives to be minimized. All these objectives are redefined in the form of fuzzy objectives to be maximized. Inequality constraints of line flow limits on transmission lines and bus voltage magnitude limits are taken into account with the inclusion of additional fuzzy objectives to be maximized. Fuzzy objectives are represented by exponential membership functions. Genetic Algorithm has been applied to simultaneously generate the non-inferior solutions in terms of active and reactive power generation schedules within the minimum and maximum capacity limits of generators. Fitness function has been achieved by implementing min–max technique. Method used here for solving multiobjective problem of allocation of active and reactive power generations to generators, is very simple and straightforward. It searches the two schedules of active and reactive power generations simultaneously. No tedious derivations of the objective functions are required to be calculated, so the technique is equally good for discontinuous functions. Convergence of solution is ensured. Optimum results obtained are sensitive to length of sub-strings used and values of crossover and mutation probabilities selected, but an expert decision-maker can easily search out the optimum values for these probabilities.

REFERENCES

Books

Arrillaga, J. and C.P. Arnold, *Computer Analysis of Power Systems*, John Wiley & Sons, Singapore, 1990.

Christensen, G.S. and S.A. Soliman, *Optimal Long-Term Operation of Electric Power Systems*, Plenum Press, New York, 1988.

Deb, K., *Optimization for Engineering Design: Algorithms and Examples*, Prentice-Hall of India, New Delhi, 1995.

Elgerd, O.I., *Electric Energy Systems Theory: An Introduction*, Tata McGraw Hill, 2nd ed., 1983.

El-Hawary, M.E. and G.S. Christensen, *Optimal Economic Operation of Power Systems*, Academic Press, New York, 1979.

Frederick Soloman, *Probability and Stochastic Processes*, Prentice Hall, New Jersey, 1987.

Kirchmayer, L.K., *Economic Operation of Power Systems*, Wiley Eastern Ltd., New Delhi, 1958.

Klir, G.J. and B.Yuan, *Fuzzy Sets and Fuzzy Logic, Theory and Applications*, Prentice Hall India, 1997.

Klir, G.J. and T.A. Folger, Fuzzy Sets, *Uncertainty and Information*, Prentice Hall India, 1993.

Kosko, B., *Neural Networks and Fuzzy Systems: Dynamical Systems Approach to Machine Intelligence*, Prentice Hall India, 1994.

Kothari D.P. and I.J. Nagrath, *Modren Power System Analysis*, 3rd ed., Tata McGraw-Hill, New Delhi, 2003.

Kusic, G.L., *Computer Aided Power Systems Analysis*, Prentice Hall of India, New Delhi, 1986.

Mahalanabis, A.K., D.P. Kothari and S.I. Ahson, *Computer Aided Power System Analysis and Control*, Tata McGraw-Hill, New Delhi, 1991.

Momoh, J.A., *Electric Power System Applications of Optimization*, Marcel Dekker, Inc., New York, 2001.

Nagrath, I.J. and D.P. Kothari, *Power System Engineering*, Tata McGraw-Hill, New Delhi, 1994.

Osyczka, A. and B.J. Davies, *Multicriterion Optimization in Engineering with FORTRAN Programs*, Ellis Horwood Ltd., 1984.

Papoulis, A., *Probability, Random Variables and Stochastic Processes*, McGraw-Hill, New Delhi, 1991.

Rao, S.S., *Optimization, Theory and Applications*, 2nd ed, Wiley Eastern Limited, New Delhi, 1987.

Sen Gupta J.K. , *Stochastic Programming*, North Holland, 1972.

Stagg, G.W. and A.H. Ei-Abiad, *Computer Methods in Power Systems Analysis*, McGraw-Hill, New Delhi, 1968.

Stevenson, W.D., *Elements of Power System Analysis*, 4th ed., McGraw-Hill, New York, 1982.

Wood, A.J. and B. Wollenberg, *Power Generation, Operation and Control*, John Wiley, New York, 1984.

Papers

Abido, M.A., A novel multiobjective evolutionary algorithm for environmental/economic power dispatch, *Electric Power Systems Research,* Vol. **65(1)**, pp. 71–81, April 2003.

Abido, M.A. and J.M. Bakhashwain, Optimal VAR dispatch using a multiobjective evolutionary algorithm, *International Journal of Electrical Power and Energy Systems*, Vol. **27(1)**, pp. 13–20, 2005.

Bala, J.L., and A. Thanikachalam, An improved second order method for optimal load flow, *IEEE Transactions on Power Apparatus and Systems*, Vol. **97(4)**, pp. 1239–1244, 1978.

Bansal, R.C., Bibliography on the fuzzy set theory applications in power systems (1994–2001), *IEEE Transactions on Power Systems*, Vol. **18(4)**, pp. 1291–1299, 2003.

Bath, S.K., J.S. Dhillon and D.P. Kotahri, Stochastic multiobjective generation dispatch, *Electric Power Components and Systems*, Vol. **32(11)**, pp. 1083–1103, 2004.

Bath, S.K., J.S. Dhillon and D.P. Kotahri, Fuzzy satisfying stochastic multiobjective generation scheduling by weightage pattern search method, *Electric Power System Research,* Vol. **69(2-3)**, pp. 311–320, 2004.

Bath, S.K., J.S. Dhillon and D.P. Kothari, Stochastic multiobjective generation allocation using pattern search method, *IEE Proceedings-Generation, Transmission and Distribution*, Vol. **153(4)**, pp. 476–484, 2006.

Bath, Sarbjeet Kaur, Jaspreet Singh Dhillon and D.P. Kothari, Security constrained stochastic multiobjective optimal power dispatch, *International Journal of Emerging Electric Power Systems*, Vol. **8(1)**, Article 7, www.bepress.com/ijeeps, January, 2007.

Bijwe, P.R., M. Hanmandlu, and V.N. Pande, Fuzzy power flow solutions with reactive limits and multiple uncertainties, *Electric Power Systems Research*, Vol. **76(1-3)**, pp. 145–152, 2005.

Brar, Y.S., J.S. Dhillon and D.P. Kothari, Multiobjective load dispatch by fuzzy logic based searching weightage pattern, *Electric Power System Research*, Vol. **63(2)**, pp. 149–160, 2002.

Brar, Y.S., J.S. Dhillon and D.P. Kothari, Fuzzy satisfying Multiobjective generation scheduling based on simplex weightage pattern search, *Inernational Journal of Electric Power and Energy Systems,* Vol. **27(7)**, pp. 518–527, 2005.

Brar, Y.S., J.S. Dhillon and D.P. Kothari, Multiobjective load dispatch by evolutionary optimisation technique based weightage pattern search method, *Electric Power Components and Systems*, Vol. **33(4)**, pp. 431–448, 2005.

Brar, Y.S., J.S. Dhillon and D.P. Kothari, Interactive fuzzy satisfying multiobjective generation scheduling based on genetic weightage pattern search, *Journal of Institute of Engineers (India)*, Vol **86**, pp. 312–318, 2006.

Brar, Y.S., J.S. Dhillon and D.P. Kothari, Genetic-fuzzy based power scheduling technique for multiobjective load dispatch problem, *International Journal of Power and Energy Systems*, Vol. **28(1)**, 2008.

Chandrasekharan, E., M.S.N. Potti, R. Sreeramakumar and K.P. Mohandas, Improved general purpose fast decoupled load flow, *International Journal of Electrical Power and Energy Systems*, Vol. **24(6)**, pp. 481–488, 2002.

Chang, Y.C., W.T. Yang and C.C. Liu, A new method for calculating loss coefficients', *IEEE Transactions on Power Systems* , Vol. **9(3)**, pp. 1665–1671, 1994.

Chen, P.-H. and H.-C. Chang, Large-scale economic dispatch by genetic algorithm, *IEEE Transactions on Power Systems*, Vol. **10(4),** pp. 1919–1926, 1995.

Chung, T.S., K.K. Li, G.J. Chen, J.D. Xie and G.Q. Tang, Multiobjective transmission network planning by hybrid GA approach with fuzzy decision analysis, *International Journal of Electrical Power and Energy Systems*, Vol. **25(3)**, pp. 187–192, 2003.

Dhillon, J.S., S.C. Parti and D.P. Kothari, Stochastic economic emission load dispatch, *Electric Power Systems Research*, Vol. **26(3)**, pp. 179–186, 1993.

Dhillon, J.S. and D.P. Kothari, The surrogate worth trade-off approach for multiobjective thermal power dispatch problem, *Electric Power Systems Research*, Vol. **56(2)**, pp 103–110, 2000.

Dhillon, J.S., S.C. Parti and D.P. Kothari, Fuzzy decision making in stochastic multiobjective short-term hydrothermal scheduling, *IEE Proceedings—Generation, Transmission and Distribution*, Vol. **149(2)**, pp. 191–200, 2002.

Dhillon, J.S., S.C. Parti and D.P. Kothari, Multiobjective decision making in stochastic economic dispatch, *Electric Machines and Power Systems*, Vol. **23(3)**, pp. 289–301, 1995.

Famideh-Vojdani, A.R. and F.D. Galiana, Economic dispatch with generation constraints, *IEEE Transactions on Automatic Control*, Vol. **25(2)**, pp. 213–217, 1980.

Hota, P.K., R. Chakrabarti and P.K. Chattopadhyay, Economic emission load dispatch through an interactive fuzzy satisfying method, *Electric Power Systems Research*, Vol. **54(3)**, pp. 151–157, 2000.

Huang, C.-M., H.-T. Yang and C.-L. Huang, Bi-objective power dispatch using fuzzy satisfaction-maximizing decision approach, *IEEE Transactions on Power Systems*, Vol. **12(4)**, pp. 1715–1721, 1997.

Leung, Y.W. and Y. Wang, U-measure: a quality measure for multiobjective programming, *IEEE Transactions on Systems, Man and Cybernetics*, Vol. **33(3)**, pp. 337–343, 2003.

Lin, C.E., S.T. Chen and C.L. Huang, A two-step sensitivity approach for real-time line flow calculation, *Electric Power Systems Research*, Vol. **21(1)**, pp. 63–69, 1991.

Ma, J.T. and L.L. Lai, Evolutionary programming approach to reactive power planning, *IEE Proceedings—Generation, Transmission and Distribution*, Vol. **143(4)**, pp. 365–370, 1996.

Momoh, J.A., M.E. El-Hawary and R. Adapa, A review of selected optimal power flow literature to 1993, *IEEE Transactions on Power Systems*, Vol. **14(1)**, pp. 96–111, 1999.

Ng, W.Y., Generalized generation distribution factors for power system security evaluations, *IEEE Transactions on Power Apparatus and Systems*, Vol. **100(3)**, pp. 1001–1005, 1981.

Niimura, T. and T. Nakashima, Multiobjective tradeoff analysis of deregulated electricity transactions, *International Journal of Electrical Power and Energy Systems*, Vol. **25(3)**, pp. 179–185, 2003.

Orero, S.O. and M.R. Irving, A genetic algorithm for generator scheduling in power systems, *International Journal of Electric Power and Energy Systems*, Vol. **18(1)**, pp. 19–26, 1996.

Rosehart, W.D., C.A. Canizares and V.H. Quintana, Multiobjective optimal power flows to evaluate voltage security costs in power networks, IEEE Transactions on Power Systems, Vol. **18(2)**, pp. 578–587, 2003.

Somasundaram, P. and K. Kuppusamy, Application of evolutionary programming to security constrained economic dispatch, *International Journal of Electrical Power and Energy Systems*, Vol. **27(5–6)**, pp. 343–351, 2005.

Talaq, J.H., F. El-Hawary and M.E. El-Hawary, Minimum emissions power flow using Newton's method and its variants, *Electric Power Systems Research*, Vol. **39(3)**, pp. 233–239, 1996.

Venkatesh, P., R. Gnanadass and N.P. Padhy, Comparison and application of evolutionary programming techniques to combined economic emission dispatch with line flow constraints, *IEEE Transactions on Power Systems*, Vol. **18(2)**, pp. 688–697, 2003.

Wong, K.P., B. Fan, C.S. Chang and A.C. Liew, Multiobjective generation dispatch using bi-criterion global optimisation, *IEEE Transactions on Power Systems*, Vol. **10(4)**, pp. 1813–1819, 1995.

Wong, K.P. and J. Yuryevich, Evolutionary programming based algorithm for environmentally constrained economic dispatch, *IEEE Transactions on Power Systems*, Vol. **13(2)**, pp. 301–306, 1998.

Yan, W., S. Lu and D.C. Yu, A novel optimal reactive power dispatch method based on an improved hybrid evolutionary programming technique, *IEEE Transactions on Power Systems*, Vol. **19(2)**, pp. 913–918, 2004.

Yang, J.B., and D. Li, Normal vector identification and interactive tradeoff analysis using minimax formulation in multiobjective optimization, *IEEE Transactions on Systems Man and Cybernetics,* Vol. **32(3)**, pp. 305–319, 2002.

Zhu, J.Z. and M.R. Irving, Combined active and reactive dispatch with multiple objectives using an analytic hierarchical process, *IEE Proceeding—Generation, Transmission and Distribution*, Vol. **143(4)**, pp. 344–352, 1996.

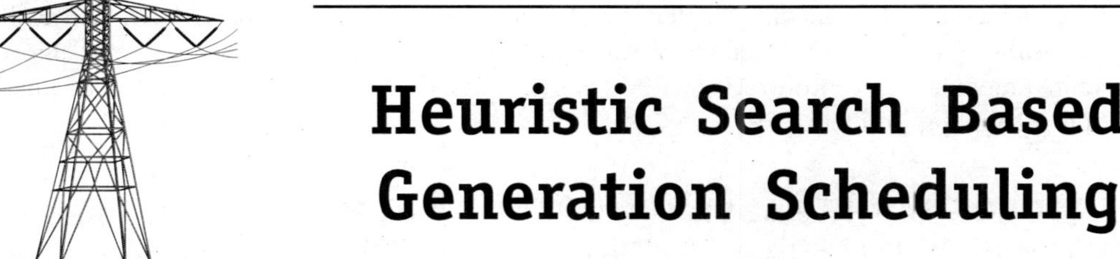

CHAPTER 9

Heuristic Search Based Generation Scheduling

9.1 INTRODUCTION

In recent years, the field of combinatorial optimization has witnessed countless "novel" meta-heuristic methods. Most of them are based on a metaphor of natural or man-made process. Several publications present a "novel" method for optimization, based on a metaphor of a process that is often seemingly completely unrelated to optimization. The jumps of frogs, the refraction of light, the flowing of water to the sea, an orchestra playing, sperm cells moving to fertilize an egg, the spiraling movements of galaxies, the colonizing behavior of empires, the behavior of bats, birds, ants, bees, flies, and virtually every other species of insects have been used as a metaphor optimization method. The authors of such papers claim that their proposed method is superior to methods previously published in the literature and gets good results.

Experiment performed is likely to be replicated to justify the achieved good results by using meta-heuristic techniques. The experiments performed may be used as a comparison in other experiments. Hence, provide as much statistical data for selected performance measures as possible like, mean values, maximum values, best values and standard deviations. Whilst replicating an experiment, the same or stricter conditions must be satisfied for instance, equal or larger number of independent runs, equal or smaller number of fitness evaluations. Performance measures must be defined appropriately. Practitioners should use the best settings for control parameters. These parameters are identified either by parameter tuning or parameter control, where some sophisticated exploration and exploitation measures can be used as well. Further additional experiments are performed to show that with an increased number of fitness evaluations; even better solutions cannot be found. Whilst replicating an experiment, all available case-studies (benchmark functions) should be used as in the original experiments. The prepared code should be made publicly available. Without an exact algorithm description, experiments cannot be replicated. Do not compare algorithms simply based on execution time. Because, different computers are used during the experiment, so it is difficult to exactly measure the time spent for running a task, and lack of confidence in the reported results. Any changes to the original experiment should be clearly discussed along with a description of the motivation for the change(s), as well as any threats to the validities of the conclusions. The algorithm stops when the number of iterations performed by the algorithm reaches a predefined value or the total number of objective function evaluations performed by the algorithm reaches a pre-set maximum number of function evaluations. Moreover, stopping criterion should be simple. Comparing algorithms based on number of fitness evaluations is safer than on the number of generations/iterations. Avoid comparison with an experiment which is poorly documented. Here in this chapter, biogeography based optimization (BBO), artificial bee colony algorithm (ABC), teacher-learner based optimization (TLBO), real coded genetic algorithm (RCG) and predator-prey optimization (PPO) are discussed to solve thermal and hydrothermal generation scheduling problems.

9.2 BIOGEOGRAPHY BASED OPTIMIZATION FOR ECONOMIC DISPATCH WITH PROHIBITED OPERATING ZONES AND RAMP RATE

The economic dispatch problem is defined by Eq. (7.52) is undertaken to solve by biogeography based optimization (BBO). BBO is a population based stochastic optimization algorithm which is inspired by the mathematics of biogeography. It deals with the geographical distribution of biological species over time and space [Simon, 2008].

Transmission losses are given by Eq. (7.53). Power balance equality constraint given by Eq. (7.26b) is taken care using heuristic and simultaneously takes care of ramp rate limits and prohibited operating zones.

9.2.1 Constraint Handling

To satisfy power balance equality constraint given by Eq. (7.26b), residual power ΔP_{Dk} is computed as:

$$\Delta P_{Dk} = \sum_{i=1}^{NG} P_{ki} - P_{Dk} - P_{Lk} \; (k = 1, 2, \ldots, \text{NP}) \tag{9.1}$$

where, NP is the population size.

The equality constraint is satisfied if residual power, ΔP_{Dk} is less than tolerance band. Otherwise out of total number of thermal generating unit those act as decision variables, one generation of slack thermal unit 's' is perturbed during each sub-interval as mentioned below:

$$P_{ki} = \begin{cases} P_{ks} - \min\{\Delta P_{Dk}, (P_{ks} - T_s^{\min})\}; \; \Delta P_{Dk} > 0 \\ P_{ks} + \min\{\Delta P_{Dk}, (T_s^{\max} - P_{ks})\}; \; \Delta P_{Dk} < 0 \end{cases} \quad (i = 1, 2, \ldots, \text{NG}; \; k = 1, 2, \ldots, \text{NP}) \tag{9.2}$$

where

$$T_s^{\min} = \max(P_s^{\min}, (P_s^0 - DR_s)) \text{ and } T_s^{\max} = \min(P_s^{\max}, (P_s^0 + UR_s))$$

P_{ks} is power generation of sth slack thermal unit for kth population.

Transmission loss is computed after evaluating the slack thermal unit power. This process continues until power balance equality constraint is satisfied. The elaborated steps are given in Algorithm 9.1.

9.2.2 Prohibited Operating Zone Constraint Handling

The prohibited operating zone (POZ) constraints on the system are specified by the Eqs. (7.52c)–(7.52e). In case the generation, P_{kj} violates the prohibited zone constraint, the violation is handled by updating the generation, P_{kj} using following equations:

$$P_{kj} = \begin{cases} P^L_{j,NZ-1} - R()\dfrac{(P^U_{j,NZ-1} - P^L_{j,NZ-1})}{P^U_{j,NZ-1}} & ;\ (P_{kj} - P^L_{j,NZ}) < (P^U_{j,NZ} - P_{kj}) \\ P^U_{j,NZ-1} + R()\dfrac{(P^U_{j,NZ-1} - P^L_{j,NZ-1})}{P^U_{j,NZ-1}} & ;\ (P_{kj} - P^L_{j,NZ}) > (P^U_{j,NZ} - P_{kj}) \quad (j = 1, 2, \ldots, \text{NG}; k = 1, 2, \ldots, \text{NP}) \\ P_{kj} & ;\ \text{else} \end{cases} \quad (9.3)$$

where $R()$ is uniform distributed random number in the range [0, 1].

Algorithm 9.1: Constraint handling

1. **FOR** (k = 1, NP)
2. **FOR** i = 1, NG
3. $T_i^{\min} = \max(P_i^{\min}, (P_i^0 - DR_i))$
4. $T_i^{\max} = \min(P_i^{\max}, (P_i^0 + UR_i))$
5. Check the limits of thermal power.
6. **IF** $(P_{ki} > T_i^{\max})\ P_{ki} = T_i^{\max}$
7. **IF** $(P_{ki} < T_i^{\min}) P_{ki} = T_i^{\max}$
8. **ENDFOR**
9. **FOR** i = 1, NG
10. Select a slack thermal unit's' randomly that has not selected so far.
11. Compute the transmission loss using Kron's formula Eq. (7.53).
12. Compute the residual power, ΔP_{Dk} as given in Eq. (9.1).
13. **IF** $\left(|\Delta P_{Dk}| \le \varepsilon\right)$ **EXIT**
14. Modify sth thermal unit power generation for members to satisfy power balance equality constraint as per Eq. (9.2).
15. Check POZ and update using Eq. (9.3)
16. **ENDFOR**
17. **ENDFOR**
18. **RETURN**

9.2.3 Biogeography Based Optimization

In BBO, each individual corresponds to a candidate solution and is called a "habitat" which is geographically isolated from other habitats. For measuring the goodness or suitability for living, each habitat has a performance index called the habitat suitability index (HSI). BBO mimics the immigration and emigration of species between habitats in a multidimensional space. Habitats that are suitable for living of biological species possess a high HSI and the less suited ones have a low HSI. Geographical features like rainfall, temperature, diversity of vegetation, topography etc. affect the HSI of a habitat. All these features that describe the suitability of the habitat are called the suitability index variables (SIVs) and mathematically composed of NG-dimensional real vector.

The movement of species among the habitats is managed by the immigration rate (λ) and the emigration rate (μ) [Simon, 2011]. The rate at which the new species enter a habitat is known as the immigration rate. The rate at which the old species become extinct from the habitat is known as the emigration rate. Habitats with smaller populations are more prone to extinction (i.e. the immigration rate is high). But, as more species occupy the habitat, the immigration rate reduces and the emigration rate increases. The greater the total numbers of species in the habitat, i.e. a high HSI, the better the solution it contains. Further, the good solutions (i.e. habitats with many species) share their features with poor solutions (i.e. habitats with fewer species), and poor solutions accept new features from good solutions. The immigration and emigration rates when there are K species in the habitat are calculated using the sinusoidal migration model [Ma, 2010, Ma and Simon, 2011] and mathematically expressed as follows:

$$\lambda_k = \frac{I}{2}\left(\cos\left(\frac{\pi K_k}{K_{\max}}\right) + 1\right) \quad (k = 1, 2, \ldots, \text{NP}) \tag{9.4}$$

$$\mu_k = \frac{E}{2}\left(-\cos\left(\frac{\pi K_k}{K_{\max}}\right) + 1\right) \quad (k = 1, 2, \ldots, \text{NP}) \tag{9.5}$$

where, I represents the maximum immigration rate, that occurs when there are zero species in the habitat. E is the maximum emigration rate, that occurs when the habitat contains the maximum number of species, $K_{\max}$.

The relationship between the fitness of habitats (a function of the number of species), emigration rate μ and immigration rate λ is shown in Figure 9.1. The point of intersection of the immigration and the emigration curve is represented by the equilibrium point, K_0. The BBO algorithm constitutes two operators, i.e., migration and mutation operators which manage its operation. The migration operation facilitates the sharing of information among the habitats while mutation operator increases the population diversity of the habitat.

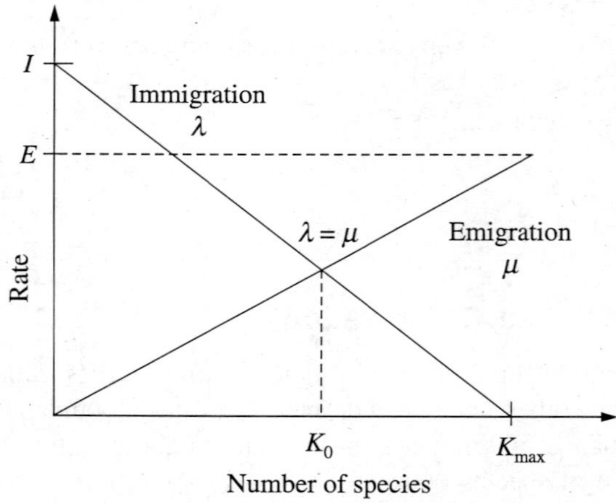

FIGURE 9.1 Species model of a single habitat.

Population Initialization

In BBO algorithm, each species in the habitat has a position made up of NG-dimensions and a fitness value. The position corresponds to the solution set and the fitness value represents the accommodation of the species to the fitness function. For a NG-dimensional optimization problem, a habitat is an 1 × NG array. And if the population consists of NP parameters, the initial population containing the candidate solutions within the solution search space is initialized as follows:

$$P_{ki}^{t} = P_i^{\min} + R(P_i^{\max} - P_i^{\min}) \quad (i = 1, 2, ..., \text{NG}; k = 1, 2, \dots, \text{NP}) \tag{9.6}$$

where, P_{ki}^{t} represents the position of the kth specie in ith dimension. $P_i^{\max}$ and $P_i^{\min}$ are the lower and the upper bounds of the kth specie and R represents a uniform random number between 0 and 1. The initially generated population must satisfy the equality and inequality constraints in order to get the feasible solutions. The random perturbation method is therefore applied to check if there occur any violations of the stability constraints.

Opposition Based Learning

Incorporating the opposition based learning strategy, the BBO method make a start with some initial random solutions that are improved with time by moving towards optimal solution. The computational time is an important parameter that is related to the distance of these initial guesses from the optimal solution. It can be improved by starting with a better solution while checking the opposite solution in the search space. The guess or its opposite guess has been chosen as initial solutions. A guess is farther from the solution than its opposite guess with 50% probability [Rahnamayan, 2008]. Therefore, starting with better guesses adjudged by its objective function has the ability to increase the convergence speed. The same approach can be applied not only to initial solutions but also continuously to each solution in the current population, during the run to reach a final optimal solution. This can be mathematically expressed as:

$$P_{k+\text{NP},i}^{t} = P_i^{\min} + P_i^{\max} - P_{ki}^{t} \quad (i = 1, 2, \dots, \text{NG}; k = 1, 2, \dots, \text{NP}) \tag{9.7}$$

where, $P_i^{\min}$ and $P_i^{\max}$ are lower and upper limits of generator outputs and are expressed as:

Fitness Evaluation

To solve the economic load dispatch problem by employing the BBO, the objective function is considered as fitness function obtained from Eq. (7.52a) and is represented by the following equation:

$$f_k = \sum_{i=1}^{\text{NG}} \left(a_i P_{ki}^2 + b_i P_{ki} + c_i + d_i \left| \sin(e_i(P_i^{\min} - P_{ki})) \right| \right) \quad (k = 1, 2, \dots, \text{NP}) \tag{9.8}$$

Migration

In BBO, the migration operation deals with the immigration and emigration rates. The migration operator modifies the selected habitat's SIVs by probabilistically sharing the information among the different habitats in the solution search space. This operation can be expressed as follows:

$$P_{ki} = P_{ei} \; (i = 1, 2, ..., \text{NG}) \tag{9.9}$$

where, P_{ki} and P_{ei} represent the immigration and the emigration habitat, respectively.

This basic migration operation in BBO possesses simple exploitation abilities of producing new solutions by simply copying the features from a constant pool by sharing the information between the habitats. This results in a lack of the exploration ability of BBO to produce new solutions from new areas in the search space. To overcome this shortcoming and to balance the exploitation and exploration features, an improved migration operator called the polyphyletic migration, is used in BBO [Xiong et al., 2014]. The sharing of information among the habitats and a habitat's SIVs modification operation is given in the following expression:

$$v_{kd} = P_{ei} + \phi_{ki}(P_{ei} - P_{ri}) \quad (i = 1, 2, \ldots, \text{NG}; k \neq e \neq r \in (1, 2, \ldots, \text{NP})) \tag{9.10}$$

where, v_{kd} is the candidate solution (habitat) and P_{ki} and P_{ei} represent the immigration and the emigration habitat, respectively. P_{ri} is a random habitat that satisfies $r \neq k \neq e$ and ϕ_{ki} is a uniform random number distributed between the range (–1, 1). The polyphyletic migration operation is illustrated in Algorithm 9.2.

The polyphyletic migration operator has three advantages. Firstly, it directly copies the information from habitat and uses the fairly good habitat as the base to produce a symmetrical perturbation. So, it fetches new information from the unexploited feasible space. Secondly, it ensures the emigration of new features from other habitats. The third advantage is that the emigration habitat mainly focuses on the exploitation while the other two habitats which are selected randomly from the current population mostly emphasize on the exploration. These three types of habitats help to attain a good sense of balance between exploitation and exploration processes.

Mutation

In BBO, the mutation operator is used to enhance the diversity of population, which helps to minimize the chances of getting trapped in the local minima [Li and Yin, 2012]. The processes of immigration and emigration can be represented mathematically by a probabilistic model. Consider the probability, p_k that a habitat contains K species. p_k changes from time t to time $t + \Delta t$ as follows:

$$p_k(t + \Delta t) = p_k(t)(1 - \lambda_k \Delta t - \mu_k \Delta t) + p_{k-1}\lambda_{k-1}\,\Delta t + p_{k+1}\mu_{k+1}\Delta t \quad (k = 1, 2, \ldots, \text{NP}) \tag{9.11}$$

where, λ_k and μ_k are the immigration and emigration rates when there are K species in the habitat. This equation holds because in order to have K species at time $t + \Delta t$, one of the following conditions must hold:

- There were NP species at time t, and no immigration or emigration occurred between t and $t + \Delta t$.
- There were (NP–1) species at time t, and single specie has immigrated.
- There were (NP+1) species at time t, and single specie has emigrated.

Equation (9.11) can be rewritten as:

$$\frac{p_k(t + \Delta t) - p_k(t)}{\Delta t} = p_k(t)(-\lambda_k - \mu_k) + p_{k-1}\lambda_{k-1} + p_{k+1}\mu_{k+1} \quad (k = 1, 2, \ldots, \text{NP})$$

Algorithm 9.2: Polyphyletic migration operation

For a target habitat P_{ki}

```
FOR k = 1 to NP
    IF (rand(0, 1) < λ_k) THEN
        Select habitat P_ei with respect to the immigration rate, μ_e
        IF (rand(0, 1) < μ_e) THEN
            Randomly select r ≠ k ≠ e
            v_ki = P_ei + φ_ki (P_ei − P_ri) (i = 1, 2, ..., NG)
        ELSE
            Randomly select s ≠ k
            v_ki = P_si  (i = 1, 2, ..., NG)
        ENDIF
    ELSE
        v_ki = P_ki  (i = 1, 2, ..., NG)
    ENDIF
ENDFOR
```

If time Δt is small enough so that the probability of more than one immigration or emigration can be ignored, then taking the limits of the above equation as $\Delta t \rightarrow 0$ gives the following equation:

$$\dot{p}_k = \begin{cases} -(\lambda_k + \mu_k)p_k + \mu_{k+1}p_{k+1}, & k = 0 \\ -(\lambda_k + \mu_k)p_k + \lambda_{k-1}p_{k-1} + \mu_{k+1}p_{k+1}, & 1 \le k \le \text{NP} - 1 \ (k = 1, 2, \ldots, \text{NP}) \\ -(\lambda_k + \mu_k)p_k + \mu_{k-1}p_{k-1}, & k = \text{NP} \end{cases} \tag{9.12}$$

where, λ_{k-1} is the immigration rate for one less than the species count of habitat i and μ_{k+1} is the emigration rate for one more than the species count of habitat k and are mathematically expressed as:

$$\lambda_{k-1} = \lambda \times \left(1 - \frac{(Spc - 1)}{\text{NP}}\right) \tag{9.13}$$

$$\mu_{k+1} = \mu\left(\frac{(Spc + 1)}{\text{NP}}\right) \tag{9.14}$$

where, Spc represents the species count of the habitat k and NP represents the maximum number of species in the habitat.

The probability is then updated as:

$$p_k = p_k + \dot{p}_k \Delta t \quad (k = 1, 2, \ldots, \text{NP}) \tag{9.15}$$

$$p_k = \frac{p_k}{\sum_{k=0}^{\text{NP}} p_k} \quad (k = 1, 2, \ldots, \text{NP}) \tag{9.16}$$

In BBO, the mutation rate, m_{rate} of a SIV in the habitat P_k is selected to be replaced by a randomly generated SIV according to a probability of existence P_k. The mutation rate, m_{rate} can be expressed as follows:

$$m_{rate} = p_{mute}\left(1 - \frac{p_k}{p_{max}}\right) \tag{9.17}$$

where, p_{mute} is a user defined parameter called the initial mutation probability and $p_{max} = \max\{p_k;$ $k = 1, 2, ..., \text{NP}\}$ The mutation operator is described in Algorithm 9.3.

Stopping Criteria

In BBO the maximum number of iterations (generations) is used as the stopping criteria. After completion of the generations the diversity of the solution is verified. If the obtained solution is found to be reliable, then the program is terminated otherwise the generation number is incremented as ($t = t + 1$) and the whole procedure is repeated until an optimal solution to the problem is obtained. Detailed algorithm is given in Algorithm 9.4.

Algorithm 9.3: Mutation operation

1. **FOR** $k = 1$ to NP
2. Compute λ_{k-1} and μ_{k+1} using Eq. (9.13) and (9.14), respectively.
3. Compute the probability p_k and time derivative of p_k, i.e., $\dot{p}_k$ using Eqn. (9.11) and Eqn. (9.12), respectively.
4. Update the probability using Eqn. (9.15) and Eqn. (9.16)
5. Compute the mutation rate, m_{rate} using Eqn. (9.17).
6. Select X_{kd} with respect to P_k
7. **IF** ($rand(0, 1) > m_{rate}$) **THEN**
8. Replace X_{kd} with a random SIV
9. **ENDIF**
10. **ENDFOR**

Algorithm 9.4: The main procedure of BBO algorithm.

1. Initialize the BBO parameters viz. Population size (NP), Maximum immigration rate (I), Maximum emigration rate (E), Mutation probability (P_{mute}), Maximum number of iterations ($MGEN$)
2. **FOR** k=1, NP
3. Initialize the population or the random set of habitats, P_{ki} using Eq. (9.6)
4. Applying opposition to find $P_{k+\text{NP},i}$ using Eq. (9.7). Each habitat corresponds to a candidate solution to the optimization problem.
5. Evaluate the fitness of 2NP habitats, i.e. the HSI for each habitat.
6. **ENDFOR**
7. Find best NP habitats and keep the overall best habitats.
8. Initialize the generation counter, t=1 and Improve=1
9. **WHILE** (Stopping criteria is not met) **DO**
10. Sort the entire population from best to worst.
11. Map the HSI to the number of species.

```
    FOR k=1,NP
        Calculate the immigration rate, λ and emigration rate, μ for each specie count using
        Eq. (9.4) and Eq. (9.5), respectively.
    ENDFOR
    Apply the Algorithm 9.2 for migration operation on the population.
    Evaluate the fitness of newly obtained migrated population, i.e. the HSI for each habitat.
    Perform the Algorithm 9.3 on the population.
    Sort them in ascending order and keep the best habitats.
    Ensure the population does not have duplicates.
    Procure the global best habitat (solution).
    IF (global best is not improved) THEN
        Improve=Improve+1
    ELSE
        Improve=1
    ENDIF
    t=t+1
ENDDO
```

9.2.4 Test System and Results

Operating cost coefficients and ramp rate limits of 6-generator system undertaken for study are given in Table 9.1. Prohibited Operating Zones (POZ) for the generators are given in Table 9.2. Ten (10) independent trial runs are given to biogeography based optimization (BBO) algorithm to justify the global solution. To justify the run time, number of function evaluations (NFE) is computed. Small standard deviation justifies the global solution achieved by performing 10 trial runs with random inputs.

TABLE 9.1 Operating cost coefficients and ramp rate limits: 6 generators units

Unit No.	a_i ₹/(MW²h)	b_i ₹/(MWh)	c_i ₹/h	P_i^{min} MW	P_i^{max} MW	UR_i MW/h	DR_i MW/h	P_i^0 MW
1.	0.0070	7.0	240	100	500	80	120	440
2.	0.0095	10.0	200	50	200	50	90	170
3.	0.0090	8.5	220	80	300	65	100	200
4.	0.0090	11.0	200	50	150	50	90	150
5.	0.0080	10.5	220	50	200	50	90	190
6	0.0075	12.0	190	50	120	50	90	110

TABLE 9.2 Prohibited operating zones: 6 generating units

Unit No.	$P_{i,1}^L$ MW	$P_{i,1}^U$ MW	$P_{i,2}^L$ MW	$P_{i,2}^U$ MW
1	210	240	350	380
2	90	110	140	160
3	150	170	210	240
4	80	90	110	120
5	90	110	140	150
6	75	85	100	105

B-coefficients' to evaluate transmission losses for 6-generating system are given below:

$$B = \begin{bmatrix} 0.0017 & 0.0012 & 0.0007 & -0.0001 & -0.0005 & -0.0002 \\ 0.0012 & 0.0014 & 0.0009 & 0.0001 & -0.0006 & -0.0001 \\ 0.0007 & 0.0009 & 0.0031 & 0.0 & -0.0010 & -0.0006 \\ -0.0001 & 0.0001 & 0.0 & 0.0024 & -0.0006 & -0.0008 \\ -0.0005 & -0.0006 & -0.0010 & -0.0006 & 0.0129 & -0.0002 \\ -0.0002 & -0.0001 & -0.0006 & -0.0008 & -0.0002 & 0.0150 \end{bmatrix} \text{MW}^{-1}$$

$$B_O = 1.0 \times 10^{-3} \begin{bmatrix} -0.3908 & -0.1297 & 0.7047 & 0.0591 & 0.2161 & -0.6635 \end{bmatrix}$$

$$B_{OO} = 0.0056 \text{ MW}$$

The parameters set to get the generation schedule and detail of results are given below for power demand, P_D = 1263 MW.

Number of species, NP = 50

Maximum number of generation, MGEN = 500.

Mutation rate, p_{mute} = 0.01

Minimum value of operating cost, $F_{\min}$ = 15443.07 ₹/*h*

Average value of operating cost, F_{avg} = 15443.07 ₹/*h*

Maximum value of operating cost, $F_{\max}$ = 15443.08 ₹/*h*

Standard deviation, SD = 5.23×10^{-3} ₹/*h*

Number of function evaluations (NFE) = 100100.

Generation schedule is given below:

P_1 = 447.8179 MW, P_2 = 174.0953 MW, P_3 = 262.9219 MW, P_4 = 138.3772 MW, P_5 = 165.251 MW, P_6 = 86.98751 MW

Transmission Loss, P_L = 12.45204 MW

Satisfaction of Equality Constraint, $\Delta P_D = 9.88 \times 10^{-4}$ MW.

9.3 MULTI-FUEL ECONOMIC DISPATCH USING ARTIFICIAL BEE COLONY ALGORITHM

The ELD problem is defined as to minimize the entire operating cost of a power system and simultaneously meet the system operational constraints. Mathematically, the multi-fuel ELD with generators' constraints and power generation limits $P_i^{\min}$ and $P_i^{\max}$ is stated as below:
Minimize the operating cost

$$f(P_i) = \sum_{i=1}^{\text{NG}} \left(a_{ji}P_i^2 + b_{ji}P_i + c_{ji} + d_{ji} \left| \sin(e_{ji}(P_i^L - P_i)) \right| \right) \tag{9.18a}$$

Subjected to:

(i) Power balance, equality constraint

$$\sum_{i=1}^{NG} P_i = P_D + P_L \tag{9.18b}$$

(ii) Generators' operating lower and upper boundaries

$$P_{ji}^L \le P_i \le P_{ji}^U \qquad (i = 1, 2, \ldots, NG; j = 1, 2, \ldots, NF) \tag{9.18c)}$$

$$P_i^{\min} \le P_i \le P_i^{\max} \qquad (i = 1, 2, \ldots, NG) \tag{9.18d}$$

where, $a_{ji}, b_{ji}, c_{ji}, d_{ji}$ and e_{ji} are cost coefficients of *j*th fuel for *i*th generator. P_{ji}^L and P_{ji}^U are minimum and maximum generation limits to use *j*th fuel of *i*th generator. $P_i^{\min}$ and $P_i^{\max}$ are minimum and maximum generating limits of *i*th unit. Transmission loss, P_L is given by Eq. (7.53). P_D is power demand is to meet. NG is number of generating units and NF is number of fuels used for a generating unit. Power equality constraint is met by following constraint handling technique given in Section 9.2.2, Algorithm 9.1.

9.3.1 Artificial Bee Colony Algorithm

The artificial bee colony (ABC) algorithm proposed by Karaboga [2005] is a novel heuristic search optimization algorithm. ABC mimics the natural foraging behavior of honey bee swarms which can be used to solve the complex optimization problems. The ABC model consists of four essential elements: the food sources, employed bees, onlooker bees and scout bees [Karaboga and Gorkemli, 2015]. Each bee in the swarm has the aim of discovering the highest quality food sources and to maximize the nectar amount stored in the beehive.

Food Sources

In ABC, each food source in the NG-dimensional search space represents a possible solution and the nectar amount of the food source is analogous to the fitness value representing the quality of the solution. The food source is selected by a forager depending on various properties like its nectar amount, taste of the nectar, difficulties in extracting the nectar and its distance from the hive.

Employed Bees

A bee is employed at each food source to explore its features, that bee is named as employed bee.

Onlooker Bees

Forager bees waiting at the hive for collecting information from the employed bees regarding a food source are known as onlooker bees.

Scout Bees

Scout bees randomly look for new food source in the vicinity of the hive if the employed and the onlooker bees fail to search a good quality food source. In order to find a best food source, like other population based optimization algorithms, the search mechanism is performed iteratively and the position of the food source is modified continuously. The different phases of ABC are explained as below:

Parameter Initialization

Firstly, the values of three important parameters of ABC algorithm: the colony size, NP, maximum number of cycles, NGEN and the limit value for each food source for the generation of scout bees are set.

Initialization of Population

Then, the NP food sources representing the candidate solutions are placed randomly on the NG-dimensional search space as follows:

$$P_{ki} = P_i^{\min} + R(\,)(P_i^{\max} - P_i^{\min}) \quad (i = 1, 2, \ldots, \text{NG}; k = 1, 2, \ldots, \text{NP}) \tag{9.19}$$

where, P_{ki} represents the position of the kth food source in ith dimension. $P_i^{\min}$ and $P_i^{\max}$ are the lower and the upper bounds of variable P_{ki}, respectively and $R(\,)$ represents a uniform random number ranging between 0 and 1.

This initially generated random population must satisfy the equality and inequality constraints in order to reach a feasible solution. The random perturbation method is therefore applied to check if there occurs any violation of the equality constraints.

Fitness Evaluation

To solve the economic load dispatch problem by employing the ABC optimization technique, the objective function is considered as fitness function obtained from Eq. (9.18a) and is represented by the following equation:

$$f_k = \sum_{i=1}^{\text{NG}} (a_{ji}P_{ki}^2 + b_{ji}P_{ki} + c_{ji} + d_{ji}\left|\sin(e_{ji}(P_i^L - P_{ki}))\right|) \quad (k = 1, 2, \ldots, \text{NP}) \tag{9.20}$$

Employed Bees' Phase

All employed bees are assigned food sources. Each employed bee then visits its associated food source, exploits its features and further tries to find new better food source randomly in the vicinity of the current food source. This is represented by the following mathematical expression:

$$V_{ki} = P_{ki} + R()(P_{ki} - P_{ji}) \quad (i = 1, 2, \ldots, \text{NG}; \; k = 1, 2, \ldots, \text{NP}) \tag{9.21}$$

where, V_{ki} represents the new food source, $R()$ is a random number between the range [–1, 1] and $j \in \{1, 2, ..., \text{NP}\}$ is randomly chosen index such that $j \neq k$.

Probabilistic Selection

As and when the employed bees have explored the entire search space, then they share the information regarding the better food source with the onlooker bees which in turn randomly choose a food source with a probability p_i expressed as follows:

$$p_k = \frac{fitness_k}{\sum_{i=1}^{\text{NP}} fitness_i} \quad (k = 1, 2, \ldots, \text{NP}) \tag{9.22}$$

where, NP represents the number of food sources and $fitness_k$, $fitness_i$ represents the fitness value of the solution P_{ki} and is mathematically defined as:

$$fitness_k = \begin{cases} \dfrac{1}{1 - f(P_{ki})} & ; f(P_{ki}) \geq 0 \\ 1 + abs(f(P_{ki})) & ; f(P_{ki}) < 0 \end{cases} \quad (k = 1, 2, \ldots, \text{NP}) \tag{9.23}$$

The fitness value of the current food source is calculated and compared with the new food source. A greedy selection procedure is used to select the food source with a better fitness. If the fitness value (nectar amount) of the new food source is more than the current food source, the employed bee memorizes the new food source, i.e. the new food source is replaced with the current food source.

Onlooker Bees' Phase

All the employed bees on finishing the search process, share the information regarding the nectar amount (fitness value) of their corresponding food source with the onlooker bees. The onlooker bees then randomly select the food sources according to the probability, p_k. Each onlooker bee then selects a food source and looks for new better food source using Eq. (9.21).

Memorize the Best Solution

After intensive search and modification, all the NP solutions are arranged according to their fitness value and the solution having the highest fitness value (nectar amount) in the current cycle is kept.

Scout Bees' Phase

Even after repeated attempts, if the employed bees and the onlooker bees fail to improve the location of the food source (solution), which is controlled by a predefined parameter limit, then the food source is abandoned or considered as exhausted and the associated employed bee becomes a scout bee. The scout bee then tries to randomly search for new better food source by using the same equation as used by the employed bee, i.e. using Eq. (9.19).

Termination/Stopping Criteria

The maximum number of iterations (generations) is used as the stopping criteria. After completion of the generations the diversity of the solution is verified. If the obtained solution is found to be reliable, then the program is terminated otherwise the generation number (t) is incremented as ($t = t + 1$) and the repeat.

Algorithm 9.5: ABC algorithm

1. Initialize control parameters: bee colony size (NP), maximum number of generations (*MGEN*), predefined parameter (*limit*)
2. **FOR** k=1, NP
3.   Initialize the population of food sources, P_{ki} using Eq. (9.19)
4.   Evaluate and Calculate the fitness of the food source, P_{ki} using Eq. (9.20)
5. **ENDFOR**
6. Find best and keep the best bee.
7. Initialize the generation counter, t = 1 and Improve = 1
8. $Trial_k = 0$ (k = 1, 2, ..., NP)
9. **WHILE** (t < *MGEN*) **DO**

```
    FOR k = 1 to NP
        Employed bees' phase: Produce new food source, V_ki from P_ki based on P_ji such
        that j≠k, using Eq. (9.21)
        Calculate the fitness of the food source, V_ki using Eq. (9.20)
        IF (new food source is better) THEN
            Select the new food source.
            Trial_k = 0
        ELSE
            Trial_k = Trial_k+1
        ENDIF
    ENDFOR
Calculate the probability values for food sources using Eq. (9.22)
FOR i=1 to NP
    IF (rand( ) < p_i
        Onlookers bees phase: Depending on P_ki, produce new food sources, V_ki from food
        source, P_ki based on P_ji such that j≠k , using Eq. (9.21)
        Calculate the fitness of the food source, V_ki using Eq. (9.20)
        IF (new food source is better) THEN
            Select the new food source
            Trial_k = 0
        ELSE
            Trial_k = Trial_k + 1
        ENDIF
    ENDIF
ENDFOR
Scout bees' phase: Big_l = max{Trial_i; i = 1, 2, ..., NP}
IF (Big_l > limit) THEN
    Replace lth food source with new generated food source using Eq. (9.19)
ENDIF
Procure the global best solution (food source) achieved so far
IF (global best is not improved) THEN
    Improve = Improve + 1
ELSE
    Improve = 1
ENDIF
t = t + 1
ENDDO
```

9.3.2 Test System and Results

Operating cost coefficients and ramp rate limits of 10-generator system undertaken for study are given in Table 9.3. The parameters set to get the generation schedule and detail of results are given below for power demand, $P_D = 2700$ MW.

Number of species, NP = 200

Maximum number of generation, MGEN = 500.

Limit, *limit* = 10

Independent 30 trial runs are performed to justify the global solution. The achieved results are given below:

Minimum value of operating cost, F_{min} = 623.8458 ₹/h
Average value of operating cost, F_{avg} = 624.0632 ₹/h
Maximum value of operating cost, F_{max} = 624.2711 ₹/h
Standard Deviation, SD = 1.1046×10^{-1}
Number of function Evaluations (NFE) = 101844.

TABLE 9.3 System data of units with multiple fuels

Unit No.	*Fuel type*	*Operating Cost Coefficients*					*Generation Limits*	
		a_i (₹/MW²–h)	b_i (₹/MWh)	c_i (₹/h)	d_i (₹/h)	e_i (rad/MW)	P_{ji}^L (MW)	P_{ji}^U (MW)
1	1	0.2176×10^{-2}	-0.3975×10^{0}	$0.2697\times10^{+2}$	0.2697×10^{-1}	$-0.3975\times10^{+1}$	100	196
	2	0.1861×10^{-2}	-0.3059×10^{0}	$0.2113\times10^{+2}$	0.2113×10^{-1}	$-0.3059\times10^{+1}$	196	250
2	1	0.4194×10^{-2}	$-0.1269\times10^{+1}$	$0.1184\times10^{+3}$	0.1184×10^{0}	$-0.1269\times10^{+2}$	157	230
	2	0.1138×10^{-2}	-0.3988×10^{-1}	$0.1865\times10^{+1}$	0.1865×10^{-2}	-0.3988×10^{0}	50	114
	3	0.1620×10^{-2}	-0.1980×10^{0}	$0.1365\times10^{+2}$	0.1365×10^{-1}	$-0.1980\times10^{+1}$	114	157
3	1	0.1457×10^{-2}	-0.3116×10^{0}	$0.3979\times10^{+2}$	0.3979×10^{-1}	$-0.3116\times10^{+1}$	200	332
	2	0.1176×10^{-4}	0.4864×10^{0}	$-0.5914\times10^{+2}$	-0.5914×10^{-1}	$0.4864\times10^{+1}$	388	500
	3	0.8035×10^{-3}	0.3389×10^{-1}	$-0.2875\times10^{+1}$	-0.2876×10^{-2}	0.3389×10^{0}	332	388
4	1	0.1049×10^{-2}	-0.3114×10^{-1}	$0.1983\times10^{+1}$	0.19830×10^{-2}	-0.3114×10^{0}	99	138
	2	0.2758×10^{-2}	-0.6348×10^{0}	$0.52.85\times10^{+2}$	0.5285×10^{-1}	$-0.6348\times10^{+1}$	138	200
	3	0.5935×10^{-2}	$-0.2338\times10^{+1}$	$0.2668\times10^{+3}$	0.2668×10^{0}	$-0.23380\times10^{+2}$	200	265
5	1	0.1066×10^{-2}	-0.8733×10^{-1}	$0.1392\times10^{+2}$	0.1392×10^{-1}	-0.8733×10^{0}	190	338
	2	0.1597×10^{-2}	-0.5206×10^{0}	$0.9976\times10^{+2}$	0.9976×10^{-1}	$-0.5206\times10^{+1}$	338	407
	3	0.1498×10^{-3}	0.4462×10^{0}	$-0.5399\times10^{+2}$	-0.5399×10^{-1}	$0.4462\times10^{+1}$	407	490
6	1	0.2758×10^{-2}	-0.6348×10^{0}	$52.85\times10^{+2}$	0.5285×10^{-1}	$-0.6348\times10^{+1}$	138	200
	2	0.1049×10^{-2}	-0.3114×10^{-1}	$1.983\times10^{+1}$	0.1983×10^{-2}	-0.3114×10^{0}	85	138
	3	0.5935×10^{-2}	$-0.2338\times10^{+1}$	$266.8\times10^{+3}$	0.2668×10^{0}	$-0.2338\times10^{+2}$	200	265
7	1	0.1107×10^{-2}	-0.1325×10^{0}	$0.1893\times10^{+2}$	0.1893×10^{-1}	$-0.1325\times10^{+1}$	200	331
	2	0.1165×10^{-2}	-0.2267×10^{0}	$0.4377\times10^{+2}$	0.4377×10^{-1}	$-0.2267\times10^{+1}$	331	391
	3	0.2454×10^{-3}	0.3559×10^{0}	$-0.4335\times10^{+2}$	-0.4335×10^{-1}	$0.3559\times10^{+1}$	391	500
8	1	0.1049×10^{-2}	-0.3114×10^{-1}	$0.1983\times10^{+1}$	0.1983×10^{-2}	-0.3114×10^{0}	99	138
	2	0.2758×10^{-2}	-0.6348×10^{0}	$0.5285\times10^{+2}$	0.5285×10^{-1}	$-0.6348\times10^{+1}$	138	200
	3	0.5935×10^{-2}	$-0.2338\times10^{+1}$	$0.2668\times10^{+3}$	0.2668×10^{0}	$-0.2338\times10^{+2}$	200	265
9	1	0.1554×10^{-2}	-0.5675×10^{0}	$0.8853\times10^{+2}$	0.8853×10^{-1}	$-0.5675\times10^{+1}$	213	370
	2	0.7033×10^{-2}	-0.4514×10^{-1}	$0.1530\times10^{+2}$	0.1423×10^{-1}	-0.1817×10^{0}	130	213
	3	0.6121×10^{-3}	-0.1817×10^{-1}	$0.1423\times10^{+2}$	0.1423×10^{-1}	-0.1817×10^{0}	370	440

Unit No.	*Fuel type*	*Operating Cost Coefficients*					*Generation Limits*	
		a_i (₹/MW²–h)	b_i (₹/MWh)	c_i (₹/h)	d_i (₹/h)	e_i (rad/MW)	P_{ji}^L (MW)	P_{ji}^U (MW)
10	1	0.1102×10^{-2}	-0.9938×10^{-1}	$0.1397\times10^{+2}$	0.1397×10^{-1}	-0.9938×10^{0}	200	362
	2	0.4164×10^{-4}	0.5084×10^{0}	$-0.6113\times10^{+2}$	-0.6113×10^{-1}	$0.5084\times10^{+1}$	362	407
	3	0.1137×10^{-2}	-0.2024×10^{0}	$0.4671\times10^{+2}$	0.4671×10^{-1}	$-0.2024\times10^{+1}$	407	490

As standard deviation is small so the ABC algorithm provides global solution. Generation schedule after performing multi-fuel generation schedule using ABC for minimum operating cost is given in Table 9.4.

TABLE 9.4 Generation scheduling

Unit No	*Fuel Type*	P_i^{min} (MW)	P_i (MW)	P_i^{max} (MW)	*Unit No.*	*Fuel Type*	P_i^{min} (MW)	P_i (MW)	P_i^{max} (MW)
1	2	100	218.1065	250	6	3	85	241.0047	265
2	1	50	210.6692	230	7	1	200	290.0925	500
3	1	200	280.6560	500	8	3	99	240.2239	265
4	3	99	239.4176	265	9	3	130	427.2079	440
5	1	190	279.9102	490	10	1	200	272.7103	490
						Operating Cost, F =	623.8458 ₹/h		
						Satisfaction of Equality Constraint, ΔP_D =	9.7656×10^{-4} MW		

9.4 COMBINED HEAT AND POWER GENERATION DISPATCH PROBLEM USING TEACHER LEARNER BASED OPTIMIZATION

Combined heat and power (CHP) generation has higher energy efficiency and less greenhouse gas emission compared with the other forms of energy supply. The main difference between CHP units and conventional condensing plant is in the type of the power obtained and the overall efficiency of each plant. In conventional condensing plants, the energy from the fuel is utilized to produce electrical power only, while in CHP systems, the energy from the fuel is utilized to produce both electrical and thermal power, thus, increasing its efficiency. The conventional condensing plant delivers power at an efficiency of 35–55%. Using efficient flue gas condensation, the total efficiency of CHP unit is found to be in the range of 80–111% (lower heating value base). The heat production depends on power generation and vice versa. This introduces complexity due to the non-separable heat in the CHP units and nature of electrical power. Economic dispatch problem of CHP plants is more complicated problem in comparison to power unit dispatch due to two-dimensional nature of the problem [Abdolomohammadi and Kazemi, 2013]. CHP systems are normally classified according to the sequence of energy use and the operating schemes adopted. On this basis CHP systems can be classified as either a topping or a bottoming cycle which are explained as below:

Topping Cycle

In a topping cycle, the fuel supplied is used to first produce power and then thermal energy, which is the by-product of the cycle and is used to satisfy process heat or other thermal requirements. Topping cycle CHP system is widely used and is the most popular method of cogeneration.

Bottoming Cycle

In a bottoming cycle, the primary fuel produces high temperature thermal energy and the heat rejected from the process is used to generate power through a recovery boiler and a turbine generator. Bottoming cycles are suitable for manufacturing processes that require heat at high temperature in furnaces and kilns, and reject heat at significantly high temperatures. Typical areas of application include cement, steel, ceramic, gas and petrochemical industries. Bottoming cycle plants are much less common than topping cycle plants.

The problem of static dispatch determines the loads of generators in a system that meets a power demand during a single scheduling period for the least cost. The conventional economic dispatch problem deals with only the electric power loads of the conventional thermal generators. The CHP economic dispatch problem on the other hand is considerably more complicated. In this problem, both the power and process heat demands must be satisfied. The power is generated by conventional thermal generators and cogeneration units whilst the heat is generated by cogeneration units and boiler units. The complexity is further increased by the non-separable nature of the power and heat generations of cogeneration units. The system under consideration has conventional thermal generators, cogeneration units, and heat-only units. Figure 9.2 shows the heat–power feasible operation region (FOR) of a combined cycle CHP unit. The feasible operation is enclosed by the boundary curve ABCD. Along the boundary curve BC, the heat capacity increases as the power generation increases. The generation declines as heat capacity increases along the curve AB and CD. The power output of the power units and the heat output of heat units are restricted by their own upper and lower limits.

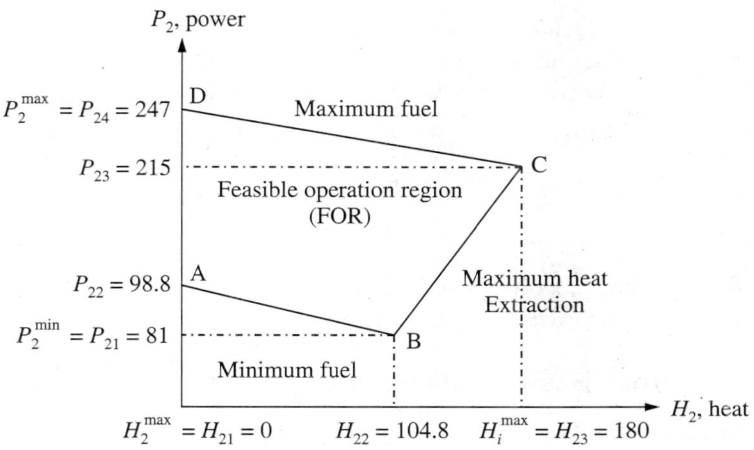

FIGURE 9.2 Feasible region of operation for cogeneration unit.

The power is generated by conventional thermal generators and cogeneration units while the heat is generated by cogeneration units and heat-only units. The CHP economic dispatch problem of a system is to determine the unit heat and power production, so that the system production cost is minimized, while the heat and power demands and other constraints are met [Subbaraj et al., 2009]. The objective function of the CHP economic dispatch problem, which is to be minimized, is:

Minimize

$$F(P_i, H_i) = \sum_{i=1}^{NG} C(P_i) + \sum_{j=1}^{NC} C(P_{j+NG}, H_j) + \sum_{i=1}^{NH} C(H_{i+NC}) \tag{9.24a}$$

Subject to: The equilibrium constraints of electricity and heat production, and the capacity limits of each unit.

$$\sum_{i=1}^{NG+NC} P_i = P_D \tag{9.24b}$$

$$\sum_{i=1}^{NC+NH} H_i = H_D \tag{9.24c}$$

$$P_i^{min} \le P_i \le P_i^{max} \quad (i = 1, 2, \ldots, NG) \tag{9.24d}$$

$$P_i^{min}(H_i) \le P_i \le P_i^{max}(H_i) \quad (i = 1, 2, \ldots, NC) \tag{9.24e}$$

$$H_j^{min}(P_j) \le H_j \le H_j^{max}(P_j) \quad (j = 1, 2, \ldots, NC) \tag{9.24f}$$

$$H_j^{min} \le H_j \le H_j^{max} \quad (j = 1, 2, \ldots, NH) \tag{9.24g}$$

where, P_i is power generation by ith thermal power unit and cogeneration unit. H_i is heat generation by jth cogeneration unit and heat only unit. P_D is total power demand. H_D is total heat demand. P_i^{min} and P_i^{max} are lower and upper bounds for power output of unit i, respectively. H_i^{min} and H_i^{max} are lower and upper bounds for heat production of unit j, respectively. NG is number of thermal power units. NC is number of cogeneration units. NH is number of heat-only units. Operating costs for thermal generation, cogeneration and heat only units are given below:

$$C(P_i) = a_{1i}P_i^2 + b_{1i}P_i + c_{1i} \ (₹/h) \tag{9.25}$$

$$C(P_i, H_i) = a_{2i}P_i^2 + b_{2i}P_i + c_{2i} + d_iH_i + e_iH_i^2 \ (₹/h) \tag{9.26}$$

$$C(H_i) = a_{3i}P_i^2 + b_{3i}P_i + c_{3i} \ (₹/h) \tag{9.27}$$

where a_{1i}, b_{1i} and c_{1i} are cost coefficients of thermal power units. a_{2i}, b_{2i}, c_{2i}, d_i and e_i are cost coefficients of cogeneration units. a_{3i}, b_{3i} and c_{3i} are cost coefficients of heat-only units.

9.4.1 Teacher Learner Based Optimization

Teacher learner based optimization (TLBO) is a nature inspired, population based algorithm that works on the process of teacher-learner. In TLBO, students in a class constitute the population and subjects offered to students represent different design variables. A set of good scores for the subject offered to students are initialized by applying opposition-based learning. The diversity of the students is maintained in TLBO by employing migration. The candidate solution comprises of design variables and the value of objective function represents the knowledge of a particular student [Roy, 2013].

The growth of every generation is greatly influenced and depends upon the quality of teachers in the society. The future and knowledge base of every child is influenced by first the parents and then the mentors/teachers. TLBO efficiently utilizes the knowledge base of a teacher to increase/improve the know-how of learners/students. A good teacher tries to impart knowledge available with him/her to his/her learners/students to best of his/her ability. The learners along with acquiring the knowledge from the teacher also interact and share the knowledge with each other. The methodology of TLBO constitute of two main phases, the Teacher phase and the Learner phase.

Class representation: Assume *NL* is the number of learners in a class (population) and each learner has been assigned number of subjects. Let there are two classes named as generation class and heat class. The decision variables of generation and heat power of CHP load dispatch are used to form generation and heat classes, respectively. The *k*th learner is represented as $P_k = [P_{k1}, P_{k2}, ..., P_{k,\mathrm{NG+NC}}]$ of generation class and $H_k = [H_{k1}, H_{k2}, ..., H_{k,\mathrm{NC+NH}}]$ of heat class, where NG is number of generators, NC is number of combined heat and power units and NH is number of heat power units.

The decision variables of the CHP load dispatch, generation and heat power are used to form the class. The generation (P_k) and heat power (H_k) of units are represented as the subjects assigned to learners in a class. For a numbers of generating unit and heat power units, NG + NC + NH, the learner is represented as a vector of length *S*.

Initialization of class: Each learner of the class is initialized with the help of random search for marks of all the subjects. The starting point is initialized by exploring the global search space and further the starting point is fine tuned in the local search space to find the best starting point. The variables P_{ki}^t and H_{ki}^t are initialized and search process is initiated using Eq. (9.28) and Eq. (9.29)

$$P_{ki}^t = P_i^{\min} + R()(P_i^{\max} - P_i^{\min}) \quad (i = 1, 2, \ldots, \mathrm{NG} + \mathrm{NC}; j = 1, 2, \ldots, \mathrm{NL}) \tag{9.28}$$

$$H_{ki}^t = H_i^{\min} + R()(H_i^{\max} - H_i^{\min}) \quad (i = 1, 2, \ldots, \mathrm{NC} + \mathrm{NH}; j = 1, 2, \ldots, \mathrm{NL}) \tag{9.29}$$

where $R()$ is a uniform generated random number ranging between {0, 1}. NG + NC denote subjects allotted to each learner. NL is number of learners in a class and *t* is the iteration counter. $P_i^{\min}$ and $P_i^{\max}$ are the maximum and minimum values of *i*th decision variable of power vector P. $H_i^{\min}$ and $H_i^{\max}$ are the maximum and minimum values of *i*th decision variable heat of vector H.

Objective Function

The objective function defined by Eq. (9.24a) is used to obtain the expected fitness function *f*. The fitness function of *k*th learner of class during *t*th iteration is given below:

$$f_k^t = \text{Minimize}(F_k(P_{ki}, H_{ki})) \qquad (k = 1, 2, \ldots, \mathrm{NL}) \tag{9.30}$$

$F_k(P_{ki}, H_{ki})$ is obtained using Eq. (9.24a) for *i*th learner of class in *t*th iteration.
At the end of first iteration the function value of the fittest learner is set as global best (f^{global}) and corresponding marks scored by him in various subjects are set as global best marks (G_j).

$$f^{\text{global}} = \min \{f_k^t; (k = 1, 2, \ldots, \mathrm{NL})\} \tag{9.31}$$

Teacher Phase

The best learner is designated as teacher for the class. The teacher acts as a role model for the rest of learners. The result of learners depends largely on the quality of teaching being imparted to the learners by the teacher. A teacher tries to increase/improve the mean score of all the learners in each of the subject allotted towards its own mean score. So, the mean fitness of the class is increased by the teacher according to his/her own capability.

The best learner is selected among all the learners in a class based upon the fitness function value calculated using Eq. (9.30) and act as teachers P_i^t and H_i^t for current iteration t. The means (m_{Pi} and m_{Hi}) for subjects allotted to the students of generation and heat classes are evaluated and a randomly weighted differential vectors (D_{Pi} and D_{Hi}) from current means and various desired mean vectors are calculated as given below:

$$D_{Pi}^t = R()\,(X_{Pi}^t - T_f m_{Pi}^t) \qquad (i = 1, 2, \ldots, \text{NG} + \text{NC}) \tag{9.32}$$

$$D_{Hi}^t = R()(X_{Hi}^t - T_f m_{Hi}^t) \qquad (i = 1, 2, \ldots, \text{NC} + \text{NH}) \tag{9.33}$$

m_{Pi}^t is mean of ith subject for all learners of a power class and m_{Hi}^t is mean of ith subject for all learners of a heat class are given below

$$m_{Pi}^t = \frac{1}{\text{NL}} \sum_{k=1}^{\text{NL}} P_{ki}^t \qquad (i = 1, 2, \ldots, \text{NG} + \text{NC}) \tag{9.34}$$

$$m_{Hi}^t = \frac{1}{\text{NL}} \sum_{k=1}^{\text{NL}} H_{ki}^t \qquad (i = 1, 2, \ldots, \text{NC} + \text{NH}) \tag{9.35}$$

where X_{Pi}^t is the score of the teacher of power class in ith subject. X_{Hi}^t is the score of the teacher of heat class in ith subject. T_f is the teaching factor. $R()$ is a uniform generated random number between {0, 1}.

The teaching factor (T_f) is one of the vital aspect that facilitates the convergence of TLBO. The value of T_f decides about the volume of effect a teacher has on the output of a learner. High T_f leads the learners to drift away from good solution, whereas too small T_f, restricts the learner's movement in limited range and leads to slow convergence. In TLBO, the value of T_f is either selected as 1 or 2 and is heuristically decided as follows:

$$T_f = \text{round } (1 + R(\,)) \tag{9.36}$$

The weighted differential vectors (D_{Pi} and D_{Hi}) generated using Eq. (9.32) and Eq. (9.33) are added to current score of learners in different subjects to generate new learners:

$$Pnew_{ki}^t = P_{ki}^t + D_{Pi}^t \qquad (i = 1, 2, \ldots, \text{NG} + \text{NC};\ k = 1, 2, \ldots, \text{NL}) \tag{9.37}$$

$$Hnew_{ki}^t = H_{ki}^t + D_{Hi}^t \qquad (i = 1, 2, \ldots, \text{NC} + \text{NH};\ k = 1, 2, \ldots, \text{NL}) \tag{9.38}$$

Newly generated learner with a better fitness value replaces the existing learner in the class.

Learner Phase

The learners in a generation or heat class not only acquire the knowledge from the teacher but also interact on regular basis among themselves and share their know-how among themselves through sharing of notes, discussions and presentations. The second phase of TLBO emulates this sharing of knowledge by learners among themselves in their respective class. Two target learners namely k and j are selected randomly such that $k \neq j$. The resultant new learners after sharing/exchange of know-how are generated as follows:

$$Pnew_{ki}^{t} = \begin{cases} P_{ki}^{t} + R()(P_{ki}^{t} - P_{ji}^{t}) & ; f_k^t < f_j^t \\ P_{ki}^{t} + R()(P_{ji}^{t} - P_{ki}^{t}) & ; f_k^t \geq f_j^t \end{cases} \quad (i = 1, 2, \ldots, \text{NG} + \text{NC}; j \neq k; k = 1, 2, \ldots, \text{NL}) \tag{9.39}$$

$$Hnew_{ki}^{t} = \begin{cases} H_{ki}^{t} + R()(H_{ki}^{t} - H_{ji}^{t}) \; ; f_k^t < f_j^t \\ H_{ki}^{t} + R()(H_{ji}^{t} - H_{ki}^{t}) \; ; f_k^t \geq f_j^t \end{cases} \quad (i = 1, 2, \ldots, \text{NC} + \text{NH}; j \neq k; k = 1, 2, \ldots, \text{NL}) \tag{9.40}$$

where P_{ki}^{t} is the score of the kth learner in ith subject. P_{ji}^{t} is the score of the jth learner in ith subject. H_{ki}^{t} is the score of the kth learner in ith subject. H_{ji}^{t} is the score of the jth learner in ith subject. $R()$ is a uniform generated random number between [0, 1].

Best Function Value

At the end of each iteration if the function value obtained by the best learner is better than the global best (f^{global}) then it replaces the global best and corresponding marks obtained by the best learner are stored as the global best marks (G_j).

$$f^{\text{global}} = \begin{cases} f_{\min}^{t} & ; f_{\min}^{t} < f^{\text{global}} \\ f^{\text{global}} & ; \text{otherwise} \end{cases} \tag{9.41}$$

where, $f_{\min}^{t} = \min\{ f_i^t \;\; (i = 1, 2, \ldots, \text{NL})\}$

Stopping Criteria

A heuristic optimization algorithm can be stopped by employing various stopping criterion. Common examples are maximum number of iterations, tolerance and number of function evaluations. TLBO employs maximum number of iterations as criteria to stop. The above procedure is repeated with incremented t value until the value of t reaches the maximum value of iterations specified. Otherwise, the best values of G_{Pi} and G_{Hi} corresponds to power and heat, respectively and f^{global} is the best/minimum value of the objective function. The search procedure of the proposed ETLBO method is as given below as Algorithm 9.6.

Algorithm 9.6: Pseudo-code for CHP economic dispatch

1. Read data; viz. Generation and heat power P and H, maximum allowed iterations, $T^{\max}$, $P_i^{\min}$ and $P_i^{\max}$ $(i = 1, 2, \ldots, \text{NG} + \text{NC})$, $H_i^{\min}$ and $H_i^{\max}$ $(i = 1, 2, .., \text{NC} + \text{NH})$, etc.
2. Generate an array of uniform random numbers.
3. Set iteration counter, $t = 0$
4. **FOR** k = 1, NL (class counter)
5.   **FOR** j = 1, NG + NC (subject (power generation) counter
6.   Generate the initial score of learners in different subjects P_{ki}^{t} using Eq.(9.28)
7.   **ENDFOR**

ENDFOR
FOR k = 1, NL (class counter)
  FOR j=1, NC+NH (subject (heat generation)) counter
    Generate the initial score of learners in different subjects H_{ki}^{t} using Eq.(9.29)
  ENDFOR
ENDFOR
FOR k = 1, NL (class counter)
  Compute f_k^t using Eq. (9.30)
ENDFOR
Arrange f_k^t (k = 1, 2, ..., NL) in ascending order
Set $f^{\text{global}} = f_1^t$ and $G_{Pi} = P_{1i}^t$ (i = 1, 2,..., NG + NC), $G_{Hi} = H_{1i}^t$ (i = 1, 2, ..., NC + NH)
WHILE ($t < T^{\max}$) *DO*
  Increment iteration counter, $t = t + 1$
  Compute $T_f = ROUND(1 + r(\ ))$
  Compute $X_{Pi}^t = P_{1i}^t$ (i = 1, NG + NC) and $X_{Hi}^t = H_{1i}^t$ (i = 1, NC + NH)
  FOR i = 1, NG + NC
    total = 0.0
    FOR k = 1, NL
      $total = total + P_{ki}^t$
    ENDFOR
    $m_{Pi}^t = \dfrac{total}{NL}$
    Compute $D_{Pi}^t = R(\ )(X_{Pi}^t - T_f m_{Pi}^t)$
  ENDFOR
  FOR i = 1, NC + NH
    total = 0.0
    FOR k = 1, NL
      $total = total + H_{ki}^t$
    ENDFOR
    $m_{Hi}^t = \dfrac{total}{NL}$
    Compute $D_{Hi}^t = R()(X_{Hi}^t - T_f m_{Hi}^t)$
  ENDFOR
  FOR k = 1, NL
    FOR i = 1, NG + NC
      $Pnew_{ki}^t = P_{ki}^t + D_{Pi}^t$
    ENDFOR
    FOR i = 1, NC + NH
      $Hnew_{ki}^t = H_{ki}^t + D_{Hi}^t$
    ENDFOR
    Compute f_k^t using Eq. (9.30) and Check limit and feasible region
    IF $(f_i^{new} < f_i^t)$ **THEN**
    Set $f_k^{new} = f_k^t$ and $P_{ki}^t = Pnew_{ki}^t$ (i = 1, *NG+NC*) and $H_{ki}^t = Hnew_{ki}^t$ (i = 1, NC+NH)
    ENDIF

ENDFOR
FOR $k = 1$, NL // Learner phase starts
Randomly select another learner j such that $j \neq k$
IF $(f_i^t < f_k^t)$ **THEN**
$Pnew_{ki}^t = P_{ki}^t + R()(P_{ki}^t - P_{ji}^t)$ $(i = 1, 2, ..., NG + NC)$
$Hnew_{ki}^t = H_{ki}^t + R()(H_{ki}^t - H_{ji}^t)$ $(i = 1, 2, ..., NC + NH)$
ELSE
$Pnew_{ki}^t = P_{ki}^t + R()(P_{ji}^t - P_{ki}^t)$ $(i = 1, 2, ..., NG + NC)$
$Hnew_{ki}^t = H_{ki}^t + R()(H_{ji}^t - H_{ki}^t)$ $(i = 1, 2, ..., NC + NH)$
ENDIF
Compute f_k^t using Eq. (9.30) and Check limit and feasible region
IF $(f_i^{new} < f_i^t)$ **THEN**
Set $f_k^{new} = f_k^t$ and $P_{ki}^t = Pnew_{ki}^t$ $(i = 1, NG+NC)$ and $H_{ki}^t = Hnew_{ki}^t$ $(i = 1, NC+NH)$
ENDIF
ENDFOR // Learner phase ends
IF $(f_{\min}^t < f^{\text{global}})$ **THEN**
Set $f^{\text{global}} = f_{\min}^t$ and $G_{Pi} = P_{\min,i}^t$ $(i = 1, NG+NC)$ and $G_{Hi} = H_{\min,i}^t$ $(i = 1, NC+NH)$
ENDIF
ENDDO
STOP

9.4.2 Test System and Results

A combined heat and power system of 4-units has been undertaken to illustrate the performance of the TLBO method to obtain heat and power generation schedule to minimize operating cost. Characteristic curves are given below for generation, cogeneration and heat only units.

Conventional power unit: $F_1 = 50P_1$ (9.42)

Co-generation units

$$F_2 = 0.0345P_2^2 + 14.5P_2 + 2650 + 4.2H_2 + 0.03H_2^2 + 0.031P_2H_2 \tag{9.43}$$

$$F_3 = 0.0435P_3^2 + 36.0P_3 + 1250 + 0.6H_3 + 0.027H_3^2 + 0.011P_3H_3 \tag{9.44}$$

Heat unit: $F_4 = 23.4H_4$ (9.45)

The power and heat demand for the system are 200 MW and 115 MWth, respectively. The heat-power feasible regions for the co-generation units are illustrated in Figures 9.2 and 9.3. Main concern of the solution is to satisfy the feasible operating region (FOR) enclosed by three lines AB, BC and DC of Figure 9.2 and lines AB, BC, CB, ED and FE of Figure 9.3.

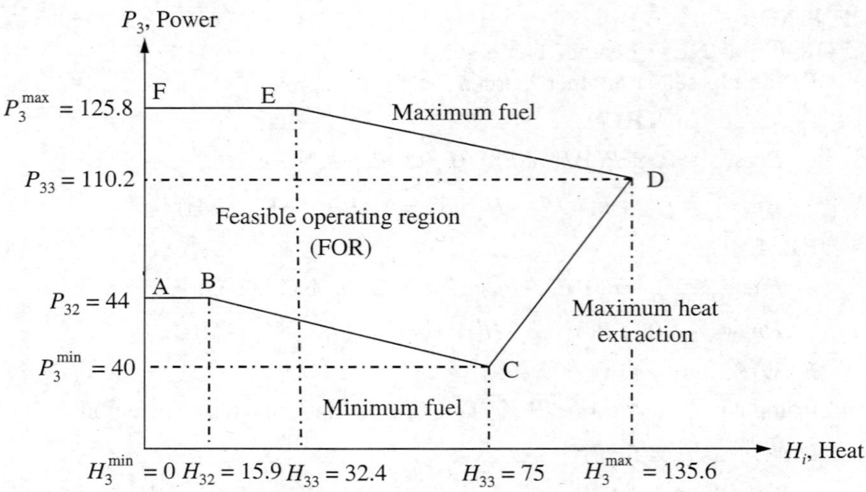

FIGURE 9.3 Feasible region of operation for cogeneration unit.

TABLE 9.5 Constraint handling for cogeneration unit for FOR given in Figure 9.2

IF$(0 \le H_2 \le 104.8)$ **THEN**
IF$(81 \le P_2 \le 98.8)$ **THEN**
IF$(104.8P_2 + 17.8H_2 - 10354.24 < 0)$ **THEN**
$P_2 = (10354.24 - 17.8H_2)/104.8$
ENDIF
ENDIF
IF$(215 \le P_2 \le 247)$ **THEN**
IF$(180P_2 + 32H_2 - 44460 > 0)$ **THEN**
$P_2 = (44460 - 32H_2)/180$
ENDIF
ENDIF
ENDIF
	IF$(104.8 < H_2 \le 180)$ **THEN**
	IF$(81 \le P_2 \le 98.8)$ **THEN**
	IF$(104.8P_2 + 17.8H_2 - 10354.24 < 0)$ **THEN**
	$P_2 = (10354.24 - 17.8H_2)/104.8$
	ENDIF
	ENDIF
	IF$(81 < P_2 \le 215)$ **THEN**
	IF$(75.2P_2 - 134H_2 + 7952 < 0)$ **THEN**
	$P_2 = (134H_2 - 7952)/75.2$
	ENDIF
	ENDIF
	IF$(215 < P_2 \le 247)$ **THEN**
	IF$(180P_2 + 32H_2 - 44460 > 0)$ **THEN**
	$P_2 = (44460 - 32H_2)/180$
	ENDIF
	ENDIF
	ENDIF

Equation for line, AB of Figure 9.2 is given below obtained from two points: (0, 98.8) and (104.8, 81):

$$P_2 - 98.8 = \frac{(81 - 98.8)}{(104.8 - 0)}(H_2 - 0)$$

On simplifying above

$$104.8P_2 + 17.8H_2 - 10354.24 = 0$$

TABLE 9.6 Constraint handling for cogeneration unit for FOR given in Figure 9.3

IF $(0 < H_3 \le 15.9)$ **THEN**
IF $(P_3 \le 44)$ $(P_3 \le 44)$
IF $(P_3 \le 125.8)$ $P_3 \le 125.8$
ENDIF
IF $(15.9 < H_3 \le 32.4)$ **THEN**
IF $(40 \le P_3 \le 44)$ **THEN**
IF $(59.1P_3 + 4H_3 - 2664 < 0)$ **THEN**
$P_3 = (2664 - 4H_3)/59.1$
ENDIF
ENDIF
IF $(P_3 \le 125.8)$ $P_3 \le 125.8$
ENDIF
	IF $(32.4 < H_3 \le 75)$ **THEN**
	IF $(40 \le P_3 \le 44)$ **THEN**
	IF $(59.1P_3 + 4H_3 - 2664 < 0)$ **THEN**
	$P_3 = (2664 - 4H_3)/59.1$
	ENDIF
	ENDIF
	IF $(110.2 \le P_3 \le 125.8)$ **THEN**
	IF $(103.2P_3 + 15.6H_3 - 13488 > 0)$ **THEN**
	$P_3 = (13488 - 15.6H_3)/103.2$
	ENDIF
	ENDIF
	ENDIF
		IF $(75 < H_3 \le 135.6)$ **THEN**
		IF $(40 \le P_2 \le 44)$ **THEN**
		IF $(59.1P_3 + 4H_3 - 2664 < 0)$ **THEN**
		$P_3 = (2664 - 4H_3)/59.1$
		ENDIF
		ENDIF
		IF $(44 < P_2 \le 110.2)$ **THEN**
		IF $(75.2P_2 - 134H_2 + 7952 < 0)$ **THEN**
		$P_2 = (134H_2 - 7952)/75.2$
		ENDIF
		ENDIF
		IF $(110.2 < P_2 \le 125.8)$ **THEN**
		IF $(103.2P_3 + 15.6H_3 - 13488 > 0)$ **THEN**
		$P_3 = (13488 - 15.6H_3)/103.2$
		ENDIF
		ENDIF
		ENDIF

Equation for line, BC of Figure 9.2 is given below obtained from two points: (104.8, 81) and (180, 215):

$$75.2P_2 - 134H_2 + 7952 = 0$$

Equation for line, DC of Figure 9.2 is given below obtained from two points: (0, 247) and (180, 215):

$$180P_2 + 32H_2 - 44460 = 0$$

Based on above lines the constraints can be satisfied by adjusting the power by checking the region of heat. Table 9.5 shows the adjustment of power for two regions of heat. Table 9.6 shows the adjustment of power for three regions of heat shown in Figure 9.3. Thirty (30) independent trial runs are given to teacher learning based optimization (TLBO) algorithm to justify the global solution. To justify the run time, number of function evaluations (NFE) is computed. Detail of results is given below.

Number of students, NP = 50
Maximum number of iterations, T^{max} =100
Minimum value of operating cost, F_{min} = 9257.078 ₹/h
Average value of operating cost, F_{avg} = 9257.136 ₹/h
Maximum value of operating cost, F_{max} = 9257.401 ₹/h
Standard Deviation, SD = 6.799031×10^{-2} ₹/h
Number of function Evaluations (NFE) = 10050.

Generation schedule giving minimum operating cost obtained by implementing TLBO is given in Table 9.7.

TABLE 9.7 Generation schedule

Unit	P_i^{min} (MW)	P_i (MW)	P_i^{max} (MW)	H_i^{min} (MWth)	H_i (MWth)	H_i^{max} (MWth)
1	0	0	150	–	–	–
2	81	160.0	247	0	40.0005	180
3	40	40.0	125.8	0	74.9995	135.6
4	–	–	–	0	0	2695.2
	Total power:	200.0		Total heat:	115.0	
					Cost =	9257.078 ₹/h

9.5 SHORT-TERM VARIABLE-HEAD HYDROTHERMAL GENERATION SCHEDULING USING REAL CODED GENETIC ALGORITHM

Consider an electric power system network having N thermal generating units and M hydro plants, where $N+M$ is the total number of generating plants. The basic problem is to find the active power generation of each plant in the system as a function of time over a time period from 0 to T.

$$\text{Minimize } J = \sum_{k=1}^{T}\left(\sum_{i=1}^{N}(a_i P_{ki}^2 + b_i P_{ki} + c_i)\right) ₹ \tag{9.46a}$$

The load demand equality constraint is expressed as

$$\sum_{i=1}^{N+M} P_{ki} = P_{Dk} + P_{Lk} \quad (k = 1, 2, \ldots, T) \tag{9.46b}$$

Each hydro plant is constrained by the amount of water available:.

$$\sum_{k=1}^{T} t_k q_{kj} = V_j \quad (j = 1, 2, \ldots, M) \tag{9.46c}$$

The generation limits are imposed as

$$P_i^{min} \le P_{ki} \le P_i^{max} \quad (i = 1, 2, \ldots, N + M;\ k = 1, 2, \ldots, T) \tag{9.46d}$$

Effective head of reservoir continuity equation can be written as

$$h_{k+1j} = h_{kj} + \frac{t_k}{S_j}[I_{kj} - q_{kj}] \quad (j = 1,2,\ldots,M;\ k = 1,2,\ldots,T) \tag{9.46e}$$

Net head of reservoir is given by:

$$h_{T+1j} = h_{0j} + \frac{t_k}{S_j}\left[\sum_{k=1}^{T} t_k I_{kj} - \sum_{k=1}^{T} t_k q_{kj}\right] \quad (j = 1, 2, \ldots, M) \tag{9.46f}$$

where, a_i, b_i and c_i are cost coefficients of ith thermal unit. P_{ki} is real power generation of ith unit during kth subinterval. P_{km} is hydro power generation of jth unit during kth subinterval.

h_{kj} is effective head of *j*th reservoir during *k*th subinterval. *N* is the number of thermal generators. M is the number of hydro generators. h_{0j} is the initial hydraulic head of *j*th hydro unit. I_{kj} is the natural inflow of water to *j*th hydro unit during *k*th subinterval. q_{kj} is the rate of water discharge at *j*th hydro unit during *k*th subinterval. S_j is the surface area of the reservoir. P_{Dk} is the power demand during *k*th subinterval. P_{Lk} is the transmission loss during time *k*th subinterval. $P_i^{\min}$ and $P_i^{\max}$ are the lower and upper limits of real thermal and hydro power, respectively.

According to Glimn-Kirchmayer model, the discharge is given by

$$q_{kj} = K_j \phi(h_{kj}) \psi(P_{km}) \quad (j = 1, 2, \ldots, M;\ m = j + N;\ k = 1, 2, \ldots, T) \tag{9.47}$$

where φ and τ are functions of head and hydro-generations, respectively. K_j is a constant of proportionality.

For a small capacity reservoir, the effective head is variable over the optimization interval. Therefore, $\phi(h_{kj})$ and $\psi(P_{km})$ are represented by quadratic equations.

$$\phi(P_{km}) = x_j P_{km}^2 + y_j P_{km} + z_j \quad (j = 1, 2, \ldots, M;\ m = j + N;\ k = 1, 2, \ldots, T) \tag{9.48}$$

$$\psi(h_{kj}) = \alpha_j h_{kj}^2 + \beta_j h_{kj} + \gamma_j \quad (j = 1, 2, \ldots, M;\ k = 1, 2, \ldots, T) \tag{9.49}$$

where, α_j, β_j and γ_j are head variation coefficients of *j*th hydro-plant. x_j, y_j and z_j are discharge coefficients of *j*th hydro-plant. V_j is predefined volume of water available for *j*th hydro-plant.

A common approach to model transmission losses in the system is to use Kron's loss formula through B-coefficients.

$$P_{Lk} = \sum_{i=1}^{N+M} \sum_{j=1}^{N+M} P_{ki} B_{ij} P_{kj} + \sum_{i=1}^{N+M} B_{io} P_{ki} + B_{oo} \tag{9.50}$$

where, B_{ij}, B_{io} and B_{oo} are B-coefficients.

Optimization problem is solved by real-coded genetic algorithm with arithmetic-average-bound-blend crossover. Wavelet mutation operator is applied to solve short-term variable-head hydro-thermal problem. Initial feasible solution has been obtained by implementing the random heuristic search. Objective and equality constraints are quantified by membership.

9.5.1 Fitness Function

The main concern of system operator is to fulfil the conflicting goals while satisfying the constraints of the system. In most decision making situations the goals, constraints and consequences of proposed alternatives are not known with precision. Much of this imprecision is neither measurable nor random. The imprecision can be due to vague, ambiguous, or fuzzy information. By taking account of the minimum and maximum values of each objective function together with the rate of increase of membership satisfaction, the power operator determines the membership function in a subjective manner. It is assumed that is a strictly monotonic linear decreasing and continuous function and is defined as:

$$\mu(J) = \begin{cases} 1 & ;\ J \le J^{\min} \\ \dfrac{J^{\max} - J}{J^{\max} - J^{\min}} & ;\ J^{\min} < J < J^{\max} \\ 0 & ;\ J \ge J^{\max} \end{cases} \tag{9.51}$$

where $J^{\min}$ and $J^{\max}$ are the minimum and maximum values of the objective function, respectively in which the solution is expected.

The hydrothermal scheduling problem is formulated as a multi-objective optimization problem by considering all the objectives simultaneously in the form of fuzzy membership functions for optimization and is defined mathematically as:

$$\text{Maximize} \qquad \mu(J) \tag{9.52}$$

Subject to the satisfaction of constraints given by Eq. (9.46b), (9.46c), (9.46d) and (9.46f).

Inequality constraint given by Eq. (9.46d) is taken care of while searching the active thermal and hydro power generations within minimum and maximum limits of generators. Slack thermal generating unit variable is created to meet the demand equality constraint given by Eq. (9.46b) during each interval and slack hydro generating unit variable is created to satisfy the available water constraint given by Eq. (9.46c) for each hydro unit. Exterior penalty method is employed to incorporate the inequality constraints of Eq. (9.46d) relating to slack thermal and slack hydro generating units while solving the above maximization problem. Real coded genetic algorithm is used to generate the set of non-inferior solutions and to select the optimum solution depending upon the maximum fitness achieved by the solution. The fitness function is defined as:

$$f = \min\left\{\mu(J), \left(\frac{\alpha}{1+Pen}\right)\right\} \tag{9.53}$$

where

$$Pen = r_p\left\{\sum_{k=1}^{T}\langle P_{sk}\rangle + \sum_{j=1}^{M}\langle P_{gh}\rangle\right\}$$

with

$$P_{ks} = \begin{cases} (P_{ks} - P_s^{\min})^2 & ; P_{ks} < P_s^{\min} \\ (P_s^{\max} - P_s)^2 & ; P_{ks} < P_s^{\max} \\ 0 & ; P_s^{\min} \le P_{ks} \le P_s^{\max} \end{cases} \qquad (k = 1, 2, \ldots, T)$$

$$P_{hm} = \begin{cases} (P_{hm} - P_m^{\min})^2 & ; P_{hm} < P_m^{\min} \\ (P_m^{\max} - P_{hm})^2 & ; P_{hm} < P_m^{\max} \\ 0 & ; P_m^{\min} \le P_{hm} \le P_m^{\max} \end{cases} \qquad (m = N+1, N+2, \ldots, N+M)$$

r_p is the penalty parameter and having higher value. s is slack thermal unit to meet power demand constraint during each interval. h is slack hydro unit to meet the available water constraint.

9.5.2 Calculation for Slack Thermal Generator

Let $N + M$ committed hydro and thermal generating units deliver the power output subject to their respective energy balance constraints Eq. (9.46b), and the capacity constraints Eq. (9.46d) during each interval of time. A dependent unit P_{ks} is selected arbitrarily from committed units to meet the equality constraints given by Eq. (9.46b). The power output of the slack thermal unit 's' is computed by rewriting the energy balance Eq. (9.46b) when transmission loss is represented by Eq. (9.50). P_{ks} is computed as below:

$$P_{ks} = \left(-Y_k \pm \sqrt{Y_k^2 - 4X_k Z_k}\right)/(2X_k) \qquad \text{where } Y_k^2 - 4X_k Z_k > 0 \tag{9.54}$$

where $X_k = B_{ss}$

$$Y_k = \sum_{j=1}^{N+M} (B_{kjs} + B_{ksj})P_{kj} + B_{so} - 1$$

$$Z_k = P_{Dk} + B_{oo} + \sum_{\substack{i=1\\ i\neq s}}^{N+M} \sum_{\substack{j=1\\ j\neq s}}^{N+M} P_{ki}B_{ij}P_{kj} + \sum_{\substack{i=1\\ i\neq s}}^{N+M} B_{io}P_{ki} - \sum_{\substack{i=1\\ i\neq s}}^{N+M} P_{ki}$$

The positive values of P_{ks} $(k = 1, 2, ..., T)$ are considered.

9.5.3 Calculation for Slack Hydro Generator

Let $N+M$ committed generating units deliver the power output subject to their respective energy balance constraints Eq. (9.46b), and the volume capacity constraints Eq. (9.46c) for whole time. Slack hydro units P_{hm} $(j = 1, 2, ..., N; m = j + N)$ are selected arbitrarily from committed units to meet the water equality constraints for the whole period. The power output of the slack hydro units during 'h' time interval is computed by rewriting the water energy balance Eq. (9.46c).

$$t_h q_{hj} + \sum_{\substack{k=1\\ k\neq h}}^{T} t_k q_{kj} = V_j \qquad (j = 1, 2, \ldots, M)$$

$$t_h(x_j P_{hm}^2 + y_j P_{hm} + z_j)\psi(h_{hj}) + \left(\sum_{\substack{k=1\\ k\neq h}}^{T} t_k q_{kj} - V_j\right) = 0 \, (j = 1, 2, \ldots, N;\, m = j + N)$$

$$P_{hm} = (-R_j \pm \sqrt{R_j^2 - 4U_j W_j})/(2U_j) \qquad \text{where, } R_j^2 - 4U_j W_j > 0 \tag{9.55}$$

where, $U_j = t_h x_j$

$R_j = t_h y_j$

$$W_j = t_h z_j + \frac{1}{\psi(h_{hj})}\left(\sum_{\substack{k=1\\ k\neq h}}^{T} t_k q_{kj} - V_j\right)$$

The positive values of P_{hm} $(j = 1, 2, ..., M; m = j + N)$ are considered.

9.5.4 Real Coded Genetic Algorithm

Real Coded Genetic Algorithm (RCGA) with genetic operators including arithmetic-average-bound-blend crossover and wavelet mutation is applied to solve the short-term fixed-head hydrothermal scheduling problem. The arithmetic-average-bound-blend crossover operator combines the arithmetic crossover, average crossover, bound and blend crossover. The arithmetic crossover operation produces some children with their parent's features; average crossover manipulates the genes of the selected parents and the minimum and maximum possible values of the genes and bound crossover is capable of moving the offspring near the domain boundary. Therefore the offspring spreads over the domain so that a higher chance of reaching the global optimum can be obtained. The wavelet mutation operation based on wavelet theory [Amjady and Nasiri-Rad, 2009] is a powerful tool for fine tuning of the genes to search the solution space locally. This property of wavelet mutation operation enhances the searching performance and provides a faster convergence than conventional real coded genetic algorithm.

Algorithm 9.7: Real coded genetic algorithm

1. Randomly generate initial population strings.
2. Evaluate fitness values of population members.
3. Is solution available among the population?
4. **IF** '*yes*' **THEN GOTO** Step 9.
5. Select highly fit member of population as parents using stochastic remainder roulette wheel selection and produce offsprings according to their fitness.
6. Create new strings by mating current offsprings. Apply crossover and mutation operators to introduce variations and form offsprings.
7. New offsprings replace existing one by applying competition and selection.
8. **GOTO** Step 3 and repeat.
9. **STOP**

Initialization

The initial population comprises combinations of only the candidate of hydro-thermal scheduling solutions, which satisfy all the constraints and are feasible solution of economic dispatch. It consists of P_{jki}^0 (i = 1, 2, ..., $N+M$; k = 1, 2, …, T, j = 1, 2, ..., NP) trial parent individuals. The elements of a parent are the combinations of power outputs of the generating units, which are chosen randomly by a random number ranging over $[P_i^{\min}, P_i^{\max}]$.

$$P_{jki}^0 = P_i^{\min} + rand(\,)(P_i^{\max} - P_i^{\min}) \;\; (i=1,2,\ldots,N+M;\, k=1,2,\ldots,T;\, j=1,2,\ldots,\text{NP}) \quad (9.56)$$

The elements of parent/offspring P_{jki}^0 may violate constraint Eq. (9.46d). This violation is corrected by fixing them either at lower or upper limits as described below:

$$P_{jki}^0 = \begin{cases} P_i^{\min} \;;\, P_{jki}^0 < P_i^{\min} \\ P_i^{\max} \;;\, P_{jki}^0 > P_i^{\max} \\ P_{jki}^0 \;;\, P_i^{min} \le P_{jki}^0 \le P_i^{\max} \end{cases} \quad (i=1,2,\ldots,N+M;\, k=1,2,\ldots,T;\, j=1,2,\ldots,\text{NP}) \quad (9.57)$$

Heuristic Random Search to Generate Feasible Solution

The heuristic random search strategy to search the hydro and thermal generations is elaborated here. Two separated search moves are performed. One move tries to meet the available water constraint for every hydro generation plant and second move tries to meet the demand constraint during each time interval. Mathematically these moves are represented below. Hydro generations are updated as:

$$P_{jkm}^{t} = \begin{cases} P_{jkm}^{t} + w_v rand()(P_m^{\max} - P_m^{\min})\Delta V_{ji}^{t}\,; (\Delta V_{ji}^{t} < 0) \\ P_{jkm}^{t} - w_v rand()(P_m^{\max} - P_m^{\min})\Delta V_{ji}^{t}\,; (\Delta V_{ji}^{t} > 0) \end{cases} \quad \begin{matrix} (i = 1, 2, \ldots, M;\ m = i + N; \\ k = 1, 2, \ldots, T; j = 1, 2, \ldots, \text{NP}) \end{matrix} \tag{9.58}$$

where

$$\Delta V_{ji}^{t} = \sum_{k=1}^{T} t_k q_{jki}^{t} - V_i \quad (i = 1, 2, \ldots, M; j = 1, 2, \ldots, \text{NP})$$

rand() is uniform random number having value between 0 and 1. w_v is scaling factor.

Thermal as well as hydro generation are updated as

$$P_{jki}^{t} = \begin{cases} P_{jki}^{t} + w_D\, rand()\,(P_m^{\max} - P_m^{\min})\,\Delta P_{Dk}^{t} \quad ; \ (\Delta P_{Dk}^{t} < 0) \\ P_{jki}^{t} - w_D\, rand()\,(P_m^{\max} - P_m^{\min})\,\Delta P_{Dk}^{t} \quad ; \ (\Delta P_{Dk}^{t} > 0) \end{cases}$$

$$(i = 1, 2, \ldots, N + M;\ k = 1, 2, \ldots, T;\ j = 1, 2, \ldots, \text{NP}) \tag{9.59}$$

where

$$\Delta P_{Dk}^{t} = P_{Dk} + P_{Lk} - \sum_{i=1}^{N+M} P_{jki}^{t} \quad (k = 1, 2, \ldots, T; j = 1, 2, \ldots, \text{NP})$$

rand() is uniform random number having value between 0 and 1. w_D is scaling factor.

The process is repeated till both the constraints are satisfied. It means for each interval $\left|\Delta P_{Dk}^{t}\right|$ (k = 1, 2, ..., T; j = 1, 2, .., NP) should meet the tolerable termination criterion 'err' and for each hydro plant $\left|\Delta V_{ji}^{t}\right|$ (i = 1, 2, ..., M; j = 1, 2, ..., NP) should meet the tolerable termination criterion 'err'. The stepwise procedure to generate feasible solution of initial generated population is given in Algorithm 9.8.

Algorithm 9.8: Heuristic random search to generate feasible solution

1. Read N, M and T, number of total generating units, number of thermal units, number of hydro units and number intervals; w_D and w_v scaling factors.
2. Initialize hydro unit counter, $j = 0$
3. **IF** ($j \ge$ M) **THEN GOTO** 12
4. Increment j, $j = j+1$
5. Set volume check counter, $v = 0$
6. Compute $\Delta V_{ji}^{t} = \sum_{k=1}^{T} t_k q_{jki}^{t} - V_i$
7. **IF** $\left|\Delta V_{ji}^{t}\right| \le err$ **THEN GOTO** 11
8. Increment check counter, $v = v + 1$
9. Update generation P_{jkm}^{t} (i = 1, 2, ..., M; $m = i + N$; k = 1, 2, ..., T) using Eq. (9.58)
10. **GOTO** 6 and Repeat
11. **GOTO** 3 and Repeat
12. Initialize time interval counter, $k = 0$

13. **IF** ($k \geq T$) **THEN GOTO** 22
14. Increment time interval counter, $k = k+1$
15. Set demand check counter, $d = 0$
16. Compute $\Delta P_{Dk}^t = P_{Dk} + P_{Lk} - \sum_{i=1}^{N+M} P_{jki}^t$
17. Check **IF** $\left|\Delta P_{Dk}^t\right| \leq err$ **THEN GOTO** 21
18. Increment demand counter, $d = d+1$
19. Update generation P_{jki}^t (i = 1, 2, ..., $N+M$) using Eq. (9.59)
20. **GOTO** 16 and Repeat.
21. **GOTO** 13 and Repeat
22. **IF** ($v \neq 0$ and $d \neq 0$) **THEN GOTO** 2 and Repeat
23. **STOP**.

Fitness Function

Genetic algorithms are suitable for solving maximization problem. Minimization problems are usually transferred into maximization problems using some suitable transformation. Fitness function f is derived from the objective function and used in successive genetic operations. The fitness function for maximization problems can be used the same way as the objective function. The fitness function used to solve short-term variable-head hydro-thermal problem is given below:

$$f_j = \min\left\{\mu(J_j), \left(\frac{\alpha}{1 + Pen_j}\right)\right\} \quad (j = 1, 2, \ldots, \text{NP}) \tag{9.60}$$

where α is the scaling factor.

Reproduction

The first genetic algorithm operator is reproduction. The reproduction genetic algorithm operator selects good members in a population and forms a mating pool. The operator is also known as selection operator. The commonly used reproduction operator is the proportionate reproduction operator where a string is selected for the mating pool with a probability proportional to its fitness. The basic roulette wheel selection method is stochastic sampling with replacement (SSR). The segment size and selection probability remain same throughout the selection phase and individuals are selected accordingly. Stochastic sampling with partial replacement (SSPR) extends upon SSR by resizing an individual's segment if it is selected. Each time an individual is selected, the size of its segment is reduced by one. If the segment size becomes negative, then it is set to zero. Remainder sampling methods involve two distinct phases. In the integral phase, the individuals are selected deterministically according to the integer part of their expected trials. The remaining individuals are then selected probabilistically from fractional part of the individual's expected values. In the study, the stochastic remainder roulette wheel selection is applied whose step-wise procedure is outlined in Chapter 5 sub-section 5.4.3.

Crossover Operators

The selection of chromosomes for the arithmetic-average-bound-blend crossover (AABBX) is based on the Roulette-wheel mechanism [Mo, 1991]. The AABBX is a combinatorial operator composed of the arithmetic crossover, average crossover, bound and blend crossover. Suppose two vectors are selected chromosomes in the *t*th generation of the RCGA execution. Each chromosome has *N*+*M* genes, which are real numbers (the generation outputs of schedulable units). The AABBX operator creates ten children from the parents P_{uki}^t and P_{vki}^t as follows:

Arithmetic crossover

$$P_{1ki}^{t+1} = w_a P_{vki}^t + (1 - w_a) P_{uki}^t (i = 1, 2, \ldots, N + M; i \neq s; k = 1, 2, \ldots, T) \quad (9.61)$$

$$P_{2ki}^{t+1} = w_a P_{uki}^t + (1 - w_a) P_{vki}^t (i = 1, 2, \ldots, N + M; i \neq s; k = 1, 2, \ldots, T) \quad (9.62)$$

$$P_{3ki}^{t+1} = \text{Min}\{P_{uki}^t, P_{vki}^t\} \; (i = 1, 2, \ldots, N + M; i \neq s; k = 1, 2, \ldots, T) \quad (9.63)$$

$$P_{4ki}^{t+1} = \text{Max}\{P_{uki}^t, P_{vki}^t\} \; (i = 1, 2, \ldots, N + M; i \neq s; k = 1, 2, \ldots, T) \quad (9.64)$$

Average crossover

$$P_{5ki}^{t+1} = 0.5(P_{uki}^t + P_{vki}^t) \quad (i = 1, 2, \ldots, N + M; i \neq s; k = 1, 2, \ldots, T) \quad (9.65)$$

$$P_{6ki}^{t+1} = 0.5[w_b(P_{uki}^t + P_{vki}^t) + (1 - w_b)(P_k^{\min} + P_k^{\max})] \quad (i = 1, 2, \ldots, N + M; i \neq s; k = 1, 2, \ldots, T) \quad (9.66)$$

Bound crossover

$$P_{7ki}^{t+1} = w_c \text{Min}\{P_{uki}^t, P_{vki}^t\} + (1 - w_c) P_k^{\min} \; (i = 1, 2, \ldots, N + M; i \neq s; k = 1, 2, \ldots, T) \quad (9.67)$$

$$P_{8ki}^{t+1} = w_c \text{Max}\{P_{uki}^t, P_{vki}^t\} + (1 - w_c) P_k^{\max} (i = 1, 2, \ldots, N + M; i \neq s; k = 1, 2, \ldots, T) \quad (9.68)$$

Blend crossover

$$P_{9ki}^{t+1} = w_d P_{vki}^t + (1 - w_d) P_{uki}^t (i = 1, 2, \ldots, N + M; i \neq s; k = 1, 2, \ldots, T) \quad (9.69)$$

$$P_{10ki}^{t+1} = w_d P_{uki}^t + (1 - w_d) P_{vki}^t \quad (i = 1, 2, \ldots, N + M; \; i \neq s; k = 1, 2, \ldots, T) \quad (9.70)$$

where

$$P_k^{\min} = \text{Min}\{P_{uki}^t, P_{vki}^t\} \quad (i = 1, 2, \ldots, N + M)$$

$$P_k^{\max} = \text{Max}\{P_{uki}^t, P_{vki}^t\} \quad (i = 1, 2, \ldots, N + M)$$

w_a, w_b and w_c are constant weights. The values are adjusted such that $0 \leq w_a, w_b, w \leq 1_c$ and w_d is also constant weight such that $1 \leq w_d \leq 2$.

Among these ten children, two children having the highest fitness values are selected as the offspring chromosomes of the crossover operation. These two offspring chromosomes are added to the previous population including the parents. So, after the execution of the crossover operator the previous population will be enlarged, which is considered for the mutation operator.

Mutation Operator

Every gene of the chromosomes should have a chance to mutate, governed by the probability of the mutation p_m. For each gene of the chromosome, a random number in the range of [0, 1] is generated. If the random number is less than p_m, that gene is selected for the mutation, otherwise it is not selected. In this work, p_m is set at 0.2. The new gene, P_{jki}^t after mutation will be as follows

$$P_{jki}^t = \begin{cases} P_{jki}^t + \Delta(\phi, t, L)\,(P_i^{\max} - P_{jki}^t) \; ; \; p_m > 0 \\ P_{jki}^t + \Delta(\phi, t, L)\,(P_{jki}^t - P_i^{min}) \; ; \; p_m < 0 \end{cases}$$

$$(i = 1, 2, \ldots, N + M;\; i \neq s;\; k = 1, 2, \ldots, T;\; j = 1, 2, \ldots, NP) \tag{9.71}$$

where

L is the maximum number of generations of the real coded genetic algorithm.
t is the current generation number.

By using the Morlet wavelet as the mother wavelet, $\Delta(\phi)$ is given by Eq. (7.124). The dilation parameter d is given by Eq. (7.125) (see Chapter 7, Section 7.16)).

Competition and Selection

Each individual P_{jki}^t in the combined population has to compete with some other individuals to have a chance to be copied to the next generation. The score for each trial vector after stochastic competition is given by Eqs. (7.72) and (7.73) (see Chapter 7, Section 7.11)). After competing, the trial 2NP solutions, including the parents and the offspring, are ranked in descending order of the score obtained in Eq. (7.72). The first NP trial solutions survive and are copied along with their objective functions into survivor set as the individuals of the next generation.

9.5.5 Test System and Results

A real-coded genetic algorithm with arithmetic-average-bound-blend crossover and wavelet mutation operator is applied to solve short-term variable-head hydro-thermal problem. Equality constraint to meet the demand requirements during each time interval is considered by creating slack thermal generating unit for each time interval. Similarly equality constraint that utilizes the available water to its maximum extent unit for the whole scheduling period by hydro units is created by introducing slack hydro units for the time interval. Initial feasible solution has been generated by implementing the random heuristic search. The search is performed within the operating generation limits of thermal as well as hydro generating units. Violation of the slack thermal and hydro units is taken care through external penalty method. The validity of the proposed technique has been illustrated on two sample test systems.

Sample Test System

The system under study comprise of two thermal plants and two hydro plants [Rashid and Nor, 1993]. A short range variable-head hydro-thermal load scheduling problem of 24h duration has been undertaken. The optimization period has been divided into 24 intervals of one hour each. Operating characteristics of thermal generators are given in Tables 9.8. Operating characteristics of hydro generators are given in Table 9.9, Table 9.10 and hydro reservoir details are given in Table 9.11. Operating limits of generators are given in Table 9.8 and Table 9.9 for thermal and hydro generating units respectively within which the optimal schedule has been searched using heuristic random search and real coded genetic algorithm. B-coefficients are given below.

$$B_{ij} = \begin{bmatrix} 0.0001400 & 0.0000100 & 0.0000150 & 0.0000150 \\ 0.0000100 & 0.0000600 & 0.0000100 & 0.0000130 \\ 0.0000150 & 0.0000100 & 0.000068 & 0.0000650 \\ 0.0000150 & 0.0000130 & 0.0000650 & 0.0000700 \end{bmatrix} \text{MW}^{-1}$$

Real coded genetic algorithm is implemented. The values of parameters like w_a, w_b, w_c and w_d are adjusted as 0.33, 0.5, 0.5 and 1.4, respectively for all the cases. To obtain the feasible solution for the scheduling period, w_v and w_D are set to 0.2 and 0.0025, respectively.

TABLE 9.8 Operating characteristics of thermal generators: 2-thermal units

Thermal Unit	*Operating cost coefficients*			*Operating limits*	
	a_i (₹/MW^{2}h)	b_i (₹/MWh)	c_i (₹/h)	P_i^{min} (MW)	P_i^{max} (MW)
1.	0.0025	3.2	25.0	100	375
2.	0.0008	3.4	30.0	175	850

TABLE 9.9 Hydro unit rate of head variation: 2-hydro units

Hydro Unit	α_i (1/ft)	β_i	γ_i (ft)	*Initial Head*, h_{oj} (ft)
1.	0.000216	0.306	0.198	300
2.	0.000360	0.612	0.936	250

TABLE 9.10 Hydro unit rate of water discharge characteristics: 2-hydro units

Hydro Unit	x_i (Mft^3/MW^{2}h)	y_i (Mft^3/MWh)	z_i (Mft^3/h)	P_i^{min} (MW)	P_i^{max} (MW)
1.	0.000216	0.306	0.198	175	550
2.	0.000360	0.612	0.936	0	250

TABLE 9.11 Hydro plant's reservoir details: 2-hydro units

Hydro Unit	*Surface area* S_j, (ft^2)	*Proportionality Constant*, K_j	*Water Inflow* I_j, (ft^3)	*Volume*, V_j (ft^3)
1.	1000	1	0	2850
2.	400	1	0	2450

The values of p_m, g and ξ are set to 2.0, 10000 and 0.5, respectively during the implantation of mutation. Results are obtained by taking population size as 70 in 70 generations. Results are shown in Tables 9.12 and 9.13 corresponding to the best solution when operating cost is searched within ₹50350 and ₹75526. Table 9.12 gives the head variation and rate of water discharge for generation schedule of 24 hours duration. Table 9.13 depicts the optimal power generation schedules obtained for hydro-thermal power systems of 24 hours duration along with transmission losses.

TABLE 9.12 Head variation and water discharge rate for the scheduling period

Interval *k*	*Head Variation*		*Water Discharge Rate*	
	h_{k1}	h_{k2}	q_{k1}	q_{k2}
1	300.0000	250.0000	84.09911	48.75573
2	299.9159	249.8781	55.52813	31.97945
3	299.8604	249.7982	56.72464	134.2820
4	299.8036	249.4624	55.16941	12.19556
5	299.7484	249.4320	93.19427	11.09622
6	299.6552	249.4042	56.13453	38.83456
7	299.5991	249.3071	88.38433	75.94097
8	299.5107	249.1173	115.22300	11.21188
9	299.3955	249.0892	178.22340	181.6925
10	299.2173	248.6350	132.86920	115.2602
11	299.0844	248.3469	169.27180	275.6708
12	298.9151	247.6577	152.63570	241.4729
13	298.7625	247.0540	145.77190	109.9694
14	298.6167	246.7791	165.15160	165.3665
15	298.4515	246.3657	111.41870	119.3768
16	298.3401	246.0672	105.84050	189.0534
17	298.2343	245.5946	134.68250	215.6756
18	298.0996	245.0554	201.46720	134.8899
19	297.8981	244.7182	202.67610	101.1441
20	297.6955	244.4653	102.16270	49.62902
21	297.5933	244.3413	118.33330	45.53951
22	297.4749	244.2274	126.15390	128.7071
23	297.3488	243.9057	100.29930	6.127791
24	297.2485	243.8903	98.58402	6.127346

TABLE 9.13 Generation schedule and transmission loss for the scheduling period

Interval *k*	*Demand* P_{Dk} (MW)	*Thermal Generation*		*Hydro Generation*		*Transmission loss* P_{Lk} (MW)
		P_{k1} (MW)	P_{k2} (MW)	P_{k3} (MW)	P_{k4} (MW)	
1	800	99.99908	417.0102	257.8075	47.7132	22.5299
2	700	170.2617	337.6592	178.5365	31.0981	17.5556
3	600	128.1658	175.0000	182.0311	128.2416	13.4385
4	600	99.99927	323.9236	177.5552	11.0831	12.5611
5	600	99.99937	220.8408	281.9272	9.9553	12.7226

(*Contd.*)

Interval k	*Demand* P_{Dk} (MW)	*Thermal Generation*		*Hydro Generation*		*Transmission loss* P_{Lk} (MW)
		P_{k1} (MW)	P_{k2} (MW)	P_{k3} (MW)	P_{k4} (MW)	
6	650	145.8594	300.4996	180.4350	38.0236	14.8175
7	800	242.9649	238.3095	269.4595	74.2712	25.0051
8	1000	243.7346	444.5429	337.8487	10.0915	36.2176
9	1330	213.4504	528.4719	483.0750	170.6555	65.6527
10	1350	334.1965	591.5360	380.7426	111.5123	67.9872
11	1450	293.8292	522.4193	463.9365	249.9693	80.1541
12	1500	370.4539	566.1710	426.9296	222.6377	86.1921
13	1300	242.3183	600.0891	411.3846	107.3855	61.1776
14	1350	256.7396	547.0097	455.4027	157.9955	67.1476
15	1350	355.7179	617.3621	329.4138	116.4949	68.9886
16	1370	110.8678	837.0598	315.5217	179.3799	72.8291
17	1450	203.7224	734.9958	386.0652	202.7740	77.5574
18	1570	283.0610	712.3875	534.1392	131.5601	91.1476
19	1430	356.9917	514.2697	536.9362	100.1687	78.3661
20	1350	331.4725	731.8283	306.7452	49.8681	69.9141
21	1270	261.4566	674.8643	347.3243	45.7713	59.4166
22	1150	337.8967	371.2675	366.4786	126.3254	51.9683
23	1000	226.1940	502.7378	302.2711	5.0	36.2029
24	900	189.9484	435.8250	297.9418	5.0	28.7152
Operating Cost			68906.34 (₹)			
Membership value			0.262935			

9.6 SHORT-TERM MULTI-CHAIN HYDROTHERMAL GENERATION SCHEDULING USING PREDATOR-PREY OPTIMIZATION

The aim of multi-objective hydrothermal optimum generation scheduling problem is to minimize the thermal operating cost and gaseous pollutant emission level of thermal plants simultaneously, while satisfying various operational constraints. The main constraints include: total generation should meet load demand in each interval. Each hydro unit is also constrained by the amount of water available for draw down in the planning period. Inequality constraints comprise of discharge, volume and power limits. A multi-chain hydro model is undertaken, in which lower reservoir volume depends on upper reservoir discharge. In this model inflow rate, spillage rate and transport delay between reservoirs has also been considered. Mathematically hydrothermal generation scheduling problem is stated below:

Minimize operating cost of thermal units:

$$J_1 = \sum_{k=1}^{T} \sum_{i=1}^{N} (a_i P_{ki}^2 + b_i P_{ki} + c_i + |d_i \sin\{e_i(P_i^{\min} - P_{ki})\}|) \text{ ₹} \tag{9.72a}$$

Minimize gaseous pollutant's emission:

$$J_2 = \sum_{k=1}^{T} \sum_{i=1}^{N} (\alpha_i P_{ki}^2 + \beta_i P_{ki} + \gamma_i + \eta_i e^{(\delta_i P_{ki})}) Ton \tag{9.72b}$$

Subjected to following set of constraints:

(i) The load demand equality constraint during each sub-interval

$$\sum_{i=1}^{N+M} P_{ki} = P_{Dk} + P_{Lk} \quad (m = j + n;\ k = 1, 2, \ldots, T) \tag{9.72c}$$

The Hydro power generation is a function of water discharge rate and reservoir storage given as

$$P_{km} = C_{1j} V_{kj}^2 + C_{2j} q_{kj}^2 + C_{3j} V_{kj} q_{kj} + C_{4j} V_{kj} + C_{5j} q_{kj} + C_{6j} \quad \begin{matrix} (j = 1, 2, \ldots, M;\ m = j + N; \\ k = 1, 2, \ldots, T) \end{matrix} \tag{9.72d}$$

(ii) The hydraulic constraint for each hydro unit at the end of planning period is given below:

$$V_{(T+1)j} = V_{1j} + \sum_{k=1}^{T} I_{kj} - \sum_{k=1}^{T} q_{kj} - \sum_{k=1}^{T} S_{kj} + \sum_{k=1}^{T} \sum_{r=1}^{U_j} (q_{r(k - tr_j)} + S_{r(k - tr_j)}) \quad (j = 1, 2, \ldots, M) \tag{9.72e}$$

(iii) The volume dynamics for the hydro reservoir network for each sub-interval

$$V_{(k+1)j} = V_{kj} + I_{kj} - q_{kj} - S_{kj} + \sum_{r=1}^{U_j} (q_{r(k - tr_j)} + S_{r(k - tr_j)}) \quad (j = 1, 2, \ldots, M;\ k = 1, 2, \ldots, T) \tag{9.72f}$$

(iv) The discharge limits are imposed on hydro unit

$$q_j^{\min} \le q_{kj} \le q_j^{\max} \quad (j = 1, 2, \ldots, M;\ k = 1, 2, \ldots, T) \tag{9.72g}$$

(v) The volume limits are imposed on hydro unit

$$V_j^{\min} \le V_{kj} \le V_j^{\max} \quad (j = 1, 2, \ldots, M;\ k = 1, 2, \ldots, T) \tag{9.72h}$$

(vi) The power limits are imposed as on hydro and thermal unit

$$P_i^{\min} \le P_{ki} \le P_i^{\max} \quad (i = 1, 2, \ldots, M + N;\ k = 1, 2, \ldots, T) \tag{9.72i}$$

where a_i, b_i, c_i, d_i and e_i are cost coefficients of ith thermal unit. P_{ki} is real power generation of ith unit during kth sub-interval. N is the number of thermal generators. T is total time duration for scheduling. α_i, β_i, γ_i, η_i and δ_i are emission curve coefficients of ith thermal unit. C_{1j}, C_{2j}, C_{3j}, C_{4j}, C_{5j} and C_{6j} are the power generation coefficients of the jth hydro unit. q_{jk} is the water discharge rate and V_{jk} is the storage volume of jth reservoir at kth time interval, respectively, M is total number of hydro units. I_{jk}, q_{jk} and S_{jk} are the water inflow, water discharge and spillage, respectively of jth hydro unit at kth subinterval. U_j is upstream reservoir; t_{rj} is representing time delay from upstream reservoir to lower reservoir; VO_{j1}, VO_{jT+1} are the initial and final reservoir storage volumes respectively. P_{Dk} is the power demand and P_{Lk} is the transmission loss during kth subinterval. $P_i^{\max}$ and $P_i^{\min}$ are the upper and lower limits of ith real power generation, respectively.

$q_i^{\max}$ and $q_i^{\min}$ are the upper and lower limits of water discharge of jth hydro unit, respectively, $V_i^{\max}$ and $V_i^{\min}$ are the upper and lower limits of jth hydro unit, respectively. Transmission losses in the system during kth sub-interval are represented by Kron's loss formula through B-coefficient is:

$$P_{Lk} = B_{oo} + \sum_{i=1}^{N+M} B_{io} P_{ki} + \sum_{i=1}^{N+M} \sum_{j=1}^{N+M} P_{ki} B_{ij} P_{kj} \quad (k = 1, 2, \ldots, T) \tag{9.73}$$

with B_{ij}, B_{io} and B_{oo} are B-coefficients.

9.6.1 Constraint Handling Technique

Variable elimination method is applied to handle the equality constraints. The variables to be eliminated are known as dependent variable and the remaining variables are known as independent variable. Dependent thermal generating unit is introduced to meet the equality constraint given by Eq. (9.72c) during each sub-interval. Water discharge rate of a sub-interval so called dependent sub-interval is computed to satisfy the available water constraint given by Eq. (9.72e) for each hydro unit. Inequality constraints are handled using exterior penalty method.

Power Balance Equality Constraint

A dependent thermal unit is selected arbitrarily from thermal units and other thermal units act as independent variables in each sub-interval. The power of the dependent thermal unit during kth sub-interval, P_{kd} is computed from Eq. (9.72c) as:

$$P_{kd} + \sum_{\substack{i=1\\i\neq d}}^{N+M} P_{ki} = \sum_{\substack{i=1\\i\neq d}}^{N+M} \sum_{\substack{j=1\\j\neq d}}^{N+M} P_{ki} B_{ij} P_{kj} + \sum_{\substack{i=1\\i\neq d}}^{N+M} P_{kj} \left(B_{jd} + B_{dj}\right) P_{kd} + B_{dd} P_{kd}^2 + \sum_{\substack{i=1\\i\neq d}}^{N+M} B_{io} P_{ki} + B_{do} P_{kd} + B_{oo}$$

The above equation is rewritten in quadratic form as:

$$X_k P_{kd}^2 + Y_k P_{kd} + Z_k = 0 \quad (k = 1, 2, \ldots, T) \tag{9.74}$$

where

$$X_k = \begin{cases} 0 & ; P_{Lk} = 0 \\ B_{dd} & ; P_{Lk} \neq 0 \end{cases}$$

$$Y_k = \begin{cases} -1 & ; P_{Lk} = 0 \\ \sum_{\substack{j=1\\j\neq d}}^{N+M} (B_{jd} + B_{dj}) + B_{do} - 1 & ; P_{Lk} \neq 0 \end{cases}$$

$$Z_k = \begin{cases} P_{Dk} - \sum_{\substack{i=1\\i\neq d}}^{N+M} P_{ki} & ; P_{Lk} = 0 \\ P_{Dk} + B_{oo} + \sum_{\substack{i=1\\i\neq d}}^{N+M} \sum_{\substack{j=1\\j\neq d}}^{N+M} P_{ki} B_{ij} P_{kj} + \sum_{\substack{i=1\\i\neq d}}^{N+M} B_{io} P_{ki} - \sum_{\substack{i=1\\i\neq d}}^{N+M} P_{ki} & ; P_{Lk} \neq 0 \end{cases}$$

The positive root of Eq. (9.74) gives the dependent generation of thermal unit at kth sub-interval.

Water Balance Equality Constraint

The dependent sub-interval 'h' is arbitrarily selected to meet the water balance equality constraint and other sub-intervals act as independent variables for each hydro unit. The water discharge rate q_{hj} for dependent sub-interval of jth hydro unit is computed as:

$$q_{hj} = V_{1j} - V_{(T+1)j} + \sum_{k=1}^{T} I_{kj} - \sum_{\substack{k=1 \\ k \neq h}}^{T} q_{kj} - \sum_{k=1}^{T} S_{kj} + \sum_{\substack{k=1 \\ k \neq h}}^{T} \sum_{r=1}^{U_j} q_{r(k-tr_j)} + \sum_{k=1}^{T} \sum_{r=1}^{U_j} S_{r(k-tr_j)} \; (j = 1, 2, \ldots, M) \tag{9.75}$$

Reservoir volume and hydro power for jth hydro unit during hth sub-interval are computed by Eq. (9.72f) and Eq. (9.72d), respectively.

Inequality Constraint Handling

Every violation of inequality constraint is penalized by exterior penalty factor r_p and cumulative effect of all inequality constraints violation is included in another objective function J_3 as:

$$J_3 = r_p \left(\sum_{k=1}^{T} \langle P_{kd} \rangle + \sum_{j=1}^{M} \langle q_{hj} \rangle + \sum_{k=1}^{T} \sum_{j=1}^{M} \langle V_{kj} \rangle + \sum_{k=1}^{T} \sum_{m=N+1}^{N+M} \langle P_{km} \rangle \right) \tag{9.76}$$

First term of Eq. (9.76) represents the total violation of inequality constraint of dependent thermal unit power in each sub-interval, where $\langle P_{kd} \rangle$ is computed as:

$$\langle P_{kd} \rangle = \begin{cases} (P_{kd} - P_d^{\min})^2 & ; \; P_{kd} < P_d^{\min} \\ (P_d^{\max} - P_{kd})^2 & ; \; P_{kd} > P_d^{max} \\ 0 & ; \; P_d^{\min} \le P_{kd} \le P_d^{\max} \end{cases}$$

Second term of Eq. (9.76) represents the total violation of inequality constraint of water discharge rate during hth dependent sub-interval for all hydro units, where $\langle q_{hj} \rangle$ is computed as:

$$\langle q_{hj} \rangle = \begin{cases} (q_{hj} - q_j^{\min})^2 & ; \; q_{hj} < q_j^{\min} \\ (q_j^{\max} - q_{hj})^2 & ; \; q_{hj} > q_j^{\max} \\ 0 & ; \; q_j^{\min} \le q_{hj} \le q_j^{\max} \end{cases}$$

Third and fourth term of Eq. (9.76) represents the total violation of inequality constraint on water storage volume and hydro power, respectively of all hydro units in each sub-interval, where $\langle V_{kj} \rangle$ and $\langle P_{km} \rangle$ are computed as:

$$\langle V_{kj} \rangle = \begin{cases} (V_{kj} - V_j^{\min})^2 & ; \; V_{kj} < V_j^{\min} \\ (V_j^{\max} - V_{kj})^2 & ; \; V_{kj} > V_j^{\max} \\ 0 & ; \; V_j^{\min} \le V_{kj} \le V_j^{\max} \end{cases}$$

$$\langle P_{km} \rangle = \begin{cases} (P_{km} - P_m^{\min})^2 & ; P_{km} < P_m^{\min} \\ (P_m^{\max} - P_{km})^2 & ; P_{kd} > P_m^{\max} \\ 0 & ; P_m^{\min} \le P_{km} \le P_m^{\max} \end{cases}$$

Decision Making

The main concern of system operator is to fulfill the conflicting goals while satisfying the constraints of the system. Considering the imprecise nature of the decision maker's judgment, it is natural to presume that the decision maker may have fuzzy or imprecise goals for each objective function. Membership functions represent the degree of membership in certain fuzzy sets using values from 0 to 1. By taking account of the minimum and maximum values of each objective function together with the rate of increase of membership satisfaction, the decision maker determines the membership function $\mu(J_i)$ in a subjective manner [Dhillon et al., 2001]. It is assumed that $\mu(J_i)$ is a strictly monotonic linear decreasing and continuous function and is defined as:

$$\mu(J_i) = \begin{cases} 1 & ; J_i < J_i^{\min} \\ \dfrac{J_i^{\max} - J_i}{J_i^{\max} - J_i^{\min}} & ; J_i^{\min} \le J_i \le J_i^{\max} \quad (i = 1, 2) \\ 0 & ; J_i > J_i^{\max} \end{cases} \tag{9.77}$$

where $J_i^{\min}$ and $J_i^{\max}$ are the minimum and maximum values of ith objective function, respectively in which the solution is expected.

Minimum values of the objectives are obtained by giving full weightage to one of the objective and neglecting others. Owing to conflicting nature of objectives, J_2 will have maximum value when J_1 has the minimum value and vice versa. Multiple objectives of hydrothermal optimization problem are clubbed to find the unified affect of both the objectives simultaneously using membership functions and are defined mathematically as:

$$\mu_D = \min\{\mu(J_1), \mu(J_2)\} \tag{9.78}$$

where μ_D is objective function value.

9.6.2 Predator-Prey Optimization (PPO)

The predator-prey optimization (PPO) technique is based on particle swarm optimization (PSO) with additional predator chasing property. PSO is a population based search technique that uses swarm intelligence such as bird flocking, fish schooling, etc. Particle changes its position with time, based on its own experience and experience of neighboring particles. The position mechanism of the particle in the search space is updated by adding the velocity vector to its position vector as given by Kennedy and Ekrhart [1995]. In PPO, the influence of predator on prey is controlled by probability fear (*pf*). Probability fear defines the probability, to change prey velocity in any available dimension due to presence of predator. It becomes difficult for prey to stay at its preferred location when chased by predator that helps to explore search area more effectively [Silva et al., 2002].

Selection of velocity limits is an important aspect in PPO. The maximum allowed velocity actually controls the maximum exploration ability of particles in the search area. In initial stage, high maximum velocity helps to search, different regions of solution space and at final stage; low maximum velocity is preferred for the final search [Amjady and Rezai, 2010]. Hence initial search is started with high maximum velocity and if no improvement is observed in objective function for same number of generations, velocity limits are squeezed by a factor α. Basic PPO algorithm with velocity squeezed method is elaborated below.

Initialization of Prey Position and Velocity

The Initial prey population is randomly initialized between upper and lower limits of decision variables. Prey velocity V_{jki}^{0} and its position P_{jki}^{0} of thermal power generating units are randomly initialized within their respective upper and lower limits.

$$P_{jki}^{0} = P_i^{\min} + R_1(P_i^{\max} - P_i^{\min}) \ (i = 1, 2, \ldots, N; i \neq d; k = 1, 2, \ldots, T; j = 1, 2, \ldots, \text{NP}) \quad (9.79)$$

$$V_{jki}^{0} = V_i^{\min} + R_2(V_i^{\max} - V_i^{\min}) \ (i = 1, 2, \ldots, N; i \neq d; k = 1, 2, \ldots, T; j = 1, 2, \ldots, \text{NP}) \quad (9.80)$$

where, $P_i^{\min}$, $P_i^{\max}$, $V_i^{\min}$ and $V_i^{\max}$, and are representing the lower and upper limit of thermal power position and velocity of ith thermal unit respectively; R_1 and R_2 are uniform random numbers having value between 0 and 1; NP is number of particles in a swarm.

Prey velocity V_{qjkm}^{0} and its position q_{mkj}^{0} representing water discharge of hydro generating units are randomly initialized within their respective upper and lower limits.

$$q_{jkm}^{0} = q_m^{\min} + R_3(q_m^{\max} - q_m^{\min}) \ (m = 1, 2, \ldots, M; k = 1, 2, \ldots, T; k \neq h; j = 1, 2, \ldots, \text{NP}) \quad (9.81)$$

$$V_{qjkm}^{0} = V_{qm}^{\min} + R_4(V_{qm}^{\max} - V_{qm}^{\min}) \ (m = 1, 2, \ldots, M; k = 1, 2, \ldots, T; k \neq h; j = 1, 2, \ldots, \text{NP}) \quad (9.82)$$

where, $q_m^{\min}$, $q_m^{\max}$, $V_{qm}^{\min}$ and $V_{qm}^{\max}$ are representing the lower and upper limit of water discharge position and velocity of mth hydro unit respectively; R_3 and R_4 are uniform random numbers having value between 0 and 1.

Initialization of Predator Position and Velocity

Single predator velocity V_{Pki}^{t} and its position P_{Pki}^{t} of thermal power generating units are randomly initialized within their respective upper and lower limits except for dth dependent thermal unit of each sub-interval.

$$P_{Pki}^{0} = P_i^{min} + R_5(P_i^{\max} - P_i^{\min}) \ (i = 1, 2, \ldots, N; i \neq d; k = 1, 2, \ldots, T) \quad (9.83)$$

$$V_{Pki}^{0} = V_i^{\min} + R_6(V_i^{\max} - V_i^{\min}) \ (i = 1, 2, \ldots, N; i \neq d; k = 1, 2, \ldots, T) \quad (9.84)$$

where R_5 and R_6 are uniform random numbers having value between 0 and 1.

Similarly, single predator velocity V_{Pqmk}^{0} and its position q_{Pmk}^{0} of hydro water discharge are randomly initialized within their respective lower and upper limits except for hth subinterval of hydro units.

$$q_{Pkm}^{0} = q_m^{\min} + R_7(q_m^{\max} - q_m^{\min}) \ (m = 1, 2, \ldots, M; k = 1, 2, \ldots, T; k \neq h) \quad (9.85)$$

$$V_{Pqkm}^{0} = V_{qm}^{\min} + R_8(V_{qm}^{\max} - V_{qm}^{\min}) \ (m = 1, 2, \ldots, M; k = 1, 2, \ldots, T; k \neq h) \quad (9.86)$$

where R_7 and R_8 are uniform random numbers having value between 0 and 1.

Feasible Solution

The heuristic random search strategy is elaborated here to find the feasible solution. Two separate search moves are performed successively to achieve the feasible solution that provides the thermal generations and water discharge by hydro units while satisfying the equality constraints. One move tries to meet the demand constraint during each time interval and second tries to meet the available water constraint for every hydro unit. Mathematically these moves are represented below. Water discharge is updated as:

$$q_{jkm}^{t} = q_{jkm}^{t} + W_v R_9 (q_m^{\max} - q_m^{\min}) \Delta V_{jm}^{t} \;\; if \Delta V_{jm}^{t} \neq 0 \quad (m = 1, 2, \ldots, M; k = 1, 2, \ldots, T; j = 1, 2, \ldots, \text{NP}) \tag{9.87}$$

where

$$\Delta V_{mj} = V_{(T+1)j} - V_{1j} + \sum_{k=1}^{T} I_{kj} - \sum_{k=1}^{T} q_{kj} - \sum_{k=1}^{T} S_{kj} + \sum_{k=1}^{T} \sum_{r=1}^{U_j} (q_{r(k-tr_j)} + S_{r(k-tr_j)})$$

and R_9 is uniform random number having value between 0 and 1, W_v is scaling factor.

On violation of limits by updated water discharge, Q_{jkm}^{t} the water discharge is set to its corresponding limit.

Thermal generation are updated as:

$$P_{jki}^{t} = P_{jki}^{t} + W_D R_{10} (P_i^{\max} - P_i^{\min}) \Delta P_{Djk}^{t} \;\; \text{if } \Delta P_{Djk}^{t} \neq 0 \quad (i = 1, 2, \ldots, N; k = 1, 2, \ldots, T; j = 1, 2, \ldots, \text{NP}) \tag{9.88}$$

where

$$\Delta P_{Djk}^{t} = P_{Dk} + P_{Lk} - \sum_{i=1}^{N+M} P_{jki}^{t} \quad (k = 1, 2, \ldots, \text{NP}; j = 1, 2, \ldots, T)$$

and R_{10} is uniform random number having value between 0 and 1, W_D is scaling factor.

On violation of limits by updated thermal power, P_{jki}^{t} the thermal power is set to its corresponding limit.

The whole process is repeated till both the constraints are satisfied. It means for each sub-interval $\left|\Delta P_{Djk}^{t}\right|$ and for each hydro unit $\left|\Delta V_{jm}^{t}\right|$ should meet the tolerable termination criterion '*err*'.

Objective Function

Predator prey algorithm is used to generate the set of non-inferior solutions and to select the optimum solution depending upon the maximum value achieved by the objective function. Inequality constraint is handled as third objective function. So overall objective function or cardinal priority ranking is defined as:

$$f_l = \min\{\mu(J_{l1}), \mu(J_{l2}), \mu(J_{l3})\} \quad (l = 1, 2, \ldots, \text{NP}) \tag{9.89}$$

where

$$\mu(J_{l3}) = \frac{1}{1 + J_{l3}} \quad \text{and} \quad J_{l3} = r_p \left(\sum_{k=1}^{T} P_{lkd} + \sum_{j=1}^{M} q_{lhj} + \sum_{k=1}^{T} \sum_{j=1}^{M} V_{lkj} + \sum_{k=1}^{T} \sum_{m=N+1}^{N+M} P_{lkm} \right)$$

where q_{lhj} and V_{lkj} are water discharge and storage volume of jth reservoir at kth time interval for lth group of swarm respectively; r_p is the penalty parameter with higher value; d is dependent thermal unit to meet power demand constraint during each interval; h is dependent time interval for hydro units to meet final volume constraint.

Predator Velocity and Position Evaluation

The predator velocity and position (thermal unit) except for dth dependent thermal unit of each sub-interval for updates as:

$$V_{Pki}^{t+1} = C_4(GPbest_{ki}^t - P_{Pki}^t) \quad (i = 1, 2, \ldots, N; i \neq d; k = 1, 2, \ldots, T) \tag{9.90}$$

$$P_{Pki}^{t+1} = P_{Pki}^t + V_{Pki}^{t+1} \quad (i = 1, 2, \ldots, N; i \neq d; k = 1, 2, \ldots, T) \tag{9.91}$$

where, $GPbest_{ki}^t$ is global best thermal power position of ith thermal unit at kth sub-interval; C_4 is the random number distributed between 0 and upper limit; V_{Pki}^{t+1} is predator velocity for kth sub-interval at (t+1)th iteration. P_{Pki}^{t+1} is predator position for kth sub-interval at (t+1)th iteration.

The predator velocity and position (water discharge) for all the subintervals except for hth dependant subinterval updates as:

$$V_{Pqkm}^{t+1} = C_4(Gqbest_{km}^t - q_{Pkm}^t) \quad (m = 1, 2, \ldots, M; k = 1, 2, \ldots, T; k \neq h) \tag{9.92}$$

$$q_{Pkm}^{t+1} = q_{Pkm}^t + V_{Pqkm}^{t+1} \quad (m = 1, 2, \ldots, M; k = 1, 2, \ldots, T; k \neq h) \tag{9.93}$$

where, $Gqbest_{km}^t$ is global best water discharge position of mth hydro unit at kth sub-interval. The elements of predator position (thermal power) P_{Pki}^t, (water discharge) q_{Pkm}^t may violates their limits. This violation is set by fixing to their corresponding limit. If Predator velocity V_{Pki}^{t+1}, V_{Pqkm}^{t+1} violates its limit, randomly initialize within limits using Eq. (9.84) and Eq. (9.86) respectively.

Prey Velocity and Position Evaluation

Predator influences the prey. Each position set is evaluated based on objective function as given by Eq. (9.89).The velocity and position of a particle representing thermal generating unit for updated as:

$$V_{jki}^{t+1} = \begin{cases} wV_{jki}^t + C_1R_{11}(Pbest_{jki}^t - P_{jki}^t) + C_2R_{12}(GPbest_{ki}^t - P_{jki}^t) & ; pf \leq pf_{\max} \\ wV_{jki}^t + C_1R_{11}(Pbest_{jki}^t - P_{jki}^t) + C_2R_{12}(GPbest_{ki}^t - P_{jki}^t) + C_3ae^{-be} & ; pf > pf_{\max} \end{cases}$$

$$(i = 1, 2, \ldots, N; i \neq d; k = 1, 2, \ldots, T; j = 1, 2, \ldots, \text{NP}) \tag{9.94}$$

where C_1 and C_2 is the acceleration constant. R_{11} and R_{12} are uniform random numbers having value between 0 and 1, w is inertia weight; V_{jki}^{t+1} is prey velocity of ith particle for kth sub-interval at (t+1)th iteration. pf is probability of fear and $Pf_{\max}$ is maximum probability of fear. e is Euclidean distance between predator and prey. Constant 'a' provides the maximum amplitude of the predator effect over a prey and 'b' allows controlling the effect. C_3 is a random number distributed between 0 and upper limit and it influences the effect of predator on prey [Silva et al., 2002].

For global exploration higher value of inertia weight and for fine tuning lower value of inertia weight is required. Inertia weight starts from its set maximum value and starts decreasing with increase of movements/iterations.

$$w = w_{\max} - \frac{(w_{\max} - w_{\min}) \times t}{T_{\text{MAX}}} \tag{9.95}$$

where $w_{\max}$ and $w_{\min}$ represent maximum and minimum value of inertia weight. t represents iteration count and T_{MAX} represent maximum number of iterations.

The position of prey representing thermal generating units, changes as:

$$P_{jki}^{t+1} = P_{jki}^{t} + V_{jki}^{t+1} \quad (i = 1, 2, \ldots, N;\ i \neq d;\ k = 1, 2, \ldots, T;\ j = 1, 2, \ldots, \text{NP}) \tag{9.96}$$

The predator influences the prey depending on probability of fear. The velocity and position of a particle representing water discharge of hydro generating units are updated as:

$$V_{qjkm}^{t+1} = \begin{cases} wV_{qjkm}^{t} + C_1R_{13}(qbest_{jkm}^{t} - q_{jkm}^{t}) + C_2R_{14}(Gqbest_{km}^{t} - q_{jkm}^{t}) & ; pf \le pf_{max} \\ wV_{qjkm}^{t} + C_1R_{13}(qbest_{jkm}^{t} - q_{jkm}^{t}) + C_2R_{14}(Gqbest_{km}^{t} - q_{jkm}^{t}) + C_3ae^{-be} & ; pf > pf_{max} \end{cases}$$

$$(i = 1, 2, \ldots, N;\ k = 1, 2, \ldots, T;\ k \neq h;\ j = 1, 2, \ldots, \text{NP}) \tag{9.97}$$

where $qbest_{jkm}^{t}$ is local best water discharge position of mth hydro unit at kth time interval and jth particle; R_{13} and R_{14} are uniform random numbers having value between 0 and 1.

The rate of water discharge position will be updated as:

$$q_{jkm}^{t+1} = q_{jkm}^{t} + V_{qjkm}^{t+1} \quad (i = 1, 2, \ldots, N;\ k = 1, 2, \ldots, T;\ k \neq h;\ j = 1, 2, \ldots, \text{NP}) \tag{9.98}$$

The prey position (thermal power) P_{jki}^{t+1} and (water discharge) q_{jkm}^{t+1} may violate its limit. This violation is set by fixing their values either at lower or upper limits. If prey velocity V_{jki}^{t+1} and V_{qjkm}^{t+1} violates its limit, randomly initialize with in limit using Eq. (9.80) and Eq. (9.82). This process is repeated for several movements until the stopping criterion is met. In the present work stopping criterion is maximum number of movements or iterations. The flow chart of the proposed Algorithm based on predator-prey optimization is shown in Algorithm 9.9.

9.6.3 Test System and Results

Electric power system consists of 4-hydro and 3-thermal units, having 24-hour scheduling period divided into 24 sub-intervals of 1 hour each. A hydraulic system network along with number of upstream reservoir and time delay between upstream and downside reservoir are shown in Figure 9.4 and Table 9.14. Hydro power generation coefficients are given Table 9.15. A set of operating characteristics of hydro unit and reservoirs inflow are given in Table 9.16 and 9.17, respectively. Thermal unit cost coefficients and gaseous pollutant emission coefficients of 3-thermal units used in the study undertaken are given in Table 9.18 and 9.19, respectively. Load demand and loss coefficient for the given study is tabulated in Table 9.20.

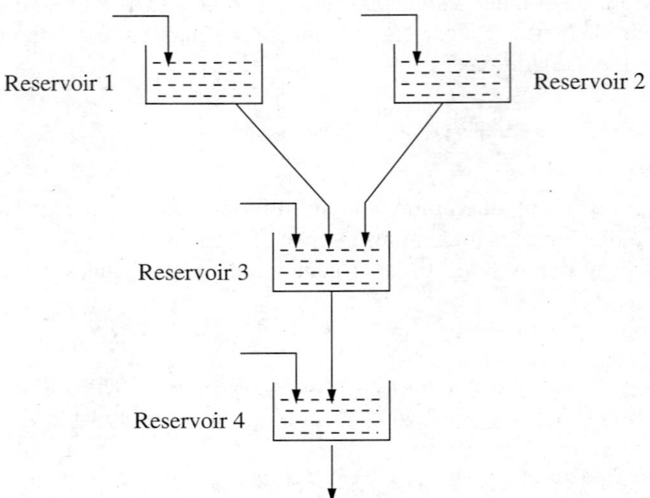

FIGURE 9.4 Hydraulic system network: Electric power system.

TABLE 9.14 Time delay between up-stream to down-stream

Plant	1	2	3	4
u_j	0	0	2	1
t_{rj}	2	3	4	0

TABLE 9.15 Hydro power generation coefficients: Electric power system

Unit (*j*)	C_{1j}	C_{2j}	C_{3j}	C_{4j}	C_{5j}	C_{6j}
1	– 0.0042	– 0.42	0.030	0.90	10.0	– 50
2	– 0.0040	– 0.30	0.015	1.14	9.5	– 70
3	– 0.0016	– 0.30	0.014	0.55	5.5	– 40
4	– 0.0030	– 0.31	0.027	1.44	14.0	– 90

Algorithm 9.9 Predator prey optimization technique for multi-objective HTGS

1. Read Input data. Set maximum and minimum values of all objectives and constant of proposed optimization technique.
2. Randomly initialize the prey positions and velocities of thermal power using Eqs. (9.79) and (9.80) and water discharge using Eq.(9.81) and (9.82), respectively.
3. Heuristic search technique is applied to satisfy equality constraints using Eq. (9.87) and Eq. (9.88) iteratively till $\left|\Delta P_{Djk}^{0}\right| \le err$ and $\left|\Delta V_{jm}^{0}\right| \le err$.
4. Compute objective f_0 using Eq. (9.89).
5. Randomly initialize the predator position and velocity of thermal power using Eq. (9.83) and (9.84) and water discharge using Eq. (9.85) and (9.86), respectively.

6. Set the movement count $t = 0$.
7. Set $IT = 50$.
8. Set the movement counter for velocity limit m=0.
   DO
   8.1 Increment the iteration count $t = t + 1$
   8.2 Compute water discharge rate for slack sub-interval as given in Eq. (9.75).
   8.3 Compute reservoir volume and hydro power for slack sub-interval by applying Eq. (9.72f) and Eq. (9.73d), respectively.
   8.4 Compute slack thermal unit power as given in Eq. (9.74).
   8.5 All inequality constraint violation is included in another objective function as given in Eq. (9.76).
   8.6 Update predator velocity and position thermal power using Eqs. (9.90) and (9.91) and water discharge using Eq. (9.92) and (9.93), respectively.
   8.7 Randomly generate probability fear between 0 and 1.
   8.8 **IF** (probability fear >maximum probability fear) ***THEN***
          Predator affects the prey.
       ENDIF
   8.9 Update prey velocities and positions thermal power using Eqs. (9.94) and (9.96) and water discharge using Eq. (9.97) and (9.98), respectively.
   8.10 Compute objective function by applying Eq. (9.89).
   8.11 Update local best position and update global best position.
   8.12 **IF** (global best position updated) ***THEN***
          $m = 0$
        ELSE
          $m = m+1$
          IF $(m > m_{max})$ **THEN**
             Squeeze the velocity limits by α factor
             $m = 0$
          ENDIF
        ENDIF
   WHILE $(t \le T_{max})$
9. Solution corresponding to maximum fuzzy membership function is the best solution.
   STOP

TABLE 9.16 Reservoir storage capacity limits, initial and final storage capacity limits, plant discharge limits, and hydro plant generation limits: Electric power system

Unit (j)	V_j^{min}	V_j^{max}	V_{j1}	V_{jF}	q_j^{min}	q_j^{max}	P_i^{min}	P_i^{max}
	($\times 10^4$ m^3)				($\times 10^4$ m^3/h)		(MW)	(MW)
1	80	150	100	120	5	15	0	500
2	60	120	80	70	6	15	0	500
3	100	240	170	170	10	30	0	500
4	70	160	120	140	6	20	0	500

TABLE 9.17 Reservoir inflows (× 10^4 m^3/h): Electric power system

k	Reservoir				k	Reservoir				k	Reservoir			
	I_{k1}	I_{k2}	I_{k3}	I_{k4}		I_{k1}	I_{k2}	I_{k3}	I_{k4}		I_{k1}	I_{k2}	I_{k3}	I_{k4}
1	10	8	8.1	2.8	9	10	8	1	0	17	9	7	2	0
2	9	8	8.2	2.4	10	11	9	1	0	18	8	6	2	0
3	8	9	4	1.6	11	12	9	1	0	19	7	7	1	0
4	7	9	2	0	12	10	8	2	0	20	6	8	1	0
5	6	8	3	0	13	11	8	4	0	21	7	9	2	0
6	7	7	4	0	14	12	9	3	0	22	8	9	2	0
7	8	6	3	0	15	11	9	3	0	23	9	8	1	0
8	9	7	2	0	16	10	8	2	0	24	10	8	0	0

TABLE 9.18 Cost coefficients and operating limits of thermal units: Electric power system

Unit (i)	a_{si} (₹/MW²h)	b_{si} (₹/MWh)	c_{si} (₹/h)	d_{si} (₹/h)	e_{si} (1/MW)	$P_i^{\min}$ (MW)	$P_i^{\max}$ (MW)
1	0.0012	2.45	100	160	0.038	20	175
2	0.0010	2.32	120	180	0.037	40	300
3	0.0015	2.10	150	200	0.035	50	500

TABLE 9.19 Emission coefficients: Electric power system

Unit (i)	α_i (lb/MW²h)	β_i (lb/MWh)	γ_i (lb/h)	η_i (lb/h)	δ_i (1/MW)
1	0.0105	−1.355	60	0.4968	0.01925
2	0.0080	−0.600	45	0.4860	0.01694
3	0.0120	−0.555	30	0.5035	0.01478

TABLE 9.20 Load demand (MW): Electric power system

k	P_{Dk} (MW)	k	P_{Dk} (MW)	k	P_{Dk} (MW)
1	750	9	1090	17	1050
2	780	10	1080	18	1120
3	700	11	1100	19	1070
4	650	12	1150	20	1050
5	670	13	1110	21	910
6	800	14	1030	22	860
7	950	15	1010	23	850
8	1010	16	1060	24	800

TABLE 9.21 Different values of PPO parameters

PPO Parameter	C_1	C_2	C_3	C_4	W_{max}	W_{min}	a	$b \times (10^{-4})$	pf_{max}
Value	2.0	2.0	1	0.1	0.9	0.4	1	0.1	0.90

TABLE 9.22 Results for test system

Technique Applied	F_1 *(cost)* $	F_2 *(emission) lb*	*CPU time* (sec)
PSO [14]	43280	17899	132.45
PPO	44111	17473	31.62

TABLE 9.23 Power generation for test system (MW)

Interval (*k*)	P_{Dk}	*Thermal generation*			*Hydro generation*			
		P_{1k}	P_{2k}	P_{3k}	P_{4k}	P_{5k}	P_{6k}	P_{7k}
1	750	99.72	208.41	135.92	84.81	62.46	26.80	131.88
2	780	175	129.14	173.22	74.33	48.95	50.33	129.03
3	700	136.15	124.85	138.9	85.23	50.12	39	125.74
4	650	20	209.27	137.06	85.5	60.86	15.69	121.62
5	670	145.04	125.83	139.69	60.61	54.16	28.85	115.82
6	800	142.15	209.52	138.42	79.89	56.43	41.44	132.14
7	950	175	220.06	229.06	84.61	65.48	34.31	141.48
8	1010	175	212.48	226.51	50.69	53.09	45.79	246.43
9	1090	175	215.25	225.50	77.65	73.15	44.35	279.1
10	1080	175	217.32	219.24	69.98	77.54	47.72	273.21
11	1100	175	209.66	213.59	81.57	72.54	41.71	305.93
12	1150	175	288	228.66	67	75.51	39.18	276.64
13	1110	175	274.35	219.93	85.19	66.06	7.94	281.53
14	1030	175	209.84	140.86	80.66	79.22	38.95	305.46
15	1010	119.73	210.25	200.21	69.73	68.39	41.2	300.49
16	1060	175	210.54	224.62	80.08	74.3	40.34	255.1
17	1050	142.06	209.44	213.62	54.98	79.21	48.94	301.74
18	1120	175	228.98	225.28	95.58	46.02	43.29	305.85
19	1070	175	211.66	200.56	84.71	46.02	48.22	303.82
20	1050	175	213.58	145.7	92.07	70.22	52.25	301.17
21	910	125.06	125.92	140.73	92.05	71.85	55.08	299.31
22	860	107.43	147.87	168.03	54.32	61.77	37.89	282.68
23	850	103.54	124.76	136.13	79.43	59.21	53.34	293.59
24	800	105.4	125.02	127.36	54.7	67.56	57.2	262.72

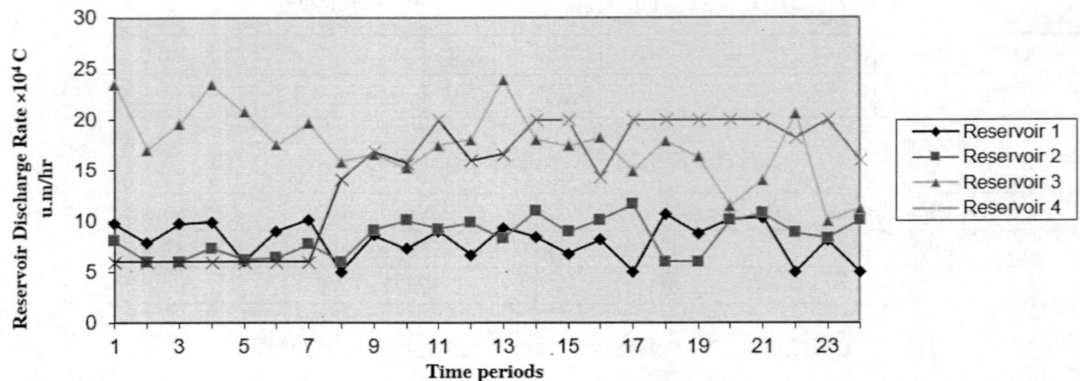

FIGURE 9.5 Hourly hydro reservoir discharge rate for test system–I.

After running a number of trials of different values of PPO parameters, such as maximum value of probability fear, the maximum amplitude of a predator over prey has been set for the best solution. The final values of PPO parameters are given in Table 9.22. For all test systems, scaling factor w_v and w_D are set to 0.05 and 0.05, respectively. Results are obtained by taking 50 preys and a single predator. Maximum and minimum velocity has been set to +1% and –1% of minimum value of the decision variable, respectively. The velocity squeezing factor has been set to 0.9. Table 9.22 gives the objectives achieved on implementation of the PPO algorithm. Table 9.23 represents power generation schedule for the given load demand. Figure 9.5 shows the hourly water discharge rate of the best compromise schedule. All equality and inequality constraints are well satisfied. The results of PPO technique are compared with interactive fuzzy method based on PSO [Mandal and Chakrabotry, 2011].

REFERENCES

Books

Arora, J.S., *Introduction to Optimum Design*, 2nd edition. Elsevier Inc., 2004.

Bergen, A.R. and Vijay Vittal, *Power System Analysis*, Pearson Education Inc., Delhi, 2004.

Davis, L., *Handbook of Genetic Algorithms*, Van Nostrand Reinhold, 1991.

Deb, K., *Optimization For Engineering Design, Algorithms and Examples*, Prentice Hall of India, 1995.

Deb, K., *Multiobjective Optimization using Evolutionary Algorithm*, John Wiley & Sons Ltd, New York, 2001.

Belegundu, A.D. and T.R. Chandrupatla, *Optimization Concepts and Applications in Engineering*, Pearson Education Inc., Delhi, 2002.

Goldberg, D.E., *Genetic Algorithm in Search, Optimization and Machine Learning*, Addison Wesley, 1989.

Frederick Soloman, *Probability and stochastic processes*, Prentice-Hall, New Jersey, 1987.

Osyczka, A. and B.J. Davies, *Multicriterion Optimization in Engineering with FORTRAN Programs*, Ellis Horwood Ltd., 1984.

Papoulis, A., Probability, *Random Variables and Stochastic Processes*, McGraw Hill, New Delhi, 1991.

Rao, S.S., *Optimization, Theory and Applications*, 2nd edition, Wiley Eastern Limited, New Delhi, 1987.

Song, Yong-Hua, *Modern Optimization Techniques in Power Systems*, Kluwer Academic Publishers, 1999.

Papers

Abido, M.A.: Multiobjective evolutionary algorithms for Electric power dispatch problem, *IEEE Transactions on Evolutionary Computation*, Vol. **10(3)**, pp. 315–329, 2006.

Abido, M.A., Optimal design of power system stabilizers using particle swarm optimization, *IEEE Transactions on Energy Conversion*, Vol. **17(3)**, pp. 406–413, 2002.

Abdolmohammadi, H.A. and Kazemi, A., A benders decomposition approach for a combined heat and power economic dispatch, *Energy Conversion and Management*, Vol. **71**, pp. 21–31, 2013.

Agarwal S. K. and O. P. Sahu, Artificial bee colony algorithm to design two–channel quadrature mirror filter banks, *Swarm and Evolutionary Computation*, Vol. **21**, pp. 24–31, 2015.

Amjady, N. and H. Nasiri-Rad, Economic dispatch using an efficient real-coded genetic algorithm, *IET Generation, Transmission, Distribution*, Vol. **3(3)**, pp. 266–278, 2009.

Amjady Nima and Soleymanpour Hassan Rezai, Daily hydrothermal generation scheduling by a new modified particle swarm optimization technique, *Electric Power Systems Research*, Vol. **80(6)**, pp.723–732, 2010.

Basu, M., An interactive fuzzy satisfying method based on evolutionary programming technique for multiobjective short-term hydrothermal scheduling, *Electric Power System Research*, Vol. **69 (2, 3)**, pp. 277–285, 2004.

Basu, M., Combined heat and power economic dispatch by using differential evolution, *Electric Power Components and Systems*, Vol. **38**, pp. 996–1004, 2010.

Bergh, F. van den and A.P. Engelbrecht, A cooperative approach to particle swarm optimization, *IEEE Transactions on Evolutionary Computation*, Vol. **8(3)**, pp. 225–239, 2004.

Bonaert, A.P., A.H. El-Abiad and A.J. Koivo, Optimal scheduling of hydro-thermal power systems, *IEEE Transactions on Power Apparatus and Systems*, Vol. **91(1)**, pp. 263–270, 1972.

Borghetti, Alberto and Claudia D'Ambrosio, Andrea Lodi, and Silvano Martello, An MILP approach for short-term hydro scheduling and unit commitment with head-dependent reservoir, *IEEE Transactions on Power Systems*, Vol. **23(3)**, pp. 1115–1124, 2008.

Catalão, J.P.S., S.J.P.S. Mariano, V.M.F. Mendes, and L.A.F.M. Ferreira, Scheduling of head-sensitive cascaded hydro systems: a nonlinear approach, *IEEE Transactions on Power Systems*, Vol. **24(1)**, pp. 337–346, 2009.

Chang, C.S. and W. FU, Stochastic multiobjective generation dispatch of combined heat and power system, *IEE Proceedings—Generation, Transmission and Distribution*, Vol. **145(5)**, pp. 583–591, 1998.

Chang, H.C. and P. H. Chen, Hydrothermal generation scheduling package: a genetic based approach, *IEE Proceedings—Generation, Transmission and Distribution*, Vol. **145(4)**, pp. 451–457, 1998.

Chiang, C.L., Improved genetic algorithm for power economic dispatch of units with valve-point effects and multiple fuels, *IEEE Transactions on Power Systems*, Vol. **20(4)**, pp. 1690–1699, 2005.

Chiang, C.L., Optimal economic emission dispatch of hydrothermal power systems, *International Journal of Electrical Power and Energy Systems*, Vol. **29 (6)**, pp. 462–469, 2007.

Chiou, Ji-Pyng: Variable scaling hybrid differential evolution for large-scale economic dispatch problems, *Electric Power Systems Research*, Vol. **77(3–4)**, pp. 212–218, 2007.

Coelho, L.S. and V.C. Mariani, Combining of chaotic differential evolution and quadratic programming for economic dispatch optimization with valve-point effect, *IEEE Transactions on Power Systems*, Vol. **21(2)**, pp. 989–996, 2006 (Erratum: Vol. **21(3)**, pp. 1465, 2006).

Coelho, L.S. and V.C. Mariani, Improved differential evolution algorithms for handling economic dispatch optimization with generator constraints, *Energy Conversion and Management*, Vol. **48(6)**, pp. 1831–1639, 2007.

Dhillon, J.S., S.C. Parti and D.P. Kothari, Fuzzy decision making in multiobjective long-term scheduling of hydrothermal system, *International Journal Electric Power Energy System*, Vol. **23(1)**, pp. 19–29, 2001.

Esmin, A.A.A., G. Lambert-Torres and A.C. Zambroni de Souza, A hybrid particle swarm optimization applied to loss power minimization, *IEEE Transactions on Power Systems*, Vol. **20(2)**, pp. 859–866, 2005.

Farag, A., S. Al-Baiyat, and T.C. Cheng, Economic load dispatch multi-objective optimization procedure using linear programming techniques, *IEEE Transactions on Power Systems*, Vol. **10(2)**, pp. 731–738, 1995.

Gaing, Zwe-Lee: Particle Swarm Optimization to Solve the Economic Dispatch Considering the Generator Constraints, *IEEE Transactions on Power Systems*, Vol. **18(3)**, pp. 1187–1195, 2003.

He, Dakuo, Fuli Wang and Zhizhong Mao, A hybrid genetic algorithm approach based on differential evolution for economic dispatch with valve-point effect, *International Journal of Electric Power and Energy Systems*, Vol. **30**, pp. 31–38, 2008.

Hota, P.K., R. Chakrabarti, and P.K. Chattopadhyay, A simulated annealing-based goal attainment method for economic emission load dispatch with non-smooth fuel cost and emission level functions, *Electric Machines and Power Systems*, Vol. **28(11)**, pp. 1037–1051, 2000.

Jayabarathi, T., G. Sadasivam and V. Ramachandran, Evolutionary programming based multiarea economic dispatch with tie line constraints, *Electric Machines and Power System*, Vol. **28(12)**, pp. 1165–1176, 2000.

Jayabarathi, T., K. Jayaprakash, D.N. Jeyakumar, and T. Raghunathan, Evolutionary Programming Techniques for Different Kinds of Economic Dispatch Problems, *Electric Power Systems Research*, Vol. **73(2)**, pp. 169–176, August 2005.

Karaboga, D., An idea based on honey bee swarm for numerical optimization (Technical Report-TR06), Engineering Faculty, Computer Engineering Department, Erciyes University, Turkey, 2005.

Karaboga, D. and B. Gorkemli, A quick artificial bee colony (qABC) algorithm and its performance on optimization problems, *Applied. Soft Computing*, Vol. **23**, pp. 227–238, 2014.

Kennedy, J. and R. Eberhart, Particle swarm optimization, *Proceedings of the IEEE International Conference on Neural Networks*, pp. 1942–1948, 1995.

Lee, F.N., and A.M. Breipohl, Reserve Constrained Economic Dispatch with Prohibited Operating Zones, *IEEE Transactions on Power Systems*, Vol. **8(1)**, pp. 246–254, 1993.

Lee, K.Y., A. Sode-Yome, and J.H. Park, Adaptive Hopfield neural networks for economic load dispatch, *IEEE Transactions on Power Systems*, Vol. **13(2)**, pp. 519–526, 1998.

Li, X. and M. Yin, Multi-operator based biogeography based optimization with mutation for global numerical optimization, *Computers and Mathematics with Applications*, Vol. **64**, pp. 2833–2844, 2012.

Liang, Z.X. and J.D. Glover, A zoom feature for a dynamic programming solution to economic dispatch including transmission losses, *IEEE Transactions on Power Systems*, Vol. **7(2)**, pp. 544–550, 1992.

Lin, W.M., F.S. Cheng and M.T. Tsay, An improved tabu search for economic dispatch with multiple minima, *IEEE Transactions on Power Systems*, Vol. **17(1)**, pp. 108–112, 2002.

Lin, C.E. and G.L. Viviani, Hierarchical economic dispatch for piecewise quadratic cost functions, *IEEE Transactions on Power Apparatus and Systems*, Vol. **103(6)**, pp. 1170–1175, 1984.

Ma, H., An analysis of the equilibrium of migration models for biogeography-based optimization, *Information Sciences*, Vol. **180(18)**, pp. 3444–3464, 2010.

Ma, H., and D. Simon, Analysis of migration models of biogeography-based optimization using Markov theory, *Engineering Applications of Artificial Intelligence*, Vol. **24**, pp. 1052–1060, 2011.

Ma, J.T. and L.L. Lai, Evolutionary programming approach to reactive power planning, *IEE Proceedings—Generation, Transmission and Distribution*, Vol. **143(4)**, pp. 365–370, 1996.

Mandal, K.K. and N. Chakrabotry, Short-term combined economic emission scheduling of hydrothermal systems with cascaded reservoirs using particle swarm optimization technique, *Applied Soft Computing*, Vol. **11(1)**, pp. 1295–1302, 2011.

Mendes, R., J. Kennedy and J. Neves, The fully informed particle swarm: simpler, maybe better, *IEEE Transactions on Evolutionary Computation*, Vol. **8(3)**, pp. 204–210, 2004.

Mo, B., J. Hegge and I. Wangensteen, Stochastic generation expansion planning by means of stochastic dynamic programming, *IEEE Transactions on Power Systems*, Vol. **6 (2)**, pp. 662–668, 1991.

Naka, S., T. Genji, T. Yura and Y. Fukuyama, A hybrid particle swarm optimization for distribution state estimation, *IEEE Transactions on Power Systems*, Vol. **18(1)**, pp. 60–68, 2003.

Nara, K., A. Shiose, M, Kitagawa and T. Shihara, Implementation of genetic algorithms for distribution systems loss minimum reconfiguration, *IEEE Transactions on Power Systems*, Vol. **7(3)**, pp. 1044–1050, 1992.

Naresh, R., J. Dubey, and J. Sharma, Two-phase neural networks based modeling framework of constrained economic load dispatch, *IEE Proceedings—Generation, Transmission and Distribution*, Vol. **151(3)**, pp. 373–378, 2004.

Orero, S.O. and M.R. Irving, Economic dispatch of generators with prohibited operating zones: a genetic algorithm approach, *IEE Proceedings—Generation, Transmission and Distribution*, Vol. **143(6)**, pp. 529–534, 1996.

Park, J.B., K.S. Lee, J.R. Shin, and K.Y. Lee, A particle swarm optimization for economic dispatch with nonsmooth cost functions, *IEEE Transactions on Power Systems*, Vol. **20(1)**, pp. 34–42, 2005.

Pereira-Neto, A., C. Unsihuay, and O.R. Saavedra, Efficient evolutionary strategy optimization procedure to solve the nonconvex economic dispatch problem with generator constraints, *IEE Proceedings—Generation, Transmission and Distribution*, Vol. **153(5)**, pp. 653–660, 2005.

Po-Hung, Chen and Chang Hong-Chan, Large-Scale Economic Dispatch by Genetic Algorithm, *IEEE Transactions on Power Systems*, Vol. **10(4)**, pp. 1919–1926, 1995.

Rahnamayan, H.R., Tizhoosh and M.A. Salama, Opposition based differential evolution, *IEEE Transactions on Evolutionary Computation*, Vol. **12(1)**, pp 64–79, 2008.

Rashid, A.H.A. and K.M. Nor, An algorithm for the optimal scheduling of variable head hydro and thermal plants, *IEEE Transactions on Power Systems*, Vol. **8 (3)**, pp. 1242–1249, 1993.

Ratnaweera, A., S.K. Halgamuge and H.C. Watson, Self-organizing hierarchical particle swarm optimizer with time-varying acceleration coefficients, *IEEE Transactions on Evolutionary Computation*, Vol. **8(3)**, pp. 240–255, 2004.

Roa-Sepulveda, C.A., M. Herrera, B. Pavez-Lazo, U.G. Knight and A.H. Coonick, Economic Dispatch Using Fuzzy Decision Trees, *Electric Power Systems Research*, Vol. **66(2)**, pp. 115–122, 2003.

Salomon, Ralf, Evolutionary algorithms and gradient search: Similarities and differences, *IEEE Transactions on Evolutionary Computation*, Vol. **2(2)**, pp. 45–55, 1998.

Selvakumar, A.I. and K. Thanushkodi, A new particle swarm optimization solution to nonconvex economic dispatch problems, *IEEE Transactions on Power Systems*, Vol. **22(1)**, pp. 42–51, 2007.

Selvakumar, A.I. and K. Thanushkodi, Anti-predatory particle swarm optimization: Solution to nonconvex economic dispatch problems, *Electric Power Systems Research*, Vol. **78(1)**, pp. 2–10, 2008.

Shaw, J.J., R.F. Gendron and D.P. Bertsekas, Optimal scheduling of large hydrothermal power systems, *IEEE Transactions on Power Apparatus and Systems*, Vol. **104(2)**, pp. 286–293, 1985.

Sifuentes, W.S. and Alberto Vargas, Hydrothermal scheduling using benders decomposition: Acceleration techniques, *IEEE Transactions on Power Systems*, Vol. **22(3)**, pp. 1351–1359, 2007.

Sheble, G.B., T.T. Maifeld, K. Brittig, G. Fahd and S. Fukurozaki-Coppinger, Unit commitment by genetic algorithm with penalty methods and a comparison of Lagrangian search and genetic algorithm—economic dispatch, *International Journal of Electric Power and Energy Systems*, Vol. **18(6)**, pp. 339–346, 1996.

Silva A., A. Nevesana and E. Costa, An Empirical comparison of Particle Swarm and Predator Prey Optimization, Lecture Notes Computer., *Science*, Vol. **2464**, pp. 103–110, 2002.

Simon, D., Biogeography-based optimization, *IEEE Transactions on Evolutionary Computation*, Vol. **12(6)**, pp. 702–713. 2008.

Simon, D.: A dynamic system model of biogeography-based optimization, *Applied Soft Computing*, Vol. **11**, pp. 5652–5661. 2011.

Sinha, N., R. Chakrabarti, and P.K. Chattopadhyay, Evolutionary Programming Techniques for Economic Load Dispatch, *IEEE Transactions on Evolutionary Computation*, Vol. **7(1)**, pp. 83–94, 2003.

Sinha, N., R. Chakrabarti and P.K. Chattopadhyay, Fast evolutionary programming techniques for short-term hydrothermal scheduling, *IEEE Transactions on Power Systems*, Vol. **18 (1)**, pp. 214–220, 2003.

Song, Y.H. and C.S.V. Chou, Advanced engineered conditioning genetic approach to power economic dispatch, *IEE Proceedings—Generation, Transmission and Distribution*, Vol. **144(3)**, pp. 285–292, 1997.

Song, Y.H., G.S. Wang, P.Y. Wang, and A.T. Johns, Environmental/economic dispatch using fuzzy logic controlled genetic algorithms, *IEE Proceedings—Generation, Transmission and Distribution*, Vol. **144(4)**, pp. 377–382, 1997.

Storn R., Differntial evolution—a simple and efficient heuristic for global optimization over continuous space, *Journal of Global Optimization*, Vol. **11(4)**, pp. 341–359, 1997.

Subbaraj, P., Rengaraj, R. and Salivahanan, S., Enhancement of combined heat and power economic dispatch using self adaptive real-coded genetic algorithm, *Applied Energy*, Vol. **86**, pp. 915–921, 2009.

Sum-Im, T., G.A. Taylor, M.R. Irving and Y.H. Song, Differential evolution algorithm for static multistage transmission expansion planning, *IET Generation, Transmission, Distribution*, Vol. **3(4)**, pp. 365–384, 2009.

Tapia, C.G. and B.A. Murtagh, Interactive fuzzy programming with preference criterion in multi-objective decision making, *Computers and Operations Research*, Vol. **18(3)**, pp. 307–316, 1991.

Varadarajan, M. and K.S. Swarup, Solving multi-objective optimal power flow using differential evolution, *IET Generation, Transmission, Distribution*, Vol. **2(5)**, pp. 720–730, 2008.

Venkatesh, P., P.S. Kannan and M. Sudhakaran, Application of computational intelligence to economic load dispatch, *Journal of Institution of Engineers, India*, Vol. **81**, pp. 39–43, 2000.

Venkatesh, P., P.S. Kannan and S. Anudevi, Application of micro Genetic algorithm to economic load dispatch, *Journal of Institution of Engineers*, India, Vol. 82, pp. 161–167, 2001.

Victorie, T.A.A. and A.E. Jeyakumar, Hybrid PSO-SQP for economic dispatch with valve-point effect, *Electric Power Systems Research*, Vol. 71(1), pp. 51–59, 2004.

Walter, D.C., and G.B. Sheble, Genetic algorithm solution of economic dispatch with valve point loading, *IEEE Transactions on Power Systems*, Vol. **8(3)**, pp.1325–1332, 1993.

Wong, K.P. and Y.W. Wong, Genetic and genetic/simulated-annealing approaches to economic dispatch, *IEE Proceedings—Generation, Transmission and Distribution*, Vol. **141(5)**, pp. 507–513, 1994.

Wong, K.P. and Y.W. Wong, Thermal Generator Scheduling Using Hybrid Genetic/Simulated Annealing Approach, *IEE Proceedings—Generation, Transmission and Distribution*, Vol. **142(4)**, pp. 372–380, 1995.

Wu, Q.H. and J.T. Ma, Power system optimal reactive power dispatch using evolutionary programming, *IEEE Transactions on Power Systems*, Vol. **10(3)**, pp. 1243–1249, 1995.

Xiong, G., D. Shi, and X. Duan, Enhancing the performance of biogeography based optimization using the polyphyletic migration operator and orthogonal learning, *Computers and Operations Research* , Vol. **41(2)**, pp. 125–139, 2014.

Yang, H.T. and P.C. Yang, C.L. Huang, Evolutionary programming based economic dispatch for units with non-smooth fuel cost functions, *IEEE Transactions on Power Systems*, Vol. **11(1)**, pp. 112–118, 1996.

Yang, P.C., H.T. Yang, C.L. Huang, Scheduling short-term hydrothermal generation using evolutionary programming techniques, *IEE Proceedings—Generation, Transmission and Distribution*, Vol. **143(4)**, pp. 371–376, 1996.

Yao, X., Yong Liu and Guangming Liu, Evolutionary programming made faster, *IEEE Transactions on Evolutionary Computation*, Vol. **3(2)**, pp. 82–102, 1999.

Zhao, B., C.X. Guo and Y.J. Cao, A multiagent-based particle swarm optimization approach for optimal reactive power dispatch, *IEEE Transactions on Power Systems*, Vol. **20(2)**, pp. 1070–1078, 2005.

CHAPTER 10

Unit Commitment

10.1 INTRODUCTION

The total load demand of power system varies throughout the day and follows different peak values from one hour to other. It is true for weekly trend of load demand. It is costly affair to run all the generators for the whole day or week. Hence, there is need to decide in advance the number of generators are to start-up and to synchronize with the power network to deliver the electric power and the number of generators are to shut-down, their order of shutting down and time duration to remain off. A generating unit is said to be committed if it is connected to the electric power network and delivers power to network. To commit a generating unit is 'to turn it on' and to de-commit a generating unit is 'to turn it off. Further, generators cannot instantly turn on and produce power. In other words Unit Commitment (UC) schedules the on and off times of the generating units, and calculates the minimum hourly operating cost of generation schedule while ensures the consideration of start-up and shut-down rates and minimum-up and -down time of units. Unit commitment is necessitated to be planned so that required generation is always available to handle system demand with an adequate reserve margin in the event that generators or transmission lines go out or load demand increases. Unit commitment handles the unit generation schedule in a power system for minimizing operating cost and satisfying prevailing physical and technological constraints such as load demand and system reserve requirements over a set of time periods [Hobbs et al., 1988].

Unit commitment is known as pre-dispatch as it fits between economic dispatch and maintenance and production scheduling in the hierarchy of generation resource management. The classic UC problem is defined as to determine the start-up and shut-down schedules of thermal units to meet forecasted demand over certain time horizons. In terms of time scales involved, UC covers the scope of hourly power system operation decisions with one day (24h) to one week. The unit commitment decisions are coupled or iteratively solved in conjunction with coordinating the use of hydro including pumped storage capabilities and ensuring system reliability using probabilistic measures. The function also includes labor constraints due to crew policy and costs. During the normal times a full operating crew is available without committing overtime costs. A foremost consideration is to adequately adopt environmental controls, such as fuel switching. System with hydro storage capability either with pumped hydro stations or with reservoirs on rivers usually require one-week horizon times. On the other hand, a system with few dynamic components can use much shorter horizon times. The plant commitment and unit-ordering schedules extend the period of optimization from a few minutes to several hours. Weekly patterns are developed from daily schedules. Likewise, monthly, seasonal, and annual schedules are prepared by taking into consideration the repetitive nature of the load demand and seasonal variations. A great deal of money is saved by turning off the units when these units are not needed for the time. The UC schedules are required for economically committing units in plant to service with the time at which individual units should be taken out from or returned to service. This problem is of importance for scheduling thermal units in a thermal plant; as for other types of generation such as hydro, their aggregate costs (such as start-up costs, operating fuel costs, and shut-down costs) are negligible so that their on-off status is not important.

The UC problem belongs to a class of combinatorial optimization problems. The methods that have been studied so far fall into roughly three categories: heuristic search, mathematical programming, and hybrid methods. Optimization techniques such as the priority list, augmented Lagrangian relaxation, dynamic programming, and the branch-and-bound algorithm have been used to solve the classic UC problem. Genetic algorithms (GA), simulated annealing (SA), analytic hierarchy process (AHP), particle swarm optimization (PSO) and meta-heuristic techniques have also been used for UC problem since the beginning of the last decade.

Unit Type

The thermal units are usually divided into the following unit types:

Must-run units

Must-run units must be on-line, if available, due to operating constraints, reliability requirements, or economic considerations. Generally in power system, some of the units are given a must-run status in order to provide voltage support for the network.

Cycling units

Cycling units can cycle on and off and are subject to minimum up and down-time constraints. Both must-run and cycling units are dispatched economically between their minimum and maximum limits.

Peak load units

Peak load units are generating units that are capable to serve the load demand in case of any sudden increase in load demand and these are able to start up in less possible time, usually gas turbines generating unit serves the motive of peak load units. The output of some units may be limited because their output would exceed the transmission capacity of the network.

Flexible units

Units are termed as flexible units in which power output can be adjusted within limits, e.g., Coal-fired, Oil-fired, Open cycle gas turbines, Combined cycle gas turbines, Hydro plants with storage. Status and power output can be optimized.

Inflexible units

Units are termed as inflexible units in which power output cannot be adjusted for technical or commercial reasons, e.g., Nuclear, Run-of-the-river hydro, renewable: wind, solar, etc., combined heat and power (CHP, cogeneration). Power output treated as given when optimization is performed.

Spinning Reserve

The spinning reserve is the amount of extra real power generation available from all synchronized units to serve the load demand in case of any fault or sudden tripping or maintenance of any generating units. Spinning reserve also compensates the loss of one or more units that does not cause a drop in system frequency. Sufficient reserves are maintained to make up for a generation-unit failure and the reserves must be allocated among fast-responding units and slow-responding units. This allows the automatic generation control system to restore frequency and interchange quickly in the event of a generating-unit outage. Beyond spinning reserve the unit commitment problem may involve various off scheduled reserves or off-line reserves. These include quick start diesel or gas-turbine units as well as most hydro-unit and pumped-storage hydro-units that can be brought on-line, synchronized, and brought up to full capacity quickly. As such, these units can be counted, in the overall reserve assessment, as long as their time to come up to full capacity is taken into account. Reserves are usually spread around the power system to avoid transmission system limitations (often called bottling of reserves) and to allow various parts of the system to run as island. It is the term used to describe the total amount of generation available from all synchronized units on the system minus the present load demand and transmission losses of the power system network being supplied. Here, the synchronized units on the system may be named units spinning on the system.

Spinning reserve obeys certain rules set by regional reliability councils. Typical rules specify that reserve should be a given percentage of forecasted peak demand, or that reserve must capable of making up the loss of the most heavily loaded unit in a given time. Further, reserve requirements are calculated as a function of the probability of not having sufficient generation to meet the load.

Operating reserves are characterized by their response speed (i.e., ramp rate and start time), response duration, frequency of use, direction of use (i.e., up or down), and type of control (i.e., control center activation, autonomous/automatic, etc.). Some operating reserves are used to respond to routine variability of the generation or the load. These variations occur on different time scales, from seconds to days. Different control strategies are required depending on the variation of the speed. Other operating reserves are needed to respond to rare unexpected events such as the tripping of a generator. The operating reserves are classified based on whether these reserves are deployed during normal conditions or event conditions. Normal conditions are based on both variability and uncertainty, but their occurrences are continuously taking place. Events can occur whether these events can be predicted or not. There is a difference in the standby costs and the deployment costs for each reserve category based on their frequent use.

Both the normal and event response categories are further sub-divided based on the needful response speed. Some events are essentially instantaneous and others take some time to occur. Different qualities of these reserves are needed for different purposes. For instance, instantaneous events require autonomous response to arrest frequency excursions. The frequency then must be corrected back to its scheduled setting and the system's ACE must be reduced to zero during instantaneous and non-instantaneous events. Further, there has to be some amount of reserves that replaces the operating reserve after these reserves are deployed to protect the system against an event. The non-event reserves have zero net energy for a particular time period. Over some time, these reserves will have equal positive deployments and negative deployments in energy.

Operating Reserve: Any capacity available for assistance in active power balance is termed as operating reserve.

Primary Reserve: Primary reserve is a portion of reserve that is automatically responsive to instantaneous active power imbalance and stabilizes system frequency.

Secondary Reserve: Secondary reserve is a portion of reserve that is not automatically responsive to the instantaneous active power imbalance and corrects frequency to nominal.

Tertiary Reserve: Tertiary reserve is a portion of reserve that is available for assistance in replacing primary and secondary reserve used during a severe instantaneous event so that they are available for a subsequent instantaneous event that occurs in the same direction.

Ramping Reserve: Ramping reserve is the capacity that is available for assistance in active power balance during infrequent events that are more severe than balancing needed during normal conditions and are used to correct non-instantaneous imbalances.

Secondary Reserve-Ramping: Secondary reserve-ramping is a portion of ramping reserve that is used to correct the imbalance of a severe non-instantaneous event and corrects the frequency to nominal.

Tertiary Reserve-Ramping: Tertiary reserve-ramping is a portion of ramping reserve that is available for assistance in replacing secondary reserve used during a severe non-instantaneous event so that eventually secondary reserve are available for a subsequent event that occurs in the same direction.

Thermal Unit Constraints

The temperature and pressure of the thermal units vary gradually and units must be synchronized before they are brought online. Thermal units usually require a crew to operate them, especially when turned on and turned off. A thermal unit can undergo only gradual temperature changes, which in turn translates into a time period of some hours that are required to bring the unit "online". As a result of such restrictions various constraints arise, in the operation of a thermal plant, such as [Pang and Chen, 1976] and are given below:

Minimum Up-time

A running unit can't be shut down immediately. In case, the unit is committed, there is a minimum time required before it can be de-committed that time is known as minimum up-time. It needs some time to reach the off state [Seki et al., 2010].

Minimum Down-time

A shut-down unit can't be started immediately. Once the unit is de-committed, there is a minimum time before it can be recommitted and that time is known as minimum down-time. It needs some time to reach the committed state.

Crew Constraints

A plant consists of two or more units. Two and more units cannot both be turned on at the same time since there are not enough crew members to attend two and more units at the start up.

Ramp Constraints

A change in a unit's generation level between any two successive periods should not exceed its ramp rate limitation. The sum of ramp limits of committed units should still allow coping with the change in the system load from one period to the next. On the system basis, the spinning reserve amount contributed by each unit is calculated by considering these ramp rate constraints. The spinning reserve calculated from the difference between maximum committed power and actual load can be large enough to meet the reserve requirement, but ramping limitations makes the actual available spinning reserve insufficient.

The sum of ramp limits of committed units should be at least sufficient to meet a change in system load from one period to the next. Ramping limitations may offset a significant part of the available spinning difference. Consequently, the ramp rate constraints link the generation variables of the previous period to that of the present, and hence introduce a dynamic characteristic in the UC model. Ramping constraints in UC cause an additional computation burden by slowing the convergence rate. Furthermore, backward economic dispatch is required to reduce the generation output from its constrained maximum to zero under the ramp down limit within one period.

The generation scheduling with relaxed ramp rate constraints is generally used since it can greatly simplify the problem. The use of ramp rate constraints to simulate the unit state and generation changes has a strong effect on optimal scheduling. Three types of ramping constraints for each unit are considered.

***Start-up Ramp Constraints*:** When an off line unit is turned on, it takes few (say T_i^{SR}) minutes to increase its generation capability from zero to its minimum level. During this period, this unit is off line but consuming fuel.

***Shut-down Ramp Constraints*:** When an online unit is turned off, it takes some time to decrease its generation capability. However, de-committing the online unit with generation output higher than minimum level is not allowed unless in a forced shutdown (generation shedding). This is because before the unit status is changed from '1' at period t to '0' at period t+1, this unit is operated at the equilibrium point where mechanical input power equals the electrical output power. At period t+1 when this unit is disconnected from the system, the electrical power of this unit is changed from its current generation to zero. This results in accelerating power on the rotor and causes transient instability, which may further affect the generator's life time and its performance. Hence, the generation output of this unit should be at the minimum level at the last committed period. It will take few (say T_i^{SDR}) minutes to decrease its generation capability from its minimum level to zero.

***Operating Ramp Constraints*:** The generation output of the current period confined by the up-ramp limit and the down-ramp limit, that is, the generation output of the current period should be less than the generation output of the previous period plus up ramp rate (MW/h) and it should be higher than the generation output of the previous period minus down ramp rate (MW/h).

1. *Generation ramp limit:* To satisfy the generation operating limit constraint, generation ramp limit are introduced.
2. *Minimum down time:* The minimum down time includes the time for the shutdown process that is from minimum generation level to zero and the time for the start-up process that is from zero to minimum generation level.
3. *Unit reserve contribution:* The up spinning reserve is required to ensure that spinning reserve is sufficient to make up for any unexpected load increase or loss of generation. Whereas, the down spinning reserve is required to ensure that a sudden outage of load won't cause the forced shutdown of thermal units. Due to the unit ramp up and ramp down limits, the up (down) spinning reserve contributed by each unit are determined by two terms: the difference between the maximum (minimum) capacity and the current generation output, and the up/down ramp capability within the specific time. The smaller of these two terms is selected as up/down spinning reserve of that unit.

4. *On/off line minimum level:* To reduce the disturbance due to an output change between zero and 'on' line generation output and to satisfy the minimum level constraints, the generation output of the shutdown period and the start-up period is limited to its minimum level.

Transmission Line Constraint

In addition to providing a sufficient reserve to make up for a generation unit failure, the reserve must be spread around the system to avoid transmission system limitations. Furthermore, the reserve must be allocated among the fast responding units and slow-responding units. This allows the automatic generation control system to restore frequency and interchange quickly in the event of a generating unit outage.

One common approach including the transmission line constraint is known as indirect method, considering the UC problem into two steps. This approach ignores transmission constraints in UC, then account for these constraints in an ED process by re-dispatching. A direct method for security constrained unit commitment was proposed in 1995. Transmission line constraints present a challenge to researchers of the UC problem since the early 1990s. Generally, a linear dc power flow was considered in the UC problem formulation for security reasons. However, optimal power flow was incorporated in the UC formulation in 1999.

Start-up Cost

A certain amount of energy is expanded to bring the unit online as the temperature and pressure of the thermal unit are required to move slowly. This energy does not result in any MW generation from the unit and is considered into the unit commitment problem as a "start-up cost." The start-up cost varies from a maximum "cold-start" value to a much smaller value, if the unit was only turned off recently and is still relatively close to operating temperature.

There are two approaches for treating a thermal unit during its down period. The first approach allows the unit's boiler to cool down and then heat back, it up to the operating temperature, in time for a scheduled turns on start-up cost when cooling is given by

$$F_t^{CS} = (1 - e^{-t/\alpha})F + C_f \tag{10.1}$$

where, F_t^{CS} is the cold-start cost for cooling model. C_f is fixed cost of generator operation including crew expense, maintenance expenses. F is fuel cost. α is thermal time constant for the unit. t is time (h) the unit was cooled.

The second approach also called banking requires that sufficient energy should be given to the boiler to just maintain operating temperature. The costs for the two are compared so that, if possible, the best approach either cooling or banking can be selected.

Start-up cost when banking is given by

$$F_t^{BS} = F_o t + C_f \tag{10.2}$$

where, F_t^{BS} is the start up cost for banking model. F_o is cost of maintaining unit at operating temperature.

Up to a certain number of hours, the cost of banking is less than the cost of cooling. Due to maintenance or unscheduled outages of various equipment in the plant the capacity limits of thermal units change frequently. It is also taken into account in unit commitment. The variation of starting cost with starting at cooling or banking is shown in Figure 10.1.

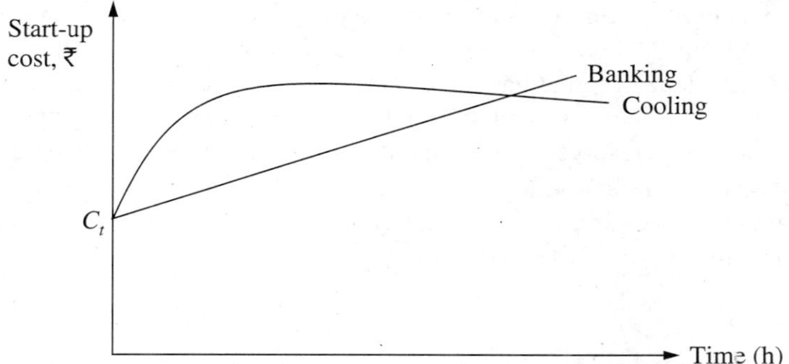

FIGURE 10.1 Startup cost of the thermal unit.

Fuel Constraints

A system in which some units have limited fuel, or else have constraints to utilize a specified amount of fuel in a given time, presents a most challenging unit commitment problem.

Hydro-constraints

The UC problem is of importance for the scheduling of thermal units. It does not mean that UC problem cannot be completely separated from the scheduling of a hydro-unit. Operation of a system having both hydro and thermal plants is, however, far more complex as hydro-plants have negligible operation costs, but are required to operate under constraints of water available for hydro-generation in a given period of time. The problem of minimizing the operating cost of a hydro-thermal system can be viewed as one of minimizing the fuel cost of thermal plants under the constraint of water availability for hydro-generation over a given period of operation.

Derating Constraint

The capacity limit of a thermal unit may change frequently according to maintenance or unscheduled outages of various equipment in the plant.

Environmental Constraints

Environmental constraints receive great attention nowadays. The Clean Air Act Amendment was passed by the United States congress in 1990. This requires utilities to limit emissions to a specific number of tons per year. The Amendments cover a countless air quality concerns including the control of the constituents of "acid rain", specifically Sulphur dioxide (SO_2) and oxides of nitrogen (NOx) in power plant emissions. To reduce power plant emissions, the use of natural gas is preferred to coal and oil-fired generating units. In addition, for existing coal power plants, either the installation of scrubbers to remove SO_2 or switching to coal with lower sulphur content are required. These and other environmental constraints were recently included in many research works on generation scheduling.

10.2 UNIT COMMITMENT PROBLEM

The unit commitment problem determines the combination of available generating units and scheduling their respective outputs to satisfy the forecasted demand with the minimum total production cost under the operating constraints enforced by the system for a specified period. The period usually varies from 24 hours to one week. The constraints to be satisfied are usually the status restriction of individual generating units, minimum up-time, minimum down-time, capacity limits, generation limit for the first and last hour, limited ramp rate, group constraint, power balance constraint, spinning reserve constraint, hydro constraint, etc. [Hobbs et al., 1988]. The high dimensionality and combinatorial nature of the unit commitment problem curtail attempts to develop any rigorous mathematical optimization method that is capable to solve the unit commitment problem for any real-size system.

Objective Function

The objective of the UC problem is to minimize the total operating costs subjected to a set of system and unit constraints over the scheduling horizon. It is assumed that the production cost is a quadratic function of the generator power output [Pappala and Erlich, 2008; Seki et al., 2010].

The total operating costs is the sum of the production costs and the start-up costs. The generator start-up cost depends on the time the unit has been switched-off prior to the start up. The start-up cost at any given time is assumed to be an exponential cost curve. Cost function is minimizing the total operating cost, subject to a number of system and unit constraints [Pappala and Erlich, 2008]. Total operating costs, C_T for the scheduling period, T is given by

$$F_T = \sum_{t=1}^{T} \sum_{i=1}^{N} [F_i(P_{ti})\, U_{ti} + F_{ti}^S U_{ti}(1 - U_{t-1,\,i})] \tag{10.3}$$

where, production cost, $F_i(P_{ti})$ of ith unit at time t is given by

$$F_i(P_{ti}) = (a_i P_{ti}^2 + b_i P_{ti} + c_i) \text{ ₹/h} \tag{10.4}$$

and starting cost, F_{ti}^S for ith unit at time t is given by

$$F_{ti}^S = \begin{cases} F_i^{HS}\ ;\ MDT_i \le T_{t-1,\,i}^{OFF} < MDT_i + T_i^{CO} \\ F_i^{CS}\ ;\ T_{t-1,\,i}^{OFF} \ge MDT_i + T_i^{CO} \end{cases} \tag{10.5}$$

where a_i, b_i and c_i are unit cost coefficients for ith thermal unit. F_i^{HS} is hot start-up cost for ith thermal unit. F_i^{CS} is cold start-up cost for ith thermal unit. MDT_i is minimum down time for ith unit. N is number of units those can be committed or de-committed. P_{ti} is generated power by ith unit at time t. T_i^{CO} is cold start up time for ith unit. T is scheduling period. U_{ti} is binary variable to indicate the on/off state of ith unit at time t and mathematically can be represented as

$$U_{ti} = \begin{cases} 1 \ ;\ \text{if unit is committed } (ON) \\ 0 \ ;\ \text{if unit is decommitted } (OFF) \end{cases} \tag{10.6}$$

Depending on the nature of power system under study, the UC problem is subjected to other constraints, the main constraints pertaining to UC problem are power balance constraints and spinning reserve constraints. The other constraints include the thermal constraints, fuel constraints, security constraints, etc. [Hobbs et al., 1988].

Power Balance Constraint

The total generated power at each hour must be equal to the load demand of the corresponding hour [Senjyu et al., 2002].

$$\sum_{i=1}^{N} U_{ti}P_{ti} = P_{Dt} \quad (t = 1, 2, \ldots, T) \tag{10.7}$$

where P_{Dt} is power demand and transmission loss at particular time t. P_{ti} is generated power at time t by ith unit. U_{ti} is binary variable to indicate the committed (on)/de-committed (off) state of ith unit at time t.

Spinning Reserve Constraint

For reliable operation, the power system has to maintain a certain megawatt capacity as spinning reserve [Nayak and Sharma, 2000].

$$\sum_{i=1}^{N} U_{ti}P_i^{\max} \geq P_{Dt} + R_t \quad (t = 1, 2, \ldots, T) \tag{10.8}$$

where R_t is spinning reserve at time t. U_{ti} is binary value to represent committed (on) or decommited (off) state of unit at time t. $P_i^{\max}$ is the maximum generation of ith units on the electric network. P_{Dt} is the load demand to meet at time t.

Generation Limit Constraint

The generation of the unit is under its minimum and maximum limit.

$$U_{ti}P_i^{\min} \leq P_{ti} \leq U_{ti}P_i^{\max} \quad (i = 1, 2, \ldots, N; t = 1, 2, \ldots, T) \tag{10.9}$$

$P_i^{\min}$ and $P_i^{\max}$ are minimum and maximum generation limits of generating units.

Minimum Up-time

A running unit can't be shut down immediately. In case, the unit is committed, there is a minimum time required before it can be de-committed. It needs some time to reach the off state [Seki et al., 2010]. This time is known as minimum up-time.

$$T_{ti}^{ON} \geq MUT_i \quad (i = 1, 2, \ldots, N; t = 1, 2, \ldots, T) \tag{10.10}$$

This action to commit the unit is stated below:

$$U_{t+1,\,i} = 1\ (ON\ state)\ ;\ \text{If } U_{ti} = 1\ (ON\ state) \text{ and } T_{ti}^{ON} < MUT_i$$

where, T_{ti}^{ON} is continuously on time duration for ith unit up to time t and MUT_i is minimum up-time for ith unit.

Minimum Down-time

A shut down unit can't be started immediately. Once the unit is de-committed, there is a minimum time before it can be recommitted. As it needs some time to reach the committed state. This time is known as minimum down-time.

$$T_{ti}^{OFF} \geq MDT_i \quad (i = 1, 2, \ldots, N; t = 1, 2, \ldots, T) \tag{10.11}$$

This action to de-commit the unit is stated below:

$$U_{t+1,i} = 0\ (OFF\ state)\ ;\ \text{If } U_{ti} = 0\ (OFF\ state) \text{ and } T_{ti}^{OFF} < MDT_i$$

where, T_{ti}^{OFF} is continuously off time duration for ith unit up to time t. MDT_i is minimum down-time for ith unit.

Ramp Rates

To avoid damaging the turbine, the electrical output of a unit cannot change by more than a certain amount over a period of time.

Maximum ramp up rate and ramp down rate constraints are given below:

$$P_{ti} - P_{t-1,i} \le UR_i \quad (i = 1, 2, \ldots, N; t = 1, 2, \ldots, T) \tag{10.12}$$

$$P_{t-1,\,i} - P_{ti} \le DR_i \quad (i = 1, 2, \ldots, N;\ t = 1, 2, \ldots, T) \tag{10.13}$$

where UR_i is ramp up limit. DR_i is ramp down limit.

Initial Status of Unit

The initial unit states at the start of the scheduling period are required to be taken into account.

10.3 ENUMERATION METHOD FOR UNIT-COMMITMENT SOLUTION

A simple but time-consuming method of finding the most economical combination of units to meet a particular load demand is to try all possible combinations of committable units that can supply this load. This load is divided optimally among the units of each combination of committed units by the use of coordination equations to find the most economical operating cost of the committed units. Then, the combination that has the least operating cost among all these is determined. Some combinations will be infeasible if the sum of all maximum MW for the units committed is less than the load or if the sum of all minimum MW for the units committed greater than the load.

Let there are N units to commit and dispatch and there are loading pattern for M periods. The M load levels and operating limits on the N committable units are such that one unit can supply the individual loads and that any combination units can also supply the loads.

Committable unit is treated as binary variable. N-committable units can be represented by N-bits. One set of N-bits combination provides 2^N possible solutions to meet the demand by committing units. For each hourly load there is need to explore $(2^N - 1)$ possible combinations for N-committable because at least one unit should be committed.

For M intervals in total period, the maximum number of possible combinations for N units is $(2^N - 1)^M$.

EXAMPLE 10.1 Let us consider a plant having 3-units. The cost characteristics as and minimum and maximum limits of power generation (MW) of each unit are given below:

$$F_1 = 0.006085P_1^2 + 10.04025P_1 + 136.9125 \text{ ₹/h}; \qquad 5 \text{ MW} \le P_1 \le 150 \text{ MW}$$

$$F_2 = 0.005915P_2^2 + 9.760576P_2 + 59.155 \text{ ₹/h}; \qquad 15 \text{ MW} \le P_2 \le 100 \text{ MW}$$

$$F_3 = 0.00525P_3^2 + 8.6625P_3 + 328.125 \text{ ₹/h}; \qquad 50 \text{ MW} \le P_3 \le 250 \text{ MW}$$

Find the combination of committable units and their generation status to supply a total load of 210 MW most economical. Up- and down-time constrains and spinning reserve constraints are relaxed.

***Solution*:** Number of units are 3, Number of combinations is 8 (2^3). Load Demand is 210 MW.

Case 1: Since all unit are off so can't meet demand.

Case 2: Unit 3 can meet demand of 210 MW.

$$F_T = (0.006085P_1^2 + 10.04025P_1 + 136.9125)\times 0 + (0.005915P_2^2 + 9.760576P_2 + 59.155)\times 0$$
$$+ (0.005250\times(210)^2 + 8.6625\times 210 + 328.125)\times 1 = 2378.775 \text{ ₹/h}$$

Case 3: Unit 2 can supply only 100 MW generation, so demand can't be met.

Case 4: Unit 1 is OFF and units 2 and 3 are ON. Units 2 and 3 supply maximum of 350 MW. Required demand is 210 MW. Perform economic dispatch to compute generation. Compute incremental cost using following relation.

$$\lambda = \frac{P_D + \sum_{i=1}^{3}\left(\frac{b_i}{2a_i}\right)U_i}{\sum_{i=1}^{3}\left(\frac{1}{2a_i}\right)U_i}$$

TABLE 10.1 Unit status for the dispatch of 210 MW

Case	*Status of Unit*			ΣP_i^{min} (MW)	ΣP_i^{max} (MW)	P_1 (MW)	P_2 (MW)	P_1 (MW)	ΣF_i (₹/h)
	U_1	U_2	U_3						
1	0	0	0	0	0	0	0	0	Infeasible
2	0	0	1	50	250	0	0	210	2378.775
3	0	1	0	15	100	0	100	0	Infeasible
4	0	1	1	65	350	0	49.5712	160.4288	2410.494
5	1	0	0	05	150	150	0	0	Infeasible
6	1	0	1	55	400	36.4909	0	173.509	2500.593
7	1	1	0	20	250	110	100	0	2409.331
8	1	1	1	70	500	17.2988	41.4370	151.2544	2544.756

$$\lambda = \frac{210 + \left(\frac{10.04025}{2\times 0.006085}\right)\times 0 + \left(\frac{9.760576}{2\times 0.005915}\right)\times 1 + \left(\frac{8.6625}{2\times 0.00525}\right)\times 1}{\left(\frac{1}{2\times 0.006085}\right)\times 0 + \left(\frac{9.760576}{2\times 0.005915}\right)\times 1 + \left(\frac{8.6625}{2\times 0.00525}\right)\times 1} = 10.347 \text{ (₹/MWh)}$$

Generation is computed using the following expression:

$$P_1 = \left(\frac{10.347 - 10.04025}{2\times 0.006085}\right)\times 0 = 0 \text{ MW}$$

$$P_2 = \left(\frac{10.347 - 9.760576}{2 \times 0.005915}\right) \times 1 = 49.5712 \text{ MW}$$

$$P_3 = \left(\frac{10.3474 - 8.6625}{2 \times 0.00525}\right) \times 1 = 160.4288 \text{ MW}$$

Total operating cost is computed as:

$$\begin{aligned} F_T &= (0.006085P_1^2 + 10.04025P_1 + 136.9125) \times 0 + (0.005915 \times (49.5712)^2 \\ &\quad + 9.760576 \times 49.5712 + 59.1550) \times 1 + (0.005250 \times (160.4288)^2 \\ &\quad + 8.6625 \times 160.4288 + 328.125) \times 1 = 2410.494 \text{ ₹/h} \end{aligned}$$

Case 5: Unit 2 can supply only 150 MW generation, so demand can't be met.

Case 6: Unit 2 is OFF and Units 1 and 3 are ON. Units 1 and 3 supply maximum of 400 MW. Required demand is 210 MW. Perform economic dispatch to compute generation. Compute incremental cost using following relation.

$$\lambda = \frac{210 + \left(\dfrac{10.04025}{2 \times 0.006085}\right) \times 1 + \left(\dfrac{9.760576}{2 \times 0.005915}\right) \times 0 + \left(\dfrac{8.6625}{2 \times 0.00525}\right) \times 1}{\left(\dfrac{1}{2 \times 0.006085}\right) \times 1 + \left(\dfrac{9.760576}{2 \times 0.005915}\right) \times 0 + \left(\dfrac{8.6625}{2 \times 0.00525}\right) \times 1} = 10.48434 \text{ (₹/MWh)}$$

$$P_1 = \left(\frac{10.48434 - 10.04025}{2 \times 0.006085}\right) \times 1 = 36.4909 \text{ MW}$$

$$P_2 = \left(\frac{10.48434 - 9.760576}{2 \times 0.005915}\right) \times 0 = 0 \text{ MW}$$

$$P_3 = \left(\frac{10.48434 - 8.6625}{2 \times 0.00525}\right) \times 1 = 173.509 \text{ MW}$$

Total operating cost is computed as:

$$\begin{aligned} F_T &= (0.006085 \times (36.4909)^2 + 10.04025 \times 36.4909 + 136.9125) \times 1 \\ &\quad + (0.005915P_2^2 + 9.760576P_2 + 59.1550) \times 0 \\ &\quad + (0.005250 \times (173.509)^2 + 8.6625 \times 173.509 + 328.125) \times 1 \\ &= 2500.593 \text{ ₹/h} \end{aligned}$$

Case 7: Unit 3 is OFF and Units 2 and 3 are ON. Units 2 and 3 supply maximum of 250 MW. Required demand is 210 MW. Perform economic dispatch to compute generation. Compute incremental cost using following relation.

$$\lambda = \frac{210 + \left(\dfrac{10.04025}{2 \times 0.006085}\right) \times 1 + \left(\dfrac{9.760576}{2 \times 0.005915}\right) \times 1 + \left(\dfrac{8.6625}{2 \times 0.00525}\right) \times 0}{\left(\dfrac{1}{2 \times 0.006085}\right) \times 1 + \left(\dfrac{9.760576}{2 \times 0.005915}\right) \times 1 + \left(\dfrac{8.6625}{2 \times 0.00525}\right) \times 0} = 11.15818 \text{ (₹/MWh)}$$

Generation is computed using the following expression:

$$P_1 = \left(\frac{10.347 - 10.04025}{2 \times 0.006085}\right) \times 1 = 91.85944 \text{ MW}$$

$$P_2 = \left(\frac{10.347 - 9.760576}{2 \times 0.005915}\right) \times 1 = 118.1406 \text{ MW} > \text{P}_2^{\text{max}} (100 \text{ MW})$$

$$P_3 = \left(\frac{10.3474 - 8.6625}{2 \times 0.00525}\right) \times 0 = 0 \text{ MW}$$

Power generation of unit 2 violates maximum limit so set it to $P_2 = P_2^{\max} = 100$ MW

$$P_1 = P_D - (P_2 + 0) = 210 - 100 = 110 \text{ MW}$$

Total operating cost is computed as:

$$\begin{aligned} F_T &= (0.006085 \times (110)^2 + 10.04025 \times 110 + 136.9125) \times 1 \\ &+ (0.005915 \times (100)^2 + 9.760576 \times 100 + 59.1550) \times 1 \\ &+ (0.005250 P_3^2 + 8.6625 P_3 + 328.125) \times 0 = 2409.331 \text{ ₹/h} \end{aligned}$$

Case 8: Units 1, 2 and 3 are ON. Units 1, 2 and 3 supply maximum of 500 MW. Required demand is 210 MW. Perform economic dispatch to compute generation. Generation values are given in Table 10.1. Incremental cost, λ is 10.25078 ₹/MWh.

It can be observed from Table 10.1, Case 2 provides minimum cost. On performing economic dispatch does not lead to economical dispatch.

10.4 PRIORITY LIST METHOD

The classic UC problem is to minimize total operational cost and is subject to minimum up- and down-time constraints, crew constraints, unit capability limits, generation constraints, and reserve constraints. Thus the objective function of UC consists of the generation cost function and start-up cost function of the generators. The former is described in Section 3.2 of Chapter 3 and Section 7.8 of Chapter 7. The latter involves the cost of the energy that brings the unit online. There are two types of startup cost model: one is bringing the unit online from a cold start. The other is bringing it from bank status, in which the unit is turned off but still close to operating temperature. The start-up cost model when cooling can be expressed by Eq. (10.1). The start-up cost model when banking can be expressed by linear function by Eq. (10.2).

The priorities of generating units are evaluated according to their average production costs. Units are committed one by one according to their priorities for each hour, until the satisfaction of power balance and security constraints. Thereafter, the minimum up- and down-time constraints are checked during this process. Further for additional cost savings, a shut down searching procedure is carried out to de-commit some inefficient units. In the optimization process, the start-up cost is not included. Only small subsets of possible schedules are examined in these methods. Hence, the resulting solution is generally far away from the optimal solution. In addition, this algorithm cannot be applied to systems consisting of fuel constrained and/or hydro units.

The simplest priority list (PL) solution method is based on startup heuristic ordering by operating cost combined with transition cost was elucidated in 1980. The pre-determined order is then used to commit units such that the system load is satisfied. Variation dynamically ranks the units sequentially in this technique. Specific guidelines are involved to establish the ranking process. To evaluate the priority commitment order the Commitment Utilization Factor (CUF) and the classical economic index Average Full Load Cost (AFLC) can be combined. Recently, to solve the problem of thermal unit commitment, the Extended Priority List has been employed. The priority list is obtained by computing the full load average production cost of each unit and sorting them in the ascending order to find a sorted set of units according to their full load average production cost.

Method-I: The average full-load cost α of a unit is defined as the cost per unit of power (₹/MW) when unit is at its full capacity.

$$\alpha_i = \frac{a_i (P_i^{\max})^2 + b_i P_i^{\max} + c_i}{P_i^{\max}} \quad (i = 1, 2, \ldots, N) \tag{10.14}$$

where, a_i, b_i and c_i are unit cost coefficients for ith thermal unit and operating cost is given by Eq. (10.4). $P_i^{\max}$ is the maximum generation limits of generating unit.

Method-II: The priority list is obtained based on the average fuel-cost of unit operating at certain fixed fraction of maximum output.

$$\alpha_i = \frac{a_i (P_i^{\text{avg}})^2 + b_i P_i^{\text{avg}} + c_i}{P_i^{\text{avg}}} \quad (i = 1, 2, \ldots, N) \tag{10.15}$$

where, $P_i^{\text{avg}} = \frac{1}{2}(P_i^{\min} + P_i^{\max})$. $P_i^{\min}$ is the minimum limit of generation of generating unit.

Method-III: Priority order in which each generator can be committed depends on its average production cost defined below:

$$\alpha_i = \frac{1}{q_i}(a_i (P_i^{\max})^2 + b_i P_i^{\max} + c_i) \quad (i = 1, 2, \ldots, N) \tag{10.16}$$

where, $q_i = \frac{P_i^{\max}}{2}\left(1 + \frac{P_i^{\min}}{P_i^{\max}}\right)$

Method-IV: The priority list is prescribed so that the unit with lower average operation cost has a higher priority to start-up. A method based on the weight between average full-load cost and maximal power output is used to determine priority list.

$$\sigma_i = \mu_1 \frac{P_i^{\max}}{\sum_{i=1}^{N_G} P_i^{\max}} + \mu_2 \frac{(1/\alpha_i^{\max})}{\sum_{i=1}^{N_G} (1/\alpha_i^{\max})} \quad (i = 1, 2, \ldots, N) \tag{10.17}$$

where, $\alpha_i = \frac{a_i (P_i^{\max})^2 + b_i P_i^{\max} + c_i}{P_i^{\max}}$ and $\mu_1 + \mu_2 = 1$

Method-V: Unit operates with maximum power generation and their increasing order is computed from incremental cost.

$$\alpha_i = 2a_i P_i^{\max} + b_i \quad (i = 1, 2, \ldots, N) \tag{10.18}$$

EXAMPLE 10.2 Let us consider a plant having 3-units. The cost characteristics and minimum and maximum limits of power generation (MW) of each unit are given in the statement of Example 10.1. Find the combination of units and their generation status to supply a total load of 210 MW most economical using priority list based on average full-load cost. Units can be started or switched off at any time. Required spinning reserve is 10% of load demand.

Solution: Number of units are 3. Number of combinations is $8(2^3)$. Load Demand is 210 MW.

TABLE 10.2 Unit commitment table and generation schedule

Unit No.	α_i (₹/MWh)	*Priority Order*	*Status of Unit,* U_i	*Generation* $P_i^{\max}$ (MW)	*Generation* P_i (MW)	
					Maximum capacity based	*Economic dispatch*
1	11.86575	3	0	0	0	0
2	10.94363	1	1	100	100	49.5712
3	11.2875	2	1	250	110	160.4288
Reserves required					21 MW	21 MW
Reserves available					140 MW	140 MW

The average full-load cost, α_i of *i*th unit is defined as the cost per unit of power (₹/MW) when unit is at its full capacity.

$$\alpha_i = \frac{a_i(P_i^{\max})^2 + b_i P_i^{\max} + c_i}{P_i^{\max}} \Rightarrow \alpha_i = a_i P_i^{\max} + b_i + \frac{c_i}{P_i^{\max}} \quad (i = 1, 2, 3)$$

$$\alpha_1 = 0.006085 \times 150 + 10.04025 + \frac{136.9125}{150} = 11.86575 \text{ ₹/MWh}$$

$$\alpha_2 = 0.005915 \times 100 + 9.760576 + \frac{59.155}{100} = 10.94363 \text{ ₹/MWh}$$

$$\alpha_3 = 0.00525 \times 250 + 8.6625 + \frac{328.125}{250} = 11.2875 \text{ ₹/MWh}$$

Priority order of units is 2, 3 and 1.

Commit unit 2 having highest priority so that $U_2 = 1$ and allocate maximum generation capacity, i.e., 100 MW.

$$P_D - P_2^{\max} = 210 - 100 = 110 \text{ MW} (> 0. \text{ It means demand is not met.})$$

Demand is still unmet then commit unit 3 having next priority so that $U_3 = 1$ and allocate maximum generation capacity, i.e. 250 MW.

$$P_D - U_2 P_2^{\max} - U_3 P_3^{\max} = 210 - 1 \times 100 - 1 \times 250 = -140 \text{ MW} (< 0. \text{ It means demand is met})$$

So, $U_1 = 0$ (OFF), $U_2 = 1$ (ON) and $U_3 = 1$ (ON)

So, Unit 1 is not committed. Unit 2 will generate $P_2 = 100$ MW and unit 3 generates $P_3 = 110$ MW.

$$F_T = \sum_{i=1}^{3} (a_i P_i^2 + b_i P_i + c_i) U_i \text{ ₹/h}$$

$$F_T = (a_1 P_1^2 + b_1 P_1 + c_1) \times 0 + (a_2 P_2^2 + b_2 P_2 + c_2) \times 1 + (a_3 P_3^2 + b_3 P_3 + c_3) \times 1$$

$$\begin{aligned} F_T = {} & (0.006085 P_1^2 + 10.04025 P_1 + 136.9125) \times 0 \\ & + (0.005915 \times (100)^2 + 9.760576 \times 100 + 59.1550) \times 1 \\ & + (0.005250 \times (110)^2 + 8.6625 \times 110 + 328.125) \times 1 \end{aligned}$$

$$F_T = 2438.888 \text{ ₹/h}$$

The priority order, status of units, maximum capacity of units, and generation allocation to units are given in Table 10.2. Available and required reserves are also mentioned in Table 10.2. On performing economic dispatch, incremental cost is computed as under:

$$\lambda = \frac{P_D + \sum_{i=1}^{3} \left(\frac{b_i}{2a_i} \right) U_i}{\sum_{i=1}^{3} \left(\frac{1}{2a_i} \right) U_i}$$

$$\lambda = \frac{210 + \left(\frac{10.04025}{2 \times 0.006085} \right) \times 0 + \left(\frac{9.760576}{2 \times 0.005915} \right) \times 1 + \left(\frac{8.6625}{2 \times 0.00525} \right) \times 1}{\left(\frac{1}{2 \times 0.006085} \right) \times 0 + \left(\frac{9.760576}{2 \times 0.005915} \right) \times 1 + \left(\frac{8.6625}{2 \times 0.00525} \right) \times 1} = 10.347 \text{ (₹/MWh)}$$

Generation is computed using the following expression:

$$P_1 = \left(\frac{10.347 - 10.04025}{2 \times 0.006085} \right) \times 0 = 0 \text{ MW}$$

$$P_2 = \left(\frac{10.347 - 9.760576}{2 \times 0.005915} \right) \times 1 = 49.5712 \text{ MW}$$

$$P_3 = \left(\frac{10.3474 - 8.6625}{2 \times 0.00525} \right) \times 1 = 160.4288 \text{ MW}$$

Total operating cost is computed as:

$$\begin{aligned} F_T = {} & (0.006085 P_1^2 + 10.04025 P_1 + 136.9125) \times 0 \\ & + (0.005915 \times (49.5712)^2 + 9.760576 \times 49.5712 + 59.1550) \times 1 \\ & + (0.005250 \times (160.4288)^2 + 8.6625 \times 160.4288 + 328.125) \times 1 \\ = {} & 2410.494 \text{ ₹/h} \quad \text{(Ans.)} \end{aligned}$$

The priority order, status of units, maximum capacity of units, and generation allocation to units by performing economic dispatch are given in Table 10.2.

EXAMPLE 10.3 Let us consider a plant having 3-units. The cost characteristics and minimum and maximum limits of power generation (MW) of each unit are given in the statement of Example 10.1. Find the combination of units and their generation status to supply a total load of 210 MW most economical using priority list based on the weight between average full-load cost and maximal power output. Units can be started or switched off at any time. Required spinning reserve is 10% of load demand.

Solution: Number of units are 3. Number of combinations is 8 (2^3). Load Demand is 210 MW. The priority list is prescribed so that the unit with lower average operation cost has a higher priority to start-up. A method based on the weight between average full-load cost and maximal power output is used to determine priority list.

$$\sigma_i = \mu_1 \frac{P_i^{\max}}{\sum_{i=1}^{N_G} P_i^{\max}} + \mu_2 \frac{(1/\alpha_i^{\max})}{\sum_{i=1}^{N_G} (1/\alpha_i^{\max})} \qquad (i = 1, 2, 3)$$

$$\alpha_1 = 0.006085 \times 150 + 10.04025 + \frac{136.9125}{150} = 11.86575 \text{ ₹/MWh}$$

$$\alpha_2 = 0.005915 \times 100 + 9.760576 + \frac{59.155}{100} = 10.94363 \text{ ₹/MWh}$$

$$\alpha_3 = 0.00525 \times 250 + 8.6625 + \frac{328.125}{250} = 11.2875 \text{ ₹/MWh}$$

$$\sigma_1 = 0.9 \times \frac{150}{150 + 100 + 250} + 0.1 \times \frac{(1/11.86575)}{(1/11.86575) + (1/10.94363) + (1/11.2875)} = 0.30189$$

$$\sigma_2 = 0.9 \times \frac{100}{150 + 100 + 250} + 0.1 \times \frac{(1/10.94363)}{(1/11.86575) + (1/10.94363) + (1/11.2875)} = 0.21458$$

$$\sigma_2 = 0.9 \times \frac{100}{150 + 100 + 250} + 0.1 \times \frac{(1/11.2875)}{(1/11.86575) + (1/10.94363) + (1/11.2875)} = 0.483527$$

Priority order of units is 2, 1 and 3.

Commit unit 2 having highest priority so that $U_2 = 1$ and allocate maximum generation capacity, i.e., 100 MW.

$$P_D - P_2^{\max} = 210 - 100 = 110 \text{ MW } (> 0 \text{. It means demand is not met})$$

Demand is still unmet then commit unit 1 having next priority so that $U_1 = 1$ and allocate maximum generation capacity, i.e., 150 MW.

$$P_D - U_2 P_2^{\max} - U_3 P_3^{\max} = 210 - 1 \times 100 - 1 \times 150 = - 40 \text{ MW } (< 0 \text{. It means demand is met})$$

So, $U_1 = 1$ (ON), $U_2 = 1$ (ON) and $U_3 = 0$ (OFF).

So Unit 1 will generate $P_1 = 110$ MW and unit 2 will generate $P_2 = 100$ MW and unit 3 is not committed.

$$F_T = \sum_{i=1}^{3} (a_i P_i^2 + b_i P_i + c_i) U_i \text{ ₹/h}$$

$$F_T = (a_1 P_1^2 + b_1 P_1 + c_1) \times 1 + (a_2 P_2^2 + b_2 P_2 + c_2) \times 1 + (a_3 P_3^2 + b_3 P_3 + c_3) \times 0$$

$$F_T = (0.006085 \times (110)^2 + 10.04025 \times 110 + 136.9125) \times 1$$

$$+ (0.005915 \times (100)^2 + 9.760576 \times 100 + 59.1550) \times 1$$

$$+ (0.005250 P_1^2 + 8.6625 P_1 + 328.125) \times 0$$

$$F_T = 2409.3310 \text{ ₹/h}$$

The priority order, status of units, maximum capacity of units, and generation allocation to units are given in Table 10.3. Available and required reserves are also mentioned in Table 10.3.

On performing economic dispatch, incremental cost is computed as under:

$$\lambda = \frac{210 + \left(\dfrac{10.04025}{2 \times 0.006085}\right) \times 1 + \left(\dfrac{9.760576}{2 \times 0.005915}\right) \times 1 + \left(\dfrac{8.6625}{2 \times 0.00525}\right) \times 0}{\left(\dfrac{1}{2 \times 0.006085}\right) \times 1 + \left(\dfrac{9.760576}{2 \times 0.005915}\right) \times 1 + \left(\dfrac{8.6625}{2 \times 0.00525}\right) \times 0} = 11.15818 \text{ (₹/MWh)}$$

TABLE 10.3 Unit commitment table and generation schedule using method-IV

Unit No.	σ_i	*Priority Order*	Status of Unit, U_i	Generation P_i^{max} (MW)	Generation P_i (MW)	
					Maximum capacity based	Economic dispatch
1	0.301893	2	1	150	110	110
2	0.21458	1	1	100	100	100
3	0.483527	3	0	0	0	0
Required Reserve					21	21
Available Reserve					40	40

Generation is computed using the following expression:

$$P_1 = \left(\frac{11.15818 - 10.04025}{2 \times 0.006085}\right) \times 1 = 91.85944 \text{ MW}$$

$$P_2 = \left(\frac{11.15818 - 9.760576}{2 \times 0.005915}\right) \times 1 = 118.1406 \text{ MW}$$

$$P_3 = \left(\frac{11.15818 - 8.6625}{2 \times 0.00525}\right) \times 0 = 0 \text{ MW}$$

Generator violates maximum generation limit, i.e., 100 MW, so set the generation to $P_1 = P_1^{max} = 100$ MW.

So, P_2 = 100 MW and $P_1 = P_D = -P_2$ = 110 MW

So total operating cost, F_T = 2409.3310 ₹/h (Ans.)

The priority order, status of units, maximum capacity of units, and generation allocation to units by performing economic dispatch are given in Table 10.3.

EXAMPLE 10.4 Let us consider a plant having 3 units. The cost characteristics as and minimum and maximum limits of power generation (MW) of each unit are given below:

$$F_1 = 0.001562P_1^2 + 7.92P_1 + 561 \text{ ₹/h}, \quad 100 \text{ MW} \le P_1 \le 600 \text{ MW}$$

$$F_2 = 0.00482P_2^2 + 7.97P_2 + 78 \text{ ₹/h}, \quad 50 \text{ MW} \le P_2 \le 200 \text{ MW}$$

$$F_3 = 0.00194P_3^2 + 7.85P_3 + 310 \text{ ₹/h}, \quad 100 \text{ MW} \le P_3 \le 400 \text{ MW}$$

Generating unit characteristics are given in Table 10.4. Determine generation schedule by committing the units to supply a five hour load pattern given in Table 10.5, which is most economical. Allocation may be done on maximum capacity allocation basis.

Solution: Number of units are 3.

TABLE 10.4 Generating unit characteristics and initial status in system

		Unit 1	*Unit 2*	*Unit 3*
Minimum Up-time, MUT_i (h)		2	2	2
Minimum Down-time MDT_i (h)		2	2	2
Initial Status: (–) hours offline, (+) hours online		–2	+2	–2
Start-up Cost	Hot, F_i^{HS} (₹)	300	100	200
	Cold, F_i^{CS} (₹)	500	200	300
	Cold start, T_i^{CO} (h)	1	1	1
Full Load average cost, α_i (₹/MWh)		9.792	9.324	9.401

TABLE 10.5 Load profile

	Time interval (h)				
	1	2	3	4	5
Load (MW)	200	540	1050	500	300

The average full-load cost α of a unit is defined as the cost per unit of power (₹/MW) when unit is at its full capacity.

$$\alpha_i = a_i P_i^{\max} + b_i + \frac{c_i}{P_i^{\max}} \quad (i = 1, 2, 3)$$

$$\alpha_1 = 0.001562 \times 600 + 7.92 + \frac{561}{600} = 9.792 \text{ ₹/MW}$$

$$\alpha_2 = 0.00482 \times 200 + 7.97 + \frac{78}{200} = 9.324 \text{ ₹/MW}$$

$$\alpha_3 = 0.00194 \times 400 + 7.85 + \frac{310}{400} = 9.401 \text{ ₹/MW}$$

As per priority order, for the 010 state, only unit-2 is committed; for the 011 state, unit-2 and unit-3 are committed; and for the 111 state, unit-2, unit-3, and unit-1 are committed.

Case 1: P_{D1} is 200 MW for 1st hour.
Commit unit 2 having highest priority so that U_{12} = 1 and allocate maximum generation capacity, i.e., 200 MW.

$$P_{D1} - U_{12}P_2^{max} = 200 - 200 = 0 \text{ MW } (= 0. \text{ It means demand is met})$$

Unit 1 is not committed, U_{11} = 0. Unit 2 remains committed, U_{12} = 1 and generates, P_{12} = 200 MW. Unit 3 is not committed, U_{13} = 0. Available reserve is nil. As unit 2 is running from 2 hours so no starting cost is involved.

$$F_{T1} = (0.001562P_{11}^2 + 7.92P_{11} + 561) \times 0 + (0.00482P_{12}^2 + 7.97P_{12} + 78) \times 1$$
$$+ (0.00194P_{13}^2 + 7.85P_{13} + 310) \times 0$$
$$F_{T1} = 1864.8 \text{ ₹/h}$$

Case 2: P_{D2} is 540 MW for 2nd hour.
Commit unit 2 having highest priority so that U_{22} = 1 and allocate maximum generation capacity, i.e., 200 MW.

$$P_{D2} - U_{22}P_2^{max} = 540 - 200 = 340 \text{ MW} (> 0. \text{ It means demand is not met.})$$

Demand is still unmet then commit unit 3 having next priority so that U_{23} = 1 and allocate maximum generation capacity, i.e., 400 MW.

$$P_{D2} - U_{22}P_2^{max} - U_{23}P_3^{max} = 540 - 1 \times 200 - 1 \times 400 = -60 \text{ MW } (< 0. \text{ It means demand is met})$$

Unit 1 is not committed, U_{21} = 0. Unit 2 is committed, U_{22} = 1 and generates, P_{22} = 200 MW. Unit 3 is committed, U_{23} = 1 and generates, P_{23} = 400 MW. 60 MW is available as reserve.

$$F_{T2} = (0.001562P_{21}^2 + 7.92P_{21} + 561) \times 0 + (0.00482P_{22}^2 + 7.97P_{22} + 78) \times 1$$
$$+ (0.00194P_{23}^2 + 7.85P_{23} + 310) \times 1$$
$$F_T = 5068.1 \text{ ₹/h}$$

Unit 2 is already running from 3hrs so starting cost is zero. However unit 3 was not running from 3 hours.

$$C_{23}^S = \begin{cases} F_3^{HS} \;;\; MDT_3 \le T_{13}^{OFF} < MDT_3 + T_3^{CO} \\ F_3^{CS} \;;\; T_{13}^{OFF} \ge MDT_3 + T_3^{CO} \end{cases}$$

Cold start cost, F_{23}^{CS} = 300 ₹ as 3h = 2h + 1h.

Case 3: P_{D3} is 1050 MW for 3rd hour.
Commit unit 2 having highest priority so that U_{32} = 1 and allocate maximum generation capacity, i.e., 200 MW.

$$P_{D3} - U_{32}P_2^{\max} = 1050 - 200 = 950 \text{ MW}(> 0. \text{ It means demand is not met.})$$

Demand is still unmet then commit unit 3 having next priority so that U_{33} = 1 and allocate maximum generation capacity, i.e. 400 MW.

$$P_{D3} - U_{32}P_2^{\max} - U_{33}P_3^{\max} = 1050 - 1\times 200 - 1\times 400 = 450 \text{ MW}(> 0. \text{ It means demand is not met})$$

Demand is still unmet then commit unit 1 having next priority so that U_{31} = 1 and allocate maximum generation capacity, i.e. 600 MW.

$$P_{D3} - U_{32}P_2^{\max} - U_{33}P_3^{\max} - U_{31}P_1^{\max} = 1050 - 1\times 200 - 1\times 400 - 1\times 600 = -150 \text{ MW } (< 0. \text{ It}$$
means demand is met)

Unit 1 is committed, U_{31} = 1 and generates, P_{31} = 450 MW. Unit 2 is committed, U_{32} = 1 and generates, P_{32} = 200 MW and unit 3 is committed, U_{33} = 1 and generates, P_{33} = 400 MW and 150 MW is available as reserve.

$$F_T = (0.001562P_{31}^2 + 7.92P_{31} + 561)\times 1 + (0.00482P_{32}^2 + 7.97P_{32} + 78)\times 1$$
$$+ (0.00194P_{33}^2 + 7.85P_{33} + 310)\times 1$$

$$F_T = 10066.5 \text{ ₹/h}$$

Units 2 and 3 are already running from 3 hrs and 1 hr, respectively, so starting cost is zero. However, unit 1 was not running from 4 hours.

$$C_{31}^S = \begin{cases} F_1^{HS} \ ; & MDT_1 \le T_{21}^{OFF} < MDT_1 + T_1^{CO} \\ F_1^{CS} \ ; & T_{21}^{OFF} \ge MDT_1 + T_1^{CO} \end{cases}$$

Cold start cost, F_1^{CS} = 500 ₹ as 4h > 2h + 1h.

Case 4: P_{D4} is 500 MW for 4th hour.
Commit unit 2 having highest priority so that U_{42} = 1 and allocate maximum generation capacity, i.e. 200 MW.

$$P_{D4} - U_{42}P_2^{\max} = 500 - 200 = 300 \text{ MW } (> 0 \text{ means demand is not met.})$$

Demand is still unmet then commit unit 3 having next priority so that U_{43} = 1 and allocate maximum generation capacity, i.e. 400 MW.

$$P_{D4} - U_{42}P_2^{\max} - U_{43}P_3^{\max} = 500 - 1\times 200 - 1\times 400 = -100 \text{ MW}(< 0. \text{ It means demand is met})$$

Unit 1 is to de-commit, U_{41} = 0. Unit 2 is committed, U_{42} = 1 and generates, P_{42} = 200 MW. Unit 3 is committed, U_{43} = 1 and generates P_{43} = 300 MW. 100 MW is available as reserve.

$$F_T = (0.001562P_{41}^2 + 7.92P_{41} + 561) \times 0 + (0.00482P_{42}^2 + 7.97P_{42} + 78) \times 1$$
$$+ (0.00194P_{43}^2 + 7.85P_{43} + 310) \times 1$$

$$F_T = 4704.4 \text{ ₹/h}$$

Unit 2 is already running from 3 hrs so starting cost is zero. Unit 1 is de-committed.

Case 5: P_{D5} is 300 MW for 5th hour.
Commit unit 2 having highest priority so that U_{52} = 1 and allocate maximum generation capacity, i.e. 200 MW.

$$P_{D5} - U_{52}P_2^{\max} = 300 - 200 = 200 \text{ MW } (> 0 \text{ means demand is not met})$$

Demand is still unmet then commit unit 3 having next priority so that U_{53} = 1 and allocate maximum generation capacity, i.e. 400 MW.

$$P_{D5} - U_{52}P_2^{\max} - U_{53}P_3^{\max} = 300 - 1 \times 200 - 1 \times 400 = -300 \text{ MW } (< 0 \text{. It means demand is met})$$

Unit 1 is to de-commit, U_{51} = 0. Unit 2 is committed, U_{52} = 1 and generates, P_{52} = 200 MW. Unit 3 is committed, U_{53} = 1 and generates, P_{53} = 100 MW. 300 MW is available as reserve.

$$F_T = (0.001562P_{51}^2 + 7.92P_{51} + 561) \times 0 + (0.00482P_{52}^2 + 7.97P_{52} + 78) \times 1$$
$$+ (0.00194P_{53}^2 + 7.85P_{53} + 310) \times 1$$

$$F_T = 2979.2 \text{ ₹/h}$$

Unit 2 is already running from 3 hrs so starting cost is zero. Unit 1 is already de-committed.

Table 10.6 details the status of units, generation schedule and cost. Table 10.7 gives the detail of on and off times and suggests modification in the status of U_{41} from 0 to 1. Table 10.8 details the revised status of units, generation schedule and cost when minimum up-down time constraints are satisfied.

$$\sum_{i=1}^{3} U_{ti}P_i^{\max} \geq P_{Dt} + R_t \quad (t = 1, 2, \ldots, 5)$$

$$U_{11}P_1^{\max} + U_{12}P_2^{\max} + U_{13}P_3^{\max} \geq P_{D1} + R_1 \quad \text{or} \quad 0 \times 600 + 1 \times 200 + 0 \times 400 < 200 + 20$$

Commit the Unit 3 having the next priority and satisfies the minimum down time, i.e. $T_{03}^{OFF} \geq MDT_3$ or $2 \geq 2$ so $U_{13} = 1$.

$$C_{13}^S = \begin{cases} F_3^{HS} & ; MDT_3 \leq T_{03}^{OFF} < MDT_3 + T_3^{CO} \\ F_3^{CS} & ; T_{03}^{OFF} \geq MDT_3 + T_3^{CO} \end{cases}$$

TABLE 10.6 Priority list solution for Example 10.4 relaxing minimum-up and down-time and spinning reserve constraints

Time, t (*h*)	Load, P_{Dt} (MW)	*Status of Units*			*Generating Power* (MW)			*Reserve* (MW)	*Start-up Cost*, C_{ti}^S (₹)	*Operating Cost* (₹)	*Overall Cost* (₹)
		U_{t1}	U_{t2}	U_{t3}	P_{t1}	P_{t2}	P_{t3}				
1	200	0	1	0	0	200	0	0	0	1864.8	1864.8
2	540	0	1	1	0	200	340	60	300	5068.1	5368.1
3	1050	1	1	1	450	200	400	150	500	10066.5	10566.5
4	500	0	1	1	0	200	300	100	0	4704.4	4704.4
5	300	0	1	1	0	200	100	300	0	2979.2	2979.2
Total Cost											25483.0

TABLE 10.7 On/off time and correction as per minimum up- and down-time

Time (*h*)	*Load* (MW)	*Status of Units*			*On time of units*			*Off time of units*			*Remarks*
		U_{t1}	U_{t2}	U_{t3}	T_{t1}^{ON}	T_{t2}^{ON}	T_{t3}^{ON}	T_{t1}^{OFF}	T_{t2}^{OFF}	T_{t3}^{OFF}	
0	—	0	1	0	0	2	0	2	0	2	
1	200	0	1	0	0	3	0	3	0	3	$T_{02}^{ON} \geq MUT_2$ or 2h = 2h so $U_{12} = 1$
2	540	0	1	1	0	4	1	4	0	0	$T_{13}^{OFF} \geq MDT_3$ or 3h ≥ 2h so $U_{23} = 1$
3	1050	1	1	1	1	5	2	0	0	0	$T_{21}^{OFF} \geq MDT_1$ or 4h ≥ 2h so $U_{31} = 1$
4	500	0/1	1	1	0	6	3	1	0	0	$T_{31}^{ON} \geq MUT_1$ or 1h < 2h so $U_{41} = 1$
5	300	0	1	1	0	7	4	2	0	0	$T_{41}^{ON} \geq MUT_1$ or 2h = 2h so $U_{51} = 0$

TABLE 10.8 Priority list solution for Example 10.2. Minimum-up and down-time constraints are considered. Spinning reserve constraints are relaxed

Time (*h*)	*Load* (MW)	*Status of Units*			*Generating Power* (MW)			*Reserve* (MW)	*Start-up Cost* (₹/h)	*Operating Cost* (₹/h)	*Overall Cost* (₹/h)
		U_1	U_2	U_3	P_1	P_2	P_3				
1	200	0	1	0	0	200	0	0	0	1864.8	1864.8
2	540	0	1	1	0	200	340	60	300	5068.1	5368.1
3	1050	1	1	1	450	200	400	150	500	10066.5	10566.5
4	500	1	1	1	100	200	200	700	0	5191.02	5191.02
5	300	0	1	1	0	200	100	300	0	2979.2	2979.2
Total Cost (₹)											25969.62

Cold start cost, $F_{23}^{HS} = 200$ ₹ as 2h ≤ 2h < 2h + 1h

$$U_{11}P_1^{\max} + U_{12}P_2^{\max} + U_{13}P_3^{\max} \geq P_{D1} + R_1 \text{ or } 0 \times 600 + 1 \times 200 + 1 \times 400 \geq 200 + 20$$

$U_{21}P_1^{\max} + U_{22}P_2^{\max} + U_{23}P_3^{\max} \geq P_{D2} + R_2$ or $0 \times 600 + 1 \times 200 + 1 \times 400 \geq 540 + 54$ (No change is required)

Unit 3 is already running so no start-up cost is involved.

$U_{31}P_1^{\max} + U_{32}P_2^{\max} + U_{33}P_3^{\max} \geq P_{D3} + R_3$ or $1 \times 600 + 1 \times 200 + 1 \times 400 \geq 1050 + 105$ (No change is required)

$U_{41}P_1^{\max} + U_{42}P_2^{\max} + U_{43}P_3^{\max} \geq P_{D4} + R_4$ or $1 \times 600 + 1 \times 200 + 1 \times 400 \geq 500 + 50$ (No change is required)

$U_{51}P_1^{\max} + U_{52}P_2^{\max} + U_{53}P_3^{\max} \geq P_{D5} + R_5$ or $0 \times 600 + 1 \times 200 + 1 \times 400 \geq 300 + 30$ (No change is required)

Revised status of units and their generation schedule, required and available reserves and cost are given in Table 10.9.

TABLE 10.9 Priority list solution for Example 10.4. Minimum-up and down-time and spinning reserve constraints are considered

Time horizon (h)	*Load* (MW)	*Status of Units*			*Generating Power* (MW)			*Reserve* (MW)		*Start-up Cost* (₹/h)	*Operating Cost* (₹/h)
		U_1	U_2	U_3	P_1	P_2	P_3	Required	Available		
1	200	0	1	1	0	100	100	20	400	200	2037.6
2	540	0	1	1	0	200	340	54	60	0	5068.1
3	1050	1	1	1	450	200	400	100	150	500	10066.5
4	500	1	1	1	100	200	100	50	700	0	5191.02
5	300	0	1	1	0	200	100	30	300	0	2979.2
Total Cost										26042.42	

Cost is increased, when minimum up- and down-constraints are applied. Cost is further increased when spinning reserves along with minimum up- and down-constraints are applied.

EXAMPLE 10.5 Let us consider a plant having 3-units. The cost characteristics and minimum and maximum limits of power generation (MW) of each unit are given in the statement of Example 10.4. Generating unit characteristics are given in Table 10.4. Determine the scheduling of units to supply a five hour load pattern given in Table 10.5, which is most economical. Allocation may be done by performing economic dispatch.

Solution: Number of units are 3. Table 10.10 gives the detail the status of units, generation schedule and cost.

Table 10.7 gives the detail of on and off times and suggests modification in the status of U_{41} from 0 to 1. Table 10.11 gives detail of the revised status of units, generation schedule and cost when minimum up-down time constraints are satisfied. Corrections are made considering spinning reserve requirement and revised status of units and their generation schedule, required and available reserves and cost are given in Table 10.12.

TABLE 10.10 Priority list solution for Example 10.5 Relaxing minimum-up and down-time and spinning reserve constraints

Time, t (h)	Load, P_{Dt} (MW)	Status of Units			Generating Power (MW)			Reserve (MW)	Start-up Cost, C_{ti}^S (₹)	Operating Cost (₹)	Overall Cost (₹)
		U_{t1}	U_{t2}	U_{t3}	P_{t1}	P_{t2}	P_{t3}				
1	200	0	1	0	0	200	0	0	0	1864.8	1864.8
2	540	0	1	1	0	146.0947	393.9055	60	300	5609.423	5909.423
3	1050	1	1	1	494.8293	155.1709	400	150	500	10053.68	10553.68
4	500	0	1	1	0	134.6154	365.3847	100	0	5236.501	5236.501
5	300	0	1	1	0	77.219	222.781	300	0	3438.293	3438.293
Total Cost										270002.697	

TABLE 10.11 Priority list solution for Example 10.5. Minimum-up and down-time constraints are considered. Spinning reserve constraints are relaxed

Time, t (h)	Load (MW)	Status of Units			Generating Power (MW)			Reserve (MW)	Start-up Cost (₹)	Operating Cost (₹)	Overall Cost (₹)
		U_1	U_2	U_3	P_1	P_2	P_3				
1	200	0	1	0	0	200	0	0	0	1864.8	1864.8
2	540	0	1	1	0	146.0947	393.9055	60	300	5609.423	5909.423
3	1050	1	1	1	494.8293	155.1709	400	150	500	10053.68	10553.68
4	500	1	1	1	228.7903	68.9565	202.2528	700	0	5082.327	5082.327
5	300	0	1	1	0	77.219	222.781	300	0	3438.293	3438.293
Total Cost (₹)										26848.523	

TABLE 10.12 Priority list solution for Example 10.5. Minimum-up and down-time and spinning reserve constraints are considered

Time, t (h)	Load (MW)	Status of Units			Generating Power (MW)			Reserve (MW)		Start-up Cost (₹)	Operating Cost (₹)
		U_1	U_2	U_3	P_1	P_2	P_3	Required	Available		
1	200	0	1	1	0	50	150	20	400	200	2580
2	540	0	1	1	0	146.0947	393.9055	60	300	0	5609.423
3	1050	1	1	1	494.8293	155.1709	400	150	500	500	10053.68
4	500	1	1	1	228.7903	68.9565	202.2528	700	0	0	5082.327
5	300	0	1	1	0	77.219	222.781	300	0	0	3438.293
Total Cost										27463.723	

It can be concluded that the allocation of generation may done on the maximum capacity of generator which gives minimum cost.

10.5 DYNAMIC PROGRAMMING

Most of practical problems forces to take decisions for a system sequentially at different points of time, at different points in space, and at different levels. The problems, in which sequential decisions are made, are called *sequential decision problems*. Since these decisions are to be taken at a number of stages that are also referred to as *multistage decision problems*. Dynamic programming is a mathematical technique for the optimization of multistage decision problems. This technique was developed by Richard Bellman in 1957.

The dynamic programming technique represents or decomposes a multistage decision problem as a sequence of single-stage decision problems when applied. Thus an N-variable problem is represented as a sequence of N single-variable problems that are solved successively. In most cases, these N sub-problems are easier to solve than the original problem. The decomposition to N sub-problems is done in such a manner that the optimal solution of the original N-variable problem can be obtained from the optimal solutions of the N one-dimensional problems.

Dynamic programming, can deal with discrete variables, non-convex, non-continuous, and non-differentiable functions. In general, it can also take into account the stochastic variability by a simple modification of the deterministic procedure. The dynamic programming technique suffers from a major drawback, known as the *curse of dimensionality*. Despite, this disadvantage, it is very suitable for the solution of a wide range of complex problems in several areas of decision making.

In a power system problem, the unit commitment table is to be arrived at for the complete load cycle. If the load is assumed to increase in small and finite steps, dynamic programming is useful to compute the unit commitment table. In such case, there is no need to solve the coordination equations while at the same time unit combinations to be tried are much reduced in number. The searching may be in forward or backward direction. For a time period, the combinations of units are known as the states. In forward dynamic programming a excellent economic schedule is obtained by commencing at the preliminary state amassing the total costs, then retracing from the combination of least accumulated cost starting at the last state and finishing at the initial state. The estimation of each and every combination is not convenient. Additionally, several of the combinations are prohibited due to insufficient existing capacity. The benefit of this method is that having the economical running N–1 units, it is simple to find out the economical running of N units. The dynamic programming approach is based on the following recursive relation.

$$F_i(x) = \min[f_i(y) + F_{i-1}(x-y)] \qquad (i = 2, 3, \ldots, N) \tag{10.19}$$

where, $F_n(x)$ is the minimum cost in ₹/h of generating x MW by N units.

$F_n(y)$ is the cost in ₹/h of generating y MW by the ith unit only.

$F_{n-1}(x-y)$ is the minimum cost in ₹/h of generating $(x-y)$ MW by $(i-1)$ units.

Let x represents the power demand P_D(MW). y represents P_i MW power generation by the ith unit. Equation (10.19) can be rewritten as

$$F_i(P_D) = \min[f_i(P_i) + F_{i-1}(P_D - P_i)] \qquad (i = 2, 3, \ldots, N) \tag{10.20}$$

The step by step procedure for dynamic programming approach is as follows:

1. Start considering two units as per priority list.
2. Compute the cost of collective output of the two units in the form of discrete load levels.
3. Determine the most economical combination of the two units for all the possible load levels. At each possible load level, the economic operation is to run either a unit or both units with a certain load sharing between the two units.
4. Obtain the more cost-effective cost curve for the two units in discrete form and it can be treated as cost curve of single equivalent unit.

5. Add the third unit and the cost curve for the combination of three units is obtained by repeating the procedure of steps 3 and 4.
6. Step 5 of procedure is repeated till all the existing units are participated.

EXAMPLE 10.6 Consider, an electric power system network having 4-thermal generating units, whose cost characteristics, and minimum and maximum generating capacities are given below:

$$f_1(P_1) = 0.385P_1^2 + 23.5P_1 \text{ ₹/h} \quad \text{and} \quad 3 \text{ MW} \le P_1 \le 12 \text{ MW}$$

$$f_2(P_2) = 0.8P_2^2 + 26.5P_2 \text{ ₹/h} \quad \text{and} \quad 3 \text{ MW} \le P_2 \le 12 \text{ MW}$$

$$f_3(P_3) = P_3^2 + 30.0P_3 \text{ ₹/h} \quad \text{and} \quad 3 \text{ MW} \le P_3 \le 12 \text{ MW}$$

$$f_4(P_4) = 1.25P_4^2 + 32P_4 \text{ ₹/h} \quad \text{and} \quad 3 \text{ MW} \le P_4 \le 12 \text{ MW}$$

Determine the most economical units to be committed for a load of 9 MW. Let the load change be in steps of 1 MW.

Solution: Generation cost $f_1(P_1)$ of generation, P_1 MW of 1st generator is computed within generator limits.

$$f_1(P_1) = 0.385P_1^2 + 23.5P_1 \text{ ₹/h, where } P_1 = 3, 4, 5, 6, 7, 8 \text{ and } 9$$

Generation cost $f_2(P_2)$ of generation, P_2 MW of 2nd generator is computed within generator limits.

$$f_2(P_2) = 0.8P_2^2 + 26.5P_2 \text{ ₹/h, where } P_2 = 3, 4, 5, 6, 7, 8 \text{ and } 9$$

Generation cost $f_3(P_3)$ of generation, P_3 MW of 3rd generator is computed within generator limits.

$$f_3(P_3) = P_3^2 + 30.0P_3 \text{ ₹/h, where } P_3 = 3, 4, 5, 6, 7, 8 \text{ and } 9$$

Generation cost $f_4(P_4)$ of generation, P_4 MW of 4th generator is computed within generator limits.

$$f_4(P_4) = 1.25P_4^2 + 32P_4 \text{ ₹/h, where } P_4 = 3, 4, 5, 6, 7, 8 \text{ and } 9$$

Generation cost for each generation value is tabulated in Table 10.13. When P_i is zero, it means that ith generating unit is OFF and when P_i is not zero, it means ith generating unit is ON.

TABLE 10.13 Generation cost of individual units for possible generation levels

$f_1(3)$ = 73.965 ₹/h	$f_2(3)$ = 86.7 ₹/h	$f_3(3)$ = 99.0 ₹/h	$f_4(3)$ = 107.25 ₹/h
$f_1(4)$ = 100.160 ₹/h	$f_2(4)$ = 118.8 ₹/h	$f_3(4)$ = 136.0 ₹/h	$f_4(4)$ = 148.0 ₹/h
$f_1(5)$ = 127.125 ₹/h	$f_2(5)$ = 152.5 ₹/h	$f_3(5)$ = 175.0 ₹/h	$f_4(5)$ = 191.25 ₹/h
$f_1(6)$ = 154.860 ₹/h	$f_2(6)$ = 187.8 ₹/h	$f_3(6)$ = 216.0 ₹/h	$f_4(6)$ = 237.0 ₹/h
$f_1(7)$ = 183.365 ₹/h	$f_2(7)$ = 224.7 ₹/h	$f_3(7)$ = 259.0 ₹/h	$f_4(7)$ = 285.25 ₹/h
$f_1(8)$ = 212.640 ₹/h	$f_2(8)$ = 263.2 ₹/h	$f_3(8)$ = 304.0 ₹/h	$f_4(8)$ = 336.0 ₹/h
$f_1(9)$ = 242.685 ₹/h	$f_2(9)$ = 303.3 ₹/h	$f_3(9)$ = 351.0 ₹/h	$f_4(9)$ = 389.25 ₹/h

Generation cost for generation P_1 to meet demand P_D when only generating unit 1 is participating is obtained as given below and is tabulated in Table 10.14.

$$F_1(P_1) = f_1(P_1), \text{ where } P_1 = 4, 5, 6, 7, 8, 9 \text{ MW}$$

To meet the demand P_D = 9 MW, the cost of generation, $F_1(9)$ comes out be 242.685 ₹/h.

Generation cost for generation ($P_1 + P_2$) when two generating units 1 and 2 are participating is obtained from recursive relation.

$$F_2(P_1 + P_2) = f_2(P_2) + F_1(P_1)$$

where, $P_1 = P_D - P_2$ and $P_2 = 0, 3, 4, 5, 6, 7, 8\ 9$ MW.

TABLE 10.14 Generating cost for unit 1

Generation Cost, $F_1(P_1)$	*Feasible P_D = 9 MW*
$F_1(3)$ = 73.965 ₹/h	Infeasible as $P_D < 9$ MW
$F_1(4)$ = 100.160 ₹/h	Infeasible as $P_D < 9$ MW
$F_1(5)$ = 127.125 ₹/h	Infeasible as $P_D < 9$ MW
$F_1(6)$ = 154.860 ₹/h	Infeasible as $P_D < 9$ MW
$F_1(7)$ = 183.365 ₹/h	Infeasible as $P_D < 9$ MW
$F_1(8)$ = 212.640 ₹/h	Infeasible as $P_D < 9$ MW
$F_1(9)$ = 242.685 ₹/h	Feasible as $P_D = 9$ MW

Minimum generation cost is computed as given below when 2 generating units are generating 9 MW (i.e., $P_1 + P_2$) power following various discrete combinations.

$$F_2(9) = \min[\{f_2(0) + F_1(9)\}, \{f_2(3) + F_1(6)\}, \{f_2(4) + F_1(5)\}, \{f_2(5) + F_1(4)\}, \{f_2(6) + F_1(3)\}, \{f_2(9) + F_1(0)\}]$$

$$F_2(9) = \min[242.685, 241.560, 245.925, 252.660, 261.765, 303.300]$$

$$F_2(9) = 241.56 \text{ ₹/h} \qquad \text{(i)}$$

Units 1 and 2 are ON for the obtained minimum cost, and are generating 3 MW and 6 MW power respectively to meet the power demand of 9 MW. Minimum generation cost is computed as given below when 2 units are generating 8 MW (i.e., $P_1 + P_2$) power for various possible discrete combinations as:

$$F_2(8) = \min[\{f_2(0) + F_1(8)\}, \{f_2(3) + F_1(5)\}, \{f_2(4) + F_1(4)\}, \{f_2(5) + F_1(3)\}, \{f_2(8) + F_1(0)\}]$$

$$F_2(8) = \min[212.640, 213.825, 218.960, 226.465, 263.200]$$

$$F_2(8) = 212.640 \text{ ₹/h}$$

Minimum generation cost is computed when 2 generating units are generating power ($P_1 + P_2$) = 7 MW for various possible discrete combinations as:

$$F_2(7) = \min[\{f_2(0) + F_1(7)\}, \{f_2(3) + F_1(4)\}, \{f_2(4) + F_1(3)\}, \{f_2(7) + F_1(0)\}]$$

$$F_2(7) = \min[183.365, 186.860, 192.765, 224.700]$$

$$F_2(7) = 183.365 \text{ ₹/h}$$

Minimum generation cost is computed when 2 generating units are generating power ($P_1 + P_2$) = 6 MW for various possible discrete combinations as:

$$F_2(6) = \min[\{f_2(0) + F_1(6)\}, \{f_2(3) + F_1(3)\}, \{f_2(6) + F_1(0)\}]$$

$$F_2(6) = \min[154.860, 160.665, 187.800]$$

$$F_2(6) = 154.860 \text{ ₹/h}$$

Minimum generation cost is computed when 2 generating units are generating power $(P_1 + P_2)$ = 5 MW for various possible discrete combinations as:

$$F_2(5) = \min[\{f_2(0) + F_1(5)\}, \{f_2(5) + F_1(0)\}]$$

$$F_2(5) = \min[127.125, 152.500]$$

$$F_2(5) = 127.125 \text{ ₹/h}$$

Minimum generation cost is computed when 2 generating units are generating power $(P_1 + P_2)$ = 4 MW for various possible discrete combinations as:

$$F_2(4) = \min[\{f_2(0) + F_1(4)\}, \{f_2(4) + F_1(0)\}]$$

$$F_2(4) = \min[100.160, 118.800]$$

$$F_2(4) = 100.160 \text{ ₹/h}$$

Minimum generation cost is computed as given below when 2 generating units are generating power $(P_1 + P_2)$ = 3 MW for various possible discrete combinations as:

$$F_2(3) = \min[\{f_2(0) + F_1(3)\}, \{f_2(3) + F_1(0)\}]$$

$$F_2(3) = \min[73.965, 86.700]$$

$$F_2(3) = 73.965 \text{ ₹/h}$$

Generation cost to generate $(P_1 + P_2 + P_3)$ when three generating units 1, 2 and 3 are working together to meet demand of 9 MW for various possible discrete combinations can be computed as:

$$F_3(P_1 + P_2 + P_3) = f_3(P_3) + F_2(P_1 + P_2)$$

where, $P_1 + P_2 = P_D - P_3$ and P_3 = 0, 3, 4, 5, 6, 7, 8, 9 MW.

Unit 3 is OFF in case P_3 is zero. Unit 3 is ON if P_3 is not zero.

$$F_3(9) = \min[\{f_3(0) + F_2(9)\}, \{f_3(3) + F_2(6)\}, \{f_3(4) + F_2(5)\}, \{f_3(5) + F_2(4)\}, \{f_3(6) + F_2(3)\}, \{f_3(9) + F_2(0)\}]$$

$$F_3(9) = \min[241.560, 253.86, 263.625, 275.160, 289.965, 351.000]$$

$$F_3(9) = 241.56 \text{ ₹/h} \qquad \text{(ii)}$$

For obtained minimum cost unit 3 remains OFF and unit 1 and unit 2 are ON.

Generation cost to generate $(P_1 + P_2 + P_3 + P_4)$ when all four generating units 1, 2, 3 and 4 are working together to meet demand of 9 MW for various possible discrete combinations can be computed as:

$$F_4(P_1 + P_2 + P_3 + P_4) = f_4(P_4) + F_3(P_1 + P_2 + P_3)$$

where, $P_1 + P_2 + P_3 = P_D - P_4$ and $P_4 = 0,\ 3, 4, 5, 6, 7, 8, 9$ MW.

Unit 4 is OFF in case P_4 is zero. Unit 4 is ON if P_4 is not zero.

$$F_4(9) = \min[\{f_4(0) + F_3(9)\}, \{f_4(3) + F_3(6)\}, \{f_4(4) + F_3(5)\}, \{f_4(5) + F_3(4)\}, \{f_4(6) + F_3(3)\}, \{f_4(9) + F_3(0)\}]$$

$$F_2(9) = \min[241.560, 261.675, 275.125, 291.410, 310.965, 389.250]$$

$$F_2(9) = 241.56 \text{ ₹/h} \qquad \text{(iii)}$$

For obtained minimum cost unit 4 remains OFF and unit 1 and unit 2 are ON and unit 3 is OFF. Summary of minimum generation cost is given in Table 10.15 when units are working together.

TABLE 10.15 Minimum generation cost when units are working together to generate various distinct possible levels.

Cost of unit 1	*Cost of units 1 and 2*	*Cost of units 1, 2 and 3*	*Cost of units 1, 2, 3 and 4*
$F_1(3)$ = 73.965 ₹/h	$F_2(3)$ = 73.965 ₹/h	$F_3(3)$ = 73.965 ₹/h	$F_4(3)$ = 73.965 ₹/h
$F_1(4)$ = 100.160 ₹/h	$F_2(4)$ = 100.160 ₹/h	$F_3(4)$ = 100.160 ₹/h	$F_4(4)$ = 100.160 ₹/h
$F_1(5)$ = 127.125 ₹/h	$F_2(5)$ = 127.125 ₹/h	$F_3(5)$ = 127.125 ₹/h	$F_4(5)$ = 127.125 ₹/h
$F_1(6)$ = 154.860 ₹/h	$F_2(6)$ = 154.860 ₹/h	$F_3(6)$ = 154.860 ₹/h	$F_4(6)$ = 154.860 ₹/h
$F_1(7)$ = 183.365 ₹/h	$F_2(7)$ = 183.365 ₹/h	$F_3(7)$ = 183.365 ₹/h	$F_4(7)$ = 183.365 ₹/h
$F_1(8)$ = 212.640 ₹/h	$F_2(8)$ = 212.640 ₹/h	$F_3(8)$ = 212.640 ₹/h	$F_4(8)$ = 212.640 ₹/h
$F_1(9)$ = 242.685 ₹/h	$F_2(9)$ = 241.560 ₹/h	$F_3(9)$ = 241.560 ₹/h	$F_4(9)$ = 241.560 ₹/h

Status of units and their generation schedule is given Table 10.16. It is concluded that Unit 1 is ON, Unit 2 is ON, unit 3 is OFF and unit 4 is OFF while meeting the demand 9 MW at minimum generation cost of 241.56 ₹/h. (Ans.)

TABLE 10.16 Unit commitment table

Unit, i	1	2	3	4
Status, U_i	ON	ON	OFF	OFF
Generation, P_i MW	3	6	—	—

Dynamic programming decomposes the main problem into a series of single stage (period or hour) decision problems and optimization is performed at each stage. For the start period, the transition cost from the given initial state plus operating cost is computed for each state as accumulated cost. Then, the states with the lowest accumulated cost are saved. The dynamic programming process moves to the next period. For each state, the cost of the transition from each saved strategy at the previous period to that state is determined to find which strategy provides the lowest accumulated cost. In this manner, strategies with the lowest cost from the initial state up to the considered period are saved until the final period is reached. Ultimately, the path with the lowest production cost is obtained and is back tracked. Dynamic programming can be performed in either forward or backward direction. Forward dynamic programming begins at the initial stage and is more appropriate for unit commitment than backward dynamic programming. Since forward dynamic programming is able to consider unit startup cost depending on off line hours and initial conditions can easily be specified.

In general, the dynamic programming has some disadvantages. While it is able to satisfy some constraints such as power balance, spinning reserve and minimum up and down time, additional constraints increase the complexity exponentially. Furthermore, the dynamic programming approach suffers from a major drawback known as the curse of dimensionality, i.e., the number of states to be handled quickly gets out of control with the number of generating units in larger systems.

In addition, only a few of the potential strategies are saved, thus the optimal solution may not be found as illustrated in the following examples. Therefore, a strict priority was imposed to reduce the number of considered states in the problem dramatically. But this still does not guarantee that the obtained solution will be optimal. It can be concluded that the allocation of generation may be done on the maximum capacity of generator which gives minimum cost.

$$F_A(t, I) = F(t, I) + \min_{L}[S_C(t-1, L \Rightarrow t, I) + F_A(t-1, L)] \tag{10.21}$$

where,

$F_A(t, I)$ is total accumulated operating cost to arrive at state I at period t.

$F(t, I)$ is the production cost of state I at period t excluding start-up and shut-down cost.

$S_C(t-1, L \Rightarrow t, I)$ is the transition cost from state L, at time period t–1 to state I at time period t.

$F_A(t-1, L)$ is total accumulated operating cost to arrive at state L at period t–1, where $L \in \Psi_S$ set of saved strategies.

Ψ_S is set of saved strategies.

For a system comprising N generating units considered over T periods, the number of states to search at each period is $2N$–1 for completed enumeration or a lower number denoted as M when a strict priority is imposed. Hence, M is less than $2N-1$. The selection procedure is illustrated in Figure 10.2, limiting the number of states to be searched I = 4 and strategies saved L = 2. At period $t-1$, 4 states (I, II, III, IV) are searched for the lowest cost. The shadow circle indicates the saved states due to their lowest cost: II, and IV. At period t, for each state (I, II, III, IV) is considered to determine which path from the two saved strategies (II, IV) from period $t-1$ will provide the lowest accumulated cost. Saved strategies for period t are states II, and III. Then, the same computation is performed for period $t+1$ to save states I and III.

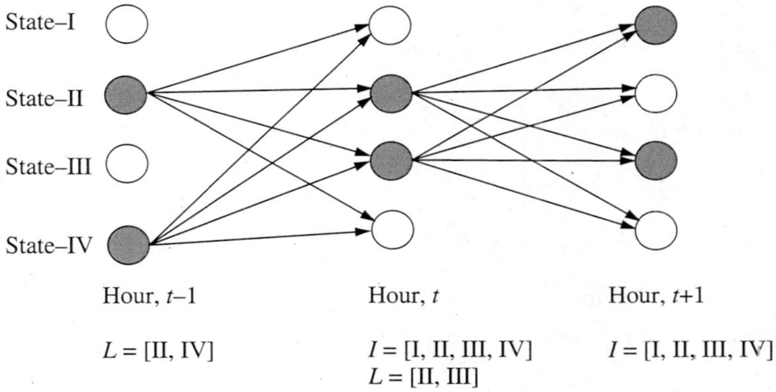

FIGURE 10.2 Search path for states, I = 4 and saved strategies, L = 2.

EXAMPLE 10.7 Consider a plant having 3 units. The cost characteristics as and minimum and maximum limits of power generation (MW) of each unit are given in the statement of Example 10.4. Generating unit characteristics are given in Table 10.4. Using dynamic programming, determine the generation scheduling by committing units to supply a five hour load pattern given in Table 10.5, that is most economical.

Solution: When t is 0 (initial sate), status of units $U = [0, 1, 0]^T$, the initial cost $F_A(0, 1) = 0.0$ ₹ $t = 1$, $P_{D1} = 200$ MW.

TABLE 10.17 Feasible states of generators for $t = 1$ and $P_{D1} = 200$ MW

Possible state I	*Status of units*			*Generation Capacity*		*Generation*			*Production cost* $F(1, I)$ (₹)
	U_{11}	U_{12}	U_{13}	ΣP_i^{min} (MW)	ΣP_i^{max} (MW)	P_{11} (MW)	P_{12} (MW)	P_{13} (MW)	
1	0	1	0	50	200	0	200	0	1864.8
2	0	1	1	150	600	0	100	100	2037.6
3	1	1	1	250	1100	$\Sigma P_{1i}^{min} > P_{D1}$			Not feasible

At time, $t = 1$, there is possibility that saved state, $L =$ [010] moves to states, $I =$ [010, 011]

Let L and I states are represented as per their serial number in Table 10.17.

There is possibility that state, $L =$ [1] moves to states, $I =$ [1, 2]

The minimum total cost when system moves from state, $L = 1$ to state, $I = 1$ at time $t = 1$ is computed as:

$$F_A(1, 1) = F(1, 1) + \min_{1}[S_C(0, L \Rightarrow 1, 1) + F_A(0, L)]$$

$$F_A(1, 1) = F(1, 1) + [S_C(0, 1 \Rightarrow 1, 1) + F_A(0, 1)]$$

$$F_A(1, 1) = 1864.8 + [0] = 1864.8 \text{ ₹}$$

$$F_A(1, 2) = F(1, 2) + \min_{1}[S_C(0, L \Rightarrow 1, 2) + F_A(0, L)]$$

The minimum total cost when system moves from state, $L = 1$ to state, $I = 2$ at time $t = 1$ is computed as

$$F_A(1, 2) = F(1, 2) + [S_C(0, 1 \Rightarrow 1, 2) + F_A(0, 1)]$$

$$F_A(1, 2) = 2037.6 + [300 + 0] = 2337.6 \text{ ₹}$$

$$F_A(1, 1) = \min\{F_A(1, 1), F_A(1, 2)\}$$

At time, $t = 2$, there is possibility that state, $L =$ [010, 011] moves to states, $I =$ [011, 111]

The L and I states are represented as per their serial number in Table 10.18.

TABLE 10.18 Possible status of generators for $t = 2$ and $P_{D2} = 540$ MW

Possible state I	*Status of units*			*Generation Capacity*		*Generation*			*Production cost* $F(2, I)$ (₹)
	U_{21}	U_{22}	U_{23}	ΣP_i^{min} (MW)	ΣP_i^{max} (MW)	P_{21} (MW)	P_{22} (MW)	P_{23} (MW)	
1	0	1	0	50	200	$\Sigma P_{2i}^{max} < P_{D2}$			Not feasible
2	0	1	1	150	600	0	200	340	5068.064
3	1	1	1	250	1100	100	200	240	4170.544

There is possibility that state, $L =$ [1, 2] moves to states, $I =$ [2, 3].

The minimum total cost when system moves from state, $L =$ [1, 2] to state, $I = 2$ at time $t = 2$ is computed as:

$$F_A(2, 2) = F(2, 2) + \min_{1,2}[S_C(1, L \Rightarrow 2, 2) + F_A(1, L)]$$

$$F_A(2, 2) = F(2, 2) + \min\begin{bmatrix} S_C(1, 1 \Rightarrow 2, 2) + F_A(1, 1) \\ S_C(1, 2 \Rightarrow 2, 2) + F_A(1, 2) \end{bmatrix}$$

$$F_A(2, 2) = 5068 + \min\begin{bmatrix} 300 + 1864 \\ 0 + 2337 \end{bmatrix} = 7232 \text{ ₹}$$

The minimum total cost when system moves from state, $L =$ [1, 2] to state, $I = 3$ at time $t = 2$ is computed as

$$F_A(2, 3) = F(2, 3) + \min_{1,2}[S_C(1, L \Rightarrow 2, 3) + F_A(1, L)]$$

$$F_A(2, 3) = F(2, 3) + \min\begin{bmatrix} S_C(1, 1 \Rightarrow 2, 3) + F_A(1, 1) \\ S_C(1, 2 \Rightarrow 2, 3) + F_A(1, 2) \end{bmatrix}$$

$$F_A(2, 3) = 4170.54 + \min\begin{bmatrix} 800 + 1864 \\ 500 + 2337 \end{bmatrix} = 6834.54 \text{ ₹}$$

TABLE 10.19 Possible status of generators for $t = 3$ and $P_{D3} = 1050$ MW

Possible state I	*Status of units*			*Generation Capacity*		*Generation*			*Production cost* $F(3, I)$ (₹)
	U_{31}	U_{32}	U_{33}	ΣP_i^{min} (MW)	ΣP_i^{max} (MW)	P_{31} (MW)	P_{32} (MW)	P_{33} (MW)	
1	0	1	0	50	200	$\Sigma P_{3i}^{max} < P_{D3}$			Not feasible
2	0	1	1	150	600	$\Sigma P_{3i}^{max} < P_{D3}$			Not feasible
3	1	1	1	250	1100	450	200	400	10066.5

At time, $t = 3$, there is possibility that state, $L =$ [011, 111] moves to states, $I =$ [111]

The L and I states are represented as per their serial number in Table 10.19.

There is possibility that state, $L =$ [2, 3] moves to states, $I =$ [3]

The minimum total cost when system moves from state, $L =$ [2, 3] to state, $I = 3$ at time $t = 3$ is computed as:

$$F_A(3, 3) = F(3, 3) + \min_{2,3}[S_C(2, L \Rightarrow 3, 3) + F_A(2, L)]$$

$$F_A(3, 3) = F(3, 3) + \min\begin{bmatrix} S_C(2, 2 \Rightarrow 3, 3) + F_A(2, 2) \\ S_C(2, 3 \Rightarrow 3, 3) + F_A(2, 3) \end{bmatrix}$$

$$F_A(3, 3) = 10066.5 + \min\begin{bmatrix} 500 + 7232 \\ 0 + 6834.54 \end{bmatrix} = 16901.04 \text{ ₹}$$

TABLE 10.20 Possible status of generators for t = 4 and P_{D4} = 500 MW

Possible state I	*Status of units*			*Generation Capacity*		*Generation*			*Production cost F(4, I)* (₹)
	U_{41}	U_{42}	U_{43}	ΣP_i^{min} (MW)	ΣP_i^{max} (MW)	P_{41} (MW)	P_{42} (MW)	P_{43} (MW)	
1	0	1	0	50	200	$\Sigma P_{4i}^{max} < P_{D2}$			Not feasible
2	0	1	1	150	600	0	200	300	4704.4
3	1	1	1	250	1100	100	200	200	5191.2

At time, t = 4, there is possibility that state, L = [111] moves to states, I = [011, 111].

The L and I states are represented as per their serial number in Table 10.20.

There is possibility that state, L = [3] moves to states, I = [2, 3].

The minimum total cost when system moves from state, L = [3] to state, I = 2 at time t = 4 is computed as:

$$F_A(4, 2) = F(4, 2) + \min_{3}[S_C(3, L \Rightarrow 4, 2) + F_A(3, L)]$$

$$F_A(4, 2) = F(4, 2) + \min[S_C(3, 3 \Rightarrow 4, 2) + F_A(3, 3)]$$

$$F_A(4, 2) = 4704.4 + [0 + 16901.04] = 21605.44 \text{ ₹}$$

The minimum total cost when system moves from state, L = [3] to state, I = 3 at time t = 4 is computed as:

$$F_A(4, 3) = F(4, 3) + \min_{3}[S_C(3, L \Rightarrow 4, 3) + F_A(3, L)]$$

$$F_A(4, 3) = F(4, 3) + \min[S_C(3, 3 \Rightarrow 4, 3) + F_A(3, 3)]$$

$$F_A(4, 3) = 5191.2 + [0 + 16901.04] = 22092.24 \text{ ₹}$$

TABLE 10.21 Possible status of generators for t = 5 and P_{D5} = 300 MW

Possible state I	*Status of units*			*Generation Capacity*		*Generation*			*Production cost F(5, I)* (₹)
	U_{51}	U_{52}	U_{53}	ΣP_i^{min} (MW)	ΣP_i^{max} (MW)	P_{51} (MW)	P_{52} (MW)	P_{53} (MW)	
1	0	1	0	50	200	$\Sigma P_{5i}^{max} < P_{D2}$			Not feasible
2	0	1	1	150	600	0	200	100	2979.2
3	1	1	1	250	1100	100	100	100	3406.22

At time, t = 5, there is possibility that state, L = [011, 111] moves to states, I = [011, 111]

The L and I states are represented as per their serial number in Table 10.21.

There is possibility that state, L = [2, 3] moves to states, I = [2, 3]

The minimum total cost when system moves from state, L = [2, 3] to state, I = 2 at time t = 5 is computed as:

$$F_A(5, 2) = F(5, 2) + \min_{2,3}[S_C(4, L \Rightarrow 5, 2) + F_A(4, L)]$$

$$F_A(5, 2) = F(5, 2) + \min\begin{bmatrix} S_C(4,\ 2 \Rightarrow 5,\ 2) + F_A(4, 2) \\ S_C(4,\ 3 \Rightarrow 5,\ 2) + F_A(4, 3) \end{bmatrix}$$

$$F_A(5, 2) = 2979.2 + \min\begin{bmatrix} 0 + 21605.49 \\ 0 + 22092.24 \end{bmatrix} = 24584.69 \text{ ₹}$$

The minimum total cost when system moves from state, L = [2, 3] to state, I = 3 at time t = 5 is computed as:

$$F_A(5, 3) = F(5, 3) + \min_{2,3}[S_C(4, L \Rightarrow 5, 2) + F_A(4,\ L)]$$

$$F_A(5, 3) = F(5, 3) + \min\begin{bmatrix} S_C(4,\ 2 \Rightarrow 5,\ 3) + F_A(4, 2) \\ S_C(4,\ 3 \Rightarrow 5,\ 3) + F_A(4, 3) \end{bmatrix}$$

$$F_A(5, 3) = 3406.22 + \min\begin{bmatrix} 500 + 21605.49 \\ 500 + 22092.24 \end{bmatrix} = 25511.71 \text{ ₹}$$

Figure 10.3 shows the path of saved sates and explored states. Figure 10.4 gives the optimal path.

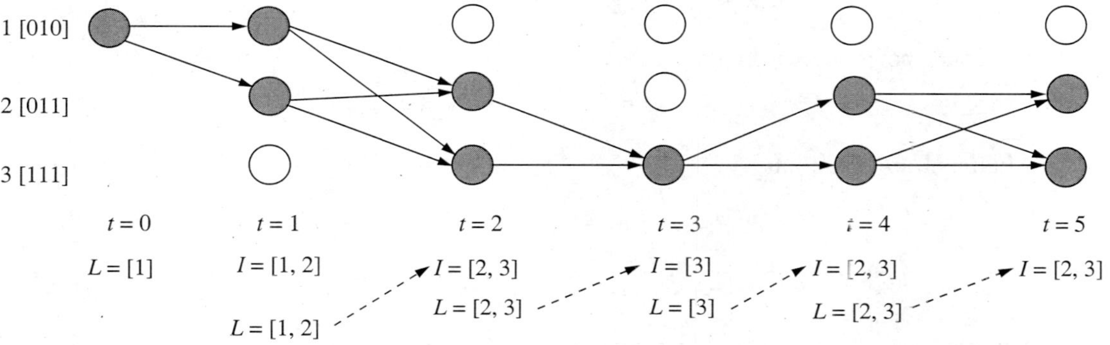

FIGURE 10.3 Search path for states, I = 4 and saved strategies, L = 2.

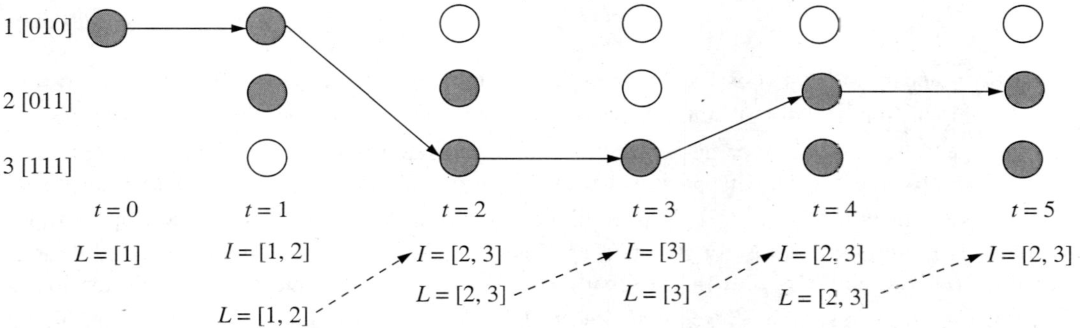

FIGURE 10.4 Search path for states, I = 4 and saved strategies, L = 2.

10.6 LAGRANGE RELAXATION METHOD

The computation burden is huge for large power systems with many generators, when the enumeration approach and dynamic programming is applied to solve unit commitment problem. The priority list is very simple, and has a fast calculation speed, but it may discard the optimum scheme. The Lagrange relaxation method overcomes the above mentioned disadvantages. The mathematical problem of the unit commitment is expressed as below.

Objective function:

$$\text{Minimize } F(P_{ti}, U_{ti}) = \sum_{t=1}^{T} \sum_{i=1}^{N} \left[(a_i P_{ti}^2 + b_i P_{ti} + c_i) U_{ti} + F_{ti}^S U_{ti}(1 - U_{t-1,i})\right] \text{₹} \qquad (10.22a)$$

Constraints:

Load balance equation

$$\sum_{i=1}^{N} P_{ti} U_{ti} = P_{Dt} \qquad (t = 1, 2, \ldots, T) \qquad (10.22b)$$

Power reserve constraints

$$\sum_{i=1}^{N} P_i^{\max} U_{ti} = P_{Dt} + P_{Rt} \qquad (i = 1, 2, \ldots, T) \qquad (10.22c)$$

Generator power output limits

$$U_{ti} P_i^{\min} \le P_{ti} \le U_{ti} P_i^{\max} \qquad (i = 1, 2, \ldots, N;\ t = 1, 2, \ldots, T) \qquad (10.22d)$$

Minimum up-/down-time

$$(T_{t-1,i}^{OFF} - MUT_i)(U_{t-1,i} - U_{ti}) \ge 0 \qquad (i = 1, 2, \ldots, N;\ t = 1, 2, \ldots, T) \qquad (10.22e)$$

$$(T_{t-1,i}^{ON} - MDT_i)(U_{ti} - U_{t-1,i}) \ge 0 \qquad (i = 1, 2, \ldots, N;\ t = 1, 2, \ldots, T) \qquad (10.22f)$$

Starting cost, F_{ti}^S for ith unit at time t is given by

$$F_{ti}^S = \begin{cases} F_i^{HS} \ ; & MDT_i \le T_{t-1,i}^{OFF} < MDT_i + T_i^{CO} \\ F_i^{CS} \ ; & T_{t-1,i}^{OFF} \ge MDT_i + T_i^{CO} \end{cases} \qquad (10.22g)$$

where a_i, b_i and c_i are unit cost coefficients for ith unit. F_i^{HS} is hot start-up cost for ith unit. F_i^{CS} is cold start-up cost for ith unit. MDT_i is minimum down time for ith unit. N is number of units those can be committed or de-committed. P_{ti} is generated power by ith unit at time t. T_i^{CO} is cold start-up time for ith unit. T is scheduling period. U_{ti} is binary variable to indicate the on/off state of ith unit at time t. T_{ti}^{ON} is continuously *ON* time duration for ith unit up to time t and MUT_i is minimum up-time for ith unit. T_{ti}^{OFF} is continuously off time duration for ith unit up to time t. MDT_i is minimum down-time for ith unit. P_{Dt} is power demand and transmission loss at particular time t. R_t is spinning reserve at time t. $P_i^{\max}$ is the maximum generation of ith units on the electric network. $P_i^{\min}$ is the minimum generation of ith units on the electric network.

The UC problem has two kinds of constraints: (i) separable and (ii) coupling constraints. Separable constraints such as capacity and minimum up- and down-time constraints are related to one single unit. On the other hand, coupling constraints involve all units. A change in one unit affects the other units. The power balance and power reserve constraints are examples of coupling constraints. The Lagrange relaxation framework relaxes the coupling constraints and incorporates them into the objective function by a dual optimization procedure. Thus the objective function can be separated into independent functions for each unit, subject to unit capacity and minimum up and down time constraints. The resulting Lagrange function of the UC problem is stated as follows:

$$L(P_{ti}, U_{ti}, \lambda_t, \mu_t) = F(P_{ti}, U_{ti}) + \sum_{t=1}^{T} \lambda_t \left(\mathrm{P_{Dt}} - \sum_{i=1}^{N} P_{ti}U_{ti} \right) + \sum_{t=1}^{T} \mu_t \left(P_{Dt} + P_{Rt} - \sum_{i=1}^{N} P_i^{\max} U_{ti} \right) \quad (10.23)$$

where, λ_t and μ_t are the Lagrange multipliers.

The unit commitment problem becomes the minimization of the Lagrange function Eq. (10.23), subject to constraints Eqs. (10.22d), (10.22e) and (10.22f). The Lagrange relaxation procedure solves the unit commitment problem by relaxing (temporarily ignoring) the coupling constraints. For dual optimization, the function to be optimized is convex and the variables are continuous, then the maximization of the dual function gives the identical result as minimizing the primal function. The application of the dual optimization method to the UC problem has been given the name "Lagrange relaxation". But, for unit commitment problem, there are 0 and 1 integer variables that indicate the status of the units, which are not continuous, or non-convex. Thus the dual theory is not exactly satisfied in UC problem. The dual procedure attempts to find the constrained optimum by maximizing the Lagrangian with respect to Lagrange multipliers while minimizing with respect to the other variables in the problem.

$$q^*(\lambda_t, \mu_t) = \max_{\lambda_t, \mu_t} (q(\lambda_t, \mu_t)) \quad (10.24)$$

where

$$q(\lambda_t, \mu_t) = \min_{P_{ti}, U_{ti}} (L(P_{ti}, U_{ti}, \lambda_t, \mu_t)) \quad (10.25)$$

Equation (10.23) is rewritten after eliminating constant terms like $\lambda_t P_{Dt}$ and $\mu_t(P_{Dt} + P_{Rt})$.

$$q(\lambda_t, \mu_t) = \text{Minimize}\left[F(P_{ti}, U_{ti}) - \sum_{t=1}^{\mathrm{T}} \left(\lambda_t \left(\sum_{i=1}^{N} P_{ti}U_{ti} \right) - \mu_t \left(\sum_{i=1}^{N} P_{\mathrm{i}}^{\max} U_{ti} \right) \right) \right] \quad (10.26)$$

Subject to the constraints given by Equations (10.22d) to (10.22f).

The following two basic steps are followed to solve the UC problem for the Lagrange procedure:

Step 1: Initializing the Lagrange multipliers with values that try to make $q(\lambda_t, \mu_t)$ larger.

Step 2: Assuming the values of the Lagrange multipliers, λ_t and μ_t in step (1) are fixed and the Lagrange function, $L(P_{ti}, U_{ti}, \lambda_t, \mu_t)$ is minimized by adjusting P_{ti} and U_{ti}.

There exists a gap between the results of the maximization of the dual function and minimizing the primal function. The aim of the Lagrange relaxation method is to reduce the duality gap by iterations. If a criterion is pre-specified, this iterative procedure continues until a duality gap criterion is met. The duality gap is also used as a measure of convergence. It is considered that the optimum has been reached, when the relative duality gap between the primal and the dual solutions is less than a specific tolerance.

Adjusting Lagrange Multipliers

To minimize the function $L(P_{ti}, U_{ti}, \lambda_t, \mu_t)$, the following necessary conditions are to be satisfied related to Lagrange multipliers λ_t and μ_t.

$$\frac{\partial L}{\partial \lambda_t} = 0 \Rightarrow P_{Dt} - \sum_{i=1}^{N} P_{ti} U_{ti} = 0$$

$$\frac{\partial L}{\partial \mu_t} = 0 \Rightarrow P_{Dt} + P_{Rt} - \sum_{i=1}^{N} \mathrm{P}_{\mathrm{i}}^{\max} U_{ti} = 0$$

By applying iterative method, the values of Lagrange multipliers λ_t and μ_t can be obtained as:

$$\lambda_t = \lambda_t + \alpha \left(P_{Dt} - \sum_{i=1}^{N} P_{ti} U_{ti} \right) \tag{10.27}$$

$$\mu_t = \mu_t + \beta \left(P_{Dt} + P_{Rt} - \sum_{i=1}^{N} P_i^{\max} U_{ti} \right) \tag{10.28}$$

where α and β are the multipliers. The values of α and β are taken for 0.01 if multiplying terms are positive. The values of α and β are taken for 0.005 if multiplying terms are negative.

Dual Problem Solution

In the conventional Lagrangian relaxation method, dynamic programming (DP) is used to solve the dual optimization. This results is decided in shorter computation time by following on/off decision criteria to obtain a dual solution instead of dynamic programming. The following objective function is minimized that depend on generation.

$$\text{Minimize } [F_i(P_{ti}) - \lambda_t P_{ti}]$$

To find the dual power of each generating unit, the optimality condition and actual dual power are:

$$\frac{d}{dP_{ti}} [F_i(P_{ti}) - \lambda_t P_{ti}] = 0 \quad (i = 1, 2, \ldots, N; t = 1, 2, \ldots, T) \tag{10.29}$$

$$\frac{d}{dP_{ti}} [F_i(P_{ti})] = \lambda_t \quad (i = 1, 2, \ldots, N; t = 1, 2, \ldots, T) \tag{10.30}$$

$$2a_i P_{ti}^{opt} + b_i = \lambda_t \quad (i = 1, 2, \ldots, N; t = 1, 2, \ldots, T) \quad \text{where, } F_i(P_{ti}) = a_i P_{ti}^2 + b_i P_{ti} + c_i$$

$$P_{ti}^{opt} = \frac{\lambda_t - b_i}{2a_i} \quad (i = 1, 2, \ldots, N; t = 1, 2, \ldots, T) \tag{10.31}$$

Adjust the generation within the limits:

$$P_{ti}^{\text{opt}} = \begin{cases} P_i^{\min} & ; P_{ti}^{\text{opt}} \le P_i^{\min} \\ P_i^{\max} & ; P_{ti}^{\text{opt}} \ge P_i^{\max} \\ P_{ti}^{\text{opt}} & ; P_i^{\max} < P_{ti}^{\text{opt}} < P_i^{\max} \end{cases} \quad (i = 1, 2, \ldots, N; t = 1, 2, \ldots, T) \tag{10.32}$$

To minimize $[F_i(P_{ti}) - \lambda_t P_{ti}]$ means to get zero value for de-committed units. So following strategy is applied to decide the ON/OFF status of units.

$$U_{ti} = \begin{cases} 0 & ; F_{ti}(P_{ti}^{\text{opt}}) - \lambda_t P_{ti}^{\text{opt}} \geq 0 \\ 1 & ; F_{ti}(P_{ti}^{\text{opt}}) - \lambda_t P_{ti}^{\text{opt}} < 0 \end{cases} \quad (i = 1, 2, \ldots, N; t = 1, 2, \ldots, T) \tag{10.33}$$

The following example explains the steps to find unit commitment solution.

EXAMPLE 10.8 The data for the 3-units, 4-hour unit commitment problem are given below. Solve the unit commitment problem using Lagrange relaxation technique.

$$F_1(P_1) = (0.002P_1^2 + 10P_1 + 500) \text{ ₹/h}$$

$$F_2(P_2) = (0.0025\, P_2^2 + 8P_2 + 300\,) \text{ ₹/h}$$

$$F_3(P_3) = (0.005\, P_3^2 + 6P_1 + 100\,)\text{₹/h}$$

$$100 \text{ MW} \leq P_1 \leq 600 \text{ MW}$$

$$100 \text{ MW} \leq P_2 \leq 400 \text{ MW}$$

$$50 \text{ MW} \leq P_3 \leq 200 \text{ MW}$$

TABLE 10.22 Hourly load data

Time, $t(h)$	1	2	3	4
Load, P_{Dt} (MW)	180	540	1050	310

Solution: The given data is tabulated in Table 10.23. Initially, the Lagrange multipliers are assumed zero and Table 10.24 gives the committed units, U_{ti}, generation, P_{ti}, mismatch of power demand, ΔP_{Dt} and scheduled generation, P_{ti}^{ed} values of iteration 1.

TABLE 10.23 Characteristics of thermal units

Unit	a_i (₹/MW²h)	b_i (₹/MWh)	c_i (₹/h)	$P_i^{\min}$ (MW)	$P_i^{\max}$ (MW)
1	0.002	10	500	100	600
2	0.0025	8	300	100	400
3	0.005	6	100	50	200

TABLE 10.24 Iteration 1

Time (*h*)	λ_t (₹/MWh)	U_{t1}	U_{t2}	U_{t3}	P_{t1} (MW)	P_{t2} (MW)	P_{t3} (MW)	ΔP_{Dt} (MW)	P_{t1}^{ea} (MW)	P_{t2}^{ed} (MW)	P_{t3}^{ed} (MW)
1	0.0	0	0	0	0	0	0	180	0	0	0
2	0.0	0	0	0	0	0	0	540	0	0	0
3	0.0	0	0	0	0	0	0	1050	0	0	0
4	0.0	0	0	0	0	0	0	310	0	0	0

Compute the dual gap by calculating $q(\lambda)$ and J^* as stated below by necessary equations.

$$q(\lambda) = \sum_{t=1}^{4}\left(\sum_{i=1}^{3}(a_i P_{ti}^2 + b_i P_{ti} + c_i)\, U_{ti}\right) + \sum_{t=1}^{4} \lambda_t\, \Delta P_{Dt}$$

where

$$\Delta P_{Dt} = \left(P_{Dt} - \sum_{i=1}^{3} P_{ti} U_{ti}\right)$$

$$J^* = \sum_{t=1}^{4} (F_t^{ed})\ ₹$$

$$F_t^{ed} = \begin{cases} 10000 & ;\ \Delta P_{Dt} > 0 \\ f_t^{ed} & ;\ \Delta P_{Dt} \le 0 \end{cases} \qquad (t = 1, 2, 3, 4)$$

$$f_t^{ed} = \sum_{i=1}^{3}(a_i (P_{ti}^{ed})^2 + b_i P_{ti}^{ed} + c_i) U_{ti}$$

$$Q(\lambda) = 0\ ₹,\ J^* = 4000\ ₹ \text{ and Duality gap (err)} = \frac{J - Q}{Q} = \text{infinite}$$

Update Lagrange multipliers

$$\lambda_t = \lambda_t + 0.01\,\Delta P_{Dt} \quad (t = 1, 2, 3, 4)$$

$$\lambda_1 = \lambda_1 + 0.01\Delta P_{D1} \Rightarrow \lambda_1 = 0 + 0.01 \times 180 \Rightarrow \lambda_1 = 1.8\ ₹/\text{MWh}$$

$$\lambda_2 = \lambda_2 + 0.01\Delta P_{D2} \Rightarrow \lambda_1 = 0 + 0.01 \times 540 \Rightarrow \lambda_2 = 5.4\ ₹/\text{MWh}$$

$$\lambda_3 = \lambda_3 + 0.01\Delta P_{D3} \Rightarrow \lambda_1 = 0 + 0.01 \times 1050 \Rightarrow \lambda_3 = 10.5\ ₹/\text{MWh}$$

$$\lambda_4 = \lambda_4 + 0.01\Delta P_{D4} \Rightarrow \lambda_4 = 0 + 0.01 \times 310 \Rightarrow \lambda_4 = 3.1\ ₹/\text{MWh}$$

Compute the optimal generation to commit the unit.

$$P_{ti}^{\text{opt}} = \frac{\lambda_t - b_i}{2a_i} \qquad (i = 1, 2, 3;\ t = 1, 2, 3, 4)$$

Check the limits of generation

$$P_{ti} = \begin{cases} P_i^{\min} & ;\ P_{ti}^{\text{opt}} < P_i^{\min} \\ P_i^{\max} & ;\ P_{ti}^{\text{opt}} < P_i^{\max} \\ \mathrm{P_{ti}^{opt}} & ;\ P_i^{\min} \le P_{ti}^{\text{opt}} \le P_i^{\max} \end{cases}$$

Compute the cost of each unit

$$F_{ti}(P_{ti}) = [(a_i P_{ti} + b_i) P_{ti} + c_i]\ ₹/\text{h} \qquad (i = 1, 2, 3;\ t = 1, 2, 3, 4)$$

Commit the units

$$U_{ti} = \begin{cases} 0 & ;\ F_{ti}(P_{ti}^{\text{opt}}) - \lambda_t P_{ti}^{\text{opt}} \ge 0 \\ 1 & ;\ F_{ti}(P_{ti}^{\text{opt}}) - \lambda_t P_{ti}^{\text{opt}} < 0 \end{cases}$$

Table 10.25 gives the λ_t, optimal computed generation, P_{t1}^{opt}, compute cost difference, $F_t(P_{t1}^{\text{opt}}) - \lambda_t P_{t1}^{\text{opt}}$ to decide the unit commitment, U_{ti} and generation, P_{t1}. Few entries in Table 10.25 are explained below:

$$P_{11}^{\text{opt}} = \frac{\lambda_1 - 10.0}{2 \times 0.002} \Rightarrow P_{11}^{\text{opt}} = \frac{1.8 - 10.0}{0.004} \Rightarrow P_{11}^{\text{opt}} = -2050 \le P_1^{\min} \Rightarrow P_{11}^{\text{opt}} = 100 \text{ MW}$$

$$F_{11}(P_{11}^{\text{opt}}) - \lambda_1 P_{11}^{\text{opt}} = [(0.002 \times 100 + 10) \times 100 + 500] - 1.8 \times 100 = 1340 \text{ ₹/h}$$

$$U_{11} = 0 \text{ and } P_{11} = 0 \text{ since } F_1(P_{11}^{\text{opt}}) - \lambda_1 P_{11}^{\text{opt}} > 0$$

$$P_{31}^{\text{opt}} = \frac{\lambda_3 - 10.0}{2 \times 0.002} \Rightarrow P_{31}^{\text{opt}} = \frac{10.5 - 10.0}{0.004} \Rightarrow P_{31}^{\text{opt}} = 125 \text{ MW}$$

$$F_{31}(P_{31}^{\text{opt}}) - \lambda_3 P_{31}^{\text{opt}} = [(0.002 \times 125 + 10) \times 125 + 500] - 10.5 \times 125 = 468.75 \text{ ₹/h}$$

$$U_{31} = 0 \text{ and } P_{31} = 0, \text{ since } F_{31}(P_{31}^{\text{opt}}) - \lambda_3 P_{31}^{\text{opt}} > 0$$

$$P_{32}^{\text{opt}} = \frac{\lambda_3 - 8.0}{2 \times 0.002} \Rightarrow P_{32}^{\text{opt}} = \frac{10.5 - 8.0}{0.005} \Rightarrow P_{32}^{\text{opt}} = 500 \text{ MW} \ge P_2^{\max} \Rightarrow P_{32}^{\text{opt}} = 400 \text{ MW}$$

$$F_{32}(P_{32}^{\text{opt}}) - \lambda_3 P_{32}^{\text{opt}} = [(0.0025 \times 400 + 8) \times 400 + 300] - 10.5 \times 400 = -300 \text{ ₹/h}$$

$$U_{32} = 1 \text{ and } P_{32} = 400 \text{ MW}, \text{ since } F_{32}(P_{32}^{\text{opt}}) - \lambda_3 P_{32}^{\text{opt}} < 0$$

TABLE 10.25 Calculations for commitment for iteration 2

Unit	t	1	2	3	4
1	λ_t	1.8	5.4	10.5	3.1
	P_{t1}^{opt} (MW)	100	100	125	100
	$F_t(P_{t1}^{\text{opt}}) - \lambda_t P_{t1}^{\text{opt}}$	1340.0	980.0	468.75	1210.0
	U_{t1}	0	0	0	0
	P_{t1}(MW)	0	0	0	0
2	P_{t2}^{opt} (MW)	100	100	400	100
	$F_t(P_{t2}^{\text{opt}}) - \lambda_t P_{t2}^{\text{opt}}$	945.0	585.0	–300	815.0
	U_{t2}	0	0	1	0
	P_{t2} (MW)	0	0	400	0
3	P_{t3}^{opt} (MW)	50.0	50.0	200	50.0
	$F_t(P_{t3}^{\text{opt}}) - \lambda_t P_{t3}^{\text{opt}}$	322.5	142.5	–600.0	257.5
	U_{t3}	0	0	1	0
	P_{t3} (MW)	0	0	200	0

Table 10.26 gives the committed units, U_{ti}, generation, P_{ti} and mismatch of power demand, ΔP_{Dt} and scheduled generation, P_{ti}^{ed} values for iteration 2.

Compute the dual gap by calculating $q(\lambda)$ and J^*.

$$Q(\lambda) = 14326.0 \text{ ₹},\ J^* = 40000 \text{ ₹} \quad \text{and} \quad \text{err} = \frac{J-Q}{Q} = 1.792126 > 0.05$$

TABLE 10.26 Iteration 2

Time (h)	λ_t (₹/MWh)	U_{t1}	U_{t2}	U_{t3}	P_{t1} (MW)	P_{t2} (MW)	P_{t3} (MW)	ΔP_{Dt} (MW)	P_{t1}^{ed} (MW)	P_{t2}^{ed} (MW)	P_{t3}^{ed} (MW)
1	1.8	0	0	0	0	0	0	180	0	0	0
2	5.4	0	0	0	0	0	0	540	0	0	0
3	10.5	0	1	1	0	400	200	450	0	0	0
4	3.1	0	0	0	0	0	0	310	0	0	0

Update Lagrange multipliers

$$\lambda_1 = \lambda_1 + 0.01\,\Delta P_{D1} \Rightarrow \lambda_1 = 1.8 + 0.01 \times 180 \Rightarrow \lambda_1 = 3.6 \text{ ₹/MWh}$$

$$\lambda_2 = \lambda_2 + 0.01\,\Delta P_{D2} \Rightarrow \lambda_1 = 5.4 + 0.01 \times 540 \Rightarrow \lambda_2 = 10.8 \text{ ₹/MWh}$$

$$\lambda_3 = \lambda_3 + 0.01\,\Delta P_{D3} \Rightarrow \lambda_1 = 10.5 + 0.01 \times 450 \Rightarrow \lambda_3 = 15.0 \text{ ₹/MWh}$$

$$\lambda_4 = \lambda_4 + 0.01\,\Delta P_{D4} \Rightarrow \lambda_4 = 3.1 + 0.01 \times 310 \Rightarrow \lambda_4 = 6.2 \text{ ₹/MWh}$$

Table 10.27 gives the λ_t, optimal computed generation, P_{ti}^{opt}, computer cost difference, $F_t(P_{ti}^{\text{opt}}) - \lambda_t P_{ti}^{\text{opt}}$ to decide the unit commitment, U_{ti} and generation, P_{ti}. Table 10.28 gives the committed units, U_{ti}, generation, P_{ti} and mismatch of power demand, ΔP_{Dt} and scheduled generation, P_{ti}^{ed} values for iteration 3.

TABLE 10.27 Calculations for commitment for iteration 3

Unit	t	1	2	3	4
1	λ_t	3.6	10.8	15.0	6.2
	P_{t1}^{opt} (MW)	100	200	600	100
	$F_t(P_{t1}^{\text{opt}}) - \lambda_t P_{t1}^{\text{opt}}$	1160.0	420.0	–1780	900.0
	U_{t1}	0	0	1	0
	P_{t1} (MW)	0	0	600	0
2	P_{t2}^{opt} (MW)	100	400	400	100
	$F_t(P_{t2}^{\text{opt}}) - \lambda_t P_{t2}^{\text{opt}}$	765.0	–420.0	–2100	505.0
	U_{t2}	0	1	1	0
	P_{t2} (MW)	0	400	400	0

Unit	*t*	1	2	3	4
3	P_{t3}^{opt} (MW)	50.0	200.0	200	50.0
	$F_t(P_{t3}^{\text{opt}}) - \lambda_t P_{t3}^{\text{opt}}$	232.5	−660.0	−1500.0	102.5
	U_{t3}	0	1	1	0
	P_{t3} (MW)	0	200	200	0

TABLE 10.28 Iteration 3

Time (*h*)	λ_t (₹/MWh)	U_{t1}	U_{t2}	U_{t3}	P_{t1} (MW)	P_{t2} (MW)	P_{t3} (MW)	ΔP_{Dt} (MW)	P_{t1}^{ed} (MW)	P_{t2}^{ed} (MW)	P_{t3}^{ed} (MW)
1	3.6	0	0	0	0	0	0	180	0	0	0
2	10.8	0	1	1	0	400	200	−60	0	340	200
3	15.0	1	1	1	600	400	200	−150	450	400	200
4	6.2	0	0	0	0	0	0	310	0	0	0

Compute the dual gap by calculating $q(\lambda)$ and J^*.

$$Q(\lambda) = 17692.0 \text{ ₹}, \; J^* = 35614 \text{ ₹} \quad \text{and} \quad \text{err} = \frac{J-Q}{Q} = 1.013 > 0.05$$

Update Lagrange multipliers

$$\lambda_1 = \lambda_1 + 0.01\,\Delta P_{D1} \Rightarrow \lambda_1 = 3.6 + 0.01 \times 180 \Rightarrow \lambda_1 = 5.4 \text{ ₹/MWh}$$

$$\lambda_2 = \lambda_2 + 0.01 \Delta P_{D2} \Rightarrow \lambda_1 = 10.8 + 0.005 \times (-60) \Rightarrow \lambda_2 = 10.5 \text{ ₹/MWh}$$

$$\lambda_3 = \lambda_3 + 0.01 \Delta P_{D3} \Rightarrow \lambda_1 = 15.0 + 0.005 \times (-150) \Rightarrow \lambda_3 = 14.25 \text{ ₹/MWh}$$

$$\lambda_4 = \lambda_4 + 0.01 \Delta P_{D4} \Rightarrow \lambda_4 = 6.2 + 0.01 \times 310 \Rightarrow \lambda_4 = 9.3 \text{ ₹/MWh}$$

Table 10.29 gives the, λ_t, optimal computed generation, P_{ti}^{opt}, computer cost difference, $F_t(P_{ti}^{\text{opt}}) - \lambda_t P_{ti}^{\text{opt}}$ to decide the unit commitment, U_{ti} and generation, P_{ti}. Table 10.30 gives the committed units, U_{ti}, generation, P_{ti} and mismatch of power demand, ΔP_{Dt} and scheduled generation, P_{ti}^{ed} values for iteration 4.

Compute the dual gap by calculating $q(\lambda)$ and J^*.

$$Q(\lambda) = 18747.5 \text{ ₹}, \; J^* = 35614 \text{ ₹}, \text{ and} \quad \text{err} = \frac{J-Q}{Q} = 0.89967 > 0.05$$

Table 10.31 gives the committed units, U_{ti}, generation, P_{ti} and mismatch of power demand, ΔP_{Dt} and scheduled generation, P_{ti}^{ed} values for iteration 12.

TABLE 10.29 Calculations for commitment for iteration 4

Unit	t	1	2	3	4
1	λ_t	5.4	10.5	14.25	9.3
	P_{t1}^{opt} (MW)	100	125	600	100
	$F_t(P_{t1}^{\text{opt}}) - \lambda_t P_{t1}^{\text{opt}}$	980.0	468.75	–1330	590.0
	U_{t1}	0	0	1	0
	P_{t1} (MW)	0	0	600	0
2	P_{t2}^{opt} (MW)	100	400	400	260
	$F_t(P_{t2}^{\text{opt}}) - \lambda_t P_{t2}^{\text{opt}}$	585.0	–300.0	–1800	130.99
	U_{t2}	0	1	1	0
	P_{t2} (MW)	0	400	400	0
3	P_{t3}^{opt} (MW)	50.0	200.0	200	200.0
	$F_t(P_{t3}^{\text{opt}}) - \lambda_t P_{t3}^{\text{opt}}$	142.5	–600.0	–1350.0	–360.0
	U_{t3}	0	1	1	1
	P_{t3} (MW)	0	200	200	200

TABLE 10.30 Iteration 3

Time (*h*)	λ_t (₹/MWh)	U_{t1}	U_{t2}	U_{t3}	P_{t1} (MW)	P_{t2} (MW)	P_{t3} (MW)	ΔP_{Dt} (MW)	P_{t1}^{ed} (MW)	P_{t2}^{ed} (MW)	P_{t3}^{ed} (MW)
1	5.4	0	0	0	0	0	0	180	0	0	0
2	10.5	0	1	1	0	400	200	–60	0	340	200
3	14.25	1	1	1	600	400	200	–150	450	400	200
4	9.3	0	0	1	0	0	200	110	0	0	0

TABLE 10.31 Iteration 12

Time (*h*)	λ_t (₹/MWh)	U_{t1}	U_{t2}	U_{t3}	P_{t1} (MW)	P_{t2} (MW)	P_{t3} (MW)	ΔP_{Dt} (MW)	P_{t1}^{ed} (MW)	P_{t2}^{ed} (MW)	P_{t3}^{ed} (MW)
1	8.4	0	0	1	0	0	200	–20	0	0	180
2	9.8	0	1	1	0	359.99	200	–19.99	0	340	200
3	14.0	1	1	1	600	400	200	–150	450	400	200
4	10.45	0	1	1	0	400	200	–290	0	110	200

Compute the dual gap by calculating $q(\lambda)$ and J^*.

$$Q(\lambda) = 19029.5 \text{ ₹}, \; J^* = 19666.25 \text{ ₹}, \text{ and } \quad \text{err} = \frac{J - Q}{Q} = 0.0334615 \leq 0.05$$

10.7 HEURISTIC METHOD

The unit commitment (UC) problem in power systems refers to the optimization problem for determining the start-up and shut-down schedule of generating units over a scheduling period so that the total production cost is minimized while satisfying several of constraints. Mathematically, the UC problem has commonly been formulated as a complex nonlinear, mixed-integer combinational optimization problem with binary (0–1) variables that represents on-off status and continuous variables that represents unit power generation and prevailing equality and inequality constraints. Unit commitment is a mixed-integer combinatorial optimization problem in a power system that involves determining a start-up and shut down schedule of the generating units to meet the hourly varying projected system demand. For each hour of a day, units are to be selected, satisfying various constraints such that the cost of supplying the load is minimum. Two types of variables are associated with each unit: (i) one to represent the selection of a generating unit, and (ii) the other to represent the output of active powers of the units. The first variable to represent the selection of the unit to contribute to meet the load and is binary in nature, and the second variable represents the power output and is real in nature. The power output of a unit is known by solving the economic load dispatch problem. Therefore, the aim of the UC problem is to properly schedule the on-off state of all the units supporting both variables in the system. The resulting schedule should minimize the system production cost during the period while simultaneously satisfying the load demand, spinning reserve, and physical operational constraints of the individual unit. On/Off status of units can be obtained by following the five steps given below.

Step 1: Randomly commit and decommit units, $U_{ti}(i = 1, 2, ..., N; t = 1, 2, ..., T)$.

Step 2: Calculate priority list of units using Eq. (10.14).

Step 3: Modify unit's status of unit combination satisfying spinning reserve constraints.

Step 4: Repair unit's status of unit combination satisfying minimum up/down time violations.

Step 5: Decommit units of each hour to reduce excessive spinning reserve constraints and repairing the minimum up/down times constraints.

Once commitment schedule of generators is available then any economic dispatch procedure can be applied. Five steps are outlined below:

Randomly Commit and De-commit Units

The ON/OFF status of units is given by binary variable, U_{ti}. Binary variable can be initialized by flipping the coin. Mathematically, flip the coin is stated below.

$$U_{ti} = \begin{cases} 0 & rand(\ ) < 0.5 \\ 1 & rand(\ ) \geq 0.5 \end{cases} \quad (i = 1, 2, \ldots, N; t = 1, 2, \ldots, T) \tag{10.34}$$

where rand( ) is uniform random number lies between 0 and 1.

Priority List for Unit-Scheduling

Priority list is created according to each unit parameters. Cost per produced unit, of a unit at its maximum output power usually is less than that at other output power levels. Priority list is based on fuel cost obtained from the average fuel cost of each unit operating at its maximum output power. The average full-load cost α_i of a unit is defined as the cost per unit of power (₹/MW) when the unit is at its full capacity. When the fuel cost of unit is given by Eq. (10.14), the units are ranked by their α_i in ascending order. Thus, the priority list of units will be formulated based on the order of α_i, in which a unit with the lowest α_i will have the highest priority to be dispatched.

Spinning Reserve Constraints

The obtained primary unit-scheduling may not satisfy the spinning reserve constraints given by Eq. (10.22c). Therefore, the spinning reserve violations are repaired by heuristic search. The algorithm for repairing the spinning reserve violations in emission constrained thermal unit commitment problem is as follows:

1. Initialize time counter, $t = 1$
2. For all uncommitted units at hour t, calculate the average full-load cost α_i using Eq. (10.14). Sort them in ascending order of α_i to obtain a unit scheduling list, $S(\alpha_i)$.
3. The amount of excessive spinning reserve at each hour is calculated by

$$\Delta P_{Dt} = \sum_{i=1}^{N} P_i^{\max} U_{ti} - P_{Dt} - P_{Rt}$$

4. IF $(\Delta P_{Dt} \le 0)$ THEN go to step 7
5. Commit an uncommitted unit in $S(\alpha_i)$ with the lowest α_i, one unit at a time
6. Go to step 3 and repeat
7. IF $(t < T)$ THEN, increment time counter, $t = t+1$ and return to step 3 ELSE, stop

Minimum Up- and Down-time Constraints

Since the obtained unit schedule may not satisfy the minimum up/down constraints, a heuristic search algorithm is required to repair any violations of these constraints. The continuously on/off times of the ith unit till time t are computed as follows:

$$T_{ti}^{ON} = (1 + T_{t-1,i}^{ON})\, U_{ti} \tag{10.35}$$

$$T_{ti}^{OFF} = (1 + T_{t-1,i}^{OFF})\,(1 - U_{ti}) \tag{10.36}$$

where, T_{ti}^{ON} is continuously on time for for ith unit upto time t. T_{ti}^{OFF} is continuously off time for ith unit upto time t.

The procedure to repair violations of the minimum up/down times constraints are as follows:

1. Calculate the duration on and off times of all units for the whole schedule time horizon using Eqs. (10.35) and (10.36).
2. Initialize time counter, $t = 1$
3. Initialize unit counter, $i = 1$
4. IF $((U_{t-1,i} = 0)$ and $(U_{ti} = 1))$ THEN

    IF $(T_{t-1,i}^{ON} < MUT_i)$ *THEN* set $U_{ti} = 1$

    $L = t + MDT_i - 1$

    IF $((L \le T)$ and $(T_{Li}^{OFF} < MDT_i)$ *THEN* set $U_{ti} = 1$

    IF $\left((L > T) \text{ and } \left(\sum_{m=t}^{T} U_{mi} > 0 \right) \right)$ *THEN* set $U_{ti} = 1$

    ENDIF

5. Update the duration on/off times for the ith unit using Eqs. (10.35) and (10.36)
6. IF ($i < N$) THEN, increment unit counter, $i = i+1$ and go to step 4 and repeat
7. IF ($t < T$) THEN, increment time counter, $t = t+1$ and go to step 2 ELSE, stop

De-commitment of Excess Units

Repairing the minimum up and down time constraints can lead to excessive spinning reserves, which is not desirable due to the high operation cost. In a heuristic search algorithm based on a priority list to de-commit the redundant units due to the minimum up and down time repairing, thereby reducing the operating cost. Starting from the committed units with the lowest priority list (the highest average operating cost), the algorithm determines units that can be de-committed without violating the minimum up and down time and spinning reserve constraints until no unit can be de-committed. So in this process, the spinning reserve and minimum up and down time constraints must be checked before de-committing a unit. Procedure for de-commitment of excessive units is as follows:

1. Initialize time counter, $t = 1$
2. Calculate the average full-load cost α_i using Eq. (10.14) of each committed unit in hour t and sort the units in the descending order of α_i to obtain a list, $S(\alpha_i)$. Let the first unit in $S(\alpha_i)$ be CU_{tj}
3. The amount of excessive spinning reserve at hour t is calculated by

$$\Delta P_{Dt} = \sum_{i=1}^{N} P_i^{\max} U_{ti} - P_{Dt} - P_{Rt}$$

4. If (ΔP_{Dt}) is less than the maximum generation power, $P_i^{\max}$ of CU_{tj} go to step 6.
5. If decommitting CU_{tj} does not violate its minimum up/down time constraint, decommit CU_{tj} and update on/off status for all units.
6. Delete CU_{tj} from $S(\alpha_i)$.
7. IF $S(\alpha_i)$ is not empty THEN, let another CU_{tj} be the first unit in $S(\alpha_i)$ and return to step 3.
8. IF ($t < T$) THEN, increment time counter, $t = t+1$ and return to step 2 and repeat ELSE, stop.

On obtaining the unit commitment schedule, any gradient or heuristic optimization technique can be applied. Chapter 8 explains the solution procedure following heuristic techniques.

REFERENCES

Books

Kothari, D.P. and I.J. Nagrath, *Modern Power System Analysis*, 3rd Edition, Tata McGraw Hill, New Delhi, 2003.

Momoh, J.A., *Electric Power System Applications of Optimization*, Marcel Dekker, Inc., New York, 2001.

Ongsakul, W. and Dieu Ngoc Vo, *Artificial Intelligence in Power System optimization*, CRC Press, New York, 2003.

Wood A.J. and B. Wollenberg, *Power Generation Operation and Control*, 2nd Edition, John Wiley, New York, 1996.

Zhu, Jizhong, *Optimization of Power System Operation*, John Wiley & Sons, Inc., Hoboken, New Jersey, 2000.

Papers

Chakraborty, S., T. Ito, T. Senjyu, and A.Y. Saber, Unit commitment strategy of thermal generators by using advanced fuzzy controlled binary particle swarm optimization algorithm, *International Journal of Electric Power and Energy Systems*, Vol. **43(1)**, pp. 1072–1080, 2012.

Chandrasekaran, K., and S,P. Simon, Network and reliability constrained unit commitment problem using binary real coded firefly algorithm, *International Journal of Electric Power and Energy Systems*, Vol. **43(1)**, pp. 921–932, 2012.

Cheng, C.P., C.W. Liu and C.C. Liu, Unit commitment by Lagrangian relaxation and genetic algorithms, *IEEE Transactions on Power Systems*, Vol. **15(2)**, pp. 707–714, 2000.

Cheng, C.P., C.W. Liu and C.C. Liu, Unit commitment by annealing genetic algorithms, *International Journal of Electric Power and Energy Systems*, Vol. **24**, pp. 149–158, 2000.

Chung, C.Y., H. Yu, and K.P. Wong, An advanced quantum-inspired evolutionary algorithm for unit commitment, *IEEE Transactions on Power Systems*, Vol. **26(2)**, pp. 847–854, 2006.

Cohen, A.I., and M. Yoshimura, A branch-and-bound algorithm for unit commitment, *IEEE Transactions on Power Apparatus and Systems*, Vol. **102(2)**, pp. 444–451, 1983.

Cohen, A.I. and S.H. Wan, A method for solving the fuel constrained unit commitment, *IEEE Transactions on Power Systems*, Vol. **1**, pp. 608–614, Feb. 1987.

Damousis, I.G., A.G. Bakirtzis and P.S. Dokopoulos, A solution to the unit commitment problem using integer-coded genetic algorithm, *IEEE Transactions on Power Systems*, Vol. **19(2)**, pp. 1165–1172, 2004

Dekrajangpetch, S., G.B. Sheble and A.J. Conejo, Auction implementation problems using Lagrangian relaxation, *IEEE Transactions on Power Systems*, Vol. **14(1)**, pp. 82–88, 1999.

Dillon, T.S., K.W. Edwin, H.D. Kochs and R.J. Taud, Integer programming approach to the problem of optimal unit commitment with probabilistic reserve determination, *IEEE Transactions on Power Apparatus and Systems*, Vol. **PAS-97(6)**, pp. 2154–2166, 1978.

Happ, H.H., J.J. Johnson and W.J. Wright, Large scale hydro-thermal unit commitment method and results, *IEEE Transactions on Power Apparatus and Systems*, Vol. **PAS-90(3)**, pp. 1373–1384, 1971.

Hobbs, W.J., G. Hermon, S. Warner and G.B. Sheble, An enhanced dynamic programming approach for unit commitment, *IEEE Power Engineering Review*, Vol. **3(3)**, pp. 1201–1205, Aug. 1988.

Kerr, R.H., J.L. Scheidt, A.J. Fontana and J.K. Wiley, Unit commitment, *IEEE Transactions on Power Apparatus and Systems*, Vol. **PAS-85(5)**, pp. 417–421, 1966.

Mantawy, A.H., Y.L. Abdel-Magid and S.Z. Selim, Unit commitment by tabu search, *IEE Proceedings: Generation, Transmission, Distribution*, Vol. **145(1)**, pp. 56–64, 1998.

Marifeld, T.T. and G.B. Sheble, Genetic based unit commitment algorithm, *IEEE Transactions on Power Systems*, Vol. **11(3)**, pp.1359–1370, 1996.

Nayak, R. and J.D. Sharma, A hybrid neural network and simulated annealing approach to the unit commitment problem, *Computers and Electrical Engineering*, Vol. **26(6)**, pp. 461–477, 2000.

Ongsakul W. and N. Petcharaks, Unit commitment by enhanced adaptive Lagrangian relaxation, *IEEE Transactions on Power Systems*, Vol. **19(1)**, pp. 620–628, 2004.

Ongsakul, W. and N. Petcharaks, Ramp rate constrained unit commitment by improved adaptive Lagrangian relaxation, *International Energy Journal*, Vol. **6(1)**, part 2, pp. 175–194, 2005.

Pang, C.K., and H.C. Chen, Optimal short-term thermal unit commitment, *IEEE Transactions on Power Apparatus and Systems*, Vol. **95(4)**, pp. 1336–1342, 1976.

Pappala,V.S. and Erlich, I., A new approach for solving the unit commitment problem by adaptive particle swarm optimization, *Power Engineering Society, IEEE General Meeting–PES*, pp. 1–6, 2008.

Salam, M.S., A.R., Hamdan, and Mohamed K. Nor, Integrating an expert system into a thermal unit-commitment algorithm, *IEE Proceedings–C*, Vol. **138**, pp. 553–559, 1991.

Seki, T., N. Yamashita and K. Kawamoto, New local search methods for improving the Lagrangian-relaxation-based unit commitment solution, *IEEE Transactions on Power Systems*, Vol. **25(1)**, pp. 272–283, 2010.

Senjyu, T., H. Yamashiro, K. Uezato, and T. Funabashi, A unit commitment problem by using genetic algorithm based on unit characteristic classification, *Proceedings IEEE/Power Engineering Society Winter Meeting*, Vol. **1**, pp. 58–63, 2002.

Senjyu,T., K. Shimabukuro, K. Uezato, and T. Funabashi, A fast technique for unit commitment problem by extended priority list. *IEEE Transactions on Power Systems*, Vol. **18(2)**, pp. 882–888, 2003.

Snyder, W.L., H.D. Powell, and C. Rayburn, Dynamic programming approach to unit commitment, *IEEE Transactions on Power Systems*, Vol. **2**, pp. 339–350, 1987.

Sriyanyong, P., and Y.H. Song, Unit commitment using particle swarm optimization combined with Lagrange relaxation. *IEEE Power Engineering Society General Meeting*, Vol **3**, June 12–16, 2005. pp 2752–2759, doi:10.1109/PES.2005.1489390.

Ting, T.O., M.V.C. Rao, and C.K. Loo, A novel approach for unit commitment problem via an effective hybrid particle swarm optimization. *IEEE Transactions on Power Systems*, Vol. **21(1)**, pp.411–418, 2006.

Valenzuela, J. and A.E. Smith, A seeded memetic algorithm for large unit commitment problems, *Journal of Heuristics*, Vol. **8**, pp.173–195, 2002.

Vemuri, S. and L. Lemonidis, Fuel constrained unit commitment, *IEEE Transactions on Power Systems*, Vol. **7,** pp. 410–415, 1992.

Wang, C. and S.M. Shahidehpour, Effects of ramp-rate limits on unit commitment and economic dispatch, *IEEE Transactions on Power Systems*, Vol. **8(3)**, pp. 1341–1350, 1993.

Wang, C. and S. M. Shahidehpour, Optimal generation scheduling with ramping costs, *IEEE Transactions on Power Systems*, Vol. **10(1)**, pp. 60–67, 1995.

Yalcinoz, T., Short, M.J., and Cory, B.J., Application of neural networks to unit commitment, *IEEE Transactions on Power Systems*, Vol. **2**, pp. 649–654, 1999. doi:10.1109/AFRCON.1999.821841.

Zhao. B., C.X. Guo, B.R. Bai and Y.J. Cao, An improved particle swarm optimization algorithm for unit commitment, *International Journal of Electric Power and Energy System*, Vol. **28(7)**, pp. 482–490, 2006.

APPENDIX A

Evaluation of the Expected Values of Functions

A.1 EXPECTED VALUE OF A FUNCTION OF SEVERAL RANDOM VARIABLES

Let $Y = g(X_1, X_2, ..., X_N)$. Expand the function g in a Taylor series about the mean values of random variables $X_1, X_2, ..., X_N$ to obtain

$$Y = g(\bar{X}_1, \bar{X}_2, ..., \bar{X}_N) + \sum_{i=1}^{N} \frac{\partial g}{\partial \bar{X}_i}(X_i - \bar{X}_i) + \frac{1}{2}\sum_{i=1}^{N}\sum_{j=1}^{N} \frac{\partial^2 g}{\partial X_i \partial X_j}(X_i - \bar{X}_i)(X_j - \bar{X}_j) + ... \quad (A.1)$$

where the derivatives are evaluated at the mean values of random variables, i.e. at $\bar{X}_1, \bar{X}_2, ..., \bar{X}_N$.

Taking expected values of (A.1) to achieve,

$$E(Y) = E[g(\bar{X}_1, \bar{X}_2, ..., \bar{X}_N)] + \sum_{i=1}^{N} \frac{\partial g}{\partial X_i} E(X_i - \bar{X}_i) + \frac{1}{2}\sum_{i=1}^{N}\sum_{j=1}^{N} \frac{\partial^2 g}{\partial X_i \partial X_j} E[(X_i - \bar{X}_i)(X_j - \bar{X}_j)] \quad (A.2)$$

where partial derivatives $\frac{\partial g}{\partial X_i}$ and $\frac{\partial^2 g}{\partial X_i \partial X_j}$ are evaluated at $\bar{X}_1, \bar{X}_2, ..., \bar{X}_N$.

On simplification,

$$E(Y) = \bar{g}(\bar{X}_1, \bar{X}_2, ..., \bar{X}_N) + \frac{1}{2}\sum_{i=1}^{N} \frac{\partial^2 g}{\partial X_i^2} \text{var}(X_i) + \frac{1}{2}\sum_{i=1}^{N}\sum_{j=1}^{N} \frac{\partial^2 g}{\partial X_i \partial X_j} \text{cov}(X_i, X_j) \quad (A.3)$$

A.2 EXPECTED COST FUNCTION

The cost function is generally described by

$$F_{1i} = F_{1i}(a_i, b_i, c_i, P_i) = a_i P_i^2 + b_i P_i + c_i \quad (A.4)$$

Expected value of the function F_{1i} can be obtained using Eq. (A.3) as discussed below.

$$F_{1i} = F_{1i}(a_i, b_i, c_i, P_i) + \frac{1}{2}\frac{\partial^2 F_{1i}}{\partial P_i^2} \text{var}(P_i) + \frac{\partial^2 F_{1i}}{\partial P_i \partial a_i} \text{cov}(a_i, P_i) + \frac{\partial^2 F_{1i}}{\partial P_i \partial b_i} \text{cov}(b_i, P_i) \quad (A.5)$$

Only the existing derivatives are considered. On substituting the values of derivatives evaluated at their mean values in Eq. (A.5),

$$F_{1i} = \overline{a}_i \overline{P}_i^2 + \overline{b}_i \overline{P}_i + \overline{c}_i + \overline{a}_i \text{ var}(P_i) + 2\overline{P}_i \text{ cov}(a_i, P_i) + \text{cov}(b_i, P_i) \tag{A.6}$$

where $\overline{a}_i, \overline{b}_i, \overline{c}_i$, and $\overline{P}_i$ are the mean values of random variables.

A.3 EXPECTED NO_x EMISSION

The NO_x emission function is described by

$$F_{2i} = F_{2i}(d_i, e_i, f_i, P_i) = d_i P_i^2 + e_i P_i + f_i \tag{A.7}$$

Expected value of the function F_{2i} can be obtained using Eq. (A.7) as discussed below.

$$F_{2i} = \overline{F}_{2i}(\overline{d}_i, \overline{e}_i, \overline{f}_i, \overline{P}_i) + \frac{1}{2}\frac{\partial^2 F_{2i}}{\partial P_i^2} \text{ var}(P_i) + \frac{\partial^2 F_{2i}}{\partial P_i \partial d_i} \text{ cov}(d_i, P_i) + \frac{\partial^2 F_{2i}}{\partial P_i \partial e_i} \text{ cov}(e_i, P_i) \tag{A.8}$$

Only the existing derivatives are considered. On substituting the values of derivatives evaluated at their mean values in Eq. (A.8),

$$F_{2i} = \overline{d}_i \overline{P}_i^2 + \overline{e}_i \overline{P}_i + \overline{f}_i + \overline{d}_i \text{ var}(P_i) + 2\overline{P}_i \text{ cov}(d_i, P_i) + \text{cov}(e_i, P_i) \tag{A.9}$$

where $\overline{d}_i, \overline{e}_i, \overline{f}_i$, and $\overline{P}_i$ are the mean values of random variables.

A.4 TRANSMISSION LOSS

The transmission loss is defined using B-coefficients,

$$P_L = \sum_{i=1}^{N}\sum_{j=1}^{N} P_i B_{ij} P_j \tag{A.10}$$

Expected value of the function P_L, can be obtained using Eq. (A.3) as below

$$\overline{P}_L = \sum_{i=1}^{N}\sum_{j=1}^{N} \overline{P}_i \overline{B}_{ij} \overline{P}_j + \sum_{i=1}^{N} \frac{\partial^2 P_L}{\partial P_i^2} \text{ var}(P_i) + \frac{1}{2}\sum_{i=1}^{N}\sum_{\substack{j=1 \\ j\neq i}}^{N} \frac{\partial^2 P_L}{\partial P_i \partial P_j} \text{ cov}(P_i, P_j)$$
$$+ \sum_{i=1}^{N} \frac{\partial^2 P_L}{\partial P_i \partial B_{ii}} \text{ cov}(P_i, B_{ii}) + \sum_{i=1}^{N}\sum_{\substack{j=1 \\ j\neq i}}^{N} \frac{\partial^2 P_L}{\partial P_i \partial B_{ij}} \text{ cov}(P_i, B_{ij}) \tag{A.11}$$

On simplification, Eq. (A.11) can be rewritten as

$$\overline{P}_L = \sum_{i=1}^{N}\sum_{j=1}^{N} \overline{P}_i \overline{B}_{ij} \overline{P}_j + \sum_{i=1}^{N} \overline{B}_{ii} \text{ var}(\overline{P}_i) + \sum_{i=1}^{N}\sum_{\substack{j=1 \\ j\neq i}}^{N} \overline{B}_{ij} \text{ cov}(\overline{P}_i, \overline{P}_j) + \sum_{i=1}^{N}\sum_{j=1}^{N} \overline{P}_j \text{ cov}(P_i, B_{ij}) \tag{A.12}$$

A.5 EXPECTED REAL POWER TRANSMISSION LOSS

Under normal operating conditions, the transmission loss is quadratic in the injected active power P_i and reactive power Q_i at all the buses, NB. Active power transmission losses, P_L is defined as:

$$P_L = \sum_{i=1}^{NB}\sum_{j=1}^{NB} [\alpha_{ij}(P_iP_j + Q_iQ_j) + \beta_{ij}(Q_iP_j - P_iQ_j)] \tag{A.13}$$

where

$P_i = P_{Gi} - P_{Di}$

$Q_i = Q_{Gi} - Q_{Di}$

and α_{ij} and β_{ij} are loss coefficients [Eq. (3.132), Chapter 3].

Active and reactive power demands and hence power generations have been considered as random that result into random nature of injected bus powers P_i and Q_i. Bus voltage magnitudes and phase angles have been treated as random which result into random nature of loss coefficients. As active power loss P_L, depends on random injected bus powers and random loss coefficients, it also becomes random. The deterministic equivalent of the power loss given by Eq. (A.13) has been obtained by expanding the function about the mean of random variables. On the basis of Eq. (A.3), expected active power loss is given by:

$$\overline{P}_L = E\{P_L(\overline{\alpha}_{ij}, \overline{\beta}_{ij}, \overline{P}_i, \overline{Q}_i)\} + \frac{1}{2}\sum_{i=1}^{NB}\frac{\partial^2 P_L}{\partial P_i^2}\text{var}(P_i) + \frac{1}{2}\sum_{i=1}^{NB}\frac{\partial^2 P_L}{\partial Q_i^2}(Q_i)$$
$$+ \sum_{i=1}^{NB}\sum_{\substack{j=1\\ j\neq i}}^{NB}\frac{\partial^2 P_L}{\partial P_i \partial P_j}\text{cov}(P_i, P_j) + \sum_{i=1}^{NB}\sum_{\substack{j=1\\ j\neq i}}^{NB}\frac{\partial^2 P_L}{\partial Q_i \partial Q_j}\text{cov}(Q_i, Q_j) + \sum_{i=1}^{NB}\sum_{j=1}^{NB}\frac{\partial^2 P_L}{\partial P_i \partial Q_j}\text{cov}(P_i, Q_j) \tag{A.14}$$

Partial derivatives of Eq. (A.13) with respect to random variables are given below which are evaluated at the mean values of random variables:

$$\frac{\partial^2 P_L}{\partial P_i^2} = 2\overline{\alpha}_{ii};\quad \frac{\partial^2 P_L}{\partial Q_i^2} = 2\overline{\alpha}_{ii};\quad \frac{\partial^2 P_L}{\partial P_i \partial P_j} = \overline{\alpha}_{ij};\quad \frac{\partial^2 P_L}{\partial Q_i \partial Q_j} = \overline{\alpha}_{ij};\quad \frac{\partial^2 P_L}{\partial P_i \partial Q_j} = 2\overline{\beta}_{ij}$$

where $\overline{\alpha}_{ij}$ and $\overline{\beta}_{ij}$ are expected loss coefficients.

Using partial derivatives which are evaluated at mean of random variables, the expected value of active power loss $\overline{P}_L$, is obtained:

$$\overline{P}_L = \sum_{i=1}^{NB}\sum_{j=1}^{NB} [\overline{\alpha}_{ij}(\overline{P}_i\overline{P}_j + \overline{Q}_i\overline{Q}_j) + \overline{\beta}_{ij}(\overline{Q}_i\overline{P}_j - \overline{P}_i\overline{Q}_j)] + \sum_{i=1}^{NB}\overline{\alpha}_{ii}[\text{var}(P_i) + \text{var}(Q_i)]$$
$$+ \sum_{i=1}^{NB}\sum_{\substack{j=1\\ j\neq i}}^{NB}\overline{\alpha}_{ij}[\text{cov}(P_i, P_j) + \text{cov}(Q_i, Q_j)] + \sum_{i=1}^{NB}\sum_{\substack{j=1\\ j\neq i}}^{NB}[(\overline{\beta}_{ji} - \overline{\beta}_{ij})\,\text{cov}(P_i, Q_j)] \tag{A.15}$$

A.6 EXPECTED REACTIVE POWER TRANSMISSION LOSS

The reactive power transmission loss Q_L is defined as:

$$Q_L = \sum_{i=1}^{NB} \sum_{j=1}^{NB} [\gamma_{ij}(P_i P_j + Q_i Q_j) + \eta_{ij}(Q_i P_j - P_i Q_j)] \tag{A.16}$$

where γ_{ij} and η_{ij} are loss coefficients [see Eq. (3.132) and Eq. (3.134), Chapter 3].

As reactive power loss also depends on random injected bus powers and random loss coefficients, γ_{ij} and η_{ij}, Q_L also becomes random. The deterministic equivalent of the reactive power loss given by Eq. (A.16) has been obtained by expanding the function about the mean of random variables.

$$\bar{Q}_L = E\{Q_L(\bar{\gamma}_{ij}, \bar{\eta}_{ij}, \bar{P}_i, \bar{Q}_i)\} + \frac{1}{2}\sum_{i=1}^{NB} \frac{\partial^2 Q_L}{\partial P_i^2} \text{var}(P_i) + \frac{1}{2}\sum_{i=1}^{NB} \frac{\partial^2 Q_L}{\partial Q_i^2} \text{var}(Q_i)$$
$$+ \sum_{i=1}^{NB} \sum_{\substack{j=1 \\ j \neq i}}^{NB} \frac{\partial^2 Q_L}{\partial P_i \partial P_j} \text{cov}(P_i, P_j) + \sum_{i=1}^{NB} \sum_{\substack{j=1 \\ j \neq i}}^{NB} \frac{\partial^2 Q_L}{\partial Q_i \partial Q_j} \text{cov}(Q_i, Q_j) + \sum_{i=1}^{NB} \sum_{j=1}^{NB} \frac{\partial^2 Q_L}{\partial P_i \partial Q_j} \text{cov}(P_i, Q_j) \tag{A.17}$$

Partial derivatives of Eq. (A.17) with respect to random variables are given below which are evaluated at mean values of random variables:

$$\frac{\partial^2 Q_L}{\partial P_i^2} = 2\bar{\gamma}_{ii}; \quad \frac{\partial^2 Q_L}{\partial Q_i^2} = 2\bar{\gamma}_{ii}; \quad \frac{\partial^2 Q_L}{\partial P_i \partial P_j} = \bar{\gamma}_{ij}; \quad \frac{\partial^2 Q_L}{\partial Q_i \partial Q_j} = \bar{\gamma}_{ij}; \quad \frac{\partial^2 Q_L}{\partial P_i \partial Q_j} = -\bar{\eta}_{ij}$$

where $\bar{\gamma}_{ij}$ and $\bar{\eta}_{ij}$ are expected loss coefficients.

Using partial derivatives, expected value of reactive power loss $\bar{Q}_L$, is obtained:

$$\bar{Q}_L = \sum_{i=1}^{NB} \sum_{j=1}^{NB} [\bar{\gamma}_{ij}(\bar{P}_i \bar{P}_j + \bar{Q}_i \bar{Q}_j) + \bar{\eta}_{ij}(\bar{Q}_i \bar{P}_j - \bar{P}_i \bar{Q}_j)] + \sum_{i=1}^{NB} \bar{\gamma}_{ii}[\text{var}(P_i) + \text{var}(Q_i)]$$
$$+ \sum_{i=1}^{NB} \sum_{\substack{j=1 \\ j \neq i}}^{NB} \bar{\gamma}_{ij}[\text{cov}(P_i, P_j) + \text{cov}(Q_i, Q_j)] + \sum_{i=1}^{NB} \sum_{\substack{j=1 \\ j \neq i}}^{NB} [(\bar{\eta}_{ji} - \bar{\eta}_{ij}) \text{cov}(P_i, Q_j)] \tag{A.18}$$

A.7 EXPECTED TRANSMISSION LOSS COEFFICIENTS

The active power loss coefficients; α_{ij} and β_{ij}, and reactive power loss coefficients; γ_{ij} and η_{ij}, between buses i and j are given as:

$$\alpha_{ij} = \frac{R_{ij}}{|V_i||V_j|} \cos(\delta_i - \delta_j) \tag{A.19}$$

$$\beta_{ij} = \frac{R_{ij}}{|V_i|\,|V_j|} \sin(\delta_i - \delta_j) \tag{A.20}$$

where R_{ij} is the real part and X_{ij} is the imaginary part of ijth element (z_{ij}) of bus impedance matrix, $|V_i|$ is the voltage magnitude at ith bus and δ_i is the voltage phase angle at ith bus.

Active and reactive power loss coefficients; α_{ij}, β_{ij}, γ_{ij} and η_{ij} are functions of bus voltage magnitudes and phase angles, which have been considered as random variables. Hence these power loss coefficients also become random variables and their expected values are derived utilizing Eq. (A.3).

Expected value of random coefficient α_{ij} is obtained by expanding, using Taylor's series expansion, about the mean of the random variables and taking its expected value. The expected value of α_{ij} is given by;

$$\bar{\alpha}_{ij} = \frac{R_{ij}}{|\bar{V}_i||\bar{V}_j|} \cos(\bar{\delta}_i - \bar{\delta}_j) + \frac{1}{2}\frac{\partial^2 \alpha_{ij}}{\partial |V_i|^2} \operatorname{var}(|V_i|) + \frac{1}{2}\frac{\partial^2 \alpha_{ij}}{\partial |V_j|^2} \operatorname{var}(|V_j|) + \frac{\partial^2 \alpha_{ij}}{\partial |V_i| \partial |V_j|} \operatorname{cov}(|V_i|, |V_j|)$$

$$+ \frac{1}{2}\frac{\partial^2 \alpha_{ij}}{\partial \delta_i^2} \operatorname{var}(\delta_i) + \frac{1}{2}\frac{\partial^2 \alpha_{ij}}{\partial \delta_j^2} \operatorname{var}(\delta_j) + \frac{\partial^2 \alpha_{ij}}{\partial \delta_i \partial \delta_j} \operatorname{cov}(\delta_i, \delta_j) \tag{A.21}$$

Partial derivatives of the above equation are evaluated at the mean values of random variables and are given below:

$$\frac{\partial^2 \alpha_{ij}}{\partial |V_i|^2} = 6\frac{R_{ij}}{|\bar{V}_i|^4} \cos(\bar{\delta}_i - \bar{\delta}_j);\ i = j$$

$$\frac{\partial^2 \alpha_{ij}}{\partial |V_i|^2} = 2\frac{R_{ij}}{|\bar{V}_i|^3 |\bar{V}_j|} \cos(\bar{\delta}_i - \bar{\delta}_j);\ i \neq j$$

$$\frac{\partial^2 \alpha_{ij}}{\partial |V_j|^2} = 2\frac{R_{ij}}{|\bar{V}_i||\bar{V}_j|^3} \cos(\bar{\delta}_i - \bar{\delta}_j);\ i \neq j$$

$$\frac{\partial^2 \alpha_{ij}}{\partial |V_i| \partial |V_j|} = \frac{R_{ij}}{|\bar{V}_i|^2 |\bar{V}_j|^2} \cos(\bar{\delta}_i - \bar{\delta}_j);\ i \neq j$$

$$\frac{\partial^2 \alpha_{ij}}{\partial \delta_i^2} = -\frac{R_{ij}}{|\bar{V}_i||\bar{V}_j|} \cos(\bar{\delta}_i - \bar{\delta}_j);\ i \neq j$$

$$\frac{\partial^2 \alpha_{ij}}{\partial \delta_j^2} = -\frac{R_{ij}}{|\bar{V}_i||\bar{V}_j|} \cos(\bar{\delta}_i - \bar{\delta}_j);\ i \neq j$$

$$\frac{\partial^2 \alpha_{ij}}{\partial \delta_i \partial \delta_j} = \frac{R_{ij}}{|\overline{V}_i||\overline{V}_j|} \cos(\overline{\delta}_i - \overline{\delta}_j);\ i \neq j$$

Substitute the derivatives in Eq. (A.21) and rearrange the terms. The expected value of loss coefficient $\overline{A}_{ij}$, is obtained as:

$$\overline{\alpha}_{ij} = \frac{\tau_{ij} R_{ij}}{|\overline{V}_i||\overline{V}_j|} \cos(\overline{\delta}_i - \overline{\delta}_j) \tag{A.22}$$

where

$$\tau_{ij} = \begin{cases} 1.0 + 3.0 \dfrac{\mathrm{var}(V_i)}{|\overline{V}_i|^2} & ;\ \text{if } i = j \\ 1.0 + \dfrac{\mathrm{var}(V_i)}{|\overline{V}_i|^2} + \dfrac{\mathrm{var}(V_j)}{|\overline{V}_j|^2} + \dfrac{\mathrm{cov}(V_i, V_j)}{|\overline{V}_i||\overline{V}_j|} - \dfrac{\mathrm{var}(\delta_i) + \mathrm{var}(\delta_j)}{2.0} + \mathrm{cov}(\delta_i, \delta_j) & ;\ \text{if } i \neq j \end{cases} \tag{A.23}$$

where

$$\mathrm{var}(V_i) = C_{Vi}^2 |\overline{V}_i|^2$$

$$\mathrm{cov}(V_i, V_j) = R_{ViVj} C_{Vi} C_{Vj} |\overline{V}_i||\overline{V}_j|$$

$$\mathrm{var}(\delta_i) = C_{\delta i}^2 \overline{\delta}_i^2$$

$$\mathrm{cov}(\delta_i, \delta_j) = R_{\delta i \delta j} C_{\delta i} C_{\delta j} \overline{\delta}_i \overline{\delta}_j$$

Similarly, expected value of random coefficient β_{ij}, is obtained by expanding the expression of Eq. (A.20) using Taylor's series expansion about the mean of the random variables as follows:

$$\overline{\beta}_{ij} = \frac{R_{ij}}{|\overline{V}_i||\overline{V}_j|} \cos(\overline{\delta}_i - \overline{\delta}_j) + \frac{1}{2} \frac{\partial^2 \beta_{ij}}{\partial |V_i|^2} \mathrm{var}(|V_i|) + \frac{1}{2} \frac{\partial^2 \beta_{ij}}{\partial |V_j|^2} \mathrm{var}(|V_j|) + \frac{\partial^2 \beta_{ij}}{\partial |V_i| \partial |V_j|} \mathrm{cov}(|V_i|, |V_j|)$$
$$+ \frac{1}{2} \frac{\partial^2 \beta_{ij}}{\partial \delta_i^2} \mathrm{var}(\delta_i) + \frac{1}{2} \frac{\partial^2 \beta_{ij}}{\partial \delta_j^2} \mathrm{var}(\delta_j) + \frac{\partial^2 \beta_{ij}}{\partial \delta_i \partial \delta_j} \mathrm{cov}(\delta_i, \delta_j) \tag{A.24}$$

Partial derivatives of the above equation are evaluated at the mean values of random variables and are given below:

$$\frac{\partial^2 \beta_{ij}}{\partial |V_i|^2} = 6 \frac{R_{ij}}{|\overline{V}_i|^4} \sin(\overline{\delta}_i - \overline{\delta}_j);\ i = j$$

$$\frac{\partial^2 \beta_{ij}}{\partial |V_i|^2} = 2 \frac{R_{ij}}{|\overline{V}_i|^3 |\overline{V}_j|} \sin(\overline{\delta}_i - \overline{\delta}_j); \; i \neq j$$

$$\frac{\partial^2 \beta_{ij}}{\partial |V_j|^2} = 2 \frac{R_{ij}}{|\overline{V}_i| |\overline{V}_j|^3} \sin(\overline{\delta}_i - \overline{\delta}_j); \; i \neq j$$

$$\frac{\partial^2 \beta_{ij}}{\partial |V_i| \partial |V_j|} = \frac{R_{ij}}{|\overline{V}_i|^2 |\overline{V}_j|^2} (\overline{\delta}_i - \overline{\delta}_j); \; i \neq j$$

$$\frac{\partial^2 \beta_{ij}}{\partial \delta_i^2} = -\frac{R_{ij}}{|\overline{V}_i| |\overline{V}_j|} \sin(\overline{\delta}_i - \overline{\delta}_j); \; i \neq j$$

$$\frac{\partial^2 \beta_{ij}}{\partial \delta_j^2} = -\frac{R_{ij}}{|\overline{V}_i| |\overline{V}_j|} \sin(\overline{\delta}_i - \overline{\delta}_j); \; i \neq j$$

$$\frac{\partial^2 \beta_{ij}}{\partial \delta_i \partial \delta_j} = \frac{R_{ij}}{|\overline{V}_i| |\overline{V}_j|} \sin(\overline{\delta}_i - \overline{\delta}_j); \; i \neq j$$

Substituting the derivatives in Eq. (A.24), the expected value of loss coefficient $\overline{\beta}_{ij}$, is obtained as:

$$\overline{\beta}_{ij} = \frac{\tau_{ij} R_{ij}}{|\overline{V}_i| |\overline{V}_j|} \sin(\overline{\delta}_i - \overline{\delta}_j) \tag{A.25}$$

where τ_{ij} is given by Eq. (A.23).

Following the same procedure as has been adopted for the evaluation of expected values of active power loss coefficients $\overline{\alpha}_{ij}$ and $\overline{\beta}_{ij}$, the expected values of reactive power loss coefficients C_{ij} and D_{ij}, can be evaluated.

A.8 EXPECTED POWER FLOW ON TRANSMISSION LINES

Active power flow P_{Tm}, on mth transmission line of a power system network connected from bus j to k, is given by:

$$P_{Tm} = g_m |V_j|^2 - |V_j| |V_k| \{g_m \cos(\delta_j - \delta_k) + b_m \sin(\delta_j - \delta_k)\} \tag{A.26}$$

where

$|V_j|$ and $|V_k|$ are the voltage magnitudes at jth and kth buses respectively.

δ_j and δ_k are the voltage phase angles at jth and kth buses respectively.

g_m is the series conductance of mth transmission line.

Applying the Taylor's series expansion, expected value of active power flow on mth transmission line is given by:

$$\overline{P}_{Tm} = E(P_{Tm}) + \frac{1}{2}\frac{\partial^2 P_{Tm}}{\partial |V_j|^2}\,\text{var}(|V_j|) + \frac{1}{2}\frac{\partial^2 P_{Tm}}{\partial |V_k|^2}\,\text{var}(|V_k|) + \frac{\partial^2 P_{Tm}}{\partial |V_j| \partial |V_k|}\,\text{cov}(|V_j|, |V_k|)$$

$$+ \frac{1}{2}\frac{\partial^2 P_{Tm}}{\partial \delta_j^2}\,\text{var}(\delta_j) + \frac{1}{2}\frac{\partial^2 P_{Tm}}{\partial \delta_k^2}\,\text{var}(\delta_k) + \frac{\partial^2 P_{Tm}}{\partial \delta_j \partial \delta_k}\,\text{var}(\delta_j, \delta_k) + \frac{\partial^2 P_{Tm}}{\partial |V_j| \partial \delta_j}\,\text{cov}(|V_j|, \delta_j)$$

$$+ \frac{\partial^2 P_{Tm}}{\partial |V_k| \partial \delta_k}\,\text{cov}(|V_k|, \delta_k) + \frac{\partial^2 P_{Tm}}{\partial |V_j| \partial \delta_k}\,\text{cov}(|V_j|, \delta_k) + \frac{\partial^2 P_{Tm}}{\partial |V_k| \partial \delta_j}\,\text{cov}(|V_k|, \delta_j) \tag{A.27}$$

Substituting the partial derivatives evaluated at the mean values of random variables, the following expression is obtained:

$$\overline{P}_{Tm} = g_m\{|\overline{V}_j|^2 + \text{var}(|\overline{V}_j|)\} - |\overline{V}_j|\,|\overline{V}_k|\,\{g_m \cos(\overline{\delta}_j - \overline{\delta}_k) + b_m \sin(\overline{\delta}_j - \overline{\delta}_k)\}\,\xi_{jk} \tag{A.28}$$

where

$$\xi_{jk} = \left[1.0 - \frac{1}{2}[\text{var}(\delta_j) + \text{var}(\delta_k)] + \text{cov}(\delta_j, \delta_k) + \frac{\text{cov}(|V_j|, |V_k|)}{|\overline{V}_j|\,|\overline{V}_k|}\right]$$

$|\overline{V}_j|$ and $|\overline{V}_k|$ are the expected voltage magnitudes at *j*th and *k*th buses respectively.
$\overline{\delta}_j$ and $\overline{\delta}_k$ are expected phase angles of voltage at *j*th and *k*th buses respectively.

REFERENCES

Papoulis, A., *Probability, Random Variables and Stochastic Processes,* McGraw-Hill, New Delhi, 1991.

Rao, S.S., *Optimization, Theory and Applications,* 2nd ed., Wiley Eastern, New Delhi, 1987.

Sen Gupta, J.K., *Stochastic Programming,* North Holland, 1972.

APPENDIX B

Evaluation of a Coefficient of a Generator Output

This appendix describes a possible method of Monte Carlo simulation to obtain sample estimate of the expected value and variance of a generator output variable via incremental production cost model incorporating random coefficients. The ratio between standard deviation and mean of the output variable is computed and interpreted in terms of coefficient of variation of the generator.

Consider an existing power system with $(N - 1)$ thermal generating units operating at equal incremental production costs to supply a system load. Knowing the value of incremental production cost, the power outputs of individual generators can be calculated. An additional Nth thermal generating unit is brought on to the system to pick up increasing load on the system. The cost coefficients are random, so the incremental production cost will be random. For a particular value of system λ, the generator output is random and is given by

$$P = \frac{\lambda - b}{2a} \tag{B.1}$$

The Monte Carlo method essentially consists of the generation of a large number of repetitive solutions of the incremental production cost model from which sample statistics of the generator output can be calculated using Eq. (B.1).

Random cost coefficients are simulated such that the uncertainty enters into these coefficients. These coefficients being estimated from experimental data are assumed to be normally distributed random variables. If only two parameters, such as mean and standard deviation, are known for normal distribution, it represents the maximum known information concerning the random variable. Each of the values of the random variables employed was computed by adding the generated random variable to a deterministic quantity as for example:

$$a = \bar{a} + \sigma_a \varepsilon_n$$

$$b = \bar{b} + \sigma_b \varepsilon_n$$

Addition of the random error ε_n having the desired distribution to the deterministic cost coefficients (a, b) yielded random variables that were used in simulation. σ_a and σ_b are ensembled standard deviations of normally distributed random variables. The characteristics of the random variables can be controlled through the values of σ_a and σ_b.

From the simulated random variables a and b, the random generation P_i, is found for a particular value of λ. The required coefficient of variation of a random variable and the coefficient of correlation of random variables pair-wise can be obtained by using the following formulae.

Coefficient of variation, $$C_X = \frac{\sigma}{\overline{X}}$$

Coefficient of correlation, $$R_{XY} = \frac{\text{cov}(X, Y)}{\sigma_X \, \sigma_Y}$$

where $$\sigma = \sqrt{\frac{1}{M}\sum_{i=1}^{M}(X_i - \overline{X})^2}$$

with X_i and Y_i as random variables. $\overline{X}$ and $\overline{Y}$ are the means of random variables X_i and Y_i respectively. σ_X and σ_Y are the standard deviations of the random variables X_i and Y_i respectively. M is the number of random numbers.

The values of coefficients of variation and those of correlation coefficients calculated using the above technique are given in Tables B.1 and B.2, respectively, for the following set of values of $\overline{a}_i$ and $\overline{b}_i$.

$\overline{a}_i$	$\overline{b}_i$
0.0050	3.6
0.0040	3.4
0.0045	3.5

Table B.1

Generator	*Mean*	*Coefficients of variation*			*Coefficients of correlation*	
no.	$\overline{P}_i$	C_{a_i}	C_{b_i}	C_{P_i}	$R_{a_iP_i}$	$R_{b_iP_i}$
1	105.636	0.0998	0.00017	0.1028	−0.9932	−0.9907
2	145.508	0.1187	0.00018	0.1233	−0.9902	−0.9958
3	123.788	0.1085	0.00018	0.1121	−0.9918	−0.9808

Table B.2

Generator		*Coefficient of correlation*
i	j	$R_{P_iP_i}$
1	2	0.9997
1	3	0.9999
2	3	0.9999

REFERENCES

Gupta, S.C. and V.K. Kapoor, *Fundamentals of Mathematical Statistics,* Sultan Chand & Sons, 1991.

Parti, S.C., *Stochastic optimal power generation scheduling,* Ph.D. (Thesis), TIET, Patiala, 1987.

Sen Gupta, J.K., *Stochastic Programming,* North Holland, 1972.

C

Kuhn-Tucker Theorem

The Kuhn-Tucker theorem makes it possible to solve the general non-linear programming problem with several variables wherein the variables are also constrained to satisfy certain equality and inequality constraints.

The minimization problem with inequality constraints for control variables can be stated as

$$\min_{u} f(x, u) \tag{C.1}$$

subject to (a) equality constraints $g(x, u, p) = 0$ (C.2)

(b) inequality constraints $u - u^{\max} \leq 0$ (C.3)

$$u^{\min} - u \leq 0 \tag{C.4}$$

where x, u, and p are vectors of variables.

The above problem is converted into an unconstrained problem using Lagrangian

$$L = f(x, u) + \lambda^T g(x, u, p) + \alpha_{\max}^T (u - u^{\max}) + \alpha_{\min}^T (u^{\min} - u) \tag{C.5}$$

where λ is Lagrangian multiplier and the multipliers $\alpha_{\min}$ and $\alpha_{\max}$ are the dual variables associated with the upper and lower limits on control variables. These are auxiliary variables similar to the Lagrangian multiplier λ for the equality constraints case.

The Kuhn-Tucker theorem [Kuhn and Tucker, 1951] gives the necessary conditions for the minimum, assuming convexity for the functions (C.1) to (C.4), as

$$\Delta L = 0 \text{ (gradient with respect to } u, x, \lambda) \tag{C.6}$$

and

$$\left\{ \begin{aligned} &\alpha_{\max}^T (u - u^{\max}) = 0 \\ &\alpha_{\min}^T (u^{\min} - u) = 0 \\ &\alpha_{\max} \geq 0, \alpha_{\min} \geq 0 \end{aligned} \right\} \tag{C.7}$$

Equations (C.7) are known as exclusion equations.

If u_i violates a limit, it can either be upper or lower limit and not both simultaneously. Thus, either inequality constraint (C.3) or (C.4) is active at a time, that is, either $\alpha_{i,\max}$ or $\alpha_{i,\min}$ exists, but never both. Equation (C.6) can be rewritten as

$$\frac{\partial L}{\partial x} = \frac{\partial f}{\partial x} + \left(\frac{\partial g}{\partial x}\right)^T \lambda = 0 \tag{C.8}$$

or

$$\frac{\partial L}{\partial u} = \frac{\partial f}{\partial u} + \left(\frac{\partial g}{\partial u}\right)^T \lambda + \alpha = 0 \tag{C.9}$$

In Eq. (C.9), the elements of α are represented as

$$\alpha_i = \alpha_{i,\,\max} \quad \text{if } u_i - u_i^{\max} < 0$$

$$\alpha_i = -\alpha_{i,\,\min} \quad \text{if } u_i^{\min} - u_i < 0$$

$$\frac{\partial L}{\partial \lambda} = g(x,\, u,\, p) = 0 \tag{C.10}$$

It is evident that α computed from Eq. (C.9) at any feasible solution, with λ from Eq. (C.8) is identical with negative gradient, i.e.

$$\alpha = -\frac{\partial L}{\partial u} = \text{negative gradient with respect to } u \tag{C.11}$$

At the optimum, α must also satisfy the exclusion equations (C.7), which state that

$$\begin{aligned} \alpha_i &= 0 && \text{if} \quad u_i^{\min} < u_i < u_i^{\max} \\ \alpha_i &= \alpha_i^{\max} \geq 0 && \text{if} \quad u_i = u_i^{\max} \\ \alpha_i &= -\alpha_i^{\min} \leq 0 && \text{if} \quad u_i = u_i^{\min} \end{aligned}$$

which can be rewritten in terms of the gradient using Eq. (C.11) as follows:

$$\begin{aligned} \frac{\partial L}{\partial u_i} &= 0 && \text{if} \quad u_i^{\min} < u_i < u_i^{\max} \\ \frac{\partial L}{\partial u_i} &\leq 0 && \text{if} \quad u_i = u_i^{\max} \\ \frac{\partial L}{\partial u_i} &\geq 0 && \text{if} \quad u_i = u_i^{\min} \end{aligned} \tag{C.12}$$

REFERENCES

Kuhn, H.W. and A.W. Tucker, Nonlinear Programming, *Proceedings of the Second Berkeley Symposium on Mathematical Statistics and Probability,* University of California Press, Berkeley, 1951.

Nagrath, I.J. and D.P. Kothari, *Modern Power System Analysis,* Tata McGraw-Hill, New Delhi, 1989.

Nagrath, I.J. and D.P. Kothari, *Power System Engineering,* Tata McGraw-Hill, New Delhi, 1994.

D

Newton-Raphson Method

The Newton–Raphson (NR) method is a powerful method of solving non-linear algebraic equations. Consider an equation

$$f(x) = 0 \tag{D.1}$$

Assume that the initial value of unknown x is x^0. Let Δx^0 be the correction to be found out, which on being added to the initial value give the actual solution. Therefore,

$$f(x^0 + \Delta x^0) = 0 \tag{D.2}$$

Expanding Eq. (D.2) around the initial value x^0 by Taylor's series,

$$f(x^0 + \Delta x^0) = f(x^0) + \Delta x^0 \left(\frac{\partial f}{\partial x}\right)^0 + \frac{1}{2!}(\Delta x^0)^2 \left(\frac{\partial^2 f}{\partial x^2}\right)^0 + \ldots = 0 \tag{D.3}$$

where $\left(\frac{\partial f}{\partial x}\right)^0$ is the derivative of f with respect to x evaluated at initial value, x^0.

Neglecting the second and higher order terms, Eq. (D.3) can be rewritten as

$$f(x^0) + \Delta x^0 \left(\frac{\partial f}{\partial x}\right)^0 = 0 \tag{D.4}$$

or

$$\Delta x^0 = \frac{-f(x^0)}{\left(\frac{\partial f}{\partial x}\right)^0} \tag{D.5}$$

Updating the value of x

$$x^1 = x^0 + \Delta x^0 \tag{D.6}$$

or

$$x^1 = x^0 + \frac{-f(x^0)}{\left(\frac{\partial f}{\partial x}\right)^0} \tag{D.7}$$

In general, for $(r + 1)$th iteration

$$x^{r+1} = x^r + \frac{-f(x^r)}{\left(\frac{\partial f}{\partial x}\right)^r} \tag{D.8}$$

The iterations are continued till the required tolerance is met, i.e.

$$|x^{r+1} - x^r| \le \varepsilon$$

where ε is the required tolerance.

A proper choice of the initial guess is very important in the NR method. If the initial choice happens to be near the root, the technique converges very fast. The initial guess leads to a divergent solution if the slope is very small or nearly equal to zero.

Algorithm D.1: Newton–Raphson Method

1. Read the initial guess x^0, tolerance ε, prescribed lower-bound for derivative of f, Δ, and the number of iterations to be allowed, N.
2. Set iteration counter, $r = 0$.
   REPEAT
3. Compute $f(x^r)$ and $\left(\frac{\partial f}{\partial x}\right)^r$
4. IF $\left|\left(\frac{\partial f}{\partial x}\right)^r\right| \le \Delta$ THEN write $f(x^r)$, $\left(\frac{\partial f}{\partial x}\right)^r$, "slope is small" and GOTO Step 8.
5. Compute x^{r+1} using Eq. (D.8), i.e.

$$x^{r+1} = x^r + \frac{-f(x^r)}{\left(\frac{\partial f}{\partial x}\right)^r}$$

6. Update counter, $r = r + 1$
   UNTIL $(|x^{r+1} - x^r| \le \varepsilon$ or $(r \ge N))$.
7. IF $(r \ge N)$ THEN write "don't converge in N iterations"

   write x^r, x^{r+1}, $f(x^r)$ and $\left(\frac{\partial f}{\partial x}\right)^r$

   ELSE write 'Convergence obtained'

   write x^{r+1}, $f(x^{r+1})$ and $\left(\frac{\partial f}{\partial x}\right)^{r+1}$

8. Stop.

Consider a system of m equations,

$$f_i(x_i, x_2, ..., x_m) = 0 \qquad (i = 1, 2, ..., m) \tag{D.9}$$

Assume that the intial values of unknowns are as x_1^0, x_2^0, ..., x_m^0. Let Δx_1^0, Δx_2^0, ..., Δx_m^0, be corrections to be found out, which on being added to the initial values give the actual solution. Therefore,

$$f(x_1^0 + \Delta x_1^0, x_2^0 + \Delta x_2^0, ..., x_m^0 + \Delta x_m^0) = 0 \tag{D.10}$$

Expanding Eq. (D.10) around initial values by Taylor's series,

$$f(x_1^0, x_2^0, ..., x_m^0) + \left[\Delta x_1^0 \left(\frac{\partial f}{\partial x_1}\right)^0 + \Delta x_2^0 \left(\frac{\partial f}{\partial x_2}\right)^0 + ... + \Delta x_m^0 \left(\frac{\partial f}{\partial x_m}\right)^0\right] + ... = 0 \tag{D.11}$$

where $\left(\frac{\partial f}{\partial x_1}\right)^0, \left(\frac{\partial f}{\partial x_2}\right)^0, ..., \left(\frac{\partial f}{\partial x_m}\right)^0$ are the derivatives of f_i with respect to x_1, x_2, ..., x_m evaluated at x_1^0, x_2^0, ..., x_m^0.

Neglecting the second and higher order terms, Eq. (D.11) can be rewritten as

$$\begin{bmatrix} f_1^0 \\ f_2^0 \\ \vdots \\ f_m^0 \end{bmatrix} + \begin{bmatrix} \left(\frac{\partial f_1}{\partial x_1}\right)^0 & \left(\frac{\partial f_1}{\partial x_2}\right)^0 & \cdots & \left(\frac{\partial f_1}{\partial x_m}\right)^0 \\ \left(\frac{\partial f_2}{\partial x_1}\right)^0 & \left(\frac{\partial f_2}{\partial x_2}\right)^0 & \cdots & \left(\frac{\partial f_2}{\partial x_m}\right)^0 \\ \vdots & \vdots & & \vdots \\ \left(\frac{\partial f_m}{\partial x_1}\right)^0 & \left(\frac{\partial f_m}{\partial x_2}\right)^0 & \cdots & \left(\frac{\partial f_m}{\partial x_m}\right)^0 \end{bmatrix} \begin{bmatrix} \Delta x_1^0 \\ \Delta x_2^0 \\ \vdots \\ \Delta x_m^0 \end{bmatrix} = \begin{bmatrix} 0 \\ 0 \\ \vdots \\ 0 \end{bmatrix} \tag{D.12}$$

or in vector matrix form

$$f^0 + J^0 \Delta x^0 = 0$$

or

$$J^0 \Delta x^0 \cong -f^0 \tag{D.13}$$

These, being a set of linear algebraic equations, can be solved for Δx^0 efficiently by triangularization and back substitution. The updated values of x are then

$$x^1 = x^0 + \Delta x^0 \tag{D.14}$$

In general, for $(r + 1)$th iteration

$$J^r \Delta x^r \cong -f^r \tag{D.15}$$

$$x^{r+1} = x^r + \Delta x^r \tag{D.16}$$

The iterations are continued till the required tolerance,

$$|x_i^{r+1} - x_i^r| \le \varepsilon \quad (i = 1, 2, ..., m) \qquad \text{or} \qquad \sqrt{\sum_{i=1}^{m} (x^{r+1} - x^r)} \le \varepsilon$$

or

$$|f_i(x^r)| \le \varepsilon \quad (i = 1, 2, ..., m) \qquad \text{or} \qquad \sqrt{\sum_{i=1}^{m} (f_i^r)} \le \varepsilon$$

where ε is the required tolerance.

Algorithm D.2: Newton–Raphson Method

1. Read the initial guess x^0, tolerance ε, prescribed lower-bound for derivative of J, Δ, and the number of iterations to be allowed, N.
2. Set iteration counter, $r = 0$
   REPEAT
3. Compute $f(x^r)$ and J^r
4. Solve Eq. (D.15) which is a set of m linear algebraic equations, using triagularization and back substitution procedure

$$J^r \Delta x^r \cong - f^r$$

5. Compute x^{r+1} using Eq. (D.16)

$$x^{r+1} = x^r + \Delta x^r$$

6. Update the iteration counter, $r = r + 1$

UNTIL $\sqrt{\sum_{i=1}^{m} (x^{r+1} - x^r)} \le \varepsilon$ or $\sqrt{\sum_{i=1}^{m} (f_i^r)} \le \varepsilon$ or $(r \ge N)$

7. IF $(r \ge N)$ THEN write "don't converge in N iterations"
   write the output
   ELSE write "Convergence obtained"
   write the output
8. Stop.

REFERENCES

Jain, M.K., S.R.K. Iyengar, and R.K. Jain, *Numerical Methods for Scientific and Engineering Computation*, Wiley Eastern, New Delhi, 1987.

Nagrath, I.J. and D.P. Kothari, *Power System Engineering*, Tata McGraw-Hill, New Delhi, 1994.

Rajaraman, V., *Computer Oriented Numerical Methods,* Prentice-Hall of India, New Delhi, 1986.

APPENDIX E

Gauss Elimination Method

The simultaneous algebraic equations are solved using two methods commonly known as the direct method and the iterative method. The direct method is based on the elimination of variables to transform the set of equations to a triangular form. The iterative method is a successive approximation process.

Consider three simultaneous equations in three unknowns, x_1, x_2, and x_3, i.e.

$$a_{11}x_1 + a_{12}x_2 + a_{13}x_3 = a_{14} \tag{E.1}$$

$$a_{21}x_1 + a_{22}x_2 + a_{23}x_3 = a_{24} \tag{E.2}$$

$$a_{31}x_1 + a_{32}x_2 + a_{33}x_3 = a_{34} \tag{E.3}$$

To eliminate $a_{21}x_1$ from Eq. (E.2), multiply Eq. (E.1) by $\dfrac{a_{21}}{a_{11}}$ (where $a_{11} \neq 0$), i.e.

$$a_{11}\frac{a_{21}}{a_{11}}x_1 + a_{12}\frac{a_{21}}{a_{11}}x_2 + a_{13}\frac{a_{21}}{a_{11}}x_3 = a_{14}\frac{a_{21}}{a_{11}} \tag{E.4}$$

Subtracting Eq. (E.4) from Eq. (E.2), we get

$$\left(a_{21} - a_{11}\frac{a_{21}}{a_{11}}\right)x_1 + \left(a_{22} - a_{12}\frac{a_{21}}{a_{11}}\right)x_2 + \left(a_{23} - a_{13}\frac{a_{21}}{a_{11}}\right)x_3 = a_{24} - a_{14}\frac{a_{21}}{a_{11}} \tag{E.5}$$

Let

$$u_{21} = \frac{a_{21}}{a_{11}}$$

Equation (E.5) can be rewritten as

$$(a_{21} - a_{11}u_{21})x_1 + (a_{22} - a_{12}u_{21})x_2 + (a_{23} - a_{13}u_{21})x_3 = a_{24} - a_{14}u_{21} \tag{E.6}$$

Eliminating $a_{31}x_1$ from Eq. (E.3), multiplying Eq. (E.1) by $\dfrac{a_{31}}{a_{11}}$ (where $a_{11} \neq 0$), we get

$$a_{11}\frac{a_{31}}{a_{11}}x_1 + a_{12}\frac{a_{31}}{a_{11}}x_2 + a_{13}\frac{a_{31}}{a_{11}}x_3 = a_{14}\frac{a_{31}}{a_{11}} \tag{E.7}$$

Subtracting Eq. (E.7) from Eq. (E.3), we obtain

$$\left(a_{31} - a_{11}\frac{a_{31}}{a_{11}}\right)x_1 + \left(a_{32} - a_{12}\frac{a_{31}}{a_{11}}\right)x_2 + \left(a_{33} - a_{13}\frac{a_{31}}{a_{11}}\right)x_3 = a_{34} - a_{14}\frac{a_{31}}{a_{11}} \tag{E.8}$$

Let

$$u_{31} = \frac{a_{31}}{a_{11}}$$

Equation (E.8) can be rewritten as

$$(a_{31} - a_{11}\,u_{31})\,x_1 + (a_{32} - a_{12}\,u_{31})\,x_2 + (a_{33} - a_{13}\,u_{31})\,x_3 = a_{34} - a_{14}\,u_{31} \tag{E.9}$$

In Eqs. (E.6) and (E.9),

$$a_{21} - a_{11}\,u_{21} = a_{31} - a_{11}\,u_{31} = 0$$

So, the reduced equations can be written as

$$a_{11}\,x_1 + a_{12}\,x_2 + a_{13}\,x_3 = a_{14} \tag{E.10}$$

$$a_{22}\,x_2 + a_{23}\,x_3 = a_{24} \tag{E.11}$$

$$a_{32}\,x_2 + a_{33}\,x_3 = a_{34} \tag{E.12}$$

where

$$a_{22} = a_{22} - a_{12}\,u_{21}, \quad a_{23} = a_{23} - a_{13}\,u_{21}, \quad a_{24} = a_{24} - a_{14}\,u_{21}$$

$$a_{32} = a_{32} - a_{12}\,u_{31}, \quad a_{33} = a_{33} - a_{13}\,u_{31}, \quad a_{34} = a_{34} - a_{14}\,u_{31}$$

$$u_{21} = \frac{a_{21}}{a_{11}}, \; u_{31} = \frac{a_{31}}{a_{11}}$$

To eliminate $a_{32}\,x_2$ from Eq. (E.12), multiply Eq. (E.11) by $u_{32} = \dfrac{a_{32}}{a_{22}}$ (where $a_{22} \neq 0$), then Eq. (E.11) becomes

$$a_{22}\,u_{32}\,x_2 + a_{23}\,u_{32}\,x_3 = a_{24}\,u_{32} \tag{E.13}$$

Subtracting Eq. (E.13) from Eq. (E.12), we get

$$(a_{32} - a_{22}\,u_{32})\,x_2 + (a_{33} - a_{23}\,u_{32})\,x_3 = a_{34} - a_{24}\,u_{32} \tag{E.14}$$

The term $(a_{32} - a_{22}\,u_{32}) = 0$.

Equations (E.10), (E.11), and (E.12) can be rewritten in reduced form as

$$a_{11}\,x_1 + a_{12}\,x_2 + a_{13}\,x_3 = a_{14} \tag{E.15}$$

$$a_{22}\,x_2 + a_{23}\,x_3 = a_{24} \tag{E.16}$$

$$a_{33}\,x_3 = a_{34} \tag{E.17}$$

where

$$a_{33} = a_{33} - a_{23}\,u_{32}, \quad a_{34} = a_{34} - a_{24}\,u_{32}$$

$$u_{32} = \frac{a_{32}}{a_{22}}$$

The procedure is known as triangularization. The triangularization process can be generalized to update the coefficients for three simultaneous equations as

$$a_{ij} = a_{ij} - ua_{kj},$$

$$\text{where} \quad u = \frac{a_{ik}}{a_{kk}} \quad (k = 1, 2;\ i = k + 1, k + 2, ..., 3 \text{ and } j = k, k + 1, ..., 4) \tag{E.18}$$

The triangularization process can further be generalized to update the coefficients for n simultaneous equations as

$$a_{ij} = a_{ij} - ua_{kj},$$

$$\text{where} \quad u = \frac{a_{ik}}{a_{kk}} \quad (k = 1, 2, ..., n - 1;\ i = k + 1, k + 2, ..., n \text{ and } j = k, k + 1, ..., n + 1) \tag{E.19}$$

The values of unknowns can be obtained from the triangular forms given in Eqs. (E.15), (E.16) and (E.17) by back substitution process. In back substitution process, x_3 is obtained from Eq. (E.17) and the obtained value is substituted in Eq. (E.16) to find x_2. Further, the obtained values of x_3 and x_2 are substituted in Eq. (E.15) to find x_1.

So, from Eq. (E.17)

$$x_3 = \frac{a_{34}}{a_{33}} \tag{E.20}$$

The value of x_3 can be utilized to evaluate x_2 from Eq. (E.16) as

$$x_2 = (a_{24} - a_{23}\, x_3)\, \frac{1}{a_{22}} \tag{E.21}$$

The values of x_3 and x_2 can be utilized to evaluate x_1 from Eq. (E.15) as

$$x_1 = (a_{14} - a_{12}\, x_2 - a_{13}\, a_3)\, \frac{1}{a_{11}} \tag{E.22}$$

Back substitution process can be generalized as

$$x_3 = \frac{a_{3m}}{a_{33}}, \text{ where } m = 3 + 1 \tag{E.23}$$

$$x_i = \left(a_{im} - \sum_{j=i+1}^{n} a_{ij}\, x_j \right) \frac{1}{a_{ii}}, \text{ where } m = 3 + 1 \text{ and } i = 3 - 1, 3 - 2 \tag{E.24}$$

For n unknowns, the back substitution process can be generalized as

$$x_n = \frac{a_{nm}}{a_{nn}}, \text{ where } m = n + 1 \tag{E.25}$$

$$x_i = \left(a_{im} - \sum_{j=i+1}^{n} a_{ij}\, x_j \right) \frac{1}{a_{ii}}, \text{ where } m = n + 1 \text{ and } i = n - 1, n - 2, ..., 1 \tag{E.26}$$

The diagonal elements in triagularized form are known as pivot elements. If the pivotal element is zero or very small then the procedure leads to no solution or undesired solution. However, the situation can be avoided by interchanging the rows. So in the elimination procedure, the elements should not be zero or a small number. For good precision the pivot element should be the largest in absolute value of all the elements below in its column. The procedure that finds the largest element as the pivot by interchanging the equation is called pivotal condensation. So the elimination process contains three steps: (a) pivotal condensation, (b) triangularization, and (c) back substitution. These three steps are elaborated in the following algorithm.

Algorithm E.1: Gauss Elimination Method

1. Input coefficients, $a_{ij}(i = 1, 2, \ldots, n; j = 1, 2, \ldots, m, m = n+1)$
2. $k = 0$
3. **REPEAT**
   - 3.1 $k = k + 1$
   - 3.2 $\max = |a_{kk}|$ and $l = k$
   - 3.3 $m = k$
   - 3.4 **REPEAT**
     - 3.4.1 $m = m + 1$
     - 3.4.2 **IF** ($| a_{kk} | > \max$) **THEN** $\max = | a_{kk} |$ and $m = k$
   - 3.5 **UNTIL** $(m < n)$
   - 3.6 **IF** ($\max < err$) **THEN** "Write ill condition" and **STOP**
   - 3.7 $p = k - 1$
   - 3.8 **REPEAT**
     - 3.8.1 $p = p + 1$, $T = a_{kp}$, $a_{kp} = a_{lp}$ and $a_{lp} = T$
   - 3.9 **UNTIL** $(p < n + 1)$
   - 3.10 $i = k$
   - 3.11 **REPEAT**
     - 3.11.1 $i = i + 1$, $u = \dfrac{a_{ik}}{a_{kk}}$ and $j = k - 1$
     - 3.11.2 **REPEAT**
       - 3.11.2.1 $j = j + 1$, $a_{ij} = a_{ij} + ua_{kj}$
     - 3.11.3 **UNTIL** $(j < n + 1)$
   - 3.12 **UNTIL** $(i < n)$
4. **UNTIL** $(k < n - 1)$
5. $x_n = \dfrac{a_{nm}}{a_{nn}}$, $m = n + 1$
6. $i = n$
7. **REPEAT**
   - 7.1 $i = i - 1$, sum $= 0$, $j = i$
   - 7.2 **REPEAT**
     - 7.2.1 $j = j + 1$, sum $=$ sum $+ a_{ij} x_j$
   - 7.3 **UNTIL** $(j < n)$
8. $x_i = \dfrac{\text{sum}}{a_{ii}}$
9. **UNTIL** $(i > 1)$
10. **STOP**

Optimal ordering

In power system studies, the matrix of coefficients is quite sparse so that the number of non-zero operations and non-zero storage required in Gauss elimination is very sensitive to the sequence in which the rows are processed. The row sequence that leads to the least number of non-zero operations is not, in general, the same as the one which yields the least storage requirement. It is believed that the absolute optimum sequence of ordering the rows of a large network matrix is too complicated and time consuming to be of any practical value. Therefore, some simple yet effective schemes have been evolved to achieve near optimal ordering with respect to both the criteria. Some of the schemes of near optimal ordering the sparse matrices, which are fully symmetrical or at least symmetric in the pattern of non-zero off-diagonal terms [Tinney and Hart, 1967] are as below.

Scheme 1

Number the matrix rows in the order of the fewest non-zero terms in each row. If more than one unnumbered row has the same number of non-zero terms, number these in any order.

Scheme 2

Number the rows in the order of the fewest non-zero in a row at each step of elimination. This scheme requires updating the count of non-zero terms after each step.

Scheme 3

Number the rows in order of the fewest non-zero off-diagonal terms generated in the remaining rows at each step of elimination. This scheme also involves an updating procedure.

The coice of a scheme is a trade-off between speed of execution and the number of times the result is to be used.

REFERENCES

Aggarwal, S.K., *Optimal Power Flow Studies,* Ph.D. Thesis, B.I.T.S., Pilani, 1970.

Jain, M.K., S.R.K. Iyengar, and R.K. Jain, *Numerical Methods for Scientific and Engineering Computation,* Wiley Eastern, New Delhi, 1987.

Nagrath, I.J. and D.P. Kothari, *Power System Engineering,* Tata McGraw-Hill, New Delhi, 1994.

Rajaraman, V., *Computer Oriented Numerical Methods,* Prentice-Hall of India, New Delhi, 1986.

Singh, L.P., *Advanced Power System Analysis and Dynamics,* 2nd ed., Wiley Eastern, New Delhi, 1986.

Tinney, W.F. and J.W. Walker, Direct solutions of sparse network equations by optimally ordered triangular factorizations, *Proc., IEEE,* **55,** pp. 1801, Nov. 1967.

Tinney, W.F. and C.E. Hart, Power flow solution by Newton's method, *IEEE Trans.,* **PAS-86(11),** pp. 1449, 1967.

APPENDIX F

Primal-Dual Interior Point Method

Since Karmarker's publication in 1984, many variants of Interior Point Methods (IPM) have been developed. Among these variants, the primal-dual IPM proves to be the best algorithm, being most elegant theoretically and the most successful computationally for linear programs. The computational efficiency of PDIPM relies on sparsity techniques. Recently, the method has also been applied to the solution of non-linear problems. The theoretical foundation of PDIPM consists of three methods: (i) Newton's method to solve non-linear equations, (ii) Lagrange's method for optimization with equality constraints, and (iii) Fiacco and McCormick's barrier method for optimization with inequality constraints.

F.1 PREDICTOR-CORRECTOR INTERIOR POINT ALGORITHM FOR LINEAR PROGRAMMING

Consider the following linear programming (LP) problem.

$$\text{Minimize} \quad \sum_{i=1}^{N} c_i\, x_i \tag{F.1a}$$

$$\text{subject to} \quad \sum_{j=1}^{N} a_{ij}\, x_j = b_i \qquad (i = 1, 2, ..., M) \tag{F.1b}$$

$$x_i^{\min} \le x_i \le x_i^{\max} \qquad (i = 1, 2, ..., N) \tag{F.1c}$$

where

c_i are cost coefficients.

x_i are unknown variables.

$x_i^{\min}$ is minimum limit of variable x_i.

$x_i^{\max}$ is maximum limit of variable x_i.

Introducing slack variables s and v to transform the bound constraints to equality constraints, the LP problem (F.1) becomes:

$$\text{Minimize} \quad \sum_{i=1}^{N} c_i\, x_i \tag{F.2a}$$

$$\text{subject to} \quad \sum_{j=1}^{N} a_{ij}\, x_j = b_i \qquad (i = 1, 2, ..., M) \tag{F.2b}$$

$$x_i - v_i = x_i^{\min} \qquad (i = 1, 2, ..., N) \tag{F.2c}$$

$$x_i + s_i = x_i^{\max} \qquad (i = 1, 2, ..., N) \tag{F.2d}$$

$$v_i \ge 0,\ s_i \ge 0 \qquad (i = 1, 2, ..., N) \tag{F.2e}$$

To eliminate the above non-negativity constraints, the objective function is appended with a logarithmic barrier term incorporating these constraints, i.e.

$$L(x_i, y_i, v_i, s_i, \alpha_i, \beta_i) = \sum_{i=1}^{N} c_i\, x_i - \mu \sum_{i=1}^{N} (\log v_i + \log s_i) - \sum_{i=1}^{M} y_i \left(\sum_{j=1}^{N} a_{ij}\, x_j - b_i \right)$$
$$+ \sum_{i=1}^{N} \alpha_i\, (-x_i + v_i + x_i^{\min}) + \sum_{i=1}^{N} \beta_i\, (x_i + s_i - x_i^{\max}) \tag{F.3}$$

where the barrier parameter $\mu > 0$ and is decreased to zero as the algorithm iteration progress. The solution of (F.3) is defined by the Karush-Kuhn-Tucker first-order necessary conditions.

Differentiating Eq. (F.3) w.r.t. x_i,

$$\frac{\partial L}{\partial x_i} = c_i - \sum_{j=1}^{M} a_{ji}\, y_j - \alpha_i + \beta_i = 0 \qquad (i = 1, 2, ..., N)$$

or

$$\sum_{j=1}^{M} a_{ji}\, y_j + \alpha_i - \beta_i = c_i \qquad (i = 1, 2, ..., N) \tag{F.4a}$$

Differentiating Eq. (F.3) w.r.t. y_i,

$$\frac{\partial L}{\partial y_i} = \sum_{j=1}^{M} a_{ij}\, x_j - b_i = 0 \qquad (i = 1, 2, ..., M)$$

or

$$\sum_{j=1}^{M} a_{ij}\, x_j = b_i \qquad (i = 1, 2, ..., M) \tag{F.4b}$$

Differentiating Eq. (F.3) w.r.t. α_i

$$\frac{\partial L}{\partial \alpha_i} = -x_i + v_i + x_i^{\min} = 0 \qquad (i = 1, 2, ..., N)$$

or

$$x_i - v_i = x_i^{\min} \qquad (i = 1, 2, ..., N) \tag{F.4c}$$

Differentiating Eq. (F.3) w.r.t. β_i

$$\frac{\partial L}{\partial \beta_i} = x_i + s_i - x_i^{\max} = 0 \qquad (i = 1, 2, ..., N)$$

or

$$x_i + s_i = x_i^{\max} \qquad (i = 1, 2, ..., N) \tag{F.4d}$$

Differentiating Eq. (F.3) w.r.t. v_i

$$\frac{\partial L}{\partial v_i} = -\frac{\mu}{v_i} + \alpha_i = 0 \qquad (i = 1, 2, ..., N)$$

or

$$v_i \alpha_i = \mu \qquad (i = 1, 2, ..., N) \tag{F.4e}$$

Differentiating Eq. (F.3) w.r.t. s_i

$$\frac{\partial L}{\partial s_i} = -\frac{\mu}{s_i} + \beta_i = 0 \qquad (i = 1, 2, ..., N)$$

$$s_i \beta_i = \mu \qquad (i = 1, 2, ..., N) \tag{F.4f}$$

The system of Kuhn-Tucker conditions formulated may be modified by introducing the following change in variables.

$$\begin{aligned}
x_i &= x_i + \Delta x_i && (i = 1, 2, ..., N)\\
y_i &= y_i + \Delta y_i && (i = 1, 2, ..., M)\\
v_i &= v_i + \Delta v_i && (i = 1, 2, ..., N)\\
s_i &= s_i + \Delta s_i && (i = 1, 2, ..., N)\\
\alpha_i &= \alpha_i + \Delta \alpha_i && (i = 1, 2, ..., N)\\
\beta_i &= \beta_i + \Delta \beta_i && (i = 1, 2, ..., N)
\end{aligned}$$

Introducing change in variables, Eq. (F.4b) can be rewritten as

$$\sum_{j=1}^{M} a_{ij}(x_j + \Delta x_j) = b_i \qquad (i = 1, 2, ..., M)$$

or

$$\sum_{j=1}^{M} a_{ij}\,\Delta x_j = b_i - \sum_{j=1}^{M} a_{ij}\, x_j \qquad (i = 1, 2, ..., M) \tag{F.5a}$$

Introducing change in variables, Eq. (F.4c) can be rewritten as

$$(x_i + \Delta x_i) - (v_i + \Delta v_i) = x_i^{\min} \qquad (i = 1, 2, ..., N)$$

or

$$\Delta x_i - \Delta v_i = x_i^{\min} - x_i + v_i \qquad (i = 1, 2, ..., N) \qquad \text{(F.5b)}$$

Introducing change in variables, Eq. (F.4d) can be rewritten as

$$(x_i + \Delta x_i) + (s_i + \Delta s_i) = x_i^{\max} \qquad (i = 1, 2, ..., N)$$

or

$$\Delta x_i + \Delta s_i = x_i^{\max} - x_i - s_i \qquad (i = 1, 2, ..., N) \qquad \text{(F.5c)}$$

Introducing change in variables, Eq. (F.4a) can be rewritten as

$$\sum_{j=1}^{M} a_{ji}(y_j + \Delta y_j) + (\alpha_i + \Delta\alpha_i) - (\beta_i + \Delta\beta_i) = c_i \qquad (i = 1, 2, ..., N)$$

or

$$\sum_{j=1}^{M} a_{ji}\,\Delta y_j + \Delta\alpha_i - \Delta\beta_i = c_i - \sum_{j=1}^{M} a_{ji}\,y_j + \beta_i - \alpha_i \qquad (i = 1, 2, ..., N) \qquad \text{(F.5d)}$$

Introducing change in variables, Eq. (F.4e) can be rewritten as

$$(v_i + \Delta v_i)\,(\alpha_i + \Delta\alpha_i) = \mu \qquad (i = 1, 2, ..., N)$$

or

$$v_i\,\Delta\alpha_i + \alpha_i\,\Delta v_i = \mu - v_i\,\alpha_i - \Delta v_i\,\Delta\alpha_i \qquad (i = 1, 2, ..., N) \qquad \text{(F.5e)}$$

Introducing change in variables, Eq. (F.4f) can be rewritten as

$$(s_i + \Delta s_i)\,(\beta_i + \Delta\beta_i) = \mu \qquad (i = 1, 2, ..., N)$$

or

$$s_i\,\Delta\beta_i + \beta_i\,\Delta s_i = \mu - s_i\,\beta_i - \Delta s_i\,\Delta\beta_i \qquad (i = 1, 2, ..., N) \qquad \text{(F.5f)}$$

The right-hand sides of Eqs. (F.5e) and (F.5f) have non-linear terms $\Delta v_i\,\Delta\alpha_i$ and $\Delta s_i\,\Delta\beta_i$. Since these non-linear terms are unknown so these can be approximately solved in two steps. The first step, "predictor", estimates the non-linear terms by solving equations (F.5a–F.5f) without μ and the non-linear terms for a primal-dual affine direction.

Equation (F.5c) can be rewritten, ignoring μ and non-linear terms, to estimate change in variables as

$$\Delta\tilde{s}_i = (x_i^{\max} - x_i - s_i) - \Delta\tilde{x}_i \qquad (i = 1, 2, ..., N)$$

or

$$\Delta\tilde{s}_i = r_i^M - \Delta\tilde{x}_i \qquad (i = 1, 2, ..., N) \qquad \text{(F.6a)}$$

where $r_i^M = x_i^{\max} - x_i - s_i$

Equation (F.5b) can be rewritten, ignoring μ and non-linear terms, to estimate change in variables as

$$\Delta\tilde{v}_i = \Delta\tilde{x}_i - (x_i^{\min} - x_i + v_i) \qquad (i = 1, 2, ..., N)$$

or

$$\Delta\tilde{v}_i = \Delta\tilde{x}_i - r_i^m \quad (i = 1, 2, ..., N) \tag{F.6b}$$

where $r_i^m = x_i^{\min} - x_i + v_i$

Ignoring μ and $\Delta s_i \Delta\beta_i$ (non-linear terms) from Eq. (F.5f), change in variables can be estimated as

$$s_i \Delta\tilde{\beta}_i + \beta_i \Delta\tilde{s}_i = -s_i\beta_i \quad (i = 1, 2, ..., N)$$

or

$$s_i \Delta\tilde{\beta}_i = -s_i \beta_i - \beta_i \Delta\tilde{s}_i \quad (i = 1, 2, ..., N)$$

Dividing by s_i,

$$\Delta\tilde{\beta}_i = -\beta_i - \frac{\beta_i}{s_i}\Delta\tilde{s}_i \quad (i = 1, 2, ..., N)$$

or

$$\Delta\tilde{\beta}_i = -\beta_i \left(1 + \frac{1}{s_i}\Delta\tilde{s}_i\right) \quad (i = 1, 2, ..., N) \tag{F.6c}$$

Ignoring μ and $\Delta v_i \Delta\alpha_i$ (non-linear terms) from Eq. (F.5e), change in variables can be estimated as

$$v_i \Delta\tilde{\alpha}_i + \alpha_i \Delta\tilde{v}_i = -v_i \alpha_i \quad (i = 1, 2, ..., N)$$

or

$$v_i \Delta\tilde{\alpha}_i = -v_i \alpha_i - \alpha_i \Delta\tilde{v}_i \quad (i = 1, 2, ..., N)$$

Dividing by v_i

$$\Delta\tilde{\alpha}_i = -\alpha_i - \frac{\alpha_i}{v_i}\Delta\tilde{v}_i \quad (i = 1, 2, ..., N)$$

or

$$\Delta\tilde{\alpha}_i = -\alpha_i \left(1 + \frac{1}{v_i}\Delta\tilde{v}_i\right) \quad (i = 1, 2, ..., N) \tag{F.6d}$$

From Eq. (F.5d), change in variables can be estimated as

$$\Delta\tilde{\beta}_i - \Delta\tilde{\alpha}_i = \sum_{j=1}^{M} a_{ji}\Delta\tilde{y}_j - \left(c_i - \sum_{j=1}^{M} a_{ji} y_j + \beta_i - \alpha_i\right) \quad (i = 1, 2, ..., N)$$

Substitute Eqs. (F.6c) and (F.6d) into the above equation,

$$-\beta_i \left(1 + \frac{1}{s_i}\Delta\tilde{s}_i\right) + \alpha_i \left(1 + \frac{1}{v_i}\Delta\tilde{v}_i\right) = \sum_{j=1}^{M} a_{ji}\Delta\tilde{y}_j - \left(c_i - \sum_{j=1}^{M} a_{ji} y_j + \beta_i - \alpha_i\right) \quad (i = 1, 2, ..., N)$$

or

$$-\frac{\beta_i}{s_i}\Delta\tilde{s}_i + \frac{\alpha_i}{v_i}\Delta\tilde{v}_i = \sum_{j=1}^{M} a_{ji}\Delta\tilde{y}_j - \left(c_i - \sum_{j=1}^{M} a_{ji} y_j\right) \qquad (i = 1, 2, ..., N)$$

Substitute Eqs. (F.6a) and (F.6b) into the above equation,

$$-\frac{\beta_i}{s_i}(r_i^m - \Delta\tilde{x}_i) + \frac{\alpha_i}{v_i}(\Delta\tilde{x}_i - r_i^m) = \sum_{j=1}^{M} a_{ji}\Delta\tilde{y}_j - \left(c_i - \sum_{j=1}^{M} a_{ji} y_j\right) \qquad (i = 1, 2, ..., N)$$

Rearranging the above equation,

$$\left(\frac{\beta_i}{s_i} + \frac{\alpha_i}{v_i}\right)\Delta\tilde{x}_i = \sum_{j=1}^{M} a_{ji}\Delta\tilde{y}_j - \left(c_i - \sum_{j=1}^{M} a_{ji} y_j - \left(\frac{\beta_i}{s_i} r_i^M + \frac{\alpha_i}{v_i} r_i^m\right)\right) \qquad (i = 1, 2, ..., N)$$

or

$$\Delta\tilde{x}_i = D_i\left(\sum_{j=1}^{M} a_{ji}\Delta\tilde{y}_j - \rho_i\right) \qquad (i = 1, 2, ..., N) \tag{F.6e}$$

where

$$\rho_i = c_i - \sum_{j=1}^{M} a_{ji} y_j - \left(\frac{\beta_i}{s_i} r_i^M + \frac{\alpha_i}{v_i} r_i^m\right)$$

$$D_i = \frac{v_i s_i}{\alpha_i s_i + \beta_i v_i}$$

Substituting Eq. (F.6e) into Eq. (F.5a),

$$\sum_{j=1}^{M} a_{ij}\left[D_j\left(\sum_{k=1}^{M} a_{kj}\Delta\tilde{y}_j - \rho_j\right)\right] = b_i - \sum_{j=1}^{M} a_{ij} x_j \qquad (i = 1, 2, ..., M)$$

or

$$\sum_{j=1}^{M} a_{ij}\left(D_j \sum_{k=1}^{M} a_{kj}\Delta\tilde{y}_j\right) = b_i - \sum_{j=1}^{M} a_{ij} x_j + \sum_{j=1}^{M} a_{ij} D_j \rho_j \quad (i = 1, 2, ..., M) \tag{F.6f}$$

The second step, "corrector", uses the affine direction to approximate the non-linear terms in the right-hand side of Eqs. (F.5e) and (F.5f) for the actual search direction.

From Eq. (F.6a), non-linear terms are approximated as

$$\Delta s_i = r_i^M - \Delta x_i \qquad (i = 1, 2, ..., N) \tag{F.7a}$$

where $r_i^M = x_i^{\max} - x_i - s_i$

From Eq. (F.6b), non-linear terms are approximated by considering estimated values as

$$\Delta v_i = \Delta x_i - r_i^m \qquad (i = 1, 2, ..., N) \tag{F.7b}$$

where $r_i^m = x_i^{\min} - x_i - v_i$

Approximation of non-linear terms can be obtained from Eq. (F.5e) by considering estimated values as

$$v_i \Delta\alpha_i + \alpha_i \Delta v_i = \mu - \Delta\tilde{v}_i \Delta\tilde{\alpha}_i - v_i \alpha_i \qquad (i = 1, 2, ..., N)$$

or

$$v_i \Delta\alpha_i = (\mu - \Delta\tilde{v}_i \Delta\tilde{\alpha}_i) - v_i \alpha_i - \alpha_i \Delta v_i \qquad (i = 1, 2, ..., N)$$

or

$$\Delta\alpha_i = \frac{1}{v_i}(\mu - \Delta\tilde{v}_i \Delta\tilde{\alpha}_i) - \alpha_i \left(1 + \frac{1}{v_i}\Delta v_i\right) \qquad (i = 1, 2, ..., N)$$

or

$$\Delta\alpha_i = \sigma_i - \alpha_i \left(1 + \frac{1}{v_i}\Delta v_i\right) \qquad (i = 1, 2, ..., N) \tag{F.7c}$$

where $\sigma_i = \frac{1}{v_i}(\mu - \Delta\tilde{v}_i \Delta\tilde{\alpha}_i)$

Approximation of non-linear terms can be obtained from Eq, (F.5f) by considering estimated values as

$$s_i \Delta\beta_i + \beta_i \Delta s_i = (\mu - \Delta\tilde{s}_i \Delta\tilde{\beta}_i) - s_i \beta_i \qquad (i = 1, 2, ..., N)$$

or

$$s_i \Delta\beta_i = (\mu - \Delta\tilde{s}_i \Delta\tilde{\beta}_i) - s_i \beta_i - \beta_i \Delta s_i \qquad (i = 1, 2, ..., N)$$

or

$$\Delta\beta_i = \frac{1}{s_i}(\mu - \Delta\tilde{s}_i \Delta\tilde{\beta}_i) - \beta_i \left(1 + \frac{1}{s_i}\Delta s_i\right) \qquad (i = 1, 2, ..., N)$$

or

$$\Delta\beta_i = \delta_i - \beta_i \left(1 + \frac{1}{s_i}\Delta s_i\right) \qquad (i = 1, 2, ..., N) \tag{F.7d}$$

where $\delta_i = \frac{1}{s_i}(\mu - \Delta\tilde{s}_i \Delta\tilde{\beta}_i)$

Considering estimated values, approximation of non-linear terms can be obtained from Eq. (F.5d) as

$$\Delta\beta_i - \Delta\alpha_i = \sum_{j=1}^{M} a_{ji} \Delta y_j - \left(c_i - \sum_{j=1}^{M} a_{ji} y_j + \beta_i - \alpha_i\right) \qquad (i = 1, 2, ..., N)$$

Substituting Eqs. (F.7c) and (F.7d) into the above equation,

$$\delta_i - \beta_i \left(1 + \frac{1}{s_i}\Delta s_i\right) - \sigma_i + \alpha_i \left(1 + \frac{1}{v_i}\Delta v_i\right)$$

$$= \sum_{j=1}^{M} a_{ji}\Delta y_j - \left(c_i - \sum_{j=1}^{M} a_{ji} y_j + \beta_i - \alpha_i\right) \qquad (i = 1, 2, ..., N)$$

or

$$-\frac{\beta_i}{s_i}\Delta s_i + \frac{\alpha_i}{v_i}\Delta v_i = \sum_{j=1}^{M} a_{ji}\Delta y_j - \left(c_i - \sum_{j=1}^{M} a_{ji} y_j\right) - \delta_i + \sigma_i \qquad (i = 1, 2, ..., N)$$

Substituting Eqs. (F.7a) and (F.7b) into the above equation,

$$-\frac{\beta_i}{s_i}(r_i^M - \Delta x_i) + \frac{\alpha_i}{v_i}(\Delta x_i - r_i^m) = \sum_{j=1}^{M} a_{ji}\Delta y_j - \left(c_i - \sum_{j=1}^{M} a_{ji} y_j + \delta_i - \sigma_i\right) \qquad (i = 1, 2, ..., N)$$

Rearranging the above equation,

$$\left(\frac{\beta_i}{s_i} + \frac{\alpha_i}{v_i}\right)\Delta x_i = \sum_{j=1}^{M} a_{ji}\Delta y_j - \left(c_i - \sum_{j=1}^{M} a_{ji} y_j - \left(\frac{\beta_i}{s_i} r_i^M + \frac{\alpha_i}{v_i} r_i^m\right) + \delta_i - \sigma_i\right) \qquad (i = 1, 2, ..., N)$$

or

$$\Delta x_i = D_i \left(\sum_{j=1}^{M} a_{ji}\Delta y_j - \eta_i\right) \qquad (i = 1, 2, ..., N) \qquad \text{(F.7e)}$$

where

$$\eta_i = c_i - \sum_{j=1}^{M} a_{ji} y_j - \left(\frac{\beta_i}{s_i} r_i^M + \frac{\alpha_i}{v_i} r_i^m\right) + \delta_i - \sigma_i$$

$$D_i = \frac{v_i s_i}{\alpha_i s_i + \beta_i v_i}$$

Substituting Eq. (F.7e) into Eq. (F.5a),

$$\sum_{j=1}^{M} a_{ij}\left[D_j \left(\sum_{k=1}^{M} a_{kj}\Delta y_j - \eta_j\right)\right] = b_i - \sum_{j=1}^{M} a_{ij} x_j \qquad (i = 1, 2, ..., M)$$

$$\sum_{j=1}^{M} a_{ij}\left(D_j \sum_{k=1}^{M} a_{kj}\Delta \tilde{y}_j\right) = b_i - \sum_{j=1}^{M} a_{ij} x_j + \sum_{j=1}^{M} a_{ij} D_j \eta_j \qquad (i = 1, 2, ..., M) \qquad \text{(F.7f)}$$

Finally, new variables are determined by updating the old variables as given below:

$$x_i = x_i + \alpha_p \Delta x_i \qquad (i = 1, 2, ..., N)$$

$$y_i = y_i + \alpha_d \Delta y_i \qquad (i = 1, 2, ..., M)$$

$$v_i = v_i + \alpha_p \Delta v_i \qquad (i = 1, 2, ..., N)$$

$$s_i = s_i + \alpha_p \Delta s_i \qquad (i = 1, 2, ..., N)$$

$$\alpha_i = \alpha_i + \alpha_d \Delta \alpha_i \qquad (i = 1, 2, ..., N)$$

$$\beta_i = \beta_i + \alpha_d \Delta \beta_i \qquad (i = 1, 2, ..., N)$$

where α_p and α_d are step sizes which are chosen to preserve the non-negativity conditions on variables with the following ratio test:

$$\tilde{\alpha}_p = \min\left\{-\frac{v_j}{\Delta v_j}, -\frac{s_j}{\Delta s_j}, \Delta v_j < 0, \Delta s_j < 0\right\} \tag{F.8a}$$

$$\tilde{\alpha}_d = \min\left\{-\frac{\alpha_j}{\Delta \alpha_j}, -\frac{\beta_j}{\Delta \beta_j}, \Delta \alpha_j < 0, \Delta \beta_j < 0\right\} \tag{F.8b}$$

and

$$\alpha_p = \min\{1, \zeta \tilde{a}_p\} \tag{F.9a}$$

$$\alpha_d = \min\{1, \zeta \tilde{a}_d\} \tag{F.9b}$$

where ζ is a factor that is chosen to prevent non-negative variables from being zero.

The Predictor-Corrector Interior-Point Algorithm (PCIPA) can be described as follows:

1. Starting with an initial point satisfying v_i, s_i, α_i, and $\beta_i \geq 0$ and with $\mu > 0$. An important feature of PCIPA is that no initial feasible point is required. But the primal and dual slack variables must be strictly positive. There are many methods to get such a starting point [McShane et al., 1989; Lustig et al, 1991; Lustig et al., 1992; Vanderbei, 1993].
2. Check convergence. The following feasibility conditions are checked for convergence:

$$\left\| \sum_{i=1}^{N} a_{ij} x_j - b_i \right\| \leq \varepsilon_1$$

$$\left\| \sum_{j=1}^{M} a_{ji} y_j + \alpha_i - \beta_i - c_i \right\| \leq \varepsilon_2$$

$$\left\| \frac{\sum_{i=1}^{N} c_i x_i - \left(\sum_{i=1}^{M} y_i b_i - \sum_{i=1}^{N} \beta_i x_i^{\min} + \sum_{i=1}^{N} \alpha_i x_i^{\max} \right)}{1 + \sum_{j=1}^{M} y_j b_j - \sum_{i=1}^{N} \beta_i x_i^{\min} + \sum_{i=1}^{N} \alpha_i x_i^{\max}} \right\| \leq \varepsilon_3$$

3. Calculate the primal-dual affine direction by solving Eqs. (F.6),

$$\sum_{j=1}^{M} a_{ij}\left(D_j \sum_{k=1}^{M} a_{kj}\Delta\tilde{y}_j\right) = b_i - \sum_{j=1}^{M} a_{ij}x_j + \sum_{k=1}^{M} a_{ij}D_j\rho_j \qquad (i = 1, 2, ..., M)$$

$$\Delta\tilde{x}_i = D_i\left(\sum_{j=1}^{M} a_{ji}\Delta\tilde{y}_j - \rho_i\right) \qquad (i = 1, 2, ..., N)$$

$$\Delta\tilde{\alpha}_i = -\alpha_i\left(1 + \frac{1}{v_i}\Delta\tilde{v}_i\right) \qquad (i = 1, 2, ..., N)$$

$$\Delta\tilde{\beta}_i = -\beta_i\left(1 + \frac{1}{s_i}\Delta\tilde{s}_i\right) \qquad (i = 1, 2, ..., N)$$

$$\Delta\tilde{v}_i = \Delta\tilde{x}_i - r_i^m \qquad (i = 1, 2, ..., N)$$

$$\Delta\tilde{s}_i = r_i^M - \Delta\tilde{x}_i \qquad (i = 1, 2, ..., N)$$

where

$$\rho_i = c_i - \sum_{j=1}^{M} a_{ji}y_j - \left(\frac{\beta_i}{s_i}r_i^M + \frac{\alpha_i}{v_i}x_i^m\right)$$

$$D_i = \frac{v_i s_i}{\alpha_i s_i + \beta_i v_i}$$

$$r_i^M = x_i^{\max} - x_i - s_i$$

$$r_i^m = x_i^{\text{mix}} - x_i + v_i$$

4. To calculate the barrier point μ, the following is used to include:

$$\mu = \begin{cases} C_g^3 \div \left[\sum_{i=1}^{N} (v_i\alpha_i + \beta_i s_i)\, n\right] & ;\text{ if feasible} \\ \sum_{i=1}^{N} (v_i\alpha_i + \beta_i s_i)\, n & ;\text{ otherwise} \end{cases}$$

where

$$C_g = \sum_{i=1}^{N}\left[(v_i + \tilde{\alpha}_p\Delta\tilde{v}_i)(\alpha_i + \tilde{\alpha}_d\Delta\tilde{\alpha}_i) + (s_i + \tilde{\alpha}_p\Delta\tilde{s}_i)(\beta_i + \tilde{\alpha}_d\Delta\tilde{\beta}_i)\right]$$

$\tilde{\alpha}_p$ and $\tilde{\alpha}_d$ are determined using Eqs. (F.8)

$$\tilde{\alpha}_p = \min\left\{-\frac{v_j}{\Delta v_j}, -\frac{s_j}{\Delta s_j}, \Delta v_j < 0, \Delta s_j < 0\right\}$$

$$\tilde{\alpha}_d = \min\left\{-\frac{\alpha_j}{\Delta\alpha_j}, -\frac{\beta_j}{\Delta\beta_j}, \Delta\alpha_j < 0, \Delta\beta_j < 0\right\}$$

5. Compute the final search direction using Eqs. (F.7),

$$\sum_{j=1}^{M} a_{ij}\left(D_j \sum_{k=1}^{M} a_{kj}\,\Delta\tilde{y}_j\right) = b_i - \sum_{j=1}^{M} a_{ij}x_j + \sum_{j=1}^{M} a_{ij}D_j\eta_j \quad (i = 1, 2, ..., M)$$

$$\Delta x_i = D_i\left(\sum_{j=1}^{M} a_{ji}\,\Delta y_j - \eta_i\right) \quad (i = 1, 2, ..., N)$$

$$\Delta\beta_i = \delta_i - \beta_i\left(1 + \frac{1}{s_i}\Delta s_i\right) \quad (i = 1, 2, ..., N)$$

$$\Delta\alpha_i = \sigma_i - \alpha_i\left(1 + \frac{1}{v_i}\Delta v_i\right) \quad (i = 1, 2, ..., N)$$

$$\Delta v_i = \Delta x_i - r_i^m \quad (i = 1, 2, ..., N)$$

$$\Delta s_i = r_i^M - \Delta x_i \quad (i = 1, 2, ..., N)$$

where

$$\eta_i = c_i - \sum_{j=1}^{M} a_{ji}y_j - \left(\frac{\beta_i}{s_i}r_i^M + \frac{\alpha_i}{v_i}r_i^m\right) + \delta_i - \sigma_i$$

$$D_i = \frac{v_i s_i}{\alpha_i s_i + \beta_i v_i}$$

$$\delta_i = \frac{1}{s_i}(\mu - \Delta s_i\,\Delta\beta_i)$$

$$\sigma_i = \frac{1}{v_i}(\mu - \Delta\tilde{v}_i\,\Delta\tilde{\alpha}_i)$$

$$r_i^m = x_i^{\min} - x_i + v_i$$

$$r_i^M = x_i^{\max} - x_i - s_i$$

6. Update the variables,

$$x_i = x_i + \alpha_p\Delta x_i \quad (i = 1, 2, ..., N)$$

$$y_i = y_i + \alpha_d\Delta y_i \quad (i = 1, 2, ..., M)$$

$$v_i = v_i + \alpha_p\Delta v_i \quad (i = 1, 2, ..., N)$$

$$s_i = s_i + \alpha_p\Delta s_i \quad (i = 1, 2, ..., N)$$

$$\alpha_i = \alpha_i + \alpha_d\Delta\alpha_i \quad (i = 1, 2, ..., N)$$

$$\beta_i = \beta_i + \alpha_d\Delta\beta_i \quad (i = 1, 2, ..., N)$$

where α_p and α_d are step sizes which are chosen to preserve the non-negativity conditions on variables with the following ratio test:

$$\tilde{\alpha}_p = \min\left\{-\frac{v_j}{\Delta v_j}, -\frac{s_j}{\Delta s_j}, \Delta v_j < 0, \Delta s_j < 0\right\}$$

$$\tilde{\alpha}_d = \min\left\{-\frac{\alpha_j}{\Delta \alpha_j}, -\frac{\beta_j}{\Delta \beta_j}, \Delta \alpha_j < 0, \Delta \beta_j < 0\right\}$$

and

$$\alpha_p = \min\{1, \zeta\tilde{\alpha}_p\}$$

$$\alpha_d = \min\{1, \zeta\tilde{\alpha}_d\}$$

F.2 PREDICTOR-CORRECTOR INTERIOR POINT ALGORITHM FOR NON-LINEAR PROGRAMMING

Consider the following non-linear programming (NLP) problem comprising both equality and inequality constraints.

Minimize $f(x)$ (F.10a)

subject to $g_i(x) = 0 \quad (i = 1, 2, ..., M)$ (F.10b)

$x_i^{\min} \le x_i \le x_i^{\max} \quad (i = 1, 2, ..., N)$ (F.10c)

where

$f(x)$ is the objective function

$g_i(x)$ are equality constraints

$x_i^{\min}$ is minimum limit of variable x_i

$x_i^{\max}$ is maximum limit of variable x_i.

Introducing slack variables s and v to transform the bound constraints to equality constraints, the NLP problem (F.10) becomes

Minimize $f(x)$ (F.11a)

subject to $g_i(x) = 0 \quad (i = 1, 2, ..., M)$ (F.11b)

$x_i - v_i = x_i^{\min} \quad (i = 1, 2, ..., N)$ (F.11c)

$x_i + s_i = x_i^{\max} \quad (i = 1, 2, ..., N)$ (F.11d)

$v_i \ge 0,\ s_i \ge 0 \quad (i = 1, 2, ..., N)$ (F.11e)

To eliminate the above non-negativity constraints, append the objective function with a logarithmic barrier term incorporating these constraints, i.e.

$$L(x_i, y_i, \lambda_j, s_i, \alpha_i, \beta_i) = f(x) - \mu\sum_{i=1}^{N}(\log v_i + \log s_i) - \sum_{j=1}^{M}\lambda_j g_j(x) + \sum_{i=1}^{N}\alpha_i(-x_i + v_i + x_i^{\min})$$

$$+ \sum_{i=1}^{N}\beta_i(x_i + s_i - x_i^{\max}) \qquad \text{(F.12)}$$

where the barrier parameter $\mu > 0$ and is decreased to zero as the algorithm iteration progresses. The solutions of (F.12) are defined by the Karush-Kuhn-Tucker first-order necessary conditions.

$$\frac{\partial L}{\partial x_i} = \frac{\partial f(x)}{\partial x_i} - \sum_{j=1}^{M} \lambda_j \frac{\partial g_j}{\partial x_i} - \alpha_i + \beta_i = 0 \qquad (i = 1, 2, ..., N) \tag{F.12a}$$

$$\frac{\partial L}{\partial \lambda_j} = g_j(x) = 0 \qquad (i = 1, 2, ..., M) \tag{F.12b}$$

$$\frac{\partial L}{\partial \beta_i} = x_i + s_i - x_i^{\max} = 0 \qquad (i = 1, 2, ..., N) \tag{F.12c}$$

$$\frac{\partial L}{\partial \alpha_i} = x_i + v_i + x_i^{\min} = 0 \qquad (i = 1, 2, ..., N) \tag{F.12d}$$

$$\frac{\partial L}{\partial v_i} = -\frac{\mu}{v_i} + \alpha_i = 0 \qquad (i = 1, 2, ..., N)$$

$$v_i \alpha_i = \mu \qquad (i = 1, 2, ..., N) \tag{F.12e}$$

$$\frac{\partial L}{\partial s_i} = -\frac{\mu}{s_i} + \beta_i = 0 \qquad (i = 1, 2, ..., N)$$

$$s_i \beta_i = \mu \qquad (i = 1, 2, ..., N) \tag{F.12f}$$

Such conditions are also sufficient if the problem is convex. The iteration of the algorithm consists in applying the Newton method to the so-obtained non-linear system of Kuhn-Tucker conditions. Expand Eqs. (F.12a) to (F.12f) using Taylor's expansion.

$$\sum_{j=1}^{N} \frac{\partial^2 L}{\partial x_i \partial x_j} \Delta x_j + \sum_{j=1}^{M} \frac{\partial^2 L}{\partial x_i \partial \lambda_j} \Delta \lambda_j + \sum_{j=1}^{N} \frac{\partial^2 L}{\partial x_i \partial \alpha_j} \Delta \alpha_j + \sum_{j=1}^{N} \frac{\partial^2 L}{\partial x_i \partial \beta_j} \Delta \beta_j + \sum_{j=1}^{N} \frac{\partial^2 L}{\partial x_i \partial v_j} \Delta v_j$$

$$+ \sum_{j=1}^{N} \frac{\partial^2 L}{\partial x_i \partial s_j} \Delta s_j = -\frac{\partial L}{\partial x_i} \tag{F.13a}$$

$$\sum_{j=1}^{N} \frac{\partial^2 L}{\partial \lambda_i \partial x_j} \Delta x_j + \sum_{j=1}^{M} \frac{\partial^2 L}{\partial \lambda_i \partial \lambda_j} \Delta \lambda_j + \sum_{j=1}^{N} \frac{\partial^2 L}{\partial \lambda_i \partial \alpha_j} \Delta \alpha_j + \sum_{j=1}^{N} \frac{\partial^2 L}{\partial \lambda_i \partial \beta_j} \Delta \beta_j + \sum_{j=1}^{N} \frac{\partial^2 L}{\partial \lambda_i \partial v_j} \Delta v_j$$

$$+ \sum_{j=1}^{N} \frac{\partial^2 L}{\partial \lambda_i \partial s_j} \Delta s_j = -\frac{\partial L}{\partial \lambda_i} \tag{F.13b}$$

$$\sum_{j=1}^{N} \frac{\partial^2 L}{\partial \alpha_i \partial x_j} \Delta x_j + \sum_{j=1}^{M} \frac{\partial^2 L}{\partial \alpha_i \partial \lambda_j} \Delta \lambda_j + \sum_{j=1}^{N} \frac{\partial^2 L}{\partial \alpha_i \partial \alpha_j} \Delta \alpha_j + \sum_{j=1}^{N} \frac{\partial^2 L}{\partial \alpha_i \partial \beta_j} \Delta \beta_j + \sum_{j=1}^{N} \frac{\partial^2 L}{\partial \alpha_i \partial v_j} \Delta v_j$$

$$+ \sum_{j=1}^{N} \frac{\partial^2 L}{\partial \alpha_i \partial s_j} \Delta s_j = -\frac{\partial L}{\partial \alpha_i} \tag{F.13c}$$

$$\sum_{j=1}^{N}\frac{\partial^2 L}{\partial\beta_i\partial x_j}\Delta x_j+\sum_{j=1}^{M}\frac{\partial^2 L}{\partial\beta_i\partial\lambda_j}\Delta\lambda_j+\sum_{j=1}^{N}\frac{\partial^2 L}{\partial\beta_i\partial\alpha_j}\Delta\alpha_j+\sum_{j=1}^{N}\frac{\partial^2 L}{\partial\beta_i\partial\beta_j}\Delta\beta_j+\sum_{j=1}^{N}\frac{\partial^2 L}{\partial\beta_i\partial v_j}\Delta v_j$$

$$+\sum_{j=1}^{N}\frac{\partial^2 L}{\partial\beta_i\partial s_j}\Delta s_j = -\frac{\partial L}{\partial\beta_i} \quad \text{(F.13d)}$$

$$\sum_{j=1}^{N}\frac{\partial^2 L}{\partial v_i\partial x_j}\Delta x_j+\sum_{j=1}^{M}\frac{\partial^2 L}{\partial v_i\partial\lambda_j}\Delta\lambda_j+\sum_{j=1}^{N}\frac{\partial^2 L}{\partial v_i\partial\alpha_j}\Delta\alpha_j+\sum_{j=1}^{N}\frac{\partial^2 L}{\partial v_i\partial\beta_j}\Delta\beta_j+\sum_{j=1}^{N}\frac{\partial^2 L}{\partial v_i\partial v_j}\Delta v_j$$

$$+\sum_{j=1}^{N}\frac{\partial^2 L}{\partial v_i\partial s_j}\Delta s_j = -\frac{\partial L}{\partial v_i} \quad \text{(F.13e)}$$

$$\sum_{j=1}^{N}\frac{\partial^2 L}{\partial s_i\partial x_j}\Delta x_j+\sum_{j=1}^{M}\frac{\partial^2 L}{\partial s_i\partial\lambda_j}\Delta\lambda_j+\sum_{j=1}^{N}\frac{\partial^2 L}{\partial s_i\partial\alpha_j}\Delta\alpha_j+\sum_{j=1}^{N}\frac{\partial^2 L}{\partial s_i\partial\beta_j}\Delta\beta_j+\sum_{j=1}^{N}\frac{\partial^2 L}{\partial s_i\partial v_j}\Delta v_j$$

$$+\sum_{j=1}^{N}\frac{\partial^2 L}{\partial s_i\partial s_j}\Delta s_j = -\frac{\partial L}{\partial s_i} \quad \text{(F.13f)}$$

Substituting the values of derivatives obtained from Eq. (F.12a) into Eq. (F.13a),

$$\sum_{j=1}^{N}A_{ij}\Delta x_j+\sum_{j=1}^{M}\frac{\partial g_j}{\partial x_i}\Delta\lambda_j+\Delta\alpha_i-\Delta\beta_i = -\left(\frac{\partial f(x)}{\partial x_i}-\sum_{j=1}^{M}\lambda_j\frac{\partial g_j}{\partial x_i}+\alpha_i-\beta_i\right) \quad \text{(F.14a)}$$

where

$$A_{ij}=\frac{\partial^2 f}{\partial x_i\partial x_j}+\sum_{j=1}^{M}\frac{\partial^2 g_j}{\partial x_i\partial x_j}$$

Substituting the values of derivatives obtained from Eq. (F.12b) into Eq. (F.13b),

$$\sum_{j=1}^{M}\frac{\partial g_i}{\partial x_j}\Delta\lambda_j = -g_j(x) \quad \text{(F.14b)}$$

Substituting the values of derivatives obtained from Eq. (F.12c) into Eq. (F.13c),

$$\Delta x_i+\Delta v_i = -(x_i+v_i-x_i^{\max}) \quad \text{(F.14c)}$$

Substituting the values of derivatives obtained from Eq. (F.12d) into Eq. (F.13d),

$$\Delta x_i-\Delta s_i = -(x_i+s_i-x_i^{\min}) \quad \text{(F.14d)}$$

Substituting the values of derivatives obtained from Eq. (F.12e) into Eq. (F.13e),

$$v_i\Delta\alpha_i+\alpha_i\Delta v_i = -(\alpha_i v_i-\mu_k) \quad \text{(F.14e)}$$

Substituting the values of derivatives obtained from Eq. (F.12f) into Eq. (F.13f),

$$s_i\Delta\beta_i+\beta_i\Delta s_i = -(\beta_i s_i-\mu_k) \quad \text{(F.14f)}$$

Substituting Eq. (F.12c) into Eq. (F.14c)

$$\Delta x_i = -\Delta v_i \tag{F.15a}$$

Substituting Eq. (F.12d) into Eq. (F.14d)

$$\Delta x_i = \Delta s_i \tag{F.15b}$$

The following equation is obtained from Eq. (F.14e).

$$v_i \Delta\alpha_i = -\alpha_i v_i + \mu_k - \alpha_i \Delta v_i$$

Substituting Eq. (F.15a) into above equation.

$$\Delta\alpha_i = -\alpha_i + \frac{\mu_k}{v_i} + \frac{\alpha_i}{v_i}\Delta x_i \tag{F.15c}$$

From Eq. (F.14f), the following equation is obtained.

$$s_i \Delta\beta_i = -\beta_i s_i + \mu_k - \beta_i \Delta s_i$$

Substituting Eq. (F.15b) into the above equation,

$$\Delta\beta_i = -\beta_i + \frac{\mu_k}{s_i} + \frac{\beta_i}{s_i}\Delta x_i \tag{F.15d}$$

Subtracting Eq. (F.15d) from Eq. (F.15c)

$$\Delta\alpha_i - \Delta\beta_i = -\alpha_i + \beta_i + \frac{\mu}{v_i} - \frac{\mu}{s_i} + \left(\frac{\alpha_i}{v_i} + \frac{\beta_i}{s_i}\right)\Delta x_i \tag{F.15e}$$

Substituting Eq. (F.15e) into Eq (F.14a),

$$\sum_{j=1}^{N} A_{ij}\,\Delta x_j + \sum_{j=1}^{M} \frac{\partial g_j}{\partial x_i}\Delta\lambda_j + \left[-\alpha_i + \beta_i + \left(\frac{\mu}{v_i} - \frac{\mu}{s_i}\right) + \left(\frac{\alpha_i}{v_i} + \frac{\beta_i}{s_i}\right)\Delta x_i\right]$$

$$= -\left(\frac{\partial f(x)}{\partial x_i} - \sum_{j=1}^{M} \lambda_j \frac{\partial g_j}{\partial x_i} + \alpha_i - \beta_i\right)$$

Rearranging the above equation,

$$\sum_{j=1}^{N} A_{ij}\,\Delta x_j + \sum_{j=1}^{M} \frac{\partial g_j}{\partial x_i}\Delta\lambda_j + \left(\frac{\alpha_i}{v_i} + \frac{\beta_i}{s_i}\right)\Delta x_i = -\left(\frac{\partial f(x)}{\partial x_i} - \sum_{j=1}^{M} \lambda_j \frac{\partial g_j}{\partial x_i}\right) - \left(\frac{\mu}{v_i} - \frac{\mu}{s_i}\right)$$

or

$$\sum_{j=1}^{N} H_{ij}\,\Delta x_j + \sum_{j=1}^{M} \frac{\partial g_j}{\partial x_i}\Delta\lambda_j = h_i \tag{F.16a}$$

where

$$H_{ii} = A_{ii} + \left(\frac{\alpha_i}{v_i} + \frac{\beta_i}{s_i}\right)$$

$$H_{ij} = A_{ij}$$

$$h_i = -\left(\frac{\partial f(x)}{\partial x_i} - \sum_{j=1}^{M} \lambda_j \frac{\partial g_j}{\partial x_i}\right) - \left(\frac{\mu}{v_i} - \frac{\mu}{s_i}\right)$$

Equations (F.14b) to (F.14d), (F.154c) and (F.15d) can be rewritten as given below.

$$\sum_{j=1}^{M} \frac{\partial g_i}{\partial x_j} \Delta\lambda_j = -g_j(x) \tag{F.16b}$$

$$\Delta\alpha_i = -\alpha_i + \frac{\mu_k}{v_i} + \frac{\alpha_i}{v_i} \Delta x_i \tag{F.16c}$$

$$\Delta\beta_i = -\beta_i + \frac{\mu_k}{s_i} + \frac{\beta_i}{s_i} \Delta x_i \tag{F.16d}$$

$$\Delta x_i + \Delta v_i = -(x_i + v_i - x_i^{\max}) \tag{F.16e}$$

$$\Delta x_i + \Delta s_i = -(x_i + s_i - x_i^{\min}) \tag{F.16f}$$

Equations (F16a) and (F16b) can be solved using Newton–Raphson method to obtain Δx_i. The rest of the variables $\Delta\alpha_i$, $\Delta\beta_i$, Δv_i and Δs_i can be obtained from Eq. (F.16c), Eq.(F.16d), Eq. (F.16e) and Eq. (F.16f), respectively and then variables are updated, with factor δ.

$$x_i = x_i + \delta\Delta x_i \qquad (i = 1, 2, ..., N)$$

$$y_i = y_i + \delta\Delta y_i \qquad (i = 1, 2, ..., M)$$

$$\lambda_i = \lambda_i + \delta\Delta\lambda_i \qquad (i = 1, 2, ..., N)$$

$$s_i = s_i + \delta\Delta s_i \qquad (i = 1, 2, ..., N)$$

$$\alpha_i = \alpha_i + \delta\Delta\alpha_i \qquad (i = 1, 2, ..., N)$$

$$\beta_i = \beta_i + \delta\Delta\beta_i \qquad (i = 1, 2, ..., N)$$

F.3 ALGORITHM

The algorithm is summarized in the following steps:

1. At the start of each iteration a solution $t^k = (x^k, \lambda^k, \alpha^k, \beta^k, v^k$, and $s^k)$ is available with: $x^k + v^k = x^{\max}$, $x^k - s^k = x^{\min}$, v^k, s^k, α^k and $\beta^k \geq 0$ and positive value of μ^k where x, λ, α, β, v, and s are vectors, and k represents the iteration number.
2. The following linear system is solved:

$$\nabla_t F(t^k, \mu_k)\Delta t = -F(t^k, \mu_k)$$

   where k represents the iteration number. Equation (F.15) is given in matrix form. Employing Gauss elimination method, search direction Δt, can be obtained.
3. The algorithm stops when Kuhn-Tucker conditions are verified with sufficient accuracy and the value of the parameter μ is sufficiently small.
4. The maximum value of step for which the solution continues to satisfy the positiveness conditions on α, β, v, and s variables are computed as

$$\delta_k = \sup\{\delta \geq 0 \mid v^k + \delta\Delta v^k \geq 0,\ \alpha^k + \delta\Delta\alpha^k \geq 0,\ s^k + \delta\Delta s^k \geq 0,\ \beta^k + \delta\Delta\beta^k \geq 0\}$$

5. The values of t^{k+1} and μ^{k+1} can be determined as follows:

$$t^{k+1} = t^k \, \xi \, \Delta t^k$$

$$\mu^{k+1} = \phi \, \frac{(\alpha^{k+1})^T \, v^k + 1 + (\beta^{k+1})^T \, s^{k+1}}{2n}$$

where $\xi = \min\,\{1, \gamma\delta^k\}$

γ is a positive constant smaller than, but very near unity (typically 0.995 or 0.999).

ϕ is a positive constant smaller than unity (typically 0.1 – 0.2).

GOTO Step 2 and repeat.

REFERENCES

Da Costa G.R.M, C.E.U. Costa, and A.M. de Souza, Comparative studies of optimization methods for the optimal power flow problem, *Electric Power Systems Research,* Vol. **56,** pp. 249–254, 2000.

Garzillo, A., M. Innorta, and M. Ricci, The problem of the active and reactive optimum power dispatching solved by utilizing a primal-dual interior point method, *J. Electrical Power &* Energy Systems, Vol. **20,** No. 6, pp. 427–434, 1998.

Lin C.H. and S.Y. Lin, A new dual-type method used in solving optimal power flow problems, *IEEE Transactions on Power Systems,* Vol. **12,** No. 4, pp. 1667–1675, November, 1997.

Lustig, I.J., R.E. Marsten, D.F. Shanno, Computational experience with a primal-dual interior point method for linear programming, *Linear Algebra and its Applications,* Vol. **152,** pp.191–222, 1991.

Lustig I.J., R.E. Marsten, D.F. Shanno, On implementing Mehrotra's predictor-corrector interior point method for linear programming, *SIAM J. Optimization,* Vol. **2,** No. 3, pp. 435–449, 1992.

McShane K.A., C.L. Monma, and D.F. Shanno, An implementation of a primal-dual interior point method for linear programming, *ORSA Journal on Computing,* Vol. **1,** pp.70–83, 1989.

Torres, G.L. and V.H. Quintana, An interior point method for nonlinear optimal power flow using voltage rectangular coordinates, *IEEE Transactions on Power Systems,* Vol. **13**, No. 4, pp. 1211–1218, November, 1998.

Vanderbei R.J., ALPO: Another Linear Program Optimizer, *ORSA Journal on Computing,* Vol. **5,** No.,2, pp.134–146, 1993.

Wei, H., H. Sasaki, J. Kubokawa, and R. Yokoyama, An interior point non-linear programming for optimal power flow problems with a noval data structure, *IEEE Transactions on Power Systems,* Vol. **13,** No. 3, pp. 870–877, August, 1998.

Wu, Y.C., Efficient two-level interior point method for optimal pumping hydro storage scheduling exploiting the non-sparse matrix structure, *IEE Proceedings-Generation, Transmission and Distribution,* Vol. **148,** No. 1, pp. 41–47, 2001.

Yan X. and V.H. Quintana, An efficient predictor-corrector interior point algorithm for security constrained economic dispatch, *IEEE Transactions on Power Systems,* Vol. **12,** No. 2, pp. 803–810, May 1997.

Index